Managing the Risks of Extreme Events and Disasters to Advance Climate Change Adaptation

Special Report of the Intergovernmental Panel on Climate Change

Extreme weather and climate events, interacting with exposed and vulnerable human and natural systems, can lead to disasters. This Special Report explores the challenge of understanding and managing the risks of climate extremes to advance climate change adaptation. Weather- and climate-related disasters have social as well as physical dimensions. As a result, changes in the frequency and severity of the physical events affect disaster risk, but so do the spatially diverse and temporally dynamic patterns of exposure and vulnerability. Some types of extreme weather and climate events have increased in frequency or magnitude, but populations and assets at risk have also increased, with consequences for disaster risk. Opportunities for managing risks of weather- and climate-related disasters exist or can be developed at any scale, local to international. Some strategies for effectively managing risks and adapting to climate change involve adjustments to current activities. Others require transformation or fundamental change.

The Intergovernmental Panel on Climate Change (IPCC) is the leading international body for the assessment of climate change, including the physical science of climate; impacts, adaptation, and vulnerability; and mitigation of climate change. The IPCC was established by the United Nations Environment Programme (UNEP) and the World Meteorological Organization (WMO) to provide the world with a comprehensive assessment of the current state of knowledge of climate change and its potential environmental and socioeconomic impacts.

Managing the Risks of Extreme Events and Disasters to Advance Climate Change Adaptation

Special Report of the Intergovernmental Panel on Climate Change

Edited by

Christopher B. Field
Co-Chair Working Group II
Carnegie Institution for Science

Vicente Barros
Co-Chair Working Group II
CIMA / Universidad de Buenos Aires

Thomas F. Stocker
Co-Chair Working Group I
University of Bern

Qin Dahe
Co-Chair Working Group I
China Meteorological Administration

David Jon Dokken **Kristie L. Ebi** **Michael D. Mastrandrea** **Katharine J. Mach**

Gian-Kasper Plattner **Simon K. Allen** **Melinda Tignor** **Pauline M. Midgley**

CAMBRIDGE
UNIVERSITY PRESS

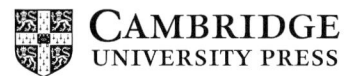

CAMBRIDGE UNIVERSITY PRESS
Cambridge, New York, Melbourne, Madrid, Cape Town,
Singapore, São Paulo, Delhi, Mexico City

Cambridge University Press
32 Avenue of the Americas, New York, NY 10013-2473, USA

www.cambridge.org
Information on this title: www.cambridge.org/9781107607804

© Intergovernmental Panel on Climate Change 2012

This publication is in copyright. Subject to statutory exception
and to the provisions of relevant collective licensing agreements,
no reproduction of any part may take place without the written
permission of Cambridge University Press.

First published 2012

Printed in the United States of America

A catalog record for this publication is available from the British Library.

ISBN 978-1-107-02506-6 Hardback
ISBN 978-1-107-60780-4 Paperback

Cambridge University Press has no responsibility for the persistence or accuracy of URLs
for external or third-party Internet Web sites referred to in this publication and does not
guarantee that any content on such Web sites is, or will remain, accurate or appropriate.

This book was printed on acid-free stock that is from SFI (Sustainable Forestry Initiative)
certified mills and distributors. It is FSC chain-of-custody certified.

Use the following reference to cite the entire volume:
IPCC, 2012: *Managing the Risks of Extreme Events and Disasters to Advance Climate
  Change Adaptation*. A Special Report of Working Groups I and II of the
  Intergovernmental Panel on Climate Change [Field, C.B., V. Barros, T.F. Stocker,
  D. Qin, D.J. Dokken, K.L. Ebi, M.D. Mastrandrea, K.J. Mach, G.-K. Plattner, S.K.
  Allen, M. Tignor, and P.M. Midgley (eds.)]. Cambridge University Press,
  Cambridge, UK, and New York, NY, USA, 582 pp.

Contents

Section I

Foreword viii

Preface ix

Section II

Summary for Policymakers 3

Section III

Chapter 1   Climate Change: New Dimensions in Disaster Risk, Exposure, Vulnerability, and Resilience 25

Chapter 2   Determinants of Risk: Exposure and Vulnerability 65

Chapter 3   Changes in Climate Extremes and their Impacts on the Natural Physical Environment 109

Chapter 4   Changes in Impacts of Climate Extremes: Human Systems and Ecosystems 231

Chapter 5   Managing the Risks from Climate Extremes at the Local Level 291

Chapter 6   National Systems for Managing the Risks from Climate Extremes and Disasters 339

Chapter 7   Managing the Risks: International Level and Integration across Scales 393

Chapter 8   Toward a Sustainable and Resilient Future 437

Chapter 9   Case Studies 487

Section IV

Annex I   Authors and Expert Reviewers 545

Annex II   Glossary of Terms 555

Annex III   Acronyms 565

Annex IV   List of Major IPCC Reports 569

Index 573

I Foreword and Preface

Foreword

This Special Report on Managing the Risks of Extreme Events and Disasters to Advance Climate Change Adaptation (SREX) has been jointly coordinated by Working Groups I (WGI) and II (WGII) of the Intergovernmental Panel on Climate Change (IPCC). The report focuses on the relationship between climate change and extreme weather and climate events, the impacts of such events, and the strategies to manage the associated risks.

The IPCC was jointly established in 1988 by the World Meteorological Organization (WMO) and the United Nations Environment Programme (UNEP), in particular to assess in a comprehensive, objective, and transparent manner all the relevant scientific, technical, and socioeconomic information to contribute in understanding the scientific basis of risk of human-induced climate change, the potential impacts, and the adaptation and mitigation options. Beginning in 1990, the IPCC has produced a series of Assessment Reports, Special Reports, Technical Papers, methodologies, and other key documents which have since become the standard references for policymakers and scientists.

This Special Report, in particular, contributes to frame the challenge of dealing with extreme weather and climate events as an issue in decisionmaking under uncertainty, analyzing response in the context of risk management. The report consists of nine chapters, covering risk management; observed and projected changes in extreme weather and climate events; exposure and vulnerability to as well as losses resulting from such events; adaptation options from the local to the international scale; the role of sustainable development in modulating risks; and insights from specific case studies.

Success in developing this report depended foremost on the knowledge, integrity, enthusiasm, and collaboration of hundreds of experts worldwide, representing a very wide range of disciplines. We would like to express our gratitude to all the Coordinating Lead Authors, Lead Authors, Contributing Authors, Review Editors, and Expert and Government Reviewers who devoted considerable expertise, time, and effort to produce this report. We are extremely grateful for their commitment to the IPCC process and we would also like to thank the staff of the WGI and WGII Technical Support Units and the IPCC Secretariat, for their unrestricted commitment to the development of such an ambitious and highly significant IPCC Special Report.

We are also very grateful to the governments which supported their scientists' participation in this task, as well as to all those that contributed to the IPCC Trust Fund, thereby facilitating the essential participation of experts from the developing world. We would also like to express our appreciation, in particular, to the governments of Australia, Panama, Switzerland, and Vietnam for hosting the drafting sessions in their respective countries, as well as to the government of Uganda for hosting in Kampala the First Joint Session of Working Groups I and II which approved the report. Our thanks are also due to the governments of Switzerland and the United States of America for funding the Technical Support Units for WGI and WGII, respectively. We also wish to acknowledge the collaboration of the government of Norway – which also provided critical support for meetings and outreach – and the United Nations International Strategy for Disaster Reduction (ISDR), in the preparation of the original report proposal.

We would especially wish to thank the IPCC Chairman, Dr. Rajendra Pachauri, for his direction and guidance of the IPCC process, as well as the Co-Chairs of Working Groups II and I, Professors Vicente Barros, Christopher Field, Qin Dahe, and Thomas Stocker, for their leadership throughout the development of this Special Report.

M. Jarraud
Secretary-General
World Meteorological Organization

A. Steiner
Executive Director
United Nations Environment Programme

Preface

This volume, Managing the Risks of Extreme Events and Disasters to Advance Climate Change Adaptation, is a Special Report of the Intergovernmental Panel on Climate Change (IPCC). The report is a collaborative effort of Working Group I (WGI) and Working Group II (WGII). The IPCC leadership team for this report also has responsibility for the IPCC Fifth Assessment Report (AR5), scheduled for completion in 2013 and 2014.

The Special Report brings together scientific communities with expertise in three very different aspects of managing risks of extreme weather and climate events. For this report, specialists in disaster recovery, disaster risk management, and disaster risk reduction, a community mostly new to the IPCC, joined forces with experts in the areas of the physical science basis of climate change (WGI) and climate change impacts, adaptation, and vulnerability (WGII). Over the course of the two-plus years invested in assessing information and writing the report, scientists from these three communities forged shared goals and products.

Extreme weather and climate events have figured prominently in past IPCC assessments. Extremes can contribute to disasters, but disaster risk is influenced by more than just the physical hazards. Disaster risk emerges from the interaction of weather or climate events, the physical contributors to disaster risk, with exposure and vulnerability, the contributors to risk from the human side. The combination of severe consequences, rarity, and human as well as physical determinants makes disasters difficult to study. Only over the last few years has the science of these events, their impacts, and options for dealing with them become mature enough to support a comprehensive assessment. This report provides a careful assessment of scientific, technical, and socioeconomic knowledge as of May 2011, the cut-off date for literature included.

The Special Report introduced some important innovations to the IPCC. One was the integration, in a single Special Report, of skills and perspectives across the disciplines covered by WGI, WGII, and the disaster risk management community. A second important innovation was the report's emphasis on adaptation and disaster risk management. A third innovation was a plan for an ambitious outreach effort. Underlying these innovations and all aspects of the report is a strong commitment to assessing science in a way that is relevant to policy but not policy prescriptive.

The Process

The Special Report represents the combined efforts of hundreds of leading experts. The Government of Norway and the United Nations International Strategy for Disaster Reduction submitted a proposal for the report to the IPCC in September 2008. This was followed by a scoping meeting to develop a candidate outline in March 2009. Following approval of the outline in April 2009, governments and observer organizations nominated experts for the author team. The team approved by the WGI and WGII Bureaux consisted of 87 Coordinating Lead Authors and Lead Authors, plus 19 Review Editors. In addition, 140 Contributing Authors submitted draft text and information to the author teams. The drafts of the report were circulated twice for formal review, first to experts and second to both experts and governments, resulting in 18,784 review comments. Author teams responded to every comment and, where scientifically appropriate, modified drafts in response to comments, with Review Editors monitoring the process. The revised report was presented for consideration at the First Joint Session of WGI and WGII, from 14 to 17 November 2011. At the joint session, delegates from over 100 countries evaluated and approved, by consensus, the Summary for Policymakers on a line-by-line basis and accepted the full report.

Structure of the Special Report

This report contains a Summary for Policymakers (SPM) plus nine chapters. References in the SPM point to the supporting sections of the technical chapters that provide a traceable account of every major finding. The first two chapters set the stage for the report. Chapter 1 frames the issue of extreme weather and climate events as a challenge in understanding and managing risk. It characterizes risk as emerging from the overlap of a triggering physical event with exposure of people and assets and their vulnerability. Chapter 2 explores the determinants of exposure and vulnerability in detail, concluding that every disaster has social as well as physical dimensions. Chapter 3, the major contribution of WGI, is an assessment of the scientific literature on observed and projected changes in extreme weather and climate events, and their attribution to causes where possible. Chapter 4 assesses observed and projected impacts, considering patterns by sector as well as region. Chapters 5 through 7 assess experience and theory in adaptation to extremes and disasters, focusing on issues and opportunities at the local scale (Chapter 5), the national scale (Chapter 6), and the international scale (Chapter 7). Chapter 8 assesses the interactions among sustainable development, vulnerability reduction, and disaster risk, considering both opportunities and constraints, as well as the kinds of transformations relevant to overcoming the constraints. Chapter 9 develops a series of case studies that illustrate the role of real life complexity but also document examples of important progress in managing risk.

Acknowledgements

We wish to express our sincere appreciation to all the Coordinating Lead Authors, Lead Authors, Contributing Authors, Review Editors, and Expert and Government Reviewers. Without their expertise, commitment, and integrity, as well as vast investments of time, a report of this quality could never have been completed. We would also like to thank the members of the WGI and WGII Bureaux for their assistance, wisdom, and good sense throughout the preparation of the report.

We would particularly like to thank the remarkable staffs of the Technical Support Units of WGI and WGII for their professionalism, creativity, and dedication. In WGI, thanks go to Gian-Kasper Plattner, Simon Allen, Pauline Midgley, Melinda Tignor, Vincent Bex, Judith Boschung, and Alexander Nauels. In WGII, which led the logistics and overall coordination, thanks go to Dave Dokken, Kristie Ebi, Michael Mastrandrea, Katharine Mach, Sandy MacCracken, Rob Genova, Yuka Estrada, Eric Kissel, Patricia Mastrandrea, Monalisa Chatterjee, and Kyle Terran. Their tireless and very capable efforts to coordinate the Special Report ensured a final product of high scientific quality, while maintaining an atmosphere of collegiality and respect.

We would also like to thank the staff of the IPCC Secretariat: Renate Christ, Gaetano Leone, Mary Jean Burer, Sophie Schlingemann, Judith Ewa, Jesbin Baidya, Joelle Fernandez, Annie Courtin, Laura Biagioni, and Amy Smith Aasdam. Thanks are also due to Francis Hayes (WMO), Tim Nuthall (European Climate Foundation), and Nick Nutall (UNEP).

Our sincere thanks go to the hosts and organizers of the scoping meeting, the four lead author meetings, and the approval session. We gratefully acknowledge the support from the host countries: Norway, Panama, Vietnam, Switzerland, Australia, and Uganda. It is a pleasure to extend special thanks to the government of Norway, which provided untiring support throughout the Special Report process.

Vicente Barros and Christopher B. Field
IPCC WGII Co-Chairs

Qin Dahe and Thomas F. Stocker
IPCC WGI Co-Chairs

II Summary for Policymakers

SPM
Summary for Policymakers

Drafting Authors:
Simon K. Allen (Switzerland), Vicente Barros (Argentina), Ian Burton (Canada), Diarmid Campbell-Lendrum (UK), Omar-Dario Cardona (Colombia), Susan L. Cutter (USA), O. Pauline Dube (Botswana), Kristie L. Ebi (USA), Christopher B. Field (USA), John W. Handmer (Australia), Padma N. Lal (Australia), Allan Lavell (Costa Rica), Katharine J. Mach (USA), Michael D. Mastrandrea (USA), Gordon A. McBean (Canada), Reinhard Mechler (Germany), Tom Mitchell (UK), Neville Nicholls (Australia), Karen L. O'Brien (Norway), Taikan Oki (Japan), Michael Oppenheimer (USA), Mark Pelling (UK), Gian-Kasper Plattner (Switzerland), Roger S. Pulwarty (USA), Sonia I. Seneviratne (Switzerland), Thomas F. Stocker (Switzerland), Maarten K. van Aalst (Netherlands), Carolina S. Vera (Argentina), Thomas J. Wilbanks (USA)

This Summary for Policymakers should be cited as:
IPCC, 2012: Summary for Policymakers. In: *Managing the Risks of Extreme Events and Disasters to Advance Climate Change Adaptation* [Field, C.B., V. Barros, T.F. Stocker, D. Qin, D.J. Dokken, K.L. Ebi, M.D. Mastrandrea, K.J. Mach, G.-K. Plattner, S.K. Allen, M. Tignor, and P.M. Midgley (eds.)]. A Special Report of Working Groups I and II of the Intergovernmental Panel on Climate Change. Cambridge University Press, Cambridge, UK, and New York, NY, USA, pp. 3-21.

Summary for Policymakers

A. Context

This Summary for Policymakers presents key findings from the Special Report on Managing the Risks of Extreme Events and Disasters to Advance Climate Change Adaptation (SREX). The SREX approaches the topic by assessing the scientific literature on issues that range from the relationship between climate change and extreme weather and climate events ('climate extremes') to the implications of these events for society and sustainable development. The assessment concerns the interaction of climatic, environmental, and human factors that can lead to impacts and disasters, options for managing the risks posed by impacts and disasters, and the important role that non-climatic factors play in determining impacts. Box SPM.1 defines concepts central to the SREX.

The character and severity of impacts from climate extremes depend not only on the extremes themselves but also on exposure and vulnerability. In this report, adverse impacts are considered disasters when they produce widespread damage and cause severe alterations in the normal functioning of communities or societies. Climate extremes, exposure, and vulnerability are influenced by a wide range of factors, including anthropogenic climate change, natural climate variability, and socioeconomic development (Figure SPM.1). Disaster risk management and adaptation to climate change focus on reducing exposure and vulnerability and increasing resilience to the potential adverse impacts of climate extremes, even though risks cannot fully be eliminated (Figure SPM.2). Although mitigation of climate change is not the focus of this report, adaptation and mitigation can complement each other and together can significantly reduce the risks of climate change. [SYR AR4, 5.3]

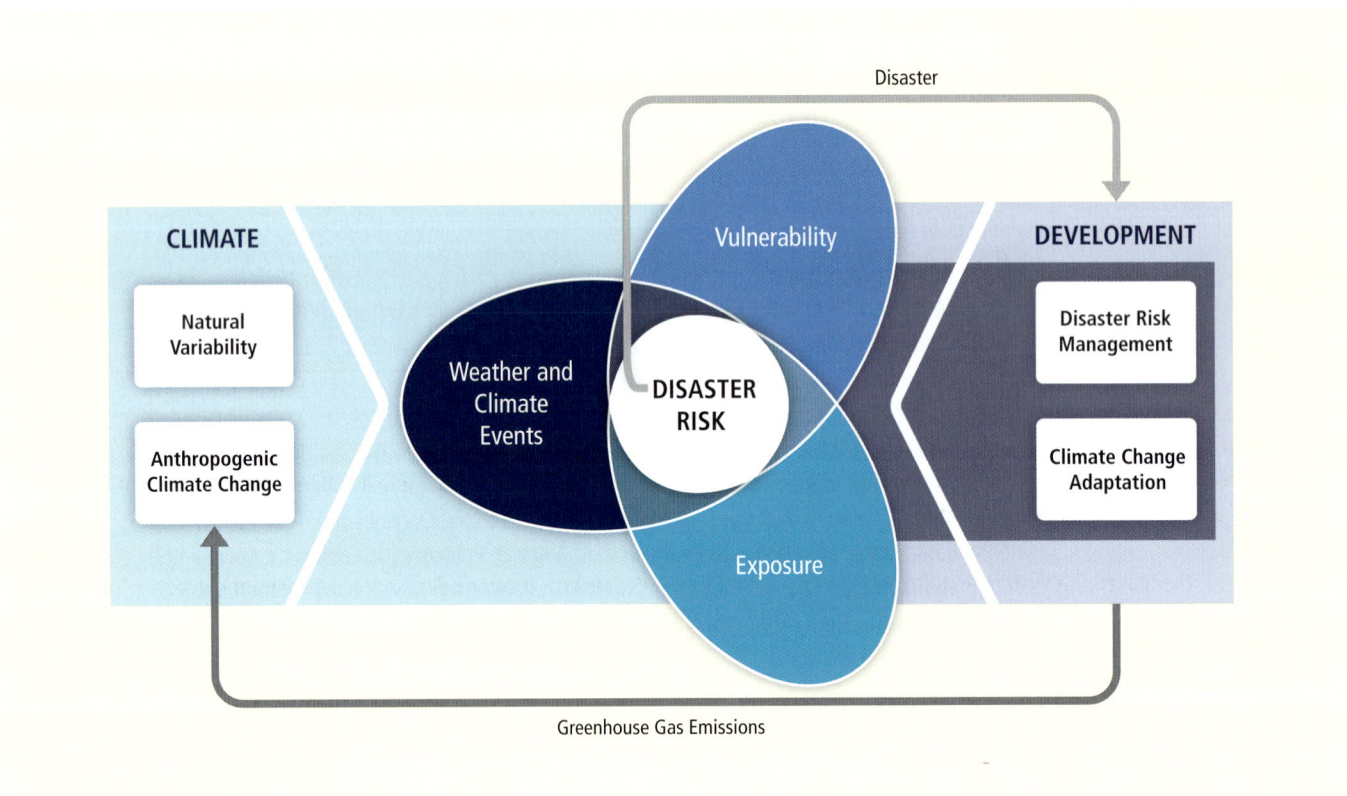

Figure SPM.1 | Illustration of the core concepts of SREX. The report assesses how exposure and vulnerability to weather and climate events determine impacts and the likelihood of disasters (disaster risk). It evaluates the influence of natural climate variability and anthropogenic climate change on climate extremes and other weather and climate events that can contribute to disasters, as well as the exposure and vulnerability of human society and natural ecosystems. It also considers the role of development in trends in exposure and vulnerability, implications for disaster risk, and interactions between disasters and development. The report examines how disaster risk management and adaptation to climate change can reduce exposure and vulnerability to weather and climate events and thus reduce disaster risk, as well as increase resilience to the risks that cannot be eliminated. Other important processes are largely outside the scope of this report, including the influence of development on greenhouse gas emissions and anthropogenic climate change, and the potential for mitigation of anthropogenic climate change. [1.1.2, Figure 1-1]

Box SPM.1 | Definitions Central to SREX

Core concepts defined in the SREX glossary[1] and used throughout the report include:

Climate Change: A change in the state of the climate that can be identified (e.g., by using statistical tests) by changes in the mean and/or the variability of its properties and that persists for an extended period, typically decades or longer. Climate change may be due to natural internal processes or external forcings, or to persistent anthropogenic changes in the composition of the atmosphere or in land use.[2]

Climate Extreme (extreme weather or climate event): The occurrence of a value of a weather or climate variable above (or below) a threshold value near the upper (or lower) ends of the range of observed values of the variable. For simplicity, both extreme weather events and extreme climate events are referred to collectively as 'climate extremes.' The full definition is provided in Section 3.1.2.

Exposure: The presence of people; livelihoods; environmental services and resources; infrastructure; or economic, social, or cultural assets in places that could be adversely affected.

Vulnerability: The propensity or predisposition to be adversely affected.

Disaster: Severe alterations in the normal functioning of a community or a society due to hazardous physical events interacting with vulnerable social conditions, leading to widespread adverse human, material, economic, or environmental effects that require immediate emergency response to satisfy critical human needs and that may require external support for recovery.

Disaster Risk: The likelihood over a specified time period of severe alterations in the normal functioning of a community or a society due to hazardous physical events interacting with vulnerable social conditions, leading to widespread adverse human, material, economic, or environmental effects that require immediate emergency response to satisfy critical human needs and that may require external support for recovery.

Disaster Risk Management: Processes for designing, implementing, and evaluating strategies, policies, and measures to improve the understanding of disaster risk, foster disaster risk reduction and transfer, and promote continuous improvement in disaster preparedness, response, and recovery practices, with the explicit purpose of increasing human security, well-being, quality of life, resilience, and sustainable development.

Adaptation: In human systems, the process of adjustment to actual or expected climate and its effects, in order to moderate harm or exploit beneficial opportunities. In natural systems, the process of adjustment to actual climate and its effects; human intervention may facilitate adjustment to expected climate.

Resilience: The ability of a system and its component parts to anticipate, absorb, accommodate, or recover from the effects of a hazardous event in a timely and efficient manner, including through ensuring the preservation, restoration, or improvement of its essential basic structures and functions.

Transformation: The altering of fundamental attributes of a system (including value systems; regulatory, legislative, or bureaucratic regimes; financial institutions; and technological or biological systems).

---

[1] Reflecting the diversity of the communities involved in this assessment and progress in science, several of the definitions used in this Special Report differ in breadth or focus from those used in the Fourth Assessment Report and other IPCC reports.

[2] This definition differs from that in the United Nations Framework Convention on Climate Change (UNFCCC), where climate change is defined as: "a change of climate which is attributed directly or indirectly to human activity that alters the composition of the global atmosphere and which is in addition to natural climate variability observed over comparable time periods." The UNFCCC thus makes a distinction between climate change attributable to human activities altering the atmospheric composition, and climate variability attributable to natural causes.

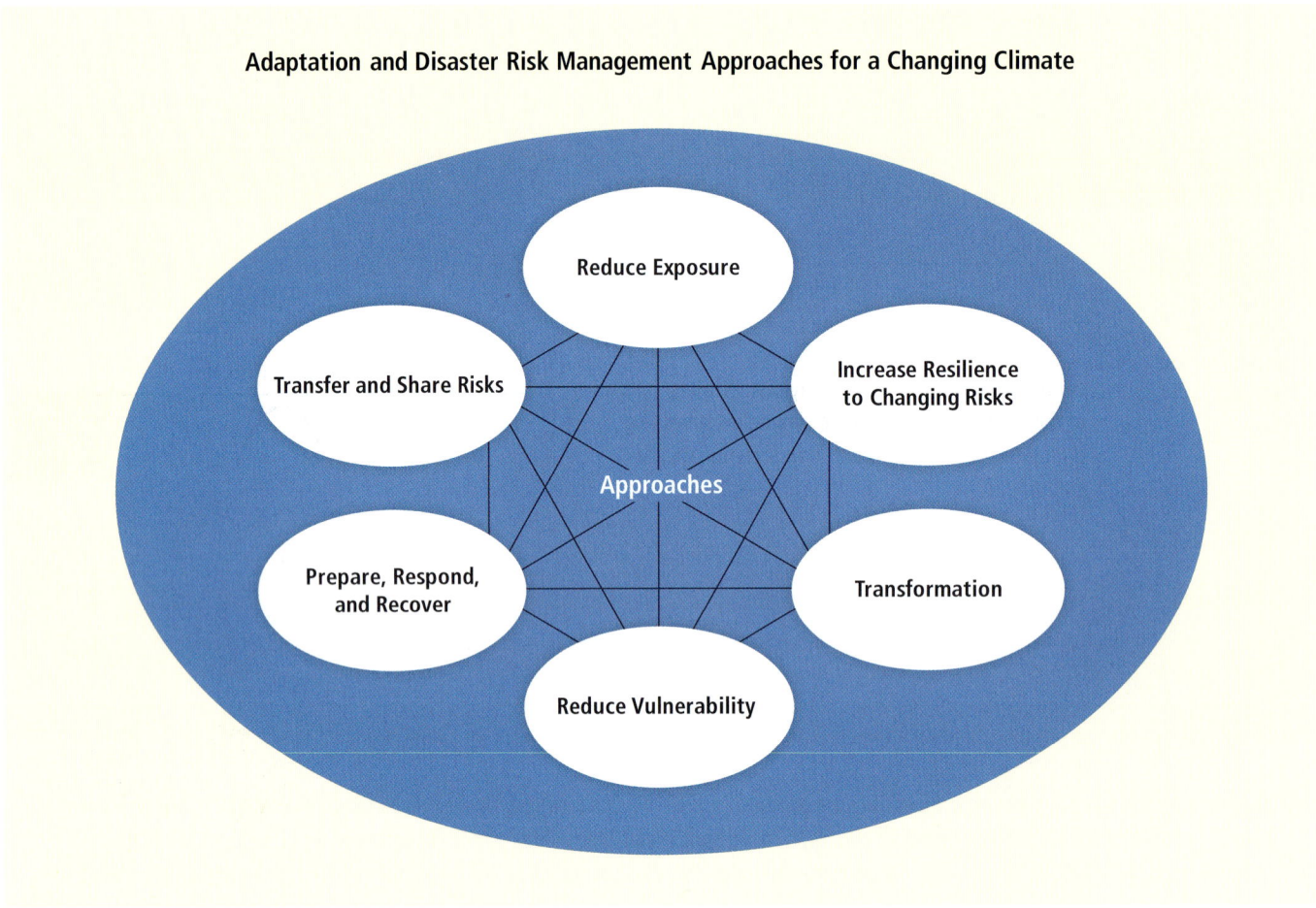

Figure SPM.2 | Adaptation and disaster risk management approaches for reducing and managing disaster risk in a changing climate. This report assesses a wide range of complementary adaptation and disaster risk management approaches that can reduce the risks of climate extremes and disasters and increase resilience to remaining risks as they change over time. These approaches can be overlapping and can be pursued simultaneously. [6.5, Figure 6-3, 8.6]

This report integrates perspectives from several historically distinct research communities studying climate science, climate impacts, adaptation to climate change, and disaster risk management. Each community brings different viewpoints, vocabularies, approaches, and goals, and all provide important insights into the status of the knowledge base and its gaps. Many of the key assessment findings come from the interfaces among these communities. These interfaces are also illustrated in Table SPM.1. To accurately convey the degree of certainty in key findings, the report relies on the consistent use of calibrated uncertainty language, introduced in Box SPM.2. The basis for substantive paragraphs in this Summary for Policymakers can be found in the chapter sections specified in square brackets.

Exposure and vulnerability are key determinants of disaster risk and of impacts when risk is realized. [1.1.2, 1.2.3, 1.3, 2.2.1, 2.3, 2.5] For example, a tropical cyclone can have very different impacts depending on where and when it makes landfall. [2.5.1, 3.1, 4.4.6] Similarly, a heat wave can have very different impacts on different populations depending on their vulnerability. [Box 4-4, 9.2.1] Extreme impacts on human, ecological, or physical systems can result from individual extreme weather or climate events. Extreme impacts can also result from non-extreme events where exposure and vulnerability are high [2.2.1, 2.3, 2.5] or from a compounding of events or their impacts. [1.1.2, 1.2.3, 3.1.3] For example, drought, coupled with extreme heat and low humidity, can increase the risk of wildfire. [Box 4-1, 9.2.2]

Extreme and non-extreme weather or climate events affect vulnerability to future extreme events by modifying resilience, coping capacity, and adaptive capacity. [2.4.3] In particular, the cumulative effects of disasters at local or sub-national levels can substantially affect livelihood options and resources and the capacity of societies and communities to prepare for and respond to future disasters. [2.2, 2.7]

A changing climate leads to changes in the frequency, intensity, spatial extent, duration, and timing of extreme weather and climate events, and can result in unprecedented extreme weather and climate events. Changes in extremes can be linked to changes in the mean, variance, or shape of probability distributions, or all of these (Figure SPM.3). Some climate extremes (e.g., droughts) may be the result of an accumulation of weather or climate events that are not extreme when considered independently. Many extreme weather and climate events continue to be the result of natural climate variability. Natural variability will be an important factor in shaping future extremes in addition to the effect of anthropogenic changes in climate. [3.1]

B. Observations of Exposure, Vulnerability, Climate Extremes, Impacts, and Disaster Losses

The impacts of climate extremes and the potential for disasters result from the climate extremes themselves and from the exposure and vulnerability of human and natural systems. Observed changes in climate extremes reflect the influence of anthropogenic climate change in addition to natural climate variability, with changes in exposure and vulnerability influenced by both climatic and non-climatic factors.

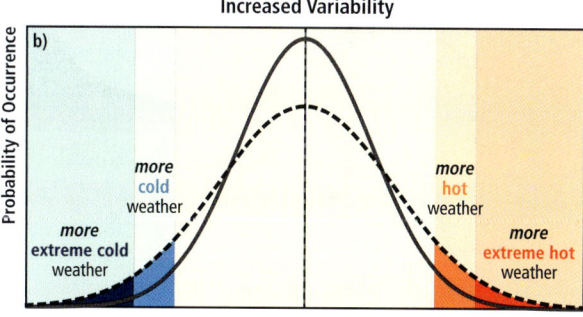
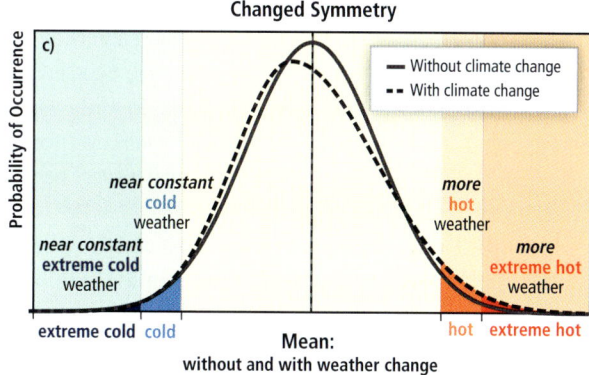

Figure SPM.3 | The effect of changes in temperature distribution on extremes. Different changes in temperature distributions between present and future climate and their effects on extreme values of the distributions: (a) effects of a simple shift of the entire distribution toward a warmer climate; (b) effects of an increase in temperature variability with no shift in the mean; (c) effects of an altered shape of the distribution, in this example a change in asymmetry toward the hotter part of the distribution. [Figure 1-2, 1.2.2]

Exposure and Vulnerability

Exposure and vulnerability are dynamic, varying across temporal and spatial scales, and depend on economic, social, geographic, demographic, cultural, institutional, governance, and environmental factors (*high confidence*). [2.2, 2.3, 2.5] Individuals and communities are differentially exposed and vulnerable based on inequalities expressed through levels of wealth and education, disability, and health status, as well as gender, age, class, and other social and cultural characteristics. [2.5]

Settlement patterns, urbanization, and changes in socioeconomic conditions have all influenced observed trends in exposure and vulnerability to climate extremes (*high confidence*). [4.2, 4.3.5] For example, coastal settlements, including in small islands and megadeltas, and mountain settlements are exposed and vulnerable to climate extremes in both developed and developing countries, but with differences among regions and countries. [4.3.5, 4.4.3, 4.4.6, 4.4.9, 4.4.10] Rapid urbanization and the growth of megacities, especially in developing countries, have led to the emergence of highly vulnerable urban communities, particularly through informal settlements and inadequate land management (*high agreement, robust evidence*). [5.5.1] See also Case Studies 9.2.8 and 9.2.9. Vulnerable populations also include refugees, internally displaced people, and those living in marginal areas. [4.2, 4.3.5]

Climate Extremes and Impacts

There is evidence from observations gathered since 1950 of change in some extremes. Confidence in observed changes in extremes depends on the quality and quantity of data and the availability of studies analyzing these data, which vary across regions and for different extremes. Assigning 'low confidence' in observed changes in a specific extreme on regional or global scales neither implies nor excludes the possibility of changes in this extreme. Extreme events are rare, which means there are few data available to make assessments regarding changes in their frequency or intensity. The more rare the event the more difficult it is to identify long-term changes. Global-scale trends in a specific extreme may be either more reliable (e.g., for temperature extremes) or less reliable (e.g., for droughts) than some regional-scale trends, depending on the geographical uniformity of the trends in the specific extreme. The following paragraphs provide further details for specific climate extremes from observations since 1950. [3.1.5, 3.1.6, 3.2.1]

It is *very likely* that there has been an overall decrease in the number of cold days and nights,[3] and an overall increase in the number of warm days and nights,[3] at the global scale, that is, for most land areas with sufficient data. It is *likely* that these changes have also occurred at the continental scale in North America, Europe, and Australia. There is *medium confidence* in a warming trend in daily temperature extremes in much of Asia. Confidence in observed trends in daily temperature extremes in Africa and South America generally varies from *low* to *medium* depending on the region. In many (but not all) regions over the globe with sufficient data, there is *medium confidence* that the length or number of warm spells or heat waves[3] has increased. [3.3.1, Table 3-2]

There have been statistically significant trends in the number of heavy precipitation events in some regions. It is *likely* that more of these regions have experienced increases than decreases, although there are strong regional and subregional variations in these trends. [3.3.2]

There is *low confidence* in any observed long-term (i.e., 40 years or more) increases in tropical cyclone activity (i.e., intensity, frequency, duration), after accounting for past changes in observing capabilities. It is *likely* that there has been a poleward shift in the main Northern and Southern Hemisphere extratropical storm tracks. There is *low confidence* in observed trends in small spatial-scale phenomena such as tornadoes and hail because of data inhomogeneities and inadequacies in monitoring systems. [3.3.2, 3.3.3, 3.4.4, 3.4.5]

There is *medium confidence* that some regions of the world have experienced more intense and longer droughts, in particular in southern Europe and West Africa, but in some regions droughts have become less frequent, less intense, or shorter, for example, in central North America and northwestern Australia. [3.5.1]

There is *limited* to *medium evidence* available to assess climate-driven observed changes in the magnitude and frequency of floods at regional scales because the available instrumental records of floods at gauge stations are limited in space and time, and because of confounding effects of changes in land use and engineering. Furthermore, there is *low agreement* in this evidence, and thus overall *low confidence* at the global scale regarding even the sign of these changes. [3.5.2]

---

[3] See SREX Glossary for definition of these terms: cold days / cold nights, warm days / warm nights, and warm spell – heat wave.

It is *likely* that there has been an increase in extreme coastal high water related to increases in mean sea level. [3.5.3]

There is evidence that some extremes have changed as a result of anthropogenic influences, including increases in atmospheric concentrations of greenhouse gases. It is *likely* that anthropogenic influences have led to warming of extreme daily minimum and maximum temperatures at the global scale. There is *medium confidence* that anthropogenic influences have contributed to intensification of extreme precipitation at the global scale. It is *likely* that there has been an anthropogenic influence on increasing extreme coastal high water due to an increase in mean sea level. The uncertainties in the historical tropical cyclone records, the incomplete understanding of the physical mechanisms linking tropical cyclone metrics to climate change, and the degree of tropical cyclone variability provide only *low confidence* for the attribution of any detectable changes in tropical cyclone activity to anthropogenic influences. Attribution of single extreme events to anthropogenic climate change is challenging. [3.2.2, 3.3.1, 3.3.2, 3.4.4, 3.5.3, Table 3-1]

Disaster Losses

Economic losses from weather- and climate-related disasters have increased, but with large spatial and interannual variability (*high confidence*, based on *high agreement*, *medium evidence*). Global weather- and climate-related disaster losses reported over the last few decades reflect mainly monetized direct damages to assets, and are unequally distributed. Estimates of annual losses have ranged since 1980 from a few US$ billion to above 200 billion (in 2010 dollars), with the highest value for 2005 (the year of Hurricane Katrina). Loss estimates are lower-bound estimates because many impacts, such as loss of human lives, cultural heritage, and ecosystem services, are difficult to value and monetize, and thus they are poorly reflected in estimates of losses. Impacts on the informal or undocumented economy as well as indirect economic effects can be very important in some areas and sectors, but are generally not counted in reported estimates of losses. [4.5.1, 4.5.3, 4.5.4]

Economic, including insured, disaster losses associated with weather, climate, and geophysical events[4] are higher in developed countries. Fatality rates and economic losses expressed as a proportion of gross domestic product (GDP) are higher in developing countries (*high confidence*). During the period from 1970 to 2008, over 95% of deaths from natural disasters occurred in developing countries. Middle-income countries with rapidly expanding asset bases have borne the largest burden. During the period from 2001 to 2006, losses amounted to about 1% of GDP for middle-income countries, while this ratio has been about 0.3% of GDP for low-income countries and less than 0.1% of GDP for high-income countries, based on *limited evidence*. In small exposed countries, particularly small island developing states, losses expressed as a percentage of GDP have been particularly high, exceeding 1% in many cases and 8% in the most extreme cases, averaged over both disaster and non-disaster years for the period from 1970 to 2010. [4.5.2, 4.5.4]

Increasing exposure of people and economic assets has been the major cause of long-term increases in economic losses from weather- and climate-related disasters (*high confidence*). Long-term trends in economic disaster losses adjusted for wealth and population increases have not been attributed to climate change, but a role for climate change has not been excluded (*high agreement, medium evidence*). These conclusions are subject to a number of limitations in studies to date. Vulnerability is a key factor in disaster losses, yet it is not well accounted for. Other limitations are: (i) data availability, as most data are available for standard economic sectors in developed countries; and (ii) type of hazards studied, as most studies focus on cyclones, where confidence in observed trends and attribution of changes to human influence is *low*. The second conclusion is subject to additional limitations: (iii) the processes used to adjust loss data over time, and (iv) record length. [4.5.3]

---

[4] Economic losses and fatalities described in this paragraph pertain to all disasters associated with weather, climate, and geophysical events.

C. Disaster Risk Management and Adaptation to Climate Change: Past Experience with Climate Extremes

Past experience with climate extremes contributes to understanding of effective disaster risk management and adaptation approaches to manage risks.

The severity of the impacts of climate extremes depends strongly on the level of the exposure and vulnerability to these extremes (*high confidence*). [2.1.1, 2.3, 2.5]

Trends in exposure and vulnerability are major drivers of changes in disaster risk (*high confidence*). [2.5] Understanding the multi-faceted nature of both exposure and vulnerability is a prerequisite for determining how weather and climate events contribute to the occurrence of disasters, and for designing and implementing effective adaptation and disaster risk management strategies. [2.2, 2.6] Vulnerability reduction is a core common element of adaptation and disaster risk management. [2.2, 2.3]

Development practice, policy, and outcomes are critical to shaping disaster risk, which may be increased by shortcomings in development (*high confidence*). [1.1.2, 1.1.3] High exposure and vulnerability are generally the outcome of skewed development processes such as those associated with environmental degradation, rapid and unplanned urbanization in hazardous areas, failures of governance, and the scarcity of livelihood options for the poor. [2.2.2, 2.5] Increasing global interconnectivity and the mutual interdependence of economic and ecological systems can have sometimes contrasting effects, reducing or amplifying vulnerability and disaster risk. [7.2.1] Countries more effectively manage disaster risk if they include considerations of disaster risk in national development and sector plans and if they adopt climate change adaptation strategies, translating these plans and strategies into actions targeting vulnerable areas and groups. [6.2, 6.5.2]

Data on disasters and disaster risk reduction are lacking at the local level, which can constrain improvements in local vulnerability reduction (*high agreement, medium evidence*). [5.7] There are few examples of national disaster risk management systems and associated risk management measures explicitly integrating knowledge of and uncertainties in projected changes in exposure, vulnerability, and climate extremes. [6.6.2, 6.6.4]

Inequalities influence local coping and adaptive capacity, and pose disaster risk management and adaptation challenges from the local to national levels (*high agreement, robust evidence*). These inequalities reflect socioeconomic, demographic, and health-related differences and differences in governance, access to livelihoods, entitlements, and other factors. [5.5.1, 6.2] Inequalities also exist across countries: developed countries are often better equipped financially and institutionally to adopt explicit measures to effectively respond and adapt to projected changes in exposure, vulnerability, and climate extremes than are developing countries. Nonetheless, all countries face challenges in assessing, understanding, and responding to such projected changes. [6.3.2, 6.6]

Humanitarian relief is often required when disaster risk reduction measures are absent or inadequate (*high agreement, robust evidence*). [5.2.1] Smaller or economically less-diversified countries face particular challenges in providing the public goods associated with disaster risk management, in absorbing the losses caused by climate extremes and disasters, and in providing relief and reconstruction assistance. [6.4.3]

Post-disaster recovery and reconstruction provide an opportunity for reducing weather- and climate-related disaster risk and for improving adaptive capacity (*high agreement, robust evidence*). An emphasis on rapidly rebuilding houses, reconstructing infrastructure, and rehabilitating livelihoods often leads to recovering in ways that recreate or even increase existing vulnerabilities, and that preclude longer-term planning and policy changes for enhancing resilience and sustainable development. [5.2.3] See also assessment in Sections 8.4.1 and 8.5.2.

Risk sharing and transfer mechanisms at local, national, regional, and global scales can increase resilience to climate extremes (*medium confidence*). Mechanisms include informal and traditional risk sharing mechanisms, micro-insurance, insurance, reinsurance, and national, regional, and global risk pools. [5.6.3, 6.4.3, 6.5.3, 7.4] These mechanisms are linked to disaster risk reduction and climate change adaptation by providing means to finance relief, recovery of livelihoods, and reconstruction; reducing vulnerability; and providing knowledge and incentives for reducing risk. [5.5.2, 6.2.2] Under certain conditions, however, such mechanisms can provide disincentives for reducing disaster risk. [5.6.3, 6.5.3, 7.4.4] Uptake of formal risk sharing and transfer mechanisms is unequally distributed across regions and hazards. [6.5.3] See also Case Study 9.2.13.

Attention to the temporal and spatial dynamics of exposure and vulnerability is particularly important given that the design and implementation of adaptation and disaster risk management strategies and policies can reduce risk in the short term, but may increase exposure and vulnerability over the longer term (*high agreement, medium evidence*). For instance, dike systems can reduce flood exposure by offering immediate protection, but also encourage settlement patterns that may increase risk in the long term. [2.4.2, 2.5.4, 2.6.2] See also assessment in Sections 1.4.3, 5.3.2, and 8.3.1.

National systems are at the core of countries' capacity to meet the challenges of observed and projected trends in exposure, vulnerability, and weather and climate extremes (*high agreement, robust evidence*). Effective national systems comprise multiple actors from national and sub-national governments, the private sector, research bodies, and civil society including community-based organizations, playing differential but complementary roles to manage risk, according to their accepted functions and capacities. [6.2]

Closer integration of disaster risk management and climate change adaptation, along with the incorporation of both into local, sub-national, national, and international development policies and practices, could provide benefits at all scales (*high agreement, medium evidence*). [5.4, 5.5, 5.6, 6.3.1, 6.3.2, 6.4.2, 6.6, 7.4] Addressing social welfare, quality of life, infrastructure, and livelihoods, and incorporating a multi-hazards approach into planning and action for disasters in the short term, facilitates adaptation to climate extremes in the longer term, as is increasingly recognized internationally. [5.4, 5.5, 5.6, 7.3] Strategies and policies are more effective when they acknowledge multiple stressors, different prioritized values, and competing policy goals. [8.2, 8.3, 8.7]

D. Future Climate Extremes, Impacts, and Disaster Losses

Future changes in exposure, vulnerability, and climate extremes resulting from natural climate variability, anthropogenic climate change, and socioeconomic development can alter the impacts of climate extremes on natural and human systems and the potential for disasters.

Climate Extremes and Impacts

Confidence in projecting changes in the direction and magnitude of climate extremes depends on many factors, including the type of extreme, the region and season, the amount and quality of observational data, the level of understanding of the underlying processes, and the reliability of their simulation in models. Projected changes in climate extremes under different emissions scenarios[5] generally do not strongly diverge in the coming two to three decades, but these signals are relatively small compared to natural climate variability over this time frame. Even the sign of projected changes in some climate extremes over this time frame is uncertain. For projected changes by the end of the 21st century, either model uncertainty or uncertainties associated with emissions scenarios used becomes dominant, depending on the extreme. Low-probability, high-impact changes associated with

---

[5] Emissions scenarios for radiatively important substances result from pathways of socioeconomic and technological development. This report uses a subset (B1, A1B, A2) of the 40 scenarios extending to the year 2100 that are described in the IPCC Special Report on Emissions Scenarios (SRES) and that did not include additional climate initiatives. These scenarios have been widely used in climate change projections and encompass a substantial range of carbon dioxide equivalent concentrations, but not the entire range of the scenarios included in the SRES.

Summary for Policymakers

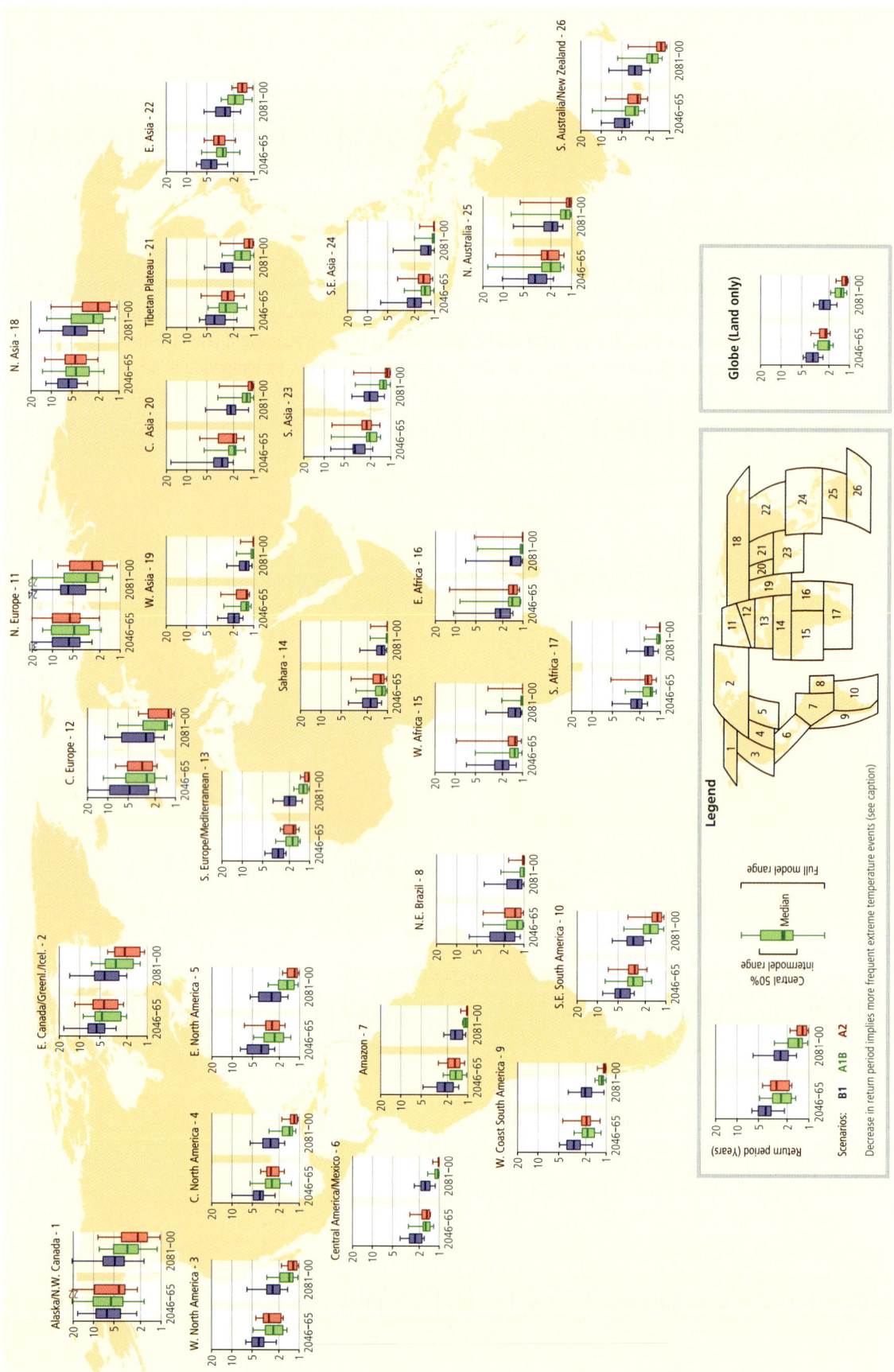

Figure SPM.4A | Projected return periods for the maximum daily temperature that was exceeded on average once during a 20-year period in the late 20th century (1981–2000). A decrease in return period implies more frequent extreme temperature events (i.e., less time between events on average). The box plots show results for regionally averaged projections for two time horizons, 2046 to 2065 and 2081 to 2100, as compared to the late 20th century, and for three different SRES emissions scenarios (B1, A1B, A2) (see legend). Results are based on 12 global climate models (GCMs) contributing to the third phase of the Coupled Model Intercomparison Project (CMIP3). The level of agreement among the models is indicated by the size of the colored boxes (in which 50% of the model projections are contained), and the length of the whiskers (indicating the maximum and minimum projections from all models). See legend for defined extent of regions. Values are computed for land points only. The 'Globe' inset box displays the values computed using all land grid points. [3.3.1, Figure 3-1, Figure 3-5]

the crossing of poorly understood climate thresholds cannot be excluded, given the transient and complex nature of the climate system. Assigning 'low confidence' for projections of a specific extreme neither implies nor excludes the possibility of changes in this extreme. The following assessments of the likelihood and/or confidence of projections are generally for the end of the 21st century and relative to the climate at the end of the 20th century. [3.1.5, 3.1.7, 3.2.3, Box 3-2]

Models project substantial warming in temperature extremes by the end of the 21st century. It is *virtually certain* that increases in the frequency and magnitude of warm daily temperature extremes and decreases in cold extremes will occur in the 21st century at the global scale. It is *very likely* that the length, frequency, and/or intensity of warm spells or heat waves will increase over most land areas. Based on the A1B and A2 emissions scenarios, a 1-in-20 year hottest day is *likely* to become a 1-in-2 year event by the end of the 21st century in most regions, except in the high latitudes of the Northern Hemisphere, where it is *likely* to become a 1-in-5 year event (see Figure SPM.4A). Under the B1 scenario, a 1-in-20 year event would *likely* become a 1-in-5 year event (and a 1-in-10 year event in Northern Hemisphere high latitudes). The 1-in-20 year extreme daily maximum temperature (i.e., a value that was exceeded on average only once during the period 1981–2000) will *likely* increase by about 1°C to 3°C by the mid-21st century and by about 2°C to 5°C by the late 21st century, depending on the region and emissions scenario (based on the B1, A1B, and A2 scenarios). [3.3.1, 3.1.6, Table 3-3, Figure 3-5]

It is *likely* that the frequency of heavy precipitation or the proportion of total rainfall from heavy falls will increase in the 21st century over many areas of the globe. This is particularly the case in the high latitudes and tropical regions, and in winter in the northern mid-latitudes. Heavy rainfalls associated with tropical cyclones are *likely* to increase with continued warming. There is *medium confidence* that, in some regions, increases in heavy precipitation will occur despite projected decreases in total precipitation in those regions. Based on a range of emissions scenarios (B1, A1B, A2), a 1-in-20 year annual maximum daily precipitation amount is *likely* to become a 1-in-5 to 1-in-15 year event by the end of the 21st century in many regions, and in most regions the higher emissions scenarios (A1B and A2) lead to a stronger projected decrease in return period. See Figure SPM.4B. [3.3.2, 3.4.4, Table 3-3, Figure 3-7]

Average tropical cyclone maximum wind speed is *likely* to increase, although increases may not occur in all ocean basins. It is *likely* that the global frequency of tropical cyclones will either decrease or remain essentially unchanged. [3.4.4]

There is *medium confidence* that there will be a reduction in the number of extratropical cyclones averaged over each hemisphere. While there is *low confidence* in the detailed geographical projections of extratropical cyclone activity, there is *medium confidence* in a projected poleward shift of extratropical storm tracks. There is *low confidence* in projections of small spatial-scale phenomena such as tornadoes and hail because competing physical processes may affect future trends and because current climate models do not simulate such phenomena. [3.3.2, 3.3.3, 3.4.5]

There is *medium confidence* that droughts will intensify in the 21st century in some seasons and areas, due to reduced precipitation and/or increased evapotranspiration. This applies to regions including southern Europe and the Mediterranean region, central Europe, central North America, Central America and Mexico, northeast Brazil, and southern Africa. Elsewhere there is overall *low confidence* because of inconsistent projections of drought changes (dependent both on model and dryness index). Definitional issues, lack of observational data, and the inability of models to include all the factors that influence droughts preclude stronger confidence than *medium* in drought projections. See Figure SPM.5. [3.5.1, Table 3-3, Box 3-3]

Projected precipitation and temperature changes imply possible changes in floods, although overall there is *low confidence* in projections of changes in fluvial floods. Confidence is *low* due to *limited evidence* and because the causes of regional changes are complex, although there are exceptions to this statement. There is *medium confidence* (based on physical reasoning) that projected increases in heavy rainfall would contribute to increases in local flooding in some catchments or regions. [3.5.2]

Summary for Policymakers

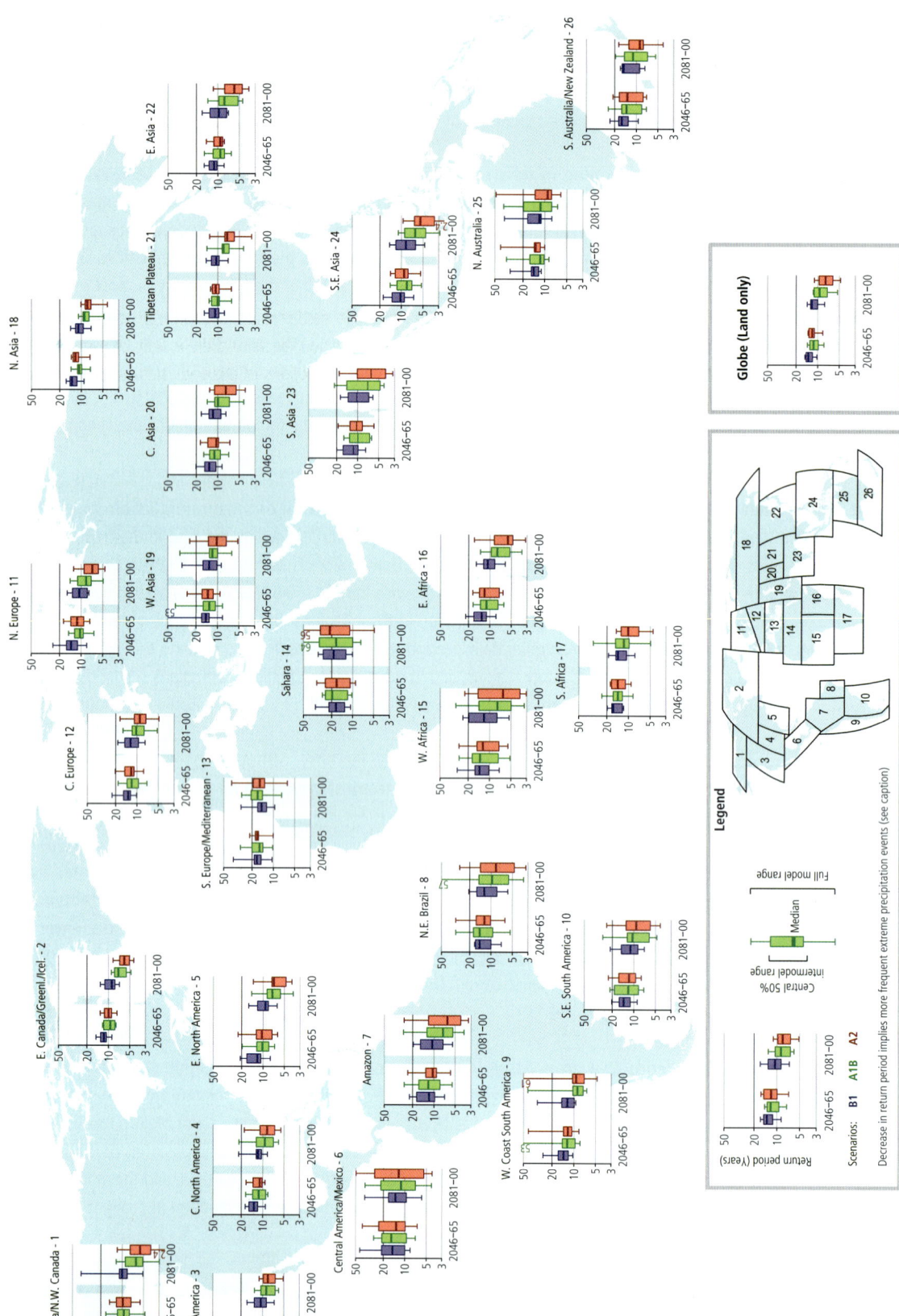

Figure SPM.4B | Projected return periods for a daily precipitation event that was exceeded in the late 20th century on average once during a 20-year period (1981–2000). A decrease in return period implies more frequent extreme precipitation events (i.e., less time between events on average). The box plots show results for regionally averaged projections for two time horizons, 2046 to 2065 and 2081 to 2100, as compared to the late 20th century, and for three different SRES emissions scenarios (B1, A1B, A2) (see legend). Results are based on 14 GCMs contributing to the CMIP3. The level of agreement among the models is indicated by the size of the colored boxes (in which 50% of the model projections are contained), and the length of the whiskers (indicating the maximum and minimum projections from all models). Values are computed for land points only. The 'Globe' inset box displays the values computed using all land grid points. See legend for defined extent of regions. [3.3.2, Figure 3-1, Figure 3-7]

It is *very likely* that mean sea level rise will contribute to upward trends in extreme coastal high water levels in the future. There is *high confidence* that locations currently experiencing adverse impacts such as coastal erosion and inundation will continue to do so in the future due to increasing sea levels, all other contributing factors being equal. The *very likely* contribution of mean sea level rise to increased extreme coastal high water levels, coupled with the *likely* increase in tropical cyclone maximum wind speed, is a specific issue for tropical small island states. [3.5.3, 3.5.5, Box 3-4]

There is *high confidence* that changes in heat waves, glacial retreat, and/or permafrost degradation will affect high mountain phenomena such as slope instabilities, movements of mass, and glacial lake outburst floods. There is also *high confidence* that changes in heavy precipitation will affect landslides in some regions. [3.5.6]

There is *low confidence* in projections of changes in large-scale patterns of natural climate variability. Confidence is *low* in projections of changes in monsoons (rainfall, circulation) because there is little consensus in climate models regarding the sign of future change in the monsoons. Model projections of changes in El Niño–Southern

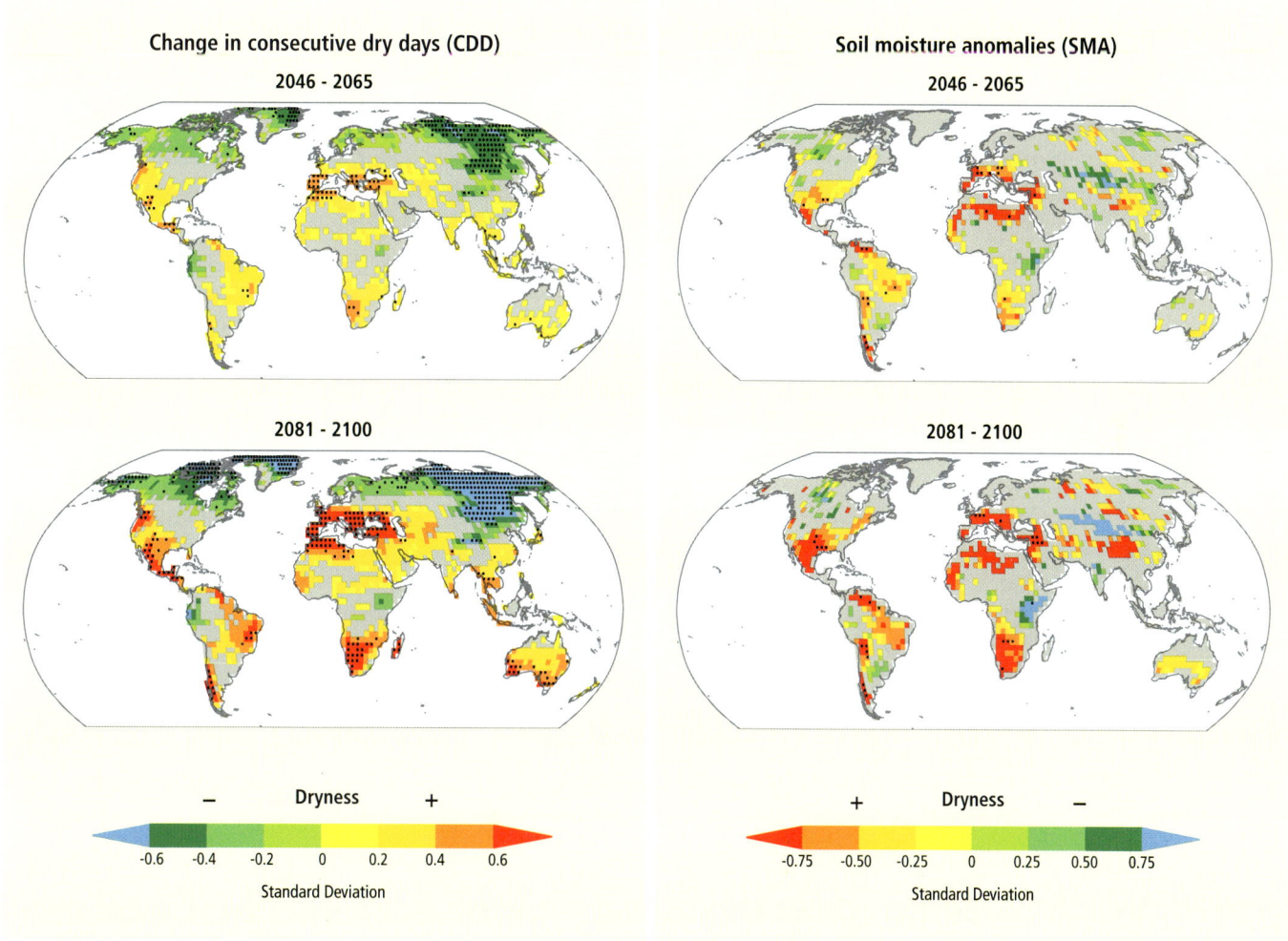

Figure SPM.5 | Projected annual changes in dryness assessed from two indices. Left column: Change in annual maximum number of consecutive dry days (CDD: days with precipitation <1 mm). Right column: Changes in soil moisture (soil moisture anomalies, SMA). Increased dryness is indicated with yellow to red colors; decreased dryness with green to blue. Projected changes are expressed in units of standard deviation of the interannual variability in the three 20-year periods 1980–1999, 2046–2065, and 2081–2100. The figures show changes for two time horizons, 2046–2065 and 2081–2100, as compared to late 20th-century values (1980–1999), based on GCM simulations under emissions scenario SRES A2 relative to corresponding simulations for the late 20th century. Results are based on 17 (CDD) and 15 (SMA) GCMs contributing to the CMIP3. Colored shading is applied for areas where at least 66% (12 out of 17 for CDD, 10 out of 15 for SMA) of the models agree on the sign of the change; stippling is added for regions where at least 90% (16 out of 17 for CDD, 14 out of 15 for SMA) of all models agree on the sign of the change. Grey shading indicates where there is insufficient model agreement (<66%). [3.5.1, Figure 3-9]

Oscillation variability and the frequency of El Niño episodes are not consistent, and so there is *low confidence* in projections of changes in this phenomenon. [3.4.1, 3.4.2, 3.4.3]

Human Impacts and Disaster Losses

Extreme events will have greater impacts on sectors with closer links to climate, such as water, agriculture and food security, forestry, health, and tourism. For example, while it is not currently possible to reliably project specific changes at the catchment scale, there is *high confidence* that changes in climate have the potential to seriously affect water management systems. However, climate change is in many instances only one of the drivers of future changes, and is not necessarily the most important driver at the local scale. Climate-related extremes are also expected to produce large impacts on infrastructure, although detailed analysis of potential and projected damages are limited to a few countries, infrastructure types, and sectors. [4.3.2, 4.3.5]

In many regions, the main drivers of future increases in economic losses due to some climate extremes will be socioeconomic in nature (*medium confidence*, based on *medium agreement, limited evidence*). Climate extremes are only one of the factors that affect risks, but few studies have specifically quantified the effects of changes in population, exposure of people and assets, and vulnerability as determinants of loss. However, the few studies available generally underline the important role of projected changes (increases) in population and capital at risk. [4.5.4]

Increases in exposure will result in higher direct economic losses from tropical cyclones. Losses will also depend on future changes in tropical cyclone frequency and intensity (*high confidence*). Overall losses due to extratropical cyclones will also increase, with possible decreases or no change in some areas (*medium confidence*). Although future flood losses in many locations will increase in the absence of additional protection measures (*high agreement, medium evidence*), the size of the estimated change is highly variable, depending on location, climate scenarios used, and methods used to assess impacts on river flow and flood occurrence. [4.5.4]

Disasters associated with climate extremes influence population mobility and relocation, affecting host and origin communities (*medium agreement, medium evidence*). If disasters occur more frequently and/or with greater magnitude, some local areas will become increasingly marginal as places to live or in which to maintain livelihoods. In such cases, migration and displacement could become permanent and could introduce new pressures in areas of relocation. For locations such as atolls, in some cases it is possible that many residents will have to relocate. [5.2.2]

E. Managing Changing Risks of Climate Extremes and Disasters

Adaptation to climate change and disaster risk management provide a range of complementary approaches for managing the risks of climate extremes and disasters (Figure SPM.2). Effectively applying and combining approaches may benefit from considering the broader challenge of sustainable development.

Measures that provide benefits under current climate and a range of future climate change scenarios, called low-regrets measures, are available starting points for addressing projected trends in exposure, vulnerability, and climate extremes. They have the potential to offer benefits now and lay the foundation for addressing projected changes (*high agreement, medium evidence*). Many of these low-regrets strategies produce co-benefits, help address other development goals, such as improvements in livelihoods, human well-being, and biodiversity conservation, and help minimize the scope for maladaptation. [6.3.1, Table 6-1]

Potential low-regrets measures include early warning systems; risk communication between decisionmakers and local citizens; sustainable land management, including land use planning; and ecosystem management and restoration.

Other low-regrets measures include improvements to health surveillance, water supply, sanitation, and irrigation and drainage systems; climate-proofing of infrastructure; development and enforcement of building codes; and better education and awareness. [5.3.1, 5.3.3, 6.3.1, 6.5.1, 6.5.2] See also Case Studies 9.2.11 and 9.2.14, and assessment in Section 7.4.3.

Effective risk management generally involves a portfolio of actions to reduce and transfer risk and to respond to events and disasters, as opposed to a singular focus on any one action or type of action (*high confidence*). [1.1.2, 1.1.4, 1.3.3] Such integrated approaches are more effective when they are informed by and customized to specific local circumstances (*high agreement, robust evidence*). [5.1] Successful strategies include a combination of hard infrastructure-based responses and soft solutions such as individual and institutional capacity building and ecosystem-based responses. [6.5.2]

Multi-hazard risk management approaches provide opportunities to reduce complex and compound hazards (*high agreement, robust evidence*). Considering multiple types of hazards reduces the likelihood that risk reduction efforts targeting one type of hazard will increase exposure and vulnerability to other hazards, in the present and future. [8.2.5, 8.5.2, 8.7]

Opportunities exist to create synergies in international finance for disaster risk management and adaptation to climate change, but these have not yet been fully realized (*high confidence*). International funding for disaster risk reduction remains relatively low as compared to the scale of spending on international humanitarian response. [7.4.2] Technology transfer and cooperation to advance disaster risk reduction and climate change adaptation are important. Coordination on technology transfer and cooperation between these two fields has been lacking, which has led to fragmented implementation. [7.4.3]

Stronger efforts at the international level do not necessarily lead to substantive and rapid results at the local level (*high confidence*). There is room for improved integration across scales from international to local. [7.6]

Integration of local knowledge with additional scientific and technical knowledge can improve disaster risk reduction and climate change adaptation (*high agreement, robust evidence*). Local populations document their experiences with the changing climate, particularly extreme weather events, in many different ways, and this self-generated knowledge can uncover existing capacity within the community and important current shortcomings. [5.4.4] Local participation supports community-based adaptation to benefit management of disaster risk and climate extremes. However, improvements in the availability of human and financial capital and of disaster risk and climate information customized for local stakeholders can enhance community-based adaptation (*medium agreement, medium evidence*). [5.6]

Appropriate and timely risk communication is critical for effective adaptation and disaster risk management (*high confidence*). Explicit characterization of uncertainty and complexity strengthens risk communication. [2.6.3] Effective risk communication builds on exchanging, sharing, and integrating knowledge about climate-related risks among all stakeholder groups. Among individual stakeholders and groups, perceptions of risk are driven by psychological and cultural factors, values, and beliefs. [1.1.4, 1.3.1, 1.4.2] See also assessment in Section 7.4.5.

An iterative process of monitoring, research, evaluation, learning, and innovation can reduce disaster risk and promote adaptive management in the context of climate extremes (*high agreement*, *robust evidence*). [8.6.3, 8.7] Adaptation efforts benefit from iterative risk management strategies because of the complexity, uncertainties, and long time frame associated with climate change (*high confidence*). [1.3.2] Addressing knowledge gaps through enhanced observation and research can reduce uncertainty and help in designing effective adaptation and risk management strategies. [3.2, 6.2.5, Table 6-3, 7.5, 8.6.3] See also assessment in Section 6.6.

Table SPM.1 presents examples of how observed and projected trends in exposure, vulnerability, and climate extremes can inform risk management and adaptation strategies, policies, and measures. The

Summary for Policymakers

Table SPM.1 | Illustrative examples of options for risk management and adaptation in the context of changes in exposure, vulnerability, and climate extremes. In each example, information is characterized at the scale directly relevant to decisionmaking. Observed and projected changes in climate extremes at global and regional scales illustrate that the direction, magnitude of, and/or degree of certainty for changes may differ across scales.

The examples were selected based on availability of evidence in the underlying chapters, including on exposure, vulnerability, climate information, and risk management and adaptation options. They are intended to reflect relevant risk management themes and scales, rather than to provide comprehensive information by region. The examples are not intended to reflect any regional differences in exposure and vulnerability, or in experience in risk management.

The confidence in projected changes in climate extremes at local scales is often more limited than the confidence in projected regional and global changes. This limited confidence in changes places a focus on low-regrets risk management options that aim to reduce exposure and vulnerability and to increase resilience and preparedness for risks that cannot be entirely eliminated. Higher-confidence projected changes in climate extremes, at a scale relevant to adaptation and risk management decisions, can inform more targeted adjustments in strategies, policies, and measures. [3.1.6, Box 3-2, 6.3.1, 6.5.2]

Example	Exposure and vulnerability at scale of risk management in the example	Information on Climate Extreme Across Spatial Scales		SCALE OF RISK MANAGEMENT Available information for the example	Options for risk management and adaptation in the example
		GLOBAL Observed (since 1950) and projected (to 2100) global changes	**REGIONAL** Observed (since 1950) and projected (to 2100) changes in the example		
Inundation related to **extreme sea levels** in tropical small island developing states	Small island states in the Pacific, Indian, and Atlantic Oceans, often with low elevation, are particularly vulnerable to rising sea levels and impacts such as erosion, inundation, shoreline change, and saltwater intrusion into coastal aquifers. These impacts can result in ecosystem disruption, decreased agricultural productivity, changes in disease patterns, economic losses such as in tourism industries, and population displacement — all of which reinforce vulnerability to extreme weather events. [3.5.5, Box 3-4, 4.3.5, 4.4.10, 9.2.9]	**Observed:** *Likely* increase in extreme coastal high water worldwide related to increases in mean sea level. **Projected:** *Very likely* that mean sea level rise will contribute to upward trends in extreme coastal high water levels. *High confidence* that locations currently experiencing coastal erosion and inundation will continue to do so due to increasing sea level, in the absence of changes in other contributing factors. *Likely* that the global frequency of tropical cyclones will either decrease or remain essentially unchanged. *Likely* increase in average tropical cyclone maximum wind speed, although increases may not occur in all ocean basins. [Table 3-1, 3.4.4, 3.5.3, 3.5.5]	**Observed:** Tides and El Niño–Southern Oscillation have contributed to the more frequent occurrence of extreme coastal high water levels and associated flooding experienced on some Pacific Islands in recent years. **Projected:** The *very likely* contribution of mean sea level rise to increased extreme coastal high water levels, coupled with the *likely* increase in tropical cyclone maximum wind speed, is a specific issue for tropical small island states. See global changes column for information on global projections for tropical cyclones. [Box 3-4, 3.4.4, 3.5.3]	Sparse regional and temporal coverage of terrestrial-based observation networks and limited in situ ocean observing network, but with improved satellite-based observations in recent decades. While changes in storminess may contribute to changes in extreme coastal high water levels, the limited geographical coverage of studies to date and the uncertainties associated with storminess changes overall mean that a general assessment of the effects of storminess changes on storm surge is not possible at this time. [Box 3-4, 3.5.3]	Low-regrets options that reduce exposure and vulnerability across a range of hazard trends: • Maintenance of drainage systems • Well technologies to limit saltwater contamination of groundwater • Improved early warning systems • Regional risk pooling • Mangrove conservation, restoration, and replanting Specific adaptation options include, for instance, rendering national economies more climate-independent and adaptive management involving iterative learning. In some cases there may be a need to consider relocation, for example, for atolls where storm surges may completely inundate them. [4.3.5, 4.4.10, 5.2.2, 6.3.2, 6.5.2, 6.6.2, 7.4.4, 9.2.9, 9.2.11, 9.2.13]
Flash floods in informal settlements in Nairobi, Kenya	Rapid expansion of poor people living in informal settlements around Nairobi has led to houses of weak building materials being constructed immediately adjacent to rivers and to blockage of natural drainage areas, increasing exposure and vulnerability. [6.4.2, Box 6-2]	**Observed:** *Low confidence* at global scale regarding (climate-driven) observed changes in the magnitude and frequency of floods. **Projected:** *Low confidence* in projections of changes in floods because of limited evidence and because the causes of regional changes are complex. However, *medium confidence* (based on physical reasoning) that projected increases in heavy precipitation will contribute to rain-generated local flooding in some catchments or regions. [Table 3-1, 3.5.2]	**Observed:** *Low confidence* regarding trends in heavy precipitation in East Africa, because of insufficient evidence. **Projected:** *Likely* increase in heavy precipitation indicators in East Africa. [Table 3-2, Table 3-3, 3.3.2]	Limited ability to provide local flash flood projections. [3.5.2]	Low-regrets options that reduce exposure and vulnerability across a range of hazard trends: • Strengthening building design and regulation • Poverty reduction schemes • City-wide drainage and sewerage improvements The Nairobi Rivers Rehabilitation and Restoration Programme includes installation of riparian buffers, canals, and drainage channels and clearance of existing channels; attention to climate variability and change in the location and design of wastewater infrastructure; and environmental monitoring for flood early warning. [6.3, 6.4.2, Box 6-6]

Continued next page →

18

Table SPM.1 (continued)

Example	Exposure and vulnerability at scale of risk management in the example	Information on Climate Extreme Across Spatial Scales		SCALE OF RISK MANAGEMENT Available information for the example	Options for risk management and adaptation in the example
		GLOBAL Observed (since 1950) and projected (to 2100) global changes	REGIONAL Observed (since 1950) and projected (to 2100) changes in the example		
Impacts of **heat waves** in urban areas in Europe	Factors affecting exposure and vulnerability include age, pre-existing health status, level of outdoor activity, socioeconomic factors including poverty and social isolation, access to and use of cooling, physiological and behavioral adaptation of the population, and urban infrastructure. [2.5.2, 4.3.5, 4.3.6, 4.4.5, 9.2.1]	**Observed:** *Medium confidence* that the length or number of warm spells or heat waves has increased since the middle of the 20th century, in many (but not all) regions over the globe. *Very likely* increase in number of warm days and nights at the global scale. **Projected:** *Very likely* increase in length, frequency, and/or intensity of warm spells or heat waves over most land areas. *Virtually certain* increase in frequency and magnitude of warm days and nights at the global scale. [Table 3-1, 3.3.1]	**Observed:** *Medium confidence* in increase in heat waves or warm spells in Europe. *Likely* overall increase in warm days and nights over most of the continent. **Projected:** *Likely* more frequent, longer, and/or more intense heat waves or warm spells in Europe. *Very likely* increase in warm days and nights. [Table 3-2, Table 3-3, 3.3.1]	Observations and projections can provide information for specific urban areas in the region, with increased heat waves expected due to regional trends and urban heat island effects. [3.3.1, 4.4.5]	Low-regrets options that reduce exposure and vulnerability across a range of hazard trends: • Early warning systems that reach particularly vulnerable groups (e.g., the elderly) • Vulnerability mapping and corresponding measures • Public information on what to do during heat waves, including behavioral advice • Use of social care networks to reach vulnerable groups Specific adjustments in strategies, policies, and measures informed by trends in heat waves include awareness raising of heat waves as a public health concern; changes in urban infrastructure and land use planning, for example, increasing urban green space; changes in approaches to cooling for public facilities; and adjustments in energy generation and transmission infrastructure. [Table 6-1, 9.2.1]
Increasing losses from **hurricanes** in the USA and the Caribbean	Exposure and vulnerability are increasing due to growth in population and increase in property values, particularly along the Gulf and Atlantic coasts of the United States. Some of this increase has been offset by improved building codes. [4.4.6]	**Observed:** *Low confidence* in any observed long-term (i.e., 40 years or more) increases in tropical cyclone activity, after accounting for past changes in observing capabilities. **Projected:** *Likely* that the global frequency of tropical cyclones will either decrease or remain essentially unchanged. *Likely* increase in average tropical cyclone maximum wind speed, although increases may not occur in all ocean basins. Heavy rainfalls associated with tropical cyclones are *likely* to increase. Projected sea level rise is expected to further compound tropical cyclone surge impacts. [Table 3-1, 3.4.4]	See global changes column for global projections.	Limited model capability to project changes relevant to specific settlements or other locations, due to the inability of global models to accurately simulate factors relevant to tropical cyclone genesis, track, and intensity evolution. [3.4.4]	Low-regrets options that reduce exposure and vulnerability across a range of hazard trends: • Adoption and enforcement of improved building codes • Improved forecasting capacity and implementation of improved early warning systems (including evacuation plans and infrastructures) • Regional risk pooling In the context of high underlying variability and uncertainty regarding trends, options can include emphasizing adaptive management involving learning and flexibility (e.g, Cayman Islands National Hurricane Committee). [5.5.3, 6.5.2, 6.6.2, Box 6-7, Table 6-1, 7.4.4, 9.2.5, 9.2.11, 9.2.13]
Droughts in the context of food security in West Africa	Less advanced agricultural practices render region vulnerable to increasing variability in seasonal rainfall, drought, and weather extremes. Vulnerability is exacerbated by population growth, degradation of ecosystems, and overuse of natural resources, as well as poor standards for health, education, and governance. [2.2.2, 2.3, 2.5, 4.4.2, 9.2.3]	**Observed:** *Medium confidence* that some regions of the world have experienced more intense and longer droughts, but in some regions droughts have become less frequent, less intense, or shorter. **Projected:** *Medium confidence* in projected intensification of drought in some seasons and areas. Elsewhere there is overall *low confidence* because of inconsistent projections. [Table 3-1, 3.5.1]	**Observed:** *Medium confidence* in an increase in dryness. Recent years characterized by greater interannual variability than previous 40 years, with the western Sahel remaining dry and the eastern Sahel returning to wetter conditions. **Projected:** *Low confidence* due to inconsistent signal in model projections. [Table 3-2, Table 3-3, 3.5.1]	Sub-seasonal, seasonal, and interannual forecasts with increasing uncertainty over longer time scales. Improved monitoring, instrumentation, and data associated with early warning systems, but with limited participation and dissemination to at-risk populations. [5.3.1, 5.5.3, 7.3.1, 9.2.3, 9.2.11]	Low-regrets options that reduce exposure and vulnerability across a range of hazard trends: • Traditional rain and groundwater harvesting and storage systems • Water demand management and improved irrigation efficiency measures • Conservation agriculture, crop rotation, and livelihood diversification • Increasing use of drought-resistant crop varieties • Early warning systems integrating seasonal forecasts with drought projections, with improved communication involving extension services • Risk pooling at the regional or national level [2.5.4, 5.3.1, 5.3.3, 6.5, Table 6-3, 9.2.3, 9.2.11]

importance of these trends for decisionmaking depends on their magnitude and degree of certainty at the temporal and spatial scale of the risk being managed and on the available capacity to implement risk management options (see Table SPM.1).

Implications for Sustainable Development

Actions that range from incremental steps to transformational changes are essential for reducing risk from climate extremes (*high agreement, robust evidence*). Incremental steps aim to improve efficiency within existing technological, governance, and value systems, whereas transformation may involve alterations of fundamental attributes of those systems. Transformations, where they are required, are also facilitated through increased emphasis on adaptive management and learning. Where vulnerability is high and adaptive capacity low, changes in climate extremes can make it difficult for systems to adapt sustainably without transformational changes. Vulnerability is often concentrated in lower-income countries or groups, although higher-income countries or groups can also be vulnerable to climate extremes. [8.6, 8.6.3, 8.7]

Social, economic, and environmental sustainability can be enhanced by disaster risk management and adaptation approaches. A prerequisite for sustainability in the context of climate change is addressing the underlying causes of vulnerability, including the structural inequalities that create and sustain poverty and constrain access to resources (*medium agreement, robust evidence*). This involves integrating disaster risk management and adaptation into all social, economic, and environmental policy domains. [8.6.2, 8.7]

The most effective adaptation and disaster risk reduction actions are those that offer development benefits in the relatively near term, as well as reductions in vulnerability over the longer term (*high agreement, medium evidence*). There are tradeoffs between current decisions and long-term goals linked to diverse values, interests, and priorities for the future. Short- and long-term perspectives on disaster risk management and adaptation to climate change thus can be difficult to reconcile. Such reconciliation involves overcoming the disconnect between local risk management practices and national institutional and legal frameworks, policy, and planning. [8.2.1, 8.3.1, 8.3.2, 8.6.1]

Progress toward resilient and sustainable development in the context of changing climate extremes can benefit from questioning assumptions and paradigms and stimulating innovation to encourage new patterns of response (*medium agreement, robust evidence*). Successfully addressing disaster risk, climate change, and other stressors often involves embracing broad participation in strategy development, the capacity to combine multiple perspectives, and contrasting ways of organizing social relations. [8.2.5, 8.6.3, 8.7]

The interactions among climate change mitigation, adaptation, and disaster risk management may have a major influence on resilient and sustainable pathways (*high agreement, limited evidence*). Interactions between the goals of mitigation and adaptation in particular will play out locally, but have global consequences. [8.2.5, 8.5.2]

There are many approaches and pathways to a sustainable and resilient future. [8.2.3, 8.4.1, 8.6.1, 8.7] However, limits to resilience are faced when thresholds or tipping points associated with social and/or natural systems are exceeded, posing severe challenges for adaptation. [8.5.1] Choices and outcomes for adaptive actions to climate events must reflect divergent capacities and resources and multiple interacting processes. Actions are framed by tradeoffs between competing prioritized values and objectives, and different visions of development that can change over time. Iterative approaches allow development pathways to integrate risk management so that diverse policy solutions can be considered, as risk and its measurement, perception, and understanding evolve over time. [8.2.3, 8.4.1, 8.6.1, 8.7]

Box SPM.2 | Treatment of Uncertainty

Based on the Guidance Note for Lead Authors of the IPCC Fifth Assessment Report on Consistent Treatment of Uncertainties,[6] this Summary for Policymakers relies on two metrics for communicating the degree of certainty in key findings, which is based on author teams' evaluations of underlying scientific understanding:
- Confidence in the validity of a finding, based on the type, amount, quality, and consistency of evidence (e.g., mechanistic understanding, theory, data, models, expert judgment) and the degree of agreement. Confidence is expressed qualitatively.
- Quantified measures of uncertainty in a finding expressed probabilistically (based on statistical analysis of observations or model results, or expert judgment).

This Guidance Note refines the guidance provided to support the IPCC Third and Fourth Assessment Reports. Direct comparisons between assessment of uncertainties in findings in this report and those in the IPCC Fourth Assessment Report are difficult if not impossible, because of the application of the revised guidance note on uncertainties, as well as the availability of new information, improved scientific understanding, continued analyses of data and models, and specific differences in methodologies applied in the assessed studies. For some extremes, different aspects have been assessed and therefore a direct comparison would be inappropriate.

Each key finding is based on an author team's evaluation of associated evidence and agreement. The confidence metric provides a qualitative synthesis of an author team's judgment about the validity of a finding, as determined through evaluation of evidence and agreement. If uncertainties can be quantified probabilistically, an author team can characterize a finding using the calibrated likelihood language or a more precise presentation of probability. Unless otherwise indicated, *high* or *very high confidence* is associated with findings for which an author team has assigned a likelihood term.

The following summary terms are used to describe the available evidence: *limited*, *medium*, or *robust*; and for the degree of agreement: *low*, *medium*, or *high*. A level of confidence is expressed using five qualifiers: *very low*, *low*, *medium*, *high*, and *very high*. The accompanying figure depicts summary statements for evidence and agreement and their relationship to confidence. There is flexibility in this relationship; for a given evidence and agreement statement, different confidence levels can be assigned, but increasing levels of evidence and degrees of agreement are correlated with increasing confidence.

A depiction of evidence and agreement statements and their relationship to confidence. Confidence increases toward the top-right corner as suggested by the increasing strength of shading. Generally, evidence is most robust when there are multiple, consistent independent lines of high-quality evidence.

The following terms indicate the assessed likelihood:

Term*	Likelihood of the Outcome
Virtually certain	99–100% probability
Very likely	90–100% probability
Likely	66–100% probability
About as likely as not	33–66% probability
Unlikely	0–33% probability
Very unlikely	0–10% probability
Exceptionally unlikely	0–1% probability

* Additional terms that were used in limited circumstances in the Fourth Assessment Report (*extremely likely*: 95–100% probability, *more likely than not*: >50–100% probability, and *extremely unlikely*: 0–5% probability) may also be used when appropriate.

---

[6] Mastrandrea, M.D., C.B. Field, T.F. Stocker, O. Edenhofer, K.L. Ebi, D.J. Frame, H. Held, E. Kriegler, K.J. Mach, P.R. Matschoss, G.-K. Plattner, G.W. Yohe, and F.W. Zwiers, 2010: *Guidance Note for Lead Authors of the IPCC Fifth Assessment Report on Consistent Treatment of Uncertainties*. Intergovernmental Panel on Climate Change (IPCC), Geneva, Switzerland, www.ipcc.ch.

III

Chapters 1 to 9

1

Climate Change: New Dimensions in Disaster Risk, Exposure, Vulnerability, and Resilience

Coordinating Lead Authors:
Allan Lavell (Costa Rica), Michael Oppenheimer (USA)

Lead Authors:
Cherif Diop (Senegal), Jeremy Hess (USA), Robert Lempert (USA), Jianping Li (China), Robert Muir-Wood (UK), Soojeong Myeong (Republic of Korea)

Review Editors:
Susanne Moser (USA), Kuniyoshi Takeuchi (Japan)

Contributing Authors:
Omar-Dario Cardona (Colombia), Stephane Hallegatte (France), Maria Lemos (USA), Christopher Little (USA), Alexander Lotsch (USA), Elke Weber (USA)

This chapter should be cited as:

Lavell, A., M. Oppenheimer, C. Diop, J. Hess, R. Lempert, J. Li, R. Muir-Wood, and S. Myeong, 2012: Climate change: new dimensions in disaster risk, exposure, vulnerability, and resilience. In: *Managing the Risks of Extreme Events and Disasters to Advance Climate Change Adaptation* [Field, C.B., V. Barros, T.F. Stocker, D. Qin, D.J. Dokken, K.L. Ebi, M.D. Mastrandrea, K.J. Mach, G.-K. Plattner, S.K. Allen, M. Tignor, and P.M. Midgley (eds.)]. A Special Report of Working Groups I and II of the Intergovernmental Panel on Climate Change (IPCC). Cambridge University Press, Cambridge, UK, and New York, NY, USA, pp. 25-64.

Table of Contents

Executive Summary 27

1.1. Introduction 29
1.1.1. Purpose and Scope of the Special Report 29
1.1.2. Key Concepts and Definitions 30
1.1.2.1. Definitions Related to General Concepts 30
1.1.2.2. Concepts and Definitions Relating to Disaster Risk Management and Adaptation to Climate Change 34
1.1.2.3. The Social Construction of Disaster Risk 36
1.1.3. Framing the Relation between Adaptation to Climate Change and Disaster Risk Management 37
1.1.4. Framing the Processes of Disaster Risk Management and Adaptation to Climate Change 38
1.1.4.1. Exceptionality, Routine, and Everyday Life 38
1.1.4.2. Territorial Scale, Disaster Risk, and Adaptation 39

1.2. Extreme Events, Extreme Impacts, and Disasters 39
1.2.1. Distinguishing Extreme Events, Extreme Impacts, and Disasters 39
1.2.2. Extreme Events Defined in Physical Terms 40
1.2.2.1. Definitions of Extremes 40
1.2.2.2. Extremes in a Changing Climate 40
1.2.2.3. The Diversity and Range of Extremes 40
1.2.3. Extreme Impacts 41
1.2.3.1. Three Classes of Impacts 41
1.2.3.2. Complex Nature of an Extreme 'Event' 42
1.2.3.3. Metrics to Quantify Social Impacts and the Management of Extremes 42
1.2.3.4. Traditional Adjustment to Extremes 43

1.3. Disaster Management, Disaster Risk Reduction, and Risk Transfer 44
1.3.1. Climate Change Will Complicate Management of Some Disaster Risks 46
1.3.1.1. Challenge of Quantitative Estimates of Changing Risks 46
1.3.1.2. Processes that Influence Judgments about Changing Risks 46
1.3.2. Adaptation to Climate Change Contributes to Disaster Risk Management 47
1.3.3. Disaster Risk Management and Adaptation to Climate Change Share Many Concepts, Goals, and Processes 48

1.4. Coping and Adapting 50
1.4.1. Definitions, Distinctions, and Relationships 51
1.4.1.1. Definitions and Distinctions 51
1.4.1.2. Relationships between Coping, Coping Capacity, Adaptive Capacity, and the Coping Range 51
1.4.2. Learning 53
1.4.3. Learning to Overcome Adaptation Barriers 54
1.4.4. 'No Regrets,' Robust Adaptation, and Learning 56

References 56

Executive Summary

Disaster signifies extreme impacts suffered when hazardous physical events interact with vulnerable social conditions to severely alter the normal functioning of a community or a society (*high confidence*). Social vulnerability and exposure are key determinants of disaster risk and help explain why non-extreme physical events and chronic hazards can also lead to extreme impacts and disasters, while some extreme events do not. Extreme impacts on human, ecological, or physical systems derive from individual extreme or non-extreme events, or a compounding of events or their impacts (for example, drought creating the conditions for wildfire, followed by heavy rain leading to landslides and soil erosion). [1.1.2.1, 1.1.2.3, 1.2.3.1, 1.3]

Management strategies based on the reduction of everyday or chronic risk factors and on the reduction of risk associated with non-extreme events, as opposed to strategies based solely on the exceptional or extreme, provide a mechanism that facilitates the reduction of disaster risk and the preparation for and response to extremes and disasters (*high confidence*). Effective adaptation to climate change requires an understanding of the diverse ways in which social processes and development pathways shape disaster risk. Disaster risk is often causally related to ongoing, chronic, or persistent environmental, economic, or social risk factors. [1.1.2.2, 1.1.3, 1.1.4.1, 1.3.2]

Development practice, policy, and outcomes are critical to shaping disaster risk (*high confidence*). Disaster risk may be increased by shortcomings in development. Reductions in the rate of depletion of ecosystem services, improvements in urban land use and territorial organization processes, the strengthening of rural livelihoods, and general and specific advances in urban and rural governance advance the composite agenda of poverty reduction, disaster risk reduction, and adaptation to climate change. [1.1.2.1, 1.1.2.2, 1.1.3, 1.3.2, 1.3.3]

Climate change will pose added challenges for the appropriate allocation of efforts to manage disaster risk (*high confidence*). The potential for changes in all characteristics of climate will complicate the evaluation, communication, and management of the resulting risk. [1.1.3.1, 1.1.3.2, 1.2.2.2, 1.3.1, 1.3.2, 1.4.3]

Risk assessment is one starting point, within the broader risk governance framework, for adaptation to climate change and disaster risk reduction and transfer (*high confidence*). The assessment and analysis process may employ a variety of tools according to management context, access to data and technology, and stakeholders involved. These tools will vary from formalized probabilistic risk analysis to local level, participatory risk and context analysis methodologies. [1.3, 1.3.1.2, 1.3.3, Box 1-2]

Risk assessment encounters difficulties in estimating the likelihood and magnitude of extreme events and their impacts (*high confidence*). Furthermore, among individual stakeholders and groups, perceptions of risk are driven by psychological and cultural factors, values, and beliefs. Effective risk communication requires exchanging, sharing, and integrating knowledge about climate-related risks among all stakeholder groups. [Box 1-1, 1.1.4.1, 1.2.2.1, 1.3.1.1, 1.3.1.2, Box 1-2, Box 1-3, 1.4.2]

Management of the risk associated with climate extremes, extreme impacts, and disasters benefits from an integrated systems approach, as opposed to separately managing individual types of risk or risk in particular locations (*high confidence*). Effective risk management generally involves a portfolio of actions to reduce and transfer risk and to respond to events and disasters, as opposed to a singular focus on any one action or type of action. [1.1.2.2, 1.1.4.1, 1.3, 1.3.3, 1.4.2]

Learning is central to adaptation to climate change. Furthermore, the concepts, goals, and processes of adaptation share much in common with disaster risk management, particularly its disaster risk reduction component (*high confidence*). Disaster risk management and adaptation to climate change offer frameworks for, and examples of, advanced learning processes that may help reduce or avoid barriers that undermine planned adaptation efforts or lead to implementation of maladaptive measures. Due to the deep uncertainty, dynamic complexity, and long timeframe associated with climate change, robust adaptation efforts would require iterative risk management strategies. [1.1.3, 1.3.2, 1.4.1.2, 1.4.2, 1.4.5, Box 1-4]

Projected trends and uncertainty in hazards, exposure, and vulnerability associated with climate change and development make return to the status quo, coping, or static resilience increasingly insufficient goals for disaster risk management and adaptation (*high confidence*). Recent approaches to resilience of social-ecological systems expand beyond these concepts to include the ability to self-organize, learn, and adapt over time. [1.1.2.1, 1.1.2.2, 1.4.1.2, 1.4.2, 1.4.4]

Given shortcomings of past disaster risk management and the new dimension of climate change, greatly improved and strengthened disaster risk management and adaptation will be needed, as part of development processes, in order to reduce future risk (*high confidence*). Efforts will be more effective when informed by the experience and success with disaster risk management in different regions during recent decades, and appropriate approaches for risk identification, reduction, transfer, and disaster management. In the future, the practices of disaster risk management and adaptation can each greatly benefit from far greater synergy and linkage in institutional, financial, policy, strategic, and practical terms. [1.1.1, 1.1.2.2, 1.1.3, 1.3.3, 1.4.2]

Community participation in planning, the determined use of local and community knowledge and capacities, and the decentralization of decisionmaking, supported by and in synergy with national and international policies and actions, are critical for disaster risk reduction (*high confidence*). The use of local level risk and context analysis methodologies, inspired by disaster risk management and now strongly accepted by many civil society and government agencies in work on adaptation at the local levels, would foster greater integration between, and greater effectiveness of, both adaptation to climate change and disaster risk management. [1.1.2.2, 1.1.4.2, 1.3.3, 1.4.2]

1.1. Introduction

1.1.1. Purpose and Scope of the Special Report

Climate change, an alteration in the state of the climate that can be identified by changes in the mean and/or the variability of its properties, and that persists for an extended period, typically decades or longer, is a fundamental reference point for framing the different management themes and challenges dealt with in this Special Report.

Climate change may be due to natural internal processes or external forcings, or to persistent anthropogenic changes in the composition of the atmosphere or in land use (see Chapter 3 for greater detail). Anthropogenic climate change is projected to continue during this century and beyond. This conclusion is robust under a wide range of scenarios for future greenhouse gas emissions, including some that anticipate a reduction in emissions (IPCC, 2007a).

While specific, local outcomes of climate change are uncertain, recent assessments project alteration in the frequency, intensity, spatial extent, or duration of weather and climate extremes, including climate and hydrometeorological events such as heat waves, heavy precipitation events, drought, and tropical cyclones (see Chapter 3). Such change, in a context of increasing vulnerability, will lead to increased stress on human and natural systems and a propensity for serious adverse effects in many places around the world (UNISDR, 2009e, 2011). At the same time, climate change is also expected to bring benefits to certain places and communities at particular times.

New, improved or strengthened processes for anticipating and dealing with the adverse effects associated with weather and climate events will be needed in many areas. This conclusion is supported by the fact that despite increasing knowledge and understanding of the factors that lead to adverse effects, and despite important advances over recent decades in the reduction of loss of life with the occurrence of hydrometeorological events (mainly attributable to important advances with early warning systems, e.g., Section 9.2.11), social intervention in the face of historical climate variability has not kept pace with the rapid increases in other adverse economic and social effects suffered during this period (ICSU, 2008) *(high confidence)*. Instead, a rapid growth in real economic losses and livelihood disruption has occurred in many parts of the world (UNISDR, 2009e, 2011). In regard to losses associated with tropical cyclones, recent analysis has shown that, with the exception of the East Asian and Pacific and South Asian regions, "both exposure and the estimated risk of economic loss are growing faster than GDP per capita. Thus the risk of losing wealth in disasters associated with tropical cyclones is increasing faster than wealth itself is increasing" (UNISDR, 2011, p. 33).

The Hyogo Framework for Action (UNISDR, 2005), adopted by 168 governments, provides a point of reference for disaster risk management and its practical implementation (see Glossary and Section 1.1.2.2 for a definition of this practice). Subsequent United Nations statements suggest the need for closer integration of disaster risk management and adaptation with climate change concerns and goals, all in the context of development and development planning (UNISDR, 2008a, 2009a,b,c). Such a concern led to the agreement between the IPCC and the United Nations International Strategy for Disaster Reduction (UNISDR), with the support of the Norwegian government, to undertake this Special Report on "Managing the Risks of Extreme Events and Disasters to Advance Climate Change Adaptation" (IPCC, 2009).

This Special Report responds to that concern by considering climate change and its effects on extreme (weather and climate) events, disaster, and disaster risk management; how human responses to extreme events and disasters (based on historical experience and evolution in practice) could contribute to adaptation objectives and processes; and how adaptation to climate change could be more closely integrated with disaster risk management practice.

The report draws on current scientific knowledge to address three specific goals:

1) To assess the relevance and utility of the concepts, methods, strategies, instruments, and experience gained from the management of climate-associated disaster risk under conditions of historical climate patterns, in order to advance adaptation to climate change and the management of extreme events and disasters in the future.
2) To assess the new perspectives and challenges that climate change brings to the disaster risk management field.
3) To assess the mutual implications of the evolution of the disaster risk management and adaptation to climate change fields, particularly with respect to the desired increases in social resilience and sustainability that adaptation implies.

The principal audience for this Special Report comprises decisionmakers and professional and technical personnel from local through to national governments, international development agencies, nongovernmental organizations, and civil society organizations. The report also has relevance for the academic community and interested laypeople.

The first section of this chapter briefly introduces the more important concepts, definitions, contexts, and management concerns needed to frame the content of this report. Later sections of the chapter expand on the subjects of extreme events and extreme impacts; disaster risk management, reduction, and transfer and their integration with climate change and adaptation processes; and the notions of coping and adaptation. The level of detail and discussion presented in this chapter is commensurate with its status as a 'scene setting' initiative. The following eight chapters provide more detailed and specific analysis.

Chapter 2 assesses the key determinants of risk, namely exposure and vulnerability in the context of climate-related hazards. A particular focus is the connection between near-term experience and long-term adaptation. Key questions addressed include whether reducing vulnerability to current hazards improves adaptation to longer-term climate change, and how near-term risk management decisions and adjustments constrain future vulnerability and enable adaptation.

Chapter 3 focuses on changes in extremes of atmospheric weather and climate variables (e.g., temperature and precipitation), large-scale phenomena that are related to these extremes or are themselves extremes (e.g., tropical and extratropical cyclones, El Niño, and monsoons), and collateral effects on the physical environment (e.g., droughts, floods, coastal impacts, landslides). The chapter builds on and updates the Fourth Assessment Report, which in some instances, due to new literature, leads to revisions of that assessment.

Chapter 4 explores how changes in climate, particularly weather and climate extremes assessed in Chapter 3, translate into extreme impacts on human and ecological systems. A key issue is the nature of both observed and expected trends in impacts, the latter resulting from trends in both physical and social conditions. The chapter assesses these questions from both a regional and a sectoral perspective, and examines the direct and indirect economic costs of such changes and their relation to development.

Chapters 5, 6, and 7 assess approaches to disaster risk management and adaptation to climate change from the perspectives of local, national, and international governance institutions, taking into consideration the roles of government, individuals, nongovernmental organizations, the private sector, and other civil society institutions and arrangements. Each chapter reviews the efficacy of current disaster risk reduction, preparedness, and response and risk transfer strategies and previous approaches to extremes and disasters in order to extract lessons for the future. Impacts, adaptation, and the cost of risk management are assessed through the prism of diverse social aggregations and means for cooperation, as well as a variety of institutional arrangements.

Chapter 5 focuses on the highly variable local contexts resulting from differences in place, social groupings, experience, management, institutions, conditions, and sets of knowledge, highlighting risk management strategies involving housing, buildings, and land use. Chapter 6 explores similar issues at the national level, where mechanisms including national budgets, development goals, planning, warning systems, and building codes may be employed to manage, for example, food security and agriculture, water resources, forests, fisheries, building practice, and public health. Chapter 7 carries this analysis to the international level, where the emphasis is on institutions, organizations, knowledge generation and sharing, legal frameworks and practices, and funding arrangements that characterize international agencies and collaborative arrangements. This chapter also discusses integration of responsibilities across all governmental scales, emphasizing the linkages among disaster risk management, climate change adaptation, and development.

Chapter 8 assesses how disaster risk reduction strategies, ranging from incremental to transformational, can advance adaptation to climate change and promote a more sustainable and resilient future. Key questions include whether an improved alignment between climate change responses and sustainable development strategies may be achieved, and whether short- and long-term perspectives may be reconciled.

Chapter 9 closes this report by presenting case studies in order to identify lessons and best practices from past responses to extreme climate-related events and extreme impacts. Cases illustrate concrete and diverse examples of disaster types as well as risk management methodologies and responses discussed in the other chapters, providing a key reference point for the entire report.

1.1.2. Key Concepts and Definitions

The concepts and definitions presented in this chapter and employed throughout the Special Report take into account a number of existing sources (IPCC, 2007c; UNISDR, 2009d; ISO, 2009) but also reflect the fact that concepts and definitions evolve as knowledge, needs, and contexts vary. Disaster risk management and adaptation to climate change are dynamic fields, and have in the past exhibited and will necessarily continue in the future to exhibit such evolution.

This chapter presents 'skeleton' definitions that are generic rather than specific. In subsequent chapters, the definitions provided here are often expanded in more detail and variants among these definitions will be examined and explained where necessary.

A glossary of the fundamental definitions used in this assessment is provided at the end of this study. Figure 1-1 provides a schematic of the relationships among many of the key concepts defined here.

1.1.2.1. Definitions Related to General Concepts

In order to delimit the central concerns of this Special Report, a distinction is made between those concepts and definitions that relate to disaster risk and adaptation to climate change generally; and, on the other hand, those that relate in particular to the options and forms of social intervention relevant to these fields. In Section 1.1.2.1, consideration is given to general concepts. In Section 1.1.2.2, key concepts relating to social intervention through 'Disaster Risk Management' and 'Climate Change Adaptation' are considered.

Extreme (weather and climate) events and disasters comprise the two central risk management concerns of this Special Report.

Extreme events comprise a facet of climate variability under stable or changing climate conditions. They are defined as the occurrence of a value of a weather or climate variable above (or below) a threshold value near the upper (or lower) ends ('tails') of the range of observed values of the variable. This definition is further discussed and amplified in Sections 1.2.2, 3.1.1, and 3.1.2.

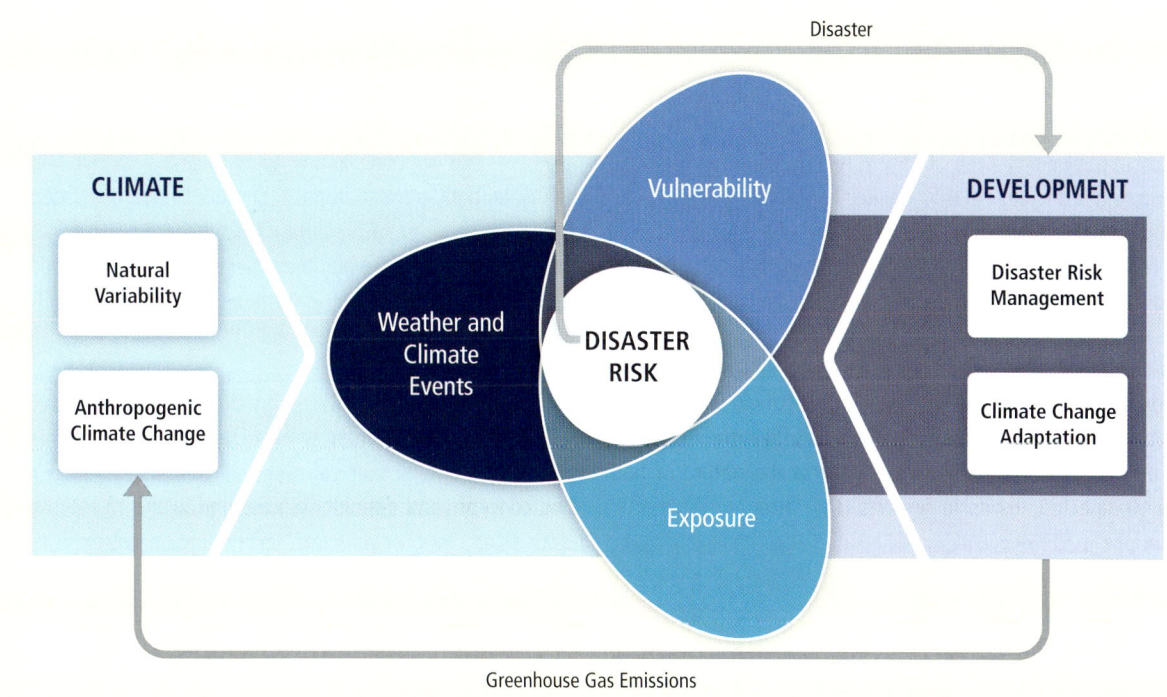

Figure 1-1 | The key concepts and scope of this report. The figure indicates schematically key concepts involved in disaster risk management and climate change adaptation, and the interaction of these with sustainable development.

Disasters are defined in this report as severe alterations in the normal functioning of a community or a society due to hazardous physical events interacting with vulnerable social conditions, leading to widespread adverse human, material, economic, or environmental effects that require immediate emergency response to satisfy critical human needs and that may require external support for recovery.

The **hazardous physical events** referred to in the definition of disaster may be of natural, socio-natural (originating in the human degradation or transformation of the physical environment), or purely anthropogenic origins (see Lavell, 1996, 1999; Smith, 1996; Tobin and Montz, 1997; Wisner et al., 2004). This Special Report emphasizes hydrometeorological and oceanographic events; a subset of a broader spectrum of physical events that may acquire the characteristic of a hazard if conditions of exposure and vulnerability convert them into a threat. These include earthquakes, volcanoes, and tsunamis, among others. Any one geographic area may be affected by one, or a combination of, such events at the same or different times. Both in this report and in the wider literature, some events (e.g., floods and droughts) are at times referred to as physical impacts (see Section 3.1.1).

Extreme events are often but not always associated with disaster. This association will depend on the particular physical, geographic, and social conditions that prevail (see this section and Chapter 2 for discussion of the conditioning circumstances associated with so-called 'exposure' and 'vulnerability') (Ball, 1975; O'Keefe et al., 1976; Timmerman, 1981; Hewitt, 1983; Maskrey, 1989; Mileti, 1999; Wisner et al., 2004).

Non-extreme physical events also can and do lead to disasters where physical or societal conditions foster such a result. In fact, a significant number of disasters registered annually in most disaster databases are associated with physical events that are not extreme as defined probabilistically, yet have important social and economic impacts on local communities and governments, both individually and in aggregate (UNISDR, 2009e, 2011) (*high confidence*).

For example, many of the 'disasters' registered in the widely consulted University of Louvaine EM-DAT database (CRED, 2010) are not initiated by statistically extreme events, but rather exhibit extreme properties expressed as severe interruptions in the functioning of local social and economic systems. This lack of connection is even more obvious in the DesInventar database (Corporación OSSO, 2010), developed first in Latin America in order to specifically register the occurrence of small- and medium-scale disasters, and which has registered tens and tens of thousands of these during the last 30 years in the 29 countries it covers to date. This database has been used by the UNISDR, the Inter-American Development Bank, and others to examine disaster occurrence, scale, and impacts in Latin America and Asia, in particular (Cardona 2005, 2008; IDEA, 2005; UNISDR, 2009e, 2011; ERN-AL, 2011). In any one place, the range of disaster-inducing events can increase if social conditions deteriorate (Wisner et al., 2004, 2011).

The occurrence of disaster is always preceded by the existence of specific physical and social conditions that are generally referred to as **disaster risk** (Hewitt, 1983; Lewis, 1999, 2009; Bankoff, 2001;

Wisner et al., 2004, 2011; ICSU, 2008; UNISDR, 2009e, 2011; ICSU-LAC, 2009).

Disaster risk is defined for the purposes of this study as the likelihood over a specified time period of severe alterations in the normal functioning of a community or a society due to hazardous physical events interacting with vulnerable social conditions, leading to widespread adverse human, material, economic, or environmental effects that require immediate emergency response to satisfy critical human needs and that may require external support for recovery. Disaster risk derives from a combination of physical hazards and the vulnerabilities of exposed elements and will signify the potential for severe interruption of the normal functioning of the affected society once it materializes as disaster. This qualitative statement will be expressed formally later in this assessment (Section 1.3 and Chapter 2).

The definitions of disaster risk and disaster posited above do not include the potential or actual impacts of climate and hydrological events on ecosystems or the physical Earth system per se. In this assessment, such impacts are considered relevant to disaster if, as is often the case, they comprise one or more of the following, at times interrelated, situations: i) they impact livelihoods negatively by seriously affecting ecosystem services and the natural resource base of communities; ii) they have consequences for food security; and/or iii) they have impacts on human health.

Extreme impacts on the physical environment are addressed in Section 3.5 and extreme impacts on ecosystems are considered in detail in Chapter 4. In excluding such impacts from the definition of 'disaster' as employed here, this chapter is in no way underestimating their broader significance (e.g., in regard to existence value) or suggesting they should not be dealt with under the rubric of adaptation concerns and management needs. Rather, we are establishing their relative position within the conceptual framework of climate-related, socially-defined 'disaster' and 'disaster risk' and the management options that are available for promoting disaster risk reduction and adaptation to climate change (see Section 1.1.2.2 and the Glossary for definitions of these terms). Thus this report draws a distinction between 'social disaster,' where extreme impacts on the physical and ecological systems may or may not play a part, and so-called 'environmental disaster,' where direct physical impacts of human activity and natural physical processes on the environment are fundamental causes (with possible direct feedback impacts on social systems).

Disaster risk cannot exist without the threat of potentially damaging physical events. However, such events, once they occur, are not in and of themselves sufficient to explain disaster or its magnitude. In the search to better understand the concept of disaster risk (thus disaster) it is important to consider the notions of hazard, vulnerability, and exposure.

When extreme and non-extreme physical events, such as tropical cyclones, floods, and drought, can affect elements of human systems in an adverse manner, they assume the characteristic of a hazard. **Hazard** is defined here as the potential occurrence of a natural or human-induced physical event that may cause loss of life, injury, or other health impacts, as well as damage and loss to property, infrastructure, livelihoods, service provision, and environmental resources. Physical events become hazards where social elements (or environmental resources that support human welfare and security) are exposed to their potentially adverse impacts and exist under conditions that could predispose them to such effects. Thus, hazard is used in this study to denote a threat or potential for adverse effects, not the physical event itself (Cardona, 1986, 1996, 2011; Smith, 1996; Tobin and Montz, 1997; Lavell, 2003; Hewitt, 2007; Wisner et al., 2004).

Exposure is employed to refer to the presence (location) of people, livelihoods, environmental services and resources, infrastructure, or economic, social, or cultural assets in places that could be adversely affected by physical events and which, thereby, are subject to potential future harm, loss, or damage. This definition subsumes physical and biological systems under the concept of 'environmental services and resources,' accepting that these are fundamental for human welfare and security (Crichton, 1999; Gasper, 2010).

Exposure may also be dictated by mediating social structures (e.g., economic and regulatory) and institutions (Sen, 1983). For example, food insecurity may result from global market changes driven by drought or flood impacts on crop production in another location. Other relevant and important interpretations and uses of exposure are discussed in Chapter 2.

Under exposed conditions, the levels and types of adverse impacts will be the result of a physical event (or events) interacting with socially constructed conditions denoted as vulnerability.

Vulnerability is defined generically in this report as the propensity or predisposition to be adversely affected. Such predisposition constitutes an internal characteristic of the affected element. In the field of disaster risk, this includes the characteristics of a person or group and their situation that influences their capacity to anticipate, cope with, resist, and recover from the adverse effects of physical events (Wisner et al., 2004).

Vulnerability is a result of diverse historical, social, economic, political, cultural, institutional, natural resource, and environmental conditions and processes.

The concept has been developed as a theme in disaster work since the 1970s (Baird et al., 1975; O'Keefe et al., 1976; Wisner et al., 1977; Lewis, 1979, 1984, 1999, 2009; Timmerman, 1981; Hewitt, 1983, 1997, 2007; Cutter, 1996; Weichselgartner, 2001; Cannon, 2006; Gaillard, 2010) and variously modified in different fields and applications in the interim (Adger, 2006; Eakin and Luers, 2006; Füssel, 2007). Vulnerability has been evaluated according to a variety of quantitative and qualitative metrics (Coburn and Spence, 2002; Schneider et al., 2007; Cardona, 2011). A detailed discussion of this notion and the drivers or root causes of vulnerability are provided in Chapter 2.

Chapter 1 — Climate Change: New Dimensions in Disaster Risk, Exposure, Vulnerability, and Resilience

The importance of vulnerability to the disaster risk management community may be appreciated in the way it has helped to highlight the role of social factors in the constitution of risk, moving away from purely physical explanations and attributions of loss and damage (see Hewitt, 1983 for an early critique of what he denominated the 'physicalist' interpretation of disaster). Differential levels of vulnerability will lead to differential levels of damage and loss under similar conditions of exposure to physical events of a given magnitude (Dow, 1992; Wisner et al., 2011).

The fundamentally social connotation and 'predictive' value of vulnerability is emphasized in the definition used here. The earlier IPCC definition of vulnerability refers, however, to "the degree to which a system is susceptible to and unable to cope with adverse effects of climate change, including climate variability and extremes. Vulnerability is a function of the character, magnitude, and rate of climate change and variation to which a system is exposed, its sensitivity, and its adaptive capacity" (IPCC, 2007c, p. 883). This definition makes physical causes and their effects an explicit aspect of vulnerability while the social context is encompassed by the notions of sensitivity and adaptive capacity (these notions are defined later). In the definition used in this report, the social context is emphasized explicitly, and vulnerability is considered independent of physical events (Hewitt, 1983, 1997, 2007; Weichselgartner, 2001; Cannon, 2006; O'Brien et al., 2007).

Vulnerability has been contrasted and complimented with the notion of **capacity**.

Capacity refers to the combination of all the strengths, attributes, and resources available to an individual, community, society, or organization that can be used to achieve established goals. This includes the conditions and characteristics that permit society at large (institutions, local groups, individuals, etc.) access to and use of social, economic, psychological, cultural, and livelihood-related natural resources, as well as access to the information and the institutions of governance necessary to reduce vulnerability and deal with the consequences of disaster. This definition extends the definition of capabilities referred to in Sen's 'capabilities approach to development' (Sen, 1983).

The lack of capacity may be seen as being one dimension of overall vulnerability, while it is also seen as a separate notion that, although contributing to an increase in vulnerability, is not part of vulnerability per se. The existence of vulnerability does not mean an absolute, but rather a relative lack of capacity.

Promoted in disaster recovery work by Anderson and Woodrow (1989) as a means, among other objectives, to shift the analytical balance from the negative aspects of vulnerability to the positive actions by people, the notion of capacity is fundamental to imagining and designing a conceptual shift favoring disaster risk reduction and adaptation to climate change. Effective **capacity building**, the notion of stimulating and providing for growth in capacity, requires a clear image of the future with clearly established goals.

Adaptive capacity comprises a specific usage of the notion of capacity and is dealt with in detail in later sections of this chapter and Chapters 2 and 8 in particular.

The existence of vulnerability and capacity and their importance for understanding the nature and extent of the adverse effects that may occur with the impact of physical events can be complemented with a consideration of the characteristics or conditions that help ameliorate or mitigate negative impacts once disaster materializes. The notions of resilience and coping are fundamental in this sense.

Coping (elaborated upon in detail in Section 1.4 and Chapter 2) is defined here generically as the use of available skills, resources, and opportunities to address, manage, and overcome adverse conditions

---

FAQ 1.1 | Is there a one-to-one relationship between extreme events and disasters?

No. Disaster entails social, economic, or environmental impacts that severely disrupt the normal functioning of affected communities. Extreme weather and climate events will lead to disaster if: 1) communities are exposed to those events; and 2) exposure to potentially damaging extreme events is accompanied by a high level of vulnerability (a predisposition for loss and damage). On the other hand, disasters are also triggered by events that are not extreme in a statistical sense. High exposure and vulnerability levels will transform even some small-scale events into disasters for some affected communities. Recurrent small- or medium-scale events affecting the same communities may lead to serious erosion of its development base and livelihood options, thus increasing vulnerability. The timing (when they occur during the day, month, or year) and sequence (similar events in succession or different events contemporaneously) of such events is often critical to their human impact. The relative importance of the underlying physical and social determinants of disaster risk varies with the scale of the event and the levels of exposure and vulnerability. Because the impact of lesser events is exacerbated by physical, ecological, and social conditions that increase exposure and vulnerability, these events disproportionately affect resource-poor communities with little access to alternatives for reducing hazard, exposure, and vulnerability. The potential negative consequences of extreme events can be moderated in important ways (but rarely eliminated completely) by implementing corrective disaster risk management strategies that are reactive, adaptive, and anticipatory, and by sustainable development.

with the aim of achieving basic functioning in the short to medium terms.

Resilience is defined as the ability of a system and its component parts to anticipate, absorb, accommodate, or recover from the effects of a potentially hazardous event in a timely and efficient manner, including through ensuring the preservation, restoration, or improvement of its essential basic structures and functions. As Gaillard (2010) points out, this term has been used in disaster studies since the 1970s (Torry, 1979) and has its origins in engineering (Gordon, 1978), ecology (Holling, 1973) and child psychology (Werner et al., 1971).

Although now widely employed in the fields of disaster risk management and adaptation, resilience has been subject to a wide range of interpretations and levels of acceptance as a concept (Timmerman, 1981; Adger, 2000; Klein et al., 2003; Berkes et al., 2004; Folke, 2006; Gallopín, 2006; Manyena, 2006; Brand and Jax, 2007; Gaillard 2007; Bosher, 2008; Cutter et al., 2008; Kelman, 2008; Lewis and Kelman, 2009; Bahadur et al., 2010; Aven, 2011). Thus, for example, the term is used by some in reference to situations at any point along the risk 'cycle' or 'continuum', that is, before, during, or after the impact of the physical event. And, in a different vein, some consider the notions of 'vulnerability' and 'capacity' as being sufficient for explaining the ranges of success or failure that are found in different recovery scenarios and are thus averse to the use of the term at all (Wisner et al., 2004, 2011). Under this latter formulation, vulnerability both potentiates original loss and damage and also impedes recovery, while capacity building can change this adverse balance and contribute to greater sustainability and reduced disaster risk.

Older conceptions of resilience, as 'bouncing back,' and its conceptual cousin, coping (see Section 1.4), have implicitly emphasized a return to a previous status quo or some other marginally acceptable level, such as 'surviving,' as opposed to generating a cyclical process that leads to continually improving conditions, as in 'bouncing forward' and/or eventually 'thriving' (Davies, 1993; Manyena, 2006). However, the dynamic and often uncertain consequences of climate change (as well as ongoing, now longstanding, development trends such as urbanization) for hazard and vulnerability profiles underscore the fact that 'bouncing back' is an increasingly insufficient goal for disaster risk management (Pelling, 2003; Vale and Campanella, 2005; Pendalla et al., 2010) (*high confidence*). Recent conceptions of resilience of social-ecological systems focus more on process than outcomes (e.g., Norris et al., 2008), including the ability to self-organize, learn, and adapt over time (see Chapter 8). Some definitions of resilience, such as that used in this report, now also include the idea of anticipation and 'improvement' of essential basic structures and functions. Section 1.4 examines the importance of learning that is emphasized within this more forward-looking application of resilience. Chapter 8 builds on the importance of learning by drawing also from literature that has explored the scope for innovation, leadership, and adaptive management. Together these strategies offer potential pathways for transforming existing development visions, goals, and practices into more sustainable and resilient futures.

Chapters 2 and 8 address the notion of resilience and its importance in discussions on sustainability, disaster risk reduction, and adaptation in greater detail.

1.1.2.2. Concepts and Definitions Relating to Disaster Risk Management and Adaptation to Climate Change

Disaster risk management is defined in this report as the processes for designing, implementing, and evaluating strategies, policies, and measures to improve the understanding of disaster risk, foster disaster risk reduction and transfer, and promote continuous improvement in disaster preparedness, response, and recovery practices, with the explicit purpose of increasing human security, well-being, quality of life, and sustainable development.

Disaster risk management is concerned with both disaster and disaster risk of differing levels and intensities. In other words, it is not restricted to a 'manual' for the management of the risk or disasters associated with extreme events, but rather includes the conceptual framework that describes and anticipates intervention in the overall and diverse patterns, scales, and levels of interaction of exposure, hazard, and vulnerability that can lead to disaster. A major recent concern of disaster risk management has been that disasters are associated more and more with lesser-scale physical phenomena that are not extreme in a physical sense (see Section 1.1.1). This is principally attributed to increases in exposure and associated vulnerability (UNISDR, 2009e, 2011).

Where the term **risk management** is employed in this chapter and report, it should be interpreted as being a synonym for disaster risk management, unless otherwise made explicit.

Disaster Risk Management can be divided to comprise two related but discrete subareas or components: **disaster risk reduction** and **disaster management**.

Disaster risk reduction denotes both a policy goal or objective, and the strategic and instrumental measures employed for anticipating future disaster risk, reducing existing exposure, hazard, or vulnerability, and improving resilience. This includes lessening the vulnerability of people, livelihoods, and assets and ensuring the appropriate sustainable management of land, water, and other components of the environment. Emphasis is on universal concepts and strategies involved in the consideration of reducing disaster risks, including actions and activities enacted pre-impact, and when recovery and reconstruction call for the anticipation of new disaster risk scenarios or conditions. A strong relationship between disaster risk and disaster risk reduction, and development and development planning has been established and validated, particularly, but not exclusively, in developing country contexts (UNEP, 1972; Cuny, 1983; Sen, 1983; Hagman, 1984; Wijkman and Timberlake, 1988; Lavell, 1999, 2003, 2009; Wisner et al., 2004, 2011; UNDP, 2004; van Niekerk, 2007; Dulal et al., 2009; UNISDR, 2009e, 2011) (*high confidence*).

Disaster management refers to social processes for designing, implementing, and evaluating strategies, policies, and measures that promote and improve disaster preparedness, response, and recovery practices at different organizational and societal levels. Disaster management processes are enacted once the immediacy of the disaster event has become evident and resources and capacities are put in place with which to respond prior to and following impact. These include the activation of early warning systems, contingency planning, emergency response (immediate post-impact support to satisfy critical human needs under conditions of severe stress), and, eventually, recovery (Alexander, 2000; Wisner et al., 2011). Disaster management is required due to the existence of 'residual' disaster risk that ongoing disaster risk reduction processes have not mitigated or reduced sufficiently or eliminated or prevented completely (IDB, 2007).

Growing disaster losses have led to rapidly increasing concerns for post-impact financing of response and recovery (UNISDR, 2009e, 2011). In this context, the concept and practice of disaster risk transfer has received increased interest and achieved greater salience. **Risk transfer** refers to the process of formally or informally shifting the financial consequences of particular risks from one party to another, whereby a household, community, enterprise, or state authority will obtain resources from the other party after a disaster occurs, in exchange for ongoing or compensatory social or financial benefits provided to that other party. Disaster risk transfer mechanisms comprise a component of both disaster management and disaster risk reduction. In the former case, financial provision is made to face up to the impacts and consequences of disaster once this materializes. In the latter case, the adequate use of insurance premiums, for example, can promote and encourage the use of disaster risk reduction measures in the insured elements. Chapters 5, 6, 7, and 9 discuss risk transfer in some detail.

Over the last two decades, the more integral notion of disaster risk management and its risk reduction and disaster management components has tended to replace the unique conception and terminology of 'disaster and emergency management' that prevailed almost unilaterally up to the beginning of the 1990s and that emphasized disaster as opposed to disaster risk as the central issue to be confronted. Disaster as such ordered the thinking on required intervention processes, whereas with disaster risk management, disaster risk now tends to assume an increasingly dominant position in thought and action in this field (see Hewitt, 1983; Blaikie et al., 1994; Smith, 1996; Hewitt, 1997; Tobin and Montz, 1997; Lavell, 2003; Wisner et al., 2004, 2011; van Niekerk, 2007; Gaillard, 2010 for background and review of some of the historical changes in favor of disaster risk management).

The notion of **disaster or disaster management cycle** was introduced and popularized in the earlier context dominated by disaster or emergency management concerns and viewpoints. The cycle, and the later 'disaster continuum' notion, depicted the sequences and components of so-called disaster management. In addition to considering preparedness, emergency response, rehabilitation, and reconstruction, it also included disaster prevention and mitigation as stated components of 'disaster management' and utilized the temporal notions of before, during, and after disaster to classify the different types of action (Lavell and Franco, 1996; van Niekerk, 2007).

The cycle notion, criticized for its mechanistic depiction of the intervention process, for insufficient consideration of the ways different components and actions merge and can act synergistically with and influence each other, and for its incorporation of disaster risk reduction considerations under the rubric of 'disaster management' (Lavell and Franco, 1996; Lewis, 1999; Wisner et al., 2004; Balamir, 2005; van Niekerk, 2007), has tended to give way over time, in many parts of the world, to the more comprehensive approach and concept of disaster risk management with its consideration of distinct risk reduction and disaster intervention components. The move toward a conception oriented in terms of disaster risk and not disaster per se has led to initiatives to develop the notion of a '**disaster risk continuum**' whereby risk is seen to evolve and change constantly, requiring different modalities of intervention over time, from pre-impact risk reduction through response to new risk conditions following disaster impacts and the need for control of new risk factors in reconstruction (see Lavell, 2003).

With regard to the influence of actions taken at one stage of the 'cycle' on other stages, much has been written, for example, on how the form and method of response to disaster itself may affect future disaster risk reduction efforts. The fostering of active community involvement, the use of existing local and community capacities and resources, and the decentralization of decisionmaking to the local level in disaster preparedness and response, among other factors, have been considered critical for also improving understanding of disaster risk and the development of future disaster risk reduction efforts (Anderson and Woodrow, 1989; Alexander, 2000; Lavell, 2003; Wisner et al., 2004) (*high confidence*). And, the methods used for, and achievements with, reconstruction clearly have important impacts on future disaster risk and on the future needs for preparedness and response.

In the following subsection, some of the major reasons that explain the transition from disaster management, with its emphasis on disaster, to disaster risk management, with its emphasis on disaster risk, are presented as a background for an introduction to the links and options for closer integration of the adaptation and disaster risk management fields.

The gradual evolution of policies that favor disaster risk reduction objectives as a component of development planning procedures (as opposed to disaster management seen as a function of civil protection, civil defense, emergency services, and ministries of public works) has inevitably placed the preexisting emergency or disaster-response-oriented institutional and organizational arrangements for disaster management under scrutiny. The prior dominance of response-based and infrastructure organizations has been complemented with the increasing incorporation of economic and social sector and territorial development agencies or organizations, as well as planning and finance ministries. Systemic, as opposed to single agency, approaches are now evolving in many places.

Synergy, collaboration, coordination, and development of multidisciplinary and multiagency schemes are increasingly seen as positive attributes for guaranteeing implementation of disaster risk reduction and disaster risk management in a sustainable development framework (see Lavell and Franco, 1996; Ramírez and Cardona, 1996; Wisner et al., 2004, 2011). Under these circumstances the notion of **national disaster risk management systems** or **structures** has emerged strongly. Such notions are discussed in detail in Chapter 6.

Adaptation to climate change, the second policy, strategic, and instrumental aspect of importance for this Special Report, is a notion that refers to both human and natural systems. Adaptation in human systems is defined here as the process of adjustment to actual or expected climate and its effects, in order to moderate harm or exploit beneficial opportunities. In natural systems, it is defined as the process of adjustment to actual climate and its effects; human intervention may facilitate adjustment to expected climate.

These definitions modify the IPCC (2007c) definition that generically speaks of the "adjustment in natural and human systems in response to actual and expected climatic stimuli, such as to moderate harm or exploit beneficial opportunities." The objective of the redefinition used in this report is to avoid the implication present in the prior IPCC definition that natural systems can adjust to expected climate stimuli. At the same time, it accepts that some forms of human intervention may provide opportunities for supporting natural system adjustment to future climate stimuli that have been anticipated by humans.

Adaptation is a key aspect of the present report and is dealt with in greater detail in Sections 1.3 and 1.4 and later chapters. The more ample introduction to disaster risk management offered above derives from the particular perspective of the present report: that adaptation is a goal to be advanced and extreme event and disaster risk management are methods for supporting and advancing that goal.

The notion of adaptation is counterposed to the notion of **mitigation** in the climate change literature and practice. **Mitigation** there refers to the reduction of the rate of climate change via the management of its causal factors (the emission of greenhouse gases from fossil fuel combustion, agriculture, land use changes, cement production, etc.) (IPCC, 2007c). However, in disaster risk reduction practice, 'mitigation' refers to the amelioration of disaster risk through the reduction of existing hazards, exposure, or vulnerability, including the use of different **disaster preparedness** measures.

Disaster preparedness measures, including early warning and the development of contingency or emergency plans, may be considered a component of, and a bridge between, disaster risk reduction and disaster management. Preparedness accepts the existence of residual, unmitigated risk, and attempts to aid society in eliminating certain of the adverse effects that could be experienced once a physical event(s) occurs (for example, by the evacuation of persons and livestock from exposed and vulnerable circumstances). At the same time, it provides for better response to adverse effects that do materialize (for example, by planning for adequate shelter and potable water supplies for the affected or destitute persons or food supplies for affected animal populations).

In order to accommodate the two differing definitions of mitigation, this report presumes that **mitigation** is a substantive action that can be applied in different contexts where attenuation of existing specified conditions is required.

Disaster mitigation is used to refer to actions that attempt to limit further adverse conditions once disaster has materialized. This refers to the avoidance of what has sometimes been called the 'second disaster' following the initial physical impacts (Alexander, 2000; Wisner et al., 2011). The 'second disaster' may be characterized, among other things, by adverse effects on health (Noji, 1997; Wisner et al., 2011) and livelihoods due to inadequate disaster response and rehabilitation plans, inadequate enactment of existing plans, or unforeseen or unforeseeable circumstances.

Disaster risk **prevention** and disaster **prevention** refer, in a strict sense, to the elimination or avoidance of the underlying causes and conditions that lead to disaster, thus precluding the possibility of either disaster risk or disaster materializing. The notion serves to concentrate attention on the fact that disaster risk is manageable and its materialization is preventable to an extent (which varies depending on the context). **Prospective (proactive) disaster risk management** and adaptation can contribute in important ways to avoiding future, and not just reducing existing, risk and disaster once they have become manifest, as is the case with **corrective** or **reactive management** (Lavell, 2003; UNISDR, 2011).

1.1.2.3. The Social Construction of Disaster Risk

The notions of hazard, exposure, vulnerability, disaster risk, capacity, resilience, and coping, and their social origins and bases, as presented above, reflect an emerging understanding that disaster risk and disaster, while potentiated by an objective, physical condition, are fundamentally a 'social construction,' the result of social choice, social constraints, and societal action and inaction (*high confidence*). The notion of social construction of risk implies that management can take into account the social variables involved and to the best of its ability work toward risk reduction, disaster management, or risk transfer through socially sustainable decisions and concerted human action (ICSU-LAC, 2009). This of course does not mean that there are not risks that may be too great to reduce significantly through human intervention, or others that the very social construction process may in fact exacerbate (see Sections 1.3.1.2 and 1.4.3). But in contrast with, for example, many natural physical events and their contribution to disaster risk, the component of risk that is socially constructed is subject to intervention in favor of risk reduction.

The contribution of physical events to disaster risk is characterized by statistical distributions in order to elucidate the options for risk reduction and adaptation (Section 1.2 and Chapter 3). But, the explicit recognition of the political, economic, social, cultural, physical, and psychological elements or determinants of risk leads to a spectrum of potential outcomes of physical events, including those captured under the notion of **extreme impacts** (Section 1.2 and Chapter 4). Accordingly, risk assessment (see Section 1.3) using both quantitative and qualitative (social and psychological) measures is required to render a more complete description of risk and risk causation processes (Section 1.3; Douglas and Wildavsky, 1982; Cardona, 2004; Wisner et al., 2004; Weber, 2006). Climate change may introduce a break with past environmental system functioning so that forecasting physical events becomes less determined by past trends. Under these conditions, the processes that cause, and the established indicators of, human vulnerability need to be reconsidered in order for risk assessment to remain an effective tool. The essential nature and structure of the characteristics that typify vulnerability can of course change without climate changing.

1.1.3. Framing the Relation between Adaptation to Climate Change and Disaster Risk Management

Adaptation to climate change and disaster risk management both seek to reduce factors and modify environmental and human contexts that contribute to climate-related risk, thus supporting and promoting sustainability in social and economic development. The promotion of adequate preparedness for disaster is also a function of disaster risk management and adaptation to climate change. And, both practices are seen to involve learning (see Section 1.4), having a corrective and prospective component dealing with existing and projected future risk.

However, the two practices have tended to follow independent paths of advance and development and have on many occasions employed different interpretations of concepts, methods, strategies, and institutional frameworks to achieve their ends. These differences should clearly be taken into account in the search for achieving greater synergy between them and will be examined in an introductory fashion in Section 1.3 and in greater detail in following chapters of this report.

Public policy and professional concepts of disaster and their approaches to disaster and disaster risk management have undergone very significant changes over the last 30 years, so that challenges that are now an explicit focus of the adaptation field are very much part of current disaster risk reduction as opposed to mainstream historical disaster management concerns (Lavell, 2010; Mercer, 2010).These changes have occurred under the stimuli of changing concepts, multidisciplinary involvement, social and economic demands, and impacts of disasters, as well as institutional changes reflected in international accords and policies such as the UN Declaration of the International Decade for Natural Disaster Reduction in the 1990s, the 2005 Hyogo Framework for Action, as well as the work of the International Strategy for Disaster Reduction since 2000.

Particularly in developing countries, this transition has been stimulated by the documented relationship between disaster risk and 'skewed' development processes (UNEP, 1972; Cuny, 1983; Sen, 1983; Hagman, 1984; Wijkman and Timberlake, 1988; Lavell, 1999, 2003; UNDP, 2004; Wisner et al., 2004, 2011; Dulal et al., 2009; UNISDR, 2009e, 2011). Significant differentiation in the distribution or allocation of gains from development and thus in the incidence of chronic or everyday risk, which disproportionately affect poorer persons and families, is a major contributor to the more specific existence of disaster risk (Hewitt, 1983, 1997; Wisner et al., 2004). Reductions in the rate of ecosystem services depletion, improvements in urban land use and territorial organization processes, the strengthening of rural livelihoods, and general and specific advances in urban and rural governance are viewed as indispensable to achieving the composite agenda of poverty reduction, disaster risk reduction, and adaptation to climate change (UNISDR, 2009e, 2011) (*high confidence*).

Climate change is at once a problem of development and also a symptom of 'skewed' development. In this context, pathways toward resilience include both incremental and transformational approaches to development (Chapter 8). Transformational strategies place emphasis on addressing risk that stems from social structures as well as social behavior and have a broader scope extending from disaster risk management into development goals, policy, and practice (Nelson et al., 2007). In this way transformation builds on a legacy of progressive, socially informed disaster risk research that has applied critical methods, including that of Hewitt (1983), Watts (1983), Maskrey (1989, 2011), Blaikie et al. (1994), and Wisner et al. (2004).

However, while there is a longstanding awareness of the role of development policy and practice in shaping disaster risk, advances in the reduction of the underlying causes – the social, political, economic, and environmental drivers of disaster risk – remain insufficient to reduce hazard, exposure, and vulnerability in many regions (UNISDR, 2009e, 2011) (*high confidence*).

The difficult transition to more comprehensive disaster risk management raises challenges for the proper allocation of efforts among disaster risk reduction, risk transfer, and disaster management efforts. Countries exhibit a wide range of acceptance or resistance to the various challenges of risk management as seen from a development perspective, due to differential access to information and education, varying levels of debate and discussion, as well as contextual, ideological, institutional, and other related factors. The introduction of disaster risk reduction concerns in established disaster response agencies may in some cases have led to a downgrading of efforts to improve disaster response, diverting scarce resources in favor of risk reduction aspects (Alexander, 2000; DFID, 2004, 2005; Twigg, 2004).

The increasing emphasis placed on considering disaster risk management as a dimension of development, and thus of development planning, as opposed to strict post-impact disaster response efforts, has been accompanied by increasing emphasis and calls for proactive, prospective disaster risk prevention as opposed to reactive, corrective disaster risk mitigation (Lavell, 2003, 2010; UNISDR, 2009e, 2011).

The more recent emergence of integrated disaster risk management reflects a shift from the notion of disaster to the notion of disaster risk as a central concept and planning concern. Disaster risk management places increased emphasis on comprehensive disaster risk reduction. This shifting emphasis to risk reduction can be seen in the increasing importance placed on developing **resistance** to the potential impacts of physical events at various social or territorial scales, and in different temporal dimensions (such as those required for corrective or prospective risk management), and to increasing the resilience of affected communities. **Resistance** refers to the ability to avoid suffering significant adverse effects.

Within this context, disaster risk reduction and adaptation to climate change are undoubtedly far closer practically than when emergency or disaster management objectives dominated the discourse and practice. The fact that many in the climate change and disaster fields have associated disaster risk management principally with disaster preparedness and response, and not with disaster risk reduction per se, contributed to the view that the two practices are essentially different, if complementary (Lavell, 2010; Mercer, 2010). Once the developmental basis of adaptation to climate change and disaster risk management are considered, along with the role of vulnerability in the constitution of risk, the temporal scale of concerns, and the corrective as well as prospective nature of disaster risk reduction, the similarities between and options for merging of concerns and practices increases commensurately.

Section 1.3 examines the current status of adaptation to climate change, as a prelude to examining in more detail the barriers and options for greater integration of the two practices. The historical frame offered in this subsection comprises an introduction to that discussion.

1.1.4. Framing the Processes of Disaster Risk Management and Adaptation to Climate Change

In this section, we explore two of the key issues that should be considered in attempting to establish the overlap or distinction between the phenomena and social processes that concern disaster risk management on the one hand, and adaptation to climate change on the other, and that influence their successful practice: 1) the degree to which the focus is on extreme events (instead of a more inclusive approach that considers the full continuum of physical events with potential for damage, the social contexts in which they occur, and the potential for such events to generate 'extreme impacts' or disasters); and 2) consideration of the appropriate social-territorial scale that should be examined (i.e., aggregations, see Schneider et al., 2007) in order to foster a deeper understanding of the causes and effects of the different actors and processes at work.

1.1.4.1. Exceptionality, Routine, and Everyday Life

Explanations of loss and damage resulting from extreme events that focus primarily or exclusively on the physical event have been referred to as 'physicalist' (Hewitt, 1983). By contrast, notions developed around the continuum of normal, everyday-life risk factors through to a linked consideration of physical and social extremes have been defined as 'comprehensive,' 'integral,' or 'holistic' insofar as they embrace the social as well as physical aspects of disaster risk and take into consideration the evolution of experience over time (Cardona, 2001; ICSU-LAC, 2009). The latter perspective has been a major contributing factor in the development of the so-called 'vulnerability paradigm' as a basis for understanding disaster (Timmerman, 1981; Hewitt, 1983, 1997; Wisner et al., 2004; Eakin and Luers, 2006; NRC, 2006).

Additionally, attention to the role of small- and medium-scale disasters (UNISDR, 2009e, 2011) highlights the need to deal integrally with the problem of cumulative disaster loss and damage, looking across the different scales of experience both in human and physical worlds, in order to advance the efficacy of disaster risk management and adaptation. The design of mechanisms and strategies based on the reduction and elimination of everyday or chronic risk factors (Sen, 1983; World Bank, 2001), as opposed to actions based solely on the 'exceptional' or 'extreme' events, is one obvious corollary of this approach. The ability to deal with risk, crisis, and change is closely related to an individual's life experience with smaller-scale, more regular physical and social occurrences (Maskrey, 1989, 2011; Lavell, 2003; Wisner et al., 2004) (*high confidence*). These concepts point toward the possibility of reducing vulnerability and increasing resilience to climate-related disaster by broadly focusing on exposure, vulnerability, and socially-determined propensity or predisposition to adverse effects across a range of risks.

As illustrated in Box 1-1, many of the extreme impacts associated with climate change, and their attendant additional risks and opportunities, will inevitably need to be understood and responded to principally at the scale of the individual, the individual household, and the community, in the framework of localities and nations and their organizational and management options, and in the context of the many other day-to-day changes, including those of an economic, political, technological, and cultural nature. As this real example illustrates, everyday life, history, and a sequence of crises can affect attitudes and ways of approaching more extreme or complex problems. In contrast, many agents and institutions of disaster risk management and climate change adaptation activities necessarily operate from a different perspective, given the still highly centralized and hierarchical authority approaches found in many parts of the world today.

Whereas disaster risk management has been modified based on the experiences of the past 30 years or more, adaptation to anthropogenic climate change is a more recent issue on most decisionmakers' policy agendas and is not informed by such a long tradition of immediate experience. However, human adaptation to prevailing climate variability and change, and climate and weather extremes in past centuries and millennia, provides a wealth of experience from which the field of adaptation to climate change, and individuals and governments, can draw.

> **Box 1-1 | One Person's Experience with Climate Variability in the Context of Other Changes**
>
> Joseph is 80 years old. He and his father and his grandfather have witnessed many changes. Their homes have shifted back and forth from the steep slopes of the South Pare Mountains at 1,500 m to the plains 20 km away, near the Pangani River at 600 m, in Tanzania. What do 'changes' (mabadiliko) mean to someone whose father saw the Germans and British fight during the First World War and whose grandfather defended against Maasai cattle raids when Victoria was still Queen?
>
> Joseph outlived the British time. He saw African Socialism come and go after Independence. A road was constructed parallel to the old German rail line. Successions of commercial crops were dominant during his long life, some grown in the lowlands on plantations (sisal, kapok, and sugar), and some in the mountains (coffee, cardamom, ginger). He has seen staple foods change as maize became more popular than cassava and bananas. Land cover has also changed. Forest retreated, but new trees were grown on farms. Pasture grasses changed as the government banned seasonal burning. The Pangani River was dammed, and the electricity company decides how much water people can take for irrigation. Hospitals and schools have been built. Insecticide-treated bed nets recently arrived for the children and pregnant mothers.
>
> Joseph has nine plots of land at different altitudes spanning the distance from mountain to plain, and he keeps in touch with his children who work them by mobile phone. What is 'climate change' (mabadiliko ya tabia nchi) to Joseph? He has suffered and benefited from many changes. He has lived through many droughts with periods of hunger, witnessed floods, and also seen landslides in the mountains. He is skilled at seizing opportunities from changes – small and large: "Mabadiliko bora kuliko mapumziko" (Change is better than resting).
>
> *The provenance of this story is an original field work interview undertaken by Ben Wisner in November 2009 in Same District, Kilimanjaro Region, Tanzania in the context of the U.S. National Science Foundation-funded research project "Linking Local Knowledge and Local Institutions for the Study of Adaptive Capacity to Climate Change: Participatory GIS in Northern Tanzania."*

The ethnographic vignette in Box 1-1 suggests the way some individuals may respond to climate change in the context of previous experience, illustrating both the possibility of drawing successfully on past experience in adapting to climate variability, or, on the other hand, failing to comprehend the nature of novel risks.

1.1.4.2. Territorial Scale, Disaster Risk, and Adaptation

Climate-related disaster risk is most adequately depicted, measured, and monitored at the local or micro level (families, communities, individual buildings or production units, etc.) where the actual interaction of hazard and vulnerability are worked out *in situ* (Hewitt, 1983, 1997; Lavell, 2003; Wisner et al., 2004; Cannon, 2006; Maskrey, 2011). At the same time, it is accepted that disaster risk construction processes are not limited to specifically local or micro processes but, rather, to diverse environmental, economic, social, and ideological influences whose sources are to be found at scales from the international through to the national, sub-national and local, each potentially in constant flux (Lavell, 2002, 2003; Wisner et al., 2004, 2011).

Changing commodity prices in international trading markets and their impacts on food security and the welfare of agricultural workers, decisions on location and cessation of agricultural production by international corporations, deforestation in the upper reaches of river basins, and land use changes in urban hinterlands are but a few of these 'extra-territorial' influences on local risk. Moreover, disasters, once materialized, have ripple effects that many times go well beyond the directly affected zones (Wisner et al., 2004; Chapter 5) Disaster risk management and adaptation policy, strategies, and institutions will only be successful where understanding and intervention is based on multi-territorial and social-scale principles and where phenomena and actions at local, sub-national, national, and international scales are construed in interacting, concatenated ways (Lavell, 2002; UNISDR, 2009e, 2011; Chapters 5 through 9).

1.2. Extreme Events, Extreme Impacts, and Disasters

1.2.1. Distinguishing Extreme Events, Extreme Impacts, and Disasters

Both the disaster risk management and climate change adaptation literature define 'extreme weather' and 'extreme climate' events and discuss their relationship with 'extreme impacts' and 'disasters.' Classification of extreme events, extreme impacts, and disasters is influenced by the measured physical attributes of weather or climatic variables (see Section 3.1.2) or the vulnerability of social systems (see Section 2.4.1).

This section explores the quantitative definitions of different classes of extreme weather events, what characteristics determine that an impact is extreme, and how climate change affects the understanding of extreme climate events and impacts.

1.2.2. Extreme Events Defined in Physical Terms

1.2.2.1. Definitions of Extremes

Some literature reserve the term 'extreme event' for initial meteorological phenomena (Easterling et al., 2000; Jentsch et al., 2007), some include the consequential physical impacts, like flooding (Young, 2002), and some the entire spectrum of outcomes for humans, society, and ecosystems (Rich et al., 2008). In this report, we use 'extreme (weather or climate) event' to refer solely to the initial and consequent physical phenomena including some (e.g., flooding) that may have human components to causation other than that related to the climate (e.g., land use or land cover change or changes in water management; see Section 3.1.2 and Glossary). The spectrum of outcomes for humans, society, and physical systems, including ecosystems, are considered 'impacts' rather than part of the definition of 'events' (see Sections 1.1.2.1 and 3.1.2 and the Glossary).

In addition to providing a long-term mean of weather, 'climate' characterizes the full spectrum of means and exceptionality associated with 'unusual' and unusually persistent weather. The World Meteorological Organization (WMO, 2010) differentiates the terms in the following way (see also FAQ 6.1): "At the simplest level the weather is what is happening to the atmosphere at any given time. Climate in a narrow sense is usually defined as the 'average weather,' or more rigorously, as the statistical description in terms of the mean and variability of relevant quantities over a period of time."

Weather and climate phenomena reflect the interaction of dynamic and thermodynamic processes over a very wide range of space and temporal scales. This complexity results in highly variable atmospheric conditions, including temperatures, motions, and precipitation, a component of which is referred to as 'extreme events.' Extreme events include the passage of an intense tornado lasting minutes and the persistence of drought conditions over decades – a span of at least seven orders of magnitude of timescales. An imprecise distinction between extreme 'weather' and 'climate' events, based on their characteristic timescales, is drawn in Section 3.1.2. Similarly, the spatial scale of extreme climate or weather varies from local to continental.

Where there is sufficient long-term recorded data to develop a statistical distribution of a key weather or climate variable, it is possible to find the probability of experiencing a value above or below different thresholds of that distribution as is required in engineering design (trends may be sought in such data to see if there is evidence that the climate has not been stationary over the sample period; Milly et al., 2008). The extremity of a weather or climate event of a given magnitude depends on geographic context (see Section 3.1.2 and Box 3-1): a month of daily temperatures corresponding to the expected spring climatological daily maximum in Chennai, India, would be termed a heat wave in France; a snow storm expected every year in New York, USA, might initiate a disaster when it occurs in southern China. Furthermore, according to the location and social context, a 1-in-10 or 1-in-20 annual probability event may not be sufficient to result in unusual consequences. Nonetheless, universal thresholds can exist – for example, a reduction in the incidence or intensity of freezing days may allow certain disease vectors to thrive (e.g., Epstein et al., 1998). These various aspects are considered in the definition of 'extreme (weather and climate) events.'

The availability of observational data is of central relevance for defining climate characteristics and for disaster risk management; and, while data for temperature and precipitation are widely available, some associated variables, such as soil moisture, are poorly monitored, or, like extreme wind speeds and other low frequency occurrences, not monitored with sufficient spatial resolution or temporal continuity (Section 3.2.1).

1.2.2.2. Extremes in a Changing Climate

An extreme event in the present climate may become more common, or more rare, under future climate conditions. When the overall distribution of the climate variable changes, what happens to mean climate may be different from what happens to the extremes at either end of the distribution (see Figure 1-2).

For example, a warmer mean climate could result from fewer cold days, leading to a reduction in the variance of temperatures, or more hot days, leading to an expansion in the variance of the temperature distribution, or both. The issue of the scaling of changes in extreme events with respect to changes in mean temperatures is addressed further in Section 3.1.6.

In general, single extreme events cannot be simply and directly attributed to *anthropogenic* climate change, as there is always a possibility the event in question might have occurred without this contribution (Hegerl et al., 2007; Section 3.2.2; FAQ 3.2). However, for certain classes of regional, long-duration extremes (of heat and rainfall) it has proved possible to argue from climate model outputs that the probability of such an extreme has changed due to anthropogenic climate forcing (Stott et al., 2004; Pall et al., 2011).

Extremes sometimes result from the interactions between two unrelated geophysical phenomena such as a moderate storm surge coinciding with an extreme spring tide, as in the most catastrophic UK storm surge flood of the past 500 years in 1607 (Horsburgh and Horritt, 2006). Climate change may alter both the frequency of extreme surges and cause gradual sea level rise, compounding such future extreme floods (see Sections 3.5.3 and 3.5.5).

1.2.2.3. The Diversity and Range of Extremes

The specification of weather and climate extremes relevant to the concerns of individuals, communities, and governments depends on the affected stakeholder, whether in agriculture, disease control, urban design, infrastructure maintenance, etc. Accordingly, the range of such extremes is very diverse and varies widely. For example, whether it falls as rain, freezing rain (rain falling through a surface layer below freezing), snow, or hail, extreme precipitation can cause significant damage (Peters et al., 2001). The absence of precipitation (McKee et al., 1993) as well as excess evapotranspiration from the soil (see Box 3-3) can be climate extremes, and lead to drought. Extreme surface winds are chiefly associated with structured storm circulations (Emanuel, 2003; Zipser et al., 2006; Leckebusch et al., 2008). Each storm type, including the most damaging tropical cyclones and mid-latitude extratropical cyclones, as well as intense convective thunderstorms, presents a spectrum of size, forward speed, and intensity. A single intense storm can combine extreme wind and extreme rainfall.

The prolonged absence of winds is a climate extreme that can also be a hazard, leading to the accumulation of urban pollution and disruptive fog (McBean, 2006).

The behavior of the atmosphere is also highly interlinked with that of the hydrosphere, cryosphere, and terrestrial environment so that extreme (or sometimes non-extreme) atmospheric events may cause (or contribute to) other rare physical events. Among the more widely documented hydroclimatic extremes are:

- Large cyclonic storms that generate wind and pressure anomalies causing coastal flooding and severe wave action (Xie et al., 2004).
- Floods, reflecting river flows in excess of the capacity of the normal channel, often influenced by human intervention and water management, resulting from intense precipitation; rapid thaw of accumulated winter snowfall; rain falling on previous snowfall (Sui and Koehler, 2001); or an outburst from an ice, landslide, moraine, or artificially dammed lake (de Jong et al., 2005). According to the scale of the catchment, river systems have characteristic response times with steep short mountain streams, desert wadis, and urban drainage systems responding to rainfall totals over a few hours, while peak flows in major continental rivers reflect regional precipitation extremes lasting weeks (Wheater, 2002).
- Long-term reductions in precipitation, or dwindling of residual summer snow and ice melt (Rees and Collins, 2006), or increased evapotranspiration from higher temperatures, often exacerbated by human groundwater extraction, reducing ground water levels and causing spring-fed rivers to disappear (Konikow and Kendy, 2005), and contributing to drought.
- Landslides (Dhakal and Sidle, 2004) when triggered by raised groundwater levels after excess rainfall or active layer detachments in thawing slopes of permafrost (Lewcowicz and Harris, 2005).

1.2.3. Extreme Impacts

1.2.3.1. Three Classes of Impacts

In this subsection we consider three classes of 'impacts': 1) changes in the natural physical environment, like beach erosion from storms and mudslides; 2) changes in ecosystems, such as the blow-down of forests in hurricanes, and 3) adverse effects (according to a variety of metrics) on human or societal conditions and assets. However, impacts are not always negative: flood-inducing rains can have beneficial effects on the following season's crops (Khan, 2011), while an intense freeze may reduce insect pests at the subsequent year's harvest (Butts et al., 1997).

An **extreme impact** reflects highly significant and typically long-lasting consequences to society, the natural physical environment, or ecosystems. Extreme impacts can be the result of a single extreme event, successive extreme or non-extreme events, including non-climatic events (e.g., wildfire, followed by heavy rain leading to landslides and soil erosion), or simply the persistence of conditions, such as those that lead to drought (see Sections 3.5.1 and 9.2.3 for discussion and examples).

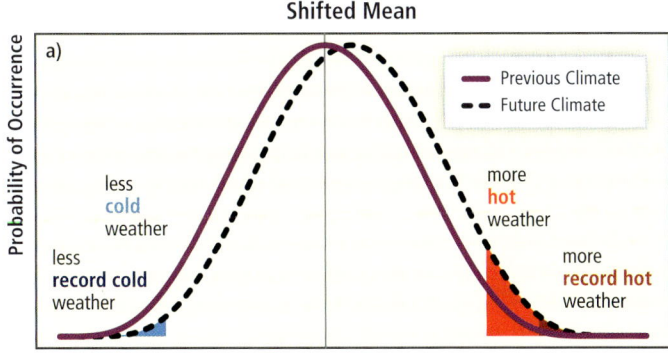
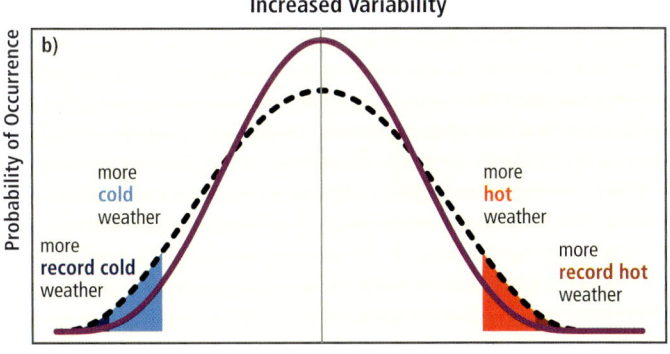
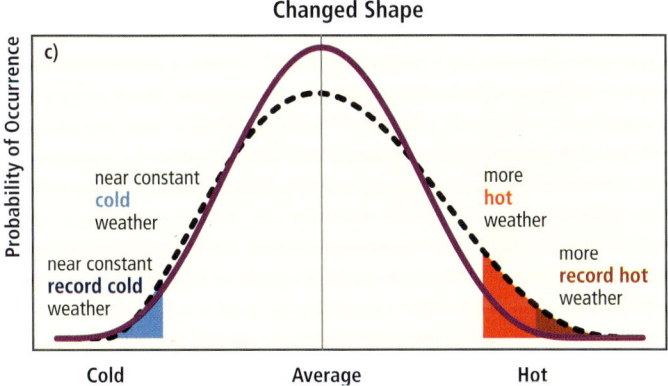

Figure 1-2 | The effect of changes in temperature distribution on extremes. Different changes in temperature distributions between present and future climate and their effects on extreme values of the distributions: a) effects of a simple shift of the entire distribution toward a warmer climate; b) effects of an increased temperature variability with no shift of the mean; and c) effects of an altered shape of the distribution, in this example an increased asymmetry toward the hotter part of the distribution.

Whether an extreme event results in extreme impacts on humans and social systems depends on the degree of exposure and vulnerability to that extreme, in addition to the magnitude of the physical event (*high confidence*). Extreme impacts on human systems may be associated with non-extreme events where vulnerability and exposure are high (Sections 1.1.2.1 and 9.2.3). A key weather parameter may cross some critical value at that location (such as that associated with heat wave-induced mortality, or frost damage to crops), so that the distribution of the impact shifts in a way that is disproportionate to physical changes (see Section 4.2). A comprehensive assessment of projected impacts of climate changes would consider how changes in atmospheric conditions (temperature, precipitation) translate to impacts on physical (e.g., droughts and floods, erosion of beaches and slopes, sea level rise), ecological (e.g., forest fires), and human systems (e.g., casualties, infrastructure damages). For example, an extreme event with a large spatial scale (as in an ice storm or windstorm) can have an exaggerated, disruptive impact due to the systemic societal dependence on electricity transmission and distribution networks (Peters et al., 2006). Links between climate events and physical impacts are addressed in Section 3.5, while links to ecosystems and human systems impacts are addressed in 4.3.

Disaster signifies extreme impacts suffered by society, which may also be associated with extreme impacts on the physical environment and on ecosystems. Building on the definition set out in Section 1.1.2.1, extreme impacts resulting from weather, climate, or hydrological events can become disasters once they surpass thresholds in at least one of three dimensions: spatial – so that damages cannot be easily restored from neighboring capacity; temporal – so that recovery becomes frustrated by further damages; and intensity of impact on the affected population – thereby undermining, although not necessarily eliminating, the capacity of the society or community to repair itself (Alexander, 1993). However, for the purposes of tabulating occurrences, some agencies only list 'disasters' when they exceed certain numbers of killed or injured or total repair costs (Below et al., 2009; CRED, 2010).

1.2.3.2. Complex Nature of an Extreme 'Event'

In considering the range of weather and climate extremes, along with their impacts, the term 'event' as used in the literature does not adequately capture the compounding of outcomes from successive physical phenomena, for example, a procession of serial storms tracking across the same region (as in January and February 1990 and December 1999 across Western Europe, Ulbrich et al., 2001). In focusing on the social context of disasters, Quarantelli (1986) proposed the use of the notion of 'disaster occurrences or occasions' in place of 'events' due to the abrupt and circumstantial nature of the connotation commonly attributed to the word 'event,' which belies the complexity and temporality of disaster, in particular because social context may precondition and extend the duration over which impacts are felt.

Sometimes locations affected by extremes within the 'same' large-scale stable atmospheric circulation can be far apart, as for example the Russian heat wave and Indus valley floods in Pakistan in the summer of 2010 (Lau and Kim, 2011). Extreme events can also be interrelated through the atmospheric teleconnections that characterize the principal drivers of oceanic equatorial sea surface temperatures and winds in the El Niño–Southern Oscillation. The relationship between modes of climate variability and extremes is discussed in greater detail in Section 3.1.1.

The aftermath of one extreme event may precondition the physical impact of successor events. High groundwater levels and river flows can persist for months, increasing the probability of a later storm causing flooding, as on the Rhine in 1995 (Fink et al., 1996). A thickness reduction in Arctic sea ice preconditions more extreme reductions in the summer ice extent (Holland et al., 2006). A variety of feedbacks and other interactions connect extreme events and physical system and ecological responses in a way that may amplify physical impacts (Sections 3.1.4 and 4.3.5). For example, reductions in soil moisture can intensify heat waves (Seneviratne et al., 2006), while droughts following rainy seasons turn vegetation into fuel that can be consumed in wildfires (Westerling and Swetman, 2003), which in turn promote soil runoff and landslides when the rains return (Cannon et al., 2001). However, extremes can also interact to reduce disaster risk. The wind-driven waves in a hurricane bring colder waters to the surface from beneath the thermocline; for the next month, any cyclone whose path follows too closely will have a reduced potential maximum intensity (Emanuel, 2001). Intense rainfall accompanying monsoons and hurricanes also brings great benefits to society and ecosystems; on many occasions it helps to fill reservoirs, sustain seasonal agriculture, and alleviate summer dry conditions in arid zones (e.g., Cavazos et al., 2008).

1.2.3.3. Metrics to Quantify Social Impacts and the Management of Extremes

Metrics to quantify social and economic impacts (thus used to define extreme impacts) may include, among others (Below et al., 2009):
- Human casualties and injuries
- Number of permanently or temporarily displaced people
- Number of directly and indirectly affected persons
- Impacts on properties, measured in terms of numbers of buildings damaged or destroyed
- Impacts on infrastructure and lifelines
- Impacts on ecosystem services
- Impacts on crops and agricultural systems
- Impacts on disease vectors
- Impacts on psychological well being and sense of security
- Financial or economic loss (including insurance loss)
- Impacts on coping capacity and need for external assistance.

All of these may be calibrated according to the magnitude, rate, duration, and degree of irreversibility of the effects (Schneider et al., 2007). These metrics may be quantified and implemented in the context of probabilistic risk analysis in order to inform policies in a variety of contexts (see Box 1-2).

Box 1-2 | Probabilistic Risk Analysis

In its simplest form, probabilistic risk analysis defines risk as the product of the probability that some event (or sequence) will occur and the adverse consequences of that event.

$$\text{Risk} = \text{Probability} \times \text{Consequence} \quad (1)$$

For instance, the risk a community faces from flooding from a nearby river might be calculated based on the likelihood that the river floods the town, inflicting casualties among inhabitants and disrupting the community's economic livelihood. This likelihood is multiplied by the value people place on those casualties and economic disruption. Equation (1) provides a quantitative representation of the qualitative definition of disaster risk given in Section 1.1. All three factors – hazard, exposure, and vulnerability – contribute to 'consequences.' Hazard and vulnerability can both contribute to the 'probability': the former to the likelihood of the physical event (e.g., the river flooding the town) and the latter to the likelihood of the consequence resulting from the event (e.g., casualties and economic disruption).

When implemented within a broader risk governance framework, probabilistic risk analysis can help allocate and evaluate efforts to manage risk. Equation (1) implies what the decision sciences literature (Morgan and Henrion, 1990) calls a decision rule – that is, a criterion for ranking alternative sets of actions by their ability to reduce overall risk. For instance, an insurance company (as part of a risk transfer effort) might set the annual price for flood insurance based on multiplying an estimate of the probability a dwelling would be flooded in any given year by an estimate of the monetary losses such flooding would cause. Ideally, the premiums collected from the residents of many dwellings would provide funds to compensate the residents of those few dwellings that are in fact flooded (and defray administrative costs). In another example, a water management agency (as part of a risk reduction effort) might invest the resources to build a reservoir of sufficient size so that, if the largest drought observed in their region over the last 100 years (or some other timeframe) occurred again in the future, the agency would nonetheless be able to maintain a reliable supply of water.

A wide variety of different expressions of the concepts in Equation (1) exist in the literature. The disaster risk management community often finds it convenient to express risk as a product of hazard, exposure, and vulnerability (e.g., UNISDR, 2009e, 2011). In addition, the decision sciences literature recognizes decision rules, useful in some circumstances, that do not depend on probability and consequence as combined in Equation (1). For instance, if the estimates of probabilities are sufficiently imprecise, decisionmakers might use a criterion that depends only on comparing estimates of potential consequences (e.g., mini-max regret, Savage, 1972).

In practice, probabilistic risk analysis is often not implemented in its pure form for reasons including data limitations; decision rules that yield satisfactory results with less effort than that required by a full probabilistic risk assessment; the irreducible imprecision of some estimates of important probabilities and consequences (see Sections 1.3.1.1 and 1.3.2); and the need to address the wide range of factors that affect judgments about risk (see Box 1-3). In the above example, the water management agency is not performing a full probabilistic risk analysis, but rather employing a hybrid decision rule in which it estimates that the consequences of running out of water would be so large as to justify any reasonable investment needed to keep the likelihood of that event below the chosen probabilistic threshold. Chapter 2 describes a variety of practical quantitative and qualitative approaches for allocating efforts to manage disaster risk.

The probabilistic risk analysis framework in its pure form is nonetheless important because its conceptual simplicity aids understanding by making assumptions explicit, and because its solid theoretical foundations and the vast empirical evidence examining its application in specific cases make it an important point of comparison for formal evaluations of the effectiveness of efforts to manage disaster risk.

Information on direct, indirect, and collateral impacts is generally available for many large-scale disasters and is systematized and provided by organizations such as the Economic Commission for Latin America, large reinsurers, and the EM-DAT database (CRED, 2010). Information on impacts of smaller, more recurrent events is far less accessible and more restricted in the number of robust variables it provides. The Desinventar database (Corporación OSSO, 2010), now available for 29 countries worldwide, and the Spatial Hazard Events and Losses Database for the United States (SHELDUS; HVRI, 2010), are attempts to satisfy this need. However, the lack of data on many impacts impedes complete knowledge of the global social and economic impacts of smaller-scale disasters (UNISDR, 2009e).

1.2.3.4. Traditional Adjustment to Extremes

Disaster risk management and climate change adaptation may be seen as attempts to duplicate, promote, or improve upon adjustments that society and nature have accomplished on many occasions spontaneously in the past, if over a different range of conditions than expected in the future.

Within the sphere of adaptation of natural systems to climate, among trees, for example, natural selection has the potential to evolve appropriate resilience to extremes (at some cost). Resistance to windthrow is strongly species-dependent, having evolved according to the climatology where that tree was indigenous (Canham et al., 2001). In their original habitat, trees typically withstand wind extremes expected every 10 to 50 years, but not extremes that lie beyond their average lifespan of 100 to 500 years (Ostertag et al., 2005).

In human systems, communities traditionally accustomed to periodic droughts employ wells, boreholes, pumps, dams, and water harvesting and irrigation systems. Those with houses exposed to high seasonal temperatures employ thick walls and narrow streets, have developed passive cooling systems, adapted lifestyles, or acquired air conditioning. In regions unaccustomed to heat waves, the absence of such systems, in particular in the houses of the most vulnerable elderly or sick, contributes to excess mortality, as in Paris, France, in August 2003 (Vandentorren et al., 2004) or California in July 2006 (Gershunov et al., 2009).

The examples given above of 'spontaneous' human system adjustment can be contrasted with explicit measures that are taken to reduce risk from an expected range of extremes. On the island of Guam, within the most active and intense zone of tropical cyclone activity on Earth, buildings are constructed to the most stringent wind design code in the world. Buildings are required to withstand peak gust wind speeds of 76 ms^{-1}, expected every few decades (International Building Codes, 2003). More generally, annual wind extremes for coastal locations will typically be highest at mid-latitudes while those expected once every century will be highest in the 10° to 25° latitude tropics (Walshaw, 2000). Consequently, indigenous building practices are less likely to be resilient close to the equator than in the windier (and storm surge affected) mid-latitudes (Minor, 1983).

While local experience provides a reservoir of knowledge from which disaster risk management and adaptation to climate change are drawing (Fouillet et al., 2008), it may not be available to other regions yet to be affected by such extremes. Thus, these experiences may not be drawn upon to provide guidance if future extremes go outside the traditional or recently observed range, as is expected for some extremes as the climate changes (see Chapter 3).

1.3. Disaster Management, Disaster Risk Reduction, and Risk Transfer

One important component of both disaster risk management and adaptation to climate change is the appropriate allocation of efforts among disaster management, disaster risk reduction, and risk transfer, as defined in Section 1.1.2.2. The current section provides a brief survey of the risk governance framework for making judgments about such an allocation, suggests why climate change may complicate effective management of disaster risks, and identifies potential synergies between disaster risk management and adaptation to climate change.

Disaster risks appear in the context of human choices that aim to satisfy human wants and needs (e.g., where to live and in what types of dwelling, what vehicles to use for transport, what crops to grow, what infrastructure to support economic activities, Hohenemser et al., 1984; Renn, 2008). Ideally, the choice of any portfolio of actions to address disaster risk would take into consideration human judgments about what constitutes risk, how to weigh such risk alongside other values and needs, and the social and economic contexts that determine whose judgments influence individuals' and societal responses to those risks.

The *risk governance* framework offers a systematic way to help situate such judgments about disaster management, risk reduction, and risk transfer within this broader context. Risk governance, under Renn's (2008) formulation, consists of four phases – pre-assessment, appraisal, characterization/evaluation, and management – in an open, cyclical, iterative, and interlinked process. Risk communication accompanies all four phases. This process is consistent with those in the UNISDR Hyogo Framework for Action (UNISDR, 2005), the best known and adhered to framework for considering disaster risk management concerns (see Chapter 7).

As one component of its broader approach, risk governance uses concepts from probabilistic risk analysis to help judge appropriate allocations in level of effort and over time and among risk reduction, risk transfer, and disaster management actions. The basic probabilistic risk analytic framework for considering such allocations regards risk as the product of the probability of an event(s) multiplied by its consequence (see Box 1-2; Bedford and Cooke, 2001). In this formulation, *risk reduction* aims to reduce exposure and vulnerability as well as the probability of occurrence of some events (e.g., those associated with landslides and forest fires induced by human intervention). *Risk transfer* efforts aim to compensate losses suffered by those who directly experience an event. *Disaster management* aims to respond to the immediate consequences and facilitate reduction of longer-term consequences (see Section 1.1).

Probabilistic risk analysis can help compare the efficacy of alternative actions to manage risk and inform judgments about the appropriate allocation of resources to reduce risk. For instance, the framework suggests that equivalent levels of risk reduction result from reducing an event's probability or by reducing its consequences by equal percentages. Probabilistic risk analysis also suggests that a series of relatively smaller, more frequent events could pose the same risk as a single, relatively less frequent, larger event. Probabilistic risk analysis can help inform decisions about alternative allocations of risk management efforts by facilitating the comparison of the increase or decrease in risk resulting from the alternative allocations (*high confidence*). Since the costs of available

> **Box 1-3 | Influence of Cognitive Processes, Culture, and Ideology on Judgments about Risk**
>
> A variety of cognitive, cultural, and social processes affect judgments about risk and about the allocation of efforts to address these risks. In addition to the processes described in Section 1.3.1.2, subjective judgments may be influenced more by emotional reactions to events (e.g., feelings of fear and loss of control) than by analytic assessments of their likelihood (Loewenstein et al., 2001). People frequently ignore predictions of extreme events if those predictions fail to elicit strong emotional reactions, but will also overreact to such forecasts when the events elicit feelings of fear or dread (Slovic et al., 1982; Slovic 1993, 2010; Weber, 2006). Even with sufficient information, everyday concerns and satisfaction of basic wants may prove a more pressing concern than attention and effort toward actions to address longer-term disaster risk (Maskrey, 1989, 2011; Wisner et al., 2004).
>
> In addition to being influenced by cognitive shortcuts (Kahneman and Tversky, 1979), the perceptions of risk and extremes and reactions to such risk and events are also shaped by motivational processes (Weber, 2010). Cultural theory combines insights from anthropology and political science to provide a conceptual framework and body of empirical studies that seek to explain societal conflict over risk (Douglas, 1992). People's worldview and political ideology guide attention toward events that threaten their desired social order (Douglas and Wildavsky, 1982). Risk in this framework is defined as the disruption of a social equilibrium. Personal beliefs also influence which sources of expert forecasts of extreme climate events will be trusted. Different cultural groups put their trust into different organizations, from national meteorological services to independent farm organizations to the IPCC; depending on their values, beliefs, and corresponding mental models, people will be receptive to different types of interventions (Dunlap and McCright, 2008; Malka and Krosnick, 2009). Judgments about the veracity of information regarding the consequences of alternative actions often depend on the perceived consistency of those actions with an individual's cultural values, so that individuals will be more willing to consider information about consequences that can be addressed with actions seen as consistent with their values (Kahan and Braman, 2006; Kahan et al., 2007).
>
> Factual information interacts with social, institutional, and cultural processes in ways that may amplify or attenuate public perceptions of risk and extreme events (Kasperson et al., 1988). The US public's estimates of the risk of nuclear power following the accident at Three Mile Island provide an example of the socio-cultural filtering of engineering safety data. Social amplification increased public perceptions of the risk of nuclear power far beyond levels that would derive only from analysis of accident statistics (Fischhoff et al., 1983). The public's transformation of expert-provided risk signals can serve as a corrective mechanism by which cultural subgroups of society augment a science-based risk analysis with psychological risk dimensions not considered in technical risk assessments (Slovic, 2000). Evidence from health, social psychology, and risk communication literature suggests that social and cultural risk amplification processes modify perceptions of risk in either direction and in ways that may generally be socially adaptive, but can also bias reactions in socially undesirable ways in specific instances (APA, 2009).

risk reduction, risk transfer, and disaster management actions will in general differ, the framework can help inform judgments about an effective mix of such actions in any particular case (see UNISDR, 2011, for efforts at stratifying different risk levels as a prelude to finding the most adequate mix of disaster risk management actions).

Probabilistic risk analysis is, however, rarely implemented in its pure form, in part because quantitative estimates of hazard and vulnerability are not always available and are not numbers that are independent of the individuals making those estimates. Rather, these estimates are determined by a combination of direct physical consequences of an event and the interaction of psychological, social, institutional, and cultural processes (see Box 1-3). For instance, perceptions of the risks of a nuclear power plant may be influenced by individuals' trust in the people operating the plant and by views about potential linkages between nuclear power and nuclear weapons proliferation – factors that may not be considered in a formal risk assessment for any given plant. Given this social construction of risk (see Section 1.1.2.2), effective allocations of efforts among risk reduction, risk transfer, and disaster management may best emerge from an integrated risk governance process, which includes the pre-assessment, appraisal, characterization/evaluation, and ongoing communications elements. Disaster risk management and adaptation to climate change each represent approaches that already use or could be improved by the use of this risk governance process, but as described in Section 1.3.1, climate change poses a particular set of additional challenges.

Together, the implications of probabilistic risk analysis and the social construction of risk reinforce the following considerations with regard to the effective allocation and implementation of efforts to manage risks in both disaster risk management and adaptation to climate change:
- As noted in Section 1.1, vulnerability, exposure, and hazard are each critical to determining disaster risk and the efficacy of actions taken to manage that risk (*high confidence*).
- Effective disaster risk management will in general require a portfolio of many types of risk reduction, risk transfer, and disaster management actions appropriately balanced in terms of resources applied over time (*high confidence*).

- Participatory and decentralized processes that are linked to higher levels of territorial governance (regions, nation) are a crucial part of all the stages of risk governance that include identification, choice, and implementation of these actions (*high confidence*).

1.3.1. Climate Change Will Complicate Management of Some Disaster Risks

Climate change will pose added challenges in many cases for attaining disaster risk management goals, and appropriately allocating efforts to manage disaster risks, for at least two sets of reasons. First, as discussed in Chapters 3 and 4, climate change is very likely to increase the occurrence and vary the location of some physical events, which in turn will affect the exposure faced by many communities, as well as their vulnerability. Increased exposure and vulnerability would contribute to an increase in disaster risk. For example, vulnerability may increase due to direct climate-related impacts on the development and development potential of the affected area, because resources otherwise available and directed towards development goals are deflected to respond to those impacts, or because long-standing institutions for allocating resources such as water no longer function as intended if climate change affects the scarcity and distribution of that resource. Second, climate change will make it more difficult to anticipate, evaluate, and communicate both probabilities and consequences that contribute to disaster risk, in particular that associated with extreme events. This set of issues, discussed in this subsection, will affect the management of these risks as discussed in Chapters 5, 6, 7, and 8 (*high confidence*).

1.3.1.1. Challenge of Quantitative Estimates of Changing Risks

Extreme events pose a particular set of challenges for implementing probabilistic approaches because their relative infrequency often makes it difficult to obtain adequate data for estimating the probabilities and consequences. Climate change exacerbates this challenge because it contributes to potential changes in the frequency and character of such events (see Section 1.2.2.2).

The likelihood of extreme events is most commonly described by the return period, the mean interval expected between one such event and its recurrence. For example, one might speak of a 100-year flood or a 50-year windstorm. More formally, these intervals are inversely proportional to the 'annual exceedance probability,' the likelihood that an event exceeding some magnitude occurs in any given year. Thus the 100-year flood has a 1% chance of occurring in any given year (which translates into a 37% chance of a century passing without at least one such flood ($(1-0.01)^{100} = 37\%$). Though statistical methods exist to estimate frequencies longer than available data time series (Milly et al., 2002), the long return period of extreme events can make it difficult, if not impossible, to reliably estimate their frequency. Paleoclimate records make clear that in many regions of the world, the last few decades of observed climate data do not represent the full natural variability of many important climate variables (Jansen et al., 2003). In addition, future climate change exacerbates the challenge of non-stationarity (Milly et al., 2008), where the statistical properties of weather events will not remain constant over time. This complicates an already difficult estimation challenge by altering frequencies and consequences of extremes in difficult-to-predict ways (Chapter 3; Meehl et al., 2007; TRB, 2008; NRC, 2009).

Estimating the likelihood of different consequences and their value is at least as challenging as estimating the likelihood of extreme events. Projecting future vulnerability and response capacity involves predicting the trends and changes in underlying causes of human vulnerability and the behavior of complex human systems under potentially stressful and novel conditions. For instance, disaster risk is endogenous in the sense that near-term actions to manage risk may affect future risk in unintended ways and near-term actions may affect perceptions of future risks (see Box 1-3). Section 1.4 describes some of the challenges such system complexity may pose for effective risk assessment. In addition, disasters affect socioeconomic systems in multiple ways so that assigning a quantitative value to the consequences of a disaster proves difficult (see Section 1.2.3.3). The literature distinguishes between direct losses, which are the immediate consequences of the disaster-related physical events, and indirect losses, which are the consequences that result from the disruption of life and activity after the immediate impacts of the event (Pelling et al., 2002; Lindell and Prater, 2003; Cochrane, 2004; Rose, 2004). Section 1.3.2 discusses some means to address these challenges.

1.3.1.2. Processes that Influence Judgments about Changing Risks

Effective risk governance engages a wide range of stakeholder groups – such as scientists, policymakers, private firms, nongovernmental organizations, media, educators, and the public – in a process of exchanging, integrating, and sharing knowledge and information. The recently emerging field of sustainability science (Kates et al., 2001) promotes interactive co-production of knowledge between experts and other actors, based on transdisciplinarity (Jasanoff, 2004; Pohl et al., 2010) and social learning (Pelling et al., 2008; Pahl-Wostl, 2009; see also Section 1.4.2). The literature on judgment and decisionmaking suggests that various cognitive behaviors involving perceptions and judgments about low-probability, high-severity events can complicate the intended functioning of such stakeholder processes (see Box 1-3). Climate change can exacerbate these challenges (*high confidence*).

The concepts of disaster, risk, and disaster risk management have very different meanings and interpretations in expert and non-expert contexts (Sjöberg, 1999a; see also Pidgeon and Fischhoff, 2011). Experts acting in formal private and public sector roles often employ quantitative estimates of both probability and consequence in making judgments about risk. In contrast, the general public, politicians, and the media tend to focus on the concrete adverse consequences of such events, paying less attention to their likelihood (Sjöberg, 1999b). As described in Box 1-3, expert estimates of probability and consequence may also not address the full range of concerns people bring to the consideration of risk. By definition (if not always in practice), expert understanding of risks associated with extreme events is based in large part on analytic tools. In particular, any estimates of changes in disaster risk due to climate change are often based on the results of complex climate models as described in Chapter 3. Non-experts, on the other hand, rely to a greater extent on more readily available and more easily processed information, such as their own experiences or vicarious experiences from the stories communicated through the news media, as well as their subjective judgment as to the importance of such events (see Box 1-1). These gaps between expert and non-expert understanding of extreme events present important communication challenges (Weber and Stern, 2011), which may adversely affect judgments about the allocation of efforts to address risk that is changing over time (*high confidence*).

Quantitative methods based on probabilistic risk analysis, such as those described in Sections 5.5 and 6.3, can allow people operating in expert contexts to use observed data, often from long time series, to make systematic and internally consistent estimates of the probability of future events. As described in Section 1.3.1.1, climate change may reduce the accuracy of such past observations as predictors for future risk. Individuals, including non-experts and experts making estimates without the use of formal methods (Barke et al., 1997), often predict the likelihood of encountering an event in the future by consulting their past experiences with such events. The 'availability' heuristic (i.e., useful shortcut) is commonly applied, in which the likelihood of an event is judged by the ease with which past instances can be brought to mind (Tversky and Kahneman, 1974). Extreme events, by definition, have a low probability of being represented in past experience and thus will be relatively unavailable. Experts and non-experts alike may essentially ignore such events until they occur, as in the case of a 100-year flood (Hertwig et al., 2004). When extreme events do occur with severe and thus memorable consequences, people's estimates of their future risks will, at least temporarily, become inflated (Weber et al., 2004).

1.3.2. Adaptation to Climate Change Contributes to Disaster Risk Management

The literature and practice of adaptation to climate change attempts to anticipate future impacts on human society and ecosystems, such as those described in Chapter 4, and respond to those already experienced. In recent years, the adaptation to climate change literature has introduced the concept of climate-related decisions (and climate proofing), which are choices by individuals or organizations, the outcomes of which can be expected to be affected by climate change and its interactions with ecological, economic, and social systems (Brown et al., 2006; McGray et al., 2007; Colls et al., 2009; Dulal et al., 2009; NRC, 2009). For instance, choosing to build in a low-lying area whose future flooding risk increases due to climate change represents a climate-related decision. Such a decision is climate-related whether or not the decisionmakers recognize it as such. The disaster risk management community may derive added impetus from the new context of a changing climate for certain of its pre-existing practices that already reflect the implementation of this concept. In many circumstances, choices about the appropriate allocation of efforts among disaster management, disaster risk reduction, and risk transfer actions will be affected by changes in the frequency and character of extreme events and other impacts of a climate change on the underlying conditions that affect exposure and vulnerability.

Much of the relevant adaptation literature addresses how expectations about future deviations from past patterns in physical, biological, and socioeconomic conditions due to climate change should affect the allocation of efforts to manage risks. While there exist differing views on the extent to which the adaptation to climate change literature has unique insights on managing changing conditions per se that it can bring to disaster risk management (Lavell, 2010; Mercer, 2010; Wisner et al., 2011), the former field's interest in anticipating and responding to the full range of consequences from changing climatic conditions can offer important new perspectives and capabilities to the latter field.

The disaster risk management community can benefit from the debates in the adaptation literature about how to best incorporate information about current and future climate into climate-related decisions. Some adaptation literature has emphasized the leading role of accurate regional climate predictions as necessary to inform such decisions (Collins, 2007; Barron, 2009; Doherty et al., 2009; Goddard et al., 2009; Shukla et al., 2009; Piao et al., 2010; Shapiro et al., 2010). This argument has been criticized on the grounds that predictions of future climate impacts are highly uncertain (Dessai and Hulme, 2004; Cox and Stephenson, 2007; Stainforth et al., 2007; Dessai et al., 2009; Hawkins and Sutton, 2009; Knutti, 2010) and that predictions are insufficient to motivate action (Fischhoff, 1994; Sarewitz et al., 2000; Cash et al., 2003, 2006; Rayner et al., 2005; Moser and Luers, 2008; Dessai et al., 2009; NRC, 2009). Other adaptation literature has emphasized that many communities do not sufficiently manage current risks and that improving this situation would go a long way toward preparing them for any future changes due to climate change (Smit and Wandel, 2006; Pielke et al., 2007). As discussed in Section 1.4, this approach will in some cases underestimate the challenges of adapting to future climate change.

To address these challenges, the adaptation literature has increasingly discussed an iterative risk management framework (Carter et al., 2007; Jones and Preston, 2011), which is consistent with risk governance as described earlier in this section. Iterative risk management recognizes that the process of anticipating and responding to climate change does not constitute a single set of judgments at some point in time, but rather an ongoing assessment, action, reassessment, and response that will continue – in the case of many climate-related decisions – indefinitely (ACC, 2010). In many cases, iterative risk management contends with conditions where the probabilities underlying estimates of future risk are imprecise and/or the structure of the models that relate events to consequences are under-determined (NRC, 2009; Morgan et al., 2009). Such deep or severe uncertainty (Lempert and Collins, 2007) can characterize not only understanding of future climatic events but also future patterns of human vulnerability and the capability to respond to such events. With many complex, poorly understood physical and socioeconomic systems, research and social learning may enrich understanding over time, but the amount of uncertainty, as measured by observers' ability to make specific, accurate predictions, may grow larger (Morgan et al., 2009, pp. 114–115; NRC, 2009, pp. 18–19; see related discussion of 'surprises' in Section 3.1.7). In addition, theory and models may change in ways that make them less, rather than more, reliable as predictive tools over time (Oppenheimer et al., 2008).

Recent literature has thus explored a variety of approaches that can help disaster risk management address such uncertainties (McGray et al., 2007; IIED 2009; Schipper, 2009), in particular approaches that help support decisions when it proves difficult or impossible to accurately estimate probabilities of events and their adverse consequences. Approaches for characterizing uncertainty include qualitative scenario methods (Parson et al., 2007); fuzzy sets (Chongfu, 1996; El-Baroudy and Simonovic, 2004; Karimi and Hullermeier, 2007; Simonovic, 2010); and the use of ranges of values or sets of distributions, rather than single values or single best-estimate distributions (Morgan et al., 2009; see also Mastrandrea et al., 2010). Others have suggested managing such uncertainty with robust policies that perform well over a wide range of plausible futures (Dessai and Hulme, 2007; Groves and Lempert, 2007; Brown, 2010; Means et al., 2010; Wilby and Dessai, 2010; Dessai and Wilby, 2011; Reeder and Ranger, 2011; also see discussion in Chapter 8). Decision rules based on the concept of robust adaptive policies go beyond 'no regrets' by suggesting how in some cases relatively low-cost, near-term actions and explicit plans to adjust those actions over time can significantly improve future ability to manage risk (World Bank, 2009; Hine and Hall, 2010; Lempert and Groves, 2010; Walker et al., 2010; Brown, 2011; Ranger and Garbett-Shiels, 2011; see also Section 1.4.5).

The resilience literature, as described in Chapter 8, also takes an interest in managing difficult-to-predict futures. Both the adaptation to climate change and vulnerability literatures often take an actor-oriented view (Wisner et al., 2004; McLaughlin and Dietz, 2007; Nelson et al., 2007; Moser 2009) that focuses on particular agents faced with a set of decisions who can make choices based on their various preferences; their institutional interests, power, and capabilities; and the information they have available. Robustness in the adaptation to climate change context often refers to a property of decisions specific actors may take (Hallegatte, 2009; Lempert and Groves, 2010; Dessai and Wilby, 2011). In contrast, the resilience literature tends to take a systems view (Olsson et al., 2006; Walker et al., 2006; Berkes, 2007; Nelson et al., 2007) that considers multi-interacting agents and their relationships in and with complex social, ecological, and geophysical systems (Miller et al., 2010). These literatures can help highlight for disaster risk management such issues as the tension between resilience to specific, known disturbances and novel and unexpected ones (sometimes referred to as the distinction between 'specified' and 'general' resilience, Miller et al., 2010), the tension between resilience at different spatial and temporal scales, and the tension between the ability of a system to persist in its current state and its ability to transform to a fundamentally new state (Section 1.4; Chapter 8; ICSU, 2002; Berkes, 2007).

Disaster risk management will find similarities to its own multi-sector approach in the adaptation literature's recent emphasis, consistent with the concept of climate-related decisions, on climate change as one of many factors affecting the management of risks. For instance, some resource management agencies now stress climate change as one of many trends such as growing demand for resources, environmental constraints, aging infrastructure, and technological change that, particularly in combination, could require changes in investment plans and business models (CCSP, 2008; Brick et al., 2010). It has become clear that many less-developed regions will have limited success in reducing overall vulnerability solely by managing climate risk because vulnerability, adaptive capacity, and exposure are critically influenced by existing structural deficits (low income and high inequality, lack of access to health and education, lack of security and political access, etc.). For example, in drought-ravaged northeastern Brazil, many vulnerable households could not take advantage of risk management interventions such as seed distribution programs because they lacked money to travel to pick up the seeds or could not afford a day's lost labor to participate in the program (Lemos, 2003). In Burkina Faso, farmers had limited ability to use seasonal forecasts (a risk management strategy) because they lacked the resources (basic agricultural technology such as plows, alternative crop varieties, fertilizers, etc.) needed to effectively respond to the projections (Ingram et al., 2002). In Bangladesh, however, despite persisting poverty, improved disaster preparedness and response and relative higher levels of household adaptive capacity have dramatically decreased the number of deaths as a result of flooding (del Ninno et al., 2002, 2003; Section 9.2.5).

Scholars have argued that building adaptive capacity in such regions requires a dialectic, two-tiered process in which climatic risk management (specific adaptive capacity) and deeper-level socioeconomic and political reform (generic adaptive capacity) iterate to shape overall vulnerability (Lemos et al., 2007; Tompkins et al., 2008). When implemented as part of a systems approach, managing climate risks can create positive synergies with development goals through participatory and transparent approaches (such as participatory vulnerability mapping or local disaster relief committees) that empower local households and institutions (e.g., Degg and Chester, 2005; Nelson, 2005).

1.3.3. Disaster Risk Management and Adaptation to Climate Change Share Many Concepts, Goals, and Processes

The efficacy of the mix of actions used by communities to reduce, transfer, and respond to current levels of disaster risk could be vastly increased. Understanding and recognition of the many development-based instruments that could be put into motion to achieve disaster risk reduction is a prerequisite for this (Lavell and Lavell, 2009; UNISDR, 2009e, 2011; Maskrey 2011; Wisner et al., 2011). At the same time, some aspects of disaster risk will increase for many communities due to climate change and other factors (Chapters 3 and 4). Exploiting the potential synergies between disaster risk management and adaptation to climate change literature and practice will improve management of both current and future risks.

Both fields share a common interest in understanding and reducing the risk created by the interactions of human with physical and biological systems. Both seek appropriate allocations of risk reduction, risk transfer, and disaster management efforts, for instance balancing pre-impact risk management or adaptation with post-impact response and recovery. Decisions in both fields may be organized according to the risk governance framework. For instance, many countries, are gaining experience in implementing cooperative, inter-sector and multi- or interdisciplinary approaches (ICSU, 2002; Brown et al., 2006; McGray et al., 2007; Lavell and Lavell, 2009). In general, disaster risk management can help those practicing adaptation to climate change to learn from addressing current impacts. Adaptation to climate change can help those practicing disaster risk management to more effectively address future conditions that will differ from those of today.

The integration of concepts and practices is made more difficult because the two fields often use different terminology, emerge from different academic communities, and may be seen as the responsibility of different government organizations. As one example, Section 1.4 will describe how the two fields use the word 'coping' with different meanings and different connotations. In general, various contexts have made it more difficult to recognize that the two fields share many concepts, goals, and processes, as well as to exploit the synergies that arise from their differences. These include differences in historical and evolutionary processes; conceptual and definitional bases; processes of social knowledge construction and the ensuing scientific compartmentalization of subject areas; institutional and organizational funding and instrumental backgrounds; scientific origins and baseline literature; conceptions of the relevant causal relations; and the relative importance of different risk factors (see Sperling and Szekely, 2005; Schipper and Pelling, 2006; Thomalla et al., 2006; Mitchell and van Aalst, 2008; Venton and La Trobe, 2008, Schipper and Burton, 2009; Lavell, 2010). These aspects will be considered in more detail in future chapters.

Potential synergies from the fields' different emphases include the following.

First, disaster risk management covers a wide range of hazardous events, including most of those of interest in the adaptation to climate change literature and practice. Thus, adaptation could benefit from experience in managing disaster risks that are analogous to the new challenges expected under climate change. For example, relocation and other responses considered when confronted with sea level change can be informed by disaster risk management responses to persistent or large-scale flooding and landslides or volcanic activity and actions with pre- or post-disaster relocation; responses to water shortages due to loss of glacial meltwater would bear similarities to shortages due to other drought stressors; and public health challenges due to modifications in disease vectors due to climate change have similarities to those associated with current climate variability, such as the occurrence of

---

FAQ 1.2 | What are effective strategies for managing disaster risk in a changing climate?

Disaster risk management has historically operated under the premise that future climate will resemble that of the past. Climate change now adds greater uncertainty to the assessment of hazards and vulnerability. This will make it more difficult to anticipate, evaluate, and communicate disaster risk. Uncertainty, however, is not a 'new' problem. Previous experience with disaster risk management under uncertainty, or where long return periods for extreme events prevail, can inform effective risk reduction, response, and preparation, as well as disaster risk management strategies in general.

Because climate variability occurs over a wide range of timescales, there is often a historical record of previous efforts to manage and adapt to climate-related risk that is relevant to risk management under climate change. These efforts provide a basis for learning via the assessment of responses, interventions, and recovery from previous impacts. Although efforts to incorporate learning into the management of weather- and climate-related risks have not always succeeded, such adaptive approaches constitute a plausible model for longer-term efforts. Learning is most effective when it leads to evaluation of disaster risk management strategies, particularly with regard to the allocation of resources and efforts between risk reduction, risk sharing, and disaster response and recovery efforts, and when it engages a wide range of stakeholder groups, particularly affected communities.

In the presence of deeply uncertain long-term changes in climate and vulnerability, disaster risk management and adaptation to climate change may be advanced by dealing adequately with the present, anticipating a wide range of potential climate changes, and promoting effective 'no-regrets' approaches to both current vulnerabilities and to predicted changes in disaster risk. A robust plan or strategy that both encompasses and looks beyond the current situation with respect to hazards and vulnerability will perform well over a wide range of plausible climate changes.

El Niño. Moreover, like disaster risk management, adaptation to climate change will often take place within a multi-hazard locational framework given that many areas affected by climate change will also be affected by other persistent and recurrent hazards (Wisner et al., 2004, 2011; Lavell, 2010; Mercer, 2010). Additionally, learning from disaster risk management can help adaptation, which to date has focused more on changes in the climate mean, increasing its focus on future changes in climate extremes and other potentially damaging events.

Second, disaster risk management has tended to encourage an expanded, bottom-up, grass roots approach, emphasizing local and community-based risk management in the framework of national management systems (see Chapters 5 and 6), while an important segment of the adaptation literature focuses on social and economic sectors and macro ecosystems over large regional scales. However, a large body of the adaptation literature – in both developed and developing countries – is very locally focused. Both fields could benefit from the body of work on the determinants of adaptive capacity that focus on the interaction of individual and collective action and institutions that frame their actions (McGray et al., 2007; Schipper, 2009).

Third, the current disaster risk management literature emphasizes the social conditioning of risk and the construction of vulnerability as a causal factor in explaining loss and damage. Early adaptation literature and some more recent output, particularly from the climate change field, prioritizes physical events and exposure, seeing vulnerability as what remains after all other factors have been considered (O'Brien et al., 2007). However, community-based adaptation work in developing countries (Beer and Hamilton, 2002; Brown et al., 2006; Lavell and Lavell, 2009; UNISDR, 2009b,c) and a growing number of studies in developed nations (Burby and Nelson, 1991; de Bruin et al., 2009; Bedsworth and Hanak, 2010; Brody et al., 2010; Corfee-Morlot et al., 2011; Moser and Eckstrom, 2011) have considered social causation. Both fields could benefit from further integration of these concepts.

Overall, the disaster risk management and adaptation to climate change literatures both now emphasize the value of a more holistic, integrated, trans-disciplinary approach to risk management (ICSU-LAC, 2009). Dividing the world up sectorally and thematically has often proven organizationally convenient in government and academia, but can undermine a thorough understanding of the complexity and interaction of the human and physical factors involved in the constitution and definition of a problem at different social, temporal, and territorial scales. A more integrated approach facilitates recognition of the complex relationships among diverse social, temporal, and spatial contexts; highlights the importance of decision processes that employ participatory methods and decentralization within a supporting hierarchy of higher levels; and emphasizes that many disaster risk management and other organizations currently face climate-related decisions whether they recognize them or not.

The following areas, some of which have been pursued by governments, civil society actors, and communities, have been recommended or proposed to foster such integration between, and greater effectiveness of, both adaptation to climate change and disaster risk management (see also WRI, 2008; Birkmann and von Teichman, 2010; Lavell, 2010):

- Development of a common lexicon and deeper understanding of the concepts and terms used in each field (Schipper and Burton, 2009)
- Implementation of government policymaking and strategy formulation that jointly considers the two topics
- Evolution of national and international organizations and institutions and their programs that merge and synchronize around the two themes, such as environmental ministries coordinating with development and planning ministries (e.g., National Environmental Planning Authority in Jamaica and Peruvian Ministries of Economy and Finance, Housing, and Environment)
- Merging and/or coordinating disaster risk management and adaptation financing mechanisms through development agencies and nongovernmental organizations
- The use of participatory, local level risk and context analysis methodologies inspired by disaster risk management that are now strongly accepted by many civil society and government agencies in work on adaptation at the local levels (IFRC, 2007; Lavell and Lavell, 2009; UNISDR, 2009 b,c)
- Implementing bottom-up approaches whereby local communities integrate adaptation to climate change, disaster risk management, and other environmental and development concerns in a single, causally dimensioned intervention framework, commensurate many times with their own integrated views of their own physical and social environments (Moench and Dixit, 2004; Lavell and Lavell, 2009).

1.4. Coping and Adapting

The discussion in this section has four goals: to clarify the relationship between adaptation and coping, particularly the notion of coping range; to highlight the role of learning in an adaptation process; to discuss barriers to successful adaptation and the issue of maladaptation; and to highlight examples of learning in the disaster risk management community that have already advanced climate change adaptation.

A key conclusion of this section is that learning is central to adaptation, and that there are abundant examples (see Section 1.4.5 and Chapter 9) of the disaster risk management community learning from prior experience and adjusting its practices to respond to a wide range of existing and evolving hazards. These cases provide the adaptation to climate change community with the opportunity not only to study the specifics of learning as outlined in these cases, but also to reflect on how another community that also addresses climate-related risk has incorporated learning into its practice over time.

As disaster risk management includes both coping and adapting, and these two concepts are central for adaptation to climate change in both scholarship and practice, it is important to start by clarifying the meanings of these terms. Without a clear conception of the distinctions between the concepts and overlaps in their meanings, it is difficult to fully understand a wide range of related issues, including those concerned with the coping range, adaptive capacity, and the role of institutional learning in promoting robust adaptation to climate change. Clarifying such distinctions carries operational significance for decisionmakers interested in promoting resilience, a process that relies on coping for immediate survival and recovery, as well as adaptation and disaster risk reduction, which entail integrating new information to moderate potential future harm.

1.4.1. Definitions, Distinctions, and Relationships

In both the disaster risk management and climate change adaptation literature, substantial differences are apparent as to the meaning and significance of coping as well as its relationship with and distinction from adaptation. Among the discrepancies, for example, some disaster risk management scholars have referred to coping as a way to engage local populations and utilize indigenous knowledge in disaster preparedness and response (Twigg, 2004), while others have critiqued this idea, concerned that it would divert attention away from addressing structural problems (Davies, 1993) and lead to a focus on 'surviving' instead of 'thriving.' There has also been persistent debate over whether coping primarily occurs before or after a disastrous event (UNISDR, 2008b,c, 2009e). This debate is not entirely resolved by the current UNISDR definition of coping, the "ability of people, organizations, and systems, using available skills and resources, to face and manage adverse conditions, emergencies or disasters" (UNISDR, 2009d). Clearly, emergencies and disasters are post facto circumstances, but 'adverse conditions' is an indeterminate concept that could include negative pre-impact livelihood conditions and disaster risk circumstances or merely post-impact effects.

The first part of this section is focused on parsing these two concepts. Once the terms are adequately distinguished, the focus shifts in the second part to important relationships between the two terms and other related concepts, which taken together have operational significance for governments and stakeholders.

1.4.1.1. Definitions and Distinctions

Despite the importance of the term coping in the fields of both disaster risk management and adaptation to climate change, there is substantial confusion regarding the term's meaning (Davies, 1996) and how it is distinguished from adaptation.

In order to clarify this aspect, it is helpful first to look outside of the disaster risk and adaptation contexts. The *Oxford English Dictionary* defines *coping* as "the action or process of overcoming a problem or difficulty" or "managing or enduring a stressful situation or condition" and *adapting* as "rendering suitable, modifying" (OED, 1989). As noted

Table 1-1 | The various dimensions of coping and adapting.

Dimension	Coping	Adapting
Exigency	Survival in the face of immediate, unusually significant stress, when resources, which may have been minimal to start with, are taxed (Wisner et al., 2004).	Reorientation in response to recent past or anticipated future change, often without specific reference to resource limitations.
Constraint	Survival is foremost and tactics are constrained by available knowledge, experience, and assets; reinvention is a secondary concern (Bankoff, 2004).	Adjustment is the focus and strategy is constrained less by current limits than by assumptions regarding future resource availability and trends.
Reactivity	Decisions are primarily tactical and made with the goal of protecting basic welfare and providing for basic human security after an event has occurred (Adger, 2000).	Decisions are strategic and focused on anticipating change and addressing this proactively (Füssel, 2007), even if spurred by recent events seen as harbingers of further change.
Orientation	Focus is on past events that shape current conditions and limitations; by extension, the focus is also on previously successful tactics (Bankoff, 2004).	Focus on future conditions and strategies; past tactics are relevant to the extent they might facilitate adjustment, though some experts believe past and future orientation can overlap and blend (Chen, 1991).

in Table 1-1, contrasting the two terms highlights several important dimensions in which they differ – exigency, constraint, reactivity, and orientation – relevant examples of which can be found in the literature cited.

Overall, coping focuses on the moment, constraint, and survival; adapting (in terms of human responses) focuses on the future, where learning and reinvention are key features and short-term survival is less in question (although it remains inclusive of changes inspired by already-modified environmental conditions).

1.4.1.2. Relationships between Coping, Coping Capacity, Adaptive Capacity, and the Coping Range

The definitions of coping and adapting used in this report reflect the dictionary definitions. As an example, a community cannot adapt its way through the aftermath of a disastrous hurricane; it must cope instead. Its coping capacity, or capacity to respond (Gallopín, 2003), is a function of currently available resources that can be used to cope, and determines the community's ability to survive the disaster intact (Bankoff, 2004; Wisner et al., 2004). Repeated use of coping mechanisms without adequate time and provisions for recovery can reduce coping capacity and shift a community into what has been termed transient poverty (Lipton and Ravallion, 1995). Rather than leaving resources for adaptation, communities forced to cope can become increasingly vulnerable to future hazards (O'Brien and Leichenko, 2000).

Adaptation in anticipation of future hurricanes, however, can limit the need for coping that may be required to survive the next storm. A community's adaptive capacity will determine the degree to which adaptation can be pursued (Smit and Pilofosova, 2003). While there is

Box 1-4 | Adaptation to Rising Levels of Risk

Before AD 1000, in the low-lying coastal floodplain of the southern North Sea and around the Rhine delta, the area that is now The Netherlands, the inhabitants lived on dwelling mounds, piled up to lie above the height of the majority of extreme storm surges. By the 10th century, with a population estimated at 300,000 people, inhabitants had begun to construct the first dikes, and within 400 years had ringed all significant areas of land above spring tide, allowing animals to graze and people to live in the protected wetlands. The expansion of habitable land encouraged a significant increase in the population exposed to catastrophic floods (Borger and Ligtendag, 1998). The weak sea dikes broke in a series of major storm surge floods through the stormy 13th and 14th centuries (in particular in 1212, 1219, 1287, and 1362), flooding enormous areas (often permanently) and causing more than 200,000 fatalities, reflecting an estimated lifetime mortality rate from floods for those living in the region in excess of 5% (assuming a 30-year average lifespan; Gottschalk, 1971, 1975, 1977).

To adapt to increasingly adverse environmental conditions (reflecting long-term delta subsidence), major improvements in the technology of dike construction and drainage engineering began in the 15th century. As the country became richer and population increased (to an estimated 950,000 by 1500 and 1.9 million by 1700), it became an imperative not only to provide better levels of protection but also to reclaim land from the sea and from the encroaching lakes, both to reduce flood hazard and expand the land available for food production (Hoeksma, 2006). Examples of the technological innovations included the development of windmills for pumping, and methods to lift water at least 4 m whether by running windmills in series or through the use of the wind-powered Archimedes screw. As important was the availability of capital to be invested in joint stock companies with the sole purpose of land reclamation. In 1607, a company was formed to reclaim the 72 km^2 Beemster Lake north of Amsterdam (12 times larger than any previous reclamation). A 50-km canal and dike ring were excavated, a total of 50 windmills installed that after five years pumped dry the Beemster polder, 3 to 4 m below the surrounding countryside, which, within 30 years, had been settled by 200 farmhouses and 2,000 people.

After the major investment in raising and strengthening flood defenses in the 17th century, there were two or three large floods, one in 1717 (when 14,000 people drowned) and two notable floods in 1825 and 1953; since that time the average flood mortality rate has been around 1,000 per century, equivalent to a lifetime mortality rate (assuming a 50-year average lifetime) of around 0.01%, 500 times lower than that which had prevailed through the Middle Ages (Van Baars and Van Kempen, 2009). This change reflects increased protection rather than any reduction in storminess. The flood hazard and attendant risk is now considered to be rising again (Bouwer and Vellinga, 2007) and plans are being developed to manage further rises, shifting the coping range in anticipation of the new hazard distribution.

some variability in how coping capacity and adaptive capacity are defined, the literature generally recognizes that adaptive capacity focuses on longer-term and more sustained adjustments (Gallopín, 2006; Smit and Wandel, 2006). However, in the same way that repeatedly invoking coping mechanisms consumes resources available for subsequent coping needs, it also consumes resources that might otherwise be available for adaptation (Adger, 1996; Risbey et al., 1999).

There is also a link between adaptation and the **coping range** – that is, a system's capacity to reactively accommodate variations in climatic conditions and their impacts (a system can range from a particular ecosystem to a society) (IPCC, 2007b). In the adaptation literature, Yohe and Tol (2002, p. 26) have used the term to refer to the range of "circumstances within which, by virtue of the underlying resilience of the system, significant consequences are not observed" in response to external stressors. Outside the coping range, communities will "feel significant effects from change and/or variability in their environments" (Yohe and Tol, 2002, p. 25). Within its coping range, a community can survive and even thrive with significant natural hazards. This is particularly the case when the historical distribution of hazard intensity is well known and relatively stable (see Section 1.2.3.4). A community's coping range is determined, in part, by prior adaptation (Hewitt and Burton, 1971; de Vries, 1985; de Freitas, 1989), and a community is most likely to survive and thrive when adaptation efforts have matched its coping range with the range of hazards it typically encounters (Smit and Pilifosova, 2003). As climate change alters future variability and the occurrence of extreme events, and as societal trends change human systems' vulnerability, adaptation is required to adjust the coping range so as to maintain societal functioning within an expected or acceptable range of risk (Moser and Luers, 2008).

Box 1-4 provides an example of this process in the region that is now The Netherlands. As this box illustrates, the process of shifting a society's coping range both depends on and facilitates further economic development (i.e., requires adaptive capacity and enhances coping capacity). The box also illustrates that the process requires continuous reassessment of risk and adjustment in response to shifting hazard distributions in order to avoid increasing, and maladaptive, hazard exposure. Successful adjustments, facilitated in part by institutional learning, can widen and shift a community's coping range, promoting resilience to a wider range of future disaster risk (Yohe and Tol, 2002), as illustrated in Box 1-4 and discussed further in Section 1.4.2 (*high confidence*).

1.4.2. Learning

Risk management decisions are made within social-ecological systems (a term referring to social systems intimately tied to and dependent on environmental resources and conditions). Some social-ecological systems are more resilient than others. The most resilient are characterized by their capacity to learn and adjust, their ability to reorganize after disruption, and their retention of fundamental structure and function in the face of system stress (Folke, 2006). The ability to cope with extreme stress and resume normal function is thus an important component of resilience, but learning, reorganizing, and changing over time are also key. As Chapter 8 highlights, transformational changes are required to achieve a future in which society's most important social-ecological systems are sustainable and resilient. Learning, along with adaptive management, innovation, and leadership, is essential to this process.

Learning related to social-ecological systems requires recognizing their complex dynamics, including delays, stock-and-flow dynamics, and feedback loops (Sterman, 2000), features that can complicate management strategies by making it difficult to perceive how a system operates. Heuristic devices and mental models can sometimes inhibit learning by obscuring a problem's full complexity (Kahneman et al., 1982; Section 1.3.1.2) and complicating policy action among both experts and lay people (Cronin et al., 2009). For instance, common heuristics (see Section 1.3.1.2) lead to misunderstanding of the relationship between greenhouse gas emission rates and their accumulation in atmospheric stocks, lending credence to a 'wait and see' approach to mitigation (Sterman, 2008). Through a variety of mechanisms, such factors can lead to paralysis and failure to engage in appropriate risk management strategies despite the availability of compelling evidence pointing to particular risk management pathways (Sterman, 2006). The resulting learning barriers thus deserve particular attention when exploring how to promote learning that will lead to effective adaptation.

Given the complex dynamics of social-ecological systems and their interaction with a changing climate, the literature on adaptation to climate change (usually referred to here, as above, simply as 'adaptation') emphasizes iterative learning and management plans that are explicitly designed to evolve as new information becomes available (Morgan et al., 2009: NRC, 2009). Unlike adaptation, the field of disaster risk management has not historically focused as explicitly on the implications of climate change and the need for iterative learning. However, the field provides several important examples of learning, including some presented in Chapter 9, that could be instructive to adaptation practitioners. Before introducing these case studies in Section 1.4.5, we will outline relevant theory of institutional learning and 'learning loops.'

Extensive literature explores both the role of learning in adaptation (Armitage et al., 2008; Moser, 2010; Pettengell, 2010) and strategies for facilitating institutional and social learning in 'complex adaptive systems' (Pahl-Wostl, 2009). Some important strategies include the use of knowledge co-production, wherein scientists, policymakers, and other actors work together to exchange, generate, and apply knowledge (van Kerkhoff and Lebel, 2006), and action research, an iterative process in which teams of researchers develop hypotheses about real-world

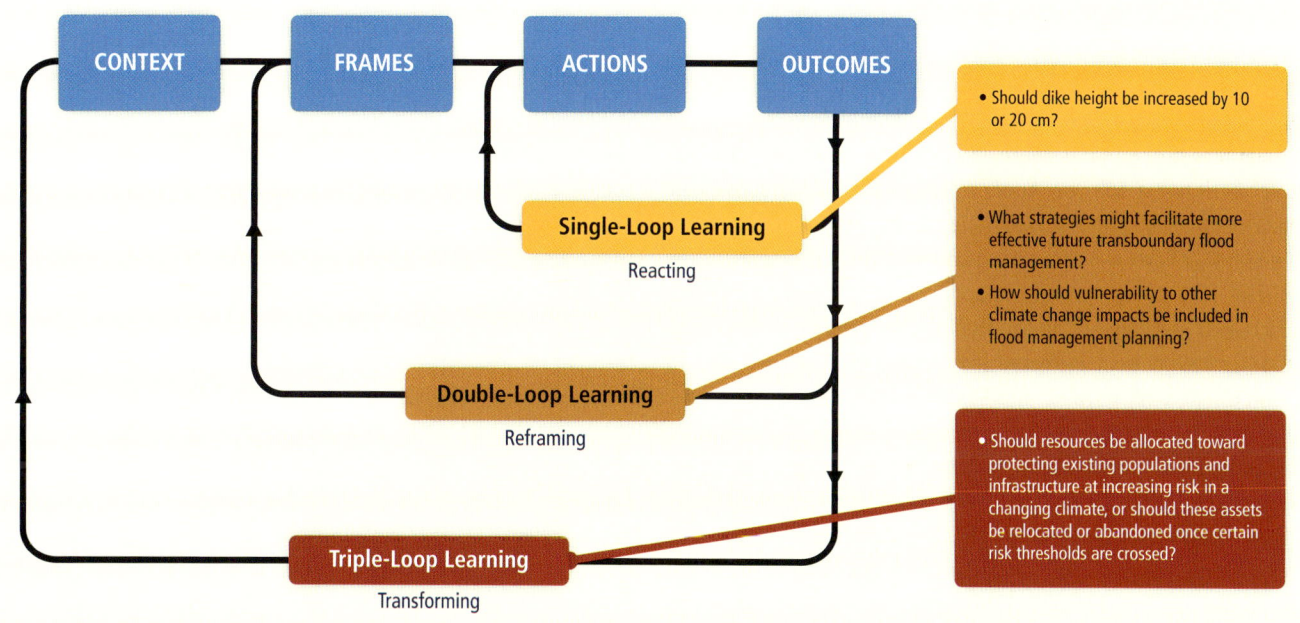

Figure 1-3 | Learning loops: pathways, outcomes, and dynamics of single-, double-, and triple-loop learning and applications to flood management. Adapted from Argyris and Schön, 1978; Hargrove, 2002; Sterman et al., 2006; Folke et al., 2009; and Pahl-Wostl, 2009.

problems and revise management strategies based on the results (List, 2006). Prior work on learning theories, for example, experiential learning (Kolb, 1984) and transformative learning (Mezirow, 1995), emphasize the importance of action-oriented problem-solving, learning-by-doing, concrete learning cycles, and how these processes result in reflection, reconsideration of meaning, and re-interpretation of value structures. The learning loop framework (Kolb and Fry, 1975; Argyris and Schön, 1978; Keen et al., 2005) integrates these theories and divides learning processes into three different loops depending on the degree to which the learning promotes transformational change in management strategies. Figure 1-3 outlines this framework and its application to the issue of flood management.

In single-loop learning processes, changes are made based on the difference between what is expected and what is observed. Single-loop learning is primarily focused on improving the efficiency of action (Pelling et al., 2008) and answering the question of "whether things are being done right" (Flood and Romm, 1996), that is, whether management tactics are appropriate or adequate to achieve identified objectives. In flood management, for example, when floodwaters threaten to breach existing flood defenses, flood managers may ask whether dike and levee heights are sufficient and make adjustments accordingly. As Figure 1-3 indicates, single-loop learning focuses primarily on actions; data are integrated and acted on but the underlying mental model used to process the data is not changed.

In double-loop learning, the evaluation is extended to assess whether actors are "doing the right things" (Flood and Romm, 1996), that is, whether management goals and strategies are appropriate. Corrective actions are made after the problem is reframed and different management goals are identified (Pelling et al., 2008); data are used to promote critical thinking and challenge underlying mental models of what works and why. Continuing with the flood management example, double-loop learning results when the goals of the current flood management regime are critically examined to determine if the regime is sustainable and resilient to anticipated shifts in hydrological extremes over a particular time period. For instance, in a floodplain protected by levees built to withstand a 500-year flood, a shift in the annual exceedance probability from 0.002 to 0.005 (equivalent to stating that the likelihood that a 500-year flood will occur in a given year has shifted to that seen historically for a 200-year event) will prompt questions about whether the increased likelihood of losses justifies different risk management decisions, ranging from increased investments in flood defenses to changed insurance policies for the vulnerable populations.

Many authors also distinguish triple-loop learning (Argyris and Schön, 1978; Hargrove, 2002; Peschl, 2007), or learning that questions deeply held underlying principles (Pelling et al., 2008). In triple-loop learning, actors question how institutional and other power relationships determine perceptions of the range of possible interventions, allowable costs, and appropriate strategies (Flood and Romm, 1996). In response to evidence that management strategies are not serving a larger agreed-upon goal, that is, they are maladaptive, triple-loop learning questions how the

social structures, cultural norms, dominant value structures, and other constructs that mediate risk and risk management (see Box 1-3) might be changed or transformed. Extending the flood control example, triple-loop learning might entail entirely new approaches to governance and participatory risk management involving additional parties, crossing cultural, institutional, national, and other boundaries that contribute significantly to flood risk, and planning aimed at robust actions instead of strategies considered optimal for particular constituents (Pahl-Wostl, 2009).

Different types of learning are more or less appropriate in given circumstances (Pahl-Wostl, 2009, p. 359). For example, overreliance on single-loop learning may be problematic in rapidly changing circumstances. Single-loop learning draws on an inventory of existing skills and memories specific to particular circumstances. As a result, rapid, abrupt, or surprising changes may confound single-loop learning processes (Batterbury, 2008). Coping mechanisms, even those that have developed over long periods of time and been tested against observation and experience, may not confer their usual survival advantage in new contexts. Double- and triple-loop learning are better suited to matching coping ranges with new hazard regimes (Yohe and Tol, 2002). Integrating double- and triple-loop learning into adaptation projects, particularly for populations exposed to multiple risks and stressors, is more effective than more narrowly planned approaches dependent on specific future climate information (McGray et al., 2007; Pettengell, 2010).

Easier said than done, triple-loop learning is analogous to what some have termed 'transformation' (Kysar, 2004; see Section 1.1.3; Chapter 8), in that it can lead to recasting social structures, institutions, and constructions that contain and mediate risk to accommodate more fundamental changes in world view (Pelling, 2010). Translating double- and triple-loop learning into policy requires not only articulation of a larger risk-benefit universe, but also mechanisms to identify, account for, and compare the costs associated with a wide range of interventions and their benefits and harms over various time horizons. Stakeholders would need also to collaborate to an unusual degree in order to collectively and cooperatively consider the wide range of risk management possibilities and their impacts.

1.4.3. Learning to Overcome Adaptation Barriers

Learning focused on barriers to adaptation can be particularly useful. Resource limitations are universally noted as a significant impediment in pursuing adaptation strategies, to a greater or lesser degree depending on the context. In addition, some recent efforts to identify and categorize adaptation barriers have focused on specific cultural factors (Nielsen and Reenberg, 2010) or issues specific to particular sectors (Huang et al., 2011), while others have discussed the topic more comprehensively (Moser and Ekstrom, 2010). Some studies identify barriers in the specific stages of the adaptation process. Moser and Ekstrom (2010), for instance, outline three phases of adaptation: understanding, planning, and management. Each phase contains several key steps, and barriers can impede progress at each. Barriers to understanding, for instance, can include difficulty recognizing a changing signal due to difficulty with its detection, perception, and appreciation; preoccupation with other pressing concerns that divert attention from the growing signal; and lack of administrative and social support for making adaptive decisions. While this study offers a diagnostic framework and avoids prescriptions about overcoming adaptation barriers, other studies, such as those mentioned above, offer more focused prescriptions relevant to particular sectors and contexts.

Research on barriers has generally focused on adaptation as a process, recognizing the difficulty in furnishing a universally acceptable *a priori* definition of successful adaptation outcomes (Adger et al., 2005). This skirts potentially important normative questions, however, and some researchers have considered whether particular activities should be considered maladaptive, defined as an "action taken ostensibly to avoid or reduce vulnerability to climate change that impacts adversely on, or increases the vulnerability of other systems, sectors, or social groups" (Barnett and O'Neill, 2009, p. 211). They identify activities that increase greenhouse gas releases, burden vulnerable populations disproportionately, and require excessive commitment to one path of action (Barnett and O'Neill, 2009). Other candidates include actions that offset one set of risks but increase others, resulting in net risk increase, for example, a dam that reduces flooding but increases the threat of zoonotic diseases, and actions that amplify risk to those who remain exposed (or are newly exposed as a result of a maladaptive action), of which there are abundant examples in the public health literature (Sterman, 2006) and other fields.

These issues have a long history in disaster risk management. For instance, in 1942, deriving from study and work in the 1930s, Gilbert White asserted that levees can provide a false sense of security and are eventually fallible, ultimately leading to increased risk, and advocated, among other 'adjustment' measures, land use planning and environmental management schemes in river basins in order to face up to flooding hazards (see Burton et al., 1978). Such findings are among the early advances in the field of 'human adjustment to hazards,' which derived from an ecological approach to human-environmental relationships. In the case of levees for example, the distinction between adaptive and maladaptive actions depends on the time period over which risks are being assessed. From a probabilistic perspective, the overall likelihood of a catastrophic flood overwhelming a levee's protective capacity is a function of time. The wrinkle that climate change introduces is that many climate-related hazards may become more frequent, shrinking the timescale over which certain decisions can be considered 'adaptive' and communities can consider themselves 'adapted' (Nelson et al., 2007).

While frameworks that help diagnose barriers to adaptation are helpful in identifying the origin of maladaptive decisions, crafting truly adaptive policies is still difficult even when the barriers are fully exposed. For instance, risk displacement is a common concern in large insurance systems when risk is not continuously reassessed, risk management strategies and mechanisms for distributing risk across populations (such as risk pricing in insurance schemes) are inadequately maintained, or if new risk management strategies are not recruited as necessary. This was the case with the levees in New Orleans prior to Hurricane Katrina, wherein the levees were built to make a hazardous area safer but paradoxically facilitated the exposure of a much larger population to a large hazard. As a result of multiple factors (Burby, 2006), inadequate levee infrastructure increased the likelihood of flooding but no other adequate risk reduction and management measures were implemented, resulting in catastrophic loss of life and property when the city was hit with the surge from a strong Category 3 storm (Comfort, 2006). Some have suggested that, as a result of the U.S. federal government's historical approach to disasters, those whose property was at risk in New Orleans anticipated that they would receive federal recovery funds in the event of a flooding disaster. This, in turn, may have distorted the risk management landscape, resulting in improper pricing of flooding risks, decreased incentives to take proper risk management actions, and exposure of a larger population to flood risk than might otherwise have been the case (Kunreuther, 2006).

This example illustrates how an adaptation barrier may have resulted in an ultimately maladaptive risk management regime, and demonstrates the importance of considering how risk, in practice, is assumed and shared. One goal of risk sharing is to properly price risk so that, in the event risk is realized, there is an adequate pool of capital available to fund recovery. When risk is improperly priced and risk sharing is not adequately regulated, as can occur when risk-sharing devices are not monitored appropriately, an adequate pool of reserves may not accumulate. When risk is realized, the responsibility for funding the recovery falls to the insurer of last resort, often the public.

The example also illustrates how an insurance system designed to motivate adaptation (by individual homeowners or flood protection agencies) can function properly only if technical rates – rates that properly reflect empirically determined levels of risk – can be established and matched with various levels of risk at a relatively high level of spatial and temporal resolution. Even in countries with free-market flood insurance systems, insurers may be reluctant to charge the full technical rate because consumers have come to assume that insurance costs should be relatively consistent in a given location. Without charging technical rates, however, it is difficult to use pricing to motivate adaptation strategies such as flood proofing or elevating the ground floor of a new development (Lamond et al., 2009), restricting where properties can be built, or justifying the construction of communal flood defenses. In such a case, barriers to adaptation (in both planning and management, in this case) can result in a strategy with maladaptive consequences in the present. In places where risk levels are rising due to climate change under prevailing negative conditions of exposure and vulnerability, reconsideration of these barriers – a process that includes double- and triple-loop learning – could promote more adaptive risk management. Otherwise, maladaptive risk management decisions may commit collective resources (public or private) to coping and recovery rather than successful adaptation and may force some segments of society to cope with disproportionate levels of risk.

1.4.4. 'No Regrets,' Robust Adaptation, and Learning

The mismatch between adaptation strategies and projected needs has been characterized as the potential for regret, that is, opportunity costs associated with decisions (and related path dependence, wherein earlier choices constrain future circumstances and decisions) that are optimal for one or a small number of possible climate futures but not necessarily robust over a wider range of scenarios (Lempert and Schlesinger, 2001). 'No regrets' adaptation refers to decisions that have net benefits over the entire range of anticipated future climate and associated impacts (Callaway and Hellmuth, 2007; Heltberg et al., 2009).

To address the challenge of risk management in the dynamically complex context of climate change and development, as well as under conditions where probabilistic estimates of future climatic conditions remain imprecise, several authors have advanced the concept of robustness (Wilby and Dessai, 2010), of which 'no regrets' adaptation is a special case (Lempert and Groves, 2010). Robustness is a property of a plan or strategy that performs well over a wide range of plausible future scenarios even if it does not perform optimally in any particular scenario. Robust adaptation plans may perform relatively well even if probabilistic assessments of risk prove wrong because they aim to address both expected and surprising changes, and may allow diverse stakeholders to agree on actions even if they disagree about values and expectations (Brown and Lall, 2006; Dessai and Hulme; 2007; Lempert and Groves, 2010; Means et al., 2010; see also Section 1.3.2).

As Section 1.4.3 highlights, currently, in many instances risks associated with extreme weather and other climate-sensitive hazards are often not well managed. To be effective, adaptation would prioritize measures that increase current as well as future resilience to threats. Robustness over time would increase if learning were a central pillar of adaptation efforts, including learning focused on addressing current vulnerabilities and enhancing current risk management (*high confidence*). Single-, double-, and triple-loop learning will all improve the efficacy of management strategies.

The case studies in Chapter 9 highlight some important examples of learning in disaster risk management relevant to a wide range of climate-sensitive threats and a variety of sectors. Section 9.2 provides examples of how single- and double-loop learning processes – enhancing public health response capacity, augmenting early warning systems, and applying known strategies for protecting health from the threat of extreme heat in new settings – had demonstrable impacts on heat-related mortality, quickly shifting a region's coping range with regard to extreme heat (Section 9.2.1). Other case studies, examining risk transfer (Section 9.2.13) and early warning systems (Section 9.2.11), provide instances of how existing methods and tools can be modified and deployed in new settings in response to changing risk profiles – examples of both double- and triple-loop learning. Similarly, the case studies on governance (Section 9.2.12) and on the limits to adaptation in small island developing states (Section 9.2.9) provide examples of third-loop learning and transformative approaches to disaster risk management.

References

A digital library of non-journal-based literature cited in this chapter that may not be readily available to the public has been compiled as part of the IPCC review and drafting process, and can be accessed via either the IPCC Secretariat or IPCC Working Group II web sites.

ACC, 2010: *Informing an Effective Response to Climate Change*. America's Climate Choices, National Academies Press, Washington, DC.

Adger, W.N., 1996: *Approaches to Vulnerability to Climate Change*. CSERGE Working Papers. University of East Anglia, Norwich, UK.

Adger, W.N., 2000: Social and ecological resilience: Are they related? *Progress in Human Geography*, **24(3)**, 347-364.

Adger, W.N., 2006: Vulnerability. *Global Environmental Change*, **16**, 268-281.

Adger, W.N., N. Arnell, and E.M. Thompkins, 2005: Successful adaptation to climate change across scales. *Global Environmental Change*, **15**, 77-86.

Alexander, D., 1993: *Natural Disasters*. UCL Press, London, 632 pp.

Alexander, D., 2000: *Confronting Catastrophe*. Oxford University Press, New York.

Anderson, M. and P. Woodrow, 1989: *Rising from the Ashes: Development Strategies in Times of Disasters*. Westview Press, Boulder, CO.

APA, 2009: *Psychology and Global Climate Change: Addressing a Multi-faceted Phenomenon and Set of Challenges*. American Psychological Association Task Force on the Interface between Psychology and Global Climate Change, American Psychological Association, Washington, DC.

Argyris, C. and D. Schön, 1978: *Organizational Learning: A Theory of Action Perspective*. Addison-Wesley, Reading, MA.

Armitage, D., M. Marschke, and R. Plummer, 2008: Adaptive co-management and the paradox of learning. *Global Environmental Change*, **18**, 86-98.

Aven, T., 2011. On some recent definitions and analysis frameworks for risk, vulnerability, and resilience. *Risk Analysis*, **31(4)**, 515-522.

Bahadur, A.V., M. Ibrahim, and T. Tanner, 2010: *The resilience renaissance? Unpacking of Resilience for Tackling Climate Change and Disasters*. Institute of Development Studies (for the Strengthening Climate Resilience (SCR) consortium), Brighton, UK.

Baird, A., P. O'Keefe, K. Westgate, and B. Wisner, 1975: *Towards an Explanation of and Reduction of Disaster Proneness*. Occasional Paper number 11, Disaster Research Unit, University of Bradford, Bradford, UK.

Balamir, M., 2005: Ways of understanding urban earthquake risks. In: *Book of Abstracts from 'Rethinking Inequalities,'* 7th Conference of the European Sociological Association, Institute of Sociology, Nicolaus Copernicus University of Torun, Poland, p. 132.

Ball, N., 1975: The myth of the natural disaster. *The Ecologist*, **5(10)**, 368-369.

Bankoff, G., 2001: Rendering the world safe: vulnerability as western discourse. *Disasters*, 25 **(10)**, 19-35.

Bankoff, G., 2004: The historical geography of disaster: "vulnerability" and "local knowledge" in western discourse. In: *Mapping Vulnerability: Disasters, Development, and People* [G. Bankoff, G. Frerks, and D. Hillhorst (eds.)]. Earthscan, London, pp. 25-36.

Barke, R, H. Jenkins-Smith, and P. Slovic, 1997: Risk perceptions of men and women scientists. *Social Science Quarterly*, **78**, 167-176.

Barnett, J. and S. O'Neill, 2009: Maladaptation. *Global Environmental Change*, **20**, 211-213.

Barron, E.J., 2009: Beyond climate science. *Science*, **326**, 643.

Batterbury, S., 2008: Anthropology and global warming: the need for environmental engagement. *Australian Journal of Anthropology*, **1**, 62-67.

Bedford, T.J. and R.M. Cooke, 2001: *Probabilistic Risk Analysis: Foundations and Methods*. Cambridge University Press, New York, NY and Cambridge, UK.

Bedsworth, L.W. and E. Hanak, 2010: Adaptation to climate change: A review of challenges and tradeoffs in six areas. *Journal of the American Planning Association*, **76(4)**, 477-495.

Beer, T. and R. Hamilton, 2002: *Natural Disaster Reduction: Safer Sustainable Communities, Making Better Decisions about Risk*. International Council for Science, ICSU Position Paper.

Below, R., A. Wirtz and D. Guha-Sapir, 2009: *Disaster Category Classification and Peril Terminology for Operational Purposes*. Common Accord Centre for Research on the Epidemiology of Disasters (CRED) and Munich Re, Brussels, Belgium and Munich, Germany, cred.be/sites/default/files/DisCatClass_264.pdf.

Berkes, F., 2007: Understanding uncertainty and reducing vulnerability: Lessons from resilience thinking. *Natural Hazards*, **41**, 283-295.

Berkes, F., J. Colding, and C. Folke (eds.), 2004: *Navigating Social-Ecological Systems: Building Resilience for Complexity and Change*. Cambridge University Press, Cambridge, UK.

Birkmann, J., and K. von Teichman, 2010: Integrating disaster risk reduction and climate change adaptation: key challenges – scales, knowledge, and norms. *Sustainability Science*, **5(2)**, 171-184.

Blaikie, P., T. Cannon, I. Davis, and B. Wisner, 1994: *At Risk: Natural Hazards, People's Vulnerability and Disasters*. 1st edition. Routledge, London.

Borger, G.J. and W.A. Ligtendag, 1998: The role of water in the development of the Netherlands – a historical perspective. *Journal of Coastal Conservation*, **4**, 109-114.

Bosher, L.S. (ed.), 2008: *Hazards and the Built Environment: Attaining Built-in Resilience*. Taylor and Francis, London.

Bouwer, L.M. and Vellinga P., 2007: On the flood risk in the Netherlands. *Advances in Natural and Technological Hazard Research*, **25**, 469-484.

Brand, F.S., and K. Jax, 2007: Focusing the meaning(s) of resilience: resilience as a descriptive concept and a boundary object. *Ecology and Society*, **12(1)**, 23, www.ecologyandsociety.org/vol12/iss21/art23/.

Brick, T., J. Kightlinger, and D. Mann, 2010: *Integrated Water Resources Plan 2010 Update*. Report No. 1373, Metropolitan Water District of Southern California, Los Angeles, CA.

Brody, S., H. Grover, E. Lindquist and A. Vedlitz, 2010: Examining climate change mitigation and adaptation behaviours among public sector organisations in the USA. *Local Environment: The International Journal of Justice and Sustainability*, **15(6)**, 591-603.

Brown, C., 2010: The end of reliability. *Journal of Water Resources Planning and Management*. **March/April**, 143-145.

Brown, C., 2011: *Decision-scaling for Robust Planning and Policy under Climate Uncertainty*. World Resources Report, World Resources Institute, Washington, DC, www.worldresourcesreport.org/files/wrr/papers/wrr_brown_uncertainty.pdf.

Brown, C. and U. Lall, 2006: Water and economic development: The role of variability and a framework for resilience. *Natural Resources Forum*, **30(4)**, 306-317.

Brown, O., A. Crawford, and A. Hammill, 2006: *Natural Disaster and Resource Rights: Building Resilience, Rebuilding Lives*. International Institute for Sustainable Development, Winnipeg, Canada.

Burby, R., 2006: Hurricane Katrina and the paradoxes of government disaster policy: Bringing about wise governmental decisions for hazardous areas. *Annals of the American Academy of Political and Social Science*, **604**, 171-191.

Burby, R.J. and A.C. Nelson, 1991: Local government and public adaptation to sea-level rise. *Journal of Urban Planning and Development*, **117(4)**, 140-163.

Burton, I., R. Kates, and G. White (eds.), 1978: *The Environment as Hazard*. Guildford, New York, NY.

Butts, R.A., R.A. Schaalje, and G. Bruce, 1997: Impact of subzero temperatures on survival, longevity, and natality of adult Russian wheat aphid (Homoptera: Aphididae). *Environmental Entomology*, **26(3)**, 661-667.

Callaway, J. and M. Hellmuth, 2007: Assessing the incremental benefits and costs of coping with development pressure and climate change: A South African case study. *Submissions from admitted non-governmental organizations on socio-economic information under the Nairobi work programme*. United Nations, Geneva, Switzerland.

Canham, C.D., M.J. Papaik, and E.F. Latty, 2001: Interspecific variation in susceptibility to windthrow as a function of tree size and storm severity for northern temperate tree species. *Canadian Journal of Forest Research*, **31(1)**, 1-10.

Cannon, S.H, R.M. Kirkham, and M. Parise, 2001: Wildfire-related debris-flow initiation processes, Storm King Mountain, CO. *Geomorphology*, **39**, 171-188.

Cannon, T., 2006: Vulnerability analysis, livelihoods and disasters. In: *Risk21. Coping with Risks due to Natural Hazards in the 21st Century* [Amman, W., S. Dannenmann, and L. Vulliet (eds.)]. Taylor and Francis Group, London, UK, pp. 41-50.

Cardona, O.D., 1986: Estudios de vulnerabilidad y evaluación del riesgo sísmico: Planificación física y urbana en áreas propensas. *Boletín Técnico de la Asociación Colombiana de Ingeniería Sísmica*, **33(2)**, 32-65.

Cardona, O.D. 1996: Manejo ambiental y prevención de desastres: dos temas asociados. In: *Ciudades en Riesgo* [Fernandez, M.A. (ed.)]. La RED-USAID, Lima, Peru, pp. 79-101.

Cardona, O.D., 2001: *Estimación Holística del Riesgo Sísmico utilizando Sistemas Dinámicos Complejos*. Doctoral dissertation, Department of Terrain Engineering, Technical University of Catalonia, Spain, www.desenredando.org/public/varios/2001/ehrisusd/index.html.

Cardona, O.D., 2004: The Need for rethinking the concepts of vulnerability and risk from a holistic perspective: A necessary review and criticism for effective risk management. In: *Mapping Vulnerability: Disasters, Development and People* [Bankoff, G., G. Frerks, and D. Hilhorst (eds.)]. Earthscan Publishers, London, UK, pp. 37-51.

Cardona, O.D., 2005. Indicators of Disaster Risk and Risk Management: Program for Latin America and the Caribbean – Summary Report. English and Spanish edition, Inter-American Development Bank, Washington, DC.

Cardona, O.D., 2008. Indicators of Disaster Risk and Risk Management: Program for Latin America and the Caribbean – Summary Report – Second Edition. English and Spanish edition, Inter-American Development Bank, Washington, DC.

Cardona, O.D., 2011: Disaster risk and vulnerability: Notions and measurement of human and environmental insecurity. In: *Coping with Global Environmental Change, Disasters and Security – Threats, Challenges, Vulnerabilities and Risks* [Brauch, H.G., U. Oswald Spring, C. Mesjasz, J. Grin, P. Kameri-Mbote, B. Chourou, P. Dunay, and J. Birkmann (eds.)]. Hexagon Series on Human and Environmental Security and Peace, Vol. 5, Springer Verlag, Heidelberg, Berlin and New York, pp. 107-121.

Carter, T.R., R.N. Jones, X. Lu, S. Bhadwal, C. Conde, L.O. Mearns, B.C. O'Neill, M.D.A. Rounsevell, and M.B. Zurek, 2007: New Assessment Methods and the Characterisation of Future Conditions. In: *Climate Change 2007: Impacts, Adaptation and Vulnerability. Contribution of Working Group II to the Fourth Assessment Report of the Intergovernmental Panel on Climate Change* [Parry, M.L., O.F. Canziani, J.P. Palutikof, P.J. van der Linden, and C.E. Hanson, (eds.)]. Cambridge University Press, Cambridge, UK, pp. 133-171.

Cash, D.W., W.C. Clark, F. Alcock, N.M. Dickson, N. Eckley, D.H. Guston, J. Jager, and R.B. Mitchell, 2003: Knowledge systems for sustainable development. *Proceedings of the National Academy of Sciences*, **100**, 8086-8091.

Cash, D.W., J.C. Borck, and A.G. Patt, 2006: Countering the loading-dock approach to linking science and decision making: Comparative analysis of El Niño/Southern Oscillation (ENSO) forecasting systems. *Science, Technology and Human Values*, **31**, 465-494.

Cavazos, T., C. Turrent, and D. P. Lettenmaier, 2008: Extreme precipitation trends associated with tropical cyclones in the core of the North American monsoon. *Geophysical Research Letters*, **35**, L21703, doi:10.1029/2008GL035832.

CCSP, 2008: *Preliminary Review of Adaptation Options for Climate-Sensitive Ecosystems and Resources* [Julius, S.H. and J.M. West (eds.)]. A Report by the U.S. Climate Change Science Program and the Subcommittee on Global Change Research, U.S. Environmental Protection Agency, Washington, DC, 873 pp.

Chen, M. (ed.), 1991: *Coping with Seasonality and Drought*. Sage Publications, New Delhi, India.

Chongfu, H., 1996: Fuzzy risk assessment of urban natural hazards. *Fuzzy Sets and Systems*, **83**, 271-282.

Coburn, A. and R. Spence, 2002: *Earthquake Protection*. 2nd edition. John Wiley and Sons, Chichester, UK.

Cochrane, H., 2004: Economic loss: Myth and measurement. *Disaster Prevention and Management*, **13**, 290-296.

Collins, M., 2007: Ensembles and probabilities: A new era in the prediction of climate change. *Philosophical Transactions of the Royal Society A*, **365**, 1957-1970.

Colls, A., N. Ash, and N. Ikkala, 2009: *Ecosystem-based Adaptation: a natural response to climate change*. IUCN, Gland, Switzerland, 16 pp.

Comfort, L., 2006: Cities at risk: Hurricane Katrina and the drowning of New Orleans. *Urban Affairs Review*, **41**(4), 501-516.

Corfee-Morlot, J., I. Cochran, S. Hallegatte, and P.-J. Teasdale, 2011: Multilevel risk governance and urban adaptation policy. *Climatic Change*, **104**(1) (Special Issue: Understanding Climate Change Impacts, Vulnerability and Adaptation at City Scale), 169-197.

Corporación OSSO, La Red de Estudios Sociales en Prevención de Desastres en América Latina, 2010: *.DesInventar online 8.1.9-2: sistema de inventario de efectos de desastres*. OSSO/La Red de Estudios Sociales en Prevención de Desastres en América Latina, Cali, Columbia, online.desinventar.org/.

Cox, P. and D. Stephenson, 2007: A changing climate for prediction. *Science*, **317**, 207-208.

CRED, 2010: *EM-DAT: The OFDA/CRED International Disaster Data Base*. Centre for Research on the Epidemiology of Disasters, Université Catholique de Louvain, Louvain, Belgium, www.emdat.be/Database/.

Cronin, M., C. Gonzalez, and J.D. Sterman, 2009: Why don't well-educated adults understand accumulation? A challenge to researchers, educators, and citizens. *Organizational Behavior and Human Decision Processes*, **108**(1), 116-130.

Cuny, F.C., 1983: *Disaster and Development*. Oxford University Press, New York, NY.

Cutter, S.L., 1996: Vulnerability to environmental hazards. *Progress in Human Geography*, **20**, 529-239.

Cutter, S. L., L. Barners, M. Berry, C. Burton, E. Evans, E. Tate, and J. Webb, 2008: A place-based model for understanding community resilience to natural disasters, *Global Environmental Change*, **18**, 598-606.

Davies, S., 1993: Are coping strategies a cop out? *Institute of Development Studies Bulletin*, **24**(4), 60-72.

Davies, S., 1996: *Adaptable Livelihoods: Coping with Food Insecurity in the Malian Sahel*. Macmillan Press, London, UK.

de Bruin, K., R. Dellink, A. Ruijs, L. Bolwidt, A. van Buuren, J. Graveland, R. de Groot, P. Kuikman, S. Reinhard, R. Roetter, V. Tassone, A. Verhagen, and E. van Ierland, 2009: Adapting to climate change in The Netherlands: an inventory of climate adaptation options and ranking of alternatives. *Climatic Change*, **95**(1), 23-45.

de Freitas, C.R., 1989: The hazard potential of drought for the population of the Sahel. In: *Population and Disaster* [Clarke, J.I., P. Curson, S.L. Kayastha, and P. Nag (eds.)]. Blackwell, Oxford, UK, pp. 98-113.

de Jong, C., D. Collins, and R. Ranzi (eds.), 2005: *Climate and Hydrology in Mountain Areas*. John Wiley and Sons, New York, NY.

de Vries, J., 1985: Analysis of historical climate-society interaction. In: *Climate Impact Assessment* [Kates, R., J. Ausubel, and M. Berberian (eds.)]. John Wiley and Sons, New York, NY.

Degg, M.R. and D.K. Chester, 2005: Seismic and volcanic hazards in Peru: changing attitudes to disaster mitigation. *The Geographical Journal*, **171**, 125-145.

del Ninno, C., P.A. Dorosh, and N. Islam, 2002: Reducing vulnerability to natural disasters – Lessons from 1998 floods in Bangladesh. *Ids Bulletin–Institute of Development Studies*, **33**, 98-107.

del Ninno, C., P.A. Dorosh, and L.C. Smith, 2003: Public policy, markets and household coping strategies in Bangladesh: Avoiding a food security crisis following the 1998 floods. *World Development*, **31**, 1221-1238.

Dessai, S., and M. Hulme, 2004: Does climate adaptation policy need probabilities? *Climate Policy*, **4**, 107-128.

Dessai, S., and M. Hulme, 2007: Assessing the robustness of adaptation decisions to climate change uncertainties: A case study on water resources management in the East of England. *Global Environmental Change*, **17**(1), 59-72.

Dessai, S. and R. Wilby, 2011: *How Can Developing Country Decision Makers Incorporate Uncertainty about Climate Risks into Existing Planning and Policymaking Processes?* World Resources Report, World Resources Institute, Washington, DC, http://www.worldresourcesreport.org/files/wrr/papers/wrr_dessai_and_wilby_uncertainty.pdf.

Dessai, S., M. Hulme, R. Lempert, and R. Pielke Jr., 2009: Climate prediction: a limit to adaptation? In: *Adapting to Climate Change: Thresholds, Values, Governance* [Adger, W.N., I. Lorenzoni, and K.L. O'Brien (eds.)]. Cambridge University Press, Cambridge, UK.

DFID, 2004: *Adaptation to Climate Change: Making Development Disaster Proof*. Key Sheet 06, UK Department for International Development, London, UK.

DFID, 2005: *Disaster Risk Reduction: A Development Concern*. Policy Briefing Document, DFID-ODG, UK Department for International Development, London, UK.

Dhakal, A.S. and R.C. Sidle, 2004: Distributed simulations of landslides for different rainfall conditions, *Hydrological Processes*, **18**(4), 757-776.

Doherty, S.J., et al., 2009: Lessons learned from IPCC AR4: Scientific developments needed to understand, predict, and respond to climate change. *Bulletin of the American Meteorological Society*, **90**, 497-513.

Douglas, M., 1992: *Risk and Blame: Essays in Cultural Theory*. Routledge, London, UK and New York, NY.

Douglas, M. and A. Wildavsky, 1982: *Risk and Culture: An Essay on the Selection of Technical and Environmental Dangers*. University of California Press, Berkeley, CA.

Dow, K., 1992: Exploring differences in our common future(s): the meaning of vulnerability to global environmental change. *Geoforum*, **23**, 417-436.

Dulal, H.B., K.U. Shah, and N. Ahmad, 2009: Social equity considerations in the implementation of Caribbean climate change adaptation policies. *Sustainability*, **1**, 363-383.

Dunlap, R.E. and A.M. McCright, 2008: A widening gap: republican and democratic views on climate change. Environment, **50**, 26-35.

Eakin, H., and A. L. Luers, 2006. Assessing the vulnerability of social-environmental systems. *Annual Review of Environment and Resources*, **31**, 365-394.

Easterling, D., G. Meehl, C. Parmesan, S. Changnon, T. Karl, and L. Mearns, 2000: Climate extremes, observations, modeling and impacts. Science, **289**, 2068-2074.

El-Baroudy, I. and S.P. Simonovic, 2004: Fuzzy criteria for the evaluation of water resources systems performance. *Water Resources Research*, **40**(10), W10503, doi:10.1029/2003WR002828.

Emanuel, K, 2001: Contribution of tropical cyclones to meridional heat transport by the oceans. *Journal of Geophysical Research*, **106**(D14), 14771-14781.

Emanuel, K., 2003: Tropical cyclones. *Annual Review of Earth and Planetary Sciences*, **31**, 75-104.

Epstein, P., H. Diaz, S. Elias, G. Grabherr, N. Graham, W. Martens, E.M. Thompson, and J. Susskind, 1998: Biological and physical signs of climate change: focus on mosquito borne diseases. *Bulletin of the American Meteorological Society*, **79**, 409-417.

ERN-AL, 2011: *Probabilistic Modelling of Natural Risk at Global Level: The Hybrid Loss Exceedance Curve – Development of a Methodology and Implementation of Case Studies, Phase 1A: Colombia, Mexico, Nepal*. Report for the GAR 2011, Consortium Evaluación de Riesgos Naturales – América Latina, Bogotá, Colombia.

Fink, A.H., U. Ulbrich, and H. Engel, 1996: Aspects of the January 1995 flood in Germany. *Weather*, **51**(2), 34-39.

Fischhoff, B., 1994: What forecasts (seem to) mean. *International Journal of Forecasting*, **10**, 387-403.

Fischhoff, B., P. Slovic, and S. Lichtenstein, 1983: The "public" vs. the "experts": Perceived vs. actual disagreement about the risks of nuclear power. In: *Analysis of Actual vs. Perceived Risks* [Covello,V., G. Flamm, J. Rodericks, and R. Tardiff (eds.)]. Plenum, New York, NY, pp. 235-249.

Flood, R. and N. Romm, 1996: *Diversity Management: Triple Loop Learning*. Wiley, Chichester, UK.

Folke, C., 2006: Resilience: The emergence of a perspective for social–ecological systems analyses. *Global Environmental Change*, **16**, 253-267.

Folke, C., F.S. Chapin III, and P. Olsson, 2009: Transformations in ecosystem stewardship. In: *Principles of Ecosystem Stewardship: Resilience-Based Natural Resource Management in a Changing World* [Chapin III, F.S., G.P. Kofinas, and C. Folke (eds.)]. Springer, New York, NY, pp. 103-125.

Fouillet, A., G. Rey, V. Wagner, K. Laaidi, P. Empereur-Bissonnet, A. Le Tertre, P. Frayssinet, P. Bessemoulin, F. Laurent, P. De Crouy-Chanel, E. Jougla, and D. Hemon, 2008: Has the impact of heat waves on mortality changed in France since the European heat wave of summer 2003? A study of the 2006 heat wave. *International Journal of Epidemiology*, **37(2)**, 309-317.

Füssel, H., 2007: Vulnerability: A generally applicable conceptual framework for climate change research. *Global Environmental Change*, **17(2)**, 155-167.

Gaillard, J.C., 2007: Resilience of traditional societies in facing natural hazard. *Disaster Prevention and Management*, **16**, 522-544.

Gaillard, J.C., 2010: Vulnerability, capacity, and resilience: Perspectives for climate and development policy. *Journal of International Development*. **22**, 218-232.

Gallopín, G.C., 2003: Box 1. A systemic synthesis of the relations between vulnerability, hazard, exposure and impact, aimed at policy identification. In: *Handbook for Estimating the Socio-Economic and Environmental Effects of Disasters*. Economic Commission for Latin American and the Caribbean, LC/MEX/G.S., Mexico, pp. 2-5.

Gallopín, G.C., 2006: Linkages between vulnerability, resilience, and adaptive capacity. *Global Environmental Change*, **16**, 293-303.

Gasper, D., 2010: The idea of human security. In: *Climate Change, Ethics and Human Security* [O'Brien, K. A.L. St. Clair, and B. Kristoffersen (eds.)]. Cambridge University Press, Cambridge, UK, pp. 23-46.

Gershunov, A., D. R. Cayan, and S. F. Iacobellis, 2009: The great 2006 heat wave over California and Nevada: Signal of an increasing trend. *Journal of Climate*, **22**, 6181-6203.

Goddard, L., W. Baethgen, B. Kirtman, and G. Meehl, 2009: The urgent need for improved climate models and predictions. *EOS*, **90(39)**, 343.

Gordon, J.E., 1978: *Structures*. Penguin Books, Harmondsworth, UK.

Gottschalk, M.K.E., 1971: *Stormvloeden en rivieroverstromingen in Nederland Deel I* (Storm surges and river floods in the Netherlands, part I). Van Gorcum, Assen, The Netherlands.

Gottschalk, M.K.E., 1975: *Stormvloeden en rivieroverstromingen in Nederland Deel II* (Storm surges and river floods in the Netherlands, part II). Van Gorcum, Assen, The Netherlands.

Gottschalk, M.K.E., 1977: *Stormvloeden en rivieroverstromingen in Nederland Deel III* (Storm surges and river floods in the Netherlands, part III). Van Gorcum, Assen, The Netherlands.

Groves, D.G. and R.J. Lempert, 2007: A new analytic method for finding policy-relevant scenarios. *Global Environmental Change*, **17**, 73-85.

Hagman, G., 1984: *Prevention Better than Cure*. Swedish Red Cross, Stockholm, Sweden.

Hallegatte, S., 2009: Strategies to adapt to an uncertain climate change. *Global Environmental Change*, **19(2)**, 240-247.

Hargrove, R., 2002. *Masterful Coaching*. Revised Edition. Jossey-Bass/Pfeiffer, Wiley, San Francisco, CA, USA.

Hawkins, E. and R. Sutton, 2009: The potential to narrow uncertainty in regional climate predictions. *Bulletin of the American Meteorological Society*, **90**, 1095-1107.

Hegerl, G.C., F.W. Zwiers, P. Braconnot, N.P. Gillett, Y. Luo, J.A. Marengo Orsini, N. Nicholls, J.E. Penner, and P.A. Stott, 2007: Understanding and attributing climate change. In: *Climate Change 2007: The Physical Science Basis. Contribution of Working Group I to the Fourth Assessment Report of the Intergovernmental Panel on Climate Change* [Solomon, S., D. Qin, M. Manning, Z. Chen, M. Marquis, K.B. Averyt, M. Tignor, and H.L. Miller (eds.)]. Cambridge University Press, Cambridge, UK and New York, NY, pp. 663-745.

Heltberg, R., P.B. Siegel, and S.L. Jorgensen, 2009: Addressing human vulnerability to climate change: Toward a "no-regrets" approach. *Global Environmental Change*, **19(1)**, 89-99.

Hertwig, R., G. Barron, E.U. Weber, and I. Erev, 2004: Decisions from experience and the effect of rare events. *Psychological Science*, **15**, 534-539.

Hewitt, K., 1983: The idea of calamity in a technocratic age. In: *Interpretations of Calamity from the Viewpoint of Human Ecology* [Hewitt, K. (ed.)]. Allen and Unwin, Boston, MA and London, UK, pp 3-32.

Hewitt, K., 1997: *Regions of Risk: A Geographical Introduction to Disasters*. Longman, Harlow, Essex, UK.

Hewitt, K., 2007: Preventable disasters: addressing social vulnerability, institutional risks and civil ethics. *Geographisches Bundschau. International Edition*, **3(1)**, 43-52.

Hewitt, K. and I. Burton, 1971: *The Hazardousness of a Place: A Regional Ecology of Damaging Events*. University of Toronto, Toronto, Canada.

Hine, D. and J.W. Hall, 2010: Information gap analysis of flood model uncertainties and regional frequency analysis. *Water Resources Research*, **46(19)**, W01514, doi:10.1029/2008WR007620.

Hoeksma, R.J., 2006: *Designed for Dry feet: Food Protection and Land Reclamation in the Netherlands*. American Society of Civil Engineers, Reson, VA.

Hohenemser, C., R.E. Kasperson, and R.W. Kates, 1984: Causal structure. In: *Perilous Progress: Managing the Hazards of Technology* [Kates, R.W., C. Hohenemser and J.X. Kasperson (eds.)]. Westview Press, Boulder, CO, pp. 25-42.

Holland, M.M., C.M. Bitz, and B. Tremblay, 2006: Future abrupt reductions in the summer Arctic sea ice. *Geophysical Research Letters*, **33**, L23503, doi:10.1029/2006GL028024.

Holling, C.S., 1973: Resilience and stability of ecological systems. *Annual Review of Ecology and Systematics*, **4**, 1-23.

Horsburgh, K.J. and M. Horritt, 2006: The Bristol Channel Floods of 1607 – reconstruction and analysis. *Weather*, **61(10)**, 272-277.

Huang, C., P. Vaneckova, X. Wang, G. FitzGerald, Y. Guo, and S. Tong, 2011: Constraints and barriers to public health adaptation to climate change. *American Journal of Preventive Medicine*, **40(2)**, 183-190.

HVRI, 2010: *The Spatial Hazard Events and Losses Database for the United States*. Version 8.0 (online database). Hazards and Vulnerability Research Institute, University of South Carolina, Columbia, SC, www.sheldus.org.

ICSU, 2002: *Resilience and Sustainable Development: Building Adaptive Capacity in a World of Transformations.* Series on Science for Sustainable Development: Resilience and Sustainable Development No. 3, International Council for Science, Paris, France, 37 pp.

ICSU, 2008: *A Science Plan for Integrated Research on Disaster Risk: addressing the challenge of natural and human induced environmental hazards*. International Council for Science, Paris, France.

ICSU-LAC, 2009: *Understanding and Managing Risk Associated with Natural Hazards: An Integrated Scientific Approach in Latin America and the Caribbean*. International Council for Science Regional Office for Latin America and the Caribbean, Rio de Janeiro, Brazil and Mexico City, Mexico.

IDB, 2007: *Disaster Risk Management Policy*. GN-2354-5, Inter-American Development Bank, Washington, DC.

IFRC, 2007. *How to Do a VCA: A Practical Step-by-Step Guide for Red Cross Red Crescent Staff and Volunteers*. International Federation of Red Cross and Red Crescent Societies.

IIED, 2009: *Natural Resilience: Healthy ecosystems as climate shock insurance*. IIED Policy Brief, International Institute for Environment and Development, London, UK.

Ingram, K.T., M.C. Roncoli, and P.H. Kirshen, 2002: Opportunities and constraints for farmers of West Africa to use seasonal precipitation forecasts with Burkina Faso as a case study. *Agricultural Systems*, **74(3)**, 331-349.

International Building Codes, 2003: International Code Council (paperback), 656 pp. ISBN 1892395568.

IPCC, 2007a: *Climate Change 2007: Synthesis Report. Contribution of Working Groups I, II and III to the Fourth Assessment Report of the Intergovernmental Panel on Climate Change* [Core Writing Team, R.K. Pachauri, and A. Reisinger (eds.)]. IPCC, Geneva, Switzerland, 104 pp.

IPCC, 2007b: *Impacts, Adaptation and Vulnerability. Contribution of Working Group II to the Fourth Assessment Report of the Intergovernmental Panel on Climate Change* [Parry, M.L., O.F. Canziani, J.P. Palutikof, P.J. van der Linden, and C.E. Hanson (eds.)]. Cambridge University Press, Cambridge, UK and New York, NY.

IPCC, 2007c: Appendix I: Glossary. In: *Climate Change 2007: Impacts, Adaptation and Vulnerability. Contribution of Working Group II to the Fourth Assessment Report of the Intergovernmental Panel on Climate Change* [Parry, M.L., O.F. Canziani, J.P. Palutikof, P.J. van der Linden, and C.E. Hanson (eds.)]. Cambridge University Press, Cambridge, UK, and New York, NY, pp. 869-883.

IPCC, 2009. *Scoping Meeting for an IPCC Special Report on Extreme Events and Disasters: Managing the Risks. Proceedings* [Barros, V., et al. (eds.)]. 23-26 March 2009, Oslo, Norway.

ISO, 2009: *Guide 73: 2009. Risk Management – Vocabulary*. International Organization for Standardization, Geneva, Switzerland.

Jansen, V., N. Stollenwerk, H.J. Jensen, M.E. Ramsay, W.J. Edmunds, and C.J. Rhodes, 2003: Measles outbreaks in a population with declining vaccine uptake. *Science*, **301**, 804.

Jasanoff, S., 2004: *States of Knowledge: The Co-Production of Science and Social Order*. Routledge, London, UK.

Jentsch, A., J. Kreyling, and C. Beierkuhnlein, 2007: A new generation of climate change experiments: events not trends. *Frontiers in Ecology and the Environment*, **6(6)**, 315-324.

Jones, R.N. and B.L. Preston, 2011: Adaptation and Risk Management. *WIREs Climate Change*, **2**, 296-308, doi:10.1002/wcc.97.

Kahan, D.M. and D. Braman, 2006: Cultural cognition and public policy. *Yale Law and Policy Review*, **24(1)**, 149-172.

Kahan, D.M., D. Braman, J. Gastil, P. Slovic, and C.K. Mertz, 2007: Culture and identity – protective cognition: Explaining the white male effect in risk perception. *Journal of Empirical Legal Studies*, **4(3)**, 465-505.

Kahneman, D. and A. Tversky, 1979: Prospect theory: An analysis of decisions under risk. *Econometrica*, **47**, 313-327.

Kahneman, D., P. Slovic, and A. Tversky (eds.), 1982: *Judgment under Uncertainty: Heuristics and Biases*. Cambridge University Press, New York, NY.

Karimi, I. and E. Hullermeier, 2007: Risk assessment system of natural hazards: A new approach based on fuzzy probability. *Fuzzy Sets and Systems*, **158**, 987-999.

Kasperson, R.E., O. Renn, P. Slovic, H.S. Brown, J.Emel, R. Goble, J.X. Kasperson, and S. Ratick, 1988: The social amplification of risk: A conceptual framework. *Risk Analysis*, **8**,177-187.

Kates, R.W., et al., 2001: Sustainability science. *Science*, **292(5517)**, 641-642.

Keen, M., V.A. Brown, and R. Dyball, 2005: *Social Learning in Environmental Management: Towards a Sustainable Future*. Earthscan, London, UK.

Kelman, I., 2008. Relocalising disaster risk reduction for urban resilience. *Urban Design and Planning*, **161(DP4)**, 197-204.

Khan, T.A., 2011: Floods increase wheat yield in KP. *Inpaper Magazine*, Pakistan, 16 May 2011.

Klein, R.J.T, R.J. Nicholls, and F. Thomalla, 2003: Resilience to natural hazards: how useful is this concept? *Global Environmental Change*, **5**, 35-45.

Knutti, R., 2010: The end of model democracy? *Climatic Change*, **102**, 395-404.

Kolb, D.A., 1984: *Experiential Learning: Experience as the Source of Learning and Development*. Prentice-Hall, Englewood Cliffs, NJ.

Kolb, D.A. and R. Fry, 1975: Toward an applied theory of experiential learning. In: *Theories of Group Process* [Cooper, C. (ed.)]. John Wiley, London, UK, pp. 33-57.

Konikow, L.F. and E. Kendy, 2005: Groundwater depletion: a global problem. *Hydrogeology Journal*, **13**, 317-320.

Kunreuther, H., 2006: Disaster mitigation and insurance: Learning from Katrina. *The Annals of the American Academy of Political and Social Science*, **604(1)**, 208-227.

Kysar, D., 2004: Climate change, cultural transformation, and comprehensive rationality. *Boston College Environmental Affairs Law Review*, **31(3)**, 555-590.

Lamond, J.E., D.G. Proverbs, and F.N. Hammond, 2009: Accessibility of flood risk insurance in the UK: confusion, competition and complacency. *Journal of Risk Research*, **12**, 825-841.

Lau, W.K.M. and K.-M. Kim, 2011: The 2010 Pakistan flood and Russian heat wave: Teleconnection of hydrometeorologic extremes. *Journal of Hydrometeorology*, doi:10.1175/JHM-D-11-016.1.

Lavell, A., 1996: Degradación ambiental, riesgo y desastre urbano. Problemas y conceptos: hacia la definición una agenda de investigación. In: *Ciudades en Riesgo* [Fernandez, M.A. (ed.)]. La RED-USAID, Lima, Peru, pp. 21-59.

Lavell, A., 1999: Environmental degradation, risks and urban disasters. Issues and concepts: Towards the definition of a research agenda. In: *Cities at Risk: Environmental Degradation, Urban Risks and Disasters in Latin America* [Fernandez, M.A. (ed.)]. La RED, US AID, Quito, Ecuador, pp. 19-58.

Lavell, A., 2002: *Riesgo y Territorio: los niveles de intervención en la Gestión del Riesgo (Risk and Territory: Levels of Intervention and Risk Management)*. Anuario Social y Político de América Latina y el Caribe. FLACSO-Nueva Sociedad, Latin American School of Social Sciences.

Lavell, A., 2003: *Local level risk management: Concept and practices*. CEPREDENAC-UNDP. Quito, Ecuador.

Lavell, A., 2009: *Technical Study in Integrating Climate Change Adaptation and Disaster Risk Management in Development Planning and Policy*. Study undertaken for the Inter-American Development Bank, Washington, DC.

Lavell, A., 2010: *Unpacking Climate Change Adaptation and Disaster Risk Management: Searching for the Links and the Differences: A Conceptual and Epistemological Critique and Proposal*. IUCN-FLACSO, International Union for Conservation of Nature - Latin American School of Social Sciences.

Lavell A. and E. Franco (eds.), 1996: *Estado, Sociedad y Gestión de los Desastres en América Latina*. Red de Estudios Sociales en Prevención de Desastres en América Latina, La RED, Tercer Mundo Editores, Bogotá, Colombia.

Lavell, A. and C. Lavell, 2009: *Local Disaster Risk Reduction Lessons from the Andes*. Series: Significant Local Development Initiatives in the Face of Disaster Risk. General Secretariat of the Andean Community, Lima, Peru.

Leckebusch, G.C., D. Renggli, and U. Ulbrich, 2008: Development and application of an objective storm severity measure for the Northeast Atlantic region. *Meteorologische Zeitschrift*, **17**, 575-587.

Lemos, M.C., 2003: A tale of two policies: the politics of seasonal climate forecast use in Ceará, Brazil. *Policy Sciences*, **32(2)**, 101-123.

Lemos, M.C., E. Boyd, E.L. Tompkins, H. Osbahr, and D. Liverman, 2007: Developing adaptation and adapting development. *Ecology and Society*, **12(2)**, 26, www.ecologyandsociety.org/vol12/iss2/art26/.

Lempert, R.J. and M. Collins, 2007: Managing the risk of uncertain threshold responses: Comparison of robust, optimum, and precautionary approaches. *Risk Analysis*, **27(4)**, 1009-1026.

Lempert, R.J. and D.G. Groves, 2010: Identifying and evaluating robust adaptive policy responses to climate change for water agencies in the American west. *Technological Forecasting and Social Change*, **77**, 960-974.

Lempert, R.J. and M. Schlesinger, 2001: Climate-change strategy needs to be robust. *Nature*, **412**, 375.

Lewcowicz, A.G. and C. Harris, 2005: Frequency and magnitude of active-layer detachment failures in discontinuous and continuous permafrost, northern Canada. *Permafrost and Periglacial Processes*, **16**,115-130.

Lewis, J., 1979: The vulnerable state: An alternative view. In: *Disaster Assistance: Appraisal, Reform and New Approaches* [Stephens, L. and S.J. Green (eds.)]. New York University Press, New York, NY, pp. 104-129.

Lewis, J., 1984: Environmental interpretations of natural disaster mitigation: The crucial need. *The Environmentalist*, **4(3)**, pp. 177-180.

Lewis, J., 1999: *Development in Disaster-prone Places: Studies in Vulnerability*. IT Publications (Practical Action), London, UK.

Lewis, J., 2009: An island characteristic: Derivative vulnerabilities to indigenous and exogenous hazards. *Shima: The International Journal of Research into Island Cultures*. **3(1)**, 3-15.

Lewis, J. and I. Kelman, 2009: Housing, flooding and risk-ecology: Thames Estuary South-Shoreland and North Kent. *Journal of Architectural and Planning Research*, **26(1)**, 14-29.

Lindell, M. and C. Prater, 2003: Assessing community impacts of natural disasters. *Natural Hazards* Review, **4(4)**, 176-185.

Lipton, M. and M. Ravallion, 1995: Poverty and policy. In: *Handbook of Development Economics, Vol. 3B* [Behrman, J.S. and T.N. Srinivasan (eds.)]. Elsevier Science, Amsterdam, New York, and Oxford, pp. 2551-2657.

List, D., 2006: Action research cycles for multiple futures perspectives. *Futures*, **38**, 673-684.

Loewenstein, G.F., E.U. Weber, C.K. Hsee, and E. Welch, 2001: Risk as feelings. *Psychological Bulletin*, **127**, 267-286.

Malka, A. and J.A. Krosnick, 2009: The association of knowledge with concern about global warming: trusted information sources shape public thinking. *Risk Analysis*, **29**, 633-647.

Manyena, S.B., 2006: The concept of resilience revisited. *Disasters*, **30(4)**, 433-450.

Maskrey, A., 1989: *Disaster Mitigation: A Community Based Approach*. Oxfam, Oxford, UK.

Maskrey, A., 2011: Revisiting community based risk management. *Environmental Hazards*, **10**, 1-11.

Mastrandrea, M.D., C.B. Field, T.F. Stocker, O. Edenhofer, K.L. Ebi, D.J. Frame, H. Held, E. Kriegler, K.J. Mach, P.R. Matschoss, G.-K. Plattner, G.W. Yohe, and F.W. Zwiers, 2010: *Guidance Note for Lead Authors of the IPCC Fifth Assessment Report on Consistent Treatment of Uncertainties*. IPCC, www.ipcc-wg2.gov/meetings/ CGCs/Uncertainties-GN_IPCCbrochure_lo.pdf.

McBean, G., 2006: An integrated approach to air pollution, climate and weather hazards. *Policy Options*, October 2006, pp. 18-24.

McGray, H., A. Hammill, and R. Bradley, 2007: *Weathering the Storm: Options for Framing Adaptation and Development*. World Resources Institute, Washington, DC.

McKee, T.B., N.J. Doesken, and J. Kleist, 1993: The duration of drought frequency and duration to time scales. In: *Eighth Conference on Applied Climatology*, 17-22 January 1993, Anaheim, California, ccc.atmos.colostate.edu/relationshipof-droughtfrequency.pdf.

McLaughlin, P. and T. Dietz, 2007: Structure, agency and environment: Toward an integrated perspective on vulnerability. *Global Environmental Change*, **39(4)**, 99-111.

Means, E., M. Laugier, J. Daw, L. Kaatz, and M. Waage, 2010: *Decision Support Planning Methods: Incorporating Climate Change Uncertainties into Water Planning*. Water Utility Climate Alliance, San Francisco, CA, www.wucaonline.org/assets/pdf/actions_whitepaper_012110.pdf.

Meehl, G.A., T.F. Stocker, W.D. Collins, P. Friedlingstein, A.T. Gaye, J.M. Gregory, A. Kitoh, R. Knutti, J.M. Murphy, A. Noda, S.C.B. Raper, Watterson, I.G., A.J. Weaver and Z.-C. Zhao, 2007: Global climate projections. In: *Climate Change 2007: The Physical Science Basis. Contribution of Working Group I to the Fourth Assessment Report of the Intergovernmental Panel on Climate Change* [Solomon, S., D. Qin, M. Manning, Z. Chen, M. Marquis, K.B. Averyt, M. Tignor, and H.L. Miller (eds.)]. Cambridge University Press, Cambridge, UK and New York, NY, pp. 747-843.

Mercer, J., 2010: Disaster risk reduction or climate change adaptation: are we reinventing the wheel? *Journal of International Development*, **22**, 247-264.

Mezirow, J., 1995: Transformation theory in adult learning. In: *In Defense of the Life World* [Welton, M.R. (ed.)]. State University of New York Press, Albany, NY, pp. 39-70.

Mileti, D., 1999: *Disasters by Design: A Reassessment of Natural Hazards in the United States*. Joseph Henry Press, Washington, DC.

Miller, F., H. Osbahr, E. Boyd, F. Thomalla, S. Bharwani, G. Ziervogel, B. Walker, J. Birkmann, S. Van der Leeuw, J. Rockström, J. Hinkel, T. Downing, C. Folke, and D. Nelson, 2010: Resilience and vulnerability: complementary or conflicting concepts? *Ecology and Society*, **15(3)**, 11.

Milly, P.C.D., R.T. Wetherald, K.A. Dunne, and T.L. Delworth, 2002: Increasing risk of great floods in a change climate. *Nature*, **415**, 514-517.

Milly, P.C.D., J. Betancourt, M. Falkenmark, R. Hirsch, Z. Kundzewicz, D. Lettenmaier, and R. Stouffer, 2008: Stationarity is dead: Whither water management? *Science*, **319(5863)**, 573-574.

Minor, J.E., 1983: Construction tradition for housing determines disaster potential from severe tropical cyclones. *Journal of Wind Engineering and Industrial Aerodynamics*, **14**, 55-66.

Mitchell, T. and M. van Aalst, 2008: *Convergence of Disaster Risk Reduction and Climate Change Adaptation*. Technical Paper, October 2008, UK Department of International Development, London, UK.

Moench, M. and A. Dixit (eds.), 2004: *Adaptive Capacity and Livelihood Resilience: Adaptive Capacities for Responding to Floods and Droughts in South Asia*. Institute for Social and Environmental Transition, Boulder, CO and Nepal.

Morgan, M.G. and M. Henrion, 1990: *Uncertainty: A Guide to Dealing with Uncertainty in Quantitative Risk and Policy Analysis*. Cambridge University Press, Cambridge, UK.

Morgan, M.G., H. Dowlatabadi, M. Henrion, D. Keith, R.J. Lempert, S. McBride, M. Small, and T. Wilbanks, 2009: *Best Practice Approaches for Characterizing, Communicating, and Incorporating Scientific Uncertainty in Decisionmaking*. Synthesis and Assessment Product 5.2 of the U.S. Climate Change Science Program, National Oceanic and Atmospheric Administration, Washington, DC.

Moser, S.C., 2009. Whether our levers are long enough and the fulcrum strong? Exploring the soft underbelly of adaptation decisions and actions. In: *Adapting to Climate Change: Thresholds, Values, Governance* [Adger, W.N., I. Loronzoni, and K. O'Brien (eds.)]. Cambridge University Press, Cambridge, UK, pp. 313-343.

Moser, S.C., 2010: Now more than ever: The need for more socially relevant research on vulnerability and adaptation to climate change. *Applied Geography*, **30(4)**, 464-474.

Moser, S.C. and J.A. Ekstrom, 2010: Barriers to climate change adaptation: A diagnostic framework. *Proceedings of the National Academy of Sciences*, **107(51)**, 22026-22031.

Moser, S.C. and J.A. Ekstrom, 2011: Taking ownership of climate change: Participatory adaptation planning in two local case studies from California. *Journal of Environmental Studies and Sciences*, **1(1)**, 63-74.

Moser, S.C. and A.L. Luers, 2008: Managing climate risks in California: The need to engage resource managers for successful adaptation to change. *Climatic Change*, **87**, S309-S322.

Nelson, D.R., 2005: *The Public and Private Sides of Persistent Vulnerability to Drought: An Applied Model for Public Planning in Ceará, Brazil*. PhD Thesis, University of Arizona, Tucson, AZ.

Nelson, D. R., N. Adger, and K. Brown, 2007: Adaptation to environmental change: contributions of a resilience framework. *The Annual Review of Environment and Resources*, **32**, 395-419.

Nielsen, J.Ø. and A. Reenberg, 2010: Cultural barriers to climate change adaptation: A case study from Northern Burkina Faso. *Global Environmental Change*, **20(1)**, 142-152.

Noji, E., 1997: *The Public Health Impacts of Disasters*. Oxford University Press, New York, NY.

Norris, F.H., S.P. Stevens, B. Pfefferbaum, K.F. Wyche, and R.L. Pfefferbaum, 2008: Community resilience as a metaphor, theory, set of capacities, and strategy for disaster readiness. *American Journal of Community Psychology*, **41**, 127-150

NRC, 2006: *Facing Hazards and Disasters: Understanding Human Dimensions*. National Research Council, National Academies Press, Washington DC.

NRC, 2009: *Informing Decisions in a Changing Climate*. Panel on Strategies and Methods for Climate-Related Decision Support, Committee on the Human Dimensions of Global Change, Division of Behavioral and Social Sciences and Education, National Research Council, The National Academies Press, Washington, DC, 188 pp.

O'Brien, K. and R. Leichenko, 2000: Double exposure: assessing the impacts of climate change within the context of economic globalization. *Global Environmental Change*, **10**, 221-232.

O'Brien, K., S. Eriksen, L.P. Nygaard, and A. Schjolden, 2007: Why different interpretations of vulnerability matter in climate change discourses. *Climate Policy*, **7(1)**, 73-88.

O'Keefe, P., K. Westgate, and B. Wisner, 1976: Taking the naturalness out of natural disasters. *Nature*, **260(5552)**, 566-577.

OED, 1989: *The Oxford English Dictionary*. Oxford University Press, Oxford, UK.

Olsson, P., L.H. Gunderson, S.R. Carpenter, P. Ryan, L. Lebel, C. Folke, and C.S. Holling, 2006: Shooting the rapids: navigating transitions to adaptive governance of social-ecological systems. *Ecology and Society*, **11(1)**, 18.

Oppenheimer, M., B.C. O'Neill, and M. Webster, 2008: Negative learning. *Climatic Change*, **89**, 155-172.

Ostertag, R., W.L. Silver, and A.E. Lugo, 2005: Factors affecting mortality and resistance to damage following hurricanes in a rehabilitated subtropical moist forest. *Biotropica*, **37(1)**, 16-24.

Pahl-Wostl, C., 2009: A conceptual framework for analysing adaptive capacity and multi-level learning processes in resource governance regimes. *Global Environmental Change*, **19(3)**, 354.

Pall, P., T. Aina, D.A. Stone, P.A. Stott, T. Nozawa, A.G. Hilberts, D. Lohmann, and M.R. Allen, 2011: Anthropogenic greenhouse gas contribution to flood risk in England and Wales in autumn 2000. *Nature*, **470**, 382-386

Parson, E., V. Burkett, K. Fisher-Vanden, D. Keith, L. Means, H. Pitcher, C. Rosenzweig, and M. Webster, 2007: *Global Change Scenarios: Their Development and Use*. Sub-report 2.1B of Synthesis and Assessment Product 2.1 by the U.S. Climate Change Science Program and the Subcommittee on Global Change Research, Department of Energy, Office of Biological and Environmental Research, Washington, DC, 106 pp.

Pelling, M., 2003: *The Vulnerability of Cities: Social Resilience and Natural Disaster*. Earthscan, London, UK.

Pelling, M., 2010: *Adaptation to Climate Change: From Resilience to Transformation*. Routledge, London, UK.

Pelling, M., A. Özerdem, and S. Barakat, 2002: The Macroeconomic impact of disasters. *Progress in Development Studies*, **2**, 283-305.

Pelling, M., C. High, J. Dearing, and D. Smith, 2008: Shadow spaces for social learning: a relational understanding of adaptive capacity to climate change within organisations. *Environment and Planning A*, **40(4)**, 867-884.

Pendalla, R., K.A. Foster, and M. Cowella, 2010: Resilience and regions: building understanding of the metaphor. *Cambridge Journal of Regions, Economy and Society*, **3(1)**, 71-84

Peschl, M.F., 2007: Triple-loop learning as foundation for profound change, individual cultivation, and radical innovation. Construction processes beyond scientific and rational knowledge. *Constructivist Foundations*, **2(2-3)**, 136-145.

Peters, G., A.M. DiGioia Jr, C. Hendrickson, and J. Apt, 2006: *Transmission Line Reliability: Climate Change and Extreme Weather*. Carnegie Mellon Electricity Industry Center Working Paper CEIC-05-06, Carnegie Mellon University, Pittsburgh, PA.

Peters, O., C. Hertlein, and K. Christensen, 2001: A complexity view of rainfall. *Physical Review Letters*, **88**, 018701, doi:10.1103/PhysRevLett.88.018701.

Pettengell, C., 2010: *Climate Change Adaptation: Enabling People Living in Poverty to Adapt*. Oxfam International Research Report, Oxfam International, Oxford, UK.

Piao, S., P. Ciais, Y. Huang, Z. Shen, S. Peng, J. Li, L. Zhou, H. Liu, Y. Ma, Y. Ding, P. Friedlingstein, C. Liu, K. Tan, Y. Yu, T. Zhang, and J. Fang, 2010: The impact of climate change on water resources and agriculture in China. *Nature*, **467**, 43-51.

Pidgeon, N. and B. Fischhoff, 2011: The role of social and decision sciences in communicating uncertain climate risks. *Nature Climate Change*, **1**, 35-41.

Pielke, Jr., R.A., 2007: Future economic damage from tropical cyclones: sensitivities to societal and climate changes. *Philosophical Transactions of the Royal Society A*, **365(1860)**, 2717-2729.

Pohl, C., S. Rist, A. Zimmermann, P. Fry, G.S. Gurung, F. Schneider, C. Ifejika Speranza, B. Kiteme, S. Boillat, E. Serrano, G. Hirsch Hadorn, and U. Wiesmann, 2010: Researchers' roles in knowledge co-production: experience from sustainability research in Kenya, Switzerland, Bolivia and Nepal. *Science and Public Policy*, **37(4)**, 267-281.

Quarantelli, E.L, 1986: *Disaster Crisis Management*. Preliminary Papers 113, University of Delaware, Disaster Research Center. Newark, DE, http://dspace.udel.edu:8080/dspace/handle/19716/487.

Ramírez, F. and O.D. y Cardona, 1996: Sistema nacional para la prevención y atención de desastres de Colombia, Estado, sociedad y gestión de los desastres en América Latina. In: *Red de Estudios Sociales en Prevención de Desastres en América Latina* [Lavell, A. and E. y Franco (eds.)]. Tercer Mundo Editores, Bogotá, Colombia.

Ranger, N. and S.-L. Garbett-Shiels, 2011: *How can Decision-Makers in Developing Countries Incorporate Uncertainty about Future Climate Risks into Existing Planning and Policymaking Processes?* Policy Paper, Centre for Climate Change Economics and Policy and Grantham Research Institute on Climate Change and the Environment in collaboration with the World Resources Report, World Resources Institute, Washington, DC, www.worldresourcesreport.org/files/wrr/papers/wrr_ranger_uncertainty.pdf.

Rayner, S., D. Lach, and H. Ingram, 2005: Weather forecasts are for wimps: Why water resource managers do not use climate forecasts. *Climatic Change*, **69**, 197-227.

Reeder, T. and N. Ranger, 2011: *How Do You Adapt In an Uncertain World? Lessons from the Thames Estuary 2100 Project*. World Resources Report, World Resources Institute, Washington, DC, www.worldresourcesreport.org/files/wrr/papers/wrr_reeder_and_ranger_uncertainty.pdf.

Rees, H.G. and D.N. Collins, 2006: Regional differences in response of flow in glacier-fed Himalayan rivers. *Hydrological Processes*, **20(10)**, 2157-2169.

Renn, O., 2008: *Risk Governance. Coping with Uncertainty in a Complex World*. Earthscan, London, UK.

Rich, P.M., D.D. Breshears, and A.B. White, 2008: Phenology of mixed woody-herbaceous ecosystems. *Ecology*, **89(2)**, 342-352.

Risbey, J., M. Kandlikar, H. Dowlatabadi, and D. Graetz, 1999: Scale, context, and decision making in agricultural adaptation to climate variability and change. *Mitigation and Adaptation Strategies for Global Change*, **4(2)**, 137-165.

Rose, A., 2004: Economic principles, issues, and research priorities in hazard loss estimation. In: *Modeling Spatial and Economic Impacts of Disasters* [Okuyama, Y. and S. Chang (eds.)]. Springer, Berlin, Germany, pp. 14–36.

Sarewitz, D., R.A. Pielke, Jr., and R. Byerly, Jr. (eds.), 2000: *Prediction: Science, Decision Making, and the Future of Nature*. Island Press, Washington, DC, 405 pp.

Savage, L.J., 1972: *Foundations of Statistics*. Dover Publications, New York, NY.

Schipper, L., 2009: Meeting at the crossroads? Exploring the linkages between climate change adaptation and disaster risk reduction. *Climate and Development*, **1**, 16-30.

Schipper, L. and I. Burton (eds.), 2009: *The Earthscan Reader on Adaptation to Climate Change*. Earthscan, London, UK.

Schipper, L. and M. Pelling, 2006: Disaster risk, climate change and international development: scope for, and challenges to, integration. *Disasters*, **30(1)**, 19-38.

Schneider, S.H., S. Semenov, A. Patwardhan, I. Burton, C.H.D. Magadza, M. Oppenheimer, A.B. Pittock, A. Rahman, J.B. Smith, A. Suarez, and F. Yamin, 2007: Assessing key vulnerabilities and the risk from climate change. In: *Climate Change 2007: Impacts, Adaptation and Vulnerability. Contribution of Working Group II to the Fourth Assessment Report of the Intergovernmental Panel on Climate Change* [Parry, M.L., O.F. Canziani, J.P. Palutikof, P.J. van der Linden, and C.E. Hanson, (eds.)]. Cambridge University Press, Cambridge, UK, pp. 779-810.

Sen, A.K., 1983: *Poverty and famines: an essay on entitlement and deprivation*. Oxford University Press, Oxford, UK.

Seneviratne, S.I., D. Lüthi, M. Litschi, and C. Schär, 2006: Land-atmosphere coupling and climate change in Europe. *Nature*, **443**, 205-209.

Shapiro, M., J. Shukla, G. Brunet, C. Nobre, M. Beland, R. Dole, K. Trenberth, R. Anthes, G. Asrar, L. Barrie, P. Bougeault, G. Brasseur, D. Burridge, A. Busalacchi, J. Caughey, D. Chen, J. Church, T. Enomoto, B. Hoskins, O. Hov, A. Laing, H. Le Treut, J. Marotzke, G. McBean, G. Meehl, M. Miller, B. Mills, J. Mitchell, M. Moncrieff, T. Nakazawa, H. Olafsson, T. Palmer, D. Parsons, D. Rogers, A. Simmons, A. Troccoli, Z. Toth, L. Uccellini, C. Velden, and J.M. Wallace, 2010: An earth-system prediction initiative for the twenty-first century. *Bulletin of the American Meteorological Society*, **91**, 1377-1388.

Shukla, J., R. Hagedorn, B. Hoskins, J. Kinter, J. Marotzke, M. Miller, T.N. Palmer, and J. Slingo, 2009: Revolution in climate prediction is both necessary and possible: A declaration at the World Modelling Summit for Climate Prediction. *Bulletin of the American Meteorological Society*, **90**, 175-178.

Simonovic, S.P., 2010: *Systems Approach to Management of Disasters: Methods and Applications*. John Wiley and Sons, New York, NY, 348 pp.

Sjöberg, L., 1999a: Risk perception by the public and by experts: A dilemma in risk management. *Human Ecology Review*, **6(2)**.

Sjöberg, L., 1999b: Consequences of perceived risk: Demand for mitigation. *Journal of Risk Research*, **(2)**, 129-149.

Slovic, P., 1993: Perceived risk, trust, and democracy. *Risk Analysis*, **13(6)**, 675-676.

Slovic, P. (ed.), 2000: *The Perception of Risk*. Earthscan, Sterling, VA.

Slovic, P. (ed.), 2010: *The Feeling of Risk: New Perspectives on Risk Perception*. Earthscan, London, UK.

Slovic, P., B. Fischhoff, and S. Lichtenstein, 1982. Facts versus fears: Understanding perceived risk. In: *Judgment Under Uncertainty: Heuristics and Biases* [Kahneman, D., P. Slovic, and A. Tversky (eds.)]. Cambridge University Press, New York, NY.

Smit, B. and O. Pilifosova, 2003. From adaptation to adaptive capacity and vulnerability reduction. In: *Climate Change, Adaptive Capacity and Development* [Smith, J.B., R.J.T. Klein, and S. Huq (eds.)]. Imperial College Press, London, UK.

Smit, B. and J. Wandel, 2006: Adaptation, adaptive capacity and vulnerability. *Global Environmental Change*, **16**, 282-292.

Smith, K., 1996: *Environmental Hazards: Assessing Risk and Reducing Disaster*. Second Edition. Routledge, London, UK.

Sperling, F. and F. Szekely, 2005: *Disaster Risk Management in a Changing Climate*. Discussion Paper prepared for the World Conference on Disaster Reduction on behalf of the Vulnerability and Adaptation Resource Group, Washington, DC.

Stainforth, D.A., M.R. Allen, E.R. Tredger, and L.A. Smith, 2007: Confidence, uncertainty, and decision-support relevance in climate predictions. *Philosophical Transactions of the Royal Society A*, **365**, 2145-2161.

Sterman, J.D., 2000: *Business Dynamics: Systems Thinking and Modeling for a Complex World*. Irwin/McGraw-Hill, Boston, MA.

Sterman, J.D., 2006: Learning from evidence in a complex world. *American Journal of Public Health*, **96**(3), 505-514.

Sterman, J.D., 2008: Risk communication on climate: mental models and mass balance. *Science*, **322**(5901), 532-533.

Stott, P.A., D.A. Stone, and M.R. Allen, 2004: Human contribution to the European heatwave of 2003. *Nature*, **432**, 610-614.

Sui, J. and G. Koehler, 2001: Rain-on-snow induced flood events in Southern Germany. *Journal of Hydrology*, **252**(1-4), 205-220.

Thomalla, F., T. Downing, E. Spanger-Siegfried, G. Han and J. Rockström, 2006: Reducing hazard vulnerability: towards a common approach between disaster risk reduction and climate adaptation. *Disasters*, **30**(1), 39-48.

Timmerman, P., 1981: *Vulnerability, Resilience and Collapse in Society*. Environmental Monograph No. 1, Institute for Environmental Studies, University of Toronto, Toronto, Canada.

Tobin, G.A. and B.E. Montz, 1997: *Natural Hazards: Explanation and Integration*. The Guildford Press, London, UK.

Tompkins, E.L., M.C. Lemos, and E. Boyd, 2008: A less disastrous disaster: Managing response to climate-driven hazards in the Cayman Islands and NE Brazil. *Global Environmental Change – Human and Policy Dimensions*, **18**, 736-745.

Torry, W., 1979: Intelligence, resilience and change in complex social systems: famine management in India. *Mass Emergencies*, **2**, 71-85.

TRB, 2008: *The Potential Impacts of Climate Change on U.S. Transportation*. Special Report 290, Transportation Research Board, National Academy of Sciences, Washington, DC. (See Appendix C: Commissioned paper: Climate Variability and Change with Implications for U.S. Transportation, Dec. 2006).

Tversky, A., and Kahneman, D., 1974: Judgment under uncertainty: Heuristics and biases. *Science*, **185**, 1124-1131.

Twigg, J., 2004: *Disaster Risk Reduction: Mitigation and Preparedness in Development and Emergency Programming*. Overseas Development Institute, London, UK.

Ulbrich, U., A.H. Fink, M. Klawa, and J.G. Pinto, 2001: Three extreme storms over Europe in December 1999. *Weather*, **56**, 70-80.

UNDP, 2004: *Reducing Disaster Risk: A Challenge for Development. A Global Report*. United Nations Development Programme, New York, NY.

UNEP, 1972: *Declaration of the United Nations Conference on the Human Environment*. United Nations Environment Programme, Nairobi, Kenya.

UNISDR, 2005: *Hyogo Framework for Action 2005-2015. Building the Resilience of Nations and Communities to Disasters*. United Nations International Strategy for Disaster Reduction, Geneva, Switzerland.

UNISDR, 2008a: *Climate Change and Disaster Risk Reduction*. Briefing Note 01, United Nations International Strategy for Disaster Reduction, Geneva, Switzerland.

UNISDR, 2008b: *Indigenous Knowledge for Disaster Risk Reduction - Good Practices and Lessons Learned from Experiences in the Asia-Pacific Region*. United Nations International Strategy for Disaster Reduction, Bangkok, Thailand.

UNISDR, 2008c: *Linking Disaster Risk Reduction and Poverty Reduction, Good Practices and Lessons Learned*. United Nations International Strategy for Disaster Reduction, UNDP, Geneva, Switzerland.

UNISDR, 2009a: *Adaptation to Climate Change by Reducing Disaster Risk: Country Practices and Lessons*. Briefing Note 02, United Nations International Strategy for Disaster Reduction, Geneva, Switzerland.

UNISDR, 2009b: *Reducing Disaster Risk through Science: Issues and Actions*. Full Report of the Scientific and Technical Committee 2009, United Nations International Strategy for Disaster Reduction, Geneva, Switzerland.

UNISDR, 2009c: *After Action Review: Second Session Global Platform for Disaster Risk Reduction*. United Nations International Strategy for Disaster Reduction, Geneva, Switzerland.

UNISDR (United Nations International Strategy for Disaster Reduction), 2009d: *Terminology: Basic Terms of Disaster Risk Reduction*. UNISDR, Geneva, Switzerland, www.unisdr.org/eng/library/lib-terminology-eng%20home.htm.

UNISDR, 2009e: *Global Assessment Report on Disaster Risk Reduction: Risk and Poverty in a Changing Climate – Invest Today for a Safer Tomorrow*. United Nations International Strategy for Disaster Reduction, Geneva, Switzerland, 207 pp.

UNISDR, 2011: *Global Assessment Report on Disaster Risk Reduction. Revealing Risk, Redefining Development*. United Nations International Strategy for Disaster Reduction, Geneva, Switzerland.

Vale, L.J. and T.J. Campanella, 2005: *The Resilient City: How Modern Cities Recover from Disaster*. Oxford University Press, Oxford, UK.

van Baars, S. and I.M. Van Kempen, 2009: The causes and mechanisms of historical dike failures in the Netherlands. *E-Water Official Publication of the European Water Association*, Hennef, Germany, 14 pp., www.dwa.de/portale/ewa/ewa.nsf/C125723B0047EC38/8428F628AB57BECFC125766C003024B6/$FILE/Historical%20Dike%20Failures.pdf.

van Kerkhoff, L. and L. Lebel, 2006: Linking knowledge and action for sustainable development. *Annual Review of Environment and Resources*, **31**, 445-477.

van Niekerk, D., 2007: Disaster risk reduction, disaster risk management and disaster management: Academic rhetoric or practical reality? *Disaster Management South Africa*, **4**(1), 6.

Vandentorren, S., F. Suzan, S. Medina, M. Pascal, A. Maulpoix, J.C. Cohen, and M. Ledrans, 2004: Mortality in 13 French cities during the August 2003 heat wave. *American Journal of Public Health*, **94**(9), 1518-1520.

Venton, P. and S. La Trobe, 2008: *Linking Climate Change Adaptation and Disaster Risk Reduction*. Technical Paper, Tearfund, Teddington, UK.

Walker, B.H., J.M. Anderies, A.P. Kinzig, and P. Ryan (eds), 2006: *Exploring Resilience in Social-Ecological Systems: Comparative Studies and Theory Development*. CSIRO Publishing, Collingwood, Australia.

Walker, W.E., A.W. Vincent, J. Marchau, and D. Swanson, 2010: Addressing deep uncertainty using adaptive policies. *Technological Forecasting and Social Change*, **77**, 917-923.

Walshaw, D., 2000: Modelling extreme wind speeds in regions prone to hurricanes. *Journal of the Royal Statistical Society, Series C (Applied Statistics)*, **49**, 51-62.

Watts, M., 1983: On the poverty of theory: natural hazards research in context. In: *Interpretations of Calamity* [Hewitt, K. (ed.)]. Allen and Unwin, Boston, MA, pp. 231-262.

Weber, E.U., 2006: Experience-based and description-based perceptions of long-term risk: why global warming does not scare us (yet). *Climatic Change*, **77**, 103-120.

Weber, E.U., 2010: What shapes perceptions of climate change? *Wiley Interdisciplinary Reviews: Climate Change*, **1**, 332-342.

Weber, E.U. and P.C. Stern, 2011: Public understanding of climate change in the United States. *American Psychologist*, **66**(4), 315-328.

Weber, E.U., S. Shafir, and A.R. Blais, 2004: Predicting risk-sensitivity in humans and lower animals: Risk as variance or coefficient of variation. *Psychological Review*, **111**, 430-445.

Weichselgartner, J., 2001: Disaster mitigation: the concept of vulnerability revisited. *Disaster Prevention and Management*, **10**(2), 85-94.

Werner, E.E., J.M. Bierman, and F.E. French, 1971: *The Children of Kauai Honolulu*. University of Hawaii Press, Honolulu, Hawaii.

Westerling, A.L. and T.W. Swetman, 2003: Interannual to decadal drought and wildfire in the Western United States. *EOS, Transactions American Geophysical Union*, **84**(49), 545-555, doi:10.1029/2003EO490001.

Wheater, H.S., 2002: Progress in and prospects for fluvial flood modeling. *Philosophical Transactions of the Royal Society A*, **360**, 1409-1431.

Wijkman, A. and L. Timberlake, 1988: *Natural Disasters: Acts of God or Acts of Man?* New Society Publishers, Philadelphia, PA.

Wilby, R.L. and S. Dessai, 2010: Robust adaptation to climate change. *Weather*, **65(7)**, 180-185.

Wisner, B., P. O'Keefe, and K. Westgate, 1977: Global systems and local disasters: the untapped power of peoples' science. *Disasters*, **1(1)**, 47-57.

Wisner, B., P. Blaikie, T. Cannon, and I. Davis, 2004: *At Risk: Natural Hazards, People's Vulnerability, and Disasters*, 2nd edition. Routledge, London, UK.

Wisner, B., J.C. Gaillard, and I. Kellman (eds.), 2011: *Handbook of Hazards and Disaster Risk Reduction*. Routledge, London, UK.

WMO, 2010: *Understanding Climate*. World Meteorological Organization, Geneva, Switzerland, www.wmo.int/pages/themes/climate/understanding_climate.php.

World Bank, 2001: *World Development Report 2000-2001*. World Bank, Washington, DC.

World Bank, 2009: *World Development Report 2010: Development and Climate Change*. World Bank, Washington, DC.

WRI (World Resources Institute), United Nations Development Programme, United Nations Environment Programme, and World Bank, 2008: *World Resources 2008: Roots of Resilience: Growing the Wealth of the Poor*. WRI, Washington, DC.

Xie, L., L.J. Pietrafesa, and M. Peng, 2004: Incorporation of a mass-conserving inundation scheme into a three dimensional storm surge model. *Journal of Coastal Research*, **20**, 1209-1233.

Yohe, G. and R.S.J. Tol, 2002: Indicators for social and economic coping capacity – moving toward a working definition of adaptive capacity. *Global Environmental Change*, **12(1)**, 25-40.

Young, P.C. 2002: Advances in real-time flood forecasting. *Philosophical Transactions of the Royal Society*, **A360**, 1433-1450.

Zipser, E.J., D.J. Cecil, C. Liu, S.W. Nesbitt and D.P. Yorty, 2006: Where are the most intense thunderstorms on Earth? *Bulletin of the American Meteorological Society*, **87**, 1057-1071.

2. Determinants of Risk: Exposure and Vulnerability

Coordinating Lead Authors:
Omar-Dario Cardona (Colombia), Maarten K. van Aalst (Netherlands)

Lead Authors:
Jörn Birkmann (Germany), Maureen Fordham (UK), Glenn McGregor (New Zealand), Rosa Perez (Philippines), Roger S. Pulwarty (USA), E. Lisa F. Schipper (Sweden), Bach Tan Sinh (Vietnam)

Review Editors:
Henri Décamps (France), Mark Keim (USA)

Contributing Authors:
Ian Davis (UK), Kristie L. Ebi (USA), Allan Lavell (Costa Rica), Reinhard Mechler (Germany), Virginia Murray (UK), Mark Pelling (UK), Jürgen Pohl (Germany), Anthony-Oliver Smith (USA), Frank Thomalla (Australia)

This chapter should be cited as:

Cardona, O.D., M.K. van Aalst, J. Birkmann, M. Fordham, G. McGregor, R. Perez, R.S. Pulwarty, E.L.F. Schipper, and B.T. Sinh, 2012: Determinants of risk: exposure and vulnerability. In: *Managing the Risks of Extreme Events and Disasters to Advance Climate Change Adaptation* [Field, C.B., V. Barros, T.F. Stocker, D. Qin, D.J. Dokken, K.L. Ebi, M.D. Mastrandrea, K.J. Mach, G.-K. Plattner, S.K. Allen, M. Tignor, and P.M. Midgley (eds.)]. A Special Report of Working Groups I and II of the Intergovernmental Panel on Climate Change (IPCC). Cambridge University Press, Cambridge, UK, and New York, NY, USA, pp. 65-108.

Table of Contents

Executive Summary 67

2.1. Introduction and Scope 69

2.2. Defining Determinants of Risk: Hazard, Exposure, and Vulnerability 69
2.2.1. Disaster Risk and Disaster 69
2.2.2. The Factors of Risk 69

2.3. The Drivers of Vulnerability 70

2.4. Coping and Adaptive Capacities 72
2.4.1. Capacity and Vulnerability 72
2.4.2. Different Capacity Needs 74
2.4.2.1. Capacity to Anticipate Risk 74
2.4.2.2. Capacity to Respond 74
2.4.2.3. Capacity to Recover and Change 75
2.4.3. Factors of Capacity: Drivers and Barriers 76

2.5. Dimensions and Trends of Vulnerability and Exposure 76
2.5.1. Environmental Dimensions 76
2.5.1.1. Physical Dimensions 77
2.5.1.2. Geography, Location, Place 77
2.5.1.3. Settlement Patterns and Development Trajectories 78
2.5.2. Social Dimensions 80
2.5.2.1. Demography 80
2.5.2.2. Education 81
2.5.2.3. Health and Well-Being 82
2.5.2.4. Cultural Dimensions 84
2.5.2.5. Institutional and Governance Dimensions 85
2.5.3. Economic Dimensions 86
2.5.4. Interactions, Cross-Cutting Themes, and Integrations 87
2.5.4.1. Intersectionality and Other Dimensions 88
2.5.4.2. Timing, Spatial, and Functional Scales 88
2.5.4.3. Science and Technology 89

2.6. Risk Identification and Assessment 89
2.6.1. Risk Identification 90
2.6.2. Vulnerability and Risk Assessment 90
2.6.3. Risk Communication 95

2.7. Risk Accumulation and the Nature of Disasters 95

References 96

Executive Summary

The severity of the impacts of extreme and non-extreme weather and climate events depends strongly on the level of vulnerability and exposure to these events (*high confidence*). [2.2.1, 2.3, 2.5] Trends in vulnerability and exposure are major drivers of changes in disaster risk, and of impacts when risk is realized (*high confidence*). [2.5] Understanding the multi-faceted nature of vulnerability and exposure is a prerequisite for determining how weather and climate events contribute to the occurrence of disasters, and for designing and implementing effective adaptation and disaster risk management strategies. [2.2, 2.6]

Vulnerability and exposure are dynamic, varying across temporal and spatial scales, and depend on economic, social, geographic, demographic, cultural, institutional, governance, and environmental factors (*high confidence*). [2.2, 2.3, 2.5] Individuals and communities are differentially exposed and vulnerable and this is based on factors such as wealth, education, race/ethnicity/religion, gender, age, class/caste, disability, and health status. [2.5] Lack of resilience and capacity to anticipate, cope with, and adapt to extremes and change are important causal factors of vulnerability. [2.4]

Extreme and non-extreme weather and climate events also affect vulnerability to future extreme events, by modifying the resilience, coping, and adaptive capacity of communities, societies, or social-ecological systems affected by such events (*high confidence*). [2.4.3] At the far end of the spectrum – low-probability, high-intensity events – the intensity of extreme climate and weather events and exposure to them tend to be more pervasive in explaining disaster loss than vulnerability in explaining the level of impact. But for less extreme events – higher probability, lower intensity – the vulnerability of exposed elements plays an increasingly important role (*high confidence*). [2.3] The cumulative effects of small- or medium-scale, recurrent disasters at the sub-national or local levels can substantially affect livelihood options and resources and the capacity of societies and communities to prepare for and respond to future disasters. [2.2.1, 2.7]

High vulnerability and exposure are generally the outcome of skewed development processes, such as those associated with environmental mismanagement, demographic changes, rapid and unplanned urbanization in hazardous areas, failed governance, and the scarcity of livelihood options for the poor (*high confidence*). [2.2.2, 2.5]

The selection of appropriate vulnerability and risk evaluation approaches depends on the decisionmaking context (*high confidence*). [2.6.1] Vulnerability and risk assessment methods range from global and national quantitative assessments to local-scale qualitative participatory approaches. The appropriateness of a specific method depends on the adaptation or risk management issue to be addressed, including for instance the time and geographic scale involved, the number and type of actors, and economic and governance aspects. Indicators, indices, and probabilistic metrics are important measures and techniques for vulnerability and risk analysis. However, quantitative approaches for assessing vulnerability need to be complemented with qualitative approaches to capture the full complexity and the various tangible and intangible aspects of vulnerability in its different dimensions. [2.6]

Appropriate and timely risk communication is critical for effective adaptation and disaster risk management (*high confidence*). Effective risk communication is built on risk assessment, and tailored to a specific audience, which may range from decisionmakers at various levels of government, to the private sector and the public at large, including local communities and specific social groups. Explicit characterization of uncertainty and complexity strengthens risk communication. Impediments to information flows and limited awareness are risk amplifiers. Beliefs, values, and norms influence risk perceptions, risk awareness, and choice of action. [2.6.3]

Adaptation and risk management policies and practices will be more successful if they take the dynamic nature of vulnerability and exposure into account, including the explicit characterization of uncertainty and complexity at each stage of planning and practice (*medium evidence, high agreement*). However, approaches to representing such dynamics quantitatively are currently underdeveloped. Projections of the impacts of climate change can be strengthened by including storylines of changing vulnerability and exposure under different development pathways. Appropriate attention to the temporal and spatial dynamics of vulnerability and exposure is particularly important given that the design and implementation of adaptation and risk management strategies and policies can reduce risk in the short term, but may increase vulnerability and exposure over the longer term. For instance, dike systems can reduce hazard exposure by offering immediate protection, but also encourage settlement patterns that may increase risk in the long term. [2.4.2.1, 2.5.4.2, 2.6.2]

Vulnerability reduction is a core common element of adaptation and disaster risk management (*high confidence*). Vulnerability reduction thus constitutes an important common ground between the two areas of policy and practice. [2.2, 2.3]

Chapter 2   Determinants of Risk: Exposure and Vulnerability

2.1. Introduction and Scope

Many climate change adaptation efforts aim to address the implications of potential changes in the frequency, intensity, and duration of weather and climate events that affect the risk of extreme impacts on human society. That risk is determined not only by the climate and weather events (the hazards) but also by the exposure and vulnerability to these hazards. Therefore, effective adaptation and disaster risk management strategies and practices also depend on a rigorous understanding of the dimensions of exposure and vulnerability, as well as a proper assessment of changes in those dimensions. This chapter aims to provide that understanding and assessment, by further detailing the determinants of risk as presented in Chapter 1.

The first sections of this chapter elucidate the concepts that are needed to define and understand risk, and show that risk originates from a combination of social processes and their interaction with the environment (Sections 2.2 and 2.3), and highlight the role of coping and adaptive capacities (Section 2.4). The following section (2.5) describes the different dimensions of vulnerability and exposure as well as trends therein. Given that exposure and vulnerability are highly context-specific, this section is by definition limited to a general overview (a more quantitative perspective on trends is provided in Chapter 4). A methodological discussion (Section 2.6) of approaches to identify and assess risk provides indications of how the dimensions of exposure and vulnerability can be explored in specific contexts, such as adaptation planning, and the central role of risk perception and risk communication. The chapter concludes with a cross-cutting discussion of risk accumulation and the nature of disasters (Section 2.7).

2.2. Defining Determinants of Risk: Hazard, Exposure, and Vulnerability

2.2.1. Disaster Risk and Disaster

Disaster risk signifies the possibility of adverse effects in the future. It derives from the interaction of social and environmental processes, from the combination of physical hazards and the vulnerabilities of exposed elements (see Chapter 1). The hazard event is not the sole driver of risk, and there is *high confidence* that the levels of adverse effects are in good part determined by the vulnerability and exposure of societies and social-ecological systems (UNDRO, 1980; Cuny, 1984; Cardona, 1986, 1993, 2011; Davis and Wall, 1992; UNISDR, 2004, 2009b; Birkmann, 2006a,b; van Aalst 2006a).

Disaster risk is not fixed but is a continuum in constant evolution. A *disaster* is one of its many 'moments' (ICSU-LAC, 2010a,b), signifying unmanaged risks that often serve to highlight skewed development problems (Westgate and O'Keefe, 1976; Wijkman and Timberlake, 1984). Disasters may also be seen as the materialization of risk and signify 'a becoming real' of this latent condition that is in itself a social construction (see below; Renn, 1992; Adam and Van Loon, 2000; Beck, 2000, 2008).

Disaster risk is associated with differing levels and types of adverse effects. The effects may assume catastrophic levels or levels commensurate with small disasters. Some have limited financial costs but very high human costs in terms of loss of life and numbers of people affected; others have very high financial costs but relatively limited human costs. Furthermore, there is *high confidence* that the cumulative effects of small disasters can affect capacities of communities, societies, or social-ecological systems to deal with future disasters at sub-national or local levels (Alexander, 1993, 2000; Quarantelli, 1998; Birkmann, 2006b; Marulanda et al., 2008b, 2010, 2011; UNISDR, 2009a).

2.2.2. The Factors of Risk

As detailed in Section 1.1, *hazard* refers to the possible, future occurrence of natural or human-induced physical events that may have adverse effects on vulnerable and exposed elements (White, 1973; UNDRO, 1980; Cardona, 1990; UNDHA, 1992; Birkmann, 2006b). Although, at times, hazard has been ascribed the same meaning as risk, currently it is widely accepted that it is a component of risk and not risk itself. The intensity or recurrence of hazard events can be partly determined by environmental degradation and human intervention in natural ecosystems. Landslides or flooding regimes associated with human-induced environmental alteration and new climate change-related hazards are examples of such socio-natural hazards (Lavell, 1996, 1999a).

Exposure refers to the inventory of elements in an area in which hazard events may occur (Cardona, 1990; UNISDR, 2004, 2009b). Hence, if population and economic resources were not located in (exposed to) potentially dangerous settings, no problem of disaster risk would exist. While the literature and common usage often mistakenly conflate exposure and vulnerability, they are distinct. Exposure is a necessary, but not sufficient, determinant of risk. It is possible to be exposed but not vulnerable (for example by living in a floodplain but having sufficient means to modify building structure and behavior to mitigate potential loss). However, to be vulnerable to an extreme event, it is necessary to also be exposed.

Land use and territorial planning are key factors in risk reduction. The environment offers resources for human development at the same time as it represents exposure to intrinsic and fluctuating hazardous conditions. Population dynamics, diverse demands for location, and the gradual decrease in the availability of safer lands mean it is almost inevitable that humans and human endeavor will be located in potentially dangerous places (Lavell, 2003). Where exposure to events is impossible to avoid, land use planning and location decisions can be accompanied by other structural or non-structural methods for preventing or mitigating risk (UNISDR, 2009a; ICSU-LAC, 2010a,b).

Vulnerability refers to the propensity of exposed elements such as human beings, their livelihoods, and assets to suffer adverse effects when impacted by hazard events (UNDRO, 1980; Cardona, 1986, 1990,

1993; Liverman, 1990; Maskrey, 1993b; Cannon, 1994, 2006; Blaikie et al., 1996; Weichselgartner, 2001; Bogardi and Birkmann, 2004; UNISDR, 2004, 2009b; Birkmann, 2006b; Janssen et al., 2006; Thywissen, 2006). Vulnerability is related to predisposition, susceptibilities, fragilities, weaknesses, deficiencies, or lack of capacities that favor adverse effects on the exposed elements. Thywissen (2006) and Manyena (2006) carried out an extensive review of the terminology. The former includes a long list of definitions used for the term vulnerability and the latter includes definitions of vulnerability and resilience and their relationship.

An early view of vulnerability in the context of disaster risk management was related to the physical resistance of engineering structures (UNDHA, 1992), but more recent views relate vulnerability to characteristics of social and environmental processes. It is directly related, in the context of climate change, to the susceptibility, sensitivity, and lack of resilience or capacities of the exposed system to cope with and adapt to extremes and non-extremes (Luers et al., 2003; Schröter et al., 2005; Brklacich and Bohle, 2006; IPCC, 2001, 2007).

While vulnerability is a key concept for both disaster risk and climate change adaptation, the term is employed in numerous other contexts, for instance to refer to epidemiological and psychological fragilities, ecosystem sensitivity, or the conditions, circumstances, and drivers that make people vulnerable to natural and economic stressors (Kasperson et al., 1988; Cutter, 1994; Wisner et al., 2004; Brklacich and Bohle, 2006; Haines et al., 2006; Villagrán de León, 2006). It is common to find blanket descriptions of the elderly, children, or women as 'vulnerable,' without any indication as to what these groups are vulnerable to (Wisner, 1993; Enarson and Morrow, 1998; Morrow, 1999; Bankoff, 2004; Cardona, 2004, 2011).

Vulnerability can be seen as situation-specific, interacting with a hazard event to generate risk (Lavell, 2003; Cannon, 2006; Cutter et al., 2008). Vulnerability to financial crisis, for example, does not infer vulnerability to climate change or natural hazards. Similarly, a population might be vulnerable to hurricanes, but not to landslides or floods. From a climate change perspective, basic environmental conditions change progressively and then induce new risk conditions for societies. For example, more frequent and intense events may introduce factors of risk into new areas, revealing underlying vulnerability. In fact, future vulnerability is embedded in the present conditions of the communities that may be exposed in the future (Patt et al., 2005, 2009); that is, new hazards in areas not previously subject to them will reveal, not necessarily create, underlying vulnerability factors (Alwang et al., 2001; Cardona et al., 2003a; Lopez-Calva and Ortiz, 2008; UNISDR, 2009a).

While vulnerability is in general hazard-specific, certain factors, such as poverty, and the lack of social networks and social support mechanisms, will aggravate or affect vulnerability levels irrespective of the type of hazard. These types of generic factors are different from the hazard-specific factors and assume a different position in the intervention actions and the nature of risk management and adaptation processes (ICSU-LAC, 2010a,b). Vulnerability of human settlements and ecosystems is intrinsically tied to different socio-cultural and environmental processes (Kasperson et al., 1988; Cutter, 1994; Adger, 2006; Cutter and Finch, 2008; Cutter et al., 2008; Williams et al., 2008; Décamps, 2010; Dawson et al., 2011). Vulnerability is linked also to deficits in risk communication, especially the lack of appropriate information that can lead to false risk perceptions (Birkmann and Fernando, 2008), which have an important influence on the motivation and perceived ability to act or to adapt to climate change and environmental stressors (Grothmann and Patt, 2005). Additionally, processes of maladaptation or unsustainable adaptation can increase vulnerability and risks (Birkmann, 2011a).

Vulnerability in the context of disaster risk management is the most palpable manifestation of the social construction of risk (Aysan, 1993; Blaikie et al., 1996; Wisner et al., 2004; ICSU-LAC 2010a,b). This notion underscores that society, in its interaction with the changing physical world, constructs disaster risk by transforming physical events into hazards of different intensities or magnitudes through social processes that increase the exposure and vulnerability of population groups, their livelihoods, production, support infrastructure, and services (Chambers, 1989; Wilches-Chaux, 1989; Cannon, 1994; Wisner et al., 2004; Wisner, 2006a; Carreño et al., 2007a; ICSU-LAC, 2010a,b). This includes:

- How human action influences the levels of exposure and vulnerability in the face of different physical events
- How human intervention in the environment leads to the creation of new hazards or an increase in the levels or damage potential of existing ones
- How human perception, understanding, and assimilation of the factors of risk influence societal reactions, prioritization, and decisionmaking processes.

There is *high agreement* and *robust evidence* that high vulnerability and exposure are mainly an outcome of skewed development processes, including those associated with environmental mismanagement, demographic changes, rapid and unplanned urbanization, and the scarcity of livelihood options for the poor (Maskrey, 1993a,b, 1994, 1998; Mansilla, 1996; Lavell, 2003; Cannon, 2006; ICSU-LAC, 2010a,b; Cardona, 2011).

Increases in disaster risk and the occurrence of disasters have been in evidence over the last five decades (Munich Re, 2011) (see Section 1.1.1). This trend may continue and may be enhanced in the future as a result of projected climate change, further demographic and socioeconomic changes, and trends in governance, unless concerted actions are enacted to reduce vulnerability and to adapt to climate change, including interventions to address disaster risks (Lavell, 1996, 1999a, 2003; ICSU-LAC, 2010a,b; UNISDR, 2011).

2.3. The Drivers of Vulnerability

In order to effectively manage risk, it is essential to understand how vulnerability is generated, how it increases, and how it builds up (Maskrey, 1989; Cardona, 1996a, 2004, 2011; Lavell, 1996, 1999a;

O'Brien et al., 2004b). Vulnerability describes a set of conditions of people that derive from the historical and prevailing cultural, social, environmental, political, and economic contexts. In this sense, vulnerable groups are not only at risk because they are exposed to a hazard but as a result of marginality, of everyday patterns of social interaction and organization, and access to resources (Watts and Bohle, 1993; Morrow, 1999; Bankoff, 2004). Thus, the effects of a disaster on any particular household result from a complex set of drivers and interacting conditions. It is important to keep in mind that people and communities are not only or mainly victims, but also active managers of vulnerability (Ribot, 1996; Pelling, 1997, 2003). Therefore, integrated and multidimensional approaches are highly important to understanding causes of vulnerability.

Some global processes are significant drivers of risk and are particularly related to vulnerability creation. There is *high confidence* that these include population growth, rapid and inappropriate urban development, international financial pressures, increases in socioeconomic inequalities, trends and failures in governance (e.g., corruption, mismanagement), and environmental degradation (Maskrey, 1993a,b, 1994, 1998; Mansilla, 1996; Cannon, 2006). Vulnerability profiles can be constructed that take into consideration sources of environmental, social, and economic marginality (Wisner, 2003). This also includes the consideration of the links between communities and specific environmental services, and the vulnerability of ecosystem components (Renaud, 2006; Williams et al., 2008; Décamps, 2010; Dawson et al., 2011). In climate change-related impact assessments, integration of underlying 'causes of vulnerability' and adaptive capacity is needed rather than focusing on technical aspects only (Ribot, 1995; O'Brien et al., 2004b).

Due to different conceptual frameworks and definitions, as well as disciplinary views, approaches to address the causes of vulnerability also differ (Burton et al., 1983; Blaikie et al., 1994; Harding et al., 2001; Twigg, 2001; Adger and Brooks, 2003, 2006; Turner et al., 2003a,b; Cardona, 2004; Schröter et al., 2005; Adger 2006; Füssel and Klein, 2006; Villagrán de León, 2006; Cutter and Finch, 2008; Cutter et al., 2008). Thomalla et al. (2006), Mitchell and van Aalst (2008), and Mitchell et al. (2010) examine commonalities and differences between the adaptation to climate change and disaster risk management communities, and identify key areas of difference and convergence. The two communities tend to perceive the nature and timescale of the threat differently: impacts due to climate change and return periods for extreme events frequently use the language of uncertainty; but considerable knowledge and certainty has been expressed regarding event characteristics and exposures related to extreme historical environmental conditions.

Four approaches to understanding vulnerability and its causes can be distinguished, rooted in political economy, social-ecology, vulnerability, and disaster risk assessment, as well as adaptation to climate change:
1) The pressure and release (PAR) model (Blaikie et al., 1994, 1996; Wisner et al., 2004) is common to social science-related vulnerability research and emphasizes the social conditions and root causes of exposure more than the hazard as generating unsafe conditions. This approach links vulnerability to unsafe conditions in a continuum that connects local vulnerability to wider national and global shifts in the political economy of resources and political power.
2) The social ecology perspective emphasizes the need to focus on coupled human-environmental systems (Hewitt and Burton, 1971; Turner et al., 2003a,b). This perspective stresses the ability of societies to transform nature and also the implications of changes in the environment for social and economic systems. It argues that the exposure and susceptibility of a system can only be adequately understood if these coupling processes and interactions are addressed.
3) Holistic perspectives on vulnerability aim to go beyond technical modeling to embrace a wider and comprehensive explanation of vulnerability. These approaches differentiate exposure, susceptibility and societal response capacities as causes or factors of vulnerability (see Cardona, 1999a, 2001, 2011; Cardona and Barbat, 2000; Cardona and Hurtado, 2000a,b; IDEA, 2005; Birkmann, 2006b; Carreño, 2006; Carreño et al.,2007a,b, 2009; Birkmann and Fernando, 2008). A core element of these approaches is the feedback loop which underlines that vulnerability is dynamic and is the main driver and determinant of current or future risk.
4) In the context of climate change adaptation, different vulnerability definitions and concepts have been developed and discussed. One of the most prominent definitions is the one reflected in the IPCC Fourth Assessment Report, which describes vulnerability as a function of exposure, sensitivity, and adaptive capacity, as also reflected by, for instance, McCarthy et al. (2001), Brooks (2003), K. O'Brien et al. (2004a), Füssel and Klein (2006), Füssel (2007), and G. O'Brien et al. (2008). This approach differs from the understanding of vulnerability in the disaster risk management perspective, as the rate and magnitude of climate change is considered. The concept of vulnerability here includes external environmental factors of shock or stress. Therefore, in this view, the magnitude and frequency of potential hazard events is to be considered in the vulnerability to climate change. This view also differs in its focus upon long-term trends and stresses rather than on current shock forecasting, something not explicitly excluded but rather rarely considered within the disaster risk management approaches.

The lack of a comprehensive conceptual framework that facilitates a common multidisciplinary risk evaluation impedes the effectiveness of disaster risk management and adaptation to climate change (Cardona, 2004). The option for anticipatory disaster risk reduction and adaptation exists precisely because risk is a latent condition, which announces potential future adverse effects (Lavell, 1996, 1999a). Understanding disaster risk management as a social process allows for a shift in focus from responding to the disaster event toward an understanding of disaster risk (Cardona and Barbat, 2000; Cardona et al., 2003a). This requires knowledge about how human interactions with the natural environment lead to the creation of new hazards, and how persons, property, infrastructure, goods, and the environment are exposed to potentially damaging events. Furthermore, it requires an understanding of the vulnerability of people and their livelihoods, including the allocation and distribution of social and economic resources that can work for or against the achievement of resistance, resilience, and security (ICSU-LAC, 2010a,b). Overall, there is *high confidence* that although hazard events are usually considered the cause of disaster risk, vulnerability and exposure are its key determining factors. Furthermore, contrary to the hazard, vulnerability and exposure can often be influenced by policy and practice, including in the short to medium term. Therefore disaster risk management and adaptation strategies have to address mainly these same risk factors (Cardona 1999a, 2011; Vogel and O'Brien, 2004; Birkmann, 2006a; Leichenko and O'Brien, 2008).

Despite various frameworks developed for defining and assessing vulnerability, it is interesting to note that at least some common causal factors of vulnerability have been identified, in both the disaster risk management and climate change adaptation communities (see Cardona, 1999b, 2001, 2011; Cardona and Barbat, 2000; Cardona and Hurtado, 2000a,b; McCarthy et al., 2001; Gallopin, 2006; Manyena, 2006; Carreño et al., 2007a, 2009; IPCC, 2007; ICSU-LAC 2010a,b; MOVE, 2010):

- *Susceptibility/fragility (in disaster risk management) or sensitivity (in climate change adaptation)*: physical predisposition of human beings, infrastructure, and environment to be affected by a dangerous phenomenon due to lack of resistance and predisposition of society and ecosystems to suffer harm as a consequence of intrinsic and context conditions making it plausible that such systems once impacted will collapse or experience major harm and damage due to the influence of a hazard event.
- *Lack of resilience (in disaster risk management) or lack of coping and adaptive capacities (in climate change adaptation)*: limitations in access to and mobilization of the resources of the human beings and their institutions, and incapacity to anticipate, adapt, and respond in absorbing the socio-ecological and economic impact.

There is *high confidence* that at the extreme end of the spectrum, the intensity of extreme climate and weather events – low-probability, high-intensity – and exposure to them tend to be more pervasive in explaining disaster loss than vulnerability itself. But as the events get less extreme – higher-probability, lower-intensity – the vulnerability of exposed elements plays an increasingly important role in explaining the level of impact. Vulnerability is a major cause of the increasing adverse effects of non-extreme events, that is, small recurrent disasters that many times are not visible at the national or sub-national level (Marulanda et al., 2008b, 2010, 2011; UNISDR, 2009a; Cardona, 2011; UNISDR, 2011).

Overall, the promotion of resilient and adaptive societies requires a paradigm shift away from the primary focus on natural hazards and extreme weather events toward the identification, assessment, and ranking of vulnerability (Maskrey, 1993a; Lavell, 2003; Birkmann, 2006a,b). Therefore, understanding vulnerability is a prerequisite for understanding risk and the development of risk reduction and adaptation strategies to extreme events in the light of climate change (ICSU-LAC, 2010a,b; MOVE, 2010; Cardona, 2011; UNISDR, 2011).

2.4. Coping and Adaptive Capacities

Capacity is an important element in most conceptual frameworks of vulnerability and risk. It refers to the positive features of people's characteristics that may reduce the risk posed by a certain hazard. Improving capacity is often identified as the target of policies and projects, based on the notion that strengthening capacity will eventually lead to reduced risk. Capacity clearly also matters for reducing the impact of climate change (e.g., Sharma and Patwardhan, 2008).

As presented in Chapter 1, coping is typically used to refer to *ex post* actions, while adaptation is normally associated with *ex ante* actions. This implies that coping capacity also refers to the ability to react to and reduce the adverse effects of experienced hazards, whereas adaptive capacity refers to the ability to anticipate and transform structure, functioning, or organization to better survive hazards (Saldaña-Zorrilla, 2007). Presence of capacity suggests that impacts will be less extreme and/or the recovery time will be shorter, but high capacity to recover does not guarantee equal levels of capacity to anticipate. In other words, the capacity to cope does not infer the capacity to adapt (Birkmann, 2011a), although coping capacity is often considered to be part of adaptive capacity (Levina and Tirpak, 2006).

2.4.1. Capacity and Vulnerability

Most risk studies prior to the 1990s focused mainly on hazards, whereas the more recent reversal of this paradigm has placed equal focus on the vulnerability side of the equation. Emphasizing that risk can be reduced through vulnerability is an acknowledgement of the power of social, political, environmental, and economic factors in driving risk. While these factors drive risk on one hand, they can on the other hand be the source of capacity to reduce it (Carreño et al., 2007a; Gaillard, 2010).

Many approaches for assessing vulnerability rely on an assessment of capacity as a baseline for understanding how vulnerable people are to a specific hazard. The relationship between capacity and vulnerability is described differently among different schools of thought, stemming from different uses in the fields of development, disaster risk management, and climate change adaptation. Gaillard (2010) notes that the concept of capacity "played a pivotal role in the progressive emergence of the vulnerability paradigm within the scientific realm." On the whole, the literature describes the relationship between vulnerability and capacity in two ways, which are not mutually exclusive (Bohle, 2001; IPCC, 2001; Moss et al., 2001; Yodmani, 2001; Downing and Patwardhan, 2004; Brooks et al., 2005; Smit and Wandel, 2006; Gaillard, 2010):

1) Vulnerability is, among other things, the result of a lack of capacity.
2) Vulnerability is the opposite of capacity, so that increasing capacity means reducing vulnerability, and high vulnerability means low capacity.

Box 2-1 | Coping and Adaptive Capacity: Different Origins and Uses

As set out in Section 1.4, there is a difference in understanding and use of the terms coping and adapting. Although coping capacity is often used interchangeably with adaptive capacity in the climate change literature, Cutter et al. (2008) point out that adaptive capacity features more frequently in global environmental change perspectives and is less prevalent in the hazards discourse.

Adaptive capacity refers to the ability of a system or individual to adapt to climate change, but it can also be used in the context of disaster risk. Because adaptive capacity is considered to determine "the ability of an individual, family, community, or other social group to adjust to changes in the environment guaranteeing survival and sustainability" (Lavell, 1999b), many believe that in the context of uncertain environmental changes, adaptive capacity will be of key significance. Dayton-Johnson (2004) defines adaptive capacity as the "vulnerability of a society before disaster strikes and its resilience after the fact." Some ways of classifying adaptive capacity include 'baseline adaptive capacity' (Dore and Etkin, 2003), which refers to the capacity that allows countries to adapt to existing climate variability, and 'socially optimal adaptive capacity,' which is determined by the norms and rules in individual locations. Another definition of adaptive capacity is the "property of a system to adjust its characteristics or behavior, in order to expand its coping range under existing climate variability, or future climate conditions" (Brooks and Adger, 2004). This links adaptive capacity to coping capacity, because coping range is synonymous with coping capacity, referring to the boundaries of systems' ability to cope (Yohe and Tol, 2002).

In simple terms, coping capacity refers to the "ability of people, organizations, and systems, using available skills and resources, to face and manage adverse conditions, emergencies, or disasters" (UNISDR, 2009b). Coping capacity is typically used in humanitarian discourse to indicate the extent to which a system can survive the impacts of an extreme event. It suggests that people can deal with some degree of destabilization, and acknowledges that at a certain point this capacity may be exceeded. Eriksen et al. (2005) link coping capacity to entitlements – the set of commodity bundles that can be commanded – during an adverse event. The ability to mobilize this capacity in an emergency is the manifestation of coping strategies (Gaillard, 2010). Furthermore, Birkmann (2011b) underscores that differences between coping and adaptation are also linked to the quality of the response process. While coping aims to maintain the system and its functions in the face of adverse conditions, adaptation involves changes and requires reorganization processes.

The capacity described by the disasters community in the past decades does not frequently distinguish between 'coping' or 'adaptive' capacities, and instead the term is used to indicate positive characteristics or circumstances that could be seen to offset vulnerability (Anderson and Woodrow, 1989). Because the approach is focused on disasters, it has been associated with the immediate-term coping needs, and contrasts from the long-term perspective generally discussed in the context of climate change, where the aim is to adapt to changes rather than to just overcome them. There has been considerable discussion throughout the vulnerability and poverty and climate change scholarly communities about whether coping strategies are a stepping stone toward adaptation, or may lead to maladaptation (Yohe and Tol, 2002; Eriksen et al., 2005) (see Chapter 1). Useful alternative terminology is to talk about 'capacity to change and adjust' (Nelson and Finan, 2009) for adaptive capacity, and 'capacity to absorb' instead of coping capacity (Cutter et al., 2008).

In the climate change community of practice, adaptive capacity has been at the forefront of thinking regarding how to respond to the impacts of climate change, but it was initially seen as a characteristic to build interventions on, and only later has been recognized as the target of interventions (Adger et al., 2004). The United Nations Framework Convention on Climate Change, for instance, states in its ultimate objective that action to reduce greenhouse gas emissions be guided by the time needed for ecosystems to adapt naturally to the impacts of climate change.

The relationship between capacity and vulnerability is interpreted differently in the climate change community of practice and the disaster risk management community of practice. Throughout the 1980s, vulnerability became a central focus of much work on disasters, in some circles overshadowing the role played by hazards in driving risk. Some have noted that the emphasis on vulnerability tended to ignore capacity, focusing too much on the negative aspects of vulnerability (Davis et al., 2004). Recognizing the role of capacity in reducing risk also indicates an acknowledgement that people are not 'helpless victims' (Bohle, 2001; Gaillard, 2010).

In many climate change-related studies, capacity was initially subsumed under vulnerability. The first handbooks and guidelines for adaptation emphasized impacts and vulnerability assessment as the necessary steps for determining adaptation options (Kate, 1985; Carter et al., 1994; Benioff et al., 1996; Feenstra et al., 1998). Climate change vulnerability was often placed in direct opposition to capacity. Vulnerability that was measured was seen as the remainder after capacity had been taken into account.

However, Davis et al. (2004), IDEA (2005), Carreño et al. (2007a,b), and Gaillard (2010) note that capacity and vulnerability are not necessarily opposites, because communities that are highly vulnerable may in fact display high capacity in certain aspects. This reflects the many elements of risk reduction and the multiple capacity needs across them. Alwang et al. (2001) also underscore that vulnerability is dynamic and determined by numerous factors, thus high capacity in the ability to respond to an extreme event does not accurately reflect low vulnerability.

2.4.2. Different Capacity Needs

The capacity necessary to anticipate and avoid being affected by an extreme event requires different assets, opportunities, social networks, and local and external institutions from capacity to deal with impacts and recover from them (Lavell, 1994; Lavell and Franco, 1996; Cardona, 2001, 2010; Carreño et al., 2007a,b; ICSU-LAC, 2010a,b; MOVE, 2010). Capacity to change relies on yet another set of factors. Importantly, however, these dimensions of capacity are not unrelated to each other: the ability to change is also necessary for risk reduction and response capacities.

Just like vulnerability, capacity is dynamic and will change depending on circumstances. The discussion in Box 2-1 indicates that there are differing perspectives on how coping and adaptive capacity relate. When coping and adapting are viewed as different, it follows that the capacity needs for each are also different (Cooper et al., 2008). This suggests that work done to understand the drivers of adaptive *ex ante* capacity (Leichenko and O'Brien, 2002; Yohe and Tol, 2002; Brenkert and Malone, 2005; Brooks et al., 2005; Haddad, 2005; Vincent, 2007; Sharma and Patwardhan, 2008; Magnan, 2010) may not be similar with the identified drivers of capacities that helped in the past (*ex post*) and are associated more closely with experienced coping processes. Many of these elements are reflected in local, national, and international contexts in Chapters 5, 6, and 7 of this Special Report.

2.4.2.1. Capacity to Anticipate Risk

Having the capacity to reduce the risk posed by hazards and changes implies that people's ability to manage is not engulfed, so they are not left significantly worse off. Reducing risk means that people do not have to devote substantial resources to dealing with a hazard as it occurs, but instead have the capacity to anticipate this sort of event. This is the type of capacity that is necessary in order to adapt to climate change, and involves conscious, planned efforts to reduce risk. The capacity to reduce risk also depends on *ex post* actions, which involve making choices after one event that reduce the impact of future events.

Capacity for risk prevention and reduction may be understood as a series of elements, measures, and tools directed toward intervention in hazards and vulnerabilities with the objective of reducing existing or controlling future possible risks (Cardona et al., 2003a). This can range from guaranteeing survival to the ability to secure future livelihoods (Batterbury, 2001; Eriksen and Silva, 2009).

Development planning, including land use and urban planning, river basin and land management, hazard-resistant building codes, and landscape design are all activities that can reduce exposure and vulnerability to hazards and change (Cardona, 2001, 2010). The ability to carry these out in an effective way is part of the capacity to reduce risk. Other activities include diversifying income sources, maintaining social networks, and collective action to avoid development that puts people at higher risk (Maskrey, 1989, 1994; Lavell, 1994, 1999b, 2003).

Up to the early 1990s, disaster preparedness and humanitarian response dominated disaster practice, and focus on capacity was limited to understanding inherent response capacity. Thus, emphasizing capacity to reduce risk was not a priority. However, in the face of growing evidence as to significant increases in disaster losses and the inevitable increase in financial and human resources dedicated to disaster response and recovery, there is an increasing recognition of the need to promote the capacity for prevention and risk reduction over time (Lavell, 1994, 1999b, 2003). Notwithstanding, different actors, stakeholders, and interests influence the capacity to anticipate a disaster. Actions to reduce exposure and vulnerability of one group of people may come at the cost of increasing it for another, for example when flood risks are shifted from upstream communities to downstream communities through large-scale upstream dike construction (Birkmann, 2011a). Consequently, it is not sufficient to evaluate the success of adaptation or capacities to reduce risk by focusing on the objectives of one group only. The evaluation of success of adaptation strategies depends on the spatial and temporal scale used (Adger et al., 2005).

2.4.2.2. Capacity to Respond

Capacity to respond is relevant both *ex post* and *ex ante*, since it encompasses everything necessary to be able to react once an extreme event takes place. Response capacity is mostly used to refer to the ability of institutions to react following a natural hazard, in particular *ex post* during emergency response. However, effective response requires substantial *ex ante* planning and investments in disaster preparedness and early warning (not only in terms of financial cost but particularly in terms of awareness raising and capacity building; IFRC, 2009). Furthermore, there are also response phases for gradual changes in ecosystems or temperature regimes caused by climate change. Responding spans everything from people's own initial reactions to a hazard upon its impact to actions to try to reduce secondary damage. It is worth noting that in climate change literature, anticipatory actions are often referred to as responses, which differs from the way this term is used in the context of disaster risk, where it only implies the actions taken once there has been an impact.

Capacity to respond is not sufficient to reduce risk. Humanitarian aid and relief interventions have been discussed in the context of their role in reinforcing or even amplifying existing vulnerabilities (Anderson and Woodrow, 1991; Wisner, 2001a; Schipper and Pelling, 2006). This does not only have implications for the capacity to respond, but also for other aspects of capacity. Wisner (2001a) shows how poorly constructed shelters, where people were placed temporarily in El Salvador following Hurricane Mitch in 1998, turned into 'permanent' housing when nongovernmental organization (NGO) support ran out. When two strong earthquakes hit in January and February 2001, the shelters collapsed, leaving the people homeless again. This example illustrates the perils associated with emergency measures that focus only on responding, rather than on the capacity to reduce risk and change. Response capacity is also differential (Chatterjee, 2010). The most effective *ex ante* risk management strategies will often include a combination of risk reduction and enhanced capacity to respond to impacts (including smarter response by better preparedness and early warning, as well risk transfer such as insurance).

2.4.2.3. Capacity to Recover and Change

Having the capacity to change is a requirement in order to adapt to climate change. Viewing adaptation as requiring transformation implies that it cannot be understood as only a set of actions that physically protect people from natural hazards (Pelling, 2010). In the context of natural hazards, the opportunity for changing is often greatest during the recovery phase, when physical infrastructure has to be rebuilt and can be improved, and behavioral patterns and habits can be contemplated (Susman et al., 1983; Renn, 1992; Comfort et al., 1999; Vogel and O'Brien, 2004; Birkmann et al., 2010a). This is an opportunity to rethink whether the crops planted are the most suited to the climate and whether it is worthwhile rebuilding hotels near the coast, taking into account what other sorts of environmental changes may occur in the area.

Capacity to recover is not only dependent on the extent of a physical impact, but also on the extent to which society has been affected, including the ability to resume livelihood activities (Hutton and Haque, 2003). This capacity is driven by numerous factors, including mental and physical ability to recover, financial and environmental viability, and political will. Because reconstruction processes often do not take people's livelihoods into account, instead focusing on their safety, new settlements are often located where people do not want to be, which brings change – but not necessarily change that leads to sustainable development. Innumerable examples indicate how people who have been resettled return back to their original location, moving into dilapidated houses or setting up new housing, even if more solid housing is available elsewhere (e.g., El Salvador after Hurricane Mitch), simply because the new location does not allow them easy access to their fields, to markets or roads, or to the sea (e.g., South and Southeast Asia after the 2004 tsunami).

Recovering to return to the conditions before a natural hazard occurs not only implies that the risk may be the same or greater, but also does not question whether the previous conditions were desirable. In fact, recovery processes are often out of sync with the evolving process of development. The recovery and reconstruction phases after a disaster provide an opportunity to rethink previous conditions and address the root causes of risk, looking to avoid reconstructing the vulnerability (IDB, 2007), but often the process is too rushed to enable effective reflection, discussion, and consensus building (Christoplos, 2006). Pushing the recovery toward transformation and change requires taking a new approach rather than returning to 'normalcy.' Several examples have shown that capacity to recover is severely limited by poverty (Chambers, 1983; Ingham, 1993; Hutton and Haque, 2003), where people are driven further down the poverty spiral, never returning to their previous conditions, however undesirable.

The various capacities to respond and to survive hazard events and changes have also been discussed within the context of the concept of resilience. While originally, the concept of resilience was strongly linked to an environmental perspective on ecosystems and their ability to maintain major functions even in times of adverse conditions and crises (Holling, 1973), the concept has undergone major shifts and has been enhanced and applied also in the field of social-ecological systems and disaster risk (Gunderson, 2000; Walker et al., 2004; UN, 2005; Abel et al., 2006). Folke (2006) differentiates three different resilience concepts that encompass an engineering resilience perspective that focuses on recovery and constancy issues, while the ecological and social resilience focus on persistence and robustness and, finally, the integrated social-ecological resilience perspective deals with adaptive capacity, transformability, learning, and innovation (Folke, 2006). In disaster risk reduction the terms resilience building and the lack of resilience have achieved a high recognition. These terms are linked to capacities of communities or societies to deal with the impact of a hazard event or crises and the ability to learn and create resilience through these experiences. Recent papers, however, also criticize the unconsidered use or the simply transfer of the concept of resilience into the wider context of adaptation (see, e.g., Cannon and Müller-Mahn, 2010). Additionally, the lack of resilience has also been used as an umbrella to examine deficiencies in capacities that communities encompass in order to deal with hazard events. Describing the lack of resilience, Cardona and Barbat (2000) identify various capacities that are often insufficient in societies that suffer heavily during disasters, such as the deficiencies regarding the capacity to anticipate, to cope with, and to adapt to changing environmental conditions and natural hazards.

Other work has argued a different view on resilience, because the very occurrence of a disaster shows that there are gaps in the development process (UNDP, 2004). Lessons learned from studying the impacts of the 2004 Indian Ocean tsunami (Thomalla et al., 2009; Thomalla and Larsen, 2010) are informative for climate-related hazards. They suggest that:
- Social vulnerability to multiple hazards, particularly rare extreme events, tends to be poorly understood.
- There is an increasing focus away from vulnerability assessment toward resilience building; however, resilience is poorly understood and a lot needs to be done to go from theory to practice.
- One of the key issues in sub-national risk reduction initiatives is a need to better define the roles and responsibilities of government and NGO actors and to improve coordination between them. Without mechanisms for joint target setting, coordination, monitoring, and

evaluation, there is much duplication of effort, competition, and tension between actors.
- Risk reduction is only meaningful and prioritized by local government authorities if it is perceived to be relevant in the context of other, more pressing day-to-day issues, such as poverty reduction, livelihood improvement, natural resource management, and community development.

2.4.3. Factors of Capacity: Drivers and Barriers

There is *high confidence* that extreme and non-extreme weather and climate events also affect vulnerability to future extreme events, by modifying the resilience, coping, and adaptive capacity of communities, societies, or social-ecological systems affected by such events. When people repeatedly have to respond to natural hazards and changes, the capitals that sustain capacity are broken down, increasing vulnerability to hazards (Wisner and Adams, 2002; Marulanda et al., 2008b, 2010, 2011; UNISDR, 2009a). Much work has gone into identifying what these factors of capacity are, to understand both what drives capacity as well as what acts as a barrier to it (Adger et al., 2004; Sharma and Padwardhan, 2008).

Drivers of capacity include: an integrated economy; urbanization; information technology; attention to human rights; agricultural capacity; strong international institutions; access to insurance; class structure; life expectancy, health, and well-being; degree of urbanization; access to public health facilities; community organizations; existing planning regulations at national and local levels; institutional and decisionmaking frameworks; existing warning and protection from natural hazards; and good governance (Cannon, 1994; Handmer et al., 1999; Klein, 2001; Barnett, 2005; Brooks et al., 2005; Bettencourt et al., 2006).

2.5. Dimensions and Trends of Vulnerability and Exposure

This section presents multiple dimensions of exposure and vulnerability to hazards, disasters, climate change, and extreme events. Some frameworks consider exposure to be a component of vulnerability (Turner et al., 2003a), and the largest body of knowledge on dimensions refers to vulnerability rather than exposure, but the distinction between them is often not made explicit. Vulnerability is: *multi-dimensional and differential* – that is, it varies across physical space and among and within social groups; *scale-dependent* with regard to space and units of analysis such as individual, household, region, or system; and *dynamic* – characteristics and driving forces of vulnerability change over time (Vogel and O'Brien, 2004). As vulnerability and exposure are not fixed, understanding the trends in vulnerability and exposure is therefore an important aspect of the discussion.

There is *high confidence* that for several hazards, changes in exposure and in some cases vulnerability are the main drivers behind observed trends in disaster losses, rather than a change in hazard character, and will continue to be essential drivers of changes in risk patterns over the coming decades (Bouwer et al., 2007; Pielke Jr. and Landsea, 1998; UNISDR, 2009a). In addition, there is *high confidence* that climate change will affect disaster risk not only through changes in the frequency, intensity, and duration of some events (see Chapter 3), but also through indirect effects on vulnerability and exposure. In most cases, it will do so not in isolation but as one of many sources of possible stress, for instance through impacts on the number of people in poverty or suffering from food and water insecurity, changing disease patterns and general health levels, and where people live. In some cases, these changes may be positive, but in many cases, they will be negative, especially for many groups and areas that are already among the most vulnerable.

Although trends in some of the determinants of risk and vulnerability are apparent (for example, accelerated urbanization), the extent to which these are altering levels of risk and vulnerability at a range of geographical and time scales is not always clear. While there is *high confidence* that these connections exist, current knowledge often does not allow us to provide specific quantifications with regional or global significance.

The multidimensional nature of vulnerability and exposure makes any organizing framework arbitrary, overlapping, and contentious to a degree. The following text is organized under three very broad headings: environmental, social, and economic dimensions. Each of these has a number of subcategories, which map out the major elements of interest.

2.5.1. Environmental Dimensions

Environmental dimensions include:
- Potentially vulnerable natural *systems* (such as low-lying islands, coastal zones, mountain regions, drylands, and Small Island Developing States (Dow, 1992; UNCED, 1992; Pelling and Uitto, 2001; Nicholls, 2004; UNISDR, 2004; Chapter 3)
- *Impacts* on systems (e.g., flooding of coastal cities and agricultural lands, or forced migration)
- The *mechanisms* causing impacts (e.g., disintegration of particular ice sheets) (Füssel and Klein, 2006; Schneider et al., 2007)
- *Responses* or *adaptations* to environmental conditions (UNEP/UNISDR, 2008).

There are important links between development, environmental management, disaster reduction, and climate adaptation (e.g., van Aalst and Burton, 2002), also including social and legal aspects such as property rights (Adger, 2000). For the purposes of vulnerability analysis in the context of climate change, it is important to acknowledge that the environment and human beings that form the socio-ecological system (Gallopin et al., 2001) behave in nonlinear ways, and are strongly coupled, complex, and evolving (Folke et al., 2002).

There are many examples of the interactions between society and environment that make people vulnerable to extreme events (Bohle et al., 1994) and highlight the vulnerability of ecosystem services (Metzger et al., 2006). As an example, vulnerabilities arising from floodplain encroachment and increased hazard exposure are typical of the intricate and finely balanced relationships within human-environment systems (Kates, 1971; White, 1974) of which we have been aware for several decades. Increasing human occupancy of floodplains increases exposure to flood hazards. It can put not only the lives and property of human beings at risk but can damage floodplain ecology and associated ecosystem services. Increased exposure of human beings comes about even in the face of actions designed to reduce the hazard. Structural responses and alleviation measures (e.g., provision of embankments, channel modification, and other physical alterations of the floodplain environment), designed ostensibly to reduce flood risk, can have the reverse result. This is variously known as the levee effect (Kates, 1971; White, 1974), the escalator effect (Parker, 1995), or the 'safe development paradox' (Burby, 2006) in which floodplain encroachment leads to increased flood risk and, ultimately, flood damages. A maladaptive policy response to such exposure provides structural flood defenses, which encourage the belief that the flood risk has been removed. This in turn encourages more floodplain encroachment and a reiteration of the cycle as the flood defenses (built to a lower design specification) are exceeded. This is typical of many maladaptive policy responses, which focus on the symptoms rather than the causes of poor environmental management.

Floodplains, even in low-lying coastal zones, have the potential to provide benefits and/or risks and it is the form of the *social* interaction (see next subsection) that determines which, and to whom. Climate variability shifts previous risk-based decisionmaking into conditions of greater uncertainty where we can be less certain of the probabilities of occurrence of any extreme event.

The environmental dimension of vulnerability also deals with the role of regulating ecosystem services and ecosystem functions, which directly impact human well-being, particularly for those social groups that heavily depend on these services and functions due to their livelihood profiles. Especially in developing countries and countries in transition, poorer rural communities often entirely depend on ecosystem services and functions to meet their livelihood needs. The importance of these ecosystem services and ecosystem functions for communities in the context of environmental vulnerability and disaster risk has been recognized by the 2009 and 2011 Global Assessment Reports on Disaster Risk Reduction (UNISDR, 2009a, 2011) as well as by the Millennium Ecosystem Assessment (MEA, 2005). The degradation of ecosystem services and functions can contribute to an exacerbation of both the natural hazard context and the vulnerability of people. The erosion of ecosystem services and functions can contribute to the decrease of coping and adaptive capacities in terms of reduced alternatives for livelihoods and income-generating activities due to the degradation of natural resources. Additionally, a worsening of environmental services and functions might also increase the costs of accessing these services, for example, in terms of the increased time and travel needed to access drinking water in rural communities affected by droughts or salinization.

Furthermore, environmental vulnerability can also mean that in the case of a hazardous event occurring, the community may lose access to the only available water resource or face a major reduction in productivity of the soil, which then also increases the risk of crop failure. For instance, Renaud (2006) underscored that the salinization of wells after the 2004 Indian Ocean tsunami had a highly negative consequence for those communities that had no alternative access to freshwater resources.

2.5.1.1. Physical Dimensions

Within the environmental dimension, physical aspects refer to a location specific context for human-environment interaction (Smithers and Smit, 1997) and to the material world (e.g., built structures).

The physical *exposure* of human beings to hazards has been partly shaped by patterns of settlement of hazard-prone landscapes for the countervailing benefits they offer (UNISDR, 2004). Furthermore, in the context of climate change, physical exposure is in many regions also increasing due to spatial extension of natural hazards, such as floods, areas affected by droughts, or delta regions affected by salinization. This does not make the inhabitants of such locations vulnerable per se because they may have capacities to resist the impacts of extreme events; this is the essential difference between exposure and vulnerability. The physical dimension of *vulnerability* begins with the recognition of a link between an extreme physical or natural phenomenon and a vulnerable human group (Westgate and O'Keefe, 1976). Physical vulnerability comprises aspects of geography, location, and place (Wilbanks, 2003); settlement patterns; and physical structures (Shah, 1995; UNISDR, 2004) including infrastructure located in hazard-prone areas or with deficiencies in resistance or susceptibility to damage (Wilches-Chaux, 1989). Further, Cutter's (1996) 'hazards of place' model of vulnerability expressly refers to the temporal dimension (see Section 2.5.4.2), which, in recognizing the dynamic nature of place vulnerability, argues for a more nuanced approach.

2.5.1.2. Geography, Location, Place

Aggregate trends in the environmental dimensions of exposure and vulnerability as they relate to geography, location, and place are given in Chapters 3 and 4, while this section deals with the more conceptual aspects.

There is a significant difference in exposure and vulnerability between developing and developed countries. While a similar (average) number of people in low and high human development countries may be exposed to hazards each year (11 and 15% respectively), the average numbers killed is very different (53 and 1% respectively) (Peduzzi, 2006).

Developing countries are recognized as facing the greater impacts and having the most vulnerable populations, in the greatest number, who are least able to easily adapt to changes in *inter alia* temperature, water resources, agricultural production, human health, and biodiversity (IPCC, 2001; McCarthy et al., 2001; Beg et al., 2002). Small Island Developing States, a number of which are also Least Developed Countries, are recognized as being highly vulnerable to external shocks including climate extremes (UN/DESA, 2010; Chapter 3). While efforts in climate change adaptation have been undertaken, progress has been limited, focusing on public awareness, research, and policy development rather than implementation (UN/DESA, 2010).

Developed countries are also vulnerable and have geographically distinct levels of vulnerability, which are masked by a predominant focus on direct impacts on biophysical systems and broad economic sectors. However, indirect and synergistic effects, differential vulnerabilities, and assumptions of relative ease of adaptation within apparently robust developed countries may lead to unforeseen vulnerabilities (O'Brien et al., 2006). Thus, development per se is not a guarantee of 'invulnerability.' Development can undermine ecosystem resilience on the one hand but create wealth that may enhance societal resilience overall if equitable (Barnett, 2001).

The importance of geography has been highlighted in an analysis of 'disaster hotspots' by Dilley et al. (2005). Hazard exposure (event incidence) is combined with historical vulnerability (measured by mortality and economic loss) in order to identify geographic regions that are at risk from a range of geophysical hazards. While flood risk is widespread across a number of regions, drought and especially cyclone risk demonstrate distinct spatial patterns with the latter closely related to the climatological pattern of cyclone tracks and landfall.

2.5.1.3. Settlement Patterns and Development Trajectories

There are specific exposure/vulnerability dimensions associated with urbanization (Hardoy and Pandiella, 2009) and rurality (Scoones, 1998; Nelson et al., 2010a,b). The major focus below is on the urban because of the increasing global trend toward urbanization and its potential for increasing exposure and vulnerability of large numbers of people.

2.5.1.3.1. The urban environment

Accelerated urbanization is an important trend in human settlement, which has implications for the consideration of exposure and vulnerability to extreme events. There has been almost a quintupling of the global urban population between 1950 and 2011 with the majority of that increase being in less developed regions (UN-HABITAT, 2011).

There is *high confidence* that rapid and unplanned urbanization processes in hazardous areas exacerbate vulnerability to disaster risk (Sánchez-Rodríguez et al., 2005). The development of megacities with high population densities (Mitchell, 1999a,b; Guha-Sapir et al., 2004) has led to greater numbers being exposed and increased vulnerability

through, *inter alia*, poor infrastructural development (Uitto, 1998) and the synergistic effects of intersecting natural, technological, and social risks (Mitchell, 1999a). Lavell (1996) identified eight contexts of cities that increase or contribute to disaster risk and vulnerability and are relevant in the context of climate change:

1) The synergic nature of the city and the interdependency of its parts
2) The lack of redundancy in its transport, energy, and drainage systems
3) Territorial concentration of key functions and density of building and population
4) Mislocation
5) Social-spatial segregation
6) Environmental degradation
7) Lack of institutional coordination
8) The contrast between the city as a unified functioning system and its administrative boundaries that many times impede coordination of actions.

The fact that urban areas are complex systems poses potential management challenges in terms of the interplay between people, infrastructure, institutions, and environmental processes (Ruth and Coelho, 2007). Alterations or trends in any of these, or additional components of the urban system such as environmental governance (Freudenberg et al., 2008) or the uptake of insurance (McLemand and Smit 2006; Lamond et al., 2009), have the potential to increase exposure and vulnerability to extreme climate events substantially.

The increasing polarization and spatial segregation of groups with different degrees of vulnerability to disaster have been identified as an emerging problem (Mitchell, 1999b). For the United States, where there is considerable regional variability, the components found to consistently increase social vulnerability (as expressed by a Social Vulnerability Index) are density (urbanization), race/ethnicity (see below), and socioeconomic status, with the level of development of the built environment, age, race/ethnicity, and gender accounting for nearly half of the variability in social vulnerability among US counties (Cutter and Finch, 2008). Social isolation, especially as it intersects with individual characteristics (see Case Study 9.2.1) and other social processes of marginalization (Duneier, 2004) also play a significant role in vulnerability creation (or, conversely, reduction).

Rapidly growing urban populations may affect the capacity of developing countries to cope with the effects of extreme events because of the inability of governments to provide the requisite urban infrastructure or for citizens to pay for essential services (UN-HABITAT, 2009). However, there is a more general concern that there has been insufficient attention to both existing needs for infrastructure maintenance and appropriate ongoing adaptation of infrastructure to meet potential climate extremes (Auld and MacIver, 2007). Further, while megacities have been associated with increasing hazard for some time (Mitchell, 1999a), small cities and rural communities are potentially more vulnerable to disasters than big cities or megacities, since megacities have considerable resources for dealing with hazards and disasters (Cross, 2001) and smaller settlements are often of lower priority for government spending.

The built environment can be both protective of, and subject to, climate extremes. Inadequate structures make victims of their occupants and, conversely, adequate structures can reduce human vulnerability. The continuing toll of deaths and injuries in unsafe schools (UNISDR, 2009a), hospitals and health facilities (PAHO/World Bank, 2004), domestic structures (Hewitt, 1997), and infrastructure more broadly (Freeman and Warner 2001) are indicative of the vulnerability of many parts of the built environment. In a changing climate, more variable and with potentially more extreme events, old certainties about the protective ability of built structures are undermined.

The increase in the number and extent of informal settlements or slums (UN-HABITAT, 2003; Utzinger and Keiser, 2006) is important because they are often located on marginal land within cities or on the periphery because of the lack of alternative locations or the fact that areas close to river systems or areas at the coast are sometimes state land that can be more easily accessed than private land. Because of their location, slums are often exposed to hydrometeorological-related hazards such as landslides (Nathan, 2008) and floods (Bertoni, 2006; Colten, 2006; Aragon-Durand, 2007; Douglas et al., 2008; Zahran et al., 2008). Vulnerability in informal settlements can also be elevated because of poor health (Sclar et al., 2005), livelihood insecurity (Kantor and Nair, 2005), lack of access to service provision and basic needs (such as clean water and good governance), and a reduction in the capacity of formal players to steer developments and adaptation initiatives in a comprehensive, preventive, and inclusive way (Birkmann et al., 2010b). Lagos, Nigeria (Adelekan, 2010), and Chittagong, Bangladesh (Rahman et al., 2010), serve as clear examples of where an upward trend in the area of slums has resulted in an increase in the exposure of slum dwellers to flooding. Despite the fact that rapidly growing informal and poor urban areas are often hotspots of hazard exposure, for a number of locations the urban poor have developed more or less successful coping and adaptation strategies to reduce their vulnerability in dealing with changing environmental conditions (e.g., Birkmann et al., 2010b).

Globally, the pressure for urban areas to expand onto flood plains and coastal strips has resulted in an increase in exposure of populations to riverine and coastal flood risk (McGranahan et al., 2007; Nicholls et al., 2011). For example, intensive and unplanned human settlements in flood-prone areas appear to have played a major role in increasing flood risk in Africa over the last few decades (Di Baldassarre et al., 2010). As urban areas have expanded, urban heat has become a management and health issue (for more on this see Section 2.5.2.3 and Chapters 3, 5, and 9). For some cities there is clear evidence of a recent trend in loss of green space (Boentje and Blinnikov, 2007; Sanli et al., 2008; Rafiee et al., 2009) due to a variety of reasons including planned and unplanned urbanization with the latter driven by internal and external migration resulting in the expansion of informal settlements. Such changes in green space may increase exposure to extreme climate events in urban areas through decreasing runoff amelioration, urban heat island mitigation effects, and alterations in biodiversity (Wilby and Perry, 2006).

While megacities have been associated with increasing hazard for some time (Mitchell, 1999a), small cities and rural communities (see next section) are potentially more vulnerable to disasters than big cities or megacities, since megacities have considerable resources for dealing with hazards and disasters (Cross, 2001) and smaller settlements are often of lower priority for government spending.

Urbanization itself is not always a driver for increased vulnerability. Instead, the type of urbanization and the context in which urbanization is embedded defines whether these processes contribute to an increase or decrease in people's vulnerability.

2.5.1.3.2. The rural environment

Many rural livelihoods are reliant to a considerable degree on the environment and natural resource base (Scoones, 1998), and extreme climate events can impact severely on the agricultural sector (Saldaña-Zorrilla, 2007). However, despite the separation here, the urban and the rural are inextricably linked. Inhabitants of rural areas are often dependent on cities for employment, as a migratory destination of last resort, and for health care and emergency services. Cities depend on rural areas for food, water, labor, ecosystem services, and other resources. All of these (and more) can be impacted by climate-related variability and extremes including changes in these associated with climate change. In either case, it is necessary to identify the many exogenous factors that affect a household's livelihood security.

Eakin's (2005) examination of rural Mexico presents empirical findings of the interactions (e.g., between neoliberalism and the opening up of agricultural markets, and the agricultural impacts of climatic extremes), which amplify or mitigate risky outcomes. The findings point to economic uncertainty over environmental risk, which most influences agricultural households' decisionmaking. However, there is not a direct and inevitable link between disaster impact and increased impoverishment of a rural population. In Nicaragua, Jakobsen (2009) found that a household's probability of being poor in the years following Hurricane Mitch was not affected by whether it was living in an area struck but by factors such as off-farm income, household size, and access to credit. Successful coping post-Hurricane Mitch resulted in poor households regaining most of their assets and resisting a decline into a state of extreme poverty. However, longer-term adaptation strategies, which might have lifted them out of the poverty category, eluded the majority and were independent of having experienced Hurricane Mitch. Thus, while poor (rural) households may cope with the impacts of a disaster in the relatively short term, their level of vulnerability, arising from a complex of environmental, social, economic, and political factors, is such that they cannot escape the poverty trap or fully reinstate development gains.

In assessing the material on exposure and vulnerability to climate extremes in urban and rural environments it is clear that there is no simple, deterministic relationship; it is not possible to show that either rural or urban environments are more vulnerable (or resilient). In either context there is the potential that climate risks can be either ameliorated or exacerbated by positive or negative adaptation processes and outcomes.

2.5.2. Social Dimensions

The social dimension is multi-faceted and cross-cutting. It focuses primarily on aspects of societal organization and collective aspects rather than individuals. However, some assessments also use the 'individual' descriptor to clarify issues of scale and units of analysis (Adger and Kelly, 1999; K. O'Brien et al., 2008). Notions of the individual are also useful when considering psychological trauma in and after disasters (e.g., Few, 2007), including that related to family breakdown and loss. The social dimension includes demography, migration, and displacement, social groups, education, health and well-being, culture, institutions, and governance aspects.

2.5.2.1. Demography

Certain population groups may be more vulnerable than others to climate variability and extremes. For example, the very young and old are more vulnerable to heat extremes than other population groups (Staffogia et al., 2006; Gosling et al., 2009). A rapidly aging population at the community to country scale bears implications for health, social isolation, economic growth, family composition, and mobility, all of which are social determinants of vulnerability. However, as discussed further below (Social Groups section), static checklists of vulnerable groups do not reflect the diversity or dynamics of people's changing conditions.

2.5.2.1.1. Migration and displacement

Trends in migration, as a component of changing population dynamics, have the potential to rise because of alterations in extreme climate event frequency. The United Nations Office for the Coordination of Humanitarian Affairs and the Internal Displacement Monitoring Centre have estimated that around 20 million people were displaced or evacuated in 2008 because of rapid onset climate-related disasters (OCHA/IDMC, 2009). Further, over the last 30 years, twice as many people have been affected by droughts (slow onset events not included in the previous point) as by storms (1.6 billion compared with approximately 718 million) (IOM, 2009). However, because of the multi-causal nature of migration, the relationship between climatic variability and change in migration is contested (Black, 2001) as are the terms environmental and climate refugees (Myers, 1993; Castles 2002; IOM, 2009). Despite an increase in the number of hydrometeorological disasters between 1990 and 2009, the International Organization on Migration reports no major impact on international migratory flows because displacement is temporary and often confined within a region, and displaced individuals do not possess the financial resources to migrate (IOM, 2009).

Although there is also a lack of clear evidence for a systematic trend in extreme climate events and migration, there are clear instances of the impact of extreme hydrometeorological events on displacement. For example, floods in Mozambique displaced 200,000 people in 2001, 163,000 people in 2007, and 102,000 more in 2008 (INGC, 2009; IOM, 2009); in Niger, large internal movements of people are due to pervasive changes related to drought and desertification trends (Afifi, 2011); in the Mekong River Delta region, changing flood patterns appear to be associated with migratory movements (White, 2002; IOM, 2009); and Hurricane Katrina, for which social vulnerability, race, and class played an important role in outward and returning migration (Elliott and Pais, 2006; Landry et al., 2007; Myers et al., 2008), resulted in the displacement of over one million people. As well as the displacement effect, there is evidence for increased vulnerability to extreme events among migrant groups because of an inability to understand extreme event-related information due to language problems, prioritization of finding employment and housing, and distrust of authorities (Enarson and Morrow, 2000; Donner and Rodriguez, 2008).

Migration can be both a condition of, and a response to, vulnerability – especially political vulnerability created through conflict, which can drive people from their homelands. Increasingly it relates to economically and environmentally displaced persons but can also refer to those who do not cross international borders but become internally displaced persons as a result of extreme events in both developed and developing countries (e.g., Myers et al., 2008).

Although data on climate change-forced displacement is incomplete, it is clear that the many outcomes of climate change processes will be seen and felt as disasters by the affected populations (Oliver-Smith, 2009). For people affected by disasters, subsequent displacement and resettlement often constitute a second disaster in their lives. As part of the Impoverishment Risks and Reconstruction approach, Cernea (1996) outlines the eight basic risks to which people are subjected by displacement: landlessness, joblessness, homelessness, marginalization, food insecurity, increased morbidity, loss of access to common property resources, and social disarticulation. When people are forced from their known environments, they become separated from the material and cultural resource base upon which they have depended for life as individuals and as communities (Altman and Low, 1992). The material losses most often associated with displacement and resettlement are losses of access to customary housing and resources. Displaced people are often distanced from their sources of livelihood, whether land, common property (water, forests, etc.), or urban markets and clientele (Koenig, 2009). Disasters and displacement may sever the identification with an environment that may once have been one of the principle features of cultural identity (Oliver-Smith, 2006). Displacement for any group can be distressing, but for indigenous peoples it can result in particularly severe impacts. The environment and ties to land are considered to be essential elements in the survival of indigenous societies and distinctive cultural identities (Colchester, 2000). The displacement and resettlement process has been consistently shown to disrupt and destroy those networks of social relationships on which the poor depend for resource access, particularly in times of stress (Cernea, 1996; Scudder, 2005).

Migration is an ancient coping mechanism in response to environmental (and other) change and does not inevitably result in negative outcomes, either for the migrants themselves or for receiving communities (Barnett and Webber, 2009). Climate variability will result in some movement of stressed people but there is *low confidence* in ability to assign direct causality to climatic impacts or to the numbers of people affected.

2.5.2.1.2. Social groups

Research evidence of the differential vulnerability of social groups is extensive and raises concerns about the disproportionate effects of climate change on identifiable, marginalized populations (Bohle et al., 1994; Kasperson and Kasperson, 2001; Thomalla et al., 2006). Particular groups and conditions have been identified as having differential exposure or vulnerability to extreme events, for example race/ethnicity (Fothergill et al., 1999; Elliott and Pais, 2006; Cutter and Finch, 2008), socioeconomic class and caste (O'Keefe et al., 1976; Peacock et al., 1997; Ray-Bennett, 2009), gender (Sen, 1981), age (both the elderly and children; Jabry, 2003; Wisner, 2006b; Bartlett, 2008), migration, and housing tenure (whether renter or owner), as among the most common social vulnerability characteristics (Cutter and Finch, 2008). Morrow (1999) extends and refines this list to include residents of group living facilities; ethnic minorities (by language); recent migrants (including immigrants); tourists and transients; physically or mentally disabled (see also McGuire et al., 2007; Peek and Stough, 2010); large households; renters; large concentrations of children and youth; poor households; the homeless (see also Wisner, 1998); and women-headed households. Generally, the state of vulnerability is defined by a specific population at a particular scale; aggregations (and generalizations) are often less meaningful and require careful interpretation (Adger and Kelly, 1999).

One of the largest bodies of research evidence, and one which can be an exemplar for the way many other marginalized groups are differentially impacted or affected by extreme events, has been on gender and disaster, and on women in particular (e.g., Neal and Phillips, 1990; Enarson and Morrow, 1998; Neumayer and Plümper, 2007). This body of literature is relatively recent, particularly in a developed world context, given the longer recognition of gender concerns in the development field (Fordham, 1998). The specific gender and climate change link including self-defined gender groups has been even more recent (e.g., Masika, 2002; Pincha and Krishna, 2009). The research evidence emphasizes the social construction of gendered vulnerability in which women and girls are often (although not always) at greater risk of dying in disasters, typically marginalized from decisionmaking fora, and discriminated and acted against in post-disaster recovery and reconstruction efforts (Houghton, 2009; Sultana, 2010).

Women or other socially marginalized or excluded groups are not vulnerable through biology (except in very particular circumstances) but are made so by societal structures and roles. For example, in the Indian Ocean tsunami of 2004, many males were out to sea in boats, fulfilling their roles as fishermen, and were thus less exposed than were many women who were on the seashore, fulfilling their roles as preparers and marketers of the fish catch. However, the women were made vulnerable not simply by their location and role but by societal norms which did not encourage survival training for girls (e.g., to swim or climb trees) and which placed the majority of the burden of child and elder care with women. Thus, escape was made more difficult for women carrying children and responsible for others (Doocy et al., 2007).

The gender and disaster/climate change literature has also recognized resilience/capacity/capability alongside vulnerability. This elaboration of the vulnerability approach makes clear that vulnerability in these identified groups is not an immutable or totalizing condition. The vulnerability 'label' can reinforce notions of passivity and helplessness, which obscure the very significant, active contributions that socially marginalized groups make in coping with and adapting to extremes. An example is provided in Box 2-2.

2.5.2.2. Education

The education dimension ranges across the vulnerability of educational building structures; issues related to access to education; and also sharing and access to disaster risk reduction and climate adaptation information and knowledge (Wisner, 2006b). Priority 3 of the Hyogo Framework for Action 2005-2015 recommends the use of knowledge, innovation, and education to build a 'culture of safety and resilience' at all levels (UNISDR, 2007a). A well-informed and motivated population can lead to disaster risk reduction but it requires the collection and dissemination of knowledge and information on hazards, vulnerabilities, and capacities. However, "It is not information per se that determines action, but how people interpret it in the context of their experience, beliefs and expectations. Perceptions of risks and hazards are culturally and socially constructed, and social groups construct different meanings for potentially hazardous situations" (McIvor and Paton, 2007). In addition to knowledge and information, explicit environmental education programs among children and adults may have benefits for public understanding of risk, vulnerability, and exposure to extreme events (UNISDR, 2004; Kobori, 2009; Nomura, 2009; Patterson et al., 2009; Kuhar et al., 2010), because they promote resilience building in socio-ecological systems through their role in stewardship of biological diversity and ecosystem services, provide the opportunity to integrate diverse forms of knowledge and participatory processes in resource management (Krasny and Tidball, 2009), and help promote action towards sustainable development (Waktola, 2009; Breiting and Wikenberg, 2010).

Many lives have been lost through the inability of education infrastructure to withstand extreme events. Where flooding is a recurrent phenomenon schools can be exposed or vulnerable to floods. For example, a survey of primary schools' flood vulnerability in the Nyando River catchment of western Kenya revealed that 40% were vulnerable, 48% were

> **Box 2-2 | Integrating Disaster Risk Reduction, Climate Adaptation, and Resilience-Building: the Garifuna Women of Honduras**
>
> The Garifuna women of Honduras could be said to show multiple vulnerability characteristics (Brondo, 2007). They are women, the gender often made vulnerable by patriarchal structures worldwide; they come from Honduras, a developing country exposed to many hazards; they belong to an ethnic group descended from African slaves, which is socially, economically, and politically marginalized; and they depend largely upon a subsistence economy, with a lack of education, health, and other resources. However, despite these markers of vulnerability, the Garifuna women have organized to reduce their communities' exposure to hazards and vulnerability to disasters through the protection and development of their livelihood opportunities (Fordham et al., 2011).
>
> The women lead the Comité de Emergencia Garifuna de Honduras, which is a grassroots, community-based group of the Afro-Indigenous Garifuna that was developed in the wake of Hurricane Mitch in 1998. After Mitch, there was a lack of external support and so the Comité women organized themselves and repaired hundreds of houses, businesses, and public buildings, in the process of which women were empowered and trained in non-traditional work. They campaigned to buy land for relocating housing to safer areas, in which the poorest families participated in the reconstruction process. Since being trained themselves in vulnerability and capacity mapping by grassroots women in Jamaica, they have in turn trained 60 trainers in five Garifuna communities to carry out mapping exercises in their communities.
>
> The Garifuna women have focused on livelihood-based activities to ensure food security by reviving and improving the production of traditional root crops, building up traditional methods of soil conservation, carrying out training in organic composting and pesticide use, and creating the first Garifuna farmers' market. In collaborative efforts, 16 towns now have established tool banks, and five have seed banks. Through reforestation, the cultivation of medicinal and artisanal plants, and the planting of wild fruit trees along the coast, they are helping to prevent erosion and reducing community vulnerability to hazards and the vagaries of climate.
>
> The Garifuna women's approach, which combines livelihood-based recovery, disaster risk reduction, and climate change adaptation, has had wide-ranging benefits. They have built up their asset base (human, social, physical, natural, financial, and political), and improved their communities' nutrition, incomes, natural resources, and risk management. They continue to partner with local, regional, and international networks for advocacy and knowledge exchange. The women and communities are still at risk (Drusine, 2005) but these strategies help reduce their socioeconomic vulnerability and dependence on external aid (Fordham et al., 2011).

marginally vulnerable, and 12% were not vulnerable; the vulnerability status was attributed to a lack of funds, poor building standards, local topography, soil types and inadequate drainage (Ochola et al., 2010). Improving education infrastructure safety can have multiple benefits. For example, the Malagasy Government initiated the Development Intervention Fund IV project to reduce cyclone risk, including safer school construction and retrofitting. In doing so, awareness and understanding of disaster issues were increased within the community (Madagascar Development Intervention Fund, 2007).

The impact of extreme events can limit the ability of parents to afford to educate their children or require them (especially girl children, whose access to education is typically prioritized less than that of boy children) to work to meet basic needs (UNDP, 2004; UNICEF, 2009).

Access to information related to early warnings, response strategies, coping and adaptation mechanisms, science and technology, and human, social, and financial capital is critical for reduction of vulnerability and increasing resilience. A range of factors may control or influence the access to information, including economic status, race (Spence et al., 2007), trust (Longstaff and Yang, 2008), and belonging to a social network (Peguero, 2006). However, the mode of information transfer or exchange must be considered because there is emerging evidence of a growing digital inequality (Rideout, 2003) that may influence trends in vulnerability as an increasing amount of information about extreme event preparedness and response is often made available via the internet (see Chapter 9). Evidence has existed for some time that people who have experienced natural hazards (and thus may have information and knowledge gained directly through that experience) are, in general, better prepared than those who have not (Kates, 1971). However, this does not necessarily translate into protective behavior because of what has been called the 'prison of experience' (Kates, 1962), in which people's response behavior is determined by the previous experience and is not based on an objective assessment of current risk. In the uncertain context of climate-related extremes, this may mean people are not appropriately educated regarding the risk.

2.5.2.3. Health and Well-Being

The health dimension of vulnerability includes differential physical, physiological, and mental health effects of extreme events in different regions and on different social groups (McMichael et al., 2003; van Lieshout et al., 2004; Haines et al., 2006; Few, 2007; Costello et al.,

2009). It also includes, in a link to the institutional dimension, health service provision (e.g., environmental health and public health issues, infrastructure and conditions; Street et al., 2005), which may be impacted by extreme events (e.g., failures in hospital/health center building structures; inability to access health services because of storms and floods). Vulnerability can also be understood in terms of functionality related to communication, medical care, maintaining independence, supervision, and transportation. In addition individuals including children, senior citizens, and pregnant women and those who may need additional response assistance including the disabled, those living in institutionalized settings, those from diverse cultures, people with limited English proficiency or are non-English speaking, those with no access to transport, have chronic medical disorders, and have pharmacological dependency can also be considered vulnerable in a health context.

Unfortunately, the health dimensions of disasters are difficult to measure because of difficulties in attributing the health condition (including mortality) directly to the extreme event because of secondary effects; in addition, some of the effects are delayed in time, which again makes attribution difficult (Bennet, 1970; Hales et al., 2003). The difficulty of collection of epidemiological data in crisis situations is also a factor, especially in low-income countries. Further understanding the post-traumatic stress disorder dimensions of extreme climate events and the psychological aspects of climate change presents a number of challenges (Amstadter et al., 2009; Kar, 2009; Mohay and Forbes, 2009; Furr et al., 2010; Doherty and Clayton, 2011).

Health vulnerability is the sum of all the risk and protective factors that determine the degree to which individuals or communities could experience adverse impacts from extreme weather events (Balbus and Malina, 2009). Vulnerabilities can arise from a wide range of institutional, geographic, environmental, socioeconomic, biological sensitivity, and other factors, which can vary spatially and temporally. Biological sensitivity can be associated with developmental stage (e.g., children are at increased mortality risk from diarrheal diseases); pre-existing medical conditions (e.g., diabetics are at increased risk during heat waves); acquired conditions (e.g., malaria immunity); and genetic factors (Balbus and Malina, 2009). Vulnerability can be viewed both from the perspective of the population groups more likely to experience adverse health outcomes and from the perspective of the public health and health care interventions required to prevent adverse health impacts during and following an extreme event.

For some extreme weather events the vulnerable population groups depend on the adverse health outcome considered. For example, in the case of heat waves socially isolated elderly people with pre-existing medical conditions are vulnerable to heat-related health effects (see Chapter 9). For floods, children are at greater risk for transmission of fecal-oral diseases, and those with mobility and cognitive constraints can be at increased risk of injuries and deaths (Ahern et al., 2005), while people on low incomes are less likely to be able to afford insurance against risks associated with flooding, such as storm and flood damage (Marmot, 2010). Flooding has been found to increase the risk of mental health problems, pre- and post-event, in both adults and children (Ginexi et al., 2000; Reacher et al., 2004; Ahern et al., 2005; Carroll et al., 2006; Tunstall et al., 2006; UK Department of Health, 2009). A UK study of over 1,200 households affected by flooding suggested that there were greater impacts on physical and mental health among more vulnerable groups and poorer households and communities (Werritty et al., 2007). However, while there is evidence for impacts on particular social groups in identified disaster types, there are some social groups that are more likely to be vulnerable whatever the hazard type; these include those at the extremes of the age range, those with underlying medical conditions, and those otherwise stressed by low socioeconomic status. The role of socioeconomic factors supports the necessity of a social, and not just a medical, model of response and adaptation.

A number of public health impacts are expected to worsen in climate-related disasters such as storms, floods, landslides, heat, drought, and wildfire. These are highly context-specific but range from worsening of existing chronic illnesses (which could be widespread), through possible toxic exposures (in air, water or food), to deaths (expected to be few to moderate but may be many in low-income countries) (Keim, 2008). Public health and health care services required for preventing adverse health impacts from an extreme weather event include surveillance and control activities for infectious diseases, access to safe water and improved sanitation, food security, maintenance of solid waste management and other critical infrastructure, maintenance of hospitals and other health care infrastructure, provision of mental health services, sufficient and safe shelter to prevent or mitigate displacement, and effective warning and informing systems (Keim, 2008). Further, it is important to consider the synergistic effects of NaTech disasters (Natural Hazard Triggering a Technological Disaster) where impacts can be considerable if only single, simple hazard events are planned for. In an increasingly urbanized world, interactions between natural disasters and simultaneous technological accidents must be given attention (Cruz et al., 2004); the combination of an earthquake, tsunami, and radiation release at the Japanese Fukushima Nuclear Power plant in March 2011 is the most recent example. Lack of provision of these services increases population vulnerability, particularly in individuals with greater biological sensitivity to an adverse health outcome. Although there is little evidence for trends in the exposure or vulnerability of public health infrastructure, the imperative for a resilient health infrastructure is widely recognized in the context of extreme climate events (Burkle and Greenough, 2008; Keim, 2008).

Deteriorating environmental conditions as a result of extremes (including land clearing, salinization, dust generation, altered ecology; Renaud, 2006; Middleton et al., 2008; Ellis and Wilcox, 2009; Hong et al., 2009; Ljung et al., 2009; Johnson et al., 2010; Tong et al., 2010) can impact key ecosystem services and exacerbate climate sensitive disease incidence (e.g., diarrheal disease; Clasen et al., 2007), particularly via deteriorating water quality and quantity.

For some health outcomes, which have direct or indirect implications for vulnerability to extreme climate events, there is evidence of trends. For example, obesity, a risk factor for cardiovascular disease, which in turn is a heat risk factor (Bouchama et al., 2007) has been noted to be on the increase in a number of developed countries (Skelton et al., 2009; Stamatakis et al., 2010). Observed trends in major public health threats such as the infectious or communicable diseases HIV/AIDS, tuberculosis, and malaria, although not directly linked to the diminution of long-term resilience of some populations, have been identified as having the potential to do so (IFRC, 2008). In addition to the diseases themselves, persistent and increasing obstacles to expanding or strengthening health systems such as inadequate human resources and poor hospital and laboratory infrastructure as observed in some countries (Vitoria et al., 2009) may also contribute indirectly to increasing vulnerability and exposure where, for example, malaria and HIV/Aids occasionally reach epidemic proportions.

However, trends in well-being and health are difficult to assess. Indicators that characterize a lack of well-being and a high degree of susceptibility are, for example, indicators of undernourishment and malnutrition. The database for the Millennium Development Goals and respective statistics of the Food and Agriculture Organization (FAO) underscore that trends in undernourishment are spatially and temporally differentiated. While, as but one example, the trend in undernourished people in Burundi shows a significant increase from 1991 to 2005, an opposite trend of a reduction in the percentage of undernourished people can be observed in Angola (see UN Statistics Division, 2011; FAOSTAT, 2011). Thus, evidence exists that trends in vulnerability, e.g., in terms of well-being and undernourishment change over time and are highly differentiated in terms of spatial patterns.

In considering health-related exposure and vulnerability to extreme events, evidence from past climate/weather-related disaster events (across a range of hazard types for which lack of space precludes coverage) makes clear the links to a range of negative outcomes for physical and mental health and health infrastructure. Furthermore, there is clear evidence (Haines et al., 2006; Confalonieri et al., 2007) that current and projected health impacts from climate change are multifarious and will affect low-income groups and low-income countries the most severely, although high-income countries are not immune.

2.5.2.4. Cultural Dimensions

The broad term 'culture' embraces a complexity of elements that can relate to a way of life, behavior, taste, ethnicity, ethics, values, beliefs, customs, ideas, institutions, art, and intellectual achievements that affect, are produced, or are shared by a particular society. In essence, all these characteristics can be summarized to describe culture as 'the expression of humankind within society' (Aysan and Oliver, 1987).

Culture is variously used to describe many aspects of extreme risks from natural disasters or climate change, including:
- Cultural aspects of risk perception
- Negative culture of danger/ vulnerability/ fear
- Culture of humanitarian concern
- Culture of organizations / institutions and their responses
- Culture of preventive actions to reduce risks, including the creation of buildings to resist extreme climatic forces
- Ways to create and maintain a 'Risk Management Culture,' a 'Safety Culture,' or an 'Adaptation Culture.'

In relation to our understanding of risk, certain cultural issues need to be noted. Typical examples are cited below:
- *Ethnicity and Culture.* Deeply rooted cultural values are a dominant factor in whether or not communities adapt to climate change. For example, recent research in Northern Burkina Faso indicates that two ethnic groups have adopted very different strategies due to cultural values and historical relations, despite their presence in the same physical environment and their shared experience of climate change (Nielsen and Reenberg, 2010).
- *Locally Based Risk Management Culture.* Wisner (2003) has argued that the point in developing a 'culture of prevention' is to build networks at the neighborhood level capable of ongoing hazard assessment and mitigation at the micro level. He has noted that while community based NGOs emerged to support recovery after the Mexico City and Northridge earthquakes, these were not sustained over time to promote risk reduction activities. This evidence confirms other widespread experience indicating that ways still need to be found to extend the agenda of Community-Based Organizations into effective action to reduce climate risks and promote adaptation to climate change.
- *Conflicting Cultures: Who Benefits, and Who Loses when Risks are Reduced?* A critical cultural conflict can arise when private actions to reduce disaster risks and adapting to climate change by one party have negative consequences on another. This regularly applies in river flood hazard management where upstream measures to reduce risks can significantly increase downstream threats to persons and property. Adger has argued that if appropriate risk reduction actions are to occur, the key players must bear all the costs and receive all the benefits from their actions (Adger, 2009). However, this can be problematic if adaptation is limited to specific local interests only.

Traditional behaviors tied to local (and wider) tradition and cultural practices can increase vulnerability – for example, unequal gender norms that put women and girls at greater risk, or traditional uses of the environment that have not adapted (or cannot adapt) to changed environmental circumstances. On the other hand, local or indigenous knowledge can reduce vulnerabilities too (Gaillard et al., 2007, 2010). Furthermore, cultural practices are often subtle and may be opaque to outsiders. The early hazards paradigm literature (White, 1974; Burton et al., 1978) referred often to culturally embedded fatalistic attitudes, which resulted in inaction in the face of disaster risk. However, Schmuck-Widmann (2000), in her social anthropological studies of char dwellers in Bangladesh, revealed how a belief that disaster occurrence and outcomes were in the hands of God did not preclude preparatory activities. Perceptions of risk (and their interpretation by others) depend on the cultural and social context (Slovic, 2000; Oppenheimer and Todorov, 2006; Schneider et al., 2007).

Research findings emphasize the importance of considering the role – and cultures – of religion and faith in the context of disaster. This includes the role of faith in the recovery process following a disaster (e.g., Davis and Wall, 1992; Massey and Sutton, 2007); religious explanations of nature (e.g., Orr, 2003; Peterson, 2005); the role of religion in influencing positions on environment and climate change policy (e.g., Kintisch, 2006; Hulme, 2009); and religion and vulnerability (Guth et al., 1995; Chester, 2005; Elliott et al., 2006; Schipper, 2010).

The cultural dimension also includes the potential vulnerability of aboriginal and native peoples in the context of climate extremes. Globally, indigenous populations are frequently dependent on primary production and the natural resource base while being subject to (relatively) poor socioeconomic conditions (including poor health, high unemployment, low levels of education, and greater poverty). This applies to groups from Canada (Turner and Clifton, 2009), to Australia (Campbell et al., 2008), to the Pacific (Mimura et al., 2007). Small island states, often with distinct cultures, typically show high vulnerability and low adaptive capacity to climate change (Nurse and Sem, 2001). However, historically, indigenous groups have had to contend with many hazards and, as a consequence, have developed capacities to cope (Campbell, 2006) such as the use of traditional knowledge systems, locally appropriate building construction with indigenous materials, and a range of other customary practices (Campbell, 2006).

Given the degree of cultural diversity identified, the importance of understanding differential risk perceptions in a cultural context is reinforced (Marris et al., 1998). Cultural Theory has contributed to an understanding of how people interpret their world and define risk according to their worldviews: hierarchical, fatalistic, individualistic, and egalitarian (Douglas and Wildavsky, 1982). Too often policies and studies focus on 'the public' in the aggregate and too little on the needs, interests, and attitudes of different social and cultural groups (see also Sections 2.5.2.1.2 and 2.5.4).

2.5.2.5. Institutional and Governance Dimensions

The institutional dimension is a key determinant of vulnerability to extreme events (Adger, 1999). Institutions have been defined in a broad sense to include "habitualized behavior and rules and norms that govern society" (Adger, 2000) and not just the more typically understood formal institutions. This view allows for a discussion of institutional structures such as property rights and land tenure issues (Toni and Holanda, 2008) that govern natural resource use and management. It forms a bridge between the social and the environmental/ecological dimensions and can induce sustainable or unsustainable exploitation (Adger, 2000). Expanding the institutional domain to include political economy (Adger, 1999) and different modes of production – feudal, capitalist, socialist (Wisner, 1978) – raises questions about the vulnerability *of* institutions and the vulnerability caused *by* institutions (including government). Institutional factors play a critical role in adaptation (Adger, 2000) as they influence the social distribution of vulnerability and shape adaptation capacity (Næss et al., 2005).

This broader understanding of the institutional dimension also takes us into a recognition of the role of social networks, community bonds and organizing structures, and processes that can buffer the impacts of extreme events (Nakagawa and Shaw, 2004) partly through increasing social cohesion but also recognizing ambiguous or negative forms (UNISDR, 2004). For example, social capital/assets (Portes, 1998; Putnam, 2000) – "the norms and networks that enable people to act collectively" (Woolcock and Narayan, 2000) – have a role in vulnerability reduction (Pelling, 1998). Social capital (or its lack) is both a cause and effect of vulnerability and thus can result in either positive benefit or negative impact; to be a part of a social group and accrue social assets is often to indicate others' exclusion. It also includes attempts to reframe climate debates by acknowledging the possibility of diverse impacts on human security, which opens up human rights discourses and rights-based approaches to disaster risk reduction (Kuwali, 2008; Mearns and Norton, 2010).

The institutional dimension includes the relationship between policy setting and policy implementation in risk and disaster management. Top-down approaches assume policies are directly translated into action on the ground; bottom-up approaches recognize the importance of other actors in shaping policy implementation (Urwin and Jordan, 2008). Twigg's categorization of the characteristics of the ideal disaster resilient community (Twigg, 2007) adopts the latter approach. This guideline document, which has been field tested by NGOs, identifies the important relations between the community and the enabling environment of governance at various scales in creating resilience, and by inference, reducing vulnerability. This set of 167 characteristics (organized under five thematic areas) also refers to institutional forms for (and processes of) engagement with risk assessment, risk management, and hazard and vulnerability mapping. These have been championed by institutions working across scales to create the Hyogo Framework for Action (UNISDR, 2007a) and associated tools (Davis et al., 2004; UNISDR, 2007b) with the goal to reduce disaster risk and vulnerability. However, linkages across scales and the inclusion of local knowledge systems are still not integrated well in formal institutions (Næss et al., 2005).

A lack of institutional interaction and integration between disaster risk reduction, climate change, and development may mean policy responses are redundant or conflicting (Schipper and Pelling, 2006; Mitchell and van Aalst, 2008; Mitchell et al., 2010). Thus, the institutional model operational in a given place and time (more or less participatory, deliberative, and democratic; integrated; or disjointed) could be an important factor in either vulnerability creation or reduction (Comfort et al., 1999). Furthermore, risk-specific policies must also be integrated (see the slippage between UK heat and cold wave policies, Wolf et al., 2010a). However, further study of the role of institutions in influencing vulnerability is called for (O'Brien et al., 2004b).

Governance is also a key topic for vulnerability and exposure. Governance is broader than governmental actions; governance can be understood as the structures of common governance arrangements and processes of steering and coordination – including markets, hierarchies, networks, and communities (Pierre and Peters, 2000). Institutionalized rule systems and habitualized behavior and norms that govern society and guide actors are representing governance structures (Adger, 2000; Biermann et al., 2009). These formal and informal governance structures also determine vulnerability, since they influence power relations, risk perceptions, and constitute the context in which vulnerability, risk reduction, and adaptation are managed.

Conflicts between formal and informal governance or governmental and nongovernmental strategies and norms can generate additional vulnerabilities for communities exposed to environmental change. An example of these conflicts of formal and informal strategies is linked to flood protection measures. While local people might expend resources to deal with increasing flood events (e.g., adapting their livelihoods and production patterns to changing flood regimes), formal adaptation strategies, particularly in developing countries, prioritize structural measures (e.g., dike systems or relocation strategies) that have severe consequences for the vulnerability of communities dependent on local ecosystem services, such as fishing and farming systems (see Birkmann, 2011a,b). These conflicts between formal and informal or governmental and nongovernmental management systems and norms are an important factor that increase vulnerability and reduce adaptive capacity of the overall system (Birkmann et al., 2010b). Countries with institutional and governance fragilities often lack the capacity to identify and reduce risks and to deal with emergencies and disasters effectively. The recent disaster and problems in coping and recovery in the aftermath of the earthquake in Haiti or the problems in terms of managing recovery and emergency management after the Pakistan floods are examples that illustrate the importance of governance as a subject of resilience and vulnerability.

In some developed countries, the last 30 years have witnessed a shift in environmental governance practices toward more integrated approaches. With the turn of the century, there has been recognition of the need to move beyond technical solutions and to deal with the patterns and drivers of unsustainable demand and consumption. This has resulted in the emergence of a more integrated approach to environmental management, a focus on prevention (UNEP, 2007), the incorporation of knowledge from the local to the global in environment policies (Karlsson, 2007), and co-management and involvement of stakeholders from all sectors in the management of natural resources (Plummer, 2006; McConnell, 2008), although some have also questioned the efficacy of this new paradigm (Armitage et al., 2007; Sandstrom, 2009).

2.5.3. Economic Dimensions

Economic vulnerability can be understood as the susceptibility of an economic system, including public and private sectors, to potential (direct) disaster damage and loss (Rose, 2004; Mechler et al., 2010) and refers to the inability of affected individuals, communities, businesses, and governments to absorb or cushion the damage (Rose, 2004).

The degree of economic vulnerability is exhibited post-event by the magnitude and duration of the indirect follow-on effects. These effects can comprise business interruption costs to firms unable to access inputs from their suppliers or service their customers, income losses of households unable to get to work, or the deterioration of the fiscal stance post-disasters as less taxes are collected and significant public relief and reconstruction expenditure is required. At a macroeconomic level, adverse impacts include effects on gross domestic product (GDP), consumption, and the fiscal position (Mechler et al., 2010). Key drivers of economic vulnerability are low levels of income and GDP, constrained tax revenue, low domestic savings, shallow financial markets, and high indebtedness with little access to external finance (OAS, 1991; Benson and Clay, 2000; Mechler, 2004).

Economic vulnerability to external shocks, including natural hazards, has been inexactly defined in the literature and conceptualizations often have overlapped with risk, resilience, or exposure. One line of research focusing on financial vulnerability, as a subset of economic vulnerability, framed the problem in terms of risk preference and aversion, a conceptualization more common to economists. Risk aversion, in this context, denotes the ability of economic agents to absorb risk financially (Arrow and Lind, 1970). There are many ways to absorb the financial burdens of disasters, with market-based insurance being one, albeit prominent, option, although more particularly in a developed country context. Households as economic agents often use informal mechanisms relying on family and relatives abroad or outside a disaster area; governments may simply rely on their tax base or international assistance. Yet, in the face of large and covariate risks, such ad hoc mechanisms often break down, particularly in developing countries (see Linnerooth-Bayer and Mechler, 2007).

Research on financial vulnerability to disasters has hitherto focused on developing countries' financial vulnerability describing financial vulnerability as a country's ability to access domestic and foreign savings for financing post-disaster relief and reconstruction needs in order to quickly recover and avoid substantial adverse ripple effects (Mechler et al., 2006; Marulanda et al., 2008a; Cardona, 2009; Cummins and Mahul, 2009). Reported and estimated substantial financial vulnerability and risk aversion in many exposed countries, as well as the emergence of novel public-private partnership instruments for pricing and transferring catastrophe risks globally, has motivated developing country governments, as well as development institutions, NGOs, and other donor organizations, to consider pre-disaster financial instruments as an important component of disaster risk management (Linnerooth-Bayer et al., 2005).

There is a distinct scale aspect to the economic dimension of exposure and vulnerability. While evidence of the economic costs of known disasters indicate impacts may be under 10% of GDP (Wilbanks et al.,

2007), at smaller and more local scales the costs can be significantly greater. A lack of good data makes it difficult to provide meaningful and specific assessments other than to acknowledge that, without investment in adaptation and resilience building measures, the intensification or increased frequency of extreme weather events is bound to impact GDP growth in the future (Wilbanks et al., 2007).

Work and Livelihoods

At the individual and community levels, work and livelihoods are an important facet of the economic dimension. These are often impacted by extreme events and by the responses to extreme events. Humanitarian/disaster relief in response to extreme events can induce dependency and weaken local economic and social systems (Dudasik, 1982) but livelihood-based relief is of growing importance (Pantuliano and Wekesa, 2008). Further, there is increasing recognition that disasters and extreme events are stresses and shocks within livelihood development processes (Cannon et al., 2003; see Kelman and Mather, 2008, for a discussion of cases applying to volcanic events).

Paavola's (2008) analysis of livelihoods, vulnerability, and adaptation to climate change in Morogoro, Tanzania, is indicative of the way extreme events impact livelihoods in specific ways. Here, rural households are found to be more vulnerable to climate variability and climate change than are those in urban environments (see also Section 2.5.1.3). This is because rural incomes and consumption levels are significantly lower, there are greater levels of poverty, and more limited access to markets and other services. More specifically, women are made more vulnerable than men because they lack access to livelihoods other than climate-sensitive agriculture. Local people have employed a range of strategies (extensification, intensification, diversification, and migration) to manage climate variability but these have sometimes had undesirable environmental outcomes, which have increased their vulnerability. In the absence of opportunities to fundamentally change their livelihood options, we see here an example of short-term coping rather than long-term climate adaptation (Paavola, 2008).

Human vulnerability to natural hazards and income poverty are largely codependent (Adger, 1999; UNISDR, 2004) but poverty does not equal vulnerability in a simple way (e.g., Blaikie et al., 1994); the determinants and dimensions of poverty are complex as well as its association with climate change (Khandlhela and May, 2006; Demetriades and Esplen, 2008; Hope, 2009). It is important to recognize that adaptation measures need to specifically target climate extremes-poverty linkages as not all poverty reduction measures reduce vulnerability to climate extremes and vice versa. Further, measures are required across scales because the drivers of poverty, although felt at a local level, may necessitate tackling political and economic issues at a larger scale (Eriksen and O'Brien, 2007; K. O'Brien et al., 2008).

Given the relationship between poverty and vulnerability, it can be argued (Tol et al., 2004) that economic growth could reduce vulnerability (with caveats). However, increasing economic growth would not necessarily decrease climate impacts because it has the potential to simultaneously increase greenhouse gas emissions. Furthermore, growth is often reliant on critical infrastructure which itself may be affected by extreme events. There are many questions still to be answered by research about the impacts of varying economic policy changes including the pursuit of narrow development trajectories and how this might shape vulnerability (Tol et al., 2004; UNDP, 2004; UNISDR, 2004)

2.5.4. Interactions, Cross-Cutting Themes, and Integrations

This section began by breaking down the vulnerability concept into its constitutive dimensions, with evidence derived from a number of discrete research and policy communities (e.g., disaster risk reduction; climate change adaptation; environmental management; and poverty reduction) that have largely worked independently (Thomalla et al., 2006). Increasingly it is recognized that collaboration and integration is necessary both to set appropriate policy agendas and to better understand the topic of interest (K. O'Brien et al., 2008), although McLaughlin and Dietz (2008) have made a critical analysis of the absence of an integrated perspective on the interrelated dynamics of social structure, human agency, and the environment.

Reviewing singular dimensions of vulnerability cannot provide an appropriate level of synthesis. Considerable conceptual advances arose from the early recognition that so-called natural disasters were not 'natural' at all (O'Keefe et al., 1976) but were the result of structural inequalities rooted in political economy. This critique required analysis of more than the hazard component (Blaikie et al., 1994). Further, it demonstrated how crossing disciplinary and other boundaries (e.g., those separating disaster and development, or developed and developing countries) can be fruitful in better understanding extremes of various kinds (see Hewitt, 1983). If we consider food security/vulnerability (as just one example), an inclusive analysis of the vulnerability of food systems (to put it broadly), must take account of aspects related to, *inter alia*: physical location in susceptible areas; political economy (Watts and Bohle, 1993); entitlements in access to resources (Sen, 1981); social capital and networks (Eriksen et al., 2005); landscape ecology (Fraser, 2006); human ecology (Bohle et al., 1994); and political ecology (Pulwarty and Riebsame, 1997; Holling, 2001; see Chapter 4 for further discussion of food systems and food security). More generally, in relation to hazards, disaster risk reduction, and climate extremes, productive advances have been made in research adopting a coupled human/social-environment systems approach (Holling, 2001; Turner et al., 2003b) which recognizes the importance of integrating often separate domains. For example, in analyzing climate change impacts, vulnerability, and adaptation in Norway, O'Brien et al. (2006) argue that a simple examination of direct climate change impacts underestimates the, perhaps more serious and larger, synergistic impacts. They use an example of projected climate change effects in the Barents Sea, which may directly impact keystone fish species. However, important as this finding is, climate change may also influence the transport sector (through reduction in ice cover); increase numbers of pollution events (through increased maritime transport of oil and other goods); may risk ecological and other damages as a result of competition from introduced species in ballast water; which, in turn, are aggravated by increases in ocean temperatures. Neither the potential level of impact nor the processes of adaptation are best represented by a singular focus on a particular sector but must consider interactions between sectors and institutional, economic, social, and cultural conditions (O'Brien et al., 2006).

2.5.4.1. Intersectionality and Other Dimensions

The dimensions discussed above generate differential effects but it is important to consider not just differences between single categories (e.g., between women and men) but the differences *within* a given category (e.g., 'women'). This refers to intersectionality, where, for example, gender may be a significant variable but only when allied with race/ethnicity or some other variable. In Hurricane Katrina, it mattered (it still matters) whether you were black or white, upper class or working class, home owner or renter, old or young, woman or man in terms of relative exposure and vulnerability factors (Cutter et al., 2006; Elliott and Pais, 2006).

Certain factors are identified as cross-cutting themes of particular importance for understanding the dynamic changes within exposure, vulnerability, and risk. In the Sphere Project's minimum standards in humanitarian response, children, older people, persons with disabilities, gender, psychosocial issues, HIV and AIDS, and environment, climate change, and disaster risk reduction are identified as cross-cutting themes and must be considered, not as separate sectors, which people may or may not select for attention, but must be integrated within each sector (Sphere Project, 2011). Exactly which topics are selected as cross-cutting themes, to be incorporated throughout an activity, is context-specific. Below, we consider just two: different timing (diachronic aspects within a single day or across longer time periods) and different spatial and functional scales.

2.5.4.2. Timing, Spatial, and Functional Scales

Cross-cutting themes of particular importance for understanding the dynamic changes within exposure, vulnerability, and risk are different timing (diachronic aspects within a single day or across longer time periods) and different spatial and functional scales.

2.5.4.2.1. Timing and timescales

Timing and timescales are important cross-cutting themes that need more attention when dealing with the identification and management of extreme climate and weather events, disasters, and adaptation strategies. The first key issue when dealing with timing and timescales is the fact that different hazards and their recurrence intervals might fundamentally change in terms of the time dimension. This implies that the identification and assessment of risk, exposure, and vulnerability needs also to deal with different time scales and in some cases might need to consider different time scales. At present most of the climate change scenarios focus on climatic change within the next 100 or 200 years, while often the projections of vulnerability just use present socioeconomic data. However, a key challenge for enhancing knowledge of exposure and vulnerability as key determinants of risk requires improved data and methods to project and identify directions and different development pathways in demographic, socioeconomic, and political trends that can adequately illustrate potential increases or decreases in vulnerability with the same time horizon as the changes in the climate system related to physical-biogeochemical projections (see Birkmann et al., 2010b).

Furthermore, the time dependency of risk analysis, particularly if the analysis is conducted at a specific point in time, has been shown to be critical. Newer research underlines that exposure – especially the exposure of different social groups – is a highly dynamic element that changes not only seasonally, but also during the day and over different days of the week (e.g., Setiadi, 2011). Disasters also exacerbate pre-disaster trends in vulnerability (Colten et al., 2008).

Consequently, time scales and dynamic changes over time have to be considered carefully when conducting risk and vulnerability assessments for extreme events and creeping changes in the context of climate change. Additionally, changes in the hazard frequency and timing of hazard occurrence during the year will have a strong impact on the ability of societies and ecosystems to cope and adapt to these changes.

The timing of events may also create 'windows of vulnerability,' periods in which the hazards are greater because of the conjunction of circumstances (Dow, 1992). Time is a cross-cutting dimension that always needs to be considered but particularly so in the case of anthropogenic climate change, which may be projected some years into the future (Füssel, 2005). In fact, this time dimension is regarded (Thomalla et al., 2006) as a key difference between the disaster management and climate change communities. To generalize somewhat, the former group typically (with obvious exceptions like slow-onset hazards such as drought or desertification) deals with fast-onset events, in discrete, even if extensive, locations, requiring immediate action. The latter group typically focuses on conditions that occur in a dispersed form over lengthy time periods and which are much more challenging in their identification and measurement (Thomalla et al., 2006). Risk perception may be reduced (Leiserowitz, 2006) for such events remote in time and/or space, such as some climate change impacts are perceived to be. Thus, in this conceptualization, different time scales are an important constraint when dealing with the link between disaster risk reduction and climate change adaptation (see Thomalla et al., 2006; Birkmann and von Teichman, 2010).

However, it is important to also acknowledge that disaster risk reduction considers risk reduction within different time frames; it encompasses short-term emergency management/response strategies and long-term risk reduction strategies, for example, building structures to resist 10,000-year earthquakes or flood barriers to resist 1,000-year storm surges. Modern prospective risk management debates involve security considerations decades ahead for production, infrastructure, houses, hospitals, etc.

2.5.4.2.2. Spatial and functional scales

Spatial and functional scales are another cross-cutting theme that is of particular relevance when dealing with the identification of exposure and vulnerability to extreme events and climate change. Leichenko and O'Brien (2002) conclude that in many areas of climate change and natural hazards societies are confronted with dynamic vulnerability, meaning that processes and factors that cause vulnerability operate simultaneously at multiple scales making traditional indicators insufficient. Leichenko and O'Brien (2002) analyze a complex mix of influences (both positive and negative) on the vulnerability, and coping and adaptive capacity of southern African farmers in dealing with climate variability. These include the impacts of globalization on national-level policies and local-level experiences (e.g., structural adjustment programs reducing local-level agricultural subsidies on the one hand, and on the other, trade liberalization measures opening up new opportunities through diversification of production in response to drought). Also Turner et al. (2003a,b) stress that vulnerability and resilience assessments need to consider the influences on vulnerability from different scales, however, the practical application and analysis of these interacting influences on vulnerability from different spatial scales is a major challenge and in most cases not sufficiently understood. Furthermore, vulnerability analysis particularly linked to the identification of institutional vulnerability has also to take into account the various functional scales that climate change, natural hazards, and vulnerability as well as administrative systems operate on. In most cases, current disaster management instruments and measures of urban or spatial planning as well as water management tools (specific plans, zoning, norms) operate on different functional scales compared to climate change. Even the various hazards that climate change may modify encompass different functional scales that cannot be sufficiently captured with one approach. For example, policy setting and management of climate change and of disaster risk reduction are usually the responsibility of different institutions or departments, thus it is a challenge to develop a coherent and integrated strategy (Birkmann and von Teichman, 2010). Consequently, functional and spatial scale mismatches might even be part of institutional vulnerabilities that limit the ability of governance system to adequately respond to hazards and changes induced by climate change.

2.5.4.3. Science and Technology

Science and technology possess the potential to assist with adaptation to extreme climate events, however there are a number of factors that determine the ultimate utility of technology for adaptation. These include an understanding of the range of technologies available, the identification of the appropriate role for technology, the process of technology transfer, and the criteria applied in selection of the technology (Klein et al., 2006). For major sectors such as water, agriculture, and health a range of possible so-called 'hard' and 'soft' technologies exist such as irrigation and crop rotation pattern (Klein et al., 2006) or the development of drought-resistant crops (IAASTD, 2009) in the case of the agricultural sector.

Although approaches alternative to pure science- and technology-based ones have been suggested for decreasing vulnerability (Haque and Etkin, 2007; Marshall and Picou, 2008), such as blending western science and technology with indigenous knowledge (Mercer et al., 2010) and ecological cautiousness and the creation of eco-technologies with a pro-nature, pro-poor, and pro-women orientation (Kesavan and Swaminathan, 2006), their efficacy in the context of risk and vulnerability reduction remain undetermined.

The increasing integration of a range of emerging weather and climate forecasting products into early warning systems (Glantz, 2003) has helped reduce exposure to extreme climate events because of an increasing improvement of forecast skill over a range of time scales (Goddard et al., 2009; Stockdale et al., 2009; van Aalst, 2009; Barnston et al., 2010; Hellmuth et al., 2011). Moreover, there is an increasing use of weather and climate information for planning and climate risk management in business (Changnon and Changnon, 2010), food security (Verdin et al., 2005), and health (Ceccato et al., 2007; Degallier et al., 2010) as well as the use of technology for the development of a range of decision support tools for climate-related disaster management (van de Walle and Turoff, 2007).

2.6. Risk Identification and Assessment

Risk accumulation, dynamic changes in vulnerabilities, and the different phases of crises and disaster situations constitute a complex environment for identifying and assessing risks and vulnerabilities, risk reduction measures, and adaptation strategies. Understanding of extreme events and disasters is a pre-requisite for the development of adaptation strategies in the context of climate change and risk reduction in the context of disaster risk management.

Current approaches to disaster risk management typically involve four distinct public policies or components (objectives) (IDEA, 2005; Carreño, 2006; IDB, 2007; Carreño et al., 2007b):
1) Risk identification (involving individual perception, evaluation of risk, and social interpretation)
2) Risk reduction (involving prevention and mitigation of hazard or vulnerability)
3) Risk transfer (related to financial protection and in public investment)
4) Disaster management (across the phases of preparedness, warnings, response, rehabilitation, and reconstruction after disasters).

The first three actions are mainly *ex ante* – that is, they take place in advance of disaster – and the fourth refers mainly to *ex post* actions, although preparedness and early warning do require *ex ante* planning (Cardona, 2004; IDB, 2007). Risk identification, through vulnerability and risk assessment can produce common understanding by the stakeholders and actors. It is the first step for risk reduction, prevention, and transfer, as well as climate adaptation in the context of extremes.

2.6.1. Risk Identification

Understanding risk factors and communicating risks due to climate change to decisionmakers and the general public are key challenges. These challenges include developing an improved understanding of underlying vulnerabilities, and societal coping and response capacities.

There is *high confidence* that the selection of appropriate vulnerability and risk evaluation approaches depends on the decisionmaking context. The promotion of a higher level of risk awareness regarding climate change-induced hazards and changes requires an improved understanding of the specific risk perceptions of different social groups and individuals, including those factors that influence and determine these perceptions, such as beliefs, values, and norms. This also requires attention for appropriate formats of communication that characterize uncertainty and complexity (see, e.g., Patt et al., 2005; Bohle and Glade, 2008; Renn, 2008, pp. 289; Birkmann et al., 2009; ICSU-LAC, 2011a,b, p. 15).

Appropriate information and knowledge are essential prerequisites for risk-aware behavior and decisions. Specific information and knowledge on the dynamic interactions of exposed and vulnerable elements include livelihoods and critical infrastructures, and potentially damaging events, such as extreme weather events or potential irreversible changes such as sea level rise. Based on the expertise of disaster risk research and findings in the climate change and climate change adaptation community, requirements for risk understanding related to climate change and extreme events particularly encompass knowledge of various elements (Kasperson et al., 2005; Patt et al., 2005; Renn and Graham, 2006; Biermann, 2007; Füssel, 2007; Bohle and Glade, 2008; Cutter and Finch, 2008; Renn, 2008; Biermann et al., 2009, Birkmann et al., 2009, 2010b; Cardona, 2010; Birkmann, 2011a; ICSU-LAC, 2011a,b), including:

- Processes by which persons, property, infrastructure, goods, and the environment itself are exposed to potentially damaging events, for example, understanding exposure in its spatial and temporal dimensions.
- Factors and processes that determine or contribute to the vulnerability of persons and their livelihoods or of socio-ecological systems. This includes an understanding of increases or decreases in susceptibility and response capacity, including the distribution of socio- and economic resources that make people more vulnerable or that increase their level of resilience.
- How climate change affects hazards, particularly regarding processes by which human activities in the natural environment or changes in socio-ecological systems lead to the creation of new hazards (e.g., NaTech hazards), irreversible changes, or increasing probabilities of hazard events occurrence.
- Different tools, methodologies, and sources of knowledge (e.g., expert/scientific knowledge, local or indigenous knowledge) that allow capturing new hazards, risk, and vulnerability profiles, as well as risk perceptions. In this context, new tools and methodologies are also needed that allow for the evaluation, for example, of new risks (sea level rise) and of current adaptation strategies.
- How risks and vulnerabilities can be modified and reconfigured through forms of governance, particularly risk governance – encompassing formal and informal rule systems and actor networks at various levels. Furthermore, it is essential to improve knowledge on how to promote adaptive governance within the framework of risk assessment and risk management.
- Adaptive capacity status and limits of adaptation. This includes the need to assess potential capacities for future hazards and for dealing with uncertainty. Additionally, more knowledge is needed on the various and socially differentiated limits of adaptation. These issues also imply an improved understanding on how different adaptation measures influence resilience and adaptive capacities.

2.6.2. Vulnerability and Risk Assessment

The development of modern risk analysis and assessments were closely linked to the establishment of scientific methodologies for identifying causal links between adverse health effects and different types of hazardous events and the mathematical theories of probability (Covello and Mumpower, 1985). Today, risk and vulnerability assessments encompass a broad and multidisciplinary research field. In this regard, vulnerability and risk assessments can have different functions and goals.

Risk and vulnerability assessment depend on the underlying understanding of the terms. In this context, two main schools of thought can be differentiated. The first school of thought defines risk as a decision by an individual or a group to act in such a way that the outcome of these decisions can be harmful (Luhmann, 2003; Dikau and Pohl, 2007). In contrast, the disaster risk research community views risk as the product of the interaction of a potentially damaging event and the vulnerable conditions of a society or element exposed (UNISDR, 2004; IPCC, 2007).

Vulnerability and risk assessment encompass various approaches and techniques ranging from indicator-based global or national assessments to qualitative participatory approaches of vulnerability and risk assessment at the local level. They serve different functions and goals (see IDEA, 2005; Birkmann, 2006a; Cardona, 2006; Dilley, 2006; Wisner, 2006a; IFRC, 2008; Peduzzi et al., 2009).

Risk assessment at the local level presents specific challenges related to a lack of data (including climate data at sufficient resolution, but also socioeconomic data at the lowest levels of aggregation) but also the highly complex and dynamic interplay between the capacities of the communities (and the way they are distributed among community members, including their power relationships) and the challenges they face (including both persistent and acute aspects of vulnerability).

To inform risk management, it is desirable that risk assessments are locally based and result in increased awareness and a sense of local ownership of the process and the options that may be employed to address the risks. Several participatory risk assessment methods, often based on participatory rural appraisal methods, have been adjusted to explicitly address changing risks in a changing climate. Examples of guidance on how to assess climate vulnerability at the community level are available from several sources (see Willows and Connell, 2003; Moench and Dixit, 2007; van Aalst et al., 2007; CARE, 2009; IISD et al., 2009; Tearfund, 2009). In integrating climate change, a balance needs to be struck between the desire for a sophisticated assessment that includes high-quality scientific inputs and rigorous analysis of the participatory findings, and the need to keep the process simple, participatory, and implementable at scale. Chapter 5 provides further details on the implementation of risk management at local levels.

The International Standards Organization defines risk assessment as a process to comprehend the nature of risk and to determine the level of risk (ISO, 2009a,b). Additionally, communication within risk assessment and management are seen as key elements of the process (Renn, 2008). More specifically, vulnerability and risk assessment deal with the identification of different facets and factors of vulnerability and risk, by means of gathering and systematizing data and information, in order to be able to identify and evaluate different levels of vulnerability and risk of societies – social groups and infrastructures – or coupled socio-ecological systems at risk. A common goal of vulnerability and risk assessment approaches is to provide information about profiles, patterns of, and changes in risk and vulnerability (see, e.g., IDEA, 2005; Birkmann, 2006a; Cardona, 2008; IFRC, 2008), in order to define priorities, select alternative strategies, or formulate new response strategies. In this context, the Hyogo Framework for Action stresses "that the starting point for reducing disaster risk and for promoting a culture of disaster resilience lies in the knowledge of the hazards and the physical, social, economic, and environmental vulnerabilities to disasters that most societies face, and of the ways in which hazards and vulnerabilities are changing in the short and long term, followed by action taken on the basis of that knowledge" (UN, 2005).

Vulnerability and risk assessments are key strategic activities that inform both disaster risk management and climate change adaptation. These require the use of reliable methodologies that allow an adequate estimation and quantification of potential losses and consequences to the human systems in a given exposure time.

Risk estimates are thus intended to be prospective, anticipating scientifically possible hazard events that may occur in the future. Usually technical risk analyses have been associated with probabilities.

Taking into account epistemic and aleatory uncertainties the probabilistic estimations of risk attempt to forecast damage or losses even where insufficient data are available on the hazards and the system being analyzed (UNDRO, 1980; Fournier d'Albe, 1985; Spence and Coburn, 1987; Blockley, 1992; Coburn and Spence, 1992; Sheldon and Golding, 1992; Woo, 1999; Grossi and Kunreuther, 2005; Cardona et al., 2008a,b; Cardona 2011). In most cases, approaches and criteria for simplification and for aggregation of different information types and sources are used, due to a lack of data or the inherent low resolution of the information. This can result in some scientific or technical and econometric characteristics, accuracy, and completeness that are desirable features when the risk evaluation is the goal of the process (Cardona et al., 2003b). Measures such as loss exceedance curves and probable maximum loss for different event return periods are of particular importance for the stratification of risk and the design of disaster risk intervention strategy considering risk reduction, prevention, and transfer (Woo, 1999; Grossi and Kunreuther, 2005; Cardona et al., 2008a,b; ERN-AL, 2011; UNISDR, 2011). However, it is also evident that more qualitative-oriented risk assessment approaches are focusing on deterministic approaches and the profiling of vulnerability using participatory methodologies (Garret, 1999).

Vulnerability and risk indicators or indices are feasible techniques for risk monitoring and may take into account both the harder aspects of risk as well as its softer aspects. The usefulness of indicators depends on how they are employed to make decisions on risk management objectives and goals (Cardona et al., 2003a; IDEA, 2005; Cardona, 2006, 2008, 2010; Carreño et al., 2007b).

However, quantitative approaches for assessing vulnerability need to be complemented with qualitative approaches to capture the full complexity and the various tangible and intangible aspects of vulnerability in its different dimensions. It is important to recognize that complex systems involve multiple variables (physical, social, cultural, economic, and environmental) that cannot be measured using the same methodology. Physical or material reality have a harder topology that allows the use of quantitative measure, while collective and historical reality have a softer topology in which the majority of the attributes are described in qualitative terms (Munda, 2000). These aspects indicate that a weighing or measurement of risk involves the integration of diverse disciplinary perspectives. An integrated and interdisciplinary focus can more consistently take into account the nonlinear relations of the parameters, the context, complexity, and dynamics of social and environmental systems, and contribute to more effective risk management by the different stakeholders involved in risk reduction or adaptation decisionmaking. Results can be verified and risk management/adaptation priorities can be established (Carreño et al., 2007a, 2009).

To ensure that risk and vulnerability assessments are also understood, the key challenges for future vulnerability and risk assessments, in the context of climate change, are, in particular, the promotion of more integrative and holistic approaches; the improvement of assessment methodologies that also account for dynamic changes in vulnerability, exposure, and risk; and the need to address the requirements of

Box 2-3 | Developing a Regional Common Operating Picture of Vulnerability in the Americas for Various Kinds of Decisionmakers

The Program of Indicators of Disaster Risk and Risk Management for the Americas of the Inter-American Development Bank (IDEA, 2005; Cardona, 2008, 2010) provides a holistic approach to relative vulnerability assessment using social, economic, and environmental indicators and a metric for sovereign fiscal vulnerability assessment taking into account that extreme impacts can generate financial deficit due to a sudden elevated need for resources to restore affected inventories or capital stock.

The Prevalent Vulnerability Index (PVI) depicts predominant vulnerability conditions of the countries over time to identify progresses and regressions. It provides a measure of direct effects (as result of exposure and susceptibility) as well as indirect and intangible effects of hazard events (as result of socioeconomic fragilities and lack of resilience). The indicators used are made up of a set of demographic, socioeconomic, and environmental national indicators that reflect situations, causes, susceptibilities, weaknesses, or relative absences of development affecting the country under study. The indicators are selected based on existing indices, figures, or rates available from reliable worldwide databases or data provided by each country. These vulnerability conditions underscore the relationship between risk and development. Figure 2-1 shows the aggregated PVI (Exposure, Social Fragility, Lack of Resilience) for 2007.

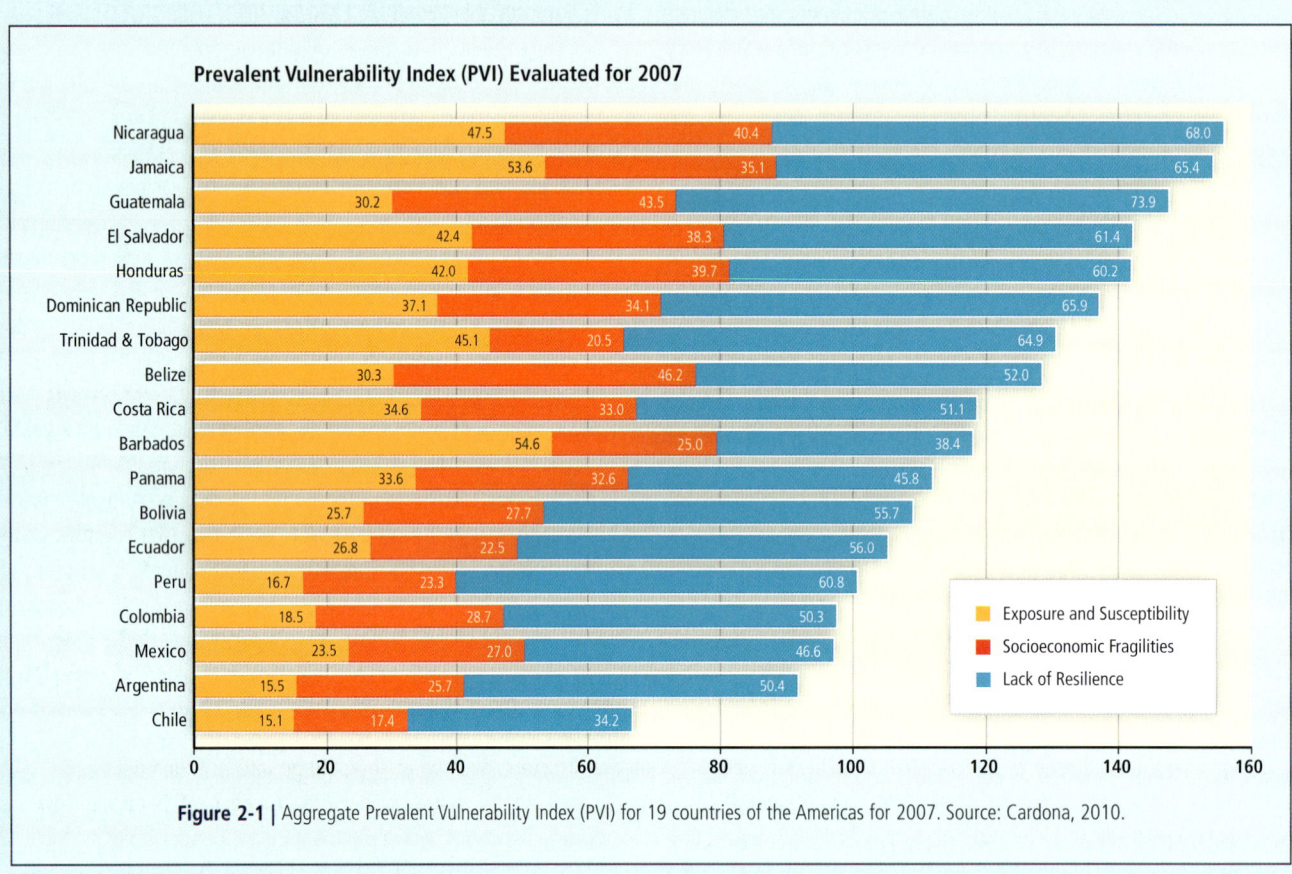

Figure 2-1 | Aggregate Prevalent Vulnerability Index (PVI) for 19 countries of the Americas for 2007. Source: Cardona, 2010.

Vulnerability and therefore risk are also the result of unsustainable economic growth and deficiencies that may be corrected by means of adequate development processes, reducing susceptibility of exposed assets, socioeconomic fragilities, and improving capacities and resilience of society (IDB, 2007). The information provided by an index such as the PVI can prove useful to ministries of housing and urban development, environment, agriculture, health and social welfare, economy, and planning. The main advantage of PVI lies in its ability to disaggregate results and identify factors that may take priority in risk management actions as corrective and prospective measures or interventions of vulnerability from a development point of view. The PVI can be used at different territorial levels, however often the indicators used by the PVI are only available at the national level; this is a limitation for its application at other sub-national scales.

On the other hand, future disasters have been identified as contingency liabilities and could be included in the balance of each nation. As pension liabilities or guaranties that the government has to assume for the credit of territorial entities or due to grants, disaster

Continued next page ⟶

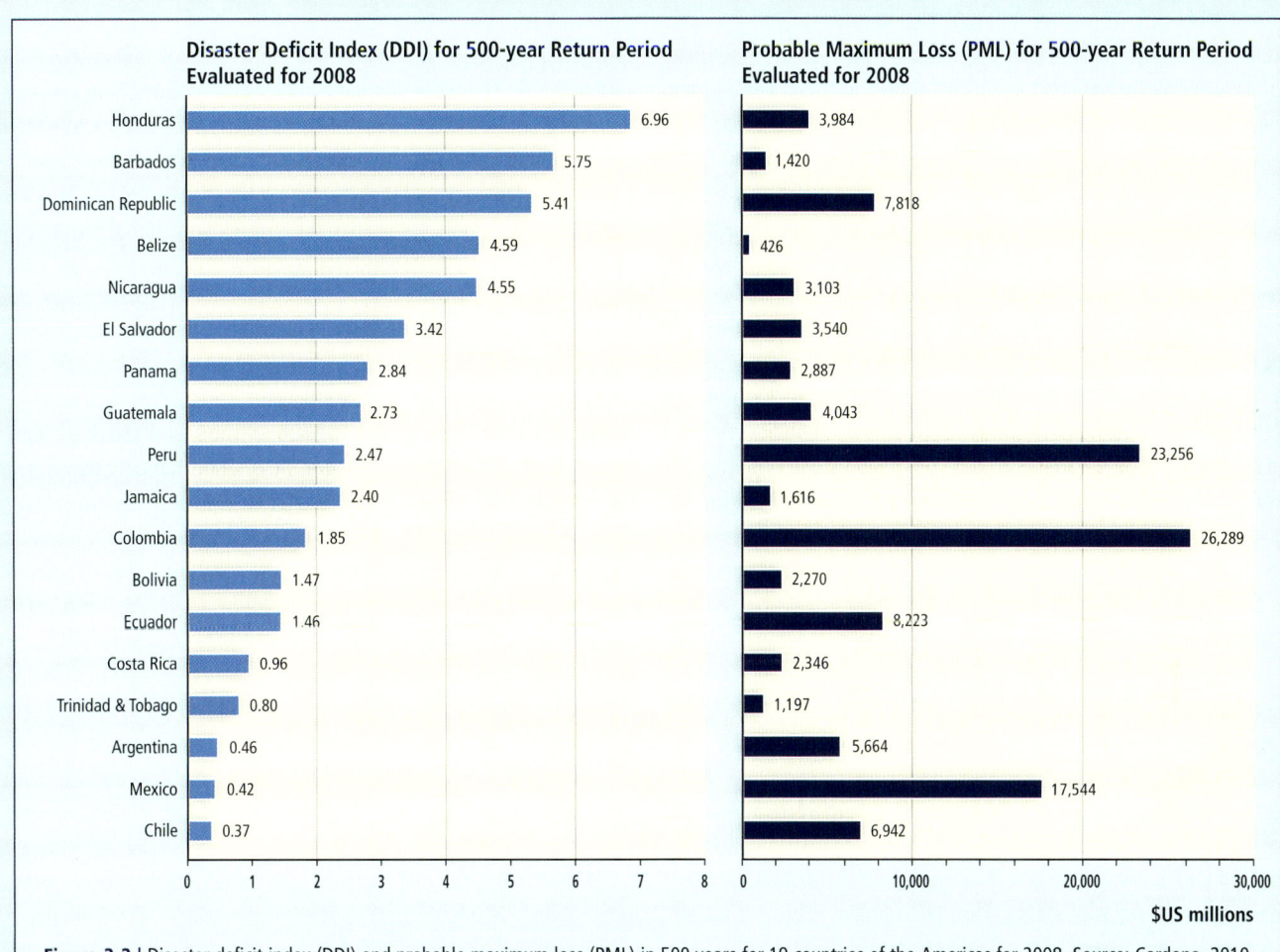

Figure 2-2 | Disaster deficit index (DDI) and probable maximum loss (PML) in 500 years for 19 countries of the Americas for 2008. Source: Cardona, 2010.

reposition costs are liabilities that become materialized when the hazard events occur. The Disaster Deficit Index (DDI) provides an estimation of the extreme impact (due to hurricane, floods, tsunami, earthquake, etc.) during a given exposure time and the financial ability to cope with such a situation. The DDI captures the relationship between the loss that the country could experience when an extreme impact occurs (demand for contingent resources) and the public sector's economic resilience – that is, the availability of funds to address the situation (restoring affected inventories). This macroeconomic risk metric underscores the relationship between extreme impacts and the capacity to cope of the government. Figure 2-2 shows the DDI for 2008.

A DDI greater than 1.0 reflects the country's inability to cope with extreme disasters, even when it would go into as much debt as possible: the greater the DDI, the greater the gap between the potential losses and the country's ability to face them. This disaster risk figure is interesting and useful for a Ministry of Finance and Economics. It is related to the potential financial sustainability problem of the country regarding the potential disasters. On the other hand, the DDI gives a compressed picture of the fiscal vulnerability of the country due to extreme impacts. The DDI has been a guide for economic risk management; the results at national and sub-national levels can be studied by economic, financial, and planning analysts, who can evaluate the potential budget problem and the need to take into account these figures in the financial planning.

decisionmakers and the general public. Many assessments still focus solely on one dimension, such as economic risk and vulnerability. Thus, they consider a very limited set of vulnerability factors and dimensions. Some approaches, e.g., at the global level, view vulnerability primarily with regard to the degree of experienced loss of life and economic damage (see Dilley et al., 2005; Dilley 2006). A more integrative and holistic perspective captures a greater range of dimensions and factors of vulnerability and disaster risk. Successful adaptation to climate change has been based on a multi-dimensional perspective, encompassing, for example, social, economic, environmental, and institutional aspects. Hence, risk and vulnerability assessments – that intend to inform these adaptation strategies – require also a multi-dimensional perspective.

Assessment frameworks with integrative and holistic perspectives have been developed by Turner et al. (2003a), Birkmann (2006b), and Cardona (2001). Key elements of these holistic views are the identification of causal linkages between factors of vulnerability and risk and the interventions (structural, non-structural) that nations, societies, and communities or individuals make to reduce their vulnerability or exposure to hazards. Turner et al. (2003a) underline the need to focus on different scales simultaneously, in order to capture the linkages between different scales (local, national, regional, etc.). The influences and linkages between different scales can be difficult to capture, especially due to their dynamic nature during and after disasters, for example, through inputs of external disaster aid (Cardona, 1999a,b; Cardona and Barbat, 2000; Turner et al., 2003a; Carreño et al., 2005, 2007a, 2009; IDEA, 2005; Birkmann, 2006b; ICSU-LAC, 2011a,b).

Several methods have been proposed to measure vulnerability from a comprehensive and multidisciplinary perspective. In some cases composite indices or indicators intend to capture favorable conditions for direct physical impacts – such as exposure and susceptibility – as well as indirect or intangible impacts of hazard events – such as socio-ecological fragilities or lack of resilience (IDEA, 2005; Cardona, 2006; Carreño et al., 2007a). In these holistic approaches, exposure and physical susceptibility are representing the 'hard' and hazard-dependent conditions of vulnerability. On the other hand, the propensity to suffer negative impacts as a result of the socio-ecological fragilities and not being able to adequately cope and anticipate future disasters can be considered 'soft' and usually non-hazard dependent conditions, that aggravate the impact. Box 2-3 describes two of these approaches, based on relative indicators, useful for monitoring vulnerability of countries over time and to communicate it to country's development and financial authorities in their own language.

To enhance disaster risk management and climate change adaptation, risk identification and vulnerability assessment may be undertaken in different phases, that is, before, during, and even after disasters occur. This includes, for instance, the evaluation of the continued viability of measures taken and the need for further or different adaptation/risk management measures. Although risk and vulnerability reduction are the primary actions to be conducted before disasters occur, it is important to acknowledge that *ex post* and forensic studies of disasters provide a laboratory in which to study risk and disasters as well as vulnerabilities revealed (see Birkmann and Fernando, 2008; ICSU-LAC, 2011a,b). Disasters draw attention to how societies and socio-ecological processes are changing and acting in crises and catastrophic situations, particularly regarding the reconfiguration of access to different assets or the role of social networks and formal organizations (see Bohle, 2008). It is noteworthy that, until today, many post-disaster processes and strategies have failed to integrate aspects of climate change adaptation and long-term risk reduction (see Birkmann et al., 2009, 2010a).

In the broader context of the assessments and evaluations, it is also crucial to improve the different methodologies to measure and evaluate hazards, vulnerability, and risks. The disaster risk research has paid more attention to sudden-onset hazards and disasters such as floods, storms, tsunamis, etc., and less on the measurement of creeping changes and integrating the issue of tipping points into these assessments (see also Section 3.1.7). Therefore, the issue of measuring vulnerability and risk, in terms of quantitative and qualitative measures also remains a challenge. Lastly, the development of appropriate assessment indicators and evaluation criteria would also be strengthened if respective integrative and consistent goals for vulnerability reduction and climate change adaptation could be defined for specific regions, such as coastal, mountain, or arid environments. Most assessments to date have based their judgment and evaluation on a relative comparison of vulnerability levels between different social groups or regions.

There is *medium evidence* (given the generally limited amount of long-term evaluations of impacts of adaptation and risk management interventions and complications associated with such assessments), but *high agreement* that adaptation and risk management policies and practices will be more successful if they take the dynamic nature of vulnerability and exposure into account, including the explicit characterization of uncertainty and complexity (Cardona 2001, 2011; Hilhorst, 2004, ICSU-LAC, 2010, Pelling, 2010). Projections of the impacts of climate change can be strengthened by including storylines of changing vulnerability and exposure under different development pathways. Appropriate attention to the dynamics of vulnerability and exposure is particularly important given that the design and implementation of adaptation and risk management strategies and policies can reduce risk in the short term, but may increase vulnerability and exposure over the longer term. For instance, dike systems can reduce hazard exposure by offering immediate protection, but also encourage settlement patterns that may increase risk in the long term. For instance, in the 40-year span between Hurricanes Betsy and Katrina, protective works – new and improved levees, drainage pumps, and canals – successfully protected New Orleans and surrounding parishes against three hurricanes in 1985, 1997, and 1998. These works were the basis for the catastrophe of Katrina, having enabled massive development of previously unprotected areas and the flooding of these areas that resulted when the works themselves were shown to be inadequate (Colten et al., 2008). For other examples, see Décamps (2010).

The design of public policy on disaster risk management is related to the method of evaluation used to orient policy formulation. If the diagnosis invites action it is much more effective than where the results are limited to identifying the simple existence of weaknesses or failures. The main quality attributes of a risk model are represented by its *applicability*, *transparency*, *presentation*, and *legitimacy* (Corral, 2000). For more details see Cardona (2004, 2011).

Several portfolio-level climate risk assessment methods for development agencies have paid specific attention to the risk of variability and extremes (see, e.g., Burton and van Aalst, 1999, 2004; Klein, 2001; van Aalst, 2006b; Klein et al., 2007; Agrawala and van Aalst, 2008; Tanner, 2009). Given the planning horizons of most development projects (typically up to about 20 years), even if the physical lifetime of the investment may be much longer, and need to combine attention to current and future risks, these tools provide linkages between adaptation to climate change and enhanced disaster risk management even in light of current hazards. For more details on the implementation of risk management at the national level, see Chapter 6.

2.6.3. Risk Communication

How people perceive a specific risk is a key issue for risk management and climate change adaptation effectiveness (e.g., Burton et al., 1993; Alexander, 2000; Kasperson and Palmlund, 2005; van Sluis and van Aalst, 2006; ICSU-LAC, 2011a,b) since responses are shaped by perception of risk (Grothmann and Patt, 2005; Wolf et al., 2010b; Morton et al., 2011).

Risk communication is a complex cross-disciplinary field that involves reaching different audiences to make risk comprehensible, understanding and respecting audience values, predicting the audience's response to the communication, and improving awareness and collective and individual decisionmaking (e.g., Cardona, 1996c; Mileti, 1996; Greiving, 2002; Renn, 2008). Risk communication failures have been revealed in past disasters, such as Hurricane Katrina in 2005 or the Pakistan floods in 2010 (DKKV, 2011). Particularly, the loss of trust in official institutions responsible for early warning and disaster management were a key factor that contributed to the increasing disaster risk. Effective and people-centered risk communication is therefore a key to improve vulnerability and risk reduction in the context of extreme events, particularly in the context of people-centered early warning (DKKV, 2011). Weak and insufficient risk communication as well as the loss of trust in government institutions in the context of early warning or climate change adaptation can be seen as a core component of institutional vulnerability.

Risk assessments and risk identification have to be linked to different types and strategies of risk communication. Risk communication or the failure of effective and people-centered risk communication can contribute to an increasing vulnerability and disaster risk. Knowledge on factors that determine how people perceive and respond to a specific risk or a set of multi-hazard risks is key for risk management and climate change adaptation (see Grothmann and Patt 2005; van Aalst et al., 2008).

Understanding the ways in which disasters are framed requires more information and communication about vulnerability factors, dynamic temporal and spatial changes of vulnerability, and the coping and response capacities of societies or social-ecological systems at risk (see Turner et al., 2003a; Birkmann, 2006a,b,c; Cardona, 2008; Cutter and Finch, 2008; ICSU-LAC, 2011a,b). 'Framing' refers to the way a particular problem is presented or viewed. Frames are shaped by knowledge of and underlying views of the world (Schon and Rein, 1994). It is related to the organization of knowledge that people have about their world in the light of their underlying attitudes toward key social values (e.g., nature, peace, freedom), their notions of agency and responsibility (e.g., individual autonomy, corporate responsibility), and their judgments about reliability, relevance, and weight of competing knowledge claims (Jasanoff and Wynne, 1997). 'Early warning' implies information interventions into an environment in which much about vulnerability is assumed. In this regard, risk communication is not solely linked to a top-down communication process, rather effective risk communication requires recognition of communication as a social process meaning that risk communication also deals with local risk perceptions and local framing of risk. Risk communication thus functions also as a tool to upscale local knowledge and needs (bottom-up approach). Therefore, effective risk communication achieves both informing people at risk about the key determinants of their particular risks and of impending disaster risk (early warning), and also engages different stakeholders in the definition of a problem and the identification of respective solutions (see van Aalst et al., 2008).

Climate change adaptation strategies as well as disaster risk reduction approaches need public interest, leadership, and acceptance. The generation and receipt of risk information occurs through a diverse array of channels. Chapter 5 and others discuss the important role of mass media and other sources (see, e.g., the case of Japan provided in Sampei and Aoyagi-Usui, 2009). Within the context of risk communication, particularly in terms of climate change and disasters, decisionmakers, scientists, and NGOs have to act in accordance with media requirements concerning news production, public discourse, and media consumption (see Carvalho and Burgess, 2005). Carvalho (2005) and Olausson (2009) underline that mass media is often closely linked to political awareness and is framed by its own journalistic norms and priorities; that means also that mass media provides little space for alternative frames of communicating climate change (Carvalho, 2005; Olausson, 2009). Boykoff and Boykoff (2007) conclude that this process might also lead to an informational bias, especially toward the presentation of events instead of a comprehensive analysis of the problem. Thus, an important aspect of improving risk communication and the respective knowledge base is the acceptance and admission of the limits of knowledge about the future (see Birkmann and von Teichman, 2010).

2.7. Risk Accumulation and the Nature of Disasters

The concept of risk accumulation describes a gradual build-up of disaster risk in specific locations, often due to a combination of processes, some persistent and/or gradual, others more erratic, often in a combination of exacerbation of inequality, marginalization, and disaster risk over time (Maskrey, 1993b; Lavell, 1994). It also reflects that the impacts of one hazard – and the response to it – can have implications for how the next hazard plays out. This is well illustrated by the example of El Salvador, where people living in temporary shelters after the 1998 Hurricane Mitch were at greater risk during the 2001 earthquakes due to the poor construction of the shelters (Wisner, 2001b). The concept of risk accumulation acknowledges the multiple causal factors of risk by the connecting development patterns and risk, as well as the links between one disaster and the next.

Risk accumulation can be driven by underlying factors such as a decline in the regulatory services provided by ecosystems, inadequate water management, land use changes, rural-urban migration, unplanned urban growth, the expansion of informal settlements in low-lying areas, and an underinvestment in drainage infrastructure. Development and governance processes that increase the marginalization of specific groups, for example, through the reduction of access to health services or the exclusion from information and power – to name just a few – can also severely increase the susceptibility of these groups and at the same time erode societal response capacities. The classic example is disaster risk in urban areas in many rapidly growing cities in developing countries (Pelling and Wisner, 2009b). In these areas, disaster risk is often very unequally distributed, with the poor facing the highest risk, for instance because they live in the most hazard-prone parts of the city, often in unplanned dense settlements with a lack of public services; where lack of waste disposal may lead to blocking of drains and increases the risk of disease outbreaks when floods occur; with limited political influence to ensure government interventions to reduce risk. The accumulation of disaster risk over time may be partly caused by a string of smaller disasters due to continued exposure to small day-to-day risks in urban areas (e.g., Pelling and Wisner, 2009a), aggravated by limited resources to cope and recover from disasters when they occur – creating a vicious cycle of poverty and disaster risk. Analysis of disaster loss data suggests that frequent low-intensity losses often highlight an accumulation of risks, which is then realized when an extreme hazard event occurs (UNISDR, 2009a). Similar accumulation of risk may occur at larger scales in hazard-prone states, especially in the context of conflict and displacement (e.g., UNDP, 2004).

A context-based understanding of these risks is essential to identify appropriate risk management strategies. This may include better collection of sub-national disaster data that allows visualization of complex patterns of local risk (UNDP, 2004), as well as locally owned processes of risk identification and reduction. Bull-Kamanga et al. (2003) suggest that one of the most effective methods to address urban disaster risk in Africa is to support community processes among the most vulnerable groups so they can identify risks and set priorities – both for community action and for action by external agencies (including local governments). Such local risk assessment processes also avoid the pitfalls of planning based on dated maps used to plan and develop large physical construction and facilities.

Disaster risk is not an autonomous or externally generated circumstance to which society reacts, adapts, or responds (as is the case with natural phenomena or events per se), but rather the result of the interaction of society and the natural or built environment. Thus disasters are often the product of parallel developments that sometimes reach a tipping point, where the cumulative effect of these parallel processes results in disaster (Dikau and Pohl, 2007; Birkmann, 2011b). After that point, recovery may be slowed by conflict between processes and goals of reconstruction (Colten et al., 2008). In addition, there is often strong pressure to restore the status quo as soon as possible after a disaster has happened, even if that status quo means continued high levels of disaster risk. Sometimes, however, disasters themselves can be a window of opportunity for addressing the determinants of disaster risk. With proactive risk assessment and reconstruction planning, more appropriate solutions can be realized while restoring essential assets and services during and after disasters (Susman et al., 1983, Renn, 1992; Comfort et al., 1999; Vogel and O'Brien, 2004).

References

A digital library of non-journal-based literature cited in this chapter that may not be readily available to the public has been compiled as part of the IPCC review and drafting process, and can be accessed via either the IPCC Secretariat or IPCC Working Group II web sites.

Abel, N., D. Cumming, and J. Anderies, 2006: Collapse and reorganization in social-ecological systems: Questions, some ideas, and policy implications. *Ecology and Society*, **11(1)**, 17-42.

Adam, B. and J. van Loon, 2000: Repositioning risk; the challenge for social theory. In: *The Risk Society and Beyond* [Adam, B., U. Beck, and J. van Loon (eds.)]. SAGE Publications, London, UK, pp. 1-31.

Adelekan, I.O., 2010: Vulnerability of poor urban coastal communities to flooding in Lagos, Nigeria. *Environment and Urbanization*, **22**, 433, doi:10.1177/0956247810380141.

Adger, W.N., 1999: Social vulnerability to climate change and extremes in coastal Vietnam. *World Development*, **27(3)**, 249-269.

Adger, W.N., 2000: Social and ecological resilience: are they related? *Progress in Human Geography*, **24(3)**, 347-364.

Adger, W.N., 2003: Social capital, collective action, and adaptation to climate change. *Economic Geography*, **79(4)**, 387-404.

Adger, W.N., 2006: Vulnerability. *Global Environmental Change*, **16**, 268-281.

Adger, W.N. (ed.), 2009: *Adapting to Climate Change, Thresholds, Values, Governance*. Cambridge University Press, Cambridge, UK.

Adger, W.N. and N. Brooks, 2003: Does global environmental change cause vulnerability to disaster? In: *Natural Disasters and Development in a Globalizing World* [Pelling, M. (ed.)]. Routledge, London, UK, pp. 19-42.

Adger, W.N., and P.M. Kelly, 1999: Social vulnerability to climate change and the architecture of entitlements. *Mitigation and Adaptation Strategies for Global Change*, **4**, 253-266.

Adger, W.N., N. Brooks, M. Kelly, S. Bentham, and S. Eriksen, 2004: *New Indicators of Vulnerability and Adaptive Capacity*. Tyndall Centre for Climate Change Research, Technical Report 7, University of East Anglia, Norwich, UK.

Adger, W.N., N.W. Arnell, and E.L. Tompkins, 2005: Successful adaptation to climate change across scales. *Global Environmental Change*, **15(2)**, 77-86.

Afifi, T., 2011: Economic or environmental migration? The push factors in Niger. *International Migration*, **49(S1)**, e95-e124.

Agrawala, S. and M.K. van Aalst, 2008: Adapting development co-operation to adapt to climate change. *Climate Policy*, **8**, 183-193.

Ahern, M., R.S. Kovats, P. Wilkinson, R. Few, and F. Matthies, 2005: Global health impacts of floods: epidemiologic evidence. *Epidemiologic Reviews*, **27**, doi:10.1093/epirev/mxi004.

Alexander, D.E., 1993: *Natural Disasters*, UCL Press Limited, London, UK.

Alexander, D.E., 2000: *Confronting Catastrophe*. Terra Publishing, Harpenden, UK.

Altman, I. and S. Low, 1992: *Place Attachment*. Plenum Press, New York, NY.

Alwang, J., P.B. Siegel, and S.L. Jorgensen, 2001: *Vulnerability: A View From Different Disciplines*. Social Protection Discussion Paper Series, No. 115, World Bank, Washington, DC.

Amstadter, A.B., R. Acierno, L.K. Richardson, D.G. Kilpatrick, D.F. Gros, R.J. Johnson, M.T. Gaboury, L.T. Trinh, T.T. Lam, T.T. Nguyen, T. Tran, T.B. La, T.H. Tran, D.C. Tran, and G. Sandro, 2009: Post typhoon prevalence of posttraumatic stress disorder, major depressive disorder, panic disorder, and generalized anxiety disorder in a Vietnamese sample. *Journal of Traumatic Stress*, **22(3)**, 180-188.

Anderson, M.B. and P.J. Woodrow, 1989: *Rising from the Ashes, Development Strategies in Times of Disaster*. 1998 ed. Lynne Rienner, London, UK, 338 pp.

Anderson, M.B. and P.J. Woodrow, 1991: Reducing vulnerability to drought and famine: Developmental approaches to relief. *Disasters*, **15(1)**, 43-54.

Aragon-Durand, F., 2007: Urbanisation and flood vulnerability in the peri-urban interface of Mexico City. *Disasters*, **31(4)**, 477-494 .

Armitage, D., B. Fikret, and N. Doubleday (eds.), 2007: *Adaptive Co-Management: Collaboration, Learning, and Multi-Level Governance*. UBC Press, Vancouver, Canada, 160 pp.

Arrow, K. and R. Lind, 1970: Uncertainty and the evaluation of public investment decisions. *The American Economic Review*, 60, 364-378.

Auld, H. and D. MacIver, 2007: *Changing Weather Patterns, Uncertainty And Infrastructure Risks: Emerging Adaptation Requirements*. Occasional Paper 9, Environment Canada, Toronto, Canada.

Aysan, Y., 1993: Vulnerability assessment. In: *Natural Disasters: Protecting Vulnerable Communities* [Merriman, P.A. and C.W.A. Browitt (eds.)]. IDNDR-Thomas Telford, London, UK.

Aysan, Y. and P. Oliver, 1987: *Housing and Culture after Earthquakes. A guide for future policy making on housing in seismic areas*. Oxford Polytechnic for the Overseas Development Administration (ODA) of the UK Government, Oxford, UK.

Balbus, J.M. and C. Malina, 2009: Identifying vulnerable subpopulations for climate change health effects in the United States. *Journal of Occupational and Environmental Medicine*, **51**, 33-37.

Bankoff, G., 2004: *The Historical Geography of Disaster: 'Vulnerability' and 'Local Knowledge' in Western Discourse*. Earthscan, London, UK.

Barnett, J., 2001: Adapting to climate change in Pacific Island countries: The problem of uncertainty. *World Development*, **29(6)**, 977-993.

Barnett, J., 2005: Titanic states? Impacts and responses to climate change in the Pacific Islands. *Journal of International Affairs*, **59**, 203-219.

Barnett, J. and M. Webber, 2009: *Accommodating Migration to Promote Adaptation to Climate Change*. A policy brief prepared for the Secretariat of the Swedish Commission on Climate Change and Development and the World Bank World Development Report 2010 team, Commission on Climate Change and Development, Stockholm, Sweden.

Barnston, A.G., S. Li, S.L. Mason, D.G. DeWitt, L. Goddard, and X. Gong, 2010: Verification of the first 11 years of iri's seasonal climate forecasts. *Journal of Applied Meteorology and Climatology*, **49**, 493-520.

Bartlett, S., 2008: Climate change and urban children: impacts and implications for adaptation in low- and middle income countries. *Environment & Urbanization*, **20(2)**, 501-519.

Batterbury, S., 2001: Landscapes of diversity: A local political ecology of livelihood diversification in south-western Niger. *Ecumene*, **8(4)**, 437-464.

Beck, U., 2000: Risk society revisited: Theory, politics and research programmes. In: *The Risk Society and Beyond* [Adam, B., U. Beck, and J. van Loon (eds.)]. SAGE Publications, London, UK, 211-229.

Beck, U., 2008: *La Sociedad del Riesgo Mundial*. Paidós, Barcelona, Spain.

Beg, N., J.C. Morlot, O. Davidson, Y. Afrane-Okesse, L. Tyani, F. Denton, Y. Sokona, J.P. Thomas, E.L. La Rovere, J.K. Parikh, K. Parikh, and A.A. Rahman, 2002: Linkages between climate change and sustainable development. *Climate Policy*, **2**, 129-144.

Benioff, R., S. Guill, and J. Lee (eds.), 1996: *Vulnerability and Adaptation Assessments: An International Handbook*. Kluwer Academic Publishers, Dordrecht, The Netherlands.

Bennet, G., 1970: Bristol floods 1968: controlled survey of effects on health of local community disaster. *British Medical Journal*, **3**, 454-458.

Benson, C. and E. Clay, 2000: Developing countries and the economic impacts of natural disasters. In *Managing disaster risk in emerging economies* [Kreimer, A. and M. Arnold (eds.)]. World Bank, Washington, DC, pp. 11-21.

Bertoni, J.C., 2006: Urban floods in Latin America: reflections on the role of risk factors. In: *Frontiers in Flood Research* [Tchiguirinskaia, I., K.N.N. Thein, and P. Hubert (eds.)]. IAHS publication 305, International Association of Hydrological Sciences, Wallingford, UK, pp. 123-141.

Bettencourt, S., R. Croad, P. Freeman, J. Hay, R. Jones, P. King, P. Lal, A. Mearns, G. Miller, I. Pswarayi-Riddihough, A. Simpson, N. Teuatabo, U. Trotz, and M.K. van Aalst, 2006: *Not If, But When. Adapting To Natural Hazards in the Pacific Islands Region*. World Bank, Washington, DC.

Biermann, F., 2007: 'Earth system governance' as a crosscutting theme of global change research. *Global Environmental Change*, **17(3-4)**, 326-337.

Biermann, F., M. Betsill, J. Gupta, N. Kanie, L. Lebel, D. Livermann, H. Schroeder, and B. Siebenhüner, 2009: *Earth System Governance – People, Places, and the Planet*. Report No. 1, Science and Implementation Plan of the Earth System Governance Project, Bonn, Germany.

Birkmann, J., 2006a: *Measuring Vulnerability to Natural Hazards – Towards Disaster Resilient Societies*. United Nations University Press, Tokyo, Japan, 450 pp.

Birkmann, J., 2006b: Measuring vulnerability to promote disaster-resilient societies: conceptual frameworks and definitions. In: *Measuring Vulnerability to Natural Hazards: Towards Disaster Resilient Societies* [Birkmann, J. (ed.)]. United Nations University Press, Tokyo, Japan, pp. 9-54.

Birkmann, J., 2006c: Conclusions and recommendations. In: *Measuring Vulnerability to Natural Hazards: Towards Disaster Resilient Societies* [Birkmann, J. (ed.)]. United Nations University Press, Tokyo, Japan, pp. 432-447.

Birkmann, J., 2011a: First and second-order adaptation to natural hazards and extreme events in the context of climate change. *Natural Hazards*, **58(2)**, 811-840, doi:10.1007/s11069-011-9806-8.

Birkmann, J., 2011b: Regulation and coupling of society and nature in the context of natural hazards. In: Coping with Global Environmental Change, Disasters and Security [Brauch, H.G., U. Oswald Spring, C. Mesjasz, J. Grin, P. Kameri-Mbote, B. Chourou, P. Dunay, J. Birkmann (eds.)]. Springer, Berlin, Germany, pp. 1103-1127.

Birkmann, J. and N. Fernando, 2008: Measuring revealed and emergent vulnerabilities of coastal communities to tsunami in Sri Lanka. *Disasters*, **32(1)**, 82-104.

Birkmann, J. and K. von Teichman, 2010: Integrating disaster risk reduction and climate change adaptation: key challenges – scales, knowledge, and norms. *Sustainability Science*, **5(2)**, 171-184.

Birkmann, J., K. von Teichman, P. Aldunce, C. Bach, N.T. Binh, M. Garschagen, S. Kanwar, N. Setiadi, and L.N. Thach, 2009: *Addressing the Challenge: Recommendations and Quality Criteria for Linking Disaster Risk Reduction and Adaptation to Climate Change* [Birkmann, J., G. Tetzlaff, and K-O. Zentel (eds.)]. DKKV Publication Series No. 38, German Committee for Disaster Reduction, Bonn, Germany.

Birkmann, J., P. Buckle, J. Jaeger, M. Pelling, N. Setiadi, M. Garschagen, N. Fernando, and J. Kropp, 2010a: Extreme events and disasters: A window of opportunity for change? Analysis of changes, formal and informal responses after mega-disasters. *Natural Hazards*, **55(3)**, 637-669.

Birkmann, J., M. Garschagen, F. Kraas, and N. Quang, 2010b: Adaptive urban governance: new challenges for the second generation of urban adaptation strategies to climate change. *Sustainability Science*, **5(2)**, 185-206.

Black, R., 2001: *Environmental Refugees: Myth or Reality?* UNHCR Working Paper 34, UN High Commissioner for Refugees, Geneva, Switzerland, 19 pp.

Blaikie, P., T. Cannon, I. Davis, and B. Wisner, 1994: *At Risk: Natural Hazards, People, Vulnerability, and Disasters*. Routledge, London, UK.

Blaikie, P., T. Cannon, I. Davis, and B. Wisner, 1996: *Vulnerabilidad, el entorno social de los desastres*. La RED-ITDG, Bogota, Colombia.

Blockley, D. (ed.), 1992: *Engineering Safety*. McGraw-Hill International Series in Civil Engineering, London, UK.

Boentje, J.P. and M.S. Blinnikov, 2007: Post-Soviet forest fragmentation and loss in the Green Belt around Moscow, Russia (1991-2001): a remote sensing perspective. *Landscape and Urban Planning*, **82(4)**, 208-221.

Bogardi, J. and J. Birkmann, 2004: Vulnerability assessment: the first step towards sustainable risk reduction. In: *Disasters and Society – From Hazard Assessment to Risk Reduction* [Malzahn, D. and T. Plapp (eds.)]. Logos Verlag, Berlin, Germany, pp. 75-82.

Bohle, H-G., 2001: Vulnerability and criticality: Perspectives from social geography. *IHDP Update*, **2/2001**, 1-7.

Bohle, H-G., 2008: Krisen, katastrophen, kollaps – Geographien von verwundbarkeit in der risikogesellschaft. In: *Umgang mit Risiken. Katastrophen – Destabilisierung – Sicherheit* [Kulke, E. and H. Popp (eds.)]. Deutscher Geographentag 2007, Bautzen, Lausitzer Druck- und Verlagshaus GmbH, Bayreuth, Germany, pp. 69-82.

Bohle, H-G., and T. Glade, 2008: Vulnerabilitätskonzepte in Sozial- und Naturwissenschaften. In: *Naturrisiken und Sozialkatastrophen* [Felgentreff, C. and T. Glade (eds.)]. Spektrum Verlag, Berlin, Germany, pp. 99-119.

Bohle, H.G., T.E. Downing, and M.J. Watts, 1994: Climate change and social vulnerability: Toward a sociology and geography of food insecurity. *Global Environmental Change*, **4(1)**, 37-48.

Bouchama, A., M. Dehbi, G. Mohamed, F. Matthies, M. Shoukri, and B. Menne, 2007: Prognostic factors in heat wave related deaths: a meta-analysis. *Archives of Internal Medicine*, **167**, 2170-2176.

Bouwer, L.M., R.P. Crompton, E. Faust, P. Hoppe, and R.A. Pielke Jr., 2007: Disaster management: Confronting disaster losses. *Science*, **318 (5851)**, 753.

Boykoff, M.T. and J. Boykoff, 2007: Climate change and journalistic norms: a case-study of US mass-media coverage. *Geoforum*, **38**, 1190-1204.

Breiting, S. and P. Wickenberg, 2010: The progressive development of environmental education in Sweden and Denmark. *Environmental Education Research*, **16(1)**, 9-37.

Brenkert, A.L. and E.L. Malone, 2005: Modeling vulnerability and resilience to climate change: A case study of India and Indian States. *Climatic Change*, **72(1-2)**, 57-102.

Brklacich, M. and H.-G. Bohle, 2006: Assessing human vulnerability to global climatic change. In: *Earth System Science in the Anthropocene* [Ehlers, E. and T. Krafft (eds.)]. Springer, Berlin, Germany, pp. 51-61.

Brondo, K.V., 2007: Land loss and Garifuna women's activism on Honduras' North Coast. *Journal of International Women's Studies*, **9(1)**, 99-116.

Brooks, N., 2003: *Vulnerability, Risk and Adaptation: A Conceptual Framework*. Tyndall Centre for Climate Change Working Paper 38, University of East Anglia, Norwich, UK.

Brooks, N. and W.N. Adger, 2004: Assessing and enhancing adaptive capacity, Technical Paper 7. In: *Adaptation Policy Frameworks for Climate Change: Developing Strategies, Policies and Measures* [Lim, B. and E. Spanger-Siegfried (eds.)]. United Nations Development Programme and Cambridge University Press, New York, NY, pp. 165-182.

Brooks, N., W.N. Adger, and M. Kelly, 2005: The determinants of vulnerability and adaptive capacity at the national level and the implications for adaptation. *Global Environmental Change Part B: Environmental Hazards*, **15**, 151-163.

Bull-Kamanga, L., K. Diagne, A. Lavell, E. Leon, F. Lerise, H. MacGregor, A. Maskrey, M. Meshack, M. Pelling, H. Reid, D. Satterthwaite, J. Songsore, K. Westgate, and A. Yitambe, 2003: From everyday hazards to disasters: the accumulation of risk in urban areas. *Environment & Urbanization*, **15(1)**, 193-203.

Burby, R.J., 2006: Hurricane Katrina and the paradoxes of government disaster policy: Bringing about wise governmental decisions for hazardous areas. *The Annals of the American Academy of Political and Social Science*, **604**, 171-191.

Burkle, F.M. and P.G. Greenough, 2008: Impact of public health emergencies on modern disaster taxonomy, planning, and response. *Disaster Medicine and Public Health Preparedness*, **2(4)**, e2-e9.

Burton, I. and M.K. van Aalst, 1999: *Come Hell or High Water: Integrating Climate Change Vulnerability and Adaptation into Bank Work*. World Bank Environment Department Papers 72, World Bank, Washington, DC.

Burton, I. and M.K. van Aalst, 2004: *Look Before You Leap. A Risk Management Approach for Climate Change Adaptation in World Bank Operations*. World Bank, Washington, DC.

Burton, I., R.W. Kates, and G.F. White, 1978: *The Environment as Hazard*. Oxford University Press, New York, NY.

Burton, I., J. Wilson, and R.E. Munn, 1983: Environmental impact assessment: national approaches and international needs. *Environmental Monitoring and Assessment*, **3**, 133-150.

Burton, I., R.W. Kates, and G.F. White, 1993: *The Environment as Hazard – Second Edition*. Oxford University Press, New York, NY.

Campbell, D., M. Stafford Smith, J. Davies, P. Kuipers, J. Wakerman, and M.J. McGregor, 2008: Responding to health impacts of climate change in the Australian desert. *Rural and Remote Health*, **8**, 1008.

Campbell, J.R. 2006. *Traditional Disaster Reduction in Pacific Island Communities*. GNS Science Report 2006/38. Institute of Geological and Nuclear Sciences, Avalon, NZ, ISBN 0-478-09961-4.

Cannon, T., 1994: Vulnerability analysis and the explanation of 'natural' disasters. In: *Disasters, Development and Environment* [Varley, A. (ed.)]. John Wiley and Sons, Chichester, UK, pp. 13-29.

Cannon, T., 2006: Vulnerability analysis, livelihoods and disasters. In: *Risk 21: Coping with Risks Due to Natural Hazards in the 21st Century* [Ammann, W.J., S. Dannenmann, and L. Vulliet (eds.)]. Taylor and Francis Group, London, UK, pp. 41-49.

Cannon, T. and D. Müller-Mahn, 2010: Vulnerability, resilience and development discourses in the context of climate change. *Natural Hazards*, **55(3)**, 621-635.

Cannon, T., J. Twigg, and J. Rowell, 2003: *Social Vulnerability, Sustainable Livelihoods and Disasters*. Conflict and Humanitarian Assistance Department and Sustainable Livelihoods Support Office, Department for International Development, London, UK, 63 pp.

Cardona, O.D., 1986: Estudios de vulnerabilidad y evaluación del riesgo sísmico: Planificación física y urbana en áreas propensas. *Boletín Técnico de la Asociación Colombiana de Ingeniería Sísmica*, **33(2)**, 32-65.

Cardona, O.D., 1990: *Terminología de Uso Común en Manejo de Riesgos*. AGID Reporte No. 13, Escuela de Administración, Finanzas, y Tecnología, Medellín, Colombia.

Cardona, O.D., 1993: Evaluación de la amenaza, la vulnerabilidad y el riesgo: Elementos para el Ordenamiento y la Planeación del Desarrollo. In: *Los Desastres No son Naturales* [Maskrey, A. (ed.)]. La RED/Tercer Mundo Editores, Bogotá, Colombia.

Cardona, O.D., 1996a: Manejo ambiental y prevención de desastres: dos temas asociados. In: *Ciudades en Riesgo* [Fernandez, M.A. (ed.)]. La RED-USAID, Lima, Peru, pp. 79-101.

Cardona, O.D., 1996b: El Manejo de riesgos y los preparativos para Desastres: Compromiso Institucional para mejorar la calidad de vida. In: *Desastres: modelo para armar* [Mansilla, E. (ed.)]. La RED, Lima, Peru, pp. 128-147.

Cardona, O.D., 1996c: Variables involucradas en el manejo de riesgos. In: *Desastres y Sociedad, Especial: Predicciones, Pronósticos, Alertas y Respuestas Sociales* [Cardona, O.D. (ed.)]. **4**, La RED, Tarea Gráfica, Lima, Peru, pp. 7-35.

Cardona, O.D., 1999a: Environmental management and disaster prevention: Two related topics: A holistic risk assessment and management approach. In: *Natural Disaster Management* [Ingleton, J. (ed.)]. IDNDR-Tudor Rose, London, UK, pp. 151-153.

Cardona, O.D., 1999b: Environmental management and disaster prevention: Two related topics. In: *Cities at Risk: Environmental Degradation, Urban Risks and Disasters in Latin America* [Fernandez, M.A. (ed.)]. A/H Editorial, La RED, US AID, Quito, Peru, pp. 79-102.

Cardona, O.D., 2001: *Estimación Holística del Riesgo Sísmico utilizando Sistemas Dinámicos Complejos*. Doctoral dissertation, Department of Terrain Engineering, Technical University of Catalonia, Spain.

Cardona, O.D., 2004: The need for rethinking the concepts of vulnerability and risk from a holistic perspective: A necessary review and criticism for effective risk management. In: *Mapping Vulnerability: Disasters, Development and People* [Bankoff, G., G. Frerks, and D. Hilhorst (eds.)]. Earthscan Publishers, London, UK, pp. 37-51.

Cardona, O.D., 2006: A system of indicators for disaster risk management in the Americas. In: *Measuring Vulnerability to Hazards of Natural Origin: Towards Disaster Resilient Societies* [Birkmann, J. (ed.)]. UNU Press, Tokyo, Japan, pp. 189-209.

Cardona, O.D., 2008: *Indicators of Disaster Risk and Risk Management: Program for Latin America and the Caribbean – Summary Report – Second Edition*. INE-08-002, Inter-American Development Bank, Washington, DC.

Cardona, O.D., 2009: *La Gestión Financiera del Riesgo de Desastres: Instrumentos Financieros de Retención y Transferencia para la Comunidad Andina*. PREDECAN, Comunidad Andina, Lima, Peru.

Cardona, O.D., 2010: *Indicators of Disaster Risk and Risk Management – Program for Latin America and the Caribbean: Summary Report*. Evaluación de Riesgos Naturales - Latino America, ERN-AL, Inter-American Development Bank, Washington, DC.

Cardona, O.D., 2011: Disaster risk and vulnerability: Notions and measurement of human and environmental insecurity. In: *Coping with Global Environmental Change, Disasters and Security – Threats, Challenges, Vulnerabilities and Risks* [Brauch, H.G., U. Oswald Spring, C. Mesjasz, J. Grin, P. Kameri-Mbote, B. Chourou, P. Dunay, J. Birkmann]. Springer Verlag, Berlin, Germany, pp. 107-122.

Cardona, O.D. and A.H. Barbat, 2000: *El Riesgo Sísmico y su Prevención*. Cuaderno Técnico 5, Calidad Siderúrgica, Madrid, Spain, 190 pp.

Cardona, O.D. and J.E. Hurtado, 2000a: Holistic seismic risk estimation of a metropolitan center. In: *Proceedings of the 12th World Conference of Earthquake Engineering*, January 30 – February 4, 2000, paper 2643. New Zealand Society for Earthquake Engineering, Auckland, New Zealand.

Cardona, O.D. and J.E. Hurtado, 2000b: Modelación numérica para la estimación holística del riesgo sísmico urbano, considerando variables técnicas, sociales y económicas. In: *Métodos Numéricos en Ciencias Sociales (MENCIS 2000)* [Oñate, E., F. García-Sicilia, and L. Ramallo (eds.)]. Centro Internacional de Métodos Numéricos en Ingeniería - Universidad Politécnica de Cataluña, Barcelona, Spain, pp. 452-466.

Cardona, O.D., J.E. Hurtado, G. Duque, A. Moreno, A.C. Chardon, L.S. Velásquez, and S.D. Prieto, 2003a: *The Notion of Disaster Risk: Conceptual Framework for Integrated Management*. National University of Colombia / Inter-American Development Bank, Washington, DC.

Cardona, O.D., J.E. Hurtado, G. Duque, A. Moreno, A.C. Chardon, L.S. Velásquez, and S.D. Prieto, 2003b: *Indicators for Risk Measurement: Methodological Fundamentals*. National University of Colombia / Inter-American Development Bank, Washington, DC.

Cardona, O.D., M.G. Ordaz, M.C. Marulanda, and A.H. Barbat, 2008a: Estimation of probabilistic seismic losses and the public economic resilience – An approach for a macroeconomic impact evaluation. *Journal of Earthquake Engineering*, **12(S2)**, 60-70.

Cardona, O.D., M.G. Ordaz, L.E. Yamín, M.C. Marulanda, and A. H. Barbat, 2008b: Earthquake loss assessment for integrated disaster risk management. *Journal of Earthquake Engineering*, **12(S2)**, 48-59.

CARE, 2009: *Climate Vulnerability and Capacity Analysis Handbook*. CARE International, Chatelaine, Switzerland.

Carreño, M.L., 2006: Técnicas innovadoras para la evaluación del riego sísmico y su gestión en centros urbanos: Acciones ex ante y ex post. Doctoral dissertation, Department of Terrain Engineering, Technical University of Catalonia, Spain.

Carreño, M.L., O.D. Cardona, and A.H. Barbat, 2005: Sistema de Indicadores para la Evaluación de Riesgos. Monografía CIMNE IS-52, Technical University of Catalonia, Barcelona, Spain, 160 pp.

Carreño, M.L., O.D. Cardona, and A.H. Barbat, 2007a: Urban seismic risk evaluation: A holistic approach. *Journal of Natural Hazards*, **40(1)**, 137-172.

Carreño, M.L., O.D. Cardona, and A.H. Barbat, 2007b: A disaster risk management performance index. *Journal of Natural Hazards*, **41(1)**, 1-20.

Carreño, M.L., O.D. Cardona, M.C. Marulanda, and A.H. Barbat, 2009: Holistic urban seismic risk evaluation of megacities: Application and robustness. In: *The 1755 Lisbon Earthquake: Revisited* [Mendes-Victor, L.A., C.S. Sousa Oliveira, J. Azevedo, and A. Ribeiro (eds.)]. Springer, Berlin, Germany.

Carroll, B., H. Morbey, R. Balogh, and G. Araoz, 2006: *Living in Fear: Health and Social Impacts of the Floods in Carlisle 2005*. Research Report, Centre for Health Research and Practice Development, St. Martins College, Carlisle, UK.

Carter, T.R., M.L. Parry, H. Harasawa, and S. Nishioka, 1994: *IPCC Technical Guidelines for Assessing Climate Change Impacts and Adaptations*. Center for Global Environmental Research, National Institute for Environmental Studies, Tsukuba, Japan.

Carvalho, A., 2005: Representing the politics of the greenhouse effect: discursive strategies in the British media. *Critical Discourse Studies*, **2(1)**, 1-29.

Carvalho, A. and J. Burgess, 2005: Cultural circuits of climate change in U.K. broadsheet newspapers, 1985-2003. *Society for Risk Analysis*, **25(6)**, 1457-1469.

Castles, S., 2002: *Environmental Change and Forced Migration: Making Sense of the Debate*. UNHCR Issues in Refugee Research, Working Paper No. 70, UN High Commissioner for Refugees, Geneva, Switzerland.

Ceccato, P., T. Ghebremeskel, M. Jaiteh, P.M. Graves, M. Levy, S. Ghebreselassie, A. Ogbamariam, A.G. Barnston, M. Bell, J. del Corral, S.J. Connor, I. Fesseha, E.P. Brantly, and M.C. Thomson, 2007: Malaria stratification, climate, and epidemic early warning in Eritrea. *American Journal of Tropical Medicine and Hygiene*, **77**, 61-68.

Cernea, M., 1996: *Eight Main Risks: Impoverishment and Social Justice in Resettlement*. World Bank Environment Department, Washington, DC.

Chambers, R., 1983: *Rural Development – Putting the Last First*. Longmans Scientific and Technical Publishers, Essex, UK, 246 pp.

Chambers, R., 1989: Vulnerability, coping and policy. *Institute of Development Studies Bulletin*, **20(2)**, 1-7.

Changnon, D. and S.A. Changnon, 2010: Major growth in some business related uses of climate information. *Journal of Applied Meteorology and Climatology*, **49**, 325-331.

Chatterjee, M., 2010: Slum dwellers response to flooding events in the megacities of India. *Adaptation and Mitigation Strategies for Global Change*, **15**, 337-353.

Chester, D.K., 2005: Theology and disaster studies: the need for dialogue. *Journal of Volcanology and Geothermal Research*, **146 (4)**, 319-328.

Christoplos, I., 2006: The elusive 'window of opportunity' for risk reduction in post-disaster recovery. Discussion Paper, *ProVention Consortium Forum*, 2-3 February 2006, Bangkok, Thailand.

Clasen, T., W.P Schmidt, T. Rabie, I. Roberts, and S. Cairncross, 2007: Interventions to improve water quality for preventing diarrhoea: systematic review and meta-analysis. *British Medical Journal*, **334(7597)**, 782-785.

Coburn, A. and R. Spence, 1992: *Earthquake Protection*. John Wiley & Sons, Chichester, UK.

Colchester, M., 2000: *Dams, Indigenous People and Vulnerable Ethnic Minorities. Thematic Review 1.2*. World Commission on Dams, Cape Town, South Africa.

Colten, C.E., 2006: Vulnerability and place: Flat land and uneven risk in New Orleans. *American Anthropologist*, **108(4)**, 731-734.

Colten, C., R. Kates, and S. Laska, 2008: *Community Resilience: Lessons for New Orleans and Hurricane Katrina*. CARRI Research Report 3, Community and Regional Resilience Institute, Oak Ridge, TN, 47 pp.

Comfort, L., B. Wisner, S. Cutter, R. Pulwarty, K. Hewitt, A. Oliver-Smith, J. Wiener, M. Fordham, W. Peacock, and F. Krimgold, 1999: Reframing disaster policy: the global evolution of vulnerable communities. *Environmental Hazards*, **1**, 39-44.

Confalonieri, U., B. Menne, R. Akhtar, K.L. Ebi, M. Hauengue, R.S. Kovats, B. Revich, and A. Woodward, 2007: Human health. In: *Climate Change 2007. Impacts, Adaptation and Vulnerability. Contribution of Working Group II to the Fourth Assessment Report of the Intergovernmental Panel on Climate Change* [Parry, M.L., O.F. Canziani, J.P. Palutikof, P.J. Van Der Linde, and C.E. Hanson (eds.)]. Cambridge University Press, Cambridge, UK, 391-431.

Cooper, P.J.M., J. Dimes, K.P.C. Rao, B. Shapiro, B. Shiferaw, and S. Twomlow, 2008: Coping better with current climatic variability in the rain-fed farming systems of sub-Saharan Africa: An essential first step in adapting to future climate change? *Agriculture, Ecosystems and Environment*, **126**, 24-35.

Corral, S., 2000: Explorando la calidad de los procesos de elaboración de políticas ambientales. In: *Métodos Numéricos en Ciencias Sociales* (MENCIS 2000) [Oñate, E., F. García-Sicilia, and L. Ramallo (eds.)]. Centro Internacional de Métodos Numéricos en Ingeniería - Universidad Politécnica de Cataluña, Barcelona, Spain, pp. 391-401.

Costello, A., M. Abbas, A. Allen, S. Ball, S. Bell, R. Bellamy, S. Friel, N. Groce, A. Johnson, M. Kett, M. Lee, C. Levy, M. Maslin, D. McCoy, B. McGuire, H. Montgomery, D. Napier, C. Pagel, J. Patel, C. Patterson, J.A. Puppim de Oliveira, N. Redclift, H. Rees, D. Rogger, J. Scott, J. Stephenson, J. Twigg, and J. Wolff, 2009: Managing the health effects of climate change. *The Lancet*, **373(16)**, 1693-1733.

Covello, V. and J. Mumpower, 1985: Risk analysis and risk management: An historical perspective. *Risk Analysis*, **5(2)**, 103-120.

Cross, J.A., 2001: Megacities and small towns: Different perspectives on hazard vulnerability. *Environmental Hazards*, **3(2)**, 63-80.

Cruz, A.M., L.J. Steinberg, A.L.V. Arellano, J.-P. Nordvik, and F. Pisano, 2004: *State of the Art in Natech Risk Management (NATECH: Natural Hazard Triggering a Technological Disaster)*. EUR 21292, European Communities, European Commission, Brussels, Belgium.

Cummins, J. and O. Mahul, 2009: *Catastrophe Risk Financing in Developing Countries: Principles for Public Intervention*. The World Bank, Washington, DC, 268 pp.

Cuny, F.C., 1984: *Disaster and Development*. Oxford University Press, New York, NY.

Cutter, S.L. (ed.), 1994: *Environmental Risks and Hazards*. Prentice Hall, Upper Saddle River, NJ.

Cutter, S.L., 1996: Vulnerability to environmental hazards. *Progress in Human Geography*, **20**, 529-539.

Cutter, S.L. and C. Finch, 2008: Temporal and spatial changes in social vulnerability to natural hazards. *Proceedings of the National Academy of Sciences*, **105(7)**, 2301-2306.

Cutter, S.L., C.T. Emrich, J.T. Mitchell, B.J. Boruff, M. Gall, M.C. Schmidtlein, C.G. Burton, and G. Melton, 2006: The long road home: Race, class, and recovery from Hurricane Katrina. *Environment*, **48(2)**, 8-20.

Cutter, S.L., L. Barnes, M. Berry, C. Burton, E. Evans, E. Tate, and J. Webb, 2008: A place-based model for understanding community resilience to natural disasters. *Global Environmental Change*, **18**, 598-606.

Davis, I. and M. Wall (eds.), 1992: *Christian Perspectives on Disaster Management: A Training Manual*. International Relief and Development Association, Middlesex, UK.

Davis, I., B. Haghebaert, and D. Peppiatt, 2004: *Social Vulnerability and Capacity Analysis Workshop Discussion paper and workshop report*. ProVention Consortium, Geneva, Switzerland.

Dawson, T.P., S.T. Jackson, J.I. House, C.I. Prentice, and G.M. Mace, 2011: Beyond predictions: Biodiversity conservation in a changing climate. *Science*, **332**, 53.

Dayton-Johnson, J., 2004: *Natural Disasters and Adaptive Capacity*. OECD Working Paper No. 237, Organisation for Economic Co-operation and Development, Paris, France.

Décamps, H. (ed.), 2010: *Événements Climatiques Extrêmes: Réduire les Vulnérabilités des Systèmes Écologiques et Sociaux*. Institut de France, Académie des Sciences, Paris, France, 194 pp.

Degallier, N., C. Favier, C. Menkes, M. Lengaigne, W.M. Ramalho, R. Souza, J. Servain, J.-P. Boulanger, 2010: Toward an early warning system for dengue prevention: modeling climate impact on dengue transmission. *Climatic Change*, **98**, 581-592.

Demetriades, J. and E. Esplen, 2008: The gender dimensions of poverty and climate change adaptation. *IDS Bulletin*, **39(4)**, 24.

Di Baldassarre, G., A. Montanari, H. Lins, D. Koutsoyiannis, L. Brandimarte, and G. Bloeschl, 2010: Flood fatalities in Africa: From diagnosis to mitigation. *Geophysical Research Letters*, **37**, L22402.

Dikau, R. and J. Pohl, 2007: "Hazards" Naturgefahren und Naturrisiken. In: *Geographie, Physische Geographie und Humangeographie* [Gebhardt, H., R. Glaser, U. Radtke, and P. Reuber (eds.)]. Spektrum Akademischer Verlag, Berlin, Germany, pp. 1029-1076.

Dilley, M., 2006: Disaster risk hotspots: A project summary. In: *Measuring Vulnerability to Natural Hazards – Towards Disaster Resilient Societies* [Birkmann, J. (ed.)]. United Nations University Press, Tokyo, Japan, pp. 182-188.

Dilley, M., R. Chen, U. Deichmann, A. Lerner-Lam, and M. Arnold, 2005: *Natural Disaster Hotspots: A Global Risk Analysis*. Columbia University and World Bank, New York, NY and Washington, DC.

DKKV, 2011: *Adaptive Disaster Risk Reduction – Enhancing Methods and Tools of Disaster Risk Reduction in the light of Climate Change*. DKKV Publication Series no. 43, German Committee for Disaster Reduction, Bonn, Germany, www.dkkv.org/de/publications/schriftenreihe.asp?h=5.

Doherty, T.J. and S. Clayton, 2011: The psychological impacts of global climate change. *American Psychologist*, **66(4)**, 265-276.

Donner, W. and H. Rodriguez, 2008: Population composition, migration and inequality: The influence of demographic changes on disaster risk and vulnerability. *Social Forces*, **87(2)**, 1089-1114.

Doocy, S., Y. Gorokhovich, G. Burnham, D. Balk, and C. Robinson, 2007: Tsunami mortality estimates and vulnerability mapping in Aceh, Indonesia. American Journal of Public Health, **97(S1)**, 146-151.

Dore, M.H.I. and D. Etkin, 2003: Natural disasters, adaptive capacity and development in the twenty-first century. In: *Natural Disasters and Development in A Globalizing World* [Pelling, M. (ed.)]. Routledge, London, UK, pp. 75-91.

Douglas, I., K. Alam, M. Maghenda, Y. McDonnell, L. McLean, and J. Campbell, 2008: Unjust waters: climate change, flooding and the urban poor in Africa. *Environment And Urbanization*, **20(1)**, 187-205.

Douglas, M. and A. Wildavsky, 1982: Risk and culture: An essay on the selection of technological and environmental dangers. University of California Press, Berkeley, CA.

Dow, K., 1992: Exploring differences in our common future(s): the meaning of vulnerability to global environmental change. *Geoforum*, **23 (3)**, 417-436.

Downing, T.E. and A. Patwardhan, 2004: Assessing vulnerability for climate adaptation, Technical Paper 3. In: *Adaptation Policy Frameworks for Climate Change: Developing Strategies, Policies and Measures* [Lim, B. and E. Spanger-Siegfried (eds.)]. United Nations Development Programme and Cambridge University Press, New York, NY, pp. 67-89.

Drusine, H., 2005: The Garifuna fight back. *Third Text*, **19(2)**, 197-202.

Dudasik, S., 1982: Unanticipated repercussions of international disaster relief. *Disasters*, **6**, 31-37.

Duneier, M., 2004: Scrutinizing the heat: On ethnic myths and the importance of shoe leather. *Contemporary Sociology*, **33 (2)**, 139-150.

Eakin, H., 2005: Institutional change, climate risk, and rural vulnerability: Cases from Central Mexico. *World Development*, **33(11)**, 1923-1938.

Elliot, L., M. Beeson, S. Akbarzadeh, G. Fealy, and S. Harris, 2006: *Religion, Faith and Global Politics*. Department of International Relations, Canberra, Australia.

Elliott, J.R. and Pais, J. 2006: Race, class, and Hurricane Katrina: Social differences in human responses to disaster. *Social Science Research*, **35**, 295-321.

Ellis, B.R. and B.A. Wilcox, 2009: The ecological dimensions of vector-borne disease research and control. *Cadernos De Saude Publica*, **25**, S155-S167.

Enarson, E. and B.H. Morrow (eds.), 1998: *The Gendered Terrain of Disaster: Through Women's Eyes*. Praeger Publishers, Westport, CT.

Enarson, E. and B.H. Morrow, 2000: A gendered perspective: The voices of women. In: *Hurricane Andrew: Ethnicity, Gender and the Sociology of Disasters* [Peacock, W.G., B.H. Morrow, and H. Gladwin (eds)]. International Hurricane Centre, Laboratory for Social and Behavioural Research, Miami, FL, pp. 116-137.

Eriksen, S.H., and K. O'Brien, 2007: Vulnerability, poverty and the need for sustainable adaptation measures. *Climate Policy*, **7(4)**, 337-352.

Eriksen, S. and J.A. Silva, 2009: The vulnerability context of a savanna area in Mozambique: household drought strategies and responses to economic change. *Environmental Science and Policy*, **12**, 33-52.

Eriksen, S., H.K. Brown, and P.M. Kelly, 2005: The dynamics of vulnerability: locating coping strategies in Kenya and Tanzania. *Geographical Journal*, **171 (4)**, 287-305.

ERN-AL, 2011: *Probabilistic Modelling of Natural Risk at Global Level: The Hybrid Loss Exceedance Curve – Development of a Methodology and Implementation of Case Studies, Phase 1A: Colombia, Mexico, Nepal*. Report for the GAR 2011, Consortium Evaluación de Riesgos Naturales – América Latina, Bogotá.

FAOSTAT, 2011: *FAOSTAT Statistical database on agricultural employment*. Food and Agriculture Organization, Rome, Italy, faostat.fao.org.

Feenstra, J.F., I. Burton, J.B. Smith, and R.S.J. Tol, 1998: *Handbook on Methods for Climate Change Impact Assessment and Adaptation Strategies*. Version 2.0, United Nations Environment Programme and Dutch National Institute for Public Health and the Environment (UNEP/RIVM), Nairobi, Kenya and Amsterdam, The Netherlands.

Few, R., 2007: Health and climatic hazards: framing social research on vulnerability, response and adaptation. *Global Environmental Change*, **17**, 281-295.

Folke, C., 2006: Resilience: The emergence of a perspective for social-ecological systems analyses. *Global Environmental Change*, **16(3)**, 253-267.

Folke, C., S. Carpenter, T. Elmqvist, L. Gunderson, C.S. Holling, and B. Walker, 2002: Resilience and sustainable development: Building adaptive capacity in a world of transformations. *AMBIO: A Journal of the Human Environment*, **31(5)**, 437-440.

Fordham, M., 1998: Making women visible in disasters: problematising the private domain. *Disasters*, **22(2)**, 126-143.

Fordham, M. and S. Gupta with S. Akerkar and M. Scharf, 2011: *Leading Resilient Development: Grassroots Women's Priorities, Practices and Innovations.* Grassroots Organizations Operating Together in Sisterhood and the UN Development Programme, New York, NY.

Fothergill, A., E.G.M. Maestas, and J.D. Darlington, 1999: Race, ethnicity and disasters in the United States: A review of the literature. *Disasters*, **23(2)**, 156-173.

Fournier d'Albe, M., 1985: The quantification of seismic hazard for the purposes of risk assessment. In: *International Conference on Reconstruction, Restoration and Urban Planning of Towns and Regions in Seismic Prone Areas*, 5-9 November 1985. Skopje, Yugoslavia, pp. 77-84.

Fraser, E.D.G., 2006: Food system vulnerability: using past famines to help understand how food systems may adapt to climate change. *Ecological Complexity*, **3**, 328-335.

Freeman, P. and K. Warner, 2001: *Vulnerability of Infrastructure to Climate Variability: How Does This Affect Infrastructure Lending Policies?* Disaster Management Facility of The World Bank and the ProVention Consortium, Washington, DC, 40 pp.

Freudenberg, W., R. Gramling, S. Laska, and K. Erickson, 2008: Organising hazards, engineering disasters? Improving the recognition of politico-economic factors in the creation of disasters. *Social Factors*, **78(2)**, 1015-1038.

Furr, J.M., J.S. Comer, J.M. Edmunds, and P.C. Kendall, 2010: Disasters and youth: A meta-analytic examination of posttraumatic stress. *Journal of Consulting and Clinical Psychology*, **78**, 765-780.

Füssel, H.-M., 2005: *Vulnerability to climate change: a comprehensive conceptual framework*. University of California International and Area Studies Breslauer Symposium Paper 6, Berkeley, CA, 36 pp.

Füssel, H.-M., 2007: Vulnerability: A generally applicable conceptual framework for climate change research. *Global Environmental Change*, **17**, 155-167.

Füssel, H.-M. and R.J.T. Klein, 2006: Climate change vulnerability assessments: an evolution of conceptual thinking. *Climatic Change*, **75**, 301-329.

Gaillard, J.C., 2007: Resilience of traditional societies in facing natural hazard. *Disaster Prevention and Management*, **16**, 522-544.

Gaillard, J.C., 2010: Vulnerability, capacity and resilience: Perspectives for climate and development policy, *Journal of International Development*, **22**, 218-232, doi:10.1002/jid.1675.

Gallopin, G.C., 2006: Linkages between vulnerability, resilience, and adaptive capacity. *Global Environmental Change*, **16(3)**, 293-303.

Gallopin, G.C., S. Funtowicz, M. O'Connor, and J. Ravetz, 2001: Science for the twenty-first century: from social contract to the scientific core. *International Social Science Journal*, **53(168)**, 219-229

Garret, M.J., 1999: *Health Futures: A Handbook for Health Professionals*. World Health Organization, Geneva. Switzerland, ISBN 92-4-154521-6.

Ginexi, E.M., K. Weihs, S.J. Simmens, and D. R. Hoyt, 2000: Natural disaster and depression: a prospective investigation of reactions to the 1993 midwest floods. *American Journal of Community Psychology*, **28**, 495-518.

Glantz, M., 2003: *Early warning systems: Dos and don'ts*. Report of Workshop on Early Warning Systems, Shanghai, China, 20-24 October 2003, ISBN 978-0756744953.

Goddard, L., Y. Aitchellouche, W. Baethgen, M. Dettinger, R. Graham, P. Hayman, M. Kadi, R. Martínez, and H. Meinke, 2009: *Providing Seasonal-to-Interannual Climate Information for Risk Management and Decision Making*. White Paper presented at the World Climate Conference 3 in Report of the World Climate Conference 3, Report No. 1048, World Meteorological Organization, Geneva, Switzerland.

Gosling, S., G. McGregor, and J. Lowe, 2009: Climate change and heat-related mortality in six cities – part 2: climate model evaluation and projected impacts from changes in the mean and variability of temperature with climate change. *International Journal of Biometeorology*, **53(1)**, 31-51.

Greiving, S., 2002: *Räumliche Planung und Risiko*. Gerling Akademischer Verlag, München, Germany, 320 pp.

Grossi, P. and H. Kunreuther (eds.), 2005: *Catastrophe Modeling: A new approach to managing risk*. Springer, New York, NY, 245 pp.

Grothmann, T. and A. Patt, 2005: Adaptive capacity and human cognition: the process of individual adaptation to climate change. *Global Environmental Change*, **15**, 199-213

Guha-Sapir, D., D. Hargitt, and P. Hoyois, 2004: *Thirty Years Of Natural Disasters 1974-2003: The Numbers*. Centre for Research on the Epidemiology of Disasters, Presses Universitaires de Louvain, Brussels, Belgium, ISBN 2930344717.

Gunderson, L.H., 2000: Resilience in theory and practice. *Annual Review of Ecology and Systematics*, **31**, 425-439.

Guth, J.L., J.C. Green, L.A. Kellstedt, and C.E. Smith, 1995: Faith and the environment: Religious beliefs and attitudes on environmental policy. *American Journal of Political Science*, **39(2)**, 364-382.

Haddad, B.M., 2005: Ranking the adaptive capacity of nations to climate change when socio-political goals are explicit. *Global Environmental Change*, **15**, 165-176.

Haines, A., R.S. Kovats, D. Campbell-Lendrum, and C. Corvalan, 2006: Climate change and human health: Impacts, vulnerability and public health. *Public Health*, **120**, 585-596

Hales, S., S. Edwards, and R.S. Kovats, 2003: Impacts on health of climate extremes. In: *Climate Change and Human Health: Risks and Responses* [McMichael, A.J., D. Campbell-Lendrum, K. Ebi, and C. Corvalan (eds.)]. World Health Organisation, Geneva, Switzerland, pp. 79-96.

Handmer, J.W., S. Dovers, and T.E. Downing, 1999: Societal vulnerability to climate change and variability. *Mitigation and Adaptation Strategies for Global Change*, **4**, 267-281.

Haque, C.E. and D. Etkin, 2007: People and community as constituent parts of hazards: the significance of societal dimensions in hazards analysis. *Natural Hazards*, **41(2)**, 271-282.

Harding, T.W., F. Romerio, J. Rossiaud, J.J. Wagner, S. Bertrand, C. Frischknecht, and J.D. Laporte, 2001: *Management des risques majeurs: des disciplines à l'interdisciplinarité.* Document de travail No 1 du Groupe de recherche Management des risques majeurs, Université de Genève, Programme plurifacultaire du Rectorat, Geneva, Switzerland.

Hardoy, J. and G. Pandiella, 2009: Urban poverty and vulnerability to climate change in Latin America. *Environment & Urbanization*, **21(1)**, 203-224.

Hellmuth, M.E., S.J. Mason, C. Vaughan, M.K. van Aalst, and R. Choularton (eds.), 2011: *A Better Climate for Disaster Risk Management.* International Research Institute for Climate and Society, Columbia University, New York, NY.

Hewitt, K., 1997: *Regions of Risk: A Geographical Introduction to Disasters*. Addison Wesley Longman, Harlow, UK, 408 pp.

Hewitt, K. (ed.), 1983: *Interpretations of Calamity*, Allen & Unwin, London, UK.

Hewitt, K. and I. Burton, 1971: *The Hazardousness of a Place; a Regional Ecology of Damaging Events*. University of Toronto Press, Toronto, Canada.

Hilhorst, D. 2004: Complexity and diversity: Unlocking social domains of disaster response. In: *Mapping Vulnerability: Disasters, Development and People* [Bankoff, G., G. Frerks, and D. Hilhorst (eds.)]. Earthscan Publishers, London, UK, pp. 52-66.

Holling, C.S., 1973: Resilience and stability of ecological systems. *Annual Reviews of Ecology and Systematics*, **4**, 1-23.

Holling, C.S., 2001: Understanding the complexity of economic, ecological, and social systems. *Ecosystems*, **4**, 390-405.

Hong, Y.C., X.C. Pan, S.Y. Kim, K. Park, E.J. Park, X. Jin, S.M. Yi, K.S. Kim, Y.H. Kim, C.H. Park, S. Song, and H. Kim, 2009: Asian dust storms and pulmonary function of school children. *Epidemiology*, **20(6)**, S121-S121.

Hope, K.R. Sr., 2009: Climate change and poverty in Africa. *International Journal of Sustainable Development and World Ecology*, **16(6)**, 451-461.

Houghton, R., 2009: Domestic violence reporting and disasters in New Zealand. *Regional Development Dialogue*, **30(1)**, 79-90.

Hulme, M., 2009: *Why We Disagree About Climate Change.* Cambridge University Press, Cambridge, UK.

Hutton, D. and C.E. Haque, 2003: Patterns of coping and adaptation among erosion-induced displacees in Bangladesh: Implications for hazard analysis and mitigation. *Natural Hazards*, **29**, 405-421.

IAASTD, 2009: *Agriculture at a Crossroads: International Assessment of Agricultural Knowledge, Science and Technology for Development*. International Assessment of Agricultural Knowledge, Science and Technology for Development, Island Press, Washington, DC.

ICSU-LAC, 2010a: *Science for a better life: Developing regional scientific programs in priority areas for Latin America and the Caribbean. Vol 2, Understanding and Managing Risk Associated with Natural Hazards: An Integrated Scientific Approach in Latin America and the Caribbean* [Cardona, O.D., J.C. Bertoni, A. Gibbs, M. Hermelin, and A. Lavell (eds.)]. ICSU Regional Office for Latin America and the Caribbean, Rio de Janeiro, Brazil.

ICSU-LAC, 2010b: *Entendimiento y gestión del riesgo asociado a las amenazas naturales: Un enfoque científico integral para América Latina y el Caribe. Ciencia para una vida mejor: Desarrollando programas científicos regionales en áreas prioritarias para América Latina y el Caribe. Vol 2* [Cardona, O.D., J.C. Bertoni, A. Gibbs, M. Hermelin, and A. Lavell (eds.)]. ICSU Regional Office for Latin America and the Caribbean, Rio de Janeiro, Brazil.

IDB, 2007: *Disaster Risk Management Policy*. GN-2354-5, Inter-American Development Bank, Washington, DC.

IDEA, 2005: *Indicators of Disaster Risk and Risk Management – Main Technical Report*. English and Spanish edition, National University of Colombia/Manizales, Institute of Environmental Studies/IDEA, Inter-American Development Bank, Washington, DC, 223 pp.

IFRC, 2008: *Vulnerability and Capacity Assessment – Guidelines*. International Federation of Red Cross and Red Crescent Societies, Geneva, Switzerland.

IFRC, 2009: *World Disasters Report 2009 – Focus on Early Warning, Early Action*. International Federation of Red Cross and Red Crescent Societies, Geneva, Switzerland.

IISD, Intercooperation, IUCN, and SEI, 2009: *CRiSTAL: Community-based Risk Screening Tool – Adaptation & Livelihoods, User's Manual, v 4.0*. IISD, Geneva, Switzerland.

INGC, 2009. Synthesis report. INGC Climate Change Report: Study on the impact of climate change on disaster risk in Mozambique [van Logchem, B. and R. Brito R (eds.)]. National Institute for Disaster Management, Mozambique.

Ingham, B., 1993: The meaning of development: interactions between 'new' and 'old' ideas. *World Development*, **21**, 1803-1821.

IOM, 2009: *Migration, Environment and Climate Change: Assessing the Evidence*. International Organization for Migration, Geneva, Switzerland, 441 pp.

IPCC, 2001: *IPCC Third Assessment Report*. Synthesis Report, Cambridge University Press, Cambridge, UK.

IPCC, 2007: *Climate Change 2007. Impacts, Adaptation and Vulnerability. Contribution of Working Group II to the Fourth Assessment Report of the Intergovernmental Panel on Climate Change* [Parry, M.L., O.F. Canziani, J.P. Palutikof, P.J. Van Der Linde, and C.E. Hanson (eds.)]. Cambridge University Press, Cambridge, UK, pp. 7-22.

ISO, 2009a: *Risk Management – Principles and guidelines*. ISO 31000, International Organization for Standardization, Geneva, Switzerland.

ISO, 2009b: *Risk Management – Vocabulary*. ISO Guide 73, International Organization for Standardization, Geneva, Switzerland.

Jabry, A., 2003: *Children in Disasters: After the Cameras Have Gone*. Plan UK, Plan International, Woking, UK, 54 pp.

Jakobsen, K.T., 2009, Views on vulnerability following hurricane Mitch in Nicaragua, IOP Conference Series: *Earth and Environmental Science*, **6**, pp. 1-2.

Janssen, M.A., M.L. Schoon, W. Ke, and K. Börner, 2006: Scholarly networks on resilience, vulnerability and adaptation within the human dimensions of global environmental change. *Global Environmental Change*, **16**(3), 240-252.

Jasanoff, S. and B. Wynne, 1997: *Handbook of Science and Technology Studies*. Sage Publications, London, UK, 828 pp., ISBN 0803940211.

Johnson, P.T.J., A.R. Townsend, C.C. Cleveland, P.M. Glibert, R.W. Howarth, V.J. McKenzie, E. Rejmankova, and M.H. Ward, 2010: Linking environmental nutrient enrichment and disease emergence in humans and wildlife. *Ecological Applications*, **20**(1), 16-29.

Kantor, P. and P. Nair, 2005: Vulnerability among slum dwellers in Lucknow, India – Implications for urban livelihood security. *International Development Planning Review*, **27**(3), 333-358.

Kar, N. 2009: Psychological impact of disasters on children: review of assessment and interventions. *World Journal of Pediatrics*, **5**(1), 5-11.

Karlsson, S., 2007: Allocating responsibilities in multi-level governance for sustainable development. *Journal of Social Economics*, **34**(1-2), 103-126.

Kasperson, R.E. and J.X. Kasperson, 2001: *Climate Change, Vulnerability, and Social Justice*. Risk and Vulnerability Programme, Stockholm Environment Institute, Stockholm, Sweden, pp. 18.

Kasperson, R. and I. Palmund, 2005: Evaluating risk communication. In: *The Social Contours of Risk. Volume I: Publics, Risk Communication & the Social Amplification of Risk* [Kasperson, J. and R. Kasperson (eds.)]. Earthscan, London, UK, pp. 51-67.

Kasperson, R.E., O. Renn, P. Slovic, H.S. Brown, J. Emel, R. Goble, J.X. Kasperson, and S. Ratick, 1988: The social amplification of risk: a conceptual framework. *Risk Analysis*, **8**(2), 177-187.

Kasperson, J., R. Kasperson, B.L. Turner, W. Hsieh, and A. Schiller, 2005: Vulnerability to global environmental change. In: *The Social Contours of Risk. Volume II: Risk Analysis, Corporations & the Globalization of Risk* [Kasperson, J. and R. Kasperson (eds.)]. Earthscan, London, UK, pp. 245-285.

Kates, R.W., 1962: *Hazard and Choice Perception in Flood Plain Management*. Dept. of Geography Research Paper No. 78, University of Chicago, Chicago, IL.

Kates, R.W., 1971: Natural hazard in human ecological perspective: Hypotheses and models. *Economic Geography*, **47**(3), pp. 438-451.

Kates, R.W., 1985: The interaction of climate and society. In: *Climate Impact Assessment* [Kates, R.W., J.H. Ausubel, and M. Berberian (eds.)]. SCOPE 27, Wiley, Chichester, UK.

Keim, M.E., 2008: Building human resilience: the role of public health preparedness and response as an adaptation to climate change. *American Journal of Preventive Medicine*, **35**, 508-516.

Kelman, I. and T.A. Mather, 2008: Living with volcanoes: the sustainable livelihoods approach for volcano-related opportunities. *Journal of Volcanology and Geothermal Research*, **172**(3-4), 189-198.

Kesavan, P.C. and M.S. Swaminathan, 2006: Managing extreme natural disasters in coastal areas. *Philosophical Transactions of The Royal Society A*, **364 (1845)**, 2191-2216.

Khandlhela, M. and J. May, 2006: Poverty, vulnerability and the impact of flooding in the Limpopo Province, South Africa. *Natural Hazards*, **39**(2), 275-287.

Kintisch, E., 2006: Evangelicals, scientists reach common ground on climate change. *Science*, **311**(5764), 1082-1083.

Klein, R.J.T., 2001: *Adaptation to Climate Change in German Official Development Assistance: An Inventory of Activities and Opportunities, with a Special Focus on Africa*. GTZ, Eschborn, Germany.

Klein, R.J.T., M. Alam, I. Burton, W.W. Dougherty, K.L. Ebi, M. Fernandes, A. Huber-Lee, A.A. Rahman, and C. Swartz, 2006: *Application of Environmentally Sound Technologies For Adaptation To Climate Change*. Technical Paper FCCC/TP/2006/2, United Nations Framework Convention on Climate Change Secretariat, Bonn, Germany.

Klein, R., S. Eriksen, L.O. Naess, A. Hammill, C. Robledo, T.M. Tanner, and K. O'Brien, 2007: Portfolio screening to support the mainstreaming of adaptation to climate change into development assistance. *Climatic Change*, **84**, 23-44.

Kobori, H., 2009: Current trends in conservation education in Japan. *Biological Conservation*, **142**(9), 1950-1957.

Koenig, D., 2009: Urban relocation and resettlement: Distinctive problems, distinctive opportunities. In: *Development and Dispossession: The Crisis of Development Forced Displacement and Resettlement* [Oliver-Smith, A. (ed.)]. SAR Press, Santa Fe, NM, pp. 119-139.

Krasny, M.E. and K.G. Tidball, 2009: Applying a resilience systems framework to urban environmental education. *Environmental Education Research*, **15**(4), 465-482.

Kuhar, C.W., T.L. Bettinger, K. Lehnhardt, O. Tracy, and D. Cox, 2010: Evaluating for long-term impact of an environmental education program at the Kalinzu Forest Reserve, Uganda. *American Journal Of Primatology*, **72**(5), 407-413.

Kuwali, D., 2008, From the West to the rest: Climate change as a challenge to human security in Africa. *African Security Review*, **17 (3)**, 18-38.

Lamond, J.E., D.G. Proverbs, and F.N. Hammond, 2009: Accessibility of flood risk insurance in the UK: confusion, competition and complacency. *Journal of Risk Research*, **12(6)**, 825-841.

Landry, C.E., O. Bin, P. Hindsley, J.C. Whitehead, and K. Wilson, 2007: *Going Home: Evacuation-Migration decisions of Hurricane Katrina Survivors*. Center for Natural Hazards Research Working Paper, Natural Hazards Center, University of Colorado at Boulder, Boulder, CO.

Lavell, A. (ed.), 1994: *Viviendo en riesgo: comunidades vulnerables y prevención de desastres en América Latina*. LA RED, Tercer Mundo Editores, Bogotá, Colombia.

Lavell, A., 1996: Degradación ambiental, riesgo y desastre urbano. Problemas y conceptos: hacia la definición de una agenda de investigación. In: *Ciudades en Riesgo* [Fernandez, M.A. (ed.)]. La RED-USAID, Lima, Peru, pp. 21-59.

Lavell, A., 1999a: Environmental degradation, risks and urban disasters. issues and concepts: Towards the definition of a research agenda. In: *Cities at Risk: Environmental Degradation, Urban Risks and Disasters in Latin America* [Fernandez, M.A. (ed.)]. A/H Editorial, La RED, US AID, Quito, Ecuador, pp. 19-58.

Lavell, A., 1999b: *Natural and Technological Disasters: Capacity Building and Human Resource Development for Disaster Management*. Concept Paper commissioned by Emergency Response Division, United Nations Development Program, Geneva Switzerland.

Lavell, A., 2003: *Local Level Risk Management: Concept and Practices*. CEPREDENAC-UNDP, Quito, Ecuador.

Lavell A. and E. Franco (eds.), 1996: *Estado, Sociedad y Gestión de los Desastres en América Latina*. Red de Estudios Sociales en Prevención de Desastres en América Latina, La RED, Tercer Mundo Editores, Bogotá, Colombia.

Leichenko, R.M. and K.L. O'Brien, 2002: The dynamics of rural vulnerability to global change: the case of southern Africa. *Mitigation and Adaptation Strategies for Global Change*, **7**, 1-18.

Leichenko, R.M. and K.L. O'Brien, 2008: *Environmental Change and Globalization, Double Exposures*. Oxford University Press, New York, NY.

Leiserowitz, A., 2006: Climate change risk perception and policy preferences: the role of affect, imagery, and values. *Climatic Change*, **77**, 45-72

Levina, E. and D. Tirpak, 2006: *Adaptation to Climate Change: Key Terms*. COM/ENV/EPOC/IEA/SLT(2006)1, OECD, Paris, France.

Linnerooth-Bayer, J. and R. Mechler, 2007: Disaster safety nets for developing countries: Extending public-private partnerships. *Environmental Hazards*, **7**, 54-61.

Linnerooth-Bayer, J., R, Mechler, and G. Pflug, 2005: Refocusing disaster aid. *Science*, **309**, 1044-1046.

Liverman, D.M., 1990: Vulnerability to global environmental change. In: *Understanding Global Environmental Change: The Contributions of Risk Analysis and Management* [Kasperson, R.E., K. Dow, D. Golding, and J.X. Kasperson (eds.)]. Clark University, Worcester, MA, pp. 27-44.

Longstaff, P.H. and S.-U. Yang, 2008: Communication management and trust: Their role in building resilience to "surprises" such as natural disasters, pandemic flu, and terrorism. *Ecology and Society*, **13 (1)**, 3.

Lopez-Calva, L.F. and E. Ortiz, 2008: *Evidence and Policy Lessons on the Link between Disaster Risk and Poverty in Latin America: Summary of Regional Studies*. RPP LAC – MDGs and Poverty – 10/2008, RBLAC-UNDP, New York, NY, USA.

Luers, A.L., D.B. Lobell, L.S. Sklar, C.L. Addams, and P.A. Matson, 2003: A method for quantifying vulnerability, applied to the Yaqui Valley, Mexico. *Global Environmental Change*, **13**, 255-267.

Luhmann, N., 2003: *Soziologie des Risikos*. Walter De Gruyterl, Berlin, Germany.

Ljung, K., F. Maley, A. Cook, and P. Weinstein, 2009: Acid sulfate soils and human health – A Millennium Ecosystem Assessment. *Environment International*, **35(8)**, 1234-1242.

Madagascar Development Intervention Fund, 2007: Disaster-Resistant Schools, a Tool for Universal Primary Education. UNISDR, Geneva, Switzerland.

Magnan, A., 2010: For a better understanding of adaptive capacity to climate change: a research framework. *IDDRI Analysis*, **2/10**, IDDRI, Paris, France.

Mansilla, E. (ed.), 1996: *Desastres: modelo para armar*. La RED, Lima, Peru.

Manyena, S.B. 2006: The concept of resilience revisited. *Disasters*, **30(4)**, 433-450.

Marmot, M., 2010: *Marmot Review final report, Fair Society, Healthy Lives*. University College London, London, UK.

Marris, C., I. Langford, T. Saunderson, and T. O'Riordan, 1998, A quantitative test of the cultural theory of risk perceptions: Comparison with the psychometric paradigm. *Risk Analysis*, **18(5)**, 635-647.

Marshall B.K. and J.S. Picou, 2008: Postnormal science, precautionary principle, and worst cases: The challenge of twenty-first century catastrophes. *Sociological Enquiry*, **78(2)**, 230-247.

Marulanda, M.C., O.D. Cardona, M.G. Ordaz, and A.H. Barbat, 2008a: *La gestión financiera del riesgo desde la perspectiva de los desastres: Evaluación de la exposición fiscal de los Estados y alternativas de instrumentos financieros de retención y transferencia del riesgo*. Monografía CIMNE IS-61, Universidad Politécnica de Cataluña, Barcelona, Spain.

Marulanda, M.C., O.D. Cardona, and A.H. Barbat, 2008b: The economic and social effects of small disasters: Revision of the Local Disaster Index and the case study of Colombia. In: *Megacities: Resilience and Social Vulnerability* [Bohle, H.G. and K. Warner (eds.)]. SOURCE No. 10, United Nations University (EHS), Munich Re Foundation, Bonn, Germany.

Marulanda, M.C., O.D. Cardona, and A.H. Barbat, 2010: Revealing the socioeconomic impact of small disasters in Colombia using the DesInventar database. *Disasters*, **34(2)**, 552-570.

Marulanda, M.C., O.D. Cardona, and A.H. Barbat, 2011: Revealing the impact of small disasters to the economic and social development. In: *Coping with Global Environmental Change, Disasters and Security - Threats, Challenges, Vulnerabilities and Risks* [Brauch, H.G., U. Oswald Spring, C. Mesjasz, J. Grin, P. Kameri-Mbote, B. Chourou, P. Dunay, and J. Birkmann (eds.)]. Springer-Verlag, Berlin, Germany.

Masika, R., 2002: Editorial. *Gender & Development*, **10(2)**, 2-9.

Maskrey, A., 1989: *Disaster Mitigation: A Community Based Approach*. Oxfam, Oxford, UK.

Maskrey, A. (Comp.), 1993a: *Los Desastres No son Naturales*. Red de Estudios Sociales en Prevención de Desastres en América Latina, LA RED, Tercer Mundo Editores, La RED, Bogotá, Colombia.

Maskrey, A., 1993b: Vulnerability accumulation in peripheral regions in Latin America: The challenge for disaster prevention and management. In: *Natural Disasters: Protecting Vulnerable Communities* [Merriman, P.A. and C.W. Browitt (eds.)]. IDNDR, Telford, London, UK, pp. 461-472.

Maskrey, A., 1994: Disaster mitigation as a crisis paradigm: Reconstructing after the Alto Mayo Earthquake, Peru. In: *Disaster, Development and Environment* [Varley, A. (ed.)]. John Wiley & Sons, Chichester, UK, pp. 109-123.

Maskrey, A. (ed.), 1998: *Navegando entre Brumas: La Aplicación de los Sistemas de Información Geográfica al Análisis de Riesgo en América Latina*. LA RED, ITDG, Lima, Peru.

Massey, K. and J. Sutton, 2007: Faith community's role in responding to disasters. *Southern Medical Journal*, **100(9)**, 944-945, doi:10.1097/SMJ.0b013e318145a847.

McCarthy, J.J., O.F. Canziani, N.A. Leary, D.J. Dokken, and K.S. White (eds.), 2001: *Climate Change 2001: Impacts, Adaptation, and Vulnerability*. Working Group II of the Intergovernmental Panel on Climate Change, Cambridge University Press, Cambridge, UK.

McConnell, W.J., 2008: Comanagement of natural resources: Local learning for poverty reduction. *Society & Natural Resources*, **21(3)**, 273-275.

McGranahan, G., D. Balk, and B. Anderson, 2007: The rising tide: assessing the risks of climate change and human settlements in low elevation coastal zones. *Environment and Urbanization*, **19(1)**, 17-37, doi:10.1177/0956247807076960.

McGuire, L.C., E.S. Ford, and C.A. Okoro, 2007: Natural disasters and older US adults with disabilities: implications for evacuation. *Disasters*, **31 (1)**, 49-56.

McIvor, D. and D. Paton, 2007: Preparing for natural hazards: normative and attitudinal influences. *Disaster Prevention and Management*, **16(1)**, 79-88.

McLaughlin, P., and T. Dietz, 2008: Structure, agency and environment: toward an integrated perspective on vulnerability. *Global Environmental Change*, **18**, 99-111.

McLeman, R. and B. Smit, 2006: Vulnerability to climate change hazards and risks: crop and flood insurance. *Canadian Geographer*, **50(2)**, 217-226.

McMichael, D.H., C.F. Campbell-Lendrum, K. Corvalan, L. Ebi, A. Githeki, J.D.Scheraga, and A. Woodward, 2003: *Climate Change and Human Health: Risks And Responses*. WMO, Geneva, Switzerland.

MEA, 2005: *Millennium Ecosystem Assessment, Ecosystems and Human Well-being: Synthesis*. Island Press, Washington, DC.

Mearns, R. and A. Norton, 2010: *The Social Dimensions of Climate Change: Equity and Vulnerability in a Warming World*. New Frontiers of Social Policy Series, World Bank, Washington, DC, ISBN 978-0-8213-7887-8.

Mechler, R., 2004: *Natural Disaster Risk Management and Financing Disaster Losses in Developing Countries*. Verlag für Versicherungswirtschaft, Karlsruhe, Germany.

Mechler, R., J. Linnerooth-Bayer, S. Hochrainer, and G. Pflug, 2006: Assessing financial vulnerability and coping capacity: The IIASA CATSIM Model. In: *Measuring Vulnerability and Coping Capacity to Hazards of Natural Origin. Concepts and Methods* [Birkmann, J. (ed.)]. United Nations University Press, Tokyo, Japan.

Mechler, R., S. Hochrainer, A. Aaheim, H. Salen and A. Wreford, 2010: Modelling economic impacts and adaptation to extreme events: Insights from European case studies. *Mitigation and Adaptation Strategies for Global Change*, **15(7)**, 737-762.

Mercer J., I. Kelman, L. Taranis, and S. Suchet-Pearson, 2010: Framework for integrating indigenous and scientific knowledge for disaster risk reduction. *Disasters*, **34(1)**, 214-239.

Metzger, M.J., M.D.A. Rounsevell, L. Acosta-Michlik, R. Leemans, and D. Schröter, 2006: The vulnerability of ecosystem services to land use change. *Agriculture, Ecosystems and Environment*, **114**, 69-85.

Mileti, D.S., 1996: Psicología social de las alertas públicas efectivas de desastres. In: *Desastres y Sociedad, Especial: Predicciones, Pronósticos, Alertas y Respuestas Sociales* [Cardona, O.D. (ed.)], **4**, La RED, Tarea Gráfica, Lima, Peru.

Mimura, N.L., L. Nurse, R.F. McLean, J. Agard, L. Briguglio, P. Lefale, R. Payet, and G. Sem, 2007: Small islands. In: *Climate Change 2007. Impacts, Adaptation and Vulnerability. Contribution of Working Group II to the Fourth Assessment Report of the Intergovernmental Panel on Climate Change* [Parry, M.L., O.F. Canziani, J.P. Palutikof, P.J. Van Der Linde, and C.E. Hanson (eds.)]. Cambridge University Press, Cambridge, UK, pp. 687-716.

Mitchell, J.K. (ed.), 1999a: *Crucibles of Hazards: Megacities and Disasters in Transition*. United Nations University Press, Tokyo, Japan.

Mitchell, J.K., 1999b: Megacities and natural disasters: A comparative analysis. *GeoJournal*, **49(2)**, 137-142.

Mitchell, T., and M.K. van Aalst, 2008: *Convergence of Disaster Risk Reduction and Climate Change Adaptation: A Review for DFID*. Department for International Development (DFID), London, UK.

Mitchell, T., M. van Aalst, and P.S. Villanueva, 2010: *Assessing Progress on Integrating Disaster Risk Reduction and Climate Change Adaptation in Development Processes*. Strengthening Climate Resilience Discussion Papers, 2, Strengthening Climate Resilience, Institute of Development Studies, Brighton, UK.

Moench, M. and A. Dixit (eds.), 2007: *Working with the Winds of Change: Towards Strategies for Responding to the Risks Associated with Climate Change and Other Hazards*. ISET, Kathmandu, Nepal, 285 pp.

Mohay, H. and N. Forbes, 2009: Reducing the risk of posttraumatic stress disorder in children following natural disasters. *Australian Journal of Guidance and Counselling*, **19(2)**, 179-195.

Morrow, B.H., 1999: Identifying and mapping community vulnerability. *Disasters*, **23(1)**, 1-18.

Morton, T.A., A. Rabinovich, D. Marshall, and P. Bretschneider, 2011: The future that may (or may not) come: How framing changes responses to uncertainty in climate change communications. *Global Environmental Change*, **21**, 103-109.

Moss, R.H., A.L. Brenkert, and E.L. Malone, 2001: *Vulnerability to Climate Change: A Quantitative Approach*. Technical Report PNNL-SA-33642, Pacific Northwest National Laboratories, Richland, WA.

MOVE, 2010: *Generic Conceptual Framework for Vulnerability Measurement*. Seven Framework Programme, Methods for the Improvement of Vulnerability Assessment in Europe, European Commission, Brussels, Belgium.

Munda, G., 2000: Multicriteria Methods and Process for Integrated Environmental Assessment. In: *Métodos Numéricos en Ciencias Sociales (MENCIS 2000)* [Oñate, E., F. García-Sicilia, and L. Ramallo (eds.)]. Centro Internacional de Métodos Numéricos en Ingeniería - Universidad Politécnica de Cataluña, Barcelona, Spain, pp. 364-375, ISBN 84-89925-71-2.

Munich Re, 2011: *TOPICS GEO, Natural Catastrophes 2010, Analyses, Assessments, Positions*. Munich Re, Munich, Germany.

Myers, C.A., T. Slack, and J. Singelmann, 2008: Social vulnerability and migration in the wake of disaster: the case of Hurricanes Katrina and Rita. *Population & Environment*, **29(6)**, 271-291.

Myers, N., 1993: Environmental refugees in a globally warmed world. *Bioscience*, **43(11)**, 752-761.

Næss, L., O. Bang, S. Eriksen, and J. Vevatne, 2005, Institutional adaptation to climate change: Flood responses at the municipal level in Norway. *Global Environmental Change*, **15**, 125-138, doi:10.1016/j.gloenvcha.2004.10.003.

Nakagawa, Y. and R. Shaw, 2004: Social capital: A missing link to disaster recovery. *International Journal of Mass Emergencies and Disasters*, **22**, 5-34.

Nathan, F., 2008: Risk perception, risk management and vulnerability to landslides in the hill slopes in the city of La Paz, Bolivia. *Disasters*, **32 (3)**, 337-357.

Neal, D.M. and B.D. Phillips, 1990: Female-dominated local movement organizations in disaster-threat situations. In: *Women and Social Protest* [West, G. and R.L. Blumberg (eds.)]. Oxford University Press, New York, NY, pp. 243-255.

Nelson, D.R. and T.J. Finan, 2009: Praying for drought: Persistent vulnerability and the politics of patronage in Ceara, Northeast Brazil. *American Anthropologist*, **111**, 302-316, doi:10.1111/j.1548-1433.2009.01134.x.

Nelson, R., P. Kokic, S. Crimp, H. Meinke, and S.M. Howden, 2010a: The vulnerability of Australian rural communities to climate variability and change, Pt. I: Conceptualising and measuring vulnerability. *Environmental Science & Policy*, **13**, 8-17.

Nelson, R., P. Kokic, S. Crimp, P. Martin, H. Meinke, S.M. Howden, P. de Voil, and U. Nidumolu, 2010b: The vulnerability of Australian rural communities to climate variability and change, Part II: Integrating impacts with adaptive capacity. *Environmental Science & Policy*, **13**, 18-27.

Neumayer, E. and T. Pluemper, 2007: The gendered nature of natural disasters: The impact of catastrophic events on the gender gap in life expectancy, 1981–2002. *Annals of the Association of American Geographers*, **97(3)**, 551-566

Nicholls, R.J., 2004: Coastal flooding and wetland loss in the 21st century: changes under the SRES climate and socio-economic scenarios. *Global Environmental Change*, **14(1)**, 69-86.

Nicholls, R.J., N. Marinova, J. Lowe, S. Brown, P. Vellinga, D. De Gusmao, J. Hinkel, and R.S. Tol, 2011: Sea-level rise and its possible impacts given a 'beyond 4 degrees C world' in the twenty-first century. *Philosophical Transactions of The Royal Society A*, **369(1934)**, 161-181.

Nielsen. J.O. and A. Reenberg, 2010: Cultural barriers to climate change adaptation: A case study from Northern Burkina Faso. *Global Environmental Change*, **20**, 142-152.

Nomura, K., 2009: A perspective on education for sustainable development: Historical development of environmental education in Indonesia. *International Journal of Educational Development*, **29(6)**, 621-627.

Nurse, L. and G. Sem, 2001: Small island states. In: *Climate Change 2001: Impacts, Adaptation & Vulnerability* [McCarthy, J., O. Canziani, N. Leary, D. Dokken, and K. White (eds.)]. Cambridge University Press, Cambridge, UK, pp. 842-875.

O'Brien, G., P. O'Keefe, H. Meena, J. Rose, and L. Wilson, 2008: Climate adaptation from a poverty perspective. *Climate Policy*, **8(2)**, 194-201.

O'Brien, K., S. Eriksen, A. Schjolen, and L. Nygaard, 2004a: *What's in a word? Conflicting interpretations of vulnerability in climate change research*. CICEROWorking Paper 2004:04, CICERO, Oslo University, Oslo, Norway.

O'Brien, K, R. Leichenko, U. Kelkar, H. Venema, G. Aandahl, H. Tompkins, A. Javed, S. Bhadwal, S. Barg, L. Nygaard, and J. West, 2004b: Mapping vulnerability to multiple stressors: climate change and globalization in India. *Global Environmental Change*, **14**, 303-313.

O'Brien, K., S. Eriksen, L. Sygna, and L.O. Naess, 2006: Questioning complacency: Climate change impacts, vulnerability, and adaptation in Norway. *Ambio*, **35(2)**, 50-56.

O'Brien, K., L. Sygna, R. Leinchenko, W.N. Adger, J. Barnett, T. Mitchell, L. Schipper, T. Tanner, C. Vogel, and C. Mortreux, 2008: *Disaster Risk Reduction, Climate Change Adaptation and Human Security.* GECHS Report 2008:3, Global Environmental Change and Human Security, Oslo, Norway.

O'Keefe, P., K. Westgate, and B. Wisner, 1976: Taking the naturalness out of natural disasters. *Nature*, **260**, 566-567.

OAS, 1991: *Primer on Natural Hazard Management in Integrated Regional Development Planning*. Organization of American States, Washington, DC.

OCHA/IDMC, 2009: *Monitoring Disaster Displacement in the Context of Climate Change*. Findings of a study by the United Nations Office for the Coordination of Humanitarian Affairs and the Internal Displacement Monitoring Centre, Geneva, Switzerland.

Ochola, S.O., B. Eitel, and D.O. Olago, 2010: Vulnerability of schools to floods in Nyando River catchment, Kenya. *Disasters*, **34(3)**, 732-754.

Olausson, U., 2009: Global warming – global responsibility? Media frames of collective action and scientific certainty. *Public Understanding of Science*, **18**, 421-436

Oliver-Smith, A., 2006: Communities after catastrophe: Reconstructing the material, reconstituting the social. In: *Community Building in the 21st Century* [Hyland, S. (ed.)]. School of American Research Press, Santa Fe, NM, pp. 45-70.

Oliver-Smith, A., 2009: Disasters and diasporas: Global climate change and population displacement in the 21st century. In: *Anthropology and Climate Change: From Encounters to Actions* [Crate, S.A. and M. Nuttall (eds.)]. Left Coast Press, Walnut Creek, CA, pp. 116-138.

Oppenheimer, M. and A. Todorov (eds.), 2006: Global warming: the psychology of long term risk. *Climatic Change*, **77(1-6)**.

Orr, M., 2003: Environmental decline and the rise of religion. *Zygon*, **38(4)**, 895-910.

Paavola, J., 2008: Livelihoods, vulnerability and adaptation to climate change in Morogoro, Tanzania. *Environmental Science & Policy*, **11**, 642-654.

PAHO/World Bank, 2004: *Guidelines for Vulnerability Reduction in the Design of New Health Facilities.* Prepared by Rubén Boroschek Krauskopf and Rodrigo Retamales Saavedra, PAHO/World Bank in collaboration with the World Health Organization, Washington, DC, 106 pp.

Pantuliano, S. and M. Wekesa, 2008: *Improving drought response in pastoral areas of Ethiopia: Somali and Afar Regions and Borena Zone of Oromiya Region*. Prepared for the CORE group (CARE, FAO, Save the Children UK and Save the Children US), Overseas Development Institute, Humanitarian Policy Group, London, UK.

Parker, D.J., 1995: Floodplain development policy in England and Wales. *Applied Geography*, **15(4)**, 341-363.

Patt, A., R. Klein, and A. Vega-Leinert, 2005: Taking the uncertainty in climate-change vulnerability seriously. *Geoscience*, **337**, 411-424.

Patt, A.G., D. Schröter, R.L.T. Klein, and A.C. Vega-Leinert, 2009: *Assessing Vulnerability to Global Environmental Change*. Earthscan Publications, London, UK.

Patterson, J., E. Linden, J.K.P. Edward, D. Wilhelmsson, and I. Lofgren, 2009: Community-based environmental education in the fishing villages of Tuticorin and its role in conservation of the environment. *Australian Journal of Adult Learning*, **49(2)**, 382-393.

Peacock, W.G., B.H. Morrow, and H. Gladwin (eds.), 1997: *Hurricane Andrew: Ethnicity, Gender and the Sociology of Disasters*. Routledge, London, UK.

Peduzzi, P., 2006: The Disaster Risk Index: Overview of a quantitative approach. In: *Measuring Vulnerability to Natural Hazards - Towards Disaster Resilient Societies* [Birkmann, J. (ed.)]. United Nations University Press, Tokyo, Japan, pp.171-181.

Peduzzi, P., H. Dao, C. Herold, and F. Mouton, 2009: Assessing global exposure and vulnerability towards natural hazards: the Disaster Risk Index. *Natural Hazards and the Earth System Science*, **9**, 1149-1159.

Peek, L. and L.M. Stough, 2010: Children with disabilities in the context of disaster: A social vulnerability perspective. *Child Development*, **81(4)**, 1260-1270.

Peguero, A.A., 2006: Latino disaster vulnerability – The dissemination of hurricane mitigation information among Florida's homeowners. *Hispanic Journal of Behavioral Sciences*, **28 (1)**, 5-22.

Pelling, M., 1997: What determines vulnerability to floods: a case study in Georgetown, Guyana. *Environment and Urbanization*, **9**, 203-226.

Pelling, M., 1998: Participation, social capital and vulnerability to urban flooding in Guyana. *Journal of International Development*, **10**, 469-486.

Pelling, M., 2003: *The Vulnerability of Cities: Natural Disasters and Social Resilience*. Earthscan Publications, London, UK.

Pelling, M., 2010: *Adaptation to Climate Change: From Resilience to Transformation*. Routledge, London, UK.

Pelling, M. and J.I. Uitto, 2001: Small island developing states: natural disaster vulnerability and global change. *Environmental Hazards*, **3**, 49-62.

Pelling, M. and B. Wisner (eds.), 2009a: *Disaster Risk Reduction, Cases From Urban Africa*. Earthscan, London, UK.

Pelling, M. and B. Wisner, 2009b: Introduction: Urbanization, human security and disaster risk in Africa. In: *Disaster Risk Reduction, cases from urban Africa* [Pelling, M. and B. Wisner (eds.)]. Earthscan, London, UK, pp. 1-16.

Pielke, Jr., R.A. and C.W. Landsea, 1998: Normalized hurricane damages in the United States: 1925-1995. *Weather and Forecasting*, **13**, 621-631.

Pierre, J. and B.G. Peters, 2000: *Governance, Politics, and the State*. St. Martin's Press, New York, NY.

Pincha, C. and N.H. Krishna, 2009: Post-disaster death ex gratia payments and their gendered impact. *Regional Development Dialogue*, **30(1)**, 95-105.

Plummer, R., 2006: Sharing the management of a river corridor: A case study of the comanagement process. *Society & Natural Resources*, **19(8)**, 709-721.

Portes, A., 1998: Social capital: its origins and applications in modern sociology. *Annual Review of Sociology*, **24**, 1-24.

Pulwarty, R.S. and W.E. Riebsame, 1997: The political ecology of vulnerability to hurricane-related hazards. In: *Hurricanes: Climate and Socio-Economic Impacts* [Diaz, H.F. and R.S. Pulwarty (eds.)]. Springer, Heidelberg, Germany, pp. 185-214.

Putnam, R.D., 2000: *Bowling Alone: The Collapse and Revival of American Community*. Simon & Schuster, New York, NY.

Quarantelli, E.L., 1998: *What is a Disaster?* Routledge, New York, NY.

Rafiee, R., A.S. Mahiny, and N. Khorasani, 2009: Assessment of changes in urban green spaces of Mashad city using satellite data. *International Journal of Applied Earth Observation and Geoinformation*, **11(6)**, 431-438.

Rahman, M.M., G. Graham Haughton, and A.E.G. Jonas, 2010: The challenges of local environmental problems facing the urban poor in Chittagong, Bangladesh: a scale-sensitive analysis. *Environment and Urbanization*, **22(2)**, 561-578, doi:10.1177/0956247810377560.

Ray-Bennett, N.S., 2009: The influence of caste, class and gender in surviving multiple disasters: A case study from Orissa, India. *Environmental Hazards-Human and Policy Dimensions*, **8(1)**, 5-22.

Reacher, M., K. McKenzie, C. Lane, T. Nichols, I. Kedge, A. Iversen, P. Hepple, T. Walter, C. Laxton, and J. Simpson, on behalf of the Lewes Flood Action Recovery Team, 2004: Health impacts of flooding in Lewes: a comparison of reported gastrointestinal and other illness and mental health in flooded and non-flooded households. *Communicable Disease and Public Health*, **7(1)**, 39-46.

Renaud, F.G., 2006: Environmental components of vulnerability. In: *Measuring Vulnerability to Natural Hazards. Towards Disaster Resilient societies* [Birkmann, J. (ed.)]. United Nations University Press, Tokyo, Japan, pp. 117-127.

Renn, O., 1992: Concepts of risk: A classification. In: *Social Theories of Risk* [Krimsky, S. and D. Golding (eds.)]. Praeger, Westport, CT, pp. 53-79.

Renn, O., 2008: *Risk Governance – Coping with Uncertainty in a Complex World.* Earthscan, London, UK.

Renn, O. and P. Graham, 2006: *Risk Governance – Towards an integrative approach*. White paper no. 1, International Risk Governance Council, Geneva, Switzerland.

Ribot, J., 1995: The causal structure of vulnerability and its application to climate impact analysis. *GeoJournal*, **35(2)**, 119-122.

Ribot, J.C., 1996: Introduction: Climate variability, climate change and vulnerability: Moving forward by looking back. In: *Climate Variability: Climate Change and Social Vulnerability in the Semi-Arid Tropics* [Ribot, J.C., A.R. Magalhaes, and S.S. Panagides (eds.)]. Cambridge University Press, Cambridge, UK.

Rideout V., 2003: Digital inequalities in eastern Canada. *Canadian Journal of Information and Library Science*, **27(2)**, 3-31.

Rose, A., 2004: Economic principles, issues, and research priorities in hazard loss estimation. In: *Modeling Spatial and Economic Impacts of Disasters* [Okuyama, Y. and S.E. Chang (eds.)]. Springer, New York, NY.

Ruth, M. and D. Coelho, 2007: Understanding and managing the complexity of urban systems under climate change. *Climate Policy*, **7**(4), 317-336.

Saldaña-Zorrilla, S.R., 2007: *Socioeconomic vulnerability to natural disasters in Mexico: rural poor, trade and public response.* CEPAL Report 92, UN-ECLAC, Disaster Evaluation Unit, Mexico, ISBN 978-92-1-121661-5.

Sampei, Y. and M. Aoyagi-Usui, 2008: Mass-media coverage, its influence on public awareness of climate-change issues, and implications for Japan's national campaign to reduce greenhouse gas emissions. *Global Environmental Change*, **19**, 203-212.

Sánchez-Rodríguez, R., K.C. Seto, D. Simon, W.D. Solecki, F. Kraas, and G. Laumann, 2005: *Science Plan Urbanization and Global Environmental Change*. IHDP Report 15, International Human Dimensions Programme on Global Environmental Change, Bonn, Germany.

Sandstrom, C., 2009: institutional dimensions of comanagement: Participation, power, and process. *Society & Natural Resources*, **22**(3), 230-244.

Sanli, F.B., F.B. Balcik, and C. Goksel, 2008: Defining temporal spatial patterns of mega city Istanbul to see the impacts of increasing population. *Environmental Monitoring and Assessment*, **146**(1-3), 267-275.

Schipper, E.L.F, 2010: Religion as an integral part of determining and reducing climate change and disaster risk: an agenda for research. In: *Climate Change: The Social Science Perspective* [Voss, M. (ed.)]. VS-Verlag, Wiesbaden, Germany, pp. 377-393.

Schipper, L. and M. Pelling, 2006: Disaster risk, climate change and international development: Scope for, and challenges to, integration. *Disasters*, **30**(1), 19-38.

Schmuck-Widmann, H., 2000: *Wissenskulturen im Vergleich. Bäuerliche und ingenieurwissenschaftliche Wahrnehmungen und Strategien zur Bewältigung der Flut in Bangladesh.* PhD Thesis, Free University Berlin, Germany. (also published in English as Schmuck-Widmann, H., 2001: *Facing the Jamuna River. Indigenous and engineering knowledge in Bangladesh.* Bangladesh Resource Centre for Indigenous Knowledge in Dhaka, Bangladesh, 242 pp.).

Schneider, S.H., S. Semenov, A. Patwardhan, I. Burton, C.H.D. Magadza, M. Oppenheimer, A.B. Pittock, A. Rahman, J.B. Smith, A. Suarez, and F. Yamin, 2007: Assessing key vulnerabilities and the risk from climate change. In: *Climate Change 2007. Impacts, Adaptation and Vulnerability. Contribution of Working Group II to the Fourth Assessment Report of the Intergovernmental Panel on Climate Change* [Parry, M.L., O.F. Canziani, J.P. Palutikof, P.J. Van Der Linde, and C.E. Hanson (eds.)]. Cambridge University Press, Cambridge, UK, pp. 779-810.

Schon, D.A. and M. Rein, 1994: Frame Reflection: Toward the Resolution of Intractable Policy Controversies. BasicBooks, New York, NY, USA, 247 pp. ISBN 0465025064.

Schröter, D., C. Polsky, and A.G. Patt, 2005: Assessing vulnerabilities to the effects of global change: an eight step approach. *Mitigation and Adaptation Strategies for Global Change*, **10**, 573-595.

Sclar, E.D., P. Garau, and G. Carolini, 2005: The 21st Century health challenge of slums and cities. *Lancet*, **365**, 901-903.

Scoones, I., 1998: *Sustainable Rural Livelihoods: A Framework For Analysis*. IDS Working Paper 72, Brighton, UK, ISBN 1-85964-224-8.

Scudder, T., 2005: *The Future of Large Dams: Dealing with Social, Environmental, Institutional and Political Costs.* Earthscan, London, UK, 408 pp.

Sen, A., 1981: *Poverty and Famines: An Essay on Entitlements and Deprivation.* Clarendon Press, Oxford, UK.

Setiadi, N., 2011: Daily mobility – excursus Padang, Indonesia. In: *Early Warning in the Context of Environmental Shocks: Demographic Change, Dynamic Exposure to Hazards, and the Role of EWS in Migration Flows and Human Displacement* [Chang Seng, D. and J. Birkmann (eds.)]. Migration and Global Environmental Change, Foresight SR4b, Vulnerability Assessment, Risk Management and Adaptive Planning Section, United Nations University for Environment and Human Security, Government Office for Science, pp. 41-45.

Shah, H.C., 1995: The increasing nature of global earthquake risk. *Global Environmental Change*, **5**(l), 65-67.

Sharma, U. and A. Patwardhan, 2008: An empirical approach to assessing generic adaptive capacity to tropical cyclone risk in coastal districts of India. *Mitigation and Adaptation Strategies to Global Change*, **13**, 819-831.

Sheldon, K. and D. Golding (eds.), 1992: *Social Theories of Risk.* Praeger, Westport, CT.

Skelton, J.A., S.R. Cook, P. Auinger, J.D. Klein, and S.E. Barlow, 2009: Prevalence and trends of severe obesity among U.S. children and adolescents. *Academic Pediatrics*, **9**(5), 322-329.

Slovic, P., 2000: *The Perception of Risk*. Earthscan, London, UK.

Smit, B. and J. Wandel, 2006: Adaptation, adaptive capacity and vulnerability. *Global Environmental Change*, **16**, 282-292.

Smithers, J. and B. Smit, 1997: Human adaptation to climatic variability and change. *Global Environmental Change*, **7**, 129-146.

Spence, P.R., K.A. Lachlan, and D.R. Griffin, 2007: Crisis communication, race, and natural disasters. *Journal of Black Studies*, **37**(4), 539-554.

Spence, R.J.S. and A.W. Coburn, 1987: Earthquake Protection: An Task for the 1990s. *The Structural Engineer*, **65A**(8), 290-296

Sphere Project, 2011: *The Sphere Project, Humanitarian Charter and Minimum Standards in Humanitarian Response*. Distributed for the Sphere Project by Practical Action Publishing, Lippincott Williams & Wilkins, Philadelphia, PA, USA, ISBN 978-1-908176-00-4.

Stafoggia M., F. Forastiere, P. Michelozzi, and C.A. Perucci, 2006: Vulnerability to heat-related mortality: a multicity, population-based, case-crossover analysis. *Epidemiology*, **17**, 315-323.

Stamatakis, E., J. Wardle, and T.J. Cole, 2010: Childhood obesity and overweight prevalence trends in England: evidence for growing socioeconomic disparities. *International Journal of Obesity*, **34**(1), 41-47.

Stockdale, T.N., O. Alves, G. Boer, M. Deque, Y. Ding, K. Kumar, W. Landman, S.J. Mason, P. Nobre, A. Scaife, O. Tomoaki, and W.-T. Yun, 2009: Understanding and predicting seasonal to interannual climate variability – the producer perspective. White Paper presented at the World Climate Conference 3 in *Report of the World Climate Conference 3*, World Meteorological Organization Report No. 1048, Geneva, Switzerland.

Street, R., A. Maarouf, and H. Jones-Otazo, 2005: Extreme weather and climate events: implications for public health. In: *Integration of Public Health with Adaptation to Climate Change: Lessons Learned and New Directions* [Ebi, K.L., J.B. Smith, and I. Burton (eds.)]. Taylor & Francis, London, UK, pp. 161-190.

Sultana, F. 2010: Living in hazardous waterscapes: Gendered vulnerabilities and experiences of floods and disasters. *Environmental Hazards-Human and Policy Dimensions*, **9**(1), 43-53.

Susman, P., P. O'Keefe, and B. Wisner, 1983: Global disasters: A radical interpretation. In: *Interpretations of Calamity* [Hewitt, K. (ed.)]. Allen & Unwin, Winchester, MA, pp. 264-283.

Tanner, T., 2009: Screening climate risks to development cooperation. *Focus*, **2.5**, IDS, Brighton, UK.

Tearfund, 2009: *Climate change and Environmental Degradation Risk and Adaptation assessment*. Tearfund, London, UK.

Thomalla, F. and R.K. Larsen, 2010: Resilience in the context of tsunami early warning systems and community disaster preparedness in the Indian Ocean region. *Environmental Hazards*, 9(3), 249-265.

Thomalla, F., T. Downing, E. Spanger-Siegfried, G. Han, and J. Rockström, 2006: Reducing hazard vulnerability: towards a common approach between disaster risk reduction and climate adaptation. *Disasters*, **30**(1), 39-48.

Thomalla, F., R.K. Larsen, F. Kanji, S. Naruchaikusol, C. Tepa, B. Ravesloot, and A.K. Ahmed, 2009: *From Knowledge to Action: Learning to Go the Last Mile. A Participatory Assessment of the Conditions for Strengthening the Technology – community Linkages of Tsunami Early Warning Systems in the Indian Ocean*, Project Report, Stockholm Environment Institute, Macquarie University, Asian Disaster Preparedness Centre, and Raks Thai Foundation, Bangkok and Sydney, Australia.

Thywissen, K., 2006: Core terminology of disaster risk reduction: A comparative glossary. In: *Measuring Vulnerability to Natural Hazards* [Birkmann, J. (ed.)]. UNU Press, Tokyo, Japan, pp. 448-496.

Tol, R.S.J., T.E. Downing, O.J. Kuik, and J.B. Smith, 2004: Distributional aspects of climate change impacts. *Global Environmental Change*, **14**, 259-272.

Tong, S.M., P. Mather, G. Fitzgerald, D. McRae, K. Verrall, and D. Walker, 2010: Assessing the vulnerability of eco-environmental health to climate change. *International Journal Of Environmental Research And Public Health*, **7(2)**, 546-564.

Toni, F. and E. Holanda, 2008: The effects of land tenure on vulnerability to droughts in Northeastern Brazil. *Global Environmental Change-Human and Policy Dimensions*, **18(4)**, 575-582.

Tunstall, S., S. Tapsell, C. Green, and P. Floyd, 2006: The health effects of flooding: social research results from England and Wales. *Journal of Water Health*, **4**, 365-380.

Turner, B.L., R.E. Kasperson, P.A. Matson, J.J. McCarthy, R.W. Corell, L. Christensen, N. Eckley, J.X. Kasperson, A. Luers, M.L. Martello, C. Polsky, A. Pulsipher, and A. Schiller, 2003a: A framework for vulnerability analysis in sustainability science. *Proceedings of the National Academy of Sciences*, **100(14)**, 8074-8079.

Turner, B.L. II, P.A. Matson, J.J. McCarthy, R.W. Corell, L. Christensen, N. Eckley, G. K. Hovelsrud-Broda, J.X. Kasperson, R.E. Kasperson, A. Luers, M.L. Martello, S. Mathiesen, R. Naylor, C. Polsky, A. Pulsipher, A. Schiller, H. Selin, and N. Tyler, 2003b: Illustrating the coupled human-environment system for vulnerability analysis: Three case studies. *Proceedings of the National Academy of Sciences*, **100(14)**, 8080-8085.

Turner, N.J. and H. Clifton, 2009: "It's so different today": Climate change and indigenous lifeways in British Columbia, Canada. *Global Environmental Change*, **19**, 180-190.

Twigg, J., 2001: *Sustainable Livelihoods and Vulnerability to Disasters*. Benfield Greig Hazard Research Centre, Disaster Management Working Paper 2/2001, BGHRC, UCL, London UK.

Twigg, J., 2007: *Characteristics of a Disaster-resilient Community. A Guidance Note. Version 1 (for field testing)*. DFID, Disaster Risk Reduction Interagency Coordination Group, Benfield, UK.

Uitto, J.I., 1998: The geography of disaster vulnerability in megacities: a theoretical framework. *Applied Geography*, **18(1)**, 7-16.

UK Department of Health, 2009: *NHS Emergency Planning Guidance: Planning for the psychosocial and mental health care of people affected by major incidents and disasters: Interim national strategic guidance*. UK DoH, London, UK.

UN, 2005: *Hyogo Framework for Action 2005-2015: Building the Resilience of Nations and Communities to Disasters*, World Conference on Disaster Reduction, 18-22 January 2005, Kobe, Japan, www.unisdr.org/wcdr/intergover/official-doc/L-docs/Hyogo-framework-for-action-english.pdf.

UN Statistics Division, 2011, *Millennium Development Goals Data Base*. UN Statistics Division, New York, NY, mdgs.un.org/unsd/mdg/Metadata.aspx.

UN/DESA, 2010: Trends in Sustainable Development in Small Island Development States. United Nations, New York, NY, USA, 40 pp. ISBN 978-92-1-104610-6.

UN-HABITAT, 2003: *Slums of the World: The Face of Urban Poverty in the New Millennium? Monitoring the Millennium Development Goal, Target 11 – World-Wide Slum Dweller Estimation*. UN-Habitat, Nairobi, Kenya.

UN-HABITAT, 2009: *Global Report on Human Settlements 2011: Planning Sustainable Cities*, United Nations Human Settlements Program, Earthscan, London, UK, ISBN 978-1-84407-898-1.

UN-HABITAT, 2011: *Global Report on Human Settlements 2011: Cities and Climate Change*. United Nations Human Settlements Program, Earthscan, London, UK.

UNCED, 1992: *The Global Partnership for Environment and Development: A Guide to Agenda 21*. United Nations Commission on Environment and Development (UNCED), Geneva, Switzerland.

UNDHA, 1992: *Internationally agreed glossary of basic terms relating to disaster management*. UNDHA, Geneva, Switzerland.

UNDP, 2004: *Reducing Disaster Risk: A Challenge for Development, A Global Report*. UNDP, New York, NY.

UNDRO, 1980: *Natural Disasters and Vulnerability Analysis*. Report of Experts Group Meeting of 9-12 July 1979, UNDRO, Geneva, Switzerland.

UNEP, 2007: *Global Environment Outlook 4*. United Nations Environment Programme, Nairobi, Kenya.

UNEP/UNISDR, 2008: *Environment and Disaster Risk Emerging Perspectives*. UNISDR/UNEP, Geneva, Switzerland.

UNICEF, 2009: *Education in Emergencies and Post-Crisis Transition Consolidated 2009*. Progress Report to the Government of The Netherlands and The European Commission, UNICEF, New York, NY.

UNISDR, 2004: *Living With Risk*. United Nations International Strategy for Disaster Reduction, Geneva, Switzerland.

UNISDR, 2007a: *Words Into Action: A Guide for Implementing the Hyogo Framework*. United Nations International Strategy for Disaster Reduction, Geneva, Switzerland.

UNISDR, 2007b: *Towards a Culture of Prevention: Disaster Risk Reduction Begins at School. Good Practices and Lessons Learned*. United Nations International Strategy for Disaster Reduction, Geneva, Switzerland.

UNISDR, 2009a: *Global Assessment Report on Disaster Risk Reduction - Risk and Poverty in a Changing Climate: Invest Today for a Safer Tomorrow*. United Nations International Strategy for Disaster Reduction, Geneva, Switzerland, 207 pp.

UNISDR, 2009b: *Terminology on Disaster Risk Reduction*. United Nations International Strategy for Disaster Reduction, Geneva, Switzerland. unisdr.org/eng/library/lib-terminology-eng.htm.

UNISDR, 2011: *Global Assessment Report on Disaster Risk Reduction: Revealing Risk, Redefining Development*. United Nations International Strategy for Disaster Reduction, Geneva, 178 pp., www.preventionweb.net/gar.

Urwin, K. and A. Jordan, 2008: Does public policy support or undermine climate change adaptation? Exploring policy interplay across different scales of governance. *Global Environmental Change*, **18**, 180-191.

Utzinger, J. and J. Keiser, 2006: Urbanization and tropical health – then and now. *Annals of Tropical Medicine and Parasitology*, **100(5-6)**, 517-533.

van Aalst, M.K., 2006a: The impacts of climate change on the risk of natural disasters. *Disasters*, **30**, 5-18.

van Aalst, M.K., 2006b: *Managing Climate Risk: Integrating Climate Change Adaptation into World Bank Operations*. World Bank, Washington, DC, USA.

van Aalst, M.K., 2009: Bridging timescales. In: *World Disasters Report 2009. Focus on Early Warning, Early Action*. IFRC, Geneva, Switzerland, pp. 68-93.

van Aalst, M.K. and I. Burton, 2002: *The Last Straw. Integrating Natural Disaster Mitigation with Environmental Management*. World Bank Disaster Risk Management Working Paper Series, 5, World Bank, Washington, DC.

van Aalst, M.K., M. Helmer, C. de Jong, F. Monasso, E. van Sluis, and P. Suarez, 2007: *Red Cross/Red Crescent Climate Guide*. Red Cross/Red Crescent Climate Centre, The Hague, The Netherlands, 144 pp.

van Aalst, M.K., T. Cannon, and I. Burton, 2008: Community level adaptation to climate change: The potential role of participatory community risk assessment. *Global Environmental Change*, **18**, 165-179.

Van de Walle, B. and M. Turoff, 2007: Emergency response information systems: Emerging trends and technologies. *Communications of the ACM*, **50(3)**, 29-31.

Van Lieshout, M., R.S. Kovats, M.T.J. Livermore, and P. Martens, 2004: Climate change and malaria: analysis of the SRES climate and socio-economic scenarios. *Global Environmental Change*, **14**, 87-99.

Van Sluis, E. and M.K. van Aalst, 2006: Climate change and disaster risk in urban environments. *Humanitarian Exchange*, **35**.

Verdin, J., C. Funk, G. Senay, and R. Choularton, 2005: Climate science and famine early warning. *Philosophical Transactions of the Royal Society B*, **360(1463)**, 2155-2168.

Villagrán de León, J.C., 2006: Vulnerability: A conceptual and methodological review. *SOURCE Publication Series of UNU-EHS*, **4**, United Nations University Institute for Environment and Human Security (UNU-EHS), Bonn, Germany.

Vincent, K., 2007: Uncertainty in adaptive capacity and the importance of scale. *Global Environmental Change*, **17(1)**, 12-24.

Vitoria, M., R. Granich, C.F. Gilks, C. Gunneberg, M. Hosseini, W. Were, M. Raviglione, and K.M. Cock, 2009: The global fight against HIV/AIDS, tuberculosis, and malaria: Current status and future perspectives. *American Journal of Clinical Pathology*, **131(6)**, 844-848.

Vogel, C. and K. O'Brien, 2004: *Vulnerability and Global Environmental Change: Rhetoric and Reality*. AVISO 13, Global Environmental Change and Human Security Project, Ottawa, Canada.

Waktola, D.K., 2009: Challenges and opportunities in mainstreaming environmental education into the curricula of teachers' colleges in Ethiopia. *Environmental Education Research*, **15(5)**, 589-605.

Walker, B., C.S. Holling, S.R. Carpenter, and A. Kinzig, 2004: Resilience, adaptability and transformability in social–ecological systems. *Ecology and Society*, **9(2)**, 5.

Watts, M.J. and H.G. Bohle, 1993: The space of vulnerability: the causal structure of hunger and famine. *Progress in Human Geography*, **17(1)**, 43-67.

Weichselgartner, J., 2001: Disaster mitigation: the concept of vulnerability revisited. *Disaster Prevention and Management*, **10(2)**, 85-94.

Werritty, A., D. Houston, T. Ball, A. Tavendale, and A. Black, 2007: *Exploring The Social Impacts Of Flood Risk And Flooding In Scotland*. Scottish Executive Social Research, Edinburgh, Scotland.

Westgate, K.N., P. O'Keefe, 1976: *Some Definitions of Disaster*. Disaster Research Unit Occasional Paper 4, Department of Geography, University of Bradford, UK.

White, G.F., 1973: Natural hazards research. In: *Directions in Geography* [Chorley, R.J. (ed.)]. Methuen and Co., London, UK, pp. 193-216.

White, G.F. (ed.), 1974: *Natural Hazards: Local, National, Global*. Oxford University Press, New York, NY.

White, I., 2002: *Water Management in the Mekong Delta: Changes, Conflicts and Opportunities*. IHP-VI Technical Papers in Hydrology No. 61, UNESCO, Paris, France.

Wijkman, A. and L. Timberlake, 1984: *Natural Disasters: Act of God or Acts of Man*. Earthscan, Washington, DC.

Wilbanks, T.J., 2003: Integrating climate change and sustainable development in a place-based context. *Climate Policy*, **3(S1)**, S147-S154.

Wilbanks, T.J., P. Romero Lankao, M. Bao, F. Berkhout, S. Cairncross, J.-P. Ceron, M. Kapshe, R. Muir-Wood, and R. Zapata-Marti, 2007: Industry, settlement and society. In: *Climate Change 2007. Impacts, Adaptation and Vulnerability. Contribution of Working Group II to the Fourth Assessment Report of the Intergovernmental Panel on Climate Change* [Parry, M.L., O.F. Canziani, J.P. Palutikof, P.J. Van Der Linde, and C.E. Hanson (eds.)]. Cambridge University Press, Cambridge, UK, 357-390.

Wilby R.L. and G. L.W. Perry, 2006: Climate change, biodiversity and the urban environment: a critical review based on London, UK. *Progress in Physical Geography*, **30(1)**, 73-98.

Wilches-Chaux, G., 1989: *Desastres, ecologismo y formación profesional*. SENA, Popayán, Colombia.

Williams, S.E, L.P. Shoo, J.L. Isaac, A.A. Hoffmann, and G. Langham, 2008: Towards an integrated framework for assessing the vulnerability of species to climate change. *PLoS Biology*, **6(12)**, e325, doi:10.1371/journal. pbio.0060325.

Willows, R.I. and R.K. Connell (eds.), 2003: *Climate Adaptation: Risk, Uncertainty and Decision-Making*. UKCIP Technical Report. UKCIP, Oxford, UK.

Wisner, B.G., 1978: An appeal for a significantly comparative method in disaster research. *Disasters*, **2**, 80-82.

Wisner, B., 1993: Disaster vulnerability: Scale, power and daily life. *GeoJournal*, **30**, 127-140.

Wisner, B., 1998: Marginality and vulnerability: why the homeless of Tokyo don't 'count' in disaster preparations. *Applied Geography*, **18(1)**, 25-33.

Wisner, B., 2001a: Capitalism and the shifting spatial and social distribution of hazard and vulnerability. *Australian Journal of Emergency Management*, **Winter 2001**, 44-50.

Wisner, B., 2001b: Risk and the Neoliberal State: Why post-Mitch lessons didn't reduce El Salvador's earthquake losses. *Disasters*, **25 (3)**, 251-268.

Wisner, B., 2003: Disaster risk reduction in megacities – Making the most of human and social capital. In: *Building Safer Cities- The Future of Disaster Risk* [Kreimer, A., M. Arnold, and C. Carlin (eds.)]. Disaster Management Facility, World Bank, Washington, DC, pp. 181-196.

Wisner, B., 2006a: Self-assessment of coping capacity: Participatory, proactive, and qualitative engagement of communities in their own risk management. In: *Measuring Vulnerability to Natural Hazards - Towards Disaster Resilient Societies* [Birkmann, J. (ed.)]. United Nations University Press, Tokyo, Japan, pp. 316-328.

Wisner, B., 2006b: *Let Our Children Teach Us! A Review of the Role of Education and Knowledge in Disaster Risk Reduction*. UNISDR System Thematic Cluster/Platform on Knowledge and Education, Geneva, Switzerland, 135 pp.

Wisner, B. and J. Adams (eds.), 2002: *Environment and Health in Emergencies and Disasters: A Practical Guide*. World Health Organization, Geneva, Switzerland.

Wisner, B., P. Blaikie, T. Cannon, and I. Davis, 2004: *At Risk, Natural Hazards, People's Vulnerability and Disasters*. Routledge, London, UK.

Wolf, J., N.W. Adger, and I. Lorenzoni, 2010a: Heat waves and cold spells: an analysis of policy response and perceptions of vulnerable populations in the UK. *Environment and Planning A*, **42**, 2721-2734, doi:10.1068/a42503.

Wolf, J., W.N. Adger, I. Lorenzoni, V. Abrahamson, and R. Raine, 2010b: Social capital, individual responses to heat waves and climate change adaptation: An empirical study of two UK cities. *Global Environmental Change*, **20**, 44-52.

Woo, G., 1999: *The Mathematics of Natural Catastrophes*. Imperial College Press, London, UK.

Woolcock, M. and D. Narayan, 2000: Social capital: Implications for development theory, research and policy. *World Bank Research Observer*, **15**, 225-249.

Yodmani, S. 2001; Disaster preparedness and management. In: *Social Protection in Asia and the Pacific* [Ortiz, I.D. (ed.)]. Asian Development Bank, Manila, pp. 481-502.

Yohe, G. and R.S.J. Tol, 2002: Indicators for social and economic coping capacity: moving toward a working definition of adaptive capacity, *Global Environmental Change*, **12**, 25-40.

3 Changes in Climate Extremes and their Impacts on the Natural Physical Environment

Coordinating Lead Authors:
Sonia I. Seneviratne (Switzerland), Neville Nicholls (Australia)

Lead Authors:
David Easterling (USA), Clare M. Goodess (United Kingdom), Shinjiro Kanae (Japan), James Kossin (USA), Yali Luo (China), Jose Marengo (Brazil), Kathleen McInnes (Australia), Mohammad Rahimi (Iran), Markus Reichstein (Germany), Asgeir Sorteberg (Norway), Carolina Vera (Argentina), Xuebin Zhang (Canada)

Review Editors:
Matilde Rusticucci (Argentina), Vladimir Semenov (Russia)

Contributing Authors:
Lisa V. Alexander (Australia), Simon Allen (Switzerland), Gerardo Benito (Spain), Tereza Cavazos (Mexico), John Clague (Canada), Declan Conway (United Kingdom), Paul M. Della-Marta (Switzerland), Markus Gerber (Switzerland), Sunling Gong (Canada), B. N. Goswami (India), Mark Hemer (Australia), Christian Huggel (Switzerland), Bart van den Hurk (Netherlands), Viatcheslav V. Kharin (Canada), Akio Kitoh (Japan), Albert M.G. Klein Tank (Netherlands), Guilong Li (Canada), Simon Mason (USA), William McGuire (United Kingdom), Geert Jan van Oldenborgh (Netherlands), Boris Orlowsky (Switzerland), Sharon Smith (Canada), Wassila Thiaw (USA), Adonis Velegrakis (Greece), Pascal Yiou (France), Tingjun Zhang (USA), Tianjun Zhou (China), Francis W. Zwiers (Canada)

This chapter should be cited as:

Seneviratne, S.I., N. Nicholls, D. Easterling, C.M. Goodess, S. Kanae, J. Kossin, Y. Luo, J. Marengo, K. McInnes, M. Rahimi, M. Reichstein, A. Sorteberg, C. Vera, and X. Zhang, 2012: Changes in climate extremes and their impacts on the natural physical environment. In: *Managing the Risks of Extreme Events and Disasters to Advance Climate Change Adaptation* [Field, C.B., V. Barros, T.F. Stocker, D. Qin, D.J. Dokken, K.L. Ebi, M.D. Mastrandrea, K.J. Mach, G.-K. Plattner, S.K. Allen, M. Tignor, and P.M. Midgley (eds.)]. A Special Report of Working Groups I and II of the Intergovernmental Panel on Climate Change (IPCC). Cambridge University Press, Cambridge, UK, and New York, NY, USA, pp. 109-230.

Table of Contents

Executive Summary 111

3.1. Weather and Climate Events Related to Disasters 115
3.1.1. Categories of Weather and Climate Events Discussed in this Chapter 115
3.1.2. Characteristics of Weather and Climate Events Relevant to Disasters 115
3.1.3. Compound (Multiple) Events 118
3.1.4. Feedbacks 118
3.1.5. Confidence and Likelihood of Assessed Changes in Extremes 120
3.1.6. Changes in Extremes and Their Relationship to Changes in Regional and Global Mean Climate 121
3.1.7. Surprises / Abrupt Climate Change 122

3.2. Requirements and Methods for Analyzing Changes in Extremes 122
3.2.1. Observed Changes 122
3.2.2. The Causes behind the Changes 125
3.2.3. Projected Long-Term Changes and Uncertainties 128

3.3. Observed and Projected Changes in Weather and Climate Extremes 133
3.3.1. Temperature 133
3.3.2. Precipitation 141
3.3.3. Wind 149

3.4. Observed and Projected Changes in Phenomena Related to Weather and Climate Extremes 152
3.4.1. Monsoons 152
3.4.2. El Niño-Southern Oscillation 155
3.4.3. Other Modes of Variability 157
3.4.4. Tropical Cyclones 158
3.4.5. Extratropical Cyclones 163

3.5. Observed and Projected Impacts on the Natural Physical Environment 167
3.5.1. Droughts 167
3.5.2. Floods 175
3.5.3. Extreme Sea Levels 178
3.5.4. Waves 180
3.5.5. Coastal Impacts 182
3.5.6. Glacier, Geomorphological, and Geological Impacts 186
3.5.7. High-latitude Changes Including Permafrost 189
3.5.8. Sand and Dust Storms 190

References 203

Boxes and Frequently Asked Questions
Box 3-1. Definition and Analysis of Climate Extremes in the Scientific Literature 116
FAQ 3.1. Is the Climate Becoming More Extreme? 124
FAQ 3.2. Has Climate Change Affected Individual Extreme Events? 127
Box 3-2. Variations in Confidence in Projections of Climate Change: Mean versus Extremes, Variables, Scale 132
Box 3-3. The Definition of Drought 167
Box 3-4. Small Island States 184

Supplementary Material
Appendix 3.A: Notes and Technical Details on Chapter 3 Figures ....Available On-Line

Executive Summary

This chapter addresses changes in weather and climate events relevant to extreme impacts and disasters. An extreme (weather or climate) event is generally defined as the occurrence of a value of a weather or climate variable above (or below) a threshold value near the upper (or lower) ends ('tails') of the range of observed values of the variable. Some climate extremes (e.g., droughts, floods) may be the result of an accumulation of weather or climate events that are, individually, not extreme themselves (though their accumulation is extreme). As well, weather or climate events, even if not extreme in a statistical sense, can still lead to extreme conditions or impacts, either by crossing a critical threshold in a social, ecological, or physical system, or by occurring simultaneously with other events. A weather system such as a tropical cyclone can have an extreme impact, depending on where and when it approaches landfall, even if the specific cyclone is not extreme relative to other tropical cyclones. Conversely, not all extremes necessarily lead to serious impacts. [3.1]

Many weather and climate extremes are the result of natural climate variability (including phenomena such as El Niño), and natural decadal or multi-decadal variations in the climate provide the backdrop for anthropogenic climate changes. Even if there were no anthropogenic changes in climate, a wide variety of natural weather and climate extremes would still occur. [3.1]

A changing climate leads to changes in the frequency, intensity, spatial extent, duration, and timing of weather and climate extremes, and can result in unprecedented extremes. Changes in extremes can also be directly related to changes in mean climate, because mean future conditions in some variables are projected to lie within the tails of present-day conditions. Nevertheless, changes in extremes of a climate or weather variable are not always related in a simple way to changes in the mean of the same variable, and in some cases can be of opposite sign to a change in the mean of the variable. Changes in phenomena such as the El Niño-Southern Oscillation or monsoons could affect the frequency and intensity of extremes in several regions simultaneously. [3.1]

Many factors affect confidence in observed and projected changes in extremes. Our confidence in observed changes in extremes depends on the quality and quantity of available data and the availability of studies analyzing these data. It consequently varies between regions and for different extremes. Similarly, our confidence in projecting changes (including the direction and magnitude of changes in extremes) varies with the type of extreme, as well as the considered region and season, depending on the amount and quality of relevant observational data and model projections, the level of understanding of the underlying processes, and the reliability of their simulation in models (assessed from expert judgment, model validation, and model agreement). Global-scale trends in a specific extreme may be either more reliable (e.g., for temperature extremes) or less reliable (e.g., for droughts) than some regional-scale trends, depending on the geographical uniformity of the trends in the specific extreme. '*Low confidence*' in observed or projected changes in a specific extreme neither implies nor excludes the possibility of changes in this extreme. [3.1.5, 3.1.6, 3.2.3; Box 3-2; Figures 3-3, 3-4, 3-5, 3-6, 3-7, 3-8, 3-10]

There is evidence from observations gathered since 1950 of change in some extremes. It is *very likely* that there has been an overall decrease in the number of cold days and nights, and an overall increase in the number of warm days and nights, at the global scale, that is, for most land areas with sufficient data. It is *likely* that these changes have also occurred at the continental scale in North America, Europe, and Australia. There is *medium confidence* of a warming trend in daily temperature extremes in much of Asia. *Confidence* in observed trends in daily temperature extremes in Africa and South America generally varies from *low* to *medium* depending on the region. Globally, in many (but not all) regions with sufficient data there is *medium confidence* that the length or number of warm spells or heat waves has increased since the middle of the 20th century. It is *likely* that there have been statistically significant increases in the number of heavy precipitation events (e.g., 95th percentile) in more regions than there have been statistically significant decreases, but there are strong regional and subregional variations in the trends. There is *low confidence* that any observed long-term (i.e., 40 years or more) increases in tropical cyclone activity are robust, after accounting for past changes in observing capabilities. It is *likely* that there has been a poleward shift in the main Northern and Southern Hemisphere extratropical storm tracks. There is *low confidence* in observed trends in small-scale phenomena such as tornadoes and hail because of data inhomogeneities and inadequacies in monitoring systems. There is *medium confidence* that since the 1950s some regions of the world have experienced a trend to more intense and longer droughts, in particular in southern Europe and West Africa, but in some regions droughts have become less frequent, less intense, or shorter, for example, in central North America and northwestern Australia. There is limited to medium evidence available to assess climate-driven observed changes in the magnitude and frequency of floods at regional scales because the available instrumental records of floods at gauge stations are limited in space and time, and because of confounding effects of changes in land use and engineering. Furthermore, there is low agreement in this evidence, and thus overall *low confidence* at the global scale regarding even the sign of these changes. It is *likely* that there has been an increase in extreme coastal high water related to increases in mean sea level in the late 20th century. [3.2.1, 3.3.1, 3.3.2, 3.3.3, 3.4.4, 3.4.5, 3.5.1, 3.5.2, 3.5.3; Tables 3-1, 3-2]

There is evidence that some extremes have changed as a result of anthropogenic influences, including increases in atmospheric concentrations of greenhouse gases. It is *likely* that anthropogenic influences have led to warming of extreme daily minimum and maximum temperatures at the global scale. There is *medium confidence* that anthropogenic influences have contributed to intensification of extreme precipitation at the global scale. It is *likely* that there has been an anthropogenic influence on increasing extreme coastal high water due to an increase in mean sea level. The uncertainties in the historical tropical cyclone records, the incomplete understanding of the physical mechanisms linking tropical cyclone metrics to climate change, and the degree of tropical cyclone variability provide only *low confidence* for the attribution of any detectable changes in tropical cyclone activity to anthropogenic influences. Attribution of single extreme events to anthropogenic climate change is challenging. [3.2.2, 3.3.1, 3.3.2, 3.4.4, 3.5.3; Table 3-1]

The following assessments of the likelihood of and/or confidence in projections are generally for the end of the 21st century and relative to the climate at the end of the 20th century. There are three main sources of uncertainty in the projections: the natural variability of climate; uncertainties in climate model parameters and structure; and projections of future emissions. Projections for differing emissions scenarios generally do not strongly diverge in the coming two to three decades, but uncertainty in the sign of change is relatively large over this time frame because climate change signals are expected to be relatively small compared to natural climate variability. For certain extremes (e.g., precipitation-related extremes), the uncertainty in projected changes by the end of the 21st century is more the result of uncertainties in climate models rather than uncertainties in future emissions. For other extremes (in particular temperature extremes at the global scale and in most regions), the emissions uncertainties are the main source of uncertainty in projections for the end of the 21st century. In the assessments provided in this chapter, uncertainties in projections from the direct evaluation of multi-model ensemble projections are modified by taking into account the past performance of models in simulating extremes (for instance, simulations of late 20th-century changes in extreme temperatures appear to overestimate the observed warming of warm extremes and underestimate the warming of cold extremes), the possibility that some important processes relevant to extremes may be missing or be poorly represented in models, and the limited number of model projections and corresponding analyses currently available of extremes. For these reasons the assessed uncertainty is generally greater than would be assessed from the model projections alone. Low-probability, high-impact changes associated with the crossing of poorly understood climate thresholds cannot be excluded, given the transient and complex nature of the climate system. Feedbacks play an important role in either damping or enhancing extremes in several climate variables. [3.1.4, 3.1.7, 3.2.3, 3.3.1, 3.3.2; Box 3-2]

Models project substantial warming in temperature extremes by the end of the 21st century. It is *virtually certain* that increases in the frequency and magnitude of warm daily temperature extremes and decreases in cold extremes will occur through the 21st century at the global scale. It is *very likely* that the length, frequency, and/or intensity of warm spells or heat waves will increase over most land areas. For the Special Report on Emissions Scenarios (SRES) A2 and A1B emission scenarios, a 1-in-20 year annual hottest day is *likely* to become a 1-in-2 year annual extreme by the end of the 21st century in most regions, except in the high latitudes of the Northern Hemisphere where it is *likely* to become a 1-in-5 year annual extreme. In terms of absolute values, 20-year extreme annual daily maximum temperature (i.e., return value) will *likely* increase by about 1 to 3°C by mid-21st century and by about 2 to 5°C by the late 21st century, depending on the region and emissions scenario (considering the B1, A1B, and A2 scenarios). Regional changes in temperature extremes will often differ from the mean global temperature change. [3.3.1; Table 3-3; Figure 3-5]

It is *likely* that the frequency of heavy precipitation or the proportion of total rainfall from heavy rainfalls will increase in the 21st century over many areas of the globe. This is particularly the case in the high latitudes and tropical regions, and in winter in the northern mid-latitudes. Heavy rainfalls associated with tropical cyclones are *likely* to increase with continued warming induced by enhanced greenhouse gas concentrations. There is *medium confidence* that, in some regions, increases in heavy precipitation will occur despite projected decreases in total precipitation. For a range of emission scenarios (SRES A2, A1B, and B1), a 1-in-20 year annual maximum 24-hour precipitation rate is *likely* to become a 1-in-5 to 1-in-15 year event by the end of the 21st century in many regions, and in most regions the higher emissions scenarios (A1B and A2) lead to a greater projected decrease in return period. Nevertheless, increases or statistically non-significant changes in return periods are projected in some regions. [3.3.2; Table 3-3; Figure 3-7]

There is generally *low confidence* in projections of changes in extreme winds because of the relatively few studies of projected extreme winds, and shortcomings in the simulation of these events. An exception is mean tropical cyclone maximum wind speed, which is *likely* to increase, although increases may not occur in all ocean basins. It is *likely* that the global frequency of tropical cyclones will either decrease or remain essentially unchanged. There is *low confidence* in projections of small-scale phenomena such as tornadoes because competing physical processes may affect future trends and because climate models do not simulate such phenomena. There is *medium confidence* that there will be a reduction in the number of mid-latitude cyclones averaged over each hemisphere due to future anthropogenic climate change. There is *low confidence* in the detailed geographical projections of mid-latitude cyclone activity. There is *medium confidence* in a projected poleward shift of mid-latitude storm tracks due to future anthropogenic forcings. [3.3.3, 3.4.4, 3.4.5]

Uncertainty in projections of changes in large-scale patterns of natural climate variability remains large. There is *low confidence* in projections of changes in monsoons (rainfall, circulation), because there is little consensus in climate models regarding the sign of future change in the monsoons. Model projections of changes in El Niño-Southern Oscillation variability and the frequency of El Niño episodes as a consequence of increased greenhouse gas concentrations are not consistent, and so there is *low confidence* in projections of changes in the phenomenon. However, most models project an increase in the relative frequency of central equatorial Pacific events (which typically exhibit different patterns of climate variations than do the classical East Pacific events). There is *low confidence* in the ability to project changes in other natural climate modes including the North Atlantic Oscillation, the Southern Annular Mode, and the Indian Ocean Dipole. [3.4.1, 3.4.2, 3.4.3]

It is *very likely* that mean sea level rise will contribute to upward trends in extreme coastal high water levels in the future. There is *high confidence* that locations currently experiencing adverse impacts such as coastal erosion and inundation will continue to do so in the future due to increasing sea levels, all other contributing factors being equal. There is *low confidence* in wave height projections because of the small number of studies, the lack of consistency of the wind projections between models, and limitations in the models' ability to simulate extreme winds. Future negative or positive changes in significant wave height are *likely* to reflect future changes in storminess and associated patterns of wind change. [3.5.3, 3.5.4, 3.5.5]

Projected precipitation and temperature changes imply possible changes in floods, although overall there is *low confidence* in projections of changes in fluvial floods. *Confidence* is *low* due to limited evidence and because the causes of regional changes are complex, although there are exceptions to this statement. There is *medium confidence* (based on physical reasoning) that projected increases in heavy rainfall would contribute to increases in local flooding, in some catchments or regions. Earlier spring peak flows in snowmelt and glacier-fed rivers are *very likely*. [3.5.2]

There is *medium confidence* that droughts will intensify in the 21st century in some seasons and areas, due to reduced precipitation and/or increased evapotranspiration. This applies to regions including southern

Europe and the Mediterranean region, central Europe, central North America, Central America and Mexico, northeast Brazil, and southern Africa. Definitional issues, lack of observational data, and the inability of models to include all the factors that influence droughts preclude stronger *confidence* than *medium* in the projections. Elsewhere there is overall *low confidence* because of inconsistent projections of drought changes (dependent both on model and dryness index). There is *low confidence* in projected future changes in dust storms although an increase could be expected where aridity increases. [3.5.1, 3.5.8; Box 3-3; Table 3-3; Figure 3-10]

There is *high confidence* that changes in heat waves, glacial retreat, and/or permafrost degradation will affect high-mountain phenomena such as slope instabilities, mass movements, and glacial lake outburst floods. There is also *high confidence* that changes in heavy precipitation will affect landslides in some regions. There is *low confidence* regarding future locations and timing of large rock avalanches, as these depend on local geological conditions and other non-climatic factors. There is *low confidence* in projections of an anthropogenic effect on phenomena such as shallow landslides in temperate and tropical regions, because these are strongly influenced by human activities such as land use practices, deforestation, and overgrazing. [3.5.6, 3.5.7]

The small land area and often low elevation of small island states make them particularly vulnerable to rising sea levels and impacts such as inundation, shoreline change, and saltwater intrusion into underground aquifers. Short record lengths and the inadequate resolution of current climate models to represent small island states limit the assessment of changes in extremes. There is insufficient evidence to assess observed trends and future projections in rainfall across the small island regions considered here. There is *medium confidence* in projected temperature increases across the Caribbean. The *very likely* contribution of mean sea level rise to increased extreme coastal high water levels, coupled with the *likely* increase in tropical cyclone maximum wind speed, is a specific issue for tropical small island states. [3.4.4, 3.5.3; Box 3-4]

This chapter does not provide assessments of projected changes in extremes at spatial scales smaller than for large regions. These large-region projections provide a wider context for national or local projections, where these exist, and where they do not exist, a first indication of expected changes, their associated uncertainties, and the evidence available. [3.2.3.1]

3.1. Weather and Climate Events Related to Disasters

A changing climate leads to changes in the frequency, intensity, spatial extent, duration, and timing of weather and climate extremes, and can result in unprecedented extremes (Sections 3.1.7, 3.3, 3.4, and 3.5). As well, weather or climate events, even if not extreme in a statistical sense, can still lead to extreme conditions or impacts, either by crossing a critical threshold in a social, ecological, or physical system, or by occurring simultaneously with other events (Sections 3.1.2, 3.1.3, 3.1.4, 3.3, 3.4, and 3.5). Some climate extremes (e.g., droughts, floods) may be the result of an accumulation of weather or climate events that are, individually, not extreme themselves (though their accumulation is extreme, e.g., Section 3.1.2). A weather system such as a tropical cyclone can have an extreme impact, depending on where and when it approaches landfall, even if the specific cyclone is not extreme relative to other tropical cyclones. Conversely, not all extremes necessarily lead to serious impacts. Changes in extremes can also be directly related to changes in mean climate, because mean future conditions in some variables are projected to lie within the tails of present-day conditions (Section 3.1.6). Hence, the definition of extreme weather and climate events is complex (Section 3.1.2 and Box 3-1) and the assessment of changes in climate that are relevant to extreme impacts and disasters needs to consider several aspects. Those related to vulnerability and exposure are addressed in Chapters 2 and 4 of this report, while we focus here on the physical dimension of these events.

Many weather and climate extremes are the result of natural climate variability (including phenomena such as El Niño), and natural decadal or multi-decadal variations in the climate provide the backdrop for anthropogenic climate changes. Even if there were no anthropogenic changes in climate, a wide variety of natural weather and climate extremes would still occur.

3.1.1. Categories of Weather and Climate Events Discussed in this Chapter

This chapter addresses changes in weather and climate events relevant to extreme impacts and disasters grouped into the following categories:
1) Extremes of atmospheric weather and climate variables (temperature, precipitation, wind)
2) Weather and climate phenomena that influence the occurrence of extremes in weather or climate variables or are extremes themselves (monsoons, El Niño and other modes of variability, tropical and extratropical cyclones)
3) Impacts on the natural physical environment (droughts, floods, extreme sea level, waves, and coastal impacts, as well as other physical impacts, including cryosphere-related impacts, landslides, and sand and dust storms).

The distinction between these three categories is somewhat arbitrary, and the categories are also related. In the case of the third category, 'impacts on the natural physical environment,' a specific distinction between these events and those considered under 'extremes of atmospheric weather and climate variables' is that they are not caused by variations in a single atmospheric weather and climate variable, but are generally the result of specific conditions in several variables, as well as of some surface properties or states. For instance, both floods and droughts are related to precipitation extremes, but are also impacted by other atmospheric and surface conditions (and are thus often better viewed as compound events, see Section 3.1.3). Most of the impacts on the natural physical environment discussed in the third category are extremes themselves, as well as often being caused or affected by atmospheric weather or climate extremes. Another arbitrary choice made here is the separate category for phenomena (or climate or weather systems) that are related to weather and climate extremes, such as monsoons, El Niño, and other modes of variability. These phenomena affect the large-scale environment that, in turn, influences extremes. For instance, El Niño episodes typically lead to droughts in some regions with, simultaneously, heavy rains and floods occurring elsewhere. This means that all occurrences of El Niño are relevant to extremes and not only extreme El Niño episodes. A change in the frequency or nature of El Niño episodes (or in their relationships with climate in specific regions) would affect extremes in many locations simultaneously. Similarly, changes in monsoon patterns could affect several countries simultaneously. This is especially important from an international disaster perspective because coping with disasters in several regions simultaneously may be challenging (see also Section 3.1.3 and Chapters 7 and 8).

This section provides background material on the characterization and definition of extreme events, the definition and analysis of compound events, the relevance of feedbacks for extremes, the approach used for the assignment of confidence and likelihood assessments in this chapter, and the possibility of 'surprises' regarding future changes in extremes. Requirements and methods for analyzing changes in climate extremes are addressed in Section 3.2. Assessments regarding changes in the climate variables, phenomena, and impacts considered in this chapter are provided in Sections 3.3 to 3.5. Table 3-1 provides summaries of these assessments for changes at the global scale. Tables 3-2 and 3-3 (found on pages 191-202) provide more regional detail on observed and projected changes in temperature extremes, heavy precipitation, and dryness (with regions as defined in Figure 3-1). Note that impacts on ecosystems (e.g., bushfires) and human systems (e.g., urban flooding) are addressed in Chapter 4.

3.1.2. Characteristics of Weather and Climate Events Relevant to Disasters

The identification and definition of weather and climate events that are relevant from a risk management perspective are complex and depend on the stakeholders involved (Chapters 1 and 2). In this chapter, we focus on the assessment of changes in 'extreme climate or weather events' (also referred to herein as 'climate extremes' see below and Glossary), which generally correspond to the 'hazards' discussed in Chapter 1.

Box 3-1 | Definition and Analysis of Climate Extremes in the Scientific Literature

This box provides some details on the definition of climate extremes in the scientific literature and on common approaches employed for their investigation.

A large amount of the available scientific literature on climate extremes is based on the use of so-called 'extreme indices,' which can either be based on the probability of occurrence of given quantities or on threshold exceedances (Section 3.1.2). Typical indices that are seen in the scientific literature include the number, percentage, or fraction of days with maximum temperature (Tmax) or minimum temperature (Tmin), below the 1st, 5th, or 10th percentile, or above the 90th, 95th, or 99th percentile, generally defined for given time frames (days, month, season, annual) with respect to the 1961-1990 reference time period. Commonly, indices for 10th and 90th percentiles of Tmax/Tmin computed on daily time frames are referred to as 'cold/warm days/nights' (e.g., Figures 3-3 and 3-4; Tables 3-1 to 3-3, and Section 3.3.1; see also Glossary). Other definitions relate to, for example, the number of days above specific absolute temperature or precipitation thresholds, or more complex definitions related to the length or persistence of climate extremes. Some advantages of using predefined extreme indices are that they allow some comparability across modelling and observational studies and across regions (although with limitations noted below). Moreover, in the case of observations, derived indices may be easier to obtain than is the case with daily temperature and precipitation data, which are not always distributed by meteorological services. Peterson and Manton (2008) discuss collaborative international efforts to monitor extremes by employing extreme indices. Typically, although not exclusively, extreme indices used in the scientific literature reflect 'moderate extremes,' for example, events occurring as often as 5 or 10% of the time. More extreme 'extremes' are often investigated using Extreme Value Theory (EVT) due to sampling issues (see below). Extreme indices are often defined for daily temperature and precipitation characteristics, and are also sometimes applied to seasonal characteristics of these variables, to other weather and climate variables, such as wind speed, humidity, or to physical impacts and phenomena. Beside analyses for temperature and precipitation indices (see Sections 3.3.1 and 3.3.2; Tables 3-2 and 3-3), other studies are, for instance, available in the literature for wind-based (Della-Marta et al., 2009) and pressure-based (Beniston, 2009a) indices, for health-relevant indices (e.g., 'heat index') combining temperature and relative humidity characteristics (e.g., Diffenbaugh et al., 2007; Fischer and Schär, 2010; Sherwood and Huber, 2010), and for a range of dryness indices (see Box 3-3).

Extreme Value Theory is an approach used for the estimation of extreme values (e.g., Coles, 2001), which aims at deriving a probability distribution of events from the tail of a probability distribution, that is, at the far end of the upper or lower ranges of the probability distributions (typically occurring less frequently than once per year or per period of interest, i.e., generally less than 1 to 5% of the considered overall sample). EVT is used to derive a complete probability distribution for such low-probability events, which can also help analyzing the probability of occurrence of events that are outside of the observed data range (with limitations). Two different approaches can be used to estimate the parameters for such probability distributions. In the *block maximum approach*, the probability distribution parameters are estimated for maximum values of consecutive blocks of a time series (e.g., years). In the second approach, instead of the block maxima the estimation is based on events that exceed a high threshold (*peaks over threshold* approach). Both approaches are used in climate research.

Continued next page ⟶

Hence, the present chapter does not directly consider the dimensions of vulnerability or exposure, which are critical in determining the human and ecosystem impacts of climate extremes (Chapters 1, 2, and 4).

This report defines an 'extreme climate or weather event' or 'climate extreme' as "the occurrence of a value of a weather or climate variable above (or below) a threshold value near the upper (or lower) ends of the range of observed values of the variable" (see Glossary). Several aspects of this definition can be clarified thus:
- Definitions of thresholds vary, but values with less than 10, 5, 1%, or even lower chance of occurrence for a given time of the year (day, month, season, whole year) during a specified *reference* period (generally 1961-1990) are often used. In some circumstances, information from sources other than observations, such as model projections, can be used as a reference.
- Absolute thresholds (rather than these relative thresholds based on the range of observed values of a variable) can also be used to identify extreme events (e.g., specific critical temperatures for health impacts).
- What is called an extreme weather or climate event will vary from place to place in an absolute sense (e.g., a hot day in the tropics will be a different temperature than a hot day in the mid-latitudes), and possibly in time given some adaptation from society (see Box 3-1).
- Some climate extremes (e.g., droughts, floods) may be the result of an accumulation of moderate weather or climate events (this accumulation being itself extreme). Compound events (see Section 3.1.3), that is, two or more events occurring simultaneously, can lead to high impacts, even if the two single events are not extreme per se (only their combination).

Recent publications have used other approaches for evaluating characteristics of extremes or changes in extremes, for instance, analyzing trends in record events or investigating whether records in observed time series are being set more or less frequently than would be expected in an unperturbed climate (Benestad, 2003, 2006; Zorita et al., 2008; Meehl et al., 2009c; Trewin and Vermont, 2010). Furthermore, besides the actual magnitude of extremes (quantified in terms of probability/return frequency or absolute threshold), other relevant aspects for the definition of climate extremes from an impact perspective include the event's duration, the spatial area affected, timing, frequency, onset date, continuity (i.e., whether there are 'breaks' within a spell), and preconditioning (e.g., rapid transition from a slowly developing meteorological drought into an agricultural drought, see Box 3-3). These aspects, together with seasonal variations in climate extremes, are not as frequently examined in climate models or observational analyses, and thus can only be partly assessed within this chapter.

As noted in the discussion of 'extreme weather or climate events' in Section 3.1.2, thresholds, percentiles, or return values used for the definition of climate extremes are generally defined with respect to a given reference period (generally historical, i.e., 1961-1990, but possibly also based on climate model data). In some cases, a transient baseline can also be considered (i.e., the baseline uses data from the period under examination and changes as the period being considered changes, rather than using a standard period such as 1961-1990). The choice of the reference period may be relevant for the magnitude of the assessed changes as highlighted, for example, in Lorenz et al. (2010). The choice of the reference period (static or transient) could also affect the assessment of the respective role of changes in mean versus changes in variability for changes in extremes discussed in Section 3.1.6. If extremes are based on the probability distribution from which they are drawn, then a simple change in the mean (and keeping the same distribution) would, strictly speaking, produce no relative change in extremes at all. The question of the choice of an appropriate reference period is tied to the notion of adaptation. Events that are considered extreme nowadays in some regions could possibly be adapted to if the vulnerability and exposure to these extremes is reduced (Chapters 1, 2, and 4 through 7). However, there are also some limits to adaptation as highlighted in Chapter 8. These considerations are difficult to include in the statistical analyses of climate scenarios because of the number of (mostly non-physical) aspects that would need to be taken into account.

To conclude, there is no precise definition of an extreme (e.g., D.B. Stephenson et al., 2008). In particular, we note limitations in the definition of both probability-based or threshold-based climate extremes and their relations to impacts, which apply independently of the chosen method of analysis:
- An event from the extreme tails of probability distributions is not necessarily extreme in terms of impact.
- Impact-related thresholds can vary in space and time, that is, single absolute thresholds (e.g., a daily rainfall exceeding 25 mm or the number of frost days) will not reflect extremes in all locations and time periods (e.g., season, decade).

As an illustration, projected patterns (in the magnitude but not the sign) of changes in annual heat wave length were shown to be highly dependent on the choice of index used for the assessment of heat wave or warm spell duration (using the mean and maximum Heat Wave Duration Indices, HWDImean and HWDImax, and the Warm Spell Duration Index, WSDI; see Orlowsky and Seneviratne, 2011), because of large geographical variations in the variability of daily temperature (Alexander et al., 2006). Similar definition issues apply to other types of extremes, especially those characterizing dryness (see Section 3.5.1 and Box 3-3).

- Not all extreme weather and climate events necessarily have extreme impacts.
- The distinction between extreme weather events and extreme climate events is not precise, but is related to their specific time scales:
  – An extreme weather event is typically associated with changing weather patterns, that is, within time frames of less than a day to a few weeks.
  – An extreme climate event happens on longer time scales. It can be the accumulation of several (extreme or non-extreme) weather events (e.g., the accumulation of moderately below-average rainy days over a season leading to substantially below-average cumulated rainfall and drought conditions).

For simplicity, we collectively refer to both extreme weather events and extreme climate events with the term 'climate extremes' in this chapter.

From this definition, it can be seen that climate extremes can be defined quantitatively in two ways:
1) Related to their probability of occurrence
2) Related to a specific (possibly impact-related) threshold.

The first type of definition can either be expressed with respect to given percentiles of the distribution functions of the variables, or with respect to specific return frequencies (e.g., '100-year event'). Compound events can be viewed as a special category of climate extremes, which result from the combination of two or more events, and which are again 'extreme' either from a statistical perspective (tails of distribution functions of climate variables) or associated with a specific threshold (Section 3.1.3.). These two definitions of climate extremes, probability-based or threshold-based, are not necessarily antithetic. Indeed, hazards for society and ecosystems are often extreme both from a probability and threshold perspective (e.g., a 40°C threshold for midday temperature in the mid-latitudes).

In the scientific literature, several aspects are considered in the definition and analysis of climate extremes (Box 3-1).

3.1.3. Compound (Multiple) Events

In climate science, compound events can be (1) two or more extreme events occurring simultaneously or successively, (2) combinations of extreme events with underlying conditions that amplify the impact of the events, or (3) combinations of events that are not themselves extremes but lead to an extreme event or impact when combined. The contributing events can be of similar (clustered multiple events) or different type(s). There are several varieties of clustered multiple events, such as tropical cyclones generated a few days apart with the same path and/or intensities, which may occur if there is a tendency for persistence in atmospheric circulation and genesis conditions. Examples of compound events resulting from events of different types are varied – for instance, high sea level coinciding with tropical cyclone landfall (Section 3.4.4), or cold and dry conditions (e.g., the Mongolian Dzud, see Case Study 9.2.4), or the impact of hot events and droughts on wildfire (Case Study 9.2.2), or a combined risk of flooding from sea level surges and precipitation-induced high river discharge (Svensson and Jones, 2002; Van den Brink et al., 2005). Compound events can even result from 'contrasting extremes', for example, the projected occurrence of both droughts and heavy precipitation events in future climate in some regions (Table 3-3).

Impacts on the physical environment (Section 3.5) are often the result of compound events. For instance, floods will more likely occur over saturated soils (Section 3.5.2), which means that both soil moisture status and precipitation intensity play a role. The wet soil may itself be the result of a number of above-average but not necessarily extreme precipitation events, or of enhanced snow melt associated with temperature anomalies in a given season. Similarly, droughts are the result of pre-existing soil moisture deficits and of the accumulation of precipitation deficits and/or evapotranspiration excesses (Box 3-3), not all (or none) of which are necessarily extreme for a particular drought event when considered in isolation. Also, impacts on human systems or ecosystems (Chapter 4) can be the results of compound events, for example, in the case of health-related impacts associated with combined temperature and humidity conditions (Box 3-1).

Although compound events can involve causally unrelated events, the following causes may lead to a correlation between the occurrence of extremes (or their impacts):
1) A common external forcing factor for changing the probability of the two events (e.g., regional warming, change in frequency or intensity of El Niño events)
2) Mutual reinforcement of one event by the other and vice versa due to system feedbacks (Section 3.1.4)
3) Conditional dependence of the occurrence or impact of one event on the occurrence of another event (e.g., extreme soil moisture levels and precipitation conditions for floods, droughts, see above).

Changes in one or more of these factors would be required for a changing climate to induce changes in the occurrence of compound events. Unfortunately, investigation of possible changes in these factors has received little attention. Also, much of the analysis of changes of extremes has, up to now, focused on individual extremes of a single variable. However, recent literature in climate research is starting to consider compound events and explore appropriate methods for their analysis (e.g., Coles, 2001; Beirlant et al., 2004; Benestad and Haugen, 2007; Renard and Lang, 2007; Schölzel and Friederichs, 2008; Beniston, 2009b; Tebaldi and Sanso, 2009; Durante and Salvadori, 2010).

3.1.4. Feedbacks

A special case of compound events is related to the presence of feedbacks within the climate system, that is, mutual interaction between several climate processes, which can either lead to a damping (negative feedback) or enhancement (positive feedback) of the initial response to a given forcing (see also 'climate feedback' in the Glossary). Feedbacks can play an important role in the development of extreme events, and in some cases two (or more) climate extremes can mutually strengthen one another. One example of positive feedback between two extremes is the possible mutual enhancement of droughts and heat waves in transitional regions between dry and wet climates. This feedback has been identified as having an influence on projected changes in temperature variability and heat wave occurrence in Central and Eastern Europe and the Mediterranean (Seneviratne et al., 2006a; Diffenbaugh et al., 2007), and possibly also in Britain, Eastern North America, the Amazon, and East Asia (Brabson et al., 2005; Clark et al., 2006). Further results also suggest that it is a relevant factor for past heat waves and temperature extremes in Europe and the United States (Durre et al., 2000; Fischer et al., 2007a,b; Hirschi et al., 2011). Two main mechanisms that have been suggested to underlie this feedback are: (1) enhanced soil drying during heat waves due to increased evapotranspiration (as a consequence of higher vapor pressure deficit and higher incoming radiation); and (2) higher relative heating of the air from sensible heat flux when soil moisture deficit starts limiting evapotranspiration/latent heat flux (e.g., Seneviratne et al., 2010). Additionally, there may also be indirect and/or non-local effects of dryness on heat waves through, for example, changes in circulation patterns or dry air advection (e.g., Fischer et al., 2007a; Vautard et al., 2007; Haarsma et al., 2009). However, the strength of these feedbacks is still uncertain in current climate models (e.g., Clark et al., 2010), in particular if additional feedbacks with precipitation (e.g., Koster et al., 2004b; Seneviratne et al., 2010) and with land use and land cover state and changes (e.g., Lobell et al., 2008; Pitman et al., 2009; Teuling et al., 2010) are considered. Also, feedbacks between trends in snow cover and changes in temperature extremes have been highlighted as being relevant for projections (e.g., Kharin et al., 2007; Orlowsky and Seneviratne, 2011). Feedbacks with soil moisture

Table 3-1 | Overview of considered extremes and summary of observed and projected changes at a global scale. Regional details on observed and projected changes in temperature and precipitation extremes are provided in Tables 3-2 and 3-3. Extremes (e.g., cold/warm days/nights, heat waves, heavy precipitation events) are defined with respect to late 20th-century climate (see also Box 3-1 for discussion of reference period).

		Observed Changes (since 1950)	Attribution of Observed Changes	Projected Changes (up to 2100) with Respect to Late 20th Century
Weather and Climate Variables	Temperature (Section 3.3.1)	*Very likely* decrease in number of unusually cold days and nights at the global scale. *Very likely* increase in number of unusually warm days and nights at the global scale. *Medium confidence* in increase in length or number of warm spells or heat waves in many (but not all) regions. *Low or medium confidence* in trends in temperature extremes in some subregions due either to lack of observations or varying signal within subregions. [Regional details in Table 3-2]	*Likely* anthropogenic influence on trends in warm/cold days/nights at the global scale. No attribution of trends at a regional scale with a few exceptions.	*Virtually certain* decrease in frequency and magnitude of unusually cold days and nights at the global scale. *Virtually certain* increase in frequency and magnitude of unusually warm days and nights at the global scale. *Very likely* increase in length, frequency, and/or intensity of warm spells or heat waves over most land areas. [Regional details in Table 3-3]
	Precipitation (Section 3.3.2)	*Likely* statistically significant increases in the number of heavy precipitation events (e.g., 95th percentile) in more regions than those with statistically significant decreases, but strong regional and subregional variations in the trends. [Regional details in Table 3-2]	*Medium confidence* that anthropogenic influences have contributed to intensification of extreme precipitation at the global scale.	*Likely* increase in frequency of heavy precipitation events or increase in proportion of total rainfall from heavy falls over many areas of the globe, in particular in the high latitudes and tropical regions, and in winter in the northern mid-latitudes. [Regional details in Table 3-3]
	Winds (Section 3.3.3)	*Low confidence* in trends due to insufficient evidence.	*Low confidence* in the causes of trends due to insufficient evidence.	*Low confidence* in projections of extreme winds (with the exception of wind extremes associated with tropical cyclones).
Phenomena Related to Weather and Climate Extremes	Monsoons (Section 3.4.1)	*Low confidence* in trends because of insufficient evidence.	*Low confidence* due to insufficient evidence.	*Low confidence* in projected changes in monsoons, because of insufficient agreement between climate models.
	El Niño and other Modes of Variability (Sections 3.4.2 and 3.4.3)	*Medium confidence* in past trends toward more frequent central equatorial Pacific El Niño-Southern Oscillation (ENSO) events. Insufficient evidence for more specific statements on ENSO trends. *Likely* trends in Southern Annular Mode (SAM).	*Likely* anthropogenic influence on identified trends in SAM.[1] Anthropogenic influence on trends in North Atlantic Oscillation (NAO) are *about as likely as not*. No attribution of changes in ENSO.	*Low confidence* in projections of changes in behavior of ENSO and other modes of variability because of insufficient agreement of model projections.
	Tropical Cyclones (Section 3.4.4)	*Low confidence* that any observed long-term (i.e., 40 years or more) increases in tropical cyclone activity are robust, after accounting for past changes in observing capabilities.	*Low confidence* in attribution of any detectable changes in tropical cyclone activity to anthropogenic influences (due to uncertainties in historical tropical cyclones record, incomplete understanding of physical mechanisms, and degree of tropical cyclone variability).	*Likely* decrease or no change in frequency of tropical cyclones. *Likely* increase in mean maximum wind speed, but possibly not in all basins. *Likely* increase in heavy rainfall associated with tropical cyclones.
	Extratropical Cyclones (Section 3.4.5)	*Likely* poleward shift in extratropical cyclones. *Low confidence* in regional changes in intensity.	*Medium confidence* in an anthropogenic influence on poleward shift.	*Likely* impacts on regional cyclone activity but *low confidence* in detailed regional projections due to only partial representation of relevant processes in current models. *Medium confidence* in a reduction in the numbers of mid-latitude storms. *Medium confidence* in projected poleward shift of mid-latitude storm tracks.
Impacts on Physical Environment	Droughts (Section 3.5.1)	*Medium confidence* that some regions of the world have experienced more intense and longer droughts, in particular in southern Europe and West Africa, but opposite trends also exist. [Regional details in Table 3-2]	*Medium confidence* that anthropogenic influence has contributed to some observed changes in drought patterns. *Low confidence* in attribution of changes in drought at the level of single regions due to inconsistent or insufficient evidence.	*Medium confidence* in projected increase in duration and intensity of droughts in some regions of the world, including southern Europe and the Mediterranean region, central Europe, central North America, Central America and Mexico, northeast Brazil, and southern Africa. Overall *low confidence* elsewhere because of insufficient agreement of projections. [Regional details in Table 3-3]
	Floods (Section 3.5.2)	Limited to medium evidence available to assess climate-driven observed changes in the magnitude and frequency of floods at regional scale. Furthermore, there is low agreement in this evidence, and thus overall *low confidence* at the global scale regarding even the sign of these changes. *High confidence* in trend toward earlier occurrence of spring peak river flows in snowmelt- and glacier-fed rivers.	*Low confidence* that anthropogenic warming has affected the magnitude or frequency of floods at a global scale. *Medium confidence* to *high confidence* in anthropogenic influence on changes in some components of the water cycle (precipitation, snowmelt) affecting floods.	*Low confidence* in global projections of changes in flood magnitude and frequency because of insufficient evidence. *Medium confidence* (based on physical reasoning) that projected increases in heavy precipitation would contribute to rain-generated local flooding in some catchments or regions. *Very likely* earlier spring peak flows in snowmelt- and glacier-fed rivers.

Continued next page →

Table 3-1 (continued)

		Observed Changes (since 1950)	Attribution of Observed Changes	Projected Changes (up to 2100) with Respect to Late 20th Century
Impacts on Physical Environment (Continued)	Extreme Sea Level and Coastal Impacts (Sections 3.5.3, 3.5.4, and 3.5.5)	*Likely* increase in extreme coastal high water worldwide related to increases in mean sea level in the late 20th century.	*Likely* anthropogenic influence via mean sea level contributions.	*Very likely* that mean sea level rise will contribute to upward trends in extreme coastal high water levels. *High confidence* that locations currently experiencing coastal erosion and inundation will continue to do so due to increasing sea level, in the absence of changes in other contributing factors.
	Other Physical Impacts (Sections 3.5.6, 3.5.7, and 3.5.8)	*Low confidence* in global trends in large landslides in some regions. *Likely* increased thawing of permafrost with *likely* resultant physical impacts.	*Likely* anthropogenic influence on thawing of permafrost. *Low confidence* of other anthropogenic influences because of insufficient evidence for trends in other physical impacts in cold regions.	*High confidence* that changes in heat waves, glacial retreat, and/or permafrost degradation will affect high mountain phenomena such as slope instabilities, mass movements, and glacial lake outburst floods. *High confidence* that changes in heavy precipitation will affect landslides in some regions. *Low confidence* in projected future changes in dust activity.

Notes: 1. Due to trends in stratospheric ozone concentrations.

and snow affect extremes in specific regions (hot extremes in transitional climate regions, and cold extremes in snow-covered regions), where they may induce significant deviations in changes in extremes versus changes in the average climate, as also discussed in Section 3.1.6. Other relevant feedbacks involving extreme events are those that can lead to impacts on the global climate, such as modification of land carbon uptake due to enhanced drought occurrence (e.g., Ciais et al., 2005; Friedlingstein et al., 2006; Reichstein et al., 2007) or carbon release due to permafrost degradation (see Section 3.5.7). These aspects are not, however, specifically considered in this chapter (but see Section 3.1.7, on projections of possible increased Amazon drought and forest dieback in this region). Chapter 4 also addresses feedback loops between droughts, fire, and climate change (Section 4.2.2.1).

3.1.5. Confidence and Likelihood of Assessed Changes in Extremes

In this chapter, all assessments regarding past or projected changes in extremes are expressed following the new IPCC Fifth Assessment Report uncertainty guidance (Mastrandrea et al., 2010). The new uncertainty guidance makes a clearer distinction between confidence and likelihood (see Box SPM.2). Its use complicates comparisons between assessments in this chapter and those in the IPCC Fourth Assessment Report (AR4), as they are not directly equivalent in terms of nomenclature. The following procedure was adopted in this chapter (see in particular the Executive Summary and Tables 3-1, 3-2, and 3-3.):

- For each assessment, the *confidence* level for the given assessment is first assessed (*low*, *medium*, or *high*), as discussed in the next paragraph.
- For assessments with *high confidence*, likelihood assessments of a direction of change are also provided (*virtually certain* for 99-100%, *very likely* for 90-100%, *likely* for 66-100%, *more likely than not* for 50-100%, *about as likely as not* for 33-66%, *unlikely* for 0-33%, *very unlikely* for 0-10%, and *exceptionally unlikely* for 0-1%). In a few cases for which there is *high confidence* (e.g., based on physical understanding) but for which there are not sufficient model projections to provide a more detailed likelihood assessment (such as '*likely*'), only the confidence assessment is provided.
- For assessments with *medium confidence*, a direction of change is provided, but without an assessment of likelihood.
- For assessments with *low confidence*, no direction of change is generally provided.

The confidence assessments are expert-based evaluations that consider the confidence in the tools and data basis (models, data, proxies) used to assess or project changes in a specific element, and the associated level of understanding. Examples of cases of *low confidence* for model projections are if models display poor performance in simulating the specific extreme in the present climate (see also Box 3-2), or if insufficient literature on model performance is available for the specific extreme, for example, due to lack of observations. Similarly for observed changes, the assessment may be of *low confidence* if the available evidence is based only on scattered data (or publications) that are insufficient to provide a robust assessment for a large region, or the observations may be of poor quality, not homogeneous, or only of an indirect nature (proxies). In cases with *low confidence* regarding past or projected changes in some extremes, we indicate whether the *low confidence* is due to lack of literature, lack of evidence (data, observations), or lack of understanding. It should be noted that there are some overlaps between these categories, as for instance a lack of evidence can be at the root of a lack of literature and understanding. Cases of changes in extremes for which confidence in the models and data is rated as '*medium*' are those where we have some confidence in the tools and evidence available to us, but there remain substantial doubts about some aspects of the quality of these tools. It should be noted that an assessment of *low* confidence in observed or projected changes or trends in a specific extreme neither implies nor excludes the possibility of changes in this extreme. Rather the assessment indicates *low confidence* in the ability to detect or project any such changes.

Changes (observed or projected) in some extremes are easier to assess than in others either due to the complexity of the underlying processes or to the amount of evidence available for their understanding. This results in differing levels of uncertainty in climate simulations and projections for different extremes (Box 3-2). Because of these issues, projections in some extremes are difficult or even impossible to provide, although projections in some other extremes have a high level of confidence. In addition, uncertainty in projections also varies over different time frames for individual extremes, because of varying contributions over time of internal climate variability, model uncertainty, and emission scenario uncertainty to the overall uncertainty (Box 3-2 and Section 3.2). Overall, we can infer that our confidence in past and future changes in extremes varies with the type of extreme, the data available, and the region, season, and time frame being considered, linked with the level of understanding and reliability of simulation of the underlying physical processes. These various aspects are addressed in more detail in Box 3-2, Section 3.2, and the subsections on specific extremes in Sections 3.3-3.5.

3.1.6. Changes in Extremes and Their Relationship to Changes in Regional and Global Mean Climate

Changes in extremes can be linked to changes in the mean, variance, or shape of probability distributions, or all of these (see, e.g., Figure 1-2). Thus a change in the frequency of occurrence of hot days (i.e., days above a certain threshold) can arise from a change in the mean daily maximum temperature, and/or from a change in the variance and/or shape of the frequency distribution of daily maximum temperatures. If changes in the frequency of occurrence of hot days were mainly linked to changes in the mean daily maximum temperature, and changes in the shape and variability of the distribution of daily maximum temperatures were of secondary importance, then it might be reasonable to use projected changes in mean temperature to estimate how changes in extreme temperatures might change in the future. If, however, changes in the shape and variability of the frequency distribution of daily maximum temperature were important, such naive extrapolation would be less appropriate or possibly even misleading (e.g., Ballester et al., 2010). The results of both empirical and model studies indicate that although in several situations changes in extremes do scale closely with changes in the mean (e.g., Griffiths et al., 2005), there are sufficient exceptions from this that changes in the variability and shape of probability distributions of weather and climate variables need to be considered as well as changes in means, if we are to project future changes in extremes (e.g., Hegerl et al., 2004; Schär et al., 2004; Caesar et al., 2006; Clark et al., 2006; Della-Marta et al., 2007a; Kharin et al., 2007; Brown et al., 2008; Ballester et al., 2010; Orlowsky and Seneviratne, 2011). This appears to be especially the case for short-duration precipitation, and for temperatures in mid- and high latitudes (but not all locations in these regions). In mid- and high latitudes stronger increases (or decreases) in some extremes are generally associated with feedbacks with soil moisture or snow cover (Section 3.1.4). Note that the respective importance of changes in mean versus changes in variability also depends on the choice of the reference period used to define the extremes (Box 3-1).

An additional relevant question is the extent to which regional changes in extremes scale with changes in global mean climate. Indeed, recent publications and the public debate have focused, for example, on global mean temperature targets (e.g., Allen et al., 2009; Meinshausen et al., 2009), however, the exact implications of these mean global changes (e.g., '2°C target') for regional extremes have not been widely assessed (e.g., Clark et al., 2010). Orlowsky and Seneviratne (2011) investigated the scaling between projected changes in the 10th and 90th percentile of Tmax on annual and seasonal (June-July-August: JJA, December-January-February: DJF) time scales with globally averaged annual mean changes in Tmax based on the whole CMIP3 ensemble (see Section 3.2.3 for discussion of the CMIP3 ensemble). The results highlight particularly large projected changes in the 10th percentile Tmax in the northern high-latitude regions in winter and the 90th percentile Tmax in Southern Europe in summer with scaling factors of about 2 in both cases (i.e., increases of about 4°C for a mean global increase of 2°C). However, in some regions and seasons, the scaling can also be below 1 (e.g., changes in 10th percentile in JJA in the high latitudes). This is also illustrated in Figure 3-5a, which compares analyses of changes in return values of annual extremes of maximum daily temperatures for the overall land and specific regions, and shows high region-to-region variability in these changes. The changes in return values at the global scale ('Globe (Land only)') for their part are almost identical to the changes in global mean daily maximum temperature, suggesting that the scaling issues are related to regional effects rather than overall differences in the changes in the tails versus the means of the distributions of daily maximum temperature. The situation is very different for precipitation (Figure 3-7a), with clearly distinct behavior between changes in mean and extreme precipitation at the global scale, highlighting the dependency of any scaling on the variable being considered. The lack of consistent scaling between regional and seasonal changes in extremes and changes in means has also been highlighted in empirical studies (e.g., Caesar et al., 2006). It should further be noted that not only do regional extremes not necessarily scale with global mean changes, but also mean global warming does not exclude the possibility of cooling in some regions and seasons, both in the recent past and in the coming decades: it has for instance been recently suggested that the decrease in sea ice caused by the mean warming could induce, although not systematically, more frequent cold winter extremes over northern continents (Petoukhov and Semenov, 2010). Also parts of central North America and the eastern United States present cooling trends in mean temperature and some temperature extremes in the spring to summer season in recent decades (Section 3.3.1). It should be noted that, independently of scaling issues for the means and extremes of the same variable, some extremes can be related to mean climate changes in other variables, such as links between mean global changes in relative humidity and some regional changes in heavy precipitation events (Sections 3.2.2.1 and 3.3.2).

Global-scale trends in a specific extreme may be either more reliable or less reliable than some regional-scale trends, depending on the geographical uniformity of the trends in the specific extreme. In particular, climate projections for some variables are not consistent, even in the sign of the projected change, everywhere across the globe (e.g., Christensen et al., 2007; Meehl et al., 2007b). For instance, projections typically include some regions with a tendency toward wetter conditions and others with a tendency toward drier conditions, with some regions displaying a shift in climate regimes (e.g., from humid to transitional or transitional to dry). Some of these regional changes will depend on how forcing changes may alter the regional atmospheric circulation, especially in coastal regions and regions with substantial orography. Hence for certain extremes such as floods and droughts, regional projections might indicate larger changes than is the case for projections of global averages (which would average the regional signals exhibiting changes of opposite signs). This also means that signals at the regional scale may be more reliable (and meaningful) in some cases than assessments at the global scale. On the other hand, temperature extremes projections, which are consistent across most regions, are thus more reliable at the global scale ('*virtually certain*') than at the regional scale (at most '*very likely*').

3.1.7. Surprises / Abrupt Climate Change

This report focuses on the most probable changes in extremes based on current knowledge. However, the possible future occurrence of low-probability, high-impact scenarios associated with the crossing of poorly understood climate thresholds cannot be excluded, given the transient and complex nature of the climate system. Such scenarios have important implications for society as highlighted in Section 8.5.1. So, an assessment that we have *low confidence* in projections of a specific extreme, or even lack of consideration of given climate changes under the categories covered in this chapter (e.g., shutdown of the meridional overturning circulation), should not be interpreted as meaning that no change is expected in this extreme or climate element (see also Section 3.1.5). Feedbacks play an important role in either damping or enhancing extremes in several climate variables (Section 3.1.4), and this can also lead to 'surprises,' that is, changes in extremes greater (or less) than might be expected with a gradual warming of the climate system. Similarly, as discussed in 3.1.3, contrasting or multiple extremes can occur but our understanding of these is insufficient to provide credible comprehensive projections of risks associated with such combinations.

One aspect that we do not address in this chapter is the existence of possible tipping points in the climate system (e.g., Meehl et al., 2007b; Lenton et al., 2008; Scheffer et al., 2009), that is, the risks of abrupt, possibly irreversible changes in the climate system. Abrupt climate change is defined as follows in the Glossary: "The nonlinearity of the climate system may lead to abrupt climate change, sometimes called rapid climate change, abrupt events, or even surprises. The term abrupt often refers to time scales faster than the typical time scale of the responsible forcing. However, not all abrupt climate changes need be externally forced. Some changes may be truly unexpected, resulting from a strong, rapidly changing forcing of a nonlinear system." Thresholds associated with tipping points may be termed 'critical thresholds,' or, in the case of the climate system, 'climate thresholds'. Scheffer et al. (2009) illustrate the possible equilibrium responses of a system to forcing. In the case of a linear response, only a large forcing can lead to a major state change in the system. However, in the presence of a critical threshold even a small change in forcing can lead to a similar

major change in the system. For systems with critical bifurcations in the equilibrium state function two alternative stable conditions may exist, whereby an induced change may be irreversible. Such critical transitions within the climate system represent typical low-probability, high-impact scenarios, which were also noted in the AR4 (Meehl et al., 2007b). Lenton et al. (2008) provided a recent review on potential tipping elements within the climate system, that is, subsystems of the Earth system that are at least subcontinental in scale and which may entail a tipping point. Some of these would be especially relevant to certain extremes [e.g., El Niño-Southern Oscillation (ENSO), the Indian summer monsoon, and the Sahara/Sahel and West African monsoon for drought and heavy precipitation, and the Greenland and West Antarctic ice sheets for sea level extremes], or are induced by changes in extremes (e.g., Amazon rainforest die-back induced by drought). For some of the identified tipping elements, the existence of bistability has been suggested by paleoclimate records, but is still debated in some cases (e.g., Brovkin et al., 2009). There is often a lack of agreement between models regarding these low-probability, high-impact scenarios, for instance, regarding a possible increased drought and consequent die-back of the Amazon rainforest (e.g., Friedlingstein et al., 2006; Poulter et al., 2010; see Table 3-3 for dryness projections in this region), the risk of an actual shutdown of the Atlantic thermohaline circulation (e.g., Rahmstorf et al., 2005; Lenton et al., 2008), or the potential irreversibility of the decrease in Arctic sea ice (Tietsche et al., 2011). For this reason, *confidence* in these scenarios is assessed as *low*.

3.2. Requirements and Methods for Analyzing Changes in Extremes

3.2.1. Observed Changes

Sections 3.3 to 3.5 of this chapter provide assessments of the literature regarding changes in extremes in the observed record published mainly since the AR4 and building on the AR4 assessment. Summaries of these assessments are provided in Table 3-1. Overviews of observed regional changes in temperature and precipitation extremes are provided in Table 3-2. In this section issues are discussed related to the data and observations used to examine observed changes in extremes.

Issues with data availability are especially critical when examining changes in extremes of given climate variables (Nicholls, 1995). Indeed, the more rare the event, the more difficult it is to identify long-term changes, simply because there are fewer cases to evaluate (Frei and Schär, 2001; Klein Tank and Können, 2003). Identification of changes in extremes is also dependent on the analysis technique employed (X. Zhang et al., 2004; Trömel and Schönwiese, 2005). Another important criterion constraining data availability for the analysis of extremes is the respective time scale on which they occur (Section 3.1.2), since this determines the required temporal resolution for their assessment (e.g., heavy hourly or daily precipitation versus multi-year drought). Longer time resolution data (e.g., monthly, seasonal, and annual values) for temperature and precipitation are available for most parts of the world starting late in the 19th to early 20th century, and allow analysis of meteorological drought (see Box 3-3) and unusually wet periods of the order of a month or longer. To examine changes in extremes occurring on short time scales, particularly of climate variables such as temperature and precipitation (or wind), normally requires the use of high-temporal resolution data, such as daily or sub-daily observations, which are generally either not available, or available only since the middle of the 20th century and in many regions only from as recently as 1970. Even where sufficient data are available, several problems can still limit their analysis. First, although the situation is changing (especially for the situation with respect to 'extreme indices,' Box 3-1), many countries still do not freely distribute their higher temporal resolution data. Second, there can be issues with the quality of measurements. A third important issue is climate data homogeneity (see below). These and other issues are discussed in detail in the AR4 (Trenberth et al., 2007). For instance, the temperature and precipitation stations considered in the daily data set used in Alexander et al. (2006) are not globally uniform. Although observations for most parts of the globe are available, measurements are lacking in Northern South America, Africa, and part of Australia. The other data set commonly used for extremes analyses is from Caesar et al. (2006; used, e.g., in Brown et al., 2008), which also has data gaps in most of South America, Africa, Eastern Europe, Mexico, the Middle East, India, and Southeast Asia. Also the study by Vose et al. (2005) has data gaps in South America, Africa, and India. It should be further noted that the regions with data coverage do not all have the same density of stations (Alexander et al., 2006; Caesar et al., 2006). While some studies are available on a country or regional basis for areas not covered in global studies (see, e.g., Tables 3-2 and 3-3), lack of data in many parts of the globe leads to limitations in our ability to assess observed changes in climate extremes for many regions.

Whether or not climate data are homogeneous is of clear relevance for an analysis of extremes, especially at smaller spatial scales. Data are defined as homogeneous when the variations and trends in a climate time series are due solely to variability and changes in the climate system. Some meteorological elements are especially vulnerable to uncertainties caused by even small changes in the exposure of the measuring equipment. For instance, erection of a small building or changes in vegetative cover near the measuring equipment can produce a bias in wind measurements (Wan et al., 2010). When a change occurs it can result in either a discontinuity in the time series (step change) or a more gradual change that can manifest itself as a false trend (Menne and Williams Jr., 2009), both of which can impact on whether a particular observation exceeds a threshold. Homogeneity detection and data adjustments have been implemented for longer averaging periods (e.g., monthly, seasonal, annual); however, techniques applicable to shorter observing periods (e.g., daily) data have only recently been developed (e.g., Vincent et al., 2002; Della-Marta and Wanner, 2006), and have not been widely implemented. Homogeneity issues also affect the monitoring of other meteorological and climate variables, for which further and more severe limitations also can exist. This is in particular the case regarding measurements of wind and relative humidity, and data required for the analysis of weather and climate phenomena (tornadoes, extratropical and tropical cyclones; Sections 3.3.3, 3.4.4, and 3.4.5), as well as impacts on the physical environment (e.g., droughts, floods, cryosphere impacts; Section 3.5).

Thunderstorms, tornadoes, and related phenomena are not well observed in many parts of the world. Tornado occurrence since 1950 in the United States, for instance, displays an increasing trend that mainly reflects increased population density and increased numbers of people in remote areas (Trenberth et al., 2007; Kunkel et al., 2008). Such trends increase the likelihood that a tornado would be observed. A similar problem occurs with thunderstorms. Changes in reporting practices, increased population density, and even changes in the ambient noise level at an observing station all have led to inconsistencies in the observed record of thunderstorms.

Studies examining changes in extratropical cyclones, which focus on changes in storm track location, intensities, and frequency, are limited in time due to a lack of suitable data prior to about 1950. Most of these studies have relied on model-based reanalyses that also incorporate observations into a hybrid model-observational data set. However, reanalyses can have homogeneity problems due to changes in the amount and type of data being assimilated, such as the introduction of satellite data in the late 1970s and other observing system changes (Trenberth et al., 2001; Bengtsson et al., 2004). Recent reanalysis efforts have attempted to produce more homogeneous reanalyses that show promise for examining changes in extratropical cyclones and other climate features (Compo et al., 2006). Results, however, are strongly dependent on the reanalysis and cyclone tracking techniques used (Ulbrich et al., 2009).

The robustness of analyses of observed changes in tropical cyclones has been hampered by a number of issues with the historical record. One of the major issues is the heterogeneity introduced by changing technology and reporting protocols within the responsible agencies (e.g., Landsea et al., 2004). Further heterogeneity is introduced when records from multiple ocean basins are combined to explore global trends, because data quality and reporting protocols vary substantially between agencies (Knapp and Kruk, 2010). Much like other weather and climate observations,

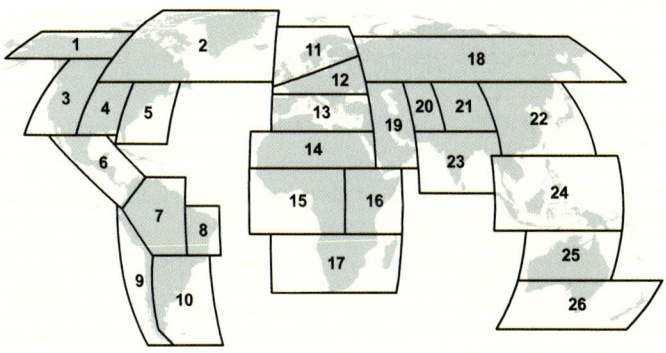

Figure 3-1 | Definitions of regions used in Tables 3-2 and 3-3, and Figures 3-5 and 3-7. Exact coordinates of the regions are provided in the on-line supplement, Appendix 3.A. Assessments and analyses are provided for land areas only.

FAQ 3.1 | Is the Climate Becoming More Extreme?

While there is evidence that increases in greenhouse gases have likely caused changes in some types of extremes, there is no simple answer to the question of whether the climate, in general, has become more or less extreme. Both the terms 'more extreme' and 'less extreme' can be defined in different ways, resulting in different characterizations of observed changes in extremes. Additionally, from a physical climate science perspective it is difficult to devise a comprehensive metric that encompasses all aspects of extreme behavior in the climate.

One approach for evaluating whether the climate is becoming more extreme would be to determine whether there have been changes in the typical range of variation of specific climate variables. For example, if there was evidence that temperature variations in a given region had become significantly larger than in the past, then it would be reasonable to conclude that temperatures in that region had become more extreme. More simply, temperature variations might be considered to be becoming more extreme if the difference between the highest and the lowest temperature observed in a year is increasing. According to this approach, daily temperature over the globe may have become less extreme because there have generally been greater increases in mean daily minimum temperatures globally than in mean daily maximum temperatures, over the second half of the 20th century. On the other hand, one might conclude that daily precipitation has become more extreme because observations suggest that the magnitude of the heaviest precipitation events has increased in many parts of the world. Another approach would be to ask whether there have been significant changes in the frequency with which climate variables cross fixed thresholds that have been associated with human or other impacts. For example, an increase in the mean temperature usually results in an increase in hot extremes and a decrease in cold extremes. Such a shift in the temperature distribution would not increase the 'extremeness' of day-to-day variations in temperature, but would be perceived as resulting in a more extreme warm temperature climate, and a less extreme cold temperature climate. So the answer to the question posed here would depend on the variable of interest, and on which specific measure of the extremeness of that variable is examined. As well, to provide a complete answer to the above question, one would also have to collate not just trends in single variables, but also indicators of change in complex extreme events resulting from a sequence of individual events, or the simultaneous occurrence of different types of extremes. So it would be difficult to comprehensively describe the full suite of phenomena of concern, or to find a way to synthesize all such indicators into a single extremeness metric that could be used to comprehensively assess whether the climate as a whole has become more extreme from a physical perspective. And to make such a metric useful to more than a specific location, one would have to combine the results at many locations, each with a different perspective on what is 'extreme.'

Continued next page →

---

tropical cyclone observations are taken to support short-term forecasting needs. Improvements in observing techniques are often implemented without any overlap or calibration against existing methods to document the impact of the changes on the climate record. Additionally, advances in technology have enabled better and more complete observations. For example, the introduction of aircraft reconnaissance in some basins in the 1940s and satellite data in the 1960s had a profound effect on our ability to accurately identify and measure tropical cyclones, particularly those that never encountered land or a ship. While aircraft reconnaissance programs have continued in the North Atlantic, they were terminated in the Western Pacific in 1987. The introduction of geostationary satellite imagery in the 1970s, and the introduction (and subsequent improvement) of new tropical cyclone analysis methods (such as the Dvorak technique for estimating storm intensity), further compromises the homogeneity of historical records of tropical cyclone activity.

Regarding impacts to the physical environment, soil moisture is a key variable for which data sets are extremely scarce (e.g., Robock et al., 2000; Seneviratne et al., 2010). This represents a critical issue for the validation and correct representation of soil moisture (agricultural) as well as hydrological drought (Box 3-3) in climate, land surface, and hydrological models, and the monitoring of ongoing changes in regional terrestrial water storage. As a consequence, these need to be inferred from simple climate indices or model-based approaches (Box 3-3). Such estimates rely in large part on precipitation observations, which have, however, inadequate spatial coverage for these applications in many regions of the world (e.g., Oki et al., 1999; Fekete et al., 2004; Koster et al., 2004a). Similarly, runoff observations are not globally available, which results in significant uncertainties in the closing of the global and some regional water budgets (Legates et al., 2005; Peel and McMahon, 2006; Dai et al., 2009; Teuling et al., 2009), as well as for the global analysis of changes in the occurrence of floods (Section 3.5.2). Additionally, ground observations of snow, which are lacking in several regions, are important for the investigation of physical impacts, particularly those related to the cryosphere and runoff generation (e.g., Essery et al., 2009; Rott et al., 2010).

All of the above-mentioned issues lead to uncertainties in observed trends in extremes. In many instances, great care has been taken to develop procedures to reduce the confounding influences of these issues on the data, which in turn helps to reduce uncertainty, and progress has been made in the last 15 years (e.g., Caesar et al., 2006;

Three types of metrics have been considered to avoid these problems, and thereby allow an answer to this question. One approach is to count the number of record-breaking events in a variable and to examine such a count for any trend. However, one would still face the problem of what to do if, for instance, hot extremes are setting new records, while cold extremes are not occurring as frequently as in the past. In such a case, counting the number of records might not indicate whether the climate was becoming more or less extreme, rather just whether there was a shift in the mean climate. Also, the question of how to combine the numbers of record-breaking events in various extremes (e.g., daily precipitation and hot temperatures) would need to be considered. Another approach is to combine indicators of a selection of important extremes into a single index, such as the Climate Extremes Index (CEI), which measures the fraction of the area of a region or country experiencing extremes in monthly mean surface temperature, daily precipitation, and drought. The CEI, however, omits many important extremes such as tropical cyclones and tornadoes, and could, therefore, not be considered a complete index of 'extremeness.' Nor does it take into account complex or multiple extremes, nor the varying thresholds that relate extremes to impacts in various sectors.

A third approach to solving this dilemma arises from the fact that extremes often have deleterious economic consequences. It may therefore be possible to measure the integrated economic effects of the occurrence of different types of extremes into a common instrument such as insurance payout to determine if there has been an increase or decrease in that instrument. This approach would have the value that it clearly takes into account those extremes with economic consequences. But trends in such an instrument will be dominated by changes in vulnerability and exposure and it will be difficult, if not impossible, to disentangle changes in the instrument caused by non-climatic changes in vulnerability or exposure in order to leave a residual that reflects only changes in climate extremes. For example, coastal development can increase the exposure of populations to hurricanes; therefore, an increase in damage in coastal regions caused by hurricane landfalls will largely reflect changes in exposure and may not be indicative of increased hurricane activity. Moreover, it may not always be possible to associate impacts such as the loss of human life or damage to an ecosystem due to climate extremes to a measurable instrument.

None of the above instruments has yet been developed sufficiently as to allow us to confidently answer the question posed here. Thus we are restricted to questions about whether specific extremes are becoming more or less common, and our confidence in the answers to such questions, including the direction and magnitude of changes in specific extremes, depends on the type of extreme, as well as on the region and season, linked with the level of understanding of the underlying processes and the reliability of their simulation in models.

Brown et al., 2008). As a consequence, more complete and homogenous information about changes is now available for at least some variables and regions (Nicholls and Alexander, 2007; Peterson and Manton, 2008). For instance, the development of global databases of daily temperature and precipitation covering up to 70% of the global land area has allowed robust analyses of extremes (see Alexander et al., 2006). In addition, analyses of temperature and precipitation extremes using higher temporal resolution data, such as that available in the Global Historical Climatology Network – Daily data set (Durre et al., 2008) have also proven robust at both a global (Alexander et al., 2006) and regional scale (Sections 3.3.1 and 3.3.2). Nonetheless, as highlighted above, for many extremes, data remain sparse and problematic, resulting in lower ability to establish changes, particularly on a global basis and for specific regions.

3.2.2. The Causes behind the Changes

This section discusses the main requirements, approaches, and considerations for the attribution of causes for observed changes in extremes. In Sections 3.3 to 3.5, the causes of observed changes in specific extremes are assessed. A global summary of these assessments is provided in Table 3-1. Climate variations and change are induced by variability internal to the climate system, and changes in external forcings, which include natural external forcings such as changes in solar irradiance and volcanism, and anthropogenic forcings such as aerosol and greenhouse gas emissions principally due to the burning of fossil fuels, and land use and land cover changes. The mean state, extremes, and variability are all related aspects of the climate, so external forcings that affect the mean climate would in general result in changes in extremes. For this reason, we provide in Section 3.2.2.1 a brief overview of human-induced changes in the mean climate to aid the understanding of changes in extremes as the literature directly addressing the causes of changes in extremes is quite limited.

3.2.2.1. Human-Induced Changes in the Mean Climate that Affect Extremes

The occurrence of extremes is usually the result of multiple factors, which can act either on the large scale or on the regional (and local) scale (see also Section 3.1.6). Some relevant large-scale impacts of external forcings affecting extremes include net increases in temperature induced by changes in radiation, enhanced moisture content of the atmosphere, and increased land-sea contrast in temperatures, which can, for example, affect circulation patterns and to some extent monsoons. At regional and local scales, additional processes can modulate the overall changes in extremes, including regional feedbacks, in particular linked to land-atmosphere interactions with, for example, soil moisture or snow (e.g., Section 3.1.4). This section briefly reviews the current understanding of the causes (i.e., in the sense of attribution to either external forcing or internal climate variability) of large-scale (and some regional) changes in the mean climate that are of relevance to extreme events, to the extent that they have been considered in detection and attribution studies.

Regarding observed increases in global average annual mean surface temperatures in the second half of the 20th century, we base our analysis on the following AR4 assessment (Hegerl et al., 2007): Most of the observed increase in global average temperatures is *very likely* due to the observed increase in anthropogenic greenhouse gas concentrations. Greenhouse gas forcing alone would *likely* have resulted in a greater warming than observed if there had not been an offsetting cooling effect from aerosol and other forcings. It is *extremely unlikely* (<5%) that the global pattern of warming can be explained without external forcing, and *very unlikely* that it is due to known natural external causes alone. Anthropogenically forced warming over the second half of the 20th century has also been detected in ocean heat content and air temperatures in all continents (Hegerl et al., 2007; Gillett et al., 2008b).

Hegerl et al. (2007) assessed literature that considered detection in temperature trends at scales as small as approximately 500 km. Recent work has provided more evidence of detection of an anthropogenic influence at increasingly smaller spatial scales and for seasonal averages (Stott et al., 2010). For instance, Min and Hense (2007) found that estimates of response to anthropogenic forcing from the multi-model Coupled Model Intercomparison Project 3 (CMIP3) ensemble (see Section 3.2.3.3) provided a better explanation for observed continental-scale seasonal temperature changes than alternative explanations such as natural external forcing or internal variability. In another study, an anthropogenic signal was detected in 20th-century summer temperatures in Northern Hemisphere subcontinental regions except central North America, although the results were more uncertain when anthropogenic and natural signals were considered together (Jones et al., 2008). An anthropogenic signal has also been detected in multi-decadal trends in a US climate extreme index (Burkholder and Karoly, 2007), in the hydrological cycle of the western United States (Barnett et al., 2008), in New Zealand temperatures (Dean and Stott, 2009), and in European temperatures (Christidis et al., 2011a).

Attribution has more stringent demands than those for the detection of an external influence in observations. Overall, attribution at scales smaller than continental has still not yet been established primarily due to the low signal-to-noise ratio and the difficulties of separately attributing effects of the wider range of possible driving processes

(either attributable to external forcing or internal climate variability) at these scales (Hegerl et al., 2007). One reason is that averaging over smaller regions reduces the internal variability less than does averaging over large regions. In addition, the small-scale details of external forcing, and the responses simulated by models, are less credible than large-scale features. For instance, temperature changes are poorly simulated by models in some regions and seasons (Dean and Stott, 2009; van Oldenborgh et al., 2009). Also the inclusion of additional forcing factors, such as land use change and aerosols that can be more important at regional scales, remains a challenge (Lohmann and Feichter, 2007; Pitman et al., 2009; Rotstayn et al., 2009).

One of the significant advances since AR4 is emerging evidence of human influence on global atmospheric moisture content and precipitation. According to the Clausius-Clapeyron relationship, the saturation vapor pressure increases approximately exponentially with temperature. It is physically plausible that relative humidity would remain roughly constant under climate change (e.g., Hegerl et al., 2007). This means that specific humidity increases about 7% for a one degree increase in temperature in the current climate. Indeed, observations indicate significant increases between 1973 and 2003 in global surface specific humidity but not in relative humidity (Willett et al., 2008), and at the largest spatial-temporal scales moistening is close to the Clausius-Clapeyron scaling of the saturated specific humidity (~7% K^{-1}; Willett et al., 2010), though relative humidity over low- and mid-latitude land areas decreased over a 10-year period prior to 2008 possibly due to a slower temperature increase in the oceans than over the land (Simmons et al., 2010). By comparing observations with model simulations, changes in the global surface specific humidity for 1973-2003 (Willett et al., 2007), and in lower tropospheric moisture content over the 1988-2006 period (Santer et al., 2007) can be attributed to anthropogenic influence.

The increase in the atmospheric moisture content would be expected to lead to an increase in extreme precipitation when other factors do not change. Min et al. (2011) detected an anthropogenic influence in annual maxima of daily precipitation over Northern Hemisphere land areas. The influence of anthropogenic forcing has been detected in the latitudinal pattern of land precipitation trends though the model-simulated magnitude of changes is smaller than that observed (X. Zhang et al., 2007). The smaller changes in model simulations may be due in part to averaging precipitation trends from different model simulations, as spatial patterns of trends simulated by different models are not exactly the same. The influence of anthropogenic greenhouse gases and aerosols on changes in precipitation over high-latitude land areas north of 55°N has also been detected (Min et al., 2008). Detection is possible there, despite limited data coverage, in part because the response to forcing is relatively strong, and because internal variability in precipitation is low in this region.

3.2.2.2. How to Attribute a Change in Extremes to Causes

The good practice guidance paper on detection and attribution (Hegerl et al., 2010) reconciles terminologies of detection and attribution used

FAQ 3.2 | Has Climate Change Affected Individual Extreme Events?

A changing climate can be expected to lead to changes in climate and weather extremes. But it is challenging to associate a single extreme event with a specific cause such as increasing greenhouse gases because a wide range of extreme events could occur even in an unchanging climate, and because extreme events are usually caused by a combination of factors. Despite this, it may be possible to make an attribution statement about a specific weather event by attributing the changed probability of its occurrence to a particular cause. For example, it has been estimated that human influences have more than doubled the probability of a very hot European summer like that of 2003.

Recent years have seen many extreme events including the extremely hot summer in parts of Europe in 2003 and 2010, and the intense North Atlantic hurricane seasons of 2004 and 2005. Can the increased atmospheric concentrations of greenhouse gases be considered the 'cause' of such extreme events? That is, could we say these events would *not* have occurred if CO_2 had remained at pre-industrial concentrations? For instance, the monthly mean November temperature averaged across the state of New South Wales in Australia for November 2009 is about 3.5 standard deviations warmer than the 1950-2008 mean, suggesting that the chance of such a temperature occurring in the 1950-2008 climate (assuming a stationary climate) is quite low. Is this event, therefore, an indication of a changing climate? In the CRUTEM3V global land surface temperature data set, about one in every 900 monthly mean temperatures observed between 1900 and 1949 lies more than 3.5 standard deviations above the corresponding monthly mean temperature for 1950-2008.[1] Since global temperature was lower in the first half of the 20th century, this clearly indicates that an extreme warm event as rare as the November 2009 temperature in any specific location could have occurred in the past, even if its occurrence in recent times is more probable.

A second complicating issue is that extreme events usually result from a combination of factors, and this will make it difficult to attribute an extreme to a single causal factor. The hot 2003 European summer was associated with a persistent high-pressure system (which led to clear skies and thus more solar energy received at the surface) and too-dry soil (which meant that less solar energy was used for evaporation, leaving more energy to heat the soil). Another example is that hurricane genesis requires weak vertical wind shear, as well as very warm sea surface temperatures. Since some factors, but not others, may be affected by a specific cause such as increasing greenhouse gas concentrations, it is difficult to separate the human influence on a single, specific extreme event from other factors influencing the extreme.

Nevertheless, climate models can sometimes be used to identify if specific factors are changing the likelihood of the occurrence of extreme events. In the case of the 2003 European heat wave, a model experiment indicated that human influences more than doubled the likelihood of having a summer in Europe as hot as that of 2003, as discussed in the AR4. The value of such a probability-based approach – "Does human influence change the likelihood of an event?" – is that it can be used to estimate the influence of external factors, such as increases in greenhouse gases, on the frequency of specific types of events, such as heat waves or cold extremes. The same likelihood-based approach has been used to examine anthropogenic greenhouse gas contribution to flood probability.

The discussion above relates to an individual, specific occurrence of an extreme event (e.g., a single heat wave). For the reasons outlined above it remains very difficult to attribute any individual event to greenhouse gas-induced warming (even if physical reasoning or model experiments suggest such an extreme may be more likely in a changed climate). On the other hand, a long-term trend in an extreme (e.g., heat wave occurrences) is a different matter. It is certainly feasible to test whether such a trend is likely to have resulted from anthropogenic influences on the climate, just as a global warming trend can be assessed to determine its likely cause.

---

[1] We used the CRUTEM3V land surface temperature data. We limit our calculation to grid points with long-term observations, requiring at least 50 non-missing values during 1950-2008 for a calendar month and a grid point to be included. A standard deviation is computed for the period 1950-2008. We then count the number of occurrences when the temperature anomaly during 1900-1949 relative to 1950-2008 mean is greater than 3.5 standard deviations, and compare it with the total number of observations for the grid and month in that period. The ratio of these two numbers is 0.00107.

---

by Working Groups I and II in the AR4. It provides detailed guidance on the procedures that include two main approaches to attribute a change in climate to causes. One is single-step attribution, which involves assessments that attribute an observed change within a system to an external forcing based on explicitly modelling the response of the variable to the external forcings. The alternate procedure is multi-step attribution, which combines an assessment that attributes an observed change in a variable of interest to a change in climate, with a separate assessment that attributes the change to external forcings. Attribution of changes in climate extremes has some unique issues. Observed data are limited in both quantity and quality (Section 3.2.1), resulting in uncertainty in the estimation of past changes; the signal-to-noise ratio may be low for many variables and insufficient data may be available to detect such weak signals. In addition, global climate models (GCMs) have several issues in simulating extremes and downscaling techniques can only partly circumvent these issues (Section 3.2.3).

Single-step attribution based on optimal detection and attribution (e.g., Hegerl et al., 2007) can in principle be applied to climate extremes. However, the difference in statistical properties between mean values and extremes needs to be carefully addressed (e.g., Zwiers et al., 2011; see also Section 3.1.6). Post-processing of climate model simulations to derive a quantity of interest that is not explicitly simulated by the models, by applying empirical methods or physically based models to the outputs from the climate models, may make it possible to directly compare observed extremes with climate model results. For example, sea level pressure simulated by multiple GCMs has been used to derive geostrophic wind to represent atmospheric storminess and to derive significant wave height on the oceans for the detection of external influence on trends in atmospheric storminess and northern oceans wave heights (X.L. Wang et al., 2009a). GCM-simulated precipitation and temperature have also been downscaled as input to hydrological and snowpack models to infer past and future changes in temperature, timing of the peak flow, and snow water equivalent for the western United States, and this enabled a detection and attribution analysis of human-induced changes in these variables (Barnett et al., 2008).

If a single-step attribution of causes to effects on extremes or physical impacts of extremes is not feasible, it might be feasible to conduct a multiple-step attribution. The assessment would then need to be based on evidence not directly derived from model simulations, that is, physical understanding and expert judgment, or their combination. For instance, in the northern high-latitude regions, spring temperature has increased, and the timing of spring peak flows in snowmelt-fed rivers has shifted toward earlier dates (Regonda et al., 2005; Knowles et al., 2006). A change in streamflow may be attributable to external influence if streamflow regime change can be attributed to a spring temperature increase and if the spring temperature increase can be attributed to external forcings (though these changes may not necessarily be linked to changes in floods; Section 3.5.2). If the chain of processes is established (e.g., in this case additionally supported by the physical understanding that snow melts earlier as spring temperature increases), the confidence in the overall assessment would be similar to, or weaker than, the lower confidence in the two steps in the assessment. In cases where the underlying physical mechanisms are less certain, such as those linking tropical cyclones and sea surface temperature (see Section 3.4.4), the confidence in multi-step attribution can be severely undermined. A necessary condition for multi-step attribution is to establish the chain of mechanisms responsible for the specific extremes being considered. Physically based process studies and sensitivity experiments that help the physical understanding (e.g., Findell and Delworth, 2005; Seneviratne et al., 2006a; Haarsma et al., 2009) can possibly play a role in developing such multi-step attributions.

Extreme events are rare, which means that there are also few data available to make assessments regarding changes in their frequency or intensity (Section 3.2.1). When a rare and high-impact meteorological extreme event occurs, a question that is often posed is whether such an event is due to anthropogenic influence. Because it is very difficult to rule out the occurrence of low-probability events in an unchanged climate and because the occurrence of such events usually involves multiple factors, it is very difficult to attribute an individual event to external forcing (Allen, 2003; Hegerl et al., 2007; Dole et al., 2011; see also FAQ 3.2). However, in this case, it may be possible to estimate the influence of external forcing on the likelihood of such an event occurring (e.g., Stott et al., 2004; Pall et al., 2011; Zwiers et al., 2011).

3.2.3. Projected Long-Term Changes and Uncertainties

In this section we discuss the requirements and methods used for preparing climate change projections, with a focus on projections of extremes and the associated uncertainties. The discussion draws on the AR4 (Christensen et al., 2007; Meehl et al., 2007b; Randall et al., 2007) with consideration of some additional issues relevant to projections of extremes in the context of risk and disaster management. More detailed assessments of projections for specific extremes are provided in Sections 3.3 to 3.5. Summaries of these assessments are provided in Table 3-1. Overviews of projected regional changes in temperature extremes, heavy precipitation, and dryness are provided in Table 3-3 (see pages 196-202).

3.2.3.1. Information Sources for Climate Change Projections

Work on the construction, assessment, and communication of climate change projections, including regional projections and of extremes, draws on information from four sources: (1) GCMs; (2) downscaling of GCM simulations; (3) physical understanding of the processes governing regional responses; and (4) recent historical climate change (Christensen et al., 2007; Knutti et al., 2010b). At the time of the AR4, GCMs were the main source of globally available regional information on the range of possible future climates including extremes (Christensen et al., 2007). This is still the case for many regions, as can be seen in Table 3-3.

The AR4 concluded that statistics of extreme events for present-day climate, especially temperature, are generally well simulated by current GCMs at the global scale (Randall et al., 2007). Precipitation extremes are, however, less well simulated (Randall et al., 2007; Box 3-2). As they continue to develop, and their spatial resolution as well as their complexity continues to improve, GCMs could become increasingly useful for investigating smaller-scale features, including changes in extreme weather events. However, when we wish to project climate and weather extremes, not all atmospheric phenomena potentially of relevance can be realistically or explicitly simulated. GCMs include a number of approximations, known as parameterizations, of processes (e.g., relating to clouds) that cannot be fully resolved in climate models. Furthermore, the assessment of climate model performance with respect to extremes (summarized in Sections 3.3 to 3.5 for specific extremes), particularly at the regional scale, is still limited by the rarity of extreme events that makes evaluation of model performance less robust than is the case for average climate. Evaluation is further hampered by incomplete data on the historical frequency and severity of extremes, particularly for variables other than temperature and precipitation, and for specific regions (Section 3.2.1; Table 3-2).

The requirement for projections of extreme events has provided one of the motivations for the development of regionalization or downscaling techniques (Carter et al., 2007). These have been specifically developed for the study of regional- and local-scale climate change, to simulate weather and climate at finer spatial resolutions than is possible with GCMs – a step that is particularly relevant for many extremes given their spatial scale. These techniques are, nonetheless, constrained by the reliability of large-scale information coming from GCMs. Recent advances in downscaling for extremes are discussed below.

As indicated in the Glossary, downscaling "is a method that derives local- to regional-scale (up to 100 km) information from larger-scale models or data analyses." Two main methods are distinguished: dynamical downscaling and empirical/statistical downscaling (Christensen et al., 2007). The dynamical method uses the output of regional climate models (RCMs), global models with variable spatial resolution, or high-resolution global models. The empirical/statistical methods develop statistical relationships that link the large-scale atmospheric variables with local/regional climate variables. In all cases, the quality of the downscaled product depends on the quality of the driving model. Dynamical and statistical downscaling techniques are briefly introduced hereafter. Specific limitations that need to be considered in the evaluation of projections are also discussed in Section 3.2.3.2.

The most common approach to dynamical downscaling uses high-resolution RCMs, currently at scales of 20 to 50 km, but in some cases down to 10 to 15 km (e.g., Dankers et al., 2007), to represent regional sub-domains, using either observed (reanalysis) or lower-resolution GCM data to provide their boundary conditions. Using non-hydrostatic mesoscale models, applications at 1- to 5-km resolution are also possible for shorter periods (typically a few months, a few full years at most) – a scale at which clouds and convection can be explicitly resolved and the diurnal cycle tends to be better resolved (e.g., Grell et al., 2000; Hay et al., 2006; Hohenegger et al., 2008; Kanada et al., 2010b). Less commonly used approaches to dynamical downscaling involve the use of stretched-grid (variable resolution) models and high-resolution 'time-slice' models (e.g., Cubasch et al., 1995; Gibelin and Deque, 2003; Coppola and Giorgi, 2005) with the latter including some simulations at 20 km globally (Kamiguchi et al., 2006; Kitoh et al., 2009; Kim et al., 2010). The main advantage of dynamical downscaling is its potential for capturing mesoscale nonlinear effects and providing information for many climate variables at a relatively high spatial resolution, although still not as high as some require. Dynamical downscaling cannot provide information at the point (i.e., weather station) scale (a scale at which the RCM and GCM parameterizations would not work). Like GCMs, RCMs provide precipitation averaged over a grid cell, which means a tendency to more days of light precipitation (Frei et al., 2003; Barring et al., 2006) and reduced magnitude of extremes (Chen and Knutson, 2008; Haylock et al., 2008) compared with point values. These scaling issues need to be considered when evaluating the ability of RCMs and GCMs to simulate precipitation and other extremes.

Statistical downscaling methods use relationships between large-scale fields (predictors) and local-scale surface variables (predictands) that have been derived from observed data, and apply these to equivalent large-scale fields simulated by climate models (Christensen et al., 2007). They may also include weather generators that provide the basis for a number of recently developed user tools that can be used to assess changes in extreme events (Kilsby et al., 2007; Burton et al., 2008; Qian et al., 2008; Semenov, 2008). Statistical downscaling has been demonstrated to have potential in a number of different regions including Europe (e.g., Schmidli et al., 2007), Africa (e.g., Hewitson and Crane, 2006), Australia (e.g., Timbal et al., 2008, 2009), South America (e.g., D'Onofrio et al., 2010) and North America (e.g., Vrac et al., 2007; Dibike et al., 2008). Statistical downscaling methods are able to access finer spatial scales than dynamical methods and can be applied to parameters that cannot be directly obtained from RCMs. Seasonal indices of extremes can, for example, be simulated directly without having to first produce daily time series (Haylock et al., 2006a), or distribution functions of extremes can be simulated (Benestad, 2007). However, statistical downscaling methods require observational data at the desired scale (e.g., the point or station scale) for a long enough period to allow the model to be well trained and validated, and in some methods can lack coherency among multiple climate variables and/or multiple sites. One specific disadvantage of some, but not all, methods based on the analog approach is that they cannot produce extreme events greater in magnitude than have been observed before (Timbal et al., 2009). Moreover, statistical downscaling does not allow for the possibility of future process-based changes in relationships between predictors and predictands (see Section 3.2.3.2). There have been few systematic intercomparisons of dynamical and statistical downscaling approaches focusing on extremes (Fowler et al., 2007b). Two examples focus on extreme precipitation for the United Kingdom (Haylock et al., 2006a) and the Alps (Schmidli et al., 2007), respectively. A few hybrid statistico-dynamical downscaling methods also exist, including a two-step approach used to downscale heavy precipitation events in southern France (Beaulant et al., 2011). A conceptually similar cascading technique has also been used to downscale tropical cyclones (Bender et al., 2010; see Section 3.4.4).

In terms of temporal resolution, while GCMs and RCMs operate at sub-daily time steps, model output at six-hourly or shorter temporal resolutions, which is desirable for some applications such as urban drainage, is less widely available than daily output. Where limited studies have been undertaken, there is evidence that at the typical spatial resolutions used (i.e., non-cloud/convection-resolving scales), RCMs do not adequately represent sub-daily precipitation and the diurnal cycle of convection (Gutowski et al., 2003; Brockhaus et al., 2008; Lenderink and Van Meijgaard, 2008). Development of sub-daily statistical downscaling methods is constrained by the availability of long observed time series for calibration and validation and this approach is not currently widely used for climate change applications, although some weather generators, for example, do provide hourly information (Maraun et al., 2010).

It is not possible in this chapter to provide assessments of projected changes in extremes at spatial scales smaller than for large regions (Table 3-3). These large-region projections provide a wider context for national or more local projections, where they exist, and, where they do not, a first indication of expected changes, their associated uncertainties, and the evidence available. Several countries, for example in Europe, North America, Australia, and some other regions, have developed national or sub-national projections (generally based on dynamical and/or statistical downscaling), including information about extremes, and a range of other high-resolution information and tools are available from national weather and hydrological services and academic institutions to assist users and decisionmakers.

3.2.3.2. Uncertainty Sources in Climate Change Projections

Uncertainty in climate change projections arises at each of the steps involved in their preparation: determination of greenhouse gas and aerosol precursor emissions (driven by socioeconomic development and represented through the use of multiple emissions scenarios), concentrations of radiatively active species, radiative forcing, and climate response including downscaling. Also, uncertainty in the estimation of the true 'signal' of climate change is introduced by both errors in the model representation of Earth system processes and by internal climate variability.

As was noted in Section 3.2.3.1, most shortcomings in GCMs and RCMs result from the fact that many important small-scale processes (e.g., representations of clouds, convection, land surface processes) are not represented explicitly (Randall et al., 2007). Some processes – particularly those involving feedbacks (Section 3.1.4), and this is especially the case for climate extremes and associated impacts – are still poorly represented and/or understood (e.g., land-atmosphere interactions, ocean-atmosphere interactions, stratospheric processes, blocking dynamics) despite some improvements in the simulations of others (see Box 3-2 and below). Therefore, limitations in computing power and in the scientific understanding of some physical processes currently restrict further global and regional climate model improvements. In addition, uncertainty due to structural or parameter errors in GCMs propagates directly from global model simulations as input to RCMs and thus to downscaled information.

These problems limit quantitative assessments of the magnitude and timing, as well as regional details, of some aspects of projected climate change. For instance, even atmospheric models with approximately 20-km horizontal resolution still do not resolve the atmospheric processes sufficiently finely to simulate the high wind speeds and low pressure centers of the most intense hurricanes (Knutson et al., 2010). Realistically capturing details of such intense hurricanes, such as the inner eyewall structure, would require models with 1-km horizontal resolution, far beyond the capabilities of current GCMs and of most current RCMs (and even global numerical weather prediction models). Extremes may also be impacted by mesoscale circulations that GCMs and even current RCMs cannot resolve, such as low-level jets and their coupling with intense precipitation (Anderson et al., 2003; Menendez et al., 2010). Another issue with small-scale processes is the lack of relevant observations, such as is the case with soil moisture and vegetation processes (Section 3.2.1) and relevant parameters (e.g., maps of soil types and associated properties, see for instance Seneviratne et al., 2006b; Anders and Rockel, 2009).

Since many extreme events, such as those associated with precipitation, occur at rather small temporal and spatial scales, where climate simulation skill is currently limited and local conditions are highly variable, projections of future changes cannot always be made with a high level of confidence (Easterling et al., 2008). The credibility of projections of changes in extremes varies with extreme type, season, and geographical region (Box 3-2). Confidence and credibility in projected changes in extremes increase when the physical mechanisms producing extremes in models are considered reliable, such as increases in specific humidity in the case of the projected increase in the proportion of summer precipitation falling as intense events in central Europe (Kendon et al., 2010). The ability of a model to capture the full distribution of variables – not just the mean – together with long-term trends in extremes, implies that some of the processes relevant to a future warming world may be captured (Alexander and Arblaster, 2009; van Oldenborgh et al., 2009). It should nonetheless be stressed that physical consistency of simulations with observed behavior provides necessary but not sufficient evidence for credible projections (Gutowski et al., 2008a).

While downscaling provides more spatial detail (Section 3.2.3.1), the added value of this step and the reliability of projections always needs to be assessed (Benestad et al., 2007; Laprise et al., 2008). A potential limitation and source of uncertainty in downscaling methods is that the calibration of statistical models and the parameterization schemes used in dynamical models are necessarily based on present (and past) climate (as well as an understanding of physical processes). Thus they may not be able to capture changes in extremes that are induced by future mechanistic changes in regional (or global) climate, that is, if used outside the range for which they were designed (Christensen et al., 2007). Spatial inhomogeneity of both land use/land cover and aerosol forcing adds to regional uncertainty. This means that the factors inducing uncertainty in the projections of extremes in different regions may differ considerably. Some specific issues inducing uncertainties in RCM projections are the interactions with the driving GCM, especially in terms of biases and climate change signal (e.g., de Elía et al., 2008; Laprise et al., 2008; Kjellström and Lind, 2009; Déqué et al., 2011) and the choice of regional domain (Wang et al., 2004; Laprise et al., 2008).

In the case of statistical downscaling, uncertainties are induced by, *inter alia*, the definition and choice of predictors (Benestad, 2001; Hewitson and Crane, 2006; Timbal et al., 2008) and the underlying assumption of stationarity (Raje and Mujumdar, 2010). In general, both approaches to downscaling are maturing and being more widely applied but are still restricted in terms of geographical coverage (Maraun et al., 2010). For many regions of the world, no downscaled information exists at all and regional projections rely only on information from GCMs (see Table 3-3).

For many user-driven applications, impact models need to be included as an additional step for projections (e.g., hydrological or ecosystem models). Because of the previously mentioned issues of scale discrepancies and overall biases, it is necessary to bias-correct RCM data before input to some impacts models (i.e., to bring the statistical properties of present-day simulations in line with observations and to use this information to correct projections). A number of bias correction methods, including quantile mapping and gamma transform, have recently been developed and exhibit promising skill for extremes of daily precipitation (Piani et al., 2010; Themeßl et al., 2011).

3.2.3.3. Ways of Exploring and Quantifying Uncertainties

Uncertainties can be explored, and quantified to some extent, through the combined use of observations and reanalyses, process understanding, a hierarchy of climate models, and ensemble simulations. Ensembles of model simulations represent a fundamental resource for studying the range of plausible climate responses to a given forcing (Meehl et al., 2007b; Randall et al., 2007). Such ensembles can be generated either by (i) collecting results from a range of models from different modelling centers (multi-model ensembles), to include the impact of structural model differences; (ii) by generating simulations with different initial conditions (intra-model ensembles) to characterize the uncertainties due to internal climate variability; or (iii) varying multiple internal model parameters within plausible ranges (perturbed and stochastic physics ensembles), with both (ii) and (iii) aiming to produce a more systematic estimate of single model uncertainty (Knutti et al., 2010b).

Many of the global models utilized for the AR4 were integrated as ensembles, permitting more robust statistical analysis than is possible if a model is only integrated to produce a single projection. Thus the available CMIP3 Multi-Model Ensemble (MME) GCM simulations reflect both inter- and intra-model variability. In advance of AR4, coordinated climate change experiments were undertaken which provided information from 23 models from around the world (Meehl et al., 2007a). The CMIP3 simulations were made available at the Program for Climate Model Diagnosis and Intercomparison (www-pcmdi.llnl.gov/ipcc/about_ipcc.php). However, the higher temporal resolution (i.e., daily) data necessary to analyze most extreme events were quite incomplete in the archive, with only four models providing daily averaged output with ensemble sizes greater than three realizations and many models not included at all. GCMs are expensive to run, thus a compromise is needed between the number of models, number of simulations, and the complexity of the models (Knutti, 2010).

Besides the uncertainty due to randomness itself, which is the canonical statistical definition, it is important to distinguish between the uncertainty due to *insufficient agreement* in the model projections, the uncertainty due to *insufficient evidence* (insufficient observational data to constrain the model projections or insufficient number of simulations from different models or insufficient understanding of the physical processes), and the uncertainty induced by *insufficient literature*, which refers to the lack of published analyses of projections. For instance, models may agree on a projected change, but if this change is controlled by processes that are not well understood and validated in the present climate, then there is an inherent uncertainty in the projections, no matter how good the model agreement may be. Similarly, available model projections may agree in a given change, but the number of available simulations may restrain the reliability of the inferred agreement (e.g., because the analyses need to be based on daily data that may not be available from all modelling groups). All these issues have been taken into account in assessing the confidence and likelihood of projected changes in extremes for this report (see Section 3.1.5).

Uncertainty analysis of the CMIP3 MME in AR4 focused essentially on the seasonal mean and inter-model standard deviation values (Christensen et al., 2007; Meehl et al., 2007b; Randall et al., 2007). In addition, confidence was assessed in the AR4 through simple quantification of the number of models that show agreement in the sign of a specific climate change (e.g., sign of the change in frequency of extremes) – assuming that the greater the number of models in agreement, the greater the robustness. However, the shortcoming of this definition of model agreement is that it does not take account of possible common biases among models. Indeed, the ensemble was strictly an 'ensemble of opportunity,' without sampling protocol, and the possible dependence of different models on one another (e.g., due to shared parameterizations) was not assessed (Knutti et al., 2010a). Furthermore, this particular metric, which assesses sign agreement only, can provide misleading conclusions in cases, for example, where the projected changes are near zero. For this reason, in our assessments of projected changes in extreme indices we consider the model agreement as a necessary but not a sufficient condition for likelihood statements [e.g., agreement of 66% of the models, as indicated with shading in several of the figures (Figures 3-3, 3-4, 3-6, 3-8, and 3-10), is a minimum but not a sufficient condition for a change being considered '*likely*'].

Post-AR4 studies have concentrated more on the use of the MME in order to better characterize uncertainty in climate change projections, including those of extremes (Kharin et al., 2007; Gutowski et al., 2008a; Perkins et al., 2009). New techniques have been developed for exploiting the full ensemble information, in some cases using observational constraints to construct probability distributions (Tebaldi and Knutti, 2007; Tebaldi and Sanso, 2009), although issues such as determining appropriate metrics for weighting models are challenging (Knutti et al., 2010a). Perturbed-physics ensembles have also become available (e.g.,

Box 3-2 | Variations in Confidence in Projections of Climate Change: Mean versus Extremes, Variables, Scale

Comparisons of observed and simulated climate demonstrate good agreement for some climate variables such as mean temperature, especially at large horizontal scales (e.g., Räisänen, 2007). For instance, Figure 9.12 of the AR4 (Hegerl et al., 2007) compares the ability of 14 climate models to simulate the temporal variations of mean temperature through the 20th century. When the models included both natural and anthropogenic forcings, they consistently reproduced the decadal variations in global mean temperature. Without the anthropogenic influences the models consistently failed to reproduce the multi-decadal temperature variations. However, when the same models' abilities to simulate the temperature variations for smaller domains were assessed, although the mean temperature produced by the ensemble generally tracked the observed temperature changes, the consistency among the models was poorer than was the case for the global mean (Figure 9.12; Hegerl et al., 2007), partly because averaging over global scales smoothes internal variability or 'noise' more than averaging over smaller domains (see also Section 3.2.2.1). We can conclude that the smaller the spatial domain for which simulations or projections are being prepared, the less confidence we should have in these projections (although in some limited cases regional-scale projections can have higher reliability than larger-scale projections; see Section 3.1.6).

This increased uncertainty at smaller scales results from larger internal variability at smaller scales or 'noise' (i.e., natural variability unrelated to external forcings) and increased model uncertainty, both of which lead to lower model consistency at these scales (Hawkins and Sutton, 2009). The latter factor is largely due to the role of unresolved processes (representations of clouds, convection, land surface processes; see also Section 3.2.3). Hawkins and Sutton (2009) also point out regional variations in these aspects: in the tropics the temperature signal expected from anthropogenic factors is large relative to the model uncertainty and the natural variability, compared with higher latitudes. Figure 9.12 from AR4 (Hegerl et al., 2007) also shows that the models are more consistent in reproducing decadal temperature variations in the tropics than at higher latitudes, even though the magnitudes of the temperature trends are larger at higher latitudes.

Uncertainty in projections also depends on the variables, phenomena, or impacts considered (Sections 3.3. to 3.5.). There is more model uncertainty for variables other than temperature, for instance precipitation (Räisänen, 2007; Hawkins and Sutton, 2011; see also Section 3.2.3). And the situation is more difficult again for extremes. For instance, climate models simulate observed changes in extreme temperatures relatively well, but the frequency, distribution, and intensity of heavy precipitation is more poorly simulated (Randall et al., 2007) as are observed changes in heavy precipitation (e.g., Alexander and Arblaster, 2009). Also, projections of changes in temperature extremes tend to be more consistent across climate models (in terms of sign) than for (wet and dry) precipitation extremes (Tebaldi et al., 2006; Orlowsky and Seneviratne, 2011; see also Figures 3-3 through 3-7 and 3-10) and significant inconsistencies are also found for projections of agricultural (soil moisture) droughts (Wang, 2005; see also Box 3-3; Figure 3-10). For some other extremes, such as tropical cyclones, differences in the regional-scale climate change projections between models can lead to marked differences in projected tropical cyclone activity associated with anthropogenic climate change (Knutson et al., 2010), and thus decrease confidence in projections of changes in that extreme.

The relative importance of various causes of uncertainties in projections is somewhat different for earlier compared with later future periods. For some variables (mean temperature, temperature extremes), the choice of emission scenario becomes more critical than model uncertainty for the second part of the 21st century (Tebaldi et al., 2006; Hawkins and Sutton, 2009, 2011) though this does not apply for mean precipitation and some precipitation-related extremes (Tebaldi et al., 2006; Hawkins and Sutton, 2009, 2011), and has in particular not been evaluated in detail for a wide range of extremes. Users need to be aware of such issues in deciding the range of uncertainties that is appropriate to consider for their particular risk or impacts assessment

In summary, confidence in climate change projections depends on the (temporal and spatial) scale and variable being considered and whether one considers extremes or mean quantities. Confidence is highest for temperature, especially at the global scale, and decreases when other variables are considered, and when we focus on smaller spatial domains (Tables 3-1 and 3-3). Confidence in projections for extremes is generally weaker than for projections of long-term averages.

Collins et al., 2006; Murphy et al., 2007) and used to examine projected changes in extremes and their uncertainties (Barnett et al., 2006; Clark et al., 2006, 2010; Burke and Brown, 2008). Advances have also been made in developing probabilistic information at regional scales from the GCM simulations, but there has been rather less development extending this to probabilistic downscaled regional information and to extremes (Fowler et al., 2007a; Fowler and Ekstrom, 2009). Perhaps the most comprehensive approach to date for quantifying the influence of the cascade of uncertainties in regional projections is that used to develop the recent United Kingdom Climate Projections (UKCP09; Murphy et al., 2009). A complex Bayesian framework is used to combine a perturbed physics ensemble exploring uncertainties in atmosphere and ocean processes, and the carbon and sulfur cycles, with structural uncertainty (represented by 12 CMIP3 models) and an 11-member RCM perturbed physics ensemble. The published projections provide probability distributions of changes in various parameters including the wettest and hottest days of each season for 25-km grid squares across the United Kingdom. These probabilities are conditional on the emissions scenario (low, medium, high) and are described as representing the "relative degree to which each climate outcome is supported by the evidence currently available, taking into account our understanding of climate science and observations, and using expert judgment" (Murphy et al., 2009).

Both statistical and dynamical downscaling methods are affected by the uncertainties that affect the global models, and a further level of uncertainty associated with the downscaling step also needs to be taken into consideration (see also Sections 3.2.3.1 and 3.2.3.2). The increasing availability of coordinated RCM simulations for different regions permits more systematic exploration of dynamical downscaling uncertainty. Such simulations are available for Europe (e.g., Christensen and Christensen, 2007; van der Linden and Mitchell, 2009) and a few other regions such as North America (Mearns et al., 2009) and West Africa (van der Linden and Mitchell, 2009; Hourdin et al., 2010). RCM intercomparisons have also been undertaken for a number of regions including Asia (Fu et al., 2005), South America (Menendez et al., 2010) and the Arctic (Inoue et al., 2006). A new series of coordinated simulations covering the globe is planned (Giorgi et al., 2009). Increasingly, RCM output from coordinated simulations is made available at the daily time scale, facilitating the analysis of some extreme events. Nevertheless, it is important to point out that ensemble runs with RCMs currently involve a limited number of driving GCMs, and hence only subsample uncertainty space. Ensuring adequate sampling of RCM simulations (both in terms of the number of considered RCMs and number of considered driving GCMs) may be more important for extremes than for changes in mean values (Frei et al., 2006; Fowler et al., 2007a). Internal variability, for example, has been shown to make a significant contribution to the spectrum of variability on at least multi-annual time scales and potentially up to multi-decadal time scales (Kendon et al., 2008; Hawkins and Sutton, 2009, 2011; Box 3-2).

3.3. Observed and Projected Changes in Weather and Climate Extremes

3.3.1. Temperature

Temperature is associated with several types of extremes, for example, heat waves and cold spells, and related impacts, for example, on human health, the physical environment, ecosystems, and energy consumption (e.g., Chapter 4, Sections 3.5.6 and 3.5.7; see also Case Studies 9.2.1 and 9.2.10). Temperature extremes often occur on weather time scales that require daily or higher time scale resolution data to accurately assess possible changes (Section 3.2.1). It is important to distinguish between daily mean, maximum (i.e., daytime), and minimum (nighttime) temperature, as well as between cold and warm extremes, due to their differing impacts. Spell lengths (e.g., duration of heat waves) are relevant for a number of impacts. Note that we do not consider here changes in diurnal temperature range or frost days, which are not typical 'climate extremes'. There is an extensive body of literature regarding the mechanisms of changes in temperature extremes (e.g., Christensen et al., 2007; Meehl et al., 2007b; Trenberth et al., 2007). Heat waves are generally caused by quasi-stationary anticyclonic circulation anomalies or atmospheric blocking (Xoplaki et al., 2003; Meehl and Tebaldi, 2004; Cassou et al., 2005; Della-Marta et al., 2007b), and/or land-atmosphere feedbacks (in transitional climate regions), whereby the latter can act as an amplifying mechanism through reduction in evaporative cooling (Section 3.1.4), but also induce enhanced persistence due to soil moisture memory (Lorenz et al., 2010). Also snow feedbacks (Section 3.1.4), and possibly changes in aerosols (Portmann et al., 2009), are relevant for temperature extremes. Trends in temperature extremes (either observed or projected) can sometimes be different for the most extreme temperatures (e.g., annual maximum/minimum daily maximum/minimum temperature) than for less extreme events [e.g., cold/warm days/nights; see, for instance, Brown et al. (2008) versus Alexander et al. (2006)]. One reason for this is that 'moderate extremes' such as warm/cold days/nights are generally computed for each day with respect to the long-term statistics for that day, thus, for example, an increase in warm days for annual analyses does not necessarily imply warming for the very warmest days of the year.

Observed Changes

Regional historical or paleoclimatic temperature reconstructions may help place the recent instrumentally observed temperature extremes in the context of a much longer period, but literature on this topic is very sparse and most regional reconstructions are for Europe. For example Dobrovolny et al. (2010) reconstructed monthly and seasonal temperature over central Europe back to 1500 using a variety of temperature proxy records. They concluded that the summer 2003 heat wave and the July 2006 heat wave exceeded the +2 standard deviation (associated with the reconstruction method) of previous monthly temperature extremes since 1500. Barriopedro et al. (2011) showed that the anomalously warm summers of 2003 in western and central Europe and 2010 in eastern Europe and Russia both broke the 500-year long seasonal temperature record over 50% of Europe. The coldest periods within the last five centuries occurred in the winter and spring of 1690. Another 500-year temperature reconstruction was recently completed for the Mediterranean basin by means of documentary data and instrumental observations (Camuffo et al., 2010). It suggests strong natural variability in the basin, possibly exceeding the recent warming, although discontinuities in the records limit the interpretation of this finding.

Changes in Climate Extremes and their Impacts on the Natural Physical Environment

The AR4 (Trenberth et al., 2007, based on Alexander et al., 2006) reported a statistically significant increase in the numbers of warm nights and a statistically significant reduction in the numbers of cold nights for 70 to 75% of the land regions with data (for the spatial coverage of the underlying data set and the definition of warm/cold days and nights, see Section 3.2.1 and Box 3-1, respectively). Changes in the numbers of warm days and cold days also showed warming, but less marked than for nights, with about 40 to 50% of the area with data showing statistically significant changes consistent with warming (Alexander et al., 2006). Less than 1% of the area with data showed statistically significant trends in cold/warm days and nights that were consistent with cooling (Alexander et al., 2006). Trenberth et al. (2007) also reported, based on Vose et al. (2005), that from 1950 to 2004, the annual trends in minimum and maximum land-surface air temperature averaged over regions with data were 0.20°C per decade and 0.14°C per decade, respectively, and that for 1979 to 2004, the corresponding linear trends for the land areas with data were 0.29°C per decade for both maximum and minimum temperature. Based on this evidence, the IPCC AR4 (SPM; IPCC, 2007b) assessed that it was *very likely* that there had been trends toward warmer and more frequent warm days and warm nights, and warmer and less frequent cold days and cold nights in most land areas.

Regions that were found to depart from this overall behavior toward more warm days and nights and fewer cold days and nights in Alexander et al. (2006) were mostly central North America, the eastern United States, southern Greenland (increase in cold days and decreases in warm days), and the southern half of South America (decrease in warm days; no data available for the northern half of the continent). In central North America and the eastern United States this partial tendency for a negative trend in extremes is also consistent with a reported mean negative trend in temperatures, mostly in the spring to summer season (also termed 'warming hole', e.g., Pan et al., 2004; Portmann et al., 2009). Several explanations have been suggested for this behavior, which seems partly associated with a change in the hydrological cycle, possibly linked to soil moisture and/or aerosol feedbacks (Pan et al., 2004; Portmann et al., 2009).

More recent analyses available since the AR4 include a global study (for annual extremes) by Brown et al. (2008) based on the data set from Caesar et al. (2006), and regional studies for North America (Peterson et al., 2008a; Meehl et al., 2009c), Central-Western Europe (since 1880; Della-Marta et al., 2007a), central and eastern Europe (Bartholy and Pongracz, 2007; Kürbis et al., 2009), the eastern Mediterranean region including Turkey (Kuglitsch et al., 2010), western Central Africa, Guinea Conakry and Zimbabwe (Aguilar et al., 2009), the Tibetan Plateau (You et al., 2008) and China (You et al., 2011), Uruguay (Rusticucci and Renom, 2008), and Australia (Alexander and Arblaster, 2009). Further references can also be found in Table 3-2. Overall, these studies are consistent with the assessment of an increase in warm days and nights and a reduction in cold days and nights on the global basis, although they do not necessarily consider trends in all four variables, and a few single studies report trends that are not statistically significant or even trends opposite to the global tendencies in some extremes, subregions,

seasons, or decades. For instance, Rusticucci and Renom (2008) found in Uruguay a reduction of cold nights, a positive but a statistically insignificant trend in warm nights, statistically insignificant decreases in cold days at most investigated stations, and inconsistent trends in warm days. Together with the previous results from Alexander et al. (2006) for southern South America (see above) and further regional studies (Table 3-2), this suggests a less consistent warming tendency in South America compared to other continents. Another notable feature is that studies for central and southeastern Europe display a marked change point in trends in temperature extremes at the end of the 1970s/beginning of 1980s (Table 3-2), which for some extremes can lead to very small and/or statistically not significant overall trends since the 1960s (e.g., Bartholy and Pongracz, 2007).

There are fewer studies available investigating changes in characteristics of cold spells and warm spells, or cold waves and heat waves, compared with studies of the intensity or frequency of warm and cold days or nights. Alexander et al. (2006) provided an analysis of trends in warm spells [based on the Warm Spell Duration Index (WSDI); see Table 3-2 and Box 3-1] mostly in the mid- and high-latitudes of the Northern Hemisphere. The analysis displays a tendency toward a higher length or number of warm spells (increase in number of days belonging to warm spells) in much of the region, with the exception of the southeastern United States and eastern Canada. Regional studies on trends in warm spells or heat waves are also listed in Table 3-2. Kunkel et al. (2008) found that the United States has experienced a general decline in cold waves over the 20th century, with a spike of more cold waves in the 1980s. Further, they report a strong increase in heat waves since 1960, although the heat waves of the 1930s associated with extreme drought conditions still dominate the 1895-2005 time series. Kuglitsch et al. (2009) reported an increase in heat wave intensity, number, and length in summer over the 1960-2006 time period in the eastern Mediterranean region. Ding et al. (2010) reported increasing numbers of heat waves over most of China for the 1961-2007 period. The record-breaking heat wave over western and central Europe in the summer of 2003 is an example of an exceptional recent extreme (Beniston, 2004; Schär and Jendritzky, 2004). That summer (June to August) was the hottest since comparable instrumental records began around 1780 and perhaps the hottest since at least 1500 (Luterbacher et al., 2004). Other examples of recent extreme heat waves include the 2006 heat wave in Europe (Rebetez et al., 2008), the 2007 heat wave in southeastern Europe (Founda and Giannakopoulos, 2009), the 2009 heat wave in southeastern Australia (National Climate Centre, 2009), and the 2010 heat wave in Russia (Barriopedro et al., 2011). Both the 2003 European heat wave (Andersen et al., 2005; Ciais et al., 2005) and the 2009 southeastern Australian heat wave were also associated with drought conditions, which can strongly enhance temperature extremes during heat waves in some regions (see also Section 3.1.4).

Some recent analyses have led to revisions of previously reported trends. For instance, Della-Marta et al. (2007a) found that mean summer maximum temperature change over Europe was +1.6 ± 0.4°C during 1880 to 2005, a somewhat greater increase than reported in earlier

Chapter 3 — Changes in Climate Extremes and their Impacts on the Natural Physical Environment

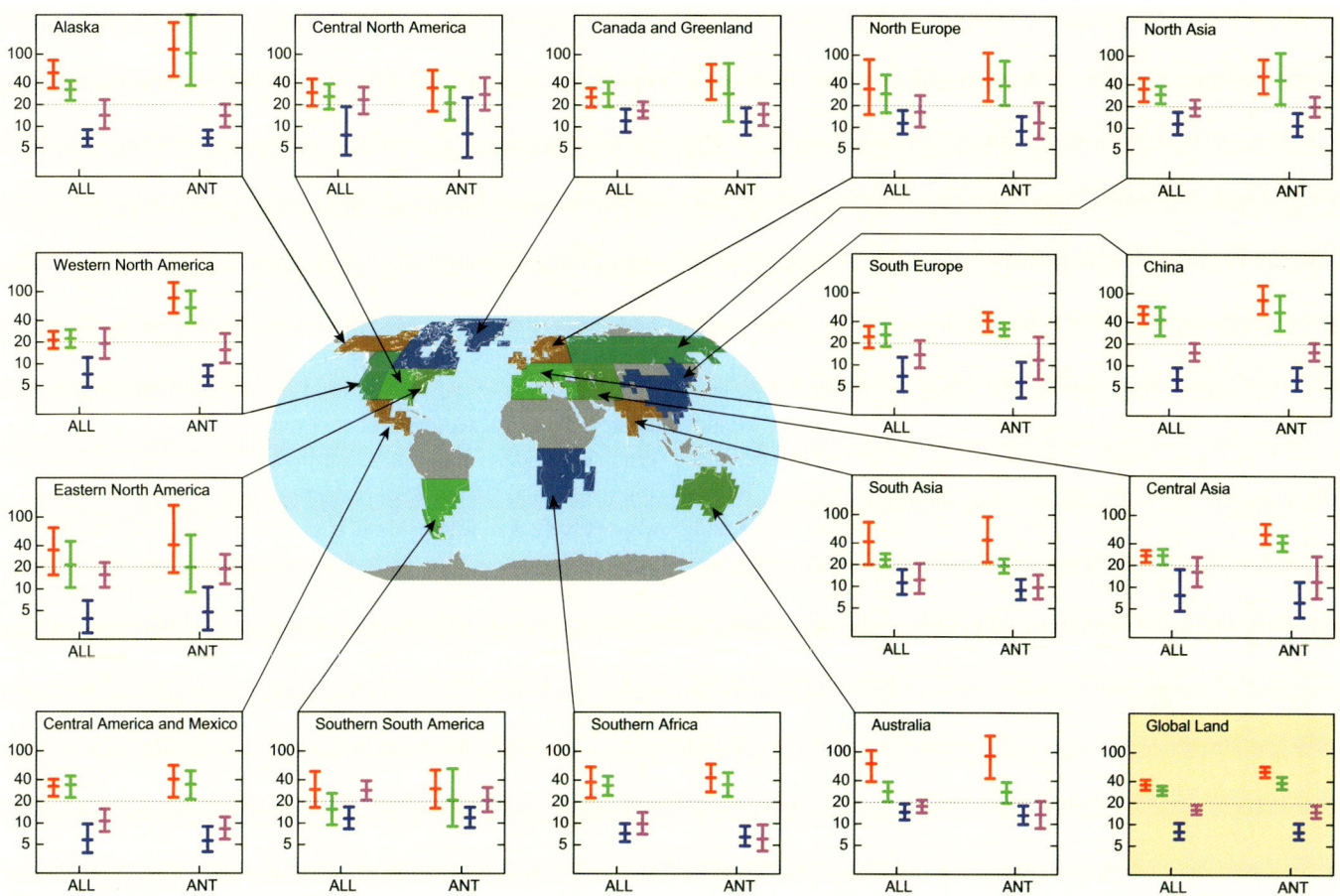

Figure 3-2 | Estimated return periods (years) and their 5 and 95% uncertainty limits for 1960s 20-year return values of annual extreme daily temperatures in the 1990s climate (see text for more details). ANT refers to model simulated responses with only anthropogenic forcing and ALL is both natural and anthropogenic forcing. Error bars are for annual minimum daily minimum temperature (red: TNn), annual minimum daily maximum temperature (green: TXn), annual maximum daily minimum temperature (blue: TNx), and annual maximum daily maximum temperature (pink: TXx), respectively. Grey areas have insufficient data. Source: Zwiers et al., (2011).

studies. Kuglitsch et al. (2009, 2010) homogenized and analyzed over 250 daily maximum and minimum temperature series in the Mediterranean region since 1960, and found that after homogenization the positive trends in the frequency of hot days and heat waves in the Eastern Mediterranean region were higher than reported in earlier studies. This was due to the correction of many warm-biased temperature data in the region during the 1960s and 1970s.

In summary, regional and global analyses of temperature extremes on land generally show recent changes consistent with a warming climate at the global scale, in agreement with the previous assessment in AR4. Only a few regions show changes in temperature extremes consistent with cooling, most notably for some extremes in central North America, the eastern United States, and also parts of South America. Based on the available evidence we conclude that it is *very likely* that there has been an overall decrease in the number of cold days and nights and *very likely* that there has been an overall increase in the number of warm days and nights in most regions, that is, for land areas with data (corresponding to about 70 to 80% of all land areas; see Table 3-2). It is *likely* that this statement applies at the continental scale in North America, Europe, and Australia (Table 3-2). However, some subregions on these continents have had warming trends in temperature extremes that were small or not statistically significant (e.g., southeastern Europe), and a few subregions have had cooling trends in some temperature extremes (e.g., central North America and eastern United States). Asia also shows trends consistent with warming in most of the continent, but which are assessed here to be of *medium confidence* because of lack of literature for several regions apart from the global study from Alexander et al. (2006). Most of Africa is insufficiently well sampled to allow an overall likelihood statement to be made at the continental scale, although most of the regions on this continent for which data are available have exhibited warming in temperature extremes (Table 3-2). In South America, both lack of data and some inconsistencies in the reported trends imply *low confidence* in the overall trends at the continental scale (Table 3-2). In many (but not all) regions with sufficient data there is *medium confidence* that the number of warm spells or heat waves has increased since the middle of the 20th century (Table 3-2).

Causes of Observed Changes

The AR4 (Hegerl et al., 2007) concluded that surface temperature extremes have *likely* been affected by anthropogenic forcing. This assessment was based on multiple lines of evidence of temperature extremes at the global scale including the reported increase in the number of warm extremes and decrease in the number of cold extremes at that scale (Alexander et al., 2006). Hegerl et al. (2007) also state that anthropogenic forcing may have substantially increased the risk of extreme temperatures (Christidis et al., 2005) and of the 2003 European heat wave (Stott et al., 2004).

Recent studies on attribution of changes in temperature extremes have tended to reaffirm the conclusions reached in the AR4. Alexander and Arblaster (2009) found that trends in warm nights over Australia could only be reproduced by a coupled model that included anthropogenic forcings. As part of the recent report of the US Climate Change Science Program (CCSP, 2008), Gutowski et al. (2008a) concluded that most of the observed changes in temperature extremes for the second half of the 20th century over the United States can be attributed to human activity. They compared observed changes in the number of frost days, the length of growing season, the number of warm nights, and the heat wave intensity with those simulated in a nine-member multi-model ensemble simulation. The decrease in frost days, an increase in growing season length, and an increase in heat wave intensity all show similar changes over the United States in 20th-century experiments that combine anthropogenic and natural forcings, though the relative contributions of each are unclear.

Results from two global coupled climate models with separate anthropogenic and natural forcing runs indicate that the observed changes are simulated with anthropogenic forcings, but not with natural forcings (even though there are some differences in the details of the forcings). Zwiers et al. (2011) compared observed annual temperature extremes including annual maximum daily maximum and minimum temperatures, and annual minimum daily maximum and minimum temperatures with those simulated responses to anthropogenic forcing or anthropogenic and natural external forcings combined by multiple GCMs. They fitted probability distributions (Box 3-1) to the observed extreme temperatures with a time-evolving pattern of location parameters as obtained from the model simulations, and found that both anthropogenic influence and the combined influence of anthropogenic and natural forcing can be detected in all four extreme temperature variables at the global scale over the land, and also over many large land areas. Globally, return periods for events that were expected to recur once every 20 years in the 1960s are now estimated to exceed 30 years for extreme annual minimum daily maximum temperature and 35 years for extreme annual minimum daily minimum temperature, although these estimates are subject to considerable uncertainty. Further, return periods were found to have decreased to less than 10 or 15 years for annual maximum daily minimum and daily maximum temperatures respectively (Figure 3-2).

However, the available detection and attribution studies for extreme maximum and minimum temperatures (Christidis et al., 2011b; Zwiers et al., 2011) suggest that the models overestimate changes in the maximum temperatures and underestimate changes in the minimum temperatures during the late 20th century.

Projected Changes and Uncertainties

Regarding projections of extreme temperatures, the AR4 (Meehl et al., 2007b) noted that cold episodes were projected to decrease significantly in a future warmer climate and considered it *very likely* that heat waves would be more intense, more frequent, and last longer in a future warmer climate. Post-AR4 studies of temperature extremes have utilized larger model ensembles (Kharin et al., 2007; Sterl et al., 2008; Orlowsky and Seneviratne, 2011) and generally confirm the conclusions of the AR4, while also providing more specific assessments both in terms of the range of considered extremes and the level of regional detail (see also Table 3-3).

There are few global analyses of multi-model projections of temperature extremes available in the literature. The study by Tebaldi et al. (2006), which provided the basis for extreme projections given in the AR4 (Figures 10.18 and 10.19 in Meehl et al., 2007b), provided global analyses of projected changes (A1B scenario) in several extremes indices based on nine GCMs (note that not all modelling groups that saved daily data also calculated the indices). For temperature extremes, analyses were provided for heat wave lengths (using only one index, see discussion in Box 3-1) and warm nights. Stippling was used where five out of nine models displayed statistically significant changes of the same sign. Orlowsky and Seneviratne (2011) recently updated the analysis from Tebaldi et al. (2006) for the full ensemble of GCMs that contributed A2 scenarios to the CMIP3, using a larger number of extreme indices [including several additional analyses of daily extremes (see Figures 3-3 and 3-4), and three heat wave indices instead of one; see also discussion of heat wave indices in Box 3-1], using other thresholds for display and stippling of the figures (no results displayed if less than 66% of the models agree on the sign of change; stippling used only for 90% model agreement), and providing seasonal analyses. This analysis confirms that strong agreement (in terms of sign of change) exists between the various GCM projections for temperature-related extremes, with projected increases in warm day occurrences (Figure 3-3) and heat wave length, and decreases in cold extremes (Figure 3-4). Temperature extremes on land are projected to warm faster than global annual mean temperature in many regions and seasons, implying large changes in extremes in some places, even for a global warming of 2 or 3°C (with scaling factors for the SRES A2 scenario ranging between 0.5 and 2 for moderate seasonal extremes; Orlowsky and Seneviratne, 2011). Based on the analyses of Tebaldi et al. (2006) and Orlowsky and Seneviratne (2011), as well as physical considerations, we assess that increases in the number of warm days and nights and decreases in the number of cold days and nights (defined with respect to present regional climate, i.e., the 1961-1990 reference period, see Box 3-1) are *virtually certain* at the global scale. Further, given the assessed changes in hot and cold days and nights and available analyses of projected changes in heat wave length in the two studies, we assess that it is *very likely* that the length, frequency, and/or intensity of heat waves will increase over most land areas.

Another global study of changes in extremes based on the CMIP3 ensemble is provided in Kharin et al. (2007), which focuses on changes in annual extremes (20-year extreme values) based on 12 GCMs for temperature extremes and 14 GCMs for precipitation extremes employing the SRES A2, A1B, and B1 emissions scenarios. This analysis projects increases in the temperature of the 1-in-20 year annual extreme hottest day of about 2 to 6°C (depending on region and scenario; Figure 3-5 adapted from Kharin et al., 2007) and strong reductions in the return periods of this extreme event by the end of the 21st century. However, as noted above, the limited number of relevant detection and attribution studies suggests that models may overestimate some changes in temperature extremes, and our assessments take this into account by reducing the level of certainty in the assessments from what would be derived by uncritical acceptance of the projections in Figure 3-5. The assessments are also weakened to reflect the possibility that some important processes relevant to extremes may be missing or be poorly represented in models, as well as the fact that the model projections considered in this study did not correspond to the full CMIP3 ensemble. Hence, we assess that in terms of absolute values, the 20-year extreme annual daily maximum temperature (i.e., return value) will *likely* increase by about 2 to 5°C by the late 21st century, and by about 1 to 3°C by mid-21st century, depending on the region and emissions scenario (considering the B1, A1B, and A2 scenarios; Figure 3-5a). Furthermore, we assess that globally under the A2 and A1B scenarios a 1-in-20 year annual extreme hot day is *likely* to become a 1-in-2 year annual extreme by the end of the 21st century in most regions, except in the high latitudes of the Northern Hemisphere where it is *likely* to become a 1-in-5 year annual extreme (Figure 3-5b, based on material from Kharin et al., 2007). Further, we assess that under the more moderate B1 scenario a current 1-in-20 year extreme would *likely* become a 1-in-5 year event (and a 1-in-10 year event in Northern Hemisphere high latitudes).

Next, regional assessments of projected changes in temperature extremes are provided. More details are found in Table 3-3. For North America, the CCSP reached the following conclusions (using IPCC AR4 likelihood terminology) regarding projected changes in temperature extremes by the end of the 21st century (Gutowski et al., 2008a):

1) Abnormally hot days and warm nights and heat waves are *very likely* to become more frequent.
2) Cold days and cold nights are *very likely* to become much less frequent.
3) For a mid-range scenario (A1B) of future greenhouse gas emissions, a day so hot that it is currently experienced only once every 20 years would occur every 3 years by the middle of the century over

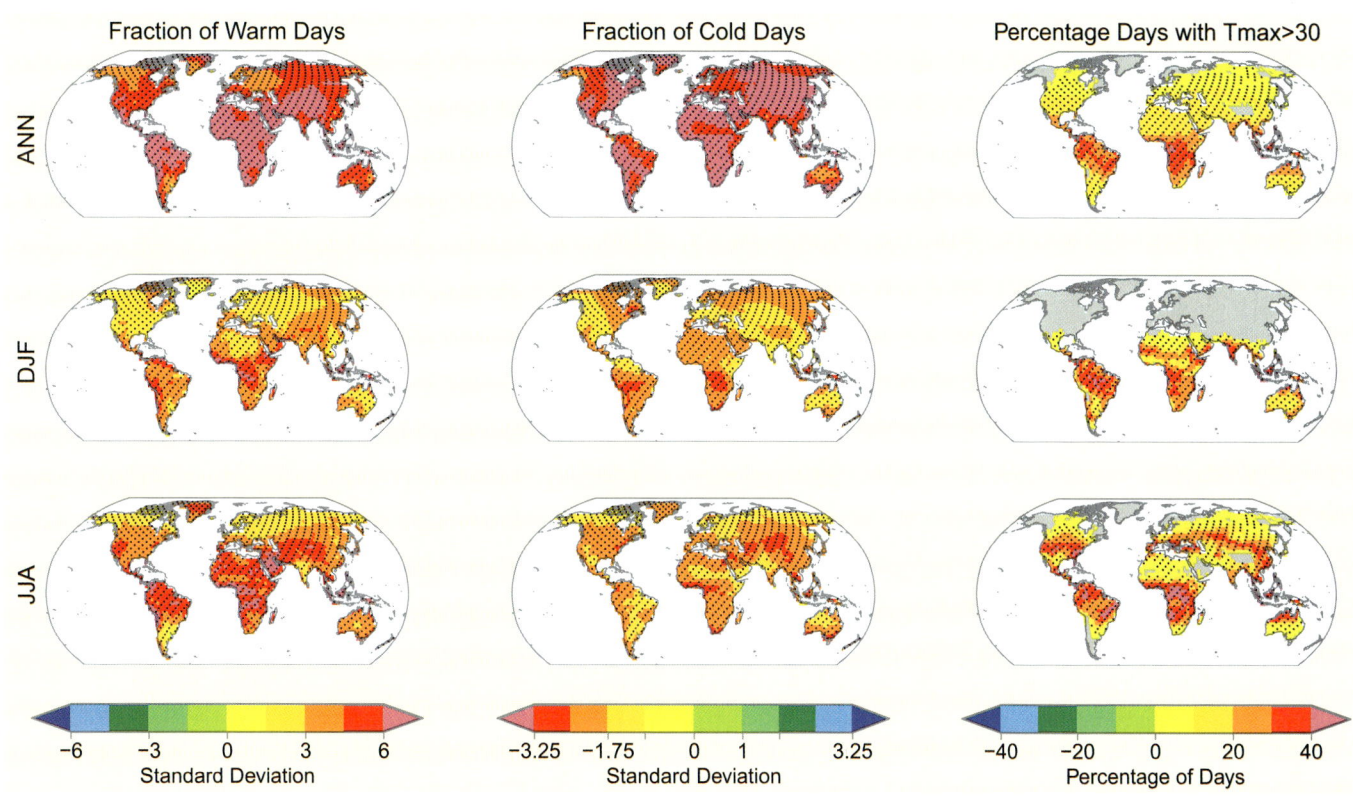

Figure 3-3 | Projected annual and seasonal changes in three indices for daily Tmax for 2081-2100 with respect to 1980-1999, based on 14 GCMs contributing to the CMIP3. Left column: fraction of warm days (days in which Tmax exceeds the 90th percentile of that day of the year, calculated from the 1961-1990 reference period); middle column: fraction of cold days (days in which Tmax is lower than the 10th percentile of that day of the year, calculated from the 1961-1990 reference period); right column: percentage of days with Tmax >30°C. The changes are computed for the annual time scale (top row) and two seasons (December-January-February, DJF, middle row, and June-July-August, JJA, bottom row) as the fractions/percentages in the 2081-2100 period (based on simulations for emission scenario SRES A2) minus the fractions/percentages of the 1980-1999 period (from corresponding simulations for the 20th century). Warm day and cold day changes are expressed in units of standard deviations, derived from detrended per year annual or seasonal estimates, respectively, from the three 20-year periods 1980-1999, 2046-2065, and 2081-2100 pooled together. Tmax >30°C changes are given directly as differences in percentage points. Color shading is only applied for areas where at least 66% (i.e., 10 out of 14) of the GCMs agree on the sign of the change; stippling is applied for regions where at least 90% (i.e., 13 out of 14) of the GCMs agree on the sign of the change. Adapted from Orlowsky and Seneviratne (2011); updating Tebaldi et al. (2006) for additional number of indices and CMIP3 models, and including seasonal time frames. For more details, see Appendix 3.A.

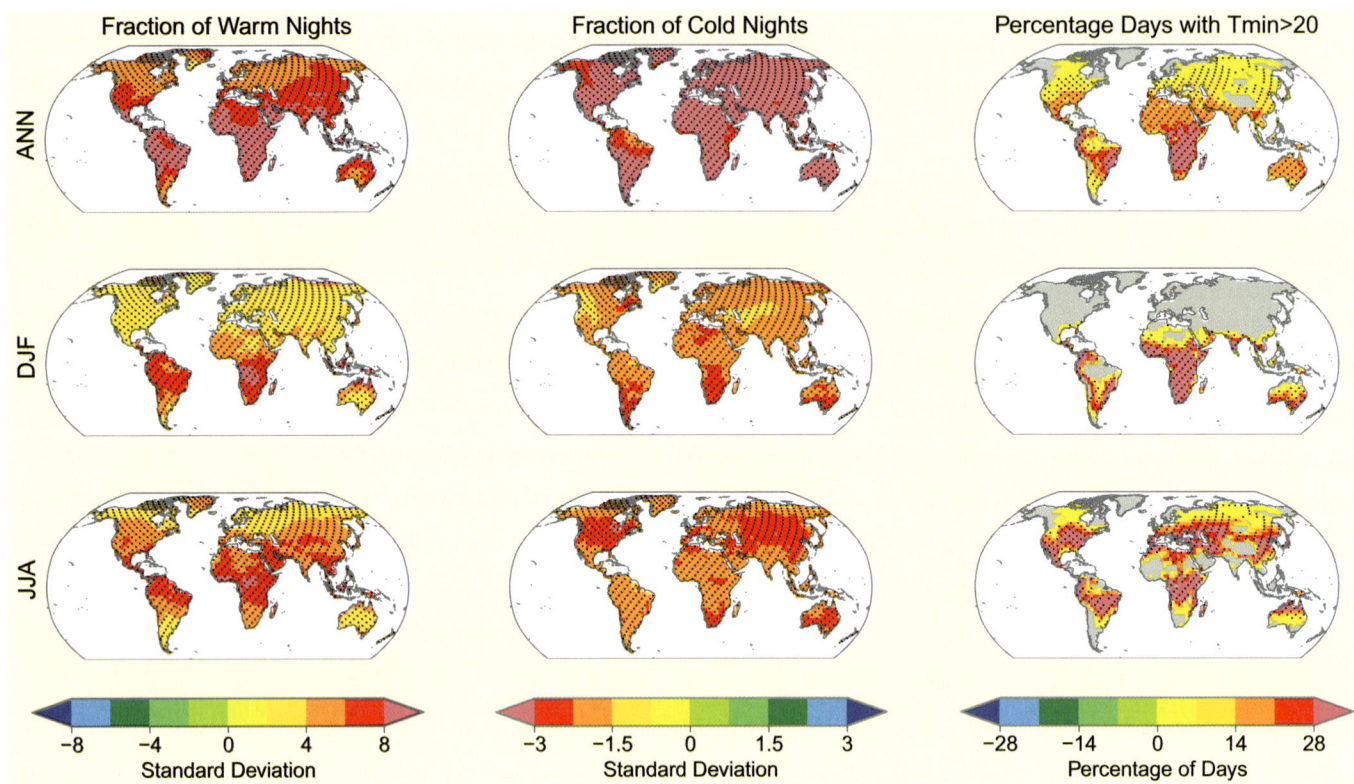

Figure 3-4 | Projected annual and seasonal changes in three indices for daily Tmin for 2081-2100 with respect to 1980-1999, based on 14 GCMs contributing to the CMIP3. Left column: fraction of warm nights (days at which Tmin exceeds the 90th percentile of that day of the year, calculated from the 1961-1990 reference period); middle column: fraction of cold nights (days at which Tmin is lower than the 10th percentile of that day of the year, calculated from the 1961-1990 reference period); right column: percentage of days with Tmin >20°C. The changes are computed for the annual time scale (top row) and two seasons (December-January-February, DJF, middle row, and June-July-August, JJA, bottom row) as the fractions/percentages in the 2081-2100 period (based on simulations under emission scenario SRES A2) minus the fractions/percentages of the 1980-1999 period (from corresponding simulations for the 20th century). Warm night and cold night changes are expressed in units of standard deviations, derived from detrended per year annual or seasonal estimates, respectively, from the three 20-year periods 1980-1999, 2046-2065, and 2081-2100 pooled together. Tmin >20°C changes are given directly as differences of percentage points. Color shading is only applied for areas where at least 66% (i.e., 10 out of 14) of the GCMs agree in the sign of the change; stippling is applied for regions where at least 90% (i.e.,13 out of 14) of the GCMs agree in the sign of the change. Adapted from Orlowsky and Seneviratne (2011); updating Tebaldi et al. (2006) for additional number of indices and CMIP3 models, and including seasonal time frames. For more details, see Appendix 3.A.

much of the continental United States and every 5 years over most of Canada; by the end of the century, it would occur every other year or more.

Meehl et al. (2009c) examined changes in record daily high and low temperatures in the United States and show that even with projected strong warming resulting in many more record highs than lows, the occasional record low is still set. For Australia, the CMIP3 ensemble projected increases in warm nights (15-40% by the end of the 21st century) and heat wave duration, together with a decrease in the number of frost days (Alexander and Arblaster, 2009). Inland regions show greater warming compared with coastal zones (Suppiah et al., 2007; Alexander and Arblaster, 2009) and large increases in the number of days above 35 or 40°C are indicated (Suppiah et al., 2007). For the entire South American region, a study with a single RCM projected more frequent warm nights and fewer cold nights (Marengo et al., 2009a). Several studies of regional and global model projections of changes in extremes are available for the European continent (see also Table 3-3). Analyses of both global and regional model outputs show major increases in warm temperature extremes across the Mediterranean region including events such as hot days (Tmax >30°C) and tropical nights (Tmin>20°C) (Giannakopoulos et al., 2009; Tolika et al., 2009).

Comparison of RCM projections using the A1B forcing scenario, with data for 2007 (the hottest summer in Greece in the instrumental record with a record daily Tmax observed value of 44.8°C) indicates that the distribution for 2007 is closer to the distribution for 2071-2100 than for the 2021-2050 period, thus 2007 might be considered a 'normal' summer of the future (Founda and Giannakopoulos, 2009; Tolika et al., 2009). Beniston et al. (2007) concluded from an analysis of RCM output that regions such as France and Hungary may experience as many days per year above 30°C as currently experienced in Spain and Sicily. In this RCM ensemble, France was the area with the largest projected warming in the uppermost percentiles of daily summer temperatures although the mean warming was greatest in the Mediterranean region (Fischer and Schär, 2009). New results from an RCM ensemble project increases in the amplitude, frequency, and duration of health-impacting heat waves, especially in southern Europe (Fischer and Schär, 2010). Overall these regional assessments are consistent with the global assessments provided above. It should be noted, however, that the assessed uncertainty is larger at the regional level than at the continental or global level (see Box 3-2). Global-scale trends in a specific extreme may be either more reliable or less reliable than regional-scale trends, depending on the geographical uniformity of the trends in the specific extreme (Section 3.1.6).

Chapter 3 Changes in Climate Extremes and their Impacts on the Natural Physical Environment

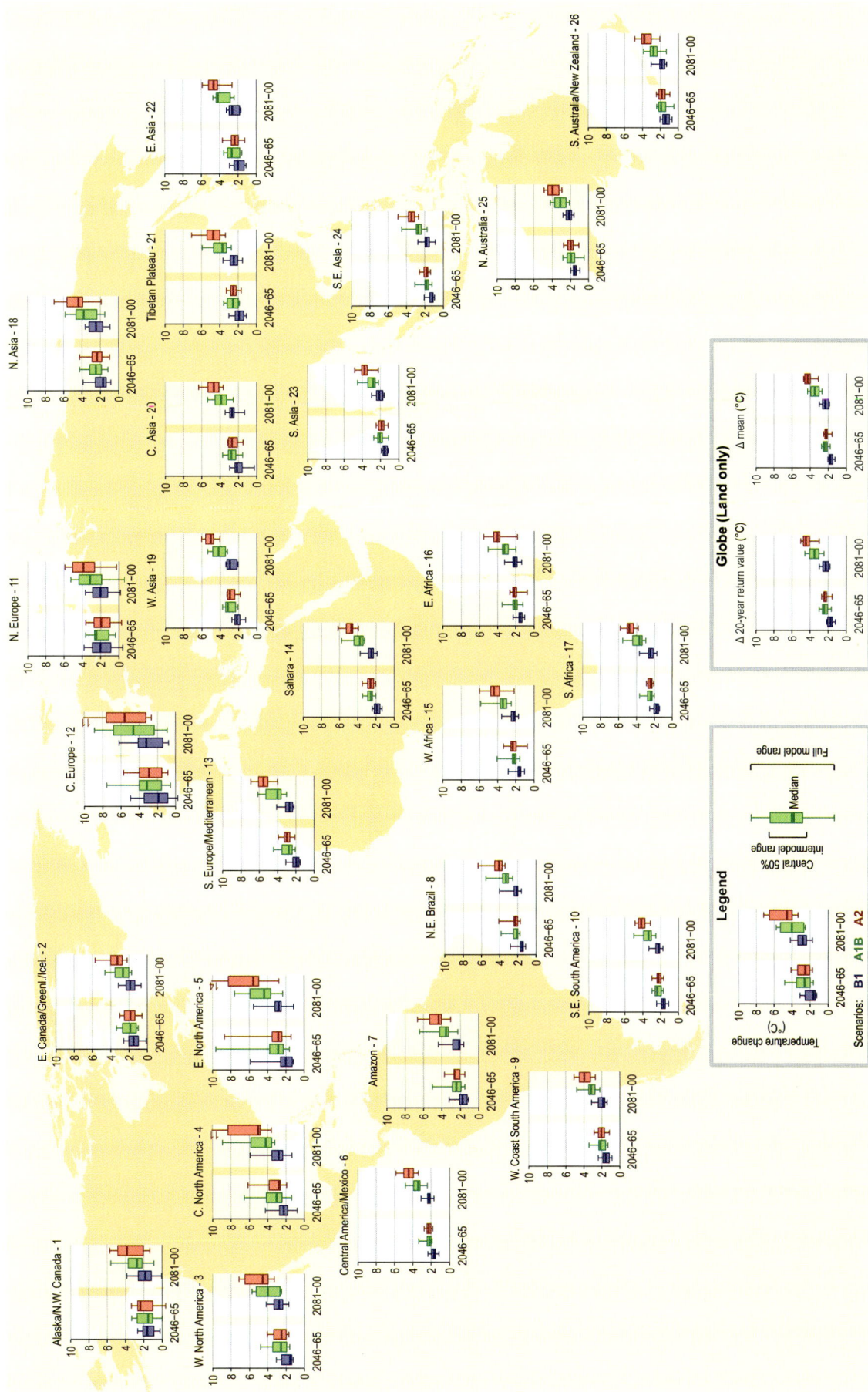

Figure 3-5a | Projected changes (in °C) in 20-year return values of the annual maximum of the daily maximum temperature. The bar plots (see legend for more information) show results for regionally averaged projections for two time horizons, 2046 to 2065 and 2081 to 2100, as compared to the late 20th century (1981-2000), and for three different SRES emission scenarios (B1, A1B, A2). Results are based on 12 GCMs contributing to the CMIP3. See Figure 3-1 for defined extent of regions. Values are computed for land points only. The 'Globe' analysis (inset box) displays the change in 20-year return values of the annual maximum of the daily maximum temperature computed using all land grid points (left), and the change in annual mean daily maximum temperature computed using all land grid points (right). Adapted from the analysis of Kharin et al. (2007). For more details, see Appendix 3.A.

139

Changes in Climate Extremes and their Impacts on the Natural Physical Environment

Chapter 3

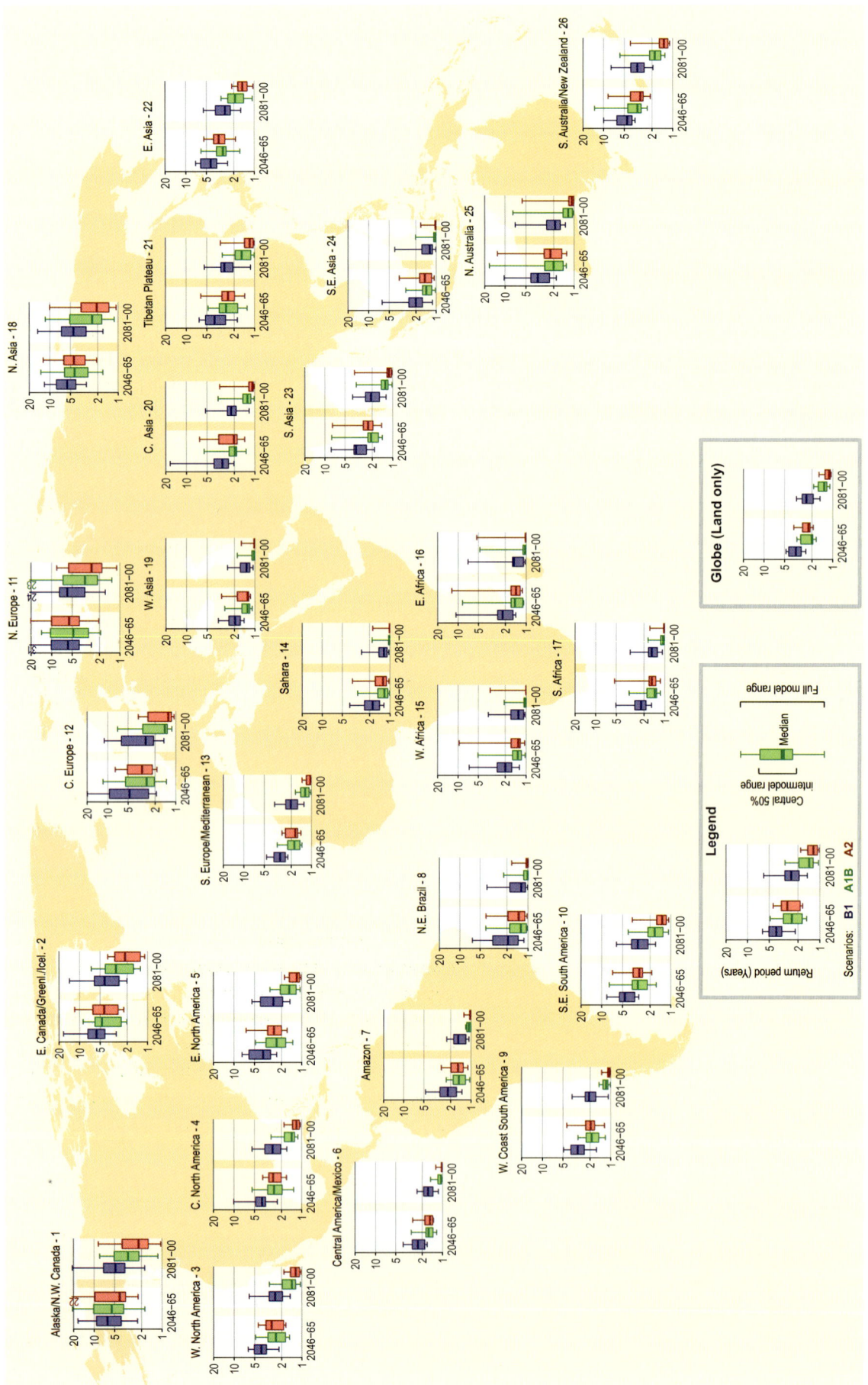

Figure 3-5b | Projected return period (in years) of late 20th-century 20-year return values of the annual maximum of the daily maximum temperature. The bar plots (see legend for more information) show results for regionally averaged projections for two time horizons, 2046 to 2065 and 2081 to 2100, as compared to the late 20th century (1981-2000), and for three different SRES emission scenarios (B1, A1B, A2). Results are based on 12 GCMs contributing to the CMIP3. See Figure 3-1 for defined extent of regions. The 'Globe' analysis (inset box) displays the projected return period (in years) of late 20th-century 20-year return values of the annual maximum of the daily maximum temperature computed using all land grid points. Adapted from the analysis of Kharin et al. (2007). For more details, see Appendix 3.A.

Temperature extremes were the type of extremes projected to change with most confidence in the AR4 (IPCC, 2007a). This is confirmed regarding the sign of change with more recent analyses (Figures 3-3 and 3-4), although there is a large spread with respect to the magnitude of changes both due to emission scenario and climate model uncertainty (Figures 3-5a,b). If changes in temperature extremes scale with changes in mean temperature (i.e., simple shifts of the probability distribution), we infer that it is *virtually certain* that hot extremes will increase and cold extremes will decrease over the 21st century with respect to the 1960-1990 climate. Changes in the tails of the temperature distributions may not scale with changes in the mean in some regions (Section 3.1.6), though in most such reported cases hot extremes tend to increase and cold extremes decrease more than mean temperature, and thus the above statement for extremes (*virtually certain* increase in hot extremes and decrease in cold extremes) still applies. Central and eastern Europe is a region where the evidence suggests that projected changes in temperature extremes result from both changes in the mean as well as from changes in the shape of the probability distributions (Schär et al., 2004). The main mechanism for the widening of the distribution is linked to the drying of the soil in this region (Sections 3.1.4 and 3.1.6). Furthermore, remote surface heating may induce circulation changes that modify the temperature distribution (Haarsma et al., 2009). Other local, mesoscale, and regional feedback mechanisms, in particular with land surface conditions (beside soil moisture, also with vegetation and snow; Section 3.1.4) and aerosol concentrations (Ruckstuhl and Norris, 2009) may enhance the uncertainties in temperature projections. Some of these processes occur at a small scale unresolved by the models (Section 3.2.3). In addition, lack of observational data (e.g., for soil moisture and snow cover; see Section 3.2.1) reduces the possibilities to evaluate climate models (e.g., Roesch, 2006; Boe and Terray, 2008; Hall et al., 2008; Brown and Mote, 2009). Because of these various processes and associated uncertainties, mean global warming does not necessarily imply warming in all regions and seasons (see also Section 3.1.6). Regarding mesoscale processes, lack of information also affects confidence in projections. One example is changes in heat waves in the Mediterranean region that are suggested to have the largest impact in coastal areas, due to the role of enhanced relative humidity in health impacts (Diffenbaugh et al., 2007; Fischer and Schär, 2010). But it is not clear how this pattern may or may not be moderated by sea breezes (Diffenbaugh et al., 2007).

In summary, since 1950 it is *very likely* that there has been an overall decrease in the number of cold days and nights and an overall increase in the number of warm days and nights at the global scale, that is, for land areas with sufficient data. It is *likely* that such changes have also occurred at the continental scale in North America, Europe, and Australia. There is *medium confidence* in a warming trend in daily temperature extremes in much of Asia. *Confidence* in historical trends in daily temperature extremes in Africa and South America generally varies from *low* to *medium* depending on the region. Globally, in many (but not all) regions with sufficient data there is *medium confidence* that the length or number of warm spells or heat waves has increased since the middle of the 20th century. It is *likely* that anthropogenic influences have led to warming of extreme daily minimum and maximum temperatures at the global scale. Models project substantial warming in temperature extremes by the end of the 21st century. It is *virtually certain* that increases in the frequency and magnitude of warm days and nights and decreases in the cold days and nights will occur through the 21st century at the global scale. This is mostly linked with mean changes in temperatures, although changes in temperature variability can play an important role in some regions. It is *very likely* that the length, frequency, and/or intensity of warm spells or heat waves (defined with respect to present regional climate) will increase over most land areas. For the SRES A2 and A1B emission scenarios a 1-in-20 year annual hottest day is *likely* to become a 1-in-2 year annual extreme by the end of the 21st century in most regions, except in the high latitudes of the Northern Hemisphere where it is *likely* to become a 1-in-5 year annual extreme. In terms of absolute values, 20-year extreme annual daily maximum temperature (i.e., return value) will *likely* increase by about 1 to 3°C by mid-21st century and by about 2 to 5°C by the late 21st century, depending on the region and emissions scenario (Figure 3-5). Moderate temperature extremes on land are projected to warm faster than global annual mean temperature in many regions and seasons. Projected changes at subcontinental scales are less certain than is the case for the global scale. Regional changes in temperature extremes will differ from the mean global temperature change. Mean global warming does not necessarily imply warming in all regions and seasons.

3.3.2. Precipitation

This section addresses changes in daily extreme or heavy precipitation events. Reductions in mean (or total) precipitation that can lead to drought (i.e., associated with lack of precipitation) are considered in Section 3.5.1. Because climates are so diverse across different parts of the world, it is difficult to provide a single definition of extreme or heavy precipitation. In general, two different approaches have been used: (1) relative thresholds such as percentiles (typically the 95th percentile) and return values; and (2) absolute thresholds [e.g., 50.8 mm (2 inches) day^{-1} of rain in the United States, and 100 mm day^{-1} of rain in China]. For more details on the respective drawbacks and advantages of these two approaches, see Section 3.1 and Box 3-1. Note that we do not distinguish between rain and snowfall (both considered as contributors to overall extreme precipitation events) as they are not treated separately in the literature, but do distinguish changes in hail from other precipitation types. Increases in public awareness and changes in reporting practices have led to inconsistencies in the record of severe thunderstorms and hail that make it difficult to detect trends in the intensity or frequency of these events (Kunkel et al., 2008). Furthermore, weather events such as hail are not well captured by current monitoring systems and, in some parts of the world, the monitoring network is very sparse (Section 3.2.1), resulting in considerable uncertainty in the estimates of extreme precipitation. There are also known biases in precipitation measurements, mostly leading to rain undercatch. Little evidence of paleoclimatic and historical changes in heavy precipitation is available to place recent variations into context.

Observed Changes

The AR4 (Trenberth et al., 2007) concluded that it was *likely* that there had been increases in the number of heavy precipitation events (e.g., 95th percentile) over the second half of the 20th century within many land regions, even in those where there had been a reduction in total precipitation amount, consistent with a warming climate and observed significant increasing amounts of water vapor in the atmosphere. Increases had also been reported for rarer precipitation events (1-in-50 year return period), but only a few regions had sufficient data to assess such trends reliably. However, the AR4 (Trenberth et al., 2007) also stated that "Many analyses indicate that the evolution of rainfall statistics through the second half of the 20th century is dominated by variations on the interannual to inter-decadal time scale and that trend estimates are spatially incoherent (Manton et al., 2001; Peterson et al., 2002; Griffiths et al., 2003; Herath and Ratnayake, 2004)". Overall, as highlighted in Alexander et al. (2006), the observed changes in precipitation extremes were found at the time to be much less spatially coherent and statistically significant compared to observed changes in temperature extremes; although statistically significant trends toward stronger precipitation extremes were generally found for a larger fraction of the land area than trends toward weaker precipitation extremes, statistically significant changes in precipitation indices for the overall land areas with data were only found for the Simple Daily Intensity index, and not for other considered indices such as Heavy Rainfall Days (Alexander et al., 2006).

Recent studies have updated the assessment of the AR4, with more regional results now available (Table 3-2). Overall, this additional evidence confirms that more locations and studies show an increase than a decrease in extreme precipitation, but that there are also wide regional and seasonal variations, and trends in many regions are not statistically significant (Table 3-2).

Recent studies on past and current changes in precipitation extremes in North America, some of which are included in the recent assessment of the CCSP report (Kunkel et al., 2008), have reported an increasing trend over the last half century. Based on station data from Canada, the United States, and Mexico, Peterson et al. (2008a) reported that heavy precipitation has been increasing over 1950-2004, as well as the average amount of precipitation falling on days with precipitation. For the contiguous United States, DeGaetano (2009) showed a 20% reduction in the return period for extreme precipitation of different return levels over 1950-2007; Gleason et al. (2008) reported an increasing trend in the area experiencing a much above-normal proportion of heavy daily precipitation from 1950 to 2006; and Pryor et al. (2009) provided evidence of increases in the intensity of events above the 95th percentile during the 20th century, with a larger magnitude of the increase at the end of the century. The largest trends toward increased annual total precipitation, number of rainy days, and intense precipitation (e.g., fraction derived from events in excess of the 90th percentile value) were focused on the Great Plains/northwestern Midwest (Pryor et al., 2009). In the core of the North American monsoon region in northwest Mexico, statistically significant positive trends were found in daily precipitation intensity and seasonal contribution of daily precipitation greater than its 95th percentile in the mountain sites for the period 1961-1998. However, no statistically significant changes were found in coastal stations (Cavazos et al., 2008). Overall, the evidence indicates a *likely* increase in observed heavy precipitation in many regions in North America, despite statistically non-significant trends and some decreases in some subregions (Table 3-2). This general increase in heavy precipitation accompanies a general increase in total precipitation in most areas of the country.

There is *low* to *medium confidence* in trends for Central and South America, where spatially varying trends in extreme rainfall events have been observed (Table 3-2). Positive trends in many areas but negative trends in some regions are evident for Central America and northern South America (Dufek and Ambrizzi, 2008; Marengo et al., 2009b; Re and Ricardo Barros, 2009; Sugahara et al., 2009). For the western coast of South America, a decrease in extreme rainfall in many areas and an increase in a few areas are observed (Haylock et al., 2006b).

There is *medium confidence* in trends in heavy precipitation in Europe, due to partly inconsistent signals across studies and regions, especially in summer (Table 3-2). Winter extreme precipitation has increased in part of the continent, in particular in central-western Europe and European Russia (Zolina et al., 2009), but the trend in summer precipitation has been weak or not spatially coherent (Moberg et al., 2006; Bartholy and Pongracz, 2007; Maraun et al., 2008; Pavan et al., 2008; Zolina et al., 2008; Costa and Soares, 2009; Kyselý, 2009; Durão et al., 2010; Rodda et al., 2010). Increasing trends in 90th, 95th, and 98th percentiles of daily winter precipitation over 1901-2000 were found (Moberg et al., 2006), which has been confirmed by more detailed country-based studies for the United Kingdom (Maraun et al., 2008), Germany (Zolina et al., 2008), and central and eastern Europe (Bartholy and Pongracz, 2007; Kyselý, 2009), while decreasing trends have been found in some regions such as northern Italy (Pavan et al., 2008), Poland (Lupikasza, 2010), and some Mediterranean coastal sites (Toreti et al., 2010). Uncertainties are overall larger in southern Europe and the Mediterranean region, where there is *low confidence* in the trends (Table 3-2). A recent study (Zolina et al., 2010) has indicated that there has been an increase of about 15 to 20% in the persistence of wet spells over most of Europe over the last 60 years, which was not associated with an increase of the total number of wet days.

There is *low* to *medium confidence* in trends in heavy precipitation in Asia, both at the continental and regional scale for most regions (Table 3-2; see also Alexander et al., 2006). A weak increase in the frequency of extreme precipitation events is observed in northern Mongolia (Nandintsetseg et al., 2007). No systematic spatially coherent trends in the frequency and duration of extreme precipitation events have been found in Eastern and Southeast Asia (Choi et al., 2009), central and south Asia (Klein Tank et al., 2006), and Western Asia (X. Zhang et al., 2005; Rahimzadeh et al., 2009). However, statistically significant positive and negative trends were observed at subregional scales within these regions. Heavy precipitation increased in Japan during 1901-2004 (Fujibe et al., 2006), and in India (Rajeevan et al., 2008; Krishnamurthy et al., 2009) especially during the monsoon seasons (Sen Roy, 2009; Pattanaik and Rajeevan, 2010). Both statistically significant increases and decreases in extreme precipitation have been found in China over the period 1951-2000 (Zhai et al., 2005) and 1978-2002 (Yao et al., 2008). In Peninsular Malaysia during 1971-2005 the intensity of extreme precipitation increased and the frequency decreased, while the trend in the proportion of extreme rainfall over total precipitation was not statistically significant (Zin et al., 2009). Heavy precipitation increased over the southern and northern Tibetan Plateau but decreased in the central Tibetan Plateau during 1961-2005 (You et al., 2008).

In southern Australia, there has been a *likely* decrease in heavy precipitation in many areas, especially where mean precipitation has decreased (Table 3-2). There were statistically significant increases in the proportion of annual/seasonal rainfall stemming from heavy rain days from 1911-2008 and 1957-2008 in northwest Australia (Gallant and Karoly, 2010). Extreme summer rainfall over the northwest of the Swan-Avon River basin in western Australia increased over 1950-2003 while extreme winter rainfall over the southwest of the basin decreased (Aryal et al., 2009). In New Zealand, the trends are positive in the western North and South Islands and negative in the east of the country (Mullan et al., 2008).

There is *low* to *medium confidence* in regional trends in heavy precipitation in Africa due to partial lack of literature and data, and due to lack of consistency in reported patterns in some regions (Table 3-2). The AR4 (Trenberth et al., 2007) reported an increase in heavy precipitation over southern Africa, but this appears to depend on the region and precipitation index examined (Kruger, 2006; New et al., 2006; Seleshi and Camberlin, 2006; Aguilar et al., 2009). Central Africa exhibited a decrease in heavy precipitation over the last half century (Aguilar et al., 2009); however, data coverage for large parts of the region was poor. Precipitation from heavy events has decreased in western central Africa, but with low spatial coherence (Aguilar et al., 2009). Rainfall intensity averaged over southern and west Africa has increased (New et al., 2006). There is a lack of literature on changes in heavy precipitation in East Africa (Table 3-2). Camberlin et al. (2009) analyzed changes in components of rainy seasons' variability over the time period 1958-1987 in this region, but did not specifically address trends in heavy precipitation. There were decreasing trends in heavy precipitation over parts of Ethiopia during the period 1965-2002 (Seleshi and Camberlin, 2006).

Changes in hail occurrence are generally difficult to quantify because hail occurrence is not well captured by monitoring systems and because of historical data inhomogeneities. Sometimes, changes in environmental conditions conducive to hail occurrence are used to infer changes in hail occurrence. However, the atmospheric conditions are typically estimated from reanalyses or from radiosonde data and the estimates are associated with high uncertainty. As a result, assessment of changes in hail frequency is difficult. For severe thunderstorms in the region east of the Rocky Mountains in the United States, Brooks and Dotzek (2008) found strong variability but no clear trend in the past 50 years. Cao (2008) identified a robust upward trend in hail frequency over Ontario, Canada. Kunz et al. (2009) found that both hail damage days and convective instability increased during 1974-2003 in a state in southwest Germany. Xie et al. (2008) identified no trend in the mean annual hail days in China from 1960 to the early 1980s but a statistically significant decreasing trend afterwards.

Causes of Observed Changes

The observed changes in heavy precipitation appear to be consistent with the expected response to anthropogenic forcing (increase due to enhanced moisture content in the atmosphere; see, e.g., Section 3.2.2.1) but a direct cause-and-effect relationship between changes in external forcing and extreme precipitation had not been established at the time of the AR4. As a result, the AR4 only concluded that it was *more likely than not* that anthropogenic influence had contributed to a global trend towards increases in the frequency of heavy precipitation events over the second half of the 20th century (Hegerl et al., 2007).

New research since the AR4 provides more evidence of anthropogenic influence on various aspects of the global hydrological cycle (Stott et al., 2010; see also Section 3.2.2), which is directly relevant to extreme precipitation changes. In particular, an anthropogenic influence on atmospheric moisture content is detectable (Santer et al., 2007; Willett et al., 2007; see also Section 3.2.2). Wang and Zhang (2008) show that winter season maximum daily precipitation in North America appears to be statistically significantly influenced by atmospheric moisture content, with an increase in moisture corresponding to an increase in maximum daily precipitation. This behavior has also been seen in model projections of extreme winter precipitation under global warming (Gutowski et al., 2008b). Climate model projections suggest that the thermodynamic constraint based on the Clausius-Clapeyron relation is a good predictor for extreme precipitation changes in a warmer world in regions where the nature of the ambient flows change little (Pall et al., 2007). This indicates that the observed increase in extreme precipitation in many regions is consistent with the expected extreme precipitation response to anthropogenic influences. However, the thermodynamic constraint may not be a good predictor in regions with circulation changes, such as mid- to higher latitudes (Meehl et al., 2005) and the tropics (Emori and Brown, 2005), and in arid regions. Additionally, changes in precipitation extremes with temperature also depend on changes in the moist-adiabatic temperature lapse rate, in the upward velocity, and in the temperature when precipitation extremes occur (O'Gorman and Schneider, 2009a,b; Sugiyama et al., 2010). This may explain why there have not been increases in precipitation extremes everywhere, although a low signal-to-noise ratio may also play a role. However, even in

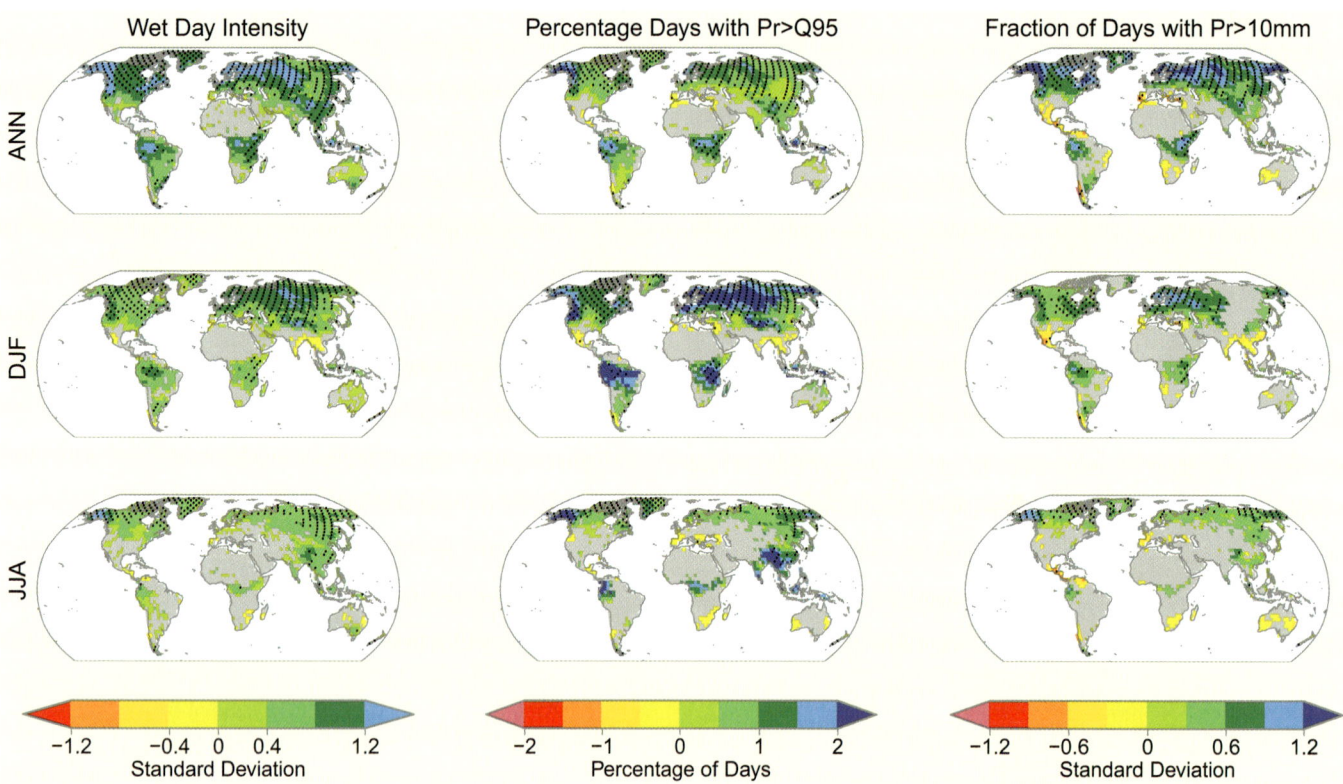

Figure 3-6 | Projected annual and seasonal changes in three indices for daily precipitation (Pr) for 2081-2100 with respect to 1980-1999, based on 17 GCMs contributing to the CMIP3. Left column: wet-day intensity; middle column: percentage of days with precipitation above the 95% quantile of daily wet day precipitation for that day of the year, calculated from the 1961-1990 reference period; right column: fraction of days with precipitation higher than 10 mm. The changes are computed for the annual time scale (top row) and two seasons (DJF, middle row, and JJA, bottom row) as the fractions/percentages in the 2081-2100 period (based on simulations under emission scenario SRES A2) minus the fractions/percentages of the 1980-1999 period (from corresponding simulations for the 20th century). Changes in wet-day intensity and in the fraction of days with Pr >10 mm are expressed in units of standard deviations, derived from detrended per year annual or seasonal estimates, respectively, from the three 20-year periods 1980-1999, 2046-2065, and 2081-2100 pooled together. Changes in percentages of days with precipitation above the 95% quantile are given directly as differences in percentage points. Color shading is only applied for areas where at least 66% (i.e., 12 out of 17) of the GCMs agree on the sign of the change; stippling is applied for regions where at least 90% (i.e., 16 out of 17) of the GCMs agree on the sign of the change. Adapted from Orlowsky and Seneviratne (2011); updating Tebaldi et al. (2006) for additional number of indices and CMIP3 models, and including seasonal time frames. For more details, see Appendix 3.A.

regions where the Clausius-Clapeyron constraint is not closely followed, it still appears to be a better predictor for future changes in extreme precipitation than the change in mean precipitation in climate model projections (Pall et al., 2007). An observational study seems also to support this thermodynamic theory. Analysis of daily precipitation from the Special Sensor Microwave Imager over the tropical oceans shows a direct link between rainfall extremes and temperature: heavy rainfall events increase during warm periods (El Niño) and decrease during cold periods (Allan and Soden, 2008). However, the observed amplification of rainfall extremes is larger than that predicted by climate models (Allan and Soden, 2008), due possibly to widely varying changes in upward velocities associated with precipitation extremes (O'Gorman and Schneider, 2008). Evidence from measurements in the Netherlands suggests that hourly precipitation extremes may in some cases increase 14% per degree of warming, which is twice as fast as what would be expected from the Clausius-Clapeyron relationship alone (Lenderink and Van Meijgaard, 2008), though this is still under debate (Haerter and Berg, 2009; Lenderink and van Meijgaard, 2009). A comparison between observed and multi-model simulated extreme precipitation using an optimal detection method suggests that the human-induced increase in greenhouse gases has contributed to the observed intensification of heavy precipitation events over large Northern Hemisphere land areas during the latter half of the 20th century (Min et al., 2011). Pall et al. (2011) linked human influence on global warming patterns with an increased risk of England and Wales flooding in autumn (September-November) 2000 that is associated with a displacement in the North Atlantic jet stream. The present assessment based on evidence from new studies and those used in the AR4 is that there is *medium confidence* that anthropogenic influence has contributed to changes in extreme precipitation at the global scale. However, this conclusion may be dependent on the season and spatial scale. For example, there is now about a 50% chance that an anthropogenic influence can be detected in UK extreme precipitation in winter, but the likelihood of the detection in other seasons is very small (Fowler and Wilby, 2010).

Projected Changes and Uncertainties

Regarding projected changes in extreme precipitation, the AR4 concluded that it was *very likely* that heavy precipitation events, that is, the frequency of heavy precipitation or proportion of total precipitation from heavy precipitation, would increase over most areas of the globe

Chapter 3 — Changes in Climate Extremes and their Impacts on the Natural Physical Environment

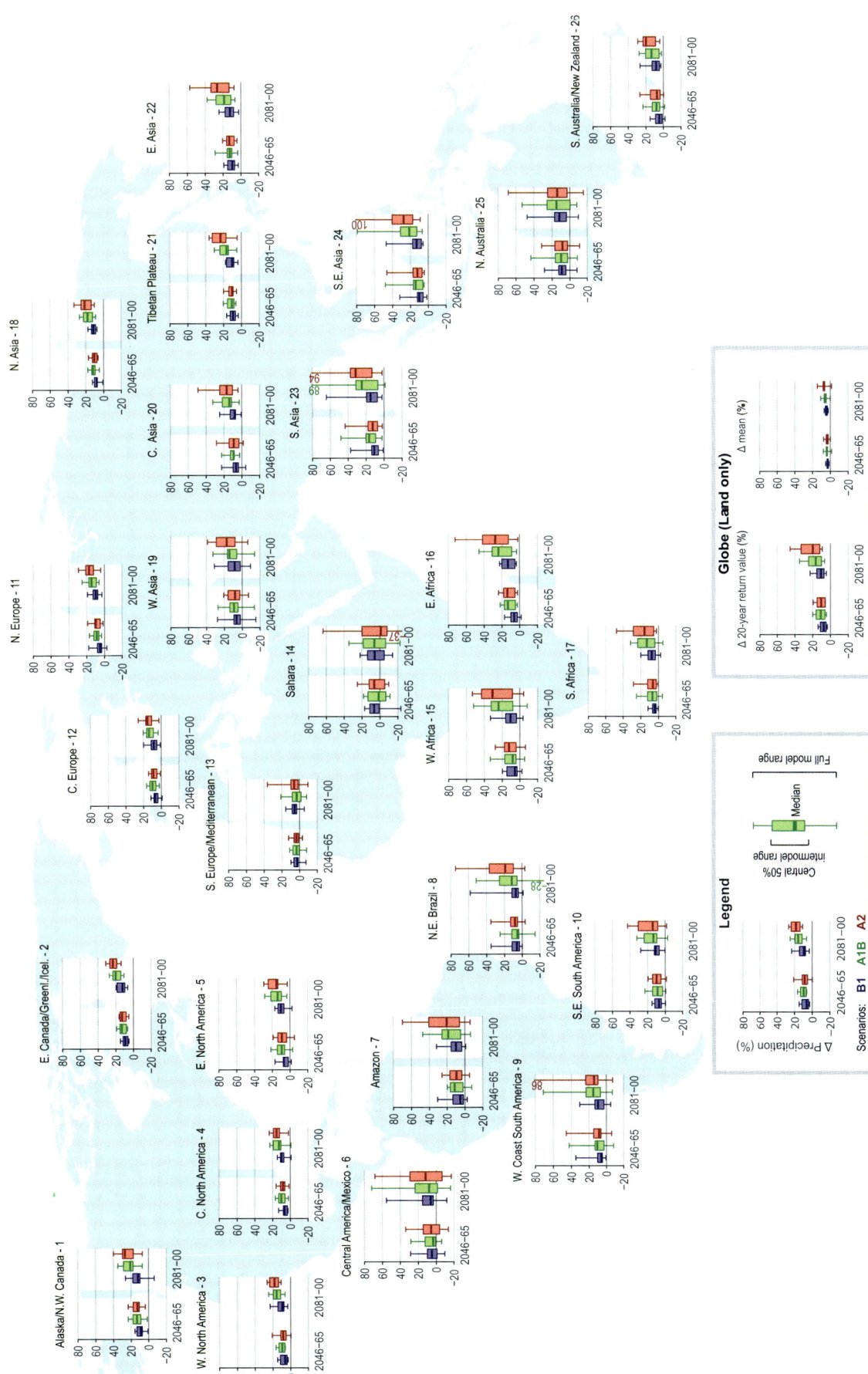

Figure 3-7a | Projected changes (%) in 20-year return values of annual maximum 24-hour precipitation rates. The bar plots (see legend for more information) show results for regionally averaged projections for two time horizons, 2046 to 2065 and 2081 to 2100, as compared to the late 20th century (1981–2000), and for three different SRES emission scenarios (B1, A1B, A2). Results are based on 14 GCMs contributing to the CMIP3. See Figure 3-1 for defined extent of regions. Values are computed for land points only. The 'Globe' analysis (inset box) displays the change in 20-year return values of the annual maximum 24-hour precipitation rates computed using all land grid points (left), and the change in annual mean 24-hour precipitation rates computed using all land grid points (right). Adapted from the analysis of Kharin et al. (2007). For more details, see Appendix 3.A.

Changes in Climate Extremes and their Impacts on the Natural Physical Environment — Chapter 3

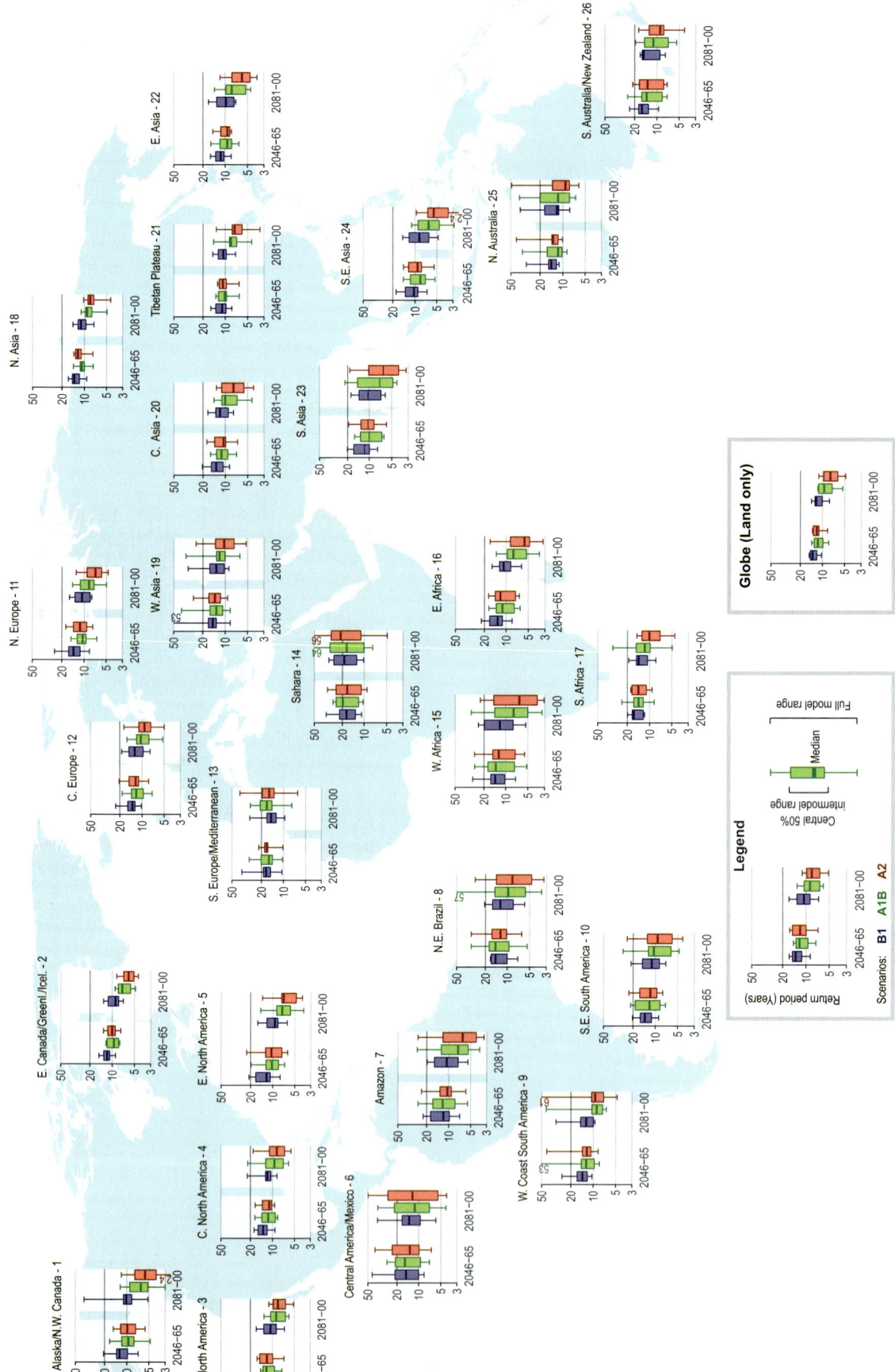

Figure 3-7b | Projected return period (in years) of late 20th century 20-year return values of annual maximum 24-hour precipitation rates. The bar plots (see legend for more information) show results for regionally averaged projections for two time horizons, 2046 to 2065 and 2081 to 2100, as compared to the late 20th century (1981-2000), and for three different SRES emission scenarios (B1, A1B, A2). Results are based on 14 GCMs contributing to the CMIP3. See Figure 3-1 for defined extent of regions. The 'Globe' analysis (inset box) displays the projected return period (in years) of late 20th-century 20-year return values of annual maximum 24-hour precipitation rates computed using all land grid points. Adapted from the analysis of Kharin et al. (2007). For more details, see Appendix 3.A.

in the 21st century (IPCC, 2007a). The tendency for an increase in heavy daily precipitation events was found in many regions, including some regions in which the total precipitation was projected to decrease.

Post-AR4 analyses of climate model simulations partly confirm this assessment but also highlight fairly large uncertainties and model biases in projections of changes in heavy precipitation in some regions (Section 3.2.3 and Table 3-3). On the other hand, more GCM and RCM ensembles have now been analyzed for some regions (Table 3-3; see also, e.g., Kharin et al., 2007; Kim et al., 2010). At the time of the AR4, Tebaldi et al. (2006) was the main global study available on projected changes in precipitation extremes (e.g., Figure 10.18 of Meehl et al., 2007b). Orlowsky and Seneviratne (2011) extended this analysis to a larger number of GCMs from the CMIP3 ensemble and for seasonal in addition to annual time frames (see also Section 3.3.1). Figure 3-6 provides corresponding analyses of projected annual and seasonal changes of the wet-day intensity, the fraction of days with precipitation above the 95% quantile of daily wet-day precipitation, and the fraction of days with precipitation above 10 mm day^{-1}. It should be noted that the 10 mm day^{-1} threshold cannot be considered extreme in several regions, but highlights differences in projections for absolute and relative thresholds (see also discussion in Box 3-1 and beginning of this section). Figure 3-6 indicates that regions with model agreement (at least 66%) with respect to changes in heavy precipitation are mostly found in the high latitudes and in the tropics, and in some mid-latitude regions of the Northern Hemisphere in the boreal winter. Regions with at least 90% model agreement are even more limited and confined to the high latitudes. Overall, model agreement in projected changes is found to be stronger in boreal winter (DJF) than summer (JJA) for most regions. Kharin et al. (2007) analyzed changes in annual maxima of 24-hour precipitation in the outputs of 14 CMIP3 models. Figure 3-7a displays the projected percentage change in the annual maximum of the 24-hour precipitation rate from the late 20th-century 20-year return values, while Figure 3-7b displays the corresponding projected return periods for late 20th-century 20-year return values of the annual maximum 24-hour precipitation rates in the mid-21st century (left) and in late 21st century (right) under three different emission scenarios (SRES B1, A1B, and A2). Between the late 20th and the late 21st century, the projected responses of extreme precipitation to future emissions show increased precipitation rates in most regions, and decreases in return periods in most regions in the high latitudes and the tropics and in some regions in the mid-latitudes consistent with projected changes in several indices related to heavy precipitation (see Figure 3-6 and Tebaldi et al., 2006), although there are increases in return periods or only small changes projected in several regions. Except for these regions, the return period for an event of annual maximum 24-hour precipitation with a 20-year return period in the late 20th century is projected to be about 5 to 15 years by the end of the 21st century. The greatest projected reductions in return period are in high latitudes and some tropical regions. The stronger CO_2 emissions scenarios (A1B and A2) lead to greater projected decreases in return period. In some regions with projected decreases in total precipitation (Christensen et al., 2007) such as southern Africa, west Asia, and the west coast of South America, heavy precipitation is nevertheless projected to increase (Figure 3-7, Table 3-3). In some other areas with projected decreases in total precipitation (e.g., Central America and northern South America), however, heavy precipitation is projected to decrease or not change. It should be noted that Figure 3-7 addresses very extreme heavy precipitation events (those expected to occur about once in 20 years) whereas Figure 3-6 addresses less extreme, but still heavy, precipitation events. Projections of changes for these differently defined extreme events may differ.

Future precipitation projected by the CMIP3 models has also been analyzed in a number of studies for various regions using different combinations of the models (see next paragraphs and Table 3-3). In general these studies confirm the findings of global-scale studies by Tebaldi et al. (2006) and Kharin et al. (2007).

By analyzing simulations with a single GCM, Khon et al. (2007) reported a projected general increase in extreme precipitation for the different regions in northern Eurasia especially for winter. Su et al. (2009) found that for the Yangtze River Basin region in 2001-2050, the 50 year heavy precipitation events become more frequent, with return periods falling to below 25 years (relative to 1951-2000 behavior). For the Indian region, the Hadley Centre coupled model HadCM3 projects increases in the magnitude of the heaviest rainfall with a doubling of atmospheric CO_2 concentration (Turner and Slingo, 2009). Simulations by 12 GCMs projected an increase in heavy precipitation intensity and mean precipitation rates in east Africa, more severe precipitation deficits in the southwest of southern Africa, and enhanced precipitation further north in Zambia, Malawi, and northern Mozambique (Shongwe et al., 2009, 2011). Rocha et al. (2008) evaluated differences in the precipitation regime over southeastern Africa simulated by two GCMs under present (1961-1990) and future (2071-2100) conditions as a result of anthropogenic greenhouse gas forcing. They found that the intensity of all episode categories of precipitation events is projected to increase practically over the whole region, whereas the number of episodes is projected to decrease in most of the region and for most episode categories. Extreme precipitation is projected to increase over Australia in 2080-2099 relative to 1980-1999 in an analysis of the CMIP3 ensemble, although there are inconsistencies between projections from different models (Alexander and Arblaster, 2009).

High spatial resolution is important for studies of extreme precipitation because the physical processes responsible for extreme precipitation require high spatial resolution to resolve them (e.g., Kim et al., 2010). Post-AR4 studies have employed three approaches to obtain high spatial resolution to project precipitation extremes: high-resolution GCMs, dynamical downscaling using RCMs, and statistical downscaling (see also Section 3.2.3.1). Based on the Meteorological Research Institute and Japan Meteorological Agency 20-km horizontal grid GCM, heavy precipitation was projected to increase substantially in south Asia, the Amazon, and west Africa, with increased dry spell persistence projected in South Africa, southern Australia, and the Amazon at the end of the 21st century (Kamiguchi et al., 2006). In the Asian monsoon region, heavy precipitation was projected to increase, notably in Bangladesh and in the Yangtze River basin due to the intensified convergence of water vapor flux in summer. Using statistical downscaling, Wang and Zhang (2008) investigated possible changes in North American extreme precipitation probability during winter from 1949-1999 to 2050-2099. Downscaled results suggested a strong increase in extreme precipitation over the south and central United States but decreases over the Canadian prairies. Projected European precipitation extremes in high-resolution studies tend to increase in northern Europe (Frei et al., 2006; Beniston et al., 2007; Schmidli et al., 2007), especially during winter (Haugen and Iversen, 2008; May, 2008), as also highlighted in Table 3-3. Fowler and Ekström (2009) project increases in both short-duration (1-day) and longer-duration (10-day) precipitation extremes across the United Kingdom during winter, spring, and autumn. In summer, model projections for the United Kingdom span the zero change line, although there is *low confidence* due to poor model performance in this season. Using daily statistics from various models, Boberg et al. (2009a,b) projected a clear increase in the contribution to total precipitation from more intense events together with a decrease in the number of days with light precipitation. This pattern of change was found to be robust for all European subregions. In double-nested model simulations with a horizontal grid spacing of 10 km, Tomassini and Jacob (2009) projected positive trends in extreme quantiles of heavy precipitation over Germany, although they are relatively small except for the high-CO_2 A2 emission scenario. For the Upper Mississippi River Basin region during October through March, the intensity of extreme precipitation is projected to increase (Gutowski et al., 2008b). Simulations with a single RCM project an increase in the intensity of extreme precipitation events over most of southeastern South America and western Amazonia in 2071-2100, whereas in northeast Brazil and eastern Amazonia smaller or no changes are projected (Marengo et al., 2009a). Outputs from another RCM indicate an increase in the magnitude of future extreme rainfall events in the Westernport region of Australia, consistent with results based on the CMIP3 ensemble (Alexander and Arblaster, 2009), and the size of this increase is greater in 2070 than in 2030 (Abbs and Rafter, 2008). When both future land use changes and increasing greenhouse gas concentrations are considered in the simulations, tropical and northern Africa are projected to experience less extreme rainfall events by 2025 during most seasons except for autumn (Paeth and Thamm, 2007). Simulations with high-resolution RCMs projected that the frequency of extreme precipitation increases in the warm climate for June through to September in Japan (Nakamura et al., 2008; Wakazuki et al., 2008; Kitoh et al., 2009). An increase in 90th-percentile values of daily precipitation on the Pacific side of the Japanese islands during July in the future climate was projected with a 5-km mesh cloud-system-resolving non-hydrostatic RCM (Kanada et al., 2010b).

Post-AR4 studies indicate that the projection of precipitation extremes is associated with large uncertainties, contributed by the uncertainties related to GCMs, RCMs, and statistical downscaling methods, and by natural variability of the climate. Kyselý and Beranova (2009) examined scenarios of change in extreme precipitation events in 24 future climate runs of 10 RCMs driven by two GCMs, focusing on a specific area of central Europe with complex orography. They demonstrated that the inter- and intra-model variability and related uncertainties in the pattern and magnitude of the change are large, although they also show that the projected trends tend to agree with those recently observed in the area, which may strengthen their credibility. May (2008) reported an unrealistically large projected precipitation change over the Baltic Sea in summer in an RCM, apparently related to an unrealistic projection of Baltic Sea warming in the driving GCM. Frei et al. (2006) found large model differences in summer when RCM formulation contributes significantly to scenario uncertainty. In exploring the ability of two statistical downscaling models to reproduce the direction of the projected changes in indices of precipitation extremes, Hundecha and Bardossy (2008) concluded that the statistical downscaling models seem to be more reliable during seasons when local climate is determined by large-scale circulation than by local convective processes. Themeßl et al. (2011) merged linear and nonlinear empirical-statistical downscaling techniques with bias correction methods, and demonstrated their ability to drastically reduce RCM error characteristics. The extent to which the natural variability of the climate affects our ability to project the anthropogenically forced component of changes in daily precipitation extremes was investigated by Kendon et al. (2008). They show that annual to multidecadal natural variability across Europe may contribute to substantial uncertainty. Also, Kiktev et al. (2009) performed an objective comparison of climatologies and historical trends of temperature and precipitation extremes using observations and 20th-century climate simulations. They did not detect significant similarity between simulated and actual patterns of the indices of precipitation extremes in most cases. Moreover, Allan and Soden (2008) used satellite observations and model simulations to examine the response of tropical precipitation events to naturally driven changes in surface temperature and atmospheric moisture content. The observed amplification of rainfall extremes was larger than that predicted by models. The underestimation of rainfall extremes by the models may be related to the coarse spatial resolution used in the model simulations – the magnitude of changes in precipitation extremes depends on spatial resolution (Kitoh et al., 2009) – suggesting that projections of future changes in rainfall extremes in response to anthropogenic global warming may be underestimated.

Confidence is still *low* for hail projections particularly due to a lack of hail-specific modelling studies, and a lack of agreement among the few available studies. There is little information in the AR4 regarding projected changes in hail events, and there has been little new literature since the AR4. Leslie et al. (2008) used coupled climate model simulations under the SRES A1B scenario to estimate future changes in hailstorms in the Sydney Basin, Australia. Their future climate simulations show an increase in the frequency and intensity of hailstorms out to 2050, and they suggest that the increase will emerge from the natural background variability within just a few decades. This result offers a different conclusion from the modelling study of Niall and Walsh (2005), which simulated Convective Available Potential Energy (CAPE) for southeastern Australia in an environment containing double the pre-industrial concentrations of equivalent CO_2. They found a statistically significant projected decrease in CAPE values and concluded that "it is possible that there will be a decrease in the frequency of hail in southeastern

Australia if current rates of CO_2 emission are sustained," assuming the strong relationship between hail incidence and the CAPE for 1980-2001 remains unchanged under enhanced greenhouse conditions.

In summary, it is *likely* that there have been statistically significant increases in the number of heavy precipitation events (e.g., 95th percentile) in more regions than there have been statistically significant decreases, but there are strong regional and subregional variations in the trends (i.e., both between and within regions considered in this report; Figure 3-1 and Tables 3-2 and 3-3). In particular, many regions present statistically non-significant or negative trends, and, where seasonal changes have been assessed, there are also variations between seasons (e.g., more consistent trends in winter than in summer in Europe). The overall most consistent trends toward heavier precipitation events are found in North America (*likely* increase over the continent). There is *low confidence* in observed trends in phenomena such as hail because of historical data inhomogeneities and inadequacies in monitoring systems. Based on evidence from new studies and those used in the AR4, there is *medium confidence* that anthropogenic influence has contributed to intensification of extreme precipitation at the global scale. There is almost no literature on the attribution of changes in hail extremes, thus no assessment can be provided for these at this point in time. Projected changes from both global and regional studies indicate that it is *likely* that the frequency of heavy precipitation or proportion of total rainfall from heavy falls will increase in the 21st century over many areas on the globe, especially in the high latitudes and tropical regions, and northern mid-latitudes in winter. Heavy precipitation is projected to increase in some (but not all) regions with projected decreases of total precipitation (*medium confidence*). For a range of emission scenarios (A2, A1B, and B1), projections indicate that it is *likely* that a 1-in-20 year annual maximum 24-hour precipitation rate will become a 1-in-5 to -15 year event by the end of 21st century in many regions. Nevertheless, increases or statistically non-significant changes in return periods are projected in some regions.

3.3.3. Wind

Extreme wind speeds pose a threat to human safety, maritime and aviation activities, and the integrity of infrastructure. As well as extreme wind speeds, other attributes of wind can cause extreme impacts. Trends in average wind speed can influence potential evaporation and in turn water availability and droughts (e.g., McVicar et al., 2008; see also Section 3.5.1 and Box 3-3). Sustained mid-latitude winds can elevate coastal sea levels (e.g., McInnes et al., 2009b), while longer-term changes in prevailing wind direction can cause changes in wave climate and coastline stability (Pirazzoli and Tomasin, 2003; see also Sections 3.5.4 and 3.5.5). Aeolian processes exert significant influence on the formation and evolution of arid and semi-arid environments, being strongly linked to soil and vegetation change (Okin et al., 2006). A rapid shift in wind direction may reposition the leading edge of a forest fire (see Section 4.2.2.2; Mills, 2005) while the fire itself may generate a local circulation response such as tornado genesis (e.g., Cunningham and Reeder, 2009). Unlike other weather and climate elements such as temperature and rainfall, extreme winds are often considered in the context of the extreme phenomena with which they are associated such as tropical and extratropical cyclones (see also Sections 3.4.4 and 3.4.5), thunderstorm downbursts, and tornadoes. Although wind is often not used to define the extreme event itself (Peterson et al., 2008b), wind speed thresholds may be used to characterize the severity of the phenomenon (e.g., the Saffir-Simpson scale for tropical cyclones). Changes in wind extremes may arise from changes in the intensity or location of their associated phenomena (e.g., a change in local convective activity) or from other changes in the climate system such as the movement of large-scale circulation patterns. Wind extremes may be defined by a range of quantities such as high percentiles, maxima over a particular time scale (e.g., daily to yearly), or storm-related highest values. Wind gusts, which are a measure of the highest winds in a short time interval (typically 3 seconds), may be evaluated in models using gust parameterizations that are applied to the maximum daily near-surface wind speed (e.g., Rockel and Woth, 2007).

Over paleoclimatic time scales, proxy data have been used to infer circulation changes across the globe from the mid-Holocene (~6000 years ago) to the beginning of the industrial revolution (Wanner et al., 2008). Over this period, there is evidence for changes in circulation patterns across the globe. The Inter-Tropical Convergence Zone (ITCZ) moved southward, leading to weaker monsoons across Asia (Haug et al., 2001). The Walker circulation strengthened and Southern Ocean westerlies moved northward and strengthened, affecting southern Australia, New Zealand, and southern South America (Shulmeister et al., 2006; Wanner et al., 2008), and an increase in ENSO variability and frequency occurred (Rein et al., 2005; Wanner et al., 2008). There is also weaker evidence for a change toward a lower Northern Atlantic Oscillation (NAO), implying weaker westerly winds over the north Atlantic (Wanner et al., 2008). While the changes in the Northern Hemisphere were attributed to changes in orbital forcing, those in the Southern Hemisphere were more complex, possibly reflecting the additional role on circulation of heat transport in the ocean. Solar variability and volcanic eruptions may also have contributed to decadal to multi-centennial fluctuations over this time period (Wanner et al., 2008).

The AR4 did not specifically address changes in extreme wind although it did report on wind changes in the context of other phenomena such as tropical and extratropical cyclones and oceanic waves and concluded that mid-latitude westerlies had increased in strength in both hemispheres (Trenberth et al., 2007). Direct investigation of changes in wind climatology has been hampered by the sparseness of long-term, high-quality wind measurements from terrestrial anemometers arising from the influence of changes in instrumentation, station location, and surrounding land use (e.g., Cherry, 1988; Pryor et al., 2007; Jakob, 2010; see also Section 3.2.1). Nevertheless, a number of recent studies report trends in mean and extreme wind speeds in different parts of the world based on wind observations and reanalyses.

Over North America, a declining trend in 50th and 90th percentile wind speeds has been reported for much of the United States over 1973 to 2005 (Pryor et al., 2007) and in 10-m hourly wind data over 1953-2006 over western and most of southern Canada (Wan et al., 2010). An increasing trend has been reported in average winds over Alaska over 1955-2001 by Lynch et al. (2004) and over the central Canadian Arctic in all seasons and in the Maritimes in spring and autumn by Wan et al. (2010) as well as in annual maximum winds in a regional reanalysis over the southern Maritimes from 1979-2003 (Hundecha et al., 2008). Over China, negative trends have been reported in 10-m monthly mean and 95th percentile winds over 1969-2005 (Guo et al., 2011), in daily maximum wind speeds over 1956-2004 by Jiang et al. (2010a), and in 2-m average winds over the Tibetan plateau from 1966-2003 (Y. Zhang et al., 2007), confirming earlier declining trends in mean and strong 10-m winds reported by Xu et al. (2006). Over Europe, Smits et al. (2005) found declining trends in extreme winds (those occurring on average 10 and 2 times per year) in 10-m anemometer data over 1962-2002. Pirazolli and Tomasin (2003) reported a generally declining trend in both annual mean and annual maximum winds from 1951 to the mid-1970s and an increasing trend since then, from observations in the central Mediterranean region. Similar to the mostly declining trends found in Northern Hemisphere studies of surface wind observations, Vautard et al. (2010) also found mostly declining trends in surface wind observations across the continental northern mid-latitudes and a stronger decline in extreme winds compared to mean winds in surface wind measurements. In the Southern Hemisphere, McVicar et al. (2008) reported declines in 2-m mean wind speed over 88% of Australia (significant over 57% of the country) over 1975-2006 and positive trends over about 12% of the mainland interior and southern and eastern coastal regions including Tasmania. In Antarctica, increasing trends in mean wind speeds have been reported over the second half of the 20th century (Turner et al., 2005). With the exception of the robust declines in wind reported over China, studies in most areas are too few in number to draw robust conclusions on wind speed change and even fewer studies have addressed extreme wind change. Some studies report opposite trends between anemometer winds and reanalysis data sets in some areas (Smits et al., 2005; McVicar et al., 2008; Vautard et al., 2010); however, comparisons of surface anemometer data at 10 m or lower with reanalysis-derived 10-m data that do not resolve complex surface features is problematic.

Trends in extreme winds have also been inferred from trends in particular phenomena. With regards to tropical cyclones (Section 3.4.4.), no statistically significant trends have been detected in the overall global annual number although a trend has been reported in the intensity of the strongest storms since 1980 [but there is *low confidence* that any observed long-term (i.e., 40 years or more) increases in tropical cyclone activity are robust, after accounting for past changes in observing capabilities; see Section 3.4.4]. In the mid-latitudes, studies have used proxies for wind such as pressure tendencies or geostrophic winds calculated from triangles of pressure (geo-winds) over Europe (e.g., Barring and von Storch, 2004; Matulla et al., 2008; Allan et al., 2009; Barring and Fortuniak, 2009; X.L. Wang et al., 2009b) and Australia (e.g., Alexander and Power, 2009; Alexander et al., 2011). For Europe, these studies suggest that storm activity was higher around 1900 and in the 1990s and lower in the 1960s and 1970s, although X.L. Wang et al. (2009b) note that seasonal trends behave differently than annual trends. In general, long-term trends differ between the different available studies as well as studies that focus on the period for which reanalysis data exist (e.g., Raible, 2007; Leckebusch et al., 2008; Della-Marta et al., 2009; Nissen et al., 2010), and strong inter-decadal variability is also often reported (e.g., Allan et al., 2009; X.L. Wang et al., 2009b; Nissen et al., 2010). Over southeast Australia, a decline in storm activity since around 1885 has been reported (Alexander and Power, 2009; Alexander et al., 2011). See Section 3.4.5 for more discussion of extratropical cyclones. Regarding other phenomena associated with extreme winds, such as thunderstorms, tornadoes, and mesoscale convective complexes, studies are too few in number to assess the effect of their changes on extreme winds. As well, historical data inhomogeneities mean that there is *low confidence* in any observed trends in these small-scale phenomena.

The AR4 reported for the mid-latitudes that trends in the Northern and Southern Annular Modes, which correspond to sea level pressure reductions over the poles, are *likely* related in part to human activity, and this in turn has affected storm tracks and wind patterns in both hemispheres (Hegerl et al., 2007). The relationship between mean and severe winds and natural modes of variability has been investigated in several post-AR4 studies. On the Canadian west coast, Abeysirigunawardena et al. (2009) found that higher extreme winds tend to occur during the negative (i.e., cold) ENSO phase. The generally increasing trend in mean wind speeds over recent decades in Antarctica is consistent with the change in the nature of the Southern Annular Mode toward its high index state (Turner et al., 2005). Donat et al. (2010b) concluded that 80% of storm days in central Europe are connected with westerly flows that occur primarily during the positive phase of the NAO. Declining trends in wind over China have mainly been linked to circulation changes due to a weaker land-sea thermal contrast (Xu et al., 2006; Jiang et al., 2010a; Guo et al., 2011). Vautard et al. (2010) attribute the slowdown in mid to high percentiles of surface winds over most of the continental northern mid-latitudes to changes in atmospheric circulation (10-50%) and an increase in surface roughness due to biomass increases (25-60%), which are supported by RCM simulations. X.L. Wang et al. (2009a) formally detected a link between external forcing and positive trends in the high northern latitudes and negative trends in the northern mid-latitudes using a proxy for wind (geostrophic wind energy) in the boreal winter. Trends in mean and annual maximum winds in the central Mediterranean region were found to be positively correlated with temperature but not with the NAO index (Pirazzoli and Tomasin, 2003). Nissen et al. (2010) used cyclone tracking to identify associated strong winds in reanalysis data from 1957 to 2002 and found a positive trend in the central Mediterranean region and southern Europe and a negative trend over the western Mediterranean region.

Projections of wind speed changes and particularly wind extremes were not specifically addressed in the AR4 although references to wind speed were made in relation to other variables and phenomena such as

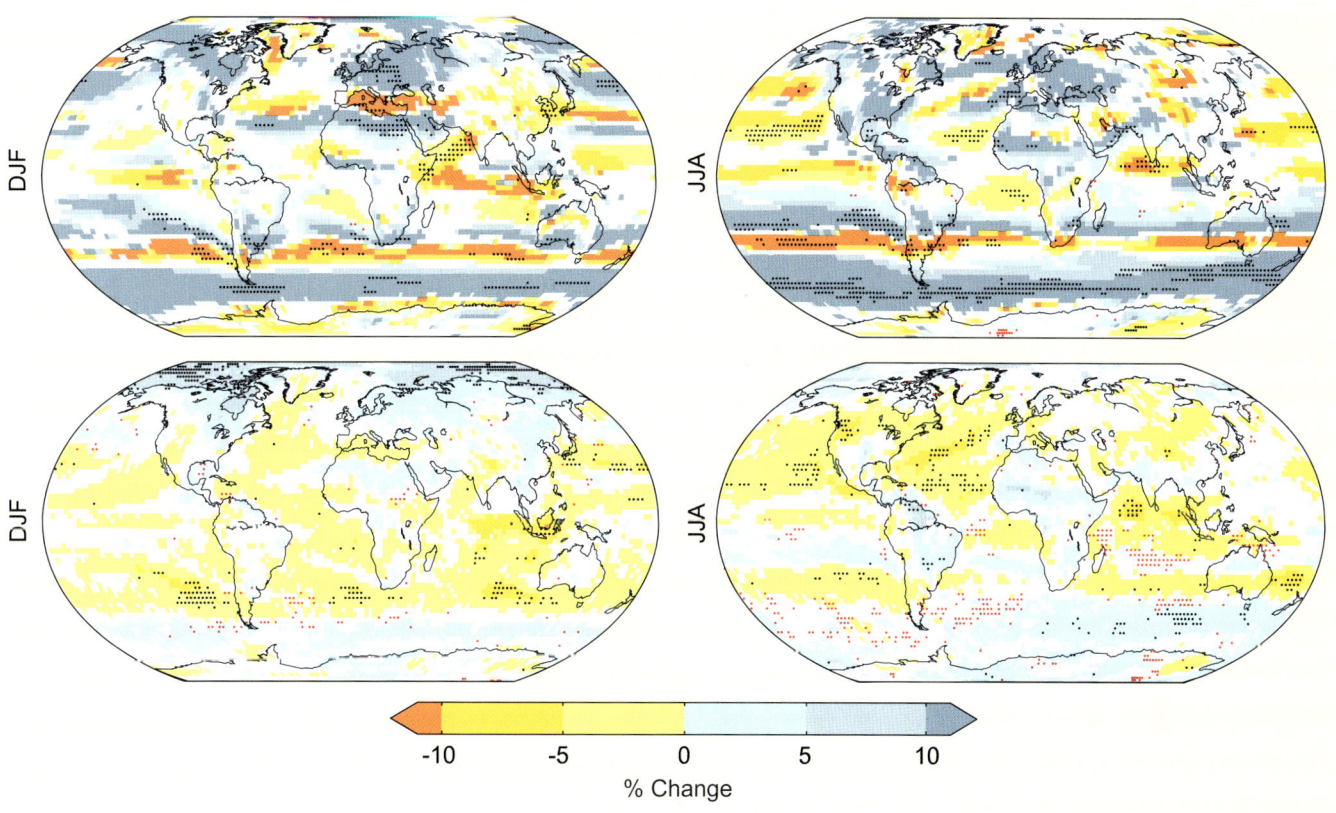

Figure 3-8 | Averaged changes from a 19-member ensemble of CMIP3 GCMs in the mean of the daily averaged 10-m wind speeds (top) and 99th percentile of the daily averaged 10-m wind speeds (bottom) for the period 2081-2100 relative to 1981-2000 (% change) for December to February (left) and June to August (right) plotted only where more than 66% of the models agree on the sign of the change. Black stippling indicates areas where more than 90% of the models agree on the sign of the change. Red stippling indicates areas where more than 66% of models agree on a small change between ±2%. Adapted from McInnes et al. (2011); for more details see Appendix 3.A.

mid-latitude storm tracks, tropical cyclones, and ocean waves (Christensen et al., 2007; Meehl et al., 2007b). Meehl et al. (2007b) projected a *likely* increase in tropical cyclone extreme winds in the future and provided more evidence for a projected poleward shift of the storm tracks and associated changes in wind patterns. Since the AR4, new studies have focused on future changes in winds. Gastineau and Soden (2009) reported a decrease in 99th-percentile winds at 850 hPa in the tropics and an increase in the extratropics in a 17-member multi-model ensemble over 2081-2100 relative to 1981-2000. McInnes et al. (2011) presented spatial maps of multi-model agreement in mean and 99th-percentile 10-m wind change between 1981-2000 and 2081-2100 in a 19-member ensemble (see Figure 3-8). These show an increase in mean winds over Europe, parts of Central and North America, the tropical South Pacific, and the Southern Ocean. Mean wind speed declines occur along the equator reflecting a slowdown in the Walker circulation (Collins et al., 2010) (and in the vicinity of the subtropical ridge in both hemispheres which, together with the strengthening of winds further poleward, reflect the contraction toward the poles of the mid-latitude storm tracks; see Section 3.4.5). Seasonal differences are also apparent with more extensive mean wind increases in the Arctic and parts of the northern Pacific in DJF and decreases over most of the northern Pacific in JJA. The 99th-percentile wind changes show declines over most ocean areas except the northern Pacific and Arctic and Southern Ocean south of 40°S in DJF, the south Pacific between about 10 and 25°S in JJA, and the Southern Ocean south of 50°S in JJA. Increases in 99th-percentile winds occur over the Arctic and large parts of the continental area in the Northern Hemisphere in DJF and in Africa, northern Australia, and Central and South America in JJA. Despite the projections displayed in Figure 3-8, the relatively few studies of projected extreme winds, combined with shortcomings in the simulation of extreme winds and the different models, regions, and methods used to develop projections of this quantity, means that we have *low confidence* in projections of changes in strong winds.

Regional increases in winter wind storm risk over Europe due to changes in storm tracks are also supported by a number of regional studies (e.g., Pinto et al., 2007b; Debernard and Roed, 2008; Leckebusch et al., 2008; Sterl et al., 2009; Donat et al., 2010a,b, 2011). However, GCMs at their current resolution are unable to resolve small-scale phenomena such as tropical cyclones, tornadoes, and mesoscale convective complexes that are associated with particularly severe winds, although as noted by McInnes et al. (2011) these winds would typically be more extreme than 99th percentile. There is evidence to suggest an increase in extreme winds from tropical cyclones in the future (see Section 3.4.4). An increase in atmospheric greenhouse gas concentrations may cause some of the atmospheric conditions conducive to tornadoes such as atmospheric instability to increase due to increasing temperature and humidity, while others such as vertical shear to decrease due to a reduced pole-to-equator temperature gradient (Diffenbaugh et al., 2008), but the literature on these phenomena is extremely limited at this time. There is thus *low confidence* in projections of changes in such small-scale systems because of limited studies, inability of climate models to resolve these phenomena, and possible competing factors affecting future changes. Confidence in the extreme wind changes is therefore lower in the regions most influenced by these phenomena irrespective of whether there is high agreement between GCMs on the sign of the wind speed change.

In addition to studies using GCMs there have also been several recent studies employing RCMs. Those focusing on Europe (e.g., Beniston et al., 2007; Rockel and Woth, 2007; Haugen and Iversen, 2008; Rauthe et al., 2010) also provide a general picture of an increasing trend in extreme winds over northern Europe despite a range of different downscaling models used, the different GCMs in which the downscaling is undertaken, and different metrics used to quantify extreme winds. Small-scale polar lows that typically form north of 60°N have been found to decline in frequency in RCM simulations downscaled from a GCM under different emission scenarios and this is related to greater stability over the region due to mid-troposphere temperatures warming faster than sea surface temperatures over the region (Zahn and von Storch, 2010). In other parts of the world there have been very few studies. Over China, Jiang et al. (2010b) projected decreases in annual and winter mean wind speed based on two RCMs that downscale two different GCMs. Over North America, statistical downscaling of winds from four GCMs over five airports in the northwest United States indicated declines in summer wind speeds and less certain changes in winter (Sailor et al., 2008).

A number of recent studies have addressed observed changes in wind speed across different parts of the globe, but due to the various shortcomings associated with anemometer data and the inconsistency in anemometer and reanalysis trends in some regions, we have *low confidence* in wind trends and their causes at this stage. We also have *low confidence* in how the observed trends in mean wind speed relate to trends in extreme winds. The few studies of projected extreme winds, combined with shortcomings in the simulation of extreme winds and the different models, regions, and methods used to develop projections of this quantity, mean that we have *low confidence* in projections of changes in extreme winds (with the exception of changes associated with tropical cyclones; Section 3.4.4). There is *low confidence* in projections of small-scale phenomena such as tornadoes because competing physical processes may affect future trends and because climate models do not simulate such phenomena.

3.4. Observed and Projected Changes in Phenomena Related to Weather and Climate Extremes

3.4.1. Monsoons

Changes in monsoon-related extreme precipitation and winds due to climate change are not well understood. Generally, precipitation is the most important variable, but it is also a variable associated with larger uncertainties in climate simulations and projections (Wang et al., 2005; Kang and Shukla, 2006). Changes in monsoons should be better depicted by large-scale dynamics, circulation, or moisture convergence more broadly than via precipitation only. However, few studies have focused on observed changes in the large-scale and regional monsoon circulations. Hence, in this section, we focus mostly on monsoon-induced changes in total and seasonal rainfall, with most discussions of intense rainfall covered in Section 3.3.2.

Modeling experiments to assess paleo-monsoons suggest that in the past, during the Holocene due to orbital forcing on a millennial time scale, there was a progressive southward shift of the Northern Hemisphere summer position of the ITCZ around 8,000 years ago. This was accompanied by a pronounced weakening of the monsoon rainfall systems in Africa and Asia and increasing dryness on both continents, while in South America the monsoon was weaker and drier than in the present, as suggested both by models and paleoclimatic indicators (Wanner et al., 2008).

The delineation of the global monsoon has been mostly performed using rainfall data or outgoing longwave radiation (OLR) fields (Kim et al., 2008). Lau and Wu (2007) identified two opposite time evolutions in the occurrence of rainfall events in the tropics: a negative trend in moderate rain events and a positive trend in heavy and light rain events. Positive trends in intense rain were located in deep convective cores of the ITCZ, South Pacific Convergence Zone, Indian Ocean, and monsoon regions.

In the Indo-Pacific region, covering the southeast Asian and north Australian monsoon, Caesar et al. (2011) found low spatial coherence in trends in precipitation extremes across the region between 1971 and 2003. In the few cases where statistically significant trends in precipitation extremes were identified, there was generally a trend towards wetter conditions, in common with the global results of Alexander et al. (2006). Liu et al. (2011) reported a decline in recorded precipitation events in China over 1960-2000, which was mainly accounted for by a decrease in light precipitation events, with intensities of 0.1-0.3 mm day^{-1}. Some of the extreme precipitation appeared to be positively correlated with a La Niña-like sea surface temperature (SST) pattern, but without suggesting the presence of a trend. With regard to wind changes, Guo et al. (2011) analyzed near-surface wind speed change in China and its monsoon regions from 1969 to 2005 and showed a statistically significant weakening in annual and seasonal mean wind.

For the Indian monsoon, Rajeevan et al. (2008) showed that extreme rain events have an increasing trend between 1901 and 2005, but the trend is much stronger after 1950. Sen Roy (2009) investigated changes in extreme hourly rainfall in India, and found widespread increases in heavy precipitation events across India, mostly in the high-elevation regions of the northwestern Himalaya as well as along the foothills of the Himalaya extending south into the Indo-Ganges basin, and particularly during the summer monsoon season during 1980-2002.

In the African monsoon region, Fontaine et al. (2011) investigated recent observed trends using high-resolution gridded precipitation (period 1979-2002), OLR, and reanalyses. Their results revealed a rainfall increase in North Africa since the mid-1990s. Over the longer term, however, Zhou et al. (2008a,b) and Wang and Ding (2006) reported an overall decreasing long-term trend in global land monsoon rainfall during the last 54 years, which was mainly caused by decreasing rainfall in the North African and South Asian monsoons.

For the American monsoon regions, Cavazos et al. (2008) reported increases in the intensity of precipitation in the mountain sites of the northwestern Mexico section of the North American monsoon over the 1961-1998 period, apparently related to an increased contribution from heavy precipitation derived from tropical cyclones. Arriaga-Ramírez and Cavazos (2010) found that total and extreme rainfall in the monsoon region of western Mexico and the US southwest presented a statistically significant increase during 1961-1998, mainly in winter. Groisman and Knight (2008) found that consecutive dry days (see Box 3-3 for definition) have significantly increased in the US southwest. On the other hand, increases in heavy precipitation during 1960-2000 in the South American monsoon have been documented by Marengo et al. (2009a,b) and Rusticucci et al. (2010). Studies using circulation fields such as 850 hPa winds or moisture flux have been performed for the South American monsoon system for assessments of the onset and end of the monsoon, and indicate that the onset exhibits a marked interannual variability linked to variations in SST anomalies in the eastern Pacific and tropical Atlantic (Gan et al., 2006; da Silva and de Carvalho, 2007; Raia and Cavalcanti, 2008; Nieto-Ferreira and Rickenbach, 2011).

Attributing the causes of changes in monsoons is difficult in part because there are substantial inter-model differences in representing Asian monsoon processes (Christensen et al., 2007). Most models simulate the general migration of seasonal tropical rain, although the observed maximum rainfall during the monsoon season along the west coast of India, the North Bay of Bengal, and adjoining northeast India is poorly simulated by many models due to limited resolution. Bollasina and Nigam (2009) show the presence of large systematic biases in coupled simulations of boreal summer precipitation, evaporation, and SST in the Indian Ocean. Many of the biases are pervasive, being common to most simulations.

The observed negative trend in global land monsoon rainfall is better reproduced by atmospheric models forced by observed historical SST than by coupled models without explicit forcing by observed ocean temperatures (Kim et al., 2008). This trend in the east Asian monsoon is strongly linked to the warming trend over the central eastern Pacific and the western tropical Indian Ocean (Zhou et al., 2008b). For the west African monsoon, Joly and Voldoire (2010) explore the role of Gulf of Guinea SSTs in its interannual variability. In most of the studied CMIP3 simulations, the interannual variability of SST is very weak in the Gulf of Guinea, especially along the Guinean Coast. As a consequence, the influence on the monsoon rainfall over the African continent is poorly reproduced. It is suggested that this may be due to the counteracting effects of the Pacific and Atlantic basins over the last decades. The decreasing long-term trend in north African summer monsoon rainfall may be due to the atmosphere response to observed SST variations (Hoerling et al., 2006; Zhou et al., 2008b; Scaife et al., 2009). A similar trend in global monsoon precipitation in land regions is reproduced in CMIP3 models' 20th-century simulations when they include anthropogenic forcing, and for some simulations natural forcing (including volcanic forcing) as well, though the trend is much weaker in general, with the exception of one model (HadCM3) capable of producing a trend of similar magnitude (Li et al., 2008). The decrease in east Asian monsoon rainfall also seems to be related to tropical SST changes (Li et al., 2008), and the less spatially coherent positive trends in precipitation extremes in the southeast Asian and north Australian monsoons appear to be positively correlated with a La Niña-like SST pattern (Caesar et al., 2011).

A variety of factors, natural and anthropogenic, have been suggested as possible causes of variations in monsoons. Changes in regional monsoons are strongly influenced by the changes in the states of dominant patterns of climate variability such as ENSO, the Pacific Decadal Oscillation (PDO), the Northern Annular Mode (NAM), the Atlantic Multi-decadal Oscillation (AMO), and the Southern Annular Mode (SAM) (see also Sections 3.4.2 and 3.4.3). Additionally, model-based evidence has suggested that land surface processes and land use changes could in some instances significantly impact regional monsoons. Tropical land cover change in Africa and southeast Asia appears to have weaker local climatic impacts than in Amazonia (Voldoire and Royer, 2004; Mabuchi et al., 2005a,b). Grimm et al. (2007) and Collini et al. (2008) explored possible feedbacks between soil moisture and precipitation during the early stages of the monsoon in South America, when the surface is not sufficiently wet, and soil moisture anomalies may thus also modulate the development of precipitation. However, the influence of historical land use on the monsoon is difficult to quantify, due both to the poor documentation of land use and difficulties in simulating the monsoon at fine scales. The impact of aerosols (black carbon and sulfate) on changes in rainfall variability jand amounts in monsoon regions has been discussed by Meehl et al. (2008), Lau et al. (2006), and Silva Dias et al. (2002). These studies suggest that there are still large uncertainties and a strong model dependency in the representation of the relevant land surface processes and the role of aerosol direct forcing, and resulting interactions (e.g., in the case of land use forcing; Pitman et al., 2009).

Regarding projections of change in the monsoons, the AR4 (Christensen et al., 2007) concluded: "There is a tendency for monsoonal circulations to result in increased precipitation due to enhanced moisture convergence, despite a tendency towards weakening of the monsoonal flows themselves. However, many aspects of tropical climatic responses remain uncertain." Held and Soden (2006) demonstrate that an increase in the hydrological cycle is accompanied by a global weakening of the large-scale circulation. As global warming is projected to lead to faster warming over land than over the oceans (e.g., Meehl et al., 2007b; Sutton et al., 2007), the continental-scale land-sea thermal contrast, a major factor affecting monsoon circulations, will become stronger in summer. Based on this projection, a simple scenario is that the summer monsoon will be stronger and the winter monsoon will be weaker in the future than now. However, model results derived from the analyses of 15 CMIP3 global models are not as straightforward as implied by this simple consideration (Tanaka et al., 2005), as they show a weakening of these tropical circulations by the late 21st century compared to the late 20th century. In turn, such changes in circulation may lead to changes in precipitation associated with monsoons. For instance, the monsoonal precipitation in Mexico and Central America is projected to decrease in association with increasing precipitation over the eastern equatorial Pacific through changes in the Walker circulation and local Hadley circulation (e.g., Lu et al., 2007). Furthermore, observations and models suggest that changes in monsoons are related at least in part to changes in observed SSTs, as noted above.

At regional scales, there is little consensus in GCM projections regarding the sign of future change in monsoon characteristics, such as circulation and rainfall. For instance, while some models project an intense drying of the Sahel under a global warming scenario, others project an intensification of the rains, and some project more frequent extreme events (Cook and Vizy, 2006). Increases in precipitation are projected in the Asian monsoon (along with an increase in interannual season-averaged precipitation variability), and in the southern part of the west African monsoon, but with some decreases in the Sahel in northern summer. In the Australian monsoon in southern summer, an analysis by Moise and Colman (2009) from the entire ensemble mean of CMIP3 simulations suggested no changes in Australian tropical rainfall during the summer and only slightly enhanced interannual variability.

A study of 19 CMIP3 global models reported a projected increase in mean south Asian summer monsoon precipitation of 8% and a possible extension of the monsoon period (Kripalani et al., 2007). A study (Ashfaq et al., 2009) from the downscaling of the National Center for Atmospheric Research (NCAR) CCSM3 global model using the RegCM3 regional model suggests a weakening of the large-scale monsoon flow and suppression of the dominant intra-seasonal oscillatory modes with overall weakening of the south Asian summer monsoon by the end of the 21st century, resulting in a decrease in summer precipitation in southeast Asia.

Kitoh and Uchiyama (2006) used 15 models under the A1B scenario to analyze the changes in intensity and duration of precipitation in the Baiu-Changma-Meiyu rain band at the end of the 21st century. They found a delay in early summer rain withdrawal over the region extending from the Taiwan province of China, and across the Ryukyu Islands to the south of Japan, contrasted with an earlier withdrawal over the Yangtze Basin. They attributed this feature to El Niño-like mean state changes over the monsoon trough and subtropical anticyclone over the western Pacific region (Meehl et al., 2007b). A southwestward extension of the subtropical anticyclone over the northwestern Pacific Ocean associated with El Niño-like mean state changes and a dry air intrusion in the mid-troposphere from the Asian continent to the northwest of Japan provides favorable conditions for intense precipitation in the Baiu season in Japan (Kanada et al., 2010a). Kitoh et al. (2009) projected changes in precipitation characteristics during the east Asian summer rainy season, using a 5-km mesh cloud-resolving model embedded in a 20-km mesh global atmospheric model with CMIP3 mean SST changes. The frequency of heavy precipitation was projected to increase at the end of the 21st century for hourly as well as daily precipitation. Further, extreme hourly precipitation was projected to increase even in the near future (2030s) when the temperature increase is still modest, even though uncertainties in the projection (and even the simulation) of hourly rainfall are still high.

Climate change scenarios for the 21st century show a weakening of the North American monsoon through a weakening and poleward expansion of the Hadley cell (Lu et al., 2007). The expansion of the Hadley cell is caused by an increase in the subtropical static stability, which pushes poleward the baroclinic instability zone and hence the outer boundary of the Hadley cell. Simple physical arguments (Held and Soden, 2006) predict a slowdown of the tropical overturning circulation under global warming. A few studies (e.g., Marengo et al., 2009a) have projected over the period 1960-2100 a weak tendency for an increase in dry spells. The projections show an increase in the frequency of rainfall extremes in southeastern South America by the end of the 21st century, possibly due to an intensification of the moisture transport from Amazonia by a more frequent/intense low-level jet east of the Andes in the A2 emissions scenario (Marengo et al., 2009a; Soares and Marengo, 2009).

There are many deficiencies in model representation of the monsoons and the processes affecting them, and this reduces confidence in their ability to project future changes. Some of the uncertainty in global and regional climate change projections in the monsoon regions results from the limits in the model representation of resolved processes (e.g., moisture advection), the parameterizations of sub-grid-scale processes (e.g., clouds, precipitation), and model simulations of feedback mechanisms at the global and regional scale (e.g., changes in land use/cover; see also Section 3.1.4). Kharin et al. (2007) made an intercomparison of precipitation extremes in the tropical region in all AR4 models with observed extremes expressed as 20-year return values. They found very large disagreement in the tropics suggesting that some physical processes associated with extreme precipitation are not well represented by the models due to model resolution and physics. Shukla (2007) noted that current climate models cannot even adequately predict the mean intensity and the seasonal variations of the Asian summer monsoon. This reduces confidence in the projected changes in extreme precipitation over the monsoon regions. Many of the important climatic effects of the Madden-Julian Oscillation (MJO, a natural mode of the climate system operating on time scales of about a month), including its impacts on rainfall variability in the monsoons, are still poorly simulated by contemporary climate models (Christensen et al., 2007).

Current GCMs still have difficulties and display a wide range of skill in simulating the subseasonal variability associated with the Asian summer monsoon (Lin et al., 2008b). Most GCMs simulate westward propagation of the coupled equatorial easterly waves, but relatively poor eastward propagation of the MJO and overly weak variances for both the easterly waves and the MJO. Most GCMs are able to reproduce the basic characteristics of the precipitation seasonal cycle associated with the South American Monsoon System (SAMS), but there are large discrepancies in the South Atlantic Convergence Zone represented by the models in both intensity and location, and in its seasonal evolution (Vera et al., 2006). In addition, models exhibit large discrepancies in the direction of the changes associated with the summer (SAMS) precipitation, which makes the projections for that region highly uncertain. Lin et al. (2008a) show that the coupled GCMs have significant problems and display a wide range of skill in simulating the North American monsoon and associated intra-seasonal variability.

Most of the models reproduce the monsoon rain belt, extending from southeast to northwest, and its gradual northward shift in early summer, but overestimate the precipitation over the core monsoon region throughout the seasonal cycle and fail to reproduce the monsoon retreat in the fall. The AR4 assessed that models fail in representing the main features of the west African monsoon although most of them do have a monsoonal climate albeit with some distortion (Christensen et al., 2007). Other major sources of uncertainty in projections of monsoon changes are the responses and feedbacks of the climate system to emissions as represented in climate models. These uncertainties are particularly related to the representation of the conversion of emissions into concentrations of radiatively active species (i.e., via atmospheric chemistry and carbon cycle models) and especially those derived from aerosol products of biomass burning, which can affect the onset of the rainy season (Silva Dias et al., 2002). The subsequent response of the physical climate system complicates the nature of future projections of monsoon precipitation. Moreover, the long-term variations of model skill in simulating monsoons and their variations represent an additional source of uncertainty for the monsoon regions, and indicate that the regional reliability of long climate model runs may depend on the time slice for which the output of the model is analyzed.

The AR4 (Hegerl et al., 2007) concluded that the current understanding of climate change in the monsoon regions remains one of considerable uncertainty with respect to circulation and precipitation. With a few exceptions in some monsoon regions, this has not changed. These conclusions have been based on very few studies, there are many issues with model representation of monsoons and the underlying processes, and there is little consensus in climate models, so there is *low confidence* in projections of changes in monsoons, even in the sign of the change. However, one common pattern is a *likely* increase in extreme precipitation in monsoon regions (see Section 3.3.2), though not necessarily induced by changes in monsoon characteristics, and not necessarily occurring in all monsoon regions.

3.4.2. El Niño-Southern Oscillation

The El Niño-Southern Oscillation (ENSO) is a natural fluctuation of the global climate system caused by equatorial ocean-atmosphere interaction in the tropical Pacific Ocean (Philander, 1990). The term 'Southern Oscillation' refers to a tendency for above-average surface atmospheric pressures in the Indian Ocean to be associated with below-average pressures in the Pacific, and vice versa. This oscillation is associated with variations in SSTs in the east equatorial Pacific. The oceanic and atmospheric variations are collectively referred to as ENSO. An El Niño episode is one phase of the ENSO phenomenon and is associated with abnormally warm central and east equatorial Pacific Ocean surface temperatures, while the opposite phase, a La Niña episode, is associated with abnormally cool ocean temperatures in this region. Both phases are associated with a characteristic spatial pattern of droughts and floods. An El Niño episode is usually accompanied by drought in southeastern Asia, India, Australia, southeastern Africa, Amazonia, and northeast Brazil, with fewer than normal tropical cyclones around Australia and in the North Atlantic. Wetter than normal conditions during El Niño episodes are observed along the west coast of tropical South America, subtropical latitudes of western North America, and southeastern America. In a La Niña episode the climate anomalies are usually the opposite of those in an El Niño. Pacific islands are strongly affected by ENSO variations. Recent research (e.g., Kenyon and Hegerl, 2008; Ropelewski and Bell, 2008; Schubert et al., 2008a; Alexander et al., 2009; Grimm and Tedeschi, 2009; Zhang et al., 2010) has demonstrated that different phases of ENSO (El Niño or La Niña episodes) also are associated with different frequencies of occurrence of short-term weather extremes such as heavy rainfall events and extreme temperatures. The relationship between ENSO and interannual variations in tropical cyclone activity is well known (e.g., Kuleshov et al., 2008). The simultaneous occurrence of a variety of climate extremes in an El Niño episode (or a La Niña episode) may provide special challenges for organizations coping with disasters induced by ENSO (see also Section 3.1.1). Monitoring and predicting ENSO can lead to disaster risk reduction through early warning (see Case Study 9.2.11).

The AR4 noted that orbital variations could affect the ENSO behavior (Jansen et al., 2007). Cane (2005) found that a relatively simple coupled model suggested that systematic changes in El Niño could be stimulated by seasonal changes in solar insolation. However, a more comprehensive model simulation (Wittenberg, 2009) has suggested that long-term changes in the behavior of the phenomenon might occur even without forcing from radiative changes. Vecchi and Wittenberg (2010) concluded that the "tropical Pacific could generate variations in ENSO frequency and intensity on its own (via chaotic behavior), respond to external radiative forcings (e.g., changes in greenhouse gases, volcanic eruptions, atmospheric aerosols, etc), or both." Meehl et al. (2009a) demonstrate that solar insolation variations related to the 11-year sunspot cycle can affect ocean temperatures associated with ENSO.

ENSO has varied in strength over the last millennium with stronger activity in the 17th century and late 14th century, and weaker activity during the 12th and 15th centuries (Cobb et al., 2003; Conroy et al., 2009). On longer time scales, there is evidence that ENSO may have changed in response to changes in the orbit of the Earth (Vecchi and Wittenberg, 2010), with the phenomenon apparently being weaker around 6,000 years ago (according to proxy measurements from corals and climate model simulations; Rein et al., 2005; Brown et al., 2006; Otto-Bliesner et al., 2009), and model simulations suggest that it was stronger at the last glacial maximum (An et al., 2004). Fossil coral evidence indicates that the phenomenon continued to operate during the last glacial interval (Tudhope et al., 2001). Thus the paleoclimatic evidence indicates that ENSO can continue to operate, although altered perhaps in intensity, in very different background climate states.

The AR4 noted that the nature of ENSO has varied substantially over the period of instrumental data, with strong events from the late 19th century through the first quarter of the 20th century and again after 1950. An apparent climate shift around 1976-1977 was associated with a shift to generally above-normal SSTs in the central and eastern Pacific and a tendency toward more prolonged and stronger El Niño episodes (Trenberth et al., 2007). Ocean temperatures in the central equatorial Pacific (the so-called NINO3 index) suggest a trend toward more frequent or stronger El Niño episodes over the past 50 to 100 years (Vecchi and Wittenberg, 2010). Vecchi et al. (2006) reported a weakening of the equatorial Pacific pressure gradient since the 1960s, with a sharp drop in the 1970s. Power and Smith (2007) proposed that the apparent dominance of El Niño during the last few decades was due in part to a change in the background state of the Southern Oscillation Index (SOI, the standardized difference in surface atmospheric pressure between Tahiti and Darwin), rather than a change in variability or a shift to more frequent El Niño events alone. Nicholls (2008) examined the behavior of the SOI and another index, the NINO3.4 index of central equatorial Pacific SSTs, but found no evidence of trends in the variability or the persistence of the indices [although Yu and Kao (2007) reported decadal variations in the persistence barrier, the tendency for weaker persistence across the Northern Hemisphere spring], nor in their seasonal patterns. There was a trend toward what might be considered more 'El Niño-like' behavior in the SOI (and more weakly in NINO3.4), but only through the period March to September and not in November to February, the season when El Niño and La Niña events typically peak. The trend in the SOI reflected only a trend in Darwin pressures, with no trend in Tahiti pressures. Apart from this trend, the temporal/seasonal nature of ENSO has been remarkably consistent through a period of strong global warming. There is evidence, however, of a tendency for recent El Niño episodes to be centered more in the central equatorial Pacific than in the east Pacific (Yeh et al., 2009), and for these central Pacific episodes to be increasing in intensity (Lee and McPhaden, 2010). In turn, these changes may explain changes that have been noted in the remote influences of the phenomenon on the climate over Australia and in the mid-latitudes (Wang and Hendon, 2007; Weng et al., 2009). For instance, Taschetto et al. (2009) demonstrated that episodes with the warming centered in the central Pacific exhibit different patterns of Australian rainfall variations relative to the east Pacific-centered El Niño events.

The possible role of increased greenhouse gases in affecting the behavior of ENSO over the past 50 to 100 years is uncertain. Yeh et al. (2009) suggested that changes in the background temperature associated with increases in greenhouse gases should affect the behavior of El Niño, such as the location of the strongest SST anomalies, because El Niño behavior is strongly related to the average ocean temperature gradients in the equatorial Pacific. Some studies (e.g., Q. Zhang et al., 2008) have suggested that increased activity might be due to increased CO_2; however, no formal attribution study has yet been completed and some other studies (e.g., Power and Smith, 2007) suggest that changes in the phenomenon are within the range of natural variability (i.e., that no change has yet been detected, let alone attributed to a specific cause).

Global warming is projected to lead to a mean reduction in the zonal mean wind across the equatorial Pacific (Vecchi and Soden, 2007b). However, this change should not be described as an 'El Niño-like' average change even though during an El Niño episode these winds also weaken, because there is only limited correspondence between these changes in the mean state of the equatorial Pacific and an El Niño episode. The AR4 determined that all models exhibited continued ENSO interannual variability in projections through the 21st century, but the projected behavior of the phenomenon differed between models, and it was concluded that "there is no consistent indication at this time of discernible changes in projected ENSO amplitude or frequency in the 21st century" (Meehl et al., 2007b). Models project a wide variety of changes in ENSO variability and the frequency of El Niño episodes as a consequence of increased greenhouse gas concentrations, with a range between a 30% reduction to a 30% increase in variability (van Oldenborgh et al., 2005). One model study even found that although ENSO activity increased when atmospheric CO_2 concentrations were doubled or quadrupled, a considerable decrease in activity occurred when CO_2 was increased by a factor of 16 times, much greater than is possible through the 21st century (Cherchi et al., 2008), suggesting a wide variety of possible ENSO changes as a result of CO_2 changes. The remote impacts, on rainfall for instance, of ENSO may change as CO_2 increases, even if the equatorial Pacific aspect of ENSO does not change substantially. For instance, regions in which rainfall increases in the future tend to show increases in interannual rainfall variability (Boer, 2009), without any strong change in the interannual variability of tropical SSTs. Also, since some long-term projected changes in response to increased greenhouse gases may resemble the climate response to an El Niño event, this may enhance or mask the response to El Niño events in the future (Lau et al., 2008b; Müller and Roeckner, 2008).

One change that models tend to project is an increasing tendency for El Niño episodes to be centered in the central equatorial Pacific, rather than the traditional location in the eastern equatorial Pacific. Yeh et al. (2009) examined the relative frequency of El Niño episodes simulated in coupled climate models with projected increases in greenhouse gas concentrations. A majority of models, especially those best able to simulate the current ratio of central Pacific locations to east Pacific locations of El Niño events, projected a further increase in the relative frequency of these central Pacific events. Such a change would also have implications for the remote influence of the phenomenon on climate away from the equatorial Pacific (e.g., Australia and India). However, even the projection that the 21st century may see an increased frequency of central Pacific El Niño episodes, relative to the frequency of events located further east (Yeh et al., 2009), is subject to considerable uncertainty. Of the 11 coupled climate model simulations examined by Yeh et al. (2009), three projected a relative decrease in the frequency of these central Pacific episodes, and only four of the models produced a statistically significant change to more frequent central Pacific events.

A caveat regarding all projections of future behavior of ENSO arises from systematic biases in the depiction of ENSO behavior through the 20th century by models (Randall et al., 2007; Guilyardi et al., 2009). Leloup et al. (2008) for instance, demonstrate that coupled climate models show wide differences in the ability to reproduce the spatial characteristics of SST variations associated with ENSO during the 20th century, and all models have failings. They concluded that it is difficult to even classify models by the quality of their reproductions of the behavior of ENSO, because models scored unevenly in their reproduction of the different phases of the phenomenon. This makes it difficult to determine which models to use to project future changes in ENSO. Moreover, most of the models are not able to reproduce the typical circulation anomalies associated with ENSO in the Southern Hemisphere (Vera and Silvestri, 2009) and the Northern Hemisphere (Joseph and Nigam, 2006).

There was no consistency in projections of changes in ENSO variability or frequency at the time of the AR4 (Meehl et al., 2007b) and this situation has not changed as a result of post-AR4 studies. The evidence is that the nature of ENSO has varied in the past apparently sometimes in response to changes in radiative forcing but also possibly due to internal climatic variability. Since radiative forcing will continue to change in the future, we can confidently expect changes in ENSO and its impacts as well, although both El Niño and La Niña episodes will continue to occur (e.g., Vecchi and Wittenberg, 2010). Our current limited understanding, however, means that it is not possible at this time to confidently predict whether ENSO activity will be enhanced or damped due to anthropogenic climate change, or even if the frequency of El Niño or La Niña episodes will change (Collins et al., 2010).

In summary, there is *medium confidence* in a recent trend toward more frequent central equatorial Pacific El Niño episodes, but insufficient evidence for more specific statements about observed trends in ENSO. Model projections of changes in ENSO variability and the frequency of El Niño episodes as a consequence of increased greenhouse gas concentrations are not consistent, and so there is *low confidence* in projections of changes in the phenomenon. However, there is *medium confidence* regarding a projected increase (projected by most GCMs) in the relative frequency of central equatorial Pacific events, which typically exhibit different patterns of climate variations than do the classical East Pacific events.

3.4.3. Other Modes of Variability

Other natural modes of variability beside ENSO (Section 3.4.2) that are relevant to extremes and disasters include the North Atlantic Oscillation (NAO), the Southern Annular Mode (SAM), and the Indian Ocean Dipole (IOD) (Trenberth et al., 2007). The NAO is a large-scale seesaw in atmospheric pressure between the subtropical high and the polar low in the North Atlantic region. The positive NAO phase has a strong subtropical high-pressure center and a deeper than normal Icelandic low. This results in a shift of winter storms crossing the Atlantic Ocean to a more northerly track, and is associated with warm and wet winters in northwestern Europe and cold and dry winters in northern Canada and Greenland. Scaife et al. (2008) discuss the relationship between the NAO and European extremes. Paleoclimatic data indicate that the NAO was persistently in its positive phase during medieval times and persistently in its negative phase during the cooler Little Ice Age (Trouet et al., 2009). The NAO is closely related to the Northern Annular Mode (NAM); for brevity we focus here on the NAO but much of what is said about the NAO also applies to the NAM. The SAM is the largest mode of Southern Hemisphere extratropical variability and refers to north-south shifts in atmospheric mass between the middle and high latitudes. It plays an important role in climate variability in these latitudes. The SAM positive phase is linked to negative sea level pressure anomalies over the polar regions and intensified westerlies. It has been associated with cooler than normal temperatures over most of Antarctica and Australia, with warm anomalies over the Antarctic Peninsula, southern South America, and southern New Zealand, and with anomalously dry conditions over southern South America, New Zealand, and Tasmania and wet anomalies over much of Australia and South Africa (e.g., Hendon et al., 2007). The IOD is a coupled ocean-atmosphere phenomenon in the Indian Ocean. A positive IOD event is associated with anomalous cooling in the southeastern equatorial Indian Ocean and anomalous warming in the western equatorial Indian Ocean. Recent work (Ummenhofer et al., 2008, 2009a,b) has implicated the IOD as a cause of droughts in Australia, and heavy rainfall in east Africa (Ummenhofer et al., 2009c). There is also evidence of modes of variability operating on multi-decadal time scales, notably the Pacific Decadal Oscillation (PDO) and the Atlantic Multi-decadal Oscillation (AMO). Variations in the PDO have been related to precipitation extremes over North America (Zhang et al., 2010).

Both the NAO and the SAM exhibited trends toward their positive phase (strengthened mid-latitude westerlies) over the last three to four decades, although the NAO has been in its negative phase in the last few years. Goodkin et al. (2008) concluded that the variability in the NAO is linked with changes in the mean temperature of the Northern Hemisphere. Dong et al. (2011) demonstrated that some of the observed late 20th-century decadal-scale changes in NAO behavior could be reproduced by increasing the CO_2 concentrations in a coupled model, and concluded that greenhouse gas concentrations may have played a role in forcing these changes. The largest observed trends in the SAM occur in December to February, and model simulations indicate that these are due mainly to stratospheric ozone changes. However it has been argued that anthropogenic circulation changes are poorly characterized by trends in the annular modes (Woollings et al., 2008). Further complicating these trends, Silvestri and Vera (2009) reported changes in the typical hemispheric circulation pattern related to the SAM and its associated impact on both temperature and precipitation anomalies, particularly over South America and Australia, between the 1960s-1970s and 1980s-1990s. The time scales of variability in modes such as the AMO and PDO are so long that it is difficult to diagnose any change in their behavior in modern data, although some evidence suggests that the PDO may be affected by anthropogenic forcing (Meehl et al., 2009b). The AR4 (Hegerl et al., 2007) concluded that trends over recent decades in the NAO and SAM are *likely* related in part to human activity. The negative NAO phase of the last few years, however, with the lack of formal attribution studies, means that attribution of changes in the NAO to human activity in recent decades now can only be considered *about as likely as not* (expert opinion). Attribution of the SAM trend to human activity is still assessed to be *likely* (expert opinion) although mainly attributable to trends in stratospheric ozone concentration (Hegerl et al., 2007).

The AR4 noted that there was considerable spread among the model projections of the NAO, leading to low confidence in NAO projected changes, but the magnitude of the increase for the SAM is generally more consistent across models (Meehl et al., 2007b). However, the ability of coupled models to simulate the observed SAM impact on climate variability in the Southern Hemisphere is limited (e.g., Miller et al., 2006; Vera and Silvestri, 2009). Variations in the longer time-scale modes of variability (AMO, PDO) might affect projections of changes in extremes associated with the various natural modes of variability and global temperatures (Keenlyside et al., 2008).

Sea level pressure is projected to increase over the subtropics and mid-latitudes, and decrease over high latitudes (Meehl et al., 2007b). This would equate to trends in the NAO and SAM, with a poleward shift of the storm tracks of several degrees latitude and a consequent increase in cyclonic circulation patterns over the Arctic and Antarctica. In the Southern Hemisphere, two opposing effects, stratospheric ozone recovery and increasing greenhouse gases, can be expected to affect the modes such as the SAM (Arblaster et al., 2011). During the 21st century, although stratospheric ozone concentrations are expected to recover, tending to lead to a weakening of the SAM, models consistently project polar vortex intensification to continue due to the increases in greenhouse gases, except in summer where the competing effects of stratospheric ozone recovery complicate this picture (Arblaster et al., 2011).

A recent study (Woollings et al., 2010) found a tendency toward a more positive NAO under anthropogenic forcing through the 21st century with one model, although they concluded that confidence in the model projections was low because of deficiencies in its simulation of current-day NAO regimes. Goodkin et al. (2008) predict continuing high variability, on multi-decadal scales, in the NAO with continued global warming. Keenlyside et al. (2008) proposed that variations associated with the multi-decadal modes of variability may offset warming due to increased greenhouse gas concentrations over the next decade or so. Conway et al. (2007) reported that model projections of future IOD behavior showed no consistency. Kay and Washington (2008) reported that under some emissions scenarios, changes in a dipole mode in the Indian Ocean could change rainfall extremes in southern Africa.

In summary, it is *likely* that there has been an anthropogenic influence on recent trends in the SAM (linked with trends in stratospheric ozone rather than changes in greenhouse gases), but it is only *about as likely as not* that there have been anthropogenic influences on observed trends in the NAO. Issues with the ability of models to simulate current behavior of these natural modes, the influence of competing factors (e.g., stratospheric ozone, greenhouse gases) on current and future mode behavior, and inconsistency between the model projections (and the seasonal dependence of these projections), means that there is *low confidence* in the ability to project changes in the modes including the NAO, SAM, and IOD. Models do, however, consistently project a strengthening of the polar vortex in the Southern Hemisphere from increasing greenhouse gases, although in summer stratospheric ozone recovery is expected to offset this intensification.

3.4.4. Tropical Cyclones

Tropical cyclones occur in most tropical oceans and pose a significant threat to coastal populations and infrastructure, and marine interests such as shipping and offshore activities. Each year, about 90 tropical cyclones occur globally, and this number has remained roughly steady over the modern period of geostationary satellites (since around the mid-1970s). While the global frequency has remained steady, there can be substantial inter-annual to multi-decadal frequency variability within individual ocean basins (e.g., Webster et al., 2005). This regional variability, particularly when combined with substantial inter-annual to multi-decadal variability in tropical cyclone tracks (e.g., Kossin et al., 2010), presents a significant challenge for disaster planning and mitigation aimed at specific regions.

Tropical cyclones are perhaps most commonly associated with extreme wind, but storm-surge and freshwater flooding from extreme rainfall generally cause the great majority of damage and loss of life (e.g., Rappaport, 2000; Webster, 2008). Related indirect factors, such as the failure of the levee system in New Orleans during the passage of Hurricane Katrina (2005), or mudslides during the landfall of Hurricane Mitch (1998) in Central America, represent important related impacts (Case Study 9.2.5). Projected sea level rise will further compound tropical cyclone surge impacts. Tropical cyclones that track poleward can undergo a transition to become extratropical cyclones. While these storms have different characteristics than their tropical progenitors, they can still be accompanied by a storm surge that can impact regions well away from the tropics (e.g., Danard et al., 2004).

Tropical cyclones are typically classified in terms of their intensity, which is a measure of near-surface wind speed (sometimes categorized according to the Saffir-Simpson scale). The strongest storms (Saffir-Simpson category 3, 4, and 5) are comparatively rare but are generally responsible for the majority of damage (e.g., Landsea, 1993; Pielke Jr. et al., 2008). Additionally, there are marked differences in the characteristics of both observed and projected tropical cyclone variability when comparing weaker and stronger tropical cyclones (e.g., Webster et al., 2005; Elsner et al., 2008; Bender et al., 2010), while records of the strongest storms are potentially less reliable than those of their weaker counterparts (Landsea et al., 2006).

In addition to intensity, the structure and areal extent of the wind field in tropical cyclones, which can be largely independent of intensity, also play an important role on potential impacts, particularly from storm surge (e.g., Irish and Resio, 2010), but measures of storm size are largely absent in historical data. Other relevant tropical cyclone measures include frequency, duration, and track. Forming robust physical links between all of the metrics briefly mentioned here and natural or human-induced changes in climate variability is a major challenge. Significant progress is being made, but substantial uncertainties still remain due largely to data quality issues (see Section 3.2.1 and below) and imperfect theoretical and modeling frameworks (see below).

Observed Changes

Detection of trends in tropical cyclone metrics such as frequency, intensity, and duration remains a significant challenge. Historical tropical cyclone records are known to be heterogeneous due to changing observing technology and reporting protocols (e.g., Landsea et al., 2004). Further heterogeneity is introduced when records from multiple ocean basins are combined to explore global trends because data quality and reporting protocols vary substantially between regions (Knapp and Kruk, 2010). Progress has been made toward a more homogeneous global record of tropical cyclone intensity using satellite data (Knapp and Kossin, 2007; Kossin et al., 2007), but these records are necessarily constrained to the satellite era and so only represent the past 30 to 40 years.

Natural variability combined with uncertainties in the historical data makes it difficult to detect trends in tropical cyclone activity. There have been no significant trends observed in global tropical cyclone frequency records, including over the present 40-year period of satellite observations (e.g., Webster et al., 2005). Regional trends in tropical cyclone frequency have been identified in the North Atlantic, but the fidelity of these trends is debated (Holland and Webster, 2007; Landsea, 2007; Mann et al., 2007a). Different methods for estimating undercounts in the earlier part of the North Atlantic tropical cyclone record provide mixed conclusions (Chang and Guo, 2007; Mann et al., 2007b; Kunkel et al., 2008; Vecchi and Knutson, 2008). Regional trends have not been detected in other oceans (Chan and Xu, 2009; Kubota and Chan, 2009; Callaghan and Power, 2011). It thus remains uncertain whether any observed increases in tropical cyclone frequency on time scales longer than about 40 years are robust, after accounting for past changes in observing capabilities (Knutson et al., 2010).

Frequency estimation requires only that a tropical cyclone be identified and reported at some point in its lifetime, whereas intensity estimation requires a series of specifically targeted measurements over the entire duration of the tropical cyclone (e.g., Landsea et al., 2006). Consequently, intensity values in the historical records are especially sensitive to changing technology and improving methodology, which heightens the challenge of detecting trends within the backdrop of natural variability. Global reanalyses of tropical cyclone intensity using a homogenous satellite record have suggested that changing technology has introduced a non-stationary bias that inflates trends in measures of intensity (Kossin et al., 2007), but a significant upward trend in the intensity of the strongest tropical cyclones remains after this bias is accounted for (Elsner et al., 2008). While these analyses are suggestive of a link between observed global tropical cyclone intensity and climate change, they are necessarily confined to a roughly 30-year period of satellite observations, and cannot provide clear evidence for a longer-term trend.

Time series of power dissipation, an aggregate compound of tropical cyclone frequency, duration, and intensity that measures total energy consumption by tropical cyclones, show upward trends in the North Atlantic and weaker upward trends in the western North Pacific over the past 25 years (Emanuel, 2007), but interpretation of longer-term trends in this quantity is again constrained by data quality concerns. The variability and trend of power dissipation can be related to SST and other local factors such as tropopause temperature and vertical wind shear (Emanuel, 2007), but it is a current topic of debate whether local SST or the difference between local SST and mean tropical SST is the more physically relevant metric (Swanson, 2008). The distinction is an important one when making projections of changes in power dissipation based on projections of SST changes, particularly in the tropical Atlantic where SST has been increasing more rapidly than in the tropics as a whole (Vecchi et al., 2008). Accumulated cyclone energy, which is an integrated metric analogous to power dissipation, has been declining globally since reaching a high point in 2005, and is presently at a 40-year low point (Maue, 2009). The present period of quiescence, as well as the period of heightened activity leading up to the high point in 2005, does not clearly represent substantial departures from past variability (Maue, 2009).

Increases in tropical water vapor and rainfall (Trenberth et al., 2005; Lau and Wu, 2007) have been identified and there is some evidence for related changes in tropical cyclone-related rainfall (Lau et al., 2008a), but a robust and consistent trend in tropical cyclone rainfall has not yet been established due to a general lack of studies. Similarly, an increase in the length of the North Atlantic hurricane season has been noted (Kossin, 2008), but the uncertainty in the amplitude of the trends and the lack of additional studies limits the utility of these results for a meaningful assessment.

Estimates of tropical cyclone variability prior to the modern instrumental historical record have been constructed using archival documents (Chenoweth and Devine, 2008), coastal marsh sediment records, and isotope markers in coral, speleothems, and tree rings, among other methods (Frappier et al., 2007a). These estimates demonstrate centennial- to millennial-scale relationships between climate and tropical cyclone activity (Donnelly and Woodruff, 2007; Frappier et al., 2007b; Nott et al., 2007; Nyberg et al., 2007; Scileppi and Donnelly, 2007; Neu, 2008; Woodruff et al., 2008a,b; Mann et al., 2009; Yu et al., 2009), but generally do not provide robust evidence that the observed post-industrial tropical cyclone activity is unprecedented.

The AR4 Summary for Policymakers concluded that it is *likely* that an increase had occurred in intense tropical cyclone activity since 1970 in some regions (IPCC, 2007b). The subsequent CCSP assessment report (Kunkel et al., 2008) concluded that it is *likely* that the frequency of tropical storms, hurricanes, and major hurricanes in the North Atlantic has increased over the past 100 years, a time in which Atlantic SSTs also increased. Kunkel et al. (2008) also concluded that the increase in Atlantic power dissipation is *likely* substantial since the 1950s. Based on research subsequent to the AR4 and Kunkel et al. (2008), which further elucidated the scope of uncertainties in the historical tropical cyclone data, the most recent assessment by the World Meteorological Organization (WMO) Expert Team on Climate Change Impacts on Tropical Cyclones (Knutson et al., 2010) concluded that it remains uncertain whether past changes in any tropical cyclone activity (frequency, intensity, rainfall) exceed the variability expected through natural causes, after accounting for changes over time in observing capabilities. The present assessment regarding observed trends in tropical cyclone activity is essentially dentical to the WMO assessment (Knutson et al., 2010): there is *low confidence* that any observed long-term (i.e., 40 years or more) increases in tropical cyclone activity are robust, after accounting for past changes in observing capabilities.

Causes of the Observed Changes

In addition to the natural variability of tropical SSTs, several studies have concluded that there is a detectable tropical SST warming trend due to increasing greenhouse gases (Karoly and Wu, 2005; Knutson et al., 2006; Santer et al., 2006; Gillett et al., 2008a). The region where this anthropogenic warming has occurred encompasses tropical cyclogenesis regions, and Kunkel et al. (2008) stated that it is *very likely* that human-caused increases in greenhouse gases have contributed to the increase in SSTs in the North Atlantic and the Northwest Pacific hurricane formation regions over the 20th century.

Changes in the mean thermodynamic state of the tropics can be directly linked to tropical cyclone variability within the theoretical framework of potential intensity theory (Bister and Emanuel, 1998). In this framework, the expected response of tropical cyclone intensity to observed climate change is relatively straightforward: if climate change causes an increase in the ambient potential intensity that tropical cyclones move through, the distribution of intensities in a representative sample of storms is expected to shift toward greater intensities (Emanuel, 2000; Wing et al., 2007). The fractional changes associated with such a shift in the distribution would be largest in the upper quantiles of the distribution as the strongest tropical cyclones become stronger (Elsner et al., 2008).

Given the evidence that SST in the tropics has increased due to increasing greenhouse gases, and the theoretical expectation that increases in potential intensity will lead to stronger storms, it is essential to fully understand the relationship between SST and potential intensity. Observations demonstrate a strong positive correlation between SST and the potential intensity. This relationship suggests that SST increases will lead to increased potential intensity, which will then ultimately lead to stronger storms (Emanuel, 2000; Wing et al., 2007). However, there is a growing body of research suggesting that local potential intensity is controlled by the difference between local SST and spatially averaged SST in the tropics (Vecchi and Soden, 2007a; Xie et al., 2010; Ramsay and Sobel, 2011). Since increases in SST due to global warming are not expected to lead to continuously increasing SST gradients, this recent research suggests that increasing SST due to global warming, by itself, does not yet have a fully understood physical link to increasingly strong tropical cyclones.

The present period of heightened tropical cyclone activity in the North Atlantic, concurrent with comparative quiescence in other ocean basins (e.g., Maue, 2009), is apparently related to differences in the rate of SST increases, as global SST has been rising steadily but at a slower rate than has the Atlantic (Holland and Webster, 2007). The present period of relatively enhanced warming in the Atlantic has been proposed to be due to internal variability (Zhang and Delworth, 2009), anthropogenic tropospheric aerosols (Mann and Emanuel, 2006), and mineral (dust) aerosols (Evan et al., 2009). None of these proposed mechanisms provide a clear expectation that North Atlantic SST will continue to increase at a greater rate than the tropical mean SST.

Changes in tropical cyclone intensity, frequency, genesis location, duration, and track contribute to what is sometimes broadly defined as 'tropical cyclone activity.' Of these metrics, intensity has the most direct physically reconcilable link to climate variability within the framework of potential intensity theory, as described above (Kossin and Vimont, 2007). Statistical correlations between necessary ambient environmental conditions (e.g., low vertical wind shear and adequate atmospheric instability and moisture) and tropical cyclogenesis frequency have been well documented (DeMaria et al., 2001) but changes in these conditions due specifically to increasing greenhouse gas concentrations do not necessarily preserve the same statistical relationships. For example, the observed minimum SST threshold for tropical cyclogenesis is roughly 26°C. This relationship might lead to an expectation that anthropogenic warming of tropical SST and the resulting increase in the areal extent of the region of 26°C SST should lead to increases in tropical cyclone frequency. However, there is a growing body of evidence that the minimum SST threshold for tropical cyclogenesis increases at about the same rate as the SST increase due solely to greenhouse gas forcing (e.g., Ryan et al., 1992; Dutton et al., 2000; Yoshimura et al., 2006; Bengtsson et al., 2007; Knutson et al., 2008; Johnson and Xie, 2010). This is because the threshold conditions for tropical cyclogenesis are controlled not just by surface temperature but also by atmospheric stability (measured from the lower boundary to the tropopause), which responds to greenhouse gas forcing in a more complex way than SST alone. That is, when the SST changes due to greenhouse warming are deconvolved from the background natural variability, that part of the SST variability, by itself, has no manifest effect on tropical cyclogenesis. In this case, the simple observed relationship between tropical cyclogenesis and SST, while robust, does not adequately capture the relevant physical mechanisms of tropical cyclogenesis in a warming world.

Another challenge to identifying causes behind observed changes in tropical cyclone activity is introduced by uncertainties in the reanalysis data used to identify environmental changes in regions where tropical cyclones develop and evolve (Bister and Emanuel, 2002; Emanuel, 2010). In particular, heterogeneity in upper-tropospheric kinematic and thermodynamic metrics complicates the interpretation of long-term changes in vertical wind shear and potential intensity, both of which are important environmental controls on tropical cyclones.

Based on a variety of model simulations, the expected long-term changes in global tropical cyclone characteristics under greenhouse warming is a decrease or little change in frequency concurrent with an increase in mean intensity. One of the challenges for identifying these changes in the existing data records is that the expected changes predicted by the models are generally small when compared with changes associated with observed short-term natural variability. Based on changes in tropical cyclone intensity predicted by idealized numerical simulations with CO_2-induced tropical SST warming, Knutson and Tuleya (2004) suggested that clearly detectable increases may not be manifest for decades to come. Their argument was based on a comparison of the amplitude of the modeled upward trend (i.e., the signal) in storm intensity with the amplitude of the interannual variability (i.e., the noise). The recent high-resolution dynamical downscaling study of Bender et al. (2010) supports this argument and suggests that the predicted increases in the frequency of the strongest Atlantic storms may not emerge as a clear statistically significant signal until the latter half of the 21st century under the SRES A1B warming scenario. Still, it should be noted that while these model projections suggest that a statistically significant signal may not emerge until some future time, the likelihood of more intense tropical cyclones is projected to continually increase throughout the 21st century.

With the exception of the North Atlantic, much of the global tropical cyclone data is confined to the period from the mid-20th century to present. In addition to the limited period of record, the uncertainties in the historical tropical cyclone data (Section 3.2.1 and this section) and the extent of tropical cyclone variability due to random processes and linkages with various climate modes such as El Niño, do not presently allow for the detection of any clear trends in tropical cyclone activity that can be attributed to greenhouse warming. As such, it remains unclear to what degree the causal phenomena described here have modulated post-industrial tropical cyclone activity.

The AR4 concluded that it is *more likely than not* that anthropogenic influence has contributed to increases in the frequency of the most intense tropical cyclones (Hegerl et al., 2007). Based on subsequent research that further elucidated the scope of uncertainties in both the historical tropical cyclone data as well as the physical mechanisms underpinning the observed relationships, no such attribution conclusion was drawn in the recent WMO assessment (Knutson et al., 2010). The present assessment regarding detection and attribution of trends in tropical cyclone activity is similar to the WMO assessment (Knutson et al., 2010): the uncertainties in the historical tropical cyclone records, the incomplete understanding of the physical mechanisms linking tropical cyclone metrics to climate change, and the degree of tropical cyclone variability – comprising random processes and linkages to various natural climate modes such as El Niño – provide only *low confidence* for the attribution of any detectable changes in tropical cyclone activity to anthropogenic influences.

Projected Changes and Uncertainties

The AR4 concluded (Meehl et al., 2007b) that a broad range of modeling studies project a *likely* increase in peak wind intensity and near-storm precipitation in future tropical cyclones. A reduction of the overall number of storms was also projected (but with lower confidence), with a greater reduction in weaker storms in most basins and an increase in the frequency of the most intense storms. Knutson et al. (2010) concluded that it is *likely* that the mean maximum wind speed and near-storm rainfall rates of tropical cyclones will increase with projected 21st-century warming, and it is *more likely than not* that the frequency of the most intense storms will increase substantially in some basins, but it is *likely* that overall global tropical cyclone frequency will decrease or remain essentially unchanged. The conclusions here are similar to those of the AR4 and Knutson et al. (2010).

The spatial resolution of some models such as the CMIP3 coupled ocean-atmosphere models used in the AR4 is generally not high enough to accurately resolve tropical cyclones, and especially to simulate their intensity (Randall et al., 2007). Higher-resolution global models have had some success in reproducing tropical cyclone-like vortices (e.g., Chauvin et al., 2006; Oouchi et al., 2006; Zhao et al., 2009), but only their coarse characteristics. Significant progress has been recently made, however, using downscaling techniques whereby high-resolution models capable of reproducing more realistic tropical cyclones are run using boundary conditions provided by either reanalysis data sets or output fields from lower-resolution climate models such as those used in the AR4 (e.g., Knutson et al., 2007; Emanuel et al., 2008; Knutson et al., 2008; Emanuel, 2010). A recent study by Bender et al. (2010) applies a cascading technique that downscales first from global to regional scale, and then uses the simulated storms from the regional model to initialize a very high-resolution hurricane forecasting model. These downscaling studies have been increasingly successful at reproducing observed tropical cyclone characteristics, which provides increased confidence in their projections, and it is expected that more progress will be made as computing resources improve. Still, awareness that limitations exist in the models used for tropical cyclone projections, particularly the ability to accurately reproduce natural climate phenomena that are known to modulate storm behavior (e.g., ENSO and MJO), is important for context when interpreting model output (Sections 3.2.3.2 and 3.4.2).

While detection of long-term past increases in tropical cyclone activity is complicated by data quality and signal-to-noise issues (as stated above), theory (Emanuel, 1987) and idealized dynamical models (Knutson and Tuleya, 2004) both predict increases in tropical cyclone intensity under greenhouse warming. Recent simulations with high-resolution dynamical models (Oouchi et al., 2006; Bengtsson et al., 2007; Gualdi et al., 2008; Knutson et al., 2008; Sugi et al., 2009; Bender et al., 2010) and statistical-dynamical models (Emanuel, 2007) consistently find that greenhouse warming causes tropical cyclone intensity to shift toward stronger storms by the end of the 21st century (2 to 11% increase in mean maximum wind speed globally). These and other models also consistently project little change or a reduction in overall tropical cyclone frequency (e.g., Gualdi et al., 2008; Sugi et al., 2009; Murakami et al., 2011), but with an accompanying substantial fractional increase in the frequency of the strongest storms and increased precipitation rates (in the models for which these metrics were examined). Current models project changes in overall global frequency ranging from a decrease of 6 to 34% by the late 21st century (Knutson et al., 2010). The downscaling experiments of Bender et al. (2010) – which use an 18-model ensemble-mean of CMIP3 simulations to nudge a high-resolution dynamical model (Knutson et al., 2008) that is then used to initialize a very high-resolution dynamical model – project a 28% reduction in the overall frequency of Atlantic storms and an 80% increase in the frequency of Saffir-Simpson category 4 and 5 Atlantic hurricanes over the next 80 years (A1B scenario).

The projected decreases in global tropical cyclone frequency may be due to increases in vertical wind shear (Vecchi and Soden, 2007c; Zhao et al., 2009; Bender et al., 2010), a weakening of the tropical circulation (Sugi et al., 2002; Bengtsson et al., 2007) associated with a decrease in the upward mass flux accompanying deep convection (Held and Soden, 2006), or an increase in the saturation deficit of the middle troposphere (Emanuel et al., 2008). For individual basins, there is much more uncertainty in projections of tropical cyclone frequency, with changes of up to ±50% or more projected by various models (Knutson et al., 2010). When projected SST changes are considered in the absence of projected radiative forcing changes, Northern Hemisphere tropical cyclone frequency has been found to increase (Wehner et al., 2010), which is congruent with the hypothesis that SST changes alone do not capture the relevant physical mechanisms controlling tropical cyclogenesis (e.g., Emanuel, 2010).

As noted above, observed changes in rainfall associated with tropical cyclones have not been clearly established. However, as water vapor in the tropics increases (Trenberth et al., 2005) there is an expectation for increased heavy rainfall associated with tropical cyclones in response to associated moisture convergence increases (Held and Soden, 2006). This increase is expected to be compounded by increases in intensity as dynamical convergence under the storm is enhanced. Models in which tropical cyclone precipitation rates have been examined are highly consistent in projecting increased rainfall within the area near the tropical cyclone center under 21st century warming, with increases of 3 to 37% (Knutson et al., 2010). Typical projected increases are near 20% within 100 km of storm centers.

Another type of projection that is sometimes inferred from the literature is based on extrapolation of an observed statistical relationship (see also Section 3.2.3). These relationships are typically constructed on past observed variability that represents a convolution of anthropogenically forced variability and natural variability across a broad range of time scales. In general, however, these relationships cannot be expected to represent all of the relevant physics that control the phenomena of interest, and their extrapolation beyond the range of the observed variability they are built on is not reliable. As an example, there is a strong observed correlation between local SST and tropical cyclone power dissipation (Emanuel, 2007). If 21st-century SST projections are applied to this relationship, power dissipation is projected to increase by about 300% in the next century (Vecchi et al., 2008; Knutson et al., 2010). Alternatively, there is a similarly strong relationship between power dissipation and relative SST, which represents the difference between local and tropical-mean SST and has been argued to serve as a proxy for local potential intensity (Vecchi and Soden, 2007a). When 21st-century projections of relative SST are considered, this latter relationship projects almost no change in power dissipation in the next century (Vecchi et al., 2006). Both of these statistical relationships can be reasonably defended based on physical arguments but it is not clear which, if either, is correct (Ramsay and Sobel, 2011).

When simulating 21st-century warming under the A1B emission scenario (or a close analog), the present models and downscaling techniques as a whole are consistent in projecting (1) decreases or no change in tropical cyclone frequency, (2) increases in intensity and fractional increases in number of most intense storms, and (3) increases in tropical cyclone-related rainfall rates. Differences in regional projections lead to lower confidence in basin-specific projections of intensity and rainfall, and confidence is particularly low for projections of frequency within individual basins. More specifically, while projections under 21st-century greenhouse warming indicate that it is *likely* that the global frequency of tropical cyclones will either decrease or remain essentially unchanged, an increase in mean tropical cyclone maximum wind speed is also *likely*, although increases may not occur in all tropical regions. This assessment is essentially identical with that of the recent WMO assessment (Knutson et al., 2010). Furthermore, while it is *likely* that overall global frequency will either decrease or remain essentially unchanged, it is *more likely than not* that the frequency of the most intense storms (e.g., Saffir-Simpson category 4 and 5) will increase substantially in some ocean basins, again agreeing with the recent WMO assessment (Knutson et al., 2010). Based on the level of consistency among models, and physical reasoning, it is *likely* that tropical cyclone-related rainfall rates will increase with greenhouse warming. Confidence in future projections for particular ocean basins is undermined by the inability of global models to reproduce accurate details at scales relevant to tropical cyclone genesis, track, and intensity evolution. Of particular concern is the limited ability of global models to accurately simulate upper-tropospheric wind (Cordero and Forster, 2006; Bender et al., 2010), which modulates vertical wind shear and tropical cyclone genesis and intensity evolution. Thus there is *low confidence* in projections of changes in tropical cyclone genesis, location, tracks, duration, or areas of impact, and existing model projections do not show dramatic large-scale changes in these features.

In summary, there is *low confidence* that any observed long-term (i.e., 40 years or more) increases in tropical cyclone activity are robust, after accounting for past changes in observing capabilities. The uncertainties in the historical tropical cyclone records, the incomplete understanding of the physical mechanisms linking tropical cyclone metrics to climate change, and the degree of tropical cyclone variability provide only *low confidence* for the attribution of any detectable changes in tropical cyclone activity to anthropogenic influences. There is *low confidence* in projections of changes in tropical cyclone genesis, location, tracks, duration, or areas of impact. Based on the level of consistency among models, and physical reasoning, it is *likely* that tropical cyclone-related rainfall rates will increase with greenhouse warming. It is *likely* that the global frequency of tropical cyclones will either decrease or remain essentially unchanged. An increase in mean tropical cyclone maximum wind speed is *likely*, although increases may not occur in all tropical regions. While it is *likely* that overall global frequency will either decrease or remain essentially unchanged, it is *more likely than not* that the frequency of the most intense storms will increase substantially in some ocean basins.

3.4.5. Extratropical Cyclones

Extratropical cyclones (synoptic-scale low-pressure systems) exist throughout the mid-latitudes in both hemispheres and mainly develop over the oceanic basins in the proximity of the upper-tropospheric jet streams, as a result of flow over mountains (lee cyclogenesis) or through conversions from tropical to extratropical systems. It should be noted that regionalized smaller-scale mid-latitude circulation phenomena such as polar lows and mesoscale cyclones are not treated in this section (but see Sections 3.3.3 and 3.4.3). Extratropical cyclones are the main poleward transporter of heat and moisture and may be accompanied by adverse weather conditions such as windstorms, the buildup of waves and storm surges, or extreme precipitation events. Thus, changes in the intensity of extratropical cyclones or a systematic shift in the geographical location of extratropical cyclone activity may have a great impact on a wide range of regional climate extremes as well as the long-term changes in temperature and precipitation. Extratropical cyclones mainly form and grow via atmospheric instabilities such as a disturbance along a zone of strong temperature contrast (baroclinic instabilities), which is a reservoir of available potential energy that can be converted into the kinetic energy associated with extratropical cyclones. Intensification of the cyclones may also take place due to processes such as release of energy due to phase changes of water (latent heat release) (Gutowski et al., 1992; Wernli et al., 2002). Why should we expect climate change to influence extratropical cyclones? A simplified line of argument would be that both the large-scale low and high level pole to equator temperature gradients may change (possibly in opposite directions) in a climate change scenario leading to a change in the atmospheric instabilities responsible for cyclone formation and growth (baroclinicity). These changes may be induced by a variety of mechanisms operating in different parts of the atmospheric column ranging from changing surface conditions (Deser et al., 2007; Bader et al., 2011) to stratospheric changes (Son et al., 2010). In addition, changes in precipitation intensities within extratropical cyclones may change the latent heat release. According to theories on wave-mean flow interaction, changes in the extratropical storm tracks are also associated with changes in the large-scale flow (Robinson, 2000; Lorenz and Hartmann, 2003). A latitudinal shift of the upper tropospheric jet would be accompanied by a latitudinal shift in the extratropical storm track. It is, however, still unclear to what extent a latitudinal shift in the jet changes the total storm track activity rather than shifting it latitudinally (Wettstein and Wallace, 2010). Even within the very simplified outline above the possible impacts of climate change on extratropical cyclone development are many and clearly not trivial.

When validated using reanalyses with similar horizontal resolution, climate models are found to represent the general structure of the storm track pattern well (Bengtsson et al., 2006; Greeves et al., 2007; Ulbrich et al., 2008; Catto et al., 2010). However, using data from five different coupled models, the rate of transfer of zonal available potential energy to eddy available potential energy in synoptic systems was found to be too large, yielding too much energy and an overactive energy cycle (Marques et al., 2011). Models tend to have excessively zonal storm tracks and some show a poor extension of the storm tracks into Europe (Pinto et al., 2006; Greeves et al., 2007; Orsolini and Sorteberg, 2009). It has also been noted that representation of cyclone activity may depend on the physics formulations and the horizontal resolution of the model (Jung et al., 2006; Greeves et al., 2007).

Paleoclimatic proxies for extratropical cyclone variability are still few, but progress is being made in using coastal dune field development and sand grain content of peat bogs as proxies for storminess. Publications covering parts of western Europe indicate enhanced sand movement in European coastal areas during the Little Ice Age (Wilson et al., 2004; de Jong et al., 2006, 2007; Clemmensen et al., 2007; Clarke and Rendell, 2009; Sjogren, 2009). It should be noted that sand influx is also influenced by sediment availability, which is controlled mainly by the degree of vegetation cover and the moisture content of the sediment (Li et al., 2004; Wiggs et al., 2004). Intense cultivation, overgrazing, and forest disturbance make soils more prone to erosion, which can lead to increased sand transport even under less windy conditions. Thus the information gained from paleoclimatic proxies to put the last 100 years of extratropical cyclone variability in context is limited.

Century-long time-series of estimates of extremes in geostrophic wind deduced from triangles of pressure stations, pressure tendencies from single stations (see Section 3.3.3 for details), or oceanic variables such as extremes in non-tide residuals are (if these are located in the vicinity of the main storm tracks) possible proxies for extratropical cyclone activity. Trend detection in extratropical cyclone variables such as number of cyclones, intensity, and activity (parameters integrating cyclone intensity, number, and possibly duration) became possible with the development of reanalyses, but remains challenging. Problems with reanalyses have been especially pronounced in the Southern Hemisphere (Hodges et al., 2003; Wang et al., 2006). Even though different reanalyses correspond well in the Northern Hemisphere (Hodges et al., 2003; Hanson et al., 2004), changes in the observing system giving artificial trends in integrated water vapor and kinetic energy (Bengtsson et al., 2004) may have influenced trends in both the number and intensity of cyclones. In addition, studies indicate that the magnitude and even the existence of the changes may depend on the choice of reanalysis (Trigo, 2006; Raible et al., 2008; Simmonds et al., 2008; Ulbrich et al., 2009) and cyclone tracking algorithm (Raible et al., 2008).

The AR4 noted a *likely* net increase in the frequency/intensity of Northern Hemisphere extreme extratropical cyclones and a poleward shift in the tracks since the 1950s (Trenberth et al., 2007; Table 3.8), and cited several papers showing increases in the number or strength of intense extratropical cyclones both over the North Pacific and the North Atlantic storm track (Trenberth et al., 2007, p. 312) during the last 50 years. Studies using reanalyses indicate a northward and eastward shift in the Atlantic cyclone activity during the last 60 years with both more frequent and more intense wintertime cyclones in the high-latitude Atlantic (Weisse et al., 2005; Wang et al., 2006; Schneidereit et al., 2007; Raible et al., 2008; Vilibic and Sepic, 2010) and fewer in the mid-latitude Atlantic (Wang et al., 2006; Raible et al., 2008). The increase in high-atitude cyclone activity was also reported in several studies of Arctic cyclone activity (X.D. Zhang et al., 2004; Sorteberg and Walsh, 2008; Sepp and Jaagus, 2011). Using ship-based trends in mean sea level pressure (MSLP) variance (which is tied to cyclone intensity), Chang (2007) found wintertime Atlantic trends to be consistent with National Centers for Environmental Prediction (NCEP) reanalysis trends in the Atlantic, but slightly weaker. There are inconsistencies among studies of extreme cyclones in reanalyses, since some show an increase in intensity and number of extreme Atlantic cyclones (Geng and Sugi, 2001; Paciorek et al., 2002; Lehmann et al., 2011) while others show a reduction (Gulev et al., 2001). These differences may in part be due to sensitivities of the identification schemes and different definitions of an extreme cyclone (Leckebusch et al., 2006; Pinto et al., 2006). New studies have confirmed that a positive NAM/NAO (see Section 3.4.3) corresponds to stronger Atlantic/European cyclone activity (e.g., Chang, 2009; Pinto et al., 2009; X.L. Wang et al., 2009b). However, studies using long historical records seem to suggest that some of these links may be statistically intermittent (Hanna et al., 2008; Matulla et al., 2008; Allan et al., 2009) due to interdecadal shifts in the location of the positions of the NAO pressure centers (Vicente-Serrano and Lopez-Moreno, 2008; X.D. Zhang et al., 2008). It is unclear to what extent the statistical intermittency implies that the underlying physical processes creating the connection act only intermittently. A possible influence of the Pacific North America (PNA) pattern on the entrance of the North Atlantic storm track (over Newfoundland) has been reported by Pinto et al. (2011). It should be noted that there is some suggestion that the reanalyses cover a time period that starts with relatively low cyclonic activity in northern coastal Europe in the 1960s and reaches a maximum in the 1990s. Long-term European storminess proxies show no clear trends over the last century (Hanna et al., 2008; Allan et al., 2009; see Section 3.3.3 for details).

Studies using reanalyses and *in situ* data for the last 50 years have noted an increase in the number and intensity of north Pacific wintertime intense extratropical cyclone systems since the 1950s (Graham and Diaz, 2001; Simmonds and Keay, 2002; Raible et al., 2008) and cyclone activity (X.D. Zhang et al., 2004), but signs of some of the trends disagreed when different tracking algorithms or reanalysis products were used (Raible et al., 2008). A slight positive trend has been found in north Pacific extreme cyclones (Geng and Sugi, 2001; Gulev et al., 2001; Paciorek et al., 2002). Using ship measurements, Chang (2007) found intensity-related wintertime trends in the Pacific to be about 20 to 60% of that found in the reanalysis. Long-term *in situ* observations of north Pacific cyclones based on observed pressure data are considerably fewer than for coastal Europe. However, using hourly tide gauge records from the western coast of the United States as a proxy for storminess, an increasing trend in the extreme winter Non-Tide Residuals (NTR) has been observed in the last decades (Bromirski et al., 2003; Menendez et al., 2008). Years having high NTR were linked to a large-scale atmospheric circulation pattern, with intense storminess associated with a broad, south-easterly displaced, deep Aleutian low that directed storm tracks toward the US West Coast. North Pacific cyclonic activity has been linked to tropical SST anomalies (NINO3.4; see Section 3.4.2) and the PNA (Eichler and Higgins, 2006; Favre and Gershunov, 2006; Seierstad et al., 2007), showing that the PNA and NINO3.4 influence storminess, in particular over the eastern North Pacific with an equatorward shift in storm tracks in the North Pacific basin, as well as an increase in storm track activity along the US East Coast during El Niño events.

Based on reanalyses, North American cyclone numbers have increased over the last 50 years, with no statistically significant change in cyclone intensity (X.D. Zhang et al., 2004). Hourly MSLP data from Canadian stations showed that winter cyclones have become significantly more frequent, longer lasting, and stronger in the lower Canadian Arctic over the last 50 years (1953-2002), but less frequent and weaker in the south, especially along the southeast and southwest Canadian coasts (Wang et al., 2006). Further south, a tendency toward weaker low-pressure systems over the past few decades was found for US East Coast winter cyclones using reanalyses, but no statistically significant trends in the frequency of occurrence of systems (Hirsch et al., 2001).

Studies on extratropical cyclone activity in northern Asia are few. Using reanalyses, a decrease in extratropical cyclone activity (X.D. Zhang et al., 2004) and intensity (X.D. Zhang et al., 2004; X. Wang et al., 2009) over the last 50 years has been reported for northern Eurasia (60-40°N) with a possible northward shift with increased cyclone frequency in the higher latitudes (50-45°N) and decrease in the lower latitudes (south of 45°N), based on a study with reanalyses. The low-latitude (south of 45°N) decrease was also noted by Zou et al. (2006), who reported a decrease in the number of severe storms for mainland China based on an analysis of extremes of observed 6-hourly pressure tendencies over the last 50 years.

Alexander and Power (2009) showed that the number of observed severe storms at Cape Otway (south-east Australia) has decreased since the mid-19th century, strengthening the evidence of a southward shift in Southern Hemisphere storm tracks previously noted using reanalyses (Fyfe, 2003; Hope et al., 2006; Wang et al., 2006). Frederiksen and Frederiksen (2007) linked the reduction in cyclogenesis at 30°S and southward shift to a decrease in the vertical mean meridional temperature gradient. Using reanalyses, both Pezza et al. (2007) and Lim and Simmonds (2009) have confirmed previous studies showing a trend toward more intense low-pressure systems. However, the trend of a decreasing number of cyclones seems to depend on the choice of reanalysis and pressure level (Lim and Simmonds, 2009), emphasizing the weaker consistency among reanalysis products for the Southern Hemisphere extratropical cyclones. Recent studies support the notion of more cyclones around Antarctica when the SAM (see Section 3.4.3) is in its positive phase and a shift of cyclones toward mid-latitudes when the SAM is in its negative phase (Pezza and Simmonds, 2008). Additionally, more intense (and fewer) cyclones seem to occur when the PDO (see Section 3.4.3) is strongly positive and vice versa (Pezza et al., 2007).

In conclusion, it is *likely* that there has been a poleward shift in the main northern and southern storm tracks during the last 50 years. There is strong agreement with respect to this change between several reanalysis products for a wide selection of cyclone parameters and cyclone identification methods and European and Australian pressure-based storminess proxies are consistent with a poleward shift over the last 50 years, which indicates that the evidence is robust. Advances have been made in documenting the observed decadal and multi-decadal variability of extratropical cyclones using proxies for storminess. So the recent poleward shift should be seen in light of new studies with longer time spans that indicate that the last 50 years coincide with relatively low cyclonic activity in northern coastal Europe in the beginning of the period. Several studies using reanalyses suggest an intensification of high-latitude cyclones, but there is still insufficient knowledge of how changes in the observational systems are influencing the cyclone intensification in reanalyses so even in cases of high agreement among the studies the evidence cannot be considered to be robust, thus we have only *low confidence* in these changes. Other regional changes in intensity and the number of cyclones have been reported. However, the level of agreement between different studies using different tracking algorithms, different reanalyses, or different cyclone parameters is still low. Thus, we have *low confidence* in the amplitude, and in some regions in the sign, of the regional changes.

Regarding possible causes of the observed poleward shift, the AR4 concluded that trends over recent decades in the Northern and Southern Annular Modes, which correspond to sea level pressure reductions over the poles, are *likely* related in part to human activity, but an anthropogenic influence on extratropical cyclones had not been formally detected, owing to large internal variability and problems due to changes in observing systems (Hegerl et al., 2007). Anthropogenic influences on these modes of variability are also discussed in Section 3.4.3.

Seasonal global sea level pressure changes have been shown to be inconsistent with simulated internal variability (Giannini et al., 2003; Gillett et al., 2005; Gillett and Stott, 2009; X.L. Wang et al., 2009a), but changes in sea level pressure in regions of extratropical cyclones (mid- and high latitudes) have not formally been attributed to anthropogenic forcings (Gillett and Stott, 2009). However, the trend pattern in atmospheric storminess as inferred from geostrophic wind energy and ocean wave heights has been found to contain a detectable response to anthropogenic and natural forcings with the effect of external forcings being strongest in the winter hemisphere (X.L. Wang et al., 2009a). Nevertheless, the models generally simulate smaller changes than observed and also appear to underestimate the internal variability, reducing the robustness of their detection results. New idealized studies have advanced the physical understanding of how storm tracks may respond to changes in the underlying surface conditions, indicating that a uniform SST increase weakens (reduced cyclone intensity or number of cyclones) and shifts the storm track poleward and strengthened SST gradients near the subtropical jet may lead to a meridional shift in the storm track either toward the poles or the equator depending on the location of the SST gradient change (Deser et al., 2007; Brayshaw et al., 2008; Semmler et al., 2008; Kodama and Iwasaki, 2009), but the average global cyclone activity is not expected to change much under moderate greenhouse gas forcing (O'Gorman and Schneider, 2008; Bengtsson et al., 2009). Studies have also emphasized the important role of stratospheric changes (induced by ozone or greenhouse gas changes) in explaining latitudinal shifts in storm tracks and several mechanisms have been proposed (Son et al., 2010). This has particularly strengthened the understanding of the Southern Hemisphere changes. According to Fogt et al. (2009) both coupled climate models and observed trends in the SAM were found to be outside the range of internal climate variability during the austral summer. This was mainly attributed to stratospheric ozone depletion (see Section 3.4.3).

In summary, there is *medium confidence* in an anthropogenic influence on the observed poleward shift in extratropical cyclone activity. It has not formally been attributed. However indirect evidence such as global anthropogenic influence on the sea level pressure distribution and trend patterns in atmospheric storminess inferred from geostrophic wind and ocean wave heights has been found. While physical understanding of how anthropogenic forcings may influence extratropical cyclone storm tracks has strengthened, the importance of the different mechanisms in the observed shifts is still unclear.

The AR4 reported that in a future warmer climate, a consistent projection from the majority of the coupled atmosphere-ocean GCMs is fewer mid-latitude storms averaged over each hemisphere (Meehl et al., 2007b) and a poleward shift of storm tracks in both hemispheres (particularly evident in the Southern Hemisphere), with greater storm activity at higher latitudes (Meehl et al., 2007b).

A poleward shift in the upper level tropospheric storm track due to increased greenhouse gas forcing is supported by post-AR4 studies (Lorenz and DeWeaver, 2007; O'Gorman, 2010; Wu et al., 2011). It should be noted that other studies indicate that the poleward shift is less clear when models including a full stratosphere or ozone recovery are used (Huebener et al., 2007; Son et al., 2008; Morgenstern et al., 2010; Scaife et al., 2011) and the strength of the poleward shift is often seen more clearly in upper-level quantities than in low-level transient parameters (Ulbrich et al., 2008). Post-AR4 single model studies support the projection of a reduction in extratropical cyclones averaged over the Northern Hemisphere during future warming (Finnis et al., 2007; Bengtsson et al., 2009; Orsolini and Sorteberg, 2009). However, neither the global changes in storm frequency or intensity were found to be statistically significant by Bengtsson et al. (2009), although they were accompanied by significant increases in total and extreme precipitation.

Models tend to project a reduction of winter cyclone activity throughout the mid-latitude North Pacific and for some models a north-eastern movement of the North Pacific storm track (Loeptien et al., 2008; Ulbrich et al., 2008; Favre and Gershunov, 2009; McDonald, 2011). However, the exact geographical pattern of cyclone frequency anomalies exhibits large variations across models (Teng et al., 2008; Favre and Gershunov, 2009; Laine et al., 2009).

Using band-passed sea level pressure data from 16 CMIP3 coupled GCMs, Ulbrich et al. (2008) showed regional increases in the storm track activity over the Eastern North Atlantic/Western European area. This eastward or southeastward extension of the storm track is also found in other studies (Ulbrich et al., 2008; Laine et al., 2009; McDonald, 2011) and may be attributed to a local minimum in ocean warming in the central North Atlantic and subsequent local changes in baroclinicity (McDonald, 2011). In line with the eastward shift, Donat et al. (2010a) projected an increase in wind storm days for central Europe by the end of the 21st century. The increase varies according to the definition of storminess and one model projects a decrease. A common deficiency among many AR4 models is a coarsely resolved stratosphere and there are still concerns that this may lead to systematic biases in the Atlantic storm track response to increased anthropogenic forcing (Scaife et al., 2011). A reduction in cyclone frequency along the Canadian east coast has been reported (Bengtsson et al., 2006; Watterson, 2006; Pinto et al., 2007a; Teng et al., 2008; Long et al., 2009). New results for Southern Hemisphere cyclones confirm the previously projected poleward shift in storm tracks under increased greenhouse gases (Lim and Simmonds, 2009). That study projected a reduction of Southern Hemisphere extratropical cyclone frequency and intensity in mid-latitudes but a slight increase at high latitudes. The poleward shift due to increased greenhouse gases may be partly opposed by ozone recovery (Son et al., 2010).

Detailed analyses of changes in physical mechanisms related to cyclone changes in coupled climate models are still few. O'Gorman (2010) showed that changes in mean available potential energy of the atmosphere can account for much of the varied response in storm-track intensity to global warming, implying that changes in storm-track intensity are sensitive to competing effects of changes in temperature gradients and static stability in different atmospheric levels. Using two coupled climate models, Laine et al. (2009) indicate that the primary cause for synoptic activity changes at the western end of the Northern Hemisphere storm tracks is related to the baroclinic conversion processes linked to mean temperature gradient changes in localized regions of the western oceanic basins. They also found downstream changes in latent heat release during the developing and mature stages of the cyclone to be of importance and indicated that changes in diabatic process may be amplified by the upstream baroclinic changes [stronger (weaker) baroclinic activity in the west gives stronger (weaker) latent heat release downstream]. Pinto et al. (2009) found that regional increases in track density and intensity of extreme cyclones close to the British Isles using a single model was associated with an eastward shift of the jet stream into Europe, more frequent extreme values of baroclinicity, and stronger upper level divergence.

The modeled reduction in Southern Hemisphere extratropical cyclone frequency and intensity in the mid-latitudes has been attributed to the tropical upper tropospheric warming enhancing static stability and decreasing baroclinicity while an increased meridional temperature gradient in the high latitudes is suggested to be responsible for the increase in cyclone activity in this region (Lim and Simmonds, 2009). In addition to details in the modeled changes in local baroclinicity and diabatic changes, the geographical pattern of modeled response in cyclone activity has been reported to be influenced by the individual model's structure of intrinsic modes of variability (Branstator and Selten, 2009) and biases in the climatology (Kidston and Gerber, 2010).

In summary it is *likely* that there has been a poleward shift in the main Northern and Southern Hemisphere extratropical storm tracks during the last 50 years. There is *medium confidence* in an anthropogenic influence on this observed poleward shift. It has not formally been attributed. There is *low confidence* in past changes in regional intensity. There is *medium confidence* that an increased anthropogenic forcing will lead to a reduction in the number of mid-latitude cyclones averaged over each hemisphere, and there is also *medium confidence* in a poleward shift of the tropospheric storm tracks due to future anthropogenic forcings. Regional changes may be substantial and CMIP3 simulations show some regions with medium agreement. However, there are still uncertainties related to how the poorly resolved stratosphere in many CMIP3 models may influence the regional results. In addition, studies using different analysis techniques, different physical quantities, different thresholds, and different atmospheric vertical levels to represent cyclone activity and storm tracks result in different projections of regional changes. This leads to *low confidence* in region-specific projections.

3.5. Observed and Projected Impacts on the Natural Physical Environment

3.5.1. Droughts

Drought is generally "a period of abnormally dry weather long enough to cause a serious hydrological imbalance" (see the Glossary and Box 3-3). While lack of precipitation (i.e., meteorological drought; Box 3-3) is often the primary cause of drought, increased potential evapotranspiration induced by enhanced radiation, wind speed, or vapor pressure deficit (itself linked to temperature and relative humidity), as well as pre-conditioning (pre-event soil moisture; lake, snow, and/or groundwater storage) can contribute to the emergence of soil moisture and hydrological drought (Box 3-3). Actual evapotranspiration is additionally controlled by soil moisture, which constitutes a limiting factor for further drying under drought conditions, and other processes that impact vegetation development and phenology (e.g., temperature) are also relevant. As noted in the AR4 (Trenberth et al., 2007), there are few direct observations of drought-related variables, in particular of soil moisture, available for a global analysis (see also Section 3.2.1). Hence, proxies for drought ('drought indices') are often used to infer changes in drought conditions. Box 3-3 provides a discussion of the issue of drought definition and a description of commonly used drought indices. In order to understand the impact of droughts (e.g., on crop yields, general ecosystem functioning, water resources, and electricity production), their timing, duration, intensity, and spatial extent need to be characterized. Several weather elements may interact to increase the impact of droughts: enhanced air temperature can indirectly lead to enhanced evaporative demand (through enhanced vapor pressure deficit), although enhanced wind speed or increased incoming radiation are generally more important factors. Moreover, climate phenomena such as monsoons (Section 3.4.1) and ENSO (Section 3.4.2) affect changes in drought occurrence in some

---

Box 3-3 | The Definition of Drought

Though a commonly used term, drought is defined in various ways, and these definitional issues make the analysis of changes in drought characteristics difficult. This explains why assessments of (past or projected) changes in drought can substantially differ between published studies or chosen indices (see Section 3.5.1). Some of these difficulties and their causes are highlighted in this box.

What is Drought or Dryness?
The Glossary defines drought as follows: "A period of abnormally dry weather long enough to cause a serious hydrological imbalance. Drought is a relative term, therefore any discussion in terms of precipitation deficit must refer to the particular precipitation-related activity that is under discussion. For example, shortage of precipitation during the growing season impinges on crop production or ecosystem function in general (due to soil moisture drought, also termed agricultural drought), and during the runoff and percolation season primarily affects water supplies (hydrological drought). Storage changes in soil moisture and groundwater are also affected by increases in actual evapotranspiration in addition to reductions in precipitation. A period with an abnormal precipitation deficit is defined as a meteorological drought. A megadrought is a very lengthy and pervasive drought, lasting much longer than normal, usually a decade or more."

As highlighted in the above definition, drought can be defined from different perspectives, depending on the stakeholders involved. The scientific literature commonly distinguishes *meteorological drought*, which refers to a deficit of precipitation, *soil moisture drought* (often called *agricultural drought*), which refers to a deficit of (mostly root zone) soil moisture, and *hydrological drought*, which refers to negative anomalies in streamflow, lake, and/or groundwater levels (e.g., Heim Jr., 2002). We use here the term 'soil moisture drought' instead of 'agricultural drought,' despite the widespread use of the latter term (e.g., Heim Jr., 2002; Wang, 2005), because soil moisture deficits have several additional effects beside those on agroecosystems, most importantly on other natural or managed ecosystems (including both forests and pastures), on building infrastructure through soil mechanical processes (e.g., Corti et al., 2009), and health through impacts on heat waves (Section 3.1.4). Water scarcity (linked to *socioeconomic drought*), which may be caused fully or in part by use from human activities, does not lie within the scope of this chapter (see Section 4.2.2); however, it should be noted that changing pressure on water resources by human uses may itself influence climate and possibly the drought conditions, for example, via declining groundwater levels, or enhanced local evapotranspiration and associated land-atmosphere feedbacks. Drought should not be confused with aridity, which describes the general characteristic of an arid climate (e.g., desert). Indeed, drought is considered a recurring feature of climate occurring in any region and is defined with respect to the average climate of the given region (e.g., Heim Jr., 2002; Dai, 2011). Nonetheless, the effects of droughts are not linear, given the existence of, for example, discrete soil moisture thresholds affecting vegetation and surface fluxes (e.g., Koster et al., 2004b; Seneviratne et al., 2010), which means that the same precipitation deficit or radiation excess relative to normal will not affect different regions equally (e.g., short-term lack of precipitation in a very humid region may not be critical for agriculture because of the ample soil moisture supply). In this chapter we often use the term 'dryness' instead of 'drought' as a more general term.

Continued next page →

Drought Drivers

For soil moisture or hydrological droughts, the main drivers are reduced precipitation and/or increased evapotranspiration (Figure 3-9). Although the role of deficits in precipitation is generally considered more prominently in the literature, several drought indicators also explicitly or indirectly consider effects of evapotranspiration. In the context of climate projections, analyses suggest that changes in simulated soil moisture drought are mostly driven by changes in precipitation, with increased evapotranspiration from higher vapor pressure deficit (often linked to increased temperature) and available radiation modulating some of the changes (e.g., Burke and Brown, 2008; Sheffield and Wood, 2008a; Orlowsky and Seneviratne, 2011). It should nonetheless be noted that under strong drought conditions, soil moisture becomes limiting for evapotranspiration, thus limits further soil moisture depletion. Other important aspects for soil moisture and hydrological droughts are persistence and pre-conditioning. Because soil moisture, groundwater, and surface waters are associated with water storage, they have a characteristic memory (e.g., Vinnikov et al., 1996; Eltahir and Yeh, 1999; Koster and Suarez, 2001; Seneviratne et al., 2006b) and thus specific response times to drought forcing (e.g., Begueria et al., 2010; Fleig et al., 2011). The memory is also a function of the atmospheric forcing and system's feedbacks (Koster and Suarez, 2001; A.H. Wang et al., 2009), and the relevant storage is dependent on soil characteristics and rooting depth of the considered ecosystems. This means that drought has a different persistence depending on the affected system, and that it is also sensitive to pre-conditioning (Figure 3-9). Effects of pre-conditioning also explain the possible occurrence of multi-year droughts, whereby soil moisture anomalies can be carried over from one year to the next (e.g., Wang, 2005). However, other features can induce drought persistence, such as persistent circulation anomalies, possibly strengthened by land-atmosphere feedbacks (Schubert et al., 2004; Rowell and Jones, 2006). The choice of variable (e.g., precipitation, soil moisture, or streamflow) and time scale can strongly affect the ranking of drought events (Vidal et al., 2010).

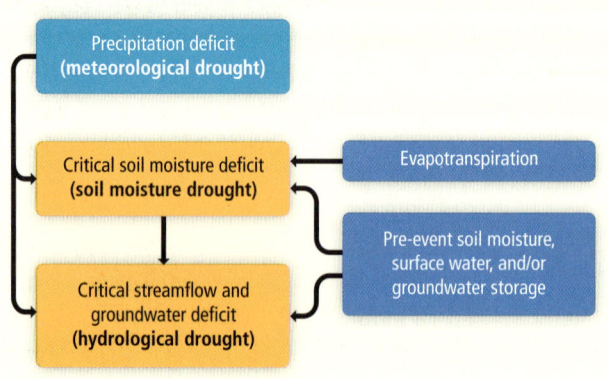

Figure 3-9 | Simplified sketch of processes and drivers relevant for meteorological, soil moisture (agricultural), and hydrological droughts.

Drought Indices

Because of the complex definition of droughts, and the lack of soil moisture observations (Section 3.2.1), several indices have been developed to characterize (meteorological, soil moisture, and hydrological) drought (see, e.g., Heim Jr., 2002; Dai, 2011). These indicators include land surface, hydrological, or climate model simulations (providing estimates of, e.g., soil moisture or runoff) and indices based on measured meteorological or hydrological variables. We provide here a brief overview of the wide range of drought indices used in the literature for the analysis of recent and projected changes. Note that information on paleoclimate proxies such as tree rings, speleothems, lake sediments, or historical evidence (e.g., harvest dates) is not detailed here.

Some indices are based solely on precipitation data. A widely used index is the Standard Precipitation Index (SPI) (McKee et al., 1993; Lloyd-Hughes and Saunders, 2002), which consists of fitting and transforming a long-term precipitation record into a normal distribution that has zero mean and unit standard deviation. SPI values of -0.5 to -1 correspond to mild droughts, -1 to -1.5 to moderate droughts, -1.5 to -2 to severe droughts, and below -2 to extreme droughts. Similarly, values from 0 to 2 correspond to mildly wet to severely wet conditions, and values above 2 to extremely wet conditions (Lloyd-Hughes and Saunders, 2002). SPI can be computed over several time scales (e.g., 3, 6, 12, or more months) and thus indirectly considers effects of accumulating precipitation deficits, which are critical for soil moisture and hydrological droughts. Another index commonly used in the analysis of climate model simulations is the Consecutive Dry Days (CDD) index, which considers the maximum consecutive number of days without rain (i.e., below a given threshold, typically 1 mm day^{-1}) within a considered period (i.e., year in general; Frich et al., 2002; Alexander et al., 2006; Tebaldi et al., 2006). For seasonal time frames, the CDD periods can either be considered to be bound to the respective seasons (e.g., Figure 3-10) or considered in their entirety (across seasons) but assigned to a specific season. Though SPI and CDD are both only based on precipitation, they do not necessarily only consider the effects of meteorological drought, since periods without rain (thus less cloud cover) are bound to have higher daytime radiation forcing and generally higher temperatures, thus possibly positive evapotranspiration anomalies (unless soil moisture conditions are too dry and limit evapotranspiration).

Some indices reflect both precipitation and estimates of actual or potential evapotranspiration, in some cases also accounting for some temporal accumulation of the forcings or persistence of the drought anomalies. These include the Palmer Drought Severity Index (PDSI)

Continued next page ⟶

(Palmer, 1965), which measures the departure of moisture balance from normal conditions using a simple water balance model (e.g., Dai, 2011), as well as other indices such as the Precipitation Potential Evaporation Anomaly (PPEA, based on the cumulative difference between precipitation and potential evapotranspiration) used in Burke and Brown (2008) and the Standardized Precipitation-Evapotranspiration Index (SPEI, which considers cumulated anomalies of precipitation and potential evapotranspiration) described in Vicente-Serrano et al. (2010). PDSI has been widely used for decades (in particular in the United States), and also in climate change analyses (e.g., Dai et al., 2004; Burke and Brown, 2008; Dai, 2011); however, it has some shortcomings for climate change monitoring and projection. PDSI was originally calibrated for the central United States, which can impair the comparability of the index across regions (and also across time periods if drought mechanisms change over time). Thus it is often of advantage to renormalize the local PDSI (Dai, 2011), which can also be done using the self-calibrated PDSI (Wells et al., 2004), but several studies do not apply these steps. Moreover, the land surface model underlying the computation of the PDSI is essentially a simple bucket-type model, which is less sophisticated than more recent land surface and hydrological models and thus implies several limitations (e.g., Dai et al., 2004; Burke et al., 2006). Another important issue is that the parameterization of potential evapotranspiration as empirically (and solely) dependent on air temperature, which is often applied for these various indices (e.g., in the study of Dai et al., 2004) can lead to biased results (e.g., Donohue et al., 2010; Milly and Dunne, 2011; Shaw and Riha, 2011). Temperature is only an indirect driver of evapotranspiration, via its effect on vapor pressure deficit and via effects on vegetation phenology. Furthermore, approaches using potential evapotranspiration as a proxy for actual evapotranspiration do not consider soil moisture and vegetation control on evapotranspiration, which are important mechanisms limiting drought development.

For the assessment of soil moisture drought, simulated soil moisture anomalies also can be considered (Wang et al., 2005; Burke and Brown, 2008; Sheffield and Wood, 2008a; A.H. Wang et al., 2009; Dai, 2011; Orlowsky and Seneviratne, 2011). Simulated soil moisture anomalies integrate the effects of precipitation forcing, simulated actual evapotranspiration (resulting from atmospheric forcing and simulated soil moisture limitation on evapotranspiration), and simulated soil moisture persistence. Although the soil moisture simulated by (land-surface, hydrological, and climate) models often exhibits strong discrepancies in absolute terms, soil moisture anomalies can be compared with simple scaling and generally match reasonably well (e.g., Koster et al., 2009; A.H. Wang et al., 2009). Soil moisture persistence is found to be an important component in projected changes in soil moisture drought, with some regions displaying year-round dryness compared to reference (late 20th or pre-industrial) conditions due to the carry-over effect of soil moisture storage from season to season, leading to year-round soil moisture deficits compared to late 20th century climate (e.g., Wang et al., 2005, Figure 3-10). However, it should be noted that some land surface and hydrological models (used offline or coupled to climate models) suffer from similar shortcomings as noted above for PDSI – that is, they use simple bucket models or simplified representations of potential evapotranspiration. The latter issue has been suggested as being particularly critical for models used in offline mode (Milly and Dunne, 2011). Nonetheless, for the assessment of soil moisture drought, using simulated soil moisture anomalies seems less problematic than many other indices for the reasons highlighted in the above paragraphs.

The indices listed above have been used in various studies analyzing drought in the context of climate change, but with a few exceptions most available studies are based only on one index, which makes their comparison difficult. Nonetheless, these studies suggest that projections can be highly dependent on the choice of drought index. For instance, one study projected changes in drought area possibly varying between a negligible impact and a 5 to 45% increase depending on the drought index considered (Burke and Brown, 2008). Other drought indices are used to quantify hydrological drought (e.g., Heim Jr., 2002; Vidal et al., 2010; Dai, 2011), but are less commonly used in climate change studies. Further analyses or indices also consider the area affected by droughts (e.g., Burke et al., 2006; Sheffield and Wood, 2008a; Dai, 2011) or additional variables (such as snow or vegetation indices from satellite measurements, e.g., Heim Jr., 2002). As for the definition of other indices (Box 3-1), the determination of the reference period is critical for the assessment of changes in drought patterns independently of the chosen index. In general, late 20th-century conditions are used as reference (e.g., Figure 3-10).

In summary, drought indices often integrate precipitation, temperature, and other variables, but may emphasize different aspects of drought and should be carefully selected with respect to the drought characteristic in mind. In particular, some indices have specific shortcomings, especially in the context of climate change. For this reason, assessments of changes in drought characteristics with climate change should consider several indices including a specific evaluation of their relevance to the addressed question to support robust conclusions. In this assessment we focus on the following indices: consecutive dry days (CDD) and simulated soil moisture anomalies (SMA), although evidence based on other indices (e.g., PDSI for present climate) is also considered (Section 3.5.1; Tables 3-2 and 3-3).

regions. Hence, drought is a complex phenomenon that is strongly affected by other extremes considered in this chapter, but that is also affected by changes in mean climate features (Section 3.1.6). In addition, via land-atmosphere interactions, drought also has the potential to impact other weather and climate elements such as temperature and precipitation and associated extremes (Koster et al., 2004b; Seneviratne et al., 2006a; Hirschi et al., 2011; see also Section 3.1.4). Case Study 9.2.3 addresses aspects related to the management of adverse consequences of droughts; while Case Study 9.2.2 considers the possible impacts of high temperatures and drought on wildfire.

Observed Changes

There are still large uncertainties regarding observed global-scale trends in droughts. The AR4 reported based on analyses using PDSI (see Box 3-3) that very dry areas had more than doubled in extent since 1970 at the global scale (Trenberth et al., 2007). This assessment was, however, largely based on the study by Dai et al. (2004) only. These trends in the PDSI proxy were found to be largely affected by changes in temperature, not precipitation (Dai et al., 2004). On the other hand, based on soil moisture simulations with an observation-driven land surface model for the time period 1950-2000, Sheffield and Wood (2008a) have inferred trends in drought duration, intensity, and severity predominantly decreasing, but with strong regional variation and including increases in some regions. They concluded that there was an overall moistening trend over the considered time period, but also a switch since the 1970s to a drying trend, globally and in many regions, especially in high northern latitudes. Some regional studies are consistent with the results from Sheffield and Wood (2008a), regarding, for example, less widespread increase (or statistically insignificant changes or decreases) in some regions compared to the study of Dai et al. (2004) (e.g., in Europe, see below). More recently, Dai (2011) by extending the record did, however, find widespread increases in drought both based on various versions of PDSI (for 1950-2008) and soil moisture output from a land surface model (for 1948-2004). Hence there are still large uncertainties with respect to global assessments of past changes in droughts. Nonetheless, there is some agreement between studies over the different time frames (i.e., since 1950 versus 1970) and using different drought indicators regarding increasing drought occurrence in some regions (e.g., southern Europe, West Africa; see below and Table 3-2), although other regions also indicate opposite trends (e.g., central North America, northwestern Australia; see below and Table 3-2). As mentioned in Section 3.1.6, spatially coherent shifts in drought regimes are expected with changing global circulation patterns. Table 3-2 provides regional and continental-scale assessments of observed trends in dryness based on different indices (Box 3-3). The following paragraphs provide more details by continent.

From a paleoclimate perspective recent droughts are not unprecedented, with severe 'megadroughts' reported in the paleoclimatic record for Europe, North America, and Australia (Jansen et al., 2007). Recent studies extend this observation to African and Indian droughts (Sinha et al., 2007; Shanahan et al., 2009): much more severe and longer droughts occurred in the past centuries with widespread ecological, political, and socioeconomic consequences. Overall, these studies confirm that in the last millennium several extreme droughts have occurred (Breda and Badeau, 2008; Kallis, 2008; Büntgen et al., 2010).

In North America, there is *medium confidence* that there has been an overall slight tendency toward less dryness (wetting trend with more soil moisture and runoff; Table 3-2), although analyses for some subregions also indicate tendencies toward increasing dryness. This assessment is based on several lines of evidence, including simulations with different hydrological models as well as PDSI and CDD estimates (Alexander et al., 2006; Andreadis and Lettenmaier, 2006; van der Schrier et al., 2006a; Kunkel et al., 2008; Sheffield and Wood, 2008a; Dai, 2011). The most severe droughts in the 20th century have occurred in the 1930s and 1950s, where the 1930s Dust Bowl was most intense and the 1950s drought most persistent (Andreadis et al., 2005) in the United States, while in Mexico the 1950s and late 1990s were the driest periods. Recent regional trends toward more severe drought conditions were identified over southern and western Canada, Alaska, and Mexico, with subregional exceptions (Dai, 2011).

In Europe, there is *medium confidence* regarding increases in dryness based on some indices in the southern part of the continent, but large inconsistencies between indices in this region, and inconsistent or statistically insignificant trends in the rest of the continent (Table 3-2). Although Dai et al. (2004) found an increase in dryness for most of the European continent based on PDSI, Lloyd-Hughes and Saunders (2002) and van der Schrier et al. (2006b) concluded, based on the analysis of SPI and self-calibrating PDSI for the 20th century (for 1901-1999 and 1901-2002, respectively), that no statistically significant changes were observed in extreme and moderate drought conditions in Europe [with the exception of the Mediterranean region in van der Schrier et al. (2006b)]. Sheffield and Wood (2008a) also found contrasting dryness trends in Europe, with increases in the southern and eastern part of the continent, but decreases elsewhere. Beniston (2009b) reported a strong increase in warm-dry conditions over all central-southern (including maritime) Europe via a quartile analysis from the middle to the end of the 20th century. Alexander et al. (2006) found trends toward increasing CDD mostly in the southern and central part of the continent. Trends of decreasing precipitation and discharge are consistent with increasing salinity in the Mediterranean Sea, indicating a trend toward freshwater deficits (Mariotti et al., 2008), but this could also be partly caused by increased human water use. In France, an analysis based on a variation of the PDSI model also reported a significant increasing trend in drought conditions, in particular from the 1990s onward (Corti et al., 2009). Stahl et al. (2010) investigated streamflow data across Europe and found negative trends (lower streamflow) in southern and eastern regions, and generally positive trends (higher streamflow) elsewhere (especially in northern latitudes). Low flows have decreased in most regions where the lowest mean monthly flow occurs in summer, but vary for catchments that have flow minima in winter and secondary low flows in summer. The exceptional 2003 summer heat wave on the European continent (see Section 3.3.1) was also associated with a major soil moisture drought, as could be inferred from satellite measurements (Andersen et al., 2005), model simulations (Fischer et al., 2007a,b), and impacts on ecosystems (Ciais et al., 2005; Reichstein et al., 2007).

There is *low confidence* in dryness trends in South America (Table 3-2), partly due to lack of data and partly due to inconsistencies. For the Amazon, repeated intense droughts have been occurring in the last decades but no particular trend has been reported. The 2005 and 2010 droughts in Amazonia are, however, considered the strongest in the last century as inferred from integrating precipitation records and water storage estimates via satellite (measurements from the Gravity Recovery and Climate Experiment; Chen et al., 2009; Lewis et al., 2011). For other parts of South America, analyses of the return intervals between droughts in the instrumental and reconstructed precipitation series indicate that the probability of drought has increased during the late 19th and 20th centuries, consistent with selected long instrumental precipitation records and with a recession of glaciers in the Chilean and Argentinean Andean Cordillera (Le Quesne et al., 2006, 2009).

Changes in drought patterns have been reported for the monsoon regions of Asia and Africa with variations at the decadal time scale (e.g., Janicot, 2009). In Asia there is overall *low confidence* in trends in dryness both at the continental and regional scale, mostly due to spatially varying trends, except in East Asia where a range of studies, based on different indices, show increasing dryness in the second half of the 20th century, leading to *medium confidence* (Table 3-2).

In the Sahel, recent years have been characterized by greater interannual variability than the previous 40 years (Ali and Lebel, 2009; Greene et al., 2009), and by a contrast between the western Sahel remaining dry and the eastern Sahel returning to wetter conditions (Ali and Lebel, 2009). Giannini et al. (2008) report a drying of the African monsoon regions, related to warming of the tropical oceans, and variability related to ENSO. In the different subregions of Africa there is overall *low* to *medium confidence* regarding regional dryness trends (Table 3-2).

For Australia, Sheffield and Wood (2008a) found very limited increases in dryness from 1950 to 2000 based on soil moisture simulated using existing climate forcing (mostly in southeastern Australia) and some marked decreases in dryness in central Australia and the northwestern part of the continent. Dai (2011), for an extended period until 2008 and using different PDSI variants as well as soil moisture output from a land surface model, found a more extended drying trend in the eastern half of the continent, but also a decrease in dryness in most of the western half. Jung et al. (2010) inferred from a combination of remote sensing and quasi-globally distributed eddy covariance flux observations that in particular the decade after 1998 became drier in Australia (and parts of Africa and South America), leading to decreased evapotranspiration, but it is not clear if this is a trend or just decadal variation.

Following the assessment of observed changes in the AR4 (Chapter 3), which was largely based on one study (Dai et al., 2004), subsequent work has drawn a more differentiated picture both regionally and temporally. There is not enough evidence at present to suggest high confidence in observed trends in dryness due to lack of direct observations, some geographical inconsistencies in the trends, and some dependencies of inferred trends on the index choice. There is *medium confidence* that since the 1950s some regions of the world have experienced more intense and longer droughts (e.g., southern Europe, west Africa) but also opposite trends exist in other regions (e.g., central North America, northwestern Australia).

Causes of the Observed Changes

The AR4 (Hegerl et al., 2007) concluded that it is *more likely than not* that anthropogenic influence has contributed to the increase in the droughts observed in the second half of the 20th century. This assessment was based on several lines of evidence, including a detection study that identified an anthropogenic fingerprint in a global PDSI data set with high significance (Burke et al., 2006), although the model trend was weaker than observed and the relative contributions of natural external forcings and anthropogenic forcings were not assessed.

There is now a better understanding of the potential role of land-atmosphere feedbacks versus SST forcing for meteorological droughts (e.g., Schubert et al., 2008a,b), and some modeling studies have also addressed potential impacts of land use changes (e.g., Deo et al., 2009), but large uncertainties remain in the field of land surface modeling and land-atmosphere interactions, in part due to lack of observations (Seneviratne et al., 2010), inter-model discrepancies (Koster et al., 2004b; Dirmeyer et al., 2006; Pitman et al., 2009), and model resolution of orographic and other effects. Nonetheless, a new set of climate modeling studies show that US drought response to SST variability is consistent with observations (Schubert et al., 2009). Inferred trends in drought are also consistent with trends in global precipitation and temperature, and the latter two are consistent with expected responses to anthropogenic forcing (Hegerl et al., 2007; X. Zhang et al., 2007). The change in the pattern of global precipitation in the observations and in model simulations is also consistent with the theoretical understanding of hydrological response to global warming that wet regions become overall wetter and dry regions drier in a warming world (Held and Soden, 2006; see also Section 3.1.6), though some regions also display shifts in climate regimes (Section 3.1.6). Nonetheless, some single events have been reported as differing from projections (Seager et al., 2009), though this is not necessarily incompatible given the superimposition of anthropogenic climate change and natural climate variability (Section 3.1). For soil moisture and hydrological drought it has been suggested that the stomatal 'antitranspirant' responses of plants to rising atmospheric CO_2 may lead to a decrease in evapotranspiration (Gedney et al., 2006). This could mean that increasing CO_2 levels alleviate soil moisture and streamflow drought, but this result is still debated (e.g., Piao et al., 2007; Gerten et al., 2008), in particular due to the uncertainty in observed runoff trends used to infer these effects (e.g., Peel and McMahon, 2006; see also Section 3.2.1).

Overall, though new studies have furthered the understanding of the mechanisms leading to drought, there is still relatively limited evidence to provide an attribution of observed changes, in particular given the issues associated with the availability of observational data (Section 3.2.1) and the definition and computation of drought indicators (Box 3-3). This latter point was mostly identified in post-AR4 studies (Box 3-3). Moreover, regions where consistent increases in drought are identified (see 'Observed Changes') are only partly consistent with those where projections indicate an enhancement of drought conditions in coming decades (see next paragraphs). We thus assess that there is *medium confidence* (see also Section 3.1.5) that anthropogenic influence has contributed to some changes in the drought patterns observed in the second half of the 20th century, based on its attributed impact on precipitation and temperature changes (though temperature can only be indirectly related to drought trends; see Box 3-3). However there is *low confidence* in the attribution of changes in droughts at the level of individual regions.

Projected Changes and Uncertainties

The AR4 assessed that projections at the time indicated an increase in droughts, in particular in subtropical and mid-latitude areas (Christensen et al., 2007). An increase in dry spell length and frequency was considered *very likely* over the Mediterranean region, southern areas of Australia, and New Zealand and *likely* over most subtropical regions, with little change over northern Europe. Continental drying and the associated risk of drought were considered *likely* to increase in summer over many mid-latitude continental interiors (e.g., central and southern Europe, the Mediterranean region), in boreal spring, and dry periods of the annual cycle over Central America.

More recent global and regional climate simulations and hydrological models mostly support the projections from the AR4, as summarized in the following paragraphs (see also Table 3-3), although we assess the overall *confidence* in drought projections as *medium* given the definitional issues associated with dryness and the partial lack of agreement in model projections when based on different dryness indices (Box 3-3). Indeed, particular care is needed in inter-comparing 'drought' projections since very many different definitions are employed (corresponding to different types of droughts), from simple climatic indices such as CDD to more complex indices of soil moisture and hydrological drought (Box 3-3). A distinction also needs to be made between short-term and longer-term events. Blenkinsop and Fowler (2007a) and Burke et al. (2010), for example, show different trend strength, and sometimes sign (Blenkinsop and Fowler, 2007a), for changes in short- and long-term droughts with RCM ensembles applied to the United Kingdom (although uncertainties in the latter projections are large; see below). These various distinctions are generally not considered and most currently available studies only assess changes in very few (most commonly one or two) dryness indices.

On the global scale, Burke and Brown (2008) provided an analysis of projected changes in drought based on four indices (SPI, PDSI, PPEA, and SMA; for definitions, see Box 3-3) using two model ensembles: one based on a GCM expressing uncertainty in parameter space, and a multi-model ensemble of 11 GCM simulations from CMIP3. Their analysis revealed that SPI, based solely on precipitation, showed little change in the proportion of the land surface in drought, and that all other indices, which include a measure of the atmospheric demand for moisture, showed a statistically significant increase with an additional 5 to 45% of the land surface in drought. This study also highlighted large uncertainties in regional changes in drought. For reasons highlighted in Box 3-3, using simulated soil moisture anomalies from the climate models avoids some shortcomings of other commonly used indices (although the quality of simulated soil moisture cannot be well evaluated due to lack of observations; Section 3.2 and Box 3-3). In the study of Burke and Brown (2008), this index showed weaker drying compared to PDSI and PPEA indices (but more pronounced drying than the SPI index). In this report, we display projected changes in soil moisture anomalies and CDD (Figure 3-10), this latter index being chosen for continuity with the AR4 (see Figure 10.18 of that report). It can be seen that the two indices partly agree on increased drought in some large regions (e.g., on the annual time scale, in Southern Europe and the Mediterranean region, central Europe, central North America, Central America and Mexico, northeast Brazil, and southern Africa), but some regions where the models show consistent increases in CDD (e.g., southeast Asia) do not show consistent decreases in soil moisture. Conversely, regions displaying a consistent decrease in CDD (e.g., in northeastern Asia) do not show a consistent increase in soil moisture. The substantial uncertainty of drought projections is particularly clear from the soil moisture projections, with, for example, no agreement among the models regarding the sign of changes in December to February over most of the globe. These results regarding changes in CDD and soil moisture are consistent with other published studies (Wang, 2005; Tebaldi et al., 2006; Burke and Brown, 2008; Sheffield and Wood, 2008b; Sillmann and Roeckner, 2008) and the areas that display consistent increasing drought tendencies for both indices have also been reported to display such tendencies for additional indices (e.g., Burke and Brown, 2008; Dai, 2011; Table 3-3). Sheffield and Wood (2008b) examined projections in drought frequency (for droughts of duration of 4 to 6 months and longer than 12 months, estimated from soil moisture anomalies) based on CMIP3 simulations with eight GCMs and the SRES scenarios A2, A1B, and B1. They concluded that drought was projected to increase in several regions under these three scenarios (mostly consistent with those displayed in Figure 3-10 for SMA), although the projections of drought intensification were stronger for the high CO_2 emissions scenarios (A2 and A1B) than for the more moderate scenario (B1). Regions showing statistically significant increases in drought frequency were found to be broadly similar for all three scenarios, despite the more moderate signal in the B1 scenario (their Figures 8 and 9). This study also highlighted the large uncertainty of scenarios for drought projections, as scenarios were found to span a large range of changes in drought frequency in most regions, from close to no change to two- to three-fold increases (their Figure 10).

Regional climate simulations and high-resolution global atmospheric model simulations over Europe also highlight the Mediterranean region

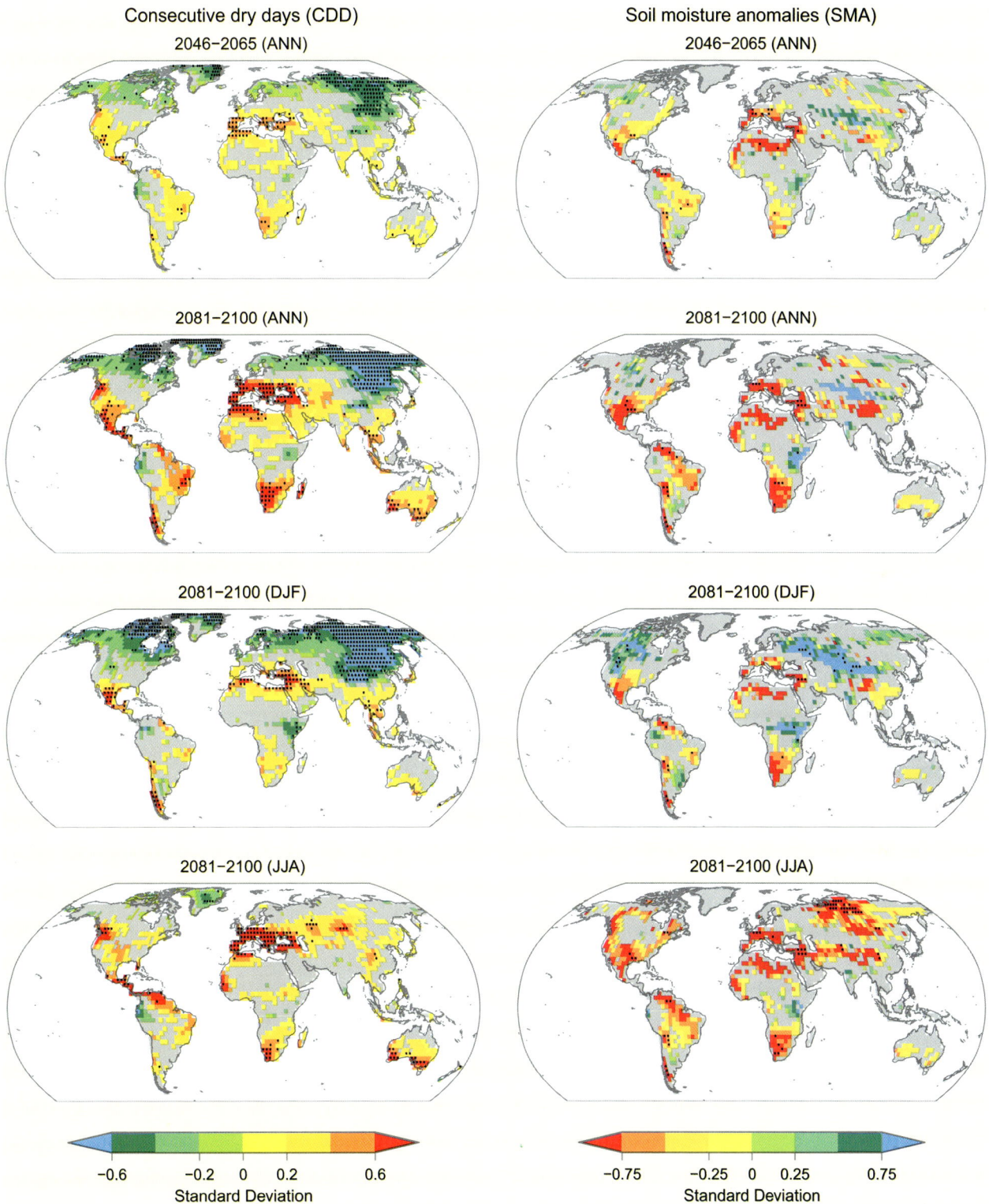

Figure 3-10 | Projected annual and seasonal changes in dryness assessed from two indices for 2081-2100 (bottom three rows, showing the annual time scale and two seasons, DJF and JJA) and 2046-2065 (top, annual time scale) with respect to 1980-1999. Left column: changes in the maximum number of CDD (days with precipitation <1 mm), based on 17 GCMs contributing to the CMIP3. Right column: changes in soil moisture (soil moisture anomalies, SMA), based on 15 GCMs contributing to the CMIP3. Increased dryness is indicated with warm colors (positive changes in CDD and negative SMA values). The maps show differences between the annual and seasonal averages over the respective 20-year periods, that is, the average of 2081-2100 or 2046-2065, respectively (based on simulations under emission scenario SRES A2), minus the average of 1980-1999 (from corresponding simulations for the 20th century). Differences are expressed in units of standard deviations, derived from detrended per year annual or seasonal estimates, respectively, from the three 20-year periods 1980-1999, 2046-2065, and 2081-2100 pooled together. Color shading is only applied for areas where at least 66% of the GCMs (12 out of 17 for CDD, 10 out of 15 for soil moisture) agree on the sign of the change; stippling is applied for regions where at least 90% of the GCMs (16 out of 17 for CDD, 14 out of 15 for soil moisture) agree on the sign of the change. Adapted from Orlowsky and Seneviratne (2011); updating Tebaldi et al. (2006) for SMA and for additional CMIP3 models, and including seasonal time frames. For more details, see Appendix 3.A.

as being affected by more severe droughts, consistent with available global projections (Table 3-3; see also Giorgi, 2006; Rowell and Jones, 2006; Beniston et al., 2007; Mariotti et al., 2008; Planton et al., 2008). Mediterranean (summer) droughts are projected to start earlier in the year and last longer. Also, increased variability during the dry and warm season is projected (Giorgi, 2006). One GCM-based study projected one to three weeks of additional dry days for the Mediterranean region by the end of the century (Giannakopoulos et al., 2009). For North America, intense and heavy episodic rainfall events with high runoff amounts are interspersed with longer relatively dry periods with increased evapotranspiration, particularly in the subtropics. There is a consensus of most climate model projections for a reduction in cool season precipitation across the US southwest and northwest Mexico (Christensen et al., 2007), with more frequent multi-year drought in the US southwest (Seager et al., 2007; Cayan et al., 2010). Reduced cool season precipitation promotes drier summer conditions by reducing the amount of soil water available for evapotranspiration in summer. For Australia, Alexander and Arblaster (2009) project increases in consecutive dry days, although consensus between models is only found in the interior of the continent. African studies indicate the possibility of relatively small-scale (500-km) heterogeneity of changes in precipitation and drought, based on climate model simulations (Funk et al., 2008; Shongwe et al., 2009). Regional climate simulations of South America project spatially coherent increases in CDD, particularly large over the Brazilian Plateau, and northern Chile and the Altiplano (Kitoh et al., 2011).

Available global and regional studies of hydrological drought (Hirabayashi et al., 2008b; Feyen and Dankers, 2009) project a higher likelihood of hydrological drought by the end of this century, with a substantial increase in the number of drought days (defined as streamflow below a specific threshold) during the last 30 years of the 21st century over North and South America, central and southern Africa, the Middle East, southern Asia from Indochina to southern China, and central and western Australia. Some regions, including eastern Europe to central Eurasia, inland China, and northern North America, project increases in drought. In contrast, wide areas over eastern Russia project a decrease in drought days. At least in Europe, hydrological drought is primarily projected to occur in the frost-free season.

Increased confidence in modeling drought stems from consistency between models and satisfactory simulation of drought indices during the past century (Sheffield and Wood, 2008a; Sillmann and Roeckner, 2008). Inter-model agreement is stronger for long-term droughts and larger spatial scales (in some regions, see above discussion), while local to regional and short-term precipitation deficits are highly spatially variable and much less consistent between models (Blenkinsop and Fowler, 2007b). Insufficient knowledge of the physical causes of meteorological droughts, and of the links to the large-scale atmospheric and ocean circulation, is still a source of uncertainty in drought simulations and projections. For example, plausible explanations have been proposed for projections of both a worsening drought and a substantial increase in rainfall in the Sahara (Biasutti et al., 2009; Burke et al., 2010). Another example is illustrated with the relationship of rainfall in southern Australia with SSTs around northern Australia. On annual time scales, low rainfall is associated with cooler than normal SSTs. Yet the warming observed in SST over the past few decades has not been associated with increased rainfall, but with a trend toward more drought-like conditions (N. Nicholls, 2010).

There are still further sources of uncertainties affecting the projections of trends in meteorological drought for the coming century. The two most important may be uncertainties in the development of the ocean circulation and feedbacks between land surface and atmospheric processes. These latter processes are related to the effects of drought on vegetation physiology and dynamics (e.g., affecting canopy conductance, albedo, and roughness), with resulting (positive or negative) feedbacks to precipitation formation (Findell and Eltahir, 2003a,b; Koster et al., 2004b; Cook et al., 2006; Hohenegger et al., 2009; Seneviratne et al., 2010; van den Hurk and van Meijgaard, 2010), and possibly – as only recently highlighted – also feedbacks between droughts, fires, and aerosols (Bevan et al., 2009). Furthermore, the development of soil moisture that results from complex interactions among precipitation, water storage as soil moisture (and snow), and evapotranspiration by vegetation is still associated with large uncertainties, in particular because of lack of observations of soil moisture and evapotranspiration (Section 3.2.1), and issues in the representation of soil moisture-evapotranspiration coupling in current climate models (Dirmeyer et al., 2006; Seneviratne et al., 2010). Uncertainties regarding soil moisture-climate interactions are also due to uncertainties regarding the behavior of plant transpiration, growth, and water use efficiency under enhanced atmospheric CO_2 concentrations, which could potentially have impacts on the hydrological cycle (Betts et al., 2007), but are not well understood yet (Hungate et al., 2003; Piao et al., 2007; Bonan, 2008; Teuling et al., 2009; see also above discussion on the causes of observed changes). The space-time development of hydrological drought as a response to a meteorological drought and the associated soil moisture drought (drought propagation, e.g., Peters et al., 2003) also needs more attention. There is some understanding of these issues at the catchment scale (e.g., Tallaksen et al., 2009), but these need to be extended to the regional and continental scales. This would lead to better understanding of the projections of hydrological droughts, which would contribute to a better identification and attribution of droughts and help to improve global hydrological models and land surface models.

In summary, there is *medium confidence* that since the 1950s some regions of the world have experienced trends toward more intense and longer droughts, in particular in southern Europe and West Africa, but in some regions droughts have become less frequent, less intense, or shorter, for example, central North America and northwestern Australia. There is *medium confidence* that anthropogenic influence has contributed to some changes in the drought patterns observed in the second half of the 20th century, based on its attributed impact on precipitation and temperature changes (though temperature can only be indirectly related to drought trends; see Box 3-3). However there is *low confidence* in the attribution of changes in droughts at the level

of single regions due to inconsistent or insufficient evidence. Post-AR4 studies indicate that there is *medium confidence* in a projected increase in duration and intensity of droughts in some regions of the world, including southern Europe and the Mediterranean region, central Europe, central North America, Central America and Mexico, northeast Brazil, and southern Africa. Elsewhere there is overall *low confidence* because of insufficient agreement of projections of drought changes (dependent both on model and dryness index). Definitional issues and lack of data preclude higher *confidence* than *medium* in observations of drought changes, while these issues plus the inability of models to include all the factors likely to influence droughts preclude stronger *confidence* than *medium* in the projections.

3.5.2. Floods

A flood is "the overflowing of the normal confines of a stream or other body of water, or the accumulation of water over areas that are not normally submerged (some specific examples are discussed in Case Study 9.2.6). Floods include river (fluvial) floods, flash floods, urban floods, pluvial floods, sewer floods, coastal floods, and glacial lake outburst floods" (see Glossary). The main causes of floods are intense and/or long-lasting precipitation, snow/ice melt, a combination of these causes, dam break (e.g., glacial lakes), reduced conveyance due to ice jams or landslides, or by a local intense storm (Smith and Ward, 1998). Floods are affected by various characteristics of precipitation, such as intensity, duration, amount, timing, and phase (rain or snow). They are also affected by drainage basin conditions such as water levels in the rivers, the presence of snow and ice, soil character and status (frozen or not, soil moisture content and vertical distribution), rate and timing of snow/ice melt, urbanization, and the existence of dikes, dams, and reservoirs (Bates et al., 2008). Along coastal areas, flooding may be associated with storm surge events (Section 3.5.5). A change in the climate physically changes many of the factors affecting floods (e.g., precipitation, snow cover, soil moisture content, sea level, glacial lake conditions, vegetation) and thus may consequently change the characteristics of floods. Engineering developments such as dikes and reservoirs regulate flow, and land use may also affect floods. Therefore the assessment of causes of changes in floods is complex and difficult. The focus in this section is on changes in floods that might be related to changes in climate (i.e., referred to as 'climate-driven'), rather than changes in engineering developments or land use. However, because of partial lack of documentation, these can be difficult to distinguish in the instrumental record.

Literature on the impact of climate change on pluvial floods (e.g., flash floods and urban floods) is scarce, although the changes in heavy precipitation discussed in Section 3.3.2 may imply changes in pluvial floods in some regions. This chapter focuses on the spatial, temporal, and seasonal changes in high flows and peak discharge in rivers related to climate change, which cause changes in fluvial (river) floods. River discharge simulation under a changing climate scenario requires a set of GCM or RCM outputs (e.g., precipitation and surface air temperature) and a hydrological model. A hydrological model may consist of a land surface model of a GCM or RCM and a river routing model. Different hydrological models may yield quantitatively different river discharge, but they may not yield different signs of the trend if the same GCM/RCM outputs are used. So the ability of models to simulate floods, in particular regarding the signs of the past and future trends, depends on the ability of the GCM or RCM to simulate precipitation changes. The ability of a GCM or RCM to simulate temperature is important for river discharge simulation in snowmelt- and glacier-fed rivers. Downscaling and/or bias-correction are frequently applied to GCM/RCM outputs before hydrological simulations are conducted, which becomes a source of uncertainty. More details on the feasibility and uncertainties in hydrological projections are described later in this section. Coastal floods are discussed in Sections 3.5.3 and 3.5.5. Glacial lake outburst floods are discussed in Section 3.5.6. The impact of floods on human society and ecosystems and related changes are discussed in Chapter 4. Case Study 9.2.6 discusses the management of floods.

Worldwide instrumental records of floods at gauge stations are limited in spatial coverage and in time, and only a limited number of gauge stations have data that span more than 50 years, and even fewer more than 100 years (Rodier and Roche, 1984; see also Section 3.2.1). However, this can be overcome partly or substantially by using pre-instrumental flood data from documentary records (archival reports, in Europe continuous over the last 500 years) (Brázdil et al., 2005), and from geological indicators of paleofloods (sedimentary and biological records over centennial to millennial scales) (Kochel and Baker, 1982). Analysis of these pre-instrumental flood records suggest that (1) flood magnitude and frequency can be sensitive to modest alterations in atmospheric circulation, with greater sensitivity for 'rare' floods (e.g., 50-year flood and higher) than for smaller and more frequent floods (e.g., 2-year floods) (Knox, 2000; Redmond et al., 2002); (2) high interannual and interdecadal variability can be found in flood occurrences both in terms of frequency and magnitude although in most cases, cyclic or clusters of flood occurrence are observed in instrumental (Robson et al., 1998), historical (Vallve and Martin-Vide, 1998; Benito et al., 2003; Llasat et al., 2005), and paleoflood records (Ely et al., 1993; Benito et al., 2008); (3) past flood records may contain analogs of unusual large floods, similar to some recorded recently, sometimes considered to be the largest on record. For example, pre-instrumental flood data show that the 2002 summer flood in the Elbe did not reach the highest flood levels recorded in 1118 and 1845 although it was higher than other disastrous floods of 1432, 1805, etc. (Brázdil et al., 2006). However, the currently available pre-instrumental flood data is also limited, particularly in spatial coverage.

The AR4 and the IPCC Technical Paper VI based on the AR4 concluded that no gauge-based evidence had been found for a climate-driven globally widespread change in the magnitude/frequency of floods during the last decades (Rosenzweig et al., 2007; Bates et al., 2008). However, the AR4 also pointed to possible changes that may imply trends in flood occurrence with climate change. For instance, Trenberth et al. (2007) highlighted a catastrophic flood that occurred along several central

European rivers in 2002, although neither flood nor mean precipitation trends could be identified in this region; however, there was a trend toward increasing precipitation variability during the last century which itself could imply an enhanced probability of flood occurrence. Kundzewicz et al. (2007) argued that climate change (i.e., observed increase in precipitation intensity and other observed climate changes) might already have had an impact on floods. Regarding the spring peak flows, the AR4 concluded with *high confidence* that abundant evidence was found for an earlier occurrence in snowmelt- and glacier-fed rivers (Rosenzweig et al., 2007; Bates et al., 2008), though we expressly note here that a change in the timing of peak flows does not necessarily imply nor preclude changes in flood magnitude or frequency in the affected regions.

Although changes in flood magnitude/frequency might be expected in regions where temperature change affects precipitation type (i.e., rain/snow separation), snowmelt, or ice cover (in particular northern high-latitude and polar regions), widespread evidence of such climate-driven changes in floods is not available. For example, there is no evidence of widespread common trends in the magnitude of floods based on the daily river discharge of 139 Russian gauge stations for the last few to several decades, though a significant shift in spring discharge to earlier dates has been found (Shiklomanov et al., 2007). Lindström and Bergström (2004) noted that it is difficult to conclude that flood levels are increasing from an analysis of runoff trends in Sweden for 1807 to 2002.

In the United States and Canada during the 20th century and in the early 21st century, there is no compelling evidence for climate-driven changes in the magnitude or frequency of floods (Lins and Slack, 1999; Douglas et al., 2000; McCabe and Wolock, 2002; Cunderlik and Ouarda, 2009; Villarini et al., 2009). There are relatively abundant studies on the changes and trends for rivers in Europe such as rivers in Germany and its neighboring regions (Mudelsee et al., 2003; Tu et al., 2005; Yiou et al., 2006; Petrow and Merz, 2009), in the Swiss Alps (Allamano et al., 2009), in France (Renard et al., 2008), in Spain (Benito et al., 2005), and in the United Kingdom (Robson et al., 1998; Hannaford and Marsh, 2008), but a continental-scale assessment of climate-driven changes in the flood magnitude and frequency for Europe is difficult to provide because geographically organized patterns are not seen in the reported changes.

Available (limited) analyses for Asia suggest the following changes: the annual flood maxima of the lower Yangtze region show an upward trend over the last 40 years (Jiang et al., 2008), the likelihood for extreme floods in the Mekong River has increased during the second half of the 20th century although the probability of an average flood has decreased (Delgado et al., 2009), and both upward and downward trends are identified over the last four decades in four selected river basins of the northwestern Himalaya (Bhutiyani et al., 2008). In the Amazon region in South America, the 2009 flood set record highs in the 106 years of data for the Rio Negro at the Manaus gauge site in July 2009 (Marengo et al., 2011). Recent increases have also been reported in flood frequency in some other river basins in South America (Camilloni and Barros, 2003; Barros et al., 2004). Conway et al. (2009) concluded that robust identification of hydrological change was severely limited by data limitations and other issues for sub-Saharan Africa. Di Baldassarre et al. (2010) found no evidence that the magnitude of African floods has increased during the 20th century. However, such analyses cover only limited parts of the world. Evidence in the scientific literature from the other parts of the world, and for other river basins, appears to be very limited.

Many river systems are not in their natural state anymore, making it difficult to separate changes in the streamflow data that are caused by the changes in climate from those caused by human regulation of the river systems. River engineering and land use may have altered flood probability. Many dams are designed to reduce flooding. Large dams have resulted in large-scale land use change and may have changed the effective rainfall in some regions (Hossain et al., 2009).

The above analysis indicates that research subsequent to the AR4 still does not show clear and widespread evidence of climate-driven observed changes in the magnitude or frequency of floods at the global level based on instrumental records, and there is thus *low confidence* regarding the magnitude and frequency and even the sign of these changes. The main reason for this lack of confidence is due to limited evidence in many regions, since available instrumental records of floods at gauge stations are limited in space and time, which limits the number of analyses. Moreover, the confounding effects of changes in land use and engineering mentioned above also make the identification of climate-driven trends difficult. There are limited regions with medium evidence, where no ubiquitous change is apparent (low agreement). Pre-instrumental flood data can provide information for longer periods, but current availability of these data is even scarcer particularly in spatial coverage. There is abundant evidence for an earlier occurrence of spring peak flows in snowmelt- and glacier-fed rivers (*high confidence*), though this feature may not necessarily be linked with changes in the magnitude of spring peak flows in the concerned regions.

The possible causes for changes in floods were discussed in the AR4 and Bates et al. (2008), but cause-and-effect between external forcing and changes in floods was not explicitly assessed. A rare example considered in Rosenzweig et al. (2007) and Bates et al. (2008) was a study by Milly et al. (2002) which, based on monthly river discharge, reported an impact of anthropogenic climate change on changes (mostly increases) in 'large' floods during the 20th century in selected extratropical river basins larger than 20,000 km^2, but they did not endorse the study because of the lack of widespread observed evidence of such trends in other studies. More recent literature has detected the influence of anthropogenically induced climate change in variables that affect floods, such as aspects of the hydrological cycle (see Section 3.2.2.2) including mean precipitation (X. Zhang et al., 2007), heavy precipitation (see Section 3.3.2), and snowpack (Barnett et al., 2008), though a direct statistical link between anthropogenic climate change and trends in the magnitude and frequency of floods is still not established.

In climates where seasonal snow storage and melting play a significant role in annual runoff, the hydrologic regime is affected by changes in temperature. In a warmer world, a smaller portion of precipitation falls as snow (Hirabayashi et al., 2008a) and the melting of winter snow occurs earlier in spring, resulting in a shift in peak river runoff to winter and early spring. This has been observed in the western United States (Regonda et al., 2005; Clow, 2010), in Canada (Zhang et al., 2001), and in other cold regions (Rosenzweig et al., 2007; Shiklomanov et al., 2007), along with an earlier breakup of river ice in Arctic rivers (Smith, 2000; Beltaos and Prowse, 2009). The observed trends toward earlier timing of snowmelt-driven streamflows in the western United States since 1950 are detectably different from natural variability (Barnett et al., 2008; Hidalgo et al., 2009). Thus, observed warming over several decades that is attributable to anthropogenic forcing has *likely* been linked to earlier spring peak flows in snowmelt- and glacier-fed rivers. It is unclear if observed warming over several decades has affected the magnitude of the snowmelt peak flows, but warming may result either in an increase in spring peak flows where winter snow depth increases (Meehl et al., 2007b) or a decrease in spring peak flows because of decreased snow cover and amounts (Hirabayashi et al., 2008b; Dankers and Feyen, 2009).

There is still a lack of studies identifying an influence of anthropogenic climate change over the past several decades on rain-generated peak streamflow trends because of availability and uncertainty in the observed streamflow data and low signal-to-noise ratio. Evidence has recently emerged that anthropogenic climate change could have increased the risk of rainfall-dominated flood occurrence in some river basins in the United Kingdom in autumn 2000 (Pall et al., 2011). Overall, there is *low confidence* (due to limited evidence) that anthropogenic climate change has affected the magnitude and frequency of floods, though it has detectably influenced several components of the hydrological cycle, such as precipitation and snowmelt, that may impact flood trends. The assessment of causes behind the changes in floods is inherently complex and difficult.

The number of studies that investigated projected flood changes in rivers especially at a regional or a continental scale was limited when the AR4 was published. Projections of flood changes at the catchment/river-basin scale were also not abundantly cited in the AR4. Nevertheless, Kundzewicz et al. (2007) and Bates et al. (2008) argued that more frequent heavy precipitation events projected over most regions would affect the risk of rain-generated floods (e.g., flash flooding and urban flooding).

The number of regional- or continental-scale studies of projected changes in floods is still limited. Recently, a few studies for Europe (Lehner et al., 2006; Dankers and Feyen, 2008, 2009) and a study for the globe (Hirabayashi et al., 2008b) have indicated changes in the frequency and/or magnitude of floods in the 21st century at large scale using daily river discharge calculated from RCM or GCM outputs and hydrological models. A notable change is projected to occur in northeastern Europe in the late 21st century because of a reduction in snow accumulation (Dankers and Feyen, 2008, 2009; Hirabayashi et al., 2008b), that is, a decrease in the probability of floods, that generally corresponds to lower flood peaks. For other parts of the world, Hirabayashi et al. (2008b) show an increase in the risk of floods in most humid Asian monsoon regions, tropical Africa, and tropical South America with a decrease in the risk of floods in non-negligible areas of the world such as most parts of northern North America.

Projections of flood changes at the catchment/river-basin scale are also not abundant in the scientific literature. Several studies have been undertaken for UK catchments (Cameron, 2006; Kay et al., 2009; Prudhomme and Davies, 2009) and catchments in continental Europe and North America (Graham et al., 2007; Thodsen, 2007; Leander et al., 2008; Raff et al., 2009; van Pelt et al., 2009). However, projections for catchments in other regions such as Asia (Asokan and Dutta, 2008; Dairaku et al., 2008), the Middle East (Fujihara et al., 2008), South America (Nakaegawa and Vergara, 2010; Kitoh et al., 2011), and Africa (Taye et al., 2011) are rare.

Uncertainty is still large in the projected changes in the magnitude and frequency of floods. It has been recently recognized that the choice of GCMs is the largest source of uncertainties in hydrological projections at the catchment/river-basin scale, and that uncertainties from emission scenarios and downscaling methods are also relevant but less important (Graham et al., 2007; Leander et al., 2008; Kay et al., 2009; Prudhomme and Davies, 2009), although, in general, hydrological projections require downscaling and/or bias-correction of GCM outputs (e.g., precipitation and temperature). Also the choice of hydrological models was found to be relevant but less important (Kay et al., 2009; Taye et al., 2011). However, the relative importance of downscaling, bias-correction, and the choice of hydrological models may depend on the selected region/catchment, the selected downscaling and bias-correction methods, and the selected hydrological models (Wilby et al., 2008). For example, the sign of the above-mentioned flood changes in northeastern Europe is affected by differences in temporal downscaling and bias-correction methods applied in the different studies (Dankers and Feyen, 2009). Chen et al. (2011) demonstrated considerable uncertainty caused by several downscaling methods in a hydrological projection for a snowmelt-dominated Canadian catchment. Downscaling (see Section 3.2.3) and bias-correction are also a major source of uncertainty in rain-dominated catchments (van Pelt et al., 2009). We also note that bias-correction and statistical downscaling tend to ignore the energy closure of the climate system, which could be a non-negligible source of uncertainty in hydrological projections (Milly and Dunne, 2011).

The number of projections of flood magnitude and frequency changes is still limited at regional and continental scales. Projections at the catchment/river-basin scale are also not abundant in the peer-reviewed scientific literature, especially for regions outside Europe and North America. In addition, considerable uncertainty remains in the projections of flood changes, especially regarding their magnitude and frequency. Therefore, our assessment is that there is *low confidence* (due to limited evidence) in future changes in flood magnitude and frequency derived from river discharge simulations. Nevertheless, as was argued by Kundzewicz et al. (2007) and Bates et al. (2008), physical reasoning suggests that projected increases in heavy rainfall in some catchments or regions would contribute to increases in rain-generated local floods (*medium confidence*). We note that heavy precipitation may be projected to increase despite a projected decrease of total precipitation depending on the regions considered (Section 3.3.2), and that changes in several variables (e.g., precipitation totals, frequency, and intensity, snow cover and snowmelt, soil moisture) are relevant for changes in floods. Confidence in change in one of these components alone may thus not be sufficient to confidently project changes in flood occurrence. Hence, *medium confidence* is attached to the above statement based on physical reasoning, although the link between increases in heavy rainfall and increases in local flooding seems apparent. The earlier shifts of spring peak flows in snowmelt- and glacier-fed rivers are robustly projected (Kundzewicz et al., 2007; Bates et al., 2008); so these are assessed as *very likely*, though this may not necessarily be relevant for flood occurrence. There is *low confidence* (due to limited evidence) in the projected magnitude of the earlier peak flows in snowmelt- and glacier-fed rivers.

In summary, there is limited to medium evidence available to assess climate-driven observed changes in the magnitude and frequency of floods at a regional scale because the available instrumental records of floods at gauge stations are limited in space and time, and because of confounding effects of changes in land use and engineering. Furthermore, there is *low agreement* in this evidence, and thus overall *low confidence* at the global scale regarding even the sign of these changes. There is *low confidence* (due to limited evidence) that anthropogenic climate change has affected the magnitude or frequency of floods, though it has detectably influenced several components of the hydrological cycle such as precipitation and snowmelt (*medium confidence* to *high confidence*), which may impact flood trends. Projected precipitation and temperature changes imply possible changes in floods, although overall there is *low confidence* in projections of changes in fluvial floods. *Confidence* is *low* due to limited evidence and because the causes of regional changes are complex, although there are exceptions to this statement. There is *medium confidence* (based on physical reasoning) that projected increases in heavy rainfall (Section 3.3.2) would contribute to increases in rain-generated local flooding, in some catchments or regions. Earlier spring peak flows in snowmelt- and glacier-fed rivers are *very likely*, but there is *low confidence* in their projected magnitude.

3.5.3. Extreme Sea Levels

Transient sea level extremes and extreme coastal high water are caused by severe weather events or tectonic disturbances that cause tsunamis. Since tsunamis are not climate-related, they are not addressed here. The drop in atmospheric pressure and strong winds that accompany severe weather events such as tropical or extratropical cyclones (Sections 3.4.4 and 3.4.5) can produce storm surges at the coast, which may be further elevated by wave setup caused by an onshore flux of momentum due to wave breaking in the surf zone. Various metrics are used to characterize extreme sea levels including storm-related highest values, annual maxima, or percentiles. Extreme sea levels may change in the future as a result of both changes in atmospheric storminess and mean sea level rise. However, neither contribution will be spatially uniform across the globe. For severe storm events such as tropical and extratropical cyclones, changes may occur in the frequency, intensity, or genesis regions of severe storms and such changes may vary between ocean basins (see Sections 3.4.4 and 3.4.5). Along some coastlines, land subsidence due to glacial isostatic adjustment (e.g., Lambeck et al., 2010) is causing a relative fall in sea levels. Variations in the rate of sea level rise can be large relative to mean sea level (Yin et al., 2010) and will occur as a result of variations in wind change (e.g., Timmermann et al., 2010), changes in atmospheric pressure and oceanic circulation (e.g., Tsimplis et al., 2008), and associated differences in water density and rates of thermal expansion (e.g., Bindoff et al., 2007; Church et al., 2010; Yin et al., 2010). In addition, if rapid melting of ice sheets occurs it would lead to non-uniform rates of sea level rise across the globe due to adjustments in the Earth's gravitational field (e.g., Mitrovica et al., 2010). On some coastlines, higher mean sea levels may alter the astronomical tidal range and the evolution of storm surges, and increase the wave height in the surf zones. As well as gradual increases in mean sea level that contribute to extreme impacts from transient extreme sea levels, rapid changes in sea level arising from, for example, collapse of ice shelves could be considered to be an extreme event with the potential to contribute to extreme impacts in the future. However, knowledge about the likelihood of such changes occurring is limited and so does not allow an assessment at this time.

Mean sea level has varied considerably over glacial time scales as the extent of ice caps and glaciers have fluctuated with global temperatures. Sea levels have risen around 120 to 130 m since the last glacial maximum 19 to 23 ka before present to around 7,000 years ago, and reached a level close to present at least 6,000 years ago (Lambeck et al., 2010). As well as the influence on sea level extremes caused by rapidly changing coastal bathymetries (Clarke and Rendell, 2009) and large-scale circulation patterns (Wanner et al., 2008), there is some evidence that changes in the behavior of severe tropical cyclones has changed on centennial time scales, which points to non-stationarity in extreme sea level events (Nott et al., 2009). Woodworth et al. (2011) use tide gauge records dating back to the 18th century, and salt marsh data, to show that sea level rise has accelerated over this time frame.

The AR4 reported that there was *high confidence* that the rate of observed sea level rise increased from the 19th to the 20th century (Bindoff et al., 2007). It also reported that the global mean sea level rose at an average rate of 1.7 (1.2 to 2.2) mm yr^{-1} over the 20th century, 1.8 (1.3 to 2.3) mm yr^{-1} over 1961 to 2003, and at a rate of 3.1 (2.4 to 3.8) mm yr^{-1} over 1993 to 2003. With updated satellite data to 2010, Church and White (2011) show that satellite-measured sea levels continue to rise at a rate close to that of the upper range of the AR4 projections. Whether the faster rate of increase during the latter period reflects decadal variability or an increase in the longer-term trend is not clear. However, there is evidence that the contribution to sea level due to mass loss from Greenland and Antarctica is accelerating (Velicogna, 2009; Rignot et al., 2011; Sørensen et al., 2011). The AR4 also reported that the rise in mean sea level and variations in regional climate led to a *likely* increase in the trend of extreme high water worldwide in the late 20th century (Bindoff et al., 2007), it was *very likely* that humans contributed to sea level rise during the latter half of the 20th century (Hegerl et al., 2007), and therefore that it was *more likely than not* that humans contributed to the trend in extreme high sea levels (IPCC, 2007a). Since the AR4, Menendez and Woodworth (2010), using data from 258 tide gauges across the globe, have confirmed the earlier conclusions of Woodworth and Blackman (2004) that there was an increasing trend in extreme sea levels globally, more pronounced since the 1970s, and that this trend was consistent with trends in mean sea level (see also Lowe et al., 2010). Additional studies at particular locations support this finding (e.g., Marcos et al., 2009; Haigh et al., 2010).

Various studies also highlight the additional influence of climate variability on extreme sea level trends. Menendez and Woodworth (2010) report that ENSO (see Section 3.4.2) has a large influence on interannual variations in extreme sea levels in the Pacific Ocean and the monsoon regions based on sea level records since the 1970s. In southern Europe, Marcos et al. (2009) report that changes in extremes are also significantly negatively correlated with the NAO (see Section 3.4.3). Ullmann et al. (2007) concluded that maximum annual sea levels in the Camargue had risen twice as fast as mean sea level during the 20th century due to an increase in southerly winds associated with a general rise in sea level pressure over central Europe (Ullmann et al., 2008). Sea level trends from two tide gauges on the north coast of British Columbia from 1939 to 2003 were twice that of mean sea level rise, the additional contribution being due to the strong positive phase of the PDO (see Section 3.4.3), which has lasted since the mid-1970s (Abeysirigunawardena and Walker, 2008). Cayan et al. (2008) reported an increase of 20-fold at San Francisco since 1915 and 30-fold at La Jolla since 1933 in the frequency of exceedance of the 99.99th percentile sea level. They also noted that positive sea level anomalies of 10 to 20 cm that often persisted for several months during El Niño events produced an increase in storm surge peaks over this time. The spatial extent of these oscillations and their influence on extreme sea levels across the Pacific has been discussed by Merrifield et al. (2007). Church et al. (2006a) examined changes in extreme sea levels before and after 1950 in two tide gauge records of approximately 100 years on the east and west coasts of Australia, respectively. At both locations a stronger positive trend was found in the sea level exceeded by 0.01% of the observations than the median sea level, suggesting that in addition to mean sea level rise, other modes of variability or climate change are contributing to the extremes. At Mar del Plata, Argentina, Fiore et al. (2009) noted an increase in the number and duration of positive storm surges in the decade 1996 to 2005 compared to previous decades, which may be due to a combination of mean sea level rise and changes in wind climatology resulting from a southward shift in the South Atlantic high.

Thus, studies since the AR4 conclude that trends in extreme sea level are generally consistent with changes in mean sea level (e.g., Marcos et al., 2009; Haigh et al., 2010; Menendez and Woodworth, 2010) although some studies note that the trends in extremes are larger than the observed trend in mean sea levels (e.g., Church et al., 2006a; Ullmann et al., 2007; Abeysirigunawardena and Walker, 2008) and may be influenced by modes of climate variability, such as the PDO on the Canadian west coast (e.g., Abeysirigunawardena and Walker, 2008). These studies are consistent with the conclusions from the AR4 that increases in extremes are related to trends in mean sea level and modes of variability in the regional climate.

The AR4 (Meehl et al., 2007b) projected sea level rise for 2090-2099 relative to 1980-1999 due to ocean thermal expansion, glaciers and ice caps, and modeled ice sheet contributions of 18 to 59 cm, which incorporates a 90% uncertainty range across all scenarios. An additional contribution to the sea level rise projections was taken into account for a possible rapid dynamic response of the Greenland and West Antarctic ice sheets, which could result in an accelerating contribution to sea level rise. This was estimated to be 10 to 20 cm of sea level rise by 2090-2099 using a simple linear relationship with projected temperature. Because of insufficient understanding of the dynamic response of ice sheets, Meehl et al. (2007b) also noted that a larger contribution could not be ruled out.

Several studies since the AR4 have developed statistical models that relate 20th-century (e.g., Rahmstorf, 2007; Horton et al., 2008) or longer (e.g., Vermeer and Rahmstorf, 2009; Grinsted et al., 2010) temperature and sea level rise to extrapolate future global mean sea level. These alternative approaches yield projections of sea level rise under a range of SRES scenarios by 2100 of 0.47 to 1.00 m (B1 to A2 scenarios; Horton et al., 2008), 0.50 to 1.40 m (B1 to A1FI scenarios; Rahmstorf, 2007), 0.75 to 1.90 m (B1 to A1FI scenarios; Vermeer and Rahmstorf, 2009), and 0.90 to 1.30 m (A1B scenario only; Grinsted et al., 2010). However, future rates of sea level rise may be less closely associated with global mean temperature if ice sheet dynamics play a larger role in the future (Cazenave and Llovel, 2010). Furthermore, Church et al. (2011) note that these models may overestimate future sea levels because non-climate related contributions to trends over the observational period such as groundwater depletion may not have been removed, and non-linear effects such as the reduction in glacier area as glaciers contract and the reduction in the efficiency of ocean heat uptake with global warming in the future are not accounted for. Pfeffer et al. (2008), using a dynamical model of glaciers, found that sea level rise of more than 2 m by 2100 is physically implausible. An estimate of 0.8 m by 2100 that included increased ice dynamics was considered most plausible.

New studies, whose focus is on quantifying the effect of storminess changes on storm surge, have been carried out over northern Europe since the AR4. Debernard and Roed (2008) used hydrodynamic models to investigate storm surge changes over Europe in four regionally downscaled GCMs including two runs with B2, one with A2, and one with an A1B emission scenario. Despite large inter-model differences, statistically significant changes between 1961-1990 and 2071-2100 consisted of decreases in the 99th percentile surge heights south of Iceland, and an 8 to 10% increase along the coastlines of the eastern North Sea and the northwest British Isles, which occurred mainly in the winter season. Wang et al. (2008) projected a significant increase in wintertime storm surges around Ireland except the south Irish coast over 2031-2060 relative to 1961-1990 using a downscaled GCM under an A1B scenario. Sterl et al. (2009) joined the output from an ensemble of 17 GCM (CMIP3) simulations using the A1B emissions scenario over the model periods 1950-2000 and 2050-2100 into a single longer time series to estimate 10,000-year return values of surge heights along the Dutch coastline. No statistically significant change in this value was projected for the 21st century because projected wind speed changes were not associated with the surge-generating northerlies but rather non-surge generating south-westerlies.

Other studies have undertaken a sensitivity approach to compare the relative impact on extreme sea levels of severe weather changes and mean sea level rise. Over southeastern Australia, McInnes et al. (2009b) found that a 10% increase in wind speeds, consistent with the upper end of the range under an A1FI scenario from a multi-model ensemble for 2070 together with an A1FI sea level rise scenario, would produce extreme sea levels that were 12 to 15% higher than those including just the A1FI sea level rise projection alone. Brown et al. (2010) also investigated the relative impact of sea level rise and wind speed change on an extreme storm surge in the eastern Irish Sea. Both studies concluded that sea level rise rather than meteorological changes has the greater potential to increase extreme sea levels in these locations in the future.

The degree to which climate models (GCM or RCM) have sufficient resolution and/or internal physics to realistically capture the meteorological forcing responsible for storm surges is regionally dependent. For example current GCMs are unable to realistically represent tropical cyclones (see Section 3.4.4). This has led to the use of alternative approaches for investigating the impact of climate change on storm surges in tropical locations whereby large numbers of cyclones are generated using statistical models that govern the cyclones' characteristics over the observed period (e.g., McInnes et al., 2003). These models are then perturbed to represent projected future cyclone characteristics and used to force a hydrodynamic model. Recent studies on the tropical east coast of Australia reported in Harper et al. (2009) that employ these approaches show a relatively small impact of a 10% increase in tropical cyclone intensity on the 1-in-100 year storm tide (the combined sea level due to the storm surge and tide), and mean sea level rise being found to produce the larger contribution to changes in future 1-in-100 year sea level extremes. However, one study that has incorporated scenarios of sea level rise in the hydrodynamic modeling of hurricane-induced sea level extremes on the Louisiana coast found that increased coastal water depths had a large impact on surge propagation over land, increasing storm surge heights by two to three times the sea level rise scenario, particularly in wetland-fronted areas (J.M. Smith et al., 2010).

To summarize, post-AR4 studies provide additional evidence that trends in extreme coastal high water across the globe reflect the increases in mean sea level, suggesting that mean sea level rise rather than changes in storminess are largely contributing to this increase (although data are sparse in many regions and this lowers the confidence in this assessment). It is therefore considered *likely* that sea level rise has led to a change in extreme coastal high water levels. It is *likely* that there has been an anthropogenic influence on increasing extreme coastal high water levels via mean sea level contributions. While changes in storminess may contribute to changes in sea level extremes, the limited geographical coverage of studies to date and the uncertainties associated with storminess changes overall (Sections 3.4.4 and 3.4.5) mean that a general assessment of the effects of storminess changes on storm surge is not possible at this time. On the basis of studies of observed trends in extreme coastal high water levels it is *very likely* that mean sea level rise will contribute to upward trends in the future.

3.5.4. Waves

Severe waves threaten the safety of coastal inhabitants and those involved in maritime activities and can damage and destroy coastal and marine infrastructure. Waves play a significant role in shaping a coastline by transporting energy from remote areas of the ocean to the coast. Energy dissipation via wave breaking contributes to beach erosion, longshore currents, and elevated coastal sea levels through wave set-up and wave run-up. Wave properties that influence these processes include wave height, the wave energy directional spectrum, and period. Studies of past and future changes in wave climate to date have tended to focus on wave height parameters such as 'Significant Wave Height' (SWH, the average height from trough to crest of the highest one-third of waves) and metrics of extreme waves, such as high percentiles or wave heights above particular thresholds, although one study (Dodet et al., 2010) also examines trends in mean wave direction and peak wave period. It should also be noted that waves may become an increasingly important factor along coastlines experiencing a decline in coastal protection afforded by sea ice (see Sections 3.5.5 and 3.5.7).

Wave climates have changed over paleoclimatic time scales. Wave modeling using paleobathymetries over the past 12,000 years indicates an increase in peak annual SWH of around 40% due to the increase in mean sea level, which redefines the location of the coastline, and hence progressively extends the fetch length in most of the shelf sea regions (Neill et al., 2009). Major circulation changes that result in changes in storminess and wind climate (see Section 3.3.3) have also affected wave climates. Evidence of enhanced storminess determined from sand drift and dune building along the western European coast indicates that enhanced storminess occurred over the period of the Little Ice Age

(1570-1900) and the mid Holocene (~8,200 years before present; Clarke and Rendell, 2009).

The AR4 reported statistically significant positive trends in SWH over the period 1950 to 2002 over most of the mid-latitudinal North Atlantic and North Pacific, as well as in the western subtropical South Atlantic, the eastern equatorial Indian Ocean and the East China and South China Sea, and declining trends around Australia, and parts of the Philippine, Coral, and Tasman Seas (Trenberth et al., 2007), based on voluntary observing ship data (e.g., Gulev and Grigorieva, 2004). Several studies that address trends in extreme wave conditions have been completed since the AR4 and the new studies generally provide more evidence for the previously reported positive trends in SWH and extreme waves in the north Atlantic and north Pacific. Global trends in 99th-percentile satellite-measured wave heights show a mostly significant positive trend of between 0.5 and 1.0% per year in the mid-latitude oceans but less clear trends over the tropical oceans from 1985 to 2008 (Young et al., 2011). X.L. Wang et al. (2009b) found that SWH increased in the boreal winter over the past half century in the high latitudes of the Northern Hemisphere (especially the northeast Atlantic), and decreased in more southerly northern latitudes based on the European Centre for Medium Range Weather Forecasts 40-year reanalysis (ERA-40). They also found that storminess around the 1880s was of similar magnitude to that in the 1990s. This is also found using the same data set by Le Cozannet et al. (2011), who relate the change in waves to the NAO pattern that is moderated by an east Atlantic pattern of climate variability during winter. A wave hindcast over the north-eastern Atlantic Ocean over the period 1953 to 2009 revealed a significant positive trend in SWH, as well as a counterclockwise shift in mean direction in the north and a slight but not significant increase in peak wave period in the northeast. In the south, no trend was found for SWH or wave period while a clockwise trend in mean direction was found (Dodet et al., 2010). In a regional North Sea hindcast, Weisse and Günther (2007) found a positive trend in 99th-percentile wave height from 1958 to the early 1990s followed by a declining trend to 2002 over the southern North Sea, except on the UK North Sea coast where negative trends occurred over much of the hindcast period.

On the North American Atlantic coast, Komar and Allan (2008) found a statistically significant trend of 0.059 m yr^{-1} in waves exceeding 3 m during the summer months over 30 years since the mid-1970s at Charleston, South Carolina, with weaker but statistically significant trends at wave buoys further north. These trends were associated with an increase in intensity and frequency of hurricanes over this period (see Section 3.4.4). In contrast, winter waves, generated by extratropical storms, were not found to have experienced a statistically significant change. In the eastern North Pacific, SWH is strongly correlated with El Niño (Section 3.4.2). However positive trends were also found in SWH and extreme wave height from the mid-1970s to 2006 in wave buoy data (Allan and Komar, 2006), for excesses of the 98th percentile SWH over 1985 to 2007 (Menendez et al., 2008) along the US west coast, and in hindcast SWH over 1948 to 1998 in the Southern Californian Bight (Adams et al., 2008). Positive though not statistically significant trends in annual mean SWH were found over south-eastern South America for *in situ* wave data over the 1996-2006 period and in satellite wave data over 1993 to 2001, while simulated wave fields using reanalysis wind forcing over the period 1971 to 2005 produced statistically significant trends in SWH (Dragani et al., 2010). Trends at particular locations may be also influenced by local factors. For example, Suursaar and Kullas (2009) reported a slight decreasing trend in mean SWHs from 1966 to 2006 in the Gulf of Riga within the Baltic Sea, while the frequency and intensity of high wave events (i.e., the difference between the maximum and 99th-percentile wave height) showed rising trends. These changes were associated with a decrease in local average wind speed, but an intensification of westerly winds and storm events occurring further to the west.

In the Southern Ocean, SWH derived from satellite observations was found to be strongly positively correlated with the SAM, particularly from March to August (Hemer et al., 2010). However, the analysis of reliable long-term trends in the Southern Hemisphere remains challenging due to limited *in situ* data and problems of temporal homogeneity in reanalysis products (Wang et al., 2006). For example, Hemer et al. (2010) also found that trends in SWH derived from satellite data over 1998-2000 relative to 1993-1996 were positive only over the Southern Ocean south of 45°S whereas trends were positive across most of the Southern Hemisphere in the Corrected ERA-40 reanalysis (C-ERA-40; Hemer, 2010). Hemer (2010) found that the frequency of wave events exceeding the 98th percentile over the period 1985 to 2002 using data from a wave buoy situated on the west coast of Tasmania showed no statistically significant trend whereas a strong positive trend was found in equivalent fields of C-ERA-40 data.

New studies have demonstrated strong links between wave climate and natural modes of climate variability (Section 3.4.3). For example, along the US west coast and the western North Pacific, SWH was found to be strongly correlated with El Niño (Allan and Komar, 2006; Sasaki and Toshiyuki, 2007) and, in the Southern Ocean, SWH was positivity correlated with the SAM (Hemer et al., 2010). On the US east coast, positive trends in summer SWH were linked to increasing numbers of hurricanes (Komar and Allan, 2008). In the northeast Atlantic, trends in SWH exhibited significant positive (negative) correlations with the NAO in the north (south) and more generally, trends in SWH, mean wave direction, and peak wave period over the period 1953 to 2009 were related to the increase in the NAO index over this time (Dodet et al., 2010). One study (X.L. Wang et al., 2009a) reported a link between external forcing (i.e., anthropogenic forcing due to greenhouse gases and aerosols, and natural forcing due to solar and volcanic forcing) and an increase in SWH in the boreal winter in the high latitudes of the Northern Hemisphere (especially the northeast North Atlantic), and a decrease in more southerly northern latitudes over the past half century.

The AR4 projected an increase in extreme wave height in many regions of the mid-latitude oceans as a result of projected increases in wind speeds associated with more intense mid-latitude storms in these regions in a future warmer climate (Meehl et al., 2007b). At the regional scale, increases in wave height were projected for most mid-latitude areas analyzed, including the North Atlantic, North Pacific, and Southern Ocean (Christensen et al., 2007) but with *low confidence* due to low confidence in projected changes in mid-latitude storm tracks and intensities (see Section 3.4.5). Several studies since then have developed wave climate projections that provide stronger evidence for future wave climate change. Global-scale projections of SWH were developed by Mori et al. (2010), using a 1.25° resolution wave model forced with projected winds from a 20-km global GCM, in which ensemble-averaged SST changes from the CMIP3 models provided the climate forcing. The spatial pattern of projected SWH change between 1979-2004 and 2075-2100 reflects the changes in the forcing winds, which are generally similar to the mean wind speed changes shown in Figure 3-8. Extreme waves (measured by a spatial and temporal average of the top 10 values over the 25-year period) were projected to exhibit large increases in the northern Pacific, particularly close to Japan due to an increase in strong tropical cyclones and also the Indian Ocean despite decreases in SWH.

A number of regional studies have also been completed since the AR4 in which forcing conditions were obtained for a few selected emission scenarios (typically B2 and A2, representing low-high ranges) from GCMs or RCMs. These studies provide additional evidence for positive projected trends in SWH and extreme waves along the western European coast (e.g., Debernard and Roed, 2008; Grabemann and Weisse, 2008) and the UK coast (Leake et al., 2007), declines in extreme wave height in the Mediterranean sea (Lionello et al., 2008) and the southeast coast of Australia (Hemer et al., 2010), and little change along the Portuguese coast (Andrade et al., 2007). However, considerable variation in projections can arise from the different climate models and scenarios used to force wave models, which lowers the confidence in the projections. For example, along the European North Sea coast, 99th-percentile wave height over the late 21st century relative to the late 20th century is projected to increase by 6 to 8% by Debernard and Roed (2008) based on wave model simulations with forcing from several GCMs under A2, B2, and A1B greenhouse gas scenarios, whereas they are projected to increase by up to 18% by Grabemann and Weisse (2008), who downscaled two GCMs under A2 and B2 emission scenarios. In one region, opposite trends in extreme waves were projected. Grabemann and Weisse (2008) project negative trends in 99th-percentile wave height along the UK North Sea coast, whereas Leake et al. (2007) downscaled the same GCM for the same emission scenarios, using a different RCM, and found positive changes in high percentile wave heights offshore of the East Anglia coastline. A wave projection study by Hemer et al. (2010) concluded that uncertainties arising from the method by which climate model winds were applied to wave model simulations (e.g., by applying bias-correction to winds or perturbing current climate winds with wind changes derived from climate models) made a larger contribution to the spread of RCM projections than the forcing from different GCMs or emission scenarios.

In summary, although post-AR4 studies are few and their regional coverage is limited, their findings generally support the evidence from earlier studies of wave climate trends. Most studies find a link between variations in waves (both SWH and extremes) and internal climate variability. There is *low confidence* that there has been an anthropogenic influence on extreme wave heights (because of insufficient literature). Despite the existence of downscaling studies for some regions such as the eastern North Sea, there is overall *low confidence* in wave height projections because of the small number of studies, the lack of consistency of the wind projections between models, and limitations in their ability to simulate extreme winds. However, the strong linkages between wave height and winds and storminess means that it is *likely* that future negative or positive changes in SWH will reflect future changes in these parameters.

3.5.5. Coastal Impacts

Severe coastal hazards such as erosion and inundation are important in the context of disaster risk management and may be affected by climate change through rising sea levels and changes in extreme events. Increasing sea levels will also increase the potential for saltwater intrusion into coastal aquifers. Coastal inundation occurs during periods of extreme sea levels due to storm surges and high waves, particularly when combined with high tides. Although tropical and extratropical cyclones (Sections 3.4.4 and 3.4.5) are the most common causes of sea level extremes, other weather events that cause persistent winds such as anticyclones and fronts can also influence coastal sea levels (Green et al., 2009; McInnes et al., 2009b). In many parts of the world, sea levels are influenced by modes of large scale variability such as ENSO (Section 3.4.2). In the western equatorial Pacific, sea levels can fluctuate up to half a meter between ENSO phases (Church et al., 2006b) and in combination with extremes of the tidal cycle, can cause extensive inundation in low-lying atoll nations even in the absence of extreme weather events (Lowe et al., 2010).

Shoreline position can change from the combined effects of various factors such as:
1) Rising mean sea levels, which cause landward recession of coastlines made up of erodible materials (e.g., Ranasinghe and Stive, 2009)
2) Changes in coastal height due to isostatic rebound (Blewitt et al., 2010; Mitrovica et al., 2010), or sediment compaction from the removal of oil, gas, and water (Syvitski et al., 2009)
3) Changes in the frequency or severity of transient storm erosion events (K.Q. Zhang et al., 2004)
4) Changes in sediment supply to the coast (Stive et al., 2003; Nicholls et al., 2007; Tamura et al., 2010)
5) Changes in wave speed due to sea level rise, which alters wave refraction, or in wave direction, which can cause realignment of shorelines (Ranasinghe et al., 2004; Bryan et al., 2008; Tamura et al., 2010)
6) The loss of natural protective structures such as coral reefs (e.g., Sheppard et al., 2005; Gravelle and Mimura, 2008) due to increased ocean temperatures (Hoegh-Guldberg, 1999) and ocean acidification (Bongaerts et al., 2010) or the reduction in permafrost

or sea ice in mid- and high latitudes, which exposes soft shores to the effects of waves and severe storms (see Section 3.5.7; Manson and Solomon, 2007).

For example, permafrost degradation and sea ice retreat may contribute to coastal erosion in Arctic regions (see Section 3.5.7).

The susceptibility of coastal regions to erosion and inundation is related to various physical (e.g., shoreline slope), and geomorphological and ecosystem attributes, and therefore may be inferred to some extent from broad coastal characterizations. These include the presence of beaches, rocky shorelines, or coasts with cliffs; deltas; back-barrier environments such as estuaries and lagoons; the presence of mangroves, salt marshes, or sea grasses; and shorelines flanked by coral reefs (e.g., Nicholls et al., 2007) or by permafrost or seasonal sea ice, each of which are characterized by different vulnerability to climate change-driven hazards. For example, deltas are low-lying and hence generally prone to inundation, while beaches are comprised of loose particles and therefore erodible. However, the degree to which these systems are impacted by erosion and inundation will also be influenced by other factors affecting disaster responses. For example, reduced protection from high waves during severe storms could occur as a result of depleted mangrove forests or the degradation of coral reefs (e.g., Gravelle and Mimura, 2008), or loss of sea ice or permafrost (e.g., Manson and Solomon, 2007); there may be a loss of ecosystem services brought about by saltwater contamination of already limited freshwater reserves due to rising sea levels and these will amplify the risks brought about by climate change (McGranahan et al., 2007), and also reduce the resilience of coastal settlements to disasters. Dynamical processes such as vertical land movement also contribute to inundation potential (Haigh et al., 2009). Coastal regions may be rising or falling due to post-glacial rebound or slumping due to aquifer drawdown (Syvitski et al., 2009). Multiple contributions to coastal flooding such as heavy rainfall and flooding in coastal catchments that coincide with elevated sea levels may also be important. Ecosystems such as coral reefs also play an important role in providing material on which atolls are formed. Large-scale oceanic changes that are particularly relevant to both coral reefs and small island countries are discussed in Box 3-4.

As discussed in Section 3.5.3, mean sea level has risen by 120 to 130 m since the end of the last glacial maximum (Jansen et al., 2007), and this has had a profound effect on coastline position around the world. Coastlines have also evolved over this time frame due to changes in the action of the ocean on the coast through changes in wave climate (Neill et al., 2009) and tides (Gehrels et al., 1995), which arise from the changing geometries of coastlines over glacial time scales and changes in storminess (e.g., Clarke and Rendell, 2009).

The AR4 (Nicholls et al., 2007) reported that coasts are experiencing the adverse consequences of impacts such as increased coastal inundation, erosion, and ecosystem losses. However, attributing these changes to sea level rise is difficult due to the multiple drivers of change over the 20th century (R.J. Nicholls, 2010) and the scarcity and fragmentary nature of data sets that contribute to the problem of identifying and attributing changes (e.g., Defeo et al., 2009). Since the AR4 there have been several new studies that examine coastline changes. In the Caribbean, the beach profiles at 200 sites across 113 beaches and eight islands were monitored on a three-monthly basis from 1985 to 2000, with most beaches found to be eroding and faster rates of erosion generally found on islands that had been impacted by a higher number of hurricanes (Cambers, 2009). However, the relative importance of anthropogenic factors, climate variability, and climate change on the eroding trends could not be separated quantitatively. In Australia, Church et al. (2008) report that despite the positive trend in sea levels during the 20th century, beaches have generally been free of chronic coastal erosion, and where it has been observed it has not been possible to unambiguously attribute it to sea level rise in the presence of other anthropogenic activities. Webb and Kench (2010) argue that the commonly held view of atoll nations being vulnerable to erosion must be reconsidered in the context of physical adjustments to the entire island shoreline, because erosion of some sectors may be balanced by progradation on other sectors. In their survey of 27 atoll islands across three central Pacific Nations (Tuvalu, Kiribati, and Federated States of Micronesia) over a 19- to 61-year period using photography and satellite imagery, they found that 43% of islands remained stable and 43% increased in area, with largest rates of increase in island area ranging from 0.1 to 5.6 ha per decade. Only 14% of islands studied exhibited a net reduction in area. On islands exhibiting either no net change or an increase in area, a larger redistribution of land area was evident in 65% of cases, consisting of mainly a shoreline recession on the ocean side and an elongation of the island or progradation of the shoreline on the lagoon side. Human settlements were present on 7 of the 27 atolls surveyed and the majority of those exhibited net accretion due in part to coastal protection works. For a coral reef island at the northern end of the Great Barrier Reef, Australia, Dawson and Smithers (2010) report a 6% increase in area and 4% increase in volume between 1967 and 2007 but with a net retreat on the east-southeast shoreline and advance on the western side. Chust et al. (2009) evaluated the relative contribution of local anthropogenic (non-climate change related) and sea level rise impacts on the coastal morphology and habitats in the Basque coast, northern Spain, for the period 1954 to 2004. They found that the impact from local anthropogenic influences was about an order of magnitude greater than that due to sea level rise over this period. Increased rates of coastal erosion have also been observed since 1935 in Canada's Gulf of St. Lawrence (Forbes et al., 2004).

The AR4 stated with *very high confidence* that the impact of climate change on coasts is exacerbated by increased pressures on the physical environment arising from human settlements in the coastal zone (Nicholls et al., 2007). The small number of studies that have been completed since the AR4 have been either unable to attribute coastline changes to specific causes in a quantitative way or else find strong evidence for non-climatic causes that are natural and/or anthropogenic.

The AR4 reported with *very high confidence* that coasts will be exposed to increasing impacts, including coastal erosion, over coming decades due to climate change and sea level rise, both of which will be

Box 3-4 | Small Island States

Small island states represent a distinct category of locations owing to their small size and highly maritime climates, which means that their concerns and information needs in relation to future climate change differ in many ways from those of the larger continental regions that are addressed in this chapter. Their small land area and often low elevation makes them particularly vulnerable to rising sea levels and impacts such as inundation, shoreline change, and saltwater intrusion into underground aquifers (Mimura, 1999). Their maritime environments lead to an additional emphasis on oceanic information to understand the impacts of climate change (see Case Study 9.2.9). Particular challenges exist for the assessment of past changes in climate given the sparse regional and temporal coverage of terrestrial-based observation networks and the limited *in situ* ocean observing network, although observations have improved somewhat in recent decades with the advent of satellite-based observations of meteorological and oceanic variables. However, the short length of these records hampers the investigation of long-term trends in the region. The resolution of GCMs is insufficient to represent small islands and few studies have been undertaken to provide projections for small islands using RCMs (Campbell et al., 2011). In regions such as the Pacific Ocean, large-scale climate features such as the South Pacific Convergence Zone ENSO (Section 3.4.2) have a substantial influence on the pattern and timing of precipitation, yet these features and processes are often poorly represented in GCMs (Collins et al., 2010). The purpose of this box is to present available information on observed trends and climate change projections that are not covered in the other sections of this chapter as well as discuss key aspects of the climate system that are particularly relevant for small islands. The *very likely* contribution of mean sea level rise to increased extreme sea levels (see Section 3.5.3), coupled with the *likely* increase in tropical cyclone maximum wind speed (see Section 3.4.4), is a specific issue for tropical small island states.

Although the underlying data sources are limited, some data for the Indian Ocean, South Pacific (Fiji), and Caribbean were available in the studies of Alexander et al. (2006) and Caesar et al. (2011). Problems of data availability and homogeneity for the Caribbean are discussed by T.S. Stephenson et al. (2008). Based on standard extremes indices, positive trends in warm days and warm nights and negative trends in cold days and cold nights[2] have occurred across the Indian Ocean and South Pacific region for the period 1971 to 2005 (Caesar et al., 2011) and the Caribbean for the period 1951 to 2003 (based on data from Alexander et al., 2006). Based on the same data sources, trends in average total wet-day precipitation were positive and statistically significant over the Indian Ocean region, negative over the South Pacific region, and weakly negative over the Caribbean. Trends in heavy and very heavy precipitation were positive over the Indian Ocean, negative over the South Pacific region, and close to zero over the Caribbean. We have *low confidence* in temperature trends over the Indian Ocean and South Pacific region due to the shorter record over which trends were assessed, whereas for the Caribbean, we have *medium confidence* in the temperature trends due to the longer records available for assessment. Because of the spatial heterogeneity exhibited in precipitation trends in general, there is insufficient evidence to assess observed rainfall trends. For the Caribbean, temperatures are projected to increase across the region by 1 to 4°C over 2071-2100 relative to 1961-1990 under the A2 and B2 scenarios and rainfall is mainly projected to decrease by 25 to 50% except in the north (Campbell et al., 2011). Based on this study and the evidence for projected temperature increases reported for other regions (see Table 3-3) we have *medium confidence* in the projected temperature increases for the Caribbean. However, due to the range of processes that contribute to rainfall change, some of which are poorly resolved by GCMs, there is insufficient evidence to assess projected rainfall changes on these small islands.

Given the low elevation of many small islands, sea level extremes are of particular relevance. Sea level extremes are strongly influenced by tidal extremes (Chowdhury et al., 2007; Merrifield et al., 2007). When the tide behavior is mostly semi-diurnal (two high and low tides per day), there will be a clustering of high spring tides around the time of the equinoxes whereas when the tide behavior is diurnal (one high and low tide per day), the clustering of high spring tides will occur around the time of the solstices. In addition, ENSO has a strong influence such that sea levels and their extremes are positively (negatively) correlated with the SOI in the tropical Pacific west (east) of 180° (Church et al., 2006b; Menendez et al., 2010). Tides and ENSO have contributed to the more frequent occurrence of sea level extremes and associated flooding experienced at some Pacific Islands such as Tuvalu in recent years, and make the task of determining the relative roles of these natural effects and mean sea level rise difficult (Lowe et al., 2010). Furthermore, the steep shelf margins that surround many islands and atolls in the Pacific support larger wave-induced contributions to sea level anomalies. Unfortunately, wave observations (including wave direction) that would facilitate more comprehensive studies of tide, surge, and wave extremes in the region are sparse, including those that are co-located with tide gauges (Lowe et al., 2010).

---

[2] Termed "cool days" and "cool nights" in that study.

Continued next page ⟶

Coral reefs are a feature of many small islands and healthy reef systems mitigate against erosion and inundation by not only providing a buffer zone for the shoreline during extreme surge and wave events but also providing a source of carbonate sand and gravel, which are delivered to the shores by storms and swell to maintain the atoll (Woodroffe, 2008; Webb and Kench, 2010). Anthropogenic oceanic changes may indirectly contribute to extreme impacts for coral atolls by affecting the health of the surrounding reef system. Such changes include: (1) warming of the surface ocean, which slows or prevents growth in temperature-sensitive species and causes more frequent coral bleaching events (e.g., Hoegh-Guldberg, 1999; see also Chapter 4); (2) ocean acidification, caused by increases in atmospheric CO_2 being absorbed into the oceans, which lowers coral growth rates (Bongaerts et al., 2010); and (3) reduction in oxygen concentration in the ocean due to a combination of changes in temperature-driven gas solubility (Whitney et al., 2007), ocean ventilation due to circulation changes, and biological cycling of organic material (Keeling et al., 2010). Quantifying these changes and understanding their impact on coral reef health will be important to understanding the impact of anthropogenic climate change on atolls.

In summary, the small land area and often low elevation of small island states make them particularly vulnerable to rising sea levels and impacts such as inundation, shoreline change, and saltwater intrusion into underground aquifers. Short record lengths and the inadequate resolution of current climate models to represent small island states limit the assessment of changes in extremes. There is insufficient evidence to assess observed trends and future projections in rainfall across the small island regions considered here. The reported increases in warm days and nights and decreases in cold days and nights are of *medium confidence* over the Caribbean and of *low confidence* over the Pacific and Indian Oceans. There is *medium confidence* in the projected temperature increases across the Caribbean. The unique situation of small islands states and their maritime environments leads to an additional emphasis on oceanic information to understand the impacts of climate change. The *very likely* contribution of mean sea level rise to increased coastal high water levels, coupled with the *likely* increase in tropical cyclone maximum wind speed, is a specific issue for tropical small island states.

exacerbated by increasing human-induced pressures (Nicholls et al., 2007). However it was also noted that since coasts are dynamic systems, adaptation to climate change required understanding of processes operating on decadal to century time scales, yet this understanding was least developed.

Because of the diverse and complex nature of coastal impacts, assessments of the future impacts of climate change have focused on a wide range of questions and employed a diverse range of methods, making direct comparison of studies difficult (R.J. Nicholls, 2010). Two types of studies are examined here: the first are assessments, typically undertaken at the country or regional scale and which combine information on physical changes with the socioeconomic implications (e.g., Nicholls and de la Vega-Leinert, 2008); the second type are studies oriented around improved scientific understanding of the impacts of climate change. In terms of coastal assessments, Aunan and Romstad (2008) reported that Norway's generally steep and resistant coastlines contribute to a low physical susceptibility to accelerated sea level rise. Nicholls and de la Vega-Leinert (2008) reported that large parts of the coasts in Great Britain (including England, Wales, and Scotland) are already experiencing widespread sediment starvation and erosion, loss/degradation of coastal ecosystems, and significant exposure to coastal flooding. Lagoons, river deltas, and estuaries are assessed as being particularly vulnerable in Poland (Pruszak and Zawadzka, 2008). In Estonia, Kont et al. (2008) reported increased beach erosion, which is believed to be the result of increased storminess in the eastern Baltic Sea since 1954, combined with a decline in sea ice cover during the winter. Sterr (2008) reported that for Germany there is a high level of reliance on hard coastal protection against extreme sea level hazards, which will increase ecological vulnerability over time. In France, the Atlantic coast Aquitaine region was considered more resilient to rising sea levels over the coming century because of the sediment storage in the extensive dune systems whereas the sandy coast regions of the Languedoc Roussillon region on the Mediterranean coast were considered more vulnerable because of narrow dune systems that are also highly urbanized (Vinchon et al., 2009). A coastal vulnerability assessment for Australia (Department of Climate Change, 2009) characterized future vulnerability in terms of coastal geomorphology, sediment type, and tide and wave characteristics, from which it concluded that the tropical northern coastline would be most sensitive to changes in tropical cyclone behavior while health of the coral reefs may also influence the tropical eastern coastline. The mid-latitude southern and eastern coastlines were expected to be most sensitive to changes in mean sea level, wave climate, and changes in storminess. A comparative study of the impact of sea level rise on coastal inundation across 84 developing countries showed that the greatest vulnerability to a 1 m sea level rise in terms of inundation of land area was located in East Asia and the Pacific, followed by South Asia, Latin America, and the Caribbean, the Middle East and North Africa, and finally sub-Saharan Africa (Dasgupta et al., 2009).

New models have been developed for the assessment of coastal vulnerability at the global to national level (Hinkel and Klein, 2009). At the local to regional scale, new techniques and approaches have also been developed to better quantify impacts from inundation due to future sea level rise. Bernier et al. (2007) evaluated spatial maps of extreme sea level for different return periods on a seasonal basis that were used to estimate seasonal risk of inundation under future sea level scenarios. McInnes et al. (2009a) developed spatial maps of storm tide and using a simple inundation model with high-resolution Light Detection and Ranging (LIDAR) data and a land subdivision database, identified the impact of inundation on several coastal towns along the southeastern Australian coastline under future sea level and wind speed scenarios. Probabilistic approaches have also been used to evaluate extreme sea level exceedance under uncertain future sea level rise scenarios. Purvis et al. (2008) constructed a probability distribution around the range of future sea level rise estimates and used Monte Carlo sampling to apply the sea level change to a two-dimensional coastal inundation model. They showed that by evaluating the possible flood-related losses in this framework they were able to represent spatially the higher losses associated with the low-frequency but high-impact inundation events instead of considering only a single midrange scenario. Hunter (2010) combined sea level extremes evaluated from observations with projections of sea level rise to 2100 and showed, for example, that planning levels in Sydney, Australia, would need to be increased substantially to cope with increased risk of flooding. Along the Portuguese coast, Andrade et al. (2007) found that projected future climate in the HadCM3 model would not affect wave height along this coastline but the projected rotation in wave direction would increase the net littoral drift and the erosional response. Along a section of the southeast coast of the United Kingdom, the effect of sea level rise, surge, and wave climate change on the inshore wave climate was evaluated and the frequency and height of extreme waves was projected to increase in the north of the domain (Chini et al., 2010). On the basis of modeling the 25-year beach response along a stretch of the Portuguese coast to various climate change scenarios, Coelho et al. (2009) concluded that the projected stormier wave climate led to higher rates of beach erosion than mean sea level rise. Modeling of the evolution of soft rock shores with rising sea levels has revealed a relatively simple relationship between sea level rise and the equilibrium cliff profile (Walkden and Dickson, 2008).

To summarize, recent observational studies that identify trends and impacts at the coast are limited in regional coverage, which means there is *low confidence*, due to insufficient evidence, that anthropogenic climate change has been a major cause of any observed changes. However, recent coastal assessments at the national and regional scale and process-based studies have provided further evidence of the vulnerability of low-lying coastlines to rising sea levels and erosion, so that in the absence of adaptation there is *high confidence* that locations currently experiencing adverse impacts such as coastal erosion and inundation will continue to do so in the future due to increasing sea levels in the absence of changes in other contributing factors.

3.5.6. Glacier, Geomorphological, and Geological Impacts

Mountains are prone to mass movements including landslides, avalanches, debris flows, and flooding that can lead to disasters. Changes in the cryosphere affect such extremes, but also water supply and hydropower generation. Many of the world's high mountain ranges are situated at the margins of tectonic plates, increasing the possibility of potentially hazardous interactions between climatic and geological processes. The principal drivers are glacier ice mass loss, mountain permafrost degradation, and possible increases in the intensity of precipitation (Liggins et al., 2010; McGuire, 2010). The possible consequences are changes in mass movement on short contemporary time scales, and modulations of seismicity and volcanic activity on longer, century to millennium time scales.

The AR4 assessed that "the late 20th century glacier wastage likely has been a response to post-1970 warming" (Lemke et al., 2007). However, the impacts of glacier retreat on the natural physical system in the context of changes in extreme events were not assessed in detail. Additionally, the AR4 did not assess geomorphological and geological impacts that might result from anthropogenic climate change. The most studied change in the high-mountain environment has been the retreat of glaciers (Paul et al., 2004; Kaser et al., 2006; Larsen et al., 2007; Rosenzweig et al., 2007). Alpine glaciers around the world were at maximum extent by the end of the Little Ice Age (~1850), and have retreated since then (Leclercq et al., 2011), with an accelerated decay during the past several decades (Zemp et al., 2007). Most glaciers have retreated since the mid-19th century (Francou et al., 2000; Cullen et al., 2006; Thompson et al., 2006; Larsen et al., 2007; Schiefer et al., 2007; Paul and Haeberli, 2008). Rates of retreat that exceed historical experience and internal (natural) variability have become apparent since the beginning of the 21st century (Reichert et al., 2002; Haeberli and Hohmann, 2008).

Outburst floods from lakes dammed by glaciers or unstable moraines [or 'glacial lake outburst floods' (GLOFs)] are commonly a result of glacier retreat and formation of lakes behind unstable natural dams (Clarke, 1982; Clague and Evans, 2000; Huggel et al., 2004; Dussaillant et al., 2010). In the past century, GLOFs have caused disasters in many high-mountain regions of the world (Rosenzweig et al., 2007), including the Andes (Reynolds et al., 1998; Carey, 2005; Hegglin and Huggel, 2008), the Caucasus and Central Asia (Narama et al., 2006; Aizen et al., 2007), the Himalayas (Vuichard and Zimmermann, 1987; Richardson and Reynolds, 2000; Xin et al., 2008; Bajracharya and Mool, 2009; Osti and Egashira, 2009), North America (Clague and Evans, 2000; Kershaw et al., 2005), and the European Alps (Haeberli, 1983; Haeberli et al., 2001; Vincent et al., 2010). However, because GLOFs are relatively rare, it is unclear whether their frequency of occurrence is changing at either the regional or global scale. Clague and Evans (2000) argue that outburst floods from moraine-dammed lakes in North America may have peaked due to a reduction in the number of the lakes since the end of the Little Ice Age. In contrast, a small but not statistically significant increase of GLOF events was observed in the Himalayas over the period 1940 to 2000 (Richardson and Reynolds, 2000), but the event documentation may not be complete. Over the past several decades, human mitigation measures at unstable glacier lakes in the Himalaya and European Alps may have prevented some potential GLOF events (Reynolds, 1998; Haeberli et al., 2001).

Evidence of degradation of mountain permafrost and attendant slope instability has emerged from recent studies in the European Alps

(Gruber and Haeberli, 2007; Huggel, 2009) and other mountain regions (Niu et al., 2005; Geertsema et al., 2006; Allen et al., 2011). This evidence includes several recent rock falls, rock slides, and rock avalanches in areas where permafrost thaw in steep bedrock is occurring. Landslides with volumes ranging up to a few million cubic meters have occurred in the Mont Blanc region (Barla et al., 2000), in Italy (Sosio et al., 2008; Huggel, 2009; Fischer et al., 2011), in Switzerland, and in British Columbia (Evans and Clague, 1998; Geertsema et al., 2006). Very large rock and ice avalanches with volumes of 30 to over 100 million m^3 include the 2002 Kolka avalanche in the Caucasus (Haeberli et al., 2004; Kotlyakov et al., 2004; Huggel et al., 2005), the 2005 Mt. Steller rock avalanche in the Alaska Range (Huggel et al., 2008), the 2007 Mt. Steele ice and rock avalanche in the St. Elias Mountains, Yukon (Lipovsky et al., 2008), and the 2010 Mt. Meager rock avalanche and debris flow in the Coast Mountains of British Columbia.

Quantification of possible trends in the frequency of landslides and ice avalanches in mountains is difficult due to incomplete documentation of past events, especially those that happened before regular satellite observations became available. Nevertheless, there has been an apparent increase in large rock slides during the past two decades, and especially during the first years of the 21st century in the European Alps (Ravanel and Deline, 2011), in the Southern Alps of New Zealand (Allen et al., 2011), and in northern British Columbia (Geertsema et al., 2006) in combination with temperature increases, glacier shrinkage, and permafrost degradation.

Research, however, has not yet provided any clear indication of a change in the frequency of debris flows due to recent deglaciation. Debris flow activity at a local site in the Swiss Alps was higher during the 19th century than today (Stoffel et al., 2005). In the French Alps no significant change in debris flow frequency has been observed since the 1950s in terrain above elevations of 2,200 m (Jomelli et al., 2004). Processes not, or not directly, driven by climate, such as sediment yield, can also be important for changes in the magnitude or frequency of alpine debris flows (Lugon and Stoffel, 2010).

Debris flows from both glaciated and unglaciated volcanoes, termed lahars, can be particularly large and hazardous. Lahars produced by volcanic eruptions on the glacier-clad Nevado del Huila volcano in Colombia in 2007 and 2008 were the largest rapid mass flows on Earth in recent years. Similarly, large mass flows occur on ice-covered active volcanoes in Iceland (Björnsson, 2003), including Eyjafjallajökull in 2010. Large rock and ice avalanches, with volumes up to 30 million m^3, have happened frequently (on average about one every four years) on the glaciated Alaskan volcano, Iliamna, and are thought to be related to elevated volcanic heat flow and possibly meteorological conditions (Huggel et al., 2007). Glacier decay on active volcanoes can lead to a reduction of lahar hazards due to less potential meltwater available for lahar generation, but it is difficult to make a general conclusion as local conditions also play important roles. In 1998, intense rainfall mobilized pyroclastic material on the flanks of Vesuvius and Campi Flegrei volcanoes, feeding approximately 150 debris flows that damaged nearby communities and resulted in 160 fatalities (Bondi and Salvatori, 2003).

In the same year, intense precipitation associated with Hurricane Mitch triggered a small flank collapse at Casita volcano in Nicaragua. This slope failure transformed into debris flows that destroyed two towns and claimed 2,500 lives (Scott et al., 2005). Following the 1991 Pinatubo eruption in the Philippines, heavy rains associated with tropical storms moved large volumes of volcanic sediment. The sediment dammed rivers, causing massive flooding across the region that continued for several years after the eruption ended (Newhall and Punongbayan, 1996).

A variety of climate and weather events can have geomorphological and geological impacts. Warming and degradation of mountain permafrost affect slope stability through a reduction in the shear strength of ice-filled rock discontinuities. For example, the 2003 European summer heat wave (Section 3.3.1) caused rapid thaw and thickening of the active layer, triggering a large number of mainly small rock falls (Gruber et al., 2004; Gruber and Haeberli, 2007). Permafrost thaw in sediment such as in talus slopes may increase both the frequency and magnitude of debris flows (Zimmermann et al., 1997; Rist and Phillips, 2005). The frost table at the base of the active layer is a barrier to groundwater infiltration and can cause the overlying non-frozen sediment to become saturated. Snow cover can also affect debris flow activity by supplying additional water to the soil, increasing pore water pressure and initiating slope failure (Kim et al., 2004). Many of the largest debris flows in the Alps in the past 20 years were triggered by intense rainfall in summer or fall when the snowline was elevated (Rickenmann and Zimmermann, 1993; Chiarle et al., 2007). Warming may increase the flow speed of frozen bodies of sediment (Kääb et al., 2007; Delaloye et al., 2008; Roer et al., 2008). Rock slopes can fail after they have been steepened by glacial erosion or unloaded (debuttressed) following glacier retreat (Augustinus, 1995). Although it may take centuries or even longer for a slope to fail following glacier retreat, recent landslides demonstrate that some slopes can respond to glacier downwasting within a few decades or less (Oppikofer et al., 2008). Twentieth-century warming may have penetrated some decameters into thawing steep rock slopes in high mountains (Haeberli et al., 1997). Case studies indicate that both small and large slope failures can be triggered by exceptionally warm periods of weeks to months prior to the events (Gruber et al., 2004; Huggel, 2009; Fischer et al., 2011).

The spatial and temporal patterns of precipitation, the intensity and duration of rainfall, and antecedent rainfall are important factors in triggering shallow landslides (Iverson, 2000; Wieczorek et al., 2005; Sidle and Ochiai, 2006). In some regions antecedent rainfall is probably a more important factor than rainfall intensity (Kim et al., 1991; Glade, 1998), whereas in other regions rainfall duration and intensity are the critical factors (Jakob and Weatherly, 2003). Landslides in temperate and tropical mountains that have no seasonal snow cover are not temperature-sensitive and may be more strongly influenced by human activities such as poor land use practices, deforestation, and overgrazing (Sidle and Ochiai, 2006).

Rock and ice avalanches on glaciated volcanoes can be triggered by heat generated by volcanic activity. Their incidence may increase with rising air and rock temperatures (Gruber and Haeberli, 2007) or during or following brief, anomalously warm events (Huggel et al., 2010) due to meltwater infiltration and shear strength reduction. Debuttressing effects due to glacier retreat can also destabilize or over-steepen slopes (Tuffen, 2010). Furthermore, on volcanoes, geothermal heat flow can enhance ice melting and thus create weak zones at the ice-bedrock interface; and hydrothermal alteration of rocks can decrease the slope stability (Huggel, 2009). On unglaciated high volcanoes in the Caribbean, Central America, Europe, Indonesia, the Philippines, and Japan, an increase in total rainfall or an increase in the frequency or magnitude of severe rainstorms (see Section 3.3.2) could cause more frequent debris flows by mobilizing unconsolidated, volcanic regolith and by raising pore-water pressures, which could lead to deep-seated slope failure. Heavy rainfall events could also influence the behavior of active volcanoes. For example, Mastin (1994) attributes the violent venting of volcanic gases at Mount St. Helens between 1989 and 1991 to slope instability or accelerated growth of cooling fractures within the lava dome following rainstorms, and Matthews et al. (2002) link episodes of intense tropical rainfall with collapses of the Soufriere Hills lava dome on Montserrat in the Caribbean. It is well established that ice mass wastage following the end of the last glaciations led to increased levels of seismicity associated with post-glacial rebound of the lithosphere (e.g., Muir-Wood, 2000; Stewart et al., 2000). There has been a large reduction in glacier cover in southern Alaska. Sauber and Molnia (2004) reported several hundred meters vertical reduction. This ice reduction may be responsible for an increase in seismicity in the region where earthquake faults are at the threshold of failure (Sauber and Molnia, 2004; Doser et al., 2007). An increase in the frequency of small earthquakes in the Icy Bay area, also in southeast Alaska, is interpreted to be a crustal response to glacier wastage between 2002 and 2006 (Sauber and Ruppert, 2008). Large-scale ice mass loss in glaciated volcanic terrain reduces the load on the crust and uppermost mantle, facilitating magma formation and its ascent into the crust (Jull and McKenzie, 1996) and allowing magma to reach the surface more easily (Sigmundsson et al., 2010). At the end of the last glaciation, this mechanism resulted in a more than 10-fold increase in the frequency of volcanic eruptions in Iceland (Sinton et al., 2005).

The AR4 projected that glaciers in mountains will lose additional mass over this century because more ice will be lost due to summer melting than is replenished by winter precipitation (Meehl et al., 2007b). The total area of glaciers in the European Alps may decrease by 20 to more than 50% by 2050 (Zemp et al., 2006; Huss et al., 2008). The projected glacier retreat in the 21st century may form new potentially unstable lakes. Probable sites of new lakes have been identified for some alpine glaciers (Frey et al., 2010). Rock slope and moraine failures may trigger damaging surge waves and outburst floods from these lakes. The temperature rise also will result in gradual degradation of mountain permafrost (Haeberli and Burn, 2002; Harris et al., 2009). The zone of warm permafrost (mean annual rock temperature approximately -2 to 0°C), which is more susceptible to slope failures than cold permafrost, may rise in elevation a few hundred meters during the next 100 years (Noetzli and Gruber, 2009). This in turn may shift the zone of enhanced instability and landslide initiation toward higher-elevation slopes that in many regions are steeper, and therefore predisposed to failure. The response of bedrock temperatures to surface warming through thermal conduction will be slow, but warming will eventually penetrate to considerable depths in steep rock slopes (Noetzli et al., 2007). Other heat transport processes such as advection, however, may induce warming of bedrock at much faster rates (Gruber and Haeberli, 2007). The response of firn and ice temperatures to an increase in air temperature is faster and nonlinear (Haeberli and Funk, 1991; Suter et al., 2001; Vincent et al., 2007). Latent heat effects from refreezing meltwater can amplify the increase in air temperature in firn and ice (Huggel, 2009; Hoelzle et al., 2010). At higher temperatures, more ice melts and the strength of the remaining ice is lower; as a result, the frequency and perhaps size of ice avalanches may increase (Huggel et al., 2004; Caplan-Auerbach and Huggel, 2007). Warm extremes can trigger large rock and ice avalanches (Huggel et al., 2010).

Current low levels of seismicity in Antarctica and Greenland may be a consequence of ice-sheet loading, and isostatic rebound associated with accelerated deglaciation of these regions may result in an increase in earthquake activity, perhaps on time scales as short as 10 to 100 years (Turpeinen et al., 2008; Hampel et al., 2010). Future ice mass loss on glaciated volcanoes, notably in Iceland, Alaska, Kamchatka, the Cascade Range in the northwest United States, and the Andes, could lead to eruptions, either as a consequence of reduced load pressures on magma chambers or through increased magma-water interaction. Reduced ice load arising from future thinning of Iceland's Vatnajökull Ice Cap is projected to result in an additional 1.4 km^3 of magma produced in the underlying mantle every century (Pagli and Sigmundsson, 2008). Ice unloading may also promote failure of shallow magma reservoirs with a potential consequence of a small perturbation of the natural eruptive cycle (Sigmundsson et al., 2010). Initially, ice thinning of 100 m or more on volcanoes with glaciers more than 150-m thick, such as Sollipulli in Chile, may cause more explosive eruptions, with increased tephra hazards (Tuffen, 2010). Additionally, the potential for edifice lateral collapse could be enhanced by loss of support previously provided by ice (Tuffen, 2010) or to elevated pore-water pressures arising from meltwater (Capra, 2006; Deeming et al., 2010). Ultimately the loss of ice cover on glaciated volcanoes may reduce opportunities for explosions arising from magma-ice interaction. The incidence of ice-sourced lahars may also eventually fall, although exposure of new surfaces of volcanic debris due to ice wastage may provide the raw material for precipitation-related lahars. The likelihood of both volcanic and non-volcanic landslides may also increase due to greater availability of water, which could destabilize slopes. Many volcanoes provide a ready source of unconsolidated debris that can be rapidly transformed into potentially hazardous lahars by extreme precipitation events. Volcanoes in coastal, near-coastal, or island locations in the tropics are particularly susceptible to torrential rainfall associated with tropical cyclones, and the rainfall rate associated with tropical cyclones is projected to increase though the number of tropical cyclones is projected to decrease or stay essentially unchanged (see Section 3.4.4). The impact of future large explosive volcanic eruptions may also be exacerbated by an increase in extreme precipitation events (see Section 3.3.2) that provide an effective means of transferring large volumes of unconsolidated ash and pyroclastic flow debris from the flanks of volcanoes into downstream areas.

Quantification of possible trends in the frequency of landslides and ice avalanches in mountains is difficult due to incomplete documentation of past events. There is *high confidence* that changes in heat waves, glacial retreat, and/or permafrost degradation will affect high mountain phenomena such as slope instabilities, mass movements, and glacial lake outburst floods, and *medium confidence* that temperature-related changes will influence bedrock stability. There is also *high confidence* that changes in heavy precipitation will affect landslides in some regions. There is *medium confidence* that high-mountain debris flows will begin earlier in the year because of earlier snowmelt, and that continued mountain permafrost degradation and glacier retreat will further decrease the stability of rock slopes. There is *low confidence* regarding future locations and timing of large rock avalanches, as these depend on local geological conditions and other non-climatic factors. There is *low confidence* in projections of an anthropogenic effect on phenomena such as shallow landslides in temperate and tropical regions, because these are strongly influenced by human activities such as poor land use practices, deforestation, and overgrazing. It is well established that ice mass wastage following the end of the last glaciations led to increased levels of seismicity, but there is *low confidence* in the nature of recent and projected future seismic responses to anthropogenic climate change.

3.5.7. High-latitude Changes Including Permafrost

Permafrost is widespread in Arctic, in subarctic, in ice-free areas of Antarctica, and in high-mountain regions, and permafrost regions occupy approximately 23 million km^2 of land area in the Northern Hemisphere (Zhang et al., 1999). Melting of massive ground ice and thawing of ice-rich permafrost can lead to subsidence of the ground surface and to the formation of uneven topography known as thermokarst, having implications for ecosystems, landscape stability, and infrastructure performance (Walsh, 2005). See also Case Study 9.2.10 for discussion of the impacts of cold events in high latitudes. The active layer (near-surface layer that thaws and freezes seasonally over permafrost) plays an important role in cold regions because most ecological, hydrological, biogeochemical, and pedogenic (soil-forming) activity takes place within it (Hinzman et al., 2005).

Observations show that permafrost temperatures have increased since the 1980s (IPCC, 2007b). Temperatures in the colder permafrost of northern Alaska, the Canadian Arctic, and Russia have increased up to 3°C near the permafrost table and up to 1 to 2°C at depths of 10 to 20 m (Osterkamp, 2007; Romanovsky et al., 2010; S.L. Smith et al., 2010) since the late 1970s/early 1980s. Temperature increases have generally been less than 1°C in the warmer permafrost of the discontinuous permafrost zone of the polar regions (Osterkamp, 2007; Romanovsky et al., 2010; S.L. Smith et al., 2010), and also in the high-altitude permafrost of Mongolia and the Tibetan Plateau (Zhao et al., 2010). When the other conditions remain constant, active layer thickness is expected to increase in response to warming. Active layer thickness has increased by about 20 cm in the Russian Arctic between the early 1960s and 2000 (T. Zhang et al., 2005) and by up to 1.0 m over the Qinghai-Tibetan Plateau since the early 1980s (Wu and Zhang, 2010), with no significant trend in the North American Arctic since the early 1990s (Shiklomanov et al., 2010). However, over extreme warm summers, active layer thickness may increase substantially (Smith et al., 2009), potentially triggering active-layer detachment failures on slopes (Lewkowicz and Harris, 2005). Extensive thermokarst development has been found in Alaska (Jorgenson z2007), and central Yakutia (Gavriliev and Efremov, 2003). Increased rates of retrogressive thaw slump activities have been reported on slopes over the Qinghai-Tibetan Plateau (Niu et al., 2005) and adjacent to tundra lakes over the Mackenzie Delta region of Canada (Lantz and Kokelj, 2008). Substantial expansion and deepening of thermokarst lakes was observed near Yakutsk with subsidence rates of 17 to 24 cm yr^{-1} from 1992 to 2001 (Fedorov and Konstantinov, 2003). Satellite remote sensing data show that thaw lake surface area has increased in continuous permafrost regions and decreased in discontinuous permafrost regions (Smith et al., 2005). Coasts with ice-bearing permafrost that are exposed to the Arctic Ocean are very sensitive to permafrost degradation. Some Arctic coasts are retreating at a rapid rate of 2 to 3 m yr^{-1} and the rate of erosion along Alaska's northeastern coastline has doubled over the past 50 years, related to declining sea ice extent, increasing sea surface temperature, rising sea level, thawing coastal permafrost, and possibly increases in storminess and waves (Jones et al., 2009; Karl et al., 2009)

Increases in air temperature are in part responsible for the observed increase in permafrost temperature over the Arctic and subarctic, but changes in snow cover also play a critical role (Osterkamp, 2005; Zhang, 2005; T. Zhang et al., 2005; S.L. Smith et al., 2010). Trends toward earlier snowfall in autumn and thicker snow cover during winter have resulted in a stronger snow insulation effect, and as a result a much warmer permafrost temperature than air temperature in the Arctic. On the other hand, permafrost temperature may decrease even if air temperature increases, if there is also a decrease in the duration and thickness of snow cover (Taylor et al., 2006). The lengthening of the thaw season and increases in summer air temperature have resulted in changes in active layer thickness. Model simulations have projected thickening of the active layer, a northward shift of the permafrost boundary, reductions in permafrost area, and an increase in permafrost temperature in the 21st century and beyond (Saito et al., 2007; Schaefer et al., 2011). The projected permafrost degradation may result in ancient carbon currently frozen in permafrost being released into the atmosphere, providing a positive feedback to the climate system (Schaefer et al., 2011). Expansion of lakes in the continuous permafrost zone may be due to thawing of ice-rich permafrost and melting of massive ground ice, while decreases in lake area in the discontinuous permafrost zone may be due to lake bottom drainage (Smith et al., 2005). Overall, increased air temperature over high latitudes is primarily responsible for the development of thermokarst terrains and thaw lakes.

In summary, it is *likely* that there has been warming of permafrost in recent decades. There is *high confidence* that permafrost temperatures will continue to increase, and that there will be increases in active layer thickness and reductions in the area of permafrost in the Arctic and subarctic.

3.5.8. Sand and Dust Storms

Sand and dust storms are widespread natural phenomena in many parts of the world. Heavy dust storms disrupt human activities. Dust aerosols in the atmosphere can cause a suite of health impacts including respiratory problems (Small et al., 2001). The long-range transport of dust can affect conditions at long distances from the dust sources, linking the biogeochemical cycles of land, atmosphere, and ocean (Martin and Gordon, 1988; Bergametti and Dulac, 1998; Kellogg and Griffin, 2006). For example, dust from the Saharan region and from Asia may reach North America and South America (McKendry et al., 2007). Some climate models have representations of dust aerosols (Textor et al., 2006). Climate variables that are most important to dust emission and transport such as soil moisture (see also Section 3.5.1), precipitation, wind, and vegetation cover are still subject to large uncertainties in climate model simulations. As a result, the sand and dust storm simulations have large uncertainties as well.

The Sahara (especially the Bodélé Depression in Chad) and east Asia have been recognized as the largest dust sources globally (Goudie, 2009). Over the few decades before the 1990s, the frequency of dust events increased in some regions such as the Sahel zone of Africa (Goudie and Middleton, 1992), and decreased in some other regions such as China (Zhang et al., 2003). There seems to be an increase in more recent years in China (Shao and Dong, 2006). Despite the importance of African dust, studies on long-term change in Sahel dust are limited. However, dust transported far away from the source region may provide some evidence of long-term changes in the Sahel region. The African dust transported to Barbados began to increase in the late 1960s and through the 1970s; transported dust reached a peak in the early 1980s but remains high into the present (Prospero and Lamb, 2003; Prospero et al., 2009).

Surface soil dust concentration during a sand and dust storm is controlled by a number of factors. The driving force for the production of dust storms is the surface wind associated with cold frontal systems sweeping across arid and semi-arid regions and lifting soil particles in the atmosphere. Dust emissions are also controlled by the surface conditions in source regions such as the desert coverage distributions, snow cover, and soil moisture. For example, in the Sahel region, the elevated high level of dust emission prior to the 1990s was related to the persistent drought during that time, and to long-term changes in the NAO (Ginoux et al., 2004; Chiapello et al., 2005; Engelstaedter et al., 2006), and perhaps to North Atlantic SST as well (Wong et al., 2008). Further evidence of the importance of climate on dust emission is that despite an increase of approximately 2 to 7% in desert areas in China over the four decades since 1960, dust storm frequency decreased in that period (Zhong, 1999). Studies on Asian soil dust production from 1960 to 2003 suggest that climatic variations have played a major role in the declining trends in dust emission and storm frequencies in China (Zhang et al., 2003; Zhou and Zhang, 2003; Zhao et al., 2004; Gong et al., 2006). Overall, changes in dust activity are affected by changes in the climate, such as wind and moisture conditions in the dust source regions. Changes in large-scale circulation play an additional role in the long-distance transport of dust. However, understanding of the physical mechanisms of the long-term trends in dust activity is not complete; for example, the relative importance of the various factors affecting dust frequency as outlined above is uncertain.

Future dust activity depends on two main factors: land use in the dust source regions, and climate both in the dust source region and large-scale circulation that affects long distance dust transport. Studies on projected future dust activity are very limited. It is difficult to project future land use. Precipitation, soil moisture, and runoff have been projected to decrease in major dust source regions (Figure 10.12 in Meehl et al., 2007b). Thomas et al. (2005) suggest that dune fields in southern Africa can become active again, and sand will become significantly exposed and move, as a consequence of 21st-century warming. A study based on simulations from two climate models also suggests increased desertification in arid and semi-arid China, especially in the second half of the 21st century (X.M. Wang et al., 2009). However, confident projected changes in wind are lacking (see Section 3.3.3).

In summary, there is *low confidence* in projecting future dust storm changes, although an increase could be expected where aridity increases. There is a lack of data and studies on past changes. There is also a lack of understanding of processes such as the relative importance of different climate variables affecting dust storms, as well as a high uncertainty in simulating important climate variables such as soil moisture, precipitation, and wind that affect dust storms.

Table 3-2 | Regional observed changes in temperature and precipitation extremes, including dryness, since 1950 unless indicated otherwise, and using late 20th-century values as reference (see Box 3-1), generally 1961-1990. See Figure 3-1 for definitions of regions. For assessments for small island states refer to Box 3-4.

A. North America and Central America

Regions	Tmax [WD = Warm Days CD = Cold Days; see Box 3-1] (using late 20th-century extreme values as reference, e.g., 90th/10th percentile)	Tmin [WN = Warm Nights CN = Cold Nights; see Box 3-1] (using late 20th-century extreme values as reference, e.g., 90th/10th percentile)	Heat Waves (HW)/ Warm Spells (WS) [WSDI = Warm Spell Duration Index, i.e., number or fraction of days belonging to spells of at least 6 days with Tmax >90th percentile] (using late 20th-century extreme values as reference)	Heavy Precipitation (HP) (using late 20th-century extreme values as reference, e.g., 90th percentile)	Dryness [CDD = Consecutive Dry Days SMA = (Simulated) Soil Moisture Anomalies PDSI = Palmer Drought Severity Index; see Box 3-3 for definitions]
All North America and Central America	*High confidence*: Likely overall increase in WD, decrease in CD (Aguilar et al., 2005; Alexander et al., 2006).	*High confidence*: Likely overall decrease in CN, increase in WN (Aguilar et al., 2005; Alexander et al., 2006).	*Medium confidence*: Overall increase since 1960 (Kunkel et al., 2008). Some areas with significant WSDI increase, others with insignificant WSDI increase or decrease (Alexander et al., 2006).	*High confidence*: Likely increase in many areas since 1950 (Aguilar et al., 2005; Alexander et al., 2006; Trenberth et al., 2007; Kunkel et al., 2008).	*Medium confidence*: Overall slight decrease in dryness (SMA, PDSI, CDD) since 1950; regional variability and 1930s drought dominate the signal (Aguilar et al., 2005; Alexander et al., 2006; Kunkel et al., 2008; Sheffield and Wood, 2008a; Dai, 2011).
W. North America (WNA, 3)	*High confidence*: Very likely large increases in WD, large decreases in CD (Robeson, 2004; Vincent and Mekis, 2006; Kunkel et al., 2008; Peterson et al., 2008a).	*High confidence*: Very likely large decreases in CN, large increases in WN (Robeson, 2004; Vincent and Mekis, 2006; Kunkel et al., 2008; Peterson et al., 2008a).	*Medium confidence*: Increase in WSDI (Alexander et al., 2006).	*Medium confidence*: Spatially varying trends. General increase, decrease in some areas (Alexander et al., 2006).	*Medium confidence*: No overall or slight decrease in dryness (SMA, PDSI, CDD) since 1950; large variability, large drought of 1930s dominates (Alexander et al., 2006; Kunkel et al., 2008; Sheffield and Wood, 2008a; Dai, 2011).
Central North America (CNA, 4)	*Medium confidence*: Spatially varying trends. Small increases in WD, decreases in CD in north CNA. Small decreases in WD, increases in CD in south CNA (Robeson, 2004; Vincent and Mekis, 2006; Kunkel et al., 2008; Peterson et al., 2008a).	*Medium confidence*: Spatially varying trends. Small decreases in CN, increases in WN in north CNA. Small increases in CN, decreases in WN in south CNA (Robeson, 2004; Vincent and Mekis, 2006; Kunkel et al., 2008; Peterson et al., 2008a).	*Medium confidence*: Spatially varying trends. Some areas with WSDI increase, others with WSDI decrease (Alexander et al., 2006).	*High confidence*: Very likely increase since 1950 (Alexander et al., 2006).	*Medium confidence*: Decrease in dryness (SMA, PDSI, CDD) and increase in mean precipitation since 1950; large variability, large drought of 1930s dominates (Alexander et al., 2006; Kunkel et al., 2008; Sheffield and Wood, 2008a; Dai, 2011).
E. North America (ENA, 5)	*Medium confidence*: Spatially varying trends. Overall increases in WD, decreases in CD; opposite or insignificant signal in a few areas (Robeson, 2004; Vincent and Mekis, 2006; Kunkel et al., 2008; Peterson et al., 2008a).	*Medium confidence*: Weak and spatially varying trends. (Robeson, 2004; Vincent and Mekis, 2006; Kunkel et al., 2008; Peterson et al., 2008a).	*Medium confidence*: Spatially varying trends. Many areas with WSDI increase, some areas with WSDI decrease (Alexander et al., 2006).	*High confidence*: Very likely increase since 1950 (Alexander et al., 2006).	*Medium confidence*: Slight decrease in dryness (SMA, PDSI, CDD) since 1950, large variability, large drought of 1930s dominates (Alexander et al., 2006; Kunkel et al., 2008; Sheffield and Wood, 2008a; Dai, 2011).
Alaska/ N.W. Canada (ALA, 1)	*High confidence*: Very likely large increases in WD, large decreases in CD (Robeson, 2004; Vincent and Mekis, 2006; Kunkel et al., 2008; Peterson et al., 2008a).	*High confidence*: Very likely large decreases in CN, large increases in WN (Robeson, 2004; Vincent and Mekis, 2006; Kunkel et al., 2008; Peterson et al., 2008a).	*Low confidence*: Insufficient evidence.	*Medium confidence*: Slight tendency for increase, in southern Alaska; no significant trend (Kunkel et al., 2008).	*Medium confidence*: Inconsistent trends; increase in dryness (SMA, PDSI, CDD) since 1950 in part of the region. (Alexander et al., 2006; Kunkel et al., 2008; Sheffield and Wood, 2008a; Dai, 2011).
E. Canada, Greenland, Iceland (CGI, 2)	*High confidence*: Likely increases in WD in some areas, decrease in others. Decreases in CD in some areas, increase in others (Robeson, 2004; Alexander et al., 2006; Vincent and Mekis, 2006; Trenberth et al., 2007; Kunkel et al., 2008; Peterson et al., 2008a).	*Medium confidence*: Small increases in unusually cold nights, decreases in WN in northeastern Canada. Small decreases in CN, increases in WN in southeastern and south central Canada. (Robeson, 2004; Vincent and Mekis, 2006; Kunkel et al., 2008; Peterson et al., 2008a).	*Medium confidence*: Some areas with WSDI increase, most others with WSDI decrease (Alexander et al., 2006).	*Medium confidence*: Increase in a few areas (Alexander et al., 2006).	*Low confidence*: Insufficient evidence.
Central America and Mexico (CAM, 6)	*Medium confidence*: Increases in WD, decreases in CD (Aguilar et al., 2005; Alexander et al., 2006).	*Medium confidence*: Decreases in CN, increases in WN (Aguilar et al., 2005; Alexander et al., 2006).	*Low confidence*: Spatially varying trends. A few areas increase, a few others decrease (Aguilar et al., 2005; Alexander et al., 2006).	*Medium confidence*: Spatially varying trends. Increase in many areas, decrease in a few areas, (Aguilar et al., 2005; Alexander et al., 2006).	*Low confidence*: Spatially varying trends, inconsistencies in trends in dryness (SMA, PDSI, CDD), (Aguilar et al., 2005; Sheffield and Wood, 2008a; Dai, 2011).

Continued next page →

Table 3-2 (continued)

B. Europe and Mediterranean Region

Regions	Tmax	Tmin	Heat Waves / Warm Spells	Heavy Precipitation	Dryness
All Europe and Mediterranean Region	*High confidence*: Overall *likely* increase in WD and *likely* decrease of CD over most of the continent since 1950. Strong increasing tendency in WD in most regions since 1976 onward; small or insignificant decrease in CD over same period (Alexander et al., 2006; see also entries for individual subregions).	*High confidence*: Overall *likely* increase in WN and *likely* decrease in CN over most of the continent since 1950. Strong increasing tendency in WN in most regions since 1976 onward; small or insignificant trends in CN over same period (Klein Tank and Können, 2003; Alexander et al., 2006; see also entries for individual subregions).	*Medium confidence*: Increase of HW since 1950. Overall consistent positive trend of WSDI across Europe, but no coherent region with significant trends (Alexander et al., 2006). Availability of a few single studies for specific regions (see below).	*Medium confidence*: Increase in part of the region, mostly in winter, insignificant or inconsistent changes elsewhere, in particular in summer. Some inconsistencies in overall patterns between studies depending on considered indices. Most consistent signal over central W. Europe and European Russia (Klein Tank and Können, 2003; Haylock and Goodess, 2004; Alexander et al., 2006; Zolina et al., 2009).	*Medium confidence*: Inconsistent trends; Increase in dryness (SMA, PDSI, CDD) in part of the region; insignificant, inconsistent, or no changes elsewhere. Most consistent signal for increase in dryness in central and S. Europe since the 1950s. No signal in N. Europe (Kiktev et al., 2003; Haylock and Goodess, 2004; Alexander et al., 2006; Sheffield and Wood, 2008a; Dai, 2011).
N. Europe (NEU, 11)	*Medium confidence*: Increase in WD and decrease in CD. Consistent signals for whole region, but generally not significant at the local scale (Alexander et al., 2006).	*Medium confidence*: Increase in WN and decrease in CN. Consistent signals over whole region but generally not significant at the local scale (Alexander et al., 2006).	*Medium confidence*: Increase in HW. Consistent tendency for increase in WSDI, but no significant trends (Alexander et al., 2006).	*Medium confidence*: Increase in winter in some areas, but often insignificant or inconsistent trends at subregional scale, in particular in summer (Fowler and Kilsby, 2003; Kiktev et al., 2003; Klein Tank and Können, 2003; Alexander et al., 2006; Maraun et al., 2008; Zolina et al., 2009).	*Medium confidence*: Spatially varying trends. Overall only slight or no increase in dryness (SMA, PDSI, CDD), slight decrease in dryness in part of the region (Kiktev et al., 2003; Alexander et al., 2006; Sheffield and Wood, 2008a; Dai, 2011).
Central Europe (CEU, 12)	*High confidence*: *Likely* overall increase in WD and *likely* decrease in CD since 1950 in most regions. Some regional and temporal variations in significance of trends. *High confidence*: *Very likely* increase in WD since 1950, 1901 and 1880 and *likely* decrease in CD since 1950 and 1901 in west Central Europe (Alexander et al., 2006; Della-Marta et al., 2007a; Laurent and Parey, 2007). *Medium confidence*: Lower confidence in trends in east Central Europe due to lack of literature, partial lack of access to observations, overall weaker signals, and change point in trends at the end of the 1970s / beginning of 1980s. Strongest increase in WD since 1976 (Alexander et al., 2006; Barthoy and Pongracz, 2007; Hirschi et al., 2011).	*High confidence*: *Likely* overall increase in WN and *likely* overall decrease in CN at the yearly time scale. Some regional and seasonal variations in significance and in a few cases also the sign of the trends. *High confidence*: *Very likely* increase in WN and *very likely* decrease in CN since 1950 and 1901 in west Central Europe (Kiktev et al., 2003; Alexander et al., 2006). *Medium confidence*: Lower confidence in trends in east Central Europe due to lack of literature, partial lack of access to observations, overall weaker signals, and change point in trends at the end of the 1970s / beginning of 1980s. (Klein Tank and Können, 2003; Alexander et al., 2006; Bartholy and Pongracz, 2007).	*Medium confidence*: Increase in heat waves. Consistent tendency for WSDI increase but no significant trends (Alexander et al., 2006). Significant increase in max HW duration since 1880 in west Central Europe in summer (JJA) (Della-Marta et al., 2007a). Less significant signal in heat wave indices in east Central Europe due to presence of change point (Bartholy and Pongracz, 2007; Hirschi et al., 2011).	*Medium confidence*: Increase in part of the domain, in particular in central W. Europe and European Russia, especially in winter. Insignificant or inconsistent trends elsewhere, in particular in summer (Kiktev et al., 2003; Klein Tank and Können, 2003; Schmidli and Frei, 2005; Alexander et al., 2006; Bartholy and Pongracz, 2007; Kyselý, 2009; Tomassini and Jacob, 2009; Zolina et al., 2009).	*Medium confidence*: Spatially varying trends. Increase in dryness (SMA, PDSI, CDD) in part of the region but some regional variation in dryness trends and dependence of trends on considered studies (index, time period) (Kiktev et al., 2003; Alexander et al., 2006; Bartholy and Pongracz, 2007; Sheffield and Wood, 2008a; Brázdil et al., 2009; Dai, 2011).
S. Europe and Mediterranean (MED, 13)	*High confidence*: *Likely* increase in WD and *likely* decrease in CD in most of the region. Some regional and temporal variations in significance of trends. *Likely* strongest and most significant trends in the Iberian Peninsula and southern France (Alexander et al., 2006; Brunet et al., 2007; Della-Marta et al., 2007a; Bartolini et al., 2008; Kuglitsch et al., 2010; Rodriguez-Puebla et al., 2010; Hirschi et al., 2011). *Medium confidence*: Smaller or less significant trends in S.E. Europe and Italy due to change point in trends at the end of the 1970s / beginning of 1980s; sometimes linked with changes in sign of trends; strongest WD increase since 1976 (Bartholy and Pongracz, 2007; Bartolini et al., 2008; Toreti and Desiato, 2008; Kuglitsch et al., 2010; Hirschi et al., 2011).	*High confidence*: *Likely* increase in WN and *likely* decrease in CN in most of the region. Some regional variations in significance of trends. *Very likely* overall increase in WN and *very likely* overall decrease in CN in S.W. Europe and W. Mediterranean; *likely* strongest signals in Spain and southern France (Kiktev et al., 2003; Klein Tank and Können, 2003; Alexander et al., 2006; Brunet et al., 2007; Rodriguez-Puebla et al., 2010). *Likely* overall tendency for increase in WN and *likely* overall tendency for decrease in CN in S.E. Europe and E. Mediterranean (Kiktev et al., 2003; Klein Tank and Können, 2003; Alexander et al., 2006).	*High confidence*: *Likely* overall increase in HW in summer (JJA). Significant increase in max HW duration since 1880 in Iberian Peninsula and west Central Europe in JJA (Della-Marta et al., 2007a). Significant increase in max HW duration in Tuscany (Italy) (Bartolini et al., 2008). Significant increase in HW indices in Turkey and to a smaller extent in S.E. Europe and Turkey in JJA (Kuglitsch et al., 2010). Less significant signal in HW indices in S.E. Europe due to presence of change point in trends (Bartholy and Pongracz, 2007; Hirschi et al., 2011).	*Low confidence*: Inconsistent trends within domain and across studies (Kiktev et al., 2003; Klein Tank and Können, 2003; Alexander et al., 2006; Garcia et al., 2007; Pavan et al., 2008; Zolina et al., 2009; Rodrigo, 2010).	*Medium confidence*: Overall increase in dryness (SMA, PDSI, CDD), but partial dependence on index and time period (Kiktev et al., 2003; Alexander et al., 2006; Sheffield and Wood, 2008a; Dai, 2011).

Continued next page →

Table 3-2 (continued)

C. Africa

Regions	Tmax	Tmin	Heat Waves / Warm Spells	Heavy Precipitation	Dryness
All Africa	*Low confidence to medium confidence*: Low confidence due to insufficient evidence (lack of literature) in many regions. *Medium confidence* in increase in frequency of WD and decrease in frequency of CD in southern part of continent (Alexander et al., 2006). See also regional assessments.	*Low confidence to medium confidence* depending on region: Low confidence due to insufficient evidence (lack of literature) in many regions. *Medium confidence* in increase in frequency of WN in northern and southern part of continent (Alexander et al., 2006). *Medium confidence* in decrease in frequency of CN in southern part of continent (Alexander et al., 2006). See also regional assessments.	*Low confidence*: Insufficient evidence (lack of literature). Some analyses for localized regions (see regional assessments).	*Low confidence*: Partial lack of data and literature and inconsistent patterns in existing studies (New et al., 2006; Aguilar et al., 2009). See also regional assessments.	*Medium confidence*: Overall increase in dryness (SMA, PDSI); regional variability, 1970s prolonged Sahel drought dominates (Sheffield and Wood, 2008a; Dai, 2011). No apparent continent-wide trends in change in rainfall over the 20th century, although there was a continent-wide drought in 1983 and 1984 (Hulme et al., 2001). Wet season arrives 9–21 days later, large inter-annual variability of wet season start, local-scale geographical variability (Kniveton et al., 2009).
W. Africa (WAF, 15)	*Medium confidence*: Significant increase in temperature of warmest day and coldest day, significant increase in frequency of WD, and significant decrease in frequency of CD in western central Africa, Guinea Conakry, Nigeria, and Gambia (New et al., 2006; Aguilar et al., 2009). *Low confidence*: Lack of literature in other parts of the region.	*Medium confidence*: Decreases in frequency of CN in western central Africa, Nigeria, and Gambia; insignificant decreases in frequency of CN in Guinea Conakry (New et al., 2006; Aguilar et al., 2009). *Low confidence*: Lack of literature on changes in CN in other parts of the region. *Medium confidence*: Increases in frequency of WN (Alexander et al., 2006; New et al., 2006; Aguilar et al., 2009).	*Low confidence*: Insufficient evidence (lack of literature) for most of the region; increases in WSDI in Nigeria and Gambia (New et al., 2006).	*Medium confidence*: Precipitation from heavy events has decreased (western central Africa, Guinea Conakry) but low spatial coherence (Aguilar et al., 2009), rainfall intensity increased (New et al., 2006).	*Medium confidence*: 1970s prolonged Sahel drought dominates, conditions are still drier (SMA, PDSI, precipitation anomalies) than during the humid 1950s (L'Hôte et al., 2002; Dai et al., 2004; Sheffield and Wood, 2008a; Dai, 2011). Dry spell duration (CDD) overall increased from 1961 to 2000 (New et al., 2006). Recent years characterized by a greater interannual variability than previous 40 years, western Sahel remaining dry and the eastern Sahel returning to wetter conditions (Ali and Lebel, 2009).
E. Africa (EAF, 16)	*Low confidence*: Lack of evidence due to lack of literature and spatially non-uniform trends. Over time period 1939–1992 spatially non-uniform trends in daytime temperature, some areas with cooling (King'uyu et al., 2000). In southern tip of domain increases in WD, decreases in CD (Alexander et al., 2006).	*Medium confidence*: Over time period 1939–1992, spatially non-uniform trends, rise of nighttime temperature at several locations, but with many coastal areas and stations near large water bodies showing a significant decrease (King'uyu et al., 2000). In southern tip of domain, decreases in CN, increases in WN (Alexander et al., 2006).	*Low confidence*: Insufficient evidence (lack of literature) for most of region; increase in WSDI in southern tip of domain (New et al., 2006).	*Low confidence*: Insufficient evidence (lack of literature) to assess trends.	*Low confidence*: Spatially varying trends in dryness (SMA, PDSI) (Sheffield and Wood, 2008a; Dai, 2011).
S. Africa (SAF, 17)	*Medium confidence*: Increases in WD, decreases in CD (Alexander et al., 2006).	*Medium confidence*: Decreases in CN, increases in WN (King'uyu et al., 2000; Alexander et al., 2006).	*Medium confidence*: Increase in WSDI (New et al., 2006).	*Low confidence*: No spatially coherent patterns of trends in precipitation extremes (Kruger, 2006; New et al., 2006; Trenberth et al., 2007).	*Medium confidence*: Slight dry spell duration increase (Alexander et al., 2006; Kruger, 2006; New et al., 2006). General increase in dryness (SMA, PDSI) (Sheffield and Wood, 2008a; Dai, 2011).
Sahara (SAH, 14)	*Low confidence*: Lack of literature.	*Medium confidence*: Increases in WN (Alexander et al., 2006). *Low confidence*: Lack of literature on trends in CN.	*Low confidence*: Insufficient evidence (lack of literature).	*Low confidence*: Insufficient evidence.	*Low confidence*: Limited data, spatial variation in the trends (Dai, 2011).

Continued next page →

Changes in Climate Extremes and their Impacts on the Natural Physical Environment — Chapter 3

Table 3-2 (continued)

D. South America

Regions	Tmax	Tmin	Heat Waves / Warm Spells	Heavy Precipitation	Dryness
All South America	*Low confidence* to *medium confidence* depending on region: *Low confidence* in trends in WD and CD due to insufficient evidence in many regions, in particular in northern half of continent. *Medium confidence* in trends in southern half of continent but often spatially varying trends. See regional assessments for details and basis for continental assessment.	*Low confidence* to *medium confidence* depending on region: *Low confidence* in trends in WN and CN due to insufficient evidence in many regions, in particular in northern half of continent. *Medium confidence* in decrease in CN and increase in WN in southern half of continent. See regional assessments for details and basis for continental assessment.	*Low confidence*: Insufficient evidence (most of continent) or lack of coherent signal (Southeast South America). See regional assessments for details and basis for continental assessment.	*Low confidence*: Insufficient evidence or spatially varying trends. See regional assessments for details and basis for continental assessment.	*Low confidence*: Spatially varying trends, inconsistencies between studies (Sheffield and Wood, 2008a; Dai, 2011).
Amazon (AMZ, 7)	*Low confidence*: Insufficient, scattered, evidence (Alexander et al., 2006; Dufek et al., 2008).	*Low confidence*: Insufficient, scattered, evidence (Alexander et al., 2006; Dufek et al., 2008).	*Low confidence*: Insufficient evidence.	*Medium confidence*: Spatially varying trends. Increase in many areas, decrease in a few areas (Alexander et al., 2006; Haylock et al., 2006b).	*Low confidence*: Spatially varying trends, mixed results. Slight decrease in CDD (Dufek et al., 2008). Tendency to decreased dryness in much of region, but some opposite trends and inconsistencies between studies (Sheffield and Wood, 2008a; Dai, 2011).
N.E. Brazil (NEB, 8)	*Medium confidence*: Increases in WD (Silva and Azevedo, 2008).	*Medium confidence*: Increases in WN (Silva and Azevedo, 2008).	*Low confidence*: Insufficient evidence.	*Medium confidence*: Increase in many areas, decrease in a few areas, (Alexander et al., 2006; Haylock et al., 2006b; Santos and Brito, 2007; Silva and Azevedo, 2008; Santos et al., 2009).	*Low confidence*: Spatially varying trends. Inconsistent trends in CDD (Santos and Brito, 2007; Dufek et al., 2008; Silva and Azevedo, 2008; Santos et al., 2009). Inconsistent trends in dryness (SMA, PDSI) between studies (Sheffield and Wood, 2008a; Dai, 2011).
S.E. South America (SSA, 10)	*Medium confidence*: Spatially varying trends. Increases in WD in some areas, decrease in others. Decreases in CD in some areas, increase in others (Rusticucci and Barrucand, 2004; Vincent et al., 2005; Alexander et al., 2006; Rusticucci and Renom, 2008; Marengo et al., 2009b).	*Medium confidence*: Decreases in CN, increases in WN (Rusticucci and Barrucand, 2004; Vincent et al., 2005; Alexander et al., 2006; Rusticucci and Renom, 2008; Marengo et al., 2009b).	*Low confidence*: Spatially varying trends. Some areas increase, others decrease (Alexander et al., 2006).	*Low confidence* to *medium confidence*, depending on subregion: *Medium confidence*: Increase in northern portion of domain (Alexander et al., 2006; Dufek et al., 2008; Sugahara et al., 2009; Penalba and Robledo, 2010). *Low confidence* in southern portion of domain: Insufficient evidence.	*Low confidence*: Slight increase in dryness, large variability (Haylock et al., 2006b; Dufek and Ambrizzi, 2008; Dufek et al., 2008; Llano and Penalba, 2011). Decrease in dryness (SMA, PDSI) in much of region (Sheffield and Wood, 2008a; Dai, 2011).
W. Coast South America (WSA, 9)	*Medium confidence*: Increases in WD in some areas, decrease in others. Decreases in CD in some areas, increase in others (Rosenbluth et al., 1997; Vincent et al., 2005; Alexander et al., 2006).	*Medium confidence*: Decreases in CN, increases in WN (Rosenbluth et al., 1997; Vincent et al., 2005; Alexander et al., 2006).	*Low confidence*: Insufficient evidence.	*Medium confidence*: Spatially varying trends. Decrease in many areas, increase in a few areas (Alexander et al., 2006; Haylock et al., 2006b).	*Low confidence*: Overall inconsistent and spatially varying signal (SMA, PDSI, CDD) (Dufek et al., 2008; Sheffield and Wood, 2008a; Dai, 2011).

E. Asia

Regions	Tmax	Tmin	Heat Waves / Warm Spells	Heavy Precipitation	Dryness
All Asia	*Low confidence* to high confidence depending on region: On continental scale, *medium confidence* in overall increase in WD and decrease in CD (Alexander et al., 2006). See also individual regional entries.	*Low confidence* to high confidence depending on region: On continental scale, *medium confidence* in overall increase in WN and decrease in CN (Alexander et al., 2006). See also individual regional entries.	*Low confidence* to *medium confidence* depending on region: *Low confidence* due to insufficient evidence in several regions; *medium confidence* in trends in other regions (Alexander et al., 2006). See also individual regional entries.	*Low confidence* to *medium confidence* depending on region: *Low confidence* due to insufficient evidence or inconsistent trends in several regions; *medium confidence* in trends in HP in a few regions (Alexander et al., 2006). See also individual regional entries.	*Low confidence* to *medium confidence* depending on region: *Low confidence* in most regions due to spatially varying trends. Some areas have consistent increases, but others display decreases in dryness indicated by different measures (SMA, PDSI, CDD) (Alexander et al., 2006; Sheffield and Wood, 2008a; Dai, 2011).
N. Asia (NAS, 18)	High confidence: Likely increases in WD, likely decreases in CD (Alexander et al., 2006).	High confidence: Likely decreases in CN, likely increases in WN (Alexander et al., 2006).	*Medium confidence*: Spatially varying trends. Overall WSDI increase, WSDI decrease in a few areas (Alexander et al., 2006).	*Low confidence*: Increase in some regions, but spatial variations (Alexander et al., 2006). Some increase in western Russia, especially in winter (DJF), 1950–2000 (Zolina et al., 2009).	*Low confidence*: Spatially varying trends. Tendency for increased dryness (SMA, PDSI, CDD) in central and northeastern N. Asia, other areas decreased dryness (Alexander et al., 2006; Sheffield and Wood, 2008a; Dai, 2011).

Continued next page →

Table 3-2 (continued)

E. Asia (Continued)

Regions	Tmax	Tmin	Heat Waves / Warm Spells	Heavy Precipitation	Dryness
Central Asia (CAS, 20)	*High confidence: Likely* increases in WD, *likely* decreases in CD (Alexander et al., 2006).	*High confidence: Likely* decreases in CN, *likely* increases in WN (Alexander et al., 2006).	*Medium confidence:* WSDI increase in a few areas, insufficient evidence elsewhere (Alexander et al., 2006).	*Low confidence:* Spatially varying trends. Increase in a few areas, decrease in a few areas (Alexander et al., 2006).	*Low confidence:* Spatially varying trends in dryness (SMA, PDSI, CDD); partial lack of coverage in some studies (Alexander et al., 2006; Sheffield and Wood, 2008a; Dai, 2011).
E. Asia (EAS, 22)	*High confidence: Likely* increases in WD, *likely* decreases in CD (Alexander et al., 2006; Ding et al., 2010).	*Medium confidence:* Decreases in CN, increases in WN (Alexander et al., 2006).	*Medium confidence:* Increase in warm season heat waves in China (Ding et al., 2010); increase in WSDI in northern China, but decline in southern China (Alexander et al., 2006).	*Low confidence:* Spatially varying trends. Increase in a few areas, decrease in a few areas (Alexander et al., 2006).	*Medium confidence:* Overall tendency for increased dryness (SMA, PDSI, CDD); few areas with opposite trends (Alexander et al., 2006; Sheffield and Wood, 2008a; Dai, 2011).
S.E. Asia (SEA, 24)	*Medium confidence:* Increases in WD, decreases in CD in northern part of domain (Alexander et al., 2006). *Low confidence:* Insufficient evidence for Malay Archipelago.	*Medium confidence:* Decreases in CN, increases in WN in northern part of domain (Alexander et al., 2006). *Low confidence:* Insufficient evidence for Malay Archipelago.	*Low confidence:* Insufficient evidence.	*Low confidence:* Spatially varying trends and partial lack of evidence. Some areas increase, some areas decrease (Alexander et al., 2006).	*Low confidence:* Spatially varying trends, inconsistent trends in dryness (SMA, PDSI) between studies (Sheffield and Wood, 2008a; Dai, 2011).
S. Asia (SAS, 23)	*Medium confidence:* Increase in WD and decrease in CD (Alexander et al., 2006).	*Medium confidence:* Decreases in CN, increases in WN (Alexander et al., 2006).	*Low confidence:* Insufficient evidence.	*Low confidence:* Mixed signal in India (Alexander et al., 2006).	*Low confidence:* Inconsistent signal for different studies and indices. Decrease in CDD over India (Alexander et al., 2006). Increased dryness (SMA, PDSI) in central India (Sheffield and Wood, 2008a; Dai, 2011).
W. Asia (WAS, 19)	*High confidence: More likely than not* decrease in CD and *very likely* increase in WD (Rahimzadeh et al., 2009; Rehman, 2010).	*High confidence: Likely* decrease in CN and *likely* increase in WN (Rehman, 2010).	*Medium confidence:* WSDI increase (Alexander et al., 2006).	*Medium confidence:* Decrease in heavy precipitation events (Kwarteng et al., 2009; Rahimzadeh et al., 2009).	*Low confidence:* Lack of studies for part of the region; mixed results (Sheffield and Wood, 2008a; Rahimzadeh et al., 2009).
Tibetan Plateau (TIB, 21)	*High confidence: Likely* increase in WD and *likely* decrease in CD (Alexander et al., 2006).	*High confidence: Likely* decreases in CN, *likely* increases in WN (Alexander et al., 2006).	*Low confidence:* Spatially varying trends (Alexander et al., 2006).	*Low confidence:* Insufficient evidence.	*Low confidence:* Lack of studies. Tendency to decreased dryness (PDSI, SMA) in Dai (2011).

F. Australia/New Zealand

Regions	Tmax	Tmin	Heat Waves / Warm Spells	Heavy Precipitation	Dryness
All Australia and New Zealand	*High confidence:* Overall *likely* increases in WD, *likely* decreases in CD. See individual regional entries for assessment basis and details.	*High confidence:* Overall *likely* decreases in CN, *likely* increases in WN. See individual regional entries for assessment basis and details.	*Low confidence to medium confidence* depending on region: See individual regional entries for assessment basis and details.	*Low confidence to high confidence* depending on region: Insufficient studies for assessment in N. Australia, *likely* decrease in HP in many areas in S. Australia. See individual regional entries for assessment basis and details.	*Medium confidence:* Some regions with dryness decreases, others with dryness increases. See individual regional entries for assessment basis and details.
N. Australia (NAU, 25)	*High confidence: Likely* increases in WD, *likely* decreases in CD. Weaker trends in northwest (Alexander et al., 2006).	*High confidence: Likely* decreases in CN, *likely* increases in WN (Alexander et al., 2006; Alexander and Arblaster, 2009).	*Low confidence:* Insufficient literature for assessment.	*Low confidence:* Insufficient studies for assessment.	*Medium confidence:* Decrease in dryness (SMA, PDSI) in northwest since mid-20th century (Sheffield and Wood, 2008a; Dai, 2011).
S. Australia/ New Zealand (SAU, 26)	*High confidence: Very likely* increases in WD, *very likely* decreases in CD (Alexander et al., 2006). NZ positive trends vary across country, related to circulation changes (Chambers and Griffiths, 2008; Mullan et al., 2008).	*High confidence: Very likely* decreases in CN, *very likely* increases in WN (Alexander et al., 2006; Alexander and Arblaster, 2009). General decrease in frosts in NZ but trends vary across country, related to circulation changes (Chambers and Griffiths, 2008; Mullan et al., 2008).	*Medium confidence:* Increase in warm spells across southern Australia (Alexander and Arblaster, 2009).	*High confidence: Likely* decrease in heavy precipitation in many areas, especially where mean precipitation has decreased (CSIRO, 2007; Gallant et al., 2007; Alexander and Arblaster, 2009). NZ trends are positive in western N. and S. Islands and negative in east of country, and are strongly correlated with changes in mean rainfall (Mullan et al., 2008).	*Medium confidence:* Increase in dryness (SMA, PDSI, CDD) in southeastern part and southwestern tip of Australia since mid-20th century. Decrease in dryness in central part of Australia (Alexander et al., 2006; Sheffield and Wood, 2008a; Dai, 2011).

Changes in Climate Extremes and their Impacts on the Natural Physical Environment — Chapter 3

Table 3-3 | Projected regional changes in temperature and precipitation (including dryness) extremes. See Figure 3-1 for definitions of regions (numbers indicated next to regions' names). For assessments for small island states, refer to Box 3-4. Projections are for the end of the 21st century versus the end of the 20th century (e.g., 1961-1990 or 1980-2000 versus 2071-2100 or 2080-2100) and for the A2/A1B emissions scenarios (except if noted otherwise). Late 20th-century extreme values (generally either 1961-1990 or ~1980-2000) are used as reference (see Box 3-1 for discussion). Codes for the source of modelling evidence: **G**: multiple GCMs; **R**: single RCM forced by single GCM; *R*: multiple RCMs forced by single GCM; **R**: multiple RCMs forced by multiple GCMs. T06 stands for Tebaldi et al. (2006), SW08b stands for Sheffield and Wood (2008b), and OS11 stands for Orlowsky and Seneviratne (2011).

A. North America and Central America

Regions	Tmax [WD = Warm Days CD = Cold Days RV20AHD: 20-year return value of annual maximum hottest day] (using late 20th-century extreme values as reference; see Box 3-1)	Tmin [WN = Warm Nights CN = Cold Nights] (using late 20th-century extreme values as reference; see Box 3-1)	Heat Waves (HW)/Warm Spells (WS) (using late 20th-century extreme values as reference; see Box 3-1)	Heavy Precipitation (HP) [HPD = Heavy Precipitation Days, e.g., precipitation >95th percentile (p) %DP10 = Percentage of Days with Precipitation >10 mm HPC = Heavy Precipitation Contribution, generally fraction from precipitation >95th percentile RV20HP = 20-year return value of annual maximum daily precipitation rates] (using late 20th-century extreme values as reference; see Box 3-1)	Dryness [CDD = Consecutive Dry Days SMA = (Simulated) Soil Moisture Anomalies; see Box 3-3 for definitions]
All North America and Central America	*High confidence*: WD *very likely* or *likely* to increase and CD *very likely* or *likely* to decrease in all regions (Christensen et al., 2007; Meehl et al., 2007b; Karl et al., 2008; Fig. 3-3). *Very likely* increase in RV20AHD in all regions except CAM (Fig. 3-5). *Medium confidence*: Largest increases in WD in summer and fall particularly over the United States; largest decrease in CD in Canada in fall and winter (OS11). G	*High confidence*: WN *very likely* or *likely* to increase and CN *very likely* or *likely* to decrease depending on subregion (T06; Christensen et al., 2007; Kharin et al., 2007; Meehl et al., 2007b; Karl et al., 2008; Fig. 3-4). *Medium confidence*: Largest increase in WN and decrease in CN in summer, particularly in the United States (OS11). G	*High confidence to medium confidence* depending on subregion. *High confidence*: *Likely* more frequent, longer, and/or more intense heat waves and warm spells over all North American subregions. *Medium confidence* in increase in warm spells in part of Central America, but lack of agreement in signal of change in heat waves (T06; Christensen et al., 2007; Karl et al., 2008; Clark et al., 2010; OS11). G	*Low confidence to high confidence* depending on region and index: *Likely* increase in HP, including HPD, HPC and RV20HP, over Canada and Alaska; *Low confidence to medium confidence* in the south (particularly CAM) due to smaller and less consistent changes, and inconsistencies between indices (T06; decreases in winter and spring) and other indices (T06; Christensen et al., 2007; Kharin et al., 2007; Meehl et al., 2007b; Karl et al., 2008; OS11; Fig. 3-6). G	*Low confidence to medium confidence* depending on region. *Medium confidence* regarding increase in CDD, and SMA drought in Texas and N. Mexico (T06; SW08b; Fig. 3-10). *Low confidence*: Inconsistent change in other regions (SMA, CDD) (T06; SW08b; Fig. 3-10). G
W. North America (WNA, 3)	*High confidence*: WD *very likely* to increase and CD *very likely* to decrease in all seasons (Christensen et al., 2007; Karl et al., 2008; Clark et al., 2010; Fig. 3-3). *Very likely* increase in RV20AHD (Fig. 3-5). *Medium confidence*: Overall weaker signal in spring and winter for both CD and WD (OS11). RCM simulations for 2030–2039 consistent with projected long-term increase in WD (Diffenbaugh and Ashfaq, 2010). G *R*	*High confidence*: WN *very likely* to increase and CN *very likely* to decrease (T06; Christensen et al., 2007; Kharin et al., 2007; Meehl et al., 2007b; Karl et al., 2008; Fig. 3-4). *Medium confidence*: Largest WN increases and CN decreases in summer (OS11). G	*High confidence*: *Likely* more frequent, longer, and/or more intense heat waves and warm spells (T06; Christensen et al., 2007; Meehl et al., 2007b; Karl et al., 2008; Clark et al., 2010; OS11). *Medium confidence*: RCM simulations for 2030–2039 and 2090–2099 consistent with projected long-term increase in frequency and/or intensity of HW (Diffenbaugh and Ashfaq, 2010; Kunkel et al., 2010). G *R*	*Low confidence to medium confidence* depending on subregion and index: *Medium confidence* in increase in HPD/HPC over northern part of domain (Canada); *low confidence* due to no signal or inconsistent signal in HPD/HPC changes over southern part of domain (T06; Fig. 3-6). *Medium confidence* in increase in RV20HP (Fig. 3-7). G	*Low confidence*: Inconsistent signal in CDD and SMA changes (T06; SW08b; Fig. 3-10). G
Central North America (CNA, 4)	*High confidence*: WD *very likely* to increase and CD *very likely* to decrease in all seasons (Christensen et al., 2007; Karl et al., 2008; Clark et al., 2010; Fig. 3-3). *Very likely* increase in RV20AHD (Fig. 3-5). *Medium confidence*: Weaker signal for CD in spring and winter (OS11). RCM simulations for 2030–2039 consistent with projected long-term increase in WD (Diffenbaugh and Ashfaq, 2010). G *R*	*High confidence*: WN *very likely* to increase and CN *very likely* to decrease (T06; Christensen et al., 2007; Kharin et al., 2007; Meehl et al., 2007b; Karl et al., 2008; Fig. 3-4). G	*High confidence*: *Likely* more frequent, longer, and/or more intense heat waves and warm spells (T06; Christensen et al., 2007; Karl et al., 2008; Clark et al., 2010; OS11). *Medium confidence*: RCM simulations for 2030–2039 and 2090–2099 consistent with projected long-term increase in frequency and/or intensity of HW (Diffenbaugh and Ashfaq, 2010; Kunkel et al., 2010). G *R*	*Low confidence to medium confidence* depending on index: *Low confidence* in changes in HPD/HPC due to inconsistent or no signal (T06; Fig. 3-6). *Medium confidence* in increase in RV20HP (Fig. 3-7). G	*Medium confidence*: Increase in CDD and decrease in SMA in southern part of the domain (SW08b; Fig. 3-10). *Low confidence*: inconsistent signal elsewhere (Fig. 3-10). G

Continued next page →

Table 3-3 (continued)

A. North America and Central America (Continued)

Regions	Tmax	Tmin	Heat Waves / Warm Spells	Heavy Precipitation	Dryness
E. North America (ENA, 5)	*High confidence*: WD *very likely* to increase and CD *very likely* to decrease in all seasons (Christensen et al., 2007; Karl et al., 2008; Clark et al., 2010; Fig. 3-3). *Very likely* increase in RV20AHD (Fig. 3-5). *Medium confidence*: Largest WD increase in summer and fall; weaker CD decrease in spring (OS11). RCM simulations for 2030–2039 consistent with projected long-term increase in WD (Diffenbaugh and Ashfaq, 2010).	*High confidence*: WN *very likely* to increase and CN *very likely* to decrease (T06; Christensen et al., 2007; Kharin et al., 2007; Meehl et al., 2007b; Karl et al., 2008; Fig. 3-4). *Medium confidence*: Largest WN increases and CN decreases in summer (OS11).	*High confidence*: *Likely* more frequent, longer, and/or more intense heat waves and warm spells (T06; Christensen et al., 2007; Meehl et al., 2007b; Karl et al., 2008; Clark et al., 2010; OS11) *Medium confidence*: RCM simulations for 2030–2039 and 2090–2099 consistent with projected long-term increase in frequency of HW (Diffenbaugh and Ashfaq, 2010; Kunkel et al., 2010).	*Medium confidence*: Increase in HPD/HPC in northern part of domain but no signal or inconsistent signal in southern part (T06; Fig. 3-6). *Medium confidence* in increase in RV20HP (Fig. 3-7).	*Low confidence*: Inconsistent signal in CDD, some consistent decrease in SMA (SW08b; Fig. 3-10).
	G R	G	G R	G	G
Alaska/ N.W. Canada (ALA, 1)	*High confidence*: WD *very likely* to increase and CD *very likely* to decrease (Christensen et al., 2007; Karl et al., 2008; Fig. 3-3). *Very likely* increase in RV20AHD (Fig. 3-5). *Medium confidence*: Strongest increase in WD in the fall (OS11).	*High confidence*: WN *very likely* to increase and CN *very likely* to decrease (T06; Christensen et al., 2007; Kharin et al., 2007; Meehl et al., 2007b; Karl et al., 2008; Fig. 3-4).	*High confidence*: *Likely* more frequent and/or longer heat waves and warm spells (T06; Christensen et al., 2007; Meehl et al., 2007b; Karl et al., 2008; OS11).	*High confidence*: *Likely* increase in HPD and HPC (T06; Fig. 3-6). *Likely* increase of RV20HP (Fig. 3-7).	*Low confidence*: Inconsistent signal in change of CDD and SMA (T06; SW08b; Fig. 3-10).
	G	G	G	G	G
E. Canada, Greenland, Iceland (CGI, 2)	*High confidence*: WD *very likely* to increase and CD *very likely* to decrease (Christensen et al., 2007; Karl et al., 2008; Fig. 3-3). *Very likely* increase in RV20AHD (Fig. 3-5). *Medium confidence*: Strongest increase of WD in fall (in summer in Greenland), weakest in spring, weaker increase of CD in summer (OS11).	*High confidence*: WN *very likely* to increase and CN *very likely* to decrease (T06; Christensen et al., 2007; Kharin et al., 2007; Meehl et al., 2007b; Fig. 3-4).	*High confidence*: *Likely* more frequent and/or longer heat waves and warm spells (T06; OS11).	*High confidence*: *Likely* increase in HPD and HPC (T06; Fig. 3-6). *Likely* increase of RV20HP (Fig. 3-7).	*Low confidence*: Inconsistent signal in CDD and/or SMA changes (T06; SW08b; Fig. 3-10).
	G	G	G	G	G
Central America and Mexico (CAM, 6)	*High confidence*: WD *likely* to increase and CD *likely* to decrease (Fig. 3-3). *Likely* increase in RV20AHD (Fig. 3-5).	*High confidence*: WN *likely* to increase (T06; Fig. 3-4) and CN *likely* to decrease (Fig. 3-4).	*Medium confidence* to *high confidence* depending on region: *Likely* more frequent, longer, and/or more intense heat waves and warm spells in most of the region; *medium confidence* in increase in warm spells in part of Central America, but lack of agreement in signal of change in heat waves (T06; OS11; Clark et al., 2010).	*Low confidence*: Lack of agreement between models and indices regarding changes in %DP10, HPC, RV20HP, and other HP indicators (Kamiguchi et al., 2006; T06; Campbell et al., 2011; Figs. 3-6 and 3-7).	*Low confidence* to *medium confidence* depending on region. *Medium confidence*: Increased dryness (CDD, SMA) in Central America and Mexico; *low confidence* in change in dryness (CDD, SMA) in the extreme south of region due to inconsistent signal (Kamiguchi et al., 2006; T06; Campbell et al., 2011; Fig. 3-10).
	G	G	G	G R	G R

Continued next page →

Table 3-3 (continued)

B. Europe and Mediterranean Region

Regions	Tmax	Tmin	Heat Waves / Warm Spells	Heavy Precipitation	Dryness
All Europe and Mediterranean Region	*High confidence*: WD *very likely* to increase – largest increases in summer and Central/S. Europe and smallest in N. Europe (Scandinavia) (Goubanova and Li, 2007; Kjellström et al., 2007; Koffi and Koffi, 2008; Fischer and Schär, 2010; Fig. 3-3) and CD *very likely* to decrease (Fig. 3-3). *Likely* increase in RV20AHD (Fig. 3-5). *Medium confidence*: Changes in higher quantiles of Tmax generally greater than changes in lower quantiles of Tmax in summer in Central Europe and Mediterranean (Diffenbaugh et al., 2007; Kjellström et al., 2007; Fischer and Schär, 2009, 2010; OS11). G <u>R</u>	*High confidence*: CN *very likely* to decrease – largest decreases in winter in E. Europe and Scandinavia (Goubanova and Li, 2007; Kjellström et al., 2007; Sillmann and Roeckner, 2008). WN *very likely* to increase (T06; Fig. 3-4).	*High confidence*: *Likely* more frequent, longer, and/or more intense heat waves and warm spells but little change over Scandinavia (Beniston et al., 2007; Koffi and Koffi, 2008; Clark et al., 2010; OS11).	*Low confidence to high confidence*, depending on region: *Likely* overall increases in HPD, %DP10, and RV20HP and decreases in return periods of long (5-day) and short (1-day) events; strong signals in N. Europe particularly in winter, but lower confidence in changes in Central Europe and in particular the Mediterranean (T06; Beniston et al., 2007; Fowler et al., 2007a; Sillmann and Roeckner, 2008; Kendon et al., 2010; Figs. 3-6 and 3-7). *Likely* increase in HPC in some regions (Boberg et al., 2009b; Kendon et al., 2010). *Likely* greater changes in extremes than mean in many regions. Increase in HP intensity (and increase in HPC) despite decrease in summer mean in some regions – e.g., Central Europe (Beniston et al., 2007; Fowler et al., 2007a; Haugen and Iversen, 2008; May, 2008; Kyselý and Beranová, 2009).	*Medium confidence*: European area affected by stronger dryness (reduced SMA and CDD) with largest and most consistent changes in Mediterranean Europe (T06; Burke and Brown, 2008; May, 2008; SW08b; Sillmann and Roeckner, 2008; Fig. 3-10).
	G <u>R</u>	G <u>R</u>	G <u>R</u>	G <u>R</u>	G
N. Europe (NEU, 11)	*High confidence*: *Very likely* increase in frequency of WD, but smaller than in Central and S. Europe (Fischer and Schär, 2010; Fig. 3-3). *Very likely* decrease in CD (Fig. 3-3). *Likely* increase in RV20AHD (Fig. 3-5). *Medium confidence*: Changes in lower quantiles of Tmax generally greater than for changes in higher quantiles of Tmax in fall, winter, and spring in Scandinavia and northeastern Europe (OS11).	*High confidence*: CN *very likely* to decrease (Kjellström et al., 2007; Sillmann and Roeckner, 2008; Fig. 3-4); WN *very likely* to increase (T06; Fig. 3-4). *Medium confidence*: Changes in lower quantiles of Tmin generally greater than changes in higher quantiles of Tmin in Scandinavia and northeastern Europe (Kjellström et al., 2007; OS11).	*High confidence*: *Likely* more frequent, longer and/or more intense heat waves and warm spells, but summer increases smaller than in S. Europe and little change over Scandinavia (Beniston et al., 2007; Koffi and Koffi, 2008; Fischer and Schär, 2010; OS11). *Medium confidence*: Some dependency of projections of changes in HW intensity on parameterization choice (Clark et al., 2010).	*High confidence*: *Very likely* increases in HP (intensity and frequency) and %DP10 north of 45°N in winter (Frei et al., 2006; T06; Beniston et al., 2007; Kendon et al., 2008; Fig. 3-6). *Likely* increase in RV20HP (Fig. 3-7).	*Medium confidence*: No major changes in dryness (CDD, SMA) in N. Europe (T06; SW08b; Sillmann and Roeckner, 2008; Fig. 3-10).
	G <u>R</u>	G <u>R</u>	G <u>R</u>	G <u>R</u>	G
Central Europe (CEU, 12)	*High confidence*: *Very likely* increase in frequency and intensity of WD (Fischer and Schär, 2010; Fig. 3-3) and decrease in frequency of CD (Fig. 3-3). *Very likely* increase in RV20AHD (Fig. 3-5). *Medium confidence*: Changes in higher quantiles of Tmax much larger than changes in lower quantiles of Tmax in summer; results in very large increase in Tmax variability (Diffenbaugh et al., 2007; Kjellström et al., 2007; Fischer and Schär, 2009, 2010; OS11).	*High confidence*: CN *very likely* to decrease (Goubanova and Li, 2007; Kjellström et al., 2007; Sillmann and Roeckner, 2008); WN *very likely* to increase (T06; Fig. 3-4).	*High confidence*: *Likely* more frequent, longer, and/or more intense heat waves and warm spells (Beniston et al., 2007; Koffi and Koffi, 2008; Clark et al., 2010; Fischer and Schär, 2010; OS11). *Medium confidence*: Some dependency of projections of changes in HW intensity on parameterization choice (Clark et al., 2010).	*High confidence*: *Likely* increases in HP (intensity and frequency) in large part of the region in winter (Frei et al., 2006; Beniston et al., 2007; Kendon et al., 2008; Kyselý and Beranová, 2009; Fig. 3-6). *Likely* increase in RV20HP (Fig. 3-7). *Medium confidence*: Inconsistent evidence for summer: increase in HP in summer evident in RCMs (Christensen and Christensen, 2003; Frei et al., 2006) versus no signal in GCMs (Fig. 3-6).	*Medium confidence*: Increase in dryness (CDD, SMA) in Central Europe (Seneviratne et al., 2006a; T06; Fig. 3-10). *Medium confidence*: Increase in short-term droughts (SW08b).
	G <u>R</u>	G <u>R</u>	G <u>R</u>	G <u>R</u>	G <u>R</u>

Continued next page →

Table 3-3 (continued)

B. Europe and Mediterranean Region (Continued)

Regions	Tmax	Tmin	Heat Waves / Warm Spells	Heavy Precipitation	Dryness
S. Europe and Mediterranean (MED, 13)	*High confidence: Very likely* increase in frequency and intensity of WD (Fischer and Schär, 2009; Fig. 3-3) and decrease in CD (Fig. 3-4). *Very likely* increase in RV20AHD (Fig. 3-5). *High confidence:* Number of days with combined hot summer days (>35°C) and tropical nights (>20°C) *very likely* to increase (Fischer and Schär, 2010). *Medium confidence:* Changes in higher quantiles of Tmax greater than changes in lower quantiles of Tmax in summer (Diffenbaugh et al., 2007; Kjellström et al., 2007; Fischer and Schär, 2009, 2010; OS11).	*High confidence:* CN *very likely* to decrease (Goubanova and Li, 2007; Kjellström et al., 2007; Sillmann and Roeckner, 2008). *High confidence:* WN *very likely* to increase (T06; Sillmann and Roeckner, 2008; Giannakopoulos et al., 2009; Fig. 3-4). *High confidence:* Tropical nights *very likely* to increase (Sillmann and Roeckner, 2008). Number of days with combined hot summer days (>35°C) and tropical nights (>20°C) *very likely* to increase (Fischer and Schär, 2010). *Medium confidence:* Changes in higher quantiles of Tmin generally greater than changes in lower quantiles of Tmin in summer in the Mediterranean (Diffenbaugh et al., 2007; OS11).	*High confidence: Likely* more frequent and/or longer heat waves and warm spells (also increases in intensity); *likely* largest increases in S.W., S., and E. (Beniston et al., 2007; Diffenbaugh et al., 2007; Koffi and Koffi, 2008; Giannakopoulos et al., 2009; Clark et al., 2010; Fischer and Schär, 2010; OS11).	*Low confidence:* Inconsistent change in HP intensity and %DP10, depends on region and season; increase in HP intensity in all seasons except summer over parts of the region, but decrease in other parts, e.g., Iberian Peninsula (T06; Goubanova and Li, 2007; Giorgi and Lionello, 2008; Giannakopoulos et al., 2009; Fig. 3-6). *Low confidence* in changes in RV20HP (Fig. 3-7).	*Medium confidence:* Increase in dryness (CDD, SMA) in Mediterranean (T06; Beniston et al., 2007; SW08b; Sillmann and Roeckner, 2008; Giannakopoulos et al., 2009; Fig. 3-10). Consistent increase in area of drought (Burke and Brown, 2008).
	G R	G R	G R	G R	G R

C. Africa

Regions	Tmax	Tmin	Heat Waves / Warm Spells	Heavy Precipitation	Dryness
All Africa	*High confidence:* WD *likely* to increase and CD *likely* to decrease in all regions (Fig. 3-3). *Likely* increase in RV20AHD in all regions (Fig. 3-5). *Medium confidence:* Increase in WD largest in summer and fall (OS11).	*High confidence:* WN *likely* to increase (T06; Kharin et al., 2007; Fig. 3-4) and CN *likely* to decrease (Fig. 3-4).	*High confidence: Likely* more frequent and/or longer heat waves and warm spells (T06; OS11).	*Low confidence to medium confidence* depending on region: Inconsistent change or no signal in HP indicators across much of continent (T06; Fig. 3-6, Fig. 3-7). Strongest and most consistent signal is *likely* increase in HP in E. Africa (T06; Figs. 3-6 and 3-7; Shongwe et al., 2011).	*Low confidence to medium confidence* depending on region: *Low confidence* of increase in dryness (CDD, SMA) in most regions, *medium confidence* of increase in dryness (CDD, SMA) in southern Africa except eastern part (T06; SW08b; Fig. 3-10).
	G	G	G	G	G
W. Africa (WAF, 15)	*High confidence:* WD *likely* to decrease (Fig. 3-3). *Likely* increase in RV20AHD (Fig. 3-5).	*High confidence:* WN *likely* to increase (T06; Fig. 3-4) and CN *likely* to decrease (Fig. 3-4).	*High confidence: Likely* more frequent and/or longer heat waves and warm spells (T06; OS11).	*Low confidence to medium confidence* depending on subregion: *Medium confidence* in slight or no change in HP indicators in most of region; *Low confidence* due to low model agreement in northern part of region (T06; Figs. 3-6 and 3-7).	*Low confidence:* Inconsistent signal of change in CDD and SMA (T06; Fig. 3-10).
	G	G	G	G	G
E. Africa (EAF, 16)	*High confidence:* WD *likely* to decrease (Fig. 3-3). *Likely* increase in RV20AHD (Fig. 3-5).	*High confidence:* WN *likely* to increase (T06; Fig. 3-4) and CN *likely* to decrease (Fig. 3-4).	*High confidence: Likely* more frequent and/or longer heat waves and warm spells (T06; OS11).	*High confidence: Likely* increase in HP indicators (T06; Fig. 3-6; Fig. 3-7; Shongwe et al., 2011).	*Medium confidence:* Decreasing dryness in large part of region, especially based on change in SMA, and partly also in CDD (T06; Fig. 3-10; Shongwe et al., 2011).
	G	G	G	G	G
S. Africa (SAF, 17)	*High confidence:* WD *likely* to decrease (Fig. 3-3). *Likely* increase in RV20AHD (Fig. 3-5).	*High confidence:* WN *likely* to increase (T06; Fig. 3-4) and CN *likely* to decrease (Fig. 3-4).	*High confidence: Likely* more frequent and/or longer heat waves and warm spells (T06; OS11).	*Low confidence:* Lack of agreement/signal in %DP10 and other HP indicators for the region as a whole (T06; Fig. 3-6). Some model agreement in increase in RV20HP (Fig. 3-7). Some evidence of increased HP intensity in the S.E. (Hewitson and Crane, 2006; Rocha et al., 2008; Shongwe et al., 2009).	*Medium confidence:* Increase in dryness (CDD, SMA), except eastern part (T06; Shongwe et al., 2009; Fig. 3-10). Consistent increase in area of drought (Burke and Brown, 2008).
	G	G	G	G	G
Sahara (SAH, 14)	*High confidence:* WD *likely* to increase and CD *likely* to decrease (Fig. 3-3). *Likely* increase in RV20AHD (Fig. 3-5).	*High confidence:* WN *likely* to increase (T06; Fig. 3-4) and CN *likely* to decrease (Fig. 3-4).	*High confidence: Likely* more frequent and/or longer heat waves and warm spells (T06; OS11).	*Low confidence:* Low agreement/no signal in %DP10, RV20HP, and other HP indicators (T06; Figs. 3-6 and 3-7).	*Low confidence:* Inconsistent signal of change in CDD and SMA (T06; Fig. 3-10).
	G	G	G	G	G

Continued next page →

Table 3-3 (continued)

D. South America

Regions	Tmax	Tmin	Heat Waves / Warm Spells	Heavy Precipitation	Dryness
All South America	*High confidence:* WD *likely* to increase and CD *likely* to decrease in all regions (Fig. 3-3). *Likely* increase in RV20AHD in all regions (Fig. 3-5).	*High confidence:* WN *likely* to increase (T06; Kharin et al., 2007; Marengo et al., 2009a; Fig. 3-4) and CN *likely* to decrease (Marengo et al., 2009a; Fig. 3-4).	*Medium confidence* to *high confidence* depending on region: *Likely* more frequent and/or longer heat waves and warm spells on annual time scale; only *medium confidence* in increase in HW duration in southeastern South America (T06; OS11).	*Low confidence* to *medium confidence* depending on region: Inconsistent sign of change in HP indicators in some regions; some regions with model agreement (T06; Figs. 3-6 and 3-7).	*Low confidence* to *medium confidence* depending on region: Inconsistent signal except for dryness increase (CDD and SMA) in northeastern Brazil (SW08b; Fig. 3-10).
	G	G R	G	G	G
Amazon (AMZ, 7)	*High confidence:* WD *likely* to increase and CD *likely* to decrease (Fig. 3-3). *Likely* increase in RV20AHD (Fig. 3-5).	*High confidence: Very likely* increase in WN (T06; Marengo et al., 2009a; Fig. 3-4) and CN *likely* to decrease (Fig. 3-4).	*High confidence: Likely* more frequent and/or longer heat waves and warm spells based on multi-model assessments (T06, OS11); non-significant signal in single-model and single-index study by Uchiyama et al. (2006).	*Medium confidence:* Tendency to increase in precip >95th p and in RV20HP but less consistent increase for some other HP indicators and single studies (Kamiguchi et al., 2006; T06; Marengo et al., 2009a; Sorensson et al., 2010; Figs. 3-6 and 3-7).	*Low confidence:* Inconsistent signal in SMA (SW08b; Fig. 3-10) and inconsistent signal in CDD (Kamiguchi et al., 2006; T06; Marengo et al., 2009a; Sorensson et al., 2010; Fig. 3-10).
	G	G R	G	G R	G R
N.E. Brazil (NEB, 8)	*High confidence:* WD *likely* to increase and CD *likely* to decrease (Fig. 3-3). *Likely* increase in RV20AHD (Fig. 3-5).	*High confidence: Likely* increase of WN (T06; Marengo et al., 2009a; Fig. 3-4) and CN *likely* to decrease (Fig. 3-4).	*High confidence: Likely* more frequent and/or longer heat waves and warm spells (T06; OS11).	*Low confidence:* Slight or no change in %DP10 and other HP indicators (Kamiguchi et al., 2006; T06; Marengo et al., 2009a; Sorensson et al., 2010; Figs. 3-6 and 3-7).	*Medium confidence:* Dryness increase (CDD, SMA) (Kamiguchi et al., 2006; T06; SW08b; Marengo et al., 2009a; Sorensson et al., 2010; Fig. 3-10).
	G	G R	G	G R	G R
S.E. South America (SSA, 10)	*High confidence:* WD *likely* to increase and CD *likely* to decrease (Fig. 3-3). *Likely* increase in RV20AHD (Fig. 3-5).	*High confidence: Very likely* increase in WN (T06; Marengo et al., 2009a; Fig. 3-4) and CN *likely* to decrease (Fig. 3-4).	*Medium confidence:* Tendency for more frequent and/or longer heat waves and warm spells on annual time scale; weak signal compared to other regions but robust sign (increase) across majority of models (T06; OS11).	*Medium confidence:* Increase of %DP10, precipitation intensity, precip >95th p, and RV20HP in the northern portion. *Low confidence* in the southern portion: inconsistent change or no signal in HP indicators (Kamiguchi et al., 2006; T06; Marengo et al., 2009a; Nunez et al., 2009; Figs. 3-6 and 3-7).	*Low confidence:* Inconsistent signal in SMA (SW08b). *Low confidence:* Inconsistent signal in CDD (Kamiguchi et al., 2006; T06; Marengo et al., 2009a; Sorensson et al., 2010; Fig. 3-10).
	G	G R	G	G R	G R
W. Coast South America (WSA, 9)	*High confidence:* WD *likely* to increase and CD *likely* to decrease (Fig. 3-3). *Likely* increase in RV20AHD (Fig. 3-5).	*High confidence: Likely* increase in WN (T06; Marengo et al., 2009a; Fig. 3-4) and CN *likely* to decrease (Fig. 3-4).	*High confidence: Likely* more frequent and/or longer heat waves and warm spells on annual time scale (T06; OS11).	*Medium confidence:* Increase in %DP10, precip >95th p and other HP indicators in the tropics. *Low confidence* for the extratropics (Kamiguchi et al., 2006; T06; Marengo et al., 2009a; Sorensson et al., 2010; Fig. 3-6). *Low confidence* in changes in RV20HP (Fig. 3-7).	*Low confidence:* Inconsistent changes in SMA across domain (Fig. 3-10), CDD decrease in the tropics and increase in the extratropics (Kamiguchi et al., 2006; T06; Marengo et al., 2009a; Sorensson et al., 2010; Fig. 3-10). *Medium confidence:* CDD increase and SMA decrease in southwest SA (SW08b; Fig. 3-10).
	G	G R	G	G R	G R

Continued next page →

Table 3-3 (continued)

E. Asia

Regions	Tmax	Tmin	Heat Waves / Warm Spells	Heavy Precipitation	Dryness
All Asia	*High confidence:* WD *likely* to increase and CD *likely* to decrease in all regions (Fig. 3-3). *Likely* increase in RV20AHD in all regions (Fig. 3-5). G	*High confidence:* WN *likely* to increase (T06; Kharin et al., 2007; Fig. 3-4) and CN *likely* to decrease (Fig. 3-4). G	*Low confidence* to *high confidence* depending on region: *Likely* more frequent and/or longer heat waves and warm spells in most regions (continental) on the annual time scale; *low confidence* due to inconsistent signal in Indonesia, Philippines, Malaysia, Papua New Guinea, and neighboring islands for some HW indices (T06; OS11). G	*Low confidence* to *high confidence* depending on region and index: *High confidence* regarding *likely* increase in HP in N. Asia; *medium confidence* regarding increase in HP in S.E. Asia, E. Asia, and Tibetan Plateau; *low confidence* regarding increase in HP in S. and W. Asia (T06; Figs. 3-6 and 3-7). G	*Low confidence:* Inconsistent change in CDD and SMA between models in large part of domain (T06; Fig. 3-10). G
N. Asia (NAS, 18)	*High confidence:* WD *likely* to increase and CD *likely* to decrease (Fig. 3-3). *Likely* increase in RV20AHD (Fig. 3-5). G	*High confidence:* WN *likely* to increase (T06; Fig. 3-4) and CN *likely* to decrease (Fig. 3-4). G	*High confidence: Likely* more frequent and/or longer heat waves and warm spells (T06; OS11). G	*High confidence: Likely* increase in HP (T06; Figs. 3-6 and 3-7) including more frequent and intense HPD over most regions (Emori and Brown, 2005; Kamiguchi et al., 2006). G	*Low confidence:* Inconsistent change in dryness (CDD, SMA) between models in large part of domain (T06; SW08b; Fig. 3-10). G
Central Asia (CAS, 20)	*High confidence:* WD *likely* to increase and CD *likely* to decrease (Fig. 3-3). *Likely* increase in RV20AHD (Fig. 3-5). G	*High confidence:* WN *likely* to increase (T06; Fig. 3-4) and CN *likely* to decrease (Fig. 3-4). G	*High confidence: Likely* more frequent, longer and/or more intense heat waves and warm spells (T06; Clark et al., 2010; OS11). G	*Low confidence:* Inconsistent signal in models regarding changes in HP (T06; Fig. 3-6). Some model consistency in projections of RV20HP (Fig. 3-7). G	*Low confidence:* Inconsistent signals across indices (CDD, SMA) (T06; SW08b; Fig. 3-10). G
E. Asia (EAS, 22)	*High confidence:* WD *likely* to increase and CD *likely* to decrease across the region (Clark et al., 2010; Fig. 3-3), including in Korea (Boo et al., 2006; Im and Kwon, 2007; Im et al., 2008, 2011; Koo et al., 2009). *Likely* increase in RV20AHD (Fig. 3-5). G R	*High confidence:* WN *likely* to increase (T06; Fig. 3-4) and CN *likely* to decrease (Fig. 3-4), including in Korea (Boo et al., 2006; Koo et al., 2009; Im et al., 2011). G R	*High confidence: Likely* more frequent, longer, and/or more intense heat waves and warm spells (T06; Clark et al., 2010; OS11). G	*Medium confidence:* Increases in HP (less consistent in %DP10 than other indicators) across the region (T06; Figs. 3-6 and 3-7), including increase in Japan and Korea (Emori and Brown, 2005; Kimoto et al., 2005; Boo et al., 2006; Kamiguchi et al., 2006; Kusunoki and Mizuta, 2008; Kitoh et al., 2009; Su et al., 2009; Kim et al., 2010; Im et al., 2011). G R	*Low confidence:* Inconsistent signal across indices (CDD, SMA) (T06; SW08b; Fig. 3-10). G
S.E. Asia (SEA, 24)	*High confidence:* WD *likely* to increase and CD *likely* to decrease (Fig. 3-3). *Likely* increase in RV20AHD (Fig. 3-5). G	*High confidence:* WN *likely* to increase (T06; Fig. 3-4) and CN *likely* to decrease (Fig. 3-4). G	*Low confidence* to *high confidence* depending on subregion and index: *Likely* more frequent and/or longer heat waves and warm spells on the annual time scale over continental areas; *Low confidence* in changes in some HW indices but model agreement of increases in a WS index in Indonesia, Philippines, Malaysia, Papua New Guinea, and neighboring islands (T06; OS11). G	*Medium confidence:* Inconsistent signal in change of %DP10 across models (T06; Fig. 3-6) but more frequent and intense HPD, including increase in FV20HP, suggested by other indicators over most regions especially non-continental parts (Emori and Brown, 2005; Kamiguchi et al., 2006; Figs. 3-6 and 3-7). G	*Low confidence:* Inconsistent signal of change in CDD and/or SMA (T06; SW08b; Fig. 3-10). G
S. Asia (SAS, 23)	*High confidence:* WD *likely* to increase and CD *likely* to decrease (Kumar et al., 2006; Rajendran and Kitoh, 2008; Fig. 3-3). *Likely* increase in RV20AHD (Fig. 3-5). G R	*High confidence:* WN *likely* to increase (T06; Fig. 3-4) and CN *likely* to decrease (Fig. 3-4). *Medium confidence:* Extreme nighttime temperature warms faster than daytime (Kumar et al., 2006). G R	*High confidence: Likely* more frequent and/or longer heat waves and warm spells on annual time scale (T06; OS11). *Medium confidence:* Some dependency of magnitude of signal on index choice (OS11). G	*Low confidence:* Slight or no increase in %DP10 (T06; Fig. 3-6). Some model consistency regarding increase in RV20HP (Fig. 3-7). *Low confidence:* More frequent and intense HPD over parts of S. Asia (Emori and Brown, 2005; Kamiguchi et al., 2006; Kharin et al., 2007; Rajendran and Kitoh, 2008). G R	*Low confidence:* Inconsistent signal of change in CDD and SMA (T06; SW08b; Fig. 3-10). G

Continued next page →

Table 3-3 (continued)
E. Asia (Continued)

Regions	Tmax	Tmin	Heat Waves / Warm Spells	Heavy Precipitation	Dryness
W. Asia (WAS, 19)	*High confidence:* WD *likely* to increase and CD *likely* to decrease (Fig. 3-3). *Likely* increase in RV20AHD (Fig. 3-5).	*High confidence:* WN *likely* to increase (T06; Fig. 3-4) and CN *likely* to decrease (Fig. 3-4).	*High confidence: Likely* more frequent, longer and/or more intense heat waves and warm spells (T06; Clark et al., 2010; OS11). *Medium confidence:* Some dependency of magnitude of signal on index choice (OS11).	*Low confidence:* Inconsistent signal of change in HP (T06; Fig. 3-6; Fig. 3-7).	*Low confidence:* Inconsistent signal of change in CDD and SMA (T06; SW08b; Fig. 3-10).
	G	G	G	G	G
Tibetan Plateau (TIB, 21)	*High confidence:* WD *likely* to decrease (Fig. 3-3). *Likely* increase in RV20AHD (Fig. 3-5).	*High confidence:* WN *likely* to increase (T06; Fig. 3-4) and CN *likely* to decrease (Fig. 3-4).	*High confidence: Likely* more frequent, longer, and/or more intense heat waves and warm spells (T06; Clark et al., 2010; OS11).	*Medium confidence:* Increase in HP (T06; Figs. 3-6 and 3-7).	*Low confidence:* Inconsistent signal of change in CDD (T06; SW08b; Fig. 3-10).
	G	G	G	G	G

F. Australia/New Zealand

Regions	Tmax	Tmin	Heat Waves / Warm Spells	Heavy Precipitation	Dryness
All Australia and New Zealand	*High confidence:* WD *very likely* to increase and CD *very likely* to decrease in all regions (CSIRO, 2007; Mullan et al., 2008; Fig. 3-3). *Very likely* increase in RV20AHD (Fig. 3-5).	*High confidence:* WN *very likely* to increase everywhere (T06; Kharin et al., 2007; Alexander and Arblaster, 2009; Fig. 3-4) and CN *very likely* to decrease (Fig. 3-4). *Medium confidence:* WN increase everywhere. Largest increases in WN in N. compared with S. and most consistent changes in inland regions (Alexander and Arblaster, 2009).	*High confidence: Likely* more frequent and/or longer heat waves and warm spells (T06; OS11; Alexander and Arblaster, 2009). *Medium confidence:* Strongest increases in HW duration in N.W. and most consistent increases inland (Alexander and Arblaster, 2009).	*Low confidence:* Lack of agreement regarding sign of change for different models and different indices, and spatial variations in signal (T06; Figs. 3-6 and 3-7). *Low confidence:* HPD tend to increase in E. and decrease in W. half of country – but considerable inter-model inconsistencies; HPC tends to increase everywhere – but considerable inter-model inconsistencies (Alexander and Arblaster, 2009).	*Low confidence* to *medium confidence* depending on region: Models agree on increase in CDD in S. Australia, but inconsistent signal over most of S. Australia in SMA; inconsistent signal in CDD and SMA in N. Australia (T06; SW08b; Fig. 3-10). Strongest CDD increases in W. half of Australia (Alexander and Arblaster, 2009). Inconsistent change in area of drought depending on index used (Burke and Brown, 2008).
	G	G	G	G	G
N. Australia (NAU, 25)	*High confidence:* WD *very likely* to increase and CD *very likely* to decrease (CSIRO, 2007; Fig. 3-3). *Very likely* increase in RV20AHD (Fig. 3-5).	*High confidence:* WN *very likely* to increase (T06; Alexander and Arblaster, 2009; Fig. 3-4) and CN *very likely* to decrease (Fig. 3-4). *Medium confidence:* Changes larger than in S. Australia (Alexander and Arblaster, 2009).	*High confidence: Likely* more frequent and/or longer heat waves and warm spells (T06; Alexander and Arblaster, 2009; OS11). *Medium confidence:* Strongest increases in N.W. and most consistent increases inland (Alexander and Arblaster, 2009).	*Low confidence:* Lack of agreement regarding sign of change for different models and different indices (T06; Fig. 3-6, Fig. 3-7).	*Low confidence:* Inconsistent signal in CDD and SMA (T06; SW08b; Fig. 3-10).
	G	G	G	G	G
S. Australia/ New Zealand (SAU, 26)	*High confidence:* WD *very likely* to decrease (CSIRO, 2007; Fig. 3-3). *Very likely* increase in RV20AHD (Fig. 3-5). *Low confidence* to *medium confidence:* Strongest New Zealand increases in WD in North Island and largest decreases in frost days in South Island (Mullan et al., 2008).	*High confidence:* WN *very likely* to increase (T06; Alexander and Arblaster, 2009; Fig. 3-4) and CN *very likely* to decrease (Fig. 3-4). *Medium confidence:* Changes smaller than in N. Australia (Alexander and Arblaster, 2009).	*High confidence: Likely* more frequent and/or longer heat waves and warm spells (T06; Alexander and Arblaster, 2009; OS11). *Medium confidence:* Most consistent increases inland (Alexander and Arblaster, 2009).	*Low confidence:* Lack of agreement regarding sign of change for different models and different indices, and spatial variations in signal (T06; Fig. 3-6, Fig. 3-7). *Low confidence* to *medium confidence:* In New Zealand, increase in HP events at most locations (Mullan et al., 2008; Carey-Smith et al., 2010).	*Medium confidence:* Models agree on increase in CDD in southern Australia including S.W. (T06; Alexander and Arblaster, 2009; Fig. 3-10), but inconsistent signal in SMA over most of the region, slight decrease in S.W. (SW08b; Fig. 3-10).
	G R	G	G	G R	G

References

A digital library of non-journal-based literature cited in this chapter that may not be readily available to the public has been compiled as part of the IPCC review and drafting process, and can be accessed via either the IPCC Secretariat or IPCC Working Group II web sites.

Abbs, D., and T. Rafter, 2008: *The Effect of Climate Change on Extreme Rainfall Events in the Western Port Region*. CSIRO Marine and Atmospheric Research, Aspendale, Australia, 23 pp.

Abeysirigunawardena, D.S. and I.J. Walker, 2008: Sea level responses to climatic variability and change in northern British Columbia. *Atmosphere-Ocean*, **46**(3), 277-296.

Abeysirigunawardena, D.S., E. Gilleland, D. Bronaugh, and W. Wong, 2009: Extreme wind regime responses to climate variability and change in the inner south coast of British Columbia, Canada. *Atmosphere-Ocean*, **47**(1), 41-62.

Adams, P.N., D.L. Imnan, and N.E. Graham, 2008: Southern California deep-water wave climate: Characterization and application to coastal processes. *Journal of Coastal Research*, **24**(4), 1022-1035.

Aguilar, E., T.C. Peterson, P. Ramírez Obando, R. Frutos, J.A. Retana, M. Solera, J. Soley, I. González García, R.M. Araujo, A. Rosa Santos, V.E. Valle, M. Brunet, L. Aguilar, L. Álvarez, M. Bautista, C. Castañón, L. Herrara, E. Ruano, J.J. Sinay, E. Sánchez, G.I. Hernández Oviedo, F. Obed, J.E. Salgado, J.L. Vázquez, M. Baca, M. Gutiérrez, C. Centella, J. Espinosa, D. Martinez, B. Olmedo, C.E. Ojeda Espinoza, R. Núñez, M. Haylock, H. Benavides, and R. Mayorga, 2005: Changes in precipitation and temperature extremes in Central America and northern South America, 1961-2003. *Journal of Geophysical Research – Atmospheres*, **110**, D23107.

Aguilar, E., A. Aziz Barry, M. Brunet, L. Ekang, A. Fernandes, M. Massoukina, J. Mbah, A. Mhanda, D.J. do Nascimento, T.C. Peterson, M. Thamba Umba, M. Tomou, and X. Zhang, 2009: Changes in temperature and precipitation extremes in western central Africa, Guinea Conakry, and Zimbabwe, 1955-2006. *Journal of Geophysical Research – Atmospheres*, **114**, D02115.

Aizen, V.B., V.A. Kuzmichenok, A.B. Surazakov, and E.M. Aizen, 2007: Glacier changes in the Tien Shan as determined from topographic and remotely sensed data. *Global and Planetary Change*, **56**(3-4), 328-340.

Alexander, L.V. and J.M. Arblaster, 2009: Assessing trends in observed and modelled climate extremes over Australia in relation to future projections. *International Journal of Climatology*, **29**(3), 417-435.

Alexander, L.V. and S. Power, 2009: Severe storms inferred from 150 years of sub-daily pressure observations along Victoria's "Shipwreck Coast." *Australian Meteorological and Oceanographic Journal*, **58**(2), 129-133.

Alexander, L.V., X. Zhang, T.C. Peterson, J. Caesar, B. Gleason, A.M.G. Klein Tank, M. Haylock, D. Collins, B. Trewin, F. Rahimzadeh, A. Tagipour, K.R. Kumar, J. Revadekar, G. Griffiths, L. Vincent, D.B. Stephenson, J. Burn, E. Aguilar, M. Brunet, M. Taylor, M. New, P. Zhai, M. Rusticucci, and J.L. Vazquez-Aguirre, 2006: Global observed changes in daily climate extremes of temperature and precipitation. *Journal of Geophysical Research – Atmospheres*, **111**, D05109.

Alexander, L.V., P. Uotila, and N. Nicholls, 2009: Influence of sea surface temperature variability on global temperature and precipitation extremes. *Journal of Geophysical Research – Atmospheres*, **114**, D18116.

Alexander, L.V., X.L. Wang, H. Wan, and B. Trewin, 2011: Significant decline in storminess over southeast Australia since the late 19th century. *Australian Meteorological and Oceanographic Journal*, **61**(1), 23-30.

Ali, A. and T. Lebel, 2009: The Sahelian standardized rainfall index revisited. *International Journal of Climatology*, **29**(12), 1705-1714.

Allamano, P., P. Claps, and F. Laio, 2009: Global warming increases flood risk in mountainous areas. *Geophysical Research Letters*, **36**, L24404.

Allan, J.C. and P.D. Komar, 2006: Climate controls on US West Coast erosion processes. *Journal of Coastal Research*, **22**(3), 511-529.

Allan, R., S. Tett, and L. Alexander, 2009: Fluctuations in autumn-winter severe storms over the British Isles: 1920 to present. *International Journal of Climatology*, **29**(3), 357-371.

Allan, R.P. and B.J. Soden, 2008: Atmospheric warming and the amplification of precipitation extremes. *Science*, **321**(5895), 1481-1484.

Allen, M., 2003: Liability for climate change: will it ever be possible to sue anyone for damaging the climate? *Nature*, **421**(6926), 891-892.

Allen, M.R., D.J. Frame, C. Huntingford, C.D. Jones, J.A. Lowe, M. Meinshausen, and N. Meinshausen, 2009: Warming caused by cumulative carbon emissions towards the trillionth tonne. *Nature*, **458**(7242), 1163-1166.

Allen, S.K., S.C. Cox, and I.F. Owens, 2011: Rock-avalanches and other landslides in the central Southern Alps of New Zealand: A regional assessment of possible climate change impacts. *Landslides*, **8**(1), 33-48.

An, S.I., A. Timmermann, L. Bejarano, F.F. Jin, F. Justino, Z. Liu, and A.W. Tudhope, 2004: Modeling evidence for enhanced El Niño-Southern Oscillation amplitude during the Last Glacial Maximum. *Paleoceanography*, **19**, PA4009.

Anders, I., and B. Rockel, 2009: The influence of prescribed soil type distribution on the representation of present climate in a regional climate model. *Climate Dynamics*, **33**(2-3), 177-186.

Andersen, O.B., S.I. Seneviratne, J. Hinderer, and P. Viterbo, 2005: GRACE-derived terrestrial water storage depletion associated with the 2003 European heat wave. *Geophysical Research Letters*, **32**, L18405.

Anderson, C.J., R.W. Arritt, E.S. Takle, Z.T. Pan, W.J. Gutowski, F.O. Otieno, R. da Silva, D. Caya, J.H. Christensen, D. Luthi, M.A. Gaertner, C. Gallardo, F. Giorgi, S.Y. Hong, C. Jones, H.M.H. Juang, J.J. Katzfey, W.M. Lapenta, R. Laprise, J.W. Larson, G.E. Liston, J.L. McGregor, R.A. Pielke, J.O. Roads, and J.A. Taylor, 2003: Hydrological processes in regional climate model simulations of the central United States flood of June-July 1993. *Journal of Hydrometeorology*, **4**(3), 584-598.

Andrade, C., H.O. Pires, R. Taborda, and M.C. Freitas, 2007: Projecting future changes in wave climate and coastal response in Portugal by the end of the 21st century. *Journal of Coastal Research,* **SI50**, 253-257.

Andreadis, K.M. and D.P. Lettenmaier, 2006: Trends in 20th century drought over the continental United States. *Geophysical Research Letters*, **33**, L10403.

Andreadis, K.M., E.A. Clark, A.W. Wood, A.F. Hamlet, and D.P. Lettenmaier, 2005: Twentieth-century drought in the conterminous United States. *Journal of Hydrometeorology*, **6**(6), 985-1001.

Arblaster, J.M., G.A. Meehl, and D.J. Karoly, 2011: Future climate change in the Southern Hemisphere: Competing effects of ozone and greenhouse gases. *Geophysical Research Letters*, **38**, L02701.

Arriaga-Ramírez, S. and T. Cavazos, 2010: Regional trends of daily precipitation indices in Northwest Mexico and the Southwest United States. *Journal of Geophysical Research – Atmospheres*, **115**, D14111.

Aryal, S.K., B.C. Bates, E.D. Campbell, Y. Li, M.J. Palmer, and N.R. Viney, 2009: Characterizing and modeling temporal and spatial trends in rainfall extremes. *Journal of Hydrometeorology*, **10**(1), 241-253.

Ashfaq, M., Y. Shi, W.W. Tung, R.J. Trapp, X. Gao, S.J. Pal, and N.S. Diffenbaugh, 2009: Suppression of south Asian summer monsoon precipitation in the 21st century. *Geophysical Research Letters*, **36**, L01704.

Asokan, S.M., and D. Dutta, 2008: Analysis of water resources in the Mahanadi River Basin, India under projected climate conditions. *Hydrological Processes*, **22**(18), 3589-3603.

Augustinus, P., 1995: Rock mass strength and the stability of some glacial valley slopes. *Zeitschrift Für Geomorphologie*, **39**(1), 55-68.

Aunan, K., and B. Romstad, 2008: Strong coasts and vulnerable communities: Potential implications of accelerated sea-level rise for Norway. *Journal of Coastal Research*, **24**(2), 403-409.

Bader, J., M.D.S. Mesquita, K.I. Hodges, N. Keenlyside, S. Østerhus, and M. Miles, 2011: A review on Northern Hemisphere sea-ice, storminess and the North Atlantic Oscillation: Observations and projected changes. *Atmospheric Research*, **101**(4), 809-834, doi:10.1016/j.atmosres.2011.1004.1007.

Bajracharya, S.R. and P. Mool, 2009: Glaciers, glacial lakes and glacial lake outburst floods in the Mount Everest region, Nepal. *Annals of Glaciology*, **50**(53), 81-86.

Ballester, J., F. Giorgi, and X. Rodo, 2010: Changes in European temperature extremes can be predicted from changes in PDF central statistics. *Climatic Change*, **98(1-2)**, 277-284.

Barla, G., F. Dutto, and G. Mortara, 2000: Brenva glacier rock avalanche of 18 January 1997 on the Mount Blanc range, northwest Italy. *Landslide News*, **13**, 2-5.

Barnett, D.N., S.J. Brown, J.M. Murphy, D.M.H. Sexton, and M.J. Webb, 2006: Quantifying uncertainty in changes in extreme event frequency in response to doubled CO_2 using a large ensemble of GCM simulations. *Climate Dynamics*, **26(5)**, 489-511.

Barnett, T.P., D.W. Pierce, H.G. Hidalgo, C. Bonfils, B.D. Santer, T. Das, G. Bala, A.W. Wood, T. Nozawa, A.A. Mirin, D.R. Cayan, and M.D. Dettinger, 2008: Human-induced changes in the hydrology of the western United States. *Science*, **319(5866)**, 1080-1083.

Barring, L. and K. Fortuniak, 2009: Multi-indices analysis of southern Scandinavian storminess 1780-2005 and links to interdecadal variations in the NW Europe-North Sea region. *International Journal of Climatology*, **29(3)**, 373-384.

Barring, L., and H. von Storch, 2004: Scandinavian storminess since about 1800. *Geophysical Research Letters*, **31**, L20202.

Barring, L., T. Holt, M.L. Linderson, M. Radziejewski, M. Moriondo, and J.P. Palutikof, 2006: Defining dry/wet spells for point observations, observed area averages, and regional climate model gridboxes in Europe. *Climate Research*, **31(1)**, 35-49.

Barriopedro, D., E.M. Fischer, J. Luterbacher, R.M. Trigo, and R. García-Herrera, 2011: The hot summer of 2010: Redrawing the temperature record map of Europe. *Science*, **332(6026)**, 220-224.

Barros, V., L. Chamorro, G. Coronel, and J. Baez, 2004: The major discharge events in the Paraguay River: Magnitudes, source regions, and climate forcings. *Journal of Hydrometeorology*, **5(6)**, 1161-1170.

Bartholy, J., and R. Pongracz, 2007: Regional analysis of extreme temperature and precipitation indices for the Carpathian Basin from 1946 to 2001. *Global and Planetary Change*, **57**, 83-95.

Bartolini, G., M. Morabito, A. Crisci, D. Grifoni, T. Torrigiani, M. Petralli, G. Maracchi, and S. Orlandini, 2008: Recent trends in Tuscany (Italy) summer temperature and indices of extremes. *International Journal of Climatology*, **28(13)**, 1751-1760.

Bates, B.C., Z.W. Kundzewics, S. Wu, and J.P. Palutikof, 2008: *Climate Change and Water. Technical Paper of the Intergovernmental Panel on Climate Change*. IPCC Secretariat, Geneva, Switzerland, 210 pp.

Beaulant, A.L., B. Joly, O. Nuissier, S. Somot, V. Ducrocq, A. Joly, F. Sevault, M. Deque, and D. Ricard, 2011: Statistico-dynamical downscaling for Mediterranean heavy precipitation. *Quarterly Journal of the Royal Meteorological Society*, **137(656)**, 736-748.

Begueria, S., S.M. Vicente-Serrano, and M. Angulo-Martinez, 2010: A multiscalar global drought dataset: The SPEIbase: A new gridded product for the analysis of drought variability and impacts. *Bulletin of the American Meteorological Society*, **91(10)**, 1351-1354.

Beirlant, J., Y. Goegebeur, J. Teugels, and J. Segers, 2004: *Statistics of Extremes: Theory and Applications*. John Wiley & Sons, Chichester, West Sussex, England, 498 pp.

Beltaos, S. and T. Prowse, 2009: River-ice hydrology in a shrinking cryosphere. *Hydrological Processes*, **23(1)**, 122-144.

Bender, M.A., T.R. Knutson, R.E. Tuleya, J.J. Sirutis, G.A. Vecchi, S.T. Garner, and I.M. Held, 2010: Modeled impact of anthropogenic warming on the frequency of intense Atlantic hurricanes. *Science*, **327(5964)**, 454-458.

Benestad, R.E., 2001: A comparison between two empirical downscaling strategies. *International Journal of Climatology*, **21(13)**, 1645-1668.

Benestad, R.E., 2003: How often can we expect a record event? *Climate Research*, **25(1)**, 3-13.

Benestad, R.E., 2006: Can we expect more extreme precipitation on the monthly time scale? *Journal of Climate*, **19(4)**, 630-637.

Benestad, R.E., 2007: Novel methods for inferring future changes in extreme rainfall over Northern Europe. *Climate Research*, **34(3)**, 195-210.

Benestad, R.E., and J.E. Haugen, 2007: On complex extremes: flood hazards and combined high spring-time precipitation and temperature in Norway. *Climatic Change*, **85(3-4)**, 381-406.

Benestad, R.E., I. Hanssen-Bauer, and E.J. Forland, 2007: An evaluation of statistical models for downscaling precipitation and their ability to capture long-term trends. *International Journal of Climatology*, **27(5)**, 649-665.

Bengtsson, L., S. Hagemann, and K.I. Hodges, 2004: Can climate trends be calculated from reanalysis data? *Journal of Geophysical Research – Atmospheres*, **109**, D11111.

Bengtsson, L., K.I. Hodges, and E. Roeckner, 2006: Storm tracks and climate change. *Journal of Climate*, **19(15)**, 3518-3543.

Bengtsson, L., K.I. Hodges, M. Esch, N. Keenlyside, L. Kornblueh, J.J. Luo, and T. Yamagata, 2007: How may tropical cyclones change in a warmer climate? *Tellus Series A – Dynamic Meteorology and Oceanography*, **59(4)**, 539-561.

Bengtsson, L., K.I. Hodges, and N. Keenlyside, 2009: Will extratropical storms intensify in a warmer climate? *Journal of Climate*, **22(9)**, 2276-2301.

Beniston, M., 2004: The 2003 heat wave in Europe: A shape of things to come? An analysis based on Swiss climatological data and model simulations. *Geophysical Research Letters*, **31**, L02202.

Beniston, M., 2009a: Decadal-scale changes in the tails of probability distribution functions of climate variables in Switzerland. *International Journal of Climatology*, **29(10)**, 1362-1368.

Beniston, M., 2009b: Trends in joint quantiles of temperature and precipitation in Europe since 1901 and projected for 2100. *Geophysical Research Letters*, **36**, L07707.

Beniston, M., D.B. Stephenson, O.B. Christensen, C.A.T. Ferro, C. Frei, S. Goyette, K. Halsnaes, T. Holt, K. Jylha, B. Koffi, J. Palutikof, R. Schoell, T. Semmler, and K. Woth, 2007: Future extreme events in European climate: an exploration of regional climate model projections. *Climatic Change*, **81**, 71-95.

Benito, G., A. Diez-Herrero, and M.F. de Villalta, 2003: Magnitude and frequency of flooding in the Tagus basin (Central Spain) over the last millennium. *Climatic Change*, **58(1-2)**, 171-192.

Benito, G., M. Barriendos, C. Llasat, M. Machado, and V. Thorndycraft, 2005: Impacts on natural hazards of climatic origin. Flood risk. In: *A Preliminary General Assessment of the Impacts in Spain Due to the Effects of Climate Change* [Moreno, J.M. (ed.)]. Ministry of Environment, Spain, pp. 507-527.

Benito, G., V.R. Thorndycraft, M. Rico, Y. Sanchez-Moya, and A. Sopena, 2008: Palaeoflood and floodplain records from Spain: Evidence for long-term climate variability and environmental changes. *Geomorphology*, **101(1-2)**, 68-77.

Bergametti, G. and F. Dulac, 1998: Mineral aerosols: Renewed interest for climate forcing and tropospheric chemistry studies. *IGACtivities Newsletter*, **11**, 13-17.

Bernier, N.B., K.R. Thompson, J. Ou, and H. Ritchie, 2007: Mapping the return periods of extreme sea levels: Allowing for short sea level records, seasonality, and climate change. *Global and Planetary Change*, **57(1-2)**, 139-150.

Betts, R.A., O. Boucher, M. Collins, P.M. Cox, P.D. Falloon, N. Gedney, D.L. Hemming, C. Huntingford, C.D. Jones, D.M.W. Sexton, and M.J. Web, 2007: Projected increase in continental runoff due to plant responses to increasing carbon dioxide. *Nature*, **448(7157)**, 1037-1041.

Bevan, S.L.N., P. R. J. North, W.M.F. Grey, S.O. Los, and S.E. Plummer, 2009: Impact of atmospheric aerosol from biomass burning on Amazon dry-season drought. *Journal of Geophysical Research – Atmospheres*, **114**, D09204.

Bhutiyani, M.R., V.S. Kale, and N.J. Pawar, 2008: Changing streamflow patterns in the rivers of northwestern Himalaya: Implications of global warming in the 20th century. *Current Science*, **95(5)**, 618-626.

Biasutti, M., A.H. Sobel, and S.J. Camargo, 2009: The role of the Sahara Low in summertime Sahel rainfall variability and change in the CMIP3 models. *Journal of Climate*, **22(21)**, 5755-5771.

Bindoff, N.L., J. Willebrand, V. Artale, A. Cazenave, J.M. Gregory, S. Gulev, K. Hanawa, C. Le Quéré, S. Levitus, Y. Nojiri, C.K. Shum, L.D. Talley, and A.S. Unnikrishnan, 2007: Observations: Oceanic climate change and sea level. In: *Climate Change 2007: The Physical Science Basis. Contribution of Working Group I to the Fourth Assessment Report of the Intergovernmental Panel on Climate Change* [Solomon, S., D. Qin, M. Manning, Z. Chen, M. Marquis, K.B. Averyt, M. Tignor and H.L. Miller (eds.)]. Cambridge University Press, Cambridge, UK, pp. 385-432.

Bister, M. and K.A. Emanuel, 1998: Dissipative heating and hurricane intensity. *Meteorology and Atmospheric Physics*, Vienna, Austria, **65(3-4)**, 233-240.

Bister, M. and K.A. Emanuel, 2002: Low frequency variability of tropical cyclone potential intensity 2. Climatology for 1982-1995. *Journal of Geophysical Research – Atmospheres*, **107(D22)**, 4621-4624.

Björnsson, H., 2003: Subglacial lakes and jökulhlaups in Iceland. *Global and Planetary Change*, **35(3-4)**, 255-271.

Blenkinsop, S. and H.J. Fowler, 2007a: Changes in drought frequency, severity and duration for the British Isles projected by the PRUDENCE regional climate models. *Journal of Hydrology*, **342**, 50-71.

Blenkinsop, S. and H.J. Fowler, 2007b: Changes in European drought characteristics projected by the PRUDENCE regional climate models. *International Journal of Climatology*, **27(12)**, 1595-1610.

Blewitt, G., Z. Altamimi, J. Davis, R. Gross, C.-Y. Kuo, F.G. Lemoine, A.W. Moore, R.E. Neilan, H.-P. Plag, M. Rothacher, C.K. Shum, M.G. Sideris, T. Schöne, P. Tregoning, and S. Zerbini, 2010: Geodetic observations and global reference frame contributions to understanding sea-level rise and variability. In: *Understanding Sea-Level Rise and Variability* [Church, J.A., P.L. Woodworth, T. Aarup, and W.S. Wilson (eds.)]. Wiley-Blackwell, Chichester, UK, pp. 256-284.

Boberg, F., P. Berg, P. Thejll, W.J. Gutowski, and J.H. Christensen, 2009a: Improved confidence in climate change projections of precipitation evaluated using daily statistics from the PRUDENCE ensemble. *Climate Dynamics*, **32(7-8)**, 1097-1106.

Boberg, F., P. Berg, P. Thejll, W.J. Gutowski, and J.H. Christensen, 2009b: Improved confidence in climate change projections of precipitation further evaluated using daily statistics from the PRUDENCE ensemble. *Climate Dynamics*, **35(7-8)**, 1509-1520.

Boe, J., and L. Terray, 2008: Uncertainties in summer evapotranspiration changes over Europe and implications for regional climate change. *Geophysical Research Letters*, **35**, L05702.

Boer, G.J., 2009: Changes in interannual variability and decadal potential predictability under global warming. *Journal of Climate*, **22**, 3098-3109.

Bollasina, M., and S. Nigam, 2009: Indian Ocean SST, evaporation, and precipitation during the South Asian summer monsoon in IPCC-AR4 coupled simulations. *Climate Dynamics*, **33(7-8)**, 1017-1032.

Bonan, G.B., 2008: Forests and climate change: Forcing, feedbacks, and the climate benefit of forests. *Science*, **320**, 1444-1449.

Bondi, F. and L. Salvatori, 2003: The 5-6 May 1998 mudflows in Campania, Italy. In: *Lessons Learnt from Landslide Disasters in Europe* [Hervas, J. (ed.)]. European Commission, Brussels, Belgium, pp. 102.

Bongaerts, P., T. Ridgway, E. Sampayo, and O. Hoegh-Guldberg, 2010: Assessing the 'deep reef refugia' hypothesis: focus on Caribbean reefs. *Coral Reefs*, **29(2)**, 309-327.

Boo, K.O., W.T. Kwon, and H.J. Baek, 2006: Change of extreme events of temperature and precipitation over Korea using regional projection of future climate change. *Geophysical Research Letters*, **33**, L01701.

Brabson, B.B., D.H. Lister, P.D. Jones, and J.P. Palutikof, 2005: Soil moisture and predicted spells of extreme temperatures in Britain. *Journal of Geophysical Research – Atmospheres*, **110**, D05104.

Branstator, G., and F. Selten, 2009: "Modes of Variability" and climate change. *Journal of Climate*, **22(10)**, 2639-2658.

Brayshaw, D.J., B. Hoskins, and M. Blackburn, 2008: The storm-track response to idealized SST perturbations in an aquaplanet GCM. *Journal of the Atmospheric Sciences*, **65(9)**, 2842-2860.

Brázdil, R., C. Pfister, H. Wanner, H. Von Storch, and J. Luterbacher, 2005: Historical climatology in Europe - The state of the art. *Climatic Change*, **70(3)**, 363-430.

Brázdil, R., Z.W. Kundzewicz, and G. Benito, 2006: Historical hydrology for studying flood risk in Europe. *Hydrological Sciences Journal*, **51(5)**, 739-764.

Brázdil, R., M. Trnka, P. Dobrovolny, K. Chromi, P. Hlavinka, and Z. Zalud, 2009: Variability of droughts in the Czech Republic, 1881-2006. *Theoretical and Applied Climatology*, **97(3-4)**, 297-315.

Breda, N., and V. Badeau, 2008: Forest tree responses to extreme drought and some biotic events: Towards a selection according to hazard tolerance? *Comptes Rendus Geoscience*, **340(9-10)**, 651-662.

Brockhaus, P., D. Luthi, and C. Schär, 2008: Aspects of the diurnal cycle in a regional climate model. *Meteorologische Zeitschrift*, **17(4)**, 433-443.

Bromirski, P.D., R.E. Flick, and D.R. Cayan, 2003: Storminess variability along the California coast: 1858-2000. *Journal of Climate*, **16(6)**, 982.

Brooks, H.E., and N. Dotzek, 2008: The spatial distribution of severe convective storms and an analysis of their secular changes. In: *Climate Extremes and Society* [Diaz, H.F., and R. Murnane (eds.)]. Cambridge University Press, Cambridge, NY, pp. 35-53.

Brovkin, V., T. Raddatz, C.H. Reick, M. Claussen, and V. Gayler, 2009: Global biogeophysical interactions between forest and climate. *Geophysical Research Letters*, **36**, L07405.

Brown, J., M. Collins, and A. Tudhope, 2006: Coupled model simulations of mid-Holocene ENSO and comparisons with coral oxygen isotope record. *Advances in Geosciences*, **6**, 29-33.

Brown, J.M., A.J. Souza, and J. Wolf, 2010: Surge modelling in the eastern Irish Sea: present and future storm impact. *Ocean Dynamics*, **60(2)**, 227-236.

Brown, R.D., and P.W. Mote, 2009: The response of Northern Hemisphere snow cover to a changing climate. *Journal of Climate*, **22(8)**, 2124-2145.

Brown, S.J., J. Caesar, and C.A.T. Ferro, 2008: Global changes in extreme daily temperature since 1950. *Journal of Geophysical Research – Atmospheres*, **113**, D05115.

Brunet, M., P.D. Jones, J. Sigró, O. Saladié, E. Aguilar, A. Moberg, P.M. Della-Marta, D. Lister, A. Walther, and D. López, 2007: Temporal and spatial temperature variability and change over Spain during 1850-2005. *Journal of Geophysical Research – Atmospheres*, **112**, D12117.

Bryan, K.R., P.S. Kench, and D.E. Hart, 2008: Multi-decadal coastal change in New Zealand: Evidence, mechanisms and implications. *New Zealand Geographer*, **64(2)**, 117-128.

Büntgen, U., V. Trouet, D. Frank, H.H. Leuschner, D. Friedrichs, J. Luterbacher, and J. Esper, 2010: Tree-ring indicators of German summer drought over the last millennium. *Quaternary Science Reviews*, **29(7-8)**, 1005-1016.

Burke, E.J., and S.J. Brown, 2008: Evaluating uncertainties in the projection of future drought. *Journal of Hydrometeorology*, **9(2)**, 292-299.

Burke, E.J., S.J. Brown, and N. Christidis, 2006: Modeling the recent evolution of global drought and projections for the twenty-first century with the Hadley Centre climate model. *Journal of Hydrometeorology*, **7(5)**, 1113-1125.

Burke, E.J., R.H.J. Perry, and S.J. Brown, 2010: An extreme value analysis of UK drought and projections of change in the future. *Journal of Hydrology*, **388(1-2)**, 131-143.

Burkholder, B.A. and D.J. Karoly, 2007: An assessment of US climate variability using the Climate Extremes Index. In: *Proceedings of the Nineteenth Conference on Climate Variability and Change*, San Antonio, TX, 15-18 January 2007, pp. 2B.9.

Burton, A., C.G. Kilsby, H.J. Fowler, P.S.P. Cowpertwait, and P.E. O'Connell, 2008: RainSim: A spatial-temporal stochastic rainfall modelling system. *Environmental Modelling & Software*, **23(12)**, 1356-1369.

Caesar, J., L. Alexander, and R. Vose, 2006: Large-scale changes in observed daily maximum and minimum temperatures: Creation and analysis of a new gridded data set. *Journal of Geophysical Research – Atmospheres*, **111**, D05101.

Caesar, J., L.V. Alexander, B. Trewin, K. Tse-ring, L. Sorany, V. Vuniyayawa, N. Keosavang, A. Shimana, M.M. Htay, J. Karmacharya, D.A. Jayasinghearachchi, J. Sakkamart, A. Soares, L.T. Hung, L.T. Thuong, C.T. Hue, N.T.T. Dung, P.V. Hung, H.D. Cuong, N.M. Cuong, and S. Sirabaha, 2011: Changes in temperature and precipitation extremes over the Indo-Pacific region from 1971 to 2005. *International Journal of Climatology*, **31(6)**, 791-801.

Callaghan, J., and S. Power, 2011: Variability and decline in the number of severe tropical cyclones making land-fall over eastern Australia since the late nineteenth century. *Climate Dynamics*, **37(3-4)**, 647-662, doi:10.1007/s00382-00010-00883-00382.

Camberlin, P., V. Moron, R. Okoola, N. Philippon, and W. Gitau, 2009: Components of rainy seasons' variability in Equatorial East Africa: onset, cessation, rainfall frequency and intensity. *Theoretical and Applied Climatology*, **98(3-4)**, 237-249.

Cambers, G., 2009: Caribbean beach changes and climate change adaptation. *Aquatic Ecosystem Health & Management*, **12(2)**, 168-176.

Cameron, D., 2006: An application of the UKCIP02 climate change scenarios to flood estimation by continuous simulation for a gauged catchment in the northeast of Scotland, UK (with uncertainty). *Journal of Hydrology*, **328(1-2)**, 212-226.

Camilloni, I.A. and V.R. Barros, 2003: Extreme discharge events in the Paraná River and their climate forcing. *Journal of Hydrology*, **278(1-4)**, 94-106.

Campbell, J.D., M.A. Taylor, T.S. Stephenson, R.A. Watson, and F.S. Whyte, 2011: Future climate of the Caribbean from a regional climate model. *International Journal of Climatology*, **31(12)**, 1866-1878, doi:10.1002/joc.2200.

Camuffo, D., C. Bertolin, M. Barriendos, F. Dominguez-Castro, C. Cocheo, S. Enzi, M. Sghedoni, A. della Valle, E. Garnier, M.J. Alcoforado, E. Xoplaki, J. Luterbacher, N. Diodato, M. Maugeri, M.F. Nunes, and R. Rodriguez, 2010: 500-year temperature reconstruction in the Mediterranean Basin by means of documentary data and instrumental observations. *Climatic Change*, **101(1-2)**, 169-199.

Cane, M.A., 2005: The evolution of El Niño, past and future. *Earth and Planetary Science Letters*, **230(3-4)**, 227-240.

Cao, Z., 2008: Severe hail frequency over Ontario, Canada: Recent trend and variability. *Geophysical Research Letters*, **35**, L14803.

Caplan-Auerbach, J., and C. Huggel, 2007: Precursory seismicity associated with frequent, large ice avalanches on Iliamna volcano, Alaska, USA. *Journal of Glaciology*, **53(180)**, 128-140.

Capra, L., 2006: Abrupt climatic changes as triggering mechanisms of massive volcanic collapses. *Journal of Volcanology and Geothermal Research*, **155(3-4)**, 329-333.

Carey, M., 2005: Living and dying with glaciers: people's historical vulnerability to avalanches and outburst floods in Peru. *Global and Planetary Change*, **47(2-4)**, 122-134.

Carey-Smith, T., S. Dean, J. Vial, and C. Thompson, 2010: Changes in precipitation extremes for New Zealand: climate model predictions. *Weather and Climate*, **30**, 23-48.

Carter, T.R., R.N. Jones, X. Lu, S. Bhadwal, C. Conde, L.O. Mearns, B.C. O'Neill, M.D.A. Rounsevell, and M.B. Zurek, 2007: New assessment methods and the characterisation of future conditions. In: *Climate Change 2007. Impacts, Adaptation and Vulnerability. Contribution of Working Group II to the Fourth Assessment Report of the Intergovernmental Panel on Climate Change* [Parry, M.L., O.F. Canziani, J.P. Palutikof, P.J. Van Der Linde, and C.E. Hanson (eds.)]. Cambridge University Press, Cambridge, UK, pp. 133-171.

Cassou, C., L. Terray, and A.S. Phillips, 2005: Tropical Atlantic influence on European heat waves. *Journal of Climate*, **18(15)**, 2805-2811.

Catto, J.L., L.C. Shaffrey, and K.I. Hodges, 2010: Can climate models capture the structure of extratropical cyclones? *Journal of Climate*, **23(7)**, 1621-1635.

Cavazos, T., C. Turrent, and D.P. Lettenmaier, 2008: Extreme precipitation trends associated with tropical cyclones in the core of the North American monsoon. *Geophysical Research Letters*, **35**, L21703.

Cayan, D.R., P.D. Bromirski, K. Hayhoe, M. Tyree, M.D. Dettinger, and R.E. Flick, 2008: Climate change projections of sea level extremes along the California coast. *Climatic Change*, **87(S1)**, 57-73.

Cayan, D.R., T. Das, D.W. Pierce, T.P. Barnett, M. Tyree, and A. Gershunov, 2010: Future dryness in the southwest US and the hydrology of the early 21st century drought. *Proceedings of the National Academy of Sciences*, **107(50)**, 21271-21276.

Cazenave, A. and W. Llovel, 2010: Contemporary sea level rise. *Annual Review of Marine Science*, **2**, 145-173.

CCSP, 2008: *Weather and Climate Extremes in a Changing Climate. Regions of Focus: North America, Hawaii, Caribbean, and U.S. Pacific Islands.* A Report by the U.S. Climate Change Science Program and the Subcommittee on Global Change Research [Karl, T.R., G.A. Meehl, D.M. Christopher, S.J. Hassol, A.M. Waple, and W.L. Murray (eds.)]. Department of Commerce, NOAA's National Climatic Data Center, Washington, DC, 164 pp.

Chambers, L.E. and G.M. Griffiths, 2008: The changing nature of temperature extremes in Australia and New Zealand. *Australian Meteorological Magazine*, **57(1)**, 13-35.

Chan, J.C.L. and M. Xu, 2009: Inter-annual and inter-decadal variations of landfalling tropical cyclones in East Asia. Part I: Time series analysis. *International Journal of Climatology*, **29(9)**, 1285-1293.

Chang, E.K.M., 2007: Assessing the Increasing Trend in Northern Hemisphere Winter Storm Track Activity Using Surface Ship Observations and a Statistical Storm Track Model. *Journal of Climate*, **20**, 5607–5628.

Chang, E.K.M., 2009: Are band-pass variance statistics useful measures of storm track activity? Re-examining storm track variability associated with the NAO using multiple storm track measures. *Climate Dynamics*, **33(2-3)**, 277-296.

Chang, E.K.M. and Y. Guo, 2007: Is the number of North Atlantic tropical cyclones significantly underestimated prior to the availability of satellite observations? *Geophysical Research Letters*, **34**, L14801.

Chauvin, F., J.F. Royer, and M. Deque, 2006: Response of hurricane-type vortices to global warming as simulated by ARPEGE-Climat at high resolution. *Climate Dynamics*, **27(4)**, 377-399.

Chen, C.-T. and T. Knutson, 2008: On the verification and comparison of extreme rainfall indices from climate models. *Journal of Climate*, **21(7)**, 1605-1621.

Chen, J., F.P. Brissette, and R. Leconte, 2011: Uncertainty of downscaling method in quantifying the impact of climate change on hydrology. *Journal of Hydrology*, **401(3-4)**, 190-202.

Chen, J.L., C.R. Wilson, B.D. Tapley, Z.L. Yang, and G.Y. Niu, 2009: 2005 drought event in the Amazon River basin as measured by GRACE and estimated by climate models. *Journal of Geophysical Research – Solid Earth*, **114**, B05404.

Chenoweth, M. and M. Devine, 2008: A document-based 318-year record of tropical cyclones in the Lesser Antilles, 1690-2007. *Geochemistry Geophysics Geosystems*, **9**, Q08013.

Cherchi, A., S. Masina, and A. Navarra, 2008: Impact of extreme CO_2 levels on tropical climate: a CGCM study. *Climate Dynamics*, **31(7-8)**, 743-758.

Cherry, N.J., 1988: Comment on long-term associations between wind speeds and the urban heat-island of Phoenix, Arizona. *Journal of Applied Meteorology*, **27(7)**, 878-880.

Chiapello, I., C. Moulin, and J.M. Prospero, 2005: Understanding the long-term variability of African dust transport across the Atlantic as recorded in both Barbados surface concentrations and large-scale Total Ozone Mapping Spectrometer (TOMS) optical thickness. *Journal of Geophysical Research – Atmospheres*, **110**, D18S10.

Chiarle, M., S. Iannotti, G. Mortara, and P. Deline, 2007: Recent debris flow occurrences associated with glaciers in the Alps. *Global and Planetary Change*, **56(1-2)**, 123-136.

Chini, N., P. Stansby, J. Leake, J. Wolf, J. Roberts-Jones, and J. Lowe, 2010: The impact of sea level rise and climate change on inshore wave climate: A case study for East Anglia (UK). *Coastal Engineering*, **57(11-12)**, 973-984.

Choi, G., D. Collins, R. Guoyu, B. Trewin, M. Baldi, Y. Fukuda, M. Afzaal, T. Pianmana, P. Gomboluudev, P.T.T. Huong, N. Lias, W.T. Kwon, K.O. Boo, Y.M. Cha, and Y. Zhou, 2009: Changes in means and extreme events of temperature and precipitation in the Asia-Pacific Network region, 1955-2007. *International Journal of Climatology*, **29(13)**, 1956-1975.

Chowdhury, M.R., P.S. Chu, and T. Schroeder, 2007: ENSO and seasonal sea-level variability – A diagnostic discussion for the US-affiliated Pacific Islands. *Theoretical and Applied Climatology*, **88(3-4)**, 213-224.

Christensen, J.H. and O.B. Christensen, 2003: Climate modeling: Severe summertime flooding in Europe. *Nature*, **421**, 805-806.

Christensen, J.H. and O.B. Christensen, 2007: A summary of the PRUDENCE model projections of changes in European climate by the end of this century. *Climatic Change*, **81(S1)**, 7-30.

Christensen, J.H., B. Hewitson, A. Busuioc, A. Chen, X. Gau, I. Held, R. Jones, R. Kolli, W. Kwon, R. Laprise, V. Magaña Rueda, L. Mearns, C. Menéndez, J. Räisänen, A. Rinke, A. Sarr, and P. Whetton, 2007: Regional climate projections. In: *Climate Change 2007: The Physical Science Basis. Contribution of Working Group I to the Fourth Assessment Report of the Intergovernmental Panel on Climate Change* [Solomon, S., D. Qin, M. Manning, Z. Chen, M. Marquis, K.B. Averyt, M. Tignor and H.L. Miller (eds.)]. Cambridge University Press, Cambridge, UK, pp. 847-940.

Christidis, N., P.A. Stott, S. Brown, G. Hegerl, and J. Caesar, 2005: Detection of changes in temperature extremes during the second half of the 20th century. *Geophysical Research Letters*, **32**, L20716.

Christidis, N., P.A. Stott, G.S. Jones, H. Shiogama, T. Nozawa, and J. Luterbacher, 2011a: Human activity and anomalously warm seasons in Europe. *International Journal of Climatology*, doi:10.1002/joc.2262.

Christidis, N., P.A. Stott, and S.J. Brown, 2011b: The role of human activity in the recent warming of extremely warm daytime temperatures. *Journal of Climate*, **24(7)**, 1922-1930.

Church, J.A. and N.J. White, 2011: Sea-level rise from the late 19th to the early 21st century. *Surveys in Geophysics*, **32(4-5)**, 585-602, doi:10.1007/s10712-10011-19119-10711.

Church, J.A., J.R. Hunter, K.L. McInnes, and N.J. White, 2006a: Sea-level rise around the Australian coastline and the changing frequency of extreme sea-level events. *Australian Meteorological Magazine*, **55(4)**, 253-260.

Church, J.A., N.J. White, and J.R. Hunter, 2006b: Sea-level rise at tropical Pacific and Indian Ocean Islands. *Global and Planetary Change*, **53(3)**, 155-168.

Church, J.A., N.J. White, J.R. Hunter, K.L. McInnes, P. J. Cowell , and S.P. O'Farrell, 2008: Sea-level rise. In: *Transitions: Pathways Towards Sustainable Urban Development in Australia* [Newton, E.P. (ed.)]. CSIRO Publishing and Springer Science, Collingwood, Australia, pp. 691.

Church, J.A., D. Roemmich, C.M. Domingues, J.K. Willis, N.J. White, J.E. Gilson, D. Stammer, A. Köhl, D.P. Chambers, F.W. Landerer, J. Marotzke, J.M. Gregory, T. Suzuki, A. Cazenave, and P.-Y. Le Traon, 2010: Ocean temperature and salinity contributions to global and regional sea-level change. In: *Understanding Sea-Level Rise and Variability* [Church, J.A., P.L. Woodworth, T. Aarup, and W.S. Wilson (eds.)]. Wiley-Blackwell, Chichester, UK, pp. 143-176.

Church, J.A., J.M. Gregory, N.J. White, S.M. Platten, and J.X. Mitrovica, 2011: Understanding and projecting sea level change. *Oceanography*, **24(2)**, 130-143.

Chust, G., A. Borja, P. Liria, I. Galparsoro, M. Marcos, A. Caballero, and R. Castro, 2009: Human impacts overwhelm the effects of sea-level rise on Basque coastal habitats (N Spain) between 1954 and 2004. *Estuarine Coastal and Shelf Science*, **84(4)**, 453-462.

Ciais, P., M. Reichstein, N. Viovy, A. Granier, J. Ogee, V. Allard, M. Aubinet, N. Buchmann, C. Bernhofer, A. Carrara, F. Chevallier, N. De Noblet, A. Friend, P. Friedlingstein, T. Grunwald, B. Heinesch, P. Keronen, A. Knohl, G. Krinner, D. Loustau, G. Manca, G. Matteucci, F. Miglietta, J. Ourcival, D. Papale, K. Pilegaard, S. Rambal, G. Seufert, J. Soussana, M. Sanz, E. Schulze, T. Vesala, and R. Valentini, 2005: Europe-wide reduction in primary productivity caused by the heat and drought in 2003. *Nature*, **437(7058)**, 529-533.

Clague, J.J. and S.G. Evans, 2000: A review of catastrophic drainage of moraine-dammed lakes in British Columbia. *Quaternary Science Reviews*, **19(17-18)**, 1763-1783.

Clark, R.T., S.J. Brown, and J.M. Murphy, 2006: Modeling northern hemisphere summer heat extreme changes and their uncertainties using a physics ensemble of climate sensitivity experiments. *Journal of Climate*, **19(17)**, 4418-4435.

Clark, R.T., J.M. Murphy, and S.J. Brown, 2010: Do global warming targets limit heatwave risk? *Geophysical Research Letters*, **37**, L17703.

Clarke, G.K.C., 1982: Glacier outburst floods from 'Hazard Lake,' Yukon Territory, and the problem of flood magnitude prediction. *Journal of Glaciology*, **28(98)**, 3-21.

Clarke, M.L., and H.M. Rendell, 2009: The impact of North Atlantic storminess on western European coasts: A review. *Quaternary International*, **195**, 31-41.

Clemmensen, L.B., M. Bjornsen, A. Murray, and K. Pedersen, 2007: Formation of aeolian dunes on Anholt, Denmark since AD 1560: A record of deforestation and increased storminess. *Sedimentary Geology*, **199(3-4)**, 171-187.

Clow, D.W., 2010: Changes in the timing of snowmelt and streamflow in Colorado: a response to recent warming. *Journal of Climate*, **23(9)**, 2293-2306.

Cobb, K.M., C.D. Charles, H. Cheng, and R.L. Edwards, 2003: El Niño/Southern Oscillation and tropical Pacific climate during the last millennium. *Nature*, **424**, 271-276.

Coelho, C., R. Silva, F. Veloso-Gomes, and F. Taveira-Pinto, 2009: Potential effects of climate change on northwest Portuguese coastal zones. *ICES Journal of Marine Science*, **66(7)**, 1497-1507.

Coles, S., 2001: *An Introduction to Statistical Modeling of Extreme Values*. Springer-Verlag, Heidelberg, Germany, 208 pp.

Collini, E.A., E.H. Berbery, V. Barros, and M. Pyle, 2008: How does soil moisture influence the early stages of the South American monsoon? *Journal of Climate*, **21(2)**, 195-213.

Collins, M., B.B.B. Booth, G.R. Harris, J.M. Murphy, D.M.H. Sexton, and M.J. Webb, 2006: Towards quantifying uncertainty in transient climate change. *Climate Dynamics*, **27(2-3)**, 127-147.

Collins, M., S.I. An, W.J. Cai, A. Ganachaud, E. Guilyardi, F.F. Jin, M. Jochum, M. Lengaigne, S. Power, A. Timmermann, G. Vecchi, and A. Wittenberg, 2010: The impact of global warming on the tropical Pacific ocean and El Nino. *Nature Geoscience*, **3(6)**, 391-397.

Compo, G.P., J.S. Whitaker, and P.D. Sardeshmukh, 2006: Feasibility of a 100-year reanalysis using only surface pressure data. *Bulletin of the American Meteorological Society*, **87(2)**, 175-190.

Conroy, J.L., A. Restrepo, J.T. Overpeck, M. Seinitz-Kannan, J.E. Cole, M.B. Bush, and P.A. Colinvaux, 2009: Unprecedented recent warming of surface temperatures in the eastern tropical Pacific Ocean. *Nature Geoscience*, **2(1)**, 46-50.

Conway, D., C.E. Hanson, R. Doherty, and A. Persechino, 2007: GCM simulations of the Indian Ocean dipole influence on East African rainfall: Present and future. *Geophysical Research Letters*, **34**, L03705.

Conway, D., A. Persechino, S. Ardoin-Bardin, H. Hamandawana, C. Dieulin, and G. Mahe, 2009: Rainfall and water resources variability in sub-Saharan Africa during the twentieth century. *Journal of Hydrometeorology*, **10(1)**, 41-59.

Cook, B.I., G.B. Bonan, and S. Levis, 2006: Soil moisture feedbacks to precipitation in southern Africa. *Journal of Climate*, **19(17)**, 4198-4206.

Cook, K.H. and E.K. Vizy, 2006: Coupled model simulations of the west African monsoon system: Twentieth- and twenty-first-century simulations. *Journal of Climate*, **19(15)**, 3681-3703.

Coppola, E. and F. Giorgi, 2005: Climate change in tropical regions from high-resolution time-slice AGCM experiments. *Quarterly Journal of the Royal Meteorological Society*, **131(612)**, 3123-3145.

Cordero, E.C. and P.M.D. Forster, 2006: Stratospheric variability and trends in models used for the IPCC AR4. *Atmospheric Chemistry and Physics*, **6(12)**, 5369-5380.

Corti, T., V. Muccione, P. Köllner-Heck, D. Bresch, and S.I. Seneviratne, 2009: Simulating past droughts and associated building damages in France. *Hydrology and Earth System Sciences*, **13(9)**, 1739-1747.

Costa, A.C., and A. Soares, 2009: Trends in extreme precipitation indices derived from a daily rainfall database for the South of Portugal. *International Journal of Climatology*, **29(13)**, 1956-1975.

CSIRO, 2007: *Climate Change in Australia*. CSIRO, Australia, 148 pp.

Cubasch, U., J. Waszkewitz, G. Hegerl, and J. Perlwitz, 1995: Regional climate changes as simulated in time-slice experiments. *Climatic Change*, **31(2-4)**, 273-304.

Cullen, N.J., T. Molg, G. Kaser, K. Hussein, K. Steffen, and D.R. Hardy, 2006: Kilimanjaro Glaciers: Recent areal extent from satellite data and new interpretation of observed 20th century retreat rates. *Geophysical Research Letters*, **33**, L16502.

Cunderlik, J.M., and T.B.M.J. Ouarda, 2009: Trends in the timing and magnitude of floods in Canada. *Journal of Hydrology*, **375(3-4)**, 471-480.

Cunningham, P., and M.J. Reeder, 2009: Severe convective storms initiated by intense wildfires: Numerical simulations of pyro-convection and pyro-tornadogenesis. *Geophysical Research Letters*, **36**, L12812.

da Silva, A.E. and L.M.V. de Carvalho, 2007: Large-scale index for South America Monsoon (LISAM). *Atmospheric Science Letters*, **8(2)**, 51-57.

Dai, A., T.T. Qian, K.E. Trenberth, and J.D. Milliman, 2009: Changes in continental freshwater discharge from 1948 to 2004. *Journal of Climate*, **22(10)**, 2773-2792.

Dai, A., 2011: Drought under global warming: a review. *Wiley Interdisciplinary Reviews: Climate Change*, **2(1)**, 45-65.

Dai, A.G., K.E. Trenberth, and T.T. Qian, 2004: A global dataset of Palmer Drought Severity Index for 1870-2002: Relationship with soil moisture and effects of surface warming. *Journal of Hydrometeorology*, **5(6)**, 1117-1130.

Dairaku, K., S. Emori, and H. Higashi, 2008: Potential changes in extreme events under global climate change. *Journal of Disaster Research*, **3(1)**, 39-50.

Danard, M.B., S.K. Dube, G. Gönnert, A. Munroe, T.S. Murty, P. Chittibabu, A.D. Rao, and P.C. Sinha, 2004: Storm surges from extra-tropical cyclones. *Natural Hazards*, **32(2)**, 177-190.

Dankers, R., O.B. Christensen, L. Feyen, M. Kalas, and A. de Roo, 2007: Evaluation of very high-resolution climate model data for simulating flood hazards in the Upper Danube Basin. *Journal of Hydrology*, **347(3-4)**, 319-331.

Dankers, R. and L. Feyen, 2008: Climate change impact on flood hazard in Europe: An assessment based on high-resolution climate simulations. *Journal of Geophysical Research – Atmospheres*, **113**, D19105.

Dankers, R. and L. Feyen, 2009: Flood hazard in Europe in an ensemble of regional climate scenarios. *Journal of Geophysical Research – Atmospheres*, **114**, D16108.

Dasgupta, S., B. Laplante, C. Meisner, D. Wheeler, and J. Yan, 2009: The impact of sea level rise on developing countries: a comparative analysis. *Climatic Change*, **93(3-4)**, 379-388.

Dawson, J.L., and S.G. Smithers, 2010: Shoreline and beach volume change between 1967 and 2007 at Raine Island, Great Barrier Reef, Australia. *Global and Planetary Change*, **72(3)**, 141-154.

de Elía, R., D. Caya, H. Côté, A. Frigon, S. Biner, M. Giguère, D. Paquin, R. Harvey, and D. Plummer, 2008: Evaluation of uncertainties in the CRCM-simulated North American climate. *Climate Dynamics*, **30(2-3)**, 113-132.

de Jong, R., S. Bjorck, L. Bjorkman, and L.B. Clemmensen, 2006: Storminess variation during the last 6500 years as reconstructed from an ombrotrophic peat bog in Halland, southwest Sweden. *Journal of Quaternary Science*, **21(8)**, 905-919.

de Jong, R., K. Schoning, and S. Bjorck, 2007: Increased aeolian activity during humidity shifts as recorded in a raised bog in south-west Sweden during the past 1700 years. *Climate of the Past*, **3(3)**, 411-422.

Dean, S.M. and P.A. Stott, 2009: The effect of local circulation variability on the detection and attribution of New Zealand temperature trends. *Journal of Climate*, **22(23)**, 6217-6229.

Debernard, J.B. and L.P. Roed, 2008: Future wind, wave and storm surge climate in the Northern Seas: a revisit. *Tellus Series A – Dynamic Meteorology and Oceanography*, **60(3)**, 427-438.

Deeming, K.R., B. McGuire, and P. Harrop, 2010: Climate forcing of volcano lateral collapse: evidence from Mount Etna, Sicily. *Philosophical Transactions of the Royal Society A: Mathematical, Physical and Engineering Sciences*, **368(1919)**, 2559-2577.

Defeo, O., A. McLachlan, D.S. Schoeman, T.A. Schlacher, J. Dugan, A. Jones, M. Lastra, and F. Scapini, 2009: Threats to sandy beach ecosystems: A review. *Estuarine Coastal and Shelf Science*, **81(1)**, 1-12.

DeGaetano, A.T., 2009: Time-dependent changes in extreme-precipitation return-period amounts in the Continental United States. *Journal of Applied Meteorology and Climatology*, **48**, 2086-2099.

Delaloye, R., T. Strozzi, C. Lambiel, E. Perruchoud, and H. Raetzo, 2008: Landslide-like development of rockglaciers detected with ERS-1/2 SAR interferometry. In: *Proceedings of the FRINGE 2007 Symposium at ESRIN* [Lacoste, H., and L. Ouwehand (eds.)], Frascati, Italy, 26-30 November 2007, ESA Communication Production Office, ESTEC, Noordwijk, The Netherlands, pp. 1-6.

Delgado, J.M., H. Apel, and B. Merz, 2009: Flood trends and variability in the Mekong River. *Hydrology and Earth System Sciences*, **6(3)**, 6691-6719.

Della-Marta, P.M. and H. Wanner, 2006: A method of homogenizing the extremes and mean of daily temperature measurements. *Journal of Climate*, **19**, 4179-4197.

Della-Marta, P.M., M.R. Haylock, J. Luterbacher, and H. Wanner, 2007a: Doubled length of western European summer heat waves since 1880. *Journal of Geophysical Research – Atmospheres*, **112**, D15103.

Della-Marta, P.M., J. Luterbacher, H. von Weissenfluh, E. Xoplaki, M. Brunet, and H. Wanner, 2007b: Summer heat waves over western Europe 1880-2003, their relationship to large-scale forcings and predictability. *Climate Dynamics*, **29(2-3)**, 251-275.

Della-Marta, P.M., H. Mathis, C. Frei, M.A. Liniger, J. Kleinn, and C. Appenzeller, 2009: The return period of wind storms over Europe. *International Journal of Climatology*, **29(3)**, 437-459.

DeMaria, M., J.A. Knaff, and B.H. Connell, 2001: A tropical cyclone genesis parameter for the tropical Atlantic. *Weather and Forecasting*, **16(2)**, 219-233.

Deo, R.C., J.I. Syktus, C.A. McAlpine, P.J. Lawrence, H.A. McGowan, and S.R. Phinn, 2009: Impact of historical land cover change on daily indices of climate extremes including droughts in eastern Australia. *Geophysical Research Letters*, **36**, L08705.

Department of Climate Change, 2009: *Climate Change Risks to Australia's Coast. A First Pass National Assessment*. Report published by the Department of Climate Change, Australian Government, Canberra, Australia, 172 pp.

Déqué, M., S. Somot, E. Sanchez-Gomez, C. Goodess, D. Jacob, G. Lenderink, and O. Christensen, 2011: The spread amongst ENSEMBLES regional scenarios: regional climate models, driving general circulation models and interannual variability. *Climate Dynamics*, doi:10.1007/s00382-00011-01053-x.

Deser, C., R.A. Tomas, and S. Peng, 2007: The transient atmospheric circulation response to North Atlantic SST and sea ice anomalies. *Journal of Climate*, **20(18)**, 4751-4767.

Di Baldassarre, G., A. Montanari, H. Lins, D. Koutsoyiannis, L. Brandimarte, and G. Blöschl, 2010: Flood fatalities in Africa: From diagnosis to mitigation. *Geophysical Research Letters*, **37**, L22402.

Dibike, Y.B., P. Gachon, A. St-Hilaire, T. Ouarda, and V.T.V. Nguyen, 2008: Uncertainty analysis of statistically downscaled temperature and precipitation regimes in Northern Canada. *Theoretical and Applied Climatology*, **91(1-4)**, 149-170.

Diffenbaugh, N.S. and M. Ashfaq, 2010: Intensification of hot extremes in the United States. *Geophysical Research Letters*, **37(15)**, L15701.

Diffenbaugh, N.S., J.S. Pal, F. Giorgi, and X. Gao, 2007: Heat stress intensification in the Mediterranean climate change hotspot. *Geophysical Research Letters*, **34**, L11706.

Diffenbaugh, N.S., R.J. Trapp, and H. Brooks, 2008: Does global warming influence tornado activity? *Eos Transactions (AGU)*, **89(53)**, 553.

Ding, T., W. Qian, and Z. Yan, 2010: Changes in hot days and heat waves in China during 1961-2007. *International Journal of Climatology*, **30(10)**, 1452-1462.

Dirmeyer, P.A., R.D. Koster, and Z. Guo, 2006: Do global models properly represent the feedback between land and atmosphere? *Journal of Hydrometeorology*, **7(6)**, 1177-1198.

Dobrovolny, P., A. Moberg, R. Brazdil, C. Pfister, R. Glaser, R. Wilson, A. van Engelen, D. Limanowka, A. Kiss, M. Halickova, J. Mackova, D. Riemann, J. Luterbacher, and R. Bohm, 2010: Monthly, seasonal and annual temperature reconstructions for Central Europe derived from documentary evidence and instrumental records since AD 1500. *Climatic Change*, **101(1-2)**, 69-107.

Dodet, G., X. Bertin, and R. Taborda, 2010: Wave climate variability in the North-East Atlantic Ocean over the last six decades. *Ocean Modelling*, **31(3-4)**, 120-131.

Dole, R., M. Hoerling, J. Perlwitz, J. Eischeid, P. Pegion, T. Zhang, X.-W. Quan, T. Xu, and D. Murray, 2011: Was there a basis for anticipating the 2010 Russian heat wave? *Geophysical Research Letters*, **38**, L06702.

Donat, M.G., G.C. Leckebusch, J.G. Pinto, and U. Ulbrich, 2010a: European storminess and associated circulation weather types: future changes deduced from a multi-model ensemble of GCM simulations. *Climate Research*, **42(1)**, 27-43.

Donat, M.G., G.C. Leckebusch, J.G. Pinto, and U. Ulbrich, 2010b: Examination of wind storms over Central Europe with respect to circulation weather types and NAO phases. *International Journal of Climatology*, **30(9)**, 1289-1300.

Donat, M.G., G.C. Leckebusch, S. Wild, and U. Ulbrich, 2011: Future changes of European winter storm losses and extreme wind speeds in multi-model GCM and RCM simulations. *Natural Hazards and Earth System Sciences*, **11(5)**, 1351-1370.

Dong, B., R. Sutton, and T. Woollings, 2011: Changes of interannual NAO variability in response to greenhouse gases forcing. *Climate Dynamics*, **37(7-8)**, 1621-1641, doi:10.1007/s00382-00010-00936-00386.

Donnelly, J.P., and J.D. Woodruff, 2007: Intense hurricane activity over the past 5,000 years controlled by El Niño and the West African monsoon. *Nature*, **447**, 465-468.

D'Onofrio, A., J.P. Boulanger, and E.C. Segura, 2010: CHAC: a weather pattern classification system for regional climate downscaling of daily precipitation. *Climatic Change*, **98(3-4)**, 405-427.

Donohue, R.J., T.R. McVicar, and M.L. Roderick, 2010: Assessing the ability of potential evaporation formulations to capture the dynamics in evaporative demand within a changing climate. *Journal of Hydrology*, **386**(1-4), 186-197.

Doser, D.I., K.R. Wiest, and J. Sauber, 2007: Seismicity of the Bering Glacier region and its relation to tectonic and glacial processes. *Tectonophysics*, **439**(1-4), 119-127.

Douglas, E.M., R.M. Vogel, and C.N. Kroll, 2000: Trends in floods and low flows in the United States: impact of spatial correlation. *Journal of Hydrology*, **240**(1-2), 90-105.

Dragani, W.C., P.B. Martin, C.G. Simionato, and M.I. Campos, 2010: Are wind wave heights increasing in south-eastern south American continental shelf between 32°S and 40°S? *Continental Shelf Research*, **30**(5), 481-490.

Dufek, A.S. and T. Ambrizzi, 2008: Precipitation variability in Sao Paulo State, Brazil. *Theoretical and Applied Climatology*, **93**(3-4), 167-178.

Dufek, A.S., T. Ambrizzi, and R.P. da Rocha, 2008: Are reanalysis data useful for calculating climate indices over South America? In: *Trends and Directions in Climate Research* [Gimeno, L., R. GarciaHerrera, and R.M. Trigo (eds.)]. Vol. 1146, Annals of the New York Academy of Sciences, New York, NY, USA, pp. 87-104.

Durante, F., and G. Salvadori, 2010: On the construction of multivariate extreme value models via copulas. *Environmetrics*, **21**(2), 143-161.

Durão, R.M., M.J. Pereira, A.C. Costa, J. Delgado, G. del Barrio, and A. Soares, 2010: Spatial-temporal dynamics of precipitation extremes in southern Portugal: a geostatistical assessment study. *International Journal of Climatology*, **30**(10), 1526-1537.

Durre, I., J.M. Wallace, and D.P. Lettenmaier, 2000: Dependence of extreme daily maximum temperatures on antecedent soil moisture in the contiguous United States during summer. *Journal of Climate*, **13**(14), 2641-2651.

Durre, I., M.J. Menne, and R.S. Vose, 2008: Strategies for evaluating quality assurance procedures. *Journal of Applied Meteorology and Climatology*, **47**(6), 1785-1791.

Dussaillant, A., G. Benito, W. Buytaert, P. Carling, C. Meier, and F. Espinoza, 2010: Repeated glacial-lake outburst floods in Patagonia: an increasing hazard? *Natural Hazards*, **54**(2), 469-481.

Dutton, J.F., C.J. Poulsen, and J.L. Evans, 2000: The effect of global climate change on the regions of tropical convection in CSM1. *Geophysical Research Letters*, **27**(19), 3049-3052.

Easterling, D.R., D.M. Anderson, S.J. Cohen, W.J. Gutowski, G.J. Holland, K.E. Kunkel, T.C. Peterson, R.S. Pulwarty, R.J. Stouffer, and M.F. Wehner, 2008: Measures to improve our understanding of weather and climate extremes. In: *Weather and Climate Extremes in a Changing Climate. Regions of focus: North America, Hawaii, Caribbean, and U.S. Pacific Islands.* [Karl, T.R., G.A. Meehl, D.M. Christopher, S.J. Hassol, A.M. Waple, and W.L. Murray (eds.)]. A Report by the U.S. Climate Change Science Program and the Subcommittee on Global Change Research, Washington, DC, pp. 117-126.

Eichler, T., and W. Higgins, 2006: Climatology and ENSO-related variability of North American extratropical cyclone activity. *Journal of Climate*, **19**(10), 2076-2093.

Elsner, J.B., J.P. Kossin, and T.H. Jagger, 2008: The increasing intensity of the strongest tropical cyclones. *Nature*, **455**(7209), 92-95.

Eltahir, E.A.B. and P.A.J.F. Yeh, 1999: On the asymmetric response of aquifer water level to floods and droughts in Illinois. *Water Resources Research*, **35**(4), 1199-1217.

Ely, L.L., Y. Enzel, V.R. Baker, and D.R. Cayan, 1993: A 5000-year record of extreme floods and climate-change in the southwestern United-States. *Science*, **262**(5132), 410-412.

Emanuel, K.A., 1987: Dependence of hurricane intensity on climate. *Nature*, **326**(61112), 483-485.

Emanuel, K.A., 2000: A statistical analysis of tropical cyclone intensity. *Monthly Weather Review*, **128**(4), 1139-1152.

Emanuel, K.A., 2007: Environmental factors affecting tropical cyclone power dissipation. *Journal of Climate*, **20**(22), 5497-5509.

Emanuel, K.A., 2010: Tropical cyclone activity downscaled from NOAA-CIRES reanalysis, 1908-1958. *Journal of Advances in Modeling Earth Systems*, **2**, 1, doi:10.3894/JAMES.2010.2.1.

Emanuel, K.A., R. Sundararajan, and J. Williams, 2008: Hurricanes and global warming: Results from downscaling IPCC AR4 simulations. *Bulletin of the American Meteorological Society*, **89**(3), 347-367.

Emori, S. and S.J. Brown, 2005: Dynamic and thermodynamic changes in mean and extreme precipitation under changed climate. *Geophysical Research Letters*, **32**, L17706.

Engelstaedter, S., I. Tegen, and R. Washington, 2006: North African dust emissions and transport. *Earth-Science Reviews*, **79**(1-2), 73-100.

Essery, R., N. Rutter, J. Pomeroy, R. Baxter, M. Stahli, D. Gustafsson, A. Barr, P. Bartlett, and K. Elder, 2009: An evaluation of forest snow process simulations. *Bulletin of the American Meteorological Society*, **90**(8), 1120-1135.

Evan, A.T., D.J. Vimont, A.K. Heidinger, J.P. Kossin, and R. Bennartz, 2009: The role of aerosols in the evolution of tropical North Atlantic Ocean temperature anomalies. *Science*, **324**(5928), 778-781.

Evans, S.G., and J.J. Clague, 1998: Rock avalanche from Mount Munday, Waddington Range, British Columbia, Canada. *Landslide News*, **11**, 23-25.

Favre, A., and A. Gershunov, 2006: Extra-tropical cyclonic/anticyclonic activity in North-Eastern Pacific and air temperature extremes in Western North America. *Climate Dynamics*, **26**(6), 617-629.

Favre, A., and A. Gershunov, 2009: North Pacific cyclonic and anticyclonic transients in a global warming context: possible consequences for Western North American daily precipitation and temperature extremes. *Climate Dynamics*, **32**(7-8), 969-987.

Fedorov, A. and P. Konstantinov, 2003: Observations of surface dynamics with thermokarst initiation, Yukechi site, central Yukutia. In: *Proceedings of the 8th International Conference on Permafrost* [Philips, M., S.M. Springman, and L.U. Arenson (eds.)]. Zurich, Switzerland, 21-25 July 2003, Balkema, Netherlands, pp. 239-243.

Fekete, B.M., C.J. Vörösmarty, J.O. Roads, and C.J. Willmott, 2004: Uncertainties in precipitation and their impacts on runoff estimates. *Journal of Climate*, **17**(2), 294-304.

Feyen, L., and R. Dankers, 2009: Impact of global warming on streamflow drought in Europe. *Journal of Geophysical Research – Atmospheres*, **114**, D17116.

Findell, K.L. and T.L. Delworth, 2005: A modeling study of dynamic and thermodynamic mechanisms for summer drying in response to global warming. *Geophysical Research Letters*, **32**, L16702.

Findell, K.L. and E.A.B. Eltahir, 2003a: Atmospheric controls on soil moisture-boundary layer interactions. Part I: Framework development. *Journal of Hydrometeorology*, **4**(3), 552-569.

Findell, K.L. and E.A.B. Eltahir, 2003b: Atmospheric controls on soil moisture-boundary layer interactions. Part II: Feedbacks within the continental United States. *Journal of Hydrometeorology*, **4**(3), 570-583.

Finnis, J., M.M. Holland, M.C. Serreze, and J.J. Cassano, 2007: Response of Northern Hemisphere extratropical cyclone activity and associated precipitation to climate change, as represented by the Community Climate System Model. *Journal of Geophysical Research – Biogeosciences*, **112**, G04S42.

Fiore, M.M.E., E.E. D'Onofrio, J.L. Pousa, E.J. Schnack, and G.R. Bertola, 2009: Storm surges and coastal impacts at Mar del Plata, Argentina. *Continental Shelf Research*, **29**(14), 1643-1649.

Fischer, E.M. and C. Schär, 2009: Future changes in daily summer temperature variability: driving processes and role for temperature extremes. *Climate Dynamics*, **33**(7-8), 917-935.

Fischer, E.M., and C. Schär, 2010: Consistent geographical patterns of changes in high-impact European heatwaves. *Nature Geoscience*, **3**(6), 398-403.

Fischer, E.M., S.I. Seneviratne, D. Lüthi, and C. Schär, 2007a: The contribution of land-atmosphere coupling to recent European summer heatwaves. *Geophysical Research Letters*, **34**, L06707.

Fischer, E.M., S.I. Seneviratne, P.L. Vidale, D. Lüthi, and C. Schär, 2007b: Soil moisture-atmosphere interactions during the 2003 European summer heatwave. *Journal of Climate*, **20**(20), 5081-5099.

Fischer, L., H. Eisenbeiss, A. Kääb, C. Huggel, and W. Haeberli, 2011: Monitoring topographic changes in a periglacial high-mountain face using high-resolution DTMs, Monte Rosa East Face, Italian Alps. *Permafrost and Periglacial Processes*, **22**(2), 140-152.

Fleig, A.K., L.M. Tallaksen, H. Hisdal, and D.M. Hannah, 2011: Regional hydrological drought in north-western Europe: linking a new Regional Drought Area Index with weather types. *Hydrological Processes*, **25(7)**, 1163-1179.

Fogt, R.L., J. Perlwitz, A.J. Monaghan, D.H. Bromwich, J.M. Jones, and G.J. Marshall, 2009: Historical SAM variability. Part II: Twentieth-century variability and trends from reconstructions, observations, and the IPCC AR4 models. *Journal of Climate*, **22(20)**, 5346-5365.

Fontaine, B., P. Roucou, M. Gaetani, and R. Marteau, 2011: Recent changes in precipitation, ITCZ convection and northern tropical circulation over North Africa (1979–2007). *International Journal of Climatology*, **31(5)**, 633-648.

Forbes, D.L., G.S. Parkes, G.K. Manson, and L.A. Ketch, 2004: Storms and shoreline retreat in the southern Gulf of St. Lawrence. *Marine Geology*, **210(1-4)**, 169-204.

Founda, D., and C. Giannakopoulos, 2009: The exceptionally hot summer of 2007 in Athens, Greece – A typical summer in the future climate? *Global and Planetary Change*, **67(3-4)**, 227-236.

Fowler, H.J., and M. Ekstrom, 2009: Multi-model ensemble estimates of climate change impacts on UK seasonal precipitation extremes. *International Journal of Climatology*, **29(3)**, 385-416.

Fowler, H.J., and C.G. Kilsby, 2003: A regional frequency analysis of United Kingdom extreme rainfall from 1961 to 2000. *Intl Journal of Climatology*, **23(11)**, 1313-1334.

Fowler, H.J., and R.L. Wilby, 2010: Detecting changes in seasonal precipitation extremes using regional climate model projections: Implications for managing fluvial flood risk. *Water Resources Research*, **46**, W0525.

Fowler, H.J., M. Ekstrom, S. Blenkinsop, and A.P. Smith, 2007a: Estimating change in extreme European precipitation using a multimodel ensemble. *Journal of Geophysical Research – Atmospheres*, **112**, D18104.

Fowler, H.J., S. Blenkinsop, and C. Tebaldi, 2007b: Linking climate change modelling to impacts studies: recent advances in downscaling techniques for hydrological modelling. *International Journal of Climatology*, **27(12)**, 1547-1578.

Francou, B., E. Ramirez, B. Caceres, and J. Mendoza, 2000: Glacier evolution in the tropical Andes during the last decades of the 20th century: Chacaltaya, Bolivia, and Antizana, Ecuador. *Ambio*, **29(7)**, 416-422.

Frappier, A.B., T. Knutson, K.B. Liu, and K. Emanuel, 2007a: Perspective: Coordinating paleoclimate research on tropical cyclones with hurricane-climate theory and modelling. *Tellus Series A – Dynamic Meteorology and Oceanography*, **59(4)**, 529-537.

Frappier, A.B., D. Sahagian, S.J. Carpenter, L.A. Gonzalez, and B.R. Frappier, 2007b: Stalagmite stable isotope record of recent tropical cyclone events. *Geology*, **35(2)**, 111-114.

Frederiksen, J.S., and C.S. Frederiksen, 2007: Interdecadal changes in southern hemisphere winter storm track modes. *Tellus Series A – Dynamic Meteorology and Oceanography*, **59(5)**, 599-617.

Frei, C. and C. Schär, 2001: Detection of probability of trends in rare events: Theory and application to heavy precipitation in the Alpine region. *Journal of Climate*, **14(7)**, 1568-1584.

Frei, C., J.H. Christensen, M. Deque, D. Jacob, R.G. Jones, and P.L. Vidale, 2003: Daily precipitation statistics in regional climate models: Evaluation and intercomparison for the European Alps. *Journal of Geophysical Research – Atmospheres*, **108**, 4124.

Frei, C., R. Schöll, S. Fukutome, J. Schmidli, and P.L. Vidale, 2006: Future change of precipitation extremes in Europe: Intercomparison of scenarios from regional climate models. *Journal of Geophysical Research – Atmospheres*, **111**, D06105.

Frey, H., W. Haeberli, A. Linsbauer, C. Huggel, and F. Paul, 2010: A multi-level strategy for anticipating future glacier lake formation and associated hazard potentials. *Natural Hazards and Earth System Sciences*, **10(5)**, 339-352.

Frich, P., L.V. Alexander, P.M. Della-Marta, B. Gleason, M. Haylock, A.M.G. Klein Tank, and T. Peterson, 2002: Observed coherent changes in climatic extremes during the second half of the twentieth century. *Climate Research*, **19(3)**, 193-212.

Friedlingstein, P., P. Cox, R. Betts, L. Bopp, W. von Bloh, V. Brovkin, P. Cadule, S. Doney, M. Eby, I. Fung, G. Bala, J. John, C. Jones, F. Joos, T. Kato, M. Kawamiya, W. Knorr, K. Lindsay, H.D. Matthews, T. Raddatz, P. Rayner, C. Reick, E. Roeckner, K.G. Schnitzler, R. Schnur, K. Strassmann, A.J. Weaver, C. Yoshikawa, and N. Zeng, 2006: Climate-carbon cycle feedback analysis: Results from the C4MIP model intercomparison. *Journal of Climate*, **19(14)**, 3337-3353.

Fu, C.B., S.Y. Wang, Z. Xiong, W.J. Gutowski, D.K. Lee, J.L. McGregor, Y. Sato, H. Kato, J.W. Kim, and M.S. Suh, 2005: Regional climate model intercomparison project for Asia. *Bulletin of the American Meteorological Society*, **86(2)**, 257-266.

Fujibe, F., N. Yamazaki, and K. Kobayashi, 2006: Long-term changes of heavy precipitation and dry weather in Japan (1901-2004). *Journal of the Meteorological Society of Japan*, **84(6)**, 1033-1046.

Fujihara, Y., K. Tanaka, T. Watanabe, T. Nagano, and T. Kojiri, 2008: Assessing the impacts of climate change on the water resources of the Seyhan River Basin in Turkey: Use of dynamically downscaled data for hydrologic simulations. *Journal of Hydrology*, **353(1-2)**, 33-48.

Funk, C., M.D. Dettinger, J.C. Michaelsen, J.P. Verdin, M.E. Brown, M. Barlow, and A. Hoell, 2008: Warming of the Indian Ocean threatens eastern and southern African food security but could be mitigated by agricultural development. *Proceedings of the National Academy of Sciences*, **105(32)**, 11081-11086.

Fyfe, J.C., 2003: Extratropical southern hemisphere cyclones: Harbingers of climate change? *Journal of Climate*, **16(17)**, 2802-2805.

Gallant, A.J.E., and D.J. Karoly, 2010: A combined climate extremes index for the Australian region. *Journal of Climate*, **23(23)**, 6153-6165.

Gallant, A.J.E., K.J. Hennessy, and J. Risbey, 2007: Trends in rainfall indices for six Australian regions: 1910-2005. *Australian Meteorological Magazine*, **56(4)**, 223-239.

Gan, M.A., V.B. Rao, and M.C.L. Moscati, 2006: South American monsoon indices. *Atmospheric Science Letters*, **6(4)**, 219-223.

García, J., M.C. Gallego, A. Serrano, and J. Vaquero, 2007: Trends in block-seasonal extreme rainfall over the Iberian Peninsula in the second half of the twentieth century. *Journal of Climate*, **20(1)**, 113-130.

Gastineau, G., and B.J. Soden, 2009: Model projected changes of extreme wind events in response to global warming. *Geophysical Research Letters*, **36**, L10810.

Gavriliev, P.P., and P.V. Efremov, 2003: Effects of cryogenic processes on Yakutian landscapes under climate warming. In: *Permafrost* [Philips, M., S.M. Springman, and L.U. Arenson (eds.)]. Proceedings of the 8th International Conference on Permafrost, Zurich, Switzerland, 21-25 July 2003, Balkema, The Netherlands, pp. 277-282.

Gedney, N., P.M. Cox, R.A. Betts, O. Boucher, C. Huntingford, and P.A. Stott, 2006: Detection of a direct carbon dioxide effect in continental river runoff records. *Nature*, **439(7078)**, 835-838.

Geertsema, M., J.J. Clague, J.W. Schwab, and S.G. Evans, 2006: An overview of recent large catastrophic landslides in northern British Columbia, Canada. *Engineering Geology*, **83(1-3)**, 120-143.

Gehrels, W.R., D.F. Belknap, B.R. Pearce, and B. Gong, 1995: Modeling the contribution of M(2) tidal amplification to the holocene rise of mean high water in the Gulf of Maine and the Bay of Fundy. *Marine Geology*, **124(1-4)**, 71-85.

Geng, Q. and M. Sugi, 2001: Variability of the North Atlantic cyclone activity in winter analyzed from NCEP–NCAR reanalysis data. *Journal of Climate*, **14(18)**, 3863-3873.

Gerten, D., S. Rost, W. von Bloh, and W. Lucht, 2008: Causes of change in 20th century global river discharge. *Geophysical Research Letters*, **35**, L20405.

Giannakopoulos, C., P. Le Sager, M. Bindi, M. Moriondo, E. Kostopoulou, and C.M. Goodess, 2009: Climatic changes and associated impacts in the Mediterranean resulting from a 2°C global warming. *Global and Planetary Change*, **68(3)**, 209-224.

Giannini, A., R. Saravanan, and P. Chang, 2003: Oceanic forcing of Sahel rainfall on interannual to interdecadal time scales. *Science*, **302(5647)**, 1027-1030.

Giannini, A., M. Biasutti, and M.M. Verstraete, 2008: A climate model-based review of drought in the Sahel: Desertification, the re-greening and climate change. *Global and Planetary Change*, **64(3-4)**, 119-128.

Gibelin, A.L., and M. Deque, 2003: Anthropogenic climate change over the Mediterranean region simulated by a global variable resolution model. *Climate Dynamics*, **20(4)**, 327-339.

Gillett, N.P. and P.A. Stott, 2009: Attribution of anthropogenic influence on seasonal sea level pressure. *Geophysical Research Letters*, **36**, L23709.

Gillett, N.P., R.J. Allan, and T.J. Ansell, 2005: Detection of external influence on sea level pressure with a multi-model ensemble. *Geophysical Research Letters*, **32**, L19714.

Gillett, N.P., P.A. Stott, and B.D. Santer, 2008a: Attribution of cyclogenesis region sea surface temperature change to anthropogenic influence. *Geophysical Research Letters*, **35**, L09707.

Gillett, N.P., D.A. Stone, P.A. Stott, T. Nozawa, A.Y. Karpechko, G.C. Hegerl, M.F. Wehner, and P.D. Jones, 2008b: Attribution of polar warming to human influence. *Nature Geoscience*, **1**, 750-754.

Ginoux, P., J.M. Prospero, O. Torres, and M. Chin, 2004: Long-term simulation of global dust distribution with the GOCART model: correlation with North Atlantic Oscillation. *Environmental Modelling & Software*, **19(2)**, 113-128.

Giorgi, F., 2006: Climate change hot-spots. *Geophysical Research Letters*, **33**, L08707.

Giorgi, F. and P. Lionello, 2008: Climate change projections for the Mediterranean region. *Global and Planetary Change*, **63(2-3)**, 90-104.

Giorgi, F., C. Jones, and G.R. Asra, 2009: Addressing climate information needs at the regional level: the CORDEX framework. *WMO Bulletin*, **58(3)**, 175-183.

Glade, T., 1998: Establishing the frequency and magnitude of landslide-triggering rainstorm events in New Zealand. *Environmental Geology*, **55(2)**, 160-174.

Gleason, K.L., J.H. Lawrimore, D.H. Levinson, and T.R. Karl, 2008: A revised U.S. climate extremes index. *Journal of Climate*, **21(10)**, 2124-2137.

Gong, S.L., X.Y. Zhang, T.L. Zhao, X.B. Zhang, L.A. Barrie, I.G. McKendry, and C.S. Zhao, 2006: A simulated climatology of Asian dust aerosol and its trans-Pacific transport. Part II: Interannual variability and climate connections. *Journal of Climate*, **19(1)**, 104-122.

Goodkin, N.F., K.A. Hughen, S.C. Doney, and W.B. Curry, 2008: Increased multi-decadal variability of the North Atlantic Oscillation since 1781. *Nature Geoscience*, **1**, 844-848.

Goubanova, K. and L. Li, 2007: Extremes in temperature and precipitation around the Mediterranean basin in an ensemble of future climate scenario simulations. *Global and Planetary Change*, **57(1-2)**, 27-42.

Goudie, A.S., 2009: Dust storms: Recent developments. *Journal of Environmental Management*, **90(1)**, 89-94.

Goudie, A.S., and N.J. Middleton, 1992: The changing frequency of dust storms through time. *Climatic Change*, **20(3)**, 197-225.

Grabemann, I., and R. Weisse, 2008: Climate change impact on extreme wave conditions in the North Sea: an ensemble study. *Ocean Dynamics*, **58(3-4)**, 199-212.

Graham, L.P., J. Andreasson, and B. Carlsson, 2007: Assessing climate change impacts on hydrology from an ensemble of regional climate models, model scales and linking methods – a case study on the Lule River basin. *Climatic Change*, **81(S1)**, 293-307.

Graham, N.E. and H.F. Diaz, 2001: Evidence for intensification of North Pacific winter cyclones since 1948. *Bulletin of the American Meteorological Society*, **82(9)**, 1869-1893.

Gravelle, G. and N. Mimura, 2008: Vulnerability assessment of sea-level rise in Viti Levu, Fiji Islands. *Sustainability Science*, **3(2)**, 171-180.

Green, D., L. Alexander, K.L. McInnes, J. Church, N. Nicholls, and W. White, 2009: An assessment of climate change impacts and adaptation for the Torres Strait Islands, Australia. *Climatic Change*, **102(3-4)**, 405-433.

Greene, A.M., A. Giannini, and S.E. Zebiak, 2009: Drought return times in the Sahel: A question of attribution. *Geophysical Research Letters*, **36**, L12701.

Greeves, C.Z., V.D. Pope, R.A. Stratton, and G.M. Martin, 2007: Representation of Northern Hemisphere winter storm tracks in climate models. *Climate Dynamics*, **28(7-8)**, 683-702.

Grell, G.A., L. Schade, R. Knoche, A. Pfeiffer, and J. Egger, 2000: Nonhydrostatic climate simulations of precipitation over complex terrain. *Journal of Geophysical Research – Atmospheres*, **105(D24)**, 29595-29608.

Griffiths, G.M., M.J. Salinger, and I. Leleu, 2003: Trends in extreme daily rainfall across the South Pacific and relationship to the South Pacific Convergence Zone. *International Journal of Climatology*, **23(8)**, 847-869.

Griffiths, G.M., L.E. Chambers, M.R. Haylock, M.J. Manton, N. Nicholls, H.J. Baek, Y. Choi, P.M. Della-Marta, A. Gosai, N. Iga, R. Lata, V. Laurent, L. Maitrepierre, H. Nakamigawa, N. Ouprasitwong, D. Solofa, L. Tahani, D.T. Thuy, L. Tibig, B. Trewin, K. Vediapan, and P. Zhai, 2005: Change in mean temperature as a predictor of extreme temperature change in the Asia-Pacific region. *International Journal of Climatology*, **25(10)**, 1301-1330.

Grimm, A.M. and R.G. Tedeschi, 2009: ENSO and extreme rainfall events in South America. *Journal of Climate*, **22(7)**, 1589-1609.

Grimm, A.M., J. Pal, and F. Giorgi, 2007: Connection between spring conditions and peak summer monsoon rainfall in South America: Role of soil moisture, surface temperature, and topography in eastern Brazil. *Journal of Climate*, **20(24)**, 5929-5945.

Grinsted, A., A.J. Moore, and S. Jevrejeva, 2010: Reconstructing sea level from paleo and projected temperatures 200 to 2100 AD. *Climate Dynamics*, **34(4)**, 461-472.

Groisman, P.Y., and R.W. Knight, 2008: Prolonged dry episodes over the conterminous United States: New tendencies emerging during the last 40 years. *Journal of Climate*, **21(9)**, 1850-1862.

Gruber, S., and W. Haeberli, 2007: Permafrost in steep bedrock slopes and its temperature-related destabilization following climate change. *Journal of Geophysical Research – Earth Surface*, **112**, F02S18.

Gruber, S., M. Hoelzle, and W. Haeberli, 2004: Permafrost thaw and destabilization of Alpine rock walls in the hot summer of 2003. *Geophysical Research Letters*, **31**, L13504.

Gualdi, S., E. Scoccimarro, and A. Navarra, 2008: Changes in tropical cyclone activity due to global warming: Results from a high-resolution coupled general circulation model. *Journal of Climate*, **21(20)**, 5204-5228.

Guilyardi, E., A. Wittenberg, A. Fedorov, M. Collins, C.Z. Wang, A. Capotondi, G.J. van Oldenborgh, and T. Stockdale, 2009: Understanding El Niño in ocean-atmosphere General Circulation Models: Progress and challenges. *Bulletin of the American Meteorological Society*, **90(3)**, 325-340.

Gulev, S.K. and V. Grigorieva, 2004: Last century changes in ocean wind wave height from global visual wave data. *Geophysical Research Letters*, **31**, L24302.

Gulev, S.K., O. Zolina, and S. Grigoriev, 2001: Extratropical cyclone variability in the Northern Hemisphere winter from the NCEP/NCAR reanalysis data. *Climate Dynamics*, **17(10)**, 795-809.

Guo, H., M. Xu, and Q. Hu, 2011: Changes in near-surface wind speed in China: 1969–2005. *International Journal of Climatology*, **31(2)**, 349-358.

Gutowski, W.J., L.E. Branscome, and D.A. Stewart, 1992: Life-cycles of moist baroclinic eddies. *Journal of the Atmospheric Sciences*, **49(4)**, 306-319.

Gutowski, W.J., S.G. Decker, R.A. Donavon, Z.T. Pan, R.W. Arritt, and E.S. Takle, 2003: Temporal-spatial scales of observed and simulated precipitation in central U.S. climate. *Journal of Climate*, **16(22)**, 3841-3847.

Gutowski, W.J., G.C. Hegerl, G.J. Holland, T.R. Knutson, L.O. Mearns, R.J. Stouffer, P.J. Webster, M.F. Wehner, and F.W. Zwiers, 2008a: Causes of observed changes in extremes and projections of future changes. In: *Weather and Climate Extremes in a Changing Climate. Regions of Focus: North America, Hawaii, Caribbean, and U.S. Pacific Islands.* [Karl, T.R., G.A. Meehl, D.M. Christopher, S.J. Hassol, A.M. Waple, and W.L. Murray (eds.)]. A Report by the U.S. Climate Change Science Program and the Subcommittee on Global Change Research, Washington, DC, pp. 81-116.

Gutowski, W.J., S.S. Willis, J.C. Patton, B.R.J. Schwedler, R.W. Arritt, and E.S. Takle, 2008b: Changes in extreme, cold-season synoptic precipitation events under global warming. *Geophysical Research Letters*, **35**, L20710.

Haarsma, R.J., F. Selten, B.V. Hurk, W. Hazeleger, and X.L. Wang, 2009: Drier Mediterranean soils due to greenhouse warming bring easterly winds over summertime central Europe. *Geophysical Research Letters*, **36**, L04705.

Haeberli, W., 1983: Frequency and characteristics of glacier floods in the Swiss Alps. *Annals of Glaciology*, **4**, 85-90.

Haeberli, W., and C.R. Burn, 2002: Natural hazards in forests: glacier and permafrost effects as related to climate change. In: *Environmental Change and Geomorphic Hazards in Forests, IUFRO Research Series 9* [Sidle, R.C. (ed.)]. CABI Publishing, Wallingford, UK, pp. 167-202.

Haeberli, W. and M. Funk, 1991: Borehole temperatures at the Colle Gnifetti core-drilling site (Monte Rosa, Swiss Alps). *Journal of Glaciology*, **37(125)**, 37-46.

Haeberli, W. and R. Hohmann, 2008: Climate, glaciers and permafrost in the Swiss Alps 2050: scenarios, consequences and recommendations. In: *Permafrost on a Warming Planet: Impacts on Ecosystems, Infrastructure and Climate* [Kane, D.L., and K.M. Hinkel (eds.)]. Proc. of the 9th Intl. Conf. on Permafrost at University of Alaska, Fairbanks, 9 June - 3 July 2008, Inst. of Northern Engineering, pp. 607-612.

Haeberli, W., M. Wegmann, and D. Vonder Mühll, 1997: Slope stability problems related to glacier shrinkage and permafrost degradation in the Alps. *Eclogae Geologicae Helvetiae*, **90(3)**, 407-414.

Haeberli, W., A. Kääb, D. Vonder Mühll, and P. Teysseire, 2001: Prevention of outburst floods from periglacial lakes at Grubengletscher, Valais, Swiss Alps. *Journal of Glaciology*, **47(156)**, 111-122.

Haeberli, W., C. Huggel, A. Kääb, S. Zgraggen-Oswald, A. Polkvoj, I. Galushkin, I. Zotikov, and N. Osokin, 2004: The Kolka-Karmadon rock/ice slide of 20 September 2002: an extraordinary event of historical dimensions in North Ossetia, Russian Caucasus. *Journal of Glaciology*, **50(171)**, 533-546.

Haerter, J.O. and P. Berg, 2009: Unexpected rise in extreme precipitation caused by a shift in rain type? *Nature Geoscience*, **2(6)**, 372-373.

Haigh, I., R. Nicholls, and N. Wells, 2009: Mean sea level trends around the English Channel over the 20th century and their wider context. *Continental Shelf Research*, **29(17)**, 2083-2098.

Haigh, I., R. Nicholls, and N. Wells, 2010: Assessing changes in extreme sea levels: Application to the English Channel, 1900–2006. *Continental Shelf Research*, **30(9)**, 1042-1055.

Hall, A., X. Qu, and J.D. Neelin, 2008: Improving predictions of summer climate change in the United States. *Geophysical Research Letters*, **35**, L01702.

Hampel, A., R. Hetzel, and G. Maniatis, 2010: Response of faults to climate-driven changes in ice and water volumes on Earth's surface. *Philosophical Transactions of the Royal Society A: Mathematical, Physical and Engineering Sciences*, **368(1919)**, 2501-2517.

Hanna, E., J. Cappelen, R. Allan, T. Jonsson, F. Le Blancq, T. Lillington, and K. Hickey, 2008: New insights into North European and North Atlantic surface pressure variability, storminess, and related climatic change since 1830. *Journal of Climate*, **21(24)**, 6739-6766.

Hannaford, J., and T.J. Marsh, 2008: High-flow and flood trends in a network of undisturbed catchments in the UK. *International Journal of Climatology*, **28(10)**, 1325-1338.

Hanson, C.E., J.P. Palutikof, and T.D. Davies, 2004: Objective cyclone climatologies of the North Atlantic – a comparison between the ECMWF and NCEP Reanalyses. *Climate Dynamics*, **22(6-7)**, 757-769.

Harper, B., T. Hardy, L. Mason, and R. Fryar, 2009: Developments in storm tide modelling and risk assessment in the Australian region. *Natural Hazards*, **51(1)**, 225-238.

Harris, C., L.U. Arenson, H.H. Christiansen, B. Etzelmüller, R. Frauenfelder, S. Gruber, W. Haeberli, C. Hauck, M. Hölzle, O. Humlum, K. Isaksen, A. Kääb, M.A. Kern-Lütschg, M. Lehning, N. Matsuoka, J.B. Murton, J. Nötzli, M. Phillips, N. Ross, M. Seppälä, S.M. Springman, and D. Vonder Mühll, 2009: Permafrost and climate in Europe: Monitoring and modelling thermal, geomorphological and geotechnical responses. *Earth-Science Reviews*, **92(3-4)**, 117-171.

Haug, G.H., K.A. Hughen, D.M. Sigman, L.C. Peterson, and U. Röhl, 2001: Southward migration of the Intertropical Convergence Zone through the Holocene. *Science*, **293(5533)**, 1304-1308.

Haugen, J.E. and T. Iversen, 2008: Response in extremes of daily precipitation and wind from a downscaled multi-model ensemble of anthropogenic global climate change scenarios. *Tellus Series A – Dynamic Meteorology and Oceanography*, **60(3)**, 411-426.

Hawkins, E. and R. Sutton, 2009: The potential to narrow uncertainty in regional climate predictions. *Bulletin of the American Meteorological Society*, **90(8)**, 1095-1107.

Hawkins, E. and R. Sutton, 2011: The potential to narrow uncertainty in projections of regional precipitation change. *Climate Dynamics*, **37(1-2)**, 407-418.

Hay, L.E., M.P. Clark, M. Pagowski, G.H. Leavesley, and W.J. Gutowski, 2006: One-way coupling of an atmospheric and a hydrologic model in Colorado. *Journal of Hydrometeorology*, **7(4)**, 569-589.

Haylock, M.R. and C.M. Goodess, 2004: Interannual variability of European extreme winter rainfall and links with mean large-scale circulation. *International Journal of Climatology*, **24(6)**, 759-776.

Haylock, M.R., G.C. Cawley, C. Harpham, R.L. Wilby, and C.M. Goodess, 2006a: Downscaling heavy precipitation over the United Kingdom: A comparison of dynamical and statistical methods and their future scenarios. *International Journal of Climatology*, **26(10)**, 1397-1415.

Haylock, M.R., T.C. Peterson, L.M. Alves, T. Ambrizzi, Y.M.T. Anunciacao, J. Baez, V.R. Barros, M.A. Berlato, M. Bidegain, G. Coronel, V. Corradi, V.J. Garcia, A.M. Grimm, D. Karoly, J.A. Marengo, M.B. Marino, D.F. Moncunill, D. Nechet, J. Quintana, E. Rebello, M. Rusticucci, J.L. Santos, I. Trebejo, and L.A. Vincent, 2006b: Trends in total and extreme South American rainfall in 1960-2000 and links with sea surface temperature. *Journal of Climate*, **19(8)**, 1490-1512.

Haylock, M.R., N. Hofstra, A. Tank, E.J. Klok, P.D. Jones, and M. New, 2008: A European daily high-resolution gridded data set of surface temperature and precipitation for 1950-2006. *Journal of Geophysical Research – Atmospheres*, **113(D20119)**.

Hegerl, G.C., F.W. Zwiers, P.A. Stott, and V.V. Kharin, 2004: Detectability of anthropogenic changes in annual temperature and precipitation extremes. *Journal of Climate*, **17(19)**, 3683-3700.

Hegerl, G.C., F.W. Zwiers, P. Braconnot, N.P. Gillett, Y. Luo, J.A. Marengo Orsini, N. Nicholls, J.E. Penner, and P.A. Stott, 2007: Understanding and Attributing Climate Change. In: *Climate Change 2007: The Physical Science Basis. Contribution of Working Group I to the Fourth Assessment Report of the Intergovernmental Panel on Climate Change* [Solomon, S., D. Qin, M. Manning, Z. Chen, M. Marquis, K.B. Averyt, M. Tignor and H.L. Miller (eds.)]. Cambridge University Press, Cambridge, UK, pp. 663-745.

Hegerl, G.C., O. Hoegh-Guldberg, G. Casassa, M.P. Hoerling, R.S. Kovats, C. Parmesan, D.W. Pierce, and P.A. Stott, 2010: Good practice guidance paper on detection and attribution related to anthropogenic climate change. In: *Meeting Report of the Intergovernmental Panel on Climate Change Expert Meeting on Detection and Attribution of Anthropogenic Climate Change* [Stocker, T.F., C.B. Field, D. Qin, V. Barros, G.-K. Plattner, M. Tignor, P.M. Midgley, and K.L. Ebi (eds.)]. IPCC Working Group I Technical Support Unit, University of Bern, Bern, Switzerland.

Hegglin, E. and C. Huggel, 2008: An integrated assessment of vulnerability to glacial hazards. *Mountain Research and Development*, **28(3)**, 299-309.

Heim Jr., R.R., 2002: A review of twentieth-century drought indices used in the United States. *Bulletin of the American Meteorological Society*, **83(8)**, 1149-1165.

Held, I.M. and B.J. Soden, 2006: Robust responses of the hydrological cycle to global warming. *Journal of Climate*, **19(21)**, 5686-5699.

Hemer, M.A., 2010: Historical trends in Southern Ocean storminess: Long-term variability of extreme wave heights at Cape Sorell, Tasmania. *Geophysical Research Letters*, **37**, L18601.

Hemer, M.A., J.A. Church, and J.R. Hunter, 2010: Variability and trends in the directional wave climate of the Southern Hemisphere. *International Journal of Climatology*, **30(4)**, 475-491.

Hendon, H.H., D.W.J. Thompson, and M.C. Wheeler, 2007: Australian rainfall and surface temperature variations associated with the Southern Hemisphere annular mode. *Journal of Climate*, **20(11)**, 2452-2467.

Herath, S. and U. Ratnayake, 2004: Monitoring rainfall trends to predict adverse impacts – a case study from Sri Lanka (1964-1993). *Global Environmental Change – Human and Policy Dimensions*, **14(1)**, 71-79.

Hewitson, B.C. and R.G. Crane, 2006: Consensus between GCM climate change projections with empirical downscaling: Precipitation downscaling over South Africa. *International Journal of Climatology*, **26(10)**, 1315-1337.

Hidalgo, H.G., T. Das, M.D. Dettinger, D.R. Cayan, D.W. Pierce, T.P. Barnett, G. Bala, A. Mirin, A.W. Wood, C. Bonfils, B.D. Santer, and T. Nozawa, 2009: Detection and attribution of streamflow timing changes to climate change in the western United States. *Journal of Climate*, **22(13)**, 3838-3855.

Hinkel, J. and R.J.T. Klein, 2009: Integrating knowledge to assess coastal vulnerability to sea-level rise: The development of the DIVA tool. *Global Environmental Change – Human and Policy Dimensions*, **19(3)**, 384-395.

Hinzman, L.D., N.D. Bettez, W.R. Bolton, F.S. Chapin, M.B. Dyurgerov, B.G. Fastie, B. Griffith, R.D. Hollister, A. Hope, H.P. Huntington, A.M. Jenson, G.J. Jia, R. Jorgenson, D. Kane, D.R. Klein, G. Kofinas, A.H. Lynch, A.H. Lloyd, D. McGuire, F.E. Nelson, W.C. Oechel, T.E. Osterkamp, C.H. Racine, V.E. Romanovsky, R.S. Stone, D.A. Stow, M. Sturm, C.E. Tweedies, G.L. Vourlitis, M.D. Walker, D.A. Walker, P.J. Webber, J.M. Welker, K.S. Winker, and K. Yoshikawa, 2005: Evidence and implications of recent climate change in northern Alaska and other arctic regions. *Climatic Change*, **72(3)**, 251-298.

Hirabayashi, Y., S. Kanae, K. Motoya, K. Masuda, and P. Doll, 2008a: A 59-year (1948-2006) global meteorological forcing data set for land surface models. Part II: Global snowfall estimation. *Hydrological Research Letters*, **2**, 65-69.

Hirabayashi, Y., S. Kanae, S. Emori, T. Oki, and M. Kimoto, 2008b: Global projections of changing risks of floods and droughts in a changing climate. *Hydrological Sciences Journal*, **53(4)**, 754-772.

Hirsch, M.E., A.T. DeGaetano, and S.J. Colucci, 2001: An East Coast winter storm climatology. *Journal of Climate*, **14(5)**, 882-899.

Hirschi, M., S.I. Seneviratne, V. Alexandrov, F. Boberg, C. Boroneant, O.B. Christensen, H. Formayer, B. Orlowsky, and P. Stepanek, 2011: Observational evidence for soil-moisture impact on hot extremes in southeastern Europe. *Nature Geoscience*, **4(1)**, 17-21.

Hodges, K.I., B.J. Hoskins, J. Boyle, and C. Thorncroft, 2003: A comparison of recent reanalysis datasets using objective feature tracking: Storm tracks and tropical easterly waves. *Monthly Weather Review*, **131(9)**, 2012-2037.

Hoegh-Guldberg, O., 1999: Climate change, coral bleaching and the future of the world's coral reefs. *Marine and Freshwater Research*, **50(8)**, 839-866.

Hoelzle, M., G. Darms, and S. Suter, 2010: Evidence of accelerated englacial warming in the Monte Rosa area, Switzerland/Italy. *The Cryosphere Discussions*, **4(4)**, 2277-2305.

Hoerling, M., J. Hurrell, J. Eischeid, and A. Phillips, 2006: Detection and attribution of twentieth-century northern and southern African rainfall change. *Journal of Climate*, **19(16)**, 3989-4008.

Hohenegger, C., P. Brockhaus, and C. Schär, 2008: Towards climate simulations at cloud-resolving scales. *Meteorology Zeitschrift*, **17(4)**, 383-394.

Hohenegger, C., P. Brockhaus, C.S. Bretherton, and C. Schar, 2009: The soil moisture-precipitation feedback in simulations with explicit and parameterized convection. *Journal of Climate*, **22(19)**, 5003-5020.

Holland, G.J. and P.J. Webster, 2007: Heightened tropical cyclone activity in the North Atlantic: Natural variability or climate trend? *Philosophical Transactions of the Royal Society A*, **365**, 2695-2716.

Hope, P.K., W. Drosdowsky, and N. Nicholls, 2006: Shifts in the synoptic systems influencing southwest Western Australia. *Climate Dynamics*, **26(7-8)**, 751-764.

Horton, R., C. Herweijer, C. Rosenzweig, J. Liu, V. Gornitz, and A.C. Ruane, 2008: Sea level rise projections for current generation CGCMs based on the semi-empirical method. *Geophysical Research Letters*, **35**, L02715.

Hossain, F., I. Jeyachandran, and R. Pielke Jr, 2009: Have large dams altered extreme precipitation patterns. *EOS Transactions (AGU)*, **90(48)**, 453.

Hourdin, F., I. Musat, F. Guichard, P.M. Ruti, F. Favot, M.A. Filiberti, M. Pham, J.Y. Grandpeix, J. Polcher, P. Marquet, A. Boone, J.P. Lafore, J.L. Redelsperger, A. Dell'aquila, T.L. Doval, A.K. Traore, and H. Gallee, 2010: AMMA-model intercomparison project. *Bulletin of the American Meteorological Society*, **91(1)**, 95-104.

Huebener, H., U. Cubasch, U. Langematz, T. Spangehl, F. Niehoerster, I. Fast, and M. Kunze, 2007: Ensemble climate simulations using a fully coupled ocean-troposphere-stratosphere general circulation model. *Philosophical Transactions of the Royal Society A*, **365(1857)**, 2089-2101.

Huggel, C., 2009: Recent extreme slope failures in glacial environments: effects of thermal perturbation. *Quaternary Science Reviews*, **28(11-12)**, 1119-1130.

Huggel, C., W. Haeberli, A. Kääb, D. Bieri, and S. Richardson, 2004: An assessment procedure for glacial hazards in the Swiss Alps. *Canadian Geotechnical Journal*, **41(6)**, 1068-1083.

Huggel, C., S. Zgraggen-Oswald, W. Haeberli, A. Kääb, A. Polkvoj, I. Galushkin, and S.G. Evans, 2005: The 2002 rock/ice avalanche at Kolka/Karmadon, Russian Caucasus: assessment of extraordinary avalanche formation and mobility, and application of QuickBird satellite imagery. *Natural Hazards and Earth System Sciences*, **5(2)**, 173-187.

Huggel, C., J. Caplan-Auerbach, C.F. Waythomas, and R.L. Wessels, 2007: Monitoring and modeling ice-rock avalanches from ice-capped volcanoes: A case study of frequent large avalanches on Iliamna Volcano, Alaska. *Journal of Volcanology and Geothermal Research*, **168(1-4)**, 114-136.

Huggel, C., J. Caplan-Auerbach, and R. Wessels, 2008: Recent extreme avalanches: Triggered by climate change. *Eos Transactions (AGU)*, **89(47)**, 469-470.

Huggel, C., N. Salzmann, S. Allen, J. Caplan-Auerbach, L. Fischer, W. Haeberli, C. Larsen, D. Schneider, and R. Wessels, 2010: Recent and future warm extreme events and high-mountain slope failures. *Philosophical Transactions of the Royal Society A*, **368**, 2435-2459.

Hulme, M., R. Doherty, T. Ngara, M. New, and D. Lister, 2001: African climate change: 1900-2100. *Climate Research*, **17(2)**, 145-168.

Hundecha, Y. and A. Bardossy, 2008: Statistical downscaling of extreme of daily precipitation and temperature and construction of their future scenarios. *International Journal of Climatology*, **28(5)**, 589-610.

Hundecha, Y., A. St-Hilaire, T.B.M.J. Ouarda, S. El Adlouni, and P. Gachon, 2008: A nonstationary extreme value analysis for the assessment of changes in extreme annual wind speed over the Gulf of St. Lawrence, Canada. *Journal of Applied Meteorology and Climatology*, **47(11)**, 2745-2759.

Hungate, B.A., J.S. Dukes, M.R.L.Y. Shaw, and C.B. Field, 2003: Nitrogen and climate change. *Science*, **302(5650)**, 1512-1513.

Hunter, J., 2010: Estimating sea-level extremes under conditions of uncertain sea-level rise. *Climatic Change*, **99(3-4)**, 331-350.

Huss, M., D. Farinotti, A. Bauder, and M. Funk, 2008: Modelling runoff from highly glacierized alpine drainage basins in a changing climate. *Hydrological Processes*, **22(19)**, 3888-3902.

Im, E.S. and W.-T. Kwon, 2007: Characteristics of extreme climate sequences over Korea using a regional climate change scenario. *SOLA*, **3**, 17-20.

Im, E.S., J.B. Ahn, W.T. Kwon, and F. Giorgi, 2008: Multi-decadal scenario simulation over Korea using a one-way double-nested regional climate model system. Part 2: Future climate projection (2021-2050). *Climate Dynamics*, **30(2-3)**, 239-254.

Im, E.S., I.W. Jung, and D.H. Bae, 2011: The temporal and spatial structures of recent and future trends in extreme indices over Korea from a regional climate projection. *International Journal of Climatology*, **31(1)**, 72-86.

Inoue, J., J.P. Liu, J.O. Pinto, and J.A. Curry, 2006: Intercomparison of Arctic Regional Climate Models: Modeling clouds and radiation for SHEBA in May 1998. *Journal of Climate*, **19(17)**, 4167-4178.

IPCC, 2007a: *Climate Change 2007: The Physical Science Basis. Contribution of Working Group I to the Fourth Assessment Report of the Intergovernmental Panel on Climate Change* [Solomon, S., D. Qin, M. Manning, Z. Chen, M. Marquis, K.B. Averyt, M. Tignor, and H.L. Miller (eds.)]. Cambridge University Press, Cambridge, UK, 996 pp.

IPCC, 2007b: Summary for Policymakers. In: *Climate Change 2007: The Physical Science Basis. Contribution of Working Group I to the Fourth Assessment Report of the Intergovernmental Panel on Climate Change* [Solomon, S., D. Qin, M. Manning, Z. Chen, M. Marquis, K.B. Averyt, M. Tignor and H.L. Miller (eds.)]. Cambridge University Press, Cambridge, UK, pp. 1-18.

Irish, J.L. and D.T. Resio, 2010: A hydrodynamics-based surge scale for hurricanes. *Ocean Engineering*, **37(1)**, 69-81.

Iverson, R.M., 2000: Landslide triggering by rain infiltration. *Water Resources Research*, **36(7)**, 1897-1910.

Jakob, D., 2010: Challenges in developing a high-quality surface wind-speed data set for Australia. *Australian Meteorological and Oceanographic Journal*, **60(4)**, 227-236.

Jakob, M. and H. Weatherly, 2003: A hydroclimatic threshold for landslide initiation on the North Shore Mountains of Vancouver, British Columbia. *Geomorphology*, **54(3-4)**, 137-156.

Janicot, S., 2009: A comparison of Indian and African monsoon variability at different time scales. *Comptes Rendus Geoscience*, **341(8-9)**, 575-590.

Jansen, E., J. Overpeck, K.R. Briffa, J.C. Duplessy, F. Joos, V. Masson-Delmotte, V. Olago, B. Otto-Bliesner, W.R. Peltier, S. Rahmstorf, R. Ramesh, D. Raynaud, D. Rind, O. Solomina, R. Villalba, and D. Zhang, 2007: Palaeoclimate. In: *Climate Change 2007: The Physical Science Basis. Contribution of Working Group I to the Fourth Assessment Report of the Intergovernmental Panel on Climate Change* [Solomon, S., D. Qin, M. Manning, Z. Chen, M. Marquis, K.B. Averyt, M. Tignor and H.L. Miller (eds.)]. Cambridge University Press, Cambridge, UK, pp. 433-497.

Jiang, T., Z.W. Kundzewicz, and B. Su, 2008: Changes in monthly precipitation and flood hazard in the Yangtze River Basin, China. *International Journal of Climatology*, **28(11)**, 1471-1481.

Jiang, Y., Y. Luo, Z.C. Zhao, and S.W. Tao, 2010a: Changes in wind speed over China during 1956-2004. *Theoretical and Applied Climatology*, **99**(3-4), 421-430.

Jiang, Y., Y. Luo, Z.C. Zhao, Y. Shi, Y.L. Xu, and J.H. Zhu, 2010b: Projections of wind changes for 21st century in China by three regional climate models. *Chinese Geographical Science*, **20**(3), 226-235.

Johnson, N.C. and S.-P. Xie, 2010: Changes in the sea surface temperature threshold for tropical convection. *Nature Geoscience*, **3**(12), 842-845.

Joly, M. and A. Voldoire, 2010: Role of the Gulf of Guinea in the inter-annual variability of the West African monsoon: what do we learn from CMIP3 coupled simulations? *International Journal of Climatology*, **30**(12), 1843-1856.

Jomelli, V., V.P. Pech, C. Chochillon, and D. Brunstein, 2004: Geomorphic variations of debris flows and recent climatic change in the French Alps. *Climatic Change*, **64**(1), 77-102.

Jones, B.M., C.D. Arp, M.T. Jorgenson, K.M. Hinkel, J.A. Schmutz, and P.L. Flint, 2009: Increase in the rate and uniformity of coastline erosion in Arctic Alaska. *Geophysical Research Letters*, **36**, L03503.

Jones, G.S., P.A. Stott, and N. Christidis, 2008: Human contribution to rapidly increasing frequency of very warm Northern Hemisphere summers. *Journal of Geophysical Research*, **113**, D02109.

Jorgenson, M.T., Y.L. Shur, and E.R. Pullman, 2006: Abrupt increase in permafrost degradation in Arctic Alaska. *Geophysical Research Letters*, **33**, L02503.

Joseph, R. and S. Nigam, 2006: ENSO evolution and teleconnections in IPCC's twentieth-century climate simulations: realistic representation? *Journal of Climate*, **19**(17), 4360-4377.

Jull, M. and D. McKenzie, 1996: The effect of deglaciation on mantle melting beneath Iceland. *Journal of Geophysical Research – Solid Earth*, **101**(B10), 21815-21828.

Jung, M., M. Reichstein, P. Ciais, S.I. Seneviratne, J. Sheffield, M.L. Goulden, G. Bonan, A. Cescatti, J.Q. Chen, R. de Jeu, A.J. Dolman, W. Eugster, D. Gerten, D. Gianelle, N. Gobron, J. Heinke, J. Kimball, B.E. Law, L. Montagnani, Q.Z. Mu, B. Mueller, K. Oleson, D. Papale, A.D. Richardson, O. Roupsard, S. Running, E. Tomelleri, N. Viovy, U. Weber, C. Williams, E. Wood, S. Zaehle, and K. Zhang, 2010: Recent decline in the global land evapotranspiration trend due to limited moisture supply. *Nature*, **467**(7318), 951-954.

Jung, T., S.K. Gulev, I. Rudeva, and V. Soloviov, 2006: Sensitivity of extratropical cyclone characteristics to horizontal resolution in the ECMWF model. *Quarterly Journal of the Royal Meteorological Society*, **132**(619), 1839-1857.

Kääb, A., R. Frauenfelder, and I. Roer, 2007: On the response of rockglacier creep to surface temperature increase. *Global and Planetary Change*, **56**(1-2), 172-187.

Kallis, G., 2008: Droughts. *Annual Review of Environment and Resources*, **33**, 85-118.

Kamiguchi, K., A. Kitoh, T. Uchiyama, R. Mizuta, and A. Noda, 2006: Changes in precipitation-based extremes indices due to global warming projected by a global 20-km-mesh atmospheric model. *SOLA*, **2**, 64-67.

Kanada, S., M. Nakano, and T. Kato, 2010a: Changes in mean atmospheric structures around Japan during July due to global warming in regional climate experiments using a cloud system resolving model. *Hydrological Research Letters*, **4**, 11-14.

Kanada, S., M. Nakano, and T. Kato, 2010b: Climatological characteristics of daily precipitation over Japan in the Kakushin regional climate experiments using a non-hydrostatic 5-km-mesh model: Comparison with an outer global 20-km-mesh atmospheric climate model. *SOLA*, **6**, 117-120.

Kang, I.S. and J. Shukla, 2006: Dynamic seasonal prediction and predictability of the monsoon. In: *The Asian Monsoon* [Wang, B. (ed.)]. Springer/Praxis, New York, NY, pp. 585-612.

Karl, T.R., G.A. Meehl, D.M. Christopher, S.J. Hassol, A.M. Waple, and W.L. Murray, 2008: *Weather and Climate Extremes in a Changing Climate. Regions of Focus: North America, Hawaii, Caribbean, and U.S. Pacific Islands*. A Report by the U.S. Climate Change Science Program and the Subcommittee on Global Change Research, Washington, DC, 164 pp.

Karl, T.R., J.M. Melillo, and T.C. Peterson (eds.), 2009: *Global Climate Change Impacts in the United States*. Cambridge University Press, New York, NY, 189 pp.

Karoly, D.J. and Q.G. Wu, 2005: Detection of regional surface temperature trends. *Journal of Climate*, **18**(21), 4337-4343.

Kaser, G., J.G. Cogley, M.B. Dyurgerov, M.F. Meier, and A. Ohmura, 2006: Mass balance of glaciers and ice caps: Consensus estimates for 1961-2004. *Geophysical Research Letters*, **33**, L19501.

Kay, A.L., H.N. Davies, V.A. Bell, and R.G. Jones, 2009: Comparison of uncertainty sources for climate change impacts: flood frequency in England. *Climatic Change*, **92**(1-2), 41-63.

Kay, G. and R. Washington, 2008: Future southern African summer rainfall variability related to a southwest Indian Ocean dipole in HadCM3. *Geophysical Research Letters*, **35**, L12701.

Keeling, R.F., A. Kortzinger, and N. Gruber, 2010: Ocean deoxygenation in a warming world. *Annual Review of Marine Science*, **2**, 199-229.

Keenlyside, N.S., M. Latif, J. Jungclaus, L. Kornblueh, and E. Roeckner, 2008: Advancing decadal-scale climate prediction in the North Atlantic sector. *Nature*, **453**, 84-88.

Kellogg, C.A. and D.W. Griffin, 2006: Aerobiology and the global transport of desert dust. *Trends in Ecology & Evolution*, **21**(11), 638-644.

Kendon, E.J., D.P. Rowell, R.G. Jones, and E. Buonomo, 2008: Robustness of future changes in local precipitation extremes. *Journal of Climate*, **21**(17), 4280-4297.

Kendon, E.J., D.P. Rowell, and R.G. Jones, 2010: Mechanisms and reliability of future projected changes in daily precipitation. *Climate Dynamics*, **35**(2-3), 489-509.

Kenyon, J., and G.C. Hegerl, 2008: Influence of modes of climate variability on global temperature extremes. *Journal of Climate*, **21**(15), 3872-3889.

Kershaw, J.A., J.J. Clague, and S.G. Evans, 2005: Geomorphic and sedimentological signature of a two-phase outburst flood from moraine-dammed Queen Bess Lake, British Columbia, Canada. *Earth Surface Processes and Landforms*, **30**(1), 1-25.

Kharin, V., F.W. Zwiers, X. Zhang, and G.C. Hegerl, 2007: Changes in temperature and precipitation extremes in the IPCC ensemble of global coupled model simulations. *Journal of Climate*, **20**(8), 1419-1444.

Khon, V.C., I.I. Mokhov, E. Roeckner, and V.A. Semenov, 2007: Regional changes of precipitation characteristics in Northern Eurasia from simulations with global climate model. *Global and Planetary Change*, **57**(1-2), 118-123.

Kidston, J., and E.P. Gerber, 2010: Intermodel variability of the poleward shift of the austral jet stream in the CMIP3 integrations linked to biases in 20th century climatology. *Geophysical Research Letters*, **37**(9), L09708.

Kiktev, D., D.M.H. Sexton, L. Alexander, and C.K. Folland, 2003: Comparison of modeled and observed trends in indices of daily climate extremes. *Journal of Climate*, **16**(22), 3560-3571.

Kiktev, D.M., J. Caesar, and L. Alexander, 2009: Temperature and precipitation extremes in the second half of the twentieth century from numerical modeling results and observational data. *Izvestiya Atmospheric and Oceanic Physics*, **45**, 284-293.

Kilsby, C., P. Jones, A. Burton, A. Ford, H. Fowler, C. Harpham, P. James, A. Smith, and R. Wilby, 2007: A daily weather generator for use in climate change studies. *Environmental Modelling and Software*, **22**(12), 1705-1719.

Kim, H.J., R.C. Sidle, R.D. Moore, and R. Hudson, 2004: Throughflow variability during snowmelt in a forested mountain catchment, coastal British Columbia, Canada. *Hydrological Processes*, **18**(7), 1219-1236.

Kim, H.J., B. Wang, and Q.H. Ding, 2008: The global monsoon variability simulated by CMIP3 coupled climate models. *Journal of Climate*, **21**(20), 5271-5294.

Kim, S., E. Nakakita, Y. Tachikawa, and K. Takara, 2010: Precipitation changes in Japan under the A1B climate change scenario. *Annual Journal of Hydraulic Engineering, JSCE*, **54**, 127-132.

Kim, S.K., W.P. Hong, and Y.M. Kim, 1991: Prediction of rainfall-triggered landslides in Korea. In: *Proceedings of the 6th International Symposium on Landslides, Christchurch, New Zealand* [Bell, D.H. (ed.)]. Balkema, Rotterdam, The Netherlands, pp. 989-994.

Kimoto, M., N. Yasutomi, C. Yokoyama, and S. Emori, 2005: Projected changes in precipitation characteristics around Japan under the global warming. *SOLA*, **1**, 85-88.

King'uyu, S.M., L.A. Ogallo, and E.K. Anyamba, 2000: Recent trends of minimum and maximum surface temperatures over eastern Africa. *Journal of Climate*, **13**(16), 2876-2886.

Kitoh, A. and T. Uchiyama, 2006: Changes in onset and withdrawal of the East Asian summer rainy season by multi-model global warming experiments. *Journal of the Meteorological Society of Japan*, **84(2)**, 247-258.

Kitoh, A., T. Ose, K. Kurihara, S. Kusunoki, M. Sugi, and KAKUSHIN Team-3 Modeling Group, 2009: Projection of changes in future weather extremes using super-high-resolution global and regional atmospheric models in the KAKUSHIN Program: Results of preliminary experiments. *Hydrological Research Letters*, **3**, 49-53.

Kitoh, A., S. Kusunoki, and T. Nakaegawa, 2011: Climate change projections over South America in the late 21st century with the 20 and 60 km mesh Meteorological Research Institute atmospheric general circulation model (MRI-AGCM). *Journal of Geophysical Research – Atmospheres*, **116**, D06105.

Kjellström, E. and P. Lind, 2009: Changes in the water budget in the Baltic Sea drainage basin in future warmer climates as simulated by the regional climate model RCA3. *Boreal Environment Research*, **14(1)**, 114-124.

Kjellström, E., L. Bärring, D. Jacob, R. Jones, G. Lenderink, and C. Schär, 2007: Modelling daily temperature extremes: recent climate and future changes over Europe. *Climatic Change*, **81(S1)**, 249-265.

Klein Tank, A.M.G., and G.P. Können, 2003: Trends in indices of daily temperature and precipitation extremes in Europe, 1946-1999. *Journal of Climate*, **16(22)**, 3665-3680.

Klein Tank, A.M.G., T.C. Peterson, D.A. Quadir, S. Dorji, X. Zou, H. Tang, K. Santhosh, U.R. Joshi, A.K. Jaswal, R.K. Kolli, A.B. Sikder, N.R. Deshpande, J.V. Revadekar, K. Yeleuova, S. Vandasheva, M. Faleyeva, P. Gomboluudev, K.P. Budhathoki, A. Hussain, M. Afzaal, L. Chandrapala, H. Anvar, D. Amanmurad, V.S. Asanova, P.D. Jones, M.G. New, and T. Specktorman, 2006: Changes in daily temperature and precipitation extremes in central and south Asia. *Journal of Geophysical Research*, **111**, D16105.

Knapp, K.R. and J.P. Kossin, 2007: A new global tropical cyclone data set from ISCCP B1 geostationary satellite observations. *Journal of Applied Remote Sensing*, **1**, 013505.

Knapp, K.R. and M.C. Kruk, 2010: Quantifying inter-agency differences in tropical cyclone best track wind speed estimates. *Monthly Weather Review*, **138(4)**, 1459-1473.

Kniveton, D.R., R. Layberry, C.J.R. Williams, and M. Peck, 2009: Trends in the start of the wet season over Africa. *International Journal of Climatology*, **29(9)**, 1216-1225.

Knowles, N., M.D. Dettinger, and D.R. Cayan, 2006: Trends in snowfall versus rainfall in the western United States. *Journal of Climate*, **19(18)**, 4545-4559.

Knox, J.C., 2000: Sensitivity of modern and Holocene floods to climate change. *Quaternary Science Reviews*, **19(1-5)**, 439-457.

Knutson, T.R. and R.E. Tuleya, 2004: Impact of CO_2-induced warming on simulated hurricane intensity and precipitation: Sensitivity to the choice of climate model and convective parameterization. *Journal of Climate*, **17(18)**, 3477-3495.

Knutson, T.R., T.L. Delworth, K.W. Dixon, I.M. Held, J. Lu, V. Ramaswamy, M.D. Schwarzkopf, G. Stenchikov, and R.J. Stouffer, 2006: Assessment of twentieth-century regional surface temperature trends using the GFDL CM2 coupled models. *Journal of Climate*, **19(9)**, 1624-1651.

Knutson, T.R., J.J. Sirutis, S.T. Garner, I.M. Held, and R.E. Tuleya, 2007: Simulation of the recent multidecadal increase of Atlantic hurricane activity using an 18-km-grid regional model. *Bulletin of the American Meteorological Society*, **88(10)**, 1549-1565.

Knutson, T.R., J.J. Sirutis, S.T. Garner, G.A. Vecchi, and I.M. Held, 2008: Simulated reduction in Atlantic hurricane frequency under twenty-first- century warming conditions. *Nature Geoscience*, **1(6)**, 359-364.

Knutson, T.R., J.L. McBride, J. Chan, K. Emanuel, G. Holland, C. Landsea, I. Held, J.P. Kossin, A.K. Srivastava, and M. Sugi, 2010: Tropical cyclones and climate change. *Nature Geoscience*, **3(3)**, 157-163.

Knutti, R., 2010: The end of model democracy? *Climatic Change*, **102(3-4)**, 395-404.

Knutti, R., R. Furrer, C. Tebaldi, J. Cermak, and G.A. Meehl, 2010a: Challenges in combining projections from multiple climate models. *Journal of Climate*, **23(10)**, 2739-2758.

Knutti, R., G. Abramowitz, M. Collins, V. Eyring, P.J. Glecker, B. Hewitson, and L. Mearns, 2010b: Good practice guidance paper on assessing and combining multi model climate projections. In: *Meeting Report of the Intergovernmental Panel on Climate Change Expert Meeting on Assessing and Combining Multi-Model Climate Projections* [Stocker, T.F., Q. Dahe, G.-K. Plattner, M. Tignor, and P.M. Midgley (eds.)]. IPCC Working Group I Technical Support Unit, University of Bern, Bern, Switzerland, pp. 1-13.

Kochel, R.C., and V.R. Baker, 1982: Paleoflood hydrology. *Science*, **215(4531)**, 353-361.

Kodama, C. and T. Iwasaki, 2009: Influence of the SST rise on baroclinic instability wave activity under an aquaplanet condition. *Journal of the Atmospheric Sciences*, **66(8)**, 2272-2287.

Koffi, B. and E. Koffi, 2008: Heat waves across Europe by the end of the 21st century: multiregional climate simulations. *Climate Research*, **36(2)**, 153-168.

Komar, P.D. and J.C. Allan, 2008: Increasing hurricane-generated wave heights along the US East Coast and their climate controls. *Journal of Coastal Research*, **24(2)**, 479-488.

Kont, A., J. Jaagus, R. Aunap, U. Ratas, and R. Rivis, 2008: Implications of sea-level rise for Estonia. *Journal of Coastal Research*, **24(2)**, 423-431.

Koo, G.S., K.O. Boo, and W.T. Kwon, 2009: Projection of temperature over Korea using an MM5 regional climate simulation. *Climate Research*, **40(2-3)**, 241-248.

Kossin, J.P., 2008: Is the North Atlantic hurricane season getting longer? *Geophysical Research Letters*, **35**, L23705.

Kossin, J.P. and D.J. Vimont, 2007: A more general framework for understanding Atlantic hurricane variability and trends. *Bulletin of the American Meteorology Society*, **88(11)**, 1767-1781.

Kossin, J.P., K.R. Knapp, D.J. Vimont, R.J. Murnane, and B.A. Harper, 2007: A globally consistent reanalysis of hurricane variability and trends. *Geophysical Research Letters*, **34**, L04815.

Kossin, J.P., S.J. Camargo, and M. Sitkowski, 2010: Climate modulation of North Atlantic hurricane tracks. *Journal of Climate*, **23(11)**, 3057-3076.

Koster, R.D. and M.J. Suarez, 2001: Soil moisture memory in climate models. *Journal of Hydrometeorology*, **2(6)**, 558-570.

Koster, R.D., M.J. Suarez, P. Liu, U. Jambor, A. Berg, M. Kistler, R. Reichle, M. Rodell, and J. Famiglietti, 2004a: Realistic initialization of land surface states: Impacts on subseasonal forecast skill. *Journal of Hydrometeorology*, **5(6)**, 1049-1063.

Koster, R.D., P.A. Dirmeyer, Z.C. Guo, G. Bonan, E. Chan, P. Cox, C.T. Gordon, S. Kanae, E. Kowalczyk, D. Lawrence, P. Liu, C.H. Lu, S. Malyshev, B. McAvaney, K. Mitchell, D. Mocko, T. Oki, K. Oleson, A. Pitman, Y.C. Sud, C.M. Taylor, D. Verseghy, R. Vasic, Y.K. Xue, and T. Yamada, 2004b: Regions of strong coupling between soil moisture and precipitation. *Science*, **305**, 1138-1140.

Koster, R.D., Z.C. Guo, R.Q. Yang, P.A. Dirmeyer, K. Mitchell, and M.J. Puma, 2009: On the nature of soil moisture in land surface models. *Journal of Climate*, **22(16)**, 4322-4335.

Kotlyakov, V.M., O.V. Rototaeva, and G.A. Nosenko, 2004: The September 2002 Kolka glacier catastrophe in North Ossetia, Russian Federation: evidence and analysis. *Mountain Research and Development*, **24(1)**, 78-83.

Kripalani, R.H., J.H. Oh, A. Kulkarni, S.S. Sabade, and H.S. Chaudhari, 2007: South Asian summer monsoon precipitation variability: Coupled climate model simulations and projections under IPCC AR4. *Theoretical and Applied Climatology*, **90(3-4)**, 133-159.

Krishnamurthy, C.K.B., U. Lall, and H.H. Kwon, 2009: Changing frequency and intensity of rainfall extremes over India from 1951 to 2003. *Journal of Climate*, **22(18)**, 4737-4746.

Kruger, A.C., 2006: Observed trends in daily precipitation indices in South Africa: 1910-2004. *International Journal of Climatology*, **26(15)**, 2275-2285.

Kubota, H. and J.C.L. Chan, 2009: Interdecadal variability of tropical cyclone landfall in the Philippines from 1902 to 2005. *Geophysical Research Letters*, **36**, L12802.

Kürbis, K., M. Mudelsee, G. Tetzlaff, and R. Brazdil, 2009: Trends in extremes of temperature, dewpoint, and precipitation from long instrumental series from central Europe. *Theoretical and Applied Climatology*, **98(1-2)**, 187-195.

Kuglitsch, F.G., A. Toreti, E. Xoplaki, P.M. Della-Marta, J. Luterbacher, and H. Wanner, 2009: Homogenization of daily maximum temperature series in the Mediterranean. *Journal of Geophysical Research – Atmospheres*, **114**, D15108.

Kuglitsch, F.G., A. Toreti, E. Xoplaki, P.M. Della-Marta, C. Zerefos, M. Türkes, and J. Luterbacher, 2010: Heat wave changes in the eastern Mediterranean since 1960. *Geophysical Research Letters*, **37**, L04802.

Kuleshov, Y., L. Qi, R. Fawcett, and D. Jones, 2008: On tropical cyclone activity in the Southern Hemisphere: Trends and the ENSO connection. *Geophysical Research Letters*, **35**, L14S08.

Kumar, K.R., A.K. Sahai, K.K. Kumar, S.K. Patwardhan, P.K. Mishra, J.V. Revadekar, K. Kamala, and G.B. Pant, 2006: High-resolution climate change scenarios for India for the 21st century. *Current Science*, **90(3)**, 334-345.

Kundzewicz, Z.W., L.J. Mata, N.W. Arnell, P. Doll, P. Kabat, B. Jimenez, K.A. Miller, T. Oki, Z. Sen, and I.A. Shiklomanov, 2007: Freshwater resources and their management. In: *Climate Change 2007. Impacts, Adaptation and Vulnerability. Contribution of Working Group II to the Fourth Assessment Report of the Intergovernmental Panel on Climate Change* [Parry, M.L., O.F. Canziani, J.P. Palutikof, P.J. Van Der Linde, and C.E. Hanson (eds.)]. Cambridge University Press, Cambridge, UK, pp. 173-210.

Kunkel, K.E., P.D. Bromirski, H.E. Brooks, T. Cavazos, A.V. Douglas, D.R. Easterling, K.A. Emanuel, P.Y. Groisman, G.J. Holland, T.R. Knutson, J.P. Kossin, P.D. Komar, D.H. Levinson, and R.L. Smith, 2008: Observed changes in weather and climate extremes. In: *Weather and Climate Extremes in a Changing Climate. Regions of Focus: North America, Hawaii, Caribbean, and U.S. Pacific Islands.* [Karl, T.R., G.A. Meehl, D.M. Christopher, S.J. Hassol, A.M. Waple, and W.L. Murray (eds.)]. A Report by the U.S. Climate Change Science Program and the Subcommittee on Global Change Research, Washington, DC, pp. 222.

Kunkel, K.E., X.-Z. Liang, and J. Zhu, 2010: Regional climate model projections and uncertainties of U.S. summer heat waves. *Journal of Climate*, **23(16)**, 4447-4458.

Kunz, M., J. Sander, and C. Kottmeier, 2009: Recent trends of thunderstorm and hailstorm frequency and their relation to atmospheric characteristics in southwest Germany. *International Journal of Climatology*, **29(15)**, 2283-2297.

Kusunoki, S. and R. Mizuta, 2008: Future changes in the Baiu rain band projected by a 20-km mesh global atmospheric model: sea surface temperature dependence. *SOLA*, **4**, 85-88.

Kwarteng, A.Y., A.S. Dorvlo, and G.T.V. Kumar, 2009: Analysis of a 27-year rainfall data (1977-2003) in the Sultanate of Oman. *International Journal of Climatology*, **29(4)**, 605-617.

Kyselý, J., 2009: Trends in heavy precipitation in the Czech Republic over 1961-2005. *International Journal of Climatology*, **29(12)**, 1745-1758.

Kyselý, J. and R. Beranová, 2009: Climate-change effects on extreme precipitation in central Europe: uncertainties of scenarios based on regional climate models. *Theoretical and Applied Climatology*, **95(3-4)**, 361-374.

Laine, A., M. Kageyama, D. Salas-Melia, G. Ramstein, S. Planton, S. Denvil, and S. Tyteca, 2009: An energetics study of wintertime Northern Hemisphere storm tracks under 4 x CO_2 conditions in two ocean-atmosphere coupled models. *Journal of Climate*, **22(3)**, 819-839.

Lambeck, K., C.D. Woodroffe, F. Antonioli, M. Anzidei, W.R. Gehrels, J. Laborel, and A.J. Wright, 2010: Paleoenvironmental records, geophysical modeling, and reconstruction of sea-level trends and variability on centennial and longer timescales. In: *Understanding Sea-Level Rise and Variability* [Church, J.A., P.L. Woodworth, T. Aarup, and W.S. Wilson (eds.)]. Wiley-Blackwell, Chichester, UK, pp. 61-121.

Landsea, C.W., 1993: A climatology of intense (or major) Atlantic hurricanes. *Monthly Weather Review*, **121(6)**, 1703-1713.

Landsea, C.W., 2007: Counting Atlantic tropical cyclones back to 1900. *EOS Transactions (AGU)*, **88(18)**, 197-202.

Landsea, C.W., C. Anderson, N. Charles, G. Clark, J. Dunion, J. Fernandez-Partagas, P. Hungerford, C. Neumann, and M. Zimmer, 2004: The Atlantic hurricane database re-analysis project: Documentation for the 1851-1910 alterations and additions to the HURDAT database. In: *Hurricanes and Typhoons: Past, Present and Future* [Murnane, R.J. and K.B. Liu (eds.)]. Columbia University Press, New York, NY, pp. 177-221.

Landsea, C.W., B.A. Harper, K. Hoarau, and J.A. Knaff, 2006: Can we detect trends in extreme tropical cyclones? *Science*, **313(5786)**, 452-454.

Lantz, T.C., and S.V. Kokelj, 2008: Increasing rates of retrogressive thaw slump activity in the Mackenzie Delta region, N.W.T., Canada. *Geophysical Research Letters*, **35**, L06502.

Laprise, R., R. de Elia, D. Caya, S. Biner, P. Lucas-Picher, E. Diaconescu, M. Leduc, A. Alexandru, L. Separovic, and C. Canadian Network Regional, 2008: Challenging some tenets of regional climate modelling. *Meteorology and Atmospheric Physics*, **100(1-4)**, 3-22.

Larsen, C.F., R.J. Motyka, A.A. Arendt, K.A. Echelmeyer, and P.E. Geissler, 2007: Glacier changes in southeast Alaska and northwest British Columbia and contribution to sea level rise. *Journal of Geophysical Research*, **112**, F01007.

Lau, K.M. and H.T. Wu, 2007: Detecting trends in tropical rainfall characteristics, 1973-2003. *International Journal of Climatology*, **27(8)**, 979-988.

Lau, K.M., M.K. Kim, and K.M. Kim, 2006: Asian summer monsoon anomalies induced by aerosol direct forcing: the role of the Tibetan Plateau. *Climate Dynamics*, **26(7-8)**, 855-864.

Lau, K.M., Y.P. Zhou, and H.T. Wu, 2008a: Have tropical cyclones been feeding more extreme rainfall? *Journal of Geophysical Research*, **113**, D23113.

Lau, N.C., A. Leetma, and M.J. Nath, 2008b: Interactions between the responses of North American climate to El Niño–La Niña and to the secular warming trend in the Indian-Western Pacific Oceans. *Journal of Climate*, **21(3)**, 476-494.

Laurent, C. and S. Parey, 2007: Estimation of 100-year return-period temperatures in France in a non-stationary climate: Results from observations and IPCC scenarios. *Global and Planetary Change*, **57(1-2)**, 177-188.

Le Cozannet, G., S. Lecacheux, E. Delvallee, N. Desramaut, C. Oliveros, and R. Pedreros, 2011: Teleconnection pattern influence on sea-wave climate in the Bay of Biscay. *Journal of Climate*, **24(3)**, 641-652.

Le Quesne, C., D.W. Stahle, M.K. Cleaveland, M.D. Therrell, J.C. Aravena, and J. Barichivich, 2006: Ancient *Austrocedrus* tree-ring chronologies used to reconstruct central Chile precipitation variability from A.D. 1200 to 2000. *Journal of Climate*, **19(22)**, 5731-5744.

Le Quesne, C., C. Acuna, J.A. Boninsegna, A. Rivera, and J. Barichivich, 2009: Long-term glacier variations in the Central Andes of Argentina and Chile, inferred from historical records and tree-ring reconstructed precipitation. *Palaeogeography, Palaeoclimatology, Palaeoecology*, **281(3-4)**, 334-344.

Leake, J., J. Wolf, J. Lowe, P. Stansby, G. Jacoub, R. Nicholls, M. Mokrech, S. Nicholson-Cole, M. Walkden, A. Watkinson, and S. Hanson, 2007: Predicted wave climate for the UK: towards an integrated model of coastal impacts of climate change. In: *Estuarine and Coastal Modeling Congress 2007* [Spaulding, M. (ed.)]. Proceedings of the Tenth International Conference on Estuarine and Coastal Modeling Congress 2007, Newport, Rhode Island, 5-7 Nov 2007, American Society of Civil Engineers, pp 393-406.

Leander, R., T.A. Buishand, B.J.J.M. van den Hurk, and M.J.M. de Wit, 2008: Estimated changes in flood quantiles of the river Meuse from resampling of regional climate model output. *Journal of Hydrology*, **351(3-4)**, 331-343.

Leckebusch, G.C., B. Koffi, U. Ulbrich, J.G. Pinto, T. Spangehl, and S. Zacharias, 2006: Analysis of frequency and intensity of European winter storm events from a multi-model perspective, at synoptic and regional scales. *Climate Research*, **31(1)**, 59-74.

Leckebusch, G.C., D. Renggli, and U. Ulbrich, 2008: Development and application of an objective storm severity measure for the Northeast Atlantic region. *Meteorologische Zeitschrift*, **17(5)**, 575-587.

Leclercq, P., J. Oerlemans, and J. Cogley, 2011: Estimating the glacier contribution to sea-level rise for the period 1800–2005. *Surveys in Geophysics*, **32(4-5)**, 519-535, doi:10.1007/s10712-10011-19121-10717.

Lee, T. and M.J. McPhaden, 2010: Increasing intensity of El Niño in the central-equatorial Pacific. *Geophysical Research Letters*, **37**, L14603.

Legates, D.R., H.F. Lins, and G.J. McCabe, 2005: Comments on "Evidence for global runoff increase related to climate warming" by Labat et al. *Advances in Water Resources*, **28(12)**, 1310-1315.

Lehmann, A., K. Getzlaff, and J. Harlass, 2011: Detailed assessment of climate variability in the Baltic Sea area for the period 1958 to 2009. *Climate Research*, **46(2)**, 185-196.

Lehner, B., P. Doll, J. Alcamo, T. Henrichs, and F. Kaspar, 2006: Estimating the impact of global change on flood and drought risks in Europe: A continental, integrated analysis. *Climatic Change*, **75(3)**, 273-299.

Leloup, J., M. Lengaigne, and J.-P. Boulanger, 2008: Twentieth century ENSO characteristics in the IPCC database. *Climate Dynamics*, **30(2-3)**, 227-291.

Lemke, P., J. Ren, R.B. Alley, I. Allison, J. Carrasco, G. Flato, Y. Fujii, G. Kaser, P. Mote, R.H. Thomas, and T. Zhang, 2007: Observations: Changes in snow, ice and frozen ground. In: *Climate Change 2007: The Physical Science Basis. Contribution of Working Group I to the Fourth Assessment Report of the Intergovernmental Panel on Climate Change* [Solomon, S., D. Qin, M. Manning, Z. Chen, M. Marquis, K.B. Averyt, M. Tignor and H.L. Miller (eds.)]. Cambridge University Press, Cambridge, UK, and New York, NY, pp. 337-383.

Lenderink, G. and E. Van Meijgaard, 2008: Increase in hourly precipitation extremes beyond expectations from temperature changes. *Nature Geoscience*, **1(8)**, 511-514.

Lenderink, G. and E. van Meijgaard, 2009: Unexpected rise in extreme precipitation caused by a shift in rain type? *Nature Geoscience*, **2(6)**, 373-373.

Lenton, T.M., H. Held, E. Kriegler, J.W. Hall, W. Lucht, S. Rahmstorf, and H.J. Schellnhuber, 2008: Tipping elements in the Earth's climate system. *Proceedings of the National Academy of Sciences*, **105(6)**, 1786-1793.

Leslie, L.M., M. Leplastrier, and B.W. Buckley, 2008: Estimating future trends in severe hailstorms over the Sydney Basin: A climate modelling study. *Atmospheric Research*, **87(1)**, 37-57.

Lewis, S.L., P.M. Brando, O.L. Phillips, G.M.F. van der Heijden, and D. Nepstad, 2011: The 2010 Amazon Drought. *Science*, **331(6017)**, 554.

Lewkowicz, A.G. and C. Harris, 2005: Morphology and geotechnique of active-layer detachment failures in discontinuous and continuous permafrost, northern Canada. *Geomorphology*, **69(1-4)**, 275-297.

L'Hôte, Y., G. Mahé, B. Somé, and J.P. Triboulet, 2002: Analysis of a Sahelian annual rainfall index from 1896 to 2000; the drought continues. *Hydrological Sciences Journal*, **47(4)**, 563-572.

Li, H., A. Dai, T. Zhou, and J. Lu, 2008: Responses of East Asian summer monsoon to historical SST and atmospheric forcing during 1950-2000. *Climate Dynamics*, **34(4)**, 501-514.

Li, X.Y., L.Y. Liu, and J.H. Wang, 2004: Wind tunnel simulation of aeolian sandy soil erodibility under human disturbance. *Geomorphology*, **59(1-4)**, 3-11.

Liggins, F., R.A. Betts, and B. McGuire, 2010: Projected future climate changes in the context of geological and geomorphological hazards. *Philosophical Transactions of the Royal Society A: Mathematical, Physical and Engineering Sciences*, **368(1919)**, 2347-2367.

Lim, E.-P., and I. Simmonds, 2009: Effect of tropospheric temperature change on the zonal mean circulation and SH winter extratropical cyclones. *Climate Dynamics*, **33(1)**, 19-32.

Lin, J.L., B.E. Mapes, K.M. Weickmann, G.N. Kiladis, S.D. Schubert, M.J. Suarez, J.T. Bacmeister, and M.I. Lee, 2008a: North American monsoon and convectively coupled equatorial waves simulated by IPCC AR4 coupled GCMs. *Journal of Climate*, **21(12)**, 2919-2937.

Lin, J.L., K.M. Weickman, G.N. Kiladis, B.E. Mapes, S.D. Schubert, M.J. Suarez, J.T. Bacmeister, and M.I. Lee, 2008b: Subseasonal variability associated with Asian summer monsoon simulated by 14 IPCC AR4 coupled GCMs. *Journal of Climate*, **21(18)**, 4541-4567.

Lindström, G. and S. Bergström, 2004: Runoff trends in Sweden 1807-2002. *Hydrological Sciences Journal*, **49(1)**, 69-83.

Lins, H.F. and J.R. Slack, 1999: Streamflow trends in the United States. *Geophysical Research Letters*, **26(2)**, 227-230.

Lionello, P., S. Cogo, M.B. Galati, and A. Sanna, 2008: The Mediterranean surface wave climate inferred from future scenario simulations. *Global and Planetary Change*, **63(2-3)**, 152-162.

Lipovsky, P., S. Evans, J. Clague, C. Hopkinson, R. Couture, P. Bobrowsky, G. Ekström, M. Demuth, K. Delaney, N. Roberts, G. Clarke, and A. Schaeffer, 2008: The July 2007 rock and ice avalanches at Mount Steele, St. Elias Mountains, Yukon, Canada. *Landslides*, **5(4)**, 445-455.

Liu, B., M. Xu, and M. Henderson, 2011: Where have all the showers gone? Regional declines in light precipitation events in China, 1960–2000. *International Journal of Climatology*, **31(8)**, 1177-1191.

Llano, M.P. and O. Penalba, 2011: A climatic analysis of dry sequences in Argentina. *International Journal of Climatology*, **31(4)**, 504-513.

Llasat, M.C., M. Barriendos, A. Barrera, and T. Rigo, 2005: Floods in Catalonia (NE Spain) since the 14th century. Climatological and meteorological aspects from historical documentary sources and old instrumental records. *Journal of Hydrology*, **313(1-2)**, 32-47.

Lloyd-Hughes, B., and M.D. Saunders, 2002: A drought climatology for Europe. *International Journal of Climatology*, **22(13)**, 1571-1592.

Lobell, D.B., C.J. Bonfils, L.M. Kueppers, and M.A. Snyder, 2008: Irrigation cooling effect on temperature and heat index extremes. *Geophysical Research Letters*, **35**, L09705.

Loeptien, U., O. Zolina, S. Gulev, M. Latif, and V. Soloviov, 2008: Cyclone life cycle characteristics over the Northern Hemisphere in coupled GCMs. *Climate Dynamics*, **31(5)**, 507-532.

Lohmann, U. and J. Feichter, 2007: Global indirect aerosol effects: a review. *Atmospheric Chemistry and Physics*, **5(3)**, 715-737.

Long, Z., W. Perrie, J. Gyakum, R. Laprise, and D. Caya, 2009: Scenario changes in the climatology of winter midlatitude cyclone activity over eastern North America and the Northwest Atlantic. *Journal of Geophysical Research – Atmospheres*, **114**, D12111.

Lorenz, D.J., and E.T. DeWeaver, 2007: Tropopause height and zonal wind response to global warming in the IPCC scenario integrations. *Journal of Geophysical Research – Atmospheres*, **112**, D10119.

Lorenz, D.J. and D.L. Hartmann, 2003: Eddy-zonal flow feedback in the Northern Hemisphere winter. *Journal of Climate*, **16(8)**, 1212-1227.

Lorenz, R., E.B. Jaeger, and S.I. Seneviratne, 2010: Persistence of heat waves and its link to soil moisture memory. *Geophysical Research Letters*, **37**, L09703.

Lowe, J.A., P.L. Woodworth, T. Knutson, R.E. McDonald, K.L. McInnes, K. Woth, H. von Storch, J. Wolf, V. Swail, N.B. Bernier, S. Gulev, K.J. Horsburgh, A.S. Unnikrishnan, J.R. Hunter, and R. Weisse, 2010: Past and future changes in extreme sea levels and waves. In: *Understanding Sea-Level Rise and Variability* [Church, J.A., P.L. Woodworth, T. Aarup, and W.S. Wilson (eds.)]. Wiley-Blackwell, Chichester, UK, pp. 326-375.

Lu, J., G.A. Vecchi, and T. Reichler, 2007: Expansion of the Hadley cell under global warming. *Geophysical Research Letters*, **34**, L06805.

Lugon, R. and M. Stoffel, 2010: Rock-glacier dynamics and magnitude-frequency relations of debris flows in a high-elevation watershed: Ritigraben, Swiss Alps. *Global and Planetary Change*, **73(3-4)**, 202-210.

Lupikasza, E., 2010: Spatial and temporal variability of extreme precipitation in Poland in the period 1951–2006. *International Journal of Climatology*, **30(7)**, 991-1007.

Luterbacher, J., D. Dietrich, E. Xoplaki, M. Grosjean, and H. Wanner, 2004: European seasonal and annual temperature variability, trends, and extremes since 1500. *Science*, **303(5663)**, 1499-1503.

Lynch, A.H., J.A. Curry, R.D. Brunner, and J.A. Maslanik, 2004: Toward an integrated assessment of the impacts of extreme wind events on Barrow, Alaska. *Bulletin of the American Meteorological Society*, **85(2)**, 209-221.

Mabuchi, K., Y. Sato, and H. Kida, 2005a: Climatic impact of vegetation change in the Asian tropical region. Part I: Case of the Northern Hemisphere summer. *Journal of Climate*, **18(3)**, 410-428.

Mabuchi, K., Y. Sato, and H. Kida, 2005b: Climatic impact of vegetation change in the Asian tropical region. Part II: Case of the Northern Hemisphere winter and impact on the extratropical circulation. *Journal of Climate*, **18(3)**, 429-446.

Mann, M.E. and K.A. Emanuel, 2006: Atlantic hurricane trends linked to climate change. *Eos Transactions (AGU)*, **87(24)**, 233-241.

Mann, M.E., K.A. Emanuel, G.J. Holland, and P.J. Webster, 2007a: Atlantic tropical cyclones revisited. *Eos Transactions (AGU)*, **88**, 349-350.

Mann, M.E., T.A. Sabbatelli, and U. Neu, 2007b: Evidence for a modest undercount bias in early historical Atlantic tropical cyclone counts. *Geophysical Research Letters*, **34**, L22707.

Mann, M.E., J.D. Woodruff, J.P. Donnelly, and Z. Zhang, 2009: Atlantic hurricanes and climate over the past 1,500 years. *Nature*, **460**, 880-883.

Manson, G.K., and S.M. Solomon, 2007: Past and future forcing of Beaufort Sea coastal change. *Atmosphere-Ocean*, **45(2)**, 107-122.

Manton, M.J., P.M. Della-Marta, M.R. Haylock, K.J. Hennessy, N. Nicholls, L.E. Chambers, D.A. Collins, G. Daw, A. Finet, D. Gunawan, K. Inape, H. Isobe, T.S. Kestin, P. Lefale, C.H. Leyu, T. Lwin, L. Maitrepierre, N. Ouprasitwong, C.M. Page, J. Pahalad, N. Plummer, M.J. Salinger, R. Suppiah, V.L. Tran, B. Trewin, I. Tibig, and D. Yee, 2001: Trends in extreme daily rainfall and temperature in Southeast Asia and the South Pacific: 1961–1998. *International Journal of Climatology*, **21(3)**, 269-284.

Maraun, D., T.J. Osborn, and N.P. Gillet, 2008: United Kingdom daily precipitation intensity: improved early data, error estimates and an update from 2000 to 2006. *International Journal of Climatology*, **28(5)**, 833-842.

Maraun, D., F. Wetterhall, A.M. Ireson, R.E. Chandler, E.J. Kendon, M. Widmann, S. Brienen, H.W. Rust, T. Sauter, M. Themessl, V.K.C. Venema, K.P. Chun, C.M. Goodess, R.G. Jones, C. Onof, M. Vrac, and I. Thiele-Eich, 2010: Precipitation downscaling under climate change. Recent developments to bridge the gap between dynamical models and the end user. *Reviews of Geophysics*, **48**, RG3003.

Marcos, M., M.N. Tsimplis, and A.G.P. Shaw, 2009: Sea level extremes in southern Europe. *Journal of Geophysical Research – Oceans*, **114**, C01007.

Marengo, J.A., R. Jones, L.M. Alves, and M.C. Valverde, 2009a: Future change of temperature and precipitation extremes in South America as derived from the PRECIS regional climate modeling system. *International Journal of Climatology*, **29(15)**, 2241-2255.

Marengo, J.A., M. Rusticucci, O. Penalba, and M. Renom, 2009b: An intercomparison of observed and simulated extreme rainfall and temperature events during the last half of the twentieth century: part 2: historical trends. *Climatic Change*, **98(3-4)**, 509-529.

Marengo, J.A., J. Tomasella, W. Soares, L. Alves, and C. Nobre, 2011: Extreme climatic events in the Amazon basin. *Theoretical and Applied Climatology*, doi:10.1007/s00704-00011-00465-00701.

Mariotti, A., N. Zeng, J.H. Yoon, V. Artale, A. Navarra, P. Alpert, and L.Z.X. Li, 2008: Mediterranean water cycle changes: transition to drier 21st century conditions in observations and CMIP3 simulations. *Environmental Research Letters*, **3(4)**, 044001.

Marques, C., A. Rocha, and J. Corte-Real, 2011: Global diagnostic energetics of five state-of-the-art climate models. *Climate Dynamics*, **36(9-10)**, 1767-1794.

Martin, J.H., and R.M. Gordon, 1988: Northeast Pacific iron distributions in relation to phytoplankton productivity. *Deep-Sea Research Part A – Oceanographic Research Papers*, **35(2)**, 177-196.

Mastin, L.G., 1994: Explosive tephra emissions at Mount St. Helens, 1989–1991: The violent escape of magmatic gas following storms? *Geological Society of America Bulletin*, **106(2)**, 175-185.

Mastrandrea, M.D., C.B. Field, T.F. Stocker, O. Edenhofer, K.L. Ebi, D.J. Frame, H. Held, E. Kriegler, K.J. Mach, P.R. Matschoss, G.-K. Plattner, G.W. Yohe, and F.W. Zwiers, 2010: *Guidance Note for Lead Authors of the IPCC Fifth Assessment Report on Consistent Treatment of Uncertainties*. Intergovernmental Panel on Climate Change (IPCC), Geneva, Switzerland.

Matthews, A.J., J. Barclay, S. Carn, G. Thompson, J. Alexander, R. Herd, and C. Williams, 2002: Rainfall-induced volcanic activity on Montserrat. *Geophysical Research Letters*, **29(13)**, 1644.

Matulla, C., W. Schoener, H. Alexandersson, H. von Storch, and X.L. Wang, 2008: European storminess: late nineteenth century to present. *Climate Dynamics*, **31(2-3)**, 125-130.

Maue, R.N., 2009: Northern Hemisphere tropical cyclone activity. *Geophysical Research Letters*, **36**, L05805.

May, W., 2008: Potential future changes in the characteristics of daily precipitation in Europe simulated by the HIRHAM regional climate model. *Climate Dynamics*, **30(6)**, 581-603.

McCabe, G.J., and D.M. Wolock, 2002: A step increase in streamflow in the conterminous United States. *Geophysical Research Letters*, **29(24)**, 2185.

McDonald, R., 2011: Understanding the impact of climate change on Northern Hemisphere extra-tropical cyclones. *Climate Dynamics*, **37(7-8)**, 1399-1425, doi:10.1007/s00382-00010-00916-x.

McGranahan, G., D. Balk, and B. Anderson, 2007: The rising tide: assessing the risks of climate change and human settlements in low elevation coastal zones. *Environment and Urbanization*, **19(1)**, 17-37.

McGuire, B., 2010: Potential for a hazardous geospheric response to projected future climate changes. *Philosophical Transactions of the Royal Society A*, **368(1919)**, 2317-2345.

McInnes, K.L., K.J.E. Walsh, G.D. Hubbert, and T. Beer, 2003: Impact of sea-level rise and storm surges on a coastal community. *Natural Hazards*, **30(2)**, 187-207.

McInnes, K.L., I. Macadam, and J.G. O'Grady, 2009a: *The Effect of Climate Change on Extreme Sea Levels along Victoria's Coast*. A Project Undertaken for the Department of Sustainability and Environment, Victoria as part of the 'Future Coasts' Program. CSIRO, Victoria, Australia, 55 pp.

McInnes, K.L., I. Macadam, G.D. Hubbert, and J.G. O'Grady, 2009b: A modelling approach for estimating the frequency of sea level extremes and the impact of climate change in southeast Australia. *Natural Hazards*, **51(1)**, 115-137.

McInnes, K.L., T.A. Erwin, and J.M. Bathols, 2011: Global Climate Model projected changes in 10 m wind speed and direction due to anthropogenic climate change. *Atmospheric Science Letters*, doi:10.1002/asl.1341.

McKee, T.B., N.J. Doesken, and J. Kleist, 1993: The relationship of drought frequency and duration to time scales. In: *Proceedings of the 8th Conference on Applied Climatology*, Anaheim, California, 17-22 Jan 1993, pp. 179-184.

McKendry, I.G., K.B. Strawbridge, N.T. O'Neill, A.M. Macdonald, P.S.K. Liu, W.R. Leaitch, K.G. Anlauf, L. Jaegle, T.D. Fairlie, and D.L. Westphal, 2007: Trans-Pacific transport of Saharan dust to western North America: A case study. *Journal of Geophysical Research – Atmospheres*, **112**, D01103.

McVicar, T.R., T.G. Van Niel, L.T. Li, M.L. Roderick, D.P. Rayner, L. Ricciardulli, and R.J. Donohue, 2008: Wind speed climatology and trends for Australia, 1975-2006: Capturing the stilling phenomenon and comparison with near-surface reanalysis output. *Geophysical Research Letters*, **35**, L20403.

Mearns, L.O., W. Gutowski, R. Jones, R. Leung, S. McGinnis, A. Nunes, and Y. Qian, 2009: A regional climate change assessment program for North America. *Eos Transactions (AGU)*, **90(36)**, 311.

Meehl, G.A. and C. Tebaldi, 2004: More intense, more frequent, and longer lasting heat waves in the 21st century. *Science*, **305(5686)**, 994-997.

Meehl, G.A., J.M. Arblaster, and C. Tebaldi, 2005: Understanding future patterns of increased precipitation intensity in climate model simulations. *Geophysical Research Letters*, **32**, L18719.

Meehl, G.A., C. Covey, T. Delworth, M. Latif, B. McAvaney, J.F.B. Mitchell, R.J. Stouffer, and K.E. Taylor, 2007a: The WCRP CMIP3 multimodel dataset – A new era in climate change research. *Bulletin of the American Meteorological Society*, **88(9)**, 1383-1394.

Meehl, G.A., T.F. Stocker, W.D. Collins, P. Friedlingstein, A.T. Gaye, J.M. Gregory, A. Kitoh, R. Knutti, J.M. Murphy, A. Noda, S.C.B. Raper, I.G. Watterson, A.J. Weaver, and Z.C. Zhao, 2007b: Global climate projections. In: *Climate Change 2007: The Physical Science Basis. Contribution of Working Group I to the Fourth Assessment Report of the Intergovernmental Panel on Climate Change* [Solomon, S., D. Qin, M. Manning, Z. Chen, M. Marquis, K.B. Averyt, M. Tignor and H.L. Miller (eds.)]. Cambridge University Press, Cambridge, UK, and New York, NY, pp. 747-845.

Meehl, G.A., J.M. Arblaster, and W.D. Collins, 2008: Effects of black carbon aerosols on the Indian monsoon. *Journal of Climate*, **21(12)**, 2869-2882.

Meehl, G.A., J.M. Arblaster, K. Matthes, F. Sassi, and H. van Loon, 2009a: Amplifying the Pacific climate system response to a small 11-year solar cycle forcing. *Science*, **325(5944)**, 1114-1118.

Meehl, G.A., A. Hu, and B.D. Santer, 2009b: The mid-1970s climate shift in the Pacific and the relative roles of forced versus inherent decadal variability. *Journal of Climate*, **22(3)**, 780-792.

Meehl, G.A., C. Tebaldi, G. Walton, D. Easterling, and L. McDaniel, 2009c: The relative increase of record high maximum temperatures compared to record low minimum temperatures in the U.S. *Geophysical Research Letters*, **36**, L23701.

Meinshausen, M., N. Meinshausen, W. Hare, S.C.B. Raper, K. Frieler, R. Knutti, D.J. Frame, and M.R. Allen, 2009: Greenhouse-gas emission targets for limiting global warming to 2°C. *Nature*, **458(7242)**, 1158-1162.

Menendez, C.G., M. de Castro, J.P. Boulanger, A. D'Onofrio, E. Sanchez, A.A. Sorensson, J. Blazquez, A. Elizalde, D. Jacob, H. Le Treut, Z.X. Li, M.N. Nunez, N. Pessacg, S. Pfeiffer, M. Rojas, A. Rolla, P. Samuelsson, S.A. Solman, and C. Teichmann, 2010: Downscaling extreme month-long anomalies in southern South America. *Climatic Change*, **98(3-4)**, 379-403.

Menendez, M. and P.L. Woodworth, 2010: Changes in extreme high water levels based on a quasi-global tide-gauge dataset. *Journal of Geophysical Research*, **115**, C10011.

Menendez, M., F.J. Mendez, I.J. Losada, and N.E. Graham, 2008: Variability of extreme wave heights in the northeast Pacific Ocean based on buoy measurements. *Geophysical Research Letters*, **35**, L22607.

Menne, M.J., and C.N. Williams Jr., 2009: Homogenization of temperature series via pairwise comparisons. *Journal of Climate*, **22(7)**, 1700-1717.

Merrifield, M.A., Y.L. Firing, and J.J. Marra, 2007: Annual climatologies of extreme water levels. In: *Aha Hulikoa: Extreme Events. Proceedings of the Hawaiian Winter Workshop*, University of Hawaii at Manoa, 23 -26 Jan 2007. SOEST, University of Hawaii, Manoa, HI, pp. 27-32.

Miller, R., G. Schmidt, and D. Shindell, 2006: Forced annular variations in the 20th century Intergovernmental Panel on Climate Change Fourth Assessment Report models. *Journal of Geophysical Research*, **111**, D18101.

Mills, G.A., 2005: A re-examination of the synoptic and mesoscale meteorology of Ash Wednesday 1983. *Australian Meteorological Magazine*, **54(1)**, 35-55.

Milly, P.C.D. and K.A. Dunne, 2011: Hydrologic adjustment of climate-model projections: The potential pitfall of potential evapotranspiration. *Earth Interactions*, **15(1)**, 1-14.

Milly, P.C.D., R.T. Wetherald, K.A. Dunne, and T.L. Delworth, 2002: Increasing risk of great floods in a changing climate. *Nature*, **415(6871)**, 514-517.

Mimura, N., 1999: Vulnerability of island countries in the South Pacific to sea level rise and climate change. *Climate Research*, **12(2-3)**, 137-143.

Min, S.-K., X. Zhang, and F. Zwiers, 2008: Human-induced arctic moistening. *Science*, **320(5875)**, 518-520.

Min, S.-K., X. Zhang, F.W. Zwiers, and G.C. Hegerl, 2011: Human contribution to more intense precipitation extremes. *Nature*, **470(7334)**, 378-381.

Min, S.K., and A. Hense, 2007: A Bayesian assessment of climate change using multimodel ensembles. Part II: Regional and seasonal mean surface temperatures. *Journal of Climate*, **20(12)**, 2769-2790.

Mitrovica, J.X., M.E. Tamisiea, E.R. Ivins, L.L.A. Vermeersen, G.A. Milne, and K. Lambeck, 2010: Surface mass loading on a dynamic earth: complexity and contamination in the geodetic analysis of global sea-level trends. In: *Understanding Sea-Level Rise and Variability* [Church, J.A., P.L. Woodworth, T. Aarup, and W.S. Wilson (eds.)]. Wiley-Blackwell, Chichester, UK, pp. 285-325.

Moberg, A., P.D. Jones, D. Lister, A. Walther, M. Brunet, J. Jacobeit, L.V. Alexander, P.M. Della-Marta, J. Luterbacher, P. Yiou, D. Chen, A.M.G.K. Tank, O. Saladie, J. Sigro, E. Aguilar, H. Alexandersson, C. Almarza, I. Auer, M. Barriendos, M. Begert, H. Bergstroem, R. Boehm, C.J. Butler, J. Caesar, A. Drebs, D. Founda, F.-W. Gerstengarbe, G. Micela, M. Maugeri, H. Osterle, K. Pandzic, M. Petrakis, L. Srnec, R. Tolasz, P. Tuomenvirta, P.C. Werner, H. Linderholm, A. Philipp, H. Wanner, and E. Xoplaki, 2006: Indices for daily temperature and precipitation extremes in Europe analyzed for the period 1901-2000. *Journal of Geophysical Research – Atmospheres*, **111**, D22106.

Moise, A.F., and R. Colman, 2009: Tropical Australia and the Australian Monsoon: general assessment and projected changes. In: *18th World IMACS Congress and MODSIM09 International Congress on Modelling and Simulation. Modelling and Simulation Society of Australia and New Zealand and International Association for Mathematics and Computers in Simulation, July 2009* [Anderssen, R.S., R.D. Braddock, and L.T.H. Newham (eds.)]. pp. 2042-2048.

Morgenstern, O., H. Akiyoshi, S. Bekki, P. Braesicke, N. Butchart, M.P. Chipperfield, D. Cugnet, M. Deushi, S.S. Dhomse, R.R. Garcia, A. Gettelman, N.P. Gillett, S.C. Hardiman, J. Jumelet, D.E. Kinnison, J.F. Lamarque, F. Lott, M. Marchand, M. Michou, T. Nakamura, D. Olivie, T. Peter, D. Plummer, J.A. Pyle, E. Rozanov, F. Saint-Martin, J.F. Scinocca, K. Shibata, M. Sigmond, D. Smale, H. Teyssedre, W. Tian, A. Voldoire, and Y. Yamashita, 2010: Anthropogenic forcing of the Northern Annular Mode in CCMVal-2 models. *Journal of Geophysical Research – Atmospheres*, **115**, D00M03.

Mori, N., T. Yasuda, H. Mase, T. Tom, and Y. Oku, 2010: Projection of extreme wave climate change under global warming. *Hydrological Research Letters*, **4**, 15-19.

Mudelsee, M., M. Borngen, G. Tetzlaff, and U. Grunewald, 2003: No upward trends in the occurrence of extreme floods in central Europe. *Nature*, **425(6954)**, 166-169.

Muir-Wood, R., 2000: Deglaciation Seismotectonics: a principal influence on intraplate seismogenesis at high latitudes. *Quaternary Science Reviews*, **19(14-15)**, 1399-1411.

Mullan, B., D. Wratt, S. Dean, M. Hollis, S. Allan, T. Williams, and G. Kenny, 2008: *Climate Change Effects and Impacts Assessment: A Guidance Manual for Local Government in New Zealand*. 2nd ed. Ministry for the Environment, Wellington, New Zealand, 149 pp.

Müller, W.A. and E. Roeckner, 2008: ENSO teleconnections in projections of future climate in ECHAM5/MPI-OM. *Climate Dynamics*, **31(5)**, 533-549.

Murakami, H., B. Wang, and A. Kitoh, 2011: Future change of western North Pacific typhoons: Projections by a 20-km-mesh global atmospheric model. *Journal of Climate*, **24(4)**, 1154-1169.

Murphy, J.M., B.B.B. Booth, M. Collins, G.R. Harris, D.M.H. Sexton, and M.J. Webb, 2007: A methodology for probabilistic predictions of regional climate change from perturbed physics ensembles. *Philosophical Transactions of the Royal Society A*, **365(1857)**, 1993-2028.

Murphy, J.M., D.M. Sexton, G.J. Jenkins, B.B.B. Booth, C.C. Brown, R.T. Clark, M. Collins, G.R. Harris, E.J. Kendon, R.A. Betts, S.J. Brown, K.A. Humphrey, M.P. McCarthy, R.E. McDonald, A. Stephens, C. Wallace, R. Warren, R. Wilby, and R.A. Wood, 2009: *UK Climate Projections Science Report: Climate change projections*. Met Office Hadley Centre, Exeter, UK.

Nakaegawa, T. and W. Vergara, 2010: First projection of climatological mean river discharges in the Magdalena River Basin, Colombia, in a changing climate during the 21st century. *Hydrological Research Letters*, **4**, 50-54.

Nakamura, M., S. Kanada, Y. Wakazuki, C. Muroi, A. Hashimoto, T. Kato, A. Noda, M. Yoshizaki, and K. Yasunaga, 2008: Effects of global warming on heavy rainfall during the Baiu season projected by a cloud-system-resolving model. *Journal of Disaster Research*, **3(1)**, 15-24.

Nandintsetseg, B., J.S. Greene, and C.E. Goulden, 2007: Trends in extreme daily precipitation and temperature near lake Hövsgöl, Mongolia. *International Journal of Climatology*, **27(3)**, 341-347.

Narama, C., Y. Shimamura, D. Nakayama, and K. Abdrakhmatov, 2006: Recent changes of glacier coverage in the western Terskey-Alatoo range, Kyrgyz Republic, using Corona and Landsat. *Annals of Glaciology*, **43(1)**, 223-229.

National Climate Centre, 2009: *The Exceptional January-February 2009 Heatwave in Southeastern Australia*. Special Climate Statement 17 (www.bom.gov.au/climate/current/statements/scs17d.pdf), Bureau of Meteorology, Melbourne, Australia.

Neill, S.P., J.D. Scourse, G.R. Bigg, and K. Uehara, 2009: Changes in wave climate over the northwest European shelf seas during the last 12,000 years. *Journal of Geophysical Research – Oceans*, **114**, C06015.

Neu, U., 2008: Is recent major hurricane activity normal? *Nature*, **451(7181)**, E5.

New, M., B. Hewitson, D.B. Stephenson, A. Tsiga, A. Kruger, A. Manhique, B. Gomez, C.A.S. Coelho, D.N. Masisi, E. Kulunaga, E. Mbambalala, F. Adesina, H. Saleh, J. Kanyanga, J. Adosi, L. Bulane, L. Fortunata, M.L. Mdoka, and R. Lajoie, 2006: Evidence of trends in daily climate extremes over southern and west Africa. *Journal of Geophysical Research*, **111**, D14102.

Newhall, C.G. and R.S. Punongbayan (eds.), 1996: *Fire and Mud: Eruptions and Lahars of Mount Pinatubo, Philippines*. University of Washington Press, Seattle, WA, 1126 pp.

Niall, S. and K. Walsh, 2005: The impact of climate change on hailstorms in southeastern Australia. *International Journal of Climatology*, **25(14)**, 1933-1952.

Nicholls, N., 1995: Long-term climate monitoring and extreme events. *Climatic Change*, **31(2-4)**, 231-245.

Nicholls, N., 2008: Recent trends in the seasonal and temporal behaviour of the El Niño Southern Oscillation. *Geophysical Research Letters*, **35**, L19703.

Nicholls, N., 2010: Local and remote causes of the southern Australian autumn-winter rainfall decline, 1958-2007. *Climate Dynamics*, **34(6)**, 835-845.

Nicholls, N. and L. Alexander, 2007: Has the climate become more variable or extreme? Progress 1992-2006. *Progress in Physical Geography*, **31**(1), 77-87.

Nicholls, R.J., 2010: Impacts of and responses to sea-level rise. In: *Understanding Sea-Level Rise and Variability* [Church, J.A., P.L. Woodworth, T. Aarup, and W.S. Wilson (eds.)]. Wiley-Blackwell, Chichester, UK, pp. 17-51.

Nicholls, R.J. and A.C. de la Vega-Leinert, 2008: Implications of sea-level rise for Europe's coasts: An introduction. *Journal of Coastal Research*, **24**(2), 285-287.

Nicholls, R.J., P.P. Wong, V.R. Burkett, J.O. Codignotto, J.E. Hay, R.F. McLean, S. Ragoonaden, and C.D. Woodroffe, 2007: Coastal systems and low-lying areas. In: *Climate Change 2007. Impacts, Adaptation and Vulnerability. Contribution of Working Group II to the Fourth Assessment Report of the Intergovernmental Panel on Climate Change* [Parry, M.L., O.F. Canziani, J.P. Palutikof, P.J. Van Der Linden, and C.E. Hanson (eds.)]. Cambridge University Press, Cambridge, UK, pp. 315-356.

Nieto-Ferreira, R. and T. Rickenbach, 2011: Regionality of monsoon onset in South America: a three stage conceptual model. *International Journal of Climatology*, **31**(9), 1309-1321.

Nissen, K.M., G.C. Leckebusch, J.G. Pinto, D. Renggli, S. Ulbrich, and U. Ulbrich, 2010: Cyclones causing wind storms in the Mediterranean: characteristics, trends and links to large-scale patterns. *Natural Hazards and Earth System Sciences*, **10**(7), 1379-1391.

Niu, F., G. Cheng, W. Ni, and D. Jin, 2005: Engineering-related slope failure in permafrost regions of the Qinghai-Tibet Plateau. *Cold Regions Science and Technology*, **42**(3), 215-225.

Noetzli, J., S. Gruber, T. Kohl, N. Salzmann, and W. Haeberli, 2007: Three-dimensional distribution and evolution of permafrost temperatures in idealized high-mountain topography. *Journal of Geophysical Research – Earth Surface*, **112**, F02S13.

Noetzli, J. and S. Gruber, 2009: Transient thermal effects in Alpine permafrost. *The Cryosphere*, **3**(1), 85-99.

Nott, J., J. Haig, H. Neil, and D. Gillieson, 2007: Greater frequency variability of landfalling tropical cyclones at centennial compared to seasonal and decadal scales. *Earth and Planetary Science Letters*, **255**(3-4), 367-372.

Nott, J., S. Smithers, K. Walsh, and E. Rhodes, 2009: Sand beach ridges record 6000 year history of extreme tropical cyclone activity in northeastern Australia. *Quaternary Science Reviews*, **28**(15-16), 1511-1520.

Nunez, M.N., S.A. Solman, and M.F. Cabre, 2009: Regional climate change experiments over southern South America. II: Climate change scenarios in the late twenty-first century. *Climate Dynamics*, **32**(7-8), 1081-1095.

Nyberg, J., B.A. Malmgren, A. Winter, M.R. Jury, K. Halimeda Kilbourne, and T.M. Quinn, 2007: Low Atlantic hurricane activity in the 1970s and 1980s compared to the past 270 years. *Nature*, **447**(7145), 698-701.

O'Gorman, P.A., 2010: Understanding the varied response of the extratropical storm tracks to climate change. *Proceedings of the National Academy of Sciences*, **107**(45), 19176-19180.

O'Gorman, P.A. and T. Schneider, 2008: Energy of midlatitude transient eddies in idealized simulations of changed climates. *Journal of Climate*, **21**(22), 5797-5806.

O'Gorman, P.A. and T. Schneider, 2009a: The physical basis for increases in precipitation extremes in simulations of 21st-century climate change. *Proceedings of the National Academy of Sciences*, **106**(35), 14773-14777.

O'Gorman, P.A., and T. Schneider, 2009b: Scaling of precipitation extremes over a wide range of climates simulated with an idealized GCM. *Journal of Climate*, **22**(21), 5676-5685.

Oki, T., T. Nishimura, and P. Dirmeyer, 1999: Assessment of annual runoff from land surface models using Total Runoff Integrating Pathways (TRIP). *Journal of the Meteorological Society of Japan*, **77**(1), 235-255.

Okin, G.S., D.A. Gillette, and J.E. Herrick, 2006: Multi-scale controls on and consequences of aeolian processes in landscape change in arid and semi-arid environments. *Journal of Arid Environments*, **65**(2), 253-275.

Oouchi, K., J. Yoshimura, H. Yoshimura, R. Mizuta, S. Kusunoki, and A. Noda, 2006: Tropical cyclone climatology in a global-warming climate as simulated in a 20 km-mesh global atmospheric model: Frequency and wind intensity analyses. *Journal of the Meteorological Society of Japan*, **84**(2), 259-276.

Oppikofer, T., M. Jaboyedoff, and H.R. Keusen, 2008: Collapse at the eastern Eiger flank in the Swiss Alps. *Nature Geoscience*, **1**(8), 531-535.

Orlowsky, B. and S.I. Seneviratne, 2011: Global changes in extremes events: Regional and seasonal dimension. *Climatic Change*, doi:10.1007/s10584-011-0122-9.

Orsolini, Y. and A. Sorteberg, 2009: Projected changes in Eurasian and Arctic Summer cyclones under global warming in the Bergen Climate Model. *Atmospheric and Oceanic Science Letters*, **2**(1), 62.

Osterkamp, T.E., 2005: The recent warming of permafrost in Alaska. *Global and Planetary Change*, **49**, 187-202.

Osterkamp, T.E., 2007: Characteristics of the recent warming of permafrost in Alaska. *Journal of Geophysical Research*, **112**, F02S02.

Osterkamp, T.E., M.T. Jorgenson, E.A.G. Schuur, Y.L. Shur, M.Z. Kanevskiy, J.G. Vogel, and V.E. Tumskoy, 2009: Physical and ecological changes associated with warming permafrost and thermokarst in Interior Alaska. *Permafrost and Periglacial Processes*, **20**(3), 235-256.

Osti, R., and S. Egashira, 2009: Hydrodynamic characteristics of the Tam Pokhari glacial lake outburst flood in the Mt. Everest region, Nepal. *Hydrological Processes*, **23**(20), 2943-2955.

Otto-Bliesner, B.L., R. Schneider, E.C. Brady, M. Kucera, A. Abe-Ouchi, E. Bard, P. Braconnot, M. Crucifix, C.D. Hewitt, M. Kageyama, O. Marti, A. Paul, A. Rosell-Melé, C. Waelbroeck, S.L. Weber, M. Weinelt, and Y. Yu, 2009: A comparison of PMIP2 model simulations and the MARGO proxy reconstruction for tropical sea surface temperatures at last glacial maximum. *Climate Dynamics*, **32**(6), 799-815.

Paciorek, C.J., J.S. Risbey, V. Ventura, and R.D. Rosen, 2002: Multiple indices of Northern Hemisphere cyclone activity, winters 1949-99. *Journal of Climate*, **15**(13), 1573-1590.

Paeth, H. and H.-P. Thamm, 2007: Regional modelling of future African climate north of 15 degrees S including greenhouse warming and land degradation. *Climatic Change*, **83**(3), 401-427.

Pagli, C. and F. Sigmundsson, 2008: Will present day glacier retreat increase volcanic activity? Stress induced by recent glacier retreat and its effect on magmatism at the Vatnajökull ice cap, Iceland. *Geophysical Research Letters*, **35**, L09304.

Pall, P., M.R. Allen, and D.A. Stone, 2007: Testing the Clausius-Clapeyron constraint on changes in extreme precipitation under CO_2 warming. *Climate Dynamics*, **28**(4), 351-363.

Pall, P., T. Aina, D.A. Stone, P.A. Stott, T. Nozawa, A.G.J. Hilberts, D. Lohmann, and M.R. Allen, 2011: Anthropogenic greenhouse gas contribution to flood risk in England and Wales in autumn 2000. *Nature*, **470**(7334), 382-385.

Palmer, W.C., 1965: *Meteorological Drought*. Report 45, US Weather Bureau, Washington, DC.

Pan, Z.T., R.W. Arritt, E.S. Takle, W.J. Gutowski, C.J. Anderson, and M. Segal, 2004: Altered hydrologic feedback in a warming climate introduces a "warming hole." *Geophysical Research Letters*, **31**, L17109.

Pattanaik, D.R. and M. Rajeevan, 2010: Variability of extreme rainfall events over India during southwest monsoon season. *Meteorological Applications*, **17**(1), 88-104.

Paul, F. and W. Haeberli, 2008: Spatial variability of glacier elevation changes in the Swiss Alps obtained from two digital elevation models. *Geophysical Research Letters*, **35**, L21502.

Paul, F., A. Kääb, M. Maisch, T. Kellenberger, and W. Haeberli, 2004: Rapid disintegration of Alpine glaciers observed with satellite data. *Geophysical Research Letters*, **31**, L21402.

Pavan, V., R. Tomozeiu, C. Cacciamani, and M. Di Lorenzo, 2008: Daily precipitation observations over Emilia-Romagna: mean values and extremes. *International Journal of Climatology*, **28**(15), 2065-2079.

Peel, M.C. and T.A. McMahon, 2006: Continental runoff: A quality-controlled global runoff data set. *Nature*, **444**(E14-E15).

Penalba, O.C. and F.A. Robledo, 2010: Spatial and temporal variability of the frequency of extreme daily rainfall regime in the La Plata Basin during the 20th century. *Climate Change*, **98**(3-4), 531-550.

Perkins, S.E., A.J. Pitman, and S.A. Sisson, 2009: Smaller projected increases in 20-year temperature returns over Australia in skill-selected climate models. *Geophysical Research Letters*, **36**, L06710.

Peters, E., P.J.J.F. Torfs, H.A.J. van Lanen, and G. Bier, 2003: Propagation of drought through groundwater – a new approach using linear reservoir theory. *Hydrological Processes*, **17(15)**, 3023-3040.

Peterson, T.C. and M.J. Manton, 2008: Monitoring changes in climate extremes – A tale of international collaboration. *Bulletin of the American Meteorological Society*, **89(9)**, 1266-1271.

Peterson, T.C., M.A. Taylor, R. Demerritte, D.L. Duncombe, S. Burton, F. Thompson, A. Porter, M. Mercedes, E. Villegas, R.S. Fils, A. Klein Tank, A. Martis, R. Warner, A. Joyette, W. Mills, L. Alexander, and B. Gleason, 2002: Recent changes in climate extremes in the Caribbean region. *Journal of Geophysical Research – Atmospheres*, **107**, 4601.

Peterson, T.C., X. Zhang, M. Brunet-India, and J.L. Vazquez-Aguirre, 2008a: Changes in North American extremes derived from daily weather data. *Journal of Geophysical Research – Atmospheres*, **113**, D07113.

Peterson, T.C., D. Anderson, S.J. Cohen, M. Cortez, R. Murname, C. Parmesan, D. Phillips, R. Pulwarty, and J. Stone, 2008b: Why weather and climate extremes matter. In: *Weather and Climate Extremes in a Changing Climate. Regions of Focus: North America, Hawaii, Caribbean, and U.S. Pacific Islands.* [Karl, T.R., G.A. Meehl, D.M. Christopher, S.J. Hassol, A.M. Waple, and W.L. Murray (eds.)]. A Report by the U.S. Climate Change Science Program and the Subcommittee on Global Change Research, Washington, DC, pp. 11-33.

Petoukhov, V. and V.A. Semenov, 2010: A link between reduced Barents-Kara sea ice and cold winter extremes over northern continents. *Journal of Geophysical Research – Atmospheres*, **115**, D21111.

Petrow, T. and B. Merz, 2009: Trends in flood magnitude, frequency and seasonality in Germany in the period 1951-2002. *Journal of Hydrology*, **371(1-4)**, 129-141.

Pezza, A.B., and I. Simmonds, 2008: Large-scale factors in tropical and extratropical cyclone transition and extreme weather events. *Trends and Directions in Climate Research: Annals of the New York Academy of Sciences*, **1146**, 189-211.

Pezza, A.B., I. Simmonds, and J.A. Renwick, 2007: Southern Hemisphere cyclones and anticyclones: Recent trends and links with decadal variability in the Pacific Ocean. *International Journal of Climatology*, **27(11)**, 1403-1419.

Pfeffer, W.T., J.T. Harper, and S. O'Neel, 2008: Kinematic constraints on glacier contributions to 21st-century sea-level rise. *Science*, **321(5894)**, 1340-1343.

Philander, S.G., 1990: *El Niño, La Niña and the Southern Oscillation*. Academic Press, San Diego, CA, 293 pp.

Piani, C., J.O. Haerter, and E. Coppola, 2010: Statistical bias correction for daily precipitation in regional climate models over Europe. *Theoretical and Applied Climatology*, **99(1-2)**, 187-192.

Piao, S., P. Friedlingstein, P. Ciais, N. de Noblet-Ducoudré, D. Labat, and S. Zaehle, 2007: Changes in climate and land use have a larger direct impact than rising CO_2 on global river runoff trends. *Proceedings of the National Academy of Sciences*, **104(39)**, 15242-15247.

Pielke Jr., R.A., J. Gratz, C.W. Landsea, D. Collins, M. Saunders, and R. Musulin, 2008: Normalized hurricane damages in the United States. *Natural Hazards Review*, **9(1)**, 29-42.

Pinto, J.G., T. Spangehl, U. Ulbrich, and P. Speth, 2006: Assessment of winter cyclone activity in a transient ECHAM4-OPYC3 GHG experiment. *Meteorologische Zeitschrift*, **15(3)**, 279-291.

Pinto, J.G., U. Ulbrich, G.C. Leckebusch, T. Spangehl, M. Reyers, and S. Zacharias, 2007a: Changes in storm track and cyclone activity in three SRES ensemble experiments with the ECHAM5/MPI-OM1 GCM. *Climate Dynamics*, **29(2-3)**, 195-210.

Pinto, J.G., E.L. Frohlich, G.C. Leckebusch, and U. Ulbrich, 2007b: Changing European storm loss potentials under modified climate conditions according to ensemble simulations of the ECHAM5/MPI-OM1 GCM. *Natural Hazards and Earth System Sciences*, **7(1)**, 165-175.

Pinto, J.G., S. Zacharias, A.H. Fink, G.C. Leckebusch, and U. Ulbrich, 2009: Factors contributing to the development of extreme North Atlantic cyclones and their relationship with the NAO. *Climate Dynamics*, **32(5)**, 711-737.

Pinto, J.G., M. Reyers, and U. Ulbrich, 2011: The variable link between PNA and NAO in observations and in multi-century CGCM simulations. *Climate Dynamics*, **36(1)**, 337-354.

Pirazzoli, P.A., and A. Tomasin, 2003: Recent near-surface wind changes in the central Mediterranean and Adriatic areas. *International Journal of Climatology*, **23(8)**, 963-973.

Pitman, A.J., N. de Noblet-Ducoudré, F.T. Cruz, E. Davin, G.B. Bonan, V. Brovkin, C. M., D. C., V. Gayler, B.J.J.M. van den Hurk, P.J. Lawrence, M.K. van der Molen, C. Müller, C.H. Reick, S.I. Seneviratne, B.J. Strengers, and A. Voldoire, 2009: Uncertainties in climate responses to past land cover change: first results from the LUCID intercomparison study. *Geophysical Research Letters*, **36**, L14814.

Planton, S., M. Deque, F. Chauvin, and L. Terray, 2008: Expected impacts of climate change on extreme climate events. *Comptes Rendus Geoscience*, **340(9-10)**, 564-574.

Portmann, R.W., S. Solomon, and G.C. Hegerl, 2009: Linkages between climate change, extreme temperature and precipitation across the United States. *Proceedings of the National Academy of Sciences*, **106(18)**, 7324-7329.

Poulter, B., F. Hattermann, E.D. Hawkins, S. Zaehle, S. Sitch, N. Restrepo-Coupe, U. Heyder, and W. Cramer, 2010: Robust dynamics of Amazon dieback to climate change with perturbed ecosystem model parameters. *Global Change Biology*, **16(9)**, 2476-2495.

Power, S.B. and I.N. Smith, 2007: Weakening of the Walker Circulation and apparent dominance of El Niño both reach record levels, but has ENSO really changed? *Geophysical Research Letters*, **34**, L18702.

Prospero, J.M. and P.J. Lamb, 2003: African droughts and dust transport to the Caribbean: Climate change implications. *Science*, **302(5647)**, 1024-1027.

Prospero, J.M., E. Blades, R. Naidu, and M.C. Lavoie, 2009: Reply to: African dust and asthma in the Caribbean-medical and statistical perspectives by M.A. Monteil and R. Antoine. *International Journal of Biometeorology*, **53(5)**, 383-385.

Prudhomme, C. and H. Davies, 2009: Assessing uncertainties in climate change impact analyses on the river flow regimes in the UK. Part 2: future climate. *Climatic Change*, **93(1-2)**, 197-222.

Pruszak, Z., and E. Zawadzka, 2008: Potential implications of sea-level rise for Poland. *Journal of Coastal Research*, **24(2)**, 410-422.

Pryor, S.C., R.J. Barthelmie, and E.S. Riley, 2007: Historical evolution of wind climates in the USA. *Journal of Physics: Conference Series*, **75**, 012065.

Pryor, S.C., J.A. Howe, and K.E. Kunkel, 2009: How spatially coherent and statistically robust are temporal changes in extreme precipitation in the contiguous USA? *International Journal of Climatology*, **29(1)**, 31-45.

Purvis, M.J., P.D. Bates, and C.M. Hayes, 2008: A probabilistic methodology to estimate future coastal flood risk due to sea level rise. *Coastal Engineering*, **55(12)**, 1062-1073.

Qian, B., S. Gameda, and H. Hayhoe, 2008: Performance of stochastic weather generators LARS-WG and AAFC-WG for reproducing daily extremes of diverse Canadian climates. *Climate Research*, **37(1)**, 17-33.

Raff, D.A., T. Pruitt, and L.D. Brekke, 2009: A framework for assessing flood frequency based on climate projection information. *Hydrology and Earth System Sciences*, **13(11)**, 2119-2136.

Rahimzadeh, F., A. Asgari, and E. Fattahi, 2009: Variability of extreme temperature and precipitation in Iran during recent decades. *International Journal of Climatology*, **29(3)**, 329-343.

Rahmstorf, S., 2007: A semi-empirical approach to projecting future sea-level rise. *Science*, **315(5810)**, 368-370.

Rahmstorf, S., M. Crucifix, A. Ganopolski, H. Goosse, I. Kamenkovich, R. Knutti, G. Lohmann, R. Marsh, L.A. Mysak, Z.M. Wang, and A.J. Weaver, 2005: Thermohaline circulation hysteresis: A model intercomparison. *Geophysical Research Letters*, **32**, L23605.

Raia, A., and I.F.D. Cavalcanti, 2008: The life cycle of the South American monsoon system. *Journal of Climate*, **21(23)**, 6227-6246.

Raible, C.C., 2007: On the relation between extremes of midlatitude cyclones and the atmospheric circulation using ERA40. *Geophysical Research Letters*, **34**, L07703.

Raible, C.C., P.M. Della-Marta, C. Schwierz, H. Wernli, and R. Blender, 2008: Northern hemisphere extratropical cyclones: A comparison of detection and tracking methods and different reanalyses. *Monthly Weather Review*, **136(3)**, 880-897.

Räisänen, J., 2007: How reliable are climate models? *Tellus Series A – Dynamic Meteorology and Oceanography*, **59**, 2-29.

Raje, D., and P.P. Mujumdar, 2010: Constraining uncertainty in regional hydrologic impacts of climate change: Nonstationarity in downscaling. *Water Resources Research*, **46**, W07543.

Rajeevan, M., J. Bhate, and A.K. Jaswal, 2008: Analysis of variability and trends of extreme rainfall events over India using 104 years of gridded daily rainfall data. *Geophysical Research Letters*, **35**, L18707.

Rajendran, K., and A. Kitoh, 2008: Indian summer monsoon in future climate projection by a super high-resolution global model. *Current Science*, **95(11)**, 1560-1569.

Ramsay, H.A., and A.H. Sobel, 2011: The effects of relative and absolute sea surface temperature on tropical cyclone potential intensity using a single column model. *Journal of Climate*, **24(1)**, 183-193.

Ranasinghe, R., and M. Stive, 2009: Rising seas and retreating coastlines. *Climatic Change*, **97(3)**, 465-468.

Ranasinghe, R., R. McLoughlin, A. Short, and G. Symonds, 2004: The Southern Oscillation Index, wave climate, and beach rotation. *Marine Geology*, **204(3-4)**, 273-287.

Randall, D.A., R.A. Wood, S. Bony, R. Colman, T. Fichefet, J. Fyfe, V. Kattsov, A. Pitman, J. Shukla, J. Srinivasan, R.J. Stouffer, A. Sumi, and K.E. Taylor, 2007: Climate models and their evaluation. In: *Climate Change 2007: The Physical Science Basis. Contribution of Working Group I to the Fourth Assessment Report of the Intergovernmental Panel on Climate Change* [Solomon, S., D. Qin, M. Manning, Z. Chen, M. Marquis, K.B. Averyt, M. Tignor and H.L. Miller (eds.)]. Cambridge University Press, Cambridge, UK, and New York, NY, pp. 589-662.

Rappaport, E.N., 2000: Loss of life in the United States associated with recent Atlantic tropical cyclones. *Bulletin of the American Meteorological Society*, **81(9)**, 2065-2073.

Rauthe, M., M. Kunz, and C. Kottmeier, 2010: Changes in wind gust extremes over Central Europe derived from a small ensemble of high resolution regional climate models. *Meteorologische Zeitschrift*, **19(3)**, 299-312.

Ravanel, L. and P. Deline, 2011: Climate influence on rockfalls in high-Alpine steep rockwalls: The north side of the Aiguilles de Chamonix (Mont Blanc massif) since the end of the 'Little Ice Age'. *The Holocene*, **21(2)**, 357-365.

Re, M., and V. Ricardo Barros, 2009: Extreme rainfalls in SE South America. *Climatic Change*, **96(1-2)**, 119-136.

Rebetez, M., O. Dupont, and M. Giroud, 2008: An analysis of the July 2006 heatwave extent in Europe compared to the record year of 2003. *Theoretical and Applied Climatology*, **95(1-2)**, 1-7.

Redmond, K.T., Y. Enzel, P.K. House, and F. Biondi, 2002: Climate impact on flood frequency at the decadal to millennial time scales. In: *Ancient Floods, Modern Hazards: Principles and Applications of Paleoflood Hydrology* [House, P.K., R.H. Webb, V.R. Baker, and D.R. Levish (eds.)]. Vol. 5. American Geophysical Union Water Science and Applications, AGU, Washington, DC, pp. 21-46.

Regonda, S.K., B. Rajagopalan, M. Clark, and J. Pitlick, 2005: Seasonal cycle shifts in hydroclimatology over the western United States. *Journal of Climate*, **18(2)**, 372-384.

Rehman, S., 2010: Temperature and rainfall variation over Dhahran, Saudi Arabia, (1970-2006). *International Journal of Climatology*, **30(3)**, 445-449.

Reichert, B.K., L. Bengtsson, and J. Oerlemans, 2002: Recent glacier retreat exceeds internal variability. *Journal of Climate*, **15(21)**, 3069-3081.

Reichstein, M., P. Ciais, D. Papale, R. Valentini, S. Running, N. Viovy, W. Cramer, A. Granier, J. Ogée, V. Allard, M. Aubinet, C. Bernhofer, N. Buchmann, A. Carrara, T. Grünwald, M. Heimann, B. Heinesch, A. Knohl, W. Kutsch, D. Loustau, G. Manca, G. Matteucci, F. Miglietta, J.M. Ourcival, K. Pilegaard, J. Pumpanen, S. Rambal, S. Schaphoff, S. Seufert, J.F. Soussana, M.J. Sanz, T. Vesala, and M. Zhao, 2007: Reduction of ecosystem productivity and respiration during the European summer 2003 climate anomaly: a joint flux tower, remote sensing and modeling analysis. *Global Change Biology*, **13(5)**, 634-651.

Rein, B., A. Lückge, L. Reinhardt, F. Sirocko, A. Wolf, and W.C. Dullo, 2005: El Niño variability off Peru during the last 20,000 years. *Paleoceanography*, **20**, PA4003.

Renard, B., and M. Lang, 2007: Use of a Gaussian copula for multivariate extreme value analysis: Some case studies in hydrology. *Advances in Water Resources*, **30(4)**, 897-912.

Renard, B., M. Lang, P. Bois, A. Dupeyrat, O. Mestre, H. Niel, E. Sauquet, C. Prudhomme, S. Parey, E. Paquet, L. Neppel, and J. Gailhard, 2008: Regional methods for trend detection: Assessing field significance and regional consistency. *Water Resources Research*, **44**, W08419.

Reynolds, J.M., 1998: High-altitude glacial lake hazard assessment and mitigation: a Himalayan perspective. *Geological Society, London, Engineering Geology Special Publications*, **15(1)**, 25-34.

Reynolds, J.M., A. Dolecki, and C. Portocarrero, 1998: The construction of a drainage tunnel as part of glacial lake hazard mitigation at Hualcan, Cordillera Blanca, Peru. *Engineering Geology Special Publications*, **15(1)**, 41-48.

Richardson, S.D. and J.M. Reynolds, 2000: An overview of glacial hazards in the Himalayas. *Quaternary International*, **65**, 31-47.

Rickenmann, D. and M. Zimmermann, 1993: The 1987 debris flows in Switzerland: documentation and analysis. *Geomorphology (Amsterdam)*, **8(2-3)**, 175-189.

Rignot, E., I. Velicogna, M.R. van den Broeke, A. Monaghan, and J. Lenaerts, 2011: Acceleration of the contribution of the Greenland and Antarctic ice sheets to sea level rise. *Geophysical Research Letters*, **38(L05503)**.

Rist, A., and M. Phillips, 2005: First results of investigations on hydrothermal processes within the active layer above alpine permafrost in steep terrain. *Norsk Geografisk Tidsskrift*, **59(2)**, 177-183.

Robeson, S., 2004: Trends in time-varying percentiles of daily minimum and maximum temperature over North America. *Geophysical Research Letters*, **31**, L04203.

Robinson, W.A., 2000: A baroclinic mechanism for the eddy feedback on the zonal index. *Journal of the Atmospheric Sciences*, **57(3)**, 415-422.

Robock, A., K.Y. Vinnikov, G. Srinivasan, J.K. Entin, S.E. Hollinger, N.A. Speranskaya, S. Liu, and A. Namkhai, 2000: The global soil moisture data bank. *Bulletin of the American Meteorological Society*, **81(6)**, 1281-1299.

Robson, A.J., T.K. Jones, D.W. Reed, and A.C. Bayliss, 1998: A study of national trend and variation in UK floods. *International Journal of Climatology*, **18(2)**, 165-182.

Rocha, A., P. Melo-Goncalves, C. Marques, J. Ferreira, and J.M. Castanheira, 2008: High-frequency precipitation changes in southeastern Africa due to anthropogenic forcing. *International Journal of Climatology*, **28(9)**, 1239-1253.

Rockel, B., and K. Woth, 2007: Extremes of near-surface wind speed over Europe and their future changes as estimated from an ensemble of RCM simulations. *Climatic Change*, **81(S1)**, 267-280.

Rodda, J.C., M.A. Little, H.J.E. Rodda, and P.E. McSharry, 2010: A comparative study of the magnitude, frequency and distribution of intense rainfall in the United Kingdom. *International Journal of Climatology*, **30(12)**, 1776-1783.

Rodier, J.A., and M. Roche, 1984: *World Catalogue of Maximum Observed Floods*. IAHS Pub. No. 143, IAHS Press, Wallingford, UK.

Rodrigo, F.S., 2010: Changes in the probability of extreme daily precipitation observed from 1951 to 2002 in the Iberian Peninsula. *International Journal of Climatology*, **30(10)**, 1512-1525.

Rodríguez-Puebla, C., A. Encinas, L. García-Casado, and S. Nieto, 2010: Trends in warm days and cold nights over the Iberian Peninsula: relationships to large-scale variables. *Climatic Change*, **100(3)**, 667-684.

Roer, I., W. Haeberli, and M. Avian, 2008: Observations and considerations on destabilizing active rock glaciers in the European Alps. In: *Permafrost on a Warming Planet: Impacts on Ecosystems, Infrastructure and Climate* [Kane, D.L. and K.M. Hinkel (eds.)]. Proceedings of the Ninth International Conference on Permafrost at University of Alaska, Fairbanks, 29 June - 03 July 2008, Institute of Northern Engineering, pp 1505-1510.

Roesch, A., 2006: Evaluation of surface albedo and snow cover in AR4 coupled climate models. *Journal of Geophysical Research*, **111**, D15111.

Romanovsky, V.E., D.S. Drozdov, N.G. Oberman, G.V. Malkova, A.L. Kholodov, S.S. Marchenko, N.G. Moskalenko, D.O. Sergeev, N.G. Ukraintseva, A.A. Abramov, D.A. Gilichinsky, and A.A. Vasiliev, 2010: Thermal state of permafrost in Russia. *Permafrost and Periglacial Processes*, **21(2)**, 136-155.

Ropelewski, C.F., and M.A. Bell, 2008: Shifts in the statistics of daily rainfall in South America conditional on ENSO phase. *Journal of Climate*, **21(5)**, 849-865.

Rosenbluth, B., H.A. Fuenzalida, and P. Aceituno, 1997: Recent temperature variations in southern South America. *International Journal of Climatology*, **17(1)**, 67-85.

Rosenzweig, C., G. Casassa, D.J. Karoly, A. Imeson, C. Liu, A. Menzel, S. Rawlins, T.L. Root, B. Seguin, and P. Tryjanowski, 2007: Assessment of Observed Changes and Responses in Natural and Managed Systems. In: *Climate Change 2007. Impacts, Adaptation and Vulnerability. Contribution of Working Group II to the Fourth Assessment Report of the Intergovernmental Panel on Climate Change* [Parry, M.L., O.F. Canziani, J.P. Palutikof, P.J. Van Der Linden, and C.E. Hanson (eds.)]. Cambridge University Press, Cambridge, UK, pp. 79-131.

Rotstayn, L.D., M.D. Keywood, B.W. Forgan, A.J. Gabric, I.E. Galbally, J.L. Gras, A.K. Luhar, G.H. McTainsh, R.M. Mitchell, and S.A. Young, 2009: Possible impacts of anthropogenic and natural aerosols on Australian climate: A review. *International Journal of Climatology*, **29(4)**, 461-479.

Rott, H., S.H. Yueh, D.W. Cline, C. Duguay, R. Essery, C. Haas, F. Heliere, M. Kern, G. Macelloni, E. Malnes, T. Nagler, J. Pulliainen, H. Rebhan, and A. Thompson, 2010: Cold regions hydrology high-resolution observatory for snow and cold land processes. *IEEE Proceedings*, **98(5)**, 752-765.

Rowell, D.P. and R.G. Jones, 2006: Causes and uncertainty of future summer drying over Europe. *Climate Dynamics*, **27(2-3)**, 281-299.

Ruckstuhl, C. and J.R. Norris, 2009: How do aerosol histories affect solar "dimming" and "brightening" over Europe? IPCC-AR4 models versus observations. *Journal of Geophysical Research – Atmospheres*, **114**, D00D04.

Rusticucci, M. and M. Barrucand, 2004: Observed trends and changes in temperature extremes over Argentina. *Journal of Climate*, **17(20)**, 4099-4107.

Rusticucci, M. and M. Renom, 2008: Variability and trends in indices of quality-controlled daily temperature extremes in Uruguay. *International Journal of Climatology*, **28(8)**, 1083-1095.

Rusticucci, M., J. Marengo, O. Penalba, and M. Renom, 2010: An intercomparison of model-simulated in extreme rainfall and temperature events during the last half of the twentieth century. Part 1: mean values and variability. *Climate Change*, **98(3-4)**, 493-508.

Ryan, B.F., I.G. Watterson, and J.L. Evans, 1992: Tropical cyclone frequencies inferred from Gray's yearly genesis parameter - Validation of GCM tropical climates. *Geophysical Research Letters*, **19(18)**, 1831-1834.

Sailor, D.J., M. Smith, and M. Hart, 2008: Climate change implications for wind power resources in the Northwest United States. *Renewable Energy*, **33(11)**, 2393-2406.

Saito, K., M. Kimoto, T. Zhang, K. Takata, and S. Emori, 2007: Evaluating a high-resolution climate model: Simulated hydrothermal regimes in frozen ground regions and their change under the global warming scenario. *Journal of Geophysical Research – Earth Surface*, **112**, F02S11.

Santer, B.D., T.M.L. Wigley, P.J. Glecker, C. Bonfils, M.F. Wehner, K. AchutaRoa, T.P. Barnett, J.S. Boyle, W. Brüggemann, M. Fiorino, N. Gillet, J.E. Hansen, P.D. Jones, S.A. Klein, G.A. Meehl, S.C.B. Raper, R.W. Reynolds, K.E. Taylor, and W.M. Washington, 2006: Forced and unforced ocean temperature changes in Atlantic and Pacific tropical cyclogenesis regions. *Proceedings of the National Academy of Sciences*, **103(38)**, 13905-13910.

Santer, B.D., C. Mears, F.J. Wentz, K.E. Taylor, P.J. Gleckler, T.M.L. Wigley, T.P. Barnett, J.S. Boyle, W. Bruggemann, N.P. Gillett, S.A. Klein, G.A. Meehl, T. Nozawa, D.W. Pierce, P.A. Stott, W.M. Washington, and M.F. Wehner, 2007: Identification of human-induced changes in atmospheric moisture content. *Proceedings of the National Academy of Sciences*, **104(39)**, 15248-15253.

Santos, C.A., and J.I.B. Brito, 2007: Análise dos índices de extremos para o semi-árido do Brasil e suas relações com TSM e IVDN. *Revista Brasileira de Meteorologia*, **22(3)**, 303-312.

Santos, C.A.C., J. Brito, I, B., T.V.R. Rao, and E.A. Meneses, 2009: Tendências dos Índices de precipitação no Estado do Ceará. *Revista Brasileira de Meteorologia*, **24(1)**, 39-47.

Sasaki, W., and H. Toshiyuki, 2007: Interannual variability and predictability of summertime significant wave heights in the western North Pacific. *Journal of Oceanography*, **63(2)**, 203-213.

Sauber, J.M., and B.F. Molnia, 2004: Glacier ice mass fluctuations and fault instability in tectonically active Southern Alaska. *Global and Planetary Change*, **42(1-4)**, 279-293.

Sauber, J.M., and N.A. Ruppert, 2008: Rapid ice mass loss: does it have an influence on earthquake occurrence in southern Alaska? In: *Active Tectonics and Seismic Potential of Alaska* [Freymueller, J.T., P.J. Haeussler, R. Wesson, and G. Ekström (eds.)]. American Geophysical Union, Washington, DC.

Scaife, A.A., C.K. Folland, L.V. Alexander, A. Moberg, and J.R. Knight, 2008: European climate extremes and the North Atlantic Oscillation. *Journal of Climate*, **21(1)**, 72-83.

Scaife, A.A., F. Kucharski, C.K. Folland, J. Kinter, S. Bronnimann, D. Fereday, A.M. Fischer, S. Grainger, E.K. Jin, I.S. Kang, J.R. Knight, S. Kusunoki, N.C. Lau, M.J. Nath, T. Nakaegawa, P. Pegion, S. Schubert, P. Sporyshev, J. Syktus, J.H. Yoon, N. Zeng, and T. Zhou, 2009: The CLIVAR C20C project: selected twentieth century climate events. *Climate Dynamics*, **33(5)**, 603-614.

Scaife, A.A., T. Spangehl, D. Fereday, U. Cubasch, U. Langematz, H. Akiyoshi, S. Bekki, P. Braesicke, N. Butchart, M. Chipperfield, A. Gettelman, S. Hardiman, M. Michou, E. Rozanov, and T. Shepherd, 2011: Climate change projections and stratosphere–troposphere interaction. *Climate Dynamics*, doi:10.1007/s00382-00011-01080-00387.

Schaefer, K., T. Zhang, L. Bruhwiler, and A.P. Barrett, 2011: Amount and timing of permafrost carbon release in response to climate warming. *Tellus Series B – Chemical and Physical Meteorology*, **63(2)**, 165-180.

Schär, C., and G. Jendritzky, 2004: Climate change: Hot news from summer 2003. *Nature*, **432(7017)**, 559-560.

Schär, C., P.L. Vidale, D. Lüthi, C. Frei, C. Häberli, M.A. Liniger, and C. Appenzeller, 2004: The role of increasing temperature variability in European summer heatwaves. *Nature*, **427(322)**, 332-336.

Scheffer, M., J. Bascompte, W.A. Brock, V. Brovkin, S.R. Carpenter, V. Dakos, H. Held, E.H. van Nes, M. Rietkerk, and G. Sugihara, 2009: Early-warning signals for critical transitions. *Nature*, **461(7260)**, 53-59.

Schiefer, E., B. Menounos, and R. Wheate, 2007: Recent volume loss of British Columbian glaciers, Canada. *Geophysical Research Letters*, **34**, L16503.

Schmidli, J., and C. Frei, 2005: Trends of heavy precipitation and wet and dry spells in Switzerland during the 20th century. *International Journal of Climatology*, **25(6)**, 753-771.

Schmidli, J., C.M. Goodess, C. Frei, M.R. Haylock, Y. Hundecha, J. Ribalaygua, and T. Schmith, 2007: Statistical and dynamical downscaling of precipitation: An evaluation and comparison of scenarios for the European Alps. *Journal of Geophysical Research – Atmospheres*, **112**, D04105.

Schneidereit, A., R. Blender, K. Fraedrich, and F. Lunkeit, 2007: Icelandic climate and north Atlantic cyclones in ERA-40 reanalyses. *Meteorologische Zeitschrift*, **16(1)**, 17-23.

Schölzel, C., and P. Friederichs, 2008: Multivariate non-normally distributed random variables in climate research - introduction to the copula approach. *Nonlinear Processes in Geophysics*, **15(5)**, 761-772.

Schubert, S.D., M.J. Suarez, P.J. Pegion, R.D. Koster, and J.T. Bacmeister, 2004: On the cause of the 1930s Dust Bowl. *Science*, **303(5665)**, 1855-1859.

Schubert, S.D., Y. Chang, M.J. Suarez, and P.J. Pegion, 2008a: ENSO and wintertime extreme precipitation events over the contiguous united states. *Journal of Climate*, **21(1)**, 22-39.

Schubert, S.D., M.J. Suarez, P.J. Pegion, R.D. Koster, and J.T. Bacmeister, 2008b: Potential predictability of long-term drought and pluvial conditions in the US Great Plains. *Journal of Climate*, **21(4)**, 802-816.

Schubert, S.D., D. Gutzler, H.L. Wang, A. Dai, T. Delworth, C. Deser, K. Findell, R. Fu, W. Higgins, M. Hoerling, B. Kirtman, R. Koster, A. Kumar, D. Legler, D. Lettenmaier, B. Lyon, V. Magana, K. Mo, S. Nigam, P. Pegion, A. Phillips, R. Pulwarty, D. Rind, A. Ruiz-Barradas, J. Schemm, R. Seager, R. Stewart, M. Suarez, J. Syktus, M.F. Ting, C.Z. Wang, S. Weaver, and N. Zeng, 2009: A US CLIVAR project to assess and compare the responses of global climate models to drought-related SST forcing patterns: Overview and results. *Journal of Climate*, **22(19)**, 5251-5272.

Scileppi, E., and J.P. Donnelly, 2007: Sedimentary evidence of hurricane strikes in western Long Island, New York. *Geochemistry Geophysics Geosystems*, **8**, Q06011.

Scott, K.M., J.W. Vallance, N. Kerle, J. Luis Macías, W. Strauch, and G. Devoli, 2005: Catastrophic precipitation-triggered lahar at Casita volcano, Nicaragua: occurrence, bulking and transformation. *Earth Surface Processes and Landforms*, **30(1)**, 59-79.

Seager, R., M.F. Ting, I. Held, Y. Kushnir, J. Lu, G. Vecchi, H.P. Huang, N. Harnik, A. Leetmaa, N.C. Lau, C.H. Li, J. Velez, and N. Naik, 2007: Model projections of an imminent transition to a more arid climate in southwestern North America. *Science*, **316(5828)**, 1181-1184.

Seager, R., A. Tzanova, and J. Nakamura, 2009: Drought in the southeastern United States: Causes, variability over the last millennium, and the potential for future hydroclimate change. *Journal of Climate*, **22(19)**, 5021-5045.

Seierstad, I.A., D.B. Stephenson, and N.G. Kvamsto, 2007: How useful are teleconnection patterns for explaining variability in extratropical storminess? *Tellus Series A – Dynamic Meteorology and Oceanography*, **59(2)**, 170-181.

Seleshi, Y. and P. Camberlin, 2006: Recent changes in dry spell and extreme rainfall events in Ethiopia. *Theoretical and Applied Climatology*, **83(1-4)**, 181-191.

Semenov, M.A., 2008: Simulation of extreme weather events by a stochastic weather generator. *Climate Research*, **35(3)**, 203-212.

Semmler, T., S. Varghese, R. McGrath, P. Nolan, S. Wang, P. Lynch, and C. O'Dowd, 2008: Regional model simulation of North Atlantic cyclones: Present climate and idealized response to increased sea surface temperature. *Journal of Geophysical Research – Atmospheres*, **113**, D02107.

Sen Roy, S., 2009: A spatial analysis of extreme hourly precipitation patterns in India. *International Journal of Climatology*, **29(3)**, 345-355.

Seneviratne, S.I., D. Lüthi, M. Litschi, and C. Schär, 2006a: Land-atmosphere coupling and climate change in Europe. *Nature*, **443(7108)**, 205-209.

Seneviratne, S.I., R.D. Koster, Z.C. Guo, P.A. Dirmeyer, E. Kowalczyk, D. Lawrence, P. Liu, C.H. Lu, D. Mocko, K.W. Oleson, and D. Verseghy, 2006b: Soil moisture memory in AGCM simulations: Analysis of global land-atmosphere coupling experiment (GLACE) data. *Journal of Hydrometeorology*, **7(5)**, 1090-1112.

Seneviratne, S.I., T. Corti, E.L. Davin, M. Hirschi, E. Jaeger, I. Lehner, B. Orlowsky, and A.J. Teuling, 2010: Investigating soil moisture-climate interactions in a changing climate: A review. *Earth Science Reviews*, **99(3-4)**, 125-161.

Sepp, M., and J. Jaagus, 2011: Changes in the activity and tracks of Arctic cyclones. *Climatic Change*, **105(3-4)**, 577-595.

Shanahan, T.M., J.T. Overpeck, K.J. Anchukaitis, J.W. Beck, J.E. Cole, D.L. Dettman, J.A. Peck, C.A. Scholz, and J.W. King, 2009: Atlantic forcing of persistent drought in West Africa. *Science*, **324(5925)**, 377-380.

Shao, Y., and C.H. Dong, 2006: A review on East Asian dust storm climate, modelling and monitoring. *Global and Planetary Change*, **52(1-4)**, 1-22.

Shaw, S.B., and S.J. Riha, 2011: Assessing temperature-based PET equations under a changing climate in temperate, deciduous forests. *Hydrological Processes*, **25(9)**, 1466-1478.

Sheffield, J., and E.F. Wood, 2008a: Global trends and variability in soil moisture and drought characteristics, 1950-2000, from observation-driven simulations of the terrestrial hydrologic cycle. *Journal of Climate*, **21(3)**, 432-458.

Sheffield, J., and E.F. Wood, 2008b: Projected changes in drought occurrence under future global warming from multi-model, multi-scenario, IPCC AR4 simulations. *Climate Dynamics*, **31(1)**, 79-105.

Sheppard, C., D.J. Dixon, M. Gourlay, A. Sheppard, and R. Payet, 2005: Coral mortality increases wave energy reaching shores protected by reef flats: Examples from the Seychelles. *Estuarine Coastal and Shelf Science*, **64(2-3)**, 223-234.

Sherwood, S.C., and M. Huber, 2010: An adaptability limit to climate change due to heat stress. *Proceedings of the National Academy of Sciences*, **107(21)**, 9552-9555.

Shiklomanov, A.I., R.B. Lammers, M.A. Rawlins, L.C. Smith, and T.M. Pavelsky, 2007: Temporal and spatial variations in maximum river discharge from a new Russian data set. *Journal of Geophysical Research – Biogeosciences*, **112**, G04S53.

Shiklomanov, N.I., D.A. Streletskiy, F.E. Nelson, R.D. Hollister, V.E. Romanovsky, C.E. Tweedie, J.G. Bockheim, and J. Brown, 2010: Decadal variations of active-layer thickness in moisture-controlled landscapes, Barrow, Alaska. *Journal of Geophysical Research – Biogeosciences*, **115**, G00I04.

Shongwe, M.E., G.J. van Oldenborgh, B.J.J.M. van den Hurk, B. de Boer, C.A.S. Coelho, and M.K. van Aalst, 2009: Projected changes in mean and extreme precipitation in Africa under global warming. Part I: Southern Africa. *Journal of Climate*, **22(13)**, 3819-3837.

Shongwe, M.E., G.J. van Oldenborgh, B.J.J.M. van den Hurk, B. de Boer, C.A.S. Coelho, and M.K. van Aalst, 2011: Projected changes in mean and extreme precipitation in Africa under global warming. Part II: East Africa. *Journal of Climate*, **24(14)**, 3718-3732.

Shukla, J., 2007: Monsoon mysteries. *Science*, **318(5848)**, 204-205.

Shulmeister, J., D.T. Rodbell, M.K. Gagan, and G.O. Seltzer, 2006: Inter-hemispheric linkages in climate change: paleo-perspectives for future climate change. *Climates of the Past*, **2(2)**, 167-185.

Sidle, R.C., and H. Ochiai, 2006: *Landslides: processes, prediction, and land use.* Water Resources Monograph 18, American Geophysical Union, Washington, DC, 312 pp.

Sigmundsson, F., V. Pinel, B. Lund, F. Albino, C. Pagli, H. Geirsson, and E. Sturkell, 2010: Climate effects on volcanism: influence on magmatic systems of loading and unloading from ice mass variations, with examples from Iceland. *Philosophical Transactions of the Royal Society A*, **368(1919)**, 2519-2534.

Sillmann, J. and E. Roeckner, 2008: Indices for extreme events in projections of anthropogenic climate change. *Climatic Change*, **86(1-2)**, 83-104.

Silva, A.G., and P. Azevedo, 2008: Índices de tendências de Mudanças Climáticas no Estado da Bahia. *Engenheiria Ambiental*, **5**, 141-151.

Silva Dias, M.A.F., S. Rutledge, P. Kabat, P.L. Silva Dias, C. Nobre, G. Fisch, A.J. Dolman, E. Zipser, M. Garstang, A.O. Manzi, J.D. Fuentes, H.R. Rocha, J. Marengo, A. Plana-Fattori, L.D.A. Sá, R.C.S. Alvalá, M.O. Andreae, P. Artaxo, R. Gielow, and L. Gatti, 2002: Cloud and rain processes in a biosphere-atmosphere interaction context in the Amazon Region. *Journal of Geophysical Research – Atmospheres*, **107(D20)**, 8072.

Silvestri, G.E. and C.S. Vera, 2009: Nonstationary impacts of the Southern Annular Mode on Southern Hemisphere climate. *Journal of Climate*, **22(22)**, 6142-6148.

Simmonds, I. and K. Keay, 2002: Surface fluxes of momentum and mechanical energy over the North Pacific and North Atlantic Oceans. *Meteorology and Atmospheric Physics*, **80(1)**, 1-18.

Simmonds, I., C. Burke, and K. Keay, 2008: Arctic climate change as manifest in cyclone behavior. *Journal of Climate*, **21(22)**, 5777-5796.

Simmons, A.J., K.M. Willett, P.D. Jones, P.W. Thorne, and D.P. Dee, 2010: Low-frequency variations in surface atmospheric humidity, temperature, and precipitation: Inferences from reanalyses and monthly gridded observational data sets. *Journal of Geophysical Research – Atmospheres*, **115**, D01110.

Sinha, A., K.G. Cannariato, L.D. Stott, H. Cheng, R.L. Edwards, M.G. Yadava, R. Ramesh, and I.B. Singh, 2007: A 900-year (600 to 1500 A. D.) record of the Indian summer monsoon precipitation from the core monsoon zone of India. *Geophysical Research Letters*, **34**, L16707.

Sinton, J., K. Grönvold, and K. Sæmundsson, 2005: Postglacial eruptive history of the Western Volcanic Zone, Iceland. *Geochemistry Geophysics Geosystems*, **6**, Q12009.

Sjogren, P., 2009: Sand mass accumulation rate as a proxy for wind regimes in the SW Barents Sea during the past 3 ka. *Holocene*, **19(4)**, 591-598.

Small, I., J. van der Meer, and R.E.G., Upshur, 2001: Acting on an environmental health disaster: The case of the Aral Sea. *Environmental Health Perspectives*, **109(6)**, 547-549.

Smith, J.M., M.A. Cialone, T.V. Wamsley, and T.O. McAlpin, 2010: Potential impact of sea level rise on coastal surges in southeast Louisiana. *Ocean Engineering*, **37(1)**, 37-47.

Smith, K., and R. Ward, 1998: *Floods. Physical Processes and Human Impacts.* John Wiley, Chichester, UK, 382 pp.

Smith, L.C., 2000: Trends in Russian Arctic river-ice formation and breakup, 1917 to 1994. *Physical Geography*, **21(1)**, 46-56.

Smith, L.C., Y. Sheng, G.M. MacDonald, and L.D. Hinzman, 2005: Disappearing Arctic lakes. *Science*, **308(5727)**, 1429.

Smith, S.L., S.A. Wolfe, D.W. Riseborough, and F.M. Nixon, 2009: Active-layer characteristics and summer climatic indices, Mackenzie Valley, Northwest Territories, Canada. *Permafrost and Periglacial Processes*, **20(2)**, 201-220.

Smith, S.L., V.E. Romanovsky, A.G. Lewkowicz, C.R. Burn, M. Allard, G.D. Clow, K. Yoshikawa, and J. Throop, 2010: Thermal state of permafrost in North America: a contribution to the international polar year. *Permafrost and Periglacial Processes*, **21(2)**, 117-135.

Smits, A., A.M.G.K. Tank, and G.P. Konnen, 2005: Trends in storminess over the Netherlands, 1962-2002. *International Journal of Climatology*, **25(10)**, 1331-1344.

Soares, W.R., and J.A. Marengo, 2009: Assessments of moisture fluxes east of the Andes in South America in a global warming scenario. *International Journal of Climatology*, **29(10)**, 1395-1414.

Son, S.W., L.M. Polvani, D.W. Waugh, H. Akiyoshi, R. Garcia, D. Kinnison, S. Pawson, E. Rozanov, T.G. Shepherd, and K. Shibata, 2008: The impact of stratospheric ozone recovery on the Southern Hemisphere westerly jet. *Science*, **320(5882)**, 1486-1489.

Son, S.W., E.P. Gerber, J. Perlwitz, L.M. Polvani, N.P. Gillett, K.H. Seo, V. Eyring, T.G. Shepherd, D. Waugh, H. Akiyoshi, J. Austin, A. Baumgaertner, S. Bekki, P. Braesicke, C. Bruhl, N. Butchart, M.P. Chipperfield, D. Cugnet, M. Dameris, S. Dhomse, S. Frith, H. Garny, R. Garcia, S.C. Hardiman, P. Jockel, J.F. Lamarque, E. Mancini, M. Marchand, M. Michou, T. Nakamura, O. Morgenstern, G. Pitari, D.A. Plummer, J. Pyle, E. Rozanov, J.F. Scinocca, K. Shibata, D. Smale, H. Teyssedre, W. Tian, and Y. Yamashita, 2010: Impact of stratospheric ozone on Southern Hemisphere circulation change: A multimodel assessment. *Journal of Geophysical Research – Atmospheres*, **115**, D00M07.

Sørensen, L.S., S.B. Simonsen, K. Nielsen, P. Lucas-Picher, G. Spada, G. Adalgeirsdottir, R. Forsberg, and C.S. Hvidberg, 2011: Mass balance of the Greenland ice sheet (2003–2008) from ICESat data – the impact of interpolation, sampling and firn density. *The Cryosphere*, **5(1)**, 173-186.

Sorensson, A.A., C.G. Menéndez, R. Ruscica, P. Alexander, P. Samuelsson, and U. Willén, 2010: Projected precipitation changes in South America: a dynamical downscaling within CLARIS. *Meteorologische Zeitschrift*, **19(4)**, 347-355.

Sorteberg, A. and J.E. Walsh, 2008: Seasonal cyclone variability at 70 degrees N and its impact on moisture transport into the Arctic. *Tellus Series A – Dynamic Meteorology and Oceanography*, **60(3)**, 570-586.

Sosio, R., G.B. Crosta, and O. Hungr, 2008: Complete dynamic modeling calibration for the Thurwieser rock avalanche (Italian Central Alps) *Engineering Geology*, **100(1-2)**, 11-26.

Stahl, K., H. Hisdal, J. Hannaford, L.M. Tallaksen, H.A.J. van Lanen, E. Sauquet, S. Demuth, M. Fendekova, and J. Jódar, 2010: Streamflow trends in Europe: evidence from a dataset of near-natural catchments. *Hydrology and Earth System Sciences Discussion*, **7(4)**, 5769-5804.

Stephenson, D.B., H.F. Diaz, and R.J. Murnane, 2008: Definition, diagnosis, and origin of extreme weather and climate events. In: *Climate Extremes and Society* [Murnane, R.J. and H.F. Diaz (eds.)]. Cambridge University Press, Cambridge, UK, pp. 11-23.

Stephenson, T.S., C.M. Goodess, M.R. Haylock, A.A. Chen, and M.A. Taylor, 2008: Detecting inhomogeneities in Caribbean and adjacent Caribbean temperature data using sea-surface temperatures. *Journal of Geophysical Research – Atmospheres*, **113**, D21116.

Sterl, A., C. Severijns, H. Dijkstra, W. Hazeleger, G.J. van Oldenborgh, M. van den Broeke, G. Burgers, B. van den Hurk, P.J. van Leeuwen, and P. van Velthoven, 2008: When can we expect extremely high surface temperatures? *Geophysical Research Letters*, **35**, L14703.

Sterl, A., H. van den Brink, H. de Vries, R. Haarsma, and E. van Meijgaard, 2009: An ensemble study of extreme storm surge related water levels in the North Sea in a changing climate. *Ocean Science*, **5(3)**, 369-378.

Sterr, H., 2008: Assessment of vulnerability and adaptation to sea-level rise for the coastal zone of Germany. *Journal of Coastal Research*, **24(2)**, 380-393.

Stewart, I.S., J. Sauber, and J. Rose, 2000: Glacio-seismotectonics: ice sheets, crustal deformation and seismicity. *Quaternary Science Reviews*, **19(14-15)**, 1367-1389.

Stive, M.J.F., Z.B. Wang, and C. Lakhan, 2003: Morphodynamic modeling of tidal basins and coastal inlets. In: *Advances in Coastal Modeling* [Lakhan, V.C. (ed.)]. Elsevier, Amsterdam, The Netherlands, pp. 367-392.

Stoffel, M., I. Lièvre, D. Conus, M.A. Grichting, H. Raetzo, H.W. Gärtner, and M. Monbaron, 2005: 400 years of debris-flow activity and triggering weather conditions: Ritigraben, Valais, Switzerland. *Arctic, Antarctic, and Alpine Research*, **37(3)**, 387-395.

Stott, P.A., D.A. Stone, and M.R. Allen, 2004: Human contribution to the European heatwave of 2003. *Nature*, **432(7017)**, 610-614.

Stott, P.A., N.P. Gillett, G.C. Hegerl, D. Karoly, D. Stone, X. Zhang, and F.W. Zwiers, 2010: Detection and attribution of climate change: a regional perspective. *Wiley Interdisciplinary Reviews: Climate Change*, **1(2)**, 192-211.

Su, B., Z.W. Kundzewicz, and T. Jiang, 2009: Simulation of extreme precipitation over the Yangtze River Basin using Wakeby distribution. *Theoretical and Applied Climatology*, **96(3-4)**, 209-219.

Sugahara, S., R.P. da Rocha, and R. Silveira, 2009: Non-stationary frequency analysis of extreme daily rainfall in Sao Paulo, Brazil. *International Journal of Climatology*, **29(9)**, 1339-1349.

Sugi, M., A. Noda, and N. Sato, 2002: Influence of the global warming on tropical cyclone climatology: An experiment with the JMA global model. *Journal of the Meteorological Society of Japan*, **80(2)**, 249-272.

Sugi, M., H. Murakami, and J. Yoshimura, 2009: A reduction in global tropical cyclone frequency due to global warming. *SOLA*, **5**, 164-167.

Sugiyama, M., H. Shiogama, and S. Emori, 2010: Precipitation extreme changes exceeding moisture content increases in MIROC and IPCC climate models. *Proceedings of the National Academy of Sciences*, **107(2)**, 571-575.

Suppiah, R., K. Hennessy, P.H. Whetton, K. McInnes, I. Macadam, J. Bathols, J. Ricketts, and C.M. Page, 2007: Australian climate change projections derived from simulations performed for the IPCC 4th Assessment Report. *Australian Meteorological Magazine*, **56(3)**, 131-152.

Suter, S., M. Laternser, W. Haeberli, R. Frauenfelder, and M. Hoelzle, 2001: Cold firn and ice of high altitude glaciers in the Alps: measurements and distribution modelling. *Journal of Glaciology*, **47(156)**, 85-96.

Sutton, R.T., B. Dong, and J.M. Gregory, 2007: Land/sea warming ratio in response to climate change: IPCC AR4 model results and comparison with observations. *Geophysical Research Letters*, **34**, L02701.

Suursaar, U., and T. Kullas, 2009: Decadal variations in wave heights off Cape Kelba, Saaremaa Island, and their relationships with changes in wind climate. *Oceanologia*, **51(1)**, 39-61.

Svensson, C., and D.A. Jones, 2002: Dependence between extreme sea surge, river flow and precipitation in eastern Britain. *International Journal of Climatology*, **22(10)**, 1149-1168.

Swanson, K.L., 2008: Non-locality of Atlantic tropical cyclone intensities. *Geochemistry Geophysics Geosystems*, **9**, Q04V01.

Syvitski, J.P.M., A.J. Kettner, I. Overeem, E.W.H. Hutton, M.T. Hannon, G.R. Brakenridge, J. Day, C. Vorosmarty, Y. Saito, L. Giosan, and R.J. Nicholls, 2009: Sinking deltas due to human activities. *Nature Geoscience*, **2(10)**, 681-686.

Tallaksen, L.M., H. Hisdal, and H.A.J. Van Lanen, 2009: Space-time modelling of catchment scale drought characteristics. *Journal of Hydrology*, **375(3-4)**, 363-372.

Tamura, T., K. Horaguchi, Y. Saito, L.N. Van, M. Tateishi, K.O.T. Thi, F. Nanayama, and K. Watanabe, 2010: Monsoon-influenced variations in morphology and sediment of a mesotidal beach on the Mekong River delta coast. *Geomorphology*, **116(1-2)**, 11-23.

Tanaka, H.L., N. Ishizaki, and D. Nohara, 2005: Intercomparison of the intensities and trends of Hadley, Walker and monsoon circulations in the global warming projections. *SOLA*, **1**, 77-80.

Taschetto, A.S., C.C. Ummenhofer, A. Sen Gupta, and M.H. England, 2009: The effect of anomalous warming in the central Pacific on the Australian monsoon. *Geophysical Research Letters*, **36**, L12704.

Taye, M.T., V. Ntegeka, N.P. Ogiramoi, and P. Willems, 2011: Assessment of climate change impact on hydrological extremes in two source regions of the Nile River Basin. *Hydrology and Earth System Sciences*, **15(1)**, 209-222.

Taylor, A.E., K. Wang, S.L. Smith, M.M. Burgess, and A.S. Judge, 2006: Canadian Arctic Permafrost Observatories: Detecting contemporary climate change through inversion of subsurface temperature time series. *Journal of Geophysical Research – Solid Earth*, **111**, B02411.

Tebaldi, C. and R. Knutti, 2007: The use of the multi-model ensemble in probabilistic climate projections. *Philosophical Transactions of the Royal Society A*, **365(1857)**, 2053-2075.

Tebaldi, C. and B. Sanso, 2009: Joint projections of temperature and precipitation change from multiple climate models: a hierarchical Bayesian approach. *Journal of the Royal Statistical Society Series A – Statistics in Society*, **172(1)**, 83-106.

Tebaldi, C., K. Hayhoe, J.M. Arblaster, and G.A. Meehl, 2006: Going to the extremes. An intercomparison of model-simulated historical and future changes in extreme events. *Climatic Change*, **79(3-4)**, 185-211.

Teng, H., W.M. Washington, and G.A. Meehl, 2008: Interannual variations and future change of wintertime extratropical cyclone activity over North America in CCSM3. *Climate Dynamics*, **30(7-8)**, 673-686.

Teuling, A.J., M. Hirschi, A. Ohmura, M. Wild, M. Reichstein, P. Ciais, N. Buchmann, C. Ammann, L. Montagnani, A.D. Richardson, G. Wohlfahrt, and S.I. Seneviratne, 2009: A regional perspective on trends in continental evaporation. *Geophysical Research Letters*, **36**, L02404.

Teuling, A.J., S.I. Seneviratne, R. Stockli, M. Reichstein, E. Moors, P. Ciais, S. Luyssaert, B. van den Hurk, C. Ammann, C. Bernhofer, E. Dellwik, D. Gianelle, B. Gielen, T. Grunwald, K. Klumpp, L. Montagnani, C. Moureaux, M. Sottocornola, and G. Wohlfahrt, 2010: Contrasting response of European forest and grassland energy exchange to heatwaves. *Nature Geoscience*, **3(10)**, 722-727.

Textor, C., M. Schulz, S. Guibert, S. Kinne, Y. Balkanski, S. Bauer, T. Berntsen, T. Berglen, O. Boucher, M. Chin, F. Dentener, T. Diehl, R. Easter, H. Feichter, D. Fillmore, S. Ghan, P. Ginoux, S. Gong, J.E. Kristjansson, M. Krol, A. Lauer, J.F. Lamarque, X. Liu, V. Montanaro, G. Myhre, J. Penner, G. Pitari, S. Reddy, O. Seland, P. Stier, T. Takemura, and X. Tie, 2006: Analysis and quantification of the diversities of aerosol life cycles within AeroCom. *Atmospheric Chemistry and Physics*, **6(7)**, 1777-1813.

Themeßl, M.J., A. Gobiet, and A. Leuprecht, 2011: Empirical-statistical downscaling and error correction of daily precipitation from regional climate models. *International Journal of Climatology*, **31(10)**, 1530-1544, doi: 10.1002/joc.2168.

Thodsen, H., 2007: The influence of climate change on stream flow in Danish rivers. *Journal of Hydrology*, **333(2-4)**, 226-238.

Thomas, D.S.G., M. Knight, and G.F.S. Wiggs, 2005: Remobilization of southern African desert dune systems by twenty-first century global warming. *Nature*, **435(7046)**, 1218-1221.

Thompson, L.G., E. Mosley-Thompson, H. Brecher, M. Davis, B. Leon, D. Les, P.N. Lin, T. Mashiotta, and K. Mountain, 2006: Abrupt tropical climate change: Past and present. *Proceedings of the National Academy of Sciences*, **103(28)**, 10536-10543.

Tietsche, S., D. Notz, J.H. Jungclaus, and J. Marotzke, 2011: Recovery mechanisms of Arctic summer sea ice. *Geophysical Research Letters*, **38**, L02707.

Timbal, B., P. Hope, and S. Charles, 2008: Evaluating the consistency between statistically downscaled and global dynamical model climate change projections. *Journal of Climate*, **21(22)**, 6052-6059.

Timbal, B., E. Fernandez, and Z. Li, 2009: Generalization of a statistical downscaling model to provide local climate change projections for Australia. *Environmental Modelling & Software*, **24(3)**, 341-358.

Timmermann, A., S. McGregor, and F.F. Jin, 2010: Wind effects on past and future regional sea level trends in the southern Indo-Pacific. *Journal of Climate*, **23(16)**, 4429-4437.

Tolika, K., P. Maheras, and I. Tegoulias, 2009: Extreme temperatures in Greece during 2007: Could this be a "return to the future"? *Geophysical Research Letters*, **36**, L10813.

Tomassini, L. and D. Jacob, 2009: Spatial analysis of trends in extreme precipitation events in high-resolution climate model results and observations for Germany. *Journal of Geophysical Research – Atmospheres*, **114**, D12113.

Toreti, A. and F. Desiato, 2008: Changes in temperature extremes over Italy in the last 44 years. *International Journal of Climatology*, **28(6)**, 733-745.

Toreti, A., E. Xoplaki, D. Maraun, F.G. Kuglitsch, H. Wanner, and J. Luterbacher, 2010: Characterisation of extreme winter precipitation in Mediterranean coastal sites and associated anomalous atmospheric circulation patterns. *Natural Hazards and Earth System Sciences*, **10(5)**, 1037-1050.

Trenberth, K.E., D. Stepaniak, J. Hurrell, and M. Fiorino, 2001: Quality of reanalyses in the tropics. *Journal of Climate*, **14(7)**, 1499-1510.

Trenberth, K.E., J. Fasullo, and L. Smith, 2005: Trends and variability in column-integrated atmospheric water vapor. *Climate Dynamics*, **24(7-8)**, 741-758.

Trenberth, K.E., P.D. Jones, P. Ambenje, R. Bojariu, D. Easterling, A. Klein Tank, D. Parker, F. Rahimzadeh, J.A. Renwick, M. Rusticucci, B. Solden, and P. Zhai, 2007: Observations: Surface and atmospheric climate change. In: *Climate Change 2007: The Physical Science Basis. Contribution of Working Group I to the Fourth Assessment Report of the Intergovernmental Panel on Climate Change* [Solomon, S., D. Qin, M. Manning, Z. Chen, M. Marquis, K.B. Averyt, M. Tignor and H.L. Miller (eds.)]. Cambridge University Press, Cambridge, UK, and New York, NY, pp. 235-336.

Trewin, B. and H. Vermont, 2010: Changes in the frequency of record temperatures in Australia, 1957-2009. *Australian Meteorological and Oceanographic Journal*, **60(2)**, 113-119.

Trigo, I.F., 2006: Climatology and interannual variability of storm-tracks in the Euro-Atlantic sector: a comparison between ERA-40 and NCEP/NCAR reanalyses. *Climate Dynamics*, **26(2-3)**, 127-143.

Trömel, S. and C.D. Schönwiese, 2005: A generalized method of time series decomposition into significant components including probability assessments of extreme events and application to observed German precipitation data. *Meteorologische Zeitschrift*, **14**, 417-427.

Trouet, V., J. Esper, N.E. Graham, A. Baker, J.D. Scourse, and D.C. Frank, 2009: Persistent positive North Atlantic Oscillation mode dominated the medieval climate anomaly. *Science*, **324(5923)**, 78-80.

Tsimplis, M.N., M. Marcos, and S. Somot, 2008: 21st century Mediterranean sea level rise: Steric and atmospheric pressure contributions from a regional model. *Global and Planetary Change*, **63(2-3)**, 105-111.

Tu, M., M.J. Hall, P.J.M. de Laat, and M.J.M. de Wit, 2005: Extreme floods in the Meuse river over the past century: aggravated by land-use changes? *Physics and Chemistry of the Earth*, **30(4-5)**, 267-276.

Tudhope, A.W., C.P. Chilcott, M.T. McCulloch, E.R. Cook, J. Chappell, R.M. Ellam, D.W. Lea, J.M. Lough, and G.B. Shimmield, 2001: Variability in the El Niño-Southern Oscillation through a glacial-interglacial cycle. *Science*, **291(5508)**, 1511-1517.

Tuffen, H., 2010: How will melting of ice affect volcanic hazards in the twenty-first century? *Philosophical Transactions of the Royal Society A*, **368(1919)**, 2535-2558.

Turner, A.G. and J.M. Slingo, 2009: Subseasonal extremes of precipitation and active-break cycles of the Indian summer monsoon in a climate-change scenario. *Quarterly Journal of the Royal Meteorological Society*, **135(640)**, 549-567.

Turner, J., S.R. Colwell, G.J. Marshall, T.A. Lachlan-Cope, A.M. Carleton, P.D. Jones, V. Lagun, P.A. Reid, and S. Iagovkina, 2005: Antarctic climate change during the last 50 years. *International Journal of Climatology*, **25(3)**, 279-294.

Turpeinen, H., A. Hampel, T. Karow, and G. Maniatis, 2008: Effect of ice sheet growth and melting on the slip evolution of thrust faults. *Earth and Planetary Science Letters*, **269(1-2)**, 230-241.

Uchiyama, T., R. Mizuta, K. Kamiguchi, A. Kitoh, and A. Noda, 2006: Changes in temperature-based extremes indices due to global warming projected by a global 20-km-mesh atmospheric model. *SOLA*, **2**, 68-71.

Ulbrich, U., J.G. Pinto, H. Kupfer, G.C. Leckebusch, T. Spangehl, and M. Reyers, 2008: Changing northern hemisphere storm tracks in an ensemble of IPCC climate change simulations. *Journal of Climate*, **21(8)**, 1669-1679.

Ulbrich, U., G.C. Leckebusch, and J.G. Pinto, 2009: Extra-tropical cyclones in the present and future climate: a review. *Theoretical and Applied Climatology*, **96(1-2)**, 117-131.

Ullmann, A., P.A. Pirazzoli, and A. Tomasin, 2007: Sea surges in Camargue: Trends over the 20th century. *Continental Shelf Research*, **27(7)**, 922-934.

Ullmann, A., P.A. Pirazzoli, and V. Moron, 2008: Sea surges around the Gulf of Lions and atmospheric conditions. *Global and Planetary Change*, **63(2-3)**, 203-214.

Ummenhofer, C.C., A. Sen Gupta, M.J. Pook, and M.H. England, 2008: Anomalous rainfall over southwest Western Australia forced by Indian Ocean sea surface temperatures. *Journal of Climate*, **21(19)**, 5113-5134.

Ummenhofer, C.C., A. Sen Gupta, M.H. England, and C.J.C. Reason, 2009a: Contributions of Indian Ocean sea surface temperatures to enhanced East African rainfall. *Journal of Climate*, **22(4)**, 993-1013.

Ummenhofer, C.C., A. Sen Gupta, A.S. Taschetto, and M.H. England, 2009b: Modulation of Australian precipitation by meridional gradients in East Indian Ocean sea surface temperature. *Journal of Climate*, **22(21)**, 5597-5610.

Ummenhofer, C.C., M.H. England, G.A. Meyers, P.C. McIntosh, M.J. Pook, J.S. Risbey, A. Sen Gupta, and A.S. Taschetto, 2009c: What causes Southeast Australia's worst droughts? *Geophysical Research Letters*, **36**, L04706.

Vallée, S. and S. Payette, 2007: Collapse of permafrost mounds along a subarctic river over the last 100 years (northern Québec). *Geomorphology*, **90(1-2)**, 162-170.

Vallve, M.B., and J. Martin-Vide, 1998: Secular climatic oscillations as indicated by catastrophic floods in the Spanish Mediterranean coastal area (14th-19th centuries). *Climatic Change*, **38(4)**, 473-491.

Van den Brink, H.W., G.P. Konnen, J.D. Opsteegh, G.J. Van Oldenborgh, and G. Burgers, 2005: Estimating return periods of extreme events from ECMWF seasonal forecast ensembles. *International Journal of Climatology*, **25(10)**, 1345-1354.

van den Hurk, B.J.J.M., and E. van Meijgaard, 2010: Diagnosing land–atmosphere interaction from a Regional Climate Model simulation over West Africa. *Journal of Hydrometeorology*, **11(2)**, 467-481.

van der Linden, P. and J.F.B. Mitchell (eds.), 2009: *ENSEMBLES: Climate Change and its Impacts: Summary of research and results from the ENSEMBLES project*. Met Office Hadley Centre, Exeter, UK, 160 pp.

van der Schrier, G., K.R. Briffa, T.J. Osborn, and E.R. Cook, 2006a: Summer moisture availability across North America. *Journal of Geophysical Research – Atmospheres*, **111**, D11102.

van der Schrier, G., K.R. Briffa, P.D. Jones, and T.J. Osborn, 2006b: Summer moisture variability across Europe. *Journal of Climate*, **19(12)**, 2818-2834.

van Oldenborgh, G.J., S. Philip, and M. Collins, 2005: El Niño in a changing climate: a multi-model study. *Ocean Science*, **1(2)**, 267-298.

van Oldenborgh, G.J., S. Drijfhout, A. van Ulden, R. Haarsma, A. Sterl, C. Severijns, W. Hazeleger, and H. Dijkstra, 2009: Western Europe is warming much faster than expected. *Climate of the Past*, **5(1)**, 1-12.

van Pelt, S.C., P. Kabat, H.W. ter Maat, B.J.J.M. van den Hurk, and A.H. Weerts, 2009: Discharge simulations performed with a hydrological model using bias corrected regional climate model input. *Hydrology and Earth System Sciences*, **13(12)**, 2387-2397.

Vautard, R., P. Yiou, F. D'Andrea, N. de Noblet, N. Viovy, C. Cassou, J. Polcher, P. Ciais, M. Kageyama, and Y. Fan, 2007: Summertime European heat and drought waves induced by wintertime Mediterranean rainfall deficit. *Geophysical Research Letters*, **34**, L07711.

Vautard, R., J. Cattiaux, P. Yiou, J.-N. Thepaut, and P. Ciais, 2010: Northern Hemisphere atmospheric stilling partly attributed to an increase in surface roughness. *Nature Geoscience*, **3(11)**, 756-761.

Vecchi, G.A. and T.R. Knutson, 2008: On estimates of historical North Atlantic tropical cyclone activity. *Journal of Climate*, **21(14)**, 3580-3600.

Vecchi, G.A. and B.J. Soden, 2007a: Effect of remote sea surface temperature change on tropical cyclone potential intensity. *Nature*, **450(7172)**, 1066-1070.

Vecchi, G.A. and B.J. Soden, 2007b: Global warming and the weakening of the tropical circulation. *Journal of Climate*, **20(17)**, 4316-4340.

Vecchi, G.A. and B.J. Soden, 2007c: Increased tropical Atlantic wind shear in model projections of global warming. *Geophysical Research Letters*, **34**, L08702.

Vecchi, G.A., and A.T. Wittenberg, 2010: El Niño and our future climate: where do we stand? *Climate Change*, **1(2)**, 260-270.

Vecchi, G.A., B.J. Soden, A.T. Wittenberg, I.M. Held, A. Leetmaa, and M.J. Harrison, 2006: Weakening of tropical Pacific atmospheric circulation due to anthropogenic forcing. *Nature*, **441**, 73-76.

Vecchi, G.A., K.L. Swanson, and B.J. Soden, 2008: Whither hurricane activity. *Science*, **322(5902)**, 687-689.

Velicogna, I., 2009: Increasing rates of ice mass loss from the Greenland and Antarctic ice sheets revealed by GRACE. *Geophysical Research Letters*, **36**, L19503.

Vera, C. and G. Silvestri, 2009: Precipitation interannual variability in South America from the WCRP-CMIP3 multi-model dataset. *Climate Dynamics*, **32(7-8)**, 1003-1014.

Vera, C., G. Silvestri, B. Liebmann, and P. Gonzalez, 2006: Climate change scenarios for seasonal precipitation in South America from IPCC-AR4 models. *Geophysical Research Letters*, **33**, L13707.

Vermeer, M. and S. Rahmstorf, 2009: Global sea level linked to global temperature. *Proceedings of the National Academy of Sciences*, **106(51)**, 21527-21532.

Vicente-Serrano, S.M. and J.I. Lopez-Moreno, 2008: Nonstationary influence of the North Atlantic Oscillation on European precipitation. *Journal of Geophysical Research – Atmospheres*, **113**, D20120.

Vicente-Serrano, S.M., S. Begueria, and J.I. Lopez-Moreno, 2010: A multiscalar drought index sensitive to global warming: The standardized precipitation evapotranspiration index. *Journal of Climate*, **23(7)**, 1696-1718.

Vidal, J.P., E. Martin, L. Franchisteguy, F. Habets, J.M. Soubeyroux, M. Blanchard, and M. Baillon, 2010: Multilevel and multiscale drought reanalysis over France with the Safran-Isba-Modcou hydrometeorological suite. *Hydrology and Earth System Sciences*, **14(3)**, 459-478.

Vilibic, I. and J. Sepic, 2010: Long-term variability and trends of sea level storminess and extremes in European Seas. *Global and Planetary Change*, **71(1-2)**, 1-12.

Villarini, G., F. Serinaldi, J.A. Smith, and W.F. Krajewski, 2009: On the stationarity of annual flood peaks in the continental United States during the 20th century. *Water Resources Research*, **45**, W08417.

Vincent, C., E. Le Meur, D. Six, M. Funk, M. Hoelzle, and S. Preunkert, 2007: Very high-elevation Mont Blanc glaciated areas not affected by the 20th century climate change. *Journal of Geophysical Research – Atmospheres*, **112**, D09120.

Vincent, C., S. Auclair, and E.L.e. Meur, 2010: Outburst flood hazard for glacier-dammed Lac de Rochemelon, France. *Journal of Glaciology*, **56(195)**, 91-100.

Vincent, L.A. and E. Mekis, 2006: Changes in daily and extreme temperature and precipitation indices for Canada over the 20th century. *Atmosphere-Ocean*, **44(2)**, 177-193.

Vincent, L.A., X. Zhang, B.R. Bonsal, and W.D. Hogg, 2002: Homogenization of daily temperatures over Canada. *Journal of Climate*, **15(11)**, 1322-1334.

Vincent, L.A., T.C. Peterson, V.R. Barros, M.B. Marino, M. Rusticucci, G. Carrasco, E. Ramirez, L.M. Alves, T. Ambrizzi, M.A. Berlato, A.M. Grimm, J.A. Marengo, J. Molion, D.F. Moncunill, E. Rebello, Y.M.T. Anunciação, J. Quintana, J.L. Santos, J. Baez, G. Coronel, J. Garcia, I. Trebejo, M. Bidegain, M.R. Haylock, and D. Karoly, 2005: Observed trends in indices of daily temperature extremes in South America 1960-2000. *Journal of Climate*, **18(23)**, 5011-5023.

Vinchon, C., S. Aubie, Y. Balouin, L. Closset, M. Garcin, D. Idier, and C. Mallet, 2009: Anticipate response of climate change on coastal risks at regional scale in Aquitaine and Languedoc Roussillon (France). *Ocean & Coastal Management*, **52(1)**, 47-56.

Vinnikov, K.Y., A. Robock, N.A. Speranskaya, and A. Schlosser, 1996: Scales of temporal and spatial variability of midlatitude soil moisture. *Journal of Geophysical Research – Atmospheres*, **101(D3)**, 7163-7174.

Voldoire, A., and J.F. Royer, 2004: Tropical deforestation and climate variability. *Climate Dynamics*, **22(8)**, 857-874.

Vose, R.S., D.R. Easterling, and B. Gleason, 2005: Maximum and minimum temperature trends for the globe: an update through 2004. *Geophysical Research Letters*, **32**, L23822.

Vrac, M., M. Stein, and K. Hayhoe, 2007: Statistical downscaling of precipitation through nonhomogeneous stochastic weather typing. *Climate Research*, **34(3)**, 169-184.

Vuichard, D. and M. Zimmermann, 1987: The 1985 catastrophic drainage of a moraine-dammed lake, Khumbu Himal, Nepal: cause and consequences. *Mountain Research and Development*, **7(2)**, 91-110.

Wakazuki, Y., M. Nakamura, S. Kanada, and C. Muroi, 2008: Climatological reproducibility evaluation and future climate projection of extreme precipitation events in the Baiu season using a high-resolution non-hydrostatic RCM in comparison with an AGCM. *Journal of the Meteorological Society of Japan*, **86(6)**, 951-967.

Walkden, M. and M. Dickson, 2008: Equilibrium erosion of soft rock shores with a shallow or absent beach under increased sea level rise. *Marine Geology*, **251(1-2)**, 75-84.

Walsh, J., 2005: Cryosphere and hydrology. In: *Arctic Climate Impact Assessment*. Cambridge Univesity Press, Cambridge, UK, pp. 183-242.

Wan, H., X.L. Wang, and V.R. Swail, 2010: Homogenization and trend analysis of Canadian near-surface wind speeds. *Journal of Climate*, **23(5)**, 1209-1225.

Wang, A.H., T.J. Bohn, S.P. Mahanama, R.D. Koster, and D.P. Lettenmaier, 2009: Multimodel ensemble reconstruction of drought over the Continental United States. *Journal of Climate*, **22(10)**, 2694-2712.

Wang, B. and Q.H. Ding, 2006: Changes in global monsoon precipitation over the past 56 years. *Geophysical Research Letters*, **33**, L06711.

Wang, B., Q.H. Ding, X.H. Fu, I.S. Kang, K. Jin, J. Shukla, and F. Doblas-Reyes, 2005: Fundamental challenge in simulation and prediction of summer monsoon rainfall. *Geophysical Research Letters*, **32**, L15711.

Wang, G. and H.H. Hendon, 2007: Sensitivity of Australian rainfall to inter-El Niño variations. *Journal of Climate*, **20(16)**, 4211-4226.

Wang, G.L., 2005: Agricultural drought in a future climate: results from 15 global climate models participating in the IPCC 4th assessment. *Climate Dynamics*, **25(7-8)**, 739-753.

Wang, J. and X. Zhang, 2008: Downscaling and projection of winter extreme daily precipitation over North America. *Journal of Climate*, **21(5)**, 923-937.

Wang, S., R. McGrath, J. Hanafin, P. Lynch, T. Semmler, and P. Nolan, 2008: The impact of climate change on storm surges over Irish waters. *Ocean Modelling*, **25(1-2)**, 83-94.

Wang, X., P. Zhai, and C. Wang, 2009: Variations in extratropical cyclone activity in northern East Asia. *Advances in Atmospheric Sciences*, **26(3)**, 471-479.

Wang, X.L., V.R. Swail, and F.W. Zwiers, 2006: Climatology and changes of extratropical cyclone activity: Comparison of ERA-40 with NCEP-NCAR reanalysis for 1958-2001. *Journal of Climate*, **19(13)**, 3145-3166.

Wang, X.L., V.R. Swail, F.W. Zwiers, X. Zhang, and Y. Feng, 2009a: Detection of external influence on trends of atmospheric storminess and northern oceans wave heights. *Climate Dynamics*, **32(2-3)**, 189-203.

Wang, X.L., F.W. Zwiers, V.R. Swail, and Y. Feng, 2009b: Trends and variability of storminess in the Northeast Atlantic region, 1874-2007. *Climate Dynamics*, **33(7-8)**, 1179-1195.

Wang, X.M., Y. Yang, Z.B. Dong, and C.X. Zhang, 2009: Responses of dune activity and desertification in China to global warming in the twenty-first century. *Global and Planetary Change*, **67(3-4)**, 167-185.

Wang, Y.Q., L.R. Leung, J.L. McGregor, D.K. Lee, W.C. Wang, Y.H. Ding, and F. Kimura, 2004: Regional climate modeling: Progress, challenges, and prospects. *Journal of the Meteorological Society of Japan*, **82(5)**, 1599-1628.

Wanner, H., J. Beer, J. Butikofer, T.J. Crowley, U. Cubasch, J. Fluckiger, H. Goosse, M. Grosjean, F. Joos, J.O. Kaplan, M. Kuttel, S.A. Muller, I.C. Prentice, O. Solomina, T.F. Stocker, P. Tarasov, M. Wagner, and M. Widmann, 2008: Mid- to Late Holocene climate change: an overview. *Quaternary Science Reviews*, **27(19-20)**, 1791-1828.

Watterson, I.G., 2006: The intensity of precipitation during extratropical cyclones in global warming simulations: a link to cyclone intensity? *Tellus Series A – Dynamic Meteorology and Oceanography*, **58(1)**, 82-97.

Webb, A.P. and P. Kench, 2010: The dynamic response of reef islands to sea level rise: Evidence from multi-decadal analysis of island change in the central Pacific. *Global and Planetary Change*, **72(3)**, 234-246.

Webster, P.J., 2008: Myanmar's deadly daffodil. *Nature Geoscience*, **1**, 488-490.

Webster, P.J., G.J. Holland, J.A. Curry, and H.R. Chang, 2005: Changes in tropical cyclone number, duration, and intensity in a warming environment. *Science*, **309(5742)**, 1844-1846.

Wehner, M.F., G. Bala, P. Duffy, A.A. Mirin, and R. Romano, 2010: Towards direct simulation of future tropical cyclone statistics in a high-resolution global atmospheric model. *Advances in Meteorology*, **2010**, Article 915303.

Weisse, R. and H. Günther, 2007: Wave climate and long-term changes for the Southern North Sea obtained from a high-resolution hindcast 1958-2002. *Ocean Dynamics*, **57(3)**, 161-172.

Weisse, R., H. Von Storch, and F. Feser, 2005: Northeast Atlantic and North Sea storminess as simulated by a regional climate model during 1958-2001 and comparison with observations. *Journal of Climate*, **18(5)**, 465-479.

Wells, N., S. Goddard, and M.J. Hayes, 2004: A self-calibrating Palmer Drought Severity Index. *Journal of Climate*, **17(12)**, 2335-2351.

Weng, H., S.K. Behera, and T. Yamagata, 2009: Anomalous winter climate conditions in the Pacific Rim during recent El Niño Modoki and El Niño events. *Climate Dynamics*, **32(5)**, 663-674.

Wernli, H., S. Dirren, M.A. Liniger, and M. Zillig, 2002: Dynamical aspects of the life cycle of the winter storm 'Lothar' (24-26 December 1999). *Quarterly Journal of the Royal Meteorological Society*, **128(580)**, 405-429.

Wettstein, J.J., and J.M. Wallace, 2010: Observed patterns of month-to-month storm-track variability and their relationship to the background flow. *Journal of the Atmospheric Sciences*, **67(5)**, 1420-1437.

Whitney, F.A., H.J. Freeland, and M. Robert, 2007: Persistently declining oxygen levels in the interior waters of the eastern subarctic Pacific. *Progress in Oceanography*, **75(2)**, 179-199.

Wieczorek, G.F., T. Glade, M. Jakob, and O. Hungr, 2005: Climatic factors influencing occurrence of debris flows. In: *Debris-Flow Hazards and Related Phenomena* [Jakob, M. and O. Hungr (eds.)]. Springer, Berlin, Germany, pp. 325-362.

Wiggs, G.F.S., A.J. Baird, and R.J. Atherton, 2004: The dynamic effects of moisture on the entrainment and transport of sand by wind. *Geomorphology*, **59(1-4)**, 13-30.

Wilby, R.L., K.J. Beven, and N.S. Reynard, 2008: Climate change and fluvial flood risk in the UK: more of the same? *Hydrological Processes*, **22(14)**, 2511-2523.

Willett, K.M., N.P. Gillett, P.D. Jones, and P.W. Thorne, 2007: Attribution of observed surface humidity changes to human influence. *Nature*, **449(7163)**, 710-712.

Willett, K.M., P.D. Jones, N.P. Gillett, and P.W. Thorne, 2008: Recent changes in surface humidity: Development of the HadCRUH dataset. *Journal of Climate*, **21(20)**, 5364-5383.

Willett, K.M., P.D. Jones, P.W. Thorne, and N.P. Gillett, 2010: A comparison of large scale changes in surface humidity over land in observations and CMIP3 general circulation models. *Environmental Research Letters*, **5(2)**, 025210.

Wilson, P., J. McGourty, and M.D. Bateman, 2004: Mid- to late-Holocene coastal dune event stratigraphy for the north coast of Northern Ireland. *Holocene*, **14(3)**, 406-416.

Wing, A.A., A.H. Sobel, and S.J. Camargo, 2007: Relationship between the potential and actual intensities of tropical cyclones on interannual time scales. *Geophysical Research Letters*, **34**, L08810.

Wittenberg, A.T., 2009: Are historical records sufficient to constrain ENSO simulations? *Geophysical Research Letters*, **36**, L12702.

Wong, S., A.E. Dessler, N.M. Mahowald, P.R. Colarco, and A. da Silva, 2008: Long-term variability in Saharan dust transport and its link to North Atlantic sea surface temperature. *Geophysical Research Letters*, **35**, L07812.

Woodroffe, C.D., 2008: Reef-island topography and the vulnerability of atolls to sea-level rise. *Global and Planetary Change*, **62(1-2)**, 77-96.

Woodruff, J.D., J.P. Donnelly, K.A. Emanuel, and P. Lane, 2008a: Assessing sedimentary records of paleohurricane activity using modeled hurricane climatology. *Geochemistry Geophysics Geosystems*, **9**, Q09V10.

Woodruff, J.D., J.P. Donnelly, D. Mohrig, and W.R. Geyer, 2008b: Reconstructing relative flooding intensities responsible for hurricane-induced deposits from Laguna Playa Grande, Vieques, Puerto Rico. *Geology*, **36(5)**, 391-394.

Woodworth, P.L. and D.L. Blackman, 2004: Evidence for systematic changes in extreme high waters since the mid-1970s. *Journal of Climate*, **17(6)**, 1190-1197.

Woodworth, P.L., M. Menendez, and W.R. Gehrels, 2011: Evidence for century-timescale acceleration in mean sea levels and for recent changes in extreme sea levels. *Surveys in Geophysics*, **32(4-5)**, 603-618, doi: 10.1007/s10712-10011-19112-10718.

Woollings, T.J., B.J. Hoskins, M. Blackburn, and P. Berrisford, 2008: A new Rossby wave-breaking interpretation of the North Atlantic Oscillation. *Journal of Atmospheric Science*, **65(2)**, 609-626.

Woollings, T.J., A. Hannachi, B. Hoskins, and A. Turner, 2010: A regime view of the North Atlantic Oscillation and its response to anthropogenic forcing. *Journal of Climate*, **23(6)**, 1291-1307.

Wu, Q., and T. Zhang, 2010: Changes in active layer thickness over the Qinghai-Tibetan Plateau from 1995-2007. *Journal of Geophysical Research*, **115**, D09107.

Wu, Y., M. Ting, R. Seager, H.-P. Huang, and M. Cane, 2011: Changes in storm tracks and energy transports in a warmer climate simulated by the GFDL CM2.1 model. *Climate Dynamics*, **37(1-2)**, 53-72.

Xie, B., Q. Zhang, and Y. Wang, 2008: Trends in hail in China during 1960–2005. *Geophysical Research Letters*, **35**, L13801.

Xie, S.-P., C. Deser, G.A. Vecchi, J. Ma, H. Teng, and A.T. Wittenberg, 2010: Global warming pattern formation: Sea surface temperature and rainfall. *Journal of Climate*, **23(4)**, 966-986.

Xin, W., L. Shiyin, G. Wanqin, and X. Junli, 2008: Assessment and simulation of glacier lake outburst floods for Longbasaba and Pida Lakes, China. *Mountain Research and Development*, **28(3)**, 310-317.

Xoplaki, E., J.F. Gonzalez-Rouco, J. Luterbacher, and H. Wanner, 2003: Mediterranean summer air temperature variability and its connection to the large-scale atmospheric circulation and SSTs. *Climate Dynamics*, **20(7-8)**, 723-739.

Xu, M., C.P. Chang, C.B. Fu, Y. Qi, A. Robock, D. Robinson, and H.M. Zhang, 2006: Steady decline of east Asian monsoon winds, 1969-2000: Evidence from direct ground measurements of wind speed. *Journal of Geophysical Research – Atmospheres*, **111**, D24111.

Yao, C., S. Yang, W. Qian, Z. Lin, and M. Wen, 2008: Regional summer precipitation events in Asia and their changes in the past decades. *Journal of Geophysical Research – Atmospheres*, **113**, D17107.

Yeh, S.W., J.S. Kug, B. Dewitte, M.H. Kwon, B.P. Kirtman, and F.F. Jin, 2009: El Niño in a changing climate. *Nature*, **461**, 511-514.

Yin, J., S.M. Griffies, and R.J. Stouffer, 2010: Spatial variability of sea level rise in twenty-first century projections. *Journal of Climate*, **23(17)**, 4585-4607.

Yiou, P., P. Ribereau, P. Naveau, M. Nogaj, and R. Brazdil, 2006: Statistical analysis of floods in Bohemia (Czech Republic) since 1825. *Hydrological Sciences Journal*, **51(5)**, 930-945.

Yoshimura, J., M. Sugi, and A. Noda, 2006: Influence of greenhouse warming on tropical cyclone frequency. *Journal of the Meteorological Society of Japan*, **84(2)**, 405-428.

You, Q., S. Kang, E. Aguilar, and Y. Yan, 2008: Changes in daily climate extremes in the eastern and central Tibetan Plateau during 1961-2005. *Journal of Geophysical Research – Atmospheres*, **113**, D07101.

You, Q., S. Kang, E. Aguilar, N. Pepin, W.A. Flügel, Y. Yan, Y. Xu, Y. Zhang, and J. Huang, 2011: Changes in daily climate extremes in China and their connection to the large scale atmospheric circulation during 1961-2003. *Climate Dynamics*, **36(11-12)**, 2399-2417.

Young, I.R., S. Zieger, and A.V. Babanin, 2011: Global trends in wind speed and wave height. *Science*, **332(6028)**, 451-455.

Yu, J.Y. and H.Y. Kao, 2007: Decadal changes of ENSO persistence barrier in SST and ocean heat content indices: 1958-2001. *Journal of Geophysical Research – Atmospheres*, **112**, D13106.

Yu, K.F., J.X. Zhao, Q. Shi, and Q.S. Meng, 2009: Reconstruction of storm/tsunami records over the last 4000 years using transported coral blocks and lagoon sediments in the southern South China Sea. *Quaternary International*, **195(1-2)**, 128-137.

Zahn, M. and H. von Storch, 2010: Decreased frequency of North Atlantic polar lows associated with future climate warming. *Nature*, **467(7313)**, 309-312.

Zemp, M., W. Haeberli, M. Hoelzle, and F. Paul, 2006: Alpine glaciers to disappear within decades? *Geophysical Research Letters*, **33**, L13504.

Zemp, M., W. Haeberli, and J. Eamer, 2007: Glaciers and ice caps. In: *Global Outlook for Ice and Snow*. UNEP/GRID, Arendal, Norway, pp. 115-152.

Zhai, P.M., X. Zhang, H. Wan, and X.H. Pan, 2005: Trends in total precipitation and frequency of daily precipitation extremes over China. *Journal of Climate*, **18(7)**, 1096-1108.

Zhang, K.Q., B.C. Douglas, and S.P. Leatherman, 2004: Global warming and coastal erosion. *Climatic Change*, **64(1-2)**, 41-58.

Zhang, Q., Y. Guan, and H. Yang, 2008: ENSO amplitude change in observation and coupled models. *Advances in Atmospheric Sciences*, **25(3)**, 361-366.

Zhang, R. and T.L. Delworth, 2009: A new method for attributing climate variations over the Atlantic Hurricane Basin's main development region. *Geophysical Research Letters*, **36**, L06701.

Zhang, T., 2005: Influence of the seasonal snow cover on the ground thermal regime: An overview. *Reviews of Geophysics*, **43**, RG4002.

Zhang, T., R.G. Barry, K. Knowles, J.A. Heginbottom, and J. Brown, 1999: Statistics and characteristics of permafrost and ground ice distribution in the Northern Hemisphere. *Polar Geography*, **23(2)**, 147-169.

Zhang, T., O.W. Frauenfeld, M.C. Serreze, A. Etringer, C. Oelke, J. McCreight, R.G. Barry, D. Gilichinsky, D. Yang, H. Ye, F. Ling, and S. Chudinova, 2005: Spatial and temporal variability of active layer thickness over the Russian Arctic drainage basin. *Journal of Geophysical Research*, **110**, D16101.

Zhang, X., K.D. Harvey, W.D. Hogg, and T.R. Yuzyk, 2001: Trends in Canadian streamflow. *Water Resources Research*, **37(4)**, 987-998.

Zhang, X., F.W. Zwiers, and G. Li, 2004: Monte Carlo experiments on the detection of trends in extreme values. *Journal of Climate*, **17(10)**, 1945-1952.

Zhang, X., E. Aguilar, S. Sensoy, H. Melkonyan, U. Tagiyeva, N. Ahmed, N. Kutaladze, F. Rahimzadeh, A. Taghipour, T.H. Hantosh, P. Albert, M. Semawi, M.K. Ali, M.H.S. Al-Shabibi, Z. Al-Oulan, T. Zatari, I.A. Khelet, S. Hamoud, R. Sagir, M. Demircan, M. Eken, M. Adiguzel, L. Alexander, T.C. Peterson, and T. Wallis, 2005: Trends in Middle East climate extreme indices from 1950 to 2003. *Journal of Geophysical Research – Atmospheres*, **110**, D22104.

Zhang, X., F.W. Zwiers, G.C. Hegerl, F.H. Lambert, N.P. Gillett, S. Solomon, P.A. Stott, and T. Nozawa, 2007: Detection of human influence on twentieth-century precipitation trends. *Nature*, **448(7152)**, 461-465.

Zhang, X., J. Wang, F.W. Zwiers, P. Ya, and P.Y. Groisman, 2010: The influence of large scale climate variability on winter maximum daily precipitation over North America. *Journal of Climate*, **23(11)**, 2902-2915.

Zhang, X.D., J.E. Walsh, J. Zhang, U.S. Bhatt, and M. Ikeda, 2004: Climatology and interannual variability of arctic cyclone activity: 1948-2002. *Journal of Climate*, **17(12)**, 2300-2317.

Zhang, X.D., A. Sorteberg, J. Zhang, R. Gerdes, and J.C. Comiso, 2008: Recent radical shifts of atmospheric circulations and rapid changes in Arctic climate system. *Geophysical Research Letters*, **35**, L22701.

Zhang, X.Y., S.L. Gong, T.L. Zhao, R. Arimoto, Y.Q. Wang, and Z.J. Zhou, 2003: Sources of Asian dust and role of climate change versus desertification in Asian dust emission. *Geophysical Research Letters*, **30(24)**, 2272.

Zhang, Y.Q., C.M. Liu, Y.H. Tang, and Y.H. Yang, 2007: Trends in pan evaporation and reference and actual evapotranspiration across the Tibetan Plateau. *Journal of Geophysical Research – Atmospheres*, **112**, D12110.

Zhao, C.S., X. Dabu, and Y. Li, 2004: Relationship between climatic factors and dust storm frequency in Inner Mongolia of China. *Geophysical Research Letters*, **31**, L01103.

Zhao, L., Q. Wu, S.S. Marchenko, and N. Sharkhuu, 2010: Thermal state of permafrost and active layer in Central Asia during the international polar year. *Permafrost and Periglacial Processes*, **21(2)**, 198-207.

Zhao, M., I.M. Held, S.J. Lin, and G.A. Vecchi, 2009: Simulations of global hurricane climatology, interannual variability, and response to global warming using a 50-km resolution GCM. *Journal of Climate*, **22(24)**, 6653-6678.

Zhong, D.C., 1999: The dynamic changes and trends of modern desert in China. *Advance in Earth Sciences*, **14(3)**, 229-234 (in Chinese).

Zhou, T.J., L.X. Zhang, and H.M. Li, 2008a: Changes in global land monsoon area and total rainfall accumulation over the last half century. *Geophysical Research Letters*, **35**, L16707.

Zhou, T.J., R.C. Yu, H.M. Li, and B. Wang, 2008b: Ocean forcing to changes in global monsoon precipitation over the recent half-century. *Journal of Climate*, **21(15)**, 3833-3852.

Zhou, Z.J. and G.C. Zhang, 2003: Typical severe dust storms in northern China during 1954-2002. *Chinese Science Bulletin*, **48(21)**, 2366-2370.

Zimmermann, M., P. Mani, and H. Romang, 1997: Magnitude-frequency aspects of alpine debris flows. *Eclogae Geologicae Helvetiae*, **90(3)**, 415-420.

Zin, W.Z.W., S. Jamaludin, S.M. Deni, and A.A. Jemain, 2009: Recent changes in extreme rainfall events in Peninsular Malaysia: 1971-2005. *Theoretical and Applied Climatology*, **99(3-4)**, 303-314.

Zolina, O., C. Simmer, A. Kapala, S. Bachner, S. Gulev, and H. Maechel, 2008: Seasonally dependent changes of precipitation extremes over Germany since 1950 from a very dense observational network. *Journal of Geophysical Research – Atmospheres*, **113**, D06110.

Zolina, O., C. Simmer, K. Belyaev, A. Kapala, and S. Gulev, 2009: Improving estimates of heavy and extreme precipitation using daily records from European rain gauges. *Journal of Hydrometeorology*, **10(3)**, 701-716.

Zolina, O., C. Simmer, S.K. Gulev, and S. Kollet, 2010: Changing structure of European precipitation: Longer wet periods leading to more abundant rainfalls. *Geophysical Research Letters*, **37**, L06704.

Zorita, E., T.F. Stocker, and H. von Storch, 2008: How unusual is the recent series of warm years? *Geophysical Research Letters*, **35**, L24706.

Zou, X., L.V. Alexander, D. Parker, and J. Caesar, 2006: Variations in severe storms over China. *Geophysical Research Letters*, **33**, L17701.

Zwiers, F.W., X. Zhang, and Y. Feng, 2011: Anthropogenic influence on long return period daily temperature extremes at regional scales. *Journal of Climate*, **24**(3), 881-892.

4 Changes in Impacts of Climate Extremes: Human Systems and Ecosystems

Coordinating Lead Authors:
John Handmer (Australia), Yasushi Honda (Japan), Zbigniew W. Kundzewicz (Poland/Germany)

Lead Authors:
Nigel Arnell (UK), Gerardo Benito (Spain), Jerry Hatfield (USA), Ismail Fadl Mohamed (Sudan), Pascal Peduzzi (Switzerland), Shaohong Wu (China), Boris Sherstyukov (Russia), Kiyoshi Takahashi (Japan), Zheng Yan (China)

Review Editors:
Sebastian Vicuna (Chile), Avelino Suarez (Cuba)

Contributing Authors:
Amjad Abdulla (Maldives), Laurens M. Bouwer (Netherlands), John Campbell (New Zealand), Masahiro Hashizume (Japan), Fred Hattermann (Germany), Robert Heilmayr (USA), Adriana Keating (Australia), Monique Ladds (Australia), Katharine J. Mach (USA), Michael D. Mastrandrea (USA), Reinhard Mechler (Germany), Carlos Nobre (Brazil), Apurva Sanghi (World Bank), James Screen (Australia), Joel Smith (USA), Adonis Velegrakis (Greece), Walter Vergara (World Bank), Anya M. Waite (Australia), Jason Westrich (USA), Joshua Whittaker (Australia), Yin Yunhe (China), Hiroya Yamano (Japan)

This chapter should be cited as:

Handmer, J., Y. Honda, Z.W. Kundzewicz, N. Arnell, G. Benito, J. Hatfield, I.F. Mohamed, P. Peduzzi, S. Wu, B. Sherstyukov, K. Takahashi, and Z. Yan, 2012: Changes in impacts of climate extremes: human systems and ecosystems. In: *Managing the Risks of Extreme Events and Disasters to Advance Climate Change Adaptation* [Field, C.B., V. Barros, T.F. Stocker, D. Qin, D.J. Dokken, K.L. Ebi, M.D. Mastrandrea, K.J. Mach, G.-K. Plattner, S.K. Allen, M. Tignor, and P.M. Midgley (eds.)]. A Special Report of Working Groups I and II of the Intergovernmental Panel on Climate Change (IPCC). Cambridge University Press, Cambridge, UK, and New York, NY, USA, pp. 231-290.

Table of Contents

Executive Summary		234
4.1.	**Introduction**	237
4.2.	**Climatic Extremes in Natural and Socioeconomic Systems**	237
4.2.1.	How Do Climate Extremes Impact on Humans and Ecosystems?	237
4.2.2.	Complex Interactions among Climate Events, Exposure, and Vulnerability	238
4.3.	**System- and Sector-Based Aspects of Vulnerability, Exposure, and Impacts**	239
4.3.1.	Introduction	239
4.3.2.	Water	241
4.3.3.	Ecosystems	244
4.3.3.1.	Heat Waves	244
4.3.3.2.	Drought	246
4.3.3.3.	Floods	246
4.3.3.4.	Other Events	246
4.3.4.	Food Systems and Food Security	246
4.3.5.	Human Settlements, Infrastructure, and Tourism	247
4.3.5.1.	Human Settlements	247
4.3.5.2.	Infrastructure	248
4.3.5.3.	Tourism	250
4.3.6.	Human Health, Well-Being, and Security	251
4.4.	**Regionally Based Aspects of Vulnerability, Exposure, and Impacts**	252
4.4.1.	Introduction	252
4.4.2	Africa	253
4.4.2.1.	Introduction	253
4.4.2.2.	Droughts and Heat Waves	253
4.4.2.3.	Extreme Rainfall Events and Floods	253
4.4.2.4.	Dust Storms	254
4.4.3.	Asia	254
4.4.3.1.	Tropical Cyclones (Typhoons or Hurricanes)	254
4.4.3.2.	Flooding	254
4.4.3.3.	Temperature Extremes	255
4.4.3.4.	Droughts	255
4.4.3.5.	Wildfires	255
4.4.4.	Central and South America	255
4.4.4.1.	Extreme Rainfalls in South America	255
4.4.4.2.	Wildfires	255
4.4.4.3.	Regional Costs	256
4.4.5.	Europe	256
4.4.5.1.	Introduction	256
4.4.5.2.	Heat Waves	256
4.4.5.3.	Droughts and Wildfires	256
4.4.5.4.	Coastal Flooding	256
4.4.5.5.	Gale Winds	257
4.4.5.6.	Flooding	258

4.4.5.7.	Landslides	258
4.4.5.8.	Snow	258
4.4.6.	**North America**	**258**
4.4.6.1.	Introduction	258
4.4.6.2.	Heat Waves	258
4.4.6.3.	Drought and Wildfire	259
4.4.6.4.	Inland Flooding	259
4.4.6.5.	Coastal Storms and Flooding	259
4.4.7.	**Oceania**	**260**
4.4.7.1.	Introduction	260
4.4.7.2.	Temperature Extremes	260
4.4.7.3.	Droughts	260
4.4.7.4.	Wildfire	261
4.4.7.5.	Intense Precipitation and Floods	261
4.4.7.6.	Storm Surges	261
4.4.8.	**Open Oceans**	**261**
4.4.9.	**Polar Regions**	**261**
4.4.9.1.	Introduction	261
4.4.9.2.	Warming Cryosphere	262
4.4.9.3.	Floods	262
4.4.9.4.	Coastal Erosion	263
4.4.10.	**Small Island States**	**263**
4.5.	**Costs of Climate Extremes and Disasters**	**264**
4.5.1.	**Framing the Costs of Extremes and Disasters**	**264**
4.5.2.	**Extreme Events, Impacts, and Development**	**265**
4.5.3.	**Methodologies for Evaluating Impact and Adaptation Costs of Extreme Events and Disasters**	**266**
4.5.3.1.	Methods and Tools for Costing Impacts	266
4.5.3.2.	Methods and Tools for Evaluating the Costs of Adaptation	267
4.5.3.3.	Attribution of Impacts to Climate Change: Observations and Limitations	268
4.5.4.	**Assessment of Impact Costs**	**269**
4.5.4.1.	Estimates of Global and Regional Costs of Disasters	269
4.5.4.2.	Potential Trends in Key Extreme Impacts	271
4.5.5.	**Assessment of Adaptation Costs**	**273**
4.5.6.	**Uncertainty in Assessing the Economic Costs of Extremes and Disasters**	**274**

References **275**

Executive Summary

Extreme impacts can result from extreme weather and climate events, but can also occur without extreme events. This chapter examines two broad categories of impacts on human and ecological systems, both of which are influenced by changes in climate, vulnerability, and exposure: first, the chapter primarily focuses on impacts that result from extreme weather and climate events, and second, it also considers extreme impacts that are triggered by less-than-extreme weather or climate events. These two categories of impacts are examined across sectors, systems, and regions. Extreme events can have positive as well as negative impacts on ecosystems and human activities.

Economic losses from weather- and climate-related disasters have increased, but with large spatial and interannual variability (*high confidence*, based on *high agreement, medium evidence*). Global weather- and climate-related disaster losses reported over the last few decades reflect mainly monetized direct damages to assets, and are unequally distributed. Estimates of annual losses have ranged since 1980 from a few US$ billion to above 200 billion (in 2010 dollars), with the highest value for 2005 (the year of Hurricane Katrina). In the period 2000 to 2008, Asia experienced the highest number of weather- and climate-related disasters. The Americas suffered the most economic loss, accounting for the highest proportion (54.6%) of total loss, followed by Asia (27.5%) and Europe (15.9%). Africa accounted for only 0.6% of global economic losses. Loss estimates are lower bound estimates because many impacts, such as loss of human lives, cultural heritage, and ecosystem services, are difficult to value and monetize, and thus they are poorly reflected in estimates of losses. Impacts on the informal or undocumented economy, as well as indirect effects, can be very important in some areas and sectors, but are generally not counted in reported estimates of losses. [4.5.1, 4.5.3.3, 4.5.4.1]

Economic, including insured, disaster losses associated with weather, climate, and geophysical events are higher in developed countries. Fatality rates and economic losses expressed as a proportion of gross domestic product (GDP) are higher in developing countries (*high confidence*). During the period from 1970 to 2008, over 95% of deaths from natural disasters occurred in developing countries. Middle-income countries with rapidly expanding asset bases have borne the largest burden. During the period from 2001 to 2006, losses amounted to about 1% of GDP for middle-income countries, while this ratio has been about 0.3% of GDP for low-income countries and less than 0.1% of GDP for high-income countries, based on *limited evidence*. In small exposed countries, particularly small island developing states, losses expressed as a percentage of GDP have been particularly high, exceeding 1% in many cases and 8% in the most extreme cases, averaged over both disaster and non-disaster years for the period from 1970 to 2010. [4.5.2, 4.5.4.1]

Increasing exposure of people and economic assets has been the major cause of long-term increases in economic losses from weather- and climate-related disasters (*high confidence*). Long-term trends in economic disaster losses adjusted for wealth and population increases have not been attributed to climate change, but a role for climate change has not been excluded (*high agreement, medium evidence*). These conclusions are subject to a number of limitations in studies to date. Vulnerability is a key factor in disaster losses, yet not well accounted. Other limitations are: (i) data availability, as most data are available for standard economic sectors in developed countries; and (ii) type of hazards studied, as most studies focus on cyclones, where confidence in observed trends and attribution of changes to human influence is *low*. The second conclusion is subject to additional limitations: the processes used to adjust loss data over time, and record length. [4.5.3.3]

Settlement patterns, urbanization, and changes in socioeconomic conditions have all influenced observed trends in exposure and vulnerability to climate extremes (*high confidence*). Settlements concentrate the exposure of humans, their assets, and their activities. The most vulnerable populations include urban poor in informal settlements, refugees, internally displaced people, and those living in marginal areas. Population growth is also a driver of changing exposure and vulnerability. [4.2.1, 4.2.2, 4.3.5.1]

In much of the developed world, societies are aging and hence can be more vulnerable to climate extremes, such as heat waves. For example, Europe currently has an aging population, with a higher population density and lower birth rate than any other continent. Nonetheless, exposure to climate extremes in Europe has increased whereas vulnerability has decreased as a result of implementation of policy, regulations, risk prevention, and risk management. Urban heat islands pose an additional risk to urban inhabitants, most affecting the elderly, ill, and socially isolated. [4.3.5.1, 4.3.6, 4.4.5]

Transportation, infrastructure, water, and tourism are sectors sensitive to climate extremes. Transport infrastructure is vulnerable to extremes in temperature, precipitation/river floods, and storm surges, which can lead to damage in road, rail, airports, and ports, and electricity transmission infrastructure is also vulnerable to extreme storm events. The tourism sector is sensitive to climate, given that climate is the principal driver of global seasonality in tourism demand. [4.3.5.2, 4.3.5.3]

Agriculture is also an economic sector exposed and vulnerable to climate extremes. The economies of many developing countries rely heavily on agriculture, dominated by small-scale and subsistence farming, and livelihoods in this sector are especially exposed to climate extremes. Droughts in Africa, especially since the end of the 1960s, have impacted agriculture, with substantial famine resulting. [4.3.4, 4.4.2]

Coastal settlements in both developed and developing countries are exposed and vulnerable to climate extremes. For example, the major factor increasing the vulnerability and exposure of North America to hurricanes is the growth in population and increase in property values, particularly along the Gulf and Atlantic coasts of the United States. Small island states are particularly vulnerable to climate extremes, especially where urban centers and/or island infrastructure predominate in coastal locations. Asia's mega-deltas are also exposed to extreme events such as flooding and have vulnerable populations in expanding urban areas. Mountain settlements are also exposed and vulnerable to climate extremes. [4.3.5.1, 4.4.3, 4.4.6, 4.4.9, 4.4.10]

In many regions, the main drivers of future increases in economic losses due to some climate extremes will be socioeconomic in nature (*medium confidence*, based on *medium agreement, limited evidence*). The frequency and intensity of extreme weather and climate events are only one factor that affects risks, but few studies have specifically quantified the effects of changes in population, exposure of people and assets, and vulnerability as determinants of loss. However, these studies generally underline the important role of projected changes (increases) in population and capital at risk. Additionally, some researchers argue that poorer developing countries and smaller economies are more likely to suffer more from future disasters than developed countries, especially in relation to extreme impacts. [4.5.2, 4.5.4.2]

Increases in exposure will result in higher direct economic losses from tropical cyclones. Losses will depend on future changes in tropical cyclone frequency and intensity (*high confidence*). Overall losses due to extratropical cyclones will also increase, with possible decreases or no change in some areas (*medium confidence*). Although future flood losses in many locations will increase in the absence of additional protection measures (*high agreement, medium evidence*), the size of the estimated change is highly variable, depending on location, climate scenarios used, and methods used to assess impacts on river flow and flood occurrence. [4.5.4.2]

Extreme events will have greater impacts on sectors with closer links to climate, such as water, agriculture and food security, forestry, health, and tourism. For example, while it is not currently possible to reliably project specific changes at the catchment scale, there is *high confidence* that changes in climate have the potential to seriously affect water management systems. However, climate change is in many instances only one of the drivers of future changes in supply reliability, and is not necessarily the most important driver at the local scale. The impacts of changes in flood characteristics are also highly dependent on how climate changes in the future, and as noted in Section 3.5.2, there is *low confidence* in projected changes in flood magnitude or frequency. However, based on the available literature, there is *high confidence* that, in some places, climate change has the potential to substantially affect flood losses. Climate-related extremes are also expected to produce large impacts on infrastructure, although detailed analysis of potential and projected damages are limited to a few countries, infrastructure types, and sectors. [4.3.2, 4.3.5.2]

Estimates of adaptation costs to climate change exhibit a large range and relate to different assessment periods. For 2030, the estimated global cost ranges from US$ 48 to 171 billion per year (in 2005 US$) with recent estimates for developing countries broadly amounting to the average of this range with annual costs of up to US$ 100 billion. Confidence in individual estimates is *low* because the estimates are derived from only three relatively independent studies. These studies have not explicitly separated costs of adapting to changes in climate extremes from other climate change impacts, do not include costs incurred by all sectors, and are based on extrapolations of bottom-up assessments and on top-down analysis lacking site-specificity. [4.5.3, 4.5.5, 4.5.6]

Chapter 4

4.1. Introduction

Chapter 3 evaluates observed and projected changes in the frequency, intensity, spatial extent, and duration of extreme weather and climate events. This physical basis provides a picture of climate change and extreme events. But it does not by itself indicate the impacts experienced by humans or ecosystems. For example, for some sectors and groups of people, severe impacts may result from relatively minor weather and climate events. To understand impacts triggered by weather and climate events, the exposure and vulnerability of humans and ecological systems need to be examined. The emphasis of this chapter is on negative impacts, in line with this report's focus on managing the risks of extreme events and disasters. Weather and climate events, however, can and often do have positive impacts for some people and ecosystems.

In this chapter, two different types of impacts on human and ecological systems are examined: (i) impacts of extreme weather and climate events; and (ii) extreme impacts triggered by less-than-extreme weather or climate events (in combination with non-climatic factors, such as high exposure and/or vulnerability). Where data are available, impacts are examined from sectoral and regional perspectives. Throughout this chapter, the term 'climate extremes' will be used to refer in brief to 'extreme weather and extreme climate events,' as defined in the Glossary and discussed more extensively in Section 3.1.2.

Activities undertaken as disaster risk reduction may also act as adaptation to trends in climate extremes resulting from climate change, and they may thereby act to reduce impacts. Strategies to reduce risk from one type of climate extreme may act to increase or decrease the risk from another. In writing this chapter, we have not considered these issues as subsequent chapters are dedicated to adaptation. Here, impacts are assessed without discussion of the specific possible adaptation or disaster risk reduction strategies or policies evaluated in subsequent chapters.

Examination of trends in impacts and disasters highlights the difficulties in attributing trends in weather- and climate-related disasters to climate change. Trends in exposure and vulnerability and their relationship with climate extremes are discussed. The chapter then examines system- and sector-based aspects of vulnerability, exposure, and impacts, both observed and projected. The same issues are examined regionally before the chapter concludes with a section on the costs of weather- and climate-related impacts, disasters, and adaptation.

4.2. Climatic Extremes in Natural and Socioeconomic Systems

4.2.1. How Do Climate Extremes Impact on Humans and Ecosystems?

The impacts of weather and climate extremes are largely determined by exposure and vulnerability. This is occurring in a context where all three components – exposure, vulnerability, and climate – are highly dynamic and subject to continuous change. Some changes in exposure and vulnerability can be considered as adaptive actions. For example, migration away from high-hazard areas (see Chapter 1 and the Glossary for a definition of the term 'hazard') reduces exposure and the chance of disaster and is also an adaptation to increasing risk from climate extremes (Adger et al., 2001; Dodman and Satterthwaite, 2008; Revi, 2008). Similar adaptive actions are reflected in changes in building regulations and livelihoods, among many other examples.

Extreme impacts on humans and ecosystems can be conceptualized as 'disasters' or 'emergencies.' Many contemporary definitions emphasize either that a disaster results when the impact is such that local capacity to cope is exceeded or such that it severely disrupts normal activities. There is a significant literature on the definitional issues, which include factors of scale and irreversibility (Quarantelli, 1998; Handmer and Dovers, 2007). Disasters result from impacts that require both exposure to the climate event and a susceptibility to harm by what is exposed. Impacts can include major destruction of assets and disruption to economic sectors, loss of human lives, mental health effects, or loss and impacts on plants, animals, and ecosystem services. The Glossary provides the definition of disaster used in this chapter.

Exposure can be conceptualized as the presence of human and ecosystem tangible and intangible assets and activities (including services) in areas affected by climate extremes (see Sections 1.1.2 and 2.2 and the Glossary for definitional discussion). Without exposure there is no impact. Temporal and spatial scales are also important. Exposure can be more or less permanent; for example, exposure can be increased by people visiting an area or decreased by evacuation of people and livestock after a warning. As human activity and settlements expand into an exposed area, more people will be subject to and affected by local climatic hazards. Population growth is predominantly in developing countries (Peduzzi et al., 2011; UNISDR, 2011). Newly occupied areas around or in urban areas were previously left vacant because they are prone to the occurrence of climatic hazards (Handmer and Dovers, 2007; Satterthwaite et al., 2007; Wilbanks et al., 2007), for example with movement of squatters to and development of informal settlements in areas prone to flooding (Huq et al., 2007) and landslides (Anderson et al., 2007). 'Informal settlements' are characterized by an absence of involvement by government in planning, building, or infrastructure and lack of secure tenure. In addition, there are affluent individuals pursuing environmental amenity through coastal canal estates, riverside, and bush locations, which are often at greater risk from floods and fires (Handmer and Dovers, 2007).

Exposure is a necessary but not sufficient condition for impacts. For exposed areas to be subjected to significant impacts from a weather or climate event there must be vulnerability. Vulnerability is composed of (i) susceptibility of what is exposed to harm (loss or damage) from the event, and (ii) its capacity to recover (Cutter and Emrich, 2006; see Sections 1.1.2 and 2.2 and the Glossary). Vulnerability is defined here as in the Glossary as the propensity or predisposition to be adversely affected. For example, those whose livelihoods are weather-dependent or whose housing offers limited protection from weather events will be particularly susceptible to harm (Dodman and Satterthwaite, 2008). Others with limited capacity to recover include those with limited personal resources for recovery or with no access to external resources such as insurance or aid after an event, and those with limited personal support networks (Handmer and Dovers, 2007). Knowledge, health, and access to services of all kinds including emergency services and political support help reduce both key aspects of vulnerability.

The concept of 'resilience' (developed in an ecological context by Holling, 1978; in a broad social sustainability context by Handmer and Dovers, 2007; and by Adger, 2000; Folke et al., 2002; see also the Glossary) emphasizes the positive components of resistance or adaptability in the face of an event and ability to cope and recover. This concept of 'resilience' is often seen as a positive way of expressing a similar concept to that contained in the term 'vulnerability' (Handmer, 2003).

Refugees, internally displaced people, and those driven into marginal areas as a result of violence can be dramatic examples of people vulnerable to the negative effects of weather and climate events, cut off from coping mechanisms and support networks (Handmer and Dovers, 2007). Reasons for the increase in vulnerability associated with warfare include destruction or abandonment of infrastructure (e.g., transport, communications, health, and education) and shelter, redirection of resources from social to military purposes, collapse of trade and commerce, abandonment of subsistence farmlands, lawlessness, and disruption of social networks (Levy and Sidel, 2000; Collier et al., 2003). The proliferation of weapons and minefields, the absence of basic health and education, and collapse of livelihoods can ensure that the effects of war on vulnerability to disasters are long lasting, although some also benefit (Korf, 2004). These areas are also characterized by an exodus of trained people and an absence of inward investment.

Many ecosystems are dependent on climate extremes for reproduction (e.g., through fire and floods), disease control, and in many cases general ecosystem health (e.g., fires or windstorms allowing new growth to replace old). How such extreme events interact with other trends and circumstances can be critical to the outcome. For example, floods that would normally be essential to river gum reproduction may carry disease and water weeds (Rogers and Ralph, 2010).

Climate extremes can cause substantial mortality of individual species and contribute to determining which species exist in ecosystems (Parmesan et al., 2000). For example, drought plays an important role in forest dynamics, as a major influence on the mortality of trees (Villalba and Veblen, 1997; Breshears and Allen, 2002; Breshears et al., 2005).

4.2.2. Complex Interactions among Climate Events, Exposure, and Vulnerability

There exist complex interactions between different climatic and non-climatic hazards, exposure, and vulnerability that have the potential of triggering complex, scale-dependent impacts. Anthropogenic changes in atmospheric systems are influencing changes in many climatic variables and the corresponding physical impacts (see Chapter 3). However, the impacts that climatic extremes have on humans and ecosystems (including those altered by humans) depend also on several other non-climatic factors (Adger, 2006). This section will explore these factors, drawing on examples of flooding and drought.

Changes in socioeconomic status are a key component of exposure; in particular, population growth is a major driver behind changing exposure and vulnerability (Downton et al., 2005; Barredo, 2009). In many regions, people have been encroaching into flood-prone areas where effective flood protection is not assured, due to human pressure and lack of more suitable and available land (McGranahan et al., 2007; Douglas et al., 2008). Urbanization, often driven by rural poverty, drives such migration (Douglas et al., 2008). In these areas, both population and wealth are accumulating, thereby increasing the flood damage potential. In many developed countries, population and wealth accumulation also occur in hazard-prone areas for reasons of lifestyle and/or lower cost (e.g., Radeloff et al., 2005). Here, a tension between climate change adaptation and development is seen; living in these areas without appropriate adaptation may be maladaptive from a climate change perspective, but this may be a risk people are willing to take, or a risk over which they have limited choice, considering their economic circumstances (Wisner et al., 2004). Furthermore, there is often a deficient risk perception present, stemming from an unjustified faith in the level of safety provided by flood protection systems and dikes in particular (Grothmann and Patt, 2005) (e.g., 2005 Hurricane Katrina in New Orleans).

Economic development and land use change can also lead to changes in natural systems. Land cover changes induce changes in rainfall-runoff patterns, which can impact on flood intensity and frequency (e.g., Kundzewicz and Schellnhuber, 2004). Deforestation, urbanization, reduction of wetlands, and river regulation (e.g., channel straightening, shortening, and embankments) change the percentage of precipitation becoming runoff by reducing the available water storage capacity (Few, 2003; Douglas et al., 2008). The proportion of impervious areas (e.g., roofs, yards, roads, pavements, parking lots, etc.) and the value of the runoff coefficient are increased. As a result, water runs off faster to rivers or the sea, and the flow hydrograph has a higher peak and a shorter time-to-peak (Cheng and Wang, 2002; Few, 2003; Douglas et al., 2008), reducing the time available for warnings and emergency action. In mountainous areas, developments extending into hilly slopes are potentially endangered by landslides and debris flows triggered by intense rains. These changes have resulted in rain that is less extreme leading to serious impacts (Crozier, 2010).

Similarly, the socioeconomic impacts of droughts may arise from the interaction between natural conditions and human water use, which can be conceptualized as a combination of supply and demand factors. Human activities (such as over-cultivation, overgrazing, deforestation) have exacerbated desertification of vulnerable areas in Africa and Asia,

> **Box 4-1 | Evolution of Climate, Exposure, and Vulnerability – The Melbourne Fires, 7 February 2009**
>
> The fires in the Australian state of Victoria, on 7 February 2009, demonstrate the evolution of risk through the relationships between the weather- and climate-related phenomena of a decade-long drought, record extreme heat, and record low humidity of 5% (Karoly, 2010; Trewin and Vermont, 2010) interacting with rapidly increasing exposure. Together the climate phenomena created the conditions for major uncontrollable wildfires (Victorian Bushfires Royal Commission, 2010).
>
> The long antecedent drought, record heat, and a 35-day period with no rain immediately before the fires turned areas normally seen as low to medium wildfire risk into very dry high-risk locations. A rapidly expanding urban-bush interface and valuable infrastructure (Berry, 2003; Burnley and Murphy, 2004; Costello, 2007, 2009) provided the values exposed and the potential for extreme impacts that was realized with the loss of 173 lives and considerable tangible and intangible damage. There was a mixture of natural and human sources of ignition, showing that human agency can trigger such fires and extreme impacts.
>
> Many people were not well-prepared physically or psychologically for the fires, and this influenced the level of loss and damage they incurred. Levels of physical and mental health also affected people's vulnerability. Many individuals with ongoing medical conditions, special needs because of their age, or other impairments struggled to cope with the extreme heat and were reliant on others to respond safely (Handmer et al., 2010). However, capacity to recover in a general sense was high for humans and human activities through insurance, government support, private donations, and nongovernmental organizations (NGOs) and was variable for the affected bush with some species and ecosystems benefitting (Lindenmayer et al., 2010; Banks et al., 2011; see also Case Study 9.2.2).
>
> Chapter 3 details projected changes in climate extremes for this region that could increase fire risk, in particular warm temperature extremes, heat waves, and dryness (see Table 3-3 for summary).

where soil and bio-productive resources became permanently degraded (Dregne, 1986). An extreme example of a human-made, pronounced hydrological drought comes from the Aral Sea basin in Central Asia. Due to excessive and non-sustainable water withdrawals from the tributaries (Syr Darya and Amu Darya), their inflow into the Aral Sea has shrunk in volume by some 75% (Micklin, 2007; Rodell et al., 2009) resulting in severe economic and ecological impacts.

The changing impacts of climate extremes on sectors, such as water and food, depend not only on changes in the characteristics of climate-related variables relevant to a given sector, but also on sector-relevant non-climatic stressors, management characteristics (including organizational and institutional aspects), and adaptive capacity (Kundzewicz, 2003).

There also may be increasing risks from possible interactions of hazards (Cruz, 2005; see Sections 3.1.3 and 3.1.4 for discussion of interactions and feedbacks). One hazard may influence other hazards or exacerbate their effects, also with dependence on scale (Buzna et al., 2006). For instance, temperature rise can lead to permafrost thaw, reduced slope stability, and damage to buildings. Another example is that intense precipitation can lead to flash flood, landslides, and infrastructure damage, for example, collapse of bridges, roads, and buildings, and interruption of power and water supplies. In the Philippines, two typhoons hitting the south of Luzon Island in 2004 caused a significant flood disaster as well as landslides on the island, leading to 900 fatalities (Pulhin et al., 2010). It is worthwhile to note that cascading system failures (e.g., among infrastructure) can happen rapidly and over large areas due to their interdependent nature.

4.3. System- and Sector-Based Aspects of Vulnerability, Exposure, and Impacts

4.3.1. Introduction

In this subsection, studies evaluating impacts and risks of extreme events are surveyed for major affected sectors and systems. Sectors and systems considered here include water; ecosystems; food systems and food security; human settlements, infrastructure, and tourism; and human health, well-being, and security. Impacts of climate extremes are determined by the climate extremes themselves as well as by exposure and vulnerability. Climate extremes, exposure, and vulnerability are characterized by uncertainty and continuous change, and shifts in any of these components of risk will have implications for the impacts of extreme events. Generally, there is limited literature on the potential future impacts of extreme events; most literature analyzes current impacts of extreme events. This focus may result in part from incomplete knowledge and uncertainties regarding future changes in some extreme events (see, for example, Section 3.2.3 and Tables 3-1 and 3-3) as well as from uncertainties regarding future exposure and vulnerabilities. Nonetheless, understanding current impacts can be important for decisionmakers preparing for future risks. Analyses of both observed and projected impacts due to extreme climate and weather events are

Box 4-2 | Observed and Projected Trends in Human Exposure: Tropical Cyclones and Floods

The International loss databases with global coverage such as EM-DAT, NatCat, and Sigma (maintained by the Centre for the Epidemiology of Disasters, Munich Re, and Swiss Re, respectively) present an increase in reported disasters through time. Although the number of reported tropical cyclone disasters, for example, has increased from a yearly average of 21.7 during the 1970s to 63 during the 2000s (see Table 4-1), one should not simply conclude that the number of disasters is increasing due to climate change. There are four factors that may individually or together explain this increase: improved access to information, higher population exposure, higher vulnerability, and higher frequency and/or intensity of hazards (Dao and Peduzzi, 2004; Peduzzi et al., 2009). Due to uncertainties in the significance of the role of each of these four possible factors (especially regarding improved access to information), a vulnerability and risk trend analysis cannot be performed based on reported losses (e.g., from EM-DAT or Munich Re). To better understand this trend, international loss databases would have to be standardized.

Here for both tropical cyclones and floods, we overview a method for better understanding these factors through calculation of past trends and future projections of human exposure at regional and global scales. Changes in population size strongly influence changes in exposure to hazards. It is estimated that currently about 1.15 billion people live in tropical cyclone-prone areas. The physical exposure (yearly average number of people exposed) to tropical cyclones is estimated to have increased from approximately 73 million in 1970 to approximately 123 million in 2010 (Figure 4-1; Peduzzi et al., 2011). The number of times that countries are hit by tropical cyclones per year is relatively steady (between 140 and 155 countries per year[1] on average; see Table 4-1 (UNISDR, 2011).

In most oceans, the frequency of tropical cyclones is *likely* to decrease or remain unchanged while mean tropical cyclone

Continued next page ⟶

Table 4-1 | Trend in tropical cyclone disasters reported versus tropical cyclones detected by satellite during the last four decades. The reported disasters as a percentage of the number of countries hit by tropical cyclones increased three-fold. Note that 'best track data' generally comprise four-times daily estimates of tropical cyclone intensity and position; these data are based on post-season reprocessing of data that were collected operationally during each storm's lifetime. Source: UNISDR, 2011.

	1970 - 79	1980 - 89	1990 - 99	2000 - 09
Number of tropical cyclones as identified in best track data (average per year)	88.4	88.2	87.2	86.5
Number of countries hit by tropical cyclones as detected by satellite (average per year)	142.1	144.0	155.0	146.3
Number of disasters triggered by tropical cyclones as reported by EM-DAT (average per year)	21.7	37.5	50.6	63.0
Reported disasters as a percentage of number of countries hit by tropical cyclones	15%	26%	33%	43%

[1] This is the number of intersections between countries and tropical cyclones. One cyclone can affect several countries, but also many tropical cyclones occur only over the oceans.

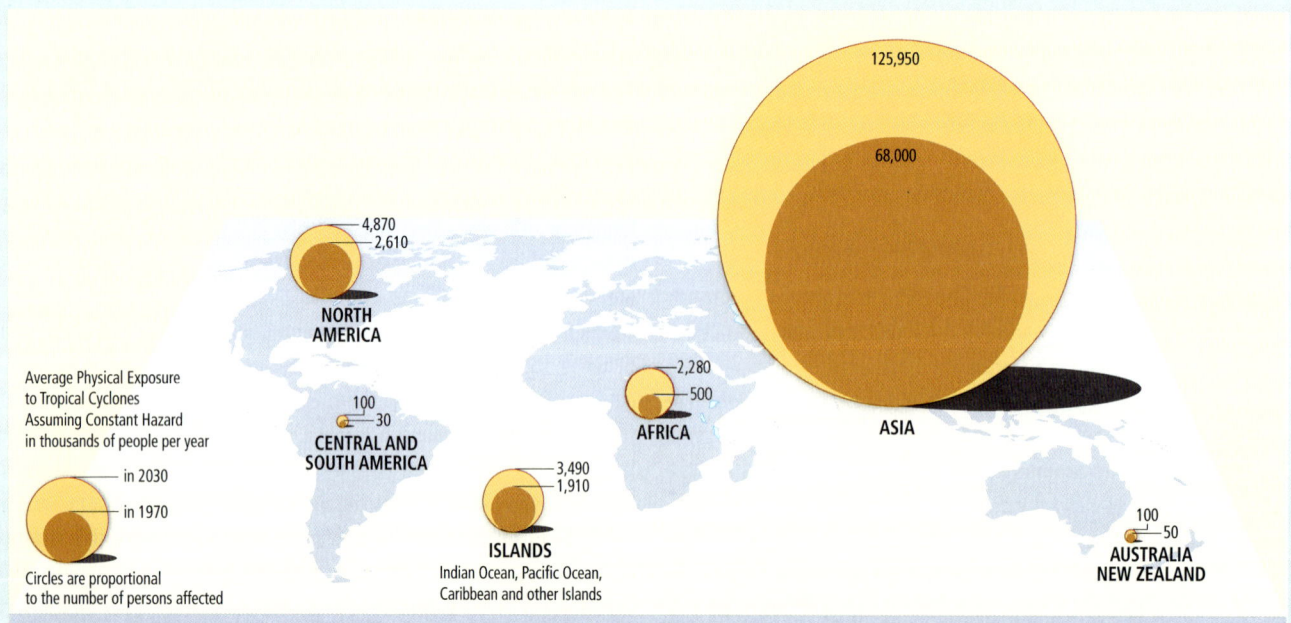

Figure 4-1 | Average physical exposure to tropical cyclones assuming constant hazard (in thousands of people per year). Data from Peduzzi et al., 2011.

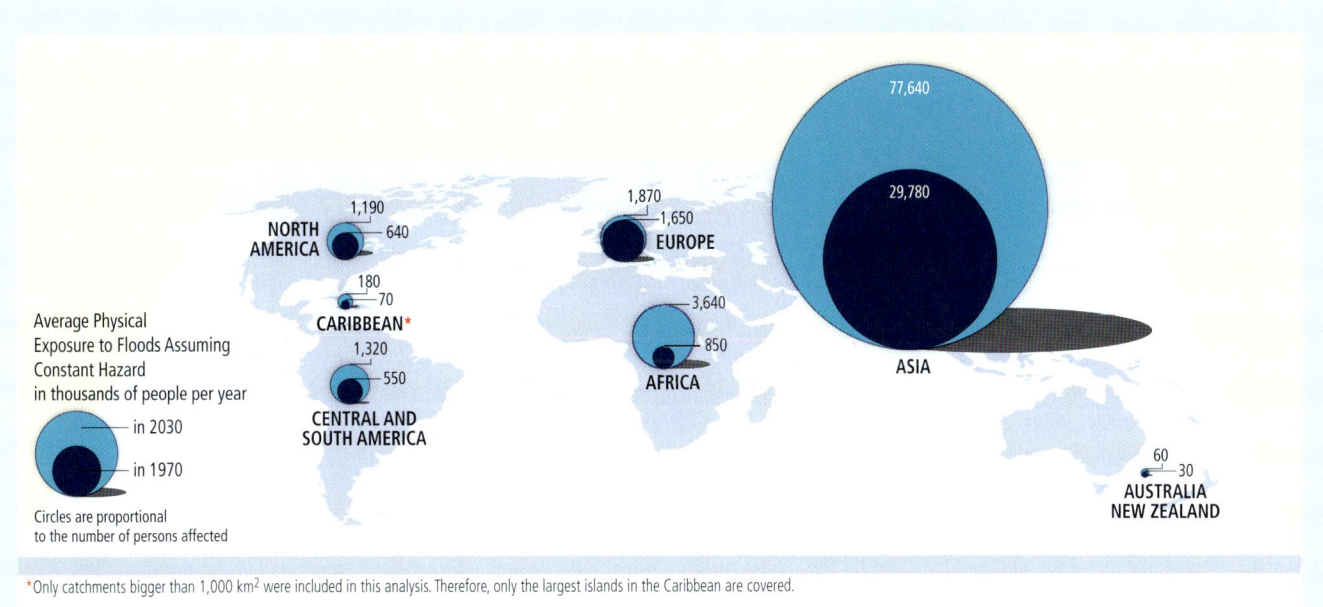

*Only catchments bigger than 1,000 km² were included in this analysis. Therefore, only the largest islands in the Caribbean are covered.

Figure 4-2 | Average physical exposure to floods assuming constant hazard (in thousands of people per year). Data from Peduzzi et al., 2011.

maximum wind speed is *likely* to increase (see Section 3.4.4). Figure 4-1 provides the modeled change in human exposure at constant hazard (without forecast of the influence of climate change on the hazard). It shows that the average number of people exposed to tropical cyclones per year globally would increase by 11.6% from 2010 to 2030 from population growth only. In relative terms, Africa has the largest percentage increase in physical exposure to tropical cyclones. In absolute terms, Asia has more than 90% of the global population exposed to tropical cyclones.

In terms of exposure to flooding, about 800 million people are currently living in flood-prone areas, and about 70 million people currently living in flood-prone areas are, on average, exposed to floods each year (UNISDR, 2011). Given the lack of complete datasets on past flood events, and the uncertainty associated with projected trends in future flood frequencies and magnitudes (see Section 3.5.2), it is difficult to estimate future flood hazards. However, using population increase in the flood-prone area, it is possible to look at trends in the number of people exposed per year on average at constant hazard (UNISDR, 2011). Figure 4-2 shows that population growth will continue to increase exposure to floods. Due to model constraints, areas north of 60°N and south of 60°S, as well as catchments smaller than 1,000 km² (typically small islands) are not modeled. The data provided in Figure 4-2 correspond to river flooding.

A number of factors underlie increases in impacts from floods and cyclones. However, trends in the population exposed to these hazards are an important factor. Population projections in tropical cyclone areas and flood-prone areas imply that impacts will almost certainly continue to increase based on this factor alone.

thus assessed. Sections 4.3.2 to 4.3.6, building on an understanding of exposure and vulnerability, evaluate knowledge of current and future risks of extreme events by sectors and systems.

4.3.2. Water

Past and future changes in exposure and vulnerability to climate extremes in the water sector are driven by both changes in the volume, timing, and quality of available water and changes in the property, lives, and systems that use the water resource or that are exposed to water-related hazards (Aggarwal and Singh, 2010). With a constant resource or physical hazard, there are two opposing drivers of change in exposure and vulnerability. On the one hand, vulnerability increases as more demands are placed on the resource (due to increased water consumption, for example, or increased discharge of polluting effluent) or exposure increases as more property, assets, and lives encounter flooding. On the other hand, vulnerability is reduced as measures are implemented to improve the management of resources and hazards and to enhance the ability to recover from extreme events. For example, enhancing water supplies, improving effluent treatment, and employing flood management measures (including the provision of insurance or disaster relief) would all lead to reductions in vulnerability in the water sector. Such measures have been widely implemented, and the runoff regime of many rivers has been considerably altered (Vörösmarty, 2002). The change in exposure and vulnerability in any place is a function of the relationship between these two opposing drivers, which also interact. Flood or water management measures may reduce vulnerability in the short term, but increased security may generate more development and ultimately lead to increased exposure and vulnerability.

Extreme events considered in this section can threaten the ability of the water supply 'system' (from highly managed systems with multiple sources to a single rural well) to supply water to users. This may be because a surplus of water affects the operation of systems, but more typically results from a shortage of water relative to demands – a drought. Water supply shortages may be triggered by a shortage of river flows and groundwater, deterioration in water quality, an increase in demand, or an increase in vulnerability to water shortage. There is *medium confidence* that since the 1950s some regions of the world have experienced more intense and longer droughts, in particular in southern Europe and West Africa (see Section 3.5.1), but it is not possible to attribute trends in the human impact of drought directly or just to these climatic trends because of the simultaneous change in the other drivers of drought impact.

There is *medium confidence* that the projected duration and intensity of hydrological drought will increase in some regions with climate change (Section 3.5.1), but other factors leading to a reduction in river flows or groundwater recharge are changes in agricultural land cover and upstream interventions. A deterioration in water quality may be driven by climate change (as shown for example by Delpla et al., 2009; Whitehead et al., 2009; Park et al., 2010), change in land cover, or upstream human interventions. An increase in demand may be driven by demographic, economic, technological, or cultural drivers as well as by climate change (see Section 2.5). An increase in vulnerability to water shortage may be caused, for example, by increasing reliance on specific sources or volumes of supply, or changes in the availability of alternatives. Indicators of hydrological and water resources drought impact include lost production (of irrigated crops, industrial products, and energy), the cost of alternative or replacement water sources, and altered human well-being, alongside consequences for freshwater ecosystems (impacts of meteorological and agricultural droughts on production of rain-fed crops are summarized in Section 4.3.4).

Few studies have so far been published on the effect of climate change on the impacts of drought in water resources terms at the local catchment scale. Virtually all of these have looked at water system supply reliability during a drought, or the change in the yield expected with a given reliability, rather than indicators such as lost production, cost, or well-being. Changes in the reliability of a given yield, or yield with a given reliability, of course vary with local hydrological and water management circumstances, the details of the climate scenarios used, and other drivers of drought risk. Some studies show large potential reductions in supply reliability due to climate change that challenge existing water management systems (e.g., Fowler et al., 2003; Kim et al., 2009; Takara et al., 2009; Vanham et al., 2009); some show relatively small reductions that can be managed – albeit at increased cost – by existing systems (e.g., Fowler et al., 2007), and some show that under some scenarios the reliability of supply increases (e.g., Kim and Kaluarachchi, 2009; Li et al., 2010). While it is not currently possible to reliably project specific changes at the catchment scale, there is *high confidence* that changes in climate have the potential to seriously affect water management systems. However, climate change is in many instances only one of the drivers of future changes in supply reliability, and is not necessarily the most important driver at the local scale. MacDonald et al. (2009), for example, demonstrate that the future reliability of small-scale rural water sources in Africa is largely determined by local demands, biological aspects of water quality, or access constraints, rather than changes in regional recharge, because domestic supply requires only 3-10 mm of recharge per year. However, they noted that up to 90 million people in low rainfall areas (200-500 mm) would be at risk if rainfall reduces to the point at which groundwater resources become nonrenewable.

There have been several continental- or global-scale assessments of potential change in hydrometeorological drought indicators (see Section 3.5.1), but relatively few on measures of water resources drought or drought impacts. This is because these impacts are very dependent on context. One published large-scale assessment (Lehner et al., 2006) used a generalized drought deficit volume indicator, calculated by comparing simulated river flows with estimated withdrawals for municipal, industrial, and agricultural uses. The indicator was calculated across Europe, using climate change projections from two climate models and assuming changes in withdrawals over time. They showed substantial changes in the return period of the drought deficit volume, comparing the 100-year return period for the 1961-1990 period with projections for the 2070s (Figure 4-3). Across large parts of Europe, the 1961-1990 100-year drought deficit volume is projected to have a return period of less than 10 years by the 2070s. Lehner et al. (2006) also demonstrated that this projected pattern of change was generally driven by changes in climate, rather than the projected changes in withdrawals of water (Figure 4-3). In southern and western Europe, changing withdrawals alone only are projected to increase deficit volumes by less than 5%, whereas the combined effect of changing withdrawals and climate change is projected to increase deficit volumes by at least 10%, and frequently by more than 25%. In eastern Europe, increasing withdrawals are projected to increase drought deficit volumes by over 5%, and more than 10% across large areas, but this is offset under both climate scenarios by increasing runoff.

Climate change has the potential to change river flood characteristics through changing the volume and timing of precipitation, by altering the proportions of precipitation falling as snow and rain, and to a lesser extent, by changing evaporation and hence accumulated soil moisture deficits. However, there is considerable uncertainty in the magnitude, frequency, and direction of change in flood characteristics (Section 3.5.2). Changes in catchment surface characteristics (such as land cover), floodplain storage, and the river network can also lead to changes in the physical characteristics of river floods (e.g., along the Rhine: Bronstert et al., 2007). The impacts of extreme flood events include direct effects on livelihoods, property, health, production, and communication, together with indirect effects of these consequences through the wider economy. There have, however, been very few studies that have looked explicitly at the human impacts of changes in flood frequency, rather than at changes in flood frequencies and magnitudes. One study has so far looked at changes in the area inundated by floods with defined return periods (Veijalainen et al., 2010), showing that the relationship between change in flood magnitude and flood extent depended strongly on local topographic conditions.

An early study in the United States (Choi and Fisher, 2003) constructed regression relationships between annual flood loss and socioeconomic and climate drivers, concluding that a 1% increase in average annual precipitation would, other things being equal, lead to an increase in annual national flood loss of around 6.5%. However, the conclusions are highly dependent on the regression methodology used, and the spatial scale of analysis. More sophisticated analyses combine estimates of current and future damage potential (as represented by a damage-magnitude relationship) with estimates of current and future flood frequency curves to estimate event damages and average annual damages (sometimes termed expected annual damage). For example, Mokrech et al. (2008) estimated damages caused by the current 10- and 75-year

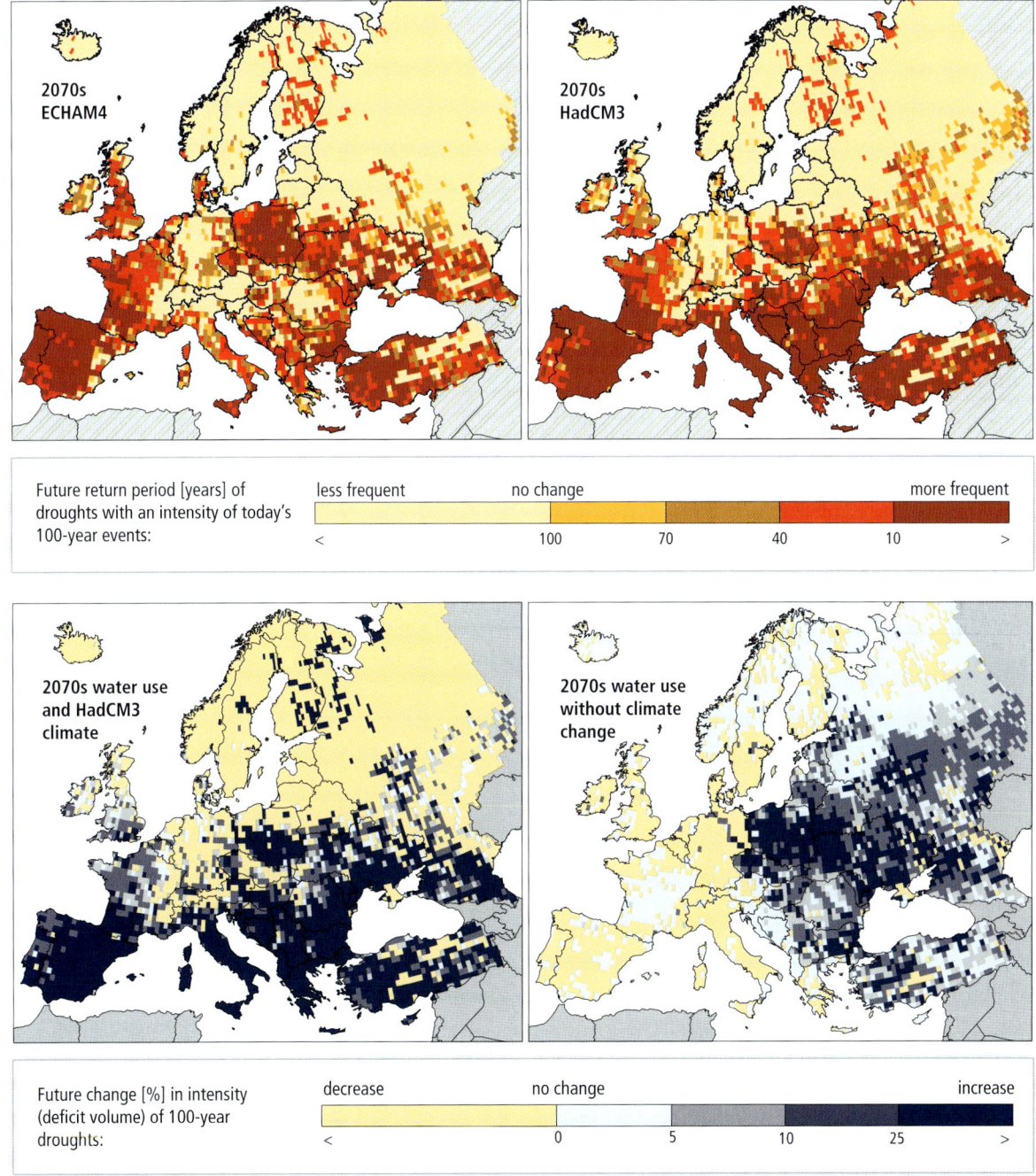

Figure 4-3 | Change in indicators of water resources drought across Europe by the 2070s. (top): projected changes in the return period of the 1961–1990 100-year drought deficit volume for the 2070s, with change in river flows and withdrawals for two climate models, ECHAM4 and HadCM3; (bottom): projected changes in the intensity (deficit volume) of 100-year droughts with changing withdrawals for the 2070s, with climate change (left, with HadCM3 climate projections) and without climate change (right). Source: Lehner et al., 2006.

events in two regions of England, combining fluvial and coastal flooding. The two main conclusions from their work were as follows. First, the percentage change in cost was greater for the rarer event than the more frequent event. Second, the absolute value of impacts, and therefore the percentage change from current impacts, was found to be highly dependent on the assumed socioeconomic change. In one region, event damage varied, in monetary terms, between four and five times across socioeconomic scenarios. An even wider range in estimated average annual damage was found in the UK Foresight Future Flooding and Coastal Defence project (Evans et al., 2004; Hall et al., 2005), which calculated average annual damage in 2080 of £1.5 billion, £5 billion, and £21 billion under similar climate scenarios but different socioeconomic futures (current average annual damage was estimated at £1 billion). The Foresight project represented the effect of climate change on flood frequency by altering the shape of the flood frequency curve using precipitation outputs from climate models and rainfall-runoff models for a sample of UK catchments. The EU-funded Projection of Economic impacts of climate change in Sectors of the European Union based on boTtom-up Analysis (PESETA) project (Ciscar, 2009; Feyen et al., 2009) used a hydrological model to simulate river flows, flooded areas, and flood frequency curves from climate scenarios derived from regional climate models, but – in contrast to the UK Foresight project – assumed no change in economic development in flood-prone areas. Figure 4-4 summarizes estimated changes in the average annual number of people flooded and average annual damage, by European region (Ciscar, 2009). There are strong regional variations in impact, with particularly large projected increases in both number of people flooded and economic damage (over 200%) in central and Eastern Europe, while in parts of North Eastern Europe, average annual flood damages decrease.

At the global scale, two studies have estimated the numbers of people affected by increases (or decreases) in flood hazard. Kleinen and Petschel-Held (2007) calculated the percentage of population living in river basins where the return period of the current 50-year event becomes shorter, for three climate models and a range of increases in global mean temperature. With an increase in global mean temperature of 2°C (above late 20th-century temperatures), between (approximately) 5 and 27% of the world's population would live in river basins where the current 50-year return period flood occurs at least twice as frequently. Hirabayashi and Kanae (2009) used a different metric, counting each year the number of people living in grid cells where the flood peak exceeded the (current) 100-year magnitude, using runoff as simulated by a high-resolution climate model fed through a river routing model. Beyond 2060, they found that at least 300 million people could be affected by substantial flooding even in years with relatively low flooding, with of the order of twice as many being flooded in flood-rich years (note that they used only one climate scenario with one climate model). This compares with a current range (using the same index) of between 20 and 300 million people. The largest part of the projected increase is due to increases in the occurrence of floods, rather than increases in population.

The impacts of changes in flood characteristics are highly dependent on how climate changes in the future, and as noted in Section 3.5.2, there is *low confidence* in projections of changes in flood magnitude or frequency. However, based on the available literature, there is *high confidence* that, in some places, climate change has the potential to substantially affect flood losses.

4.3.3. Ecosystems

Available information shows that high temperature extremes (i.e., heat wave), drought, and floods substantially affect ecosystems. Increasing gaps and overall contraction of the distribution range for species habitat could result from increases in the frequency of large-scale disturbances due to extreme weather and climate events (Opdam and Wascher, 2004). Fischlin et al. (2007), from assessment of 19 studies, found that 20 to 30% of studied plant and animal species may be at an increased risk of extinction if warming exceeds 2 to 3°C above the preindustrial level. Changes due to climate extremes could also entail shifts of ecosystems to less-desired states (Scheffer et al., 2001; Chapin et al., 2004; Folke et al., 2004) through, for example, the exceedance of critical temperature thresholds, with potential loss of ecosystem services dependent on the previous state (Reid et al., 2005; see also Fischlin et al., 2007).

4.3.3.1. Heat Waves

Heat waves can directly impact ecosystems by, for example, constraining carbon and nitrogen cycling and reducing water availability, with the result of potentially decreasing production or even causing species mortality.

Warming can decrease net ecosystem carbon dioxide (CO_2) exchange by inducing drought that suppresses net primary productivity. More frequent warm years may lead to a sustained decrease in CO_2 uptake by terrestrial ecosystems (Arnone et al., 2008). Extreme temperature conditions can shift forest ecosystems from being a net carbon sink to being a net carbon source. For example, tall-grass prairie net ecosystem CO_2 exchange levels decreased in both an extreme warming year (2003) and the following year in grassland monoliths from central Oklahoma, United States (Arnone et al., 2008). A 30% reduction in gross primary productivity together with decreased ecosystem respiration over Europe during the heat wave in 2003 resulted in a strong net source of CO_2 (0.5 Pg C yr^{-1}) to the atmosphere and reversed the effect of four years of net ecosystem carbon sequestration. Such a reduction in Europe's primary productivity is unprecedented during the last century (Ciais et al., 2005).

Impacts are determined not only by the magnitude of warming but also by organisms' physiological sensitivity to that warming and by their ability to compensate behaviorally and physiologically. For example, warming may affect tropical forest lizards' physiological performance in summer, as well as their ability to compete with warm-adapted, open-habitat competitors (Huey et al., 2009). Projected increases in maximum

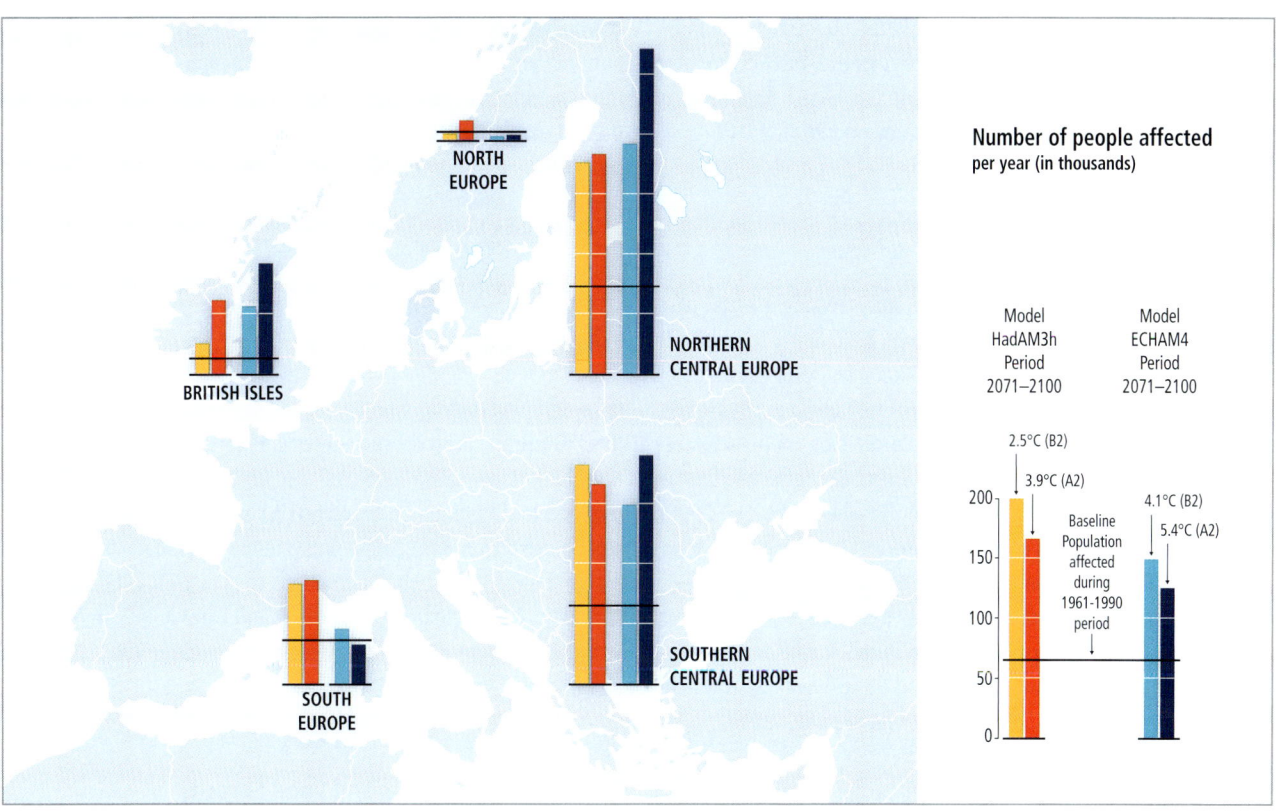
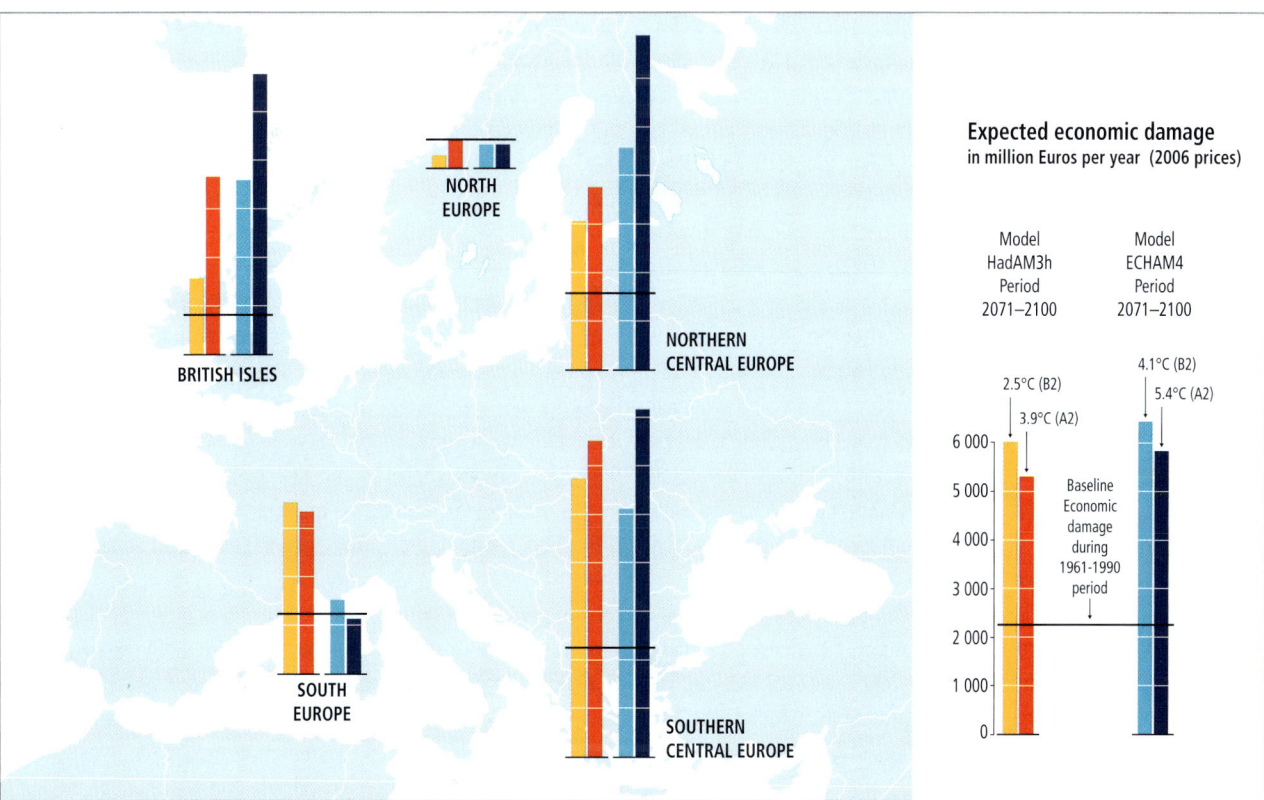

Figure 4-4 | Impact of climate change by 2071–2100 on flood risk in Europe. Note that the numbers assume no change in population or development in flood-prone areas. As illustrated in the legend on the right of each panel, projections are given for two Special Report on Emissions Scenarios (SRES) scenarios (A2 and B2) and for two global climate models (HadAM3h and ECHAM4). Projected mean temperature increase in the European region for the period 2071–2100 compared with 1961–1990 is indicated for each scenario and model combination. (top): For each region, baseline simulated population affected over 1961–1990 (thousands per year) and expected population affected (thousands per year) for 2071–2100 for each scenario and model combination. (bottom): For each region, baseline simulated economic damage over 1961–1990 (million € per year, 2006 prices) and expected economic damage (million € per year, 2006 prices) for 2071–2100 for each scenario and model combination. Data from Ciscar, 2009.

air temperatures may increase evaporative water requirements in birds, thus influencing survival during extreme heat events (McKechnie and Wolf, 2010). Heat waves could also cause increased likelihood of catastrophic avian mortality events (McKechnie and Wolf, 2010).

4.3.3.2. Drought

A rapid, drought-induced die-off of overstory woody plants at a subcontinental scale was triggered by the 2000–2003 drought in southwestern North America. Following 15 months of diminished soil water content, more than 90% of the dominant tree species, *Pinus edulis*, died. Limited observations indicate that die-off was more extensive than during the previous drought of the 1950s, also affecting wetter sites within the tree species' distribution (Breshears et al., 2005). Regional-scale piñon pine mortality was observed following an extended drought (2000–2004) in northern New Mexico (Rich et al., 2008). Dominant plant species from diverse habitat types (i.e., riparian, chaparral, and low-to high-elevation forests) exhibited significant mortality during a drought in the southwestern United States; average mortality among dominant species was 3.3 to 41.4% (Gitlin et al., 2006).

Evergreen coniferous species mortality caused by the coupling of drought and higher temperatures from winter to spring has been observed in the Republic of Korea (Lim et al., 2010). In 1998, 2002, 2007, and 2009, years of high winter-spring temperatures and lower precipitation, *P. densiflora* and *P. koraiensis* were affected by droughts, with many dying in the crown layer, while deciduous species survived. Similarly, *Abies koreana*, an endemic species in Korea, at high elevation has declined following a rise in winter temperatures since the late 1990s (Lim et al., 2010). Beech crown condition was observed to decline following severe drought in 1976 (Power, 1994), 1989 (Innes, 1992), and 1990 (Stribley and Ashmore, 2002). Similarly, the percentage of moderately or severely damaged trees displayed an upward trend after the 1989 drought in Central Italy, especially for *P. pinea* and *Fagus sylvatica* (Bussotti et al., 1995). As final examples, defoliation and mortality in Scots pine observed in each year during 1996 to 2002 was related to the precipitation deficit and hot conditions of the previous year in the largest inner-alpine valley of Switzerland (Valais) (Rebetez and Dobbertin, 2004), and both gross primary production and total ecosystem respiration decreased in 2003 in many regions of Europe (Granier et al., 2007).

In a shallow temperate southern European estuary, the Mondego Estuary in Portugal, the severe drought in 2004–2005 was responsible for spatial shifts in the estuary's zooplankton community, with an increase in abundance and diversity during the period of low freshwater flow (Marques et al., 2007).

4.3.3.3. Floods

Floods also impact ecosystems. Floods can cause population- and community-level changes superimposed on a background of more gradual trends (Thibault and Brown, 2008). As an example, an extreme flood event affected a desert rodent community (that had been monitored for 30 years) by inducing a large mortality rate, eliminating the advantage of previously dominant species, resetting long-term population and community trends, altering competitive and meta-population dynamics, and reorganizing the community (Thibault and Brown, 2008).

4.3.3.4. Other Events

Other events, such as hurricanes and storms, can also impact ecosystems. Hurricanes can cause widespread mortality of wild birds, and their aftermath may cause declines due to the birds' loss of resources required for foraging and breeding (Wiley and Wunderle, 1994). Winter storms can also impact forest ecosystems, particularly in pre-alpine and alpine areas (Faccio, 2003; Schelhaas et al., 2003; Fuhrer et al., 2006). In addition, saltmarshes, mangroves, and coral reefs can be vulnerable to climate extremes (e.g., Bertness and Ewanchuk, 2002; Hughes et al., 2003; Fischlin et al., 2007).

4.3.4. Food Systems and Food Security

Food systems and food security can be affected by extreme events that impair food production and food storage and delivery systems (food logistics). Impacts transmitted through an increase in the price of food can be especially challenging for the urban poor in developing countries (FAO, 2008). Global food price increases are borne disproportionally by low-income countries, where people spend more of their income on food (OECD-FAO, 2008).

When agricultural production is not consumed where it is produced, it must be transported and often processed and stored. This process involves complex interdependent supply chains exposed to multiple hazards. At every step of the process, transport and associated infrastructure such as roads, railways, bridges, warehouses, airports, ports, and tunnels can be at risk of direct damage from climate events, making the processing and delivery chain as a whole at risk of disruption resulting from damage or blockages at any point in the chain.

The economies of many developing countries rely heavily on agriculture, dominated by small-scale and subsistence farming. People's livelihoods in this sector are especially exposed to weather extremes (Easterling and Apps, 2005; Easterling et al., 2007). Subsistence farmers can be severely impacted by climate and weather events. For example, the majority of households produce maize in many African countries, but only a modest proportion sells it – the great majority eat all they produce. In Kenya for example, nearly all households grow maize, but only 36% sell it, with 20% accounting for the majority of sales (FAO, 2009). Both such famers and their governments have limited capacity for recovery (Easterling and Apps, 2005).

Evidence that the current warming trends around the world have already begun to impact agriculture is reported by Lobell et al. (2011). They show that crop yields have already declined due to warmer conditions compared to the expected yields without warming. Both Schlenker and Roberts (2009) and Muller et al. (2011), after their evaluation of projected temperature effects on crops in the United States and Africa, concluded that climate change would have negative impacts on crop yields. These effects were based on temperature trends and an expected increase in the probability of extremes during the growing season; however, there is also the potential occurrence of extreme events after the crop is grown, which could affect harvest and grain quality. Fallon and Betts (2010) stated that increasing flooding and drought risks could affect agricultural production and require the adoption of robust management practices to offset these negative impacts. Their analysis for Europe showed a probable increase in crop productivity in northern regions but a decrease in the southern regions, leading to a greater disparity in production.

In a recent evaluation of high temperature as a component of climate trends, Battisti and Naylor (2009) concluded that future growing season temperatures are *very likely* to exceed the most extreme temperatures observed from 1900 to 2006, for both tropical and subtropical regions, with substantial potential implications for food systems around the world.

The effects of temperature extremes on a number of different crop species have been summarized in Hatfield et al. (2011). Many crops are especially sensitive to extreme temperatures that occur just prior to or during the critical pollination phase of crop growth (Wheeler et al., 2000; Hatfield et al., 2008, 2011). Crop sensitivity and ability to compensate during later improved weather will depend on the length of time for anthesis in each crop.

Extreme temperatures can negatively impact grain yield (Kim et al., 1996; Prasad et al., 2006). For example, Tian et al. (2010) observed in rice that high temperatures (>35°C) coupled with high humidity and low wind speed caused panicle temperatures to be as much as 4°C higher than air temperature, inducing floret sterility. Impacts of temperature extremes may not be limited to daytime events. Mohammed and Tarpley (2009) observed that rice yields were reduced by 90% when night temperatures were increased from 27 to 32°C. An additional impact of extremes has been found in the quality of the grain. Kettlewell et al. (1999) found that wheat quality in the United Kingdom was related to the North Atlantic Oscillation and probably caused by variation in rainfall during the grain-filling period. In a more recent study, Hurkman et al. (2009) observed that high-temperature events during grain-filling of wheat altered the protein content of the grain, and these responses were dependent upon whether the exposure was imposed early or midway through the grain-filling period. Skylas et al. (2002) observed that high temperature during grain-filling was one of the most significant factors affecting both yield and flour quality in wheat.

Drought causes yield variation, and an example from Europe demonstrates that historical yield records show that drought has been a primary cause of interannual yield variation (Hlavinka et al., 2009; Hatfield, 2010). Water supply for agricultural production will be critical to sustain production and even more important to provide the increase in food production required to sustain the world's growing population. With glaciers retreating due to global warming and El Niño episodes, the Andean region faces increasing threats to its water supply (Mark and Seltzer, 2003; Cadier et al., 2007). With precipitation limited to only a few months of the year, melt from glaciers is the only significant source of water during the dry season (Mark and Seltzer, 2003). Glacier recession reduces the buffering role of glaciers, hence inducing more floods during the rainy season and more water shortages during the dry season. Cadier et al. (2007) found that warm anomalies of the El Niño-Southern Oscillation (ENSO) corresponded to an increase in melting four months later.

Food security is linked to our ability to adapt agricultural systems to extreme events using our understanding of the complex system of production, logistics, utilization of the produce, and the socioeconomic structure of the community. The spatial variability and context sensitivity of each of these factors points to the value of downscaled scenarios of climate change and extreme events.

4.3.5. Human Settlements, Infrastructure, and Tourism

4.3.5.1. Human Settlements

Settlements concentrate the exposure of humans, their assets, and their activities. In the case of very large cities, these concentrations can represent a significant proportion of national wealth and may result in additional forms of vulnerability (Mitchell, 1998). Flooding, landslides, storms, heat waves, and wildfires have produced historically important damages in human settlements, and the characteristics of these events and their underlying climate drivers are projected to change (see Chapter 3; Kovats and Akhtar, 2008; Satterthwaite, 2008). The concentration of economic assets and people creates the possibility of large impacts, but also the capacity for recovery (Cutter et al., 2008). Coastal settlements are especially at risk with sea level rise and changes in coastal storm activity (see Sections 3.4.4 and 3.4.5 and Case Study 9.2.8).

At very high risk of impacts are the urban poor in informal settlements (Satterthwaite, 2008; Douglas, 2009). Worldwide, about one billion people live in informal settlements, and informal settlements are growing faster than formal settlements (UN-HABITAT, 2008; UNISDR, 2011). Informal settlements are also found in developed countries; for example, there are about 50 million people in such areas in Europe (UNECE, 2009). Occupants of informal settlements are typically more exposed to climate events with no or limited hazard-reducing infrastructure. The vulnerability is high due to very low-quality housing and limited capacity to cope due to a lack of assets, insurance, and marginal livelihoods, with less state support and limited legal protection (Dodman and Satterthwaite, 2008).

The number and size of coastal settlements and their associated infrastructure have increased significantly over recent decades (McGranahan et al., 2007; Hanson et al., 2011; see also Case Study 9.2.8). In many cases these settlements have affected the ability of natural coastal systems to respond effectively to extreme climate events by, for example, removing the protection provided by sand dunes and mangroves. Small island states, particularly small island developing states (see Case Study 9.2.9), may face substantial impacts from climate change-related extremes.

Urbanization exacerbates the negative effects of flooding through greatly increased runoff concentration, peak, and volume, the increased occupation of flood plains, and often inadequate drainage planning (McGranahan et al., 2007; Douglas et al., 2008). These urbanization issues are universal but often at their worst in informal settlements, which are generally the most exposed to flooding and usually do not have the capacity to deal with the issues (Hardoy et al., 2001). Flooding regularly disrupts cities, and urban food production can be severely affected by flooding, undermining local food security in poor communities (Douglas, 2009; Aggarwal and Singh, 2010). A further concern for low- and middle-income cities as a result of flooding, particularly in developing countries, is human waste, as most of these cities are not served by proper water services such as sewers, drains, or solid waste collection services (Hardoy et al., 2001).

Slope failure can affect settlements in tropical mountainous areas, particularly in deforested areas (e.g., Vanacker et al., 2003)and hilly areas (Loveridge et al., 2010), and especially following heavy prolonged rain (e.g., see Case Study 9.2.5). Informal settlements are often exposed to potential slope failure as they are often located on unstable land with no engineering or drainage works (Alexander, 2005; Anderson et al., 2007). Informal settlements have been disproportionately badly impacted by landslides in Colombia and Venezuela in the past (e.g., Takahashi et al., 2001; Ojeda and Donnelly, 2006) and were similarly affected in 2010 during unusual heavy rains associated with the La Niña weather phenomenon (NCDC, 2011). Densely settled regions in the Alps (Crosta et al., 2004) and Himalayas have been similarly impacted (Petley et al., 2007).

Cities can substantially increase local temperatures and reduce temperature drop at night (e.g., see Case Study 9.2.1). This is the urban heat island effect resulting from the large amount of heat-absorbing material, building characteristics, and emissions of anthropogenic heat from air conditioning units and vehicles (e.g., Rizwan et al., 2008; for a critical review of heat island research, see Stewart, 2011). Heat waves combined with urban heat islands (Basara et al., 2010; Tan et al., 2010) can result in large death tolls with the elderly, the unwell, the socially isolated, and outdoor workers (Maloney and Forbes, 2011) being especially vulnerable, although acclimatization and heat health-warning systems can substantially reduce excess deaths (Fouillet et al., 2008). Heat waves thus pose a future challenge for major cities (e.g., Endlicher et al., 2008; Bacciniet al., 2011; for London, Wilby, 2003). In urban areas, heat waves also have negative effects on air quality and the number of days with high pollutants, ground level ozone, and suspended particle concentrations (Casimiro and Calheiros, 2002; Sanderson et al., 2003; Langner et al., 2005).

The largest impacts from coastal inundation due to sea level rise (and/or relative sea level rise) in low-elevation coastal zones (i.e., coastal areas with an elevation less than 10 m above present mean sea level; see McGranahan et al., 2007) are thought to be associated with extreme sea levels due to tropical and extratropical storms (e.g., Ebersole et al., 2010; Mozumder et al., 2011) that will be superimposed upon the long-term sea level rise (e.g., Frazier et al., 2010). An increase in the mean maximum wind speed of tropical cyclones is *likely* over the 21st century, but possibly not in all ocean basins (see Table 3-1). The destructive potential of tropical cyclones may increase in some regions as a result of this projected increase in intensity of mean maximum wind speed and tropical cyclone-related rainfall rates (see Section 3.4.4). Storms generally result in considerable disruption and local destruction, but cyclones and their associated storm surges have in some cases caused very substantial destruction in modern cities (e.g., New Orleans and Darwin; see also Case Study 9.2.5). The impacts are considered to be more severe for large urban centers built on deltas and small island states (McGranahan et al., 2007; Love et al., 2010; Wardekker et al., 2010), particularly for those at the low end of the international income distribution (Dasgupta et al., 2009). The details of exposure will be controlled by the natural or human-induced characteristics of the system, for example, the occurrence/distribution of protecting barrier islands and/or coastal wetlands that may attenuate surges (see, e.g., Irish et al., 2010; Wamsley et al., 2010) or changes such as land reclamation (Guo et al., 2009). Recent studies (Nicholls et al., 2008; Hanson et al., 2011) have assessed the asset exposure of port cities with more than one million inhabitants (in 2005). They demonstrated that large populations are already exposed to coastal inundation (~40 million people or 0.6% of the global population) by a 1-in-100-year extreme event, while the total value of exposed assets was estimated at US$ 3,000 billion (~ 5% of the global GDP in 2005). By the 2070s, population exposure was estimated to triple, whereas asset exposure could grow ten-fold to some US$ 35,000 billion; these estimates, however, do not account for the potential construction of effective coastal protection schemes (see also Dawson et al., 2005), with the exposure growth rate being more rapid in developing countries (e.g., Adamo, 2010). Lenton et al. (2009) estimated a substantial increase in the exposure of coastal populations to inundation (see Figure 4-5).

4.3.5.2. Infrastructure

Weather- and climate-related extremes are expected to produce large impacts on infrastructure, although detailed analyses of potential and projected damages are limited to a few countries (e.g., Australia, Canada, the United States; Holper et al., 2007), infrastructure types (e.g., power lines), and sectors (e.g., transport, tourism). Inadequate infrastructure design may increase the impacts of climate and weather extremes, and some infrastructure may become inadequate where climate

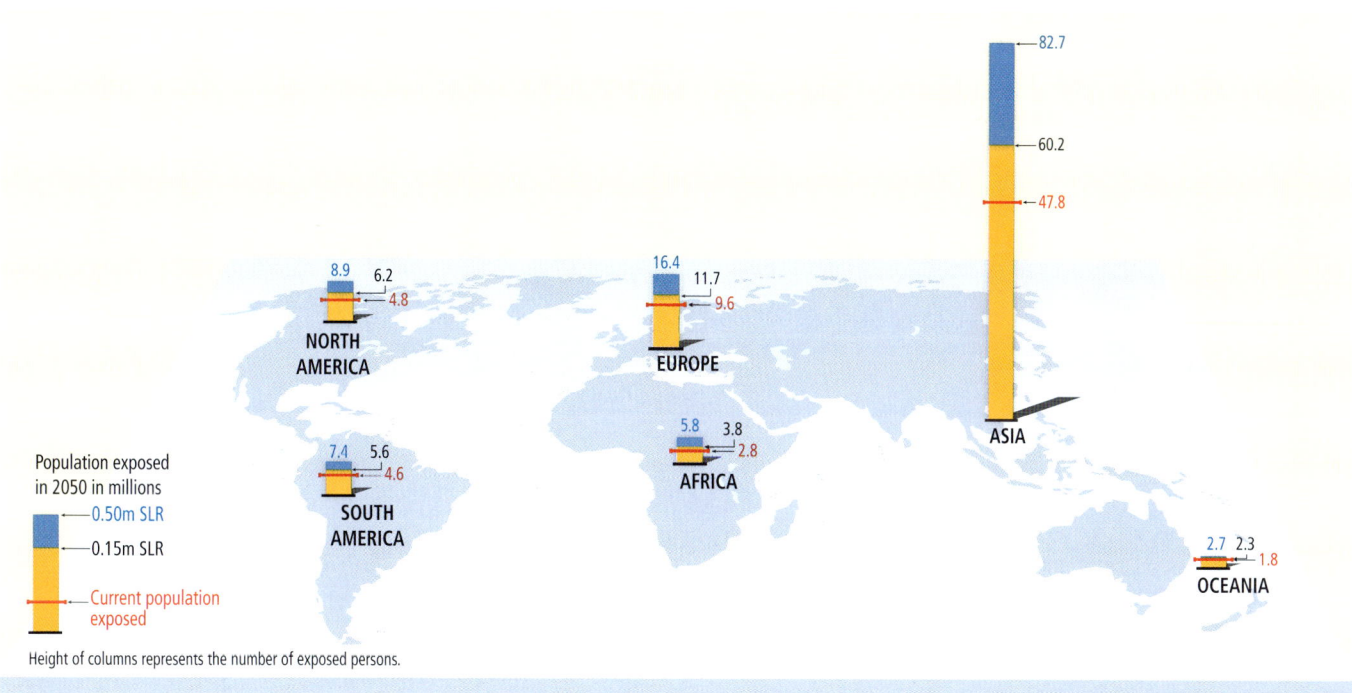

Figure 4-5 | For low-elevation coastal areas, current and future (2050) population exposure to inundation in the case of the 1-in-100-year extreme storm for sea level rise of 0.15 m and for sea level rise of 0.50 m due to the partial melting of the Greenland and West Antarctic Ice Sheets. Data from Lenton et al., 2009.

change alters the frequency and severity of extremes, for example, an increase in heavy rainfalls may affect the capacity and maintenance of storm water, drainage, and sewerage infrastructure (Douglas et al., 2008). In some infrastructure, secondary risks in case of extreme weather may cause additional hazards (e.g., extreme rainfall can damage dams). The same is true for industrial and mining installations containing hazardous substances (e.g., heavy rainfall is the main cause of tailings dam failure, accounting for 25% of incidents worldwide and 35% in Europe; Rico et al., 2008).

In many parts of the world, including Central Asia and parts of Europe, aging infrastructure, high operating costs, low responsiveness to customers, and poor access to capital markets may limit the operability of sewerage systems (Evans and Webster, 2008). Moreover, most urban centers in sub-Saharan Africa and in Asia have no sewers (Hardoy et al., 2001). Current problems of pollution and flooding will be exacerbated by an increase in climatic and weather extremes (e.g., intense rainfall; see Table 3-3 for projected regional changes).

Major settlements are dependent on lengthy infrastructure networks for water, power, telecommunications, transport, and trade, which are exposed to a wide range of extreme events (e.g., heavy precipitation and snow, gale winds). Modern logistics systems are intended to minimize slack and redundancies and as a result are particularly vulnerable to disruption by extreme events (Love et al., 2010).

Transport infrastructure is vulnerable to extremes in temperature, precipitation/river floods, and storm surges, which can lead to damage in roads, rail, airports, and ports. Impacts on coastal infrastructure, on services, and particularly on ports, key nodes of international supply chains, are expected (e.g., Oh and Reuveny, 2010). This may have far-reaching implications for international trade, as more than 80% of global trade in goods (by volume) is carried by sea (UNCTAD, 2009). All coastal modes of transportation are considered vulnerable, but exposure and impacts will vary, for example, by region, mode of transportation, location/elevation, and condition of transport infrastructure (NRC, 2008; UNCTAD, 2009). Coastal inundation due to storm surges and river floods can affect terminals, intermodal facilities, freight villages, storage areas, and cargo and disrupt intermodal supply chains and transport connectivity (see Figure 4-6). These effects would be of particular concern to small island states, whose transportation facilities are mostly located in low-elevation coastal zones (UNCTAD, 2009; for further examples, see Love et al., 2010).

Regarding road infrastructure, Meyer (2008) pointed to bridges and culverts as vulnerable elements in areas with projected increases in heavy precipitation. Moreover, the lifetime of these rigid structures is longer than average road surfaces and they are costly to repair or replace. Increased temperatures could reduce the lifetime of asphalt on road surfaces (Meizhu et al., 2010). Extreme temperature may cause expansion and increased movement of concrete joints, protective cladding, coatings, and sealants on bridges and airport infrastructure, impose stresses in the steel in bridges, and disrupt rail travel (e.g., Arkell and Darch, 2006). Nevertheless, roads and railways are typically replaced every 20 years and can accommodate climate change at the time of replacement (Meyer, 2008).

Electricity transmission infrastructure is also vulnerable to extreme storm events, particularly wind and lightning, and in some cases heat

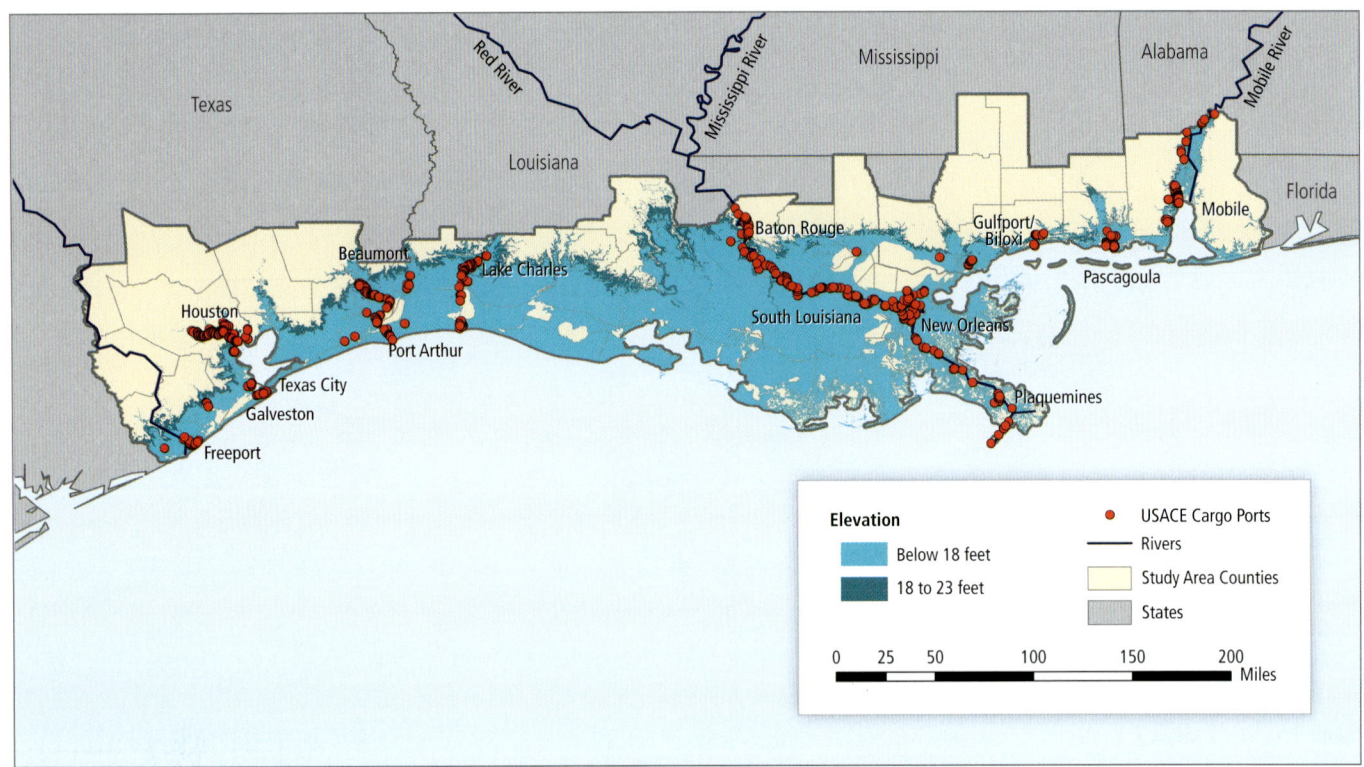

Figure 4-6 | Freight-handling port facilities at risk from storm surge of 5.5 and 7 m on the US Gulf Coast. Adapted from CCSP, 2008.

waves (McGregor et al., 2007). The passage of the Lothar and Martin storms across France in 1999 caused the greatest devastation to an electricity supply network ever seen in a developed country, as 120 high-voltage transmission pylons were toppled, and 36 high-tension transmission lines (one-quarter of the total lines in France) were lost (Abraham et al., 2000). Severe droughts may also affect the supply of cooling water to power plants, disrupting the ongoing supply of power (see Box 4-4; Rübbelke and Vögele, 2011).

Buildings and urban facilities may be vulnerable to increasing frequency of heavy precipitation events (see Section 3.3.2). Those close to the coast are particularly at risk when storm surges are combined with sea level rise. In commercial buildings, vulnerable elements are lightweight roofs commonly used for warehouses, causing water spoilage to stored goods and equipment. During the Lothar and Martin storms, the most vulnerable public facilities were schools, particularly those built in the 1960s and 1970s and during the 1990s with the use of lightweight architectural elements of metal, plastic, and glass in walls and roofs (Abraham et al., 2000).

4.3.5.3. Tourism

The tourism sector is highly sensitive to climate, since climate is the principal driver of global seasonality in tourism demand (Lise and Tol, 2002; Becken and Hay, 2007). Approximately 10% of global GDP is spent on recreation and tourism, constituting a major source of income and foreign currency in many developing countries (Berrittella et al., 2006). Extreme events may play an important role in tourist decisions (e.g., Hein et al., 2009; Yu et al., 2009).

There are three broad categories of impacts of climate extremes that can affect tourism destinations, competitiveness, and sustainability (Scott et al., 2008): (1) direct impacts on tourist infrastructure (hotels, access roads, etc.), on operating costs (heating/cooling, snowmaking, irrigation, food and water supply, evacuation, and insurance costs), on emergency preparedness requirements, and on business disruption (e.g., sun-and-sea or winter sports holidays); (2) indirect environmental change impacts of extreme events on biodiversity and landscape change (e.g., coastal erosion), which may negatively affect the quality and attractiveness of tourism destinations; and (3) tourism-adverse perception of particular touristic regions after occurrence of the extreme event itself. For example, adverse weather conditions or the occurrence of an extreme event can reduce a touristic region's popularity among tourists during the following season.

Apart from extreme events, large impacts on some tourist destinations may be produced by medium-term projected climate change effects (e.g., Bigano et al., 2008). Salinization of the groundwater resources due to sea level rise, land reclamation, and overexploitation of coastal aquifers (e.g., Alpa, 2009) as well as changing weather extreme patterns (Hein et al., 2009) will pose additional stresses for the industry. Nevertheless, the potential impacts on the tourist industry will depend also on tourists' perceptions of the coastal destinations (e.g., of destinations experiencing beach erosion) that, however, cannot be easily predicted (Buzinde et al., 2009). Capacity to recover is related to the degree of dependence on tourism, with diversified economies being more robust (Ehmer and Heymann, 2008). However, low-lying coastal areas and areas currently on the edge of the snow limit may have limited alternatives. Some ski resorts will be able to adapt using snowmaking, which has become an integral component of the ski industry in Europe (Elsasser and Bürki, 2002), although at the expense of high water and energy consumption.

In some regions, the main impact of extreme events on tourism is decline in revenue, with loss of livelihoods for those working in the sector (Hamilton et al., 2005; Scott et al., 2008; Hein et al., 2009). Quantitative regional climate projections of the frequency or magnitude of certain weather and climate extremes (e.g., heat waves and droughts; see, for example, Table 3-3) inform qualitative understanding of regional impacts on tourism activities (see Box 4-3). The vulnerable hotspot regions in terms of extreme impacts of climate change on tourism include the Mediterranean, Caribbean, small islands of the Indian and Pacific Oceans, and Australia and New Zealand (Scott et al., 2008). Direct and indirect effects of extremes in these regions will vary greatly with location (Gössling and Hall, 2006a,b; Wilbanks et al., 2007).

Box 4-3 points out a number of potential climate extreme impacts on tourism regions and activities.

4.3.6. Human Health, Well-Being, and Security

Climate extremes, such as heat waves, floods, droughts, and cyclones, influence human health, well-being, and security.

Heat waves have affected developed countries, as exemplified by the 2003 European heat wave (see Case Study 9.2.1 and Box 4-4). In much

---

Box 4-3 | Regional Examples of Potential Impacts of Climate Extremes on Tourism

Tropics
Projections indicate a *likely* increase in mean maximum wind speed (but not in all basins) and in tropical cyclone-related rainfall rates (see Table 3-1). In the Caribbean, tourist activities may be reduced where beaches erode with sea level rise and where coral is bleached, impacting snorkelers and divers (Uyarra et al., 2005).

Small island states are dependent on tourism, and the tourism infrastructure that lies on the coast is threatened by climate change (Berrittella et al., 2006). Sea level rise in the 20th century – with an average rate of 1.7 mm yr^{-1} and a significantly higher rate (3.1 mm yr^{-1}) for the period 1993–2003 (Bindoff et al., 2007) – poses risks for many touristic resorts of small islands in the Pacific and Indian Oceans (Becken and Hay, 2007; Scott et al., 2008).

Alpine Regions
Warming temperatures will raise the snowline elevation (Elsasser and Bürki, 2002; Scott et al., 2006). In Switzerland, only 44% of ski resorts are projected to be above the 'snow-reliable' altitude (snow for 100 days per season) by approximately 2030, as opposed to 85% today (Elsasser and Bürki, 2002). In Austria, 83% of ski resorts are currently snow-reliable but an increase in temperature of 1 and 2°C is projected to reduce this number to 67 and 50%, respectively (Abegg et al., 2007). Ski season simulations show that snowmaking technology can maintain snow-reliable conditions in Austria until the 2040s (A1B) to the 2050s (B1), but by the end of the century the required production in snow volume is projected to increase by up to 330% (Steiger, 2010). This artificial snow production will increase vulnerability to water shortage and local water conflicts, in particular in the French Alps (EEA, 2009).

Mediterranean Countries
More frequent heat waves and tropical nights (>20°C) in summer (see Table 3-3) may lead to exceedance of comfortable temperature levels and reduce the touristic flow by 2060 (Hein et al., 2009). Tourism occupancy may increase during spring and autumn and decrease in summer (Perry, 2003; Esteban-Talaya et al., 2005). Northern European countries are expected to become relatively more attractive, closing their gap with the currently popular southern European countries (Hamilton et al., 2005).

There are major regional gaps in understanding how climate change may affect the natural and cultural resources in Africa and South America, preventing further insight on corresponding impacts for tourism activities (Scott et al., 2008).

In many regions, some types of tourism will benefit from, or be unaffected by, climate extremes (Scott et al., 2008). When an area is impacted directly by an extreme event, tourists will often go to another destination with the result that one area's loss becomes another's gain. The impacted area may also gain in the longer term through the provision of new infrastructure. City and cultural tourism is generally seen as relatively unaffected by climate and weather events (Scott et al., 2008).

of the developed world, societies are aging and hence can be more sensitive to climate extremes, such as heat waves (Hennessy et al., 2007). Heat extremes can claim casualties even in tropical countries, where people are acclimatized to the hot climate; McMichael et al. (2008) evaluated the relation between daily temperature and mortality in middle- and low-income countries, and reported that higher mortality was observed on very hot days in most of the cities, including tropical cities, such as Bangkok, Thailand; Delhi, India; and Salvador, Brazil.

Floods can cause deaths and injuries and can be followed by infectious diseases (such as diarrhea) and malnutrition due to crop damage (see Section 4.4.2.3). In Dhaka, Bangladesh, the severe flood in 1998 was associated with an increase in diarrhea during and after the flood, and the risk of non-cholera diarrhea was higher among those from a lower socioeconomic group and not using tap water (Hashizume et al., 2008). Floods may also lead to a geographical shift of malaria epidemic regions by changing breeding sites for vector mosquitoes. Outbreaks of malaria were associated with changes in habitat after the 1991 floods in Costa Rica's Atlantic region (Saenz et al., 1995; for another example, see Case Study 9.2.6). Malaria epidemics can also occur when people with little immunity move into endemic regions, although the displacement of large populations has rarely occurred as a result of acute natural disasters (Toole, 1997).

Drought can affect water security, as well as food security through reduction of agricultural production (MacDonald, 2010), and it can be a factor contributing to human-ignited forest fires, which can lead to widespread deforestation and carbon emissions (D'Almeida et al., 2007; van der Werf et al., 2008; Field et al., 2009; Phillips et al., 2009; Costa and Pires, 2010). Also, drought can increase or decrease the prevalence of mosquito-borne infectious diseases such as malaria, depending on the local conditions (Githeko et al., 2000), and is associated with meningitis (Molesworth et al., 2003). Studies indicate that there is a climate signal in forest fires throughout the American West and Canada and that there is a projected increase in severe wildfires in many areas (Gillett et al., 2004; Westerling et al., 2006; Westerling and Bryant, 2008). As described by McMichael et al. (2003a), the direct effects of fire on human health can include burns and smoke inhalation, with indirect health impacts potentially resulting from loss of vegetation on slopes, increased soil erosion, and resulting increased risk of landslides.

Evaluation of how impacts of climate extremes affect human health tend to focus on the direct, immediate effects of the event, using parameters that are often easier to obtain and quantify like death statistics or hospitalizations. These direct observable outcomes are used to demonstrate the extremity of an event and as a comparison metric to measure against other extreme events. However, indirect health impacts are not often reported, because they are one step removed from the event. Because indirect impacts are hard to monitor and are often temporally separated from the event, they are effectively removed from the cause-and-effect linkage to that event. Examples of indirect health impacts from extreme weather events include illnesses or injury resulting from disruption of human infrastructure built to deal with basic needs like medical services; exposure to infectious or toxic agents after an extreme event like cyclones or flooding (Schmid et al., 2005); stress, anxiety, and mental illness after evacuation or geographical displacement (Fritze et al., 2008) as well as increased susceptibility to infection (Yee et al., 2007); and disruption of socioeconomic structures and food production that leads to increases of malnutrition that might not manifest until months after an extreme event (Haines et al., 2006; McMichael et al., 2006). Indirect health impacts are therefore a potentially large but under-examined outcome of extreme weather events that lead to a substantial underestimation of the total health burden.

There is a growing body of evidence that the mental health impact from extreme events is substantial (Neria et al., 2008; Berry et al., 2010). Often overshadowed by the physical health outcomes of an event, the psychological effects can be long lasting and can affect a large portion of a population (Morrissey and Reser, 2007). An extreme event may affect mental health directly from acute traumatic stress from an event, with common outcomes of anxiety and depression. It can also have indirect impacts during the recovery period associated with the stress and challenges of loss, disruption, and displacement. Furthermore, indirect mental health impacts could even affect individuals not directly associated with an event, like grieving friends and family of those who die from an event or the rescue and aid workers who suffer post-traumatic stress disorder (PTSD) after their aid efforts. Long-term mental health impacts are not often adequately monitored, but the body of research conducted after natural disasters in the past three decades suggests that the burden of PTSD among persons exposed to disasters is substantial (Neria et al., 2008). A range of other stress-related problems such as grief, depression, anxiety disorders, somatoform disorders, and drug and alcohol abuse (Fritze et al., 2008) have lasting effects, long after the causative event.

There remain large limitations in evaluating health impacts of climate change. The largest research gap is a lack of information on impact outcomes themselves in developing countries in general. This includes the mortality/morbidity data and information on other contributing factors such as nutritional status or access to safe water and medical facilities.

4.4. Regionally Based Aspects of Vulnerability, Exposure, and Impacts

4.4.1. Introduction

The regional subsections presented here discuss the impacts of extreme weather and climate events within the context of other issues and trends. Regional perspective, in social and economic dimensions, is important especially since decisionmaking often has a strong regional context. For a comprehensive assessment of observed and projected regional changes in climate extremes, see Sections 3.3 to 3.5 and Tables 3-2 and 3-3.

For various climate extremes, the following aspects are considered on a regional basis: exposure of humans and their activities to given climate extremes; the vulnerability of what is exposed to the climate extreme; and the resulting impacts. The individual sections below are structured as is most logical for the trends relevant to each region.

4.4.2. Africa

4.4.2.1. Introduction

Climate extremes exert a significant control on the day-to-day economic development of Africa, particularly in traditional rain-fed agriculture and pastoralism, and water resources, at all scales. Floods and droughts can cause major human and environmental impacts on and disruptions to the economies of African countries, thus exacerbating vulnerability (AMCEN/UNEP, 2002; Scholes and Biggs, 2004; Washington et al., 2004; Thornton et al., 2006). There is still limited scientific information available on observed frequency and projections of many extreme events in Africa (e.g., see Tables 3-2 and 3-3), despite frequent reporting of such events, including their impacts.

Agriculture as an economic sector is most vulnerable and most exposed to climate extremes in Africa. It contributes approximately 50% to Africa's total export value and approximately 21% of its total GDP (Mendelsohn et al., 2000; PACJA, 2009). In particular, with an inefficient agriculture industry, sub-Saharan Africa is extremely vulnerable to climate extremes. This vulnerability is exacerbated by poor health, education, and governance standards (Brooks et al., 2005). Reid et al. (2007) project climate impacts on Namibia's natural resources that would cause annual losses of 1 to 6% of GDP, of which livestock production, traditional agriculture, and fishing are expected to be hardest hit, with a combined loss of US$ 461 to 2,045 million per year by 2050.

4.4.2.2. Droughts and Heat Waves

An overall increase in dryness in Africa has been observed (*medium confidence*), with prolonged Sahel drought, but regional variability is observed (see Table 3-2). Droughts have affected the Sahel, the Horn of Africa, and Southern Africa particularly since the end of the 1960s (Richard et al., 2001; L'Hôte et al., 2002; Brooks, 2004; Christensen et al., 2007; Trenberth et al., 2007). One of the main consequences of multi-year drought periods is severe famine, such as the one associated with the drought in the Sahel in the 1980s, causing many casualties and important socioeconomic losses. The people in Africa who live in drought-prone areas are vulnerable to the direct impacts of droughts (e.g., famine, death of cattle, soil salinization), as well as indirect impacts (e.g., illnesses such as cholera and malaria) (Few et al., 2004).

The water sector is strongly influenced by, and sensitive to, periods of prolonged drought conditions in a continent with limited water storage infrastructure. Natural water reservoirs such as lakes experience a marked interannual water level fluctuation related to rainfall interannual variability (Nicholson et al., 2000; Verschuren et al., 2000).

Large changes in hydrology and water resources linked to climate variability have led to water stress conditions in human and ecological systems in a number of African countries (Schulze et al., 2001; New, 2002; Legesse et al., 2003; Eriksen et al., 2005; de Wit and Stankiewicz, 2006; Nkomo and Bernard, 2006). Twenty-five percent of the contemporary African population has limited water availability and thus constitutes a drought-sensitive population, whereas 69% of the population experiences relative water abundance (Vörösmarty et al., 2005). Even for this latter part of the population, however, relative abundance does not necessarily correspond to access to safe drinking water and sanitation, and this effective reduction of the quantity of freshwater available for human use negatively affects vulnerability. Despite the considerable improvements in access to freshwater in the 1990s, only about 62% of the African population had access to improved water supplies in 2000 (WHO/ UNICEF, 2000). As water demand increases, the population exposed to different drought conditions (agricultural, climate, urban) is expected to increase as well.

Increasing drought risk may cause a decline in tourism, fisheries, and cropping (UNWTO, 2003). This could reduce the revenue available to governments, enterprises, and individuals, and hence further deteriorate the capacity for adaptation investment. For example, the 2003-2004 drought cost the Namibian Government N$ 275 million (US$ 43-48 million) in provision of emergency relief (Reid et al., 2007). Cameroon's economy is highly dependent on rain-fed agriculture; a 14% reduction in rainfall is projected to cause significant losses, of up to US$ 4.56 billion (Molua and Lambi, 2006).

4.4.2.3. Extreme Rainfall Events and Floods

There are inconsistent patterns of change in heavy precipitation in Africa and partial lack of data; hence there is *low confidence* in observed precipitation trends (see Table 3-2). Heavy precipitation may induce landslides and debris flows in tropical mountain regions (Thomas and Thorp, 2003) with potential impacts for human settlements. In the arid and semi-arid areas of countries of the Horn of Africa, extreme rainfall events are often associated with a higher risk of the vector and epidemic diseases of malaria, dengue fever, cholera, Rift Valley fever, and hantavirus pulmonary syndrome (Anyamba et al., 2006; McMichael et al., 2006).

The periods of extreme rainfall and recurrent floods seem to correlate with the El Niño phase (Reason and Kiebel, 2004; Reason et al., 2005; Washington and Preston, 2006; Christensen et al., 2007) of ENSO events (e.g., 1982-1983, 1997-1998, 2006-2007). When such events occur, important economic and human losses result. In 2000, floods in Mozambique (see Case Study 9.2.6), particularly along the valleys of the rivers Limpopo, Save, and Zambezi, resulted in 700 reported deaths and about half a million homeless. The floods had a devastating effect on livelihoods, destroying agricultural crops, disrupting electricity supplies, and demolishing basic infrastructure (Osman-Elasha et al., 2006). However, floods can be highly beneficial in African drylands (e.g.,

Sahara and Namib Deserts) since the floodwaters infiltrate and recharge alluvial aquifers along ephemeral river pathways, extending water availability to dry seasons and drought years (Morin et al., 2009; Benito et al., 2010), and supporting riparian systems and human communities (e.g., Walvis Bay in Namibia).

Damage to African port cities from flooding, storm surge, and high winds might increase due to climate change. For instance, it is indicated that in Alexandria, US$ 563.28 billion worth of assets could suffer damage or be lost because of coastal flooding alone by 2070 (Nicholls et al., 2008).

4.4.2.4. Dust Storms

Atmospheric dust is a major element of the Saharan and Sahelian environments. The Sahara Desert is the world's largest source of airborne mineral dust, which is transported over large distances, traversing northern Africa and adjacent regions and depositing dust on other continents (Osman-Elasha et al., 2006; Moulin et al., 1997). Dust storms have negative impacts on agriculture, health, and structures. They erode fertile soil; uproot young plants; bury water canals, homes, and properties; and cause respiratory problems. Meningitis transmission is associated with dust in semi-arid conditions and overcrowded living conditions. The frequency of dust events has increased in the Sahel zone, but studies of observations and in particular studies of projections of dust activity are limited (see Section 3.5.8).

4.4.3. Asia

Asia includes mega-deltas, which are susceptible to extreme impacts due to a combination of the following factors: high-hazard rivers, coastal flooding, and increased population exposure from expanding urban areas with large proportions of high vulnerability groups (Nicholls et al., 2007). Asia can also expect changes in the frequency and magnitude of extreme weather and climate events, such as heat waves and heavy precipitation (see, e.g., Table 3-3). Such changes may have ramifications not only for physical and natural systems but also for human systems.

4.4.3.1. Tropical Cyclones (Typhoons or Hurricanes)

Damage due to storm surge is sensitive to any changes in the magnitude of tropical cyclones (Xiao and Xiao, 2010). For example, changes in storm surge and associated damage were projected for the inner parts of three major bays (Tokyo, Ise, and Osaka) in Japan (Suzuki, 2009). The projections were based on calculations of inundations for different sea levels and different strengths of typhoons, using a spatial model with information on topography and levees. The research indicated that a typhoon that is 1.3 times as strong as the design standard with a sea level rise of 60 cm would cause damage costs of about US$ 3, 40, and 27 billion, respectively, in the investigated bays.

Awareness, improved governance, and development are essential in coping with extreme tropical cyclone and typhoon events in developing Asian countries (Cruz et al., 2007). For example, two cyclones in the Indian Ocean (Sidr and Nargis) of similar magnitude and strength caused a significantly different number of fatalities. A comparison is presented in Case Study 9.2.5.

For the period from 1983 to 2006, the direct economic losses in China increased, but there is no trend if the losses are normalized by annual total GDP and GDP per capita, suggesting Chinese economic development contributed to the upward trend. This hypothesis is consistent with data on tropical cyclone casualties, which showed no significant trend over the 24 years (Zhang et al., 2009). Similarly, normalized losses from typhoons on the Indian southeast coast since 1977 show no increases (Raghavan and Rajesh, 2003).

4.4.3.2. Flooding

The geographical distribution of flood risk is heavily concentrated in India, Bangladesh, and China, causing high human and material losses (Brouwer et al., 2007; Dash et al., 2007; Shen et al., 2008). Regarding the occurrence of the extreme events themselves, different flooding trends have been detected and projected in various catchments, but the evidence for broader regional trends is limited (see Section 3.5.2).

In July 2005, severe flooding occurred in Mumbai, India, after 944 mm of rain fell in a 24-hour period (Kshirsagar et al., 2006). The consequent flooding affected households, even in more affluent neighborhoods. Poor urban drainage systems in many parts of India can be easily blocked. Ranger et al. (2011) analyzed risk from heavy rainfall in the city of Mumbai, concluding that total losses (direct plus indirect) for a 1-in-100 year event could triple by the 2080s compared with the present (increasing from US$ 700 to 2,305 million), and that adaptation could help reduce future damages.

As noted in the final report for the Ministry of Environment and Forest (2005) of the People's Republic of Bangladesh, flooding in Bangladesh is a normal, frequently recurrent, phenomenon. Bangladesh experiences four types of floods: flash floods from the overflowing of hilly rivers; rain floods due to poor drainage; monsoon floods in the flood plains of major rivers; and coastal floods following storm surge. In a normal year, river spills and drainage congestions cause inundation of 20 to 25% of the country's area. Inundation areas for 10-, 50-, and 100-year floods constitute 37, 52, and 60% of the country's area, respectively. In 1987, 1988, and 1998, floods inundated more than 60% of the country. The 1998 flood alone led to 1,100 deaths, caused inundation of nearly 100,000 km^2, left 30 million people homeless, and substantially damaged infrastructure.

There have been increases in flood impacts associated with changes in surrounding environments. Flooding has increased over the past few decades in the Poyang Lake, South China, due to levee construction protecting a large rural population (Shankman et al., 2006). Such levees reduce the area for floodwater storage, leading to higher lake stages during the summer flood season and then levee failures. The most extreme floods occurred during or immediately following El Niño events (Shankman et al., 2006). Fengqing et al. (2005) analyzed losses from flooding in the Xinjiang autonomous region of China, and found an increase that seems to be linked to changes in rainfall and flash floods since 1987.

Heavy rainfall and flooding also affect environmental health in urban areas because surface water can be quickly contaminated. Urban poor populations in low- and middle-income countries can experience higher rates of infectious disease after floods, such as cholera, cryptosporidiosis, and typhoid fever (Kovats and Akhtar, 2008).

4.4.3.3. Temperature Extremes

Increases in warm days/nights and heat wave duration, frequency, and/or intensity are observed and projected in Asia (see Tables 3-2 and 3-3), with adverse impacts on both human and natural systems. In 2002, a heat wave was reported to have killed 622 people in the southern Indian state of Andhra Pradesh. Persons living in informal settlements and structures are more exposed to high temperatures (Kovats and Akhtar, 2008).

Agriculture is also affected directly by temperature extremes. For example, rice, the staple food in many parts of Asia, is adversely affected by extremely high temperature, especially prior to or during critical pollination phases (see Section 4.3.4).

4.4.3.4. Droughts

Asia has a long history of drought, which has been linked with other climate extremes. Spatially varying trends have been observed during the second half of the 20th century, with increasing dryness noted in some areas, particularly in East Asia (see Table 3-2), adversely affecting socioeconomic, agricultural, and environmental conditions. Drought causes water shortages, crop failures, starvation, and wildfire.

In Southeast Asia, El Niño is associated with comparatively dry conditions: 93% of droughts in Indonesia between 1830 and 1953 occurred during El Niño years (Quinn et al., 1978). In four El Niño years between 1973 and 1992, the average annual rainfall amounted to only around 67% of the 20-year average in two major rice growing areas in Java, Indonesia, causing a yield decline of approximately 50% (Amien et al., 1996).

During drought, severe water scarcity results from one of, or a combination of, the following mechanisms: insufficient precipitation; high evapotranspiration; and over-exploitation of water resources (Bhuiyan et al., 2006).

About 15% (23 million ha) of Asian rice areas experience frequent yield loss due to drought (Widawsky and O'Toole, 1990). The problem is particularly pertinent to eastern India, where the area of drought-prone fields exceeds more than 10 million ha (Pandey et al., 2000). Even when the total rainfall is adequate, shortages during critical periods reduce yield (Kumar et al., 2007). Lowland rice production in the Mekong region is generally reduced because crops are cultivated under rain-fed conditions, rather than irrigated, and often exposed to drought. In Cambodia, severe drought that affects grain yield mostly occurs late in the growing season, and longer-duration genotypes are more likely to encounter drought during grain filling (Tsubo et al., 2009).

Asian wetlands provide resources to people in inundation areas, who are susceptible to droughts. For achieving the benefits from fertilization for inundation agriculture in Cambodia, wide areas along the rivers need to be flooded (Kazama et al., 2009). Flood protection in this area needs to consider this benefit of inundation.

4.4.3.5. Wildfires

Grassland fire disaster is a critical problem in China (Su and Liu, 2004; Zhang et al., 2006), especially in northwestern and northeastern China due to expansive territory and complex physiognomy. Statistical analysis of historical grassland fire disaster data has suggested a gradual increase in grassland fire disasters with economic development and population growth in 12 northern provinces of China between 1991 and 2006 (Liu et al., 2006).

In tropical Asia, although humans are igniting the fires, droughts are predisposing factors for fire occurrence (Field et al., 2009). Drought episodes, forest fires, drainage of rice fields, and oil palm plantations are drying peatlands, which are then more susceptible to fires (van der Werf et al., 2008). Peatland fires are an important issue given the difficulties of extinguishing them and their potential effects on climate.

4.4.4. Central and South America

4.4.4.1. Extreme Rainfalls in South America

Extreme rainfall episodes have caused disasters in parts of South America, with hundreds to thousands of fatalities in mudslides and landslides, as typified, for example, by the December 1999 incident in Venezuela (Lyon, 2003). However, there is *low* to *medium confidence* in observed (Table 3-2) and in projected (Table 3-3) changes in heavy precipitation in the region.

4.4.4.2. Wildfires

There is a *low* to *medium confidence* in projections of trends in dryness in South America (see Table 3-3). Magrin et al. (2007) indicated that more frequent wildfires are probable (an increase in frequency of 60% for a temperature increase of 3°C) in much of South America. In most of central and northern Mexico, the semi-arid vegetation could be replaced by the vegetation of arid regions (Villers and Trejo, 2004). Due to the interrelated nature of forest fires, deforestation, drought, and climate change, isolating one of the processes fails to describe the complexity of the interconnected whole.

4.4.4.3. Regional Costs

Climatic disasters account for the majority of natural disasters in Central America, with most of its territory located in tropical and equatorial areas. Low-lying states are especially vulnerable to hurricanes and tropical storms. In October 1998, Hurricane Mitch, one of the most powerful hurricanes of the tropical Atlantic Basin of the 20th century, caused direct and indirect damages to Honduras of US$ 5 billion, equivalent to 95% of Honduras' 1998 GDP (Cardemil et al., 2000). Some literature indicates that hurricane losses, when corrected for population and wealth in Latin America and the Caribbean, have not increased since the 1940s (Pielke Jr. et al., 2003); and that increasing population and assets at risk are the main reason for increasing impacts.

4.4.5. Europe

4.4.5.1. Introduction

This section assesses vulnerability and exposure to climate extremes in Europe, evaluating observed and projected impacts, disasters, and risks. Europe has a higher population density and lower birth rate than any other continent. It currently has an aging population; life expectancy is high and increasing, and child mortality is low and decreasing (Eurostat, 2010). European exposure to climate- and weather-related hazards has increased whereas vulnerability has decreased as a result of implementation of policy, regulations, and risk prevention and management (EEA, 2008; UNISDR, 2009).

4.4.5.2. Heat Waves

Summer heat waves have increased in frequency and duration in most of Europe (Section 3.3.1 and Table 3-2) and have affected vulnerable segments of European society. During the 2003 heat wave, several tens of thousands of additional heat-related deaths were recorded (see Case Study 9.2.1 and Box 4-4). Urban heat islands pose an additional risk to urban inhabitants. Those most affected are the elderly, ill, and socially isolated (Kunst et al., 1993; Laschewski and Jendritzky, 2002; see Case Study 9.2.1). There are mounting concerns about increasing heat intensity in major European cities (Wilby, 2003) because of the large population that inhabits urban areas. Building characteristics, emissions of anthropogenic heat from air conditioning units and vehicles, as well as lack of green open areas in some parts of the cities, may exacerbate heat load during heat waves (e.g., Stedman, 2004; Wilby, 2007). However, as high summer temperatures and urban heat waves become more common, populations are able to adapt to such 'expected' temperature conditions, decreasing mortality during subsequent heat waves (Fouillet et al., 2008).

4.4.5.3. Droughts and Wildfires

Drought risk is a function of the frequency, severity, and spatial and temporal extent of dry spells and of the vulnerability and exposure of a population and its economic activity (Lehner et al., 2006). In Mediterranean countries, droughts can lead to economic damages larger than floods or earthquakes (e.g., the drought in Spain in 1990 affected 6 million people and caused material losses of US$ 4.5 billion; after CRED, 2010). The most severe human consequences of droughts are often found in semiarid regions where water availability is already low under normal conditions, water demand is close to, or exceeds, natural availability, and/or society lacks the capacity to mitigate or adapt to drought (Iglesias et al., 2009). Direct drought impacts affect all forms of water supply (municipal, industrial, and agricultural). Other sectors and systems affected by drought occurrence are hydropower generation, tourism, forestry, and terrestrial and aquatic ecosystems.

Forest fire danger (length of season, frequency, and severity) depends on the occurrence of drought. There is *medium confidence* in observed changes in drought in Europe (Table 3-2). Projections indicate increasing dryness in central Europe and the Mediterranean, with no major change in Northern Europe (*medium confidence*) (see Table 3-3). In the Mediterranean, an increase in dryness may lead to increased dominance of shrubs over trees (Mouillot et al., 2002); however, it does not translate directly into increased fire occurrence or changes in vegetation (Thonicke and Cramer, 2006). Analysis of post-fire forest resilience contributes to identifying 'risk hotspots' where post-fire management measures should be applied as a priority (Arianoutsou et al., 2011).

4.4.5.4. Coastal Flooding

Coastal flooding is an important natural disaster, since many Europeans live near the coasts. Storm surges can be activated as a result of wind-driven waves and winter storms (Smith et al., 2000), whereas long-term processes are linked to global mean sea level rise (Woodworth et al., 2005). Locations currently experiencing adverse impacts such as coastal erosion and inundation will continue to do so in the future (see Section 3.5.5). Expected sea level rise is projected to have impacts on Europe's coastal areas including land loss, groundwater and soil salinization, and damage to property and infrastructure (Devoy, 2008). Hinkel et al. (2010) found that the total monetary damage in coastal areas of the Member Countries of the European Union caused by flooding, salinity intrusion, land erosion, and migration is projected to rise without adaptation by 2100 to roughly € 17 billion per year under the A2 and

> **Box 4-4 | Extraordinary Heat Wave in Europe, Summer 2003**
>
> The extraordinarily severe heat wave over large parts of the European continent in the summer of 2003 produced record-breaking temperatures particularly during June and August (Beniston, 2004; Schär et al., 2004). Average summer (June to August) temperatures were by up to five standard deviations above the long-term mean, implying that this was an extremely unusual event (Schär and Jendritzky, 2004). Regional climate model simulations suggest the 2003 heat wave bears resemblance to summer temperatures in the late 21st century under the A2 scenario (Beniston, 2004).
>
> Electricity demand increased with the high heat levels. Additionally, drought conditions created stress on health, water supplies, food storage, and energy systems; for example, reduced river flows reduced the cooling efficiency of thermal power plants (conventional and nuclear), and six power plants were shut down completely (Létard et al., 2004). Many major rivers (e.g., the Po, Rhine, Loire, and Danube) were at record low levels, resulting in disruption of inland navigation and irrigation, as well as power plant cooling (Beniston and Díaz, 2004; Zebisch et al., 2005). In France, electricity became scarce, construction productivity fell, and the cold storage systems of approximately 20 to 30% of all food-related establishments were found to be inadequate (Létard et al., 2004). The (uninsured) economic losses for the agriculture sector in the European Union were estimated at € 13 billion (Sénat, 2004). A record drop in crop yield of 36% occurred in Italy for maize grown in the Po valley, where extremely high temperatures prevailed (Ciais et al., 2005). The hot and dry conditions led to many very large wildfires. Glacier melting in the Alps prevented even lower river flows in the Danube and Rhine (Fink et al., 2004).
>
> Health and health service-related impacts of the heat wave were dramatic, with excess deaths of about 35,000 (Kosatsky, 2005). Elderly people were among those most affected (WHO, 2003; Borrell et al., 2006; Kovats and Ebi, 2006), but deaths were also associated with housing and social conditions, for example, being socially isolated or living on the top floor (Vandentorren et al., 2006). The high mortality during the 2003 heat wave marked an inflexion point in public awareness of the dangers of high temperatures, conducive to increasing the preventive measures set up by health institutions and authorities (Koppe et al., 2004; Pascal et al., 2006).
>
> During the July 2006 heat wave, about 2,000 excess deaths occurred in France (Rey et al., 2007). The excess mortality during the 2006 heat wave was markedly lower than that predicted by Fouillet et al. (2008) based on the quantitative association between temperature and mortality observed during 1975-2003. Fouillet et al. (2008) interpreted this mortality reduction (~4,400 deaths) as a decrease in the population's vulnerability to heat, together with increased awareness of the risk related to extreme temperatures, preventive measures, and the warning system established after the 2003 heat wave.

B1 emission scenarios. The Netherlands is an example of a country that is highly susceptible to both sea level rise and coastal flooding, with damage costs relative to GDP of up to 0.3% of GDP under the A2 scenario (Hinkel et al., 2010). By 2100, adaptation can reduce the number of people flooded by two orders of magnitude and the total damage costs by a factor of seven to nine (Hinkel et al., 2010).

4.4.5.5. Gale Winds

Storms have been one of the most important climate hazards for the insurance industry in Europe (Munich Re NatCatSERVICE data cited in EEA, 2008). In the most severe extratropical windstorm month, December 1999, when three events struck Europe (Anatol – December 3, Denmark; Lothar – December 26, France, Germany, and Switzerland; and Martin – December 28, France, Spain, and Italy), insured damage was in excess of US$ 12 billion (Schwierz et al., 2010). Typical economic losses were generated by gale winds via effects on electrical distribution systems, transportation, and communication lines; by damage to vulnerable elements of buildings (e.g., lightweight roofs); and by trees falling on houses. Some researchers have found no contribution from climate change to trends in the economic losses from floods in Europe since the 1970s (Barredo, 2009). Some studies have found evidence of increasing damages to forests in Sweden and Switzerland (Nilsson et al., 2004; Usbeck et al., 2010). Still other studies assert that increases in forest disturbances in Europe are mostly due to changes in forest management (e.g., Schelhaas et al., 2003).

There is *medium confidence* in projected poleward shifts of mid-latitude storm tracks but *low confidence* in detailed regional projections (see Section 3.4.5). According to a study by Swiss Re (2009), if by the end of this century once-in-a-millennium storm surge events strike northern Europe every 30 years, this could potentially result in a disproportionate increase in annual expected losses from a current € 0.6 to 2.6 billion by end of the century. Similar results are obtained from global and regional climate models run under the IPCC SRES A1B emission scenario (Donat et al., 2010). Adaptation to the changing wind climate may reduce by half the estimated losses (Leckebusch et al., 2007; Donat et al., 2010), indicating that adaptation through adequate sea defenses and the management of residual risk is beneficial.

4.4.5.6. Flooding

Flooding is the most frequent natural disaster in Europe (EEA, 2008). Economic losses from flood hazards in Europe have increased considerably over previous decades (Lugeri et al., 2010), and increasing exposure of people and economic assets is probably the major cause of the long-term changes in economic disaster losses (Barredo, 2009). Exposure is influenced by socioeconomic development, urbanization, and infrastructure construction on flood-prone areas. Large flood impacts have been caused by a few individual flood events (e.g., the 1997 floods in Poland and Czech Republic, the 2002 floods in much of Europe, and the 2007 summer floods in the United Kingdom). The projected increase in frequency and intensity of heavy precipitation over large parts of Europe (Table 3-3) may increase the probability of flash floods, which pose the highest risk of fatality (EEA, 2004). Particularly vulnerable are new urban developments and tourist facilities, such as camping and recreation areas (e.g., a large flash flood in 1996 in the Spanish Pyrenees, conveying a large amount of water and debris to a camping site, resulted in 86 fatalities; Benito et al., 1998). Apart from new developed urban areas, linear infrastructure, such as roads, railroads, and underground rails with inadequate drainage, will probably suffer flood damage (DEFRA, 2004; Arkell and Darch, 2006). Increased runoff volumes may increase risk of dam failure (small water reservoirs and tailings dams) with high environmental and socioeconomic damages as evidenced by historical records (Rico et al., 2008).

In glaciated areas of Europe, glacial lake outburst floods, although infrequent, have the potential to produce immense socioeconomic and environmental impacts. Glacial lakes dammed by young, unstable, and unconsolidated moraines, and lakes in contact with the active ice body of a glacier, increase the potential of triggering an event (e.g., Huggel et al., 2004). Intense lake level and dam stability monitoring on most glacial lakes in Europe helps prevent major breach catastrophes. In case of flooding, major impacts are expected on infrastructure and settlements even at long distances downstream from the hazard source area (Haeberli et al., 2001; Huggel et al., 2004).

4.4.5.7. Landslides

There is a general lack of information on trends in landslide activity, and for regions with reasonably well-established databases (e.g., Switzerland), significant trends have not been found in the number of events and impacts (Hilker et al., 2009). Reactivation of large movements usually occurs in areas with groundwater flow and river erosion. In southern Europe the risk is reduced through revegetation on scree slopes, which enhances cohesion and slope stability coupled with improved hazard mitigation (Corominas, 2005; Clarke and Rendell, 2006).

4.4.5.8. Snow

Snow avalanches are an ever-present hazard with the potential for loss of life, property damage, and disruption of transportation. Due to an increased use of mountainous areas for recreation and tourism, there is increased exposure for the population leading to an increased rate of mortality due to snow avalanches. During the period 1983 to 2003, avalanche fatalities have averaged about 25 per year in Switzerland (McClung and Schaerer, 2006). In economic terms, direct losses related to avalanches are small (Voigt et al., 2010), although short-term reactions by tourists may result in a reduction in overnight stays one year after a disaster (Nöthiger and Elsasser, 2004). Increased winter precipitation may result in higher than average snow depth or duration of snow cover, which could contribute to avalanche formation (Schneebeli et al., 1997). Climate change impacts on snow cover also include decreases in its duration, depth, and extent and a possible altitudinal shift of the snow/rain limit (Beniston et al., 2003), with adverse consequences to winter tourism. Increased avalanche occurrence would have a negative impact on humans (loss of life and infrastructure) but could have a positive result in mountain forests due to higher biodiversity within the affected areas (Bebi et al., 2009).

4.4.6. North America

4.4.6.1. Introduction

North America (Canada, Mexico, and the United States) is relatively well developed, although differentiation in living standards exists across and within countries. This differentiation in adaptive capacity, combined with a decentralized and essentially reactive response capability, underlies the region's vulnerability (Field et al., 2007). Furthermore, population trends within the region have increased vulnerability by heightening exposure of people and property in areas that are affected by extreme events. For example, population in coastline regions of the Gulf of Mexico region in the United States increased by 150% from 1960 to 2008, while total US population increased by 70% (U.S. Census Bureau, 2010).

4.4.6.2. Heat Waves

For North America, there is *medium confidence* in observations (Table 3-2) and *high confidence* in projections (Table 3-3) of increasing trends in heat wave frequency and duration.

Heat waves have impacts on many sectors, most notably on human health, agriculture, forestry and natural ecosystems, and energy infrastructure. One of the most significant concerns is human health, in particular mortality and morbidity. In 2006 in California, at least 140 deaths and more than 1,000 hospitalizations were recorded during a severe heat wave (CDHS, 2007; Knowlton et al., 2008). In 1995 in Chicago, more than 700 people died during a severe heat wave. Following that 1995 event, the city developed a series of response measures through an extreme heat program. In 1999, the city experienced another extreme heat event but far fewer lives were lost. While conditions in the 1999 event were somewhat less severe, the city's response measures were also credited with contributing to the lower mortality (Palecki et al., 2001).

While heat waves are projected to increase in intensity and duration (Table 3-3), their net effect on human health is uncertain, largely because of uncertainties about the structure of cities in the future, adaptation measures, and access to cooling (Ebi and Meehl, 2007). Many cities have installed heat watch warning systems. Several studies show that the sensitivity of the population of large US cities to extreme heat events has been declining over time (e.g., Davis et al., 2003; Kalkstein et al., 2011).

Heat waves have other effects. There is increased likelihood of disruption of electricity supplies during heat waves (Wilbanks et al., 2008). Air quality can be reduced, particularly if stagnant high-pressure systems increase in frequency and intensity (Wang and Angell, 1999). Additionally, extreme heat can reduce yields of grain crops such as corn and increase stress on livestock (Karl et al., 2009).

4.4.6.3. Drought and Wildfire

There is *medium confidence* in an overall slight decrease in dryness since 1950 across the continent, with regional variability (Table 3-2). For some regions of North America, there is *medium confidence* in projections of increasing dryness (Table 3-3).

Droughts are currently the third most costly category of natural disaster in the United States (Carter et al., 2008). The effects of drought include reduced water quantity and quality, lower streamflows, decreased crop production, ecosystem shifts, and increased wildfire risk. The severity of impacts of drought is related to the exposure and vulnerability of affected regions.

From 2000 to 2010, excluding 2003, crop losses accounted for nearly all direct damages resulting from US droughts (NWS, 2011). Similarly, drought has had regular recurring impacts on agricultural activities in Northern Mexico (Endfield and Tejedo, 2006). In addition to impacts on crops and pastures, droughts have been identified as causes of regional-scale ecosystem shifts throughout southwestern North America (Allen and Breshears, 1998; Breshears et al., 2005; Rehfeldt et al., 2006).

Drought also has multiple indirect impacts in North America, although they are more difficult to quantify. Droughts pose a risk to North American power supplies due to associated reliance on sufficient water supplies for hydropower generation and cooling of nuclear, coal, and natural gas generation facilities (Goldstein, 2003; Wilbanks et al., 2008). Studies of water availability in heavily contested reservoir systems such as the Colorado River Basin indicate that climate change is projected to reduce states' abilities to meet existing agreements (Christensen et al., 2004). The effects of climate change on the reliability of the water supply have been thoroughly explored by Barnett and Pierce (2008, 2009).

Additionally, droughts and dry conditions more generally have been linked to increases in wildfire activity in North America. Westerling et al. (2006) found that wildfire activity in the western United States increased substantially in the late 20th century and that the increase is caused by higher temperatures and earlier snowmelt. Similarly, increases in wildfire activity in Alaska from 1950 to 2003 have been linked to increased temperatures (Karl et al., 2009). Anthropogenic warming was identified as a contributor to increases in Canadian wildfires (Gillett et al., 2004).

In Canada, forest fires are responsible for one-third of all particulate emissions, leading to heightened incidence of respiratory and cardiac illnesses as well as mortality (Rittmaster et al., 2006). Wildfires not only cause direct mortality, but the air pollution produces increases in eye and respiratory illnesses (Ebi et al., 2008). The principal economic costs of wildfires include timber losses, property destruction, fire suppression, and reductions in the tourism sector (Butry et al., 2001; Morton et al., 2003).

4.4.6.4. Inland Flooding

There has been a *likely* increase in heavy precipitation in many areas of North America since 1950 (Table 3-2), with projections suggesting further increases in heavy precipitation in some regions (Table 3-3). Flooding and heavy precipitation events have a variety of significant direct and indirect human health impacts (Ebi et al., 2008). Heavy precipitation events are strongly correlated with the outbreak of waterborne illnesses in the United States – 51% of waterborne disease outbreaks were preceded by precipitation events in the top decile (Curriero et al., 2001). In addition, heavy precipitation events have been linked to North American outbreaks of vector-borne diseases such as Hantavirus and plague (Engelthaler et al., 1999; Parmenter et al., 1999; Hjelle and Glass, 2000).

Beyond direct destruction of property, flooding has important negative impacts on a variety of economic sectors including transportation and agriculture. Heavy precipitation and field flooding in agricultural systems delays spring planting, increases soil compaction, and causes crop losses through anoxia and root diseases; variation in precipitation is responsible for the majority of the crop losses (Mendelsohn, 2007). In 1993, heavy precipitation flooded 8.2 million acres (~3.3 million ha) of American Midwest soybean and corn fields, leading to a 50% decrease in corn yields in Iowa, Minnesota, and Missouri, and a 20 to 30% decrease in Illinois, Indiana, and Wisconsin (Changnon, 1996). Furthermore, flood impacts include temporary damage or permanent destruction of infrastructure for most modes of transportation (Zimmerman and Faris, 2010). For example, heavy precipitation events are a very costly weather condition facing US rail transportation (Changnon, 2006).

4.4.6.5. Coastal Storms and Flooding

Global observed and projected changes in coastal storms and flooding are complex. Since 1950, there has been a *likely* increase in extreme sea level, related to trends in mean sea level. With upward trends in sea level *very likely* to continue (Section 3.5.3), there is *high confidence* that locations currently experiencing coastal erosion and inundation will continue to do so in the future (Section 3.5.5).

North America is exposed to coastal storms, and in particular, hurricanes. 2005 was a particularly severe year with 14 hurricanes (out of 27 named storms) in the Atlantic (NCDC, 2005). There were more than 2,000 deaths during 2005 (Karl et al., 2009) and widespread destruction on the Gulf Coast and in New Orleans in particular. Property damages exceeded US$ 100 billion (Beven et al., 2008; Pielke Jr. et al., 2008). Hurricanes Katrina and Rita destroyed more than 100 oil and gas platforms in the Gulf and damaged 558 pipelines, halted all oil and gas production in the Gulf, and disrupted 20% of US refining capacity (Karl et al., 2009). It is reported that the direct overall losses of Hurricane Katrina were about US$ 138 billion in 2007 dollars (Spranger, 2008). However, 2005 may be an outlier for a variety of reasons – the year saw storms of higher than average frequency, with greater than average intensity, which made more frequent landfall, including in the most vulnerable region of the country (Nordhaus, 2010). The major factor increasing the vulnerability and exposure of North America to hurricanes is the growth in population (see, e.g., Pielke Jr. et al., 2008) and increase in property values, particularly along the Gulf and Atlantic coasts of the United States. While some of this increase has been offset by adaptation and improved building codes, Nordhaus (2010) suggests the ratio of hurricane damages to national GDP has increased by 1.5% per year over the past half-century. However, the choice of start and end dates influences this figure.

Future sea level rise and potential increases in storm surge could increase inundation and property damage in coastal areas. Hoffman et al. (2010) assumed no acceleration in the current rate of sea level rise through 2030 and found that property damage from hurricanes would increase by 20%. Frey et al. (2010) simulated the combined effects of sea level rise and more powerful hurricanes on storm surge in southern Texas in the 2080s. They found that the area inundated by storm surge could increase from 6-25% to 60-230% across scenarios evaluated. No adaptation measures were assumed in either study. Globally, uncertainties associated with changes in tropical and extratropical cyclones mean that a general assessment of the projected effects of storminess on future storm surge is not currently possible (Section 3.5.3).

4.4.7. Oceania

The region of Oceania consists of Australia, New Zealand, and several small island states that are considered separately in Section 4.4.10.

4.4.7.1. Introduction

Extreme events have severe impacts in both Australia and New Zealand. In Australia, weather- and climate-related events cause around 87% of economic damage due to natural disasters (storms, floods, cyclones, earthquakes, fires, and landslides; BTE, 2001). In New Zealand, floods and droughts are the most costly climate disasters (Hennessy et al., 2007). Economic damage from extreme weather is projected to increase and provide challenges for adaptation (Hennessy et al., 2007).

Observed and projected trends in temperature and precipitation extremes for the region are extensively covered in Chapter 3 (e.g., Tables 3-2 and 3-3). ENSO is a strong driver of climate variability in this region (see Section 3.4.2).

4.4.7.2. Temperature Extremes

During the Eastern Australian heat wave, in February 2004, temperatures reached 48.5°C in western New South Wales. About two-thirds of continental Australia recorded maximum temperatures over 39°C. Due to heat-related stresses, the Queensland ambulance service recorded a 53% increase in ambulance call-outs (Steffen et al., 2006). A week-long heat wave in Victoria in 2009 corresponded with a sharp increase in deaths in the state. For the week of the heat wave a total of 606 deaths were expected and there were a total of 980 deaths, representing a 62% increase (DHS, 2009).

An increase in heat-related deaths is projected given a warming climate (Hennessy et al., 2007). In Australian temperate cities, the number of deaths is projected to more than double in 2020 from 1,115 per year at present and to increase to between 4,300 and 6,300 per year by 2050 for all emission scenarios, including demographic change (McMichael et al., 2003b). In Auckland and Christchurch, a total of 14 heat-related deaths occur per year in people aged over 65, but this number is projected to rise approximately two-, three-, and six-fold for warming of 1, 2, and 3°C, respectively (McMichael et al., 2003b). An aging society in Australia and New Zealand would amplify these figures. For example, it has been projected that, by 2100, the Australian annual death rate in people aged over 65 would increase from a 1999 baseline of 82 per 100,000 to a range of between 131 and 246 per 100,000 in 2100 for the scenarios examined (SRES B2 and A2, with stabilization of atmospheric CO_2 at 450 ppm; Woodruff et al., 2005). In Australia, cities with a temperate climate are expected to experience more heat-related deaths than those with a tropical climate (McMichael et al., 2003b).

4.4.7.3. Droughts

There is a complex pattern of observed and projected changes in dryness over the region, with increasing dryness in some areas, and decreasing dryness or inconsistent signals in others (Tables 3-2 and 3-3). However, several high-impact drought events have been recorded (OCDESC, 2007).

In Australia, the damages due to the droughts of 1982-1983, 1991-1995, and 2002-2003 were US$ 2.3, 3.8, and 7.6 billion, respectively (Hennessy et al., 2007). Droughts have a negative impact on water security in the

Murray-Darling Basin in Australia, as it accounts for most of the water for irrigated crops and pastures in the country.

New Zealand has a high level of economic dependence on agriculture, and drought can cause significant disruption for this industry. The 1997-1998 El Niño resulted in severe drought conditions across large areas of New Zealand with losses estimated at NZ$ 750 million (2006 values) or 0.9% of GDP (OCDESC, 2007). Severe drought in two consecutive summers, 2007-2009, affected a large area of New Zealand and caused on-farm net income to drop by NZ$ 1.9 billion (Butcher and Ford, 2009). Drought conditions also have a serious impact on electricity production in New Zealand where around two-thirds of supply is from hydroelectricity and low precipitation periods result in increased use of fossil fuel for electricity generation, a maladaptation to climate change. Auckland, New Zealand's largest city, suffered from significant water shortages in the early 1990s, but has since established a pipeline to the Waikato River to guarantee supply (OCDESC, 2007).

Climate change may cause land use change in southern Australia. Cropping could become non-viable at the dry margins if rainfall substantially decreases, even though yield increases from elevated CO_2 partly offset this effect (Luo et al., 2003).

4.4.7.4. Wildfire

Wildfires around Canberra in January 2003 caused AUS$ 400 million damage (Lavorel and Steffen, 2004), with about 500 houses destroyed, four people killed, and hundreds injured. Three of the city's four water storage reservoirs were contaminated for several months by sediment-laden runoff (Hennessy et al., 2007). The 2009 fire in the state of Victoria caused immense damage (see Box 4-1 and Case Study 9.2.2).

An increase in fire danger in Australia is associated with a reduced interval between fire events, increased fire intensity, a decrease in fire extinguishments, and faster fire spread (Hennessy et al., 2007). In southeast Australia, the frequency of very high and extreme fire danger days is expected to rise 15 to 70% by 2050 (Hennessy et al., 2006). By the 2080s, the number of days with very high and extreme fire danger are projected to increase by 10 to 50% in eastern areas of New Zealand, the Bay of Plenty, Wellington, and Nelson regions (Pearce et al., 2005), with even higher increases (up to 60%) in some western areas. In both Australia and New Zealand, the fire season length is expected to be extended, with the window of opportunity for fuel reduction burning shifting toward winter (Hennessy et al., 2007).

4.4.7.5. Intense Precipitation and Floods

There has been a *likely* decrease in heavy precipitation in many parts of southern Australia and New Zealand (Table 3-2), while there is generally *low* to *medium confidence* in projections due to a lack of consistency between models (Table 3-3).

Floods are New Zealand's most frequently experienced hazard (OCDESC, 2007) affecting both agricultural and urban areas. Being long and narrow, New Zealand is characterized by small river catchments and accordingly shorter time-to-peak and shorter flood warning times, posing a difficult preparedness challenge. Projected increases in heavy precipitation events across most parts of New Zealand (Table 3-3) is expected to cause greater erosion of land surfaces, more landslides, and a decrease in the protection afforded by levees (Hennessy et al., 2007).

4.4.7.6. Storm Surges

Over 80% of the Australian population lives in the coastal zone, and outside of the major capital cities is also where the largest population growth occurs (Harvey and Caton, 2003; ABS, 2010). Over 500,000 addresses are within 3 km of the coast and less than 5 m above sea level (Chen and McAneney, 2006). As a result of being so close to sea level, the risk of inundation from sea level rise and large storm surges increases with climate change (Hennessy et al., 2007). The risk of a 1-in-100 year storm surge in Cairns is expected to more than double by 2050 (McInnes et al., 2003). Projected changes in coastal hazards from sea level rise and storm surge are also an issue for New Zealand (e.g., Ministry for the Environment, 2008).

4.4.8. Open Oceans

The ocean's huge mass in comparison to the atmosphere gives it a crucial role in global heat budgets and chemical budgets. Possible extreme impacts can be triggered by (1) warming of the surface ocean, with a major cascade of physical effects, (2) ocean acidification induced by increases in atmospheric CO_2, and (3) reduction in oxygen concentration in the ocean due to a temperature-driven change in gas solubility and physical impacts from (1). All have potentially nonlinear multiplicative impacts on biodiversity and ecosystem function, and each may increase the vulnerability of ocean systems, triggering an extreme impact (Kaplan et al., 2010; Griffith et al., 2011). Surface warming of the oceans can itself directly impact biodiversity by slowing or preventing growth in temperature-sensitive species. One of the most well-known biological impacts of warming is coral bleaching, but ocean acidification can also affect coral growth rates (Bongaerts et al., 2010). The seasonal sea ice cycle affects biological habitats. Such species of Arctic mammals as polar bears, seals, and walruses depend on sea ice for habitat, hunting, feeding, and breeding. Declining sea ice can decrease polar bear numbers (Stirling and Parkinson, 2006).

4.4.9. Polar Regions

4.4.9.1. Introduction

The polar regions consist of the Arctic and the Antarctic, including associated water bodies. The Arctic region consists of a vast treeless permafrost territory (parts of northern Europe, northern Asia, and North America, and several islands including Greenland). Delimitation of the Arctic may differ according to different disciplinary and political definitions (ACIA, 2004). Population density in the polar regions is low, so that impacts of climate change and extremes on humans may not be as noticeable as elsewhere throughout the world. The territory of the Russian Arctic is more populated than other polar regions, hence impacts of climate change are most noticeable there as they affect human activities. Specific impacts of climate extremes on the natural physical environment in polar regions are discussed in Section 3.5.7.

4.4.9.2. Warming Cryosphere

Polar regions have experienced significant warming in recent decades. Warming has been most pronounced across the Arctic Ocean Basin and along the Antarctic Peninsula, with significant decreases in the extent and seasonal duration of sea ice, while in contrast, temperatures over mainland Antarctica have not warmed over recent decades (Lemke et al., 2007; Trenberth et al., 2007). Sea ice serves as primary habitat for marine organisms central to the food webs of these regions. Changes in the timing and extent of sea ice can impose temporal and spatial mismatches between energy requirements and food availability for many higher trophic levels, leading to decreased reproductive success, lower abundances, and changes in distribution (Moline et al., 2008).

Warming in the Arctic may be leading to a shift of vegetation zones (e.g., Sturm et al., 2001; Tape et al., 2006; Truong et al., 2006), bringing wide-ranging impacts and changes in species diversity and distribution.

In the Russian North, the seasonal soil thawing depth has increased overall over the past four decades (Sherstyukov, 2009). As frozen ground thaws, many existing buildings, roads, pipelines, airports, and industrial facilities are destabilized. In the 1990s, the number of damaged buildings increased by 42 to 90% in comparison with the 1980s in the north of western Siberia (Anisimov and Belolutskaya, 2002; Anisimov and Lavrov, 2004). Arctic infrastructure faces increased risks of damage due to changes in the cryosphere, particularly the loss of permafrost and land-fast sea ice (SWIPA, 2011).

An apartment building collapsed in the upper part of the Kolyma River Basin, and over 300 buildings were severely damaged in Yakutsk as a result of retreating permafrost (Anisimov and Belolutskaya, 2002; Anisimov and Lavrov, 2004). Changes in permafrost damage the foundations of buildings and disrupt the operation of vital infrastructure in human settlements (Anisimov et al., 2004; see also Case Study 9.2.10). Transport options and access to resources are altered by differences in the distribution and seasonal occurrence of snow, water, ice, and permafrost in the Arctic. This affects both daily living and commercial activities (SWIPA, 2011).

In conditions of land impassability, frozen rivers are often used as transport ways. In the conditions of climate warming, rivers freeze later and melt earlier than before, and the duration of operation of transport routes to the far north of Russia decreases with the increase in air temperature in winter and spring (Mirvis, 1999).

Ice cover does not allow ship navigation. Navigation in the Arctic Ocean is only possible during the ice-free period off the northern coasts of Eurasia and North America. During periods of low ice concentration, ships navigate toward ice-free passages, away from multi-year ice that has accumulated over several years. Regional warming provides favorable conditions for sea transport going through the Northern Sea Route along the Eurasian coasts and through the Northwest Passage in the north of Canada and Alaska (ACIA, 2004).

In September 2007, when the Arctic Sea ice area was extremely low, the Northwest Passage was opened up. In Russia, this enabled service to ports of the Arctic region and remote northern regions (import of fuel, equipment, food, timber, and export of timber, oil, and gas). However, owing to deglaciation in Greenland, New Land, and Northern Land, the number of icebergs may increase, creating navigation hazards (Roshydromet, 2005, 2008; Rignot et al., 2010, Straneo et al., 2010).

4.4.9.3. Floods

From the mid-1960s to the beginning of the 1990s, winter runoff in the three largest rivers of Siberia (Yenisei, Lena, and Ob; jointly contributing approximately 70% of the global river runoff to the Arctic Ocean) increased by 165 km^3 (Savelieva et al., 2004).

Rivers in Arctic Russia experience floods, but their frequency, stage, and incidence are different across the region, depending on flood formation conditions. Floods on the Siberian rivers can be produced by a high peak of the spring flood, by rare heavy rain, or by a combination of snow and rain, as well as by ice jams, hanging dams, and combinations of factors (Semyonov and Korshunov, 2006).

Maximum river discharge was found to decrease from the mid-20th century through 1980 in Western Siberia and the Far East (except for the Yenisei and the Lena rivers). However, since 1980, maximum streamflow values began to increase over much of Russia (Semyonov and Korshunov, 2006).

Snowmelt and rain continue to be the most frequent cause of hazardous floods on the rivers in the Russian Arctic (85% of all hazardous floods in the past 15 years). Hazardous floods produced by ice jams and wind tides make up 10 and 5% of the total number of hazardous floods, respectively. For the early 21st century, Pomeranets (2005) suggests that the probability of catastrophic wind tide-related floods and ice jam-related floods increased. The damage from floods depends not only on their level, but also on the duration of exposure. On average, a flood lasts 5 to 10 days, but sometimes high water marks have been recorded to persist longer, for example, for 20 days or more (Semyonov and Korshunov, 2006).

4.4.9.4. Coastal Erosion

Coastal erosion is a significant problem in the Arctic, where coastlines are highly variable due to environmental forcing (wind, waves, sea level changes, sea ice, etc.), geology, permafrost, and other elements (Rachold et al., 2005). For example, the amount of coastal erosion along a 60 km stretch of Alaska's Beaufort Sea doubled between 2002 and 2007. Jones et al. (2009) considered contributing factors to be melting sea ice, increasing summer sea surface temperature, sea level rise, and increases in storm power and associated stronger ocean waves.

Increasing coastal retreat will have further ramifications for Arctic landscapes, including losses in freshwater and terrestrial wildlife habitats, in subsistence grounds for local communities, and in disappearing cultural sites, as well as adverse impacts on coastal villages and towns. In addition, oil test wells may be impacted (Jones et al., 2009). Coastal erosion has also become a problem for residents of Inupiat and on the island of Sarichev (Russian Federation) (Revich, 2008).

Permafrost degradation along the coast of the Kara Sea may lead to intensified coastal erosion, driving the coastline back by up to 2 to 4 m per year (Anisimov and Belolutskaya, 2002; Anisimov and Lavrov, 2004). Coastline retreat poses considerable risks for coastal population centers in Yamal and Taymyr and other littoral lowland areas.

4.4.10. Small Island States

Small Island States (SIS) in the Pacific, Indian, and Atlantic Oceans have been identified as being among the most vulnerable to climate change and climate extremes (e.g., UNFCCC, 1992; DSD, 1994; UNISDR, 2005). In the light of current experience and model-based projections, SIS, with high vulnerability and low adaptive capacity, have substantial future risks (Mimura et al., 2007). Smallness renders island countries at risk of high proportionate losses when impacted by a climate extreme (Pelling and Uitto, 2001; see also Case Study 9.2.9 and Box 3-4).

Sea level rise could lead to a reduction in island size (FitzGerald et al., 2008). Island infrastructure, including international airports, roads, and capital cities, tends to predominate in coastal locations (Hess et al., 2008). Sea level rise exacerbates inundation, erosion, and other coastal hazards; threatens vital infrastructure, settlements, and facilities; and thus compromises the socioeconomic well-being of island communities and states (Hess et al., 2008).

In 2005, regionally averaged temperatures were the warmest in the western Caribbean for more than 150 years (Eakin et al., 2010). These extreme temperatures caused the most severe coral bleaching ever recorded in the Caribbean: more than 80% of the corals surveyed were bleached, and at many sites more than 40% died. Recovery from such large-scale coral mortality is influenced by the extent to which coral reef health has been compromised and the frequency and severity of subsequent stresses to the system.

Since the early 1950s, when the quality of disaster monitoring and reporting improved in the Pacific Islands region, there has been a general increasing trend in the number of disasters reported annually (Hay and Mimura, 2010).

Pacific Island Countries and Territories (PICs) exhibit a variety of characteristics rendering generalization difficult (see Table 4-2; Campbell, 2006). One form of PICs is large inter-plate boundary islands formed by subduction and found in the southwest Pacific Ocean. These may be compared to the Oceanic (or intra-plate) islands which were, or are being, formed over 'hot spots' in the Earth's mantle into volcanic high islands. Some of these are still being formed and some are heavily eroded with steep slopes and barrier reefs. Another form of PICs is atolls that consist of coral built on submerging former volcanic high islands, through raised limestone islands (former atolls stranded above contemporary sea levels). Each island type has specific characteristics in relation to disaster risk reduction, with atolls being particularly vulnerable to tropical cyclones, where storm surges can completely inundate them and there is no high ground to which people may escape. In contrast, the inter-plate islands are characterized by large river systems and fertile flood plains in addition to deltas, both of which tend to be heavily populated. Fatalities in many of the worst weather- and climate-related disasters in the region have been mostly from river flooding (AusAID, 2005). Raised atolls are often saved from the storm surge effects of

Table 4-2 | Pacific Island type and exposure to climate extremes. Adapted from Campbell, 2006.

Island Type	Exposure to climate risks
Plate-Boundary Islands • Large land area • High elevations • High biodiversity • Well-developed soils • River flood plains • Orographic rainfall	These islands are located in the western Pacific. River flooding is more likely to be a problem than in other island types. In Papua New Guinea, high elevations expose areas to frost (extreme during El Niño). Most major settlements are on the coast and exposed to storm damage.
Intra-Plate (Oceanic) Islands **Volcanic High Islands** • Steep slopes • Different stages of erosion • Barrier reefs • Relatively small land area • Less well-developed river systems • Orographic rainfall	Because of size these islands have substantial exposure to tropical cyclones, which cause the most damage in coastal areas and catchments. Streams and rivers are subject to flash flooding. Most islands are exposed to drought. Barrier reefs may ameliorate storm surge and tsunamis.
Atolls • Very small land areas • Very low elevations • No or minimal soil • Small islets surround a lagoon • Shore platform on windward side • Larger islets on windward side • No surface (fresh) water • Ghyben Herzberg (freshwater) lens • Convectional rainfall	These islands are exposed to storm surge, 'king' tides, and high waves. They are exposed to freshwater shortages and drought. Freshwater limitations may lead to health problems.
Raised Limestone Islands • Steep outer slopes • Concave inner basin • Sharp karst topography • Narrow coastal plains • No surface water • No or minimal soil	Depending on height these islands may be exposed to storm surges and wave damage during cyclones and storms. They are exposed to freshwater shortages and drought. Freshwater problems may lead to health problems.

tropical cyclones, but during Cyclone Heta that struck Niue in 2004, the cliffs were unable to provide protection.

Drought is a hazard of considerable importance in SIS. In particular, atolls have very limited water resources, being dependent on their freshwater lens, whose thickness decreases with sea level rise (e.g., Kundzewicz et al., 2007), floating above sea water in the pervious coral, and is replenished by convectional rainfall. During drought events, water shortages in SIS become acute on atolls in particular, resulting in rationing in some cases (Campbell, 2006).

The main impacts from climatic extremes in PICs are damage to structures, infrastructure, and crops during tropical cyclones and crop damage and water supply shortages during drought events. On atolls, salinization of the freshwater lens and garden areas is a serious problem following storm surges, high wave events, and 'king' tides (Campbell, 2006).

4.5. Costs of Climate Extremes and Disasters

The following section focuses on the economic costs imposed by climate extremes and disasters on humans, societies, and ecosystems and the costs of adapting to the impacts. Cost estimates are composed of observed and projected economic impacts, including economic losses, future trends in extreme events and disasters in key regions, and the costs of adaptation. The section stands at the interface between chapters, using the conceptual framework of Chapters 1 and 2 and the scientific foundation of Chapter 3 and earlier subsections in this chapter, and leading into the following Chapters 5 through 9.

4.5.1. Framing the Costs of Extremes and Disasters

The economic costs associated with climate extremes and disasters can be subdivided into impact or damage costs (or simply losses) and adaptation costs. Costs arise due to economic, social, and environmental impacts of a climate extreme or disaster and adaptation to those impacts in key sectors. Residual damage costs are the impact and damage costs after all desirable and practical adaptation actions have been implemented. Conceptually, comparing costs of adaptation with damages before adaptation and residual damages can help in assessing the economic efficiency of adaptation (Parry et al., 2009).

The *impact of climate extremes and disasters* on economies, societies, and ecosystems can be measured as the *damage costs* and *losses* of economic assets or stocks, as well as consequential indirect effects on economic flows, such as on GDP or consumption. In line with general definitions in Chapters 1 and 2, economic disaster *risk* may be defined as a probability distribution indicating *potential* economic damage costs and associated return periods. The cost categories of direct, indirect, and intangible are rarely fully exclusive, and items or activities can have elements in all categories.

Direct damage costs or losses are often defined as those that are a direct consequence of the weather or climate event (e.g., floods, windstorms, or droughts). They refer to the costing of the physical impacts of climate extremes and disasters – on the lives and health of directly affected persons; on all types of tangible assets, including private dwellings, and agricultural, commercial, and industrial stocks and facilities; on infrastructure (e.g., transport facilities such as roads, bridges, and ports, energy and water supply lines, and telecommunications); on public facilities (e.g., hospitals, schools); and on natural resources (ECLAC, 2003; World Bank, 2010).

Indirect impact costs generally arise due to the disruption of the flows of goods and services (and therefore economic activity) because of a disaster, and are sometimes termed consequential or secondary impacts as the losses typically flow from the direct impact of a climate event (ECLAC, 2003; World Bank, 2010). Indirect damages may be caused by the direct damages to physical infrastructure or sources of livelihoods, or because reconstruction pulls resources away from production. Indirect damages include additional costs incurred from the need to use alternative and potentially inferior means of production and/or distribution of normal goods and services (Cavallo and Noy, 2010). For example, electricity transmission lines may be destroyed by wind, a direct impact, causing a key source of employment to cease operation, putting many people out of work, and in turn creating other problems that can be classified as indirect impacts. These impacts can emerge later in the affected location, as well as outside the directly affected location (Pelling et al., 2002; ECLAC, 2003; Cavallo and Noy, 2010). Indirect impacts include both negative and positive factors – for example, transport disruption, mental illness or bereavement resulting from disaster shock, rehabilitation, health costs, and reconstruction and disaster-proof investment, which can include changes in employment in a disaster-hit area (due to reconstruction and other recovery activity) or additional demand for goods produced outside of a disaster-affected area (ECLAC, 2003; World Bank, 2010). As another example, long-running droughts can induce indirect losses such as local economic decline, out-migration, famine, the partial collapse of irrigation areas, or loss of livelihoods dependent on hydroelectricity or rain-fed agriculture. It is important to note that impacts on the informal or undocumented economy may be very important in some areas and sectors, but are generally not counted in reported estimates of losses.

Many impacts, such as loss of human lives, cultural heritage, and ecosystem services, are difficult to measure as they are not normally given monetary values or bought and sold, and thus they are also poorly reflected in estimates of losses. These items are often referred to as *intangibles* in contrast to tangibles such as tradable assets, structures, and infrastructure (Handmer et al., 2002; Pelling et al., 2002; Benson and Clay, 2003; ECLAC, 2003; Cavallo and Noy, 2010; World Bank, 2010).

Adaptation costs are those associated with adaptation and facilitation in terms of planning (e.g., developing appropriate processes including key stakeholders), actual adaptation (e.g., risk prevention, preparedness, and risk financing), reactive adaptation (e.g., emergency disaster responses, rehabilitation, and reconstruction), and finally the implementation of adaptation measures (including transition costs) (Smit et al., 2001; also see the Glossary). The benefits of adaptation can generally be assessed as the value of avoided impacts and damages as well as the co-benefits generated by the implementation of adaptation measures (Smit et al., 2001). The value of all avoidable damage can be taken as the gross (or theoretically maximum) benefit of adaptation and risk management, which may be feasible to adapt to but not necessarily economically efficient (Pearce et al., 1996; Tol, 2001; Parry et al., 2009). The *adaptation deficit* is identified as the gap between current and optimal levels of adaptation to climate change (Burton and May, 2004). However, it is difficult to assess the optimal adaptation level due to the uncertainties inherent in climate scenarios, the future patterns of exposure and vulnerability to climate events, and debates over methodological issues such as discount rates. In addition, as social values and technologies change, what is considered avoidable also changes, adding additional uncertainty to future projections.

4.5.2 Extreme Events, Impacts, and Development

The relationship between socioeconomic development and disasters, including those triggered by climatic events, has been explored by a number of researchers over the last few years using statistical techniques and numerical modeling approaches. It has been suggested that natural disasters exert adverse impacts on the pace and nature of economic development (Benson and Clay, 1998, 2003; Kellenberg and Mobarak, 2008). (The 'poverty trap' created by disasters is discussed in Chapter 8.) A growing literature has emerged that identifies these important adverse macroeconomic and developmental impacts of natural disasters (Cuny, 1983; Otero and Marti, 1995; Benson and Clay, 1998, 2000, 2003, 2004; Charveriat, 2000; Crowards, 2000; ECLAC, 2003; Mechler, 2004; Raddatz, 2009; Noy, 2009; Okuyama and Sahin, 2009; Cavallo and Noy, 2010). Yet, confidence in the adverse economic impacts of natural disasters is only *medium*, as, although the bulk of studies identify negative effects of disasters on shorter-term economic growth (up to three years after an event), others find positive effects (Albala-Bertrand, 1993; Skidmore and Toya, 2002; Caselli and Malhotra, 2004; see Section 4.2). Differences can be partly explained by the lack of a robust counterfactual in some studies (e.g., what would GDP have been if a disaster had not occurred?), failure to account for the informal sector, varying ways of accounting for insurance and aid flows, different patterns of impacts resulting from, for example, earthquakes versus floods, and the fact that national accounting does not record the destruction of assets, but reports relief and reconstruction as additions to GDP (World Bank and UN, 2010). In terms of longer-run economic growth (beyond three years after events), there are mixed findings with the exception of very severe disasters, which have been found to set back development (World Bank and UN, 2010).

In terms of the nexus between development and disaster vulnerability, researchers argue that poorer developing countries and smaller economies are more likely to suffer more from future disasters than developed countries, especially in relation to extreme impacts (Hallegatte et al., 2007; Heger et al., 2008; Hallegatte and Dumas, 2009; Loayza et al., 2009; Raddatz, 2009). In general, the observed or modeled relationship between development and disaster impacts indicates that a wealthier country is better equipped to manage the consequences of extreme events by reducing the risk of impacts and by managing the impacts when they occur. This is due (*inter alia*) to higher income levels, more governance capacity, higher levels of expertise, amassed climate-proof investments, and improved insurance systems that can act to transfer costs in space and time (Wildavsky, 1988; Albala-Bertrand, 1993; Burton et al., 1993; Tol and Leek, 1999; Mechler, 2004; Rasmussen, 2004; Brooks et al., 2005; Kahn, 2005; Toya and Skidmore, 2007; Raschky, 2008; Noy, 2009). While the countries with highest income account for most of the total economic and insured losses from disasters (Swiss Re, 2010), in developing countries there are higher fatality rates and the impacts consume a greater proportion of GDP. This in turn imposes a greater burden on governments and individuals in developing countries. For example, during the period from 1970 to 2008 over 95% of deaths from natural disasters occurred in developing countries (Cavallo and Noy, 2010; CRED, 2010). From 1975 to 2007, Organisation for Economic Co-operation and Development (OECD) countries accounted for 71.2% of global total economic losses from tropical cyclones, but only suffered 0.13% of estimated annual loss of GDP (UNISDR, 2009).

There is general consensus that, as compared to developed countries, developing countries are more economically vulnerable to climate extremes largely because: (i) developing countries have less resilient economies that depend more on natural capital and climate-sensitive activities (cropping, fishing, etc.; Parry et al., 2007); (ii) they are often poorly prepared to deal with the climate variability and physical hazards they currently face (World Bank, 2000); (iii) more damages are caused by maladaptation due to the absence of financing, information, and techniques in risk management, as well as weak governance systems; (iv) there is generally little consideration of climate-proof investment in regions with a fast-growing population and asset stocks (such as in coastal areas) (IPCC, 2001; Nicholls et al., 2008); (v) there is an adaptation deficit resulting from the low level of economic development (World Bank, 2007) and a lack of ability to transfer costs through insurance and fiscal mechanisms; and vi) they have large informal sectors. However, in some cases like Hurricane Katrina in New Orleans, United States, developed countries also suffer severe disasters because of social vulnerability and inadequate disaster protection (Birch and Wachter, 2006; Cutter and Finch, 2008).

While some literature has found that the relationship between income and some natural disaster consequences is nonlinear (Kellenberg and Mobarak, 2008; Patt et al., 2010), much empirical evidence supports a negative relationship between the relative share of GDP and fatalities, with fatalities from hydrometeorological extreme events falling with rising level of income (Kahn, 2005; Toya and Skidmore, 2007; World Bank and UN, 2010). Some emerging developing countries, such as China, India, and Thailand, are projected to face increased future exposure to extremes, especially in highly urbanized areas, as a result of the rapid urbanization and economic growth in those countries (Bouwer et al., 2007; Nicholls et al., 2008).

It should also be noted that in a small country, a disaster can directly affect much of the country and therefore the magnitude of losses and recovery demands can be extremely high relative to GDP and public financial resources (Mechler, 2004). This is particularly the case in the event of multiple and/or consecutive disasters in short periods. For example, in Fiji, consecutive natural disasters have resulted in reduced national GDP as well as decreased socioeconomic development as captured by the Human Development Index (HDI) (Lal, 2010). In Mexico, natural disasters resulted in the HDI regressing by approximately two years and in an increase in poverty levels (Rodriguez-Oreggia et al., 2010). Patt et al. (2010) indicated that vulnerability in the least-developed countries will rise most quickly, which implies an urgent need for international assistance.

Costs and impacts not only vary among developing and developed countries, but also between and within countries, regions, local areas, sectors, systems, and individuals due to the heterogeneity of vulnerability and resilience (see Chapter 2). Some individuals, sectors, and systems would be less affected, or may even benefit, while other individuals, sectors, and systems may suffer significant losses in the same event. In general, the poorest and those who are socially or economically marginalized will be the most at risk in terms of being exposed and vulnerable (Wisner et al., 2004). For example, women and children are found to be more vulnerable to disasters in many countries, with larger disasters having an especially unequal impact (Neumayer and Plümper, 2007).

4.5.3. Methodologies for Evaluating Impact and Adaptation Costs of Extreme Events and Disasters

4.5.3.1. Methods and Tools for Costing Impacts

Direct, tangible impacts are comparatively easy to measure, but costing approaches are not necessarily standardized and assessments are often incomplete, which can make aggregation and comparability across the literature difficult. In some countries, flood impact assessment has long been standardized, for example, in Britain and parts of the United States (e.g., Handmer et al., 2002). Intangible losses can generally be estimated using valuation techniques such as loss of life/morbidity (usually estimated using value of statistical life benchmarks), replacement value, benefits transfer, contingent evaluation, travel cost, hedonic pricing methods, and so on (there is a vast literature on this subject, e.g., Handmer et al., 2002; Carson et al., 2003; Pagiola et al., 2004; Ready and Navrud, 2006; TEEB, 2009). Yet, assessing the intangible impacts of extremes and disasters in the social, cultural, and environmental fields is more difficult, and there is little agreement on methodologies (Albala-Bertrand, 1993; Tol, 1995; Hall et al., 2003; Huigen and Jens, 2006; Schmidt et al., 2009).

Studies and reports on the economic impacts of extremes, such as insurance or post-disaster reports, have mostly focused on direct and tangible losses, such as on impacts on produced capital and economic activity. Intangibles such as loss of life and impacts on the natural environment are generally not considered using monetary metrics (Parry et al., 2009). Loss of life due to natural disasters, including future changes, is accounted for in some studies (e.g., BTE, 2001; Jonkman, 2007; Jonkman et al., 2008; Maaskant et al., 2009). Estimates of impacts that account for tangibles and intangibles are expected to be much larger than those that consider tangible impacts only (Handmer et al., 2002; Parry et al., 2009). Potential impacts include all direct, indirect, and intangible costs, including the losses from public goods and natural capital (in particular ecosystem services), as well as the longer-term economic impact of disasters. Indirect impacts and intangible impacts can outweigh those of direct impacts. There will therefore often be a large gap between potential impacts and the estimates from studies that consider only direct impacts.

Indirect economic loss assessment methodologies exist but produce uncertain and method-dependent results. Such assessments at national, regional, and global levels fall into two categories: a 'top down' approach that uses models of the whole economy under study, and a bottom-up or partial equilibrium approach that identifies and values changes in specific parts of an economy (van der Veen, 2004).

The top-down approach is grounded in macroeconomics under which the economy is described as an ensemble of interacting economic sectors. Most studies have focused on impact assessment remodeling actual events in the past and aim to estimate the various, often hidden follow-on impacts of disasters (e.g., Ellson et al., 1984; Yezer and Rubin, 1987; Guimaraes et al., 1993; West and Lenze, 1994; Brookshire et al., 1997; Hallegatte et al., 2007; Rose 2007). Existing macroeconomic or top-down approaches utilize a range of models such as the Input-Output, Social Accounting Matrix multiplier, Computable General Equilibrium models, economic growth frameworks, and simultaneous-equation econometric models. These models attempt to capture the impact of the extreme event as it is felt throughout the whole economy. Only a few models have aimed at representing extremes in a risk-based framework in order to assess the potential impacts of events and their probabilities using a stochastic approach, which is desirable given the fact that extreme events are non-normally distributed and the tails of the distribution matter (Freeman et al., 2002; Mechler, 2004; Hallegatte and Ghil, 2007; Hallegatte, 2008).

The bottom-up approach, derived from microeconomics, scales up data from sectors at the regional or local level to aggregate an assessment of disaster costs and impacts (see van der Veen, 2004). The bottom-up approach to disaster impact assessment attempts to evaluate the impact of an actual or potential disaster on consumers' willingness to pay (or willingness to accept). This approach values direct loss of or damage to property, as well as that of the interruption to the economy, impacts on health and well-being, and impacts on environmental amenity and ecosystem services. In short, it attempts to value the impact of the disaster on society.

Overall, measuring the many effects of disasters is problematic, prone to both overestimation (for example, double counting) and underestimation (because it is difficult to value loss of life or damage to the environment). Both over- and underestimation can be issues in different parts of the same impact assessment, for example, ecological and quality of life impacts may be ignored, while double counting occurs in the measurement of indirect impacts. As discussed earlier in this section, most large-scale estimates leave out significant areas of cost and are therefore underestimates. Biases also affect the accuracy of estimates; for example, the prospect of aid may create incentives to inflate losses. How disaster impacts are evaluated depends on numerous factors, such as the types of impacts being evaluated, the objective of the evaluation, the spatial and temporal scale under consideration, and importantly, the information, expertise, and data available. In practice, the great majority of post-disaster impact assessments are undertaken pragmatically using whatever data and expertise are available. Many studies utilize both partial and general equilibrium analysis in an 'integrated assessment' that attempts to capture both the bottom-up and the economy-wide impacts of disasters (Ciscar, 2009; World Bank, 2010).

4.5.3.2. Methods and Tools for Evaluating the Costs of Adaptation

Over the last few years, a wide range of methodologies using different metrics, time periods, and assumptions has been developed and applied for assessing adaptation costs and benefits. However, much of the literature remains focused on gradual changes such as sea level rise and effects on agriculture (IPCC, 2007). Extreme events are generally represented in an ad hoc manner using add-on damage functions based on averages of past impacts and contingent on gradual temperature increase (see comment in Nordhaus and Boyer, 2000). In a review of existing literature, Markandya and Watkiss (2009) identify the following types of analyses: investment and financial flows; impact assessments (scenario-based assessments); vulnerability assessments; adaptation assessments; risk management assessments; economic integrated assessment models; multi-criteria analysis; computable general equilibrium models; cost-benefit analysis; cost effectiveness analysis; and portfolio/real options analysis.

Global and regional assessments of adaptation costs, the focus of this section, have essentially used two approaches: (1) determining the pure *financial costs*, that is, outlays necessary for specific adaptation interventions (known as *investment and financial flow analyses*); and (2) *economic costs* involving estimating the wider overall costs and benefits to society and comparing this to mitigation, often using Integrated Assessment Models (IAMs). The IAM approach leads to a broader estimate of costs (and benefits) over long time scales, but requires detailed models of the economies under study (UNFCCC, 2007). One way of measuring the costs of adaptation involves first establishing a baseline development path (for a country or all countries) with no climate change, and then altering the baseline to take into account the impacts of climate change (World Bank, 2010). Then the potential effects of various adaptation strategies on development or growth can be examined. Adaptation cost estimates are based on various assumptions about the baseline scenario and the effectiveness of adaptation measures. The difference between these assumptions makes it very difficult to compare or aggregate results (Yohe et al., 1995, 1996; West et al., 2001).

An example illustrating methodological challenges comes from agriculture, where estimates have been made using various assumptions about adaptation behavior (Schneider et al., 2000). These assumptions about behavior range from the farmers who do not react to observed changes in climate conditions (especially in studies that use crop yield sensitivity to weather variability) (Deschênes and Greenstone, 2007; Lobell et al., 2008; Schlenker and Lobell, 2010), to the introduction of selected adaptation measures within crop yield models (Rosenzweig and Parry, 1994), to the assumption of 'perfect' adaptation – that is, farmers have complete or 'perfect' knowledge and apply that knowledge in ways that ensure outcomes align exactly with theoretical predictions (Kurukulasuriya and Mendelsohn, 2008a,b; Seo and Mendelsohn, 2008). Realistic assessments fall between these extremes, and a realistic representation of future adaptation patterns depends on the in-due-time detection of the climate change signal (Schneider et al., 2000; Hallegatte, 2009); the inertia in adoption of new technologies (Reilly and Schimmelpfennig, 2000); the existence of price signals (Fankhauser et al., 1999); and assessments of plausible behavior by farmers.

Cost-benefit analysis (CBA) is an established tool for determining the economic efficiency of development interventions. CBA compares the costs of conducting such projects with their benefits and calculates the net benefits or economic efficiency (Benson and Twigg, 2004). Ideally CBA accounts for all costs and benefits to society including environmental impacts, not just financial impacts on individual businesses. All costs and benefits are monetized so that tradeoffs can be compared with a common measure. The fact that intangibles and other items that are difficult to value are often left out is one of the major criticisms of the approach (Gowdy, 2007). In the case of disaster risk reduction (DRR) and adaptation interventions, CBA weighs the costs of the DRR project against the disaster damage costs avoided. While the benefits created by development interventions are the additional benefits due to, for example, improvements in physical or social infrastructure, in DRR the benefits are mostly the avoided or reduced potential damages and losses (Smyth et al., 2004). The net benefit can be calculated in terms of net present value, the rate of return, or the benefit-cost ratio. OECD countries such as the United Kingdom and the United States, as well as international financial institutions such as the World Bank, Asian Development Bank, and Inter-American Development Bank, have used CBA for evaluating disaster risk management (DRM) in the context of development assistance (Venton and Venton, 2004; Ghesquiere et al., 2006) and use it routinely for assessing engineering DRM strategies domestically. CBA can be, and has been, applied at any level from the global to local (see Kramer, 1995; Benson and Twigg, 2004; Venton and

Venton, 2004; UNFCCC, 2007; Mechler, 2008). Because the chance of occurrence of a disaster event can be expressed as a probability, it follows that the benefits of reducing the impact of that event can be expressed in probabilistic terms. Costs and benefits should be calculated by multiplying probability by consequences; this leads to risk estimates that account for hazard intensity and frequency, vulnerability, and exposure (Smyth et al., 2004; Ghesquiere et al., 2006).

National-level studies of adaptation effectiveness in the European Union, the United Kingdom, Finland, and The Netherlands, as well as in a larger number of developing countries using the National Adaptation Programme of Action approach, have been conducted or are underway (Ministry of Agriculture and Forestry, 2005; DEFRA, 2006; Lemmen et al., 2008; de Bruin et al., 2009; Parry et al., 2009). Yet the evidence base on the economic aspects including economic efficiency of adaptation remains limited and fragmented (Adger et al., 2007; Moench et al., 2007; Agrawala and Fankhauser, 2008; Parry et al., 2009). As noted at the start of Section 4.5.3.2, many adaptation studies focus on gradual change, especially for agriculture. Those studies considering extreme events, and finding or reporting net benefits over a number of key options (Agrawala and Fankhauser, 2008; Parry et al., 2009), do so by treating extreme events similarly to gradual onset phenomena and using deterministic impact metrics, which is problematic for disaster risk. A recent, risk-focused study (ECA, 2009) concentrating on national and sub-national levels went so far as to suggest an adaptation cost curve, which organizes relevant adaptation options around their cost-benefit ratios. However, given available data including future projections of risk and the effectiveness of options, this is probably at most heuristic rather than a basis for policy.

There are several complexities and uncertainties inherent in the estimates required for a CBA of DRR. As these are compounded by climate change, CBA's utility in evaluating adaptation may be reduced. These include difficulties in handling intangibles and, as is particularly important for extremes, in the discounting of future impacts; CBA does not account for the distribution of costs and benefits or the associated equity issues. Moench et al. (2007) argue that CBA is most useful as a decision support tool that helps the policymaker categorize, organize, assess, and present information on the costs and benefits of a potential project, rather than give a definite answer. Overall, the applicability of rigorous CBAs for evaluations of adaptation is thus limited based on *limited evidence* and *medium agreement*.

4.5.3.3. Attribution of Impacts to Climate Change: Observations and Limitations

Attribution of the impacts of climate change can be defined and used in a way that parallels the well-developed applications for the physical climate system (IPCC, 2010). Detection is the process of demonstrating that a system affected by climate has changed in some defined statistical sense, without providing a reason for that change. Attribution is the process of establishing the most probable causes, natural or anthropogenic, for the detected change with some defined level of confidence.

The IPCC Working Group II Fourth Assessment Report found, with *very high confidence*, that observational evidence shows that biological systems on all continents and in most oceans are already being affected by recent climate changes, particularly regional temperature increases (Rosenzweig et al., 2007).

Attribution of changes in individual weather and climate events to anthropogenic forcing is complicated because any such event might have occurred by chance in an unmodified climate as a result of natural climate variability (see FAQ 3.2). An approach that addresses this problem is to look at the likelihood of such an event occurring, rather than the occurrence of the event itself (Stone and Allen, 2005). For example, human-induced changes in mean temperature have been shown to increase the likelihood of extreme heat waves (Meehl and Tebaldi, 2004; Stott et al., 2004). For a large region of continental Europe, Stott et al. (2004) showed that anthropogenic climate change *very likely* doubled the probability of surpassing a mean summer temperature not exceeded since advent of the instrumental record in 1851, but which was by the 2003 event in Europe. More recent work provides further support for such a linkage (Barriopedro et al., 2011; see Section 3.3.1).

Most published studies on the attribution of impacts of extremes to natural and anthropogenic climate change have focused on long-term records of disaster losses, or examine the likelihood of the event occurring. Most published effort has gone into the analysis of long-term disaster loss records.

There is *high confidence*, based on *high agreement* and *medium evidence*, that economic losses from weather- and climate-related disasters have increased (Cutter and Emrich, 2005; Peduzzi et al., 2009, 2011; UNISDR, 2009; Mechler and Kundzewicz, 2010; Swiss Re 2010; Munich Re, 2011). A key question concerns whether trends in such losses, or losses from specific events, can be attributed to climate change. In this context, changes in losses over time need to be controlled for exposure and vulnerability. Most studies of long-term disaster loss records attribute these increases in losses to increasing exposure of people and assets in at-risk areas (Miller et al., 2008; Bouwer, 2011), and to underlying societal trends – demographic, economic, political, and social – that shape vulnerability to impacts (Pielke Jr. et al., 2005; Bouwer et al., 2007). Some authors suggest that a (natural or anthropogenic) climate change signal can be found in the records of disaster losses (e.g., Mills, 2005; Höppe and Grimm, 2009), but their work is in the nature of reviews and commentary rather than empirical research. Attempts have been made to normalize loss records for changes in exposure and wealth. There is *medium evidence* and *high agreement* that long-term trends in normalized losses have not been attributed to natural or anthropogenic climate change (Choi and Fisher, 2003; Crompton and McAneney, 2008; Miller et al., 2008; Neumayer and Barthel, 2011). The evidence is *medium* because of the issues set out toward the end of this section.

The statement about the absence of trends in impacts attributable to natural or anthropogenic climate change holds for tropical and extratropical storms and tornados (Boruff et al., 2003; Pielke Jr. et al., 2003, 2008; Raghavan and Rajesh, 2003; Miller et al 2008; Schmidt et al., 2009; Zhang et al., 2009; see also Box 4-2). Most studies related increases found in normalized hurricane losses in the United States since the 1970s (Miller et al., 2008; Schmidt et al., 2009; Nordhaus, 2010) to the natural variability observed since that time (Miller et al., 2008; Pielke Jr. et al., 2008). Bouwer and Botzen (2011) demonstrated that other normalized records of total economic and insured losses for the same series of hurricanes exhibit no significant trends in losses since 1900.

The absence of an attributable climate change signal in losses also holds for flood losses (Pielke Jr. and Downton, 2000; Downton et al., 2005; Barredo, 2009; Hilker et al., 2009), although some studies did find recent increases in flood losses related in part to changes in intense rainfall events (Fengqing et al., 2005; Chang et al., 2009). For precipitation-related events (intense rainfall, hail, and flash floods), the picture is more diverse. Some studies suggest an increase in damages related to a changing incidence in extreme precipitation (Changnon, 2001, 2009), although no trends were found for normalized losses from flash floods and landslides in Switzerland (Hilker et al., 2009). Similarly, a study of normalized damages from bushfires in Australia also shows that increases are due to increasing exposure and wealth (Crompton et al., 2010).

Increasing exposure of people and economic assets has been the major cause of long-term increases in economic losses from weather- and climate-related disasters (*high confidence*). The attribution of economic disaster losses is subject to a number of limitations in studies to date: data availability (most data are available for standard economic sectors in developed countries); type of hazards studied (most studies focus on cyclones, where confidence in observed trends and attribution of changes to human influence is *low*; Section 3.4.4); and the processes used to normalize loss data over time. Different studies use different approaches to normalization, and most normalization approaches take account of changes in exposure of people and assets, but use only limited, if any, measures of vulnerability trends, which is questionable. Different approaches are also used to handle variations in the quality and completeness of data on impacts over time. Finding a trend or 'signal' in a system characterized by large variability or 'noise' is difficult and requires lengthy records. These are all areas of potential weakness in the methods and conclusions of longitudinal loss studies and more empirical and conceptual efforts are needed. Nevertheless, the results of the studies mentioned above are strengthened as they show similar results, although they have applied different data sets and methodologies.

A general area of uncertainty in the studies concerns the impacts of weather and climate events on the livelihoods and people of informal settlements and economic sectors, especially in developing countries. Some one billion people live in informal settlements (UNISDR, 2011), and over half the economy in some developing countries is informal (Schneider et al., 2010). These impacts have not been systematically documented, with the result that they are largely excluded from both longitudinal impact analysis and attribution to defined weather episodes.

Another general area of uncertainty comes from confounding factors that can be identified but are difficult to quantify, and relates to the usual assumption of constant vulnerability in studies of loss trends. These include factors that would be expected to increase resilience (Chapters 2 and 5 of this report) and thereby mask the influence of climate change, and those that could act to increase the impact of climate change. Those that could mask the effects of change include gradual improvements in warnings and emergency management (Adger et al., 2005), building regulations (Crichton, 2007), and changing lifestyles (such as the use of air conditioning), and the almost instant media coverage of any major weather extreme that may help reduce losses. In the other direction are changes that may be increasing risk, such as the movement of people in many countries to coastal areas prone to cyclones (Pompe and Rinehart, 2008) and sea level rise.

4.5.4. Assessment of Impact Costs

Much work has been conducted on the analysis of direct economic losses from natural disasters. The examples mentioned below mainly focus on national and regional economic losses from particular climate extremes and disasters, and also discuss uncertainty issues related to the assessment of economic impacts.

4.5.4.1. Estimates of Global and Regional Costs of Disasters

Observed trends in extreme impacts: Data on global weather- and climate-related disaster losses reported since the 1960s reflect mainly monetized direct damages to assets, and are unequally distributed. Estimates of annual losses have ranged since 1980 from a few US$ billion to above 200 billion (in 2010 dollars) for 2005 (the year of Hurricane Katrina) (UNISDR, 2009; Swiss Re 2010; Munich Re, 2011). These estimates do not include indirect and intangible losses.

On a global scale, annual material damage from large weather and climate events has been found to have increased eight-fold between the 1960s and the 1990s, while the insured damage has been found to have increased by 17-fold in the same interval, in inflation-adjusted monetary units (Mechler and Kundzewicz, 2010). Between 1980 and 2004, the total costs of extreme weather events totaled US$ 1.4 trillion, of which only one-quarter was insured (Mills, 2005). Material damages caused by natural disasters, mostly weather- and water-related, have increased more rapidly than population or economic growth, so that these factors alone may not fully explain the observed increase in damage. The loss of life has been brought down considerably (Mills, 2005; UNISDR, 2011).

Developing regions are vulnerable both because of exposure to weather- and climate-related extremes and their status as developing economies. However, disaster impacts are unevenly distributed by type of disaster, region, country, and the exposure and vulnerability of different communities and sectors.

Percentage of direct economic losses by regions: The concentration of information on disaster risk generally is skewed toward developed countries and the Northern Hemisphere (World Bank and UN, 2010). Some global databases, however, do allow a regional breakdown of disaster impacts. The unequal distribution of the human impact of natural disasters is reflected in the number of disasters and losses across regions (Figure 4-7). In the period 2000 to 2008, Asia experienced the highest number of weather- and climate-related disasters. The Americas suffered the most economic loss, accounting for the highest proportion (54.6%) of total loss, followed by Asia (27.5%) and Europe (15.9%). Africa accounted for only 0.6% of global economic losses, but economic damages from natural disasters are underreported in these data compared to other regions (Vos et al., 2010). Although reporting biases exist, they are judged to provide *robust evidence* of the regional distribution of the number of disasters and of direct economic losses for this recent period 2000 to 2008, and there is *high agreement* regarding this distribution among different databases collected by independent organizations (Guha-Sapir et al., 2011; Munich Re, 2011; Swiss Re, 2011).

Damage losses in percentage of GDP by regions: The relative economic burden in terms of direct loss expressed as a percentage of GDP has been substantially higher for developing states. Middle-income countries with rapidly expanding asset bases have borne the largest burden, where during the period from 2001 to 2006 losses amounted to about 1% of GDP, while this ratio has been about 0.3% of GDP for low-income countries and less than 0.1% of GDP for high-income countries, based on *limited evidence* (Cummins and Mahul, 2009). In small exposed countries, particularly small island developing states, these wealth losses expressed as a percentage of GDP and averaged over both disaster and non-disaster years can be considerably higher, exceeding 1% in many cases and 8% in the most extreme cases over the period from 1970 to 2010 (World Bank and UN, 2010), and individual events may consume more than the annual GDP (McKenzie et al., 2005). This indicates a far higher vulnerability of the economic infrastructure in developing countries (Cavallo and Noy 2009; UNISDR, 2009).

Increasing weather- and climate-related disasters: The number of reported weather- and climate-related disasters and their direct financial costs have increased over the past decades. Figure 4-8 illustrates an increasing trend (coupled with large interannual variability) in losses based on data for large weather-and climate-related disasters over the period 1980 to 2010, for which data have been gathered consistently and systematically (see Neumayer and Barthel, 2011).

This increase in affected population and direct economic losses is also coupled with the increasing numbers of reported weather- and climate-related disasters (UNISDR, 2009; Munich Re, 2011; Swiss Re 2011).

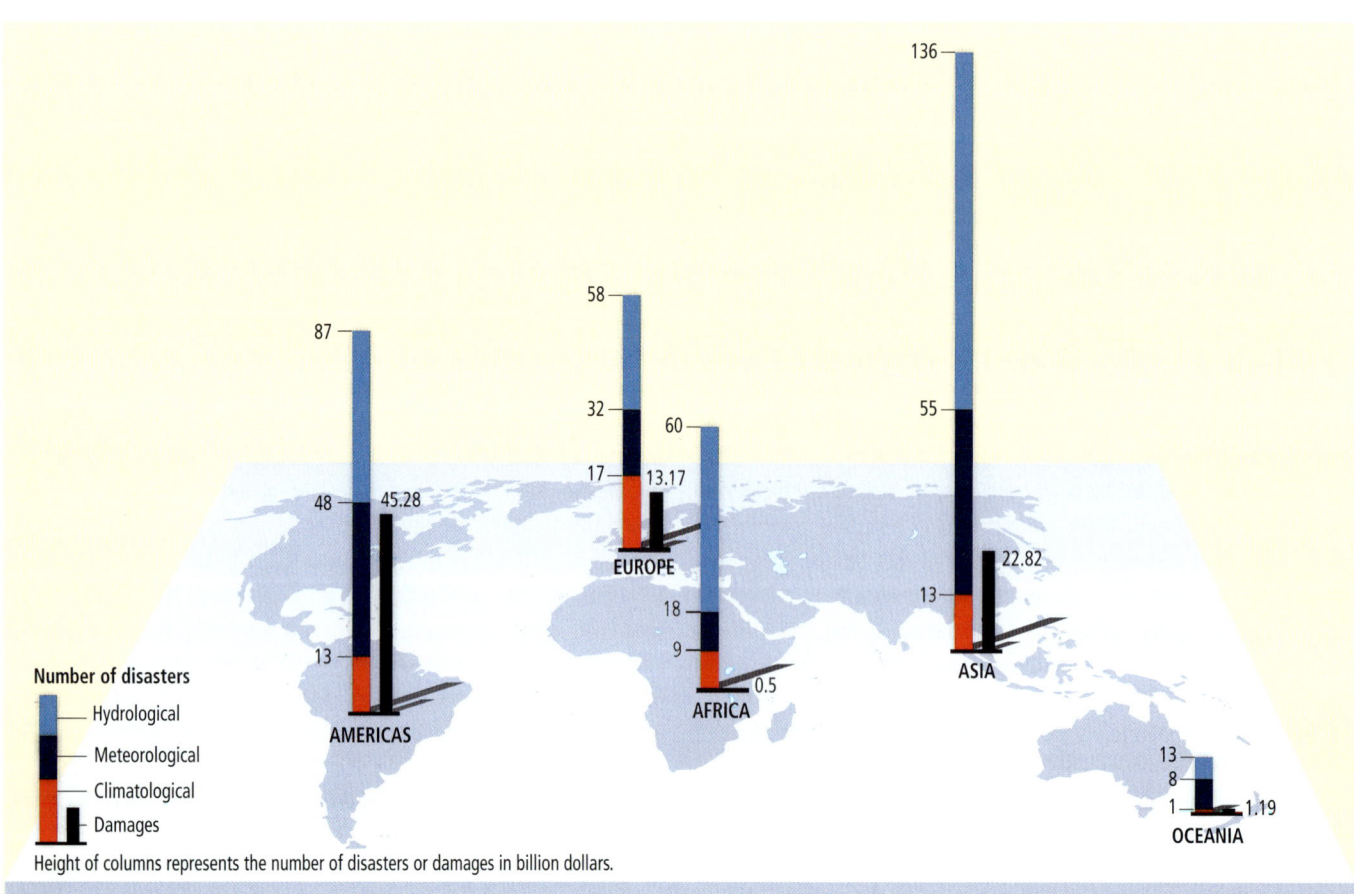

Figure 4-7 | Weather- and climate-related disaster occurrence and regional average impacts from 2000 to 2008. The number of climatological (e.g., extreme temperature, drought, wildfire), meteorological (e.g., storm), and hydrological (e.g., flood, landslides) disasters is given for each region, along with damages (2009 US$ billion). Data from Vos et al., 2010.

These statistics imply the increasing cost of such disasters to society, regardless of cause. It is also important to note that the number of weather- and climate-related disasters has increased more rapidly than losses from non-weather disasters (Mills, 2005; Munich Re, 2011; Swiss Re, 2011). This could indicate a change in climate extremes, but there are other possible explanations (Bouwer, 2011). Drought and flood losses may have grown due to a number of non-climatic factors, such as increasing water withdrawals effectively exacerbating the impact of droughts, decrease in storage capacity in catchments (urbanization, deforestation, sealing surfaces, channelization) adversely affecting both flood and drought preparedness, increase in runoff coefficients, and growing settlements in floodplains around urban areas (see Section 4.2.2; Field et al., 2009).

4.5.4.2. Potential Trends in Key Extreme Impacts

As indicated in Sections 3.3 to 3.5 and Tables 3-1 and 3-3, climate extremes may have different trends in the future; some such as heat waves are projected to increase over most areas in length, frequency, and intensity, while projected changes in some other extremes are given with less confidence. However, uncertainty is a key aspect of disaster/climate change trend analysis due to attribution issues discussed above, incomparability of methods, changes in exposure and vulnerability over time, and other non-climatic factors such as mitigation and adaptation. A challenge is ensuring that the projections of losses from future changes in extreme events are examined not for current populations and economies, but for scenarios of possible future socioeconomic development. See Box 4-2 for a discussion of this with respect to cyclones.

It is *more likely than not* that the frequency of the most intense tropical cyclones will increase substantially in some ocean basins (Section 3.4.4). Many studies have investigated impacts from tropical cyclones (e.g., ABI, 2005a, 2009; Hallegatte, 2007; Pielke Jr., 2007; Narita et al., 2009; Bender et al., 2010; Nordhaus, 2010; Crompton et al., 2011). Table 4-3 presents the projected percentage increase in direct economic losses from tropical cyclones from a number of these studies, scaled to the year 2040 relative to a common baseline (year 2000). There is *high confidence* that increases in exposure will result in higher direct economic losses from tropical cyclones and that losses will also depend on future changes in tropical cyclone frequency and intensity. One study, building on global climate model results from Bender et al. (2010), found that to attribute increased losses to increased tropical cyclone activity in the United States with a high degree of certainty would take another 260 years of records, due to the high natural variability of storms and their impacts

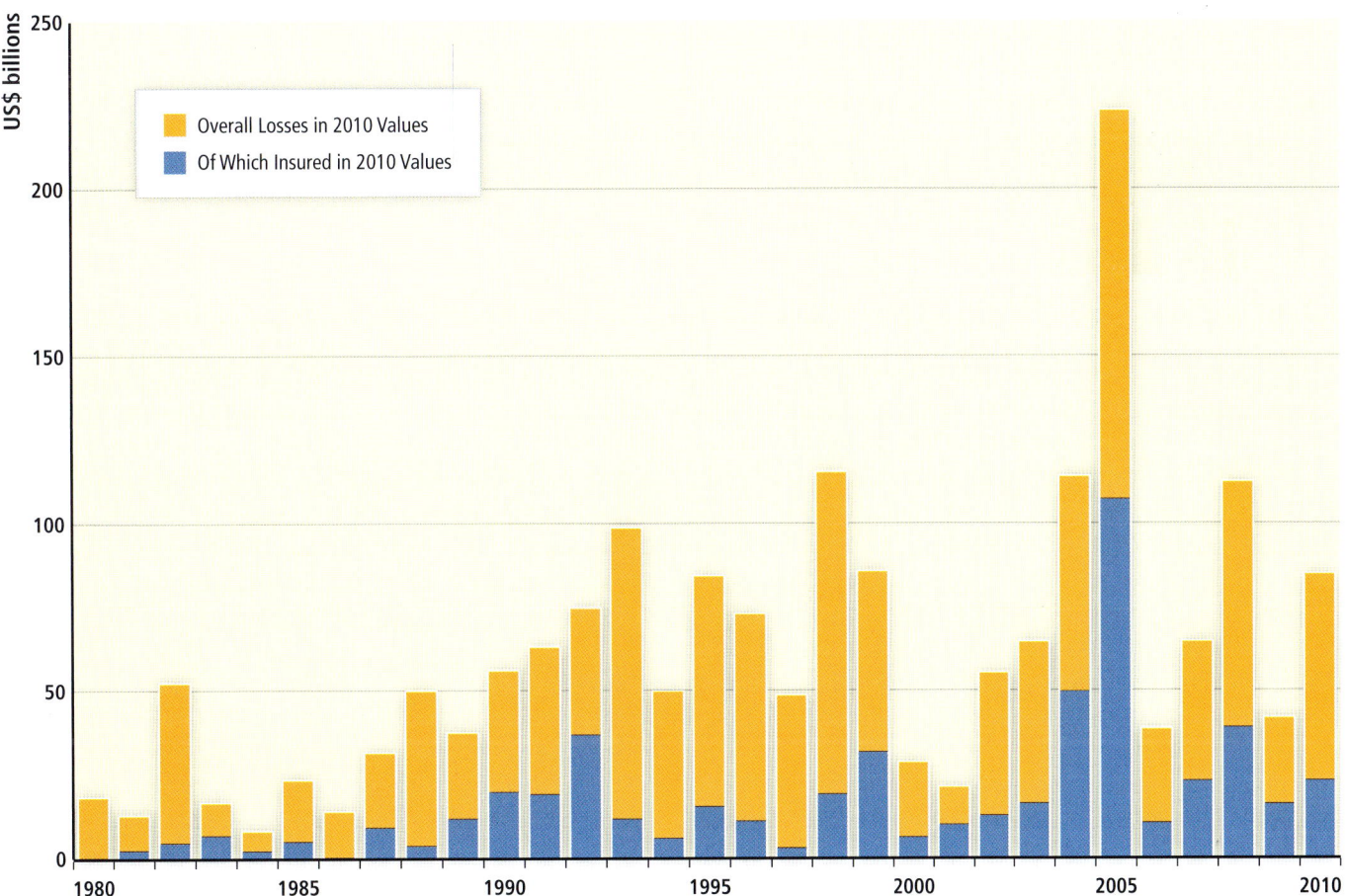

Figure 4-8 | The overall losses and insured losses from weather- and climate-related disasters worldwide (in 2010 US$). These data for weather- and climate-related 'great' and 'devastating' natural catastrophes are plotted without inclusion of losses from geophysical events. A catastrophe in this data set is considered 'great' if the number of fatalities exceeds 2,000, the number of homeless exceeds 200,000, the country's GDP is severely hit, and/or the country is dependent on international aid. A catastrophe is considered 'devastating' if the number of fatalities exceeds 500 and/or the overall loss exceeds US$ 650 million (in 2010 values). Data from Munich Re, 2011.

Table 4-3 | Estimated change in disaster losses in 2040 under projected climate change and exposure change, relative to 2000, from 21 impact studies including median estimates by type of weather hazard. Source: Bouwer, 2010.

A. Impact of projected climate change

Study	Hazard type	Region	Estimated loss change [%] in 2040			Median
			Min	Max	Mean	
Pielke (2007)	Tropical storm	Atlantic	58	1,365	417	
Nordhaus (2010)	Tropical storm	United States	12	92	47	
Narita et al. (2009)	Tropical storm	Global	23	130	46	
Hallegatte (2007)	Tropical storm	United States	-	-	22	
ABI (2005a,b)	Tropical storm	United States, Caribbean	19	46	32	30
ABI (2005a,b)	Tropical storm	Japan	20	45	30	
ABI (2009)	Tropical storm	China	9	19	14	
Schmidt et al. (2009)	Tropical storm	United States	-	-	9	
Bender et al. (2010)	Tropical storm	United States	-27	36	14	
Narita et al. (2010)	Extra-tropical storm	High latitude	-11	62	22	
Schwierz et al. (2010)	Extra-tropical storm	Europe	6	25	16	
Leckebusch et al. (2007)	Extra-tropical storm	United Kingdom, Germany	-6	32	11	15
ABI (2005a,b)	Extra-tropical storm	Europe	-	-	14	
ABI (2009)	Extra-tropical storm	United Kingdom	-33	67	15	
Dorland et al. (1999)	Extra-tropical storm	Netherlands	80	160	120	
Bouwer et al. (2010)	River flooding	Netherlands	46	201	124	
Feyen et al. (2009)	River flooding	Europe	-	-	83	
ABI (2009)	River flooding	United Kingdom	3	11	7	65
Feyen et al. (2009)	River flooding	Spain (Madrid)	-	-	36	
Schreider et al. (2000)	Local flooding	Australia	67	514	361	
Hoes (2007)	Local flooding	Netherlands	16	70	47	

B. Impact of projected exposure change

Study	Hazard type	Region	Estimated loss change [%] in 2040			Median
			Min	Max	Mean	
Pielke (2007)	Tropical storm	Atlantic	164	545	355	
Schmidt et al. (2009)	Tropical storm	United States	-	-	240	
Dorland et al. (1999)	Extra-tropical storm	Netherlands	12	93	50	172
Bouwer et al. (2010)	River flooding	Netherlands	35	172	104	
Feyen et al. (2009)	River flooding	Spain (Mad)	-	-	349	
Hoes (2007)	Local flooding	Netherlands	-4	72	29	

(Crompton et al., 2011). See Section 4.5.3.3 on attribution and the use of a risk-based approach to cope with this issue. Other studies have investigated impacts from increases in the frequency and intensity of extratropical cyclones at high latitudes (Dorland et al., 1999; ABI, 2005a, 2009; Narita et al., 2010; Schwierz et al., 2010; Donat et al., 2011). In general there is *medium confidence* that increases in losses due to extratropical cyclones will occur with climate change, with possible decreases or no change in some areas. Projected increases generally are slightly lower than increases in tropical cyclone losses (see Table 4-3). Patt et al. (2010) projected future losses due to weather- and climate-related extremes in least-developed countries.

Many studies have addressed future economic losses from river floods, most of which are focused on Europe, including the United Kingdom (Hall et al., 2003, 2005; ABI, 2009), Spain (Feyen et al., 2009), and The Netherlands (Bouwer et al., 2010) (see Table 4-3). Maaskant et al. (2009) is one of the few studies that addresses future loss of life from flooding, and projects up to a four-fold increase in potential flood victims in The Netherlands by the year 2040, when population growth is accounted for. Some studies are available on future coastal flood risks (Hall et al., 2005; Mokrech et al., 2008; Nicholls et al., 2008; Dawson et al., 2009; Hallegatte et al., 2010). Although future flood losses in many locations will increase in the absence of additional protection measures (*high agreement, medium evidence*), the size of the estimated change is highly variable, depending on location, climate scenarios used, and methods used to assess impacts on river flow and flood occurrence (see Table 4-3 for a comparison of some regional studies) (Bouwer, 2010).

Some studies have addressed economic losses from other types of weather extremes, often smaller-scale compared to river floods and cyclones. These include hail damage, for which mixed results are found: McMaster (1999) and Niall and Walsh (2005) found no significant effect on hailstorm losses for Australia, while Botzen et al. (2010) find a significant increase (up to 200% by 2050) for damages in the agricultural sector in The Netherlands, although the approaches used vary considerably. Rosenzweig et al. (2002) report on a possible doubling of losses to crops due to excess soil moisture caused by more intense rainfall. Hoes (2007), Hoes and Schuurmans (2006), and Hoes et al.

(2005) estimated increases in damages due to extreme rainfall in The Netherlands by mid-century.

It is well known that the frequency and intensity of extreme weather and climate events are only one factor that affects risks, as changes in population, exposure of people and assets, and vulnerability determine loss potentials (see Sections 4.2 to 4.4). Few studies have specifically quantified these factors. However, the ones that do generally underline the important role of projected changes (increases) in population and capital at risk. Some studies indicate that the expected changes in exposure are much larger than the effects of climate change (see Table 4-3), which is particularly true for tropical and extratropical storms (Pielke Jr., 2007; Feyen et al., 2009; Schmidt et al., 2009). Other studies show that the effect of increasing exposure is about as large as the effect of climate change (Hall et al., 2003; Maaskant et al., 2009; Bouwer et al., 2010), or estimate that these are generally smaller (Dorland et al., 1999; Hoes, 2007). There is therefore *medium confidence* that, for some climate extremes in many regions, the main driver for future increasing losses in many regions will be socioeconomic in nature (based on *medium agreement* and *limited evidence*). Finally, many studies underline that both factors need to be taken into account, as the factors do in fact amplify each other, and therefore need to be studied jointly when expected losses from climate change are concerned (Hall et al., 2003; Bouwer et al., 2007, 2010; Pielke Jr., 2007; Feyen et al., 2009).

4.5.5. Assessment of Adaptation Costs

The World Bank (2006) estimated the cost of climate-proofing foreign direct investments, gross domestic investments, and Official Development Assistance, which was taken up and modified by Stern (2007), Oxfam (2007), and UNDP (2007). The second source of adaptation cost estimates is UNFCCC (2007), which calculated the value of existing and planned investment and financial flows required for the international community to effectively and appropriately respond to climate change impacts. The third source is World Bank (2010), which also conducted a number of country-level studies to complement the global assessment, following UNFCCC (2007), but aimed at improving upon this by assessing the climate-proofing of existing and new infrastructure, using more precise unit cost estimates and including the costs of maintenance as well as those of port upgrading and the risks from sea level rise and storm surges. Also, the investment in education necessary to neutralize impacts of extreme weather is calculated. Estimates of costs to adapt to climate change (rather than simply to extremes and disasters), which have mostly been made for developing countries, exhibit a large range and relate to different assessment periods (such as today, 2015, or 2030). For 2030, the estimated global cost from UNFCCC (2007) ranges from US$ 48 to 171 billion per year for developed and developing countries, and US$ 28 to 67 billion per year for developing countries (in 2005 dollars). Recent estimates from World Bank (2010) for developing countries lead to higher projected costs and broadly amount to the average of this range with annual costs of up to US$ 100 billion (in 2005 US$) (see Table 4-4). Confidence in individual global estimates is *low* because, as mentioned above and discussed by Parry et al. (2009), the estimates are derived from only three relatively independent studies, which explains the seeming convergence of the estimates in latter studies. As well, Parry et al. (2009) consider the estimates a significant underestimation by at least a factor of two to three and possibly more if the costs incurred by other sectors were included, such as ecosystem services, energy, manufacturing, retailing, and tourism. The adaptation cost estimates are also based mostly on low levels of investment due to an existing adaptation deficit in many regions. Unavoidable residual damages remain absent from these analyses.

In terms of regional costs and as reported in the World Bank (2010) study, the largest absolute adaptation costs would arise in East Asia and the Pacific, followed by the Latin American and Caribbean region as well as sub-Saharan Africa. This pattern held for the two scenarios assessed

Table 4-4 | Estimates of global costs of adaptation to climate change. Source: Extended based on Agrawala and Fankhauser (2008) and Parry et al. (2009).

Study	Results (billion US$ yr^{-1})	Time Frame and Coverage	Sectors	Methodology and Comment
World Bank (2006)	9–41[1]	Present, developing countries	Unspecified	Cost of climate-proofing foreign direct investments, gross domestic investments, and Official Development Assistance
Stern (2007)	4–37[1]	Present, developing countries	Unspecified	Update of World Bank (2006)
Oxfam (2007)	>50[1]	Present, developing countries	Unspecified	World Bank (2006) plus extrapolation of cost estimates from National Adaptation Programmes of Action and NGO projects
UNDP (2007)	86–109[2]	In 2015, developing countries	Unspecified	World Bank (2006) plus costing of targets for adapting poverty reduction programs and strengthening disaster response systems
UNFCCC (2007)	48–171 (28–67 for developing countries)[2]	In 2030, developed and developing countries	Agriculture, forestry, and fisheries; water supply; human health; coastal zones; infrastructure; ecosystems (but no estimate for 2030 for ecosystem adaptation)	Additional investment and financial flows needed for adaptation in 2030
World Bank (2010)	70–100[2]	Annual from 2010 to 2050, developing countries	Agriculture, forestry, and fisheries; water supply and flood protection; human health; coastal zones; infrastructure; extreme weather events	Impact costs linked to adaptation costs, improvement upon UNFCCC (2007): climate-proofing existing and new infrastructure, more precise unit cost, inclusion of cost of maintenance and port upgrading, risks from sea level rise and storm surges, riverine flood protection, education investment to neutralize impacts of extreme weather events

Notes: 1. in 2000 US$; 2. in 2005 US$.

Table 4-5 | Range of regionalized annual costs of adaptation for wet and dry scenarios (in 2005 US$ billion). Reflecting the full range of estimated costs, the wet scenario costs do not include benefits from climate change while the dry scenario costs include benefits from climate change within and across countries. Source: World Bank, 2010.

Scenario/Region	East Asia & Pacific	Europe & Central Asia	Latin America & Caribbean	Middle East & North America	South Asia	sub-Saharan Africa	Total
Wet	25.7	12.6	21.3	3.6	17.1	17.1	97.5
Dry	17.7	6.5	14.5	2.4	14.6	13.8	69.6

in the study, which were a scenario with the most precipitation ('wet') and one with the least precipitation ('dry') among all scenarios chosen for the study, which employ socioeconomic driver information from IPCC's SRES A2 scenario (see Table 4-5).

Taking Africa as an example, based on various estimates the potential additional costs of adaptation investment range from US$ 3 to 10 billion per year by 2030 (UNFCCC, 2007; PACJA, 2009). However, this could be also an underestimate considering the desirability of improving Africa's resilience to climate extremes as well as the flows of international humanitarian aid in the aftermath of disasters.

4.5.6. Uncertainty in Assessing the Economic Costs of Extremes and Disasters

Upon reviewing the estimates to date, the costing of weather- and climate-related disasters and estimating adaptation costs is still preliminary, incomplete, and subject to a number of assumptions with the result that there is considerable uncertainty (Agrawala and Fankhauser, 2008; Parry et al., 2009). This is largely due to modeling uncertainties in climate change and damage estimates, limited data availability, and methodological shortcomings in analyzing disaster damage statistics. Such costing is further limited by the interaction between numerous adaptation options and assumptions about future exposure and vulnerabilities, social preferences, and technology, as well as levels of resilience in specific societies. Additionally the following challenges can be identified.

Risk assessment methods: Technical challenges remain in developing robust risk assessment and damage costing methods. Study results can vary significantly between top-down and bottom-up approaches. Risk-based approaches are utilized for assessing and projecting disaster risk (Jones, 2004; Carter et al., 2007), for which input from both climate and social scenarios is required. All climatic phenomena are subject to the limitation that historically based relationships between damages and disasters cannot be used with confidence to deduce future risk of extreme events under changing characteristics of frequency and intensity (UNDP, 2004). Yet climate models are today challenged when reproducing spatially explicit climate extremes, due to coarse resolution and physical understanding of the relevant process, as well as challenges in modeling low-probability, high-impact events (see Section 3.2.3). Therefore, projections of future extreme event risk involve uncertainties that can limit understanding of sudden onset risk, such as flood risk. Future socioeconomic development is also inherently uncertain. A uniform set of assumptions can help to provide a coherent global picture and comparison and extrapolation between regions.

Data availability and consistency: Lack of data and robust information increases the uncertainty of costing when scaling up to global levels from a very limited (and often very local) evidence base. There are double-counting problems and issues of incompatibility between types of impacts in the process of multi-sectoral and cross-scale analyses, especially for the efforts to add both market and non-market values (e.g., ecosystem services) (Downton and Pielke Jr., 2005; Pielke Jr. et al., 2008; Parry et al., 2009). Moreover the full impacts of weather- and climate-related extremes in developing countries are not fully understood, and a lack of comprehensive studies on damage, adaptation, and residual costs indicates that the full costs are underestimated.

Information on future vulnerability: Apart from climate change, vulnerability and exposure will also change over time, and the interaction of these aspects should be considered (see, e.g., Hallegatte, 2008; Hochrainer and Mechler, 2011). This has been recognized and assessments of climate change impacts, vulnerability, and risk are changing in focus, leading to more integration across questions. While initial studies focused on an analysis of the problem, the field proceeded to assess potential impacts and risks, and now more recently started to combine such assessments with the consideration of specific risk management methods (Carter et al., 2007).

Some studies have suggested incorporating an analysis of the ongoing or chronic economic impact of disasters into the adaptation planning process (Freeman, 2000). A fuller assessment of disaster cost at varying spatial and temporal scales and costs related to impacts on human, social, built, and natural capital, and their associated services at different levels can set the stage for comparisons of post-disaster development strategies. This would make disaster risk reduction planning and preparedness investment more cost-effective (Gaddis et al., 2007). For example, there is consensus on the important role of ecosystems in risk reduction and well-being, which would make the value of ecosystem services an integral part of key policy decisions associated with adaptation (Tallis and Kareiva, 2006; Costanza and Farley, 2007).

References

A digital library of non-journal-based literature cited in this chapter that may not be readily available to the public has been compiled as part of the IPCC review and drafting process, and can be accessed via either the IPCC Secretariat or IPCC Working Group II web sites.

Abegg, B., S. Agrawala, F. Crick, and A. Montfalcon, 2007: Climate change impacts and adaptation in winter tourism. In: *Climate Change in the European Alps: Adapting Winter Tourism and Natural Hazards Management* [S. Agrawala (ed.)]. Organisation for Economic Co-operation and Development, Paris, France, pp. 25-60.

ABI, 2005a: *Financial Risks of Climate Change: Summary Report*. Association of British Insurers, London, UK, 129 pp.

ABI, 2005b: *Financial Risks of Climate Change: Technical Annexes*. Climate Risk Management Ltd. for the Association of British Insurers, London, UK, 125 pp.

ABI, 2009. *The Financial Risk of Climate Change*. Research Paper No. 19, Association of British Insurers, London, UK, 107 pp.

Abraham, J., F. Bendimerad, A. Berger, A. Boissonnade, O. Collignon, E. Couchmann, F. Grandjean, S. McKay, C. Miller, C. Mortgat, R. Muir-Wood, B. Page, T. Shah, S. Smith, P. Wiart, and Y. Xien, 2000: *Windstorms Lothar and Martin. December 26-28, 1999*. Risk Management Solutions, Newark, CA, 20 pp.

ABS, 2010: *Regional Population Growth, Australia, 2008-09*. Australian Bureau of Statistics, Canberra, Australia, abs.gov.au/ausstats/abs@.nsf/Products/3218.0~2008-09~Main+Features~Main+Features?OpenDocument#PARALINK5.

ACIA, 2004: *Impacts of a Warming Arctic*. Overview Report of the Arctic Climate Impact Assessment, Cambridge University Press, Cambridge, UK, 139 pp.

Adamo, S.B., 2010: Environmental migration and cities in the context of global environmental change. *Current Opinion in Environmental Sustainability*, **2(3)**, 161-165, doi:10.1016/j.cosust.2010.06.005.

Adger, W.N., 2000. Social and ecological resilience: Are they related? *Progress in Human Geography*, **24**, 347-364.

Adger, W.N., 2006: Vulnerability. *Global Environmental Change*, **16(3)**, 268-281.

Adger, W.N., T.A. Benjaminsen, K. Brown, and H. Svarstad, 2001: Advancing a political ecology of global environmental discourses. *Development and Change*, **32(4)**, 681-715. doi:10.1111/1467-7660.00222.

Adger, W.N., T.P. Hughes, C. Folke, S.R. Carpenter, and J. Rockstrom, 2005: Social-ecological resilience to coastal disasters. *Science*, **309(5737)**, 1036-1039.

Adger, W.N., S. Agrawala, M.M.Q. Mirza, C. Conde, K. O'Brien, J. Pulhin, R. Pulwarty, B. Smit, and K. Takahashi, 2007: Assessment of adaptation practices, options, constraints and capacity. In: *Climate Change 2007: Impacts, Adaptation and Vulnerability. Contribution of Working Group II to the Fourth Assessment Report of the Intergovernmental Panel on Climate Change* [Parry, M.L., O.F. Canziani, J.P. Palutikof, P.J. van der Linden and C.E. Hanson (eds.)]. Cambridge University Press, Cambridge, UK, pp. 717-743.

Aggarwal, P.K. and A.K. Singh, 2010: Implications of global climatic change on water and food security. In: *Global Change: Impacts on Water and Food Security* [Ringler, C., A.K. Biswas, and S. Kline (eds.)]. Springer-Verlag, Berlin and Heidelberg, Germany, pp. 49-63.

Agrawala, S. and S. Fankhauser (eds.), 2008: *Economic Aspects of Adaptation to Climate Change: Costs, Benefits and Policy Instruments*. Organisation for Economic Co-operation and Development, Paris, France, 133 pp.

Albala-Bertrand, J., 1993: *Political Economy of Large Natural Disasters*. Oxford University Press, New York, NY, 272 pp.

Alexander, D., 2005: Vulnerability to landslides. In: *Landslide Hazards and Risk* [Glade, T., M. Anderson, and M. Crozier (eds.)]. Wiley & Sons, Sussex, UK, pp.175-198.

Allen, C.D. and D.D. Breshears, 1998: Drought-induced shift of a forest–woodland ecotone: Rapid landscape response to climate variation. *Proceedings of the National Academy of Sciences*, **95(25)**, 14839 -14842.

Alpa, B., 2009: Vulnerability of Turkish coasts to accelerated sea-level rise. *Geomorphology*, **107**, 58-63.

AMCEN/UNEP, 2002: Africa environment outlook: past, present and future perspectives. United Nations Environmental Programme and the African Ministerial Conference on the Environment, Earthprint, Stevenage, UK, 422 pp.

Amien, I., P. Rejekiningrum, A. Pramudia, and E. Susanti, 1996: Effects of interannual climate variability and climate change on rice yields in Java, Indonesia. *Water, Air &Soil Pollution*, **92**, 29-39.

Anderson, M.G., L. Holcombe, and J.-P. Renaud, 2007: Assessing slope stability in unplanned settlements in developing countries. *Journal of Environmental Management*, **85(1)**, 101-111.

Anisimov, O.A. and M.A. Belolutskaya, 2002: Assessment of the impact of climate change and permafrost degradation on infrastructure in northern regions of Russia. *Meteorology and Hydrology*, **6**, 15-22.

Anisimov, O.A. and S.A. Lavrov, 2004: Global warming and permafrost melting: assessment of risks for energy-sector industrial facilities. *Tekhnologii TEK*, **3**, 78-83.

Anisimov, O.A., A.A. Velichko, P.F. Demchenko, A.V. Yeliseyev, I.I. Mokhov, and V.P. Nechaev, 2004: Impact of climate change on permafrost in the past, present and future. *Physics of Atmosphere and Oceans*, **38(1)**, 25-39.

Anyamba, A., J-P. Chretien, J. Small, C.J. Tucker, and K.J. Linthicum, 2006: Developing global climate anomalies suggest potential disease risks for 2006-2007. *International Journal of Health Geographics*, **5(60)**, doi:10.1186/1476-072X-5-60.

Arianoutsou, M., S. Koukoulas, and D. Kazanis, 2011: Evaluating post-fire forest resilience using GIS and multi-criteria analysis: an example from Cape Sounion National Park, Greece. *Journal of Environmental Management*, **47**, 384-397.

Arkell, B.P. and G.J.C. Darch, 2006: Impact of climate change on London's transport network. *Proceedings of the Institute of Civil Engineers – Municipal Engineer*, **159(4)**, 231-237.

Arnone, III, J.A., P.S.J. Verburg, D.W. Johnson, J.D. Larsen, R.L. Jasoni, L.J. Annmarie, C.M. Batts, C. von Nagy, W.G. Coulombe, D.E. Schorran, P.E. Buck, B.H. Braswell, J.S. Coleman, R.A. Sherry, L.L. Wallace, Y. Luo, and D.S. Schimel, 2008: Prolonged suppression of ecosystem carbon dioxide uptake after an anomalously warm year. *Nature*, **455**, 383-386.

AusAID, 2005: *Economic Impact of Natural Disasters on Development in the Pacific*. Volume 1: Research Report, Australian Agency for International Development, Canberra, Australia, 92 pp., ausaid.gov.au/publications/pdf/impact_pacific_report.pdf.

Baccini, M., T. Kosatsky, A. Analitis, H.R. Anderson, M. D'Ovidio, B. Menne, P. Michelozzi, A. Biggeri, and the PHEWE Collaborative Group, 2011: Impact of heat on mortality in 15 European cities: attributable deaths under different weather scenarios. *Journal of Epidemiology and Community Health*,**65(1)**, 64-70, doi:10.1136/jech.2008.085639.

Banks, S., M. Blyton, D. Blair, L. McBurney, and D. Lindenmayer, 2011: Adaptive responses and disruptive effects: how major wildfire influences kinship-based social interactions in a forest marsupial. *Molecular Ecology*, doi:10.1111/j.1365-294X.2011.05282.x.

Barnett, T.P. and D.W. Pierce, 2008: When will Lake Mead run dry? *Journal of Water Resources Research*, **44**, W03201, doi:10.1029/2007WR006704.

Barnett, T.P. and D.W. Pierce, 2009: Sustainable water deliveries from the Colorado River in a changing climate. *Proceedings of the National Academy of Sciences*, **106(18)**, 7334-7338, doi:10.1073/pnas.0812762106.

Barredo, J.I., 2009: Normalised flood losses in Europe: 1970-2006. *Natural Hazards and Earth System Sciences*, **9**, 97-104.

Barriopedro, D., E.M. Fischer, J. Luterbacher, R.M. Trigo, and R. Garcia-Herrara, 2011: The hot summer of 2010: redrawing the temperature record map of Europe. *Science*, **332**, 220-224.

Basara, J.B., H.G. Basara, B.G. Illston, and K.C. Crawford, 2010: The impact of the urban heat island during an intense heat wave in Oklahoma City. *Advances in Meteorology*, **(2010)**, 230365, doi:10.1155/2010/230365.

Battisti, D.S. and R.L. Naylor, 2009: Historical warnings of future food insecurity with unprecedented seasonal heat. *Science*, **323**, 240-244.

Bebi, P., D. Kulakowski, and C. Rixen, 2009: Snow avalanche disturbances in forest ecosystems – State of research and implications for management. *Forest Ecology and Management*, **257(9)**, 1883-1892.

Becken, S. and J.E. Hay, 2007: *Tourism and climate change: risks and opportunities*. Climate Change, Economies and Society Series, Channel View Publications, Clevedon, UK, 352 pp.

Bender, M.A., T.R. Knutson, R.E. Tuleya, J.J. Sirutis, G.A. Vecchi, S.T. Garner, and I.M. Held, 2010: Modeled impact of anthropogenic warming on the frequency of intense Atlantic hurricanes. *Science*, **327**, 454-458.

Beniston, M., 2004: The 2003 heatwave in Europe: A shape of things to come? An analysis based on Swiss climatological data and model simulations. *Geophysical Research Letters*, **31**, L02202.

Beniston, M. and H. Diaz, 2004: The 2003 heatwave as an example of summers in a greenhouse climate? Observations and climate model simulations for Basel, Switzerland. *Global and Planetary Change*, **44(1-4)**, 73-81.

Beniston, M., F. Keller, and S. Goyette, 2003: Snow pack in the Swiss Alps under changing climatic conditions: An empirical approach for climate impact studies. *Theoretical and Applied Climatology*, **74**, 19-31.

Benito, G., T. Grodek, and Y. Enzel, 1998: The geomorphic and hydrologic impacts of the catastrophic failure of flood-control-dams during the 1996-Biescas flood (Central Pyrennes, Spain). *Zeitschrift für Geomorphologie*, **42(4)**, 417-437.

Benito, G., R.F. Rohde, M. Seely, C. Kulls, O. Dahan, Y. Enzel, S. Todd, B. Botero, and T. Grodek, 2010: Management of alluvial aquifers in two southern African ephemeral rivers: implications for IWRM. *Water Resources Management*, **24**, 641-667.

Benson, C. and E. Clay, 1998: *The Impact of Drought on Sub-Saharan African Economies: A Preliminary Examination*. World Bank Technical Paper Series No. 401, World Bank, Washington, DC, 80 pp.

Benson, C. and E.J. Clay, 2000: Developing countries and the economic impacts of natural disasters. In: *Managing disaster risk in emerging economies* [Kreimer, A. and A. Margaret (eds.)]. Disaster Risk Management Series No. 2, World Bank, Washington, DC, pp. 11-21.

Benson, C. and E. Clay, 2003: Disasters, vulnerability, and global economy. In: *Building Safer Cities: The Future of Disaster Risk* [Kreimer, A., M. Arnold, and A. Carlin (eds.)]. Disaster Risk Management Series No. 3, World Bank, Washington, DC, pp. 3-31.

Benson, C. and E. Clay, 2004: *Understanding the Economic and Financial Impacts of Natural Disasters*. Disaster Risk Management Series, No. 4, World Bank, Washington, DC, 119 pp..

Benson, C. and J. Twigg, 2004. *Measuring Mitigation: Methodologies for assessing natural hazard risks and the net benefits of mitigation — A Scoping Study*. The International Federation of Red Cross and Red Crescent Societies/The ProVention Consortium, Geneva, Switzerland, 149 pp.

Berrittella M., A. Bigano, and R.S.J. Tol, 2006: A general equilibrium analysis of climate change impacts on tourism. *Tourism Management*, **27(5)**, 913-924.

Berry, M., 2003: Why is it important to boost the supply of affordable housing in Australia - and how can we do it? *Urban Policy and Research*, **21**, 413-435.

Berry, H., K. Bowen, and T. Kjellstrom, 2010: Climate change and mental health: a causal pathways framework. *International Journal of Public Health*, **55**, 123-132.

Bertness, M.D. and P.J. Ewanchuk, 2002: Latitudinal and climate-driven variation in the strength and nature of biological interactions in New England salt marshes. *Oecologia*, **132**, 392-401.

Beven, J.L., L.A. Avila, E.S. Blake, D.P. Brown, J.L. Franklin, R.D. Knabb, R.J. Pasch, J.R. Rhome, and S.R. Stewart, 2008: Atlantic hurricane season of 2005. *Monthly Weather Review*, **136**, 1109-1173.

Bhuiyan, C., R.P. Singh, and F.N. Kogan, 2006: Monitoring drought dynamics in the Aravalli region (India) using different indices based on ground and remote sensing data. *International Journal of Applied Earth Observation and Geoinformation*, **8(4)**, 289-302.

Bigano, A., F. Bosello, R. Roson, R.S.J. Tol, 2008: Economy-wide impacts of climate change: a joint analysis for sea level rise and tourism. *Mitigation and Adaption Strategies for Global Change*, **13**, 765-791.

Bindoff, N., J. Willebrand, V. Artale, A. Cazenave, J. Gregory, S. Gulev, K. Hanawa, C. Le Quéré, S. Levitus, Y. Nojiri, C.K. Shum, L.D. Talley and A. Unnikrishnan, 2007: Observations: Oceanic climate change and sea level. In: *Climate Change 2007: The Physical Science Basis, Contribution of Working Group I to the Fourth Assessment Report of the Intergovernmental Panel on Climate Change* [Solomon, S., D. Qin, M. Manning, Z. Chen, M. Marquis, K.B. Averyt, M. Tignor, and H.L. Miller (eds.)]. Cambridge University Press, Cambridge, UK, pp. 385-432.

Birch, E.L. and S.M. Wachter (eds.), 2006: *Rebuilding Urban Places after Disaster: Lessons from Hurricane Katrina*. University of Pennsylvania Press, Philadelphia, PA, 375 pp.

Bongaerts, P., T. Ridgeway, E.M. Sampayo, and O. Hoegh-Guldberg, 2010: Assessing the 'deep reef refugia' hypothesis: focus on Caribbean reefs. *Coral Reefs*, **29**, 309-327.

Borrell, C, M. Marí-Dell'Olmo, M. Rodríguez-Sanz, P.Garcia-Olalla, J.A. Caylà, J. Benach, and C. Muntaner, 2006: Socioeconomic position and excess mortality during the heat wave of 2003 in Barcelona. *European Journal of Epidemiology*, **21(9)**, 633-640.

Boruff, B.J., J.A. Easoz, S.D. Jones, H.R. Landry, J.D. Mitchem, and S.L. Cutter, 2003: Tornado hazards in the United States. *Climate Research*, **24**, 103-117.

Botzen, W.J.W., L.M. Bouwer, and J.C.J.M. van den Bergh, 2010: Climate change and hailstorm damage: Empirical evidence and implications for agriculture and insurance. *Resource and Energy Economics*, **32**, 341-362.

Bouwer, L.M., 2010: *Disasters and climate change: analysis and methods for projecting future losses from extreme weather*. Ph.D. thesis, Vrije Universiteit, Amsterdam, The Netherlands, 141 pp., dare.ubvu.vu.nl/bit-stream/1871/16355/1/ dissertation.pdf.

Bouwer, L.M., 2011: Have disaster losses increased due to anthropogenic climate change? *Bulletin of the American Meteorological Society*, **92**, 39-46.

Bouwer, L.M. and W.J.W. Botzen, 2011: How sensitive are US hurricane damages to climate? Comment on a paper by W.D. Nordhaus. *Climate Change Economics*, **2**, 1-7.

Bouwer, L.M., R.P. Crompton, E. Faust, P. Höppe, and R.A. Pielke Jr., 2007: Confronting disaster losses. *Science*, **318**, 753.

Bouwer, L.M., P. Bubeck, and J.C.J.H. Aerts, 2010: Changes in future flood risk due to climate and development in a Dutch polder area. *Global Environmental Change*, **20**, 463-471.

Breshears, D.D. and C.D. Allen, 2002: The importance of rapid, disturbance-induced losses in carbon management and sequestration. *Global Ecology and Biogeography*, **11**, 1-5.

Breshears, D.D., N.S. Cobb, P.M. Rich, K.P. Price, C.D. Allen, R.G. Balice, W.H. Romme, J.H. Kastens, M.L. Floyd, J. Belnap, J.J. Anderson, O.B. Myers, and C.W. Meyer, 2005: Regional vegetation die-off in response to global-change-type drought. *Proceedings of the National Academy of Sciences*, **102(42)**, 15144-15148.

Bronstert, A., A. Bárdossy, C. Bismuth, H. Buiteveld, M. Disse, H. Engel, U. Fritsch, Y. Hundecha, R. Lammersen, D. Niehoff, and N. Ritter, 2007: Multi-scale modelling of land-use change and river training effects on floods in the Rhine basin. *River Research and Applications*, **23(10)**, 1102-1125.

Brooks, N., 2004: *Drought in the Africa Sahel: long-term perspectives and future prospects*. Working paper 61, Tyndall Centre for Climate Research, University of East Anglia, Norwich, UK, 31 pp.

Brooks, N., W.N. Adger, and P.M. Kelly, 2005: The determinants of vulnerability and adaptive capacity at the national level and the implications for adaptation. *Global Environmental Change*, **15**, 151-163.

Brookshire, D.S., S.E. Chang, H. Cochrane, R.A. Olson, A. Rose, and J. Steenson, 1997: Direct and indirect economic losses from earthquake damage. *Earthquake Spectra*, **13**, 683-701.

Brouwer, R., S. Akter, L. Brander, and E. Haque, 2007: Socioeconomic vulnerability and adaptation to environmental risk: acase study of climate change and flooding in Bangladesh. *Risk Analysis*, **27**, 313-326.

BTE, 2001: *Economic costs of natural disasters in Australia*. Report 103, Bureau of Transport Economics, Canberra, Australia, 170 pp.

Burnley, I. and P. Murphy, 2004: *Sea Change: Movement from Metropolitan to Arcadian Australia*. UNSW Press, Sydney, Australia, 272 pp.

Burton, I. and E. May, 2004: The adaptation deficit in water resources management. *IDS Bulletin*, **35(3)**, 31-37.

Burton, K., R. Kates, and G. White, 1993: *The Environment as Hazard*. 2nd ed. Guilford Press, New York, NY, 290 pp.

Bussotti, F., E. Cenni, M. Ferretti, A. Cozzi, L. Brogi, and A. Mecci, 1995: Forest condition in Tuscany (Central Italy) - Field surveys 1987-1991. *Forestry*, **68(1)**, 11-24.

Butcher, G.V. and S. Ford, 2009: Modeling the regional economic impacts of the 2007/08 drought: results and lessons. Paper presented at the *2009 NZARES Conference*, Nelson, New Zealand, 27-28 August 2009, 11 pp., ageconsearch.umn.edu/bitstream/97167/2/2009_22_Economic%20Impact%20of%202007-08%20Drought_ButcherG.pdf.

Butry, D., E. Mercer, J. Prestemon, J. Pye, and T. Holmes, 2001: What is the price of catastrophic wildfire? *Journal of Forestry*, **99(11)**, 9-17.

Buzinde, C.N., D. Manuel-Navarrete, E.E. Yoo, and D. Morais, 2010: Tourists' perceptions in a climate of change: Eroding destinations. *Annals of Tourism Research*, **37(2)**, 333-354.

Buzna, L., K. Peters, and D. Helbing, 2006: Modelling the dynamics of disaster spreading in networks. *Physica A: Statistical Mechanics and its Applications*, **363(1)**, 132-140.

Cadier, E., M. Villacis, A. Graces, P. Lhuissier, L. Maisincho, R. Laval, D. Paredes, B. Caceres, and B. Francou, 2007: Variation of a low latitude Andean glacier according to global and local climate variations: first results. In: *Glacier Mass Balances and Meltwater Discharge* [Ginot, P. and J.E. Sicart (eds.)]. IAHS Pub. No. 318, International Association of Hydrological Sciences, Wallingford, UK, pp. 66-74.

Campbell, J.R., 2006: *Traditional Disaster Reduction in Pacific Island Communities*. GNS Science Report 2006/38, Institute of Geological and Nuclear Sciences, Lower Hutt, New Zealand, 46 pp.

Cardemil, L., J.C. Di Tata, and F. Frantischek, 2000: Central America: adjustment and reforms in the 1990s. *Finance and Development*, **37(1)**, 34-37, www.imf.org/external/pubs/ft/fandd/2000/03/cardemil.htm.

Carson, R., R. Mitchell, M. Hanemann, R. Kopp, S. Presser, and P. Ruud, 2003: Contingent valuation and lost passive use: damages from the Exxon Valdez oil spill. *Environmental and Resource Economics*, **25**, 257-286.

Carter, L.M., S.J. Cohen, N.B. Grimm, and J.L. Hatfield, 2008: *Weather and Climate Extremes in a Changing Climate, Regions of Focus: North America, Hawaii, Caribbean and U.S. Pacific Islands*. Synthesis and Assessment Product 3.3, Report by the U.S. Climate Change Science Program and the Subcommittee on Global Change, Washington, DC, 162 pp.

Carter, M.R., P.D. Little, T. Mogues, and W. Negatu, 2007: Poverty traps and natural disasters in Ethiopia and Honduras. *World Development*, **35(5)**, 835-856, doi:10.1016/j.worlddev.2006.09.010.

Caselli, F. and P. Malhotra, 2004: *Natural Disasters and Growth: from Thought Experiment to Natural Experiment*, International Monetary Fund, Washington, DC, 29 pp.

Casimiro, E. and J.M. Calheiros, 2002: Human health. In: *Climate Change in Portugal. Scenarios, Impacts and Adaptation Measures—SIAM Project* [Santos F.D., K. Forbes, and R. Moita (eds.)]. Gradiva Publishers, Lisbon, Portugal, pp. 241-300.

Cavallo, E. and I. Noy, 2009: *The Economics of Natural Disasters: A Survey*. IDB Working Paper 09-19, Inter-American Development Bank, Washington, DC, 48 pp.

Cavallo, E. and I. Noy, 2010: *The Economics of Natural Disasters: A Survey*. IDB Working Paper Series, No. IDB-WP-124, Inter-American Development Bank, Washington, DC, 50 pp., www.iadb.org/res/publications/pubfiles/pubIDB-WP-124.pdf.

CCSP, 2008: *Impacts of Climate Change and Variability on Transportation Systems and Infrastructure: Gulf Coast Study, Phase I*. A Report by the U.S. Climate Change Science Program and the Subcommittee on Global Change Research [Savonis, M.J., V.R. Burkett, and J.R. Potter (eds.)]. Department of Transportation, Washington, DC, 445 pp.

CDHS, 2007: *Review of July 2006 Heatwave Related Fatalities in California*. CDHS Report, Epidemiology and Prevention for Injury Control Branch, California Department of Health Services, Sacramento, CA, 10 pp.

Chang, H., J. Franczyk, and C. Kim, 2009: What is responsible for increasing flood risks? The case of Gangwon Province, Korea. *Natural Hazards*, **48**, 399-354.

Changnon, S. (ed.), 1996: *The Great Flood of 1993*. Westview Press, Boulder, CO, 321 pp.

Changnon, S.A., 2001: Damaging thunderstorm activity in the United States. *Bulletin of the American Meteorological Society*, **82**, 597-608.

Changnon, S.A., 2006: *Railroads and Weather: From Fogs to Floods and Heat to Hurricanes, the Impacts of Weather and Climate on American Railroading*. American Meteorological Society, Boston, MA, 136 pp.

Changnon, S.A., 2009: Increasing major hail losses in the U.S. *Climatic Change*, **96**, 161-166.

Chapin, F.S., T.V. Callaghan, Y. Bergeron, M. Fukuda, J.F. Johnstone, G. Juday, and S.A. Zimov, 2004: Global change and the boreal forest: thresholds, shifting states or gradual change? *Ambio*, **33**, 361-365.

Charveriat, C., 2000: *Natural Disasters in Latin America and the Caribbean: An Overview of Risk*. IDB Working Paper 434, Inter-American Development Bank, Washington, DC, 104 pp.

Chen, K. and J. McAneney, 2006: High-resolution estimates of Australia's coastal population with validations of global population, shoreline and elevation datasets. *Geophysical Research Letters*, **33**, L16601, doi:10.1029/2006GL026981.

Cheng, S. and R. Wang, 2002: An approach for evaluating the hydrological effects of urbanization and its application. *Hydrological Processes*, **16(7)**, 1403-1418.

Choi, O. and A. Fisher, 2003: The impacts of socioeconomic development and climate change on severe weather catastrophe losses: Mid-Atlantic Region (MAR) and the US. *Climatic Change*, **58(1-2)**, 149-170.

Christensen, J.H., B. Hewitson, A. Busuioc, A. Chen, X. Gao, I. Held, R. Jones, R.K. Kolli, W. Kwon, R. Laprise, V.M. Rueda, L. Mearns, C.G. Menéndez, J. Räisänen, A. Rinke, A. Sarr, and P. Whetton, 2007: Regional climate projections. In: *Climate Change 2007: The Physical Science Basis. Contribution of Working Group I to the Fourth Assessment Report of the Intergovernmental Panel on Climate Change* [Solomon, S., D. Qin, M. Manning, Z. Chen, M. Marquis, K.B. Averyt, M. Tignor, and H.L. Miller (eds.)]. Cambridge University Press, Cambridge, UK, pp. 847-933.

Christensen, N.S., A.W. Wood, N. Voisin, D.P. Lettenmaier, and R.N. Palmer, 2004: The effects of climate change on the hydrology and water resources of the Colorado River basin. *Climatic Change*, **62(1)**, 337-363.

Ciais, P., M. Reichstein, N. Viovy, A. Granier, J. Ogée, V. Allard, M. Aubinet, N. Buchmann, Chr. Bernhofer, A. Carrara, F. Chevallier, N. De Noblet, A.D. Friend, P. Friedlingstein, T. Grünwald, B. Heinesch, P. Keronen, A. Knohl, G. Krinner, D. Loustau, G. Manca, G. Matteucci, F. Miglietta, J.M. Ourcival, D. Papale, K. Pilegaard, S. Rambal, G. Seufert, J.F. Soussana, M.J. Sanz, E.D. Schulze, T. Vesala, and R. Valentini, 2005: Europe-wide reduction in primary productivity caused by the heat and drought in 2003. *Nature*, **437(7058)**, 529-533.

Ciscar, J.C. (ed.), 2009: *Climate Change Impacts in Europe: Final Report of the PESETA research project*. EUR 24093 EN-2009, European Commission, Joint Research Centre, Institute for Prospective Technological Studies, EU Publications, Luxembourg, Luxembourg, 113 pp.

Clarke, M.L. and H.M. Rendell, 2006: Hindcasting extreme events: the occurrence and expression of damaging floods and landslides in Southern Italy. *Land Degradation and Development*, **17**(4), 365-380.

Collier, P., V. Elliot, H. Hegre, A. Hoeffler, M. Reynal-Querol, and N. Sambanis, 2003: *Breaking the Conflict Trap: Civil War and Development Policy*. World Bank, Washington, DC.

Corominas, J., 2005: Impacto sobre los riesgos naturales de origen climático: inestabilidad de laderas. In: *Proyecto ECCE. Evaluación Preliminar de los impactos en España por efecto del Cambio Climático* [Moreno, J.M. (ed.)]. Ministerio de Medio Ambiente, Madrid, Spain, pp. 549-579.

Costa, M.H. and G.F. Pires, 2010: Effects of Amazon and Central Brazil deforestation scenarios on the duration of the dry season in the arc of deforestation. *International Journal of Climatology*, **30(13)**, 1970-1979.

Costanza, R. and J. Farley. 2007. Ecological economics of coastal disasters: Introduction to the Special Issue. *Ecological Economics*, **63**, 249-253.

Costello, L., 2007: Going bush: the implications of urban-rural migration. *Geographical Research*, **45**, 85-94.

Costello, L., 2009: Urban-rural migration: housing availability and affordability. *Australian Geographer*, **40**, 219-233.

CRED, 2010: *EM-DAT: The OFDA/CRED International Disaster Data Base*. Centre for Research on the Epidemiology of Disasters, Université Catholique de Louvain, Louvain, Belgium, www.emdat.be/Database/.

Crichton, D., 2007: What can cities do to increase resilience? *Philosophical Transactions of the Royal Society A*, **365(1860)**, 2731-2739.

Crompton, R.P. and K.J. McAneney, 2008: Normalised Australian insured losses from meteorological hazards: 1967-2006. *Environmental Science and Policy*, **11**, 371-378.

Crompton, R.P., K.J. McAneney, and K. Chen, 2010: Influence of location, population, and climate on building damage and fatalities due to Australian bushfire: 1925–2009. *Weather, Climate, and Society*, **2**, 300-310.

Crompton, R., R.A. Pielke Jr., and J. McAneney, 2011: Emergence timescales for detection of anthropogenic climate change in US tropical cyclone loss data. *Environmental Research Letters*, **6(1)**, doi:10.1088/1748-9326/6/1/014003.

Crosta, G.B., H. Chen, and C.F. Lee, 2004: Replay of the 1987 Val Pola Landslide, Italian Alps. *Geomorphology*, **60(2, 3)**, 127-146.

Crowards, T., 2000: *Comparative Vulnerability to Natural Disasters in the Caribbean*. Staff Working Paper No. 1/00, Caribbean Development Bank, Charleston, SC, 21 pp.

Crozier, M.J., 2010: Deciphering the effect of climate change on landslide activity: A review. *Geomorphology*, **124(3-4)**, 260-267, doi:10.1016/j.geomorph.2010.04.009.

Cruz, A., 2005: NaTech disasters: a review of practices, lessons learned and future research needs. Paper presented at the 5th Annual IIASA-DPRI Forum, *Integrated Disaster Risk Management: Innovations in Science and Policy*, Beijing, China, 14-18 September 2005, 10 pp.

Cruz, R.V., H. Harasawa, M. Lal, S. Wu, Y. Anokhin, B. Punsalmaa, Y. Honda, M. Jafari, C. Li and N. Huu Ninh, 2007: Asia. In: *Climate Change 2007: Impacts, Adaptation and Vulnerability. Contribution of Working Group II to the Fourth Assessment Report of the Intergovernmental Panel on Climate Change* [Parry, M.L., O.F. Canziani, J.P. Palutikof, P.J. van der Linden and C.E. Hanson (eds.)]. Cambridge University Press, Cambridge, UK, pp. 469-506.

Cummins, J., and O. Mahul, 2009: *Catastrophe Risk Financing in Developing Countries: Principles for Public Intervention.* World Bank, Washington, DC.

Cuny, F.C., 1983: *Disasters and Development.* Oxford University Press, Oxford, UK, 278 pp.

Curriero, F.C., J.A. Patz, J.B. Rose, and S. Lele, 2001: The association between extreme precipitation and waterborne disease outbreaks in the United States, 1948-1994. *American Journal of Public Health*, **91(8)**, 1194-1199, doi:10.2105/AJPH.91.8.1194.

Cutter, S. and C. Emrich, 2005: Are natural hazards and disaster losses in the U.S. increasing? *EOS*, **86(41)**, 381-396.

Cutter, S.L. and C.T. Emrich, 2006: Moral hazard, social catastrophe: the changing face of vulnerability along the hurricane coasts. *Annals of the American Academy of Political and Social Science*, **604**, 102-112.

Cutter, S.L. and C. Finch, 2008: Temporal and spatial changes in social vulnerability to natural hazards. *Proceedings of the National Academy of Sciences*, **105(7)**, 2301-2306.

Cutter, S.L., L. Barnes, M. Berry, C. Burton, E. Evans, E. Tate, and J. Webb, 2008: A place-based model for understanding community resilience to natural disasters. *Global Environmental Change*, **18**, 598-606.

D'Almeida, C., C.J. Vorosmarty, G.C. Hurtt, J.A. Marengo, S.L. Dingman, and B.D. Keim, 2007: The effects of deforestation on the hydrological cycle in Amazonia: a review on scale and resolution. *International Journal of Climatology*, **27**, 633-648.

Dao, H. and P. Peduzzi, 2004: Global evaluation of human risk and vulnerability to natural hazards In: *Proceedings of ENVIROINFO 2004, 18th International Conference Informatics for Environmental Protection*, Vol. 1, CERN, Geneva, Switzerland, 21-23 October 2004, Sh@ring, Editions du Tricorne, Geneva, Switzerland, pp. 435-446.

Dasgupta, S., B. Laplante, S. Murray, and D. Wheeler, 2009: *Sea-Level Rise and Storm Surges: A Comparative Analysis of Impacts in Developing Countries*. Policy Research Working Paper 4901, World Bank Development Research Group, Environment and Energy Team, Washington, DC, 41 pp.

Dash, S., R. Jenamani, S. Kalsi, and S. Panda, 2007: Some evidence of climate change in twentieth-century India, *Climatic Change*, **85**, 299-321.

Davis, R.E, P.C. Knappenberger, P.J. Michaels, and W.M. Novicoff, 2003: Changing heat-related mortality in the United States. *Environmental Health Perspectives*, **111**, 1712-1718.

Dawson, R. J., J.W. Hall, P.D. Bates, and R.J. Nicholls, 2005: Quantified analysis of the probability of flooding in the Thames Estuary under imaginable worst-case sea level rise scenarios. *International Journal of Water Resources Development*, **21(4)**, 577-591, doi:10.1080/07900620500258380.

Dawson, R.J., M.E. Dickson, R.J. Nicholls, J.W. Hall, M.J.A. Walkden, P.K. Stansby, M. Mokrech, J. Richards, J. Zhou, J. Milligan, A. Jordan, S. Pearson, J. Rees, P.D. Bates, S. Koukoulas, and A.R. Watkinson, 2009: Integrated analysis of risks of coastal flooding and cliff erosion under scenarios of long term change. *Climatic Change*, **95**, 249-288.

de Bruin, K., R.B.Dellink, A.Ruijs, L. Bolwidt, A. vanBuuren, J. Graveland, R.S. deGroot, P.J.Kuikman, S. Reinhard, R.P. Roetter, V.C. Tassone, A. Verhagen, and E.C. van Ierland, 2009: Adapting to climate change in The Netherlands: an inventory of climate adaptation options and ranking of alternatives. *Climatic Change*, **95(1-2)**, 23-45.

de Wit, M. and J. Stankiewicz, 2006: Changes in water supply across Africa with predicted climate change. *Science*, **311**, 1917-1921.

DEFRA, 2004: *Scientific and Technical Aspects of Climate Change, including Impacts and Adaptation and Associated Costs*. Department for Environment, Food and Rural Affairs, London, UK, 19 pp., www.newforestnpa.gov.uk/__data/assets/pdf_file/0016/20473/152-07_annex_3_climate_change.pdf.

DEFRA, 2006: *The UK's Fourth National Communication under the United Nations Framework Convention on Climate Change*. United Kingdom Department for Environment, Food and Rural Affairs, London, UK, 135 pp., unfccc.int/resource/docs/natc/uknc4.pdf.

Delpla, I., A.-V. Jung, E. Baures, M. Clement, and O. Thomas, 2009: Impacts of climate change on surface water quality in relation to drinking water production. *Environment International*, **35(8)**, 1225-1233.

Deschênes, O. and M. Greenstone, 2007: The economic impacts of climate change: Evidence from agricultural output and random fluctuations in weather. *The American Economic Review*, **97(1)**, 354-385.

Devoy, R.J.N., 2008: Coastal vulnerability and the implications of sea-level rise for Ireland. *Journal of Coastal Research*, **24(2)**, 325-341.

DHS, 2009: *January 2009 Heatwave in Victoria: an Assessment of Health Impacts*. Victorian Govt. Department of Human Services, Melbourne, Australia, 16 pp.

Dodman, D. and D. Satterthwaite, 2008: Institutional capacity, climate change adaptation and the urban poor. *Institute of Development Studies Bulletin*, **39(4)**, 67-74.

Donat, M.G., G.C. Leckebusch, J.G. Pinto, and U. Ulbrich, 2010: European storminess and associated circulation weather types: future changes deduced from a multi-model ensemble of GCM simulations. *Climate Research*, **42**, 27-43.

Donat, M.G., G.C. Leckebusch, S. Wild, and U. Ulbrich, 2011: Future changes in European winter storm losses and extreme wind speeds inferred from GCM and RCM multi-model simulations. *Natural Hazards and Earth System Sciences*, **11**, 1351-1370.

Dorland, C., R.S.J. Tol, and J.P. Palutikof, 1999: Vulnerability of the Netherlands and Northwest Europe to storm damage under climate change. *Climatic Change*, **43**, 513-535.

Douglas, I., 2009: Climate change, flooding and food security in south Asia, *Food Security*, **1(2)**, 127-136.

Douglas, I., K. Alam, M. Maghenda, Y. Mcdonnell, L. Mclean, and J. Campbell, 2008: Unjust waters: climate change, flooding and the urban poor in Africa, *Environment and Urbanization*, **20 (1)**, 187-205.

Downton, M. and R.A. Pielke, Jr., 2005: How accurate are disaster loss data? The case of U.S. flood damage. *Natural Hazards*, **352**, 211-228.

Downton, M.W., J.Z.B. Miller, and R.A. Pielke, 2005: Reanalysis of US National Weather Service flood loss database. *Natural Hazards Review*, **6**, 13-22.

Dregne, H.E., 1986: Desertification of arid lands. In: *Physics of Desertification* [El-Baz, F. and M.H.A. Hassan (eds.)]. Martinus Nijhoff, Dordrecht, The Netherlands, pp. 4-34.

DSD,1994: Barbados Declaration. *A/CONF.167/9,I, Annex I. Report of the Global Conference on the Sustainable Development of Small Island Developing States*, Bridgetown, Barbados, 25 April-6 May 1994, United Nations Division for Sustainable Development, Department of Economic and Social Affairs, New York, NY, USA, 77 pp., islands.unep.ch/dbardecl.htm.

Eakin, C.M., J. A. Morgan, S. F. Heron, T. B. Smith, G. Liu, et al., 2010: Caribbean corals in crisis: record thermal stress, bleaching, and mortality in 2005. *Public Library of Science ONE*, **5(11)**, e13969, doi:10.1371/journal.pone.0013969.

Easterling, W. and M. Apps, 2005: Assessing the consequences of climate change for food and forest resources: A view from the IPCC. *Climate Change*, **70**, 165-189.

Easterling, W.E., P.K. Aggarwal, P. Batima, K.M. Brander, L. Erda, S.M. Howden, A. Kirilenko, J. Morton, J.-F. Soussana, J. Schmidhuber, and F.N. Tubiello, 2007: Food, fibre and forest products. In: *Climate Change 2007: Impacts, Adaptation and Vulnerability. Contribution of Working Group II to the Fourth Assessment Report of the Intergovernmental Panel on Climate Change* [Parry, M.L., O.F. Canziani, J.P. Palutikof, P.J. van der Linden and C.E. Hanson (eds.)]. Cambridge University Press, Cambridge, UK, pp. 273-313.

Ebersole, B.A., J.J. Westerink, S. Bunya, J.C. Dietrich, and M.A. Cialone, 2010: Development of storm surge which led to flooding in St. Bernard Polder during Hurricane Katrina. *Ocean Engineering*, **37**, 91-103.

Ebi, K.L. and G.A. Meehl, 2007: *The Heat is On: Climate Change & Heatwaves in the Midwest*. Pew Center on Global Climate Change, Arlington, VA, 14 pp., www.pewclimate.org/docUploads/Regional-Impacts-Midwest.pdf.

Ebi, K.L., J. Balbus, P.L. Kinney, E. Lipp, D. Mills, M.S. O'Neill, and M. Wilson, 2008: Effects of global change on human health. In: *Analyses of the Effects of Global Change on Human Health and Welfare and Human Systems* [Gamble, J.L. (ed.)]. A Report by the U.S. Climate Change Science Program and the Subcommittee on Global Change Research. U.S. Environmental Protection Agency, Washington, DC, pp. 2-1 to 2-78.

ECA, 2009: *Shaping Climate-Resilient Development: A Framework for Decision-Making Study*. Report of the Economics of Climate Adaptation Working Group, New York, NY, 159 pp.

ECLAC, 2003: *Handbook for Estimating the Socio-economic and Environmental Effects of Disasters*. LC/MEX/G.5, LC/L 1874, Economic Commission for Latin America and the Caribbean, Santiago, Chile, 354 pp., www.undp.org/cpr/disred/documents/publications/eclac_handbook.pdf.

EEA, 2004: *Impacts of Europe's changing climate. An indicator-based assessment*. EEA Report No. 2/2004, European Environmental Agency, Copenhagen, Denmark, 100 pp.

EEA, 2008: *Impacts of Europe's changing climate- 2008 indicator-based assessment*. EEA Report No.4/2008, European Environmental Agency, Copenhagen, Denmark, 19 pp., doi:10.2800/48117.

EEA, 2009: *Regional Climate Change and Adaptation: The Alps Facing the Challenge of Changing Water Resources*. EEA Report No. 8, European Environmental Agency, Copenhagen, Denmark, 143 pp.

Ehmer, P. and E. Heymann, 2008: *Climate change and tourism: Where will the journey lead?* Deutsche Bank Research, Frankfurt am Main, Germany, 28 pp.

Ellson, R.W., J.W. Milliman, and R.B. Roberts, 1984: Measuring the regional economic effects of earthquakes and earthquake predictions. *Journal of Regional Science*, **24(4)**, 559-579.

Elsasser, H. and R. Bürki, 2002: Climate change as a threat to tourism in the Alps. *Climate Research*, **20**, 253-257.

Emanuel, K., 2005: Increasing destructiveness of tropical cyclones over the past 30 years. *Nature*, **436(7051)**, 686-688.

Endfield, G.H. and I.F. Tejedo, 2006: Decades of drought, years of hunger: archival investigations of multiple year droughts in late colonial Chihuahua. *Climatic Change*, **75(4)**, 391-419, doi:10.1007/s10584-006-3492-7.

Endlicher, W., G. Jendritzky, J. Fischer, and J.-P. Redlich, 2008: Heat waves, urban climate and human health. In *Urban Ecology* [Marzluff, J.M., E. Shulenberger, W. Endlicher, M. Alberti, G. Bradley, C. Ryan, U. Simon, and C. ZumBrunnen (eds.)]. Springer, New York, NY, pp. 269-278.

Engelthaler, D.M., D.G. Mosley, J.E. Cheek, C.E. Levy, K.K. Komatsu, P. Ettestad, T. Davis, D.T. Tanda, L. Miller, J.W. Frampton, R. Porter, and R.T. Bryan, 1999: Climatic and environmental patterns associated with hantavirus pulmonary syndrome, Four Corners region, United States. *Emerging Infectious Diseases*, **5(1)**, 87-94.

Eriksen, S.H., K. Brown, and P.M. Kelly, 2005: The dynamics of vulnerability: locating coping strategies in Kenya and Tanzania. *Geographical Journal*, **171**, 287-305.

Esteban-Talaya, A., F. López Palomeque, and E. Aguiló Pérez, 2005: Impacts in the touristic sector. In: *A Preliminary Assessment of the Impacts in Spain due to the Effect of Climate Change* [Moreno, J.M. (ed.)]. Ministry of Environment, Madrid, Spain, pp. 653-690.

Eurostat, 2010: *Ageing in the European Union: where exactly? Rural areas are losing the young generation quicker than urban areas*. Eurostat Statistics in Focus, 26/2010, Eurostat, European Commission, Luxembourg, 16 pp.

Evans, B. and M. Webster, 2008: *Adapting to Climate Change in Europe and Central Asia. Background Paper on Water Supply and Sanitation*. World Bank Sustainable Development Department, Europe and Central Asia Region, Washington, DC, 34 pp.

Evans, E.P., R. Ashley, J.W. Hall, E.C. Penning-Rowsell, A. Saul, P.B. Sayers, C.R. Thorne, and A. Watkinson, 2004. *Foresight Flood and Coastal Defence Project: Scientific Summary: Volume I, Future risks and their drivers*. Office of Science and Technology, London, UK, 366 pp., www.bis.gov.uk/foresight/our-work/projects/published-projects/flood-and-coastal-defence/project-outputs/volume-1.

Faccio, S.D., 2003: Effects of ice storm-created gaps on forest breeding bird communities in central Vermont. *Forest Ecology and Management*, **186(1-3)**, 133-145.

Fankhauser, S., J.B. Smith, and R.S.J. Tol, 1999. Weathering climate change: some simple rules to guide adaptation decisions. *Ecological Economics*, **30(1)**, 67-78.

FAO, 2008: *The State of Food Insecurity in the World 2008, High food prices and food security – threats and opportunities*. Food and Agriculture Organization of the United Nations, Rome, Italy, 56 pp.

FAO, 2009: *The State of Agricultural Commodity Markets 2009*. Food and Agriculture Organization of the United Nations, Rome, Italy, 63 pp.

Fallon, P. and R. Betts. 2010: Climate impacts on European agriculture and water management in the context of adaptation and mitigation – the importance of an integrated approach. *Science of Total Environment*, **408**, 5667-5687.

Fengqing, J., Z. Cheng, M. Guijin, H. Ruji, and M. Qingxia, 2005: Magnification of flood disasters and its relation to regional precipitation and local human activities since the 1980s in Xinxiang, Northwestern China. *Natural Hazards*, **36**, 307-330.

Few, R., 2003: Flooding, vulnerability and coping strategies: local responses to a global threat. *Progress in Development Studies*, **3(43)**, 43-58.

Few, R., M. Ahern, F. Matthies, and S. Kovats, 2004: *Floods, Health and Climate Change: A Strategic Review*. Working Paper 63, Tyndall Centre for Climate Change Research, University of East Anglia, Norwich, UK, 138 pp.

Feyen, L., J.I. Barredo, and R. Dankers, 2009: Implications of global warming and urban land use change on flooding in Europe. Water and urban development paradigms – towards an integration of engineering. In: *Design and Management Approaches* [Feyen, J., K. Shannon, and M. Neville (eds.)]. CRC Press, Boca Raton, FL, pp. 217-225.

Field, C.B., L.D. Mortsch, M. Brklacich, D.L. Forbes, P. Kovacs, J.A. Patz, S.W. Running, and M.J. Scott, 2007: North America. In: *Climate Change 2007: Impacts, Adaptation and Vulnerability. Contribution of Working Group II to the Fourth Assessment Report of the Intergovernmental Panel on Climate Change* [Parry, M.L., O.F. Canziani, J.P. Palutikof, P.J. van der Linden and C.E. Hanson (eds.)]. Cambridge University Press, Cambridge, UK, pp. 617-652.

Field, R.D., G.R. Van Der Werf, and S.S.P. Shen, 2009: Human amplification of drought-induced biomass burning in Indonesia since 1960. *Nature Geoscience*, **2**, 185-188.

Fink, A., T. Brücher, A. Krüger, G. Leckebusch, J. Pinto, and U. Ulbrich, 2004: The 2003 European summer heatwaves and drought – synoptic diagnosis and impacts. *Weather*, **59**, 209-216.

Fischlin, A., G.F. Midgley, J.T. Price, R. Leemans, B. Gopal, C. Turley, M.D.A. Rounsevell, O.P. Dube, J. Tarazona, and A.A. Velichko, 2007: Ecosystems, their properties, goods, and services. In: *Climate Change 2007: Impacts, Adaptation and Vulnerability. Contribution of Working Group II to the Fourth Assessment Report of the Intergovernmental Panel on Climate Change* [Parry, M.L., O.F. Canziani, J.P. Palutikof, P.J. van der Linden and C.E. Hanson (eds.)]. Cambridge University Press, Cambridge, UK, pp. 211-272.

FitzGerald, D.M., M.S. Fenster, B.A. Argow, and I.V. Buynevich, 2008: Coastal impacts due to sea-level rise, *Annual Review of Earth and Planetary Sciences*, **36(1)**, 601-647.

Folke, C., S.R. Carpenter, T. Elmquist, L.H. Gunderson, C.S. Holling, and B.H. Walker, 2002: Resilience and sustainable development: Building adaptive capacity in a world of transformations. *Ambio*, **31**, 437-440.

Folke, C., S. Carpenter, B. Walker, M. Scheffer, T. Elmqvist, L. Gunderson, and C.S. Holling, 2004: Regime shifts, resilience, and biodiversity in ecosystem management. *Annual Review of Ecology, Evolution, and Systematics*, **35**, 557-581.

Fouillet, A, G. Rey, V. Wagner, K. Laaidi, P. Empereur-Bissonnet, A. Le Tertre, P. Frayssinet, P. Bessemoulin, F. Laurent, P. De Crouy-Chanel, E. Jougla, D. Hémon, 2008: Has the impact of heat waves on mortality changed in France since the European heat wave of summer 2003? A study of the 2006 heat wave. *International Journal of Epidemiology*, **37(2)**, 309-317.

Fowler, H.J., C.G. Kilsby, and P.E. O'Connell, 2003: Modeling the impacts of climatic change and variability on the reliability, resilience, and vulnerability of a water resource system. *Water Resources Research*, **39(8)**, 1222-1232.

Fowler, H.J., C.G. Kilsby, and J. Stunell, 2007: Modelling the impacts of projected future climate change on water resources in north-west England. *Hydrology and Earth System Sciences*, **11(3)**, 1115-1124.

Frazier, T.G., N. Wood, B. Yarnal, and D.H. Bauer, 2010: Influence of potential sea level rise on societal vulnerability to hurricane storm-surge hazards, Sarasota County, Florida. *Applied Geography*, **30(4)**, 490-505, doi:10.1016/j.apgeog.2010.05.005.

Freeman, P.K., 2000: Estimating chronic risk from natural disasters in developing countries: A case study in Honduras. Presented at the *Annual Bank Conference on Development Economics – Europe Development Thinking at the Millennium*, Paris, France, 26-28June, 2000, 10 pp.

Freeman, P.K., L. Martin, R. Mechler, K. Warner, with P. Hausman, 2002: *Catastrophes and Development, Integrating Natural Catastrophes into Development Planning*. World Bank Disaster Risk Management Working Paper Series No.4, World Bank, Washington, DC, 64 pp.

Frey, A.E., F. Olivera, J.L. Irish, L.M. Dunkin, J.M. Kaihatu, C.M. Ferreira, and B.L. Edge, 2010: The impact of climate change on hurricane flooding inundation, population affected, and property damages. *Journal of the American Water Resources Association*, **46(5)**, 1049-1059.

Fritze, J., G. Blashki, S. Burke, and J. Wiseman, 2008: Hope, despair and transformation: climate change and the promotion of mental health and wellbeing. *International Journal of Mental Health Systems*, **2**,13.

Fuhrer, J., M. Beniston, A. Fischlin, C. Frei, S. Goyette, K. Jasper, and C. Pfister, 2006: Climate risks and their impact on agriculture and forests in Switzerland. *Climatic Change*, **79(1-2)**, 79-102.

Gaddis, E., B. Miles, S. Morse, and D. Lewis, 2007: Full-cost accounting of coastal disasters in the United States: Implications for planning and preparedness. *Ecological Economics*, **63**, 307-318.

Ghesquiere, F., O. Mahul, M. Forni, and R. Gartley, 2006: *Caribbean Catastrophe Risk Insurance Facility: A solution to the short-term liquidity needs of small island states in the aftermath of natural disasters*. IAT03-13/3, World Bank, Washington, DC, 4 pp., siteresources.worldbank.org/PROJECTS/Resources/Catastrophicriskinsurancefacility.pdf.

Gillett, N.P., A.J. Weaver, F.W. Zwiers, and M.D. Flannigan, 2004: Detecting the effect of climate change on Canadian forest fires. *Geophysical Research Letters*, **31(18)**, L18211, doi:10.1029/2004GL020876.

Githeko, A.K., S.W. Lindsay, U.E. Confalonieri, and J.A. Patz, 2000: Climate change and vector-borne diseases: a regional analysis. *Bulletin of the World Health Organization*, **78**, 1136-1147.

Gitlin, A.R., C.M. Sthultz, M.A. Bowker, S. Stumpof, K.L. Paxton, K.A. Kennedy, A. Munoz, J.K. Bailey, and T.G. Whitham, 2006: Mortality gradients within and among dominant plant populations as barometers of ecosystem change during extreme drought. *Conservation Biology*, **20(5)**, 1477-1486.

Goldstein, R., 2003: *A Survey of Water Use and Sustainability in the United States with a Focus on Power Generation*. Electric Power Research Institute, Palo Alto, CA, 64 pp.

Gössling, S. and C.M. Hall, 2006a: *Tourism and Global Environmental Change. Ecological, Social, Economic and Political Interrelationships*. Routledge, London, UK, 330 pp.

Gössling, S. and C.M. Hall, 2006b: Uncertainties in predicting tourist flows under scenarios of climate change. *Climatic Change*, **79(3-4)**, 163-173.

Gowdy, J., 2007: Toward an experimental foundation for benefit-cost analysis. *Ecological Economics*, **63**, 649-655.

Granier, A., M. Reichstein, N. Bréda, I.A. Janssens, E. Falge, P. Ciais, T. Grünwald, M. Aubinet, P. Berbigier, C. Bernhofer, N. Buchmann, O. Facini, G. Grassi, B. Heinesch, H. Ilvesniemi, P. Keronen, A. Knohlc, B. Köstner, F. Lagergren, A. Lindroth, B. Longdoz, D. Loustau, J. Mateusq, L. Montagnanir, C. Nyst, E. Moors, D. Papale, M. Peiffer, K. Pilegaard, G. Pita, J. Pumpanen, S. Rambal, C. Rebmann, A. Rodrigues, G. Seufert, J. Tenhunen, T. Vesala, and Q. Wang, 2007: Evidence for soil water control on carbon and water dynamics in European forests during the extremely dry year: 2003. *Agricultural and Forest Meteorology*, **143(1-2)**, 123-145.

Griffith, G.P., E.A. Fulton, and A.J. Richardson, 2011: Effects of fishing and acidification-related benthic mortality on the southeast Australian marine ecosystem. *Global Change Biology*, **17(10)**, 3058-3074, doi:10.1111/j.1365-2486.2011.02453.x.

Grothmann, T. and A. Patt, 2005: Adaptive capacity and human cognition: The process of individual adaptation to climate change. *Global Environmental Change, Part A: Human and Policy Dimensions*, **3(15)**, 199-213.

Guha-Sapir, D., F. Vos, R. Below, with S. Ponserre, 2011: *Annual Disaster Statistical Review 2010: The numbers and trends*. Centre for Research on the Epidemiology of Disasters, Université Catholique de Louvain, Brussels, Belgium, 42 pp.,cred.be/sites/default/files/ADSR_2010.pdf.

Guimaraes, P., F.L. Hefner, and D.P. Woodward, 1993: Wealth and income effects of natural disasters: An econometric analysis of Hurricane Hugo. *Review of Regional Studies*, **23**, 97-114.

Guo, Y., J. Zhang, L. Zhang, and Y. Shen, 2009: Computational investigation of typhoon-induced storm surge in Hangzhou Bay, China. *Estuarine, Coastal and Shelf Science*, **85**, 530-553.

Haeberli, W., A. Kääb, D. Vonder Mühll, and P. Teysseire, 2001: Prevention of outburst floods from periglacial lakes at Grubengletscher, Valais, Swiss Alps. *Journal of Glaciology*, **47**, 111-122.

Haines, A., R.S. Kovats, D. Campbell-Lendrum, and C. Corvalan, 2006: Climate change and human health: impacts, vulnerability, and mitigation. *The Lancet*,**367**, 2101-2109.

Hall, J.W., E.P. Evans, E.C. Penning-Rowsell, P.B. Sayers, C.R. Thorne, and A.J. Saul, 2003: Quantified scenarios analysis of drivers and impacts of changing flood risk in England and Wales: 2030-2100. *Environmental Hazards*, **5**, 51-65.

Hall, J.W., P.B. Sayers, and R.J. Dawson, 2005: National-scale assessment of current and future flood risk in England and Wales. *Natural Hazards*, **36(1-2)**, 147-164.

Hallegatte, S., 2007: The use of synthetic hurricane tracks in risk analysis and climate change damage assessment. *Journal of Applied Meteorology and Climatology*, **46**, 1956-1966.

Hallegatte, S., 2008: A roadmap to assess the economic cost of climate change with an application to hurricanes in the United States. In: *Hurricanes and Climate Change* [Elsner, J.B. and T.H. Jagger (eds.)]. Springer-Verlag, Berlin and Heidelberg, Germany, pp. 361-386.

Hallegatte, S., 2009: Strategies to adapt to an uncertain climate change. *Global Environmental Change*, **19(2)**, 240-247.

Hallegatte, S. and P. Dumas, 2009: Can natural disasters have positive consequences? Investigating the role of embodied technical change. *Ecological Economics*, **68**, 777-786.

Hallegatte, S. and M. Ghil, 2007: *Endogenous Business Cycles and the Economic Response to Exogenous Shocks*. Working Papers 2007.20, Fondazione Eni Enrico Mattei, Milan, Italy, 20 pp.

Hallegatte, S., J.C. Hourcade, and P. Dumas, 2007: Why economic dynamics matter in assessing climate change damages: illustration on extreme events. *Ecological Economics*, **62**, 330-340.

Hallegatte, S., F. Henriet, A. Patwardhan, K. Narayanan, S. Ghosh, S. Karmakar, U. Patnaik, A. Abhayankar, S. Pohit, J. Corfee-Morlot, C. Herweijer, N. Ranger, S. Bhattacharya, M. Bachu, S. Priya, K. Dhore, F. Rafique, P. Mathur, and N. Naville, 2010: *Flood Risks, Climate Change Impacts and Adaptation Benefits in Mumbai: An Initial Assessment of Socio-Economic Consequences of Present and Climate Change Induced Flood Risks and of Possible Adaptation Options*. OECD Environment Working Papers No. 27, Organisation for Economic Co-operation and Development Publishing, Paris, France, 61 pp. dx.doi.org/10.1787/5km4hv6wb434-en.

Hamilton, J.M., D.J. Maddison, and R.S.J. Tol, 2005: Climate change and international tourism: a simulation study. *Global Environmental Change*, **15**, 253-266.

Handmer, J., 2003: We are all vulnerable. *Australian Journal of Emergency Management*, **18(3)**, 55-60.

Handmer, J. and S. Dovers, 2007: *The Handbook of Disaster and Emergency Policy and Institutions*. Earthscan, London, UK, 192 pp.

Handmer, J., C. Reed, and O. Percovich, 2002: *Disaster Loss Assessment Guidelines*. Department of Emergency Services, State of Queensland, Emergency Management Australia, Commonwealth Australia, 90 pp.

Handmer, J., S. O'Neil, and D. Killalea, 2010: *Review of Fatalities in the February 7, 2009, Bushfires*. Victorian Bushfires Royal Commission, Melbourne, Australia.

Hanson, S., R. Nicholls, N. Ranger, S. Hallegatte, J. Corfee-Morlot, C. Herweijer, and J. Chateau, 2011: A global ranking of port cities with high exposure to climate extremes. *Climatic Change*, **104**, 89-111.

Hardoy, J.E., D. Mitlin, and D. Satterthwaite, 2001: *Environmental Problems in an Urbanizing World: Finding Solutions for Cities in Africa, Asia and Latin America*. Earthscan, London, UK, 448 pp.

Harvey, N. and B. Caton, 2003: *Coastal Management in Australia*. Oxford University Press, Melbourne, Australia, 342 pp.

Hashizume, M., Y. Wagatsuma, A.S. Faruque, T. Hayashi, P.R. Hunter, B.G. Armstrong, D.A. Sack, 2008: Factors determining vulnerability to diarrhoea during and after severe floods in Bangladesh. *Journal of Water and Health*, **6(3)**, 323-332.

Hatfield, J.L., 2010: Climate impacts on agriculture in the United States: The value of past observations. In: *Handbook of Climate Change and Agroecosystems: Impact, Adaptation and Mitigation* [Hillel, D. and C. Rosenzweig(eds.)]. Imperial College Press, London, UK, pp. 239-253.

Hatfield, J.L., K.J. Boote, P. Fay, L. Hahn, C. Izaurralde, B.A. Kimball, T. Mader, J. Morgan, D. Ort, W. Polley, A. Thomson, and D. Wolfe, 2008: Agriculture. In: *The Effects of Climate Change on Agriculture, Land Resources, Water Resources, and Biodiversity in the United States*. A report by the U.S. Climate Change Science Program and the Subcommittee on Global Change Research, Washington, DC, pp. 21-74.

Hatfield, J.L., K.J. Boote, B.A. Kimball, L. Ziska, R.C. Izaurralde, D. Ort, A.M. Thomson, and D.A. Wolfe, 2011: Climate impacts on agriculture: Implications for crop production. *Agronomy Journal*, **103(2)**, 351-370.

Hay, J. and N. Mimura, 2010: The changing nature of extreme weather and climate events: risks to sustainable development. *Geomatics, Natural Hazards and Risk*, **1(1)**, 3-18.

Heger, M., A. Julca, and O. Paddison, 2008: *Analysing the Impact of Natural Hazards in Small Economies: The Caribbean Case*. UNU/WIDER Research Paper 2008/25, The World Institute for Development Economics Research, Helsinki, Finland, 27 pp.

Hein, L., M.J. Metzger, and A. Moren, 2009: Potential impacts of climate change on tourism; a case study for Spain. *Current Opinion in Environmental Sustainability*, **1**, 170-178.

Hennessy, K., C. Lucas, N. Nicholls, J. Bathols, R. Suppiah, and J. Ricketts, 2006: *Climate change impacts on fire-weather in south-east Australia*. CSIRO Marine and Atmospheric Research, Bushfire CRC and Australia Bureau of Meteorology, Clayton South Victoria, Australia, 88 pp.

Hennessy, K., B. Fitzharris, B.C. Bates, N. Harvey, S.M. Howden, L. Hughes, J. Salinger, and R. Warrick, 2007: Australia and New Zealand. In: *Climate Change 2007. Impacts, Adaptation and Vulnerability. Contribution of Working Group II to the Fourth Assessment Report of the Intergovernmental Panel on Climate Change* [Parry, M.L., O.F. Canziani, J.P. Palutikof, P.J. Van Der Linde, and C.E. Hanson (eds.)]. Cambridge University Press, Cambridge, UK, pp. 507-540.

Hess, J.J., J.N. Malilay, and A.J. Parkinson, 2008: Climate change: The importance of place. *American Journal of Preventive Medicine*, **35(5)**, 468-478.

Hilker, N., A. Badoux, and C. Hegg, 2009: The Swiss flood and landslide damage database 1972-2007. *Natural Hazards and Earth System Sciences*, **9**, 913-925.

Hinkel, J., R.J. Nicholls, A.T. Vafeidis, R.S.J. Tol, and T. Avagianou, 2010: Assessing risk of and adaptation to sea-level rise in the European Union: an application of DIVA. *Mitigation and Adaptation Strategies for Global Change*, **15**, 703-719.

Hirabayashi, Y. and S. Kanae, 2009: First estimate of the future global population at risk of flooding. *Hydrological Research Letters*, **3**, 6-9.

Hjelle, B. and G.E. Glass, 2000: Outbreak of Hantavirus infection in the Four Corners region of the United States in the wake of the 1997-1998 El Niño-Southern Oscillation. *Journal of Infectious Diseases*, **181(5)**, 1569-1573.

Hlavinka, P., M. Trnka, D. Semeradova, M. Dubrovsky, Z. Zalud, and M. Mozny, 2009: Effect of drought on yield variability of key crops in Czech Republic. *Agricultural and Forest Meteorology*, **149**, 431-442.

Hochrainer, S., and R. Mechler, 2011: Natural disaster risk in Asian megacities. A case for risk pooling? *Cities*, **28(1)**, 53-61.

Hoes, O.A.C. 2007: *Aanpak wateroverlast in polders op basis van risicobeheer*. Ph.D. thesis, Delft University of Technology, Delft, The Netherlands, 188 pp., repository.tudelft.nl/assets/uuid:69de94f9-385c-4c50-beff-8cc7b955beba/ceg_hoes_20070319.pdf.

Hoes, O., and W. Schuurmans, 2006: Flood standards or risk analyses for polder management in The Netherlands. *Irrigation and Drainage*, **55**, S113-S119.

Hoes, O.A.C., W. Schuurmans, and J. Strijker, 2005: Water systems and risk analysis. *Water Science and Technology*, **51**, 105-112.

Hoffman, R.N., P. Dailey, S. Hopsch, R.M. Ponte, K. Quinn, E.M. Hill, and B. Zachry, 2010: An estimate of increases in storm surge risk to property from sea level rise in the first half of the twenty-first century. *Weather, Climate and Society*, **2**, 271-293.

Holling, C.S. (ed.), 1978: *Adaptive environmental assessment and management*. John Wiley, New York, NY, 377 pp.

Holper, P.N., S. Lucy, M. Nolan, C. Senese, and K. Hennessy, 2007: *Infrastructure and Climate Change Risk Assessment for Victoria*. Consultancy Report to the Victorian Government prepared by CSIRO, Maunsell Australia, and Phillips Fox, Aspendale, Victoria, Australia, 84 pp.

Höppe, P. and T. Grimm, 2009: Rising natural catastrophe losses: What is the role of climate change? In: *Economics and Management of Climate Change: Risks, Mitigation and Adaptation* [Hansjürgens, B. and R. Antes (eds.)]. Springer-Verlag, Heidelberg, Germany, pp. 13-22.

Huey, C.A.D., J.J. Tewksbury, L.J. Vitt, P.E. Hertz, H.J.A. Pérez, and T. Garland, Jr., 2009: Why tropical forest lizards are vulnerable to climate warming. *Proceedings of the Royal Society, Biology*, **276**, 1939-1948.

Huggel, C., W. Haeberli, A. Kaab, D. Bieri, and S. Richardson, 2004: An assessment procedure for glacial hazards in the Swiss Alps. *Canadian Geotechnical Journal*, **41**, 1068-1083.

Hughes, T.P., A.H. Baird, D.R. Bellwood, M. Card, S.R. Connolly, C. Folke, R. Grosberg, O. Hoegh-Guldberg, J.B.C. Jackson, J. Kleypas, J.M. Lough, P. Marshall, M. Nystrom, S.R. Palumbi, J.M. Pandolfi, B. Rosen, and J. Roughgarden, 2003: Climate change, human impacts, and the resilience of coral reefs. *Science*, **301**, 929-933.

Huigen, M.G.A. and I.C. Jens, 2006: Socio-economic impact of Super Typhoon Harurot in San Mariano, Isabela, the Philippines. *World Development*, **34(12)**, 2116-2136.

Huq, S., S. Kovats, H. Reid, and D. Satterthwaite, 2007: Editorial: Reducing risks to cities from disasters and climate change. *Environment and Urbanization*, **19(1)**, 3-15.

Hurkman, W.J., W.H. Vensel, C.K. Tanaka, L. Whitehead, and S.B. Altenbach, 2009: Effect of high temperature on albumin and globulin accumulation in the endosperm proteome of the developing wheat grain. *Journal of Cereal Science*, **49**, 12-23.

Iglesias, A., L. Garrote, and F. Martín-Carrasco, 2009: Drought risk management in Mediterranean river basins. *Integrated Environmental Assessment and Management*, **5**, 11-16, doi:10.1897/IEAM_2008-044.1.

Innes, J.L., 1992: Observations on the condition of beech (*Fagus sylvatica* L.) in Britain in 1990. *Forestry*, **65(1)**, 35-60.

IPCC, 2001: *Climate Change 2001: Impacts, Adaptation and Vulnerability. Contribution of Working Group II to the Third Assessment Report of the Intergovernmental Panel on Climate Change* [McCarthy, J.J., O.F. Canziani, N.A. Leary, D.J. Dokken, and K. S. White (eds.)]. Cambridge University Press, Cambridge, UK, 1032 pp.

IPCC, 2007: *Climate Change 2007: Impacts, Adaptation and Vulnerability. Contribution of Working Group II to the Fourth Assessment Report of the Intergovernmental Panel on Climate Change* [Parry, M.L., O.F. Canziani, J.P. Palutikof, P.J. van der Linden, and C.E. Hanson (eds.)]. Cambridge University Press, Cambridge, UK, 976 pp.

IPCC, 2010: *Meeting Report of the Intergovernmental Panel on Climate Change Expert Meeting on Detection and Attribution Related to Anthropogenic Climate Change* [Stocker, T.F., C.B. Field, D. Qin, V. Barros, G.-K. Plattner, M. Tignor, P.M. Midgley, and K.L. Ebi (eds.)]. IPCC Working Group I Technical Support Unit, University of Bern, Bern, Switzerland, 55 pp.

Irish, J.L., A.E. Frey, J.D. Rosati, F. Olivera, L.M. Dunkin, J.M. Kaihatu, C.M. Ferreira, and B.L. Edge, 2010: Potential implications of global warming and barrier island degradation on future hurricane inundation, property damages, and population impacted. *Ocean & Coastal Management*, **53(10)**, 645-657, doi:10.1016/j.ocecoaman.2010.08.001.

Jones, B.M., C.D. Arp, M.T. Jorgenson, K.M. Hinkel, J.A. Schmutz, and P.L. Flint, 2009: Increase in the rate and uniformity of coastline erosion in Arctic Alaska. *Geophysical Research Letters*, **36**, L03503, doi:10.1029/2008GL036205.

Jones, R.N., 2004: Managing climate change risks. In: *The Benefits of Climate Change Policies: Analytical and Framework Issues* [Agrawal, S. and J. Corfee-Morlot (eds.)]. Organisation for Economic Co-operation and Development, Paris, France, pp. 249-297.

Jonkman, S.N., 2007: *Loss of Life Estimation in Flood Risk Assessment: Theory And Applications*. Ph.D. Thesis, Delft University of Technology, Delft, The Netherlands, 354 pp.

Jonkman, S.N., M. Bočkarjova, M. Kok, and P. Bernardini, 2008: Integrated hydrodynamic and economic modelling of flood damage in The Netherlands. *Ecological Economics*, **66**, 77-90.

Kahn, M.E., 2005: The death toll from natural disasters: The role of income, geography and institutions. *The Review of Economics and Statistics*, **87(2)**, 271-284.

Kalkstein, L., S. Greene, D. Mills, and J. Samenow, 2011: An evaluation of the progress in reducing heat-related human mortality in major U.S. cities. *Natural Hazards*, **56(1)**, 113-129, doi:10.1007/s11069-010-9552-3.

Kaplan, I.C., P.S. Levin, M. Burden, and E.A. Fulton, 2010: Fishing catch shares in the face of global change: a framework for integrating cumulative impacts and single species management. *Canadian Journal of Fisheries and Aquatic Sciences*, **67**, 1968-1982.

Karl, T.R., J.M. Melillo, and T.C. Peterson (eds.), 2009: *Global Climate Change Impacts in the United States*. U.S. Global Change Research Program, Cambridge University Press, New York, NY, 188 pp., downloads.globalchange.gov/usimpacts/pdfs/climate-impacts-report.pdf.

Karoly, D., 2010: The recent bushfires and extreme heat wave in southeast Australia. *Bulletin of Australian Meteorological and Oceanographical Societies*, **22**, 10-13.

Kazama, S., T. Kono, K. Kakiuchi, and M. Sawamoto, 2009: Evaluation of flood control and inundation conservation in Cambodia using flood and economic growth models. *Hydrological Processes*, **23**, 623-632.

Kellenberg, D.K. and A.M. Mobarak, 2008: Does rising income increase or decrease damage risk from natural disasters? *Journal of Urban Economics*, **63(3)**, 788-802.

Kettlewell, P.S., R.B. Sothern, and W.L. Koukkari, 1999: U.K. wheat quality and economic value are dependent on the North Atlantic Oscillation. *Journal of Cereal Science*, **29**, 205-209.

Kim, H.Y., T. Horie, H. Nakagawa, and K. Wada, 1996: Effects of elevated CO_2 concentration and high temperature on growth and yield of rice. II. The effect of yield and its component of Akihikari rice. *Japanese Journal of Crop Science*, **65**, 644-651.

Kim, S., Y. Tachikawa, E. Nakakita, and K. Takara, 2009: Reconsideration of reservoir operations under climate change: case study with Yagisawa Dam, Japan. *Annual Journal of Hydraulic Engineering, JSCE*, **53**, 597-611.

Kim, U. and J.J. Kaluarachchi, 2009: Climate change impacts on water resources in the upper blue Nile river basin, Ethiopia. *Journal of the American Water Resources Association*, **45(6)**, 1361-1378.

Kleinen, T. and G. Petschel-Held, 2007: Integrated assessment of changes in flooding probabilities due to climate change. *Climatic Change*, **81(3)**, 283-312.

Knowlton, K., M. Rotkin-Ellman, G. King, H.G. Margolis, D. Smith, G. Solomon, R. Trent, and P. English, 2008: The 2006 California heatwave: Impacts on hospitalizations and emergency department visits. *Environmental Health Perspectives*, **117(1)**, 61-67.

Koppe, C., G. Jendritzky, R.S. Kovats, and B. Menne, 2004: *Heat Waves: Risks and Responses*. Health and Global Environmental Change, Series No. 2, World Health Organization, WHO Regional Office for Europe, Copenhagen, Denmark, 123 pp., www.euro.who.int/__data/assets/pdf_file/0008/96965/E82629.pdf.

Korf, B., 2004: War, livelihoods and vulnerability in Sri Lanka. *Development and Change*, **35(2)**, 275-295.

Kosatsky, T., 2005: The 2003 European heatwaves. *Euro Surveillance*, **10(7)**, 148-149.

Kovats, R.S. and R. Akhtar, 2008: Climate, climate change and human health in Asian cities. *Environment and Urbanization*, **20(1)**, 165-175.

Kovats, R.S. and K. Ebi, 2006: Heatwaves and public health in Europe. *European Journal of Public Health*, **16(6)**, 592-599.

Kramer, R.A. 1995. Advantages and limitations of benefit-cost analysis for evaluating investments in natural disaster mitigation. In: *Disaster Prevention for Sustainable Development: Economic and Policy Issues* [Munasinghe, M. and C. Clarke (eds.)]. World Bank, Washington, DC, pp. 61-76.

Kshirsagar, N., R. Shinde, and S. Mehta, 2006: Floods in Mumbai: Impact of public health service by hospital staff and medical students. *Journal of Postgraduate Medicine*, **52(4)**, 312.

Kumar, R., R. Venuprasad, and G.N. Atlin, 2007: Genetic analysis of rainfed lowland rice drought tolerance under naturally-occurring stress in eastern India: Heritability and QTL effects. *Field Crops Research*, **103**, 42-52.

Kundzewicz, Z., 2003: Water and climate – The IPCC TAR perspective. *Nordic Hydrology*, **34(5)**, 387-398.

Kundzewicz, Z.W. and H.-J. Schellnhuber, 2004: Floods in the IPCC TAR perspective. *Natural Hazards*, **31**, 111-128.

Kundzewicz, Z.W., L.J. Mata, N. Arnell, P. Döll, P. Kabat, B. Jiménez, K. Miller, T. Oki, Z. Sen, and I. Shiklomanov, 2007: Freshwater resources and their management. *Climate Change 2007. Impacts, Adaptation and Vulnerability. Contribution of Working Group II to the Fourth Assessment Report of the Intergovernmental Panel on Climate Change* [Parry, M.L., O.F. Canziani, J.P. Palutikof, P.J. van der Linden, and C.E. Hanson (eds.)]. Cambridge University Press, Cambridge, UK, pp.173-210.

Kunst, A.E., C.W.N. Looman, and J.P. Mackenbath, 1993: Outdoor air temperature and mortality in The Netherlands: a time-series analysis. *American Journal of Epidemiology*, **137**, 331-341.

Kurukulasuriya, P. and R. Mendelsohn, 2008a: Crop switching as a strategy for adapting to climate change. *African Journal of Agricultural and Resource Economics*, **2(1)**, 105-126.

Kurukulasuriya, P. and R. Mendelsohn, 2008b: *How Will Climate Change Shift Agro-Ecological Zones and Impact African Agriculture?* Policy Research Working Paper Series 4717, World Bank, Washington, DC, 31 pp.

L'Hôte, Y., G. Mahé, B. Some, and J.P. Triboulet, 2002: Analysis of a Sahelian annual rainfall index from 1896 to 2000: the drought continues. *Hydrological Science Journal*, **47**, 563-572.

Lal, P.N., 2010. Vulnerability to natural disasters: an economic analysis of the impact of the 2009 floods on the Fijian sugar belt. *Pacific Economic Bulletin*, **25(2)**, 62-77.

Langner, J., R. Bergström, and V. Foltescu, 2005: Impact of climate change on surface ozone and deposition of sulphur and nitrogen in Europe. *Atmospheric Environment*, **39(6)**, 1129-1141.

Laschewski, G. and G. Jendritzky, 2002: Effects of the thermal environment on human health: an investigation of 30 years of daily mortality data from southwest Germany. *Climate Research*, **21**, 91-103.

Lavorel, S. and W. Steffen, 2004: Cascading impacts of land use through time: the Canberra bushfire disaster. In: *Global Change and the Earth System: A Planet Under Pressure* [Steffen, W., A. Sanderson, P.D. Tyson, J. Jäger, P.A. Matson, B. Moore III, F. Oldfield, K. Richardson, H.J. Schellnhuber, B.L. Turner II, and R.J. Wasson (eds.)]. Springer-Verlag, Berlin, Germany, pp. 186-188.

Leckebusch, G.C., U. Ulbrich, L. Fröhlich, and J.G. Pinto, 2007: Property loss potentials for European midlatitude storms in a changing climate. *Geophysical Research Letters*, **34**, L05703, doi:10.1029/2006GL027663.

Legesse, D., C. Vallet-Coulomb, and F. Gasse, 2003: Hydrological response of a catchment to climate and land use changes in Tropical Africa: case study South Central Ethiopia. *Journal of Hydrology*, **275**, 67-85.

Lehner, B., P. Doell, J. Alcamo, T. Henrichs, and F. Kaspar, 2006: Estimating the impact of global change on flood and drought risks in Europe: A continental, integrated analysis. *Climatic Change*, **75(3)**, 273-299.

Lemke, P., J. Ren, R.B. Alley, I. Allison, J. Carrasco, G. Flato, Y. Fujii, G. Kaser, P. Mote, R.H. Thomas, and T. Zhang, 2007: Observations: Changes in snow, ice and frozen ground. In: *Climate Change 2007: The Physical Science Basis. Contribution of Working Group I to the Fourth Assessment Report of the Intergovernmental Panel on Climate Change* [Solomon, S., D. Qin, M. Manning, Z. Chen, M. Marquis, K.B. Averyt, M. Tignor and H.L. Miller (eds.)]. Cambridge University Press, Cambridge, UK, pp. 337-383

Lemmen, D., F. Warren, J. Laeroix, and E. Bush (eds.), 2008: *From Impacts to Adaptation: Canada in a Changing Climate 2007*. Climate Change Impacts and Adaptation Division, Earth Sciences Sector, Natural Resources Canada, Ottawa, Canada, 448 pp.

Lenton, T., A. Footitt, and A. Dlugolecki, 2009: *Major Tipping Points in the Earth's Climate System and Consequences for the Insurance Sector*. World Wide Fund for Nature, Gland, Switzerland and Allianz SE, Munich, Germany, 89 pp.

Létard, V., H. Flandre, and S. Lepeltier, 2004: *France and the French response to the heatwave: Lessons from a crisis*. In: Information report of the Senate number 195(2003-2004), Paris, France, 391 pp. (in French).

Levy, B.S. and V.W. Sidel (eds.), 2000: *War and Public Health*. American Public Health Association, Washington DC, 436 pp.

Li, L.H., H.G. Xu, X. Chen, and S.P. Simonovic, 2010: Streamflow forecast and reservoir operation performance assessment under climate change. *Water Resources Management*, **24(1)**, 83-104.

Lim, J.H., J.H. Chun, and J.H. Shin, 2010: Global warming related dieback of evergreen coniferous forests in Korea due to high temperature and drought stress in winter season. Proceedings of the XXIII IUFRO 2010 World Congress, 23-28 August, Seoul. *The International Forestry Review*, **12(5)**, 44-44.

Lindenmayer, D., D. Blair, L. McBurney, and S. Banks, 2010: *Forest Phoenix: How A Great Forest Recovers After Wildfire*. CSIRO, Melbourne, Australia.

Lise, W. and R.S.J. Tol, 2002: Impact of climate on tourism demand. *Climatic Change*, **55(4)**, 429-449.

Liu, X.P., J.Q. Zhang, D.W. Zhou, Z.S. Song, and X.T. Wuet, 2006: Study on grassland fire risk dynamic distribution characteristic and management policy. *Chinese Journal of Grassland*, **28(6)**, 77-83 (in Chinese).

Loayza, N., O. Eduardo, R. Jamele, and L. Christiaensen, 2009: *Natural Disasters and Growth - Going Beyond the Averages*. Policy Research Working Paper Series 4980, World Bank, Washington, DC, 40 pp.

Lobell, D.B., M.B. Burke, C. Tebaldi, M.D. Mastrandrea, W.P. Falcon, and R.L. Naylor, 2008: Prioritizing climate change adaptation needs forfood security in 2030. *Science*, **319**, 607.

Lobell, D.B., W. Schlenker, and J. Costa-Roberts, 2011: Climate trends and global crop production since 1980. *Science*, **333(6042)**, 616-620, doi:10.1126/science.1204531.

Love, G., A. Soares, and H. Püempel, 2010: Climate change, climate variability and transportation. *Procedia Environmental Sciences*, **1**, 130-145.

Loveridge, F.A., T.W. Spink, A.S. O'Brien, K.M. Briggs, and D. Butcher, 2010: The impact of climate and climate change on infrastructure slopes, with particular reference to southern England. *Quarterly Journal of Engineering Geology and Hydrogeology*, **43(4)**, 461-472.

Lugeri, N., Z.W. Kundzewicz, E. Genovese, S. Hochrainer, and M. Radziejewski, 2010: River flood risk and adaptation in Europe – assessment of the present status. *Mitigation and Adaptation Strategies for Global Change*, **15(7)**, 621-639, doi:10.1007/s11027-009-9211-8.

Luo, Q., M.A.J. Williams, W. Bellotti, and B. Bryan, 2003: Quantitative and visual assessments of climate change impacts on South Australian wheat production. *Agricultural Systems*, **77**, 173-186.

Lyon, B., 2003: Enhanced seasonal rainfall in northern Venezuela and the extreme events of December 1999. *Journal of Climate*, **16**, 2302-2306.

Maaskant, B., S.N. Jonkman, and L.M. Bouwer, 2009: Future risk of flooding: an analysis of changes in potential loss of life in South Holland (The Netherlands). *Environmental Science and Policy*, **12**, 157-169.

MacDonald, A.M., R.C. Calow, D.M.J. MacDonald, W.G. Darling, and B.E.O. Dochortaigh, 2009: What impact will climate change have on rural groundwater supplies in Africa? *Hydrological Sciences Journal*, **54(4)**, 690-703.

MacDonald, G.M., 2010: Climate change and water in Southwestern North America special feature: water, climate change, and sustainability in the southwest. *Proceedings of the National Academy of Sciences*, **107**, 21256-21262.

Magrin, G., C. Gay García, D. Cruz Choque, J.C. Giménez, A.R. Moreno, G.J. Nagy, C. Nobre, and A. Villamizar, 2007: Latin America. In: *Climate Change 2007. Impacts, Adaptation and Vulnerability. Contribution of Working Group II to the Fourth Assessment Report of the Intergovernmental Panel on Climate Change* [Parry, M.L., O.F. Canziani, J.P. Palutikof, P.J. Van Der Linde, and C.E. Hanson (eds.)]. Cambridge University Press, Cambridge, UK, pp. 581-615.

Maloney, S.K. and C.F. Forbes, 2011: What effect will a few degrees of climate change have on human heat balance? Implications for human activity. *International Journal of Biometeorology*, **55(2)**, 147-160.

Mark, B.G. and G.O. Seltzer, 2003: Tropical glacier meltwater contribution to stream discharge: a case study in the Cordillera Blanca, Peru. *Journal of Glaciology*, **49**, 271-281.

Markandya, A. and P. Watkiss, 2009: *Potential Costs and Benefits of Adaptation Options: A Review of Existing Literature*. Technical paper prepared for the United Nations Framework Convention on Climate Change, UNFCCC, Bonn, Germany.

Marques, S.C., U.M. Azeiteiro, F. Martinhoa, and M.A. Pardala, 2007: Climate variability and planktonic communities: The effect of an extreme event (severe drought) in a southern European estuary. *Estuarine Coastal and Shelf Science*, **73(3-4)**, 725-734.

McClung, D.M. and Schaerer, P., 2006: *The Avalanche Handbook*. The Mountaineers Books, Seattle, WA, 342 pp.

McGranahan, G., D. Balk, and B. Anderson, 2007: The rising tide: assessing the risks of climate change and human settlements in low elevation coastal zones. *Environment and Urbanization*, **19(1)**, 17-37.

McGregor, G., M. Pelling, T. Wolf, and S. Gosling, 2007: *The Social Impacts of Heat Waves*. Science Report SC20061/SR6, Environment Agency, Bristol, UK.

McInnes, K.L., K.J.E. Walsh, G.D. Hubbert, and T. Beer, 2003: Impact of sea-level rise and storm surges on a coastal community. *Natural Hazards*, **30**, 187-207.

McKechnie, A.E. and B.O. Wolf, 2010: Climate change increases the likelihood of catastrophic avian mortality events during extreme heatwaves. *Biology Letters*, **6(2)**, 253-256.

McKenzie, E., B. Prasad, and A. Kaloumaira, 2005: *Economic Impact of Natural Disasters on Development in the Pacific*. Volume I, South Pacific Applied Geoscience Commission and University of the South Pacific, Suva, Fiji.

McMaster, H.J., 1999: The potential impact of global warming on hail losses to winter cereal crops in New South Wales. *Climatic Change*, **43**, 455-476.

McMichael, A.J., D.H. Campbell-Lendrum, C.F. Corvalán, K.L. Ebi, A.K. Githeko, J.D. Scheraga, and A. Woodward (eds.), 2003a: *Climate Change and Human Health: Risks and Responses*. World Health Organization, Geneva, Switzerland, 322 pp.

McMichael, A., R. Woodruff, P. Whetton, K. Hennessy, N. Nicholls, S. Hales, A. Woodward, and T. Kjellstrom, 2003b: *Human Health and Climate Change in Oceania: A Risk Assessment 2002*. Commonwealth Department of Health and Ageing, Canberra, Australia, 128 pp.

McMichael, A.J., R.E. Woodruff, and S. Hales, 2006: Climate change and human health: present and future risks. *The Lancet*, **367**, 859-869.

McMichael, A.J., P. Wilkinson, R.S. Kovats, S. Pattenden, B. Armstrong, N. Vajanapoom, E.M. Niciu, H. Mahomed, C. Kingkeow, M. Kosnik, M.S. O'Neill, I. Romieu, M. Ramirez-Aguilar, M.L. Barreto, N. Gouveia, and B. Nikiforov, 2008: International study of temperature, heat and urban mortality: the 'ISOTHURM' project. *International Journal of Epidemiology*, **37**, 1121-1131.

Mechler, R., 2004: *Natural Disaster Risk Management and Financing Disaster Losses in Developing Countries*. Verlag Versicherungswirtsch, Karlsruhe, Germany, 235 pp.

Mechler, R., 2008: *From Risk to Resilience. The Cost-Benefit Analysis Methodology*. Working Paper 1, Provention Consortium, Geneva, Switzerland.

Mechler, R. and Z.W. Kundzewicz, 2010: Assessing adaptation to extreme weather events in Europe. *Mitigation and Adaptation Strategies for Global Change*, **15(7)**, 611-620.

Meehl, G.A. and C. Tebaldi, 2004: More intense, more frequent, and longer lasting heat waves in the 21st century. *Science*, **305(5686)**, 994-997.

Meizhu, C., X. Guangji, W. Shaopeng, and Z. Shaoping, 2010: High-temperature hazards and prevention measurements for asphalt pavement. In: *Proceedings of the International Conference on Mechanic Automation and Control Engineering (MACE)*, Wuhan, China, 26-28 June 2010.Institute of Electrical and Electronics Engineers, New York, NY, pp. 1341-1344, doi:10.1109/MACE.2010.5536275.

Mendelsohn, R., 2007: What causes crop failure? *Climatic Change*, **81(1)**, 61-70.

Mendelsohn, R., A. Dinar, A. Dafelt, 2000: *Climate Change Impacts on African Agriculture*. World Bank, Washington, DC, 25 pp., www.ceepa.co.za/Climate_Change/pdf/%285-22-01%29afrbckgrnd-impact.pdf.

Meyer, M.D., 2008: *Design Standards for U.S. Transportation Infrastructure. The Implications of Climate Change*. Accession number 01104982, Transportation Research Board of the National Academies, Washington, DC, 30 pp., onlinepubs.trb.org/onlinepubs/sr/sr290Meyer.pdf.

Micklin, P., 2007: The Aral Sea disaster. *Annual Review of Earth and Planetary Sciences*, **35**, 47-72.

Miller, S., R. Muir-Wood, and A. Boissonnade, 2008: An exploration of trends in normalised weather-related catastrophe losses. In: *Climate Extremes and Society* [Diaz, H.F. and R.J. Murnane (eds.)]. Cambridge University Press, Cambridge, UK, pp. 225-247.

Mills, E., 2005: Insurance in a climate of change. *Science*, **309**, 1040-1044.

Mimura, N., L. Nurse, R.F. McLean, J. Agard, L. Briguglio, P. Lefale, R. Payet, and G. Sem, 2007: Small islands. In: *Climate Change 2007. Impacts, Adaptation and Vulnerability. Contribution of Working Group II to the Fourth Assessment Report of the Intergovernmental Panel on Climate Change* [Parry, M.L., O.F. Canziani, J.P. Palutikof, P.J. Van Der Linde, and C.E. Hanson (eds.)]. Cambridge University Press, Cambridge, UK, pp. 687-716.

Ministry for the Environment, 2008: *Coastal Hazards and Climate Change. A Guidance Manual for Local Government in New Zealand*. 2nd ed. Revised by Ramsay, D. and R. Bell, Ministry of Agriculture and Forestry, Government of New Zealand, Wellington, New Zealand, 129 pp., mfe.govt.nz/publications/climate/coastal-hazards-climate-change-guidance-manual/coastal-hazards-climate-change-guidance-manual.pdf .

Ministry of Agriculture and Forestry, 2005: *Finland's National Strategy for Adaptation to Climate Change*. Ministry of Agriculture and Forestry of Finland, Helsinki, Finland, 281 pp.

Ministry of Environment and Forest, 2005: *National Adaptation Programme of Action (NAPA), Final Report*. Government of the People's Republic of Bangladesh, Dhaka, Bangladesh.

Mirvis, V.M., 1999: Regularities of a change in the air temperature regime in Russia. In: *Investigations in the Main Geophysical Observatory*. Gidrometeoizdat, St. Petersburg, Russia.

Mitchell, J.K., 1998: Introduction: Hazards in changing cities. *Applied Geography*, **18**, 1-6.

Moench, M., R. Mechler, and S. Stapleton, 2007. Guidance note on the costs and benefits of disaster risk reduction. In: *UNISDR High Level Platform on Disaster Risk Reduction*, Geneva, Switzerland, 4-7 June 2007. International Strategy for Disaster Reduction, Geneva, Switzerland, 29 pp.

Mohammed, A.R. and L. Tarpley, 2009: High nighttime temperatures affect rice productivity through altered pollen germination and spikelet fertility. *Agricultural and Forest Meteorology*, **149**, 999-1008.

Mokrech, M., R.J. Nicholls, J.A. Richards, C. Henriques, I.P. Holman, and S. Shackley, 2008: Regional impact assessment of flooding under future climate and socioeconomic scenarios for East Anglia and North West England. *Climatic Change*, **90**, 31-55.

Molesworth, A.M., L.E. Cuevas, S.J. Connor, A.P. Morse, and M.C. Thomson, 2003: Environmental risk and meningitis epidemics in Africa. *Emerging Infectious Diseases*, **9**, 1287-1293.

Moline, M.A., N.J. Karnovsky, Z. Brown, G.J. Divoky, T.K. Frazer, C.A. Jacoby, J.J. Torres, and W.R. Fraser, 2008: High latitude changes in ice dynamics and their impact on polar marine ecosystems. *Annals of the New York Academy of Sciences*, **1134**, 267-319, doi:10.1196/annals.1439.010.

Molua, E.L. and C.M. Lambi, 2006: *The Economic Impact of Climate Change on Agriculture in Cameroon*. CEEPA Discussion Paper No. 17, Centre for Environmental Economics and Policy in Africa, Pretoria, South Africa, 33 pp.

Morin, E., T. Grodek, O. Dahan, G. Benito, C. Kulls, Y. Jacoby, G. Van Langenhove, M. Seely, and Y. Enzel, 2009: Flood routing and alluvial aquifer recharge along the ephemeral arid Kuiseb River, Namibia. *Journal of Hydrology*, **368**, 262-275.

Morrissey, S.A. and J.P. Reser, 2007: Natural disasters, climate change and mental health considerations for rural Australia. *Australian Journal of Rural Health*, **15**, 120-125.

Morton, D.C., M.E. Roessing, A.E. Camp, and M.L. Tyrrell, 2003: *Assessing the Environmental, Social, and Economic Impacts of Wildfire*. Research Paper 001, Yale University Global Institute of Sustainable Forestry, New Haven, CT, 54 pp.

Mouillot, F., S. Rambal, and R. Joffre, 2002: Simulating climate change impacts on fire frequency and vegetation dynamics in a Mediterranean-type ecosystem. *Global Change Biology*, **8**, 423-437.

Moulin, C., C.E. Lambert, F. Dulac, and U. Dayan, 1997: Control of atmospheric export of dust from North Africa by the North Atlantic Oscillation. *Nature*, **387**, 691-694.

Mozumder, P., E. Flugman, and T. Randhir, 2011: Adaptation behavior in the face of global climate change: Survey responses from experts and decision makers serving the Florida Keys. *Ocean & Coastal Management*, **54(1)**, 37-44, doi:10.1016/j.ocecoaman.2010.10.008.

Muller, C., W. Cramer, W.L. Hare, and H. Lotze-Campen, 2011: Climate change risks for African agriculture. *Proceedings of the National Academy of Sciences*, **108**, 4313-4315.

Munich Re, 2011: *TOPICS GEO, Natural catastrophes 2010, Analyses, assessments, positions*. Munich Reinsurance Company, Munich, Germany, www.munichre.com/publications/302-06735_en.pdf

Narita, D., R.S.J. Tol, and D. Anthoff, 2009: Damage costs of climate change through intensification of tropical cyclone activities: an application of FUND. *Climate Research*, **39**, 87-97.

Narita, D., R.S.J. Tol, and D. Anthoff, 2010: Economic costs of extratropical storms under climate change: an application of FUND. *Journal of Environmental Planning and Management*, **53**, 371-384.

NCDC, 2005: *State of the Climate: Hurricanes & Tropical Storms for Annual 2005*. National Oceanic and Atmospheric Administration, National Climatic Data Center, Washington, DC, 7 pp., www.ncdc.noaa.gov/sotc/tropical-cyclones/2005/13.

NCDC, 2011: *State of the Climate: Global Hazards for December 2010*. National Oceanic and Atmospheric Administration, National Climatic Data Center, Washington, DC, lwf.ncdc.noaa.gov/sotc/hazards/2010/12.

Neria, Y., A. Nandi, and S. Galea, 2008. Post-traumatic stress disorder following disasters: a systematic review. *Psychological Medicine*, **38**, 467-480.

Neumayer, E. and F. Barthel, 2011: Normalizing economic loss from natural disasters: a global analysis. *Global Environmental Change*, **22(1)**, 13-24.

Neumayer, E. and T. Plümper, 2007: The gendered nature of natural disasters: The impact of catastrophic events on the gender gap in life expectancy, 1981-2002. *Annals of the Association of American Geographers*, **97(3)**, 551-566.

New, M., 2002: Climate change and water resources in the southwestern Cape, South Africa. *South African Journal of Science*, **98**, 369-373.

Niall, S. and K. Walsh, 2005: The impact of climate change on hailstorms in southeastern Australia. *International Journal of Climatology*, **25**, 1933-1952.

Nicholls, R.J., P.P. Wong, V.R. Burkett, J.O. Codignotto, J.E. Hay, R.F. McLean, S. Ragoonaden, and C.D. Woodroffe, 2007: Coastal systems and low-lying areas. In: *Climate Change 2007. Impacts, Adaptation and Vulnerability. Contribution of Working Group II to the Fourth Assessment Report of the Intergovernmental Panel on Climate Change* [Parry, M.L., O.F. Canziani, J.P. Palutikof, P.J. van der Linden, and C.E. Hanson (eds.)]. Cambridge University Press, Cambridge, UK, pp. 315-356.

Nicholls, R.J., S. Hanson, C. Herweijer, N. Patmore, S. Hallegatte, J. Corfee-Morlot, J. Château, and R. Muir-Wood, 2008: *Ranking Port Cities With High Exposure And Vulnerability to Climate Extremes: Exposure Estimates*. OECD ENV/WKP 2007-1, Organisation for Economic Co-operation and Development, Paris, France, 62 pp.

Nicholson, S.E., X. Yin, and M.B. Ba, 2000: On the feasibility of using a lake water balance model to infer rainfall: An example from Lake Victoria. *Hydrological Sciences Journal*, **45**, 75-95.

Nilsson, C., I. Stjerquist, L. Bärring, P. Schlyter, A.M. Jönsson, and H. Samuelson, 2004: Recorded storm damage in Swedish forests 1901-2000. *Forest Ecology and Management*, **199**, 165-173.

Nkomo, J.C. and G. Bernard, 2006: *Estimating and Comparing Costs and Benefits of Adaptation Projects: Case Studies in South Africa and The Gambia*. AIACC Project No. AF 47, Assessments of Impacts and Adaptations to Climate Change, International START Secretariat, Washington, DC, 137 pp.

Nordhaus, W.D., 2010: The economics of hurricanes and implications of global warming. *Climate Change Economics*, **1**, 1-20.

Nordhaus, W.D. and J. Boyer, 2000: *Warming the World: Economic Models of Global Warming*. MIT Press, Cambridge, MA, 232 pp.

Nöthiger, C. and H. Elsasser, 2004: Natural hazards and tourism: new findings on the European Alps. *Mountain Research and Development*, **24(1)**, 24-27.

Noy, I., 2009: The macroeconomic consequences of disasters. *Journal of Development Economics*, **88(2)**, 221-231.

NRC, 2008: *Potential Impacts of Climate Change on U.S. Transportation*. National Research Council, Transportation Research Board Special Report 290, National Academy of Sciences, Washington, DC.

NWS, 2011: *National Weather Service Natural Hazard Statistics*. National Oceanic and Atmospheric Administration, National Weather Service, Office of Climate, Water, and Weather Services, Silver Spring, MD, www.nws.noaa.gov/om/hazstats.shtml.

OCDESC, 2007: *National Hazardscape Report*. Officials Committee for Domestic and External Security Coordination, Department of the Prime Minister and Cabinet, Wellington, New Zealand, 139 pp.

OECD-FAO, 2008: *OECD-FAO Agricultural Outlook 2008-2017, Highlights*. Organisation for Economic Co-operation and Development and Food and Agriculture Organization, Paris, France, 72 pp., www.fao.org/es/ESC/common/ecg/550/en/AgOut2017E.pdf.

Oh, C.H. and R. Reuveny, 2010: Climatic natural disasters, political risk, and international trade. *Global Environmental Change*, **20**, 243-254.

Ojeda, J. and L. Donnelly, 2006: *Landslides in Colombia and their Impact on Towns and Cities*. IAEG2006 Paper 112, Geological Society of London, London, UK, 13 pp., iaeg.info/iaeg2006/papers/iaeg_112.pdf.

Okuyama, Y. and S. Sahin, 2009: *Impact Estimation of Disasters: A Global Aggregate for 1960 to 2007*. World Bank Policy Research Working Paper 4963, World Bank, Washington, DC, 40 pp.

Opdam, P. and D. Wascher, 2004: Climate change meets habitat fragmentation: linking landscape and biogeographical scale levels in research and conservation. *Biological Conservation*, **117**, 285-297.

Osman-Elasha, B., M. Medany, I. Niang-Diop, T. Nyong, R. Tabo, and C. Vogel, 2006: *Impacts, Vulnerability and Adaptation to Climate Change in Africa*. Background paper for the African Workshop on Adaptation, Implementation of Decision 1/CP.10 of the UNFCCC Convention, Accra, Ghana, 21-23 September 2006, United Nations Framework Convention on Climate Change, Bonn, Germany, 54 pp.

Otero, R.C. and R.Z. Marti, 1995. The impacts of natural disasters on developing economies: Implications for the international development and disaster community. In: *Disaster Prevention for Sustainable Development: Economic and Policy Issues* [Munasinghe, M. and C. Clarke (eds.)]. World Bank, Washington, DC, pp.11-40.

Oxfam, 2007: *Adapting to Climate Change. What is Needed in Poor Countries and Who Should Pay?* Oxfam Briefing Paper 104, Oxfam International, Oxford, UK, 47 pp.

PACJA, 2009: *The Economic Cost of Climate Change in Africa*. The Pan African Climate Justice Alliance, Nairobi, Kenya, 49 pp.

Pagiola, S., K. von Ritter, and J. Bishop, 2004: *Assessing the Economic Value of Ecosystem Conservation*. World Bank Environment Department Paper No.101, World Bank, Washington, DC, 58 pp.

Palecki, M.A., S.A. Changnon, and K.E. Kunkel, 2001: The nature and impacts of the July 1999 heatwave in the midwestern United States: Learning from the lessons of 1995. *Bulletin of the American Meteorological Society*, **82(7)**, 1353-1368.

Pandey, S., D.D. Behura, R. Villano, and D. Naik, 2000: *Economic Costs of Drought and Farmers' Coping Mechanisms: A Study of Rainfed Rice Systems in Eastern India*. IRRI Discussion Paper Series 39, International Rice Research Institute, Los Banos, Philippines, 35 pp.

Park, J.H., L. Duan, B. Kim, M.J. Mitchell, and H. Shibata, 2010: Potential effects of climate change and variability on watershed biogeochemical processes and water quality in Northeast Asia. *Environment International*, **36(2)**, 212-225.

Parmenter, R., E. Yadav, C. Parmenter, P. Ettestad, and K. Gage,1999: Incidence of plague associated with increased winter-spring precipitation in New Mexico. *American Journal of Tropical Medicine and Hygiene*, **61(5)**, 814-821.

Parmesan, C., T.L. Root, and M.R. Willig, 2000: Impacts of extreme weather and climate on terrestrial biota. *Bulletin of the American Meteorological Society*, **81**, 443-450.

Parry, M.L., O.F. Canziani, J.P. Palutikof and co-authors, 2007. Technical summary. *Climate Change 2007. Impacts, Adaptation and Vulnerability. Contribution of Working Group II to the Fourth Assessment Report of the Intergovernmental Panel on Climate Change* [Parry, M.L., O.F. Canziani, J.P. Palutikof, P.J. Van Der Linde, and C.E. Hanson (eds.)]. Cambridge University Press, Cambridge, UK, pp. 23-78.

Parry, M., N. Arnell, P. Berry, D. Dodman, S. Fankhauser, C. Hope, S. Kovats, R. Nichollas, D. Satterthwaite, R. Tiffin, and R. Wheeler, 2009: *Assessing the Costs of Adaptation to Climate Change: A review of the UNFCCC and other recent estimates*. International Institute for Environment and Development and Grantham Institute for Climate Change, London, UK, 111 pp.

Pascal, M., K. Laaidi, M. Ledrans, E. Baffert, C. Caserio-Schönemann, A. Le Tertre, J. Manach, S. Medina, J. Rudant, and P. Empereur-Bissonnet, 2006: France's heat health watch warning system. *International Journal of Biometeorology*, **50**, 144-153.

Patt, A.G., M. Tadross, P. Nussbaumer, K. Asante, M. Metzger, J. Rafael, A. Goujon, and G. Brundrit, 2010: Estimating least-developed countries' vulnerability to climate-related extreme events over the next 50 years. *Proceedings of the National Academy of Sciences*, **107(4)**, 1333-1337.

Pearce, D.W., W.R. Cline, A.N. Achanta, S.R. Fankhauser, R.K. Pachauri, R.S.J. Tol, and P. Vellinga, 1996: The social costs of climate change: Greenhouse damage and the benefits of control. In: *Intergovernmental Panel on Climate Change. Working Group III. Climate Change 1995: Economic and Social Dimensions of Climate Change* [Bruce, J.P., H. Yi, and E.F. Haites (eds.)]. Cambridge University Press, Cambridge, UK, pp. 179-224.

Pearce, G., A.B. Mullan, M.J. Salinger, T.W. Opperman, D. Woods, and J.R. Moore, 2005: *Impact of Climate Change on Long-Term Fire Danger*. Research Report Number 50, New Zealand Fire Service Commission, Wellington, New Zealand, 75 pp.

Peduzzi, P., H. Dao, C. Herold, and F. Mouton, 2009: Assessing global exposure and vulnerability towards natural hazards: the Disaster Risk Index. *Natural Hazards and Earth System Sciences*, **9**, 1149-1159.

Peduzzi, P., B. Chatenoux, H. Dao, C. Herold, and G. Giuliani, 2011: *Preview Global Risk Data Platform*. UNEP/GRID and UNISDR, Geneva, Switzerland, preview.grid.unep.ch/index.php?preview=tools&cat=1&lang=eng.

Pelling, M. and J.I. Uitto, 2001: Small island developing states: natural disaster vulnerability and global change. *Global Environmental Change Part B: Environmental Hazards*, **3(2)**, 49-62.

Pelling, M., A. Özerdem, and S. Barakat, 2002: The macro-economic impact of disasters. *Progress in Development Studies*, **2(4)**, 283-305.

Perry, A., 2003: Impacts of climate change on tourism in the Mediterranean. In: *Climate Change in the Mediterranean: socio-economic perspectives of impacts, vulnerability and adaptation* [Gioponi, C., M. Schechter, and E. Elgar (eds.)]. Edward Elgar Publishing, Cheltenham, UK, pp. 279-289.

Petley, D.N., G.J. Hearn, A. Hart, N.J. Rosser, S.A. Dunning, K. Oven, and W.A. Mitchell, 2007: Trends in landslide occurrence in Nepal. *Natural Hazards*, **43(1)**, 23-44, doi:10.1007/s11069-006-9100-3.

Phillips, O.L., L. Aragao, S.L. Lewis, J.B. Fisher, J. Lloyd, G. Lopez-Gonzalez, Y. Malhi, A. Monteagudo, J. Peacock, and C.A. Quesada, 2009: Drought sensitivity of the Amazon rainforest. *Science*, **323**, 1344-1347.

Pielke Jr., R.A. 2007: Future economic damage from tropical cyclones: sensitivities to societal and climate changes. *Philosophical Transactions of the Royal Society A*, **365**, 2717-2729.

Pielke Jr., R.A. and Downton, M.W., 2000: Precipitation and damaging floods: trends in the United States, 1932-1997. *Journal of Climate*, **13(20)**, 3625-3637.

Pielke Jr., R.A., J. Rubiera, C. Landsea, M.L. Fernandez, and R. Klein, 2003: Hurricane vulnerability in Latin America and the Caribbean: normalized damage and loss potentials. *Natural Hazards Review*, **4**, 101-114.

Pielke Jr., R.A., S. Agrawala, L.M. Bouwer, I. Burton, S. Changnon, M.H. Glantz, W.H. Hooke, R.J.T. Klein, K. Kunkel, D. Mileti, D. Sarewitz, E.L. Thompkins, N. Stehr, and H. von Storch, 2005: Clarifying the attribution of recent disaster losses: a response to Epstein and McCarthy. *Bulletin of the American Meteorological Society*, **86**, 1481-1483.

Pielke Jr., R.A., J. Gratz, C.W. Landsea, D. Collins, M. Saunders, and R. Musulin, 2008: Normalized hurricane damages in the United States: 1900-2005. *Natural Hazards Review*, **9**, 29-42.

Pomeranets, K.S., 2005: *Three Centuries of Floods in St-Petersburg*. Iskusstvo, St. Petersburg, Russia, 214 pp.

Pompe, J.J. and J.R. Rinehart, 2008: Mitigating damage costs from hurricane strikes along the southeastern U.S. Coast: A role for insurance markets. *Ocean & Coastal Management*, **51(12)**, 782-788.

Power, S.A., 1994: Temporal trends in twig growth of *Fagus sylvatica* L and their relationships with environmental factors. *Forestry*, **67(1)**, 13 30.

Prasad, P.V.V., K.J. Boote, L.H. Allen, Jr., J.E. Sheehy, and J.M.G. Thomas, 2006: Species, ecotype and cultivar differences in spikelet fertility and harvest index of rice in response to high temperature stress. *Field Crops Research*, **95**, 398-411.

Pulhin, J., M. Tapia, and R. Perez, 2010: Integrating disaster risk reduction and climate change adaptation: Initiatives and challenges in the Philippines. Climate Change Adaptation and Disaster Risk Reduction: An Asian Perspective. *Community, Environment and Disaster Risk Management*, **5**, 217-235.

Quarantelli, E.L. (ed.), 1998: *What is a Disaster? Perspectives on the Question*. Routledge, London, UK, 312 pp.

Quinn, W.H., D.O. Zopf, K.S. Short, R.T.W. Kuo Yang, 1978: Historical trends and statistics of the Southern Oscillation, El Niño, and Indonesian droughts. *Fishery Bulletin*, **76**, 663-678.

Rachold, V., H. Lantuit, N. Couture, and W. Pollard (eds.), 2005: *Arctic Coastal Dynamics*. Report of the 5th International Workshop, McGill University, Montreal, Canada, 13-16 October 2004. Reports on Polar and Marine Research 506, Alfred Wegener Institute for Polar and Marine Research, Bremerhaven, Germany, 131 pp.

Raddatz, C., 2009: *The Wrath of God: Macroeconomic Costs of Natural Disasters*. Policy Research Working Paper 5039, Development Research Group, Macroeconomics and Growth Team, World Bank, Washington, DC, 35 pp.

Radeloff, V.C., R.B. Hammer, S.I. Stewart, J.S. Fried, S.S. Holcomb, and J.F. Mckeefry, 2005: The wildland–urban interface in the United States. *Ecological Applications*, **15(3)**, 799-805.

Raghavan, S. and S. Rajesh, 2003: Trends in tropical cyclone impact: a study in Andhra Pradesh, India. *Bulletin of the American Meteorological Society*, **84**, 635-644.

Ranger, N., S. Hallegatte, S. Bhattacharya, M. Bachu, S. Priya, K. Dhore, F. Rafique, P. Mathur, N. Naville, F. Henriet, C. Herweijer, S. Pohit, and J. Corfee-Morlot, 2011: An assessment of the potential impact of climate change on flood risk in Mumbai. *Climatic Change*, **104**, 139-167.

Raschky, P.A. 2008: Institutions and the losses from natural disasters. *Natural Hazards Earth Systems Science*, **8**, 627-634.

Rasmussen, T.N., 2004: *Macroeconomic Implications of Natural Disasters in the Caribbean*. IMF Working Papers WP/04/224, International Monetary Fund, Washington, DC.

Ready, R. and S. Navrud, 2006: International benefit transfer methods and validity tests. *Ecological Economics*, **60**, 429-434.

Reason, C.J.C. and A. Keibel, 2004: Tropical cyclone Eline and its unusual penetration and impacts over the Southern African mainland. *Weather Forecast*, **19**, 789-805.

Reason, C.J.C., S. Hachigonta, and R.F. Phaladi, 2005: Interannual variability in rainy season characteristics over the Limpopo region of southern Africa. *International Journal of Climatology*, **25**, 1835-1853.

Rebetez, M. and M. Dobbertin, 2004: Climate change may already threaten Scots pine stands in the Swiss Alps. *Theoretical and Applied Climatology*, **79**, 1-9.

Rehfeldt, G.E., N.L. Crookston, M.V. Warwell, and J.S. Evans, 2006: Empirical analyses of plant-climate relationships for the western United States. *International Journal of Plant Sciences*, **167(6)**, 1123-1150.

Reid, H., L. Sahlén, J. MacGregor, and J. Stage, 2007: *The economic impact of climate change in Namibia: How climate change will affect the contribution of Namibia's natural resources to its economy*. Environmental Economics Programme Discussion Paper 07-02, International Institute for Environment and Development, London, UK, 46 pp.

Reid, W.V., H.A. Mooney, A. Cropper, D. Capistrano, S.R. Carpenter, K. Chopra, P. Dasgupta, T. Dietz, A.K. Duraiappah, R. Hassan, R. Kasperson, R. Leemans, R.M. May, A.J. McMichael, P. Pingali, C. Samper, R. Scholes, R.T. Watson, A.H. Zakri, Z. Shidong, N.J. Ash, E. Bennett, P. Kumar, M.J. Lee, C. Raudsepp-Hearne, H. Simons, J. Thonell, and M.B. Zurek (eds.), 2005: *Ecosystems and Human Well-being: Synthesis*. Island Press, Washington, DC, 155 pp.

Reilly, J. and D. Schimmelpfennig, 2000: Irreversibility, uncertainty, and learning: portraits of adaptation to long term climate change. *Climatic Change*, **45(1)**, 253-278.

Revi, A., 2008: Climate change risk: an adaptation and mitigation agenda for Indian cities. *Environment and Urbanization*, **20(1)**, 207-229, doi:10.1177/0956247808089157.

Revich, B., 2008: *Climate Change Impact on Public Health in the Russian Arctic*. United Nations in the Russian Federation, Moscow, Russia, 24 pp.

Rey, G., A. Fouillet, É. Jougla, and D. Hémon, 2007: Heat waves, ordinary temperature fluctuations and mortality in France since 1971. *Population (English Edition)*, **62**, 457-485, doi:10.3917/pope.703.0457,cairn.info/revue-population-english-2007-3-page-457.htm.

Rich, P.M., D.D. Breshears, and A.B. White, 2008: Phenology of mixed woody-herbaceous ecosystems following extreme events: net and differential responses. *Ecology*, **89(2)**, 342-352.

Richard, Y., N. Fauchereau, I. Poccard, M. Rouault, and S. Trzaska, 2001: 20th century droughts in Southern Africa: spatial and temporal variability, teleconnections with oceanic and atmospheric conditions. *International Journal of Climatology*, **21**, 873-885.

Rico, M., G. Benito, A.R. Salgueiro, A. Díez-Herrero, and H. Pereira, 2008: Reported tailings dam failures. A review of the European incidents in the worldwide context. *Journal of Hazardous Materials*, **152**, 846-852.

Rignot, E., M. Koppes, and I. Velicogna, 2010: Rapid submarine melting of the calving faces of West Greenland glaciers. *Nature Geoscience*, **3**, 187-191, doi:10.1038/ngeo765.

Rittmaster, R., W. Adamowicz, B. Amiro, and R.T. Pelletier, 2006: Economic analysis of health effects from forest fires. *Canadian Journal of Forest Research*, **36(4)**, 868-877.

Rizwan, A.M., L. Dennis, and C. Liu, 2008: A review on the generation, determination, and mitigation of urban heat island. *Journal of Environmental Sciences*, **20(1)**, 120-128.

Rodell, M., I. Velicogna, and J. Famiglietti, 2009: Satellite-based estimates of groundwater depletion in India. *Nature*, **460**, 999-1002.

Rodriguez-Oreggia, E., A. de la Fuente, R. de la Torre, H. Moreno, and C. Rodriguez, 2010: *The Impact of Natural Disasters on Human Development and Poverty at the Municipal Level in Mexico*. CID Working Paper No. 43, Center for International Development at Harvard University, Cambridge, MA, USA, 34 pp.

Rogers, K. and T. Ralph (eds.), 2010: *Floodplain Wetland Biota in the Murry-Darling Basin: Water and Habitat Requirements*. CSIRO Publishing, Collingwood, Victoria, Australia, 348 pp.

Rose, A., 2007: Economic resilience to natural and man-made disasters: Multidisciplinary origins and contextual dimensions. *Environmental Hazards*, **7(4)**, 383-398.

Rosenzweig, C. and M.L. Parry, 1994: Potential impact of climate change on world food supply. *Nature*, **367**, 133-138.

Rosenzweig, C., F.N. Tubiello, R. Goldberg, E. Mills, and J. Bloomfield, 2002: Increased crop damage in the US from excess precipitation under climate change. *Global Environmental Change*, **12**, 197-202.

Rosenzweig, C., G. Casassa, D.J. Karoly, A. Imeson, C. Liu, A. Menzel, S. Rawlins, T.L. Root, B. Seguin, and P. Tryjanowski, 2007: Assessment of observed changes and responses in natural and managed systems. In: *Climate Change 2007. Impacts, Adaptation and Vulnerability. Contribution of Working Group II to the Fourth Assessment Report of the Intergovernmental Panel on Climate Change* [Parry, M.L., O.F. Canziani, J.P.Palutikof, P.J. Van Der Linde, and C.E. Hanson (eds.)]. Cambridge University Press, Cambridge, UK, pp. 79-131.

Roshydromet, 2005: *Strategic Prediction for the Period of up to 2010—2015 of Climate Change Expected in Russia and Its Impact on Sectors of the Russian National Economy*. Federal Service for Hydrometeorology and Environmental Monitoring (Roshydromet), Moscow, Russia, 24pp.

Roshydromet, 2008: *Assessment Report on Climate Change and Its Impacts on the Territory of the Russian Federation*. Federal Service for Hydrometeorology and Environmental Monitoring (Roshydromet),Vol.II, Climate Change Impacts, RIHMI-WDC, Obninsk, Kaluga Region, Russia, 288pp.

Rübbelke, D. and S. Vogele, 2011: Impacts of climate change on European critical infrastructures: The case of the power sector. *Environmental Science & Policy*, **14**, 53-63.

Saenz, R., R.A. Bissell, and F. Paniagua, 1995: Post-disaster malaria in Costa Rica. *Prehospital Disaster Medicine*, **10(3)**, 154-160.

Sanderson, M.G., C.D. Jones, W.J. Collins, C.E. Johnson, and R.G. Derwent, 2003: Effect of climate change on isoprene emissions and surface ozone levels. *Geophysical Research Letters*, **30(18)**, 1936, doi:10.1029/2003GL017642.

Satterthwaite, D., 2008: *Climate Change and Urbanization: Effects and Implications for Urban Governance*. Paper presented at the United Nations Expert Meeting on Population Distribution, Urbanization, Internal Migration and Development, New York, NY, 21-23 January 2008, United Nations Population Division, New York, NY, 29 pp., un.org/esa/population/meetings/EGM_PopDist/P16_Satterthwaite.pdf.

Satterthwaite, D., S. Huq, M. Pelling, H. Reid, and P. Romero-Lankao, 2007: *Adapting to Climate Change in Urban Areas: The Possibilities and Constraints in Low- and Middle-income Nations*. Human Settlements Working Paper Series, Climate Change and Cities No.1, International Institute for Environment and Development, London, UK, 107 pp., pubs.iied.org/pdfs/10549IIED.pdf.

Savelieva, N.I., I.P. Semiletov, G.E. Weller, and L.N. Vasilevskaya, 2004: Climate change in the northern Asia in the second half of the 20th century. *Pacific Oceanography*, **2(1-2)**, 74-84.

Schär, C. and G. Jendritzky, 2004: Climate change: Hot news from summer 2003. *Nature*, **432**, 559-560.

Schär, C., P.L. Vidale, D. Luthi, C. Frei, C. Haberli, M.A. Liniger, and C. Appenzeller, 2004: The role of increasing temperature variability in European summer heatwaves. *Nature*, **427**, 332-336.

Scheffer, M., S. Carpenter, J.A. Foley, C. Folke, and B. Walker, 2001: Catastrophic shifts in ecosystems. *Nature*, **413**, 591-596.

Schelhaas, M.J., G.J. Nabuurs, and A. Schuck, 2003: Natural disturbances in the European forests in the 19th and 20th centuries. *Global Change Biology*, **9(11)**, 1620-1633.

Schlenker, W. and D. Lobell, 2010: Robust negative impacts of climate change on African agriculture. *Environmental Research Letters*, **5(1)**, doi:10.1088/1748-9326/5/1/014010.

Schlenker, W. and M.J. Roberts, 2009: Nonlinear temperature effects indicate severe damages to U.S. crop yields under climate change. *Proceedings of the National Academy of Sciences*, **106**, 15594-15598.

Schmid, D., I. Lederer, P. Much, A. Pichler, and F. Allerberger, 2005: Outbreak of norovirus infection associated with contaminated flood water, Salzburg, 2005. *Eurosurveillance*, **10(24)**, E050616.

Schmidt, S., C. Kemfert, and E. Faust, 2009: *Simulation of Economic Losses from Tropical Cyclones in the Years 2015 and 2050: The Effects of Anthropogenic Climate Change and Growing Wealth*. Discussion paper 914, German Institute for Economic Research, Berlin, Germany, 29 pp.

Schneebeli, M., M. Laternser and W. Ammann, 1997: Destructive snow avalanches and climate change in the Swiss Alps. *Eclogae Geologicae Helvetiae*, **90**, 457-461.

Schneider, F., A. Buehn, and C.E. Montenegro, 2010. *Shadow Economies All over the World: New Estimates for 162 Countries from 1999 to 2007*. Policy Research Working Paper 5356, World Bank Development Research Group, Poverty and Inequality Team and Europe and Central Asia Region, Human Development Economics Unit, Washington, DC, 52 pp. www-wds.worldbank.org/external/default/WDSContentServer/IW3P/IB/2010/10/14/000158349_20101014160704/Rendered/PDF/WPS5356.pdf.

Schneider, S.H., W.E. Easterling, and L.O. Mearns, 2000: Adaptation: Sensitivity to natural variability, agent assumptions and dynamic climate changes. *Climatic Change*, **45**, 203-221.

Scholes, R.J. and R. Biggs (eds.), 2004: *Ecosystem Services in Southern Africa: A Regional Assessment*. A contribution to the Millennium Ecosystem Assessment, CSIR, Pretoria, South Africa, 76 pp., www.icp-confluence-sadc.org/sites/default/files/SAfMA_Regional_Report_-_final.pdf.

Schreider, S.Y., D.I. Smith, and A.J. Jakeman, 2000: Climate change impacts on urban flooding. *Climatic Change*, **47**, 91-115.

Schulze, R., J. Meigh, and M. Horan, 2001: Present and potential future vulnerability of eastern and southern Africa's hydrology and water resources. *South African Journal of Science*, **97**, 150-160.

Schwierz, C., P. Köllner-Heck, E. Zenklusen Mutter, D.N. Bresch, P.L. Vidale, M. Wild, and C. Schär, 2010: Modelling European winter wind storm losses in current and future climate. *Climatic Change*, **101**, 485-514.

Scott, D., G. McBoyle, B. Mills, and A. Minogue, 2006: Climate change and the sustainability of ski-based tourism in eastern North America. *Journal of Sustainable Tourism*, **14(4)**, 367-375

Scott, D., B. Amelung, S. Becken, J.P. Ceron, G. Dubois, S. Gossling, P. Peeters, and M.C. Simpson, 2008: *Climate Change and Tourism: Responding to Global Challenges*. United Nations World Tourism Organization, Madrid, Spain, 256 pp.

Semyonov, V.A. and A.A. Korshunov, 2006: Floods on the Russian rivers in the late 20th century and the early 21st century. *Issues of Geography and Geo-ecology*, **5**, 6-12.

Sénat. 2004. *France and the French face the canicule: The lessons of a crisis*. Information Report no. 195, Sénat, Paris, France, pp. 59-62 (in French), www.senat.fr/rap/r03-195/r03-19510.html.

Seo, N. and R. Mendelsohn, 2008: An analysis of crop choice: Adapting to climate change in Latin American farms. *Ecological Economics*, **67**, 109-116.

Shankman, D., B. D. Keim, and J. Song, 2006: Flood frequency in China's Poyang Lake Region: trends and teleconnections. *International Journal of Climatology*, **26**, 1255-1266.

Shen, C., W.-C. Wang, Z. Hao, and W. Gong, 2008: Characteristics of anomalous precipitation events over eastern China during the past five centuries. *Climate Dynamics*, **31(4)**, 463-476, doi:10.1007/s00382-007-0323-0.

Sherstyukov, A.B., 2009: *Climate Change and Its Impact in the Russian Permafrost Zone*. RIHMI-WDC, Obninsk, Russia, 127 pp.

Skidmore, M. and H. Toya, 2002: Do natural disasters promote long-run growth? *Economic Inquiry*, **40(4)**, 664-687.

Skylas, D.J., S.J. Cordwell, P.G. Hains, M.R. Larsen, D.J. Basseal, B.J. Walsh, C. Blumenthal, W. Rathmell, L. Copeland, and C.W. Wrigley, 2002: Heat shock of wheat during grain filling: proteins associated with heat-intolerance. *Journal of Cereal Science*, **35**, 175-188.

Smit, B., O. Pilifosova, I. Burton, B. Challenger, S. Huq, R. Klein, and G. Yohe, 2001: Adaptation to climate change in the context of sustainable development and equity. In: *Climate Change 2001: Impacts, Adaptation, and Vulnerability. Contribution of Working Group II to the Third Assessment Report of the Intergovernmental Panel on Climate Change* [McCarthy, J.J., O.F. Canziani, N.A. Leary, D.J. Dokken, and K.S. White (eds.)]. Cambridge University Press, Cambridge, UK, pp. 879-906.

Smith, D.E, S.B. Raper, S. Zerbini, and A. Sánchez-Arcilla (eds.), 2000: *Sea Level Change and Coastal Processes: Implications for Europe*. Proceedings of the Climate change and Sea level research in Europe: State of the art and future research needs Conference, Mataro, Spain, April 1997. Office for Official Publications of the European Communities, Luxembourg, 247 pp.

Smith, O., 2006: Trends in Transportation Maintenance Related to Climate Change. Paper presented to the *Climate Change Workshop*, Center for Transportation and Environment, Washington, DC, 29 March, 2006.

Smyth, A.W., G. Altay, G. Deodatis, M. Erdik, G. Franco, P. Gulkan, H. Kunreuther, H. Lus, E. Mete, N. Seeber, and O. Yuzugullu, 2004: Probabilistic benefit-cost analysis for earthquake damage mitigation: Evaluating measures for apartment houses in Turkey. *Earthquake Spectra*, 20(1), 171-203.

Spranger, M., 2008: Special nature of disaster risk in megacities. In: *Natural Catastrophe Risk Insurance Mechanisms for Asia and the Pacific*, Tokyo, Japan, 4-5 November 2008. Munich Reinsurance Company, Munich, Germany, 11 pp.

Stedman, J.R., 2004: The predicted number of air pollution related deaths in the UK during the August 2003 heatwave. *Atmospheric Environment*, **38**, 1083-1085.

Steffen, W., G. Love, and P.H. Whetton, 2006: Approaches to defining dangerous climate change: an Australian perspective. In: *Avoiding Dangerous Climate Change* [Schellnhuber, H.J., W. Cramer, N. Nakicenovic, T. Wigley, and G. Yohe (eds.)]. Cambridge University Press, Cambridge, UK, 392 pp.

Steiger, R., 2010: The impact of climate change on ski season length and snowmaking requirements in Tyrol, Austria. *Climate Research*, **43**, 251-262.

Stern, N., 2007: *The Economics of Climate Change: The Stern Review*. Cambridge University Press, Cambridge, UK.

Stewart, I.D., 2011. A systematic review and scientific critique of methodology in modern urban heat island literature. *International Journal of Climatology*, **31(2)**, 200-217.

Stirling, I. and C.L. Parkinson, 2006: Possible effects of climate warming on selected populations of polar bears In the Canadian Arctic. *Arctic*, **59(3)**, 261-275.

Stone, D.A. and M.R. Allen, 2005: The end-to-end attribution problem: from emissions to impacts. *Climatic Change*, **71**, 303-318.

Stott, P.A., D.A. Stone, and M.R. Allen, 2004: Human contribution to the European heatwave of 2003. *Nature*, **432**, 610-614.

Straneo, F., G.S. Hamilton, D.A. Sutherland, L.A. Stearns, F. Davidson, M.O. Hammill, G.B. Stenson and A. Rosing-Asvid, 2010: Rapid circulation of warm subtropical waters in a major glacial fjord in East Greenland. *Nature Geoscience*, **3**, 182-186, doi:10.1038/ngeo764.

Stribley, G.H. and M.R. Ashmore, 2002: Quantitative changes in twig growth pattern of young woodland beech (*Fagus sylvatica* L.) in relation to climate and ozone pollution over 10 years. *Forest Ecology and Management*, **157(1-3)**, 191-204.

Sturm, M., C. Racine, and K. Tape, 2001: Climate change: Increasing shrub abundance in the Arctic. *Nature*, **411**, 546-547.

Su, H. and G.X. Liu, 2004: Elementary analyses on the progress of grassland fire disaster information management technique. *Grassland of China*, **26(3)**, 69-71 (in Chinese).

Suzuki, T., 2009: Estimation of inundation damage caused by global warming in three major bays and western parts of Japan. *Global Environmental Research*, **14(2)**, 237-246 (in Japanese).

Swenson, S. and J. Wahr, 2009: Monitoring the water balance of Lake Victoria, East Africa, from space. *Journal of Hydrology*, **370**, 163-176.

SWIPA, 2011: *SWIPA 2011 Assessment Executive Summary: Snow, Water, Ice and Permafrost in the Arctic*. Snow, Water, Ice and Permafrost in the Arctic (SWIPA), Arctic Monitoring and Assessment Programme Secretariat, Oslo, Norway, 15pp, amap.no/swipa/SWIPA2011ExecutiveSummaryV2.pdf.

Swiss Re, 2009: *The Effects of Climate Change: An Increase in Coastal Flood Damage in Northern Europe*. Focus Report, Swiss Reinsurance Company, Zurich, Switzerland, 4 pp., media.swissre.com/documents/the_effects_of_climate_change_an_increase_in_coastal_flood_damage_in_northern_europe.pdf.

Swiss Re, 2010: *Natural Catastrophes and Man-Made Disasters in 2009: Catastrophe Claims Few Victims, Insured Losses Fall*. Sigma, No 1/2010, Swiss Reinsurance Company, Zurich, Switzerland, 36 pp.

Swiss Re, 2011: *Natural Catastrophes and Man-Made Disasters in 2010: A Year of Devastating and Costly Events*. Sigma, No 1/2011, Swiss Reinsurance Company, Zurich, Switzerland, 36 pp.

Takahashi, T., H. Nakagawa, Y. Satofuka, and K. Kawaike, 2001: Flood and sediment disasters triggered by 1999 rainfall in Venezuela; a river restoration plan for an alluvial fan. *Journal of Natural Disaster Science*, **23(2)**, 65-82.

Takara, K., S. Kim, Y. Tachikawa, and E. Nakita, 2009: Assessing climate change impact on water resources in the Tone River Basin, Japan, using super-high-resolution atmospheric model output. *Journal of Disaster Research*, **4(1)**, 12-23.

Tallis, H.M. and P. Kareiva, 2006: Shaping global environmental decisions using socio-ecological models. *Trends in Ecology & Evolution*, **21(10)**, 562-568.

Tan, J., Y. Zheng, X. Tang, C. Gup, L. Li, G. Song, X. Zhen, D. Yuan, A. Kalkstein, and F. Li, 2010: The urban heat island and its impact on heat waves and human health in Shanghai. *Journal of Biometeorology*, **54(1)**,

Tape, K., M. Sturm, and C. Racine, 2006: The evidence for shrub expansion in Northern Alaska and the Pan-Arctic. *Global Change Biology*, **12**, 686-702.

TEEB, 2009: *The Economics of Ecosystems and Biodiversity for National and International Policy Makers – Summary: responding to the value of nature*. The Economics of Ecosystems and Biodiversity, UNEP, Geneva, Switzerland, 39 pp.

Thibault, K.M. and J.H. Brown, 2008: Impact of an extreme climatic event on community assembly. *Proceedings of the National Academy of Sciences*, **105(9)**, 3410-3415.

Thomas, M.F. and M.B. Thorp, 2003: Palaeohydrological reconstructions for tropical Africa – evidence and problems. In: *Palaeohydrology: Understanding Global Change* [Benito, G. and K.J. Gregory (eds.)]. John Wiley and Sons, Chichester, UK, pp. 167-192.

Thonicke, K. and W. Cramer, 2006: Long-term trends in vegetation dynamics and forest fires in Brandenburg (Germany) under a changing climate. *Natural Hazards*, **38**, 283-300.

Thornton, P.K., P.G. Jones, T. Owiyo, R.L. Kruska, M. Herrero, P. Kristjanson, A. Notenbaert, N. Bekele, and A. Omolo, with contributions from V. Orindi, B. Otiende, A. Ochieng, S. Bhadwal, K. Anantram, S. Nair, V. Kumar, and U. Kulkar, 2006: *Mapping Climate Vulnerability and Poverty in Africa*. ILRI Report to the Department for International Development, International Livestock Research Institute, Nairobi, Kenya, 171 pp.

Tian, X, T. Matsui, S. Li, M. Yoshimoto, K. Kobaysai, and T. Hasegawa. 2010: Heat-induced floret sterility of hybrid rice (Oryza sativa L.) cultivars under humid and low wind conditions in the field of Jainghan basin, China. *Plant Production Science*, **13(3)**, 243-251.

Tol, R.S.J., 1995: The damage costs of climate change: Toward more comprehensive calculations. *Environmental and Resource Economics*, **5**, 353-374.

Tol, R.S.J., 2001: Estimates of the damage costs of climate change. *Environmental and Resource Economics*, **21**, 47-73.

Tol, R.S.J. and F.P.M. Leek, 1999: Economic analysis of natural disasters. In: *Climate, Change and Risk*. [Downing, T.E., A.A. Olsthoorn, and R.S.J. Tol (eds.)]. Routledge, London, UK, pp. 308-327.

Toole, M.J., 1997: Communicable diseases and disease control. In: *The Public Health Consequences of Disasters*. [Noji, E.K. (ed.)]. Oxford University Press, New York, NY, pp. 79-100.

Toya, H. and M. Skidmore, 2007: Economic development and the impacts of natural disasters. *Economics Letters*, **94**, 20-25.

Trenberth, K.E., P.D. Jones, P. Ambenje, R. Bojariu, D. Easterling, A. Klein Tank, D. Parker, F. Rahimzadeh, J.A. Renwick, M. Rusticucci, B. Soden and P. Zhai, 2007: Observations: Surface and atmospheric climate change. In: *Climate Change 2007: The Physical Science Basis. Contribution of Working Group I to the Fourth Assessment Report of the Intergovernmental Panel on Climate Change* [Solomon, S., D. Qin, M. Manning, Z. Chen, M. Marquis, K.B. Averyt, M. Tignor, and H.L. Miller (eds.)]. Cambridge University Press, Cambridge, UK, pp. 237-336.

Trewin, B. and H. Vermont, 2010. Changes in the frequency of record temperatures in Australia, 1957-2009. *Australian Meteorological and Oceanographic Journal*, **60**, 113-119.

Truong, G., A.E. Palm, and F. Felber, 2006: Recent invasion of the mountain birch *Betula pubescens ssp. tortuosa* above the treeline due to climate change: genetic and ecological study in northern Sweden. *Journal of Evolutionary Biology*, **20(1)**, 369-380.

Tsubo, M., S. Fukai, J. Basnayake, M. Ouk, 2009: Frequency of occurrence of various drought types and its impact on performance of photoperiod-sensitive and insensitive rice genotypes in rainfed lowland conditions in Cambodia. *Field Crops Research*, **113**, 287-296.

U.S. Census Bureau, 2010: *Coastline Population Trends in the United States: 1960 to 2008*. U.S. Department of Commerce, Economics and Statistics Administration, U.S. Census Bureau, Washington, DC, 27 pp., www.census.gov/ prod/2010pubs/p25-1139.pdf.

UN-HABITAT, 2008: *State of the World's Cities 2008/2009: Harmonious Cities*. United Nations Human Settlements Programme, Earthscan, London, UK, 259 pp.

UNCTAD, 2009: *Multi-year Expert Meeting on Transport and Trade Facilitation: Maritime Transport and the Climate Change Challenge, 16-18 February 2009*. Summary of Proceedings, UNCTAD/SDTE/TLB/2009/1, UN Conference on Trade and Development, Geneva, Switzerland, 47 pp.

UNDP, 2004: *Reducing Disaster Risk: A Challenge for Development*. United Nations Development Programme, Bureau for Crisis Prevention and Recovery, New York, NY, 146 pp.

UNDP, 2007: *Fighting Climate Change: Human Solidarity in a Divided World* [Watkins, K. (ed.)]. Human Development Report 2007/2008 of the United Nations Development Programme, Palgrave Macmillan, Basingstoke, UK, 384 pp.

UNECE, 2009: *Self-Made Cities: In search of sustainable solutions for informal settlements in the United Nations Economic Commission for Europe region*. United Nations Economic Commission for Europe, United Nations, Geneva, Switzerland, 113 pp.

UNFCCC, 1992: *United Nations Framework Convention on Climate Change*. United Nations, FCCC/INFORMAL/84 GE.05-62220 (E) 200705, Secretariat of the United Nations Framework Convention on Climate Change, Bonn, Germany, 24 pp., unfccc.int/resource/docs/convkp/conveng.pdf.

UNFCCC, 2007: *Investment and Financial Flows to Address Climate Change*. Secretariat of the United Nations Framework Convention on Climate Change, Bonn, Germany, 272 pp.

UNHCR, 2010: *2009 Global Trends: Refugees, Asylum-seekers, Returnees, Internally Displaced and Stateless Persons*. United Nations High Commissioner for Refugees, Division of Programme Support and Management, Geneva, Switzerland, 29 pp.

UNISDR, 2005: *Framework for Action 2005-2015. Building the Resilience of Nations and Communities to Disasters*. United Nations International Strategy for Disaster Reduction, Geneva, Switzerland, 22 pp., http://www.unisdr.org/2005/wcdr/intergover/official-doc/L-docs/Hyogo-declaration-english.pdf.

UNISDR, 2009, *Global Assessment Report on Disaster Risk Reduction: Risk and Poverty in a Changing Climate – Invest Today for a Safer Tomorrow*. United Nations International Strategy for Disaster Reduction Secretariat, Geneva, Switzerland, Oriental Press, Manama, Kingdom of Bahrain, 207 pp., www.preventionweb.net/english/hyogo/gar/report/index.php?id=1130&pid:34&pih:2.

UNISDR, 2011: *Global Assessment Report on Disaster Risk Reduction: Revealing Risk, Redefining Development*. United Nations International Strategy for Disaster Reduction Secretariat, Geneva, Switzerland, Information Press, Oxford, UK, 178 pp.

UNWTO, 2003. *Tourism, Peace, and Sustainable Development for Africa: Luanda, Angola, 29-30 May 2003*. United Nations World Tourism Organization, Madrid, Spain, 276 pp.

Usbeck, T., T. Wohlgemuth, M. Dobbertin, C. Pfister, A. Bürgi, and M. Rebetez, 2010: Increasing storm damage to forests in Switzerland from 1858 to 2007. *Agricultural and Forest Meteorology*, **150**, 47-55.

Uyarra, M.C., I.M. Côté, J.A. Gill, R.T.T. Tinch, D. Viner, and A.R. Watkinson, 2005: Island-specific preferences of tourists for environmental features: implications of climate change for tourism-dependent states. *Environmental Conservation*, **32(1)**, 11-19.

van der Veen, A., 2004: Disasters and economic damage: macro, meso and micro approaches. *Disaster Prevention and Management*, **13(4)**, 274-279.

van der Werf, G.R., J. Dempewolf, S.N. Trigg, J.T. Randerson, P.S. Kasibhatla, L. Giglio, D. Murdiyarso, W. Peters, D.C. Morton, G.J. Collatz, A.J. Dolman, and R.S. DeFries, 2008: Climate regulation of fire emissions and deforestation in equatorial Asia. *Proceeding of the National Academy of Sciences*, **105(51)**, 20350-20355.

Vanacker, V., M. Vanderschaeghe, G. Govers, E. Willems, J. Poesen, J. Deckers, and B. De Bievre, 2003: Linking hydrological, infinite slope stability and land use change models through GIS for assessing the impact of deforestation on landslide susceptibility in High Andean watersheds. *Geomorphology*, **52**, 299-315.

Vandentorren, S., P. Bretin, A. Zeghnoun, L. Mandereau-Bruno, A. Croisier, C. Cochet, J. Ribéron, I. Siberan, B. Declercq, and M. Ledrans, 2003: August 2003 heat wave in France: risk factors for death of elderly people living at home. *European Journal of Public Health*, **16(6)**, 583-591.

Vanham, D., E. Fleischhacker, and W. Rauch, 2009: Impact of an extreme dry and hot summer on water supply security in an alpine region. *Water Science and Technology*, **59(3)**, 469-477.

Veijalainen, N., E. Lotsari, P. Alho, B. Vehviläinen, and J. Käyhkö, 2010: National scale assessment of climate change impacts on flooding in Finland. *Journal of Hydrology*, **391(3-4)**, 333-350.

Venton, C., and P. Venton, 2004: *Disaster Preparedness Programmes in India: A Cost Benefit Analysis*. Network Paper 49, Humanitarian Practice Network at Overseas Development Institute, London, UK, 21 pp., odihpn.org/documents/networkpaper049.pdf.

Verschuren, D., K.R. Laird, and B.F. Cumming, 2000: Rainfall and drought in equatorial east Africa during the past 1,100 years. *Nature*, **403**, 410-414.

Victorian Bushfires Royal Commission, 2010: *The 2009 Victorian Bushfires Royal Commission Final Report*. State Government of Victoria, Melbourne, Australia, 42 pp., royalcommission.vic.gov.au/Commission-Reports/Final-Report.

Villalba, R. and T.T. Veblen, 1997: Regional patterns of tree population age structures in northern Patagonia: climatic and disturbance influences. *Journal of Ecology*, **85**, 113-124.

Villers, L. and I. Trejo, 2004: Evaluación de la vulnerabilidad en los sistemas forestales. Cambio Climático. In: *Una Visión desde México* [Martínez, J. and A. Fernández Bremauntz (eds.)]. Secretaría de Medio Ambiente y Recursos Naturalesand Instituto Nacional de Ecologia, Tlalpan, Mexico, pp. 239-254.

Voigt, T., H.-M. Füssel, I. Gärtner-Roer, C. Huggel, M. Zemp, and C. Marty, 2010: *Impacts of Climate Change On Snow, Ice, and Permafrost in Europe: Observed Trends, Future Projections, and Socio-Economic Relevance*. ETC/ACC Technical Paper 2010/13, The European Topic Centre on Air and Climate Change, Bilthoven, The Netherlands, 117 pp.

Vörösmarty, C.J., 2002: Global change, the water cycle, and our search for Mauna Loa. *Hydrological Processes*, **16**, 135-139, doi:10.1002/hyp.527.

Vörösmarty, C.J., E.M. Douglas, P.A. Green, and C. Revenga, 2005: Geospatial indicators of emerging water stress: an application to Africa. *Ambio*, **34**, 230-236.

Vos, F., J. Rodriguez, R. Below, and D. Guha-Sapir, 2010: *Annual Disaster Statistical Review 2009: The Numbers and Trends*. Centre for Research on the Epidemiology of Disasters, Université Catholique de Louvain, Brussels, Belgium, 38 pp.

Wamsley, T.V., M.A. Cialone, J.M. Smith, J.H. Atkinson, and J.D. Rosati, 2010: The potential of wetlands in reducing storm surge. *Ocean Engineering*, **37**, 59-68.

Wang, J.X.L. and J.K. Angell, 1999: *Air Stagnation Climatology for the United States (1948-1998)*. NOAA/Air Resources Laboratory ATLAS, No.1, Silver Spring, MD, 73 pp.

Wardekker, J.A., A. de Jong, J.M. Knoop, and J.P. van der Sluijs, 2010: Operationalising a resilience approach to adapting an urban delta to uncertain climate changes. *Technological Forecasting & Social Change*, 77, 987-998.

Washington, R. and A. Preston, 2006: Extreme wet years over southern Africa: role of Indian Ocean Sea surface temperatures. *Journal of Geophysical Research*, **111**, D15104, doi:10./029/2005 JD006724.

Washington, R., M. Harrison, and D. Conway, 2004: *African Climate Report*. Report commissioned by the UK Government to review African climate science, policy and options for action, UK Department for Environment Food and Rural Affairs and the Department for International Development, London, UK, 45 pp.

West, C.T. and D.G. Lenze, 1994: Modeling the regional impact of natural disaster and recovery: A general framework and an application to Hurricane Andrew. *International Regional Science Review*, **17(2)**, 121-150.

West, J.J., M.J. Small, and H. Dowlatabadi, 2001: Storms, investor decisions, and the economic impacts of sea level rise. *Climatic Change*, **48**, 317-342.

Westerling, A.L. and B.P. Bryant, 2008: Climate change and wildfire in California. *Climatic Change*, **87**, 231-249.

Westerling, A.L., H.G. Hidalgo, D.R. Cayan, and T.W. Swetnam, 2006: Warming and earlier spring increases western U.S. forest wildfire activity. *Science*, **313**, 940-943.

Wheeler T.R., P.Q. Craufurd, R.H. Ellis, J.R. Porter, and P.V. Vara Prasad, 2000: Temperature variability and the yield of annual crops. *Agriculture, Ecosystems and Environment*, **82**, 159-167.

Whitehead, P.G., R.L. Wilby, R.W. Battarbee, M. Kernan, and A.J. Wade, 2009: A review of the potential impacts of climate change on surface water quality. *Hydrological Sciences Journal*, **54(1)**, 101-123.

WHO, 2003: *The Health Impacts of 2003 Summer Heat-Waves*. Briefing note for the delegations of the fifty-third session of the World Health Organization Regional Committee for Europe, Vienna, Austria, 8-11 September 2003, WHO Regional Committee for Europe, Copenhagen, Denmark, 12 pp.

WHO/UNICEF, 2000: *Global Water Supply and Sanitation Assessment: 2000 Report*. World Health Organization, Geneva, 87 pp., who.int/entity/water_sanitation_health/monitoring/jmp2000.pdf.

Widawsky, D.A., and J.C. O'Toole, 1990: Prioritizing the rice biotechnology research agenda for Eastern India. Rockefeller Foundation Press, New York, NY, 86 pp.

Wilbanks, T.J., P. Romero Lankao, M. Bao, F. Berkhout, S. Cairncross, J.-P. Ceron, M. Kapshe, R. Muir-Wood and R. Zapata-Marti, 2007: Industry, settlement and society. In: *Climate Change 2007: Impacts, Adaptation and Vulnerability. Contribution of Working Group II to the Fourth Assessment Report of the Intergovernmental Panel on Climate Change* [Parry, M.L., O.F. Canziani, J.P. Palutikof, P.J. van der Linden, and C.E. Hanson (eds.)]. Cambridge University Press, Cambridge, UK, pp. 357-390.

Wilbanks, T., V. Bhatt, D. Bilello, S. Bull, J. Ekmann, W. Horak, J. Huang, M. Levine, M. Sale, D. Schmalzer, and M. Scott, 2008. *Effects of Climate Change on Energy Production and Use in the United States*. U.S. Climate Change Science Program Synthesis and Assessment Product 4.5, Report by the U.S. Climate Change Science Program and the Subcommittee on Global Change Research, Washington, DC, 84 pp.

Wilby, R.L., 2003: Past and projected trends in London's urban heat island. *Weather*, **58**, 251-260.

Wilby, R.L., 2007: A review of climate change impacts on the built environment. *Built Environment*, **33(1)**, 31-45.

Wildavsky, A. 1988: *Searching for Safety*. Transaction Books, New Brunswick, NJ, 253 pp.

Wiley, J.W. and J.M. Wunderle, Jr., 1994: The effects of hurricanes on birds, with special reference to Caribbean islands. *Bird Conservation International*, **3**, 319-349.

Wisner, B., P. Blaikie, T. Cannon, and I. Davis, 2004: *At Risk: Natural Hazards, People's Vulnerability and Disasters*. Routledge, London, UK, 471 pp.

Woodruff, R.E., S. Hales, C. Butler, and A.J. McMichael, 2005: *Climate Change Health Impacts in Australia: Effects of Dramatic CO_2 Emission Reductions*. Australian Conservation Foundation and the Australian Medical Association, Canberra, Australia, 44 pp.

Woodworth, P.L., J.M. Gregory, and R.J. Nicholls, 2005: Long term sea level changes and their impacts. In: *The Global Coastal Ocean: Multiscale Interdisciplinary Processes* [Robinson, A.R. and K.H. Brink (eds.)]. Harvard University Press, Cambridge, MA, pp. 715-753.

World Bank, 2000: *Cities, Seas and Storms: Managing Change in Pacific Island Economies*. Volume IV, Adapting to Climate Change, Papua New Guinea and Pacific Islands Country Unit, International Bank for Reconstruction and Development/World Bank, Washington, DC, 48 pp., web.worldbank.org/WBSITE/EXTERNAL/COUNTRIES/EASTASIAPACIFICEXT/PACIFICISLANDSEXTN/0,,contentMDK:20218394~pagePK:141137~piPK:217854~theSitePK:441883,00.html.

World Bank, 2006: *Clean Energy and Development: Towards an Investment Framework*. World Bank, Washington, DC, 146 pp.

World Bank, 2007: *Disasters, Climate Change, and Economic Development in Sub-Saharan Africa: Lessons and Future Directions*. Evaluation Brief 3, Independent Evaluation Group, World Bank, Washington, DC, 31 pp.

World Bank, 2010: *Economics of Adaptation to Climate Change: Synthesis Report*. World Bank, Washington, DC, 101 pp.

World Bank and UN, 2010: *Natural Hazards, UnNatural Disasters: The Economics of Effective Prevention*. International Bank for Reconstruction and Development/World Bank, Washington, DC, 254 pp., gfdrr.org/gfdrr/nhud-home.

Xiao, F. and Z. Xiao, 2010: Characteristics of tropical cyclones in China and their impacts analysis. *Natural Hazards*, **54**, 827-837

Yee, E.L., H. Palacio, R.L. Atmar, U. Shah, C. Kilborn, M. Faul, T.E. Gavagan, R.D. Feigin, J. Versalovic, and F.H. Neill, 2007: Widespread outbreak of norovirus gastroenteritis among evacuees of Hurricane Katrina residing in a large "megashelter" in Houston, Texas: lessons learned for prevention. *Clinical Infectious Diseases*, **44**, 1032-1039.

Yezer, A.M. and C.B. Rubin, 1987: *The Local Economic Effects of Natural Disasters*. National Hazard Research, George Washington University, Washington, DC.

Yohe, G., J. Neumann, and H. Ameden, 1995: Assessing the economic cost of greenhouse induced sea level rise: Methods and applications in support of a national survey. *Journal of Environmental Economics and Management*, **29**, S78-S97.

Yohe, G., J. Neumann, P. Marshall, and H. Ameden, 1996: The economic cost of greenhouse induced sea level rise in the United States. *Climatic Change*, **32**, 387-410.

Yu, G., Z. Schwartz, and J.E. Walsh, 2009: A weather-resolving index for assessing the impact of climate change on tourism related climate resources. *Climatic Change*, **95**, 551-573.

Zebisch, M., T. Grothmann, D. Schröter, C. Hasse, U. Fritsch, and W. Cramer, 2005: *Climate change in Germany–Vulnerability and adaptation of climate sensitive sectors*. Report commissioned by the Federal Environmental Agency, Germany (UFOPLAN 201 41 253), Potsdam Institute of Climate Impact Research, Potsdam, Germany, 205 pp.

Zhang, J.Q., D.W. Zhou, Z.S. Song, and Z.J. Tong, 2006: A new perception on risk assessment and risk assessment of grassland fire disaster. *Journal of Basic Science and Engineering*, **14**, 56-62 (in Chinese).

Zhang, Q., L. Wu, and Q. Liu, 2009: Tropical cyclone damages in China: 1983-2006. *Bulletin of the American Meteorological Society*, **90**, 489-495.

Zimmerman, R. and C. Faris, 2010: Infrastructure impacts and adaptation challenges. In: *Annals of the New York Academy of Sciences. 2010*. Blackwell Publishing Inc., New York, NY, pp. 63-86.

5

Managing the Risks from Climate Extremes at the Local Level

Coordinating Lead Authors:
Susan Cutter (USA), Balgis Osman-Elasha (Sudan)

Lead Authors:
John Campbell (New Zealand), So-Min Cheong (Republic of Korea), Sabrina McCormick (USA), Roger Pulwarty (USA), Seree Supratid (Thailand), Gina Ziervogel (South Africa)

Review Editors:
Eduardo Calvo (Peru), Khamaldin Daud Mutabazi (Tanzania)

Contributing Authors:
Alex Arnall (UK), Margaret Arnold (World Bank), Joanne Linnerooth Bayer (USA), Hans-Georg Bohle (Germany), Christopher Emrich (USA), Stephane Hallegatte (France), Bettina Koelle (South Africa), Noel Oettle (South Africa), Emily Polack (UK), Nicola Ranger (UK), Stephan Rist (Switzerland), Pablo Suarez (Argentina), Gustavo Wilches-Chaux (Colombia)

This chapter should be cited as:

Cutter, S., B. Osman-Elasha, J. Campbell, S.-M. Cheong, S. McCormick, R. Pulwarty, S. Supratid, and G. Ziervogel, 2012: Managing the risks from climate extremes at the local level. In: *Managing the Risks of Extreme Events and Disasters to Advance Climate Change Adaptation* [Field, C.B., V. Barros, T.F. Stocker, D. Qin, D.J. Dokken, K.L. Ebi, M.D. Mastrandrea, K.J. Mach, G.-K. Plattner, S.K. Allen, M. Tignor, and P.M. Midgley (eds.)]. A Special Report of Working Groups I and II of the Intergovernmental Panel on Climate Change (IPCC). Cambridge University Press, Cambridge, UK, and New York, NY, USA, pp. 291-338.

Table of Contents

Executive Summary 293

5.1. Introduction: Why the Local is Important 296

5.2. How Local Places Currently Cope with Disaster Risk 298
5.2.1. Emergency Assistance and Disaster Relief 299
5.2.2. Population Movements 300
5.2.3. Recovery and Reconstruction 301

5.3. Anticipating and Responding to Future Disaster Risk 302
5.3.1. Communicating Risk 302
5.3.1.1. Message Design 302
5.3.1.2. Modes and Timing of Risk Communication 302
5.3.1.3. Warnings and Warning Systems 303
5.3.2. Structural Measures 304
5.3.3. Land Use and Ecosystem Protection 306
5.3.4. Storage and Rationing of Resources 307

5.4. Building Capacity at the Local Level for Risk Management in a Changing Climate 308
5.4.1. Proactive Behaviors and Protective Actions 308
5.4.2. Empowerment for Local Decisionmaking 310
5.4.3. Social Drivers 310
5.4.4. Integrating Local Knowledge 311
5.4.5. Local Government and Nongovernment Initiatives and Practices 312

5.5. Challenges and Opportunities 313
5.5.1. Differences in Coping and Risk Management 313
5.5.1.1. Gender, Age, and Wealth 313
5.5.1.2. Livelihoods and Entitlements 314
5.5.1.3. Health and Disability 316
5.5.1.4. Human Settlements 316
5.5.2. Costs of Managing Disaster Risk and Risk from Climate Extremes 317
5.5.2.1. Costs of Impacts, Costs of Post-Event Responses 317
5.5.2.2. Adaptation and Risk Management – Present and Future 318
5.5.2.3. Consistency and Reliability of Cost and Loss Estimations at the Local Level 318
5.5.3. Limits to Local Adaptation 319
5.5.4. Advancing Social and Environmental Justice 320

5.6. Management Strategies 320
5.6.1. Basics of Planning in a Changing Climate 320
5.6.2. Community-Based Adaptation 321
5.6.3. Risk Sharing and Transfer at the Local Level 321
5.6.4. A Transformative Framework for Management Strategies 323

5.7. Information, Data, and Research Gaps at the Local Level 323

5.8. Summary 325

References 326

Executive Summary

Disasters are most acutely experienced at the local level (*high agreement, robust evidence*). The reality of disasters in terms of loss of life and property occurs in local places and to local people. These localized impacts can then cascade to have national and international consequences. In this chapter, local refers to a range of places, social groupings, experience, management, institutions, conditions, and sets of knowledge that exist at a sub-national scale. [5.1]

Developing strategies for disaster risk management in the context of climate change requires a range of approaches, informed by and customized to specific local circumstances (*high agreement, robust evidence*). These differences and the context (national to global, urban to rural) in which they are situated shape local vulnerability and local impacts. [5.1]

The impacts of climate extremes and weather events may threaten human security at the local level (*high agreement, medium evidence*). Vulnerability at the local level is attributed to social, political, and economic conditions and drivers including localized environmental degradation and climate change. Addressing disaster risk and climate extremes at the local level requires attention to much wider issues relating to sustainable development. [5.1]

While structural measures provide some protection from disasters, they may also create a false sense of safety (*high agreement, robust evidence*). Such measures result in increased property development, heightened population density, and more disaster exposure. Current regulations and design levels for structural measures may be inadequate under conditions of climate change. [5.3.2]

Sustainable land management is an effective disaster risk reduction tool (*high agreement, robust evidence*). Land management includes land use, planning, zoning, conservation zones, buffer zones, or land acquisition. Often it is difficult for local jurisdictions to implement such measures as a result of political and economic pressures for development. However, such measures are often less disruptive to the environment and more sustainable at the local level than structural measures. [5.3.3]

Humanitarian relief is often required when disaster risk reduction measures are absent or inadequate (*high agreement, robust evidence*). Such assistance is more effective when it takes local social, cultural, and economic conditions into account, acknowledges local agency in disaster response, and recognizes that the initial assistance during and immediately after disasters is nearly always locally generated. [5.2.1]

Post-disaster recovery and reconstruction provide an opportunity for reducing weather- and climate-related disaster risk and for improving adaptive capacity (*high agreement, robust evidence*). An emphasis on rapidly rebuilding houses, reconstructing infrastructure, and rehabilitating livelihoods often leads to recovering in ways that recreate or even increase existing vulnerabilities, and that preclude longer-term planning and policy changes for enhancing resilience and sustainable development. Including local actors benefits the recovery process. [5.2.3]

Disasters associated with climate extremes influence population mobility and relocation affecting host and origin communities (*medium agreement, medium evidence*). Most people return and participate in the post-disaster recovery in their local areas. If disasters occur more frequently and/or with greater magnitude, some local areas will become increasingly marginal as places to live or in which to maintain livelihoods. In such cases, migration and displacement could become permanent and could introduce new pressures in areas of relocation. For locations such as atolls, in some cases it is possible that many residents will have to relocate. In other cases, migration is an adaptation to climate change, with remittances supporting community members who remain at home. [5.2.2]

Integration of local knowledge with additional scientific and technical knowledge can improve disaster risk reduction and climate change adaptation (*high agreement, robust evidence*). Local populations document their experiences with the changing climate, particularly extreme weather events, in many different ways, and this type of self-generated knowledge induces discussions of proactive adaptation strategies and can uncover existing capacity within the community and important current shortcomings. [5.4.4]

Effectively communicating risk involves multiple pathway exchanges between decisionmakers and local citizens (*high agreement, medium evidence*). Viewing risk communication as a social process allows for effective participatory approaches, relationship building, and the production of visual, compelling, and engaging information for use by local stakeholders. [5.3.1]

Inequalities influence local coping and adaptive capacity and pose disaster risk management and adaptation challenges (*high agreement, robust evidence*). These inequalities reflect differences in gender, age, wealth, class, ethnicity, health, and disability. They may also be reflected in differences in access to livelihoods and entitlements. Understanding and increasing the awareness of coping mechanisms in the context of local-level livelihood is important to climate change adaptation planning and risk management. This signifies the need for the identification and accommodation of these differences to enhance opportunities arising from their incorporation into adaptation planning and disaster response. [5.5.1]

Ecosystem management and restoration activities that focus on addressing deteriorating environmental conditions are essential to protecting and sustaining people's livelihoods in the face of climate extremes (*high agreement, robust evidence*). Such activities include, among others, watershed rehabilitation, agro-ecology, and forest landscape restoration. Moreover, provision of better access to and control of resources will improve people's livelihoods, and build long-term adaptive capacity. Such approaches have been recommended in the past, but have not been incorporated into capacity building to date. [5.3.3]

Local-level institutions and self-organization are critical for social learning, innovations, and action; all are essential elements for local risk management and adaptation (*high agreement, medium evidence*). Adaptive capacities are not created in a vacuum – local institutions provide the enabling environment for community-based adaptation planning and implementation. Local participation (community-based organizations, development committees) contributes to empowering the most vulnerable and strengthening innovations. Addressing political and cultural issues at the local levels are fundamental to the development of any strategy aiming at sustained disaster risk management and adaptation. [5.4]

The rapid urbanization of the sub-national populations and the growth of megacities, especially in developing countries, have led to the emergence of highly vulnerable urban communities, particularly through informal settlements and inadequate land management, presenting challenges to disaster management (*high agreement, robust evidence*). Addressing these critical vulnerabilities means consideration of the social, political, and economic driving forces, including rural-to-urban migration, changing livelihoods, and wealth inequalities as key inputs into decisionmaking. [5.5.1]

Effective local adaptation strategy requires addressing a number of factors that limit the ability of local people to undertake necessary measures to protect themselves against climate extremes and disasters (*high agreement, robust evidence*). Closing the information gap is critical to reducing vulnerability of natural-resource dependent communities. Maintaining the ability of a community to ensure equitable access and entitlement to key resources and assets is essential to building local adaptive capacity in a changing climate. Moreover, capacity building and development of new skills for diversifying local livelihoods are key to flexibility in disaster reduction, improving local adaptation, and managing disasters. [5.5]

Comprehensive assessments of local disaster risk are lacking in many places (*high agreement, medium evidence*). As a foundation for management options, the methodology for locally based vulnerability assessments (exposure and sensitivity) and potential costs needs more development and testing for applications to the local context. [5.6]

Insurance is a risk transfer mechanism used at the local level (*medium agreement, medium evidence*). Risk sharing (formal insurance, micro-insurance, crop insurance) can be a tool for risk reduction and for recovering livelihoods after a disaster. Under certain conditions such tools can provide disincentives for reducing disaster risk at the local level through the transfer of the risk spatially (to other places) or temporally (to the future). [5.6.3]

Local participation supports community-based adaptation to benefit management of disaster risk and climate extremes (*medium agreement, medium evidence*). However, improvements in the availability of human and financial capital and of disaster risk and climate information customized for local stakeholders can enhance community-based adaptation. [5.6]

Data on natural disasters and disaster risk reduction are lacking at the local level, which can constrain improvements in local vulnerability reduction (*high agreement, medium evidence*). This is the case in all areas but especially so in developing countries. Local knowledge systems are often neglected in disaster risk management. There is considerable potential for adapting geographic information systems to include local-level knowledge to support disaster management activities. [5.7]

Disaster loss estimates are inconsistent and highly dependent on the scale of the analysis, and result in wide variations among community, state, province, and sub-national regions (*high agreement, robust evidence*). Indirect losses are increasingly taken into account as significant factors in precipitating negative economic impacts. Adaptation costs, though hard to estimate, can be reduced if climate change adaptation is integrated into existing disaster risk management and disaster risk management is in turn embedded in development strategies and decisionmaking. [5.5]

Mainstreaming disaster risk management into policies and practices provides key lessons that apply to climate change adaptation at the local level (*high agreement, medium evidence*). Addressing social welfare, quality of life, infrastructure, and livelihoods, and incorporating a multi-hazards approach into planning and action for disasters in the short term, facilitates adaptation to climate extremes in the longer term. [5.4, 5.5, 5.6]

The main challenge for local adaptation to climate extremes is to apply a balanced portfolio of approaches as a one-size-fits-all strategy may prove limiting for some places and stakeholders (*high confidence, medium evidence*). Successful measures simultaneously address fundamental issues related to the enhancement of local collective actions, and the creation of approaches at national and international scales that complement, support, and legitimize such local actions. [5.4, 5.6]

5.1. Introduction: Why the Local is Important

Disasters occur first at the local level and affect local people. These localized impacts can then cascade to have national and international ramifications. As a result, the responsibility for managing such risks requires the linkage of local, national, and global scales (Figure 5-1). Some disaster risk management options are bottom-up strategies, designed by and for local places, while other management options are products of global negotiations (Chapter 7) that are then implemented through national institutions (Chapter 6) to local levels. Institutions, actors, governance, and geographic units of analysis are not uniform across these scales. Even within each scale there are differences. While some communities are able to cope with disaster risks, others have limited disaster resilience and capacity to cope with present disaster risk let alone adapt to climate variability and extremes. This is the topic of this chapter: to present evidence on where disasters are experienced, how disaster risks are managed at present, and the variability in coping mechanisms and capacity in the face of climate variability and change, all from the perspective of local places and local actors. The chapter explores three themes: how disaster risks are managed at present; how the impact of climate extremes threatens human security at the local level; and the role of scale and context in shaping variability in vulnerability, coping, adaptive capacity, and the management of disaster risks and climate extremes at the local level.

The idea of local has many connotations. For the purposes of this report, local refers to a range of places, management structures, institutions, social groupings, conditions, and sets of experiences and knowledge that exist at a scale below the national level. As administrative units, local can range from villages, districts, suburbs, cities, and metropolitan areas, through to regions, states, and provinces. The conception of local includes the set of institutions (public and private) that maintain and protect local people as well as those that have some administrative control over space and resources. In these places, choices and actions for disaster risk management and adaptation to climate extremes can be initially independent of national interventions. At the local level there is traditional knowledge about disaster risk and grassroots actions to manage it. Functional or physical units such as watersheds, ecological zones, or economic regions operate at the local level, including the private and public institutions that govern their use and management. Each of the differing connotations of local means that there are differing approaches to and contents of disaster risk management practice, differing stakeholders and interest groups, and more significantly, differing relations with the national and international levels (Adger et al., 2005). We recognize that states and provinces in many countries are large complex entities with similar powers as smaller nations. Where we discuss states and provinces and similar administrative structures in this chapter, we refer to them as sub-national for clarification purposes.

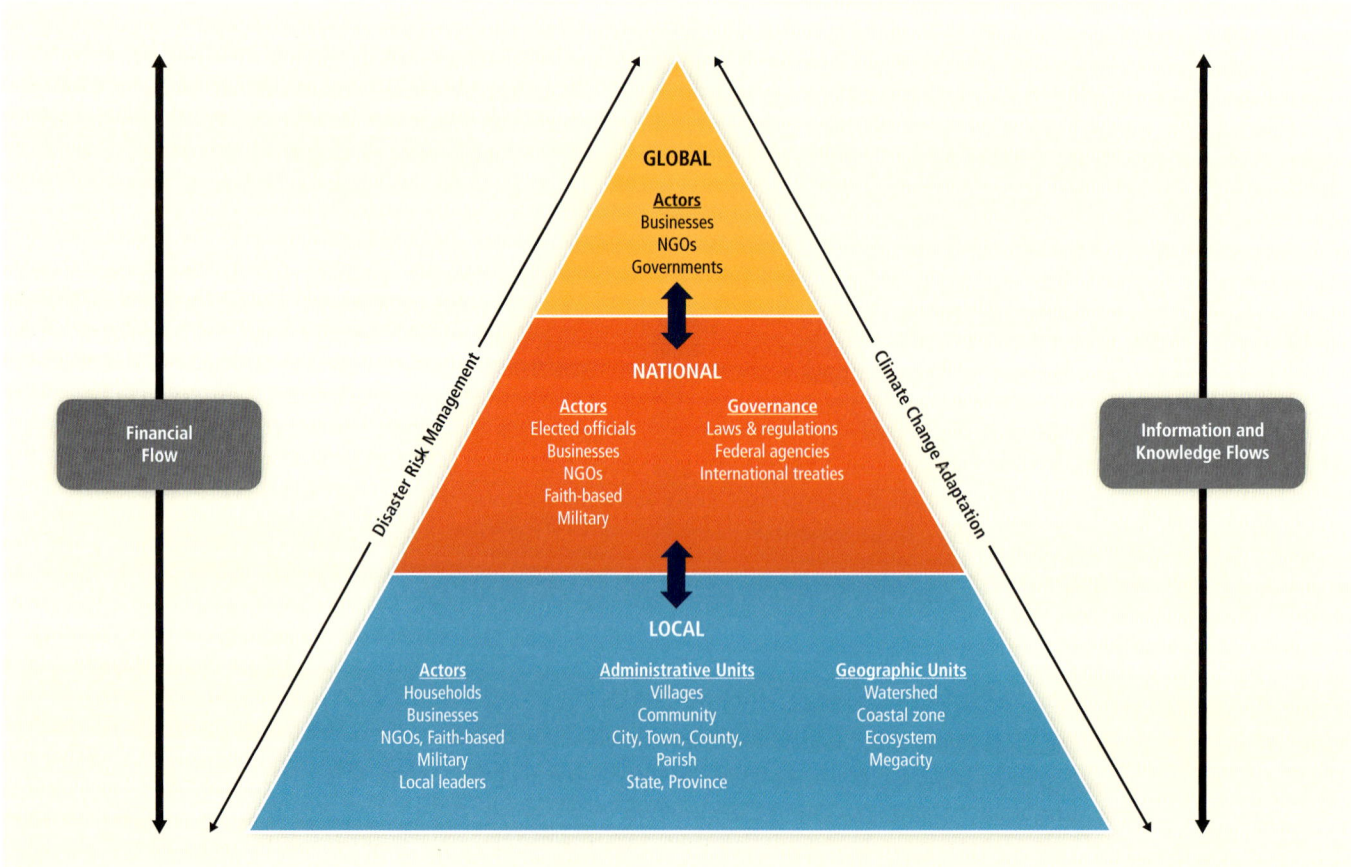

Figure 5-1 | Linking local to global actors and responsibilities.

Box 5-1 | Climate Change and Violent Conflict

Linking climate change and violent conflict is controversial. The conceptual debate links climate change to resource scarcity (or those essential resources to support livelihoods), which in turn leads to human insecurity. At the local scale, there are two distinct outcomes: armed conflict or migration, with the latter potentially leading to increased conflict in the receiving locality (Barnett and Adger, 2007; Nordås and Gleditsch, 2007). For example, some research suggests that environmental stresses feed the tensions between localities as they compete for land to support their livelihoods (Kates, 2000; Barnett, 2003; Osman-Elasha and El Sanjak, 2009). Extreme events such as droughts and heat waves could increase these tensions in areas already facing situations of water scarcity and environmental degradation, giving rise to conflicts resulting in the dislocation of large numbers of refugees and people within and across borders. However, there is limited agreement and evidence to support the link between climate change and violent conflict, especially in Africa (Burke et al., 2009; Buhaug, 2010). While the causal chain suggested in the literature (climate change increases the risk of violent conflict) has found currency within the policy community, it has not been adequately substantiated in the scientific literature (*low agreement* and *limited evidence*). Where empirical studies exist, they are methodologically flawed in a number of ways: not controlling for population size; focusing only on conflict cases and not all migration instances; using aggregated, not disaggregated climate data at sub-national scales; and having inherent inconsistencies in the time frames used (short-term variability in violent conflict; longer-term variability in climate). More research on the local climate-conflict nexus is warranted in order to determine if a causal linkage exists.

Local places vary in their disaster experience, who and what is at risk, the potential geographical extent of the potential impact and responses, and in stakeholders and decisionmakers. Local places have considerable experience with short-term coping responses and adjustments to disaster risk (UNISDR, 2004), as well as with longer-term adjustments such as the establishment of local flood defenses, the selection of drought resistant crops, or seasonal or longer migration by one or more family members. For example, the use of remittances is a substantial source of post-disaster income and is regularly used as a means for diversifying livelihoods to enhance resilience and to proactively cope with extremes (Adger et al., 2002).

Climate-sensitive hazards such as flooding, tropical cyclones, drought, heat, and wildfires regularly affect many localities with frequent, yet low-level, losses (UNISDR, 2009). Because of their frequent occurrence, localities have developed extensive reactive disaster risk management practices. However, disaster risk management also entails the day-to-day struggle to improve livelihoods, social services, and environmental services. Local response and long-term adaptation to climate extremes will require disaster risk management that acknowledges the role of climate variability. This can mean a modification and expansion of local disaster risk management principles and experience through innovative organizational, institutional, and governmental measures at all jurisdictional levels (local, national, international). Institutionally-driven arrangements may constrain or impede local actions and ultimately limit the coping capacity and adaptation of local places.

Local communities routinely experience hazard impacts, with many resulting from extreme weather and climate events (see Chapter 3). The significance of discussing these from the local perspective is that extreme weather and climate events will vary from place to place and not all places have the same experience with that particular initiating event. Research demonstrates that disaster experience influences proactive behaviors in preparing for and responding to subsequent events (see Section 5.4.1). In the context of climate change, some localities could be experiencing certain types of hazards for the first time and not have the existing capacities for preparedness and response. For example, Hurricane Catarina, the first South Atlantic hurricane that made landfall as a category 1 storm just north of Porto Alegre, Brazil, in March 2004 (McTaggart-Cowan et al., 2006), was the region's first local experience with a hurricane. However, there is *low confidence* in attribution of any long-term increases in hurricane formation in this ocean region where tropical cyclones had not been previously recorded (Table 3-2). Finally, not all of the extreme events become severe enough to cause a disaster of national or international magnitude, yet they will create ongoing problems for local disaster risk management.

The second theme of the chapter examines how climate extremes could threaten the human security of local populations. Because these risks often affect the basic functioning of society, it is increasingly recognized that climate change adaptation and disaster risk management should be integral components of development planning and implementation to increase sustainability (Thomalla et al., 2006; see Box 5-1). In other words, both have to be mainstreamed into national development plans, poverty reduction strategies, sectoral policies, and other development tools and techniques (UNDP, 2007). For example, rural communities in many world regions face greater risks of livelihood loss resulting from flooding of low-lying coastal areas, water scarcity and drought, decline in agricultural yields and fisheries resources, and loss of biological resources (Osman-Elasha and Downing, 2007). In some African countries where recurrent floods are closely linked with El Niño-Southern Oscillation (ENSO) events (Ward et al., 2010), the result is major economic and human loss seen in places such as Mozambique (Mirza, 2003; Obasi, 2005) and Somalia. For such communities, with less developed infrastructure and health services, the impacts of floods are often further exacerbated by health problems associated with water scarcity and quality, such as malnutrition, diarrhea, cholera, and malaria (Kabat et al., 2002).

In order to develop preparedness measures for disaster risk management and climate adaptation, the vast contextual differences of localities will have to be considered. They include differences in population characteristics that influence vulnerability, differences in settlement patterns ranging from urban to rural, differences in administrative units from municipalities to provincial governments, and differences within developing and developed country contexts. Given the wide disparities, it is clear that single solutions for disaster risk management are not possible. For example, there are differences between urban and rural communities in terms of disaster and climate change vulnerability and disaster risk and adaptation options. Given the rapid pace of urbanization and diffusion of communication and transportation networks into distant areas, the sharp distinction between urban and rural is less visible in many areas. In its place is a continuum with local places exhibiting both rural and urban characteristics with a mix of vulnerabilities and jurisdictional issues that are neither totally urban nor rural (McGregor et al., 2006; Aragon-Durand, 2007).

Scalar considerations must also be emphasized in planning. Efforts to forge greater and more equitable capacity at the local scale have to be supported by policies at the national level to increase the ability of local institutions and communities to cope with present and future risks from climate-sensitive hazards. To effectively reduce vulnerabilities to hazards associated with climate change, coordination across different levels and sectors is required, in addition to the involvement of a broad range of stakeholders beginning at the local level (UNISDR, 2004; DFID, 2006; Tearfund., 2006; Devereux and Coll-Black, 2007; Davies, 2009). The larger global context within which a locality is situated affects outcomes. It is possible that the history of resource exploitation, globalization, and the processes of development as currently practiced may be increasing, rather than reducing disaster vulnerability at the local level (see Chapter 2). Those choosing strategies for reducing disaster risk and adapting to climate change, especially in developing countries, need to take these processes into account (UNISDR, 2009).

These contextual factors are critical to planning for climate extremes. They suggest the need for strengthening coordination between climate change adaptation and disaster risk management locally that will in turn improve the implementation of plans (Mitchell and van Aalst, 2008). Such coordination is also needed in order to avoid any negative impacts across different sectors or scales that could potentially result from fragmented adaptation and development plans. This is evident in the implementation of some of the adaptation strategies, such as large-scale agriculture, irrigation, and hydroelectric development, that may benefit large groups or the national interests but may also harm local, indigenous, and poor populations (Kates, 2000; Rojas Blanco, 2006). Some sources believe that it is essential that any new disaster risk reduction or climate change adaptation strategies must be built on strengthening local actors and enhancing their livelihoods (Osman-Elasha, 2006a). Moreover, a key aspect of planning for adaptation at the local level is the identification of the differentiated social impacts of climate change based on gender, age, disability, ethnicity, geographical location, livelihood, and migrant status (Tanner and Mitchell, 2008). Emphasis

> **FAQ 5.1 | Why is the local context important in climate change adaptation and disaster risk management?**
>
> In the context of this report, the local refers to a range of places (community, city, province, region, state), management structures, institutions, social groupings, conditions, and sets of experiences and knowledge that exist at a scale below the national level. It also includes the set of institutions (public and private) that maintain and protect social relations as well as those that have some administrative control over space and resources. The definition of the local influences the context for disaster risk management, the experience of disasters, and conditions, actions, and adaptation to climate changes. Local is important because locals respond and experience disasters at first hand, they retain local and traditional knowledge valuable for disaster reduction and adaptation plans, and lastly they implement adaptation plans.

needs to be given to identifying the adaptation measures that serve the most vulnerable groups, address their urgent needs, and increase their resilience. This often means using a more coordinated and integrated management approach with the involvement of diverse stakeholder groups (Sperling and Szekely, 2005), which may assist in avoiding maladaptation across sectors or scales and provide for win-win solutions.

5.2. How Local Places Currently Cope with Disaster Risk

Local people everywhere have developed skills, knowledge, and management systems that enable them to interact with their environment. Often these interactions are beneficial and provide the livelihoods that people living in local places depend on. At the same time communities have developed ways of responding to disruptive environmental events. These coping mechanisms include measures that seek to modify the impacts of disruptive events, modify some of the attributes or environmental aspects of the events themselves, and/or actions to share or reduce the disaster risk burdens (Burton et al., 1993). By the same token, some actions taken at local levels (e.g., deforestation and coral mining) may also increase disaster risks. It is important to acknowledge that while climate change may alter the magnitude and/or frequency of some climatic extremes (see Chapter 3), other environmental, social, political, or economic processes (many of them also global in scale) are affecting the abilities of communities to cope with disaster risks and climate-sensitive hazards (Wisner et al., 2004; Adger and Brown, 2009). Accordingly, disaster losses have increased significantly in recent decades (UNDP, 2004; UNISDR, 2004). These social, economic, and political processes are complex and deep seated and present major obstacles to reducing disaster risk, and may constrain efforts to reduce community vulnerabilities to extreme events under conditions of climate change. In Section 5.2 we outline three common local-level coping strategies: emergency assistance and disaster relief, population movements, and recovery and reconstruction.

5.2.1. Emergency Assistance and Disaster Relief

Humanitarian assistance is often required when other measures to reduce disasters have been unsuccessful, and plays a critical role in helping local people cope with the effects of disasters. Such relief often helps to offset distress and suffering at the local level and to assist in recovery and rehabilitation. Sometimes external relief is unnecessary or inappropriate because the local people affected by disasters often are not completely helpless or passive and are capable of helping themselves (Cuny, 1983; De Ville de Groyet, 2000). This view is sustained by commonplace definitions of disasters as situations where communities or even countries cannot cope without external assistance (Cuny, 1983; Quarantelli, 1998).

It is important to realize that the first actors providing assistance during and after disasters are members of the affected community (De Ville de Groyet, 2000) who provide relief through local charities, kinship networks, or local governments. In isolated communities such as those in the outer islands of small-island developing states, external assistance may be subject to considerable delay and self-help is an essential element of response, especially in the period before assistance arrives. Typically, emergency assistance and disaster relief in developed countries comes in the form of assistance from national and state/provincial level governments to local communities. The provision of international relief is usually from members of the Organisation for Economic Co-operation and Development to developing countries (Development Initiatives, 2009). The international provision of disaster relief to local places has become highly sophisticated and much broader in scope over the past two decades, involving both development and humanitarian organizations, with the increasing recognition that external relief providers make use of local knowledge in planning their relief efforts (Morgan, 1994; Darcy and Hofmann, 2003; Méheux et al., 2010). The relief itself includes such things as assistance in post-disaster assessment, food provision, water and sanitation, medical assistance and health services, household goods, temporary shelter, transport, tools and equipment, security, logistics, communications, and community services (Bynander et al., 2005; Cahill, 2007). Many of these activities are organized into clusters of specialists from multilateral organizations and nongovernmental organizations (NGOs), among others, coordinated by the United Nations.

While much of the relief tends to be organized at more of a national and international scale than local scale, the distribution and use of relief occur at the local level. From this perspective it is vital to understand what is locally appropriate in terms of the type of relief provided, and how it is distributed (Darcy and Hofmann, 2003; Kovác and Spens, 2007). Similarly, local resources and capacities should be utilized as much as possible (Beamon and Balcik, 2008). There has also been a trend towards international humanitarian organizations working with local partners, although on occasion this can result in the imposition of external cultural values resulting in resentment or resistance (Hillhorst, 2002).

Relief, nevertheless, is often a critically important strategy for coping. Relief organizations have built capacity based on experience in recent years, have become increasingly accountable, and are obliged to follow humanitarian principles. Despite these improvements, some problems remain. Relief cannot cover all losses, most of which are borne locally. Relief can undermine local coping capacities and reduce resilience and sustainability (Susman et al., 1983; Waddell, 1989), reinforce the status quo that was characterized by vulnerability (O'Keefe et al., 1976), and in some cases, serve to remove independence or autonomy from disaster 'victims' so that ownership of the event and control over the recovery phase is lost at the local level (Hillhorst, 2002). Relief is often inequitably distributed and in some disasters there is insufficient relief. Corruption is also a factor in some disaster relief operations with local elites often benefiting more than others (Pelling and Dill, 2010). Humanitarian organizations are increasingly aware of these concerns and many are addressing them through coordination of activities, addressing gendered inequalities, and working in partnership with local organizations in disaster relief. There is also a growing recognition of the need for accountability in humanitarian work (Humanitarian Accountability Partnership International, 2011).

Not all disasters engender the same response, as local communities receive different levels of assistance. For example, those people most affected by a small event can suffer just as much as a globally publicized big event but are sometimes overlooked by relief agencies. Fast onset and unusual disasters such as tsunamis generate considerably more public interest and contributions from governments, NGOs, and the public, sometimes referred to as the CNN factor (Olsen et al., 2003). Disasters that are overshadowed by other newsworthy or media events, such as coverage of major sporting events, are often characterized by lower levels of relief support (Eisensee and Stromberg, 2007). Where there is widespread media coverage, NGOs and governments are often pressured to respond quickly with the possibility of an oversupply of relief and personnel. This has worsened in recent times when reporters are 'parachuted' into disaster sites often in advance of relief teams but they have little understanding of the contextual factors that often underlie vulnerability to disasters (Silk, 2000). Such media coverage often perpetrates disaster myths such as the prevalence of looting, helplessness, and social collapse, putting pressure on interveners to select military options for relief when humanitarian assistance would be more helpful (Tierney et al., 2006).

Relief is politically more appealing than disaster risk management (Seck, 2007) and it often gains much greater political support and funding than measures that would help offset the need for it in the first place. Providing relief reflects well on politicians (both in donor and recipient countries) who are seen to be caring, taking action, and responding to public demand (Eisensee and Stromberg, 2007).

> **FAQ 5.2 | What lessons have been learned about effective disaster management and climate change adaptation at the local scales?**
>
> In fostering sustainable and disaster-resilient areas, local response to climate extremes will require disaster risk management that acknowledges the role of climate variability and change and the associated uncertainties and that will contribute to long-term adaptation. In order to anticipate the risks and uncertainties associated with climate change there are a number of emerging approaches and responses at the local level. One set of responses focus on integrating information about changing climate risks into disaster planning and scenario assessments of the future. Setting up plans in advance, for example, enabled communication systems to be strengthened before the extreme event struck. Another is community-based adaptation (CBA), which helps to define solutions for managing risks while considering climate change. CBA responses provide increased participation by locals and recognition of the local context and the access to adaptation resources and promote adaptive capacity within communities. A critical factor in community-based actions is that community members are empowered to take control of the processes involved. Scaling up community-based approaches poses a challenge as well as integrating climate information and other interventions such as ecosystem management and restoration, watershed rehabilitation, agroecology, and forest landscape restoration. These types of interventions protect and enhance natural resources at the local scale, improve local capacities to adapt to future climate, and may also address immediate development needs.

Major shares of the costs of disaster relief and recovery still fall on the governments of disaster-affected countries. Bilateral relief is limited to materials from donor countries and most relief is subject to relatively strict criteria to reduce perceived levels of corruption. In both cases, flexibility is heavily restricted. Relief can also produce local economic distortions such as causing shops to lose business as the market becomes flooded with relief supplies. These problems can be overcome by directly transferring cash to local people to buy building materials, seed, and the like. Such programs have performed well where local supplies are available (Farrington and Slater, 2009). At the same time, there is the view that disaster relief can create a culture of dependency and expectation at the local level (Burby, 2006), where disaster relief becomes viewed as an entitlement program as local communities are not forced to bear the responsibility for their own locational choices, land use, and lack of mitigation practices.

5.2.2. Population Movements

A second coping strategy is population movements. Natural disasters are linked with population movements in a number of ways (Hunter, 2005; Perch-Nielson et al., 2008; Warner et al., 2010). Evacuations occur before, during, and after some disaster events. Longer-term relocation of affected communities sometimes occurs. Relocations can be temporary or permanent. These different forms of population movements have variable social, psychological, health, and financial implications for the communities concerned. Population movements may also be differentiated on the basis of whether the mobility is voluntary or forced (displacement) and whether or not international borders are crossed. Most contemporary research views population mobility as a continuum from completely voluntary movements to completely forced migrations (Laczko and Aghazarm, 2009). The United Nations Office for the Coordination of Humanitarian Affairs and the Internal Displacement Monitoring Centre estimated that at least 36 million people were displaced by natural disasters in 2008. While these displaced people would come from and arrive at local origins and destinations there is little information on the local implications and time frames of the displacement (UNOCHA and IDMC, 2009).

Where climate change increases the marginality of livelihoods and settlements beyond a sustainable level, communities may be forced to migrate or be displaced (McLeman and Smit, 2006). While migration typically has many causes, of which the environment (including climate) is just one factor, extremes often serve as precipitating events (Hugo, 1996). Furthermore, a number of researchers consider that climate-related migration, other than forced displacement, may not necessarily be a problem and indeed may be a positive adaptive response, with people who remain at the place of origin benefitting from remittances (Barnett and Webber, 2009; Tacoli, 2009). Nomadic pastoralists migrate as part of their livelihoods but often respond to disruptive events by modifying their patterns of mobility (Anderson et al., 2010). Migration is highly gendered in terms of both drivers and impacts, which differ between men and women, although it is not clear how these differences might be played out in the context of climate change (Hugo, 2010).

Global estimations provide little insight into the local implications of such large scale migratory patterns. Migration will have local effects, not only for the communities generating the migrants, but those communities where they may settle. Barnett and Webber (2009) also note that the less voluntary the migration choice is, the more disruptive it will become. In the context of dam construction, for example Hwang et al. (2007) found that communities anticipating forced migration experienced stress. Hwang et al. (2010) also found that forced migration directly led to increased levels of depression and the weakening of social safeguards in the relocation process. Much post-disaster relocation is temporary, which is also associated with psychological and social effects such as disruption of social networks and trauma (Neria et al., 2009).

One outcome of climate change is that entire communities could be required to relocate and in some cases, such as those living in atoll countries, the relocation will be international. Such relocation can have significant social, cultural, and psychological impacts (Campbell, 2010b). Community relocation schemes are those in which whole communities are relocated to a new non-exposed site. Perry and Lindell (1997) examined one such instance in Allenville, Arizona. They developed a set of five principles for achieving positive outcomes in relocation projects: 1) the community to be relocated should be organized; 2) all potential relocatees should be involved in the relocation decisionmaking process; 3) citizens must understand the multi-organizational context in which the relocation is to be conducted; 4) special attention should be given to the social and personal needs of the relocatees; and 5) social networks need to be preserved. For many communities relocation is difficult, especially in those communities with communal land ownership. In the Pacific Islands, for example, relocation within one's own lands is least disruptive but leaving it completely is much more difficult, as is making land available for people who have been relocated (Campbell, 2010b).

5.2.3. Recovery and Reconstruction

Recovery and reconstruction include actions that seek to establish or re-establish the everyday life of the locality affected by disaster (Hewitt, 1997). Often reconstruction enables communities and businesses to return to the same conditions that existed prior to the disaster, and in so doing create the potential for further similar losses, thus reproducing the same exposure that resulted in disaster in the first place (Jha et al., 2010). Recovery and reconstruction (especially housing rehabilitation and rebuilding) are among the more contentious elements of disaster response. One of the major issues surrounding recovery is the lack of clarity between recovery as a process and recovery as an outcome. The former emphasizes betterment processes where pre-existing vulnerability issues are addressed. The latter focuses on the material manifestation of recovery such as building houses or infrastructure. Often following large disasters, top-down programs result in rebuilding houses but fail to provide homes (Petal et al., 2008). Moreover, haste in reconstruction, while achieving short-term objectives, often results in unsustainable outcomes and increasing vulnerability (Ingram et al., 2006). As seen in the aftermath of Hurricane Katrina, there are measureable local disparities in recovery, leading to questions of recovery for whom and recovery to what (Curtis et al., 2010; Finch et al., 2010; Stevenson et al., 2010). There are a number of obstacles to effective and timely reconstruction including lack of labor, lack of capacity among local construction companies, material shortages, resolution of land tenure considerations, and insufficiency of funds (Keraminiyage et al., 2008). While there is urgency to have people re-housed and livelihoods re-established, long-term benefits may be gained through carefully implemented reconstruction (Hallegatte, 2008; Hallegatte and Dumas, 2009) in order to achieve greater disaster resilience.

Most research on recovery and reconstruction has tended to focus on housing and the so-called lifelines of infrastructure: electricity, water supply, and transport links. Less is published on the equally important, if indeed not more so, rehabilitation of livelihoods, and addressing the problems of power inequities that often include land and resource grabbing by the economic and politically powerful after disaster in both developed and developing countries. Agricultural rehabilitation (e.g., the provision of seeds, planting material, fertilizers, and stock, and the remediation of land) is particularly important where local livelihoods are directly affected such as in subsistence or semi-subsistence societies (Dorosh et al., 2010). In addition, some climate-related disaster events, such as droughts, do not always directly destroy the built environment infrastructure (like flooding or tropical cyclones) so the rehabilitation of livelihoods, in particular sustainable livelihoods, becomes an important aspect of disaster risk reduction and development (Nakagawa and Shaw, 2004).

As with relief, major problems can occur where planning and implementation of recovery and reconstruction is taken out of the hands of the local communities concerned. In addition, the use of inappropriate (culturally, socially, or environmentally) materials and techniques may render rebuilt houses unsuitable for their occupants (Jha et al., 2010). As Davidson et al. (2007) found, this is often the case and results in local community members having little involvement in decisionmaking for the recovery process; instead they are used to provide labor. It is also important to acknowledge that post-disaster recovery often does not reach all community members and in many recovery programs, the most vulnerable, those who have suffered the greatest losses, often do not recover from disasters and endure long-term hardship (Wisner et al., 2004). In this context, it is important to take into account the diversity of livelihoods in many local areas and to work with local residents and stakeholders to develop strategies that are potentially more resilient in the face of future events (Pomeroy et al., 2006).

During the post-recovery phase, reconstruction requires weighing, prioritizing, and sequencing of policy programming, given the multiple, and sometimes competing agendas for most decisionmakers and operational actors. Often there are opportunities for change in policy directions and agenda setting at local to national levels at this time (Birkland, 1997). The post-event lobbying for action and resources requires a balance between short-term needs and long-term goals. The most significant is the pressure to quickly return to conditions prior to the event rather than incorporate longer-term and more sustainable development policies (Christoplos, 2006; Kates et al., 2006). How long such a window will stay open or precisely what factors will make it close under a given set of conditions is not well known even though three to six months has been recognized in specific cases (Kates et al., 2006).

The most often used strategies for coping with present disaster risk at the local level are emergency assistance (including disaster relief), population movements, and recovery and reconstruction. As illustrated above, there is considerable variability among and between local places in how these actions are implemented and the impacts of their use.

5.3. Anticipating and Responding to Future Disaster Risk

This section examines how local places anticipate future risks and how they respond to them. In addition to enhanced communication, other approaches to anticipating and responding to future risks include structural interventions such as dikes or dams, natural resources planning and ecosystem protection, and storage and rationing of resources.

5.3.1. Communicating Risk

Effective communication is necessary across the full cycle of disaster management: reduction, preparedness, response, and recovery, especially at the local level where communications face particular constraints and possibilities. A burgeoning field of research explores the barriers to communicating the impacts of climate change to motivate constructive behaviors and policy choices (Frumkin and McMichael, 2008). Communicating the likelihood of extreme impacts of climate change also presents an important and difficult challenge (Moser and Dilling, 2007). Research on climate communications addresses how information can be designed, and the mechanisms and timing of its distribution.

5.3.1.1. Message Design

As used here, the term *risk communication* refers to intentional efforts on the part of one or more sources (e.g., international agencies, national governments, local government) to provide information about hazards and hazard adjustments through a variety of channels to different audience segments (e.g., the general public, specific at-risk communities). The characteristics of messages that have a significant impact on local adoption of adjustments involve information quality (specificity, consistency, and source certainty), information reinforcement (number of warnings and repetition) (Mileti and O'Brien, 1992; O'Brien and Mileti, 1992; Mileti and Fitzpatrick, 1993), and the ways in which information is designed. Messages targeted to specific audiences are more readily received (Maibach et al., 2008) than those which are not. Targeting threats to future generations may generate more concern than overt actions to reduce contemporary climate change impacts (Maibach et al., 2008). In addition, communication is more effective when the information regarding risk does not exceed the capacity for coping and therefore galvanizes resilience (Fritze et al., 2008). Some research suggests that a focus on personal risk of specific damages of climate change can be a central element in motivating interest and behavior change (Leiserowitz, 2007). Risk messages vary in threat specificity, guidance specificity, repetition, consistency, certainty, clarity, accuracy, and sufficiency (Mileti and Sorensen, 1990; Mileti and Peek, 2002).

Communications that include social, interpersonal, physical, environmental, and policy factors can foster civic engagement and social change fundamental to reducing risk (Brulle, 2010). A participatory approach highlights the need for multiple pathways of communication that engenders credibility, trust, and cooperation (NRC, 1989; Frumkin and McMichael, 2008), which are especially important in high-stress situations such as those associated with climate extremes. For example, participatory video production is effective in communicating the extreme impacts of climate change (Suarez et al., 2008; Baumhardt et al., 2009). Participatory video involves a community or group creating their own videos through storyboarding and production (Lunch and Lunch, 2006). Such projects are traditionally used in contexts, such as poor communities, where there are constraints to accessing accurate climate information (Patt and Gwata, 2002; Patt and Schröter, 2008). Engaging with community leaders or opinion leaders in accessing social networks through which to distribute information is another approach, traditionally used by health educators but also applicable to the translation of climate risks in a community context (Maibach et al., 2008). Another approach used in health communications that is relevant to climate education is the 'community drama' in which community members engage in plays to communicate health risks (Middlekoop et al., 2006). These types of communication projects can motivate community action necessary to promote preparedness (Semenza, 2005; Jacobs et al., 2009).

Visualizing methods such as mapping, cartographic animations, and graphic representations are also used to engage with stakeholders who may be impacted by extreme events (McCall, 2008; A. Shaw et al., 2009). Many programs are developing ways to use visualizations to help decisionmakers adapt to a changing environment, suggesting that such tools can increase climate literacy (Niepold et al., 2008). Visualizations can be powerful tools, but issues of validity, subjectivity, and interpretation must be seriously considered in such work (Nicholson-Cole, 2004). These communications are most effective when they take local experiences or points of view and locally relevant places into account (O'Neill and Ebi, 2009). Little evaluation has been done of visualization projects, therefore leaving a gap in understanding of how to most effectively communicate future risks of extreme events.

5.3.1.2. Modes and Timing of Risk Communication

The generation and receipt of risk information occurs through a diverse array of channels. They include: interpersonal contact with particular researchers; planning and conceptual foresight; outside consultation on the planning process; user-oriented transformation of information; and individual and organizational leadership (NRC, 2006). Researchers have long recognized a variety of informal information source vehicles including peers (friends, relatives, neighbors, and coworkers) and news media (Drabek, 1986). These sources systematically differ in terms of such characteristics as perceived expertise, trustworthiness, and protection responsibility (Lindell and Perry, 1992; Lindell and Whitney, 2000; Pulwarty, 2007). Risk-area residents use information channels for different purposes: the internet, radio, and television are useful for immediate updates; meetings are useful for clarifying questions; and newspapers and brochures are useful for retaining information that might be needed later. In addition, within-community discussions on risks to livelihoods,

Box 5-2 | Successful Communication of Local Risk-Based Climate Information

The following questions have been identified as shaping the successful communication of risk-based climate information (Ascher, 1978; Fischhoff, 1992; Pulwarty, 2003):

- What do people *know* and *believe* about the risks being posed?
- What is the *past experience*/outcomes of information use?
- Is the new information *relevant* for decisions in the particular community?
- Are the sources/providers of information *credible* to the intended user?
- Are practitioners (e.g., farmers) *receptive* to the information and to research?
- Is the information *accessible* to the decisionmaker?
- Is the information *compatible* with existing decision models (e.g., for farming practice)?
- Does the community (or individuals in the community) have the *capacity* to use information?

such as during droughts, act as mechanisms for risk communication and response actions (Dekens, 2007).

Policies and actions affecting communications and advanced warning have a major impact on the adaptive capacity and resilience of livelihoods. The collection and transmittal of weather- (and climate-) related information is often a governmental function and timely issuance remains a key weakness in climate information systems, especially for communication passed on to communities from the national early warning units (UNISDR, 2006). There are other localized forms of communication that can be used rapidly, such as neighborhood watch systems (Lichterman, 2000). Some private communication methods, such as text messaging, Facebook, and Twitter, may reach affected populations before government directives (Palen et al., 2007). However, some research shows that there has been too much reliance on one-way devices for communication (such as the radio), which were felt to be inadequate for agricultural applications (for example, farmers are not able to ask further questions regarding the information provided) (Ziervogel, 2004). Within many rural communities, low bandwidth and poor computing infrastructure pose serious constraints to risk-message receipt. Such gaps are evident in developed as well as lesser-developed regions.

The degree of acceptability of information and trust in the providers dictates the context of communicating disaster and climate information (see Box 5-2). Pre-decisional processes (reception, attention, and comprehension) influence the disaster message's effectiveness (Lindell and Perry, 2004). Several studies have identified the characteristics of pre-decisional practices that lead to effective communication over the long-term (Fischhoff, 1992; Cutter, 2001; Pulwarty, 2007). These include: 1) understanding the goals, objectives, and constraints of communities in the target system; 2) mapping practical pathways to different outcomes carried out as joint problem definition and fact-finding strategies among research, extension, and farmer communities; 3) bringing the delivery persons (e.g., extension personnel, research community, etc.) to an understanding of what has to be done to translate current information into usable information; 4) interacting with actual and potential users to better understand informational needs, desired formats of information, and timeliness of delivery; 5) assessing impediments and opportunities to the flow of information including issues of credibility, legitimacy, compatibility (appropriate scale, content, match with existing practice), and acceptability; and 6) relying on existing stakeholders' networks and organizations to disseminate and assess climate information and forecasts.

Much research has yet to be done regarding risk communication on climate change. There has been little systematic investigation, for example, on message effectiveness in prompting local action based on differing characteristics such as the precision of message dissemination, penetration into normal activities, message specificity, message distortion, rate of dissemination over time, receiver characteristics, sender requirements, and feedback (Lindell and Perry, 1992; NRC, 2006). Little research attention has been devoted to how information can be distributed within a family, although the existing research does show there are emotional, social, and structural barriers to such distribution (Norgaard, 2009).

5.3.1.3. Warnings and Warning Systems

The disaster research community has shown that warnings of impending hazards need to be complemented by information on the risks actually posed by the hazards as well as the potential strategies and pathways to mitigate the damage in the particular context in which they arise (Drabek, 1999; UNISDR, 2006). Local-level early warnings based on traditional knowledge (e.g., water turning a different color, winds shifting) are frequently used. The use of radios, megaphones, and cell phones are also used at the local level to warn.

Effective early warning implies information interventions into an environment where vulnerability is assumed (Olson, 2000). This backdrop is reinforced through significant lessons that have been identified from the use of seasonal climate forecasts over the past 15 years (Podestá et al., 2002; Pulwarty, 2007). It is now widely accepted that the existence of predictable climate variability and impacts are necessary but not sufficient to achieve effective use of climate information, including seasonal forecasts. The practical obstacles to using information about future conditions at the local scale are diverse. They include: limitations in modeling the climate system's complexities (e.g., projections having coarse spatial and temporal resolution; limited predictability of some relevant variables; and forecast skill characterization; see Chapter 3); procedural, institutional, and cognitive barriers in receiving or

> **Box 5-3 | Large Dams in Brazil: Scalar Challenges to Climate Adaptation**
>
> Effective climate adaptation requires consideration of cross-scale management concerns. Any project or impact that crosses jurisdictions from local to regional to national to transnational is best planned using a perspective that takes into account all levels of management (Adger et al., 2005). The planned or built large dams in Amazonia, Brazil (McCormick, 2011), exemplify these issues. These dams are related to water management and would cross local, regional, and national boundaries. At the national level, these dams would provide large-scale energy needs and serve major urban centers and industrial sectors across the country. At the regional level, the large Amazonian dams could both generate energy and assist in drought management through storage of hydrological resources (Postel et al., 1996). Because of the expansive range and impacts of large dams, their planning and management raise a variety of scalar concerns about climate adaptation. While on one level a dam may present benefits regionally and nationally, it may also cause serious environmental and social problems locally (McCormick, 2009). For example, dams upstream may lead to erosion and inundation of deltas (Yang et al., 2011).
>
> While there are many environmental benefits of hydroelectric power and large-scale water management, the uncertainty of climate change could alter such benefits at local to global scales and influence the social and environmental ramifications of these projects. For example, the flooding caused by the construction of reservoirs could result in migration of locally affected communities, thereby increasing community fragmentation, poverty, and ill health of humans and biota (Kingsford, 2000). This becomes a local and regional impact of dam construction that may increase vulnerability to climate change in many localities. Changing rainfall patterns that affect reservoir levels could impact the availability of energy generation at the national level (DeLucena et al., 2009). Degradation of flora and fauna also results in additional greenhouse gas emissions at local to global scales (Fearnside, 1995).

understanding climatic information; and the capacity and willingness of decisionmakers to modify actions (Kasperson et al., 1988; Stern and Easterling, 1999; Roncoli et al., 2001; Patt and Gwata, 2002; Marx et al., 2007). In addition, functional, structural, and social factors inhibit joint problem identification and collaborative knowledge production between providers and users. These include divergent objectives, needs, scope, and priorities; different institutional settings and standards; as well as differing cultural values, understanding, and mistrust (Pulwarty et al., 2004; Rayner et al., 2005; Weichselgartner and Kasperson, 2010).

Significant advances in warning systems in terms of improved monitoring, instrumentation, and data collection have occurred (Case Study 9.2.11), but the management of the information and its dissemination to at-risk populations is still problematic (Sorensen, 2000). Researchers have identified several aspects of information communication, such as stakeholder awareness, key relationships, and language and terminology, that are socially contingent in addition to the nature of the predictions themselves. More is known about the effects of these message characteristics on warning recipients than is known about the degree to which generators and providers of information including hazards researchers address them in their risk communication messages. For example, warnings may be activated (such as the tsunami early warning system), yet fail to reach potentially affected communities (Oloruntoba, 2005). Similarly, many communities do not have access to climate-sensitive hazard warning systems such as tone alert radio, emergency alert system, emergency phone numbers (reverse 911 in the United States, but in other parts of the world you call 110, 112, or other numbers), and thus never hear the warning message, let alone act upon the information (Sorensen, 2000). On the other hand, Valdes (1997) demonstrated that flood warning systems based on community operation and participation in Costa Rica make a difference as to whether early warnings are acted upon to save lives and property. Implementing community early warning systems (such as the correlations of rain data and water levels among monitoring stations along the river) serves to encourage communities to become more proactive in their hazard mitigation approaches.

Part of the research gap regarding risk communication stems from the lack of projects that can be tested and shown to affect preparedness. On the most basic level, there is considerable understanding of the information needed for preparing for disasters, but less-specific understanding of what information and trusted communication processes are necessary to generate local confidence and preparedness for climate change (Fischhoff, 2007). The very discussion of climate forecasts and projections within potentially impacted communities has served as a vehicle for democratizing the drought discourse in Ceará in Northeast Brazil (Finan and Nelson, 2001). Developing a seamless continuum across emergency responses, preparedness, and coping and adaptation requires insight into the demands that different types of disasters will place upon the local area and the need to perform basic emergency functions – pre-event assessments, proactive hazards mitigation, and incident management (Lindell and Perry, 1996). As noted in previous IPCC reports (IPCC, 2007), preparing for short-term disasters enhances the capacity to adapt to longer-term climate change.

5.3.2. Structural Measures

Structural measures may be used to reduce the effects of climate-related events such as floods, droughts, coastal erosion, and heat waves. Structural interventions to reduce the effects of extreme events often employ engineering works to provide protection from flooding such as dikes, embankments, seawalls, river channel modification, flood gates, and reservoirs. However, structural measures also include those that strengthen buildings (during construction and retrofitting), that enhance water collection in drought-prone areas (e.g., roof catchments, water tanks, wells), and that reduce the effects of heat waves (e.g., insulation and cooling systems). Although many of these structural interventions can achieve success in reducing disaster impacts, they can also fail due to lack of maintenance, age, or due to extreme events that exceed the engineering design level (Galloway, 2007; Doyle et al., 2008; Galloway et al., 2009). In the event that the frequency and magnitude of extreme events increase as a result of climate change, new design levels may be necessary. Technical considerations should also include local social, cultural, and environmental considerations (WMO, 2003; Opperman et al., 2009).

Implementing structural measures from planning through implementation that involve participatory approaches with local residents who are proactively involved often leads to increased local ownership and more sustainable outcomes. Such an example is a program of building, managing, and maintaining cyclone shelters in Bangladesh (Zimmermann and Stössel, 2011). One of the key reasons why sub-national structural projects are often ineffective is that they are approved on the basis of technical information alone, rather than based on both technical information and local knowledge (ActionAid, 2005; Prabhakar et al., 2009; see also Section 5.4.4). In addition, national legislation has important influences on the choice of disaster risk reduction strategies at the local level as local and national institutional arrangements that often favor structural responses over other non-structural approaches (Burby, 2006; Galloway, 2009). Technological responses alone may also have unintended geomorphologic and social consequences, including increasing flood hazard in downstream locations, increasing costs of long-term flood protection works, or increasing coastal erosion in areas deprived of sediments by coastal protection works (Adger et al., 2005; Hudson et al., 2008; Box 5-3).

The method of protecting an entire area by building a dike has been in use for thousands of years and is still being applied by communities in flood-prone countries. Embankments, dikes, levees, and floodwalls are all designed to protect areas from flooding by confining the water to a river channel, thus protecting the areas immediately behind them. Building dikes is one of the most economical means of flood control (Asian Disaster Preparedness Centre, 2005). Dikes built by communities normally involve low technology and traditional knowledge (such as earth embankments). Sand bagging is also a very common form of flood-proofing. Generally, structures that are built of earth are highly susceptible to erosion, leading to channel siltation and reduced water conveyance on the wet side and slope instability and failure on the dry side. Slopes can be stabilized by various methods, including turfing by planting vegetation such as Catkin grass and Vetiver grass in Bangladesh and Thailand, respectively. However there is a continuing debate in the region as to whether the grass strips prevent erosion, whether erosion is in fact the main problem instead of soil fertility, and whether farmers still need slope stabilization (Forsyth and Walker, 2008).

Decisionmaking for large-scale structural measures is often based on cost-benefit analyses and technical approaches. In many cases, particularly in developed countries, structural measures are subsidized by national governments and local governments and communities are required to cover only partial costs. In New Zealand, this led to a preponderance of structural measures despite planning legislation that enabled non-structural measures. As a result, the potential for major disasters was increased and development intensified in areas with structural measures only to be seriously devastated by events that exceeded the engineering design level (Ericksen, 1986). Reduction of centralized subsidies in the mid-1980s and changes in legislation saw greater responsibility for the costs of disaster risk management falling on the communities affected and a move toward more integrated disaster risk reduction processes within New Zealand (Ericksen et al., 2000). Similar trends have been observed in relation to coastal protection, where structural measures are often favored over non-structural options (Titus et al., 2009; Titus, 2011).

Building codes closely align with engineering and architectural structural approaches to disaster risk reduction (Petal et al., 2008; Kang et al., 2009). This is accompanied by the elevation of buildings and ground floor standards in the case of flooding (Aerts et al., 2009; Kang et al., 2009). Though building code regulations exist, non-adoption, especially in developing countries, is problematic (Spence, 2004). Damages to the structure occur not only because of noncompliance with the codes, but also through a lack of inspections, the ownership status of the structure, and the political context and mechanisms of local governance (May and Burby, 1998). Insurance arrangements can provide incentives to local governments and households to implement building codes (Botzen et al., 2009).

Short-term risk reduction strategies can actually produce greater vulnerability to future events as shown in diverse contexts such as ENSO-related impacts in parts of Latin America, induced development below dams or levees in the United States, and flooding in the United Kingdom (Bowden et al., 1981; Pulwarty et al., 2004; Berube and Katz, 2005; Penning-Rowsell et al., 2006). While locally based protection works often enable areas to be productively used and will continue to be needed for areas that are already densely settled, they are commonly misperceived as providing complete protection, and actually increase development – and thus vulnerability – in hazard-prone areas, resulting in the so-called 'levee effect' (Tobin, 1995; Montz and Tobin, 2008). A more general statement of this proposition is found in the safe development paradox in which increased safety measures such as dams or levees induce increased development, which in turn leads to increased exposure and ultimately losses (Burby, 2006). The conflicting policy goals of rapid recovery, safety, betterment, and equity and their relative strengths and weaknesses largely reflect experience with large disasters in other places and times. The actual decisions and rebuilding undertaken to date clearly demonstrate the rush by government at all levels and the residents themselves to rebuild the familiar or increase risks in new locations through displacement (Kates et al., 2006). Similarly, in drought-prone areas, provision of assured water supplies encourages the development of intensive agricultural systems – and for that matter, domestic water use habits – that are poorly suited to the inherent variability of supply and will be even more so in areas projected to become increasingly arid in a changing climate (Chapter 3).

5.3.3. Land Use and Ecosystem Protection

Changes in land use not only contribute to global climate change but they are equally reflective of adaptation to the varying signals of economic, policy, and environmental change (Lambin et al., 2001). Local land use planning embedded in zoning, local comprehensive plans, and retreat and relocation policies is a useful approach to disaster risk management to keep people and property away from locations exposed to risk (Burby, 1998). However, some countries and rural areas may not have formal land use regulations that restrict development or settlement. As land use management regulates the movement of people and industries in hazard-prone zones, such policies face strong opposition from development pressures, real estate interests accompanied by property rights, and local resistance against land acquisition (Burby, 2000; Thomson, 2007). Buffer zones, setback lines in coastal zones, and inundation zones based on flood and sea level rise projections can result in controversies and lack of enforcement that bring temporary resettlement, land speculation, and creation of new vulnerabilities (Ingram et al., 2006; Jha et al., 2010). The government of Sri Lanka, for example, created buffer zones after the Indian Ocean tsunami of 2004, and relocated people to safer locations. Distance from people's coastal livelihoods and social disruptions led to the revision of buffers and new resettlement (Ingram et al., 2006). In the United States, coastal retreat measures are difficult to implement as coastal property carries high value and wealthy property owners can exert political pressure to build along the coast (Ruppert, 2008). Shorefront property owners and realtors especially oppose setback regulations because they consider the regulation to deter growth (NOAA, 2007).

Land use planning and the application of spatial hazard information is another avenue for disaster risk management, especially hazard mitigation. Berke and Beatley (1992) examined a range of hazard mitigation measures and ranked them according to effectiveness and ease of enforcement. The most effective measures include land acquisition, density reduction, clustering of development, building codes for new construction, and mandatory retrofit of existing structures. The high cost of land acquisition programs can make them unattractive to small communities. There has been limited systematic scientific characterization of the ways in which different hazard agents vary in their threats and characteristics, thus require different pre-impact interventions and post-impact responses by households, businesses, and community hazard management organizations. However, Burby et al. (1997) have found evidence for some communities that previous occurrence of a disaster did not have a strong effect on the number of hazard mitigation techniques subsequently employed.

Formal approaches to land use planning as a means of disaster risk management are often less appropriate for many rural areas in developing countries where traditional practices and land tenure systems operate. Systems of land tenure often are very complex and flexible and can contribute to vulnerability reduction. In the case of pastoralists in dryland environments, sharing of land for grazing and of access to water are important drought responses (Anderson et al., 2010). This is not always the case and some land tenure systems marginalize certain groups and increase their vulnerability (Clot and Carter, 2009; Robledo et al., 2011). There are also restrictions on land use planning in regards to slums and squatter settlements. Poverty and the lack of infrastructure and services increase the vulnerability of urban poor to adverse impacts from disasters, and national governments and international agencies have had little success in reversing such trends. As a result, successful efforts to reduce hazard exposure have been locally led, built upon successful initiatives, and composed of informal measures rather than those imposed by governments at the local level (Satterthwaite et al., 2007; Zimmermann and Stössel, 2011).

Land acquisition is another means of protecting property and people by relocating them away from hazardous areas (Olshansky and Kartez, 1998). Many jurisdictions have the power of eminent domain to purchase property but this is rarely used as a form of disaster risk management (Godschalk et al., 2000) or climate change adaptation. Voluntary acquisition of land, for example, requires local authorities to purchase exposed properties, which in turn enables households to obtain less risky real estate elsewhere without suffering large economic losses in the process (Handmer, 1987), but this is rarely used in developing countries because of lack of resources and political support. Given the rapid population growth in coastal areas and in flood plains in many parts of the world, and the large number and high value of exposed properties in coastal zones in developed countries such as the United States and Australia, this buy out strategy is cost-prohibitive and thus rarely used (Anning and Dominey-Howes, 2009). Similarly, voluntary acquisition schemes for developing countries are equally fraught with problems as people have strong ties to the land, and land is held communally in places like the Pacific Islands where community identity cannot be separated from the land to which its members belong (Campbell, 2010b). Land use planning alone, therefore, may not be successful as a singular strategy but when coupled with related policies such as tax incentives or disincentives, insurance, and drainage and sewage systems it could be effective (Yohe et al., 1995; Cheong, 2011b). However, if sea level rise adversely affects local coastal areas some form of relocation may become necessary in all exposed jurisdictions. In the United States, some state and local governments have adopted rolling easement policies, which allow construction in vulnerable areas subject to the requirement that the structures will be removed if and when the landward edge of a wetland or beach encroaches (Titus, 2011).

Ecosystem conservation offers long-term protection from climate extremes. The mitigation of soil erosion, landslides, waves, and storm surges are some of the ecosystem services that protect people and infrastructure from extreme events and disasters (Sudmeier-Rieux et al.,

2006). The 2004 Indian Ocean tsunami attests to the utility of mangroves, coral reefs, and sand dunes in alleviating the influx of large waves to the shore (Das and Vincent, 2009). The use of dune management districts to protect property along developed shorelines has achieved success in many places along the US eastern shore and elsewhere (Nordstrom, 2000, 2008). Carbon sequestration is another benefit of ecosystem-based adaptation based on sustainable watershed and community forest management (McCall, 2010). While the extent of their protective ecosystem functions is still debated (Gedan et al., 2011), the merits of ecosystem services in general are proven, and development of quantified models of the services is well under way (Barbier et al., 2008; Nelson et al., 2009). These nonstructural measures are considered to be less intrusive and more sustainable, and when integrated with engineering responses provide mechanisms for adapting to disasters and climate extremes (Galloway, 2007; Opperman et al., 2009; Cheong, 2011a).

5.3.4. Storage and Rationing of Resources

Communities may take a range of approaches to cope with disaster-induced shortages of resources, including producing surpluses and storing them. If the surpluses are not available, rationing of food may occur. Many localities produce food surpluses that enable them to manage during periods of seasonal or disaster-initiated disruptions to their food supplies, although such practices were more prevalent in pre-capitalist societies. In Pacific Island communities, for example, food crops such as taro and breadfruit were often stored for periods up to and exceeding a year by fermentation in leaf-lined pits. Yams could be stored for several years in dry locations, and most communities maintained famine foods such as wild yams, swamp taro, and sago, which were only harvested during times of food shortage (Campbell, 2006). The provision of disaster relief, among other factors, has seen these practices decline (Campbell, 2010a). In Mali, women store part of their harvest as a hedge against drought (Intercooperation, 2008). Stockpiling and prepositioning of emergency response equipment, materials, foods, and pharmaceuticals and medical equipment is also an important form of disaster preparedness at the local level, especially for many indigenous communities.

Rationing may be seen as the initial response to food shortages at or near the onset of a food crisis. However, in many cases rationing is needed on a seasonal basis. Rationing at the local level is often self-rationing instituted at the level of households – particularly poor ones without the ability to accumulate wealth or surpluses. Often, rationing initially occurs among women and children (Hyder et al., 2005; Ramachandran, 2006). Most rationing takes place in response to food shortages and is, for most poor communities, the first response to the disruption of livelihoods (Walker, 1989; Barrett, 2002; Devereux and Sabates-Wheeler, 2004; Baro and Deubel, 2006). In many cases increases in food prices force those with insufficient incomes to ration as well.

When the food shortage becomes too severe, households may reduce future security by eating seeds or selling livestock, followed by severe illness, migration, starvation, and death if the shortages persist. While climate change may alter the frequency and severity of droughts (see Section 3.5.1), the causes of food crises are multi-faceted and often lie in social, economic, and political processes in addition to climatic variability (Sen, 1981; Corbett, 1988; Bohle et al., 1994; Wisner et al., 2004).

Food rationing is unusual in developed countries where most communities are not based on subsistence production. Welfare systems and NGO agencies respond to needs of those with livelihood deficits in these countries. However, other forms of rationing do exist particularly in response to drought events. Reductions in water use are achieved through a number of measures including: metering, rationing (fixed amounts, proportional reductions, or voluntary reductions), pressure reduction, leakage reduction, conservation devices, education, plumbing codes, market mechanisms (e.g., transferable quotas, tariffs, pricing), and water use restrictions (Lund and Reed, 1995; Froukh, 2001).

Electricity supplies may also be disrupted by disaster events resulting in partial or total blackouts. While a number of countries have national electrical grids, decisions on responses to shortages are often made at local levels causing considerable disruption to other services, domestic customers, and to businesses. Rose et al. (2007) show that many American businesses can be quite resilient in such circumstances, adapting a variety of strategies including conserving energy, using alternative forms of energy, using alternative forms of generation, rescheduling activities to a future date, or focusing on the low- or no-energy elements of the business operation. Rose and Liao (2005) had similar findings for water supply disruption. Electricity storage (in advance) and rationing may also be required when low precipitation reduces hydroelectricity production, a possible scenario in some places under climate projections (Vörösmarty et al., 2000; Boyd and Ibarrarán, 2009). In some cases there may be competition among a range of sectors including industry, agriculture, electricity production, and domestic water supply (Vörösmarty et al., 2000) that may have to be addressed through rationing and other measures such as those listed above. Clear rules outlining which consumers have priority in using water or electricity is important.

Other elements that may be rationed as a result of natural hazards or disasters include prioritization of medical and health services where disasters may simultaneously cause a large spike in numbers requiring medical assistance and a reduction in medical facilities, equipment, pharmaceuticals, and personnel. This may require classifying patients and giving precedence to those with the greatest need and the highest likelihood of a positive outcome. This approach seeks to achieve the best results for the largest number of people (Alexander, 2002; Iserson and Moskop, 2007).

Responding to future disaster risk will entail multiple approaches at the local level. Starting with risk communication and warning information, the following dominate the range of adjustments local areas presently undertake in responding to future risks: structural measures, land use planning, ecosystem protection, and storage and rationing of resources.

5.4. Building Capacity at the Local Level for Risk Management in a Changing Climate

Local risk management has traditionally dealt with extreme events without considering the climate change context. This section provides examples of adaptations to disaster risk and how such proactive behaviors at the community level by local government and NGOs provide guidance for reducing the longer-term impacts of climate change. Although reacting to extreme events and their impacts is important, it is crucial to focus on building the resilience of communities, cities, and sectors in order to ameliorate the impacts of extreme events now and into the future.

5.4.1. Proactive Behaviors and Protective Actions

Researchers have identified some of the physical and social characteristics that allow for the adoption of effective partnerships and implementation practices during events (Birkland, 1997; Pulwarty and Melis, 2001). These include the occurrence of previous strong focusing events (such as catastrophic extreme events) that generate significant public interest and the personal attention of key leaders, a social basis for cooperation including close inter-jurisdictional partnerships, and the existence of a supported collaborative framework between research and management.

Factors conditioning this outcome have been summed up by White et al. (2001) as "knowing better and losing even more." In this context, knowing better indicates the accumulation of readily available knowledge on drivers of impacts and effective risk management practices. For instance, researchers had understood the consequences of a major hurricane hitting New Orleans prior to Hurricane Katrina with fairly detailed understanding of planning and response needs. This knowledge appears to have been ignored at all levels of government including the local level after Hurricane Katrina (Kates et al., 2006). White et al. (2001) offer four explanations for why such conditions exist from an information standpoint: 1) knowledge continues to be flawed by areas of ignorance; 2) knowledge is available but not used effectively; 3) knowledge is used effectively but takes a long time to have an impact; and 4) knowledge is used effectively in some respects but is overwhelmed by increases in vulnerability and in population, wealth, and poverty. Another possibility is that some individuals or communities choose to take the risk. For example, there is some evidence that the value of living near the coast, especially in developed countries, pays back the cost of the structure in a few years due to increases in housing values (Kunreuther et al., 2009), so the risk is worth taking by the individual and the community. Finally, knowledge is often discounted.

Individuals can make choices to reduce their risk but social relations, context, and certain structural features of the society in which they live and work mediate these choices and their effects. The recognition that dealing with risk and insecurity is a central part of how poor people develop their livelihood strategies is giving rise to prioritizing disaster mitigation and preparedness as important components of many poverty alleviation agendas (Cuny, 1983; Olshansky and Kartez, 1998; UNISDR, 2009). A number of long-standing challenges remain as the larger and looser coalitions of interests that sometimes emerge after disasters rarely last long enough to sustain the kind of efforts needed to reduce present disaster risk, let alone climate extremes in a climate change context.

At the household and community level, individuals often engage in protective actions to minimize the impact of extreme events on themselves, their families, and their friends and neighbors. In some cases individuals ignore the warning messages and choose to stay in places of risk. The range and choice of actions are often event specific and time dependent, but they are also constrained by location, adequate infrastructure, socioeconomic characteristics, access to disaster risk information, and risk perception (Tierney et al., 2001). For example, evacuation is used when there is sufficient warning to temporarily relocate out of harm's way such as for tropical storms, flooding, and wildfires. Collective evacuations are not always possible given the location, population size, transportation networks, and the rapid onset of the event. At the same time, individual evacuation may be constrained by a host of factors ranging from access to transportation, monetary resources, health impairment, job responsibilities, gender, and the reluctance to leave home. There is a consistent body of literature on hurricane evacuations in the United States, for example, that finds: 1) individuals tend to evacuate as family units, but they often use more than one private vehicle to do so; 2) social influences (neighbors, family, friends) are key to individual and households evacuation decision-making; if neighbors are leaving then the individual is more inclined to evacuate and vice versa; 3) risk perception, especially the personalization of risk by individuals, is a more significant factor in prompting evacuation than prior adverse experience with hurricanes; 4) pets and concerns about property safety reduce household willingness to evacuate; and 5) social and demographic factors (age, presence of children, elderly, or pets in households, gender, income, disability, and race or ethnicity) either constrain or motivate evacuation depending on the particular context (Perry and Lindell, 1991; Dow and Cutter, 1998, 2000, 2002; Whitehead et al., 2000; Bateman and Edwards, 2002; Van Willigen et al., 2002; Sorensen et al., 2004; Lindell et al., 2005; Dash and Gladwin, 2007; McGuire et al., 2007; Sorensen and Sorensen, 2007; Edmonds and Cutter, 2008; Adeola, 2009). Culture also plays an important role in evacuation decisionmaking (Clot and Carter, 2009). For example, recent studies in Bangladesh have shown that there are high rates of non-evacuation despite improvements in warning systems and the construction of shelters. While there are a variety of reasons for this, gender issues (e.g., shelters were dominated by males, shelters didn't have separate spaces for males and females) have a major influence upon females not evacuating (Paul and Dutt, 2010; Paul et al., 2010).

A different protective action, shelter-in-place, occurs when there is little time to act in response to an extreme event or when leaving the community would place individuals more at risk (Sorensen et al., 2004). Seeking higher ground or moving to higher floors in residential structures to get out of rising waters is one example. Another is the movement into interior spaces within buildings to seek refuge from strong winds. In the

Box 5-4 | Collective Behavior and the Moral Economy at Work

A variety of socio-political networks that were used to offset disaster losses existed throughout the Pacific region prior to colonization (Sahlins, 1962; Paulson, 1993). One example of such a system is the *Suqe*, or graded society, which existed in northern Vanuatu, a small island nation in the South West Pacific Ocean. In the *Suqe* 'big men' achieved the highest status by accumulating surpluses of valued goods such as shell money, specially woven mats, and pigs. Men increased their grade within the system by making payments of these goods to men of higher rank. In accumulating the items men would also accumulate obligations to those they had borrowed from. Accordingly, networks and alliances emerged among the islands of northern Vanuatu. When tropical cyclones destroyed crops, the obligations could be called in and assistance given from members of the networks who lived in islands that escaped damage (Campbell, 1990). A number of processes associated with colonialism (changes to the socio-political order), the introduction of the cash economy (the replacement of shell money), and religious conversion that resulted in the banning of the *Suqe*), as well as the provision of post-disaster relief, have caused a number of elements of the moral economy to fall into disuse (Campbell, 2006).

In many traditional or pre-capitalist societies it appears that mechanisms existed that protected community members from periodic shocks such as natural hazards. These mechanisms, sometimes referred to as the *moral economy*, were underpinned by reciprocity and limiting exploitation so that everyone had basic security. The mechanisms are often linked to kinship networks, and serve to redistribute resources to reduce the impacts on those who had sustained severe losses, and have been identified in Southeast Asia (Scott, 1976), Western Africa (Watts, 1983), and the Pacific Islands (Paulson, 1993). The moral economy incorporated social, cultural, political, and religious arrangements, which ensured that all community members had a minimal level of subsistence (see Box 5-4). For example, traditional political systems in the semiarid Limpopo Basin in northern South Africa enabled chiefs to reallocate surpluses during bad years, but this practice has declined under contemporary systems where surpluses are sold (Dube and Sekhwela, 2008). In Northern Kenya, social security networks existed among some groups of nomadic pastoralists that enabled food and livestock stock to be redistributed following drought events, but these are also breaking down with the monetization of the local economy, among other factors (Oba, 2001).

Although the concept of moral economy is generally associated with pre-capitalist societies and those in transition to capitalism (in the past), significant features of moral economy, such as reciprocity, barter, crop sharing, and other forms of cooperation among families and communities or community-based management of agricultural lands, waters, or woods are still part of the social reality of developing countries that cannot be considered anymore as pre-capitalist. Many studies show that moral economy-based social relationships are still present such as traditional institutions regulating access, use, and on-going redistribution of community-owned land (Sundar and Jeffery, 1999; Rist, 2000; Hughes, 2001; Trawick, 2001; Rist et al., 2003). The revitalization, enhancement, and innovation of such moral economy-based knowledge, technologies, and forms of cooperation and interfamily organization represent an important and still existing source of fostering collective action that serves as an enabling condition for preventing and coping with hazards related to natural resource management. While aspects of the traditional moral economy have declined in many societies, informal networks remain important in disaster risk reduction (see Section 5.4.3).

The notion of the moral economy does not recognize the inequalities in some of the social systems that enabled such practices to be sustained (e.g., gender-based power relationships) and tended to perhaps provide an unrealistic notion of a less risky past. In addition, kinship-based sharing networks may foster freeloading among some members (diFalco and Bulte, 2009). Nevertheless, a reduction in traditional coping mechanisms including the moral economy is reflected in growing disaster losses and increasing dependency on relief (Campbell, 2006).

case of wildfires, shelter-in-place becomes a back-up strategy when evacuation routes are restricted because of the fire and include protecting the structure with garden hoses or finding a safe area such as a water body (lake or backyard swimming pool) as temporary shelter (Cova et al., 2009). In Australia, the shelter-in-place action is slightly different. Here the local community engagement with wildfire risks has two options: stay and defend or leave early. In this context, the decisions to remain are based on social networks, prior experience with wildfires, gender (males will remain to protect and guard property), and involvement with the local fire brigade (McGee and Russell, 2003). The study also found that rural residents were more self-reliant and prepared to defend then suburban residents (McGee and Russell, 2003).

The social organization of societies dictates the flexibility in the choice of protective actions – some are engaged in voluntarily (such as in the United States, Australia, and Europe), while other protective actions for individuals or households are coordinated by centralized authorities such as Cuba and China. Planning for disasters is a way of life for Cuba, where everyone is taught at an early age to mobilize quickly in the case of a natural disaster (Sims and Vogelmann, 2002; Bermejo, 2006). The organization of civil defense committees at block, neighborhood, and community levels working in conjunction with centralized governmental authority makes the Cuban experience unique (Sims and Vogelmann, 2002; Bermejo, 2006). Recent experience with hurricanes affecting Cuba suggests that such efforts are successful because there has been little loss of life.

Collective action to prepare for or respond to disaster risk and extreme climate impacts can also be driven by localized organizations and social movements. Many such groups represent networks or first responders for climate-sensitive disasters. However, there are many constraints that these movements face in building effective coalitions including the need to connect with other movement organizations and frame the problem in an accessible way (McCormick, 2010). One means of mobilizing collective responses at the local level is through participatory approaches to disaster risk reduction such as Community-Based Disaster Reduction or Community-Based Disaster Preparedness (see Section 5.6.2). Such approaches build on local needs and priorities, knowledge, and social structures and are increasingly being used in relation to climate change adaptation (Reid et al., 2009).

5.4.2. Empowerment for Local Decisionmaking

A critical factor in community-based disaster risk reduction is that community members are empowered to take control of the processes involved. Marginalization (Mustafa, 1998; Adger and Kelly, 1999; Polack, 2008) and disempowerment (Hewitt, 1997) are critical factors in creating vulnerability and efforts to reduce these characteristics play an important role in building resilient communities. In this chapter, *empowerment* refers to giving community members control over their lives with support from outside (Sagala et al., 2009). This requires external facilitators to respect community structures and traditional and local knowledge systems, to assist but not take a dominating role, to share knowledge, and to learn from community members (Petal et al., 2008). A key element in empowering communities is building trust between the community and the external facilitators (Sagala et al., 2009). In the Philippines, for example, Allen (2006) found that many aspects of community disaster preparedness such as building on local institutions and structures, building local capacity to act independently, and building confidence through achieving project outcomes were already present. She also found that where agencies focused on the physical hazard as the cause of disasters and neglected the underlying causes of the social vulnerability within these small specific projects, the result was disempowerment. It is also important to note that communities have choices from a range of disaster management options (Mercer et al., 2008). Empowerment in community-based disaster risk management may also be applied to groups within communities whose voice may otherwise not be heard or who are in greater positions of vulnerability (Wisner et al., 2004). These include women (Wiest et al., 1994; Bari, 1998; Clifton and Gell, 2001; Polack, 2008) and disabled people (Wisner, 2002).

Another key element of empowerment is ownership of or responsibility for disaster management (Buvinić et al., 1999). This applies to all aspects of the disaster, from the ownership of a disaster itself so that the community has control of relief and reconstruction, to a local project to improve preparedness. Empowerment and ownership ensure that local needs are met, that community cohesion is sustained, and a greater chance of success of the disaster management process. Empowerment and ownership of the disaster impacts may be particularly important in achieving useful (for the locality) post-disaster assessments (Pelling, 2007). It is important for external actors to identify those voices who speak for the local constituencies. Also, accountability and governance of disaster and climate management issues is growing in importance.

5.4.3. Social Drivers

Similar to empowerment is the role of localized social norms, social capital, and social networks as these also shape behaviors and actions before, during, and after extreme events. Each of these factors operates on their own and in some cases also intersects with the others. As vulnerability to disasters and climate change is socially constructed (Sections 2.4 and 2.5.2), the breakdown of collective action often leads to increased vulnerability. Norms regarding gender also play a role in determining outcomes. For example, women were more prone to drowning than men during the Asian tsunami because they were less able to swim and because they were attempting to save their children (Rofi et al., 2006; Section 2.5.8).

Social norms are rules and patterns of behavior that reflect expectations of a particular social group (Horne, 2001). Norms structure many different kinds of action regarding climate change (Pettenger, 2007). Norms are embedded in formal institutional responses, as well as informal groups that encounter disasters (Raschky, 2008). Norms of reciprocity, trust, and associations that bridge social divisions are a central part of social cohesion that fosters community capacity (Kawachi and Berkman, 2000). A number of types of groups drive norms and, consequently, vulnerability, including religious, neighborhood, cultural, and familial groups (Brenkert and Malone, 2005). In the occurrence of extreme events, affected groups interact with one another in an attempt to develop a set of norms appropriate to the situation, otherwise known as the emergent norm theory of collective behavior (NRC, 2006). This is true of those first affected at the local level whose norms and related social capital affect capacity for response (Dolan and Walker, 2004).

Social capital is a multifaceted concept that captures a variety of social engagements within the community that bonds people and generates a positive collective value. It is also an important element in disaster preparedness, coping, and response. It is an important element in the face of climate extremes because community social resources such as networks, social obligations, trust, and shared expectations create social capital to prevent, prepare, and cope with disasters (Dynes, 2006). In climate change adaptation, scholars and policymakers increasingly promote social capital as a long-term adaptation strategy (Adger, 2003; Pelling and High, 2005). Although often positive, social capital can have some negative outcomes. Internal social networks are oftentimes self-referential and insular (Portes and Landolt, 1996; Dale and Newman, 2010). This results in a closed society that lacks innovation and diversity, which are essential elements for climate change adaptation. Disaster itself is overwhelming, and can lead to the erosion of social capital and the demise of the community (Ritchie and Gill, 2007). This invites external engagement beyond local-level treatment of the disaster and extreme events (Brondizio et al., 2009; Cheong, 2010). The inflow of external aids, expertise, and the emergence of new groups to cope with disaster are indicative of bridging and linking social capital beyond local boundaries.

Social capital is embedded in social networks (Lin, 2001), or the social structure composed of individuals and organizations, through multiple types of dependency, such as kinship, financial exchange, or prestige (Wellman and Berkowitz, 1988). Social networks provide a diversity of functions, such as facilitating sharing of expertise and resources across stakeholders (Crabbé and Robin, 2006). Networks can function to promote messages within communities through preventive advocacy, or the engagement of advocates in promoting preventive behavior (Weibel, 1988). Information about health risks has often been effectively distributed through a social network structure using opinion leaders as a guide (Valente and Davis, 1999; Valente et al., 2003), and has promising application for changing behavior regarding climate adaptation (Maibach et al., 2008). Such opinion leaders may span a range of types, from formally elected officials, celebrities, and well-known leaders, to local community members who are well-embedded in local social networks. It is important to note that more potential has been shown in influencing behavior through community-level interventions than through individual-level directives at the population level (Kawachi and Berkman, 2000). Local and international networks can support the development of policies and practices that result in greater preparedness (Tompkins, 2005). Local resilience in the face of climate change can be fostered by strong social networks that support effective responses (Ford et al., 2006). For example, networks facilitate the transmission of information about risks (Berkes and Jolly, 2001). Therefore, communities with stronger social networks appear to be better prepared for extreme climate impacts because of access to information and social support (Buckland and Rahman, 1999).

At the same time, it is important to note that social networks may not always be sufficient to foster effective adaptation to extreme events. Some social networks actually discourage people from moving away from high risk zones, such as has been the case in storms and floods when residents have not wanted to leave (Eisenman et al., 2007). The impacts of climate change itself may also change the structure and utility of social networks. As people migrate away from climate and other risks or are pulled toward alternative locations for social or ecological resources, those left behind can experience fragmented or weakened social networks. The utilization of social networks can also be prevented by the status of particular social groups, such as illegal and legal settlers or immigrants (Wisner et al., 2004). Other social and environmental contextual factors must be considered when conceptualizing the role of social networks in managing extreme events. For example, strong social networks have facilitated adaptability in Inuit communities, but are being undermined by the dissolution of traditional ways of life (Ford et al., 2006).

5.4.4. Integrating Local Knowledge

Local and traditional knowledge is increasingly valued as important information to include when preparing for disasters (McAdoo et al., 2009; R. Shaw et al., 2009). It is embedded in local culture and social interactions and transmitted orally over generations (Berkes, 2008). Place-based memory of vulnerable areas, know-how for responding to recurrent extreme events, and detection of abnormal environmental conditions manifest the power of local knowledge. Because local knowledge is often tacit and invisible to outsiders, community participation in disaster management is essential to tap this information as it can offer alternative perspectives and approaches to problem-solving (Battista and Baas, 2004; Turner and Clifton, 2009).

Within a climate change context, indigenous people as well as long-term residents often conserved their resources *in situ*, providing important information about changing environmental conditions as well as actively adapting to the changes (Salick and Byg, 2007; Macchi et al., 2008; Salick and Ross, 2009; Turner and Clifton, 2009). Research is emerging that helps to document changes that local people are experiencing (Ensor and Berger, 2009; Salick and Ross, 2009). Although this evidence might be similar to scientific observations from external researchers, the fact that local communities are observing it is initiating discussions about existing and potential adaptation to climate changes from within the community.

The following example is illustrative. In six villages in eastern Tibet, near Mt. Khawa Karpo, local documentation of warmer temperatures, less snow, and glacial retreat across areas were consistent, whereas other observations were more varied, including those for river levels and landslide incidences (Byg and Salick, 2009). In Gitga'at (Coast Tsimshian) Nation of Hartley Bay, British Columbia, indigenous people observe the decline of some species but also new appearances of others, anomalies in weather patterns, and declining health of forests and grasslands that have affected their ability to harvest food (Turner and Clifton, 2009). The Alaska Native Tribal Health Consortium generated climate change and health impact assessment reports from observations, data, and traditional ecological knowledge (ANTHC, 2011). Other than knowledge from indigenous groups, local knowledge associated with contemporary societies and cities exists though more research is needed in this area (Hordijk and Baud, 2011).

Integration of local knowledge with external scientific, global, and technical knowledge is an important dimension of climate change adaptation and disaster management. Experiences in environmental management and integrated assessment suggest mechanisms for such knowledge transfers from the bottom up and from the top down (Burton et al., 2007; Prabhakar et al., 2009). For example, communities set up trusted intermediaries to transfer and communicate external knowledge such as technology-based early warning systems and innovative and sustainable farming techniques that incorporate the local knowledge system (Bamdad, 2005; Kristjanson et al., 2009). Another example is the re-engineering of local practices to adapt to climate change as shown in the conversion of traditional dry-climate adobe construction to more stabilized earth construction built to withstand regular rainfall. The utilization of participatory methods to draw in the perspectives of local stakeholders for subsequent input into hazards vulnerability assessments or climate change modeling or scenario development is well documented. Stakeholder interactions and related workshops using participatory or mediated modeling elicit discussions of model assumptions, local impacts, consistencies of observed and modeled patterns, and adaptation strategies (Cabrera et al., 2008; Langsdale et al., 2009).

Obstacles to utilizing local knowledge as part of adaptation strategies exist. Climate-induced biodiversity change threatens historical coping strategies of indigenous people as they depend on the variety of wild plants, crops, and their environments particularly in times of disaster (Turner and Clifton, 2009). In dryland areas such as in Namibia and Botswana, one of the indigenous strategies best adapted to frequent droughts is livestock herding, including nomadic pastoralism (Ericksen et al., 2008). Decreased access to water sources through fencing and privatization has inhibited this robust strategy. Also in Botswana, it has been suggested that government policies have weakened traditional institutions and practices, as they have not adequately engaged with local community institutions and therefore the mechanisms for redistributing resources have not been strengthened sufficiently (Dube and Sekhwela, 2008).

5.4.5. Local Government and Nongovernment Initiatives and Practices

Governance structures are pivotal to addressing disaster risk and informing responses as they help shape efficiency, effectiveness, equity, and legitimacy (Adger et al., 2003; UNISDR, 2009). Some places centralize climate change management practices at the national level (see Chapter 6). This may be due to the ways in which many climate extremes affect environmental systems that cross political boundaries resulting in discordance if solely locally managed (Cash and Moser, 2000) but could also be based on old practices of operations. In other places, actions are more decentralized, emerging at the local level and tailored to local contexts (Bizikova et al., 2008). If multiple levels of planning are to be implemented, mechanisms for facilitation and guidance at the local level are needed in order that fairness is guaranteed during the implementation of national policies at the local scale (Thomas and Twyman, 2005). Local governments play an important role as they are responsible for providing infrastructure, preparing and responding to disasters, developing and enforcing planning, and connecting national government programs with local communities (Huq et al., 2007; UNISDR, 2009). The quality and provision of these services have an impact on disaster and climate risk (Tanner et al., 2009). Effective localized planning, for example, can minimize both the causes and consequences of climate change (Bulkeley, 2006).

Though local government-led climate adaptation policies and initiatives are less pronounced than climate change mitigation measures, a growing number of cities and sub-national entities are developing adaptation plans, but few have implemented their strategies (Heinrichs et al., 2009; Birkmann et al., 2010). The Greater London Authority (2010), for example, has prepared a Public Consultation Draft of their climate change adaptation strategy for London. The focus of this is on the changing risk of flood, drought, and heat waves through the century and actions for managing them. Some of the actions include improvement in managing surface water flood risk, an urban greening program to buffer the impacts from floods and hot weather, and retrofitting homes to improve water and energy efficiency. ICLEI, a non-profit network of more than 1,200 local government members across the globe, provides web-based information (www.iclei.org) in support of local sustainability efforts using customized tools and case studies on assessing climate resilience and climate change adaptation.

Some assessments of urban adaptations exist. For example, adaptation efforts in eight cities (Bogotá, Cape Town, Delhi, Pearl River Delta, Pune, Santiago, Sao Paulo, and Singapore) tend to support existing disaster management strategies (Heinrichs et al., 2009). Another study comparing both formal adaptation plans and less formal adaptation studies in nine cities including Boston, Cape Town, Halifax, Ho Chi Minh City, London, New York, Rotterdam, Singapore, and Toronto demonstrates that the focus is mostly on risk reduction and the protection of citizens and infrastructure, with Rotterdam seeing adaptation as opportunity for transformation (Birkmann et al., 2010). These nine cities have focused more on expected biophysical impacts than on socioeconomic impacts and have not had a strong focus on vulnerability and the associated susceptibility or coping capacity. Despite the intention that city adaptation responses aim at an integrated approach, they tend to have sectoral responses, with limited integration of local voices. There is a good understanding of the impacts, but the implementation of policy and outcomes on the ground are harder to see (Bulkeley, 2006; Burch and Robinson, 2007).

In these adaptation strategies, the size of the local government is important, and it varies depending on the population and location. Primate and large cities exert more independence, whereas smaller municipalities depend more on higher levels of the government units, and often form associations to pool their resources (Lundqvist and Borgstede, 2008). In the latter case, state-mandated programs and state-generated grants are the main incentives to formulate mitigation policies (Aall et al., 2007) and can be applicable to adaptation policies. Lack of resources and capabilities has led to outsourcing of local adaptation plans, and can generate insensitive and unrefined local solutions and more reliance on technological fixes (Crabbé and Robin, 2006).

The history and process of decentralization are significant in the capacity of the local government to formulate and implement adaptation policies. Aligning local climate adaptation policies with the state/provincial and national/federal units is a significant challenge for local governments (Roberts, 2008; van Aalst et al., 2008). The case of decentralization in climate change adaptation is relatively new, and we can draw some lessons from decentralized natural resource management and crisis management. One of the problems of decentralization has been the complexity and uniqueness of each locality that policy planners often failed to take into account because of the lack of understanding and consultation with the local community, and this could result in recentralizing the entire process in some instances (Ribot et al., 2006; Geiser and Rist, 2009). Some remedies include working with local institutions, ensuring appropriate transfer of various rights and access, and providing sufficient time for the process (Ribot, 2003). The crisis management literature also points out that there has been a lack of coordination and integration between central and local governments

(Waugh and Streib, 2006; Schneider, 2008). Moynihan (2009) suggests a networked collaboration as a solution and posits that even a hierarchical disaster management structure such as the incident command system in the United States operates on the network principles of negotiation, trust, and reciprocity.

Although government actors play a key role, it is evident that partnerships between public, civic, and private actors are crucial in addressing climate hazards-related adaptation (Agrawal, 2010). While international agencies, the private sector, and NGOs play a norm-setting agenda at provincial, state, and national levels, community-based organizations (CBOs) often have greater capacity to mobilize at the local scale (Milbert, 2006). NGO and CBO networks play a critical role in capturing the realities of local livelihoods, facilitating sharing information, and identifying the role of local institutions that lead to strengthened local capacity (Bull-Kamanga et al., 2003). Strong city-wide initiatives are often based on strategic alliances, and local community organizations are essential to making city planning operational (Hasan, 2007). This can be seen in the case of the New York City Panel on Climate Change that acted as a scientific advisory group to both Mayor Bloomberg's Office of Long-term Planning and Sustainability and the New York City Climate Change Adaptation Task Force, a stakeholder group of approximately 40 public agencies and private sector organizations that manage the critical infrastructure of the region (Rosenzweig et al., 2011). The Panel and stakeholders separated functions between scientists (knowledge provision) and stakeholders (planning and action) and communicated climate change uncertainties with coordination by the Mayor's office (Rosenzweig et al., 2011).

Many nongovernment actors charged with managing climate risks use community risk assessment tools to engage communities in risk reduction efforts and influence planning at district and sub-national levels (van Aalst, 2006; Twigg, 2007). NGO engagement in risk management activities ranges from demonstration projects, training and awareness-raising, legal assistance, alliance building, small-scale infrastructure, and socioeconomic projects, to mainstreaming and advocacy work (Luna, 2001; Shaw, 2006). Bridging citizen-government gaps is a recognized role of civil society organizations and NGOs often act as social catalysts or social capital, an essential for risk management in cities (Wisner, 2003). Conversely, the potential benefits of social capital are not always maximized due to mistrust, poor communications, or lack of functioning either within municipalities or nongovernment agencies. This has major implications for risk reduction (Wisner, 2003) and participation of the most vulnerable in nongovernment initiatives at municipal or subnational level is not guaranteed (Tanner et al., 2009).

This section highlighted mechanisms for building capacity for local adaptation to climate extremes ranging from empowerment for decisionmaking to utilization of social networks. A balanced portfolio of approaches that capture local knowledge, proactive behaviors, and governmental and nongovernmental initiatives and practices will prove most successful in managing the risk of climate extremes at the local level.

5.5. Challenges and Opportunities

As illustrated earlier in the chapter, disaster risk management actions increase the coping capacity of local places to disasters in the short term and benefit a community's resilience in the long term. Differences in coping, risk management, and adaptation along with the costs of managing disaster risk at the local level present challenges and opportunities for adaptation to climate extremes. They not only influence human security, but the scale and context of the differences highlight opportunities for proactive actions for risk reduction and climate change adaptation, and also identify constraints to such actions.

5.5.1. Differences in Coping and Risk Management

There are significant differences among localities and population groups in the ability to prepare for, respond to, recover from, and adapt to disasters and climate extremes. During the last century, social science researchers have examined those factors that influence coping responses by households and local entities through post-disaster field investigations as well as pre-disaster assessments (Mileti, 1999; NRC, 2006). Among the most significant individual characteristics are gender, age, wealth, ethnicity, livelihoods, entitlements, health, and settlements. However, it is not only these characteristics operating individually, but also their synergistic effects that give rise to variability in coping and managing risks at the local level.

5.5.1.1. Gender, Age, and Wealth

The literature suggests that at the local level gender makes a difference in vulnerability (Section 2.4) and in the differential mortality from disasters (Neumayer and Plümper, 2007). The evidence is *robust* with *high agreement*. In disasters, women tend to have different coping strategies and constraints on actions than men (Fothergill, 1996; Morrow and Enarson, 1996; Peacock et al., 1997). These are due to socialized gender factors such as social position (class), marital status, education, wealth, and caregiver roles, as well as physical differences in stature and endurance. At the local level for example, women's lack of mobility, access to resources, lack of power and legal protection, and social isolation found in many places across the globe tend to augment disaster risk, and vulnerability (Schroeder, 1987; IFRC, 1991; Mutton and Haque, 2004; UNIFEM, 2011). Relief and recovery operations are often insensitive to gender issues (Hamilton and Halvorson, 2007), and so the provision of such supplies and services also influences the differential capacities to cope (Enarson, 2000; Ariyabandu, 2006; Wachtendorf et al., 2006; Fulu, 2007), especially at the local level. However, the active participation of women has been shown to increase the effectiveness of prevention, disaster relief, recovery, and reconstruction, thereby improving disaster management (Enarson and Morrow, 1997, 1998; Fothergill, 1999, 2004; Hamilton and Halvorson, 2007; Enarson, 2010; see Box 5-5).

Age acts as an important factor in coping with disaster risk (Cherry, 2009). Older people are more prone to ill health, isolation, disabilities, and immobility (Dershem and Gzirishvili, 1999; Ngo, 2001), which negatively influence their coping capacities in response to extreme events (see Case Study 9.2.1). In North America, for example, retired people often choose to live in hazardous locations such as Florida or Baja California because of warmer weather and lifestyles, which in turn increases their potential exposure to climate-sensitive hazards. Often because of hearing loss, mental capabilities, or mobility, older persons are less able to receive warning messages or take protective actions, and are more reluctant to evacuate (O'Brien and Mileti, 1992; Hewitt, 1997). However, older people have more experience and wisdom with accumulated know-how on specific disasters/extreme events as well as the enhanced ability to transfer their coping strategies arising from life experiences.

Children have their own knowledge of hazards, hazardous places, and vulnerability that is often different than adults (Plush, 2009; Gaillard and Pangilinan, 2010). Research has shown significant diminishment of coping skills (and increases in post-traumatic stress disorder and other psychosocial effects) among younger children following Hurricane Katrina (Barrett et al., 2008; Weems and Overstreet, 2008). In addition to physical impacts and safety (Lauten and Lietz, 2008; Weissbecker et al., 2008), research also suggests that emotional distress caused by fear of separation from the family, and increased workloads following disasters affects coping responses of children (Babugura, 2008; Ensor, 2008). However, the research also suggests that children are quite resilient and can adapt to environmental changes thereby enhancing the adaptive capacity of households and communities (Bartlett, 2008; Manyena et al., 2008; Mitchell et al., 2008; Pfefferbaum et al., 2008; Ronan et al., 2008; Williams et al., 2008).

Wealth, especially at the local level affects the ability of a households or localities to prepare for, respond to, and rebound from disaster events (Cutter et al., 2003; Masozera et al., 2007). Wealthier places have a greater potential for large monetary losses, but at the same time, they have the resources (insurance, income, political cache) to cope with the impacts and recover from extreme events, and they are less socially vulnerable. In Asia, for example, wealth shifted construction practices from wood to masonry, which made many of the cities more vulnerable and less able to cope with disaster risk (Bankoff, 2007), especially in seismic regions. Poorer localities and populations often live in cheaper hazard-prone locations, and face challenges not only in responding to the event, but also recovering from it. Poverty also enhances disaster risk (Carter et al., 2007). In some instances, it is neither the poor no the rich that face recovery challenges, but rather localities that are in between, such as those not wealthy enough to cope with the disaster risk on their own, but not poor enough to receive full federal or international assistance.

In some localities, it is not just wealth or poverty that influence coping strategies and disaster risk management, but rather the interaction between wealth, power, and status, that through time and across space has led to a complicated system of social stratification (Heinz Center,

> ### Box 5-5 | The Role of Women in Proactive Behavior
>
> Women's involvement in running shelters and processing food was crucial to the recovery of families and communities after Hurricane Mitch hit Honduras. One-third of the shelters were run by women, and this figure rose to 42% in the capital. The municipality of La Masica in Honduras, with a mostly rural population of 24,336 people, stands out in the aftermath of Mitch because, unlike other municipalities in the northern Atlanta Department, it reported no mortality. Some attributed this outcome to a process of community emergency preparedness that began about six months prior to the disaster. Gender lectures were given and, consequently, the community decided that men and women should participate equally in all hazard management activities. When Mitch struck, the municipality was prepared and vacated the area promptly, thus avoiding deaths. Women participated actively in all relief operations. They went on rescue missions, rehabilitated local infrastructure (such as schools), and along with men, distributed food. They also took over from men who had abandoned the task of continuous monitoring of the early warning system. This case study illustrates the more general finding that the active incorporation of women into disaster preparedness and response activities helps to ensure success in reducing the impacts of disasters (Buvinić et al., 1999; Cupples, 2007; Enarson, 2009).

2002). One of the best examples of this is the human experience with Hurricane Katrina (see Box 5-6).

5.5.1.2. Livelihoods and Entitlements

Adaptive capacity is influenced to a large extent by the institutional rules and behavioral norms that govern individual responses to hazards (Dulal et al., 2010). It is also socially differentiated along the lines of age, ethnicity, class, religion, and gender (Adger et al., 2007). Local institutions regulate the access to adaptation resources: those that ensure equitable opportunities for access to resources promote adaptive capacity within communities and other local entities (Jones et al., 2010). Institutions, as purveyors of the rules of the game (North, 1990), mediate the socially differential command over livelihood assets, thus determining protection or loss of entitlements.

Livelihood is the generic term for all the capabilities, assets, and activities required for a means of living. Livelihood influences how families and communities cope with and recover from stresses and shocks (Carney, 1998). Another definition of livelihoods gives more emphasis to access to assets and activities that is influenced by social relations (gender, class, kin, and belief systems) and institutions (Ellis, 2000). Understanding

Box 5-6 | Race, Class, Age, and Gender: Hurricane Katrina Recovery and Reconstruction

The intersection of race, class, age, and gender influenced differential decisionmaking; the uneven distribution of vulnerability and exposure; and variable access to post-event aid, recovery, and reconstruction in New Orleans before, during, and after Hurricane Katrina (Elliott and Pais, 2006; Hartman and Squires, 2006; Tierney, 2006). Evacuation can protect people from injury and death, but there are inequalities in who can evacuate and when, with the elderly, poor, and minority residents least able to leave without assistance (Cutter and Smith, 2009). Extended evacuations (or temporary displacements lasting weeks to months) produce negative effects. Prolonged periods of evacuation can result in a number of physical and mental health problems (Curtis et al., 2007; Mills et al., 2007). Furthermore, separation from family and community members and not knowing when a return home will be possible also adds to stress among evacuees (Curtis et al., 2007). DeSalvo et al. (2007) found that long periods of displacement were among the key causes of post-traumatic stress disorder in a study of New Orleans workers. These temporary displacements can also lead to permanent outmigration by specific social groups as shown by the depopulation of New Orleans five years after Hurricane Katrina (Myers et al., 2008). In terms of longer-term recovery, New Orleans is progressing, however large losses in population, housing, and employment suggest a pattern of only partial recovery for the city with significant differences in the location and the timing at the neighborhood or community level (Finch et al., 2010).

how natural resource-dependent people cope with climate change in the context of wider livelihood influences is critical to formulating valid adaptation frameworks.

Local people's livelihoods and their access to and control of resources can be affected by events largely beyond their control such as climatic extremes (e.g., floods, droughts), conflict, or agricultural problems such as pests and disease and economic shocks that can largely impact their livelihoods (Chambers and Conway, 1992; Jones et al., 2010). For poor communities living on fragile and degraded lands such as steep hillsides, dry lands, and floodplains, climate extremes present additional threats to their livelihoods that could be lost completely if exposed to repeated disastrous events within short intervals leaving insufficient time for recovery. Actions aiming at improving local adaptive capacity focus more on addressing the deteriorating environmental conditions. A central element in their adaptation strategies involves ecosystem management and restoration activities such as watershed rehabilitation, agroecology, and forest landscape restoration (Ellis, 2000; Ellis and Allison, 2004; Osman-Elasha, 2006b). As some suggest (Spanger-Siegfried et al., 2005) these types of interventions often protect and enhance natural resources at the local scale and address immediate development priorities, but can also improve local capacities to adapt to future climate change. The buffering capacities of local people's livelihoods and their institutions are critical for their adaptation to extreme climate stress. They often rest on the ability of communities to generate potentials for self-organization and for social learning and innovations (Adger et al., 2006).

A number of studies indicated that sustainable strategies for disaster reduction help improve livelihoods (UNISDR, 2004), while social capital and community networks support adaptation and disaster risk reduction by diminishing the need for emergency relief in times of drought and/or crop failure (Devereux and Coll-Black, 2007; see Section 5.2.1). A research study in South Asia suggests that adaptive capacity and livelihood resilience depend on social capital at the household level (i.e., education and other factors that enable individuals to function within a wider economy), the presence or absence of local enabling institutions (local cooperatives, banks, self-help groups), and the larger physical and social infrastructure that enables goods, information, services, and people to flow. Interventions to catalyze effective adaptation are important at all these multiple levels (Moench and Dixit, 2004). Diversification within and beyond agriculture is a widely recognized strategy for reducing risk and increasing well-being in many developing countries (Ellis, 2000; Ellis and Allison, 2004).

Entitlements are assets of the individuals and household. Assets are broadly defined and include not only physical assets such as land, but also human capital such as education and training. At the local scale assets include institutional assets such as technical assistance or credit; social capital such as mutual assistance networks; public assets such as basic infrastructure like water and sanitation; and environmental assets such as access to resources and ownership of them (Leach et al., 1999). The link between disaster risk, access to resources, and adaptation has been widely documented in the literature (Sen, 1981; Adger, 2000; Brooks, 2003). Extreme climate events generally lead to entitlement decline in terms of the rights and opportunities that local people have to access and command the livelihood resources that enable them to deal with and adapt to climate stress.

Assessment of livelihoods provides the explanation as to the differences in responses based on the understanding of endowments, entitlements, and capabilities, within the organizational structure and power relations of individuals, households, communities, and other local entities (Scoones, 1998). Access to assets and entitlements is an important element in improving the ability of localities to lessen their vulnerability and to cope with and respond to disasters and environmental change. In some instances, this may not be true. For example, if a disaster affects a household asset, but the household is still paying off its debt regarding the initial cost of the asset and assuming that the asset is not protected or insured against hazards, the asset loss coupled with the need to pay off the loan renders the household more vulnerable, not less (Twigg, 2001). Entitlement protection thus requires adaptive types of institutions and patterns of behavior (Meehl et al., 2007), with a focus on local people's agency within specific configurations of power relations. The challenge is, therefore, to empower the most vulnerable to pursue livelihood options that strengthen their entitlements and protect what they themselves consider the social sources of adaptation and resilience in the face of extreme climate stress. Better management of disaster risk also maximizes use of available resources for adapting to climate change (Kryspin-Watson et al., 2006).

5.5.1.3. Health and Disability

The changes in extreme events and impacts of climate change influence the morbidity and mortality of many populations now, and even more so in the future (Campbell-Lendrum et al., 2003). The extreme impacts of climate change (see Sections 3.1.1 and 4.4.6) directly or indirectly affect the health of many populations and these will be felt first at the local level. Heat waves lead to heatstroke and cardiovascular disease, while shifts in air pollution concentrations such as ozone that often increase with higher temperatures cause morbidity from other diseases (Bernard et al., 2001). Heat waves differentially affect populations based on their ethnicity, gender, age (Díaz et al., 2002), and medical and socioeconomic status (O'Neill and Ebi, 2009), consequently raising concerns about health inequalities (see Case Study 9.2.1), especially at the local scale. Health inequalities are of concern in extreme impacts of climate change more generally, as those with the least resources often have the least ability to adapt, making the poor and disenfranchised most vulnerable to climate-related illnesses (McMichael et al., 2008). For extreme events, pre-existing health conditions that characterize vulnerable populations can exacerbate the impact of disaster events since these populations are more susceptible to additional injuries from disaster impacts (Brauer, 1999; Brown, 1999; Parati et al., 2001). Chronic health conditions/disabilities can also lead to subsequent communicable diseases and illnesses in the short term, to lasting chronic illnesses, and to longer-term mental health conditions (Shoaf and Rottmann, 2000; Bourque et al., 2006; Few and Matthies, 2006).

A range of vector-borne illnesses has been linked to climate, including malaria, dengue, Hantavirus, Bluetongue, Ross River Virus, and cholera (Patz et al., 2005). Cholera, for example, has seasonal variability that may be directly affected by climate change (Koelle et al., 2004). Vector-borne illnesses have been projected to increase in geographic reach and severity as temperatures increase (McMichael et al., 2006), but these changes depend on a variety of human interventions like deforestation and land use. The areas of habitation by mosquitoes and other vectors are moving to areas previously free from such vectors of transmission (Lafferty, 2009). Pools of standing water that are breeding grounds for mosquitoes promise to expand, therefore increasing illness exposure (Depradine and Lovell, 2004; Meehl et al., 2007). At the same time, some literature shows that illnesses like malaria are less prone to increase than originally thought (Gething et al., 2010). Much of the nuance of this literature is due to the location-specific nature of these outcomes. Therefore, vector control programs will be best suited to the local characteristics of changing risks. Some programs, like those geared toward surveillance, need common characteristics to support national programs and also need to be coordinated across scales from local to national and between local places. In addition, there are a variety of social factors that have the potential to influence disease rates that are most suitably managed at the sub-national level or urban scale. For instance, certain types of population growth or change may increase risk and affect disease rates (Patz et al., 2005). Increased population and related land use changes can also increase disease rates. Vector control programs generally implemented at the local level also have the potential to influence health outcomes (Tanser et al., 2003). Infectious disease patterns also have the potential to change dramatically, necessitating improved prevention on the part of local providers who have knowledge of local environmental change (Parkinson and Butler, 2005).

There is concern regarding the mental health impacts of storms and floods that lead to destruction of livelihoods and displacement, especially for vulnerable populations (Balaban, 2006). In some hurricanes, the mental health of residents in affected communities is extremely negatively impacted over an extended period of time (Weisler et al., 2006). Policy responses to the event were insufficient to manage these impacts, and provide a lesson for future events where greater mental health services may be necessary (Lambrew and Shalala, 2006). Managing public health and disability is important in the response to disasters (Shoaf and Rottmann, 2000).

Human health is at risk from many extreme events linked to climate change. While resources from scales above the local are often necessary, the direction and application of those resources by local actors who know how to best apply them could make significant differences in human morbidity and mortality linked to climate extremes.

5.5.1.4. Human Settlements

Settlement patterns are another factor that influences disaster risk management and coping with extremes. Human settlements differ in their physical and governance structures, population growth patterns, as well as in the types, drivers, impacts, and responses to disasters. As noted earlier (see Section 5.5.1.2), rural livelihoods and poverty are drivers of disaster risk, but not the only ones. Poverty, resource scarcity, access to resources, as well as inaccessibility constrain disaster risk management. When these are coupled with climate variability, conflict, and health issues they reduce the coping capacity of rural places (UNISDR, 2009). At the other extreme are the concentrated settlements of towns and cities where the disaster risks are magnified because of population densities, poor living conditions including overcrowded and substandard housing, lack of sanitation and clean water, and health impairments from pollution and lack of adequate medical care (Bull-Kamanga et al., 2003; De Sherbinin et al., 2007). Strengthening local capacity in terms of housing, infrastructure, and disaster preparedness is one mechanism shown to improve urban resilience and the adaptive capacity of cities to climate-sensitive hazards (Pelling, 2003). It is also instructive to note how communities with differing capacities address similar problems (Walker and Sydneysmith, 2008).

One important locality receiving considerable research and policy attention is megacities (see Case Study 9.2.8) due to the density of infrastructure, the population at risk, the growing number and location of informal settlements, complex governance, and disaster risk management (Mitchell, 1999). Given the rapid rate of growth in the largest of these world's cities and the increasing urbanization, the disaster risks will increase in the next decade, placing more people in harm's way with billions of dollars in infrastructure located in highly exposed areas (Munich Re Group, 2004; Kraas et al., 2005; Wenzel et al., 2007).

For many regions, the ability to limit urban exposure has already been achieved through building codes, land management, and disaster risk mitigation, yet losses keep increasing. For disaster reduction to become more effective, megacities will need to address their societal vulnerability and the driving forces that produce it (rural to urban migration, livelihood pattern changes, wealth inequities, informal settlements) (Wisner and Uitto, 2009). Many megacities are seriously compromised in their ability to prepare for and respond to present disasters, let alone adapt to future ones influenced by climate change (Fuchs, 2009; Heinrichs et al., 2009; Prasad et al., 2009).

However, it is not only the megacities that pose challenges, but the overall growth in urban populations. Currently more than half of the global population lives in urban areas with an increasing population exposed to multiple risk factors (UNFPA, 2009). Risk is increasing in urban agglomerations of different size due to unplanned urbanization and accelerated migration from rural areas or smaller cities (UN-HABITAT, 2007). The 2009 Global Assessment Report on Disaster Risk Reduction (UNISDR, 2009) lists unplanned urbanization and poor urban governance as two main underlying factors accelerating disaster risk. It highlighted that the increase in global urban growth of informal settlements in hazard-prone areas reached 900 million in informal settlements, increasing by 25 million per year (UNISDR, 2009). Urban hazards exacerbate disaster risk by the lack of investment in infrastructure as well as poor environmental management, thus limiting the adaptive capacity of these areas.

5.5.2. Costs of Managing Disaster Risk and Risk from Climate Extremes

5.5.2.1. Costs of Impacts, Costs of Post-Event Responses

It is extremely difficult to assess the total cost of a large disaster, such as Hurricane Katrina, especially at the local scale since most economic data are only available at the national scale. Direct losses consist of direct market losses and direct non-market losses (intangible losses). The latter include health impacts, loss of lives, natural asset damages and ecosystem losses, and damages to historical and cultural assets. Indirect losses (also labeled higher-order losses (Rose, 2004) or hidden costs (Heinz Center, 1999)) include all losses that are not provoked by the disaster itself, but by its consequences. Measuring indirect losses is important as it evaluates the overall economic impact of the disaster on society. Another difficulty with the measurement of economic losses at the local level has to do with the boundary delineation for local analyses. For example, local losses can be compensated from various inflows of goods, workers, capital, and governmental or foreign aid from outside the affected area (Eisensee and Stromberg, 2007). Local disasters also provide ripple effects and influence world markets, for instance, oil prices in the case of Hurricane Katrina due to the temporary shutdown of oil rigs. It is important to consider tradeoffs at different spatial scales especially when estimating indirect losses at the local level. Disaster loss estimates are, therefore, highly dependent on the scale of the analysis, and result in wide variations among community, state, province, and sub-national regions.

Despite the difficulties in assessing local economic impact, several studies exist. For example, Strobl (2008) provided an econometric analysis of the impact of the hurricane landfall on county-level economic growth in the United States. This analysis showed that a county struck by at least one hurricane over a year led to a decline in economic growth on average by 0.79% and an increase by 0.22% the following year. The economic impact of the 1993 Mississippi flooding in the United States showed significant spatial variability within the affected regions. In particular, states with a strong dependence on the agricultural sector had a disproportionate loss of wealth compared to states that had a more diversified economy (Hewings and Mahidhara, 1996). Noy and Vu (2010) investigated the impact of disasters on economic growth in Vietnam at the provincial level, and found that fatal disasters decreased economic production while costly disasters increased short-term growth. Rodriguez-Oreggia et al. (2010) focused on poverty and the World Bank's Human Development Index at the municipality level in Mexico, and demonstrated that municipalities affected by disasters saw an increase in poverty of 1.5 to 3.6%. Studies also found that regional indirect losses increase nonlinearly with direct losses (Hallegatte, 2008), and can be compensated by importing the means for reconstruction (workers, equipment, finance) from outside the affected area.

The U.S. Bureau of Labor Statistics (2006) also provided a detailed analysis of the labor market consequences of Hurricane Katrina within Louisiana and found a marked economic and employment loss for Louisiana businesses, high unemployment rates immediately after the disaster, and continued unemployment into 2006 for returning evacuees. At the household level, Smith and McCarty (2006) show that households are more often forced to move outside the affected area due to infrastructure problems than due to structural damages to their home.

Modeling approaches are also used to assess disaster indirect losses at sub-national levels. These approaches include input-output models (Okuyama, 2004; Haimes et al., 2005; Hallegatte, 2008) and Computable General Equilibrium models (Rose et al., 1997; Rose and Liao, 2005; Tsuchiya et al., 2007). Most of the published analyses were carried out in developed countries. In the United States, West and Lenze (1994), for example, discuss the merits of combining different impact models to triangulate, obtaining better primary data to reduce uncertainty, and developing tools for estimating the impact of Hurricane Andrew on Florida using reconstruction scenarios. The lack of research on disaster loss estimates in developing countries creates problems of underreported economic losses or overestimation of disaster losses depending on political or other interests. This is a big research gap.

5.5.2.2. Adaptation and Risk Management – Present and Future

Studies on the costs of local disaster risk management are scarce, fragmented, and conducted mostly in rural areas. One study estimated the benefit/cost ratio of disaster management and preparedness programs in the villages of Bihar and Andra Pradesh, India to be 3.76 and 13.38, respectively (Venton and Venton, 2004), suggesting higher benefits than costs. Research undertaken by the Institute for Social and Environmental Transition on a number of cases in India, Nepal, and Pakistan demonstrated that benefits exceed the costs for local interventions (Dixit et al., 2008; Moench and Risk to Resilience Study Team, 2008). For example, they note that return rates are particularly robust for lower-cost interventions (e.g., raising house plinths and fodder storage units, community-based early warning, establishing community grain or seed banks, and local maintenance of key drainage points), when compared to embankment infrastructure strategies that require capital investment (Moench and Risk to Resilience Study Team, 2008). The studies demonstrated a sharp difference in the effectiveness of the two approaches, concluding that the embankments historically have not had an economically satisfactory performance in that study area. In contrast, the benefit/cost ratio for the local-level strategies indicated economic efficiency over time and for all climate change scenarios (Dixit et al., 2008). In developed countries, there are cost differences in adaptation strategies between urban and rural areas. For example, in Japan disaster damage is several hundred times more costly in urban than in rural areas, necessitating different disaster risk management strategies depending on the benefit to cost analysis (Kazama et al., 2009).

Though disaster risk management and adaptation policies are closely linked, few integrated cost analyses of risk management and adaptation are available at the local level. One example draws from recent studies of the cost of city-scale adaptation. Rosenzweig et al. (2007, 2011) developed a sophisticated analytical response to a projected fall in water availability in New York. This frames adaptation assessment within a step-wise decision analysis by identifying and quantifying impact risks before identifying adaptation options that are then screened, evaluated, and finally implemented. Another series of studies used simplified catastrophe risk assessment to calculate the direct costs of storm surges under scenarios of sea level rise coupled with an economic input-output model for Copenhagen and Mumbai (Hallegatte et al., 2008a,b, 2011; Ranger et al., 2011). The output is an assessment of the direct and indirect economic impacts of storm surge under climate change including production, job losses, reconstruction time, and the benefits of investment in upgraded coastal defenses. Results show that the consideration of adaptation is an important element in the economic assessment of

> **FAQ 5.3 | Is it possible to estimate the cost of risk management and adaptation at the local scale?**
>
> Studies on the costs of local disaster risk management are scarce, fragmented, and conducted mostly in rural areas. Most economic data (e.g., input-output table, income data) are available at the national scale. Moreover, there is a clear lack of research on disaster estimates in developing countries, which presents a big gap in need of further research. In developed countries, there are cost differences in adaptation strategies between urban and rural areas. The reliability of disaster economic loss estimates is especially problematic at the local level due to factors associated with the global nature of spatial coverage and resolution. In addition there is some ambiguity on impact and adaptation costs that affect local-level economic analyses, such as the lack of consensus on physical impacts of climate change and adaptive capacity and on the evaluation of non-market costs (e.g., biodiversity or cultural heritage), which creates some uncertainty about local impact and adaptation costs.

extreme disaster risks related to climate change (Hallegatte et al., 2011). Ranger et al. (2011) show that by improving the drainage system in Mumbai, losses associated with a 1-in-100 year flood event could be reduced by as much as 70%. This means that the annual losses could be reduced in absolute terms compared with the current level, even with climate change. Full insurance coverage of flooding could also cut the indirect cost by half. These analyses highlight the fact that adaptation to extreme events and climate change can focus on reducing the direct losses (e.g., through the upgrade of coastal defenses) or indirect losses by making the economy more robust, utilizing insurance schemes, or enacting public policies to support small businesses after the disaster.

5.5.2.3. Consistency and Reliability of Cost and Loss Estimations at the Local Level

There are inconsistencies in disaster-related economic loss data at all levels – local, national, global – that ultimately influence the accuracy of such estimates (Guha-Sapir and Below, 2002; Downton and Pielke Jr., 2005; Pielke Jr. et al., 2008). The reliability of disaster economic loss estimates is especially problematic at the local level. First, the spatial coverage and resolution of databases are global in coverage, but only have data that represent the entire country, not sub-units within it such as provinces, states, or counties. Second, thresholds for inclusion, where only large economically significant disasters are included, bias the data toward singular events with large losses, rather than multiple, smaller events with fewer losses. Third, what gets counted varies between databases (e.g., insured versus uninsured losses; direct versus indirect; Gall et al., 2009). Moreover, disaster loss estimates have various purposes (e.g., assessment of foreign aid needs; cost-benefit analysis of protection investments; World Bank, 2010). Depending on the purpose, spatial and conceptual gaps exist depending on the inclusion of loss-only data or a combination of loss and gain estimates as well as the calculation of non-market losses.

Similarly, there is some ambiguity about impact and adaptation costs that affect local-level economic analyses. The lack of consensus on physical impacts of climate change and adaptive capacity (see Section 4.5) is one issue. Another is the discount rate (Heal, 1997; Tol, 2003; Nordhaus, 2007; Stern, 2007; Weitzman, 2007) and the evaluation of non-market costs, especially the value of biodiversity or cultural heritage (Pearce and Moran, 1994), the latter contributing some uncertainty about local impact and adaptation costs. Finally, the possibility of low-probability high-consequence climate change is not fully included in most analyses (Stern, 2007; Weitzman, 2007; Lonsdale et al., 2008; Nicholls et al., 2008).

5.5.3. Limits to Local Adaptation

Local adaptation is set within larger spatial and temporal scales (Adger et al., 2005), which influence the range of actors involved and the types of potential barriers to the adaptation process (Moser and Ekstrom, 2010; see Sections 6.3 and 7.6). At the local scale, limits and barriers to local adaptation generally fall into three interconnected categories: ecological and physical; human informational related to knowledge, technology, economics, and finances; and psychological, behavioral, and socio-cultural barriers (ICIMOD, 2009; Adger et al., 2010). The social and cultural limits to adaptation are not well researched, with little attention within the climate change literature devoted to this thus far.

Lack of access to information by local people has restricted improvements in knowledge, understanding, and skills – needed elements in helping localities undertake improved measures to protect themselves against disasters and climate change impacts (Agrawal et al., 2008). The information gap is particularly evident in many developing countries with limited capacity to collect, analyze, and use scientific data on mortality and demographic trends as well as evolving environmental conditions (IDRC, 2002; Carraro et al., 2003; NRC, 2007). Based on Fischer et al. (2002), closing the information gap is critical to reducing climate change-related threats to rural livelihoods and food security in Africa.

Lack of capacities and skills, particularly for women, also has been identified as a limiting factor for effective local adaptation actions (Osman-Elasha et al., 2006). For example, localities in areas prone to climate extremes such as frequent drought have developed certain coping responses that assist them in surviving harsh conditions. Over time, such coping responses proved inadequate due to the magnitude of the problem (Ziervogel et al., 2006). For example, in Mali, one initiative involves empowering women and giving them the skills to diversify their livelihoods, thus linking environmental management, disaster risk reduction, and the position of women as key resource managers (UNISDR and UNOCHA, 2008).

In financial terms, microfinance services typically do not reach the poorest and most vulnerable groups at local levels who have urgent and immediate needs to be addressed (Amin et al., 2001; Helms, 2006). The ability of a community to ensure equitable access and entitlement to key resources and assets is a key factor in building local adaptive capacity.

In developed countries, household decisions regarding disaster risk reduction and adaptation are often guided by factors other than cost. For example, Kunreuther et al. (2009) found that most individuals underestimate the risk and do not make cost-benefit tradeoffs in their decisions to purchase hazard insurance and/or have adequate coverage. They also found empirical evidence to suggest that the hazard insurance purchase decision was driven not only by the need to protect assets, but also to reduce anxiety, satisfy mortgage requirements, and social norms. For other types of disaster mitigation activities, households do

---

FAQ 5.4 | What are the limits to adaptation at the local level?

Traditionally, local risk management strategies focused only on short-term climatic events without considering the long-term trajectories presented by a changing climate. Although reacting to climate extreme events and their impacts is important, it is more crucial now to focus on building the resilience of communities, cities, and sectors in order to ameliorate the impacts of future climatic changes. The range and choice of actions that can be taken at the levels of individual or households are often event-specific and time-dependent. They are also constrained by location, adequate infrastructure, socioeconomic characteristics, and access to disaster risk information. For example, the increased urban vulnerability due to urbanization and rising population exacerbates disaster risk by the lack of investment in infrastructure as well as poor environmental management, and can have spillover effects to rural areas.

The obstacles to information transfer and communications are diverse, ranging from limitations in modeling the climate system to procedural, institutional, and cognitive barriers in receiving or understanding climatic information and advance warnings and the capacity and willingness of decisionmakers to modify action. Within many rural communities, low bandwidth and poor computing infrastructure pose serious constraints to risk message receipt. Such gaps are evident in developed as well as lesser-developed regions. Constraints exist in locally-organized collective action because of the difficulties of building effective coalitions with other organizations.

not voluntarily invest in cost-effective mitigation because of underestimating the risk, taking a short-term rather than long-term view, and not learning from previous experience. However, they found social norms significant: if homeowners in the neighborhood installed hurricane shutters, most would follow suit; the same was true of purchasing insurance (Kunreuther et al., 2009). For municipal governments, adoption of building codes in hurricane-prone areas reduces damages by US$ 108 per square meter for homes built from 1996 to 2004 in Florida (Kunreuther et al., 2009). However, enforcement of building codes by municipalities is highly variable and becomes a limiting factor in disaster risk management and adaptation.

Local-level adaptation actions in many cases are portrayed as reactive and short term, unlike the higher-level national or regional plans that are considered anticipatory and involve formulation of policies and programs (Bohle, 2001; Burton et al., 2003). Poverty, increased urbanization, and extreme climate events limit the capacity to initiate planned livelihood adaptations at the local scale. If extreme events happen more frequently or with greater intensity or magnitude some locations may be uninhabitable for lengthy and repeated periods, rendering sustainable development impossible. In such a situation, not all places will be able to adapt without considerable disruption and costs (economic, social, cultural, and psychological). In some cases forced migration may be the only alternative (Brown, 2008).

As the above shows, the main challenge for local adaptation to climate extremes is to find a good balance of measures that simultaneously address fundamental issues related to the enhancement of local collective actions, and the creation of subsidiary structures at national and international scales that complement such local actions. This means that the localized expression of the type, frequency, and extremeness of climate-sensitive hazards will be set within these national and international contexts.

5.5.4. Advancing Social and Environmental Justice

One of the key issues in examining outcomes of local strategies for disaster risk management and climate change adaptation is the principle of fairness and equity. There is a burgeoning research literature on climate justice looking at the differential impacts of adaptation policies (Kasperson and Kasperson, 2001; Adger et al., 2006) at local, national, and global scales. The primary considerations at the local level are the differential impacts of policies on communities, subpopulations, and regions from present management actions (or inactions) (Thomas and Twyman, 2005). There is also concern regarding the impact of present management (or inactions) in transferring the vulnerability of disaster risk from one local place to another (spatial inequity) or from one generation to another (intergenerational equity) (Cooper and McKenna, 2008). There is less research on the mechanisms or practical actions needed for advancing social and environmental justice at the local scale, independent of the larger issues of accountability and governance at all scales. This is an important gap in the literature.

5.6. Management Strategies

5.6.1. Basics of Planning in a Changing Climate

Prior to the development and implementation of management strategies and adaptation alternatives, local entities need baseline assessments on disaster risk and the potential impacts of climate extremes. The assessment of local disaster risk includes three distinct elements: 1) exposure hazard assessment, or the identification of hazards and their potential magnitudes/severities as they relate to specific local places (see below); 2) vulnerability assessments that identify the sensitivity of the population to such exposures and the capacity of the population to cope with and recover from them (see below and Sections 2.6.2 and 4.4); and 3) damage assessments that determine direct and indirect losses from particular events (either ex-post in real events or ex-ante through modeling of hypothetical events) (described in Section 5.5.2; see Section 4.5.1). Each of these plays a part in understanding the hazard vulnerability of a particular locale or characterizing not only who is at risk but also the driving forces behind the differences in disaster vulnerabilities in local places.

There are numerous examples of exposure and vulnerability assessment methodologies and metrics (Birkmann, 2006; see Chapter 2). Of particular note are those studies focused on assessing the sub-national exposure to coastal hazards (Gornitz et al., 1994; Hammar-Klose and Thieler, 2001), drought (Wilhelmi and Wiilhite, 2002; Alcamo et al., 2008; Kallis, 2008), or multiple hazards such as the Federal Emergency Management Agency's multi-hazard assessment for the United States (FEMA, 1997).

Vulnerability assessments highlight the interactive nature of disaster risk exposure and societal vulnerability. While many of them are qualitatively based (Bankoff et al., 2004; Birkmann, 2006), there is an emergent literature on quantitative metrics in the form of vulnerability indices. The most prevalent vulnerability indices, however, are national in scale (SOPAC and UNEP, 2005; Cardona, 2007) and compare countries to one another, not places at sub-national geographies. The exceptions are the empirically-based Social Vulnerability Index (Cutter et al., 2003) and extensions of it (Fekete, 2009).

Vulnerability assessments are normally hazard-specific and many have focused on climate-sensitive threats such as extreme storms in Revere, Massachusetts (Clark et al., 1998), sea level rise in Cape May, New Jersey (Wu et al., 2002), or flooding in Germany (Fekete, 2009) and the United States (Burton and Cutter, 2008; Zahran et al., 2008). Research focused on multi-hazard impact assessments ranges from locally based county-level assessments for all hazards (Cutter et al., 2000) to sub-national studies such as those involving all hazards for Barbados and St. Vincent (Boruff and Cutter, 2007) to those involving a smaller subset of climate-related threats (O'Brien et al., 2004; Brenkert and Malone, 2005; Alcamo et al., 2008). The intersection of local exposure to climate-sensitive hazards and social vulnerability was recently assessed for the northeast (Cox et al., 2007) and southern region of the United States (Oxfam, 2009).

However, the full integration of hazard exposure and social vulnerability into a comprehensive vulnerability assessment for the local area or region of concern is often lacking for many places. Part of this is a function of the bifurcation of the science inputs (e.g., natural scientists provide most of the relevant data and models for exposure assessments while social scientists provide the inputs for the populations at risk). It also relates to the difficulties of working across disciplinary or knowledge boundaries.

The development of methodologies and metrics for climate adaptation assessments is emerging and mostly derivative of the methodologies employed in vulnerability assessments noted above. For example, some are extensions or modifications of community vulnerability assessment methodologies and employ community participatory approaches such as those used by World Vision (Greene, n.d.), the Red Cross (van Aalst et al., 2008), and others. Still others begin with livelihood or risk assessment frameworks and use a wide range of techniques including multi-criteria decision analyses (Eakin and Bojorquez-Tapia, 2008); index construction (Vescovi et al., 2009); segmentation and regional to global comparisons (Torresan et al., 2008); and scenarios (Wilby et al., 2009).

5.6.2. Community-Based Adaptation

Community-based adaptation (CBA) empowers communities to decide how they want to prepare for climate risks and coordinate community action to achieve adaptation to climate change (Ebi, 2008). Part of this entails community risk assessment for climate change adaptation that assesses the hazards, vulnerabilities, and capacities of the community (van Aalst et al., 2008), which has also been called community-based disaster preparedness among other names. The intention is to foster active participation in collecting information that is rooted in the communities and enables affected people to participate in their own assessment of risk and identify responses than can enhance resilience by strengthening social-institutional measures including social relations (Allen, 2006; Patiño and Gauthier, 2009). In assessing short- and long-term risks, the needs of vulnerable groups are often excluded (Douglas et al., 2009). The tools for engaging vulnerable groups in the process include transect walks and risk maps that capture the climate-related hazards and risks and storylines about possible future climate change impacts (Ebi, 2008; van Aalst et al., 2008; Patiño and Gauthier, 2009), although these tools often require input from participants external to the community who have long-term climate information.

The challenges in using community-based adaptation approaches include the challenge of scaling up information (Burton et al., 2007), the fact that it is resource-intensive (van Aalst et al., 2008), and recognizing that disempowerment occurs when local stories are distorted or not valued sufficiently (Allen, 2006). The integration of climate change information increases this challenge as it introduces an additional layer of uncertainty and may conflict with the principle of keeping CBA simple. There is little evidence that secondary data on climate change has been used in CBA, partly because of the challenge of limited access to downscaled climate change scenarios relevant at the local level (Ziervogel and Zermoglio, 2009) and because of the uncertainty of projections.

Examples of community-based approaches illustrate some of the processes involved. In northern Bangladesh, a flooding adaptation project helped to establish early warning committees within villages that linked to organizations outside the community, with which they did not usually interact and that had historically blocked collective action and resource distribution (Ensor and Berger, 2009). Through this revised governance structure, the building of small roads, digging culverts, and planting trees to alleviate flood impacts was facilitated. In Portland, Oregon, another project involved a range of actors to reduce the impact of urban heat islands through engaging neighborhoods and linking them to experts to install green roofs, urban vegetation, and fountains that led to an increased sense of ownership in the improvements (Ebi, 2008). In the Philippines, the community-based approach enabled a deeper understanding of locally specific vulnerability than in previous disaster management contexts (Allen, 2006). While individually important, these community-based approaches should be viewed as part of a wider system that recognizes the drivers at multiple scales, including the municipalities and national levels.

CBA responses provide increased participation and recognition of the local context, which is important when adapting to climate change (see Box 5-7). The need for coordinated collective action was seen in Kampala, where land cover change and changing climate are increasing the frequency and severity of urban flooding and existing response activities are uncoordinated and consist of clearing drainage channels (Douglas et al., 2009). However, residents felt more could be done to adapt to frequent flooding, including increasing awareness of roles and responsibilities in averting floods, improving the drainage system, improving garbage and solid waste disposal, strengthening the building inspection unit, and enforcing bylaws on the construction of houses and sanitation facilities. Similarly, in Accra, residents felt that municipal laws on planning and urban design need to be enforced, suggesting that strong links are needed between community responses and municipal responses (Douglas et al., 2009).

5.6.3. Risk Sharing and Transfer at the Local Level

Risk transfer and risk sharing are pre-disaster financing arrangements that shift economic risk from one party to another and are more fully discussed in Chapter 2, Section 7.4.4, and Case Study 9.2.13. Informal risk sharing practices are common and important for post-disaster relief and reconstruction. In the absence of more formal mechanisms like insurance, those incurring losses may employ diverse non-insurance financial coping strategies, such as relying on the solidarity of international aid, remittances, selling and pawning fungible assets, and borrowing from moneylenders. Traditional livestock loans are one example (Oba, 2001). At-risk individuals in low-income countries rely extensively on reciprocal exchange, kinship ties, and community self-help. For example, women in high-risk areas often engage in innovative ways to access post-disaster

Box 5-7 | Taking Collective Action to Improve Livelihoods Strategies: Small-Scale Farmers Adapting to Climate Change in Northern Cape, South Africa

The Northern Cape Province, South Africa, is a harsh landscape, with frequent and severe droughts and extreme conditions for the people, animals, and plants living there. This has long had a negative impact on small-scale rooibos farmers living in some of the more marginal production areas. Rooibos is an indigenous crop that is well adapted to the prevailing hot, dry summer conditions, but is sensitive to prolonged drought. Rooibos tea has become well-accepted on world markets, but this success has brought little improvement to marginalized small-scale producers.

In 2001, a small group of farmers decided to take collaborative action to improve their livelihoods and founded the Heiveld Co-operative Ltd. Initially established as a trading cooperative to help the farmers produce and market their tea jointly, it subsequently became apparent that the local organization was also an important vehicle for social change in the wider community (Oettlé et al., 2004). The Heiveld became a repository and source of local and scientific knowledge related to sustainable rooibos production. Following a severe drought (2003-2005) and a perceived increase in weather variability, the Heiveld farmers decided to monitor the local climate and to discuss seasonal forecasts and possible strategies in quarterly climate change preparedness workshops. These workshops are facilitated in collaboration with two local NGOs (Indigo and Environmental Monitoring Group). They are also supported by scientists to address farmers' questions in a participatory action research approach – to ensure that local knowledge and scientific input can be combined to increase the resilience of local livelihoods. The Heiveld Co-operative has been an important organizational vehicle for this learning process, strongly supported by their long-term partners, with the focus on supporting the development of possible adaptation strategies through a joint learning approach to respond to and prepare for climate variability and change.

The extension of social, participatory, and organizational learning to climate change adaptation illustrated in this case study emphasizes the significance of identifiable climate change signals, informal networks, and boundary organizations to enhance the preparation of people and organizations for the changing climate (Berkhout et al., 2006; Pelling et al., 2008). Participatory learning is especially emphasized (Berkhout, 2002; A. Shaw et al., 2009; R. Shaw et al., 2009). Focusing on what can be learned from managing current climate risk is a good starting point, particularly for poor and marginalized communities (Someshwar, 2008).

capital by joining informal risk-hedging schemes, becoming clients of multiple microfinance institutions, or maintaining reciprocal social relationships. Combined analysis of multiple surveys suggests that about 40% of households in low- and lower-middle income countries are involved in private transfers in a given year as recipients or donors (Davies and Leavy, 2007).

Households in disaster-prone slum areas in El Salvador spend an average of 9.2% of their yearly income on risk management, including financing emergency relief and recovery (Wamsler, 2007). A particularly important informal risk-sharing mechanism is remittances, or transfers of money from foreign workers to their home countries (discussed further in Section 7.4.4). Household savings can be accessed from a bank, but they can also be in the form of stockpiles of food, grains, seeds, and fungible assets. Small savings institutions can be directly impacted by catastrophes, which can result in insufficient liquidity to handle a run on their accounts, as occurred during the 1998 floods in Bangladesh (Kull, 2006). Lacking sufficient savings, many disaster victims take out loans to cover their post-disaster expenses. The interest rate (18-60%) charged on formal microcredit, while relatively high, generally is far below the rate (120-300%) charged by local moneylenders (Linnerooth-Bayer and Mechler, 2009).

Insurance, including micro-insurance, is the most common formal risk transfer mechanism at the local level. An insurance contract spreads stochastic losses geographically and temporally, and can assure timely liquidity for the recovery and reconstruction process. As such, it is an effective disaster risk reduction tool especially when combined with other risk management measures. For example, in most industrialized countries, insurance is utilized in combination with early warning systems, risk information, disaster preparation, and disaster mitigation. Where insurance is applied without adequate risk reduction, it can be a disincentive for adaptation, as individuals may rely on insurance to manage their risks and are left overly exposed to impacts (Rao and Hess, 2009; see Section 5.4.1). Furthermore, insurance can provide the necessary financial security to take on productive but risky investments (Höppe and Gurenko, 2006). Examples include a pilot project in Malawi where micro-insurance is bundled with loans that enable farmers to access agricultural inputs that increase their productivity (Hess and Syroka, 2005), and a project in Mongolia that protects herders' livestock from extreme winter weather to reduce livestock losses (Skees et al., 2008).

Micro-insurance is a financial arrangement to protect low-income people against specific perils in exchange for regular premium payments (Churchill, 2006, 2007). Several pilot projects have yielded promising outcomes, yet experience is too short to judge if micro-insurance schemes are viable in the long run for local places. Many of the ongoing micro-insurance initiatives are index-based – a relatively new approach whereby the insurance contract is not against the loss itself, but against an event that causes loss, such as insufficient rainfall during critical stages of plant growth (Turvey, 2001). Weather index insurance is largely at a pilot stage, with several projects operating around the globe, including in Mongolia, Kenya, Malawi, Rwanda, and Tanzania (Hellmuth et al., 2009). Index insurance for agriculture is more developed in India, where the Agricultural Insurance Company of India has extended coverage against inadequate rainfall to 700,000 farmers (Hellmuth et al., 2009).

Index-based contracts as an alternative to traditional crop insurance have the advantages of greatly limiting transaction costs (from reduced claims handling) and in improving emergency response (Chantarat et al., 2007). A disadvantage is the potential of a mismatch between yield and payout, a critical issue given the current lack of density of meteorological stations in vulnerable regions – a challenge remote sensing may help address (Skees and Barnett, 2006). Participants' understanding of how insurance operates, as well as their trust in the product and the stakeholders involved, may also be a problem for scaling up index insurance pilots, although simulation games and other innovative communication approaches are yielding promising results (Patt et al., 2009). Affordability can also be a problem. Disasters can affect whole communities or regions (co-variant risks), and because of this insurers must be prepared for meeting large claims all at once, with the cost of requisite backup capital potentially raising the premium far above the client's expected losses – or budget. While valuable in reducing the long-term effects on poverty and development, insurance instruments, particularly if left entirely to the market, are not appropriate in all contexts (Linnerooth-Bayer et al., 2010).

The insurance industry itself is vulnerable to climate change. The continuing exit of private insurers in some market areas is seen with the increasingly catastrophic local losses in the United States (Lecomte and Gahagan, 1998), United Kingdom (Priest et al., 2005), and Germany (Thieken et al., 2006; Botzen and van den Bergh, 2008), which in turn reduces disaster management options at the sub-national scale. Climate change could be particularly problematic for this sector at the local scale (Vellinga et al., 2001), including the probable maximum loss and pressures from regulators responding to changing prices and coverage (Kunreuther et al., 2009).

One response to rising levels and volatility of risk has been to increase insurance and reinsurance capacity through new alternative risk transfer instruments, such as index-linked securities (including catastrophe bonds and weather derivatives) (Vellinga et al., 2001). These tools could play an increasingly important role in a new era of elevated catastrophe risks (Kunreuther et al., 2009). Another approach is to reduce risks through societal adaptation (Herweijer et al., 2009), and risk communication and financial incentives from insurers (Ward et al., 2008). For example, Lloyds of London (2008) demonstrated that in exposed coastal regions, increases in average annual losses and extreme losses due to sea level rise in 2030 could be offset through investing in property-level resilience to flooding or sea walls. Similarly, RMS (2009) shows that wind-related losses in Florida could be significantly reduced through strengthening buildings.

Risk transfer is broader than shifting the economic burden from one party to another. It also entails the transfer of risks from one generation (intergenerational equity) to the next. Risk transfer also has a spatial element in shifting the risk burdens from one geographic location to another. Both of these larger transfer mechanisms are significant for disaster risk management and climate change adaptation at the local scale, but more research is required to assess the localized effects. Spatial and intergenerational equity are considered in Chapter 8.

5.6.4. A Transformative Framework for Management Strategies

Management strategies need to consider adaptation as a process rather than measures and actions for a particular event or time period. Experience in planning and implementing adaptation to climate change as well as disaster response reveals that socio-institutional processes are important in bringing together a set of intertwined elements (Tschakert and Dietrich, 2010; see Chapter 8). O'Brien et al. (2011) suggest an adaptation continuum (see Figure 5-2), where the goal is to move toward partnerships that enable social transformations and increased resilience.

A key component of the disaster risk management and adaptation process is the ability to learn (Pahl-Wostl et al., 2007; Armitage et al., 2008; Lonsdale et al., 2008). This focus on learning partly derives from the fields of social-ecological resilience and sustainability science (Berkes, 2009; Kristjanson et al., 2009). As scenarios combine quantitative indicators of climate, demographic, biophysical, and economic change, as well as qualitative storylines of socio-cultural changes at the local level, the participation of local stakeholders is essential to generate values and understandings of climate extremes.

Adaptation is a process rather than an endpoint and requires a focus on the institutions and policies that enable or hinder this process (Inderberg and Eikeland, 2009) as well as the acknowledgment that there are often competing stakeholder goals (Ziervogel and Ericksen, 2010). Fostering better adaptive capacity for disaster and climate risk will help to accelerate future adaptation (Inderberg and Eikeland, 2009; Moser, 2009; Patt, 2009). However, there are barriers, including lack of coordination between actors, the complexity of the policy field (Mukheibir and Ziervogel, 2007; Winsvold et al., 2009), and limited human capacity to implement policies (Ziervogel et al., 2010). Lastly, individual, sector, and institutional perceptions of risk and adaptive capacity can determine whether adaptation responses are initiated or not (Grothmann and Patt, 2005).

5.7. Information, Data, and Research Gaps at the Local Level

The causal processes by which disasters produce systemic effects over time and across space is reasonably well-known (Kreps, 1985; Cutter, 1996; Lindell and Prater, 2003; NRC, 2006). Yet, local emergency

Managing the Risks from Climate Extremes at the Local Level — Chapter 5

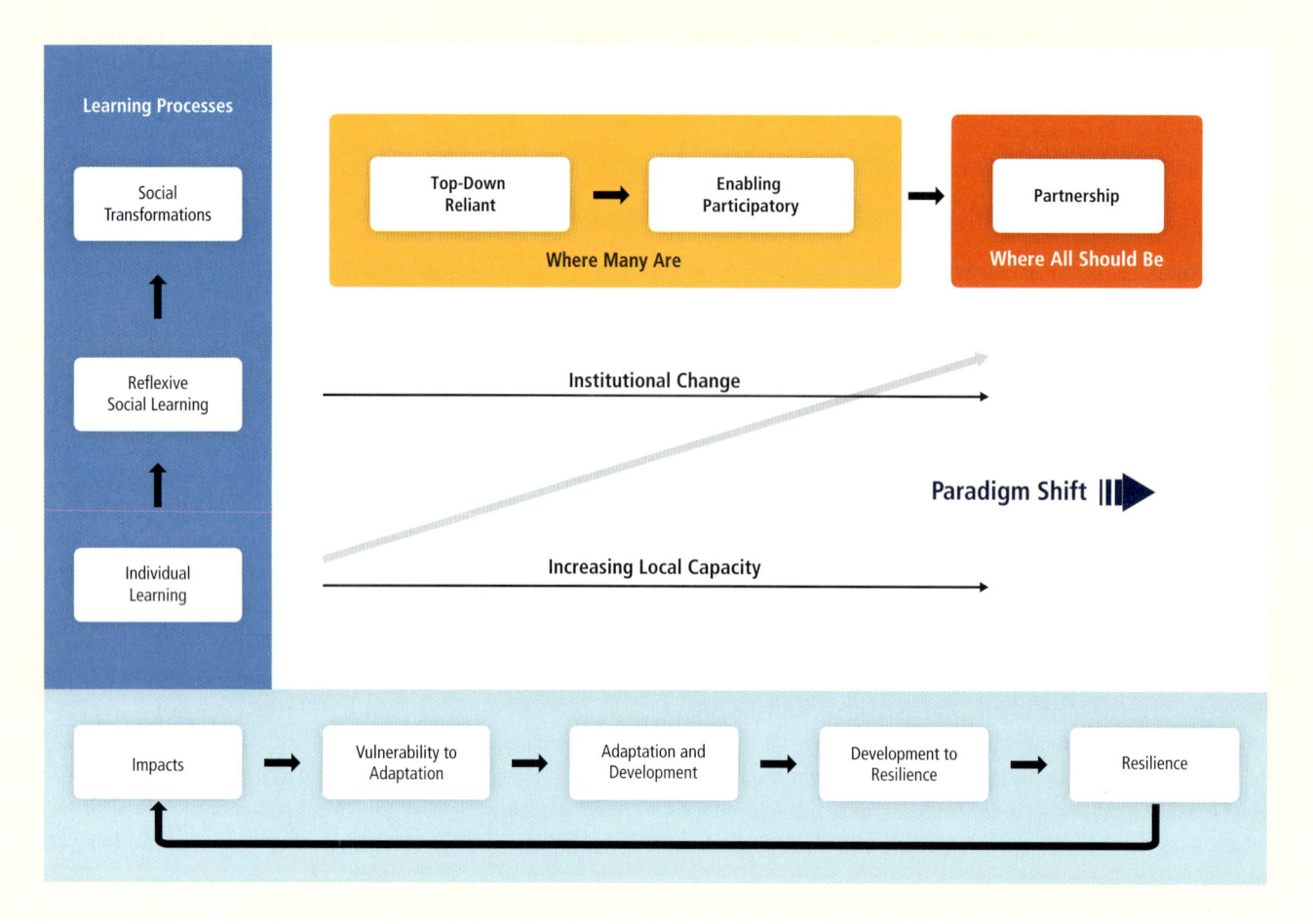

Figure 5-2 | Learning and transformation. Throughout the adaptation process, learning is expected to increase along with institutional change leading to the potential for paradigmatic transformation – the community moves away from an impact-focus perspective to a resilience-centric one where there is an expectation of risk and where good governance and key partnerships are the norm. Source: adapted from O'Brien et al., 2011.

management communities have by and large paid little attention to the links between climate change and natural hazards (Bullock et al., 2009). As a result, state and local disaster mitigation plans, even when required by law, usually fail to include climate change, sea level rise, or climate extreme events in hazard assessments or do so in entirely deterministic ways. Decisions about development, hazard mitigation, and emergency preparedness in the context of climate change give rise to critical questions about social and economic adaptation, and the information and data to support it, especially at the local scale (Mileti, 1999; Cutter, 2001; Mileti and Peek, 2002). For example: How do cumulative impacts of smaller events over time compare to single high-impact events for localities? Do increased levels of hazard mitigation and disaster preparedness increase local risk-taking by individuals and social systems? How do short-term adjustments or coping strategies enable or constrain long-term vulnerabilities in localities? What are the tradeoffs among decision acceptability versus decision quality, especially within local contexts (Comfort et al., 1999; Travis, 2010)?

For many of these questions, sufficient empirical information is lacking, especially at the sub-national scale (see also Section 5.4.2.3). Two recent all-hazards studies for the United States found that from 1970 to 2004, climate-sensitive hazards accounted for the majority of recorded fatalities from natural hazards (Borden and Cutter, 2008; Thacker et al., 2008). Yet, these are the only databases for monitoring mortality from natural hazards at the local level and suffer from lack of consistency and completeness.

The hurricane recovery process includes ample evidence of how efforts to ensure that the rush to return to normal have also led to depletion of natural resources and increased risk. How decisions regarding the right to migrate (even temporarily), the right to organize, and the right of access to information are made will, as a result, have major implications for the ability of different groups to adapt successfully to floods, droughts, and storms. The idea of linking place-based recovery, preparedness, and resilience to adaptation is intuitively appealing. However, the constituency that supports improved disaster risk management has historically proven too small to bring about many of the changes that have been recommended by researchers, especially those that focus on strengthening the social fabric to decrease vulnerability. Behind the specific questions of the transparency of risk, are broader questions about the public sphere. What public goods will be provided by governments at all levels (and how will they be funded), what public goods will be provided by private or organizations in civil society, what will be provided by market actors, and what will not? How will these influence local-level disaster risk management, especially of climate-sensitive hazards (Mitchell, 1988, 1999; Thomalla et al., 2006; van Aalst et al., 2008)?

While there has been increasing focus on the processes by which knowledge has been produced, less time has been spent examining the capacity of local communities to critically assess knowledge claims made by others for their reliability and relevance to those communities (Fischhoff, 2007; Pulwarty, 2007). There is the need to move beyond the integration of physical and societal impacts to focus on practice and evaluation. How are impediments to the flow of information created? Is a focus on communication adequate to ensure effective response? How are these nodes defined among differentially vulnerable groups, for example, based on economic class, race, or gender? However, there is little research on the extent to which local jurisdictions have adopted policy options and practice and the ways in which they are being implemented. Most of the studies to date have addressed factors that lead to policy adoption and not necessarily successful implementation.

Beyond infrastructure and retrofitting concerns, successful adaptation strategies integrate urban planning, water management, early warning systems, and preparedness. One widely acknowledged goal is to address, directly, the problem of an inadequate fit between what the research community knows about the physical and social dimensions of uncertain environmental hazards and what society chooses to do with that knowledge. An even larger challenge is to consider how different systems of knowledge about the physical environment and competing systems of action can be brought together in pursuit of diverse goals that we wish to pursue (Mitchell, 2003). Several sources (Comfort et al., 1999; Bullock et al., 2009; McKinsey Group, 2009) have identified key research and data requirements for addressing these challenges:

1) Multi-way information exchange systems – effective adaptation will always be locally driven. Communities need reliable measurements and assessment tools, integrated information about risks that those tools reveal, and best approaches to minimize those risks. The goal is to improve the assessment and transparency of risk in a geographic place-based approach for vulnerable regions. Improving the collection and quality control of locally based data on economic losses, disaster and adaptation costs, and human losses (fatalities) will ensure improved empirically based baseline assessments.

2) Maps of the decision processes for disaster mitigation, preparedness, response, and recovery and guidance for using such decision support tools are needed. Hazard maps developed through collaboration between researchers and affected communities are the simplest and often most powerful form of risk information. They capture the likelihood and impact of a peril and are important for informing many aspects of disaster risk management including disaster risk reduction, risk-based pooling of resources, and risk transfer. Such devices would identify segments of threatened social systems that could suffer disproportionate disaster impacts; critical actors at each jurisdictional level; their risk assumptions; their different types of information needs; and the design of an information infrastructure that would support decisions at critical entry points (Comfort, 1993).

3) People who face hazards often need assistance to manage their own environments over the long term and develop systematic actions to improve resilience in vulnerable localities. Research is needed on how local governments and institutions can support, provide incentives, and legitimize successful approaches to increasing capacity and action.

4) Methodologies, indicators, and measurement of progress in reducing vulnerability and enhancing community capacity at the local level are under-researched at present. Locally based risk management, cost-effectiveness methodologies and analyses, quantification of societal impacts of catastrophic events at local to national scales, and research on implementation and evaluation of risk management and mitigation programs are needed. Similarly, there is a critical need for the assessment and coordination of multi-jurisdictional and multi-sectoral efforts to help avoid the unintended consequences of actions and interventions especially at the local scale.

5) Underserved people require access to the social and economic security that comes from sharing risk, through financial risk transfer mechanisms such as insurance. There is a paucity of studies at the local level to assess the efficacy of alternative risk reduction, risk-based resource pooling and transfer methods, analysis of benefits and costs to various stakeholder groups, analysis of complementary roles of mitigation and insurance, and analysis of safeguards against insurance industry insolvency.

Interdisciplinary collaboration is clearly needed to prioritize and address these research needs. Situating the scientific understanding of hazards, disaster risk, and climate change adaptation within a broader discourse about different forms of knowledge will increase the likelihood of public actions that are better grounded in scientific knowledge and customized for the local context.

5.8. Summary

This chapter presented evidence on how climate extremes affect local places; how local places currently cope with disasters such as emergency assistance and disaster relief; and how they anticipate and plan for future disaster risk using improved communication, structures such as dams and levees, land use management and ecosystem protection, and storage of resources. The role of scale and context shapes variability in building adaptive capacity at the local level. Differences in coping and risk management also are scale-dependent and context-specific, and could affect or limit adaptations to climate extremes at the local level. Lastly, climate extremes threaten human security at the local level. Localized vulnerability attributed to social, economic, environmental, and climate change drivers at a variety of scales heightens the impacts of climate extremes on local places. While some places have considerable experience with disasters and some inherent capacity to cope with climate extremes, others do not. These differences in coping and management necessitate a range of approaches for disaster risk management and climate change adaptation, thus attention to a broader set of national and international contexts relating to social welfare, quality of life, and sustainable livelihoods.

References

A digital library of non-journal-based literature cited in this chapter that may not be readily available to the public has been compiled as part of the IPCC review and drafting process, and can be accessed via either the IPCC Secretariat or IPCC Working Group II web sites.

Aall, C., K. Groven, and G. Lindseth, 2007: The scope of action for local climate policy: The case of Norway. *Global Environmental Politics*, **7(2)**, 83-101.

ActionAid, 2005: *People-Centered Governance: Reducing Disaster for Poor and Excluded People*. ActionAid International, Johannesburg, South Africa.

Adeola, F.O., 2009: Katrina cataclysm: Does duration of residency and prior experience affect impacts, evacuation, and adaptation behavior among survivors? *Environment and Behavior*, **41(4)**, 459-489.

Adger, W.N., 2000: Social and ecological resilience: Are they related? *Progress in Human Geography*, **24(3)**, 347-364.

Adger, W.N., 2003: Social capital, collective action, and adaptation to climate change. *Economic Geography*, **79(4)**, 387-404.

Adger, W.N. and K. Brown, 2009: Adaptation, vulnerability and resilience: Ecological and social perspectives. In: *Companion to Environmental Geography* [Castree, N., D. Demeritt, D. Liverman, and B. Rhoads (eds.)]. Wiley-Blackwell, Chichester, UK, pp. 109-122.

Adger, W.N. and P.M. Kelly, 1999: Social vulnerability to climate change and the architecture of entitlements. *Mitigation and Adaptation Strategies for Global Change*, **4(3-4)**, 253-266.

Adger, W.N., P.M. Kelly, A. Winkels, L.Q. Huy, and C. Locke, 2002: Migration, remittances, livelihood trajectories, and social resilience. *Ambio*, **31(4)**, 358-366.

Adger, W.N., K. Brown, J. Fairbrass, A. Jordan, J. Paavola, S. Rosendo, and G. Seyfang, 2003: Governance for sustainability: Towards a 'thick' analysis of environmental decision-making. *Environment and Planning A*, **35**, 1095-1110.

Adger, W.N., N.W. Arnell, and E.L. Tompkins, 2005: Successful adaptation to climate change across scales. *Global Environmental Change*, **15(1)**, 77-86.

Adger, W.N., J. Paavola, S. Huq, and M.J. Mace (eds.), 2006: *Fairness in Adaptation to Climate Change*. MIT Press, Cambridge, MA.

Adger, W.N., S. Agrawala, M.M.Q. Mirza, C. Conde, K. O'Brien, J. Pulhin, R. Pulwarty, B. Smit, and K. Takahashi, 2007: Assessment of adaptation practices, options, constraints and capacity. In: *Climate Change 2007. Impacts, Adaptation and Vulnerability. Contribution of Working Group II to the Fourth Assessment Report of the Intergovernmental Panel on Climate Change* [Parry, M.L., O.F. Canziani, J.P. Palutikof, P.J. Van Der Linde, and C.E. Hanson (eds.)]. Cambridge University Press, Cambridge, UK, pp. 717-743.

Adger, W.N., S. Dessai, M. Goulden, M. Hulme, I. Lorenzoni, D.R. Nelson, L.O. Naess, J. Wolf, and A. Wreford, 2010: Are there social limits to adaptation to climate change? *Climatic Change*, **93(3-4)**, 335-354.

Aerts, J., D. Major, M. Bowman, and P. Dircke, 2009: *Connecting Delta Cities. Coastal Cities, Flood Risk Management and Adaptation to Climate*. VU University Press, Amsterdam, The Netherlands, 90 pp.

Agrawal, A., 2010: Local institutions and adaptation to climate change. In: *Social Dimensions of Climate Change: Equity and Vulnerability in a Warming World* [Mearns, R. and A. Norton (eds.)]. World Bank, Washington, DC, pp. 173-198.

Agrawal, A., C. McSweeney, and N. Perrin, 2008: *Social Development Notes. Local Institutions and Climate Change Adaptation - the Social Dimensions of Climate Change*. World Bank, Washington, DC.

Alcamo, J., L. Acosta-Michlik, A. Carius, F. Eierdanz, R. Klein, D. Krömker, and D. Tänzler, 2008: A new approach to quantifying and comparing vulnerability to drought. *Regional Environmental Change*, **8(4)**, 137-149.

Alexander, D., 2002: *Principles of Emergency Planning and Management*. Oxford University Press, Oxford, UK.

Allen, K., 2006: Community-based disaster preparedness and climate adaptation: Local capacity-building in the Philippines. *Disasters*, **30(1)**, 81-101.

Amin, S., A.S. Rai, and G. Topa, 2001: *Does Microcredit Reach and Poor and Vulnerable? Evidence from Northern Bangladesh*. Center for International Development, Harvard University, Cambridge, MA.

Anderson, S., J. Morton, and C. Toulmin, 2010: Climate change for agrarian societies in drylands: Implications and future pathways. In: *Social Dimensions of Climate Change: Equity and Vulnerability in a Warming World* [Mearns, R. and A. Norton (eds.)]. World Bank, Washington, DC, pp. 199-230.

Anning, D. and D. Dominey-Howes, 2009: Valuing climate change impacts on Sydney beaches to inform coastal management decisions: A research outline. *Management of Environmental Quality*, **20(4)**, 408-421.

ANTHC, 2011: *Climate Change & Health Impact Assessment Reports*. Alaska Native Tribal Health Consortium, www.anthctoday.org/community/reports.html.

Aragon-Durand, F., 2007: Urbanization and flood vulnerability in the peri-urban interface of Mexico City. *Disasters*, **31(4)**, 477-494.

Ariyabandu, M.M., 2006: Gender issues in recovery from the December 2004 Indian ocean tsunami: The case of Sri Lanka. *Earthquake Spectra*, **22(3)**, 759-775.

Armitage, D., M. Marschke, and R. Plummer, 2008: Adaptive co-management and the paradox of learning. *Global Environmental Change*, **18**, 86-98.

Ascher, W., 1978: *Forecasting: An appraisal for policy-makers and planners*. Johns Hopkins Press, Baltimore, MD, 256 pp.

Asian Disaster Preparedness Centre, 2005: *A Primer. Integrated Flood Risk Management in South East Asia*. Asian Disaster Preparedness Centre, Bangkok, Thailand.

Babugura, A.A., 2008: Vulnerability of children and youth in drought disasters: A case study of Botswana. *Children, Youth and Environments*, **18(1)**, 126-157.

Balaban, V., 2006: Psychological assessment of children in disasters and emergencies. *Disasters*, **30**, 178-198.

Bamdad, N., 2005: *The Role of Community Knowledge in Disaster Management: The Bam Earthquake Lesson in Iran*. Institute of Management and Planning Studies, Tehran, Iran.

Bankoff, B., 2007: Living with risk: Coping with disasters. *Education about Asia*, **12(2)**, 26-29.

Bankoff, G., G. Frerks, and D. Hilhorst (eds.), 2004: *Mapping vulnerability: Disasters, Development & People*. Earthscan, London, UK.

Barbier, E.B., E.W. Koch, B.R. Silliman, S.D. Hacker, E. Wolanski, J. Primavera, E.F. Granek, S. Polasky, S. Aswani, L.A. Cramer, D.M. Stoms, C.J. Kennedy, D. Bael, C.V. Kappel, G.M.E. Perillo, and D.J. Reed, 2008: Coastal ecosystem-based management with nonlinear ecological functions and values. *Science*, **319**, 321-323.

Bari, F., 1998: Gender, disaster and empowerment: A case study from Pakistan. In: *The Gendered Terrain of Disaster: Through Women's Eyes* [Enarson, E.P. and B.H. Morrow (eds.)]. Praeger, Westport, CT, pp. 125-132.

Barnett, J., 2003: Security and climate change. *Global Environmental Change*, **13(1)**, 7-17.

Barnett, J. and W.N. Adger, 2007: Climate change, human security and violent conflict. *Political Geography*, **26**, 639-655.

Barnett, J. and M. Webber, 2009: *Accommodating Migration to Promote Adaptation to Climate Change*. Commission on Climate Change and Development, Stockholm, Sweden.

Baro, M. and T. Deubel, 2006: Persistent hunger: Perspectives on vulnerability, famine, and food security in sub-Saharan Africa. *Annual Review of Anthropology*, **35**, 521-538.

Barrett, C.B., 2002: Food security and food assistance programs. In: *Handbook of Agricultural Economics* [Gardner, B. and G. Rausser (eds.)]. Elsevier Science, Amsterdam, The Netherlands.

Barrett, E.J., M. Martinez-Cosio, and C.Y.B. Ausbrooks, 2008: The school as a source of support for Katrina-evacuated youth. *Children, Youth and Environments*, **18(1)**, 202-236.

Bartlett, S., 2008: The implications of climate change for children in lower-income countries. *Children, Youth and Environments*, **18(1)**, 71-98.

Bateman, J.M. and B. Edwards, 2002: Gender and evacuation: A closer look at why women are more likely to evacuate for hurricanes. *Natural Hazards Review*, **3(3)**, 107-117.

Battista, F. and S. Baas, 2004: *Consolidated Report on Case Studies and Workshop Findings. The Role of Local Institutions in Reducing Vulnerability to Recurrent Natural Disasters and in Sustainable Livelihoods Development*. Rural Institutions and Participation Service, Food and Agriculture Organization, Rome, Italy.

Baumhardt, F., R. Lasage, P. Suarez, and C. Chadza, 2009: Farmers become filmmakers: Climate change adaptation in Malawi. *Participatory Learning and Action*, **60**, 129-138.

Beamon, B.M. and B. Balcik, 2008: Performance measurement in humanitarian supply chains. *International Journal of Public Sector Management*, **21(1)**, 4-25.

Berke, P.R. and T. Beatley, 1992: A national assessment of earthquake mitigation: Implications for land use planning and public policy. *Earthquake Spectra*, **8**, 1-17.

Berkes, F., 2008: *Sacred Ecology*. 2nd ed. Routledge, New York, NY.

Berkes, F., 2009: Evolution of co-management: Role of knowledge generation, bridging organizations and social learning. *Journal of Environmental Management*, **90(5)**, 1692-1702.

Berkes, F. and D. Jolly, 2001: Adapting to climate change: Social-ecological resilience in a Canadian western Arctic community. *Ecology and Society*, **5(2)**, 18.

Berkhout, F., 2002: Technological regimes, path dependency and the environment. *Global Environmental Change, Part A: Human and Policy Dimensions*, **12(1)**, 1-4.

Berkhout, F., J. Hertin, and D. Gann, 2006: Learning to adapt: Organisational adaptation to climate change impacts. *Climatic Change*, **78(1)**, 135-156.

Bermejo, P.M., 2006: Preparation and response in case of natural disasters: Cuban programs and experience. *Journal of Public Health Policy*, **27(1)**, 13-21.

Bernard, S.M., J.M. Samet, A. Grambsch, K.L. Ebi, and I. Romieu, 2001: The potential impacts of climate variability and change on air pollution-related health effects in the United States. *Environmental Health Perspectives*, **109**, 199-209.

Berube, A. and B. Katz, 2005: *Katrina's Window: Confronting Concentrated Poverty Across America*. The Brookings Institution Metropolitan Policy Program, Washington, DC.

Birkland, T.A., 1997: *After disaster: Agenda Setting, Public Policy, and Focusing Events*. Georgetown University Press, Washington, DC.

Birkmann, J., 2006: *Measuring Vulnerability to Natural Hazards: Towards Disaster Resilient Societies*. United Nations Publications, New York, NY.

Birkmann, J., M.A. Garschagen, F. Krass, and N. Quang, 2010: Adaptive urban governance–new challenges for the second generation of urban adaptation strategies to climate change. *Sustainability Science*, **5(2)**, 185-206.

Bizikova, L., T. Neale, and I. Burton, 2008: *Canadian Communities' Guidebook for Adaptation to Climate Change. Including an Approach to Generate Mitigation Co-Benefits in the Context of Sustainable Development*. 1st ed. Environment Canada and University of British Columbia, Vancouver, Canada, 100 pp.

Bohle, H.-G., 2001: Vulnerability and criticality: Perspectives from social geography. *Newsletter of the International Human Dimensions Programme on Global Environmental Change*, **2/2001**.

Bohle, H., T. Downing, and M. Watts, 1994: Climate change and social vulnerability: Toward a sociology and geography of food insecurity. *Global Environmental Change*, **4(1)**, 371-396.

Borden, K.A. and S.L. Cutter, 2008: Spatial patterns of natural hazards mortality in the United States. *International Journal of Health Geographics*, **7(64)**.

Boruff, B.J. and S.L. Cutter, 2007: The environmental vulnerability of Caribbean island nations. *Geographical Review*, **97(1)**, 24-45.

Botzen, W.J.W. and J.C.J.M. van den Bergh, 2008: The insurance against climate change and flooding in The Netherlands: Present, future, and comparison with other countries. *Risk Analysis*, **28(2)**, 413-426.

Botzen, W., J. Aerts, and J. van den Bergh, 2009: Willingness of homeowners to mitigate climate risk through insurance. *Ecological Economics*, **68**, 2265-2277.

Bourque, L., J. Siegel, M. Kano, and M. Wood, 2006: Weathering the storm: The impact of hurricanes on physical and mental health. *The Annals of the American Academy of Political and Social Science*, **604**, 129-151.

Bowden, M., R.W. Kates, P.A. Kay, W.E. Riebsame, R.A. Warrick, D.L. Johnson, H.A. Gould, and D. Weiner, 1981: The effect of climatic fluctuations on human populations: Two hypotheses. In: *Climate and History: Studies in Past Climates and their Impact on Man* [Wigley, T.M., M.J. Ingrams, and G. Farmer (eds.)]. Cambridge University Press, Cambridge, UK, pp. 479-513.

Boyd, R. and M.E. Ibarrarán, 2009: Extreme climate events and adaptation: An exploratory analysis of drought in Mexico. *Environment and Development Economics*, **14(3)**, 371-396.

Brauer, M., 1999: Health impacts of biomass air pollution. In: *Health Guidelines for Vegetation Fire Events* [Goh, K.-T., D.H. Schwela, J.G. Goldammer, and O. Simpson (eds.)]. World Health Organization, Singapore.

Brenkert, A.L. and E.L. Malone, 2005: Modeling vulnerability and resilience to climate change: A case study of India and Indian states. *Climatic Change*, **72(1-2)**, 57-102.

Brondizio, E., E. Ostrom, and O. Young, 2009: Connectivity and the governance of multilevel social-ecological systems: The role of social capital. *Annual Review of Environment and Resources*, **34**, 253-278.

Brooks, N., 2003: *Vulnerability, Risk and Adaptation: A Conceptual Framework*. Tyndall Centre for Climate Change Research, University of East Anglia, Norwich, UK.

Brown, D.L., 1999: Disparate effects of the 1989 Loma Prieta and 1994 Northridge earthquakes on hospital admissions for acute myocardial infarction: Importance of superimposition of triggers. *American Heart Journal*, **7(5)**, 830-836.

Brown, O., 2008: *Migration and Climate Change*. International Organization for Migration, Geneva, Switzerland.

Brulle, R., 2010: From environmental campaigns to advancing the public dialogues: Environmental communication for civic engagement. *Environmental Communication*, **4(1)**, 82-98.

Buckland, J. and M. Rahman, 1999: Community-based disaster management during the 1997 Red River flood in Canada. *Disasters*, **23(3)**, 174-191.

Buhaug, H., 2010: Climate not to blame for African civil wars. *Proceedings of the National Academy of Sciences*, **107(38)**, 16477-16482.

Bulkeley, H., 2006: A changing climate for spatial planning? *Planning Theory and Practice*, **7(2)**, 203-214.

Bull-Kamanga, L., K. Diagne, A. Lavell, E. Leon, F. Lerise, H. MacGregor, A. Maskrey, M. Meshack, M. Pelling, H. Reid, D. Satterthwaite, J. Songsore, K. Westgate, and A. Yitambe, 2003: From everyday hazards to disasters: The accumulation of risk in urban areas. *Environment and Urbanization*, **15(1)**, 193-203.

Bullock, J., G. Haddow, and K. Haddow, 2009: *Global Warming, Natural Hazards, and Emergency Management*. CRC Press, Boca Raton, FL, 282 pp.

Burby, R.J. (ed.), 1998: *Cooperating with Nature: Confronting Natural Hazards with Land-Use Planning for Sustainable Communities*. Joseph Henry Press/National Academy of Sciences, Washington, DC, 356 pp.

Burby, R.J., 2000: Creating hazard resilient communities through land-use planning. *Natural Hazards Review*, **1(2)**, 99-106.

Burby, R.J., 2006: Hurricane Katrina and the paradoxes of government disaster policy: Bringing about wise government decisions for hazardous areas. *Annals of the American Academy of Political and Social Science*, **604(1)**, 171-191.

Burby, R.J., May, P.J. with Berke, P. R., L.C. Dalton, S.P. French, and E.J. Kaiser, 1997: *Making Government Plan: State Experiments in Managing Land Use*. Johns Hopkins University Press, Baltimore, MD.

Burch, S. and J. Robinson, 2007: A framework for explaining the links between capacity and action in response to global climate change. *Climate Policy*, **7(4)**, 304-316.

Burke, M.B., E. Miguel, S. Satyanath, J.A. Dykema, and D.B. Lobell, 2009: Warming increases the risk of civil war in Africa. *Proceedings of the National Academy of Sciences*, **106(49)**, 20670-20674.

Burton, C. and S.L. Cutter, 2008: Levee failures and social vulnerability in the Sacramento–San Joaquin delta area, California. *Natural Hazards Review*, **9(3)**, 136-149.

Burton, I., R.W. Kates, and G.F. White, 1993: *The Environment as Hazard*. 2nd ed. Guilford Press, New York, NY.

Burton, I., J. Soussan, and A. Hammill, 2003: *Livelihoods and Climate Change, Combining Disaster Risk Reduction, Natural Resource Management and Climate Change Adaptation in a New Approach to the Reduction of Vulnerability and Poverty*. A Conceptual Framework Paper Prepared by the Task Force on Climate Change, Vulnerable Communities and Adaptation, International Institute for Sustainable Development, Winnipeg, Canada.

Burton, I., L. Bizikova, T. Dickinson, and Y. Howard, 2007: Integrating adaptation into policy: Upscaling evidence from local to global. *Climate Policy*, **7(4)**, 371-376.

Buvinić, M., G. Vega, M. Bertrand, A. Urban, R. Grynspan, and G. Truitt, 1999: *Hurricane Mitch: Women's Needs and Contributions*. Sustainable Development Department Technical Papers Series, Inter-American Development Bank, Washington, DC.

Byg, A. and J. Salick, 2009: Local perspectives on a global phenomenon – climate change in eastern Tibetan villages. *Global Environmental Change*, **19**, 156-166.

Bynander, F., L. Newlove, and B. Ramberg, 2005: *Sida and the Tsunami of 2004 – A Study of Organizational Crisis Response*. Swedish International Development Cooperation Agency, Stockholm, Sweden.

Cabrera, V.E., N.E. Breuer, and P.E. Hildebrand, 2008: Participatory modeling in dairy farm systems: A method for building consensual environmental sustainability using seasonal climate forecasts. *Climatic Change*, **89**, 395-409.

Cahill, K., 2007: *The Pulse of Humanitarian Assistance*. Fordham University Press, New York, NY.

Campbell, J.R., 1990: Disasters and development in historical context: Tropical cyclone response in the Banks Islands of northern Vanuatu. *International Journal of Mass Emergencies and Disasters*, **8(3)**, 401-424.

Campbell, J.R., 2006: *Traditional Disaster Reduction in Pacific Island Communities*. Institute of Geological and Nuclear Sciences Ltd., Avalon, New Zealand.

Campbell, J.R., 2010a: An overview of natural hazard planning in the Pacific island region. *The Australasian Journal of Disaster and Trauma Studies*, **2010(1)**.

Campbell, J.R., 2010b: Climate-induced community relocation in the Pacific: The meaning and importance of land. In: *Climate Change and Displacement: Multidisciplinary Perspectives* [McAdam, J. (ed.)]. Hart Publishing, Oxford, UK, pp. 57-79.

Campbell-Lendrum, D., A. Pruss-Ustun, and C. Corvalan, 2003: How much disease could climate change cause? In: *Climate Change and Health: Risks and Responses* [Campbell-Lendrum, D., C. Corvalan, K. Ebi, A. Githeko, and J. Scheraga (eds.)]. World Health Organization, Geneva, Switzerland, pp. 133-158.

Cardona, O.D., 2007: Indicators of Disaster Risk and Risk Management: Program for Latin America and the Caribbean: Summary Report. Inter-American Development Bank, Washington, DC.

Carney, E. (ed.), 1998: *Sustainable Rural Livelihoods: What Contributions Can We Make?* Department for International Development, London, UK.

Carraro, L., S. Khan, S. Hunt, G. Rawle, M. Robinson, M. Antoninis, and L. Street, 2003: *Monitoring the Millennium Development Goals: Current Weaknesses and Possible Improvements*. Oxford Policy Management, Department for International Development, Glasgow, Scotland.

Carter, M.R., P.D. Little, T. Mogues, and W. Negatu, 2007: Poverty traps and natural disasters in Ethiopia and Honduras. *World Development*, **35(5)**, 835-856.

Cash, D.W. and S.C. Moser, 2000: Linking global and local scales: Designing dynamic assessment and management processes. *Global Environmental Change*, **10(2)**, 109-120.

Chambers, R. and G. Conway, 1992: *Sustainable Rural Livelihoods: Practical Concepts for the 21st Century*. Institute of Development Studies, Brighton, UK.

Chantarat, S., C.B. Barrett, A.G. Mude, and C.G. Turvey, 2007: Using weather index insurance to improve drought response for famine prevention. *American Journal of Agricultural Economics*, **89(5)**, 1262-1268.

Cheong, S., 2010: Initial responses to the MT Hebei-Spirit oil spill. *Marine Technology Society Journal*, **44(1)**, 69-74.

Cheong, S., 2011a: Guest editorial: Coastal adaptation. *Climatic Change*, **106(1)**, 1-4.

Cheong, S., 2011b: Policy solutions in the U.S. *Climatic Change*, **106(1)**, 57-70.

Cherry, K.E. (ed.), 2009: *Lifespan Perspectives on Natural Disasters: Coping with Katrina, Rita, and Other Storms.* Springer, New York, NY.

Christoplos, I., 2006: The Elusive "Window of Opportunity" For Risk Reduction in Post-Disaster Recovery. In: *ProVention Consortium Forum 2006*, Bangkok, Thailand, 2-3 Feb 2006.

Churchill, C. (ed.), 2006: *Protecting the Poor: A Microinsurance Compendium*. International Labor Organization, Geneva, Switzerland.

Churchill, C., 2007: Insuring the low-income market: Challenges and solutions for commercial insurers. *Geneva Papers on Risk and Insurance–Issues and Practice*, **32(3)**, 401-412.

Clark, G., S. Moser, S. Ratick, K. Dow, W. Meyer, S. Emani, W. Jin, J. Kasperson, R. Kasperson, and H. Schwartz, 1998: Assessing the vulnerability of coastal communities to extreme storms: The case of Revere, MA, USA. *Mitigation and Adaptation Strategies for Global Change*, **3(1)**, 59-82.

Clifton, D. and F. Gell, 2001: Saving and protecting lives by empowering women. *Gender and Development*, **9(1)**, 8-18.

Clot, N. and J. Carter, 2009: *Disaster Risk Reduction: A Gender and Livelihood Perspective*. Swiss Agency for Cooperation and Development, Berne, Switzerland, 13 pp.

Comfort, L.K., 1993: Integrating information technology into international crisis management and policy. *Journal of Contingencies and Crisis Management*, **1**, 15-26.

Comfort, L., B. Wisner, S. Cutter, R. Pulwarty, K. Hewitt, A. Oliver-Smith, J. Wiener, M. Fordham, W. Peacock, and F. Krimgold, 1999: Reframing disaster policy: The global evolution of vulnerable communities. *Global Environmental Change Part B: Environmental Hazards*, **1**, 39-44.

Cooper, J.A.G. and J. McKenna, 2008: Social justice in coastal erosion management: The temporal and spatial dimensions. *Geoforum*, **39**, 294-306.

Corbett, J., 1988: Famine and household coping strategies. *World Development*, **16(9)**, 1099-1112.

Cova, T.J., F.A. Drews, L.K. Siebeneck, and A. Musters, 2009: Protective actions in wildfires: Evacuate or shelter in place? *Natural Hazards Review*, **10(4)**, 151-162.

Cox, J.R., C. Rosenzweig, W.D. Solecki, R. Goldberg, and P.L. Kinney, 2007: *Social Vulnerability to Climate Change: A Neighborhood Analysis of the Northeast U.S. Megaregion*. Union of Concerned Scientists, Cambridge, MA.

Crabbé, P. and M. Robin, 2006: Institutional adaptation of water resource infrastructures to climate change in eastern Ontario. *Climatic Change*, **78(1)**, 103-133.

Cuny, F., 1983: *Disasters and Development*. Oxford University Press, New York, NY.

Cupples, J., 2007: Gender and hurricane Mitch: Reconstructing subjectivities after disaster. *Disasters*, **21(2)**, 155-175.

Curtis, A., J.W. Mills, and M. Leitner, 2007: Katrina and vulnerability: The geography of stress. *Journal of Health Care for the Poor and Underserved*, **18(2)**, 315-330.

Curtis, A., A. Duval-Diop, and J. Novak, 2010: Identifying spatial patterns of recovery and abandonment in the post-Katrina holy cross neighborhood of New Orleans. *Cartography and Geographic Information Science*, **37(1)**, 45-56.

Cutter, S.L., 1996: Vulnerability to environmental hazards. *Progress in Human Geography*, **20**, 529-539.

Cutter, S.L. (ed.), 2001: *American Hazardscapes: The Regionalization of Hazards and Disasters.* Joseph Henry Press/National Academies of Science, Washington, DC.

Cutter, S.L. and M.M. Smith, 2009: Fleeing from the hurricane's wrath: Evacuation and the two Americas. *Environment*, **51(2)**, 26-36.

Cutter, S.L., J.T. Mitchell, and M.S. Scott, 2000: Revealing the vulnerability of people and places: A case study of Georgetown County, South Carolina. *Annals of the Association of American Geographers*, **90(4)**, 713-737.

Cutter, S.L., B.J. Boruff, and W.L. Shirley, 2003: Social vulnerability to environmental hazards. *Social Science Quarterly*, **84(1)**, 242-261.

Dale, A. and L. Newman, 2010: Social capital: A necessary and sufficient condition for sustainable community development? *Community Development Journal*, **45(1)**, 5-21.

Darcy, J. and C. Hofmann, 2003: *Humanitarian Needs Assessment and Decision-Making*. HPG Report, Overseas Development Institute (ODI), London, UK, 1-74 pp.

Das, S. and J. Vincent, 2009: Mangroves protected villages and reduced death toll during Indian super cyclone. *Proceedings of the National Academy of Sciences*, **106(18)**, 7357-7360.

Dash, N. and H. Gladwin, 2007: Evacuation decision making and behavioral responses: Individual and household. *Natural Hazards Review*, **8(3)**, 69-77.

Davidson, C.H., C. Johnson, L. Gonzalo, N. Dikmen, and A. Sliwinski, 2007: Truths and myths about community participation in post-disaster housing projects. *Habitat International*, **31(1)**, 100-115.

Davies, M., 2009: *DFID Social Transfers Evaluation Summary Report*. Department for International Development, London, UK.

Davies, M. and J. Leavy, 2007: Connecting social protection and climate change adaptation. *IDS in Focus*, **2**.

De Sherbinin, A., A. Schiller, and A. Pulsipher, 2007: The vulnerability of global cities to climate hazards. *Environment and Urbanization*, **19(1)**, 39-64.

De Ville de Groyet, C., 2000: Stop propagating disaster myths. *Lancet*, **356(9231)**, 762-764.

Dekens, J., 2007: *Local Knowledge for Disaster Preparedness: A Literature Review*. Hillside Press, Kathmandu, Nepal, 85 pp.

DeLucena, A.F.P., A.S. Szklo, R. Schaeffer, R.R. de Souza, B. S. M. C. Borba, I. V. L. da Costa, A.O.P. Júnior, and S. H. F. da Cunha, 2009: The vulnerability of renewable energy to climate change in Brazil. *Energy Policy*, **37(3)**, 879-888.

Depradine, C.A. and E.H. Lovell, 2004: Climatological variables and the incidence of dengue fever in Barbados. *International Journal of Dengue Fever in Barbados*, **14**, 429-441.

Dershem, L. and D. Gzirishvili, 1999: Informal social support networks and household vulnerability: Empirical findings from Georgia. *World Development*, **26(10)**, 1827-1838.

DeSalvo, K., D. Hyre, D. Ompad, A. Menke, L. Tynes, and P. Muntner, 2007: Symptoms of posttraumatic stress disorder in a New Orleans workforce following Hurricane Katrina. *Journal of Urban Health: Bulletin of the New York Academy of Medicine*, **84(2)**, 142-152.

Development Initiatives, 2009: *Global Humanitarian Assistance*. Development Initiatives, Wells, Somerset, UK.

Devereux, S. and S. Coll-Black, 2007: *DFID Social Transfers Evaluation. Review of Evidence and Evidence Gaps on the Effectiveness and Impacts of DFID-Supported Pilot Social Transfer Schemes*. Institute of Development Studies, Brighton, UK.

Devereux, S. and R. Sabates-Wheeler, 2004: *Transformative Social Protection*. Institute of Development Studies, Brighton, UK.

DFID, 2006: *Eliminating World Poverty: Making Governance Work for the Poor: A White Paper on International Development*. Department for International Development, London, UK.

Díaz, J., C. López, J.C. Alberdi, A. Jordán, R. García, E. Hernández, and A. Otero, 2002: Heat waves in Madrid 1986-1997: Effects on the health of the elderly. *International Archives of Occupational & Environmental Health*, **75(163-170)**.

diFalco, S. and E. Bulte, 2009: *Social Capital and Weather Shocks in Ethiopia: Climate Change and Culturally-Induced Poverty Traps*. London School of Economics, London, UK.

Dixit, A., A. Pokhrel, and M. Moench, 2008: *From Risk to Resilience: Costs and Benefits of Flood Mitigation in the Lower Bagmati Basin: Case of Nepal Terai and North Bihar*. ISET-Nepal and ProVention, Kathmandu, Nepal.

Dolan, A.H. and I.J. Walker, 2004: Understanding vulnerability of coastal communities to climate change related risks. In: Proceedings of the 8th International Coastal Symposium, Itajai, Brazil, 2004. *Journal of Coastal Research*, **S39**.

Dorosh, P., S. Malik, and M. Krausova, 2010: *Rehabilitating Agriculture and Promoting Food Security Following the 2010 Pakistan Floods Insights from South Asian Experience*. IFPRI Discussion Paper, World Bank/International Food Policy Research Institute, Washington, DC, 26 pp.

Douglas, I., K. Alam, M. Maghenda, Y. Mcdonnell, L. Mclean, and J. Campbell, 2009: Unjust waters: Climate change, flooding and the urban poor in Africa. In: *Adapting Cities to Climate Change* [Bicknell, J., D. Dodman, and D. Satterthwaite (eds.)]. Earthscan, London, UK, pp. 201-224.

Dow, K. and S.L. Cutter, 1998: Crying wolf: Repeat responses to hurricane evacuation orders. *Coastal Management*, **26(4)**, 237-252.

Dow, K. and S.L. Cutter, 2000: Public orders and personal opinions: Household strategies for hurricane risk assessment. *Environmental Hazards*, **2(4)**, 143-155.

Dow, K. and S.L. Cutter, 2002: Emerging hurricane evacuation issues: Hurricane Floyd and South Carolina. *Natural Hazards Review*, **3(1)**, 12-18.

Downton, M.W. and R.A. Pielke Jr., 2005: How accurate are disaster loss data? The case of U.S. flood damage. *Natural Hazards*, **35**, 211-228.

Doyle, M.W., E.H. Stanley, D.G. Havlick, M.J. Kaiser, G. Steinbach, W.L. Graf, G.E. Galloway, and J.A. Riggsbee, 2008: Aging infrastructure and ecosystem restoration. *Science*, **319(5861)**, 286-287.

Drabek, T.E., 1986: *Human System Responses to Disaster: An Inventory of Sociological Findings*. Springer Verlag, New York, NY.

Drabek, T.E., 1999: Understanding disaster warning responses. *Social Science Journal*, **36(3)**, 515-523.

Dube, O.P. and S. Sekhwela, 2008: Indigenous knowledge, institutions and practices for coping with variable climate in the Limpopo Basin of Botswana. In: *Climate change and adaptation* [Leary, N., J. Adejuwon, and V. Barros (eds.)]. Earthscan, Sterling, UK, pp. 71-89.

Dulal, H., G. Brodnig, G. Onoriose, and H. Thakur, 2010: *Capitalising on Assets: Vulnerability and Adaptation to Climate Change in Nepal*. World Bank, Washington, DC.

Dynes, R., 2006: Social capital: Dealing with community emergencies. *Homeland Security Affairs*, **2(2)**, 1-26.

Eakin, H. and L.A. Bojorquez-Tapia, 2008: Insights into the composition of household vulnerability from multicriteria decision analysis. *Global Environmental Change*, **18(1)**, 112-127.

Ebi, K.L., 2008: Community-based adaptation to the health impacts of climate change. *American Journal of Preventive Medicine*, **35(5)**, 501-507.

Edmonds, A. and S.L. Cutter, 2008: Planning for pet evacuations during disasters. *Journal of Homeland Security and Emergency Management*, **5(1)**, 33.

Eisenman, D.P., K.M. Cordasco, S. Asch, J.F. Golden, and D. Glik, 2007: Disaster planning and risk communication with vulnerable communities: Lessons from hurricane Katrina. *American Journal of Public Health*, **97(S1)**, S109-S115.

Eisensee, T. and D. Stromberg, 2007: News droughts, news floods, and U.S. disaster relief. *The Quarterly Journal of Economics*, **122(2)**, 693-728.

Elliott, J.R. and J. Pais, 2006: Race, class, and Hurricane Katrina: Social differences in human responses to disaster. *Social Science Research*, **35**, 295-321.

Ellis, F., 2000: The determinants of rural livelihood diversification in developing countries. *Journal of Agricultural Economics*, **51(2)**, 289-302.

Ellis, F. and E. Allison, 2004: *Livelihood Diversification and Natural Resource Access*. Food and Agriculture Organization, Rome, Italy.

Enarson, E., 2000: We will make meaning out this: Women's cultural responses to the Red River valley flood. *International Journal of Mass Emergencies and Disasters*, **18(1)**, 39-64.

Enarson, E., 2009: *Gender Mainstreaming in Emergency Management*. Prairie Women's Health Centre of Excellence, Toronto, Canada.

Enarson, E., 2010: Gender. In: *Social Vulnerability to Disasters* [Phillips, B.D., D.S.K. Thomas, A. Fothergill, and L. Blinn-Pike (eds.)]. CRC Press, Boca Raton, FL, pp. 123-154.

Enarson, E. and B.H. Morrow, 1997: A gendered perspective: The voices of women. In: *Hurricane Andrew: Ethnology, Gender, and the Sociology of Disasters* [Peacock, W.G., B.H. Morrow, and H. Gladwin (eds.)]. Routledge, London, UK, pp. 116-140.

Enarson, E. and B.H. Morrow (eds.), 1998: *The Gendered Terrain of Disaster: Through Women's Eyes*. Praeger, Westport, CT.

Ensor, J. and R. Berger, 2009: *Understanding Climate Change Adaptation: Lesson from Community-Based Approaches*. Practical Action Publishing, Bourton-on-Dunsmore, UK.

Ensor, M.O., 2008: Displaced once again: Honduran migrant children in the path of Katrina. *Children, Youth and Environments*, **18(1)**, 280-302.

Ericksen, N.J., 1986: *Creating Flood Disasters? New Zealand's Need for a New Approach to Urban Flood Hazard*. Ministry of Works and Development, Wellington, NZ.

Ericksen, N.J., J. Dixon, and P. Berke, 2000: Managing natural hazards under the resource management act. In: *Environmental Planning and Management in New Zealand* [Memon, A. and Perkins, H. (ed.)]. Dunmore Press, Palmerston North, UK, pp. 123-132.

Ericksen, N.J., K. O'Brien, and L. Rosentrater, 2008: *Climate Change in Eastern and Southern Africa: Impacts, Vulnerability and Adaptation*. GECHS Report. University of Oslo, Oslo, Norway.

Farrington, J. and R. Slater, 2009: *ODI Social Protection Cash Transfers Series. Lump Sum Cash Transfers in Developmental and Post-Emergency Contexts: How Well have they Performed?* Overseas Development Institute, London, UK, 1-26 pp.

Fearnside, P., 1995: Hydroelectric dams in the Brazilian Amazon as sources of "greenhouse" gases. *Environmental Conservation*, **22**, 7-19.

Fekete, A., 2009: Validation of a social vulnerability index in context to river-floods in Germany. *Natural Hazards and Earth Systems Science*, **9**, 393403.

FEMA, 1997: *Multi Hazard Identification and Risk Assessment*. Government Printing Office, Washington, DC.

Few, R. and F. Matthies, 2006: *Flood hazards and health.* Earthscan, London, UK.

Finan, T.J. and D.R. Nelson, 2001: Making rain, making roads, making do: Public and private responses to drought in Ceará, Brazil. *Climate Research*, **19**, 97-108.

Finch, C., C.T. Emrich, and S.L. Cutter, 2010: Disaster disparities and differential recovery in New Orleans. *Population and Environment*, **31**, 179-202.

Fischer, G., H. van Velthuizen, M. Shah, and F.O. Nachtergaele, 2002: *Global Agro-Ecological Assessment for Agriculture in the 21st Century: Methodology and Results.* International Institute for Applied Systems Analysis, Laxenburg, Austria.

Fischhoff, B., 1992: What forecasts (seem to) mean. *International Journal of Forecasting*, **10**, 387-403.

Fischhoff, B., 2007: Nonpersuasive communication about matters of greatest urgency: Climate change. *Environmental Science and Technology*, **41(21)**, 7204-7208.

Ford, J., B. Smit, and J. Wandel, 2006: Vulnerability to climate change in the Arctic: A case study from Arctic Bay, Canada. *Global Environmental Change*, **16(2)**, 145-160.

Forsyth, T. and A. Walker, 2008: *Forest Guardians, Forest Destroyers: The Politics of Environmental Knowledge in Northern Thailand.* University of Washington Press, Seattle, WA, pp. 302.

Fothergill, A., 1996: Gender, risk, and disaster. *International Journal of Mass Emergencies and Disasters*, **14(1)**, 33-56.

Fothergill, A., 1999: Women's role in disaster. *American Behavioral Science Review*, **7(2)**, 125-143.

Fothergill, A., 2004: *Heads above Water: Gender, Class and Family in the Grand Forks Flood.* SUNY Press, Albany, NY.

Fritze, J., G. Blashki, S. Burke, and J. Wieseman, 2008: Hope, despair and transformation: Climate change and the promotion of mental health and well-being. *International Journal of Mental Health Systems*, **2**, 1-12.

Froukh, M.L., 2001: Decision-support system for domestic water demand forecasting and management. *Water Resources Management*, **15(5)**, 363-382.

Frumkin, H. and A.J. McMichael, 2008: Climate change and public health: Thinking, communicating, acting. *American Journal of Preventive Medicine*, **35(5)**, 403-410.

Fuchs, R., 2009: *Cities at Risk: Developing Adaptive Capacity for Climate Change in Asia's Coastal Megacities.* International START Secretariat and the East West Center, Washington, DC and Honolulu, HI.

Fulu, E., 2007: Gender, vulnerability, and the experts: Responding to the Maldives tsunami. *Development and Change*, **38(5)**, 843-864.

Gaillard, J.C. and M.L.C.J.D. Pangilinan, 2010: Participatory mapping for raising disaster risk awareness among the youth. *Journal of Contingencies and Crisis Management*, **18(3)**, 175-179.

Gall, M., K.A. Borden, and S.L. Cutter, 2009: When do losses count? Six fallacies of natural hazards loss data. *Bulletin of the American Meteorological Society*, **90(6)**, 799-809.

Galloway, G.E.J., 2007: New directions in floodplain management. *Journal of the American Water Resources Association*, **31(3)**, 351-357.

Galloway, G.E.J., 2009: Corps of Engineers responses to the changing national approach to the floodplain management since the 1993 Midwest flood. *Journal of Contemporary Water Research & Education*, **130(1)**, 5-12.

Galloway, G.E., D.F. Boesch, and R.R. Twilley, 2009: Restoring and protecting coastal Louisiana. *Issues in Science and Technology*, **25(2)**.

Gedan, K.B., M.L. Kirwan, E. Wolanski, E.B. Barbier, and B.R. Silliman, 2011: The present and future role of coastal wetland vegetation in protecting shorelines: Answering recent challenges to the paradigm. *Climatic Change*, **106(1)**, 7-29.

Geiser, U. and S. Rist, 2009: Decentralization meets local complexity: Conceptual entry points, field-level findings, and insights gained. In: *Decentralization Meets Local Complexity: Local Struggles, State Decentralization and Access to Natural Resources In South Asia And Latin America* [Geiser, U. and S. Rist (eds.)]. Geographica Bernesia/Swiss National Centre of Competence in Research (NCCR) North-South, Berne, Switzerland, pp. 15-55.

Gething, P.W., D.L. Smith, A.P. Patil, A.J. Tatem, R.W. Snow, and S.I. Hay, 2010: Climate change and the global malaria recession. *Nature*, **465**, 342-345.

Godschalk, D.R., R. Norton, C. Richardson, and D. Salvesen, 2000: Avoiding coastal hazard areas: Best state mitigation practices. *Environmental Geosciences*, **7(1)**, 13-22.

Gornitz, V.N., R.C. Daniels, T.W. White, and K.R. Birdwell, 1994: The development of a coastal assessment database: Vulnerability to sea-level rise in the U.S. southeast. *Journal of Coastal Research*, **12**, 327-338.

Greater London Authority, 2010: *The Draft Climate Change Adaptation Strategy for London: Public Consultation Draft.* Greater London Authority, London, UK.

Greene, S., n.d.: *Community Owned Vulnerability and Capacity Assessment.* World Vision, Nairobi, Kenya, .

Grothmann, T. and A. Patt, 2005: Adaptive capacity and human cognition: The process of individual adaptation to climate change. *Global Environmental Change, Part A: Human and Policy Dimensions*, **15(3)**, 199-213.

Guha-Sapir, D. and R. Below, 2002: *The Quality and Accuracy of Disaster Data: A Comparative Analysis of Three Global Data Sets*, ProVention Consortium, Geneva, Switzerland.

Haimes, Y., B. Horowitz, J. Lambert, J. Santos, C. Lian, and K. Crowther, 2005: Inoperability input-output model for interdependent infrastructure sectors. I: Theory and methodology. *Journal of Infrastructure Systems*, **11**, 67-79.

Hallegatte, S., 2008: An adaptive regional input-output model and its application to the assessment of the economic cost of Katrina. *Risk Analysis*, **28(3)**, 779-799.

Hallegatte, S. and P. Dumas, 2009: Can natural disasters have positive consequences? Investigating the role of embodied technical change. *Ecological Economics*, **68(3)**, 777-786.

Hallegatte, S., F. Henriet, and J. Corfee-Morlot, 2008a: *The Economics of Climate Change Impacts and Policy Benefits at City Scale: A Conceptual Framework.* OECD, Paris, France.

Hallegatte, S., N. Patmore, O. Mestre, P. Dumas, J. Corfee-Morlot, C. Herweijer, and R.M. Wood, 2008b: *Assessing Climate Change Impacts, Sea Level Rise and Storm Surge Risk in Port Cities: A Case Study on Copenhagen.* OECD, Paris, France.

Hallegatte, S., N. Ranger, O. Mestre, P. Dumas, J. Corfee-Morlot, C. Herweijer, and R. Muir Wood, 2011: Assessing climate change impacts, sea level rise and storm surge risk in port cities: A case study on Copenhagen. *Climatic Change*, **104(1)**, 113-137.

Hamilton, J.P. and S.J. Halvorson, 2007: The 2005 Kashmir earthquake. *Mountain Research and Development*, **27(4)**, 296-301.

Hammar-Klose, E. and R. Thieler, 2001: *Coastal Vulnerability to Sea-Level Rise: A Preliminary Database for the US Atlantic, Pacific and Gulf Of Mexico Coasts.* USGS, Woods Hole, MA.

Handmer, J.W., 1987: Guidelines for floodplain acquisition. *Applied Geography*, **7(3)**, 203-221.

Hartman, C. and G.D. Squires (eds.), 2006: *There is no such thing as a natural disaster: Race, class, and Hurricane Katrina.* Routledge, New York, NY.

Hasan, A., 2007: The urban resource center, Karachi. *Environment and Urbanization*, **19(1)**, 275-292.

Heal, G., 1997: Discounting and climate change; an editorial comment. *Climatic Change*, **37(2)**, 335-343.

Heinrichs, D., R. Aggarwal, J. Barton, E. Bharucha, C. Butsch, M. Fragkias, P. Johnston, F. Kraas, K. Krellenberg, A. Lampis, and O.G. Ling, 2009: *Adapting Cities to Climate Change: Opportunities and Constraints (Findings from Eight Cities).* Paper presented at Fifth Urban Research Symposium, Marseilles, France, 28-30 June 2009. World Bank, Washington DC.

Heinz Center, 1999: *The Hidden Costs of Coastal Hazards: Implications for Risk Assessment and Mitigation.* Island Press, Washington, DC.

Heinz Center, 2002: *Human Links to Coastal Disasters.* The H. John Heinz III Center for Science, Economics and the Environment, Washington, DC.

Hellmuth, M.E., D.E. Osgood, U. Hess, A. Moorhead, and H. Bhojwani (eds.), 2009: *Index Insurance And Climate Risk: Prospects for Development and Disaster Management.* Climate and Society No. 2, International Research Institute for Climate and Society (IRI), Columbia University, New York, NY.

Helms, B., 2006: *Access For All: Building Inclusive Financial Systems.* World Bank, Washington, DC.

Herweijer, C., N. Ranger, and R.E.T. Ward, 2009: Adaptation to climate change: Threats and opportunities for the insurance industry. *The Geneva Papers*, **34(3)**, 360-380.

Hess, U. and J. Syroka, 2005: *Weather-Based Insurance in Southern Africa: The Case of Malawi*. World Bank, Washington, DC.

Hewings, G.J.D. and R. Mahidhara, 1996: Economic impacts: Lost income, ripple effects, and recovery. In: *The Great Flood of 1993: Causes, Impacts, Responses* [Changnon, S.A. (ed.)]. Westview, Boulder, CO, pp. 205-217.

Hewitt, K., 1997: *Regions of risk: A geographical introduction to disasters*. Longman, Harlow, UK, 389 pp.

Hillhorst, D., 2002: Being good or doing good? Quality and accountability of humanitarian NGOs. *Disasters*, **26(3)**, 193-212.

Höppe, P. and E. Gurenko, 2006: Scientific and economic rationales for innovative climate insurance solutions. *Climate Policy*, **6(6)**, 607-620.

Hordijk, M. and I. Baud, 2011: Inclusive adaptation: Linking participatory learning and knowledge management to urban resilience. *Local Sustainability*, **1(3)**, 111-121.

Horne, C., 2001: Sociological perspectives on the emergence of norms. In: *Social Norms* [Hechter, M. and K.-. Opp (eds.)]. Russell Sage Foundation, New York, NY, pp. 3-34.

Hudson, P.F., H. Middelkoop, and E. Stouthamer, 2008: Flood management along the lower Mississippi and Rhine rivers (the Netherlands) and the continuum of geomorphic adjustment. *Geomorphology*, **101(2)**, 209-236.

Hughes, D.M., 2001: Cadastral politics: The making of community-based resource management in Zimbabwe and Mozambique. *Development and Change*, **32(4)**, 741-768.

Hugo, G., 1996: Environmental concerns and international migration. *The International Migration Review*, **30(1)**, 105-131.

Hugo, G., 2010: Climate change-induced mobility and the existing migration regime in Asia and the Pacific. In: *Climate Change and Displacement. Multidisciplinary Perspectives* [McAdam, J. (ed.)]. Hart, Oxford, UK, pp. 9-35.

Humanitarian Accountability Partnership International, 2011: 2010 Humanitarian Accountability Report. Humanitarian Accountability Partnership International, Geneva, Switzerland, 1 pp.

Hunter, L.M., 2005: Migration and environmental hazards. *Population and Environment*, **26(4)**, 273-302.

Huq, S., S. Kovats, H. Reid, and D. Satterthwaite, 2007: Reducing risks to cities from disasters and climate change. *Environment and Urbanization*, **19(1)**, 39-64.

Hwang, S., J. Xi, Y. Cao, X. Feng, and X. Qia, 2007: Anticipation of migration and psychological stress and the Three Gorges dam project, China. *Social Science and Medicine*, **65(5)**, 1012-1024.

Hwang, S., Y. Cao, and J. Xi, 2010: Project-induced migration and depression: A panel analysis. *Social Science and Medicine*, **70(11)**, 1765-1772.

Hyder, A.A., S. Maman, J.E. Nyoni, S.A. Khasiani, N. Teoh, Z. Premji, and S. Sohani, 2005: The pervasive triad of food security, gender inequity and women's health: Exploratory research from Sub-Saharan Africa. *African Health Sciences*, **5(4)**, 328-334.

ICIMOD, 2009: *Biodiversity and Climate Change*. International Centre for Integrated Mountain Development, Kathmandu, Nepal.

IDRC, 2002: *Population and Health in Developing Countries–Volume 1: Population, Health and Survival at INDEPTH Sites*. International Development Research Centre, Ottawa, Canada.

IFRC, 1991: *Working with Women in Emergency Relief and Rehabilitation Programmes*. International Federation of Red Cross and Red Crescent Societies, Geneva, Switzerland.

Inderberg, T.H. and P.O. Eikeland, 2009: Limits to adaptation: Analysing institutional constraints. In: *Adapting to Climate Change: Thresholds, Values and Governance* [Adger, W.N., I. Lorenzoni, and K. O'Brien (eds.)]. Cambridge University Press, Cambridge, UK, pp. 433-447.

Ingram, J., G. Franco, C.R. Rio, and B. Khazai, 2006: Post-disaster recovery dilemmas: Challenges in balancing short-term and long-term needs for vulnerability reduction. *Environmental Science and Policy*, **9(7-8)**, 607-613.

Intercooperation, 2008: Mali: Highlighting local coping strategies for drought. In: *Gender Perspectives: Integrating Disaster Risk Reduction into Climate Change Adaptation Good Practices and Lessons Learned*. United Nations Secretariat of the International Strategy for Disaster Reduction, Geneva, Switzerland, pp. 57-61.

IPCC, 2007: *Climate Change 2007: The Physical Science Basis. Contribution of Working Group I to the Fourth Assessment Report of the Intergovernmental Panel on Climate Change* [Solomon, S., D. Qin, M. Manning, Z. Chen, M. Marquis, K.B. Averyt, M. Tignor and H.L. Miller (eds.)]. Cambridge University Press, Cambridge, UK.

Iserson, K.V. and J.C. Moskop, 2007: Triage in medicine, part I: Concept, history, and types. *Annals of Emergency Medicine*, **49(3)**, 275-281.

Jacobs, L., F. Cook, and M. Carpini, 2009: *Talking Together: Public Deliberation and Political Participation in America*. University of Chicago Press, Chicago, IL.

Jha, A.K., J.D. Barenstein, P.M. Phelps, D. Pittet, and S. Sena, 2010: *Safer Homes, Stronger Communities: A Handbook for Reconstructing After Natural Disasters*. World Bank, Washington, DC, 370 pp.

Jones, L., S. Jaspars, S. Pavanello, E. Ludi, R. Slater, A. Arnall, N. Grist, and S. Mtisi, 2010: *Responding to a Changing Climate: Exploring how Disaster Risk Reduction, Social Protection and Livelihoods Approaches Promote Features of Adaptive Capacity*. ODI, London, UK.

Kabat, P., R.E. Schulze, M.E. Hellmuth, and J.A. Veraart (eds.), 2002: *Coping with Impacts of Climate Variability and Climate Change in Water Management: A Scoping Paper*. DWC-Report no. DWCSSO-01(2002), International Secretariat of the Dialogue on Water and Climate, Wageningen, The Netherlands, 114 pp.

Kallis, G., 2008: Droughts. *Annual Review of Environment and Resources*, **33**, 85-118.

Kang, S.J., S.J. Lee, and K.H. Lee, 2009: A study on the implementation of non-structural measures to reduce urban flood damage. *Journal of Asian Architecture and Building Engineering*, **8(2)**, 385-392.

Kasperson, R.E. and J.X. Kasperson, 2001: *Climate Change, Vulnerability, and Social Justice*. Stockholm Environment Institute, Stockholm, Sweden.

Kasperson, R.E., O. Renn, P. Slovic, H.S. Brown, J. Emel, R. Goble, J.X. Kasperson, and S. Ratick, 1988: The social amplification of risk: A conceptual framework. *Risk Analysis*, **2**, 177-187.

Kates, R.W., 2000: Cautionary tales: Adaptation and the global poor. *Climatic Change*, **45**, 5-17.

Kates, R.W., C.E. Colten, S. Laska, and S.P. Leatherman, 2006: Reconstruction of New Orleans after Hurricane Katrina: A research perspective. *Proceedings of the National Academy of Sciences*, **103**, 14653-14660.

Kawachi, I. and L. Berkman, 2000: Social cohesion, social capital, and health. In: *Social Epidemiology* [Kawachi, I. and L. Berkman (eds.)]. Oxford University Press, New York, NY, pp. 174-190.

Kazama, S., A. Sato, and S. Kawagoe, 2009: Evaluating the cost of flood damage based on changes in extreme rainfall in Japan. *Sustainability Science*, **4(1)**, 61-69.

Keraminiyage, K., S. Jayasena, R. Haigh, and D. Amaratunga (eds.), 2008: *Post Disaster Recovery Challenges in Sri Lanka*. School of the Built Environment, University of Salford, UK.

Kingsford, R.T., 2000: Ecological impacts of dams, water diversions and river management on floodplain wetlands in Australia. *Austral Ecology*, **25(2)**, 109-127.

Koelle, K., M. Pascual, and M.D. Yunus, 2004: Pathogen adaptation to seasonal forcing and climate change. *Proceedings of the Royal Society*, **272(1566)**, 971-977.

Kovác, G. and K.M. Spens, 2007: Humanitarian logistics in disaster relief operations. *International Journal of Physical Distribution & Logistics Management*, **37(2)**, 99-114.

Kraas, F., S. Aggarwal, M. Coy, G. Heiken, E. de Mulder, B. Marker, K. Nenonen, and W. Yu, 2005: *Megacities–our Global Urban Future*. Earth Sciences for Society Foundation, International Year for Planet Earth, Leiden, The Netherlands.

Kreps, G.A., 1985: Disaster and the social order. *Sociological Theory*, **3**, 49-64.

Kristjanson, P., R. Reid, N. Dickson, W. Clark, D. Romney, R. Puskur, S. MacMillan, and D. Grace, 2009: Linking international agricultural research knowledge with action for sustainable development. *Proceedings of the National Academy of Sciences*, **106(13)**, 5047-5052.

Kryspin-Watson, J., J. Arkedis, and W. Zakout, 2006: *Mainstreaming Hazard Risk Management into Rural Projects. Disaster Risk Management*. World Bank, Washington, DC.

Kull, D., 2006: Financial services for disaster risk management for the poor. In: *MicroFinance and Disaster Risk Reduction* [Bhatt, M. and P.G.D. Chakrabarti (eds.)]. Knowledge World, Delhi, India, pp. 39-64.

Kunreuther, H.C., E.O Michel-Kerjan, with N.A. Doherty, M.F. Grace, R.W. Klein, and M.V. Pauly, 2009: *At War With the Weather: Managing Large-Scale Risks in a New Era of Catastrophes.* MIT Press, Cambridge, MA.

Laczko, F. and C. Aghazarm (eds.), 2009: *Migration, Environment and Climate Change: Assessing the Evidence.* International Organization for Migration, Geneva, Switzerland.

Lafferty, K.D., 2009: The ecology of climate change and infectious diseases. *Ecology*, **90**, 888-900.

Lambin, E.F., B.L. Turner, H.J. Geist, S.B. Agbola, A. Angelsen, J.W. Bruce, O.T. Coomes, R. Dirzo, G. Fischer, C. Folke, P.S. George, K. Homewood, J. Imbernon, R. Leemans, X. Li, E.F. Moran, M. Mortimore, P.S. Ramakrishnan, J.F. Richards, H. Skånes, W. Steffen, G.D. Stone, U.S. Svedin, T.A. Veldkamp, C. Vogel, and J. Xu, 2001: The causes of land-use and land-cover change: Moving beyond the myths. *Global Environmental Change*, **11(4)**, 261-269.

Lambrew, J.M. and D.E. Shalala, 2006: Federal health policy response to Hurricane Katrina: What it was and what it should have been. *Journal of the American Medical Association*, **296**, 1394-1397.

Langsdale, S.M., A. Beall, J. Carmichael, S.J. Cohen, C.B. Forster, and T. Neale, 2009: Exploring the implications of climate change on water resources through participatory modeling: Case study of the Okanagan Basin, British Columbia. *Journal of Water Resources Planning and Management*, **135(5)**, 373-382.

Lauten, A.W. and K. Lietz, 2008: A look at standards gap: Comparing child protection responses in the aftermath of Hurricane Katrina and the Indian Ocean tsunami. *Children, Youth and Environments*, **18(1)**, 158-201.

Leach, M., R. Mearns, and I. Scoones, 1999: Environmental entitlements: Dynamics and institutions in community-based natural resource management. *World Development*, **27(2)**, 225-247.

Lecomte, E. and K. Gahagan, 1998: Hurricane insurance protection in Florida. In: *Paying the Price: The Status and Role of Insurance Against Natural Disasters in the United Sates* [Kunreuther, H. and R. Roth Sr. (eds.)]. Joseph Henry Press, Washington, DC, pp. 97-124.

Leiserowitz, A., 2007: Communicating the risks of global warming: American risk perceptions, affective images and interpretive communities. In: *Creating a Climate for Change* [Moser, S. and L. Dilling, L. (eds.)]. Cambridge University Press, New York, NY, pp. 44-63.

Lichterman, J.D., 2000: A "community as resource" strategy for disaster response. *Public Health Reports*, **115(2-3)**, 262-265.

Lin, N., 2001: Building a network theory of social capital. In: *Social capital: Theory and research* [Lin, N., K.S. Cook, and R.S. Burt (eds.)]. Transaction Press, New Brunswick, NJ, pp. 3-30.

Lindell, M.K. and R.W. Perry, 1992: *Behavioral Foundations of Community Emergency Management.* Hemisphere Publishing Corp., Washington, DC.

Lindell, M.K. and R.W. Perry, 1996: Identifying and managing conjoint threats. *Journal of Hazardous Materials*, **50**, 31-46.

Lindell, M.K. and R.W. Perry, 2004: *Communicating Environmental Risk in Multiethnic Communities.* Sage Publications, Thousand Oaks, CA.

Lindell, M.K. and C.S. Prater, 2003: Assessing community impacts of natural disasters. *Natural Hazards Review*, **4**, 176-185.

Lindell, M.K. and D.J. Whitney, 2000: Correlates of household seismic hazard adjustment adoption. *Risk Analysis*, **20**, 13-25.

Lindell, M.K., J. Lu, and C.S. Prater, 2005: Household decision making and evacuation in response to Hurricane Lili. *Natural Hazards Review*, **6(4)**, 171-179.

Linnerooth-Bayer, J. and R. Mechler, 2009: *Insurance Against Losses from Natural Disasters in Developing Countries.* United Nations Department of Economic and Social Affairs, New York, NY and Geneva, Switzerland.

Linnerooth-Bayer, J., C. Bals, and R. Mechler, 2010: Insurance as part of a climate adaptation strategy. In: *Making Climate Change Work for Us* [Hulme, M. and H. Neufeldt (eds.)]. Cambridge University Press, Cambridge, UK, pp. 340-366.

Lloyds of London, 2008: *Coastal Communities and Climate Change: Maintaining Future Insurability.* 360 Degree Risk Project Report, Lloyd's of London, London, UK.

Lonsdale, K.G., T. Downing, R. Nicholls, D. Parker, M.C. Uyarra, R. Dawson, and J. Hall, 2008: Plausible responses to the threat of rapid sea-level rise for the Thames estuary. *Climatic Change*, **91(1-2)**, 145-169.

Luna, E., 2001: Disaster mitigation and preparedness: The case of NGOs in the Philippines. *Disasters*, **25(3)**, 216-226.

Lunch, N. and C. Lunch, 2006: *Insights into Participatory Video: A Handbook for The Field.* Insight, Oxford, UK.

Lund, J.R. and R.U. Reed, 1995: Drought water rationing and transferable rations. *Journal of Water Resources Planning and Management*, **121(6)**, 429-437.

Lundqvist, L.J. and C.V. Borgstede, 2008: Whose responsibility? Swedish local decision makers and the scale of climate change abatement. *Urban Affairs Review*, **43(3)**, 299-324.

Macchi, M., G. Oveido, S. Gotheil, K. Cross, A. Boedhihartono, C. Wolfangel, and M. Howell, 2008: *Indigenous and Traditional Peoples and Climate Change.* IUCN Issues Paper, IUCN, Gland, Switzerland.

Maibach, E., C. Roser-Renouf, and A. Leiserowitz, 2008: Communication and marketing as climate change - intervention assets: A public health perspective. *American Journal of Preventive Medicine*, **35(5)**, 488-500.

Manyena, S.B., M. Fordham, and A. Collins, 2008: Disaster resilience and children: Managing food security in Zimbabwe's Binga District. *Children, Youth and Environments*, **18(1)**, 303-331.

Marx, S., E. Wever, B. Orlove, A. Leiserowitz, D. Krantz, C. Roncoli, and J. Phillips, 2007: Communication and mental processes: Experiential and analytic processing of uncertain climate information. *Global Environmental Change*, **17**, 47-58.

Masozera, M., M. Bailey, and C. Kerchner, 2007: Distribution of impacts of natural disasters across income groups: A case study of New Orleans. *Ecological Economics*, **63(2-3)**, 299-306.

May, P. and R. Burby, 1998: Making sense out of regulatory reform. *Law and Policy*, **20(2)**, 158-182.

McAdoo, B., A. Moore, and J. Baumwoll, 2009: Indigenous knowledge and the near field population. *Natural Hazards*, **48**, 73-82.

McCall, M., 2008: *Participatory Mapping and Participatory GIS (PGIS) for CRA, Community DRR and Hazard Assessment.* ProVention Consortium, CRA Toolkit, Participation Resources, Geneva, Switzerland.

McCall, M.K., 2010: Local participation in mapping, measuring and monitoring for community carbon forestry. In: *Community Forest monitoring for the Carbon Market: Opportunities Under REDD* [Skutch, M. (ed.)]. Routledge, London, UK, pp. 31-44.

McCormick, S., 2009: From "politico-scientists" to democratizing science movements: The changing climate of citizens and science. *Organization and Environment*, **22(1)**, 34-51.

McCormick, S., 2010: Hot or not? Obstacles to the emergence of climate-induced illness movements. In: *Social Movements and Health Care in the United States* [Zald, J.B.-.a.S.L.M. (ed.)]. Oxford University Press, Oxford, UK.

McCormick, S., 2011: Damming the Amazon: Local movements and transnational struggles over water. *Society & Natural Resources*, **24(1)**, 34-48.

McGee, T.K. and S. Russell, 2003: "It's just a natural way of life..." an investigation of wildfire preparedness in rural Australia. *Environmental Hazards*, **5(1-2)**, 1-12.

McGregor, D., D. Simon, and D. Thompson (eds.), 2006: *The Peri-Urban Interface: Approaches to Sustainable Natural and Human Resource Use.* Earthscan, London, UK.

McGuire, L., E. Ford, and C. Okoro, 2007: Natural disasters and older US adults with disabilities: implications for evacuation. *Disasters*, **31(1)**, 49-56.

McKinsey Group, 2009: *Shaping Climate Resilient Development: A Framework for Decision Making.* Climate Works Foundation, Global Environment Facility, European Commission, McKinsey & Company, Rockefeller Foundation, and Swiss Re, 164 pp.

McLeman, R. and B. Smit, 2006: Migration as an adaptation to climate change. *Climatic Change*, **76**, 31-53.

McMichael, A.J., R.E. Woodruff, and S. Hales, 2006: Climate change and human health: Present and future risks. *Lancet*, **367**, 859-869.

McMichael, A.J., S. Friel, A. Nyong, and C. Corvalan, 2008: Global environmental change and health: Impacts, inequalities, and the health sector. *British Medical Journal*, **336**, 191-194.

McTaggart-Cowan, R., L.F. Brosart, C.A. Davis, E.H. Atallah, J.R. Gyakum, and K.A. Emanuel, 2006: Analysis of Hurricane Catarina (2004). *Monthly Weather Review*, **134(11)**, 3029-3053.

Meehl, G.A., T.F. Stocker, W.D. Collins, P. Friedlingstein, A.T. Gaye, J.M. Gregory, A. Kitoh, R. Knutti, J.M. Murphy, A. Noda, S.C.B. Raper, I.G. Watterson, A.J. Weaver, and Z-. Zhao, 2007: Global climate projections. In: *Climate Change 2007: The Physical Science Basis. Contribution of Working Group I to the Fourth Assessment Report of the Intergovernmental Panel on Climate Change* [Solomon, S., D. Qin, M. Manning, Z. Chen, M. Marquis, K.B. Averyt, M. Tignor and H.L. Miller (eds.)]. Cambridge University Press, Cambridge, UK, and New York, NY, pp. 747-845.

Méheux, K., D. Dominey-Howes, and K. Lloyd, 2010: Operational challenges to community participation in post-disaster damage assessments: Observations from Fiji. *Disasters*, **34(4)**, 1102-1122.

Mercer, J., I. Kelman, K. Lloyd, and S. Suchet-Pearson, 2008: Reflections on use of participatory research for disaster risk reduction. *Area*, **40(2)**, 172-183.

Middlekoop, K., L. Meyer, J. Smit, R. Wood, and L. Bekker, 2006: Design and evaluation of a drama-based intervention to promote voluntary counseling and HIV testing in a South African community. *Sexually Transmitted Diseases*, **33(8)**, 524-526.

Milbert, E., 2006: Slums, slum dwellers and multilevel governance. *European Journal of Development Research*, **18(2)**, 299-318.

Mileti, D.S., 1999: *Disasters by Design: A Reassessment of Natural Hazards In the United States.* Joseph Henry Press, Washington, DC.

Mileti, D.S. and C. Fitzpatrick, 1993: *The Great Earthquake Experiment: Risk Communication and Public Action.* Westview Press, Boulder, CO.

Mileti, D.S. and P.W. O'Brien, 1992: Warnings during disaster: Normalizing communicated risk. *Social Problems*, **39**, 40-57.

Mileti, D.S. and L.A. Peek, 2002: Understanding individual and social characteristics in the promotion of household disaster preparedness. In: *New Tools for Environmental Protection: Education, Information, and Voluntary Measures* [Dietz, T. and P.C. Stern (eds.)]. The National Academies Press, Washington, DC, pp. 125-140.

Mileti, D.S. and J.H. Sorensen, 1990: *Communication of Emergency Public Warnings: A Social Science Perspective and State-Of-The-Art Assessment.* Oak Ridge National Laboratory, Oak Ridge, TN.

Mills, M.A., D. Edmonson, and C.L. Park, 2007: Trauma and stress response among Hurricane Katrina evacuees. *American Journal of Public Health*, **97(S1)**, S116-S123.

Mirza, M.M.Q., 2003: Climate change and extreme weather events: Can developing countries adapt? *Climate Policy*, **3**, 233-248.

Mitchell, J.K., 1988: Confronting natural disasters: An international decade for natural hazard reduction. *Environment*, **30(2)**, 25-29.

Mitchell, J.K. (ed.), 1999: *Crucibles of Hazard: Mega-cities and Disasters in Transition.* United Nations University Press, Tokyo, Japan.

Mitchell, J.K., 2003: European river floods in a change world. *Risk Analysis*, **23**, 567-574.

Mitchell, T. and M. van Aalst, 2008: *Convergence of Disaster Risk Reduction and Climate Change Adaptation*. UK Department for International Development, London, UK.

Mitchell, T., K. Haynes, W. Choong, N. Hall, and K. Oven, 2008: The role of children and youth in communicating disaster risk. *Children, Youth and Environments*, **18(1)**, 254-279.

Moench, M. and A. Dixit (eds.), 2004: *Adaptive Capacity and Livelihood Resilience: Adaptive Strategies for Responding To Floods and Droughts in South Asia.* Institute for Social and Environmental Transition, Kathmandu, Nepal.

Moench, M. and Risk to Resilience Study Team, 2008: *From Risk to Resilience: Understanding the Costs and Benefits of Disaster Risk Reduction Under Changing Climatic Conditions*. Working Paper 9, ProVention and ISET, Geneva, Switzerland.

Montz, B.E. and G.A. Tobin, 2008: Living large with levees: Lessons learned and lost. *Natural Hazards Review*, **9(3)**, 150-157.

Morgan, J., 1994: Sudanese refugees in Koboko. *Gender and Development*, **2(1)**, 41-44.

Morrow, B.H. and E. Enarson, 1996: Hurricane Andrew through women's eyes: Issues and recommendations. *International Journal of Mass Emergencies and Disasters*, **14(1)**, 5-22.

Moser, S.C., 2009: Whether our are levers are long enough and the fulcrum strong? Exploring the soft underbelly of adaptation decisions and actions. In: *Adapting to Climate Change: Thresholds, Values, Governance* [Adger, W.N., I. Lorenzoni, and K. O'Brien (eds.)]. Cambridge University Press, Cambridge, UK, pp. 313-334.

Moser, S. and L. Dilling, 2007: *Creating a Climate for Change: Communicating Climate Change and Facilitating Social Change.* Cambridge University Press, Cambridge, UK.

Moser, S.C. and J.A. Ekstrom, 2010: A framework to diagnose barriers to climate change adaptation. *Proceedings of the National Academy of Sciences*, **107(51)**, 22026-22031.

Moynihan, D.P., 2009: The network governance of crisis response: Case studies of incident command systems. *Public Administration Review*, **19(4)**, 895-915.

Mukheibir, P. and G. Ziervogel, 2007: Developing a municipal adaptation plan (MAP) for climate change: The city of Cape Town. *Environment and Urbanization*, **19(1)**, 143-158.

Munich Re Group, 2004: *Megacities–Megarisks: Trends and Challenges for Insurance and Risk Management*. Munich Re Group, Munich, Germany.

Mustafa, D., 1998: Structural causes of vulnerability to flood hazard in Pakistan. *Economic Geography*, **74(3)**, 289-305.

Mutton, D. and C.E. Haque, 2004: Human vulnerability, dislocation and resettlement: Adaptation processes of river-bank erosion-induced displacees in Bangladesh. *Disasters*, **28(1)**, 41-62.

Myers, C.A., T. Slack, and J. Singlemann, 2008: Social vulnerability and migration in the wake of disaster: The case of Hurricanes Katrina and Rita. *Population and Environment*, **29**, 271-291.

Nakagawa, Y. and R. Shaw, 2004: Social capital: A missing link to disaster recovery. *International Journal of Mass Emergencies and Disasters*, **22(1)**, 5-34.

Nelson, E., G. Mendoza, J. Regetz, S. Polasky, H. Tallis, D.R. Cameron, K.M.A. Chan, G.C. Daily, J. Goldstein, P.M. Kareiva, E. Lonsdorf, R. Naidoo, T.H. Ricketts, and M.R. Shaw, 2009: Modeling multiple ecosystem services, biodiversity conservation, commodity production, and tradeoffs at landscape scales. *Frontiers in Ecology and Environment*, **7(1)**, 4-11.

Neria, Y., S. Galea, and F.H. Norris (eds.), 2009: *Mental Health and Disasters.* Cambridge University Press, Cambridge, UK.

Neumayer, E. and T. Plümper, 2007: The gendered nature of natural disasters: The impact of catastrophic events on the gender gap in life expectancy, 1981-2002. *Annals of the Association of American Geographers*, **97(3)**, 551-566.

Ngo, E.B., 2001: When disasters and age collide: Reviewing vulnerability of the elderly. *Natural Hazards Review*, **2(2)**, 80-89.

Nicholls, R., R.S.J. Tol, and A. Vafeidis, 2008: Global estimates of the impact of a collapse of the west Antarctic ice sheet: An application of FUND. *Climatic Change*, **91(1-2)**, 171-190.

Nicholson-Cole, S., 2004: Representing climate change futures: A critique on the use of images for visual communication. *Computers, Environment and Urban Systems*, **29**, 255-273.

Niepold, F., D. McConville, and D. and Herring, 2008: The role of narrative and geospatial visualization in fostering climate literate citizens. *Physical Geography*, **29**, 255-273.

NOAA, 2007: *Construction Setbacks*. National Oceanic and Atmospheric Administration, Office of Ocean and Coastal Resource Management, Washington, DC.

Nordås, R. and N.P. Gleditsch, 2007: Climate change and conflict. *Political Geography*, **26**, 627-638.

Nordhaus, W.D., 2007: Review: A review of the "Stern review on the economics of climate change." *Journal of Economic Literature*, **45(3)**, 686-702.

Nordstrom, K.F., 2000: *Beaches and Dunes of Developed Coasts.* Cambridge University Press, Cambridge, UK.

Nordstrom, K.F., 2008: *Beach and Dune Restoration.* Cambridge University Press, Cambridge, UK.

Norgaard, K., 2009: *Cognitive and Behavioral Challenges in Responding to Climate Change*. The World Bank, Washington, DC.

North, D.C., 1990: *Institutions, Institutional Change and Economic Performance.* Cambridge University Press, Cambridge, UK.

Noy, I. and T. Vu, 2010: The economics of natural disasters in a developing country: The case of Vietnam. *Journal of Asian Economics*, **21**, 345-354.

NRC, 1989: *Improving Risk Communication*. National Research Council, National Academies Press, Washington, DC.

NRC, 2006: *Facing Hazards and Disasters: Understanding Human Dimensions*. National Research Council, National Academy Press, Washington, DC.

NRC, 2007: *Tools and Methods for Estimating Populations at Risk from Natural Disasters and Complex Humanitarian Crises*. National Research Council, National Academies Press, Washington, DC.

Oba, G., 2001: The importance of pastoralists' indigenous coping strategies for planning drought management in the arid zone of Kenya. *Nomadic Peoples*, **5(1)**, 89-119.

Obasi, G.O.P., 2005: The impacts of ENSO in Africa. In: *Climate Change and Africa* [Low, P.S. (ed.)]. Cambridge University Press, Cambridge, UK, pp. 218-230.

O'Brien, G., T. Devisscher, and P. O'Keefe, 2011: *The Adaptation Continuum: Groundwork for the Future*. Lambert Academic Publishing, Germany.

O'Brien, K., R. Leichenko, U. Kelkar, H. Venema, G. Aandahl, H. Tompkins, A. Javed, S. Bhadwal, S. Barg, L. Nygaard, and J. West, 2004: Mapping vulnerability to multiple stressors: Climate change and globalization in India. *Global Environmental Change*, **14(4)**, 303-313.

O'Brien, P. and D.S. Mileti, 1992: Citizen participation in emergency response following the Loma Prieta earthquake. *International Journal of Mass Emergencies and Disasters*, **10(1)**, 71-89.

Oettlé, N., A. Arendse, B. Koelle, and A. Van Der Poll, 2004: Community exchange and training in the Suid Bokkeveld: A UNCCD pilot project to enhance livelihoods and natural resource management. *Environmental Monitoring and Assessment*, **99(1-3)**, 115-125.

O'Keefe, P., K. Westgate, and B. Wisner, 1976: Taking the naturalness out of natural disasters. *Nature*, **260(5552)**, 566-567.

Okuyama, Y., 2004: Modeling spatial economic impacts of an earthquake: Input-output approaches. *Disaster Prevention and Management*, **13**, 297-306.

Oloruntoba, R., 2005: A wave of destruction and the waves of relief: Issues, challenges and strategies. *Disaster Prevention and Management*, **14(4)**, 506-521.

Olsen, G.R., N. Carstensen, and K. Høyen, 2003: Humanitarian crises: What determines the level of emergency assistance? Media coverage, donor interests and the aid business. *Disasters*, **27(2)**, 109-126.

Olshansky, R.B. and J.D. Kartez, 1998: Managing land use to build resilience. In: *Cooperating with Nature: Confronting Natural Hazards with Land Use Planning for Sustainable Communities* [Burby, R. (ed.)]. Joseph Henry Press, Washington, DC, pp. 167-201.

Olson, R., 2000: Towards a politics of disaster: Losses, values, agendas and blame. *International Journal of Mass Emergencies and Disasters*, **18**, 265-287.

O'Neill, M.S. and K.L. Ebi, 2009: Temperature extremes and health: Impacts of climate variability and change in the United States. *Journal of Occupational and Environmental Medicine*, **51(1)**, 13-25.

Opperman, J.J., G.E. Galloway, J. Fargione, J.F. Mount, B.D. Richter, and S. Secchi, 2009: Sustainable floodplains through large-scale reconnection to rivers. *Science*, **326**, 1487-1488.

Osman-Elasha, B., 2006a: *Project AF14,-Assessments of Impacts and Adaptations to Climate Change. Environmental Strategies to Increase Human Resilience to Climate Change: Lessons for Eastern and Northern Africa, Final Report*. International START Secretariat, Washington, DC.

Osman-Elasha, B., 2006b: *Human Resilience to Climate Change: Lessons for Eastern and Northern Africa*. International START Secretariat, Washington, DC.

Osman-Elasha, B. and T.E. Downing, 2007: *Lessons Learned in Preparing National Adaptation Programmes of Action in Eastern and Southern Africa*. SEI Oxford Working Paper, Stockholm Environment Institute, Oxford, UK.

Osman-Elasha, B. and A. El Sanjak, 2009: Global climate changes: Impacts on water resources and human security in africa. In: *Environment and Conflict in Africa: Reflections on Darfur* [Leroy, M. (ed.)]. University for Peace/Africa Programme, Addis Ababa, Ethiopia, pp. 406-427.

Osman-Elasha, B., N. Goutbi, E. Spanger-Siegfried, W. Dougherty, A. Hanafi, S. Zakieldeen, A. Sanjak, H.A. Atti, and H.M. Elhassan, 2006: *Adaptation Strategies to Increase Human Resilience Against Climate Variability and Change: Lessons from the Arid Regions of Sudan*. AIACC, International START Secretariat, Washington, DC.

Oxfam, 2009: *Exposed: Social Vulnerability and Climate Change in the US Southeast*. Oxfam America, Boston, MA.

Pahl-Wostl, C., J. Sendzimir, P. Jeffrey, J. Aerts, G. Berkamp, and K. Cross, 2007: Managing change toward adaptive water management through social learning. *Ecology and Society*, **12(2)**, 23/06/2010-30.

Palen, L., S. Vieweg, J. Sutton, S.B. Liu, and A. Hughes, 2007: Crisis informatics: Studying crises in a networked world. In: *Third International Conference on e-Social Science*, Ann Arbor, MI, 7-9 October 2007.

Parati, G., R. Antonicelli, F. Guazzarotti, E. Paciaroni, and G. Mancia, 2001: Cardiovascular effects of an earthquake: Direct evidence by ambulatory blood pressure monitoring. *Hypertension*, **38(5)**, 1093-1095.

Parkinson, A.J. and J.C. Butler, 2005: Potential impacts of climate change on infectious diseases in the Arctic. *International Journal of Circumpolar Health*, **64**, 478-486.

Patiño, L. and D. Gauthier, 2009: Integrating local perspectives into climate change decision making in rural areas of the Canadian prairies. *International Journal of Climate Change Strategies and Management*, **1(2)**, 179-196.

Patt, A.G., 2009: Learning to crawl: How to use seasonal climate forecasts to build adaptive capacity. In: *Adapting to Climate Change: Thresholds, Values, Governance* [Adger, W.N., I. Lorenzoni, and K. O'Brien (eds.)]. Cambridge University Press, Cambridge, UK.

Patt, A.G. and C. Gwata, 2002: Effective seasonal forecast applications: Examining constraints for subsistence farming in Zimbabwe. *Global Environmental Change–Human and Policy Dimensions*, **12**, 185-195.

Patt, A. and D. Schröter, 2008: Perceptions of climate risk in Mozambique: Implications for the success of adaptation and coping strategies. *Global Environmental Change*, **18**, 458-467.

Patt, A.G., N. Peterson, M. Carter, M. Velez, U. Hess, and P. Suarez, 2009: Making index insurance attractive to farmers. *Mitigation and Adaptation Strategies for Global Change*, **14**, 737-757.

Patz, J.A., D. Campbell-Lendrum, T. Holloway, and J.A. Foley, 2005: Impact of regional climate change on human health. *Nature*, **438**, 310-317.

Paul, B.K. and S. Dutt, 2010: Hazard warnings and responses to evacuation orders: The case of Bangladesh's Cyclone Sidr. *Geographical Review*, **100(3)**, 336-355.

Paul, B.K., H. Rashid, M.S. Islam, and L.M. Hunt, 2010: Cyclone evacuation in Bangladesh: Tropical Cyclones Gorky (1991) vs. Sidr (2007). *Environmental Hazards*, **9(1)**, 89-101.

Paulson, D.D., 1993: Hurricane hazard in western Samoa. *Geographical Review*, **83(1)**, 43-53.

Peacock, W., B.H. Morrow, and H. Gladwin (eds.), 1997: *Hurricane Andrew: Ethnicity, Gender, and the Sociology of Disasters*. Routledge, London, UK.

Pearce, D. and D. Moran, 1994: *The economic value of biodiversity*. Earthscan Publications Ltd., London, UK, 172 pp.

Pelling, M., 2003: *The Vulnerability of Cities: Natural Disasters and Social Resilience*. Earthscan, London, UK.

Pelling, M., 2007: Learning from others: The scope and challenges for participatory disaster risk assessment. *Disasters*, **31(4)**, 373-385.

Pelling, M. and K. Dill, 2010: Disaster politics: Tipping points for change in the adaptation of sociopolitical regimes. *Progress in Human Geography*, **34(1)**, 21-37.

Pelling, M. and C. High, 2005: Understanding adaptation: What can social capital offer assessments of adaptive capacity? *Global Environmental Change*, **15**, 308-319.

Pelling, M., C. High, J. Dearing, and D. Smith, 2008: Shadow spaces for social learning: A relational understanding of adaptive capacity to climate change within organisations. *Environment and Planning A*, **40(4)**, 867-884.

Penning-Rowsell, E.C., C. Johnson, and S.M. Tunstall, 2006: Signals from pre-crisis discourse: Lessons from UK flooding for global environmental policy change? *Global Environmental Change*, **16**, 323-339.

Perch-Nielson, S.L., M.B. Battig, and D. Imboden, 2008: Exploring the link between climate change and migration. *Climatic Change*, **91(3-4)**, 375-393.

Perry, R.W. and M.K. Lindell, 1991: The effects of ethnicity on evacuation decision making. *International Journal of Mass Emergencies and Disasters*, **9(1)**, 47-68.

Perry, R.W. and M.K. Lindell, 1997: Principles for managing community relocation as a hazard mitigation measure. *Journal of Contingencies and Crisis Management*, **5(1)**, 49-59.

Petal, M., R. Green, I. Kelman, R. Shaw, and A. Dixit, 2008: Community-based construction for disaster risk reduction. In: *Hazards and the Built Environment: Attaining Built-In Resilience* [Bosher, L. (ed.)]. Routledge, London, UK, pp. 191-217.

Pettenger, M.E., 2007: *The Social Construction of Climate Change: Power, Knowledge, Norms, Discourses.* Ashgate Publishers, Hampshire, UK.

Pfefferbaum, B., J.B. Houston, K.F. Wyche, R.L. Van Horn, G. Reyes, H. Jeon-Slaughter, and C.S. North, 2008: Children displaced by Hurricane Katrina: A focus group study. *Journal of Loss & Trauma*, **13(4)**, 303-318.

Pielke Jr., R.A., J. Gratz, C.W. Landsea, D. Collins, M.A. Saunders, and R. Musulin, 2008: Normalized hurricane damage in the United States: 1900-2005. *Natural Hazards Review*, **9**, 29-42.

Plush, T., 2009: Amplifying children's voices on climate change: The role of participatory video. *Participatory Learning and Action*, **60**, 119-128.

Podestá, G., D. Letson, C. Messina, F. Royce, R.A. Ferreyra, J. Jones, J. Hansen, I. Llovet, M. Grondona, and J.J. O'Brien, 2002: Use of ENSO-related climate information in agricultural decision making in Argentina: A pilot experience. *Agricultural Systems*, **74**, 371-392.

Polack, E., 2008: A right to adaptation: Securing the participation of marginalized groups. *IDS Bulletin*, **39(4)**, 16-23.

Pomeroy, R.S., B.D. Ratner, S.J. Hall, J. Pimoljind, and V. Vivekanandan, 2006: Coping with disaster: Rehabilitating coastal livelihoods and communities. *Marine Policy*, **30(6)**, 786-793.

Portes, A. and P. Landolt, 1996: The downside of social capital. *The American Prospect*, **26**, 18-21.

Postel, S.L., G.C. Daily, and P.R. Ehrlich, 1996: Human appropriation of renewable fresh water. *Science*, **271(5250)**, 785-788.

Prabhakar, S.V.R.K., A. Srinivasan, and R. Shaw, 2009: Climate change and local level disaster reduction planning: Need, opportunities and challenges. *Mitigation and Adaptation Strategies for Global Change*, **14**, 7-33.

Prasad, N., F. Ranghieri, F. Shah, Z. Trohanis, E. Kessler, and R. Sinha, 2009: *Climate Resilient Cities: A Primer On Reducing Vulnerabilities.* World Bank, Washington, DC.

Priest, S.J., M.J. Clark, and E.J. Treby, 2005: Flood insurance: The challenge of the uninsured. *Area*, **37(3)**, 295-302.

Pulwarty, R.S., 2003: Climate and water in the west: Science, information and decision making. *Water Resources*, **124**, 4-12.

Pulwarty, R.S., 2007: *Communicating Agroclimatological Information, Including Forecasts for Agricultural Decision. Guide to Agrometeorological Practices.* World Meteorological Organization, Geneva, Switzerland.

Pulwarty, R.S. and T. Melis, 2001: Climate extremes and adaptive management on the Colorado River. *Journal of Environmental Management*, **63**, 307-324.

Pulwarty, R., T. Broad, and T. Finan, 2004: ENSO forecasts and decision making in Peru and Brazil. In: *Mapping Vulnerability: Disasters, Development and People* [Bankoff, G., G. Frerkes, and T. Hilhorst (eds.)]. Earthscan, London, UK, pp. 83-98.

Quarantelli, E.L. (ed.), 1998: *What is a Disaster? A Dozen Perspectives on the Question.* Routledge, London, UK.

Ramachandran, N., 2006: *Women and Food Security in South Asia. Current Issues and Emerging Concerns.* WIDER Research Paper, UN University, World Institute for Development Economic Research, Helsinki, Finland, 19 pp.

Ranger, N., S. Hallegatte, S. Bhattacharya, M. Bachu, S. Priya, K. Dhore, F. Rafique, P. Mathur, N. Naville, F. Henriet, C. Herweijer, S. Pohit, and J. Corfee-Morlot, 2011: An assessment of the potential impact of climate change on flood risk in Mumbai. *Climatic Change*, **104(1)**, 139-167.

Rao, K. and U. Hess, 2009: Scaling up with India: The public sector. In: *Index Insurance and Climate Risk: Prospects for Development and Disaster Management* [Hellmuth, M.E., D.E. Osgood, U. Hess, A. Moorhead, and H. Bhojwani (eds.)]. International Research Institute for Climate and Society (IRI), New York, NY, pp. 87-89.

Raschky, P.A., 2008: Institutions and the losses from natural disasters. *Natural Hazards Earth Systems Science*, **8**, 627-634.

Rayner, S., H. Ingram, and D. Lach, 2005: Weather forecasts are for wimps: Why water resource managers do not use climate forecasts. *Climatic Change*, **69**, 197-227.

Reid, H., M. Alam, R. Berger, T. Cannon, S. Huq, and A. Milligan, 2009: Community-based adaptation to climate change: An overview. *Participatory Learning and Action*, **60**, 11-38.

Ribot, J.C., 2003: Democratic decentralization of natural resources: Institutional choice and discretionary power transfers in sub-Saharan Africa. *Public Administration and Development*, **23(1)**, 53-65.

Ribot, J.C., A. Agrawal, and A.M. Larson, 2006: Recentralizing while decentralizing: How national governments reappropriate forest resources. *World Development*, **34(11)**, 1864-1886.

Rist, S., 2000: Linking ethics and market – Campesino economic strategies in the Bolivian Andes. *Mountain Research and Development*, **20(4)**, 310-315.

Rist, S., F. Delgado, and U. Wiesmann, 2003: The role of social learning processes in the emergence and development of Aymara land use systems. *Mountain Research and Development*, **23(3)**, 263-270.

Ritchie, L. and D. Gill, 2007: Social capital theory as an integrating theoretical framework in technological disaster research. *Sociological Spectrum*, **27(1)**, 103-129.

RMS, 2009: *Analyzing the Effects of the 'my Safe Florida Home' Program on Florida Insurance Risk.* RMS Special Report, Summary of Analysis Prepared for the Florida Department of Financial Services, Risk Management Solutions, Newark, CA.

Roberts, S., 2008: Effects of climate change on the built environment. *Energy Policy*, **36(12)**, 4552-4557.

Robledo, C., N. Clot, A. Hammill, and B. Riche, 2011: The role of forest ecosystems in community-based coping strategies to climate hazards: Three examples from rural areas in Africa. *Forest Policy and Economics*, doi:10.1016/j.forpol.2011.04.006.

Rodriguez-Oreggia, E., A.d.l. Fuente, R.d.l. Torre, H. Moreno, and C. Rodriguez, 2010: *The Impact of Natural Disasters on Human Development and Poverty at the Municipal Level in Mexico.* CID Working paper 43, Center for International Development, Harvard University, Cambridge, MA.

Rofi, A., S. Doocy, and C. Robinson, 2006: Tsunami mortality and displacement in Aceh Province, Indonesia. *Disasters*, **30(3)**, 340-350.

Rojas Blanco, A.V., 2006: Local initiatives and adaptation to climate change. *Disasters*, **30(1)**, 140-147.

Ronan, K.R., K. Crellin, D.M. Johnston, J. Becker, K. Finnis, and D. Paton, 2008: Promoting child and family resilience to disasters: Effects, interventions and prevention effectiveness. *Children, Youth and Environments*, **18(1)**, 332-353.

Roncoli, C., K. Ingram, and P. Kirshen, 2001: The costs and risks of coping with drought: Livelihood impacts and farmers' responses in Burkina Faso. *Climate Research*, **19**, 119-132.

Rose, A., 2004: Economic principles, issues, and research priorities in hazard loss estimation. In: *Modeling Spatial and Economic Impacts of Disasters* [Okuyama, Y. and S. Chang (eds.)]. Springer, Berlin, Germany, pp. 14-36.

Rose, A. and S. Liao, 2005: Modeling regional economic resilience to disasters: A computable general equilibrium analysis of water service disruptions. *Journal of Regional Science*, **45**, 75-112.

Rose, A., J. Benavides, S. Chang, P. Szczesnjak, and D. Lim, 1997: The regional economic impact of an earthquake: Direct and indirect effects of electricity lifeline disruptions. *Journal of Regional Science*, **37**, 437-458.

Rose, A., G. Oladosu, and S. Liao, 2007: Business interruption impacts of a terrorist attack on the electric power system of Los Angeles: Customer resilience to a total blackout. *Risk Analysis*, **27(3)**, 513-531.

Rosenzweig, C., D.C. Major, K. Demong, C. Stanton, R. Horton, and C. Stults, 2007: Managing climate change risks in New York City's water system: Assessment and adaptation planning. *Mitigation and Adaptation Strategies for Global Change*, **12(8)**, 1391-1409.

Rosenzweig, C., W.D. Solecki, R. Blake, M. Bowman, C. Faris, V. Gornitz, R. Horton, K. Jacob, A. LeBlanc, R. Leichenko, M. Linkin, D. Major, M. O'Grady, L. Patrick, E. Sussman, G. Yohe, and R. Zimmerman, 2011: Developing coastal adaptation to climate change in New York City infrastructure-shed: Process, approach, tools, and strategies. *Climatic Change*, **106(1)**, 93-127.

Ruppert, T.K., 2008: Eroding long-term prospects for Florida's beaches: Florida's coastal construction control line program. *Sea Grant Law and Policy Journal*, **1(1)**, 65-98.

Sagala, S., N. Okada, and D. Paton, 2009: Predictors of intention to prepare for volcanic risks in Mt. Merapi, Indonesia. *Journal of Pacific Rim Psychology*, **3(2)**, 47-54.

Sahlins, M.D., 1962: *Moala: Culture and Nature on a Fijian Island*. University of Michigan Press, Ann Arbor, MI.

Salick, J. and A. Byg (eds.), 2007: *Indigenous Peoples and Climate Change*. Tyndall Centre for Climate Change Research, Oxford, UK.

Salick, J. and N. Ross, 2009: Traditional peoples and climate change. *Global Environmental Change*, **19**, 137-139.

Satterthwaite, D., S. Huq, M. Pelling, H. Reid, and P. Lankao, 2007: *Adapting to Climate Change in Urban Areas: The Possibilities and Constraints in Low-and Middle-Income Nations*. International Institute for Environment and Development, London, UK.

Schneider, S., 2008: Who's to blame? (mis)perceptions of the intergovernmental response to disasters. *Disasters*, **38(4)**, 715-738.

Schroeder, R.A., 1987: *Gender Vulnerability to Drought: A Case Study of the Hausa Social Environment*. Natural Hazards Research, University of Colorado, Boulder, CO.

Scoones, I., 1998: *Sustainable Rural Livelihoods: A Framework for Analysis*. IDS, University of Sussex, Brighton, UK.

Scott, J.C., 1976: *The Moral Economy of the Peasant Rebellion and Subsistence in Southeast Asia*. Yale University Press, New Haven, CT.

Seck, P., 2007: *Links between Natural Disasters, Humanitarian Assistance and Disaster Risk Reduction: A Critical Perspective*. United Nations Development Programme, New York, NY.

Semenza, J.C., 2005: Building healthy cities: A focus on interventions. In: *Handbook of Urban Health* [Galea, S. and D. Vlahov (eds.)]. Spring Science and Business Media, New York, NY, pp. 459-502.

Sen, A., 1981: *Poverty and Famines: An Essay on Entitlement and Deprivation*. Clarendon, Oxford, UK.

Shaw, A., S. Sheppard, S. Burch, D. Flanders, A. Wiek, J. Carmichael, J. Robinson, and S. Cohen, 2009: Making local futures tangible–Synthesizing, downscaling, and visualizing climate change scenarios for participatory capacity building. *Global Environmental Change*, **19**, 447-463.

Shaw, R., 2006: Community-based climate change adaptation in Vietnam: Inter-linkages of environment, disaster, and human security. In: *Multiple Dimensions of Global Environmental Changes* [Sonack, S. (ed.)]. TERI Publication, Energy and Resources Institute, New Delhi, India, pp. 521-547.

Shaw, R., A. Sharma, and Y. Takeuchi, 2009: *Indigenous Knowledge and Disaster Risk Reduction: From Practice to Policy*. Nova Science Publishers, Hauppauge, NY, pp. 409.

Shoaf, K. and S. Rottmann, 2000: The public health impact of disasters. *Australian Journal of Emergency Management*, **15(3)**, 58-63.

Silk, J., 2000: Caring at a distance: (im)partiality, moral motivation, and the ethics of representation – introduction. *Ethics, Place & Environment*, **3(3)**, 303-309.

Sims, H. and K. Vogelmann, 2002: Popular mobilization and disaster management in Cuba. *Public Administration and Development*, **22**, 389-400.

Skees, J.R. and B.J. Barnett, 2006: Enhancing microfinance using index-based risk-transfer products. *Agricultural Finance Review*, **66(2)**, 235-250.

Skees, J.R., B.J. Barnett, and A.G. Murphy, 2008: Creating insurance markets for natural disaster risk in lower income countries: The potential role for securitization. *Agricultural Finance Review*, **68**, 151-157.

Smith, S.K. and C. McCarty, 2006: Florida's 2004 hurricane season: Demographic response and recovery. In: *Proceedings of Southern demographic association meeting*, Durham, NC, 2-4 Nov 2006.

Someshwar, S., 2008: Adaptation as 'Climate-smart' development. *Development*, **51**, 366-374.

SOPAC and UNEP, 2005: *Building Resilience in SIDS: The Environmental Vulnerability Index*. South Pacific Applied Geoscience Commission, Suva, Fiji.

Sorensen, J.H., 2000: Hazard warning systems: Review of 20 years of progress. *Natural Hazards Review*, **1(2)**, 119-125.

Sorensen, J.H. and B.V. Sorensen, 2007: Community processes: Warning and evacuation. In: *Handbook of Disaster Research* [Rodriquez, H., E.L. Quarantelli, and R.R. Dynes (eds.)]. Springer, New York, NY, pp. 183-199.

Sorensen, J.H., B.L. Shumpert, and B.M. Vogt, 2004: Planning for protective action decision making: Evacuate or shelter-in-place. *Journal of Hazardous Materials*, **109(1-3)**, 1-11.

Spanger-Siegfried, E., W. Dougherty, and B. Osman-Elasha, 2005: *Methodological Framework an Internal Scoping Report of the Project Strategies for Increasing Human Resilience in Sudan: Lessons for Climate Change Adaptation in North and East Africa*. AIACC Working Paper No. 18, Assessments of Impacts and Adaptations to Climate Change, Washington, DC.

Spence, R., 2004: Risk and regulation: Can improved government action reduce the impacts of natural disasters? *Building Research & Information*, **32(5)**, 391-405.

Sperling, F. and F. Szekely, 2005: *Disaster Risk Management in a Changing Climate*. Vulnerability and Adaptation Resource Group, Washington, DC.

Stern, N., 2007: *The Economics of Climate Change: The Stern Review*. Cambridge University Press, Cambridge, UK.

Stern, P. and W. Easterling (eds.), 1999: *Making Climate Forecasts Matter*. National Academies Press, Washington DC, 175 pp.

Stevenson, J.R., C.T. Emrich, J.T. Mitchtell, and S.L. Cutter, 2010: Using building permits to monitor disaster recovery: A spatio-temporal case study of coastal Mississippi following Hurricane Katrina. *Cartography and Geographic Information Science*, **37(1)**, 57-68.

Strobl, E., 2008: *The Economic Growth Impact of Hurricanes: Evidence from US Coastal Counties*. IZA Discussion Papers Series, Institute for the Study of Labor, Bonn, Germany.

Suarez, P., F. Ching, G. Ziervogel, I. Lemaire, D. Turnquest, J.M. de Suarez, and B. Wisner, 2008: *Video-Mediated Approaches for Community-Level Climate Adaptation*. Institute of Development Studies, University of Sussex, Brighton, UK.

Sudmeier-Rieux, K., H. Masundire, A. Rizvi, and S. Rietbergen, 2006: *Ecosystems, Livelihoods And Disasters: An Integrated Approach to Disaster Risk Management*. International Union for Conservation of Nature and Natural Resources, Gland, Switzerland, 56 pp.

Sundar, N. and R. Jeffery, 1999: *A New Moral Economy for Indian Forests? Discourses of Community and Participation*. Sage, New Delhi, India, 304 pp.

Susman, P., P. O'Keefe, and B. Wisner, 1983: Global disasters, a radical reinterpretation. In: *Interpretations of Calamity from the Viewpoint of Human Ecology* [Hewitt, K. (ed.)]. Allen and Unwin, Boston, MA, pp. 263-283.

Tacoli, C., 2009: Crisis or adaptation? Migration and climate change in a context of high mobility. *Environment and Urbanization*, **21(2)**, 513-525.

Tanner, T.M. and T. Mitchell, 2008: Poverty in a changing climate. *IDS Bulletin*, **39(4)**, 1-5.

Tanner, T., T. Mitchell, E. Polack, and B. Guenther, 2009: *Urban Governance for Adaptation: Assessing Climate Change for Resilience in Ten Asian Cities*. Institute of Development Studies, Brighton, UK.

Tanser, F.C., B. Sharp, and D. le Sueur, 2003: Potential effect of climate change on malaria transmission in Africa. *The Lancet*, **362(9398)**, 1792-1798.

Tearfund, 2006: *Adapting to Climate Change: Challenges and Opportunities for the Development Community*. Institute of Development Studies, Brighton, UK.

Thacker, M.T.F., R. Lee, R.I. Sabogal, and A. Henderson, 2008: Overview of deaths associated with natural events, United States, 1979-2004. *Disasters*, **32(2)**, 303-315.

Thieken, A.H., T. Petrow, and H. Kreibich, and B. Merz, 2006: Insurability and mitigation of flood losses in private households in Germany. *Risk Analysis*, **26(2)**, 383-395.

Thomalla, F., T. Downing, E. Spanger-Siegfried, G. Han, and J. Rockström, 2006: Reducing hazard vulnerability: Towards a common approach between disaster risk reduction and climate adaptation. *Disasters*, **30(1)**, 39-48.

Thomas, D.S.G. and C. Twyman, 2005: Equity and justice in climate change adaptation among natural-resource-dependent societies. *Global Environmental Change*, **15**, 115-124.

Thomson, R., 2007: Cultural models and shoreline social conflict. *Coastal Management*, **35**, 211-237.

Tierney, K., 2006: Social inequality, hazards, and disasters. In: *On Risk and Disaster: Lessons from Hurricane Katrina* [Daniels, R.J., D.F. Kettl, and H. Kunreuther (eds.)]. University of Pennsylvania Press, Philadelphia, PA, pp. 109-128.

Tierney, K.J., M.K. Lindell, and R.W. Perry, 2001: *Facing the Unexpected: Disaster Preparedness and Response in the United States*. National Academies Press, Washington, DC.

Tierney, K., C. Bevc, and E. Kuligowski, 2006: Metaphors matter: Disaster myths, media frames, and their consequences in hurricane Katrina. *Annals of the American Academy of Political and Social Science*, **604(1)**, 57-81.

Titus, J.G., 2011: *Rolling Easements*. United States Environmental Protection Agency, Washington, DC, 166 pp.

Titus, J., D. Hudgens, D. Trescott, M. Craghan, W. Nuckols, C. Hershner, J. Kassakian, C. Linn, P. Merritt, J. McCue, J. O'Connell, J. Tanski, and J. Wang, 2009: State and local governments' plan for development of most land vulnerable to rising sea level along the US Atlantic coast. *Environmental Research Letters*, **4(4)**, 044008.

Tobin, G.A., 1995: The levee love affair: A stormy relationship. *Water Resources Bulletin*, **31**, 359-367.

Tol, R.S.J., 2003: Is the uncertainty about climate change too large for expected cost-benefit analysis? *Climatic Change*, **56(3)**, 265-289.

Tompkins, E.L., 2005: Planning for climate change in small islands: Insights from national hurricane preparedness in the Cayman Islands. *Global Environmental Change A*, **15(2)**, 139-149.

Torresan, S., A. Critto, M.D. Valle, N. Harvey, and A. Marcomini, 2008: Assessing coastal vulnerability to climate change: Comparing segmentation at global and regional scales. *Sustainability Science*, **3(1)**, 45-65.

Travis, W., 2010: Going to extremes: Propositions on the social response to severe climate change. *Climatic Change*, **98**, 1-19.

Trawick, P., 2001: The moral economy of water: Equity and antiquity in the Andean commons. *American Anthropologist*, **103(2)**, 361-379.

Tschakert, P. and K. Dietrich, 2010: Anticipatory learning for climate change adaptation and resilience. *Ecology and Society*, **15(2)**, 11.

Tsuchiya, S., H. Tatano, and N. Okada, 2007: Economic loss assessment due to railroad and highway disruptions. *Economic Systems Research*, **19(2)**, 147-162.

Turner, N.J. and H. Clifton, 2009: "It's so different today": Climate change and indigenous lifeways in British Columbia, Canada. *Global Environmental Change*, **19(2)**, 180-190.

Turvey, C., 2001: Weather insurance and specific event risks in agriculture. *Review of Agricultural Economics*, **23(2)**, 333-351.

Twigg, J., 2001: *Sustainable Livelihoods and Vulnerability to Disasters*. Benfield Greig Hazard Research Centre, Disaster Management Working Paper 2/2001, University College London, London, UK.

Twigg, J., 2007: *Characteristics of a Disaster-Resilient Community: A Guidance Note*. DFID Disaster Risk Reduction Interagency Coordination Group, Benfield, UK.

U.S. Bureau of Labor Statistics, 2006: The labor market impact of Hurricane Katrina: An overview. *Monthly Labor Review*, **August 2006**, 3.

UNDP, 2004: *Reducing Disaster Risk: A Challenge for Development*. United Nations Development Programme, New York, NY.

UNDP, 2007: *Human Development Report 2007/2008: Fighting Climate Change: Human Solidarity in a Divided World*. United Nations Development Programme, Palgrave Macmillan, Hampshire, UK.

UNFPA, 2009: Facing a Changing World: Women, Population and Climate. United Nations Population Fund, New York, NY.

UN-HABITAT, 2007: *Global Report on Human Settlements, 2007: Enhancing Urban Safety and Security*. Earthscan, London, UK.

UNIFEM, 2011: *Pakistan Floods 2010: Rapid Gender Needs Assessment of Flood Affected Communities*. UN Development Fund for Women, Geneva, Switzerland.

UNISDR, 2004: *Living With Risk: A Global Review of Disaster Reduction Initiatives*. United Nations International Strategy for Disaster Reduction, Geneva, Switzerland.

UNISDR, 2006: *Global Survey of Early Warning Systems: An Assessment of Capacities, Gaps and Opportunities Toward Building a Comprehensive Global Early Warning System for all Natural Hazards*. United Nations International Strategy for Disaster Reduction, Geneva, Switzerland, 56 pp.

UNISDR, 2009: *Global Assessment Report on Disaster Risk Reduction: Risk and Poverty in a Changing Climate*. United Nations International Strategy for Disaster Reduction, Geneva, Switzerland.

UNISDR and UNOCHA, 2008: *Disaster Preparedness for Effective Response, Guidance and Indicator Package for Implementing Priority Five of the Hyogo Framework*. United Nations Secretariat of the International Strategy for Disaster Reduction (UNISDR) and the United Nations Office for Coordination of Humanitarian Affairs (UNOCHA), Geneva, Switzerland, 51 pp.

UNOCHA and IDMC, 2009: *Monitoring Disaster Displacement in the Context of Climate Change*. Findings of a Study by the United Nations Office for the Coordination of Humanitarian Affairs and the Internal Displacement Monitoring Centre, Norwegian Refugee Council, Internal Displacement Monitoring Centre, Geneva, Switzerland, 32 pp.

Valdes, H.M., 1997: Community-operated early warning system for floods. In: *UN-IDNDR & QUIPUNET Internet conference on Floods, Drought: Issues for the 21st Century*. 22 September – 24 October 1997.

Valente, T.W. and R.L. Davis, 1999: Accelerating the diffusion of innovations using opinion leaders. *Annals of the American Academy of Political and Social Science*, **566**, 55-67.

Valente, T.W., B.R. Hoffman, A. Ritt-Olson, K. Lichtman, and C.A. Johnson, 2003: Effects of a social-network method for group assignment strategies on peer-led tobacco prevention programs in schools. *American Journal of Public Health*, **93**, 1837-1843.

van Aalst, M.K., 2006: The impacts of climate change on the risk of natural disasters. *Disasters*, **30(1)**, 5-18.

van Aalst, M., K. Maarten, T. Cannon, and I. Burton, 2008: Community level adaptation to climate change: The potential role of participatory community risk assessment. *Global Environmental Change: Human and Policy Dimensions*, **18(1)**, 165-179.

Van Willigen, M., T. Edwards, B. Edwards, and S. Hessee, 2002: Riding out the storm: Experiences of the physically disabled during Hurricanes Bonnie, Dennis, and Floyd. *Natural Hazards Review*, **3(3)**, 98-106.

Vellinga, P., E. Mills, G. Berz, L. Bouwer, S. Huq, L.A. Kozak, J. Palutikof, B. Schanzenbacher, and G. Soler, 2001: Insurance and other financial services. In: *Impacts, Adaptation and Vulnerability. Contribution of Working Group II to the Third Assessment Report of the Intergovernmental Panel on Climate Change* [McCarthy, J.J., O.F. Canziani, N.A. Leary, D.J. Dokken, and K.S. White (eds.)]. Cambridge University Press, Cambridge, UK, pp. 417-450.

Venton, C.C. and P. Venton, 2004: *Disaster Preparedness Programmes in India: A Cost Benefit Analysis*. HPN Network Paper No. 49, Overseas Development Institute, London, UK.

Vescovi, L., A. Bourque, G. Simonet, and A. Musy, 2009: Transfer of climate knowledge via a regional climate-change management body to support vulnerability, impact assessments and adaptation measures. *Climate Research*, **40(2-3)**, 163-173.

Vörösmarty, C., G. Green, J. Salisbury, and R. Lammers, 2000: Global water resources: Vulnerability from climate change and population growth. *Science*, **289(5477)**, 284-289.

Wachtendorf, T., J.M. Kendra, H. Rodriguez, and J. Trainor, 2006: The social impacts and consequences of the December 2004 Indian ocean tsunami: Observations from India and Sri Lanka. *Earthquake Spectra*, **22(3)**, 693-714.

Waddell, E., 1989: Observations on the 1972 frosts and subsequent relief programme among the Enga of the western highlands. *Mountain Research and Development*, **9(3)**, 210-223.

Walker, I.J. and R. Sydneysmith, 2008: British Columbia. In: *From Impacts to Adaptation: Canada in a Changing Climate 2007* [Lemmen, D.S., F.J. Warren, J. Lacroix, and E. Bush (eds.)]. Government of Canada, Ottawa, Canada, pp. 358-362.

Walker, P., 1989: *Famine Early Warning Systems: Victims and Destitution*. Earthscan Publications, London, UK.

Wamsler, C., 2007: Bridging the gaps: Stakeholder-based strategies for risk reduction and financing for the urban poor. *Environment and Urbanization*, **19(1)**, 115-142.

Ward, P.J., W. Beets, L.M. Bouwer, J.C.J.H. Aerts, and H. Renssen, 2010: Sensitivity of river discharge to ENSO. *Geophysical Research Letters*, **37(L12402)**, doi:10.1029/2010GL043215.

Ward, R.E.T., C. Herweijer, N. Patmore, and R. Muir-Wood, 2008: The role of insurers in promoting adaptation to the impacts of climate change. *The Geneva Papers*, **33**, 133-139.

Warner, K., M. Hamza, A. Oliver-Smith, F. Renaud, and A. Julca, 2010: Climate change, environmental degradation and migration. *Natural Hazards*, **55(3)**, 689-715.

Watts, M., 1983: Hazards and crises: A political economy of drought and famine in northern Nigeria. *Antipode*, **15(1)**, 24-34.

Waugh, W.L.J. and G. Streib, 2006: Collaboration and leadership for effective emergency management. *Public Administration Review*, **66**, 131-140.

Weems, C.F. and S. Overstreet, 2008: Child and adolescent mental health research in the context of Hurricane Katrina: A ecological needs-based perspective and introduction to the special section. *Journal of Clinical Child and Adolescent Psychology*, **37(3)**, 487-494.

Weibel, W.W., 1988: Combining ethnographic and epidemiologic methods in targeted AIDS interventions: The Chicago model. In: *Needle Sharing among Intravenous Drug Abusers: National and International Perspectives* [Battjes, R.J. and R.W. Pickens (eds.)]. NIDA Research Monograph 80, National Institute on Drug Abuse, Rockville, MD, pp. 137-150.

Weichselgartner, J. and R.E. Kasperson, 2010: Barriers in the science-policy-practice interface: Toward a knowledge-action-system in global environmental change research. *Global Environmental Change*, **20**, 266-277.

Weisler, R.H., J.G. Barbee IV, and M.H. Townsend, 2006: Mental health and recovery in the Gulf Coast after Hurricanes Katrina and Rita. *Journal of the American Medical Association*, **296**, 585-588.

Weissbecker, I., S.E. Sephton, M.B. Martin, and D.M. Simpson, 2008: Psychological and physiological correlates of stress in children exposed to disaster: Current research and recommendations for intervention. *Children, Youth and Environments*, **18(1)**, 30-70.

Weitzman, M.L., 2007: A review of the Stern Review on the economics of climate change. *Journal of Economic Literature*, **45**, 703-724.

Wellman, B. and S.D. Berkowitz, 1988: *Social Structures: A Network Approach*. Cambridge University Press, Cambridge, UK.

Wenzel, F., F. Bendimerad, and R. Sinha, 2007: Megacities–megarisks. *Natural Hazards*, **42(3)**, 481-491.

West, C.T. and D.G. Lenze, 1994: Modeling the regional impact of natural disasters and recovery: A general framework and an application to Hurricane Andrew. *International Regional Science Review*, **17**, 121-150.

White, G.F., R.W. Kates, and I. Burton, 2001: Knowing better and losing even more: The use of knowledge in hazards management. *Global Environmental Change Part B: Environmental Hazards*, **3**, 81-92.

Whitehead, J.C., B. Edwards, M. Van Willigen, J.R. Maiolo, K. Wilson, and K.T. Smith, 2000: Heading for higher ground: Factors affecting real and hypothetical hurricane evacuation behavior. *Environmental Hazards*, **2(4)**, 133-142.

Wiest, R.E., J.S.P. Mocellin, and D.T. Motsisi, 1994: *The Needs of Women in Disasters and Emergencies*. Prepared for the Disaster Management Training Programme of the United Nations Development Programme and the Office of the United Nations Disaster Relief Coordinator, Disaster Research Unit, University of Manitoba, Winnipeg, Canada.

Wilby, R.L., J. Troni, Y. Biot, L. Tedd, B.C. Hewitson, D.C. Smith, and R.T. Sutton, 2009: A review of climate risk information for adaptation and development planning. *International Journal of Climatology*, **29(9)**, 1193-1215.

Wilhelmi, O.V. and D.A. Wiilhite, 2002: Assessing vulnerability to agricultural drought: A Nebraska case study. *Natural Hazards*, **25(1)**, 37-58.

Williams, R., D.A. Alexander, D. Bolsover, and F.K. Bakke, 2008: Children, resilience and disasters: Recent evidence that should influence a model of psychosocial care. *Current Opinion in Psychiatry*, **21(4)**, 338-344.

Winsvold, M., K.B. Stokke, J.E. Klausen, and I. Saglie, 2009: Organizational learning and governance in adaptation in urban development. In: *Adapting to Climate Change: Thresholds, Values and Governance* [Adger, W.N., I. Lorenzoni, and K. O'Brien (eds.)]. Cambridge University Press, Cambridge, UK, pp. 476-490.

Wisner, B., 2002: Disability and disaster: Victimhood and agency in earthquake risk reduction. In: *Earthquakes..* [Rodrigue, C. and E. Rovai (eds.)]. Routledge, London, UK.

Wisner, B., 2003: Disaster risk reduction in megacities: Making the most of human and social capital. In: *Building Safer Cities: The Future of Disaster Risk* [Kreimer, A., M. Arnold, and A. Carlin (eds.)]. World Bank Books, Washington, DC, pp. 181-196.

Wisner, B. and J. Uitto, 2009: Life on the edge: Urban social vulnerability and decentralized, citizen-based disaster risk reduction in four large cities of the Pacific Rim. *Facing Global Environmental Change: Hexagon Series on Human and Environmental Security and Peace*, **4(II)**, 215-231.

Wisner, B., P. Blaike, T. Cannon, and I. Davis, 2004: *At Risk: Natural Hazards, People's Vulnerability and Disasters*. Routledge, London, UK.

WMO, 2003: *Integrated Flood Plain Management Case Study: Bangladesh Flood Management*. World Meteorological Organization, Geneva, Switzerland.

World Bank, 2010: *Natural Hazards, Unnatural Disasters, the Economics of Effective Prevention*. World Bank, Washington, DC.

Wu, S., B. Yarnal, and A. Fisher, 2002: Vulnerability of coastal communities to sea-level rise: A case study of Cape May county, New Jersey, USA. *Climate Research*, **22**, 255-270.

Yang, S.L., J.D. Milliman, P. Li, and K. Xu, 2011: 50,000 dams later: Erosion of the Yangtze River and its delta. *Global and Planetary Change*, **75(1)**, 14-20.

Yohe, G., J. Neumann, and H. Ameden, 1995: Assessing the economic cost of greenhouse induced sea level rise: Methods and applications in support of a national survey. *Journal of Environmental Economics and Management*, **29**, 78-97.

Zahran, S., S.D. Brody, W.G. Peacock, A. Vedlitz, and H. Gover, 2008: Social vulnerability and the natural and built environment: A model of flood casualties in Texas. *Disasters*, **32(4)**, 537-560.

Ziervogel, G., 2004: Targeting seasonal climate forecasts for integration into household level decisions: The case of smallholder farmers in Lesotho. *The Geographical Journal*, **170**, 6-21.

Ziervogel, G. and P. Ericksen, 2010: Adapting to climate change to sustain food security. *Wiley Interdisciplinary Reviews: Climate Change*, **1(4)**, doi:10.1002/wcc.56.

Ziervogel, G. and F. Zermoglio, 2009: Climate-change scenarios and the development of adaptation strategies in Africa: Challenges and opportunities. *Climate Research*, **40(2-3)**, 133-146.

Ziervogel, G., A.O. Nyong, B. Osman-Elasha, C. Conde, S. Cortés, and T. Downing, 2006: *Climate Variability and Change: Implications for Household Food Security*. AIACC Working Paper No. 20, Assessments of Impacts and Adaptations to Climate Change, Washington, DC.

Ziervogel, G., M. Shale, and M. Du, 2010: Climate change adaptation in a developing country context: The case of urban water supply in Cape Town. *Climate and Development*, **2**, 94-110.

Zimmermann, M. and F. Stössel, 2011: *Disaster Risk Reduction in International Cooperation: Switzerland's Contribution to the Protection of Lives and Livelihoods*. Swiss Agency for Development and Cooperation, Berne, Switzerland, 23 pp.

National Systems for Managing the Risks from Climate Extremes and Disasters

6

Coordinating Lead Authors:
Padma Narsey Lal (Australia), Tom Mitchell (UK)

Lead Authors:
Paulina Aldunce (Chile), Heather Auld (Canada), Reinhard Mechler (Germany), Alimullah Miyan (Bangladesh), Luis Ernesto Romano (El Salvador), Salmah Zakaria (Malaysia)

Review Editors:
Andrew Dlugolecki (UK), Takuo Masumoto (Japan)

Contributing Authors:
Neville Ash (Switzerland), Stefan Hochrainer (Austria), Robert Hodgson (UK), Tarik ul Islam (Canada), Sabrina McCormick (USA), Carolina Neri (Mexico), Roger Pulwarty (USA), Ataur Rahman (Bangladesh), Ben Ramalingam (UK), Karen Sudmeier-Reiux (France), Emma Tompkins (UK), John Twigg (UK), Robert Wilby (UK)

This chapter should be cited as:

Lal, P.N., T. Mitchell, P. Aldunce, H. Auld, R. Mechler, A. Miyan, L.E. Romano, and S. Zakaria, 2012: National systems for managing the risks from climate extremes and disasters. In: *Managing the Risks of Extreme Events and Disasters to Advance Climate Change Adaptation* [Field, C.B., V. Barros, T.F. Stocker, D. Qin, D.J. Dokken, K.L. Ebi, M.D. Mastrandrea, K.J. Mach, G.-K. Plattner, S.K. Allen, M. Tignor, and P.M. Midgley (eds.)]. A Special Report of Working Groups I and II of the Intergovernmental Panel on Climate Change (IPCC). Cambridge University Press, Cambridge, UK, and New York, NY, USA, pp. 339-392.

Table of Contents

Executive Summary 341

6.1. Introduction 344

6.2. National Systems and Actors for Managing the Risks from Climate Extremes and Disasters 345
- 6.2.1. National and Sub-National Governments 346
- 6.2.2. Private Sector Organizations 347
- 6.2.3. Civil Society and Community-Based Organizations 348
- 6.2.4. Bilateral and Multilateral Agencies 348
- 6.2.5. Research and Communication 349

6.3. Planning and Policies for Integrated Risk Management, Adaptation, and Development Approaches 349
- 6.3.1. Developing and Supporting National Planning and Policy Processes 349
- 6.3.2. Mainstreaming Disaster Risk Management and Climate Change Adaptation into Sectors and Organizations 355
- 6.3.3. Sector-Based Risk Management and Adaptation 357

6.4. Strategies including Legislation, Institutions, and Finance 357
- 6.4.1. Legislation and Compliance Mechanisms 358
- 6.4.2. Coordinating Mechanisms and Linking across Scales 358
- 6.4.3. Finance and Budget Allocation 360

6.5. Practices including Methods and Tools 362
- 6.5.1. Building a Culture of Safety 362
  - 6.5.1.1. Assessing Risks and Maintaining Information Systems 363
  - 6.5.1.2. Preparedness: Risk Awareness, Education, and Early Warning Systems 364
- 6.5.2. Reducing Climate-Related Disaster Risk 366
  - 6.5.2.1. Applying Technological and Infrastructure-Based Approaches 366
  - 6.5.2.2. Human Development and Vulnerability Reduction 368
  - 6.5.2.3. Investing in Natural Capital and Ecosystem-Based Adaptation 370
- 6.5.3. Transferring and Sharing 'Residual' Risks 371
- 6.5.4. Managing the Impacts 373

6.6. Aligning National Disaster Risk Management Systems with the Challenges of Climate Change 375
- 6.6.1. Assessing the Effectiveness of Disaster Risk Management in a Changing Climate 375
- 6.6.2. Managing Uncertainties and Adaptive Management in National Systems 377
- 6.6.3. Tackling the Underlying Drivers of Vulnerability 379
- 6.6.4. Approaching Disaster Risk, Adaptation, and Development Holistically 379

References 381

Executive Summary

This chapter assesses how countries are managing current and projected disaster risks, given knowledge of how risks are changing with observations and projections of weather and climate extremes [Table 3-2, 3.3], vulnerability and exposure [4.3], and impacts [4.4]. It focuses on the design of national systems for managing such risks, the roles played by actors involved in the system, and the functions they perform, acknowledging that complementary actions to manage risks are also taken at local and international level as described in Chapters 5 and 7.

National systems are at the core of countries' capacity to meet the challenges of observed and projected trends in exposure, vulnerability, and weather and climate extremes *(high agreement, robust evidence)*. Effective national systems comprise multiple actors from national and sub-national governments, private sector, research bodies, and civil society, including community-based organizations, playing differential but complementary roles to manage risk according to their accepted functions and capacities. These actors work in partnership across temporal, spatial, administrative, and social scales, supported by relevant scientific and traditional knowledge. Specific characteristics of national systems vary between countries and across scales depending on their socio-cultural, political, and administrative environments and development status. [6.2]

The national level plays a key role in governing and managing disaster risks because national government is central to providing risk management-related public goods as it commonly maintains financial and organizational authority in planning and implementing these goods *(high agreement, robust evidence)*. National governments are charged with the provision of public goods such as ensuring the economic and social well-being, safety, and security of their citizens from disasters, including the protection of the poorest and most vulnerable citizens. They also control budgetary allocations as well as creating legislative frameworks to guide actions by other actors. Often, national governments are considered to be the 'insurer of last resort'. In line with the delivery of public goods, national governments and public authorities 'own' a large part of current and future disaster risks (public infrastructure, public assets, and relief spending). In terms of managing risk, national governments act as risk aggregators and by pooling risk, hold a large portfolio of public liabilities. This provides governments responsibility to accurately quantify and manage risks associated with this portfolio – functions that are expected to become more important given projected impacts of climate change and trends in vulnerability and exposure. [6.2.1]

In providing such public goods, governments choose to manage disaster risk by enabling national systems to guide and support stakeholders to reduce risk where possible, transfer risk where feasible, and manage residual risk, recognizing that risks can never be totally eliminated *(high agreement, robust evidence)*. The balance between reducing risk and other disaster risk management strategies is influenced by a range of factors, including financial and technical capacity of stakeholders, robustness of risk assessment information, and cultural elements involving risk tolerance. [6.2.1, 6.2-6.5]

The ability of governments to implement disaster risk management responsibilities differs significantly across countries, depending on their capacity and resource constraints *(high agreement, robust evidence)*. Smaller or economically less-diversified countries face particular challenges in providing the public goods associated with disaster risk management, in absorbing the losses caused by climate extremes and disasters, and in providing relief and reconstruction assistance. [6.4.3] However, there is *limited evidence* to suggest any correlation between the type of governance system in a country (e.g., centralized or decentralized; unitary or federal) and the effectiveness of disaster risk management efforts. There is *robust evidence* and *high agreement* to suggest that actions generated within and managed by communities with supporting government policies are generally most effective since they are specific and tailored to local environments. [6.4.2]

In the majority of countries, national systems have been strengthened by applying the principles of the Hyogo Framework for Action to mainstream risk considerations across society and sectors, although greater efforts are required to address the underlying drivers of risk and generate the political will to invest in disaster risk reduction *(high agreement, robust evidence)*. The Hyogo Framework for Action has encouraged countries to develop and implement a systematic disaster risk management approach, and in some cases has led to strategic shifts in the management of disaster risks, with governments and other actors placing greater attention on disaster risk reduction compared to more reactive measures. This has included improvements in coordination between actors, enhanced early warning and preparedness, more rigorous risk assessments, and increased awareness. However, there is *limited evidence* and *low agreement* to suggest improvements in integration between efforts to implement the Hyogo Framework for Action, the United Nations Framework Convention on Climate Change, and broader development and environmental policy frameworks. [6.4.2]

A set of factors can be identified that make efforts to systematically manage current disaster risks more successful (all *high agreement, robust evidence*). Systems to manage current disaster risk are more successful if:

- Risks are recognized as dynamic and are mainstreamed and integrated into development policies, strategies, and actions, and into environmental management. [6.3.1]
- Legislation for managing disaster risks is supported by clear regulations that are effectively enforced across scales and complemented by other sectoral development and management legislations where risk considerations are explicitly integrated. [6.4.1]
- Disaster risk management functions are coordinated across sectors and scales and led by organizations at the highest political level. [6.4.2]
- They include considerations of disaster risk in national development and sector plans, and, if they adopt climate change adaptation strategies, translating these plans and strategies into actions targeting vulnerable areas and groups. [6.5.2]
- Risk is quantified and factored into national budgetary processes, and a range of measures including budgeting for relief expenditure, reserve funds, and other forms of risk financing have been considered or implemented. [6.4.3]
- Decisions are informed by comprehensive information about observed changes in weather, climate, and vulnerability and exposure, and historic disaster losses, using a diversity of readily available tools and guidelines. [6.5.1]
- Early warning systems deliver timely, relevant, and accurate predictions of hazards, and are developed and made operational in partnership with the public and trigger effective response actions. [6.5.1]
- Strategies include a combination of hard infrastructure-based options responses and soft solutions such as individual and institutional capacity building and ecosystem-based responses, including conservation measures associated with, for example, forestry, river catchments, coastal wetlands, and biodiversity. [6.5.2]

While there is robust evidence and high agreement on efforts to tackle current disaster risks, the assessment found *limited evidence* **of national disaster risk management systems and associated risk management measures explicitly integrating knowledge of and uncertainties in projected changes in exposure, vulnerability, and climate extremes.** The effectiveness of efforts to manage projected disaster risks at the national level are dependent on a range of factors, including the effectiveness of the system for managing current risks, the ability of the system to flexibly respond to new knowledge, the availability of suitable data, and the resources available to invest in longer-term risk reduction and adaptation measures. Developed countries are better equipped financially and institutionally to adopt explicit measures to effectively respond and adapt to projected changes in exposure, vulnerability, and climate extremes than developing countries. Nonetheless, all countries face challenges in assessing, understanding, and then responding to such projected changes. [6.3.2, 6.6]

Measures that provide benefits under current climate and a range of future climate change scenarios, called low-regrets measures, are available starting points for addressing projected trends in exposure, vulnerability, and climate extremes. They have the potential to offer benefits now and lay the foundation for addressing projected changes (*high agreement, medium evidence***).** The assessment considered such 'low' regrets options across a range of key sectors, with some of the most commonly cited measures associated with improvements to early warning systems, health surveillance, water supply, sanitation, and drainage systems; climate proofing of major infrastructure and enforcement of building codes; better education and awareness; and restoration of degraded ecosystems and nature conservation. Many of these low-regrets strategies produce co-benefits; help address other development goals, such as improvements in livelihoods, human well-being, and biodiversity conservation; and help minimize the scope for maladaptation. [6.3.1, Table 6-1]

Ecosystem-based solutions in the context of changing climate risks can offer 'triple-win' solutions, as they can provide cost-effective risk reduction, support biodiversity conservation, and enable improvements in economic livelihoods and human well-being, particularly for the poor and vulnerable *(high agreement, robust evidence)*. The assessment found that such ecosystem-based adaptation strategies, including mangrove conservation and rehabilitation, integrated catchment management, and sustainable forest and fisheries management, also minimize the scope for maladaptation in developed and developing countries. In choosing amongst ecosystem-based adaptation options, decisionmakers may need to make tradeoffs between particular climate risk reduction strategies and other valued ecosystem services. [6.5.2]

Insurance-related instruments are key mechanisms for helping households, business, and governments absorb the losses from disasters; but their uptake is unequally distributed across regions and hazards, and often public-private partnerships are required *(high agreement, robust evidence)*. Disaster insurance and other risk transfer instruments covered about 20% of reported weather-related losses over the period 1980 to 2003. Distribution, though, is uneven, with about 40% of the losses insured in high-income as compared to 4% of losses in low-income countries. Existing national insurance systems differ widely as to whether policies are compulsory or voluntary, and importantly in how systems allocate liability and responsibility for disaster risks across society. With changing weather and extreme events, vulnerability, and exposure, extended and innovative private-public sector partnerships are required to better estimate and price risk as well as to develop robust insurance-related products, which may be supported in developing countries by development partner funds. [6.5.3]

Pooling of risk by and between national governments contributes to reducing the fiscal and socioeconomic consequences of disasters *(medium agreement, medium evidence)*. As national governments hold a large portfolio of public liabilities (infrastructure, public assets, and the provision of disaster relief), risk aggregation and pooling are expected to become more important given projected impacts of climate change and trends in vulnerability and exposure. In addition, particularly for small, low-income, and highly exposed countries, risk transfer of public sector assets and relief expenditure recently have become a cornerstone of disaster risk reduction. Key innovative and promising applications recently implemented comprise sovereign insurance for hurricane risk, insurance for humanitarian assistance following droughts, and intergovernmental risk pooling. [6.4.3, 6.5.3]

Flexible and adaptive national systems are better suited to manage projected trends and associated uncertainties in exposure, vulnerability, and weather and climate extremes than static and rigid national systems *(high agreement, limited evidence)*. Adaptive management brings together different scientific, social, and economic information, experiences, and traditional knowledge into decisionmaking through 'learning by doing.' Multi-criteria analysis, scenario planning, and flexible decision paths offer options for taking action when faced with large uncertainties or incomplete information. National systems for managing disaster risk can adapt to climate change and shifting exposure and vulnerability by (i) frequently assessing and mainstreaming knowledge of dynamic risks; (ii) adopting 'low regrets' strategies; (iii) improving learning and feedback across disaster, climate, and development organizations at all scales; (iv) addressing the root causes of poverty and vulnerability; (v) screening investments for climate change-related impacts and risks to minimize scope for maladaptation; and (vi) increasing standing capacity for emergency response as climatic conditions change over time. [6.6.1, 6.6.2, 6.6.4]

6.1. Introduction

The socioeconomic impacts of disaster events can be significant in all countries, but low- and middle-income countries are especially vulnerable, and experience higher fatalities even when exposed to hazards of similar magnitude (O'Brien et al., 2006; Thomalla et al., 2006; Ibarraran et al., 2009; IFRC, 2010). The number of deaths per cyclone event in the last several decades, for example, was highest in low-income countries even though a higher proportion of population exposed to cyclones lives in countries with higher income; 11% of the people exposed to hazards live in low human development countries, but they account for more than 53% of the total recorded deaths resulting from disasters (UNDP, 2004a). At the same time, while in absolute terms the direct economic losses from disasters are far greater in high-income countries, middle- and low-income states bear the heaviest burden of these costs in terms of damage relative to annual gross domestic product (GDP: UNDP, 2004a; DFID, 2005; O'Brien et al., 2006; Kellenberg et al., 2008; Pelham et al., 2011). This burden has been increasing in the middle-income countries, where the asset base is rapidly expanding and losses over the period from 2001 to 2006 amounted to about 1% of GDP. For the low-income group, losses totaled an average of 0.3% and for the high-income countries amounted to less than 0.1% of GDP (Cummins and Mahul, 2009). In some particularly exposed countries, including many small island developing states, these wealth losses expressed as a percentage of GDP can be considerably higher, with the average costs over disaster and non-disaster years close to 10%, such as reported for Grenada and St. Lucia (World Bank and UN, 2010). In extreme cases, the costs of individual events can be as high as 200% of the annual GDP as experienced in the Polynesian island nation of Niue following cyclone Heta in 2004, or in the Hurricane Ivan event affecting Grenada in 2004 (McKenzie et al., 2005).

In terms of the macroeconomic and developmental consequences of high exposure to disaster risk, a growing body of literature has shown significant adverse effects in developing countries (Otero and Marti, 1995; Charveriat, 2000; Crowards, 2000; Murlidharan and Shah, 2001; ECLAC, 2002, 2003; Mechler, 2004; Hochrainer, 2006; Noy, 2009). These include reduced direct and indirect tax revenue, dampened investment, and reduced long-term economic growth through their negative effect on a country's credit rating and an increase in interest rates for external borrowing. Among the reasons behind limited coping capacity of individuals, communities, and governments are reduced tax bases and high levels of indebtedness, combined with limited household income and savings, a lack of disaster risk transfer and other financing instruments, few capital assets, and limited social insurance.

This body of evidence emphasizes that disasters can cause a setback for development, and even a reversal of recent development gains in the short- to medium-term, emphasizing the point that disaster risk management is a development issue as much as a humanitarian one. Poor development status of communities and countries increases their sensitivity to disasters. Disaster impacts can also force households to fall below the basic needs poverty line, further increasing their vulnerability to other shocks (Owens et al., 2003; Lal, 2010). Consequently, disasters are seen as barriers for development, requiring ex-ante disaster risk reduction policies that also target poverty and development (del Ninno et al., 2003; Owens et al., 2003; Skoufias, 2003; Benson and Clay, 2004; Hallegatte et al., 2007; Raddatz, 2007; Cardona et al., 2010; IFRC, 2010). However, some literature suggests that disasters may not always have a negative effect on economic growth and development and for some countries disasters may be regarded as a problem *of*, and not *for* development (Albala-Bertrand, 1993; Skidmore and Toya, 2002; Caselli and Malthotra, 2004; Hallegatte and Ghil, 2007). Disasters have also been considered to increase economic growth in the short term as well as spur positive economic growth and technological renewal in the longer term, depending on the domestic capacity of nations to rebuild and the inflow of international assistance (Skidmore and Toya, 2002). This observation may be partially attributable to national accounting practices, which positively record reconstruction efforts but do not account for the immediate destruction of assets and wealth in some cases (Skidmore and Toya, 2002).

To better respond to the impacts of disasters on human livelihoods, environment, and economies, national disaster risk management systems have evolved in recent years, guided in some cases by international instruments, particularly the Hyogo Framework for Action (HFA) 2005-2015 and more recently as part of the adaptation agenda under the United Nations Framework Convention on Climate Change (UNFCCC; see Section 7.3). Increasing knowledge, understanding, and experiences in dealing with disaster risks have gradually contributed to a paradigm shift globally that recognizes the importance of reducing risks by addressing underlying drivers of vulnerability and exposure, such as targeting poverty, improving human well-being, better environmental management, and adaptation to climate change as well as responding to and rebuilding after disaster events (Yodmani, 2001; IFRC, 2004, 2010; Thomalla et al., 2006; UNISDR, 2008a; Venton and LaTrobe, 2008; Pelham et al., 2011). While governments cannot act alone, the majority are well placed and equipped to support communities and the private sector to address disaster risks. Yet recent reported experiences suggest that countries vary considerably in their responses, and concerns remain about the lack of integration of disaster risk management into sustainable development policies and planning as well as insufficient implementation at different levels (CCCD, 2009; UNFCCC, 2008b).

It is at the national level that overarching development policies and legislative frameworks are formulated and implemented to create appropriate enabling environments to guide other stakeholders to reduce, share, and transfer risks, albeit in different ways (Carter, 1992; Freeman et al., 2003). National-level governments in developed countries are often the de facto 'insurers of last resort' and used to be considered the most effective insurance instruments of society (Priest, 1996). Governments also have the ability to mainstream risks associated with climate variability and change into existing disaster risk management and sectoral development, policies, and plans, albeit to differing degrees depending on their capacity. These include initiatives to assess risks and uncertainties, manage these across sectors, share and transfer risks, and establish baseline information and research priorities (Freeman et al., 2003; Mechler, 2004; Prabhakar et al., 2009). Ideally, national-level institutions are best able to respond to the challenges of climate extremes, particularly given that when disasters occur they often surpass people and businesses' coping capacity (OAS, 1991; Otero and Marti, 1995; Benson and Clay, 2002a,b). National governments are also better placed to appreciate key uncertainties and risks and take strategic actions, particularly based on their power of taxation (see Sections 6.4.3 and 6.5.3), although particularly exposed developing countries may be financially challenged to attend to the risks and liabilities imposed by natural disasters (Mechler, 2004; Cummins and Mahul, 2009; UNISDR, 2011a).

Changes in weather and climate extremes and related impacts pose new challenges for national disaster risk management systems, which in many instances remain poorly adapted to the risks posed by existing climatic variability and extremes (Lavell, 1998; McGray et al., 2007; Venton and La Trobe, 2008; Mitchell et al., 2010b). Nonetheless, valuable lessons for advancing adaptation to climate change can be drawn from existing national disaster risk management systems (McGray et al., 2007; Mitchell et al., 2010b). Such national systems are comprised of actors operating across scales, fulfilling a range of roles and functions, guided by an enabling environment of institutions, international agreements, and experience of previous disasters (Carter, 1992; Freeman et al., 2003). These systems vary considerably between countries in terms of their capacities and effectiveness and in the way responsibilities are distributed between actors. Countries also put differential emphasis on integration of disaster risk management with development processes and tackling vulnerability and exposure, compared with preparing for and responding to extreme events and disasters (Cardona et al., 2010).

Recent global assessments of disaster risk management point to a general lack of integration of disaster risk management into sustainable development policies and planning across countries and regions, although progress has been made especially in terms of passing legislation, in setting up early warning systems, and in strengthening disaster preparedness and response (Amendola et al., 2008; UNISDR, 2011b; Wisner, 2011). Closing the gap between current provision and what is needed for tackling even current climate variability and disaster risk is a priority for national risk management systems and is also a crucial aspect of countries' responses to projected climate change. With a history of managing climatic extremes, involving a large number of experienced actors across scales and levels of government and widespread instances of supporting legislation and cross-sectoral coordinating bodies (Section 6.4.2), national disaster risk management systems offer a promising avenue for supporting adaptation to climate change and reducing projected climate-related disaster risks.

Accordingly, this chapter assesses the literature on national systems for managing disaster risks and climate extremes, particularly the design of such systems of functions, actors, and roles they play, emphasizing the importance of government and governance for improved adaptation to climate extremes and variability. Focusing particularly on developing country challenges, the assessment reflects on the adequacy of existing knowledge, policies, and practices globally and considers the extent to which the current disaster risk management systems may need to evolve to deal with the uncertainties associated with and the effects of climate change on disaster risks. Section 6.2 characterizes national systems for managing existing climate extremes and disaster risk by focusing on the actors that help create the system – national and sub-national government agencies, bilateral and multilateral organizations, the private sector, research agencies, civil society, and community-based organizations. Drawing on a range of examples from developed and developing countries, Sections 6.3 through 6.5 describe what is known about the status of managing current and future risk, what is desirable in an effective national system for adapting to climate change, and what gaps in knowledge exist. The latter parts of the chapter are organized by the set of functions undertaken by the actors discussed in Section 6.2. The functions are divided into three main categories – those associated with planning and policies (Section 6.3), strategies (Section 6.4), and practices, including methods and tools (Section 6.5), for reducing climatic risks. Section 6.6 reflects on how national systems for managing climate extremes and disaster risk can become more closely aligned to the challenges posed by climate change and development – particularly those associated with uncertainty, changing patterns of risk and exposure, and the impacts of climate change on vulnerability and poverty. Aspects of Section 6.6 are further elaborated in Chapter 8.

6.2. National Systems and Actors for Managing the Risks from Climate Extremes and Disasters

Managing climate-related disaster risks is a concern of multiple actors, working across scales from international, national, and sub-national and community levels, and often in partnership, to ultimately help individuals, households, communities, and societies to reduce their risks (Twigg, 2004; Schipper, 2009; Wisner, 2011). Comprising national and sub-national governments, the private sector, research bodies, civil society, and community-based organizations and communities, effective national systems would ideally have each actor performing to their accepted functions and capacities. Each actor would play differential but complementary roles across spatial and temporal scales (UNISDR, 2008a; Schipper, 2009; Miller et al., 2010) and would draw on a mixture of scientific and local knowledge to shape their actions and their appreciation of the dynamic nature of risk (see Figure 6-1). Given that national systems are at the core of a country's capacity to meet the challenges of observed and projected trends in exposure, vulnerability, and weather and climate extremes, this section assesses the literature on the roles played by different actors working within such national systems.

Figure 6-1 encapsulates the discussions to follow on the interface and interaction between different levels of actors, roles, and functions, with the centrality of national organizations and institutions engaging at the international level and creating enabling environments to support

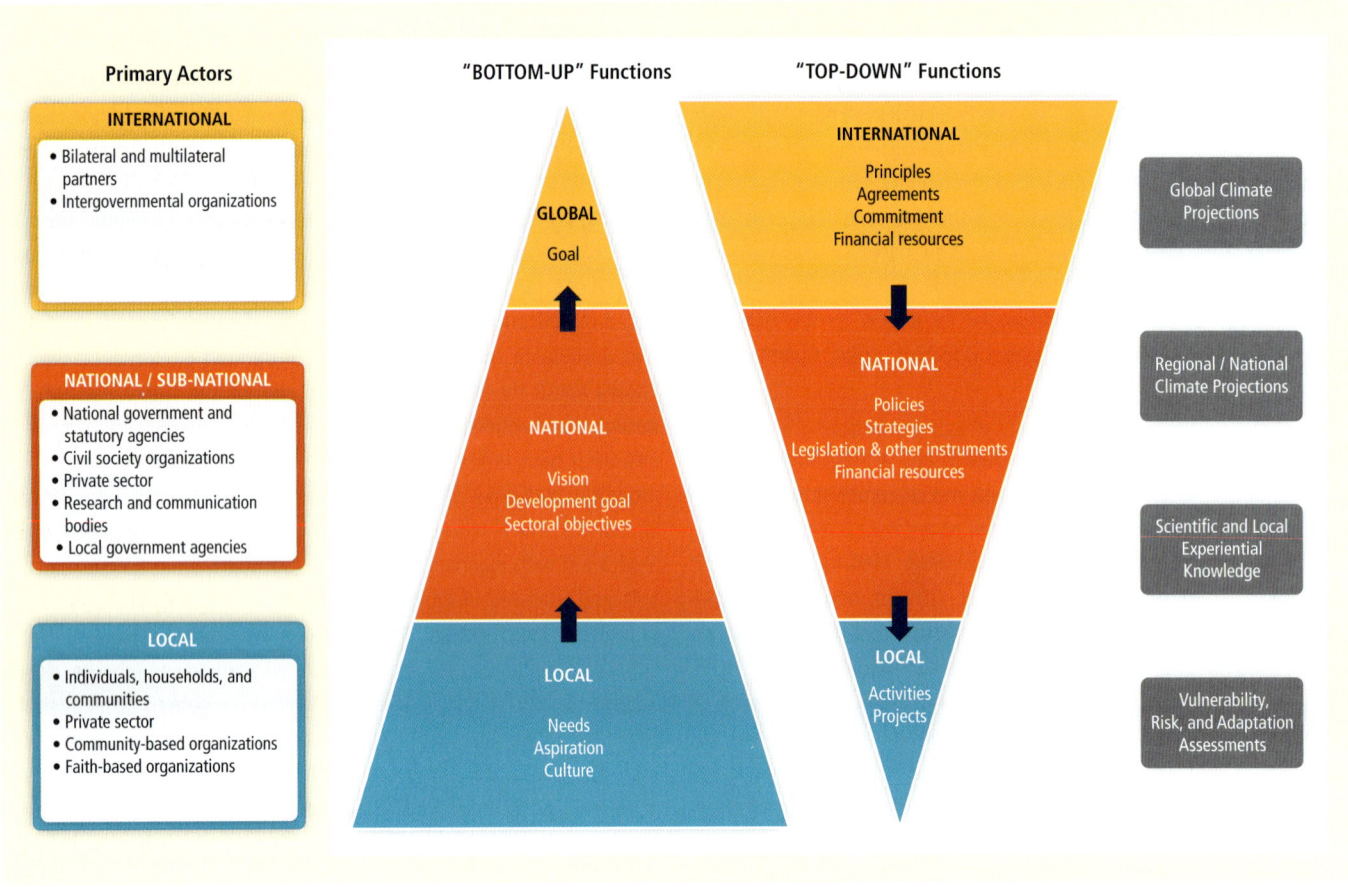

Figure 6-1 | National system of actors and functions for managing disaster risk and adapting to climate change.

actions across the country, supported by scientific information and traditional knowledge.

6.2.1. National and Sub-National Governments

The national level plays a key role in governing and managing disaster risks because national governments are central to providing risk management-related public goods as they maintain organizational and financial authority in planning and providing such goods. National governments have the moral and legal responsibility to ensure economic and social well-being, including safety and security of their citizens from disasters (UNISDR, 2004). It is also argued that it is government's responsibility to protect the poorest and most vulnerable citizens from disasters, and to implement disaster risk management that reaches all (McBean, 2008; O'Brien et al., 2008; CCCD, 2009). In terms of risk ownership, government and public disaster authorities 'own' a large part of current and future extreme event risks and are expected to govern and regulate risks borne by other parts of society (Mechler, 2004). Various normative literature sources support this. As one example, literature on economic welfare theory suggests that national governments are exposed to natural disaster risk and potential losses due to their three main functions: provision of public goods and services (e.g., education, clean environment, and security); the redistribution of income; and stabilizing the economy (Musgrave, 1959; Twigg, 2004; White et al., 2004; McBean, 2008; Shaw et al., 2009). The risks faced by governments include losing public infrastructure, assets, and national reserves. National-level governments also redistribute income across members of society and thus are called upon when those are in need (Linnerooth-Bayer and Amendola, 2000), or when members of society are in danger of becoming poor, and in need of relief payments to sustain a basic standard of living, especially in countries with low per capita income and/or that have large proportions of the population in poverty (Cummins and Mahul, 2009). Finally, it can be argued that governments are expected to stabilize the economy, for example, by supply-side interventions when the economy is in disequilibrium.

National-level governments are often called 'insurers of last resort' as the governments are often the final entity that private households and firms turn to in case of need, although the degree of compliance and ability to honor those responsibilities by governments differs significantly across countries. Nonetheless, in the context of a changing climate, it is argued that governments have a particularly critical role to play in relation to not only addressing the current gaps in disaster risk management but also in response to uncertainties and changing needs due to increases in the frequency, magnitude, and duration of some climate extremes (Katz and Brown, 1992; Meehl et al., 2000; Christensen et al., 2007; also refer to Chapter 3).

Different levels of governments – national, sub-national, and local level – as well as respective sectoral agencies play multiple roles in addressing drivers of vulnerability and managing the risk of extreme events, although their effectiveness varies within a country as well as across countries. They are well placed to create multi-sectoral platforms to guide, build, and develop policy, regulatory, and institutional frameworks that prioritize risk management (Handmer and Dovers, 2007; UNISDR, 2008b; OECD, 2009); integrate disaster risk management with other policy domains like development or environmental management, which often are separated in different ministries (UNISDR 2004, 2009c; White et al., 2004; Tompkins et al., 2008); and address drivers of vulnerability and assist the most vulnerable populations (McBean, 2008; CCCD, 2009). Governments across sectors and levels also provide many public goods and services that help address drivers of vulnerability as well as those that support disaster risk management (White et al., 2004; Shaw et al., 2009) through education, training, and research (Twigg, 2004; McBean, 2008; Shaw et al., 2009).

Governments also allocate financial and administrative resources for disaster risk management, as well as provide political authority (Spence, 2004; Twigg, 2004; Handmer and Dovers, 2007; CCCD, 2009). Evidence suggests that successful disaster risk management is partly contingent on resources being made available at all administration levels, but to date, insufficient policy and institutional commitments have been made to disaster risk management in many countries, particularly at the local government level (Twigg, 2004; UNISDR, 2009d). It is argued that governments also have an important role to guide and support the private sector, civil society organizations, and other development partners in playing their differential roles in managing disaster risk (O'Brien et al., 2008; Prabhakar et al., 2009).

6.2.2. Private Sector Organizations

The private sector plays a small, but increasingly important role in disaster risk management and adaptation, and some aspects of disaster risk management may be suitable for nongovernmental stakeholders to implement, albeit this would often effectively be coordinated within a framework created and enabled by governments. Three avenues for private sector engagement may be identified: (1) corporate social responsibility (CSR); (2) public-private partnerships (PPP); and (3) businesses model approaches. CSR involves voluntary advocacy and raising awareness by businesses for disaster risk reduction as well as involving funding support and the contribution of volunteers and expertise to implement risk management measures. PPPs focus on enhancing the provision of public goods for disaster risk reduction in joint undertakings between public and private sector players. The business model approach pursues the integration and alignment of disaster risk reduction with operational and strategic goals of an enterprise (Warhurst, 2006; Roeth, 2009). While CSR and PPP have received substantial attention, business model approaches remain rather untouched areas, one very important exception being the insurance industry as a supplier of tools for transferring and sharing disaster risks and losses.

In terms of business model approaches, insurance is a key sector. In exchange for pre-disaster premium payments, disaster insurance and other risk transfer instruments in 2010 covered about 30% of disaster losses overall (Munich Re, 2011). In terms of weather-related events, for the period 1980 to 2003, insurance overall covered about 20% of the losses, yet the distribution according to country income groups is uneven, with about 40% of the losses insured in high-income as compared to 4% in low-income countries (Mills, 2007). In developing countries, despite complexities and uncertainties involved in both supply and demand for risk transfer, risk financing mechanisms have been found to demonstrate substantial potential for absorbing the financial burden of disasters (Pollner, 2000; Andersen, 2001; Varangis et al., 2002; Auffret, 2003; Dercon, 2005; Hess and Syroka, 2005; Linnerooth-Bayer et al., 2005; Skees et al., 2005; World Bank, 2007; Cummins and Mahul, 2009; Hazell and Hess, 2010). There is, though, some uncertainty as to the extent to which the private sector would continue to play this role in the context of a changing environment due to uncertainty and imperfect information, missing and misaligned markets, and financial constraints (Smit et al., 2001; Aakre et al., 2010). Private insurers are concerned about changes in risks and associated risk ambiguity, that is, the uncertainty about the changes induced by climate change in terms of potentially modified extreme event intensity and frequency. Accordingly, as climate change, and other drivers such as changes in vulnerability and exposure (see Chapters 1, 2, and 3), are projected to lead to changes in frequency and intensity of some weather risks and extremes, insurers may be less prepared to underwrite insurance for extreme event risks. Innovative private-public sector partnerships may thus be required to better estimate and price risk as well as develop robust insurance-related products, which may be supported in developing countries by development partner funds as well (see Section 6.5.3 and Case Study 9.2.13).

Professional societies (such as builders and architects) and trade associations also play a key role in developing and implementing standards and practices for disaster risk reduction. These practices may include national and international standards and model building codes that are adopted in the regulations of local, state, and national governments. Although the potential for private sector players in disaster risk reduction in sectors such as engineering and construction, information communication technology, media and communication, as well as utilities and transportation seems large, *limited evidence* of successful private sector activity has been documented, owing to a number of reasons (Roeth, 2009). The business case for private sector involvement in disaster risk reduction remains unclear, hampering private sector engagement. Companies may also be averse to reporting activities that are fundamental to their business; and, in more community-focused projects, companies often work with local nongovernmental organizations (NGOs) and do not often report such efforts. Considering climate variability and change within the business model, companies may be an important entry point for disaster risk reduction, particularly in terms of guaranteeing global value chains in the presence of potentially large-scale disruptions triggered by climate-related disasters. For example, the economic viability of the Chinese coastal zone – the economic heartland of China and home to many multinational companies producing a large share of consumer goods globally – is highly exposed to typhoon risk and will increasingly depend on well-implemented disaster risk reduction mechanisms (Roeth, 2009).

6.2.3. Civil Society and Community-Based Organizations

At the national level, civil society organizations (CSOs) and community-based organizations (CBOs) play a significant role in developing initiatives to respond to disasters, reduce the risk of disasters, and, recently, adapt to climate-related hazards (see Section 5.1 for a discussion of 'local' and 'community' and Section 5.4.1 for the role of CBOs at the local level). CSOs and CBOs are referred to here as the wide range of associations around which society voluntarily organizes itself, with CBO referring to those associations primarily concerned with local interests and ties. CSO and CBO initiatives in the field of disaster risk management, which may usually begin as a humanitarian concern, often evolve to also embrace the broader challenge of disaster risk reduction following community-focused risk assessment, including specific activities targeting education and advocacy; environmental management; sustainable agriculture; infrastructure construction; and increased livelihood diversification (McGray et al., 2007; CARE International, 2008; Oxfam America, 2008; Practical Action Bangladesh, 2008; SEEDS India, 2008; Tearfund, 2008; World Vision, 2008).

Recently in some high-risk regions there has been rapid development of national platforms of CSOs and CBOs that have been working together in order to push for the transformation of policies and practices related to disaster risk reduction. This is true in the case of Central America, where at least four platforms are functioning in the same number of countries, involving more than 120 CSOs and CBOs (CRGR, 2007a). The efforts of these platforms have been aimed at advocacy, training, research, and capacity building in disaster risk reduction. In Central America, the experience is that advocacy on climate policy construction has become a new feature of such platforms since 2007 (CRGR, 2009). While beyond the scope of this chapter, on balance the majority of CSOs and CBOs focus efforts at the local level, trying to link disaster risk management with local development goals associated with water, sanitation, education, and health, for example (GNDR 2009; Lavell, 2009). Faith-based organizations are also influential in assisting local communities in disaster risk management, not only providing pastoral care in times of disasters but also playing an important role in raising awareness and training, with many international development partners often working with local church groups to build community resilience (see, for example, ADPC 2007; Gero et al., 2011; Tearfund, 2011).

In several countries in Latin America, CSOs and CBOs are considered, by law, as part of national systems for civil protection (Lavell and Franco, 1996; CRGR, 2007b) though participation, with the exception of National Red Cross/Red Crescent Societies, remains patchy (UNISDR, 2008c). In some countries where governments are not able or willing to fulfill certain disaster risk management functions, such as training, supporting food security, providing adequate housing, and preparedness, CSOs and CBOs have stepped in (Benson et al., 2001). While CSOs often face challenges in securing resources for replicating successful initiatives and scaling out geographically (CARE International, 2008; Oxfam America, 2008; Practical Action Bangladesh, 2008; SEEDS India, 2008; Tearfund, 2008; World Vision, 2008); sustaining commitment to work with local governments and stakeholders over the long term and maintaining partnerships with local authorities (Oxfam America, 2008); and coordinating and linking local-level efforts with sub-national government initiatives and national plans during the specific project implementation (SEEDS India, 2008), they are particularly well positioned to draw links between disaster risk reduction and climate change adaptation given that such organizations are currently among the few to combine such expertise (Mitchell et al., 2010b).

6.2.4. Bilateral and Multilateral Agencies

In developing countries, particularly where the government is weak and has limited resources, bilateral and multilateral agencies play a significant role in supplying financial, technical, and in some cases strategic support to government and nongovernment agencies to tackle the multifaceted challenges of disaster risk management and climate change adaptation in the context of national development goals (e.g., AusAid, 2009; DFID, 2011). Multilateral agencies are referred to here as international institutions with governmental membership that have a significant focus on development and aid recipient countries. Such agencies can include United Nations agencies, regional groupings (e.g., some European Union agencies), and multilateral development banks (e.g., World Bank, Asian Development Bank). Bilateral agencies (e.g., United Kingdom Department for International Development) are taken here as national institutions that focus on the relationship between one government and another. In the development sphere, this is often in the context of a richer government providing support to a poorer government. The role of international institutions, including bilateral and multilateral agencies, is discussed extensively in Section 7.3.

Bilateral and multilateral agencies have been key actors in advancing mainstreaming of disaster risk reduction and climate change adaptation into development planning (Eriksen and Næss, 2003; Klein et al., 2007; see Section 6.3). This has primarily been driven by a concern that development investments are increasingly exposed to climate- and disaster-related risks and that climate change poses security concerns (Harris, 2009; Persson and Klein, 2009). As a result, such agencies are influencing development policy and implementation at a national level as they require disaster and climate risk assessments and environmental screening to be conducted at different points in the project approval process and in some cases retrospectively when projects are already underway (Klein et al., 2007; OECD, 2009; Hammill and Tanner, 2010). A range of tools and methods have been developed, primarily by bilateral and multilateral agencies, to support such processes (Klein et al., 2007; Hammill and Tanner, 2010).

While significant progress has been made in developing appropriate tools and methods for assessing and screening risk, many bilateral and multilateral agencies continue to address disaster risk management and climate change adaptation separately, and link with respective regional and national agencies in the context of distinct international instruments (Mitchell and Van Aalst, 2008; Mitchell et al., 2010b; Gero et al., 2011). However, recent assessments suggest that the situation is improving, partially attributable to the process of authoring this Special Report and in the focus on risk management in the text of the Bali Action Plan (2007) and Cancun Agreement (2010) (Mitchell et al., 2010b; see Section 7.3.2.2 for more detail).

The diversity of national contexts requires bilateral and multilateral agencies to adopt different modalities to maximize the effectiveness of technical, financial, and strategic support. For example, in the Pacific and the Caribbean, regional bodies (e.g., the Caribbean Disaster Emergency Management Agency) commonly operate as an intermediary, channeling resources to island countries where it is not efficient for international agencies to establish a permanent adaptation or risk management-focused presence (Hay, 2009; Gero et al., 2011). In countries with weak national institutions, bilateral and multilateral agencies commonly choose to channel resources through civil society organizations with the intention of ensuring that resources reach the poorest and most vulnerable (Wickham et al., 2009). In such situations, coordination between agencies can be challenging and in certain circumstances can further reduce the risk management capacity of government organizations (Wickham et al., 2009). However, the broad trend is to maximize the support to national governments by seeking to improve national ownership of risk management and adaptation processes and in that respect support national governments to lead national systems (GFDRR, 2010; DFID, 2011).

6.2.5. Research and Communication

The effectiveness of national systems for managing climate extremes and disaster risks is highly dependent on the availability and communication of robust and timely scientific data and information (Sperling and Szekely, 2005; Thomalla et al., 2006; CACCA, 2010) and traditional knowledge (Mercer et al., 2007; Kelman et al., 2011; see Box 5-7) to inform not only community-based decisions and policymakers who manage national approaches to disaster risk and climate change adaptation, but also researchers who provide further analytical information to support such decisions.

Scientific and research organizations range from specialized research centers and universities, to regional organizations, to national research agencies, multilateral agencies, and CSOs playing differential roles, but generally continue to divide into disaster risk management or climate change adaptation communities. Scientific research bodies play important roles in managing climate extremes and disaster risks by: (a) supporting thematic programs to study the evolution and consequences of past hazard events, such as cyclones, droughts, sandstorms, and floods; (b) analyzing time-and-space dependency in patterns of weather-related risks; (c) building cooperative networks for early warning systems, modeling, and long-term prediction; (d) actively engaging in technical capacity building and training; (e) translating scientific evidence into adaptation practice; (f) collating traditional knowledge and lessons learned for wider dissemination; and (g) translating scientific information into user-friendly forms for community consumption (Sperling and Szekely, 2005; Thomalla et al., 2006; Aldunce and González, 2009).

Disaster practitioners largely focus on making use of short-term weather forecasting and effective dissemination and communication of hazard information and responses (Thomalla et al., 2006). Such climate change expertise can typically be found in meteorological agencies, environment or energy departments, and in academic institutions (Sperling and Szekely, 2005), while disaster risk assessments have been at the core of many multilateral and civil society organizations and national disaster management authorities (Sperling and Szekely, 2005; Thomalla et al., 2006). Although progress has been reported in the communication and availability of scientific information, there is still a lack of, for example, sufficient local or sub-national data on hazards and risk assessments to underpin area-specific disaster risk management (Chung, 2009; UNISDR, 2009c).

6.3. Planning and Policies for Integrated Risk Management, Adaptation, and Development Approaches

Given that learning will come from doing and in spite of differences, there are many ways that countries can learn from each other in prioritizing their climate and disaster risks; in mainstreaming climate change adaptation and disaster risk management into plans, policies, and processes for development; and in securing additional financial and human resources needed to meet increasing demands (UNDP, 2002; Thomalla et al., 2006; Schipper, 2009). This subsection will address frameworks for national disaster risk management and climate change adaptation planning and policies (Section 6.3.1), the mainstreaming of plans and policies nationally (Section 6.3.2), and the various sectoral disaster risk management and climate change adaptation options available for national systems (Section 6.3.3), recognizing the range of actors engaged in these processes as described in Section 6.2.

6.3.1. Developing and Supporting National Planning and Policy Processes

National and sub-national government and statutory agencies have a range of planning and policy options to help create the enabling environments for departments, public service agencies, the private sector, and individuals to act (UNDP, 2002; Heltberg et al., 2009; OECD, 2009; ONERC, 2009; Hammill and Tanner, 2010). When considering disaster risk management and adaptation to climate change actions, it is often the scale of the potential climate and disaster risks and impacts, the capacity of the governments or agencies to act, the level of certainty about future changes, the timeframes within which these future impacts and disasters will occur, and the costs and consequences of decisions that play an important role in their prioritization and adoption (Heltberg et al., 2008; World Bank, 2008; Wilby and Dessai, 2010).

The complexity and diversity of adaptation to climate change situations implies that there can be no single recommended approach for assessing, planning, and implementing adaptation options (Füssel, 2007; Hammill and Tanner, 2010; Lu, 2011). When the planning horizons are short and adaptation decisions only impact the next one or two decades, adaptation to recent climate variability and observed trends may be sufficient (Hallegatte, 2009; Wilby and Dessai, 2010; Lu, 2011). For long-lasting risks and decisions, the timing and sequencing of adaptation options and incorporation of climate change scenarios become increasingly important (Hallegatte, 2009; OECD, 2009; Wilby and Dessai, 2010). Studies suggest that the most pragmatic adaptation and disaster risk management options depend on the timeframes under consideration and the adaptive capacity and ability of the country or sectoral agencies to effectively integrate information on climate change and its uncertainties (McGray et al., 2007; Biesbroek et al., 2010; Krysanova et al., 2010; Wilby and Dessai, 2010; Juhola and Westerhoff, 2011). Given the various uncertainties at decisionmaking scales, studies suggest that adaptation actions based on information on the observed climate and its trends may be preferable in some cases while, in other cases with long-term irreversible decisions, climate change scenario-guided adaptation actions will be required (Auld, 2008b; Hallegatte, 2009; OECD, 2009; Krysanova et al., 2010; Wilby and Dessai, 2010). Climate change scenarios provide needed guidance for adaptation options when the direction of the climate change impacts are known and when the decisions involve long-term building infrastructure, development plans, and actions to avoid catastrophic impacts from more intense extreme events (Haasnoot et al., 2009; Hallegatte, 2009; Wilby and Dessai, 2010).

In dealing with climate change and disaster risk uncertainties, many national studies identify gradations or categories of adaptation and disaster risk management planning and policy options (Dessai and Hulme, 2007; Auld, 2008b; Hallegatte, 2009; Kwadijk et al., 2010; Mastrandrea et al., 2010; Wilby and Dessai, 2010). These gradations in options range from climate vulnerability or resilience approaches, sometimes described as 'bottom-up'; vulnerability, tipping point, critical threshold, or policy-first approaches to climate modeling, impact-based approaches, sometimes described as 'top-down'; model or impacts-first; science-first; or classical approaches (as illustrated in Figure 6-2 and outlined in the sectoral option headings of Table 6-1 and described in Section 6.3.3). Although the bottom-up and top-down terms sometimes refer to scale, subject matter, or policy (e.g., national versus local, physical to socioeconomic systems), the terms are used here to describe the sequences or steps needed to develop adaptation and disaster risk management plans and policies at the national level. When dealing with long-term future climate change risks, the main differences between the scenarios-impacts-first and vulnerability-thresholds-first approaches lie in the timing or sequencing of the stages of the analyses, as shown in Figure 6-2 (Kwadijk et al., 2010; Ranger et al., 2010). Although this difference appears subtle, it has significant implications for the management of uncertainty, the timing of adaptation options, and the efficiency of the policymaking (Dessai and Hulme, 2007; Auld, 2008b; Kwadijk et al., 2010; Wilby and Dessai, 2010; Lu, 2011). For example,

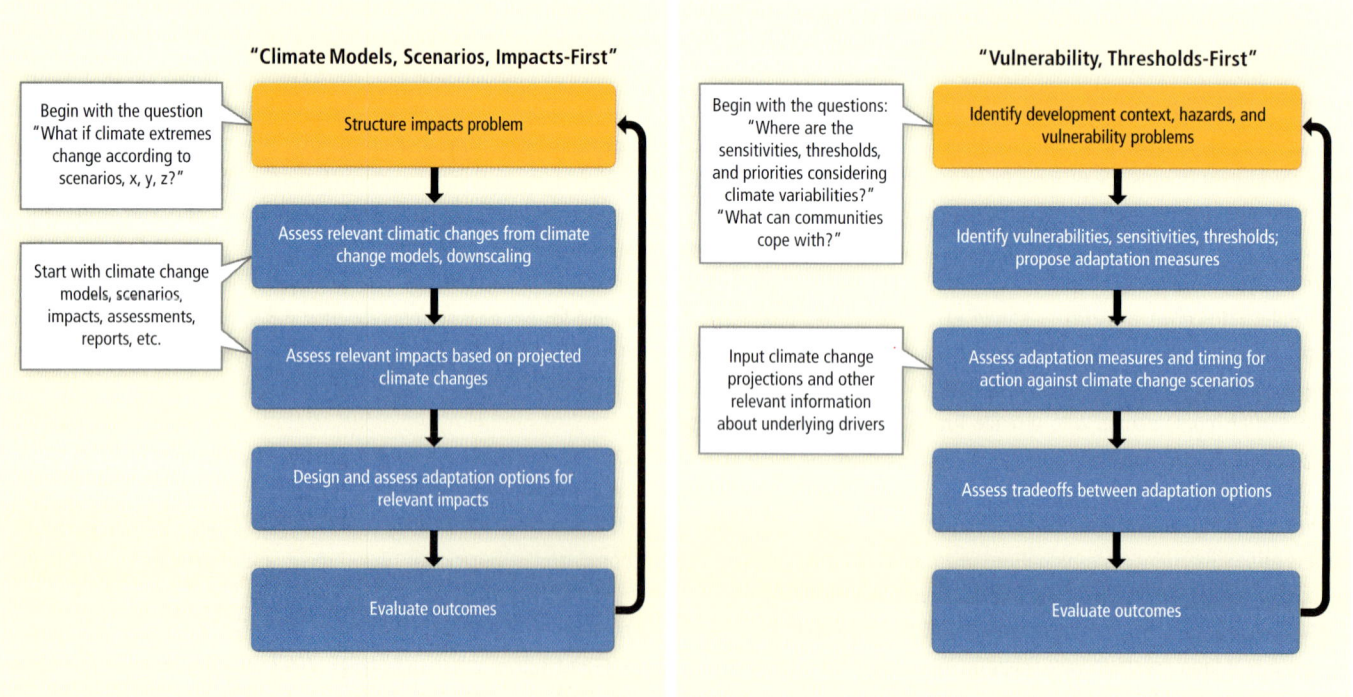

Figure 6-2 | Top-down scenario, impacts-first approach (left panel) and bottom-up vulnerability, thresholds-first approach (right panel) – comparison of stages involved in identifying and evaluating adaptation options under changing climate conditions. Adapted from Kwadijk et al. (2010) and Ranger et al. (2010).

when the lifespan of a decision, policy, or measure has implications for multiple decades or the decision is irreversible and sensitive to climate, the performance of adaptation and risk reduction options across a range of climate change scenarios becomes critical (Auld, 2008b; Kwadijk et al., 2010; Wilby and Dessai, 2010).

Vulnerability thresholds-based approaches start at the level of the decisionmaker, identify desired system objectives and constraints, consider how resilient or robust a system or sector is to changes in climate, assess adaptive capacity and critical 'tipping points' or threshold points, then identify the viable adaptation strategies that would be required to improve resilience and robustness under future climate scenarios (Auld, 2008b; Urwin and Jordan, 2008; Hallegatte, 2009; Kwadijk et al., 2010; Mastrandrea et al., 2010; Wilby and Dessai, 2010). Vulnerability-thresholds approaches can be independent of any specific future climate condition.

Options that are known as 'no regrets' and 'low regrets' provide benefits under any range of climate change scenarios, although they may not be optimal for every future scenario, and are recommended when uncertainties over future climate change directions and impacts are high (Dessai and Hulme, 2007; Auld, 2008b; Hallegatte, 2009; Kwadijk et al., 2010). These 'low regrets' adaptation options typically include improvements to coping strategies or reductions in exposure to known threats (Auld, 2008b; Kwadijk et al., 2010; Wilby and Dessai, 2010), such as better forecasting and warning systems, use of climate information to better manage agriculture in drought-prone regions, flood-proofing of homesteads, or interventions to ensure up-to-date climatic design information for engineering projects. The vulnerability-thresholds-first approaches are particularly useful for identifying priority areas for action now, assessing the effectiveness of specific interventions when current climate-related risks are not satisfactorily controlled, when climatic stress factors are closely intertwined with non-climatic factors, planning horizons are short, resources are very limited (i.e., expertise, data, time, and money), or uncertainties about future climate impacts are very large (Agrawala and van Aalst, 2008; Hallegatte, 2009; Prabhakar et al., 2009; Wilby and Dessai, 2010).

Vulnerability-thresholds-first approaches have sometimes been critiqued for the time required to complete a vulnerability assessment, for their reliance on experts, and for their largely qualitative results and limited comparability across regions (Patt et al., 2005; Kwadijk et al., 2010). Vulnerability-thresholds approaches can sometimes prove less suited for guiding future adaptation decisions if coping thresholds change, or if climate change risks emerge that are outside the range of recent experiences (e.g., successive drought years could progressively reduce coping thresholds of the rural poor by increasing indebtedness) (McGray et al., 2007; Agrawala and van Aalst, 2008; Auld, 2008b; Hallegatte, 2009; Prabhakar et al., 2009; Wilby and Dessai, 2010).

The scenarios-impact-first approaches typically start with several climate change modeling scenarios and socioeconomic scenarios, evaluate the expected impacts of climate change, and subsequently identify adaptation and risk reduction options to reduce projected risks (Kwadijk et al., 2010; Mastrandrea et al., 2010; Wilby and Dessai, 2010). The scenarios-impacts-first approaches are most useful to raise awareness of the problem, to explore possible adaptation strategies, and to identify research priorities, especially when current climate and disaster risks can be effectively controlled, when sufficient data and resources are available to produce state-of-the-art climate scenarios at the spatial resolutions relevant for adaptation, and when future climate impacts can be projected reliably (Kwadijk et al., 2010; Wilby and Dessai, 2010). Scenarios-impacts approaches depend strongly on the chosen climate change scenarios and downscaling techniques, as well as the assumptions about scientific and socioeconomic uncertainties (OECD, 2009; Kwadijk et al., 2010). Pure scenarios-impacts approaches may not be available at the spatial scales relevant to the decisionmaker, may not be applicable for the purpose of the decisionmaker, and usually give less consideration to current risks from natural climate variability, to non-climatic stressors, and to key uncertainties along with their implications for robust adaptation policies (Füssel, 2007; Wilby and Dessai, 2010). In practice, there are very limited examples of actual adaptation policies being developed and planned adaptation decisions being implemented based on scenarios-impacts approaches only (Füssel, 2007; Biesbroek et al., 2010; Wilby and Dessai, 2010).

Increasingly, studies are recognizing that the scenarios-impacts and vulnerability-thresholds approaches are complementary and need to be integrated and that both can benefit from the addition of stakeholder and scientific input to determine critical thresholds for climate change vulnerabilities (Auld, 2008b; Haasnoot et al., 2009; Kwadijk et al., 2010; Mastrandrea et al., 2010; Wilby and Dessai, 2010). Critical thresholds (or adaptation tipping points) help in answering the basic adaptation questions of decision- and policymakers – namely, what are the first priority issues that need to be addressed as a result of increasing disaster risks under climate change and when might these critical thresholds be reached (Auld, 2008b; Haasnoot et al., 2009; Kwadijk et al., 2010; Mastrandrea et al., 2010). The integration of scenarios-impacts and vulnerability-thresholds approaches provides guidance on the sensitivity of sectors and durability of options under different climate change scenarios (Haasnoot et al., 2009; Kwadijk et al., 2010; Mastrandrea et al., 2010). Integrated approaches that link changes in climate variables to decisions and policies and express uncertainties in terms of timeframes over which a policy or plan may be effective (i.e., roughly when will the critical threshold be reached) also provide valuable information for plans and policies and their implementation (Haasnoot et al., 2009; Kwadijk et al., 2010; Mastrandrea et al., 2010).

Regardless of the approaches used, it is important that uncertainty over future climate change risks not become a barrier to climate change risk reduction actions (Auld, 2008b; Hallegatte, 2009; Krysanova et al., 2010; Wilby and Dessai, 2010). In cases where climate change uncertainties remain high, countries may choose to increase or build on their capacity to cope with uncertainty, rather than risk maladaptation from use of ambiguous impact studies or no action (McGray et al., 2007; Hallegatte, 2009; Wilby and Dessai, 2010). In order to reduce the risk of maladaptation

Table 6-1 | National policies, plans, and programs: a selection of disaster risk reduction and adaptation to climate change options by selected sectors.

Sector/ Response	'No regrets' and 'low regrets' actions for current and future risks	('No/low regrets' options plus...) Preparing for climate change risks by reducing uncertainties (building capacity)	("Preparing for climate change" risks plus...) Reduce risks from future climate change	Risk transfer	Accept and deal with increased and unavoidable (residual) risks	'Win-win' synergies for GHG reduction, adaptation, risk reduction, and development benefits
Natural Ecosystems and Forestry	• Use of Ecosystem-based Adaptation (EbA) or 'soft engineering'; integrate disaster risk reduction and climate into integrated coastal zone and water resources management, forest management, and land use management; conserve, enhance resilience of ecosystems; restore protective ecosystem services [1] • Adaptive forest management; forest fire management, controlled burns; agroforestry; biodiversity [2] • Reduce forest degradation, unsustainable harvests, and provide incentives for alternate livelihoods, eco-tourism [6]	• Synergies between UNFCCC and Rio Conventions; avoid actions that interfere with goals of other UN conventions [3] • Research on climate change–ecosystem–forest links, climate and ecosystem prediction systems, climate change projections; monitor ecosystem and climate trends [3] • Incorporate ecosystem management into National Adaptation Programmes of Action and disaster risk reduction plans [3]	• Adaptation to climate change interventions to maintain ecosystem resilience; corridors, assisted migrations; plan EbA for climate change [4] • Seed, genetic banks; new genetics; tree species improvements to maintain ecosystem services in future; adaptive agroforestry [4] • Changed timber harvest management, new technologies for adaptation to climate change, new uses to conserve forest ecosystem services [4]	• Micro-finance and insurance to compensate for lost livelihoods [5] • Investments in additional insurance, government reserve funds for increased risks due to loss of protective ecosystem services [5]	• Replace lost ecosystem services through additional hard engineering, health measures [6] • Restore loss of damaged ecosystems [6]	• Sustainable afforestation (for robust forests), reforestation, conservation of forests, wetlands and peatlands, sustainable and increased biomass; land use, land-use change, and forestry; reducing emissions from deforestation [7] • Incentives for sustainable sequestration of carbon; sustainable bio-energy; energy self-sufficiency [7]
Agriculture and Food Security	• Food security via sustainable land and water management; training; efficient water use, storage; agro-forestry; protection shelters, crop and livestock diversification; improved supply of climate stress tolerant seeds; integrated pest, disease management [8] • Climate monitoring; improved weather predictions; disaster management, crop yield and distribution models and predictions [9]	• Increased agriculture-climate research and development [10] • Research on climate tolerant crops, livestock; agrobiodiversity for genetics [10] • Integration of climate change scenarios into national agronomic assessments [11] • Diversification of rural economies for sensitive agricultural practices [10]	• Adaptive agricultural and agroforestry practices for new climates, extremes [12] • New and enhanced agricultural weather, climate prediction services [11] • Food emergency planning; distribution and infrastructure networks [12] • Diversify rural economies [12]	• Improved access to crop, livestock, and income loss insurance (e.g., weather derivatives) [13] • Micro-financing and micro-insurance [13] • Subsidies, tax credits [13]	• Changed livelihoods and relocations in regions with climate sensitive practices [12] • Secure emergency stock and improve distribution of food and water for emergencies [12]	• Energy efficient and sustainable carbon sequestering practices; training; reduced use of chemical fertilizers [14] • Use of bio-gas from agricultural waste and animal excreta [14] • Agroforestry [14]
Coastal Zone and Fisheries	• EbA; integrated coastal zone management (ICZM); combat salinity; alternate drinking water availability; soft and hard engineering [15] • Strengthen institutional, regulatory, and legal instruments; setbacks; tourism development planning [16] • Marine protected areas, monitoring fish stocks, alter catch quantities, effort, timing; salt-tolerant fish species [17] • Climate risk reduction planning; hazard delineation; improve weather forecasts, warnings, environmental prediction [16]	• Climate change projections for coastal management planning; develop modeling capacity for coastal zone-climate links; climate-linked ecological and resource predictions; improved monitoring, geographic and other databases for coastal management [18] • Monitor fisheries; selective breeding for aquaculture, fish genetic stocks; research on salt-tolerant crop varieties [19]	• Incorporate adaptation to climate change, sea level rise into ICZM, coastal defenses [18] • Hard and 'soft' engineering for adaptation to climate change; sustainable tourism development planning; resilient vessels and coastal facilities [16] • Manage for changed fisheries, invasives [20] • Inland lakes: alter transportation and industrial practices, soft and hard engineering [20]	• Enhance insurance for coastal regions and resources; fisheries insurance [21] • Government reserve funds [21]	• Enhance emergency preparedness measures for more frequent and intense extremes, including more evacuations [16] • Relocations of communities, infrastructure [16] • Exit fishing; provide alternate livelihoods [19]	• Use of sustainable renewable energy; conservation, energy self-sufficiency (especially for islands, coastal regions) [22] • Offshore renewable energy for alternate incomes and aquaculture habitat [22]

Continued next page →

Table 6-1 (continued)

Sector/ Response	'No regrets' and 'low regrets' actions for current and future risks	('No/low regrets' options plus...) Preparing for climate change risks by reducing uncertainties (building capacity)	("Preparing for climate change" risks plus...) Reduce risks from future climate change	Risk transfer	Accept and deal with increased and unavoidable (residual) risks	'Win-win' synergies for GHG reduction, adaptation, risk reduction, and development benefits
Water resources	• Implement Integrated Water Resource Management (IWRM), national water efficiency, storage plans [20] • Effective surveillance, prediction, warning and emergency response systems; better disease and vector control, detection and prediction systems; better sanitation; awareness and training on public health [24] • Adequate funding, capacity for resilient water infrastructure and water resource management; Improved institutional arrangements, negotiations for water allocations, joint river basin management [23]	• Develop prediction, climate projection, and early warning systems for flood events and low water flow conditions; research and downscaling for hydrological basins [24] • Multi-sectoral planning for water; selective decentralization of water resource management (e.g., catchments and river basins); joint river basin management (e.g., bi-national) [23]	• National water policy frameworks, robust integrated and adaptive water resource management for adaptation to climate change [25] • Investments in hard and soft infrastructure considering changed climate; river restoration [25] • Improved weather, climate, hydrology-hydraulics, water quality forecasts for new conditions [24]	• Public-private partnerships; Economics for water allocations beyond basic needs [26] • Mobilize financial resources and capacity for technology and EbA [26] • Insurance for infrastructure [26]	• Enhance national preparedness and evacuation plans for greater risks [24] • Enhance health infrastructure for more failures [24] • Alter transport, engineering; increases to temporary consumable water taking permits [24] • Enhance food , water distribution for emergencies, plan for alternate livelihoods [24]	• Integrated and sustainable water efficiency and renewable hydro power for adaptation to climate change [23]
Infra-structure, Housing, Cities, Transportation, Energy	• Building codes, standards with updated climatic values; climate- resilient infrastructure (and energy) designs; training, capacity, inspection, enforcement; monitoring for priority retrofits (e.g., permafrost); maintenance [27] • Legal alternatives to informal settlements, sanitation [27] • Strengthen early warning systems, hazard awareness; improved weather warning systems; disaster-resilient building components (rooms) in high-risk areas; tourism development planning; heat-health responses [28] • Integrate urban planning, engineering, maintenance [27] • Diversified energy systems; maintenance; self-sufficiency, clean energy technologies for national energy plans, international agreement goals (biogas, solar cooker); use of renewable energy in remote and vulnerable regions, use of appropriate energy mixes nationally [29] • Energy security; distributed energy generation and distribution [29]	• Improved downscaling of climate change information; maintain climate data networks, update climatic design information; increased safety/uncertainty factors in codes and standards; develop adaptation to climate change tools [28] • Research on climate, energy, coastal, and built environment interface, including flexible designs, redundancy; forensic studies of failures (adaptation learning); improved maintenance [27] • Investments for sustainable energy development; cooperation on trans-boundary energy supplies (e.g., wind energy at times of peak wind velocity) [29]	• Codes, standards for changed extremes [30] • Publicly funded infrastructure, coastal development and post-disaster reconstruction to include adaptation to climate change [30] • New materials, engineering approaches; flexible design and use structures; asset management for adaptation to climate change [30] • Hazard mapping: zoning and avoidance; prioritized retrofits, abandon the most vulnerable; soft engineering services [30] • Design energy generation, distribution systems for adaptation; switch to less risky energy systems, mixes; embed sustainable energy in disaster risk reduction and adaptation to climate change planning [29]	• Infrastructure insurance and financial risk management [29] • Insurance for energy facilities, interruption [29] • Innovative risk sharing instruments [29] • Government reserve funds [29]	• More relocations [28] • Enhance evacuation, transportation, and energy contingency planning for increases in extreme events [28] • Increase climate-resilient shelter construction [28]	• Implement energy- and water-efficient GHG reductions, disaster risk reduction and adaptation to climate change synergies [29] • Scale up, market penetration for sustainable renewable energy production: increased hydroelectric potential; sustainable biomass; 'greener' distributed community energy systems [29]

National Systems for Managing the Risks from Climate Extremes and Disasters

Table 6-1 (continued)

Sector/Response	'No regrets' and 'low regrets' actions for current and future risks	('No/low regrets' options plus...) Preparing for climate change risks by reducing uncertainties (building capacity)	("Preparing for climate change" risks plus...) Reduce risks from future climate change	Risk transfer	Accept and deal with increased and unavoidable (residual) risks	'Win-win' synergies for GHG reduction, adaptation, risk reduction, and development benefits
Health	• Community/urban and coastal zone planning, building standards and guidelines; cooling shelters; safe health facilities; retrofits for vulnerable structures; health facilities designed using updated climate information [31] • Strengthen surveillance, health preparedness; early warning weather-climate-health systems, heat alerts and responses; capacity for response to early warnings; prioritize disaster risks; disaster prevention and preparedness; public education campaigns; food security [31] • Strengthen disease surveillance and controls; improve health care services, personal health protection; improve water treatment/sanitation; water quality regulations; vaccinations, drugs, repellants; development of rapid diagnostic tests [31] • Monitor air and water quality; regulations; urban planning [31] • Better land and water use management to reduce health risks [31]	• Research on climate-health linkages and adaptation to climate change options; develop new health prediction systems for emerging risks; research on landscape changes, new diseases, and climate; urban weather-health modeling [31] • Education, disaster prevention and preparedness [31]	• New food and water security, distribution systems; air quality regulations, alternate fuels [32] • New warning and response systems; predict and manage health risks from landscape changes; target services for most at risk populations [32] • Climate proofing, refurbish/maintain national health facilities and services [32] • Address needs for additional health facilities and services [32]	• Extend and expand health insurance coverage to include new and changed weather and climate risks [33] • Government reserve funds [33]	• National plan for heat and extremes emergencies [32] • New disease detection and management systems [32] • Enhanced prediction and warning systems for new risks [32]	• Use of clean and sustainable renewable energy and water sources; increase energy efficiency; air quality regulations; clean energy technologies to reduce harmful air emissions (e.g., cooking stoves) [34] • Design sustainable infrastructure for climate change and health [30]

References: 1. Adger et al., 2005; Barbier, 2009; Colls et al., 2009; FAO, 2008a; MEA, 2005; SCBD, 2009; Shepherd, 2004, 2008; UNEP, 2009; UNISDR, 2009d; World Bank, 2010. 2. FAO, 2007; Neufeldt et al., 2009; Shugart et al., 2003; Spittlehouse and Stewart, 2003; Weih, 2004. 3. Colls et al., 2009; FAO, 2008a; OECD, 2009; Rahel and Olden, 2008; Robledo et al., 2005; SCBD, 2009; UNEP, 2009; UNFCCC, 2006a. 4. Berry, 2007; FAO, 2007, 2008a,b; Leslie and McLeod, 2007; OECD, 2009; SCBD, 2009. 5. CCCD, 2009; Colls et al., 2009; FAO, 2008b; ProAct Network, 2008; UNFCCC, 2006a. 6. Chhatre and Agrawal, 2009; FAO, 2008b; Mansourian et al., 2009; Reid and Huq, 2005; SCBD, 2009; UNEP, 2006; Venter et al., 2009. 8. Arnell, 2004; Branco et al., 2005; Campbell et al., 2008; Easterling et al., 2007; FAO, 2008a, 2009; Howden et al., 2007; McGray et al., 2007; Neufeldt et al., 2009; SCBD, 2009; UNISDR, 2009d; World Bank, 2009. 9. Easterling et al., 2007; FAO, 2007, 2010; Hammer et al., 2003; McCarl, 2007; UNFCCC, FAO, 2006a; World Bank, 2009. 10. Campbell et al., 2008; CCCD, 2009; Easterling et al., 2007; FAO, 2007, 2010; World Bank, 2009. 11. Easterling et al., 2007; FAO, 2007, 2008a; World Bank, 2009. 12. Butler and Oluoch-Kosura, 2006; Butt et al., 2005; CCCD, 2009; Davis, 2004; FAO, 2006, 2008a; Howden et al., 2007; McCarl, 2007; World Bank, 2009. 13. CCCD, 2009; FAO, 2007; UNISDR, 2009d; World Bank, 2009. 14. FAO, 2007, 2008a; Rosenzweig and Tubiello, 2007. 15. Adger et al., 2005; FAO, 2008c; Kay and Adler, 2005; Kesavan and Swaminathan, 2006. 16. Adger et al., 2005; FAO, 2008c; Kesavan and Swaminathan, 2006; Klein et al., 2001; Nicholls, 2007; Nicholls et al., 2008; Romieu et al., 2010; UNFCCC, 2006a. 17. FAO, 2007, 2008c; Rahel and Olden, 2008; UNFCCC, 2006a. 18. Adger et al., 2005; Dolan and Walker, 2006; FAO, 2008b; Nicholls, 2007; Thorne et al., 2006; UNFCCC, 2006b; World Bank, 2010. 19. FAO, 2008c; Kesavan and Swaminathan, 2006; Rahel and Olden, 2008. 20. FAO, 2007, 2008c; IIED, 2009; Romieu et al., 2010. 21. FAO, 2007, 2008c; Nicholls, 2007. 22. FAO, 2008c; UNFCCC, 2006a. 23. Branco et al., 2005; CCCD, 2009; Hedger and Cacouris, 2008; UNFCCC, 2006c; Hedger and Cacouris, 2008; ICHARM, 2009; Kijin et al., 2004; Krysanova et al., 2010; Mills, 2007; Olsen, 2006; Rahaman and Varis, 2005; World Bank, 2009; WSSD, 2002; WWAP, 2009. 24. Arnell and Delaney, 2006; Auld et al., 2004; CCCD, 2009; DaSilva et al., 2004; Hedger and Cacouris, 2008; Rahaman and Varis, 2005; WWAP, 2009. 26. Few et al., 2006; Kirshen, 2007; Mills, 2007; Muller, 2007; Rahaman and Varis, 2005; World Bank, 2009; WWC, 2009. 25. CCCD, 2009; Crabbé and Robin, 2006; Hedger and Cacouris, 2008; Krysanova et al., 2010; Rahaman and Varis, 2005; WWAP, 2009. 26. Few et al., 2006; Kirshen, 2007; Mills, 2007; Muller, 2007; Rahaman and Varis, 2005; WWAP, 2009. 27. Auld, 2008b; Haasnoot et al., 2009; Hallegatte, 2009; Hodgson and Carter, 1999; Lowe, 2003; Mills, 2007; Nicholls et al., 2008; NRTEE, 2009; ProVention, 2009; Rossetto, 2007; Wamsler, 2004; Wilby and Dessai, 2010; World Bank, 2000, 2008; WWC, 2009. 28. Auld, 2008a,b; Haasnoot et al., 2009; Maréchal, 2007; Mills, 2007; Neumann, 2009; Robledo et al., 2005; Van Buskirk, 2006; Warner et al., 2009; Younger et al., 2008. 29. Auld, 2008b; Islam and Ferdousi, 2007; Kagiannas et al., 2003; Maréchal, 2007; Mills, 2007; Neumann, 2009; Robledo et al., 2005; Van Buskirk, 2006; Warner et al., 2009; Younger et al., 2008. 30. Auld, 2008b; Freeman and Warner, 2001; Mills, 2007; Neumann, 2009; NRTEE, 2009; ProVention, 2009; Stevens, 2008; Younger et al., 2008. 31. Auld et al., 2004; Auld, 2008a; CCCD, 2009; Curriero et al., 2001; DaSilva et al., 2004; Ebi et al., 2006; Haines et al., 2006; FAO, 2008b; Nicholls, 2007; FAO, 2008b; Nicholls, 2007; UNFCCC, 2006b; WHO, 2003; WWAP, 2009; WWC, 2009. 25. CCCD, 2009; Crabbé and Robin, 2006; Hedger and Cacouris, 2008; Krysanova et al., 2010; Rahaman and Varis, 2005; WWAP, 2009. 26. Few et al., 2006; Patz et al., 2000, 2005; UNFCCC, 2006a; WHO, 2003, 2005; Younger et al., 2008. 33. Mills, 2005, 2006. 34. Haines et al., 2006; Younger et al., 2008.

into the future, some studies recommend the use of pro-adaptation and robust options to deal with climate change uncertainties (Auld, 2008b; Hallegatte, 2009; Wilby and Dessai, 2010). These robust options include actions that are reversible, flexible, less sensitive to future climate conditions (i.e., no and low regret), and can incorporate safety margins (e.g., infrastructure investments), employ 'soft' solutions (e.g., ecosystem services), and are mindful of actions being taken by others to either reduce greenhouse gases (GHGs) or adapt to climate change in other sectors (Hallegatte, 2009; Wilby and Dessai, 2010). Flexible options are those that provide benefits under a variety of climate conditions or reduce stress on affected systems to increase their flexibility (e.g., reducing pollution or demand on resources) (Auld, 2008b; Hallegatte, 2009; Wilby and Dessai, 2010).

Options that allow for incremental changes in, for example, infrastructure over time, or allow incorporation of future change, for example, support more flexible systems (Auld, 2008b; Hallegatte, 2009; OECD, 2009). Uncertainties over future risks can also be accounted for through 'safety margin' or over-design strategies to reduce vulnerability and increase resiliency at low and sometimes null costs (Auld, 2008b; Hallegatte, 2009). These safety margin strategies have been used to manage future risks for sea level rise and coastal defenses, for water drainage management, and for investments in other infrastructure (Hallegatte, 2009). Given uncertainties, national policies may need to become more adaptable and flexible, particularly where national plans and policies currently operate within a limited range of conditions and are based on certainty (McGray et al., 2007; Wilby and Dessai, 2010). Without flexibility, rigid national policies may become disconnected from evolving climate risks and bring unintended consequences or maladaptation (Sperling and Szekely, 2005; Hallegatte, 2009). Rigid plans and policies that are irreversible and based on a specific climate scenario that does not materialize can result in future maladaptation and imply wasted investments or harm to people and ecosystems that can prove unnecessary.

Several studies indicate that national plans and policies for adaptation to climate change and disaster risk management tend to favor options that deal with the current or near-term climate risks and 'win-win' options that satisfy multiple synergies for GHG reduction, disaster risk management, climate change adaptation, and development issues (World Bank, 2008; Heltberg et al., 2009; Ribeiro et al., 2009; Fankhauser, 2010; Mitchell and Maxwell, 2010). Many of these 'win-win' options include ecosystem-based adaptation actions, sustainable land and water use planning, carbon sequestration, energy efficiency, and energy and food self-sufficiency. For example, the ecosystem management practices of afforestation, reforestation, and conservation of forests offer co-benefits for disaster risk reduction from floods, landslides, avalanches, coastal storms, and drought while contributing to adaptation to future climates, economic opportunities, increased biomass and carbon sequestration, energy efficiency, energy savings, as well as energy and food self sufficiency (Thompson et al., 2009).

Disaster risk transfer options offer a viable adaptation response to current and future climate risks and include instruments such as insurance, micro-insurance, and micro-financing; government disaster reserve funds; government-private partnerships involving risk sharing; and new, innovative insurance mechanisms (Linnerooth-Bayer and Mechler, 2006; EC, 2009; World Bank, 2010). Risk transfer options can provide much needed, immediate liquidity after a disaster, allow for more effective government response, provide some relief from the fiscal burden placed on governments due to disaster impacts, and constitute critical steps in promoting more proactive risk management strategies and responses (Arnold, 2008). Case Study 9.2.13 and Section 6.5.3 provide more detail on risk transfer options.

Even with risk transfer instruments and adaptation to climate change options in place, residual losses can be realized when extreme events – well beyond those typically expected – result in high impacts. In spite of the evidence, decisions to ignore increasing future risks and even current risks remain common, particularly when uncertainties over the directions of future climate change impacts are high, when capacity is initially very limited, adaptation options are not available, or when the risks of future impacts are considered to be very low (Linnerooth-Bayer and Mechler, 2006; Heltberg et al., 2009; World Bank, 2010).The losses from deferring adaptation and disaster risk reduction actions are borne by all actors.

Table 6-1 outlines some of the adaptation to climate change and disaster risk management policy and planning options available nationally for selected sectors and described in the literature. Many of these options are incremental actions that complement and reinforce each other. The actions are organized using the gradations of planning and policy options described in this section.

6.3.2. Mainstreaming Disaster Risk Management and Climate Change Adaptation into Sectors and Organizations

National adaptation to climate change will involve stand-alone adaptation policies and plans as well as the integration or mainstreaming of adaptation measures into existing activities (OECD, 2009). Mainstreaming of adaptation and disaster risk management actions implies that national, sub-national, and local authorities adopt, expand, and enhance measures that factor disaster and climate risks into their normal plans, policies, strategies, programs, sectors, and organizations (Few et al., 2006; UNISDR, 2008a; OECD, 2009; Biesbroek et al., 2010; CACCA, 2010).

In reality, it can be challenging to provide clear pictures of what mainstreaming is, let alone how it can be made operational, supported, and strengthened at the various national and sub-national levels (Olhoff and Schaer, 2010). Some studies indicate that the real challenge to mainstreaming adaptation is not planning but implementation (Biesbroek et al., 2010; Krysanova et al., 2010; Tompkins et al., 2010). Some of the barriers to implementation include lack of funding, limited budget flexibility, lack of relevant information or expertise, lack of political will or support, and institutional silos (Krysanova et al., 2010;

Preston et al., 2011). Studies indicate that effective plans, policies, and programs for adaptation to climate change and disaster risk management need to go beyond identifying potential options to include better inventories of existing assets and liabilities for managing risk and specific actions for overcoming adaptation barriers (Haasnoot et al., 2009; Preston et al., 2011).

Recent studies investigating the success of existing adaptation plans and policies for Australia, the United States, countries in Europe, and major river basins in Africa and Asia, for example, indicate that there is a need for mainstreaming of adaptation into existing national policies and plans and a priority for capitalizing on 'win-win' or options that take advantage of synergies with other national objectives (Biesbroek et al., 2010; Tompkins et al., 2010; Preston et al., 2011). The studies found that many strategies and institutions were focused to a greater extent on lower-risk actions dealing with science and outreach (knowledge acquisition) and capacity building rather than moving forward on specific, more costly and difficult to implement adaptation and disaster risk management actions and managing at-risk public goods (Tompkins et al., 2010; Preston et al., 2011).

Preston et al. (2011) found in their studies from Australia, the United States, and the United Kingdom that most national adaptation strategies were based on vulnerability assessments informed by broad international and national climate change guidance, rather than any consistent or systematic use of scenarios, and favored bottom-up approaches for coordination across sectors and multiple government scales. Biesbroek et al. (2010) noted similar results for nine countries in Europe. Tompkins et al. (2010) and Krysanova et al. (2010) found that the sectors with the highest levels of adaptation implementation in the United Kingdom were those that tended to be most affected by current weather variability and extremes and that specific government initiatives had been successful in stimulating adaptation and disaster risk reduction (e.g., mandatory planning for flood-prone areas, ISO 14001). Tompkins et al. (2010) also found that successful implementation frequently resulted from multiple triggers, that few of these adaptation actions were solely initiated in response to climate change, and that the relative impact of weather on core business and organizational culture encouraged an ability and willingness to proactively act on climate change information.

Adaptation to climate change and disaster risk management needs to typically identify more adaptation options than most countries can reasonably implement in the short term due to resource constraints, requiring that actions be prioritized (OECD, 2009; Krysanova et al., 2010). Initially, actions that remove the existing barriers to managing disaster risks from today's climate variability can help to reduce the even greater barriers to managing future climate risks (UNDP, 2002, 2004a; CCCD, 2009; Prabhakar et al., 2009; Tompkins et al., 2010). As a result, a key challenge, and an opportunity for mainstreaming adaptation and disaster risk management, lies in building bridges between current disaster risk management actions for existing climate vulnerabilities and the additional revised efforts needed for future vulnerabilities (Few et al., 2006; Krysanova et al., 2010; Olhoff and Schaer, 2010; Wilby and Dessai, 2010).

An important prerequisite for informed decisions on adaptation to climate change and disaster risk management is that they should be based upon the best available information (OECD, 2009; Biesbroek et al., 2010; Lu, 2011). Preston et al. (2011) noted that many of the specific adaptation plans from Australia, the United States, and the United Kingdom indicated a need for improved gathering and sharing of climate and climate change science information prior to or in conjunction with the delivery of adaptation actions, perhaps reflecting a preference for delaying adaptation actions until greater certainty or better information on different adaptation actions was known. As noted in Chapter 3 (Section 3.2.3 and Box 3-2), many extreme events occur at small temporal and spatial scales, where climate change models, even when downscaled, cannot provide simulations at such spatial and temporal resolutions. A number of studies also contend that increased and better information on climate change scenarios and projections and potential impacts will accomplish little on their own to mainstream and alter on-the-ground decisions, policies, and plans unless the information provided can directly meet decisionmakers' needs (Stainforth et al., 2007; Auld, 2008b; Haasnoot et al., 2009; Krysanova et al., 2010; Mastrandrea et al., 2010; Wilby and Dessai, 2010). Users require relevant climate risk information that is accessible, can be explained in understandable language, provides straightforward estimates of uncertainties, and is relevant or tailored to their management functions (Stainforth et al., 2007; Mastrandrea et al., 2010; Lu, 2011). Increasingly, studies are showing that this is best accomplished through sustained interactions between scientists and stakeholders and policymakers, usually maintained through years of relationship- and trust-building (Mastrandrea et al., 2010; Wilby and Dessai, 2010; Lu, 2011).

Studies generally indicate that the most essential means for effectively mainstreaming both adaptation and disaster risk management nationally involve 'whole of government' coordination across different levels and sectors of governance, including the involvement of a broad range of stakeholders (Few et al., 2006; Thomalla et al., 2006; OECD, 2009; also Section 6.4.2). In spite of the strong interdependencies, governments have tended to manage these issues in their 'silos' with environment or energy authorities and scientific institutions typically responsible for climate change adaptation while disaster risk management authorities may reside in a variety of national government departments and national disaster management offices (Sperling and Szekely, 2005; Thomalla et al., 2006; Prabhakar et al., 2009). Progress in planning for adaptation and developing and implementing strategies within government agencies usually depends on political commitment, institutional capacity, and, in some cases, on enabling legislation, regulations, and financial support (Few et al., 2006; OECD, 2009; Krysanova et al., 2010; see Section 6.4). Nationally, studies indicate that it may be important to clearly identify a lead for disaster and climate risk reduction efforts where that lead has influence on budgeting and planning processes (Few et al., 2006; OECD, 2009). In some cases, countries and regions may be able to build on phases of raised awareness and increased attention to disaster risk in order to develop and strengthen their responsible institutions (Few et al., 2006; Krysanova et al., 2010).

While developed countries may be more financially equipped to meet many of the challenges of mainstreaming adaptation and disaster risk reduction into national plans and policies, the situation is often more challenging in developing countries (Krysanova et al., 2010). Nonetheless, there are examples from developing countries where adaptation to climate change and disaster risk management mainstreaming issues have been priorities for many years and significant progress in mainstreaming has been noted (e.g., the Caribbean Mainstreaming Adaptation to Climate Change project, which was implemented from 2004 to 2007; Case Studies 9.2.9 and 9.2.12). In other cases, international funding mechanisms such as the Least Developed Countries (LDC) Fund, the Special Climate Change Fund, the Multi-donor Trust Fund on Climate Change, and the Pilot Programme for Climate Resilience under the Climate Investment Fund are making funding and resources available to developing countries to pilot and mainstream changing climate risks and resilience into core development and as an incentive for scaled-up action and transformational change, although needs exceed availability of funds (O'Brien et al., 2008; Krysanova et al., 2010; see Sections 7.4.3.3 and 7.4.2 for additional discussion).

6.3.3. Sector-Based Risk Management and Adaptation

The challenge for countries is to manage short-term climate variability while also ensuring that different sectors and systems remain resilient and adaptable to changing extremes and risks over the long term (Füssel, 2007; Wilby and Dessai, 2010). The requirement is to balance the short-term and the longer-term actions needed to resolve the underlying causes of vulnerability and to understand the nature of changing climate hazards (UNFCCC, 2008a; OECD, 2009). Achieving adaptation and disaster risk management objectives while attaining human development goals requires a number of cross-cutting, interlinked sectoral and development processes, as well as effective strategies within sectors and coordination between sectors (Few et al., 2006; Thomalla et al., 2006; Biesbroek et al., 2010). Climate change is far too big a challenge for any single ministry of a national government to undertake (CCCD, 2009; Biesbroek et al., 2010).

Sector-based organizations and departments play a central role in national decisionmaking and are a logical focus for adaptation actions (McGray et al., 2007; Biesbroek et al., 2010). The impacts of changing climate risks in one sector, such as tourism, can affect other sectors and scales significantly, especially since sectoral linkages operate both vertically and horizontally. Sector plans, policies, and programs are linked vertically from national to local levels within the same sector as well as horizontally across different sectors at the same level (Urwin and Jordan, 2008; UNFCCC, 2008b; CCCD, 2009; Biesbroek et al., 2010). While the case and need for integration within sectors and levels may be clear, the issue of how to integrate or mainstream nationally across multiple sectors and multiple levels still remains challenging, requiring governance mechanisms and coordination that can cut across governments and sectoral organizations (UNISDR, 2005; UNFCCC, 2008b; CCCD, 2009; ONERC, 2009; Biesbroek et al., 2010). Typically, multi-sector integration tends to deal with the broader national scale (e.g., entire economy or system) and aims to be as comprehensive as possible in covering several affected sectors, regions, and issues (UNFCCC, 2008b). Studies from organizations and academia indicate that effective adaptation and risk reduction coordination between all sectors may only be realized if all areas of government are coordinated from the highest political and organizational level (Schipper and Pelling, 2006; UNFCCC, 2008b; CCCD, 2009; Prabhakar, 2009). Even when 'political champions' at the highest levels encourage mainstreaming across sectors and departments, competing national priorities will remain an impediment to progress.

Table 6-1 (Section 6.3.1) outlines adaptation to climate change and disaster risk management options for several selected sectors. As the table indicates, adaptation and disaster risk management approaches for many development sectors benefit jointly from ecosystem-based adaptation and integrated land, water, and coastal zone management actions. For example, conservation and sustainable management of ecosystems, forests, land use, and biodiversity have the potential to create win-win disaster risk protection services for agriculture, infrastructure, cities, water resource management, and food security. They can also create synergies between climate change adaptation and mitigation measures (SCBD, 2009; CCCD, 2009), as well as produce many co-benefits that address other development goals, including improvements in livelihoods and human well being, particularly for the poor and vulnerable, and biodiversity conservation, and are discussed further in Section 6.5.2.3 and in Case Studies 9.2.3, 9.2.4, 9.2.5, 9.2.7, 9.2.8, and 9.2.9. Likewise, water resource, land, and coastal zone management options deal with many sectors and issues and jointly provide disaster risk management and adaptation solutions, as mentioned in Case Studies 9.2.6 and 9.2.8 (WHO, 2003; Urwin and Jordan, 2008; UNFCCC, 2008b; CCCD, 2009; WWAP, 2009). Human health is a cross-cutting issue impacted by actions taken in many sectors, as indicated in Table 6-1 and discussed in Case Studies 9.2.2 and 9.2.7.

6.4. Strategies including Legislation, Institutions, and Finance

National systems for managing the risks of extreme events and disasters are shaped by legislative provision, compliance mechanisms, the nature of cross-stakeholder bodies, and financial and budgetary processes that allocate resources to actors working at different scales. These elements help to create the technical architecture of national systems and are often led by national government agencies. However non-technical dimensions of good governance, such as the distribution and decentralization of power and resources, processes for decisionmaking, transparency, and accountability are woven into the technical architecture and are significant factors in determining the effectiveness of risk management systems and actions (UNDP, 2004b, 2009). These technical and non-technical aspects of risk governance vary between countries as governance capacity varies (and, as detailed in Section 6.3, are critical in shaping investment in particular adaptation and disaster risk management options). Accordingly, risks can be addressed through both formal and informal governance modes and institutions in all countries (Jaspars and Maxwell, 2009), but a clear correlation between particular risk governance models and specific political-administrative contexts is difficult to identify (UNISDR, 2011a). The balance between formal or informal, or technical and non-technical, risk governance strategies depends on the economic, political, and environmental contexts of individual countries or scales within countries, and the culture of managing risks (Menkhaus, 2007; Kelman, 2008).

6.4.1. Legislation and Compliance Mechanisms

Disaster risk management legislation commonly establishes organizations and their mandates, clarifies budgets, provides (dis)incentives, and develops compliance and accountability mechanisms (UNDP, 2004b; Llosa and Zodrow, 2011). Creating and improving legislation for disaster risk reduction was included as a priority area in the HFA (UNISDR, 2005) and the majority of countries – in excess of 80% – now have some form of disaster risk management legislation (UNISDR, 2005; Bhavnani et al., 2008). Legislation continues to be considered as an important component of effective national disaster risk management systems (UNDP, 2004b; UNISDR, 2011a) as it creates the legal context of the enabling environment in which others, working at different scales, can act, and it helps to define people's rights to protection from disasters, assistance, and compensation (Pelling and Holloway, 2006). Multi-stakeholder, cross-sector bodies for coordinating disaster risk management actions and implementing the HFA, known commonly as National Platforms, are seen as key advocacy routes for achieving new and improved legislation (UNISDR, 2005, 2007b). Where National Platforms are less prevalent or less well organized, literature suggests that regional disaster management bodies are viewed as responsible for advancing legislation (Pelling and Holloway, 2006; UNISDR, 2007b). With new information on the impacts of climate change, legislation on managing disaster risk may need to be modified and strengthened to reflect changing rights and responsibilities and to support the uptake of adaptation options (UNDP, 2009; Llosa and Zodrow, 2011; see Case Study 9.2.12 on legislation).

There have been few detailed cross-comparative studies that assess the extent to which legislation in different countries is oriented toward managing uncertainty and reducing disaster risk compared with disaster response (Llosa and Zodrow, 2011). Limited evidence suggests that legislation in some countries (such as the United Kingdom, the United States, and Indonesia) has led to a focus on building institutional capacity to help create resilience to disasters at different scales, but even in such cases a strongly reactive culture is retained when observing the system as a whole (O'Brien and Read, 2005; O'Brien, 2006, 2008; UNDP, 2009; O'Brien and O'Keefe, 2010). This has been attributed to lack of political will and insufficient financial and human resources for disaster risk reduction (O'Brien 2006, 2008). Additionally, few studies have assessed whether disaster risk management legislation includes provision for the impact of climate change on disaster risk or whether aspects of managing disaster risk are included in other complementary pieces of legislation

(Case Study 9.2.12; Llosa and Zodrow, 2011), though there are also a very limited number of normative studies on these aspects (Llosa and Zodrow, 2011). However, where reforms of disaster management legislation have occurred, they have tended to: (a) demonstrate a transition from emergency response to a broader treatment of managing disaster risk; (b) recognize that protecting people from disaster risk is at least partly the responsibility of governments; and (c) promote the view that reducing disaster risk is everyone's responsibility (Case Study 9.2.12; UNDP, 2004b; Llosa and Zodrow, 2011).

Vietnam has taken steps to integrate disaster risk management into legislation across key development sectors, including its Land Use Law and Law on Forest Protection. Vietnam's Poverty Reduction Strategy Paper also included a commitment to reduce by 50% those falling back into poverty as a result of disasters and other risks (Pelling and Holloway, 2006). Case Study 9.2.12, in examining legislation development processes in the Philippines and South Africa, highlights a number of components of effective disaster risk management legislation. An act needs to be: (a) comprehensive and overarching; (b) establish management structures and secure links with development processes at different scales; and (c) establish participation and accountability mechanisms that are based on information provision and effective public awareness and education. Box 6-1 supplements these cases with reflections on the process that led to the creation of disaster risk management legislation in Indonesia.

Where risk management dimensions are a feature of national legislation, positive changes are not always guaranteed (UNDP, 2004b). A lack of financial, human, or technical resources and capacity constraints present significant obstacles to full implementation, especially as experience suggests that legislation should be implemented continuously from the national to local level and is contingent on strong monitoring and enforcement frameworks and adequate decentralization of responsibilities and human and financial resources at every scale (UNDP, 2004b; Pelling and Holloway, 2006). In some countries, building codes, for instance, are often not implemented properly because of a lack of technical capacity and political will of officials concerned (UNDP, 2004b). Where enforcement is unfeasible, accountability for disaster risk management actions is extremely challenging; this supports the need for an inclusive, consultative process for discussing and drafting the legislation (UNDP, 2004b; UNISDR, 2007b). Effective legislation includes benchmarks for action, a procedure for evaluating actions, integrated planning to assist coordination across geographical or sectoral areas of responsibility, and a feedback system to monitor risk reduction activities and their outcomes (UNISDR, 2005; Pelling and Holloway, 2006).

6.4.2. Coordinating Mechanisms and Linking across Scales

As the task of managing the risks of changing climate conditions and climate extremes and disasters cuts across the majority of sectors and involves a wide range of actors, multi-stakeholder and cross-government mechanisms are commonly cited as preferred way to 'organize' disaster

> ## Box 6-1 | Enabling Disaster Risk Management Legislation in Indonesia
>
> *Indonesia: Disaster Management Law (24/2007)*
>
> The legislative reform process in Indonesia that resulted in the passing of the 2007 Disaster Management Law (24/2007) created a stronger association between disaster risk management and development planning processes. The process was considered successful due to the following factors:
>
> - **Strong, visible professional networks** – Professional networks born out of previous disasters meant a high level of trust and willingness to coordinate and became pillars of the legal reform process. The political and intellectual capital in these networks, along with leadership from the MPBI (The Indonesian Society for Disaster Management), was instrumental in convincing the lawmakers about the importance of disaster management reform.
> - **Civil society leading the advocacy** – Civil society leading the advocacy for reform has resulted in CSOs being recognized by the Law as key actors in implementing disaster risk management in Indonesia.
> - The impact of the 2004 South Asian tsunami helping to create a supportive **political environment** – The reform process was initiated in the aftermath of the tsunami that highlighted major deficiencies in disaster management. However, the direction of the reform (from emergency management toward disaster risk reduction) was influenced by the international focus, through the HFA, on disaster risk reduction.
> - An **inclusive drafting process** – Consultations on the new Disaster Management Law were inclusive of practitioners and civil society, but were not so far-reaching as to delay or lose focus on the timetable for reform.
> - Consensus that **passing an imperfect law is better than no law at all** – An imperfect law can be supplemented by additional regulations, which helps to maintain interest and focus.
>
> Source: UNDP (2004b, 2009); Pelling and Holloway (2006).

risk management systems at the national level (UNISDR, 2005, 2007b; see Section 6.3.3), as well as for addressing the challenges associated with adaptation to climate change (ONERC, 2009). The HFA terms these 'National Platforms,' defined as a "generic term for national mechanisms for coordination and policy guidance on disaster risk reduction that are multi-sectoral and inter-disciplinary in nature, with public, private and civil society participation involving all concerned entities within a country" (UNISDR, 2005). In some countries such coordinating mechanisms are referred to by other names (Hay, 2009; Gero et al., 2011) but essentially perform the same function. Guidelines on establishing National Platforms suggest that they need to be built on existing relevant systems and should include participation from different levels of government, key line ministries, disaster management authorities, scientific and academic institutions, civil society, the Red Cross/Red Crescent, the private sector, opinion shapers, and other relevant sectors associated with disaster risk management (UNISDR, 2007b). Evaluations and reflections on the effectiveness of National Platforms for delivering results on the HFA and on disaster risk management more broadly indicate widely varying results (GTZ/DKKV, 2007; UNISDR, 2007c, 2008c; UNISDR/DKKV/Council of Europe, 2008; Sharma, 2009). An assessment in Asia found National Platforms struggling to obtain the legal mandate to secure full participation of stakeholders, particularly NGOs, difficulty in obtaining sustainable funding sources, and challenges associated with translating intent into implementation (Sharma, 2009). On the other hand, pockets of evidence exist where National Platforms have succeeded in generating senior political commitment for disaster risk reduction, in strengthening integration of disaster risk reduction into national policy and development plans, and in establishing institutions and programs on disaster risk management with engagement from academia, media, and the private sector (UNISDR, 2008b; Sharma, 2009). This assessment found only a limited number of genuinely independent studies on the effectiveness of National Platforms, with evidence particularly weak in Africa and elsewhere.

While the evidence again suggests significant differences between countries, on balance, national coordination mechanisms for adaptation to climate change and disaster risk management remain largely disconnected, although evidence suggests that the trajectory is one of improvement (National Platform for Kenya, 2009; Mitchell et al., 2010b; discussed in Chapter 1). Benefits of improved coordination between adaptation to climate change and disaster risk management bodies, and development and disaster management agencies, include the ability to (i) explore common tradeoffs between present and future action, including addressing human development issues and reducing sensitivity to disasters versus addressing post-disaster vulnerability; (ii) identify synergies to make best use of available funds for short- to longer-term adaptation to climate risks as well as to tap into additional funding sources; (iii) share human, information, technical, and practice resources; (iv) make best use of past and present experience to address emerging risks; (v) avoid duplication of project activities; and (vi) collaborate on reporting requirements (Mitchell and Van Aalst, 2008). Barriers to integrating disaster risk management and adaptation coordination mechanisms include the underdevelopment of the 'preventative' component of disaster risk management, the paucity of projects that

> **Box 6-2 | National and Sub-National Coordination for Managing Disaster Risk in a Changing Climate: Kenya**
>
> Kenya's National Platform is situated under the Office of the President and has made significant achievements in coordinating multiple stakeholders, but is constrained by limited resources and lack of budgets for disaster risk reduction in line ministries (National Platform for Kenya, 2009). Some key constraints of the national system are recognized as being difficulties in integrating disaster risk reduction in planning processes in urban and rural areas and lack of data on risks and vulnerabilities at different scales (Few et al., 2006). In this regard, Nairobi has experienced periods of drought and heavy rains in the last decade, prompting action to reduce exposure and vulnerability to what is perceived as changing hazard trends (ActionAid, 2006). Increasing exposure and vulnerability has resulted from a rapid expansion of poor people living in informal settlements around Nairobi, leading to houses of weak building materials being constructed immediately adjacent to rivers and blocking natural drainage areas. While data and coordination systems are still lacking, the Government of Kenya has established the Nairobi Rivers Rehabilitation and Restoration Programme (African Development Bank Group, 2010), designed to install riparian buffers, canals, and drainage channels, while also clearing existing channels. The Programme also targets the urban poor with improved water and sanitation, paying attention to climate variability and change in the location and design of wastewater infrastructure and environment monitoring for flood early warning (African Development Bank Group, 2010). This demonstrates the kind of options for investments that can be achieved in the absence of a fully fledged nationally coordinated disaster management system and in the absence of complete multi-hazard, exposure, and vulnerability data sets.

integrate climate change in the context of disaster risk management, disconnects between different levels of government, and the weakness of both disaster risk management and adaptation to climate change in national planning and budgetary processes (Few et al., 2006; Mitchell and Van Aalst 2008; Mitchell et al., 2010b) (see Box 6-2).

While national level coordination is important and the majority of risks associated with disasters and climate extremes are owned by national governments and are managed centrally (see Section 6.2.1), sources suggest that decentralization can be an effective risk management strategy, especially in support of community-based disaster risk management processes (Mitchell and Van Aalst, 2008; GNDR, 2009; Scott and Tarazona, 2011). However, there are few studies that critically examine the effectiveness of decentralization of disaster risk management in detail (Twigg, 2004; Tompkins et al., 2008; Scott and Tarazona, 2011). One such study of four countries – Colombia, Mozambique, Indonesia, and South Africa – found that effective decentralization of disaster risk reduction can be constrained by (a) low capacity at the local level; (b) funds dedicated to disaster risk reduction often being channeled elsewhere; (c) the fact that decentralization does not automatically lead to more inclusive decisionmaking processes; (d) an appreciation that decentralized systems face significant communications challenges; and (e) knowledge that robust measures for ensuring accountability and transparency are vital for effective disaster risk management but are often missing (Scott and Tarazona, 2011). It appears that motivation for management at a particular scale promises to influence how well the impacts of disasters and climate change are managed, and therefore affect disaster outcomes (Tsing et al., 1999). Decisions made at one scale may have unintended consequences for another (Brooks and Adger, 2005), meaning that governance decisions will have ramifications across scale and contexts. In all cases, the selection of a framework for governance of disasters and climate change-related risks may be issue- or context-specific (Sabatier, 1986).

6.4.3. Finance and Budget Allocation

Governments in the past have ignored catastrophic risks in decisionmaking, implicitly or explicitly exhibiting risk-neutrality (Mechler, 2004). This is consistent with the Arrow Lind theorem (Arrow and Lind, 1970), according to which a government may be well equipped to efficiently (i) pool risks as it possesses a large number of independent assets and infrastructure so that the aggregate risk converges to zero, and/or (ii) spread risk across the taxpaying population base, so that per capita risk accruing to risk-averse households converges to zero. In line with this theorem, due to their ability to spread and diversify risks, governments are sometimes termed 'the most effective insurance instrument of society' (Priest 1996). Accordingly, it has been deduced that, although individuals are risk-averse [to disasters risk], governments can take a risk-neutral approach. However, the experiences of highly exposed countries suggest otherwise and have led to a recent paradigm shift, with governments changing from being 'risk neutral' to being risk averse and managing disaster risks. Many highly exposed developing and developed countries (especially in the wake of the recent financial crisis) have very limited economic means, rely on small and exhausted tax bases, have high levels of indebtedness, and are unable to raise sufficient and timely capital to replace or repair damaged assets and restore livelihoods following major disasters. This can lead to increased impacts of disaster shocks on poverty and development (OAS, 1991; Linnerooth-Bayer et al., 2005; Hochrainer, 2006; Mahul and Ghesquiere, 2007; Cummins and Mahul, 2009). Exposed countries thus have had to rely on donors to 'bail' them out after events, although ex-post assistance usually only provides partial relief and reconstruction funding, and such assistance is also often associated with substantial time lags (Pollner, 2000; Mechler, 2004).

Furthermore, extreme events that are associated with large losses may lead to important downstream economic effects (see Section 4.5), causing depressed incomes and reduced ability to share the losses.

Table 6-2 | Government liabilities and disaster risk. Modified from Polackova Brixi and Mody (2002).

Liabilities	Direct: obligation in any event	Contingent: obligation if a particular event occurs
Explicit: Government liability recognized by law or contract	Foreign and domestic sovereign borrowing, expenditures by budget law and budget expenditures	State guarantees for non-sovereign borrowing and public and private sector entities, reconstruction of public infrastructure
Implicit: A 'moral' obligation of the government	Future recurrent costs of public investment projects, pensions, and health care expenditure	Default of sub-national government as public or private entities provide disaster relief

Consequently, a risk-neutral stance in dealing with catastrophic risk (implying that the consideration of risk broadly in terms of means – the statistical expectation – is sufficient) may not be suitable for exposed developing countries with limited diversification of their economies or small tax bases. Accordingly, assessing and managing risks over the whole spectrum of probabilities is gaining momentum (Cardenas et al., 2007; Cummins and Mahul, 2009). As the Organization of American States suggests: "Government decisions should be based on the opportunity costs to society of the resources invested in the project and on the loss of economic assets, functions and products. In view of the responsibility vested in the public sector for the administration of scarce resources, and considering issues such as fiscal debt, trade balances, income distribution, and a wide range of other economic and social, and political concerns, governments should not act risk-neutral" (OAS, 1991). Also, in more developed economies, less-pronounced but still considerable effects imposed by events linked to climate variability can be identified. This has been shown by the Austrian political and fiscal crisis in the aftermath of large-scale flooding that led to billions of Euros in losses in 2002 (Mechler et al., 2010).

Budget and resource planning for extremes is not an easy proposition. Governments commonly plan and budget for *direct* liabilities, that is, liabilities that manifest themselves through certain and annually recurrent events. Those liabilities are of explicit (as recognized by law or contract), or implicit nature (moral obligations) (see Table 6-2). Yet, governments are not good at planning for contingencies even for probable events, let alone improbable events. Explicit, contingent liabilities deal with the reconstruction of infrastructure destroyed by events, whereas implicit obligations are associated with providing relief – commonly considered as a moral liability for governments (Polackova Brixi and Mody, 2002). In many countries, governments do not explicitly plan for contingent liabilities, and rely on reallocating their resources following disasters and raising capital from domestic and international donations to meet infrastructure reconstruction needs and costs.

More recently, some developing and transition countries that face large contingent liabilities in the aftermath of extreme events and associated financial gaps have begun to plan for and consider contingent natural events (also see Section 6.5.3). Mexico, Colombia, and many Caribbean countries now include contingent liabilities in their budgetary process and eventually even transfer some of these risks (Cardenas et al., 2007; Linnerooth-Bayer and Mechler, 2007; Cummins and Mahul, 2009; see Box 6-3). Similarly, many countries have also started to focus on improving human development conditions as an adaptation strategy for climate change and extreme events, particularly with the help of international agencies such as the World Bank. These deliberations are

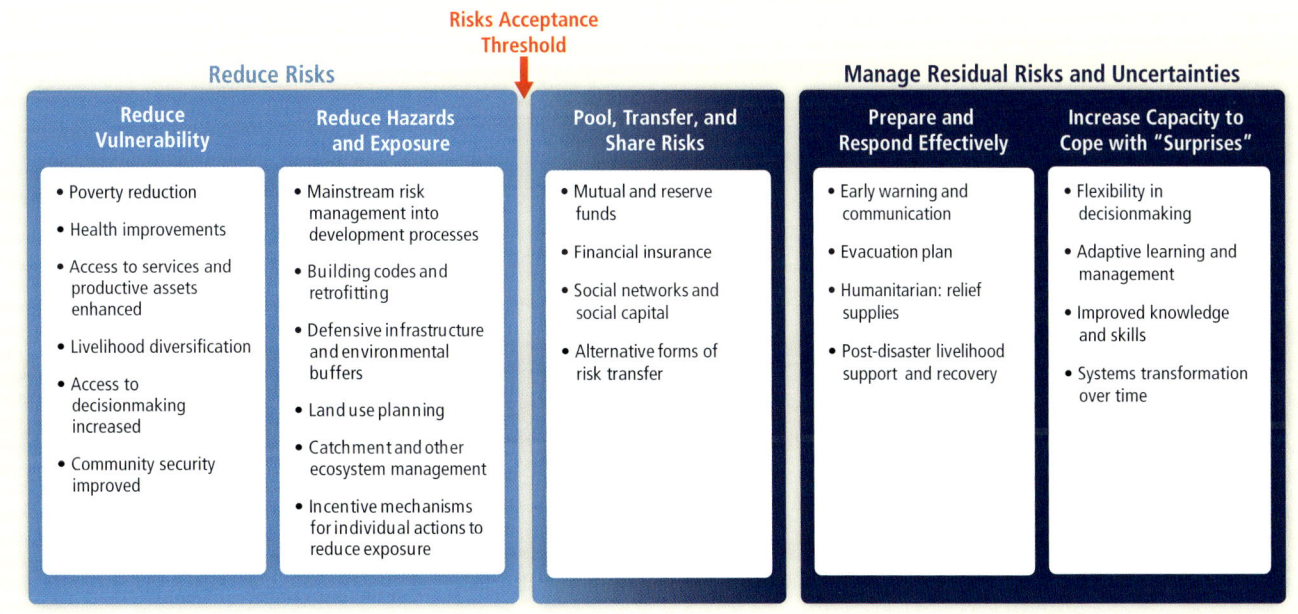

Figure 6-3 | Complementary response measures for observed and projected disaster risks supported by respective institutional and individual capacity for making informed decisions.

Box 6-3 | Mexico's Fund for Natural Disasters, FONDEN

Mexico is exposed to natural hazards due to its location within one of the world's most active seismic regions and in the path of hurricanes and tropical storms originating in the Caribbean Sea and Atlantic and Pacific Oceans. There have been many disaster events and in the past severe hurricane disasters (in addition to earthquakes) have created large fiscal liabilities and imbalances (Cardenas et al., 2007). Given high perceived financial vulnerability to disasters, in 1994, the Mexican Government passed a law that required federal, state, and municipal public assets to be insured. This was intended to relieve the central government of the obligation of having to pay for the reconstruction of public infrastructure, although the adequate level of insurance, particularly for very large events remained a concern. As a next step, in 1996, the national government established a system of allocating resources for disaster spending (FONDEN) in order to enhance the country's financial preparedness for disaster losses and prevent imbalances in the federal government finances derived from outlays caused by catastrophes. FONDEN serves as last-resort funding for uninsurable losses, such as emergency response and disaster relief expenditures. In addition to this budgetary program, in 1999, a reserve fund was created that accumulates the surplus of the previous year's FONDEN budget item (Cardenas et al., 2007).

After the initial phase that was characterized by spending in line with requirements caused by disaster events, one concern for the disaster management authorities became the regular demands on the funds in non-disaster years. As a consequence, budgeted FONDEN resources were declining while demands on FONDEN's resources were becoming more volatile, and outlays often exceeded budgeted funds causing the reserve fund to decline. In 2005, after the severe hurricane season affecting large parts of coastal Mexico, the fund was finally exhausted. This forced the Mexican Government to look at alternative risk financing strategies, which included hedging against disaster shocks, government agencies at all levels providing their insurance protection independent of FONDEN, and that FONDEN itself should only indemnify losses that exceed the financial capacity of the federal, local, or municipal government agencies. In 2006, Mexico became the first transition country to transfer part of its public sector catastrophe risk to the international reinsurance and capital markets, and, in 2009, the transaction was renewed for another three years, covering both hurricane and earthquake risk (Cardenas et al., 2007).

in line with the described 'no' and 'low regrets' strategies discussed in Section 6.3.1.

6.5. Practices including Methods and Tools

With some success and with many challenges, countries are increasingly adopting a diverse range of approaches, methods, and tools to manage disaster risk and adapt to a changing climate, with the intention of building a safe, secure society. This section discusses efforts made in building a culture of safety (Section 6.5.1), which includes methods associated with assessing and communicating risk; reducing climate-related disaster risks (Section 6.5.2); transferring and sharing residual risks (Section 6.5.3); and managing the impacts of disasters holistically (Section 6.5.4), as disaster risks can never be reduced to zero. Accordingly, it is important to recognize that the approaches, methods, and tools discussed here are complementary, often overlapping, and can be pursued simultaneously. Whereas the Summary for Policymakers includes a visual representation of the range of such approaches (see Figure SPM-2), Figure 6-3 on the previous page is tailored to incremental action at the national level.

Figure 6-3 characterizes the range of risk management and adaptation options open to stakeholders involved in national systems for managing disaster risk. Such options exist along a continuum of action, with choices between different options being dependent on the quality of information and how it is communicated, the findings of risk assessments, the culture of risk management/acceptability of risk, and on capacities and resources. In practice, different options will likely be pursued simultaneously and will have a high degree of co-dependence.

6.5.1. Building a Culture of Safety

Building a culture of safety involves several strategies and activities that start with the assessment of risk factors and development of information systems that provide relevant information for critical decisionmaking. It also involves understanding the large variety of beliefs and core value systems, which will help determine decisions made by different actors and stakeholders. A key ingredient is appropriate public education and awareness raising, and as such, early warning systems play an important role in managing residual risk as they can provide timely warnings to exposed communities and thus can promote action for a quick response. Time series empirical data used to generate risk assessments, including those contributing to early warnings, are also critical for long-term planning because of their relevance in generating appropriate information about adequate land use planning, for example, to reduce climatic risks. As examples, in the same sense, analyzed information about climate-adapted infrastructure, enhanced human development, ecosystems protection, risks transfer, and sharing and managing the impacts of climate-related disasters can play a fundamental role in building a culture and practice of human safety.

6.5.1.1. Assessing Risks and Maintaining Information Systems

As discussed widely in Chapter 1, the first key step in managing risk is to assess and characterize it. In terms of risk factors, disaster risk is commonly defined by three elements: the hazard, exposure of elements, and vulnerability (Swiss Re, 2000; Kuzak, 2004; Grossi and Kunreuther, 2005; CACCA, 2010). Thus, understanding risk involves observing and recording hazards and hazard analysis, studying exposure and drivers of vulnerability, and vulnerability assessment. Responding to risks is dependent on the way risk-based information is framed in the context of public perception and risk management needs.

Given the 'public good' nature of much of disaster-related information (Benson and Clay, 2004), governments have a fundamental role in providing good-quality and context-specific risk information about, for example, the geographical distribution of people, assets, hazards, risks, and disaster impacts and vulnerability to support disaster risk management (McBean, 2008). Good baseline information and robust time series information are key for long-term risk monitoring and assessments, not only for hazards, but also for evaluating the evolution of vulnerability and exposure (McEntire and Myers, 2004; Aldunce and León, 2007). Regular updating of information about hazards, exposure, and vulnerability are also necessary because of the dynamic nature of disaster risk, especially due to the effects of climate change and the associated uncertainty this creates (UNISDR, 2004; Prabhakar et al., 2009; CACCA, 2010).

A key component in the risk assessment process is to determine exposed elements at risk. This may relate to persons, buildings, infrastructure (e.g., water and sewer facilities, roads, and bridges), agricultural assets, livelihoods, ecosystems, natural infrastructure, and ecosystem services in harm's way that can be impacted in case of a disaster event. For national level assessments, their aggregate values are of interest. Ideally, this would be based on national asset inventories, national population census, and other national information. In practice, collecting an inventory on assets and their values often proves very difficult and expensive due to the heterogeneity and sheer number of the examined elements (see Cummins and Mahul, 2009). In addition, risk management processes require identifying those elements of the social process that also contribute to vulnerability – such as organizational and economic capacities, human development status of communities at risk, and capacity to respond to disasters (Lavell, 1996; Cardona et al., 2010) – as well as assessing the impacts following disaster events (ECLAC, 2003; Benson and Clay, 2004). Considerable progress has been made in the generation and use of such information including in some developing countries (Benson and Clay, 2004; UNISDR 2009c). Nevertheless, in many countries this is not a regular practice and efforts to document

Table 6-3 | Information requirements for selected disaster risk management and adaptation to climate change activities. Adapted from Wilby (2009).

	Activities	Examples of information needs
Cross-cutting	Climate change modeling	Time series information on climate variables – air and sea surface temperatures, rainfall and precipitation measures, wind, air circulation patterns, and greenhouse gas levels
	Hazard zoning and 'hot spot' mapping	Georeferenced inventories of landslide, flood, drought, and cyclone occurrence and impacts at local, sub-national and national levels
	Human development indicators	Geospatial distribution of poverty, livelihood sources, access to water and sanitation
	Disbursement of relief payments	Household surveys of resource access, social well-being, and income levels
	Seasonal outlooks for preparedness planning	Seasonal climate forecasts; sea surface temperatures; remotely sensed and *in situ* measurements of snow cover/depth, soil moisture, and vegetation growth; rainfall-runoff; crop yields; epidemiology
	A system of risk indicators reflecting macro and financial health of nation, social and environmental risks, human vulnerability conditions, and strength of governance (Cardona et al., 2010)	Macroeconomic and financial indicators (Disaster Deficit Index) Measures of social and environmental risks Measures of vulnerability conditions reflected by exposure in disaster-prone areas, socioeconomic fragility, and lack of social resilience in general Measures of organizational, development, and institutional strengths
Flood risk management	Early warning systems for fluvial, glacial, and tidal hazards	Real-time meteorology and water-level telemetry; rainfall, stream flow, and storm surge; remotely sensed snow, ice, and lake areas; rainfall-runoff model and time series; probabilistic information on extreme wind velocities and storm surges
	Flooding hot spots, and structural and non-structural flood controls	Rainfall data, rainfall-runoff, stream flow, floods, and flood inundation maps Inventories of pumps, stream gauges, drainage and defense works; land use maps for hazard zoning; post-disaster plan; climate change allowances for structures; floodplain elevations
	Artificial draining of proglacial lakes	Satellite surveys of lake areas and glacier velocities; inventories of lake properties and infrastructure at risk; local hydro-meteorology
Drought management	Traditional rain and groundwater harvesting, and storage systems	Inventories of system properties including condition, reliable yield, economics, ownership; soil and geological maps of areas suitable for enhanced groundwater recharge; water quality monitoring; evidence of deep-well impacts
	Long-range reservoir inflow forecasts	Seasonal climate forecast model; sea surface temperatures; remotely sensed snow cover; *in situ* snow depths; multi-decadal rainfall-runoff series
	Water demand management and efficiency measures	Integrated climate and river basin water monitoring; data on existing systems' water use efficiency; data on current and future demand metering and survey effectiveness of demand management

impacts are started only after major disasters (Prabhakar et al., 2009). Regular monitoring of vulnerability is also at a nascent stage (Dilley, 2006; Cardona et al., 2010). Table 6-3 on the previous page shows a sample of the kinds of information required for effective disaster risk management and adaptation to climate change activities.

Country- and context-specific information on disaster impacts and losses, including baseline data about observations (different types of losses, weather data) from past events, is often very limited and of mixed quality (see Embrechts et al., 1997; Carter et al., 2007). Data records at best may date back several decades, and thus often would provide only one reference data point for extreme events, such as a 100-year event (see Section 3.2.1). Data on losses from extremes can also be systematically biased due to high media attention (Guha-Sapir and Below, 2002). At times the data on losses are incomplete, as in the Pacific small island developing states, because of limited capacity to systematically collect information at the time of disaster, or because of inconsistent methodologies and the costs of measures used (Chung, 2009; Lal, 2010).

International disaster impact databases are available, such as the EM-DAT database of the Centre for the Epidemiology of Disasters (CRED) in Brussels, Desinventar maintained by a network of scientists involved in studying disasters in Latin America (Red de Estudios Sociales en Prevención de Desastres en América Latina – LA RED), as well as databases of reinsurers such as Munich Re. Comparisons of international and national disaster loss databases have shown significant variations in documented losses due to inconsistencies in the definition of key parameters and estimation methods used. This emphasizes the need to standardize parameter definitions and estimation methods (Guha-Sapir and Below, 2002; ECLAC, 2003; Tschoegl et al., 2006). For some countries, a reasonable quality and quantity of information may exist on the direct impacts, particularly where the reinsurance industry, consulting firms, and multi-lateral financial institutions have worked together with the research communities. Limited information is generally available on socially relevant effects of disasters, such as the incidence of health effects after a disaster as well as the impacts on ecosystems, which have not been well studied (Benson and Twigg 2004). Furthermore, the assessment of indirect disaster impacts on social or economic systems, such as on income-generating sectors and national savings, needs greater attention (ECLAC, 2003; Benson and Clay, 2004). Such information can often also be very useful in order to assess risks by using statistical estimation techniques (Embrechts et al., 1997) or catastrophe modeling approaches (Grossi and Kunreuther, 2005).

6.5.1.2. Preparedness: Risk Awareness, Education, and Early Warning Systems

National governments create the environment and communication channels to develop and disseminate different kinds of information on, for example, the hazards that affect different populations and preparedness for disaster response. Numerous studies indicate that up-to-date and robust early warning systems play a critical role in reducing the impacts of potential disasters and enable populations to protect lives and some property and infrastructure (White et al., 2004; Aldunce and León, 2007; McBean, 2008; Rogers and Tsirkunov, 2010), and as illustrated in Case Study 9.2.11.

Traditionally, early warning systems have been interpreted narrowly as technological instruments for detecting and forecasting impending hazard events and for issuing alerts (Rogers and Tsirkunov, 2010). However, this interpretation does not clarify whether warning information is received by or helpful to the population it serves or is actually used to reduce risks (Basher, 2006; Rogers and Tsirkunov, 2010). As noted in Case Study 9.2.11, the HFA 2005-2015 stated a need for more than just accurate predictions, stressing that early warning systems should be "people centered" and that warnings need to be "timely and understandable to those at risk" and include "guidance on how to act upon warnings" (UNISDR, 2005).

Governments also maintain early warning systems to warn their citizens and themselves about impending creeping climate- and weather-related hazards. For example, 'early warnings' of potentially poor seasons have been successful at informing key actions for agricultural planning on longer time scales and for producing proactive responses (Meinke et al., 2006; Vogel and O'Brien, 2006). Case Study 9.2.11 provides examples of early warning systems for short-response hazards as well as for creeping hazards operating on time scales from weeks to seasonal. This case study also highlights the possibility of using weather and climate predictions for timeframes longer than a few days to provide advanced warning of extreme conditions, which has been only a very recent development. Studies indicate that successful early warning systems are reliant on close inter-institutional collaboration between national meteorological and hydrological services and the agencies that directly intervene in rural areas, such as extension services, development projects, and civil society organizations (Meinke et al., 2006; Vogel and O'Brien, 2006; Rogers and Tsirkunov, 2010).

An effective early warning system delivers accurate, timely, and meaningful information, with its success dependent on whether the warnings trigger effective responses (UNISDR, 2005; Basher, 2006; Gwimbi, 2007; Auld, 2008a; van Aalst, 2009; Rogers and Tsirkunov, 2010). Warnings fail in both developing and developed countries for a number of reasons, including inaccurate weather and climate forecasting, public ignorance of prevailing conditions of vulnerability, failure to communicate the threat clearly or in time, lack of local organization, and failure of the recipients to understand or believe in the warning or to take suitable action (UNISDR, 2006; Auld, 2008a; Rogers and Tsirkunov, 2010). To be effective and complete, an early warning system typically is composed of four interacting elements (Basher, 2006; UNISDR, 2006): (1) generation of risk knowledge including monitoring and forecasting; (2) surveillance and warning services; (3) dissemination and communication; and (4) response capability. Warnings are received and understood by the target audience and are most relevant when the communications have meaning that is shared between those who issue the forecasts, local knowledge, and the decisionmakers they are intended to inform (Basher, 2006; UNISDR, 2006; Auld, 2008a; Case Study 9.2.2). Because emergency responders, the media, and the public often are unable to translate the scientific information on forecast hazards in warnings into risk levels and responses, early warning systems are most effective when their users can identify and interpret the general warning messages into simple and relevant local impacts and actions (e.g., flash flood warning and the need to evacuate areas at risk), prioritize the most dangerous hazards, assess potential contributions from cumulative and sequential events to risks, and identify thresholds linked to escalating risks for infrastructure, communities, and disaster response (UNISDR, 2006; Auld, 2008a).

Different hazards and different sectors often require unique preparedness, warnings, and response strategies (Basher, 2006; UNISDR, 2006). For example, the needs and responses behind a warning of a drought, a tornado, a cyclone, or a fire are very different. Some hazards may represent singular extreme events, sequences, or compound combinations of hazards while other hazards can be described as 'creeping' or accumulations of events (or non-events). For example, the World Meteorological Organization (WMO), national meteorological and hydrological services, the World Health Organization (WHO), the Food and Agriculture Organization (FAO), and others recognize that combinations of weather and climate hazards can result in complex emergency response situations and are working to establish multi-hazard early warning systems for complex risks such as heat waves and vector-borne diseases (UNISDR, 2006; WMO, 2007) and early warnings of pests and food safety threats and disease outbreaks (e.g., prediction of a potential desert locust crisis) (WMO, 2004, 2007; FAO, 2010). Other 'creeping' hazards can evolve over a period of days to months; floods and droughts, for example, can result from cumulative or sequential multi-hazard events, especially when accompanied by an already existing vulnerability, while other hazards such as accumulated precipitation can lead to critical infrastructure failure (Basher, 2006; Auld, 2008a; Rogers and Tsirkunov, 2010). Section 3.1.3 provides more detail on compound, multiple, and creeping hazards.

Studies indicate that an understanding by the public and community organizations of their risks and vulnerabilities are critical but insufficient for risk management and that early warning systems need to be complemented by preparedness programs as well as public education and awareness programs (Basher, 2006; UNISDR, 2006; Gwimbi, 2007; Rogers and Tsirkunov, 2010). This requires systematic linkages and integration between early warning systems and contingency planning processes (Pelham et al., 2011). For example, a significant long-term social protection program known as the Productive Safety Net Programme (PSNP) was implemented in Ethiopia in 2007 in response to experiences from a series of drought-related disaster responses during the late 1990s and early 2000s (Pierro and Desai, 2008; Conway and Schipper, 2011). The aim of the PSNP was to shift institutional approaches away from just emergency responses and into more sustainable livelihood approaches involving asset protection and food security. Under this program, millions of people in 'chronically' food-insecure households in rural Ethiopia received resources from the PSNP through cash transfers or food payments for their participation in labor-intensive public works projects with a particular focus on environmental rehabilitation (Conway and Schipper, 2011). The case study on drought (Case Study 9.2.3) also emphasizes the importance of proactive steps in the form of drought preparedness and mitigation, and improved monitoring and early warning systems.

Some studies indicate that public awareness and support for disaster prevention and preparedness are often high immediately after a major disaster event and that such moments can be capitalized on to strengthen and secure the sustainability of, for example, early warning systems (Basher, 2006; Rossetto, 2007). It should be noted that such windows require the pre-existence of a social basis for cooperation that, in turn, supports a collaborative framework between research and management (Rossetto, 2007; Tompkins et al., 2008; Pelham et al., 2011).

The timing and form of climatic information (including forecasts and projections), and access to trusted guidance to help interpret and implement the information and projections in decisionmaking processes, may be more important to individual users than improved reliability and forecast skill (Pulwarty and Redmond, 1997; Rayner et al., 2005; Gwimbi, 2007; Rogers and Tsirkunov, 2010). Decisionmakers typically manage risks holistically, while scientific information is generally derived using reductionist approaches (Meinke et al., 2006). The net outcome can be a 'disconnect' between scientists and decisionmakers with the result that climate and hydro-meteorological information can be developed that, although scientifically sound, may lack relevance to the decisionmaker (Cash and Buizer, 2005; Meinke et al., 2006; Vogel and O'Brien, 2006; Averyt, 2010). Perceptions of irrelevance, inconsistency, confusion, or doubt can delay action (NRC, 2009). Some studies (Lowe, 2003; Glantz, 2005; Meinke et al., 2006; Feldman and Ingram, 2009) advise scientists and practitioners to work together to produce trustworthy knowledge that combines scientific excellence with social relevance, a point also emphasized in Case Study 9.2.2 on fire. These studies suggest that decision support activities should be driven by users' needs, not by scientific research priorities, and that these user needs are not always known in advance, but should be identified collaboratively and iteratively in ongoing two-way communication between knowledge producers and decisionmakers (Cash and Buizer, 2005; NRC, 2009). It has been suggested that this ongoing interaction, two-way communication, and collaboration allows scientists and decisionmakers to get to know each other, to develop an understanding of what decisionmakers need to know and what science can provide, to build trust, and, over time, develop highly productive relationships as the basis for effective decision support (Feldman and Ingram, 2009; NRC, 2009; Averyt, 2010).

Since early warning information systems are multi-jurisdictional and multidisciplinary, they usually require anticipatory coordination across a spectrum of technical and non-technical actors. National governments can play an important role in setting the high-level policies and supporting frameworks involving multiple organizations, in adopting multi-hazard and multi-stakeholder approaches, and in promoting community-based early warning systems (Pulwarty et al., 2004; Basher, 2006, UNISDR, 2010). National governments can also interact with regional and international governments and agencies to strengthen early warning capacities and to ensure that warnings and related responses are directed toward the most vulnerable populations (Basher, 2006; UNISDR, 2010). At the same time, national governments can also play an important role in supporting regions and sub-national governments in developing operational and local response capabilities (Basher, 2006; UNISDR, 2010; see Section 6.5.4). In Japan and the Mekong region, for example, in addition to using an early warning system based on extensive flood modeling exercise, the emergency basin-level management relies on the flood mitigation capacity of paddy fields (Masumoto et al., 2006, 2008).

6.5.2. Reducing Climate-Related Disaster Risk

National climate-related disaster risk reduction activities include a broad range of options that vary from safe infrastructure and building codes to those aimed to enhance and protect natural ecosystems, support human development and even 'build back better' following a disaster. Each of these strategies can prove minimally effective in isolation but highly effective in combination. These and other different options, along with their limitations (e.g. lack of information and understanding, human resource capacity, scientific requirements, financing) are addressed in the following subsections, noting how risk reduction measures are increasingly being considered as good practices to promote adaptation to climate change.

6.5.2.1. Applying Technological and Infrastructure-Based Approaches

Climate change has the potential to directly and indirectly impact the safety of existing infrastructure and to alter engineering and maintenance practices, and will require changes in building codes and standards where they exist (Bourrelier et al., 2000; Füssel, 2007; Wilby, 2007; Auld, 2008b; Stevens, 2008; Hallegatte, 2009). The changing climate also has the potential regionally to increase premature deterioration and weathering impacts on the built environment, exacerbating vulnerabilities to climate extremes and disasters and negatively impacting the expected and useful life spans of structures (Auld, 2008b; Larsen et al., 2008; Stewart et al., 2011). As noted in Case Study 9.2.8, people living with un-adapted and inadequate infrastructure and housing will be more at risk from climate change.

With projected increases in the magnitude and/or frequency of some extreme events in many regions (see Chapter 3), small increases in climate extremes above thresholds or regional infrastructure 'tipping points' have the potential to result in large increases in damages to all forms of existing infrastructure nationally and to increase disaster risks (Coleman, 2002; Munich Re, 2005; Auld, 2008b; Larsen et al., 2008; Kwadijk et al., 2010; Mastrandrea et al., 2010). Since infrastructure systems, such as buildings, water supply, flood control, and transportation networks often function as a whole or not at all, an extreme event that exceeds an infrastructure design or 'tipping point' can sometimes result in widespread failure and a potential disaster (Ruth and Coelho, 2007; Haasnoot et al., 2009). For example, a break in a water main, dike, or bridge can impact other systems and sectors and render the regional system incapable of providing needed services (Ruth and Coelho, 2007). These infrastructure thresholds or adaptation 'tipping points' become important when considering sensitivities to climate change and adaptation and disaster risk reduction options for the future (see Section 6.6.1 for further discussion on thresholds and management of climate change uncertainties). Infrastructure thresholds refer here to the critical climate conditions where acceptable technical, economic, spatial, or societal limits are exceeded and the current built environment system is no longer "future climate proof" (i.e., it fails, requiring proactive adaptation actions and changes in infrastructure codes, standards, and management processes) (Auld, 2008b; Haasnoot et al., 2009; Kwadijk et al., 2010; Mastrandrea et al., 2010).

The need to address the risk of climate extremes and disasters in the built environment and urban areas, particularly for low- and middle-income countries, is one that is not always fully appreciated by many national governments and development and disaster specialists (Rossetto, 2007; Moser and Satterthwaite, 2008). Low- and middle-income countries, which account for close to three-quarters of the world's urban populations, are at greatest risk from extreme events and also have far less capacity than do high-income countries, largely due to backlogs in protective infrastructure and services and limitations in urban government (Satterthwaite et al., 2007; Moser and Satterthwaite, 2008). Rapid growth and expansion in urban areas, particularly in developing countries, can outpace infrastructure development and lead to a lack of infrastructure services for housing, sewer systems, effective transportation, and emergency response and increased vulnerability to weather and climate extremes (Satterthwaite et al., 2007; Birkmann et al., 2011). These impacts from the changing climate will be particularly severe for populations living in poor-quality housing on illegally occupied land, where there is little incentive for investments in more resilient buildings or infrastructure and service provision (Freeman and Warner, 2001; Satterthwaite et al., 2007; Birkmann et al., 2011). Case Study 9.2.8 provides further discussion on the best adaptation and risk management practices for cities and their built environment.

An inevitable result of potentially increased damages to infrastructure will be a dramatic increase in the national resources needed to restore infrastructure and assist the poor affected by damaged infrastructure (Freeman and Warner, 2001). A study by the Australian Academy of Technological Sciences and Engineering concluded that national retrofit measures will be needed to safeguard existing infrastructure in Australia and new adaptation approaches and national codes and standards will be required for construction of new infrastructure (Stevens, 2008). Recommendations reported from this study call for research to fill gaps on the future climate risks, comprehensive risk assessments for existing critical climate-sensitive infrastructure, development of information and supporting tools (e.g., non-stationary extreme value analysis methods) about future climate change events, investigation of the links between soft and hard engineering solutions, and strengthened research efforts to improve the modeling of small-scale climate events (Wilby, 2007; Auld, 2008b; Stevens, 2008).

The recommended national adaptation options to deal with projected impacts to the built environment range from deferral of actions pending development of new climate change information to modification of infrastructure components according to national guidance, acceptance of residual losses, reliance on insurance and other risk transfer instruments, formalized asset management and maintenance, mainstreaming into environmental assessments, new structural materials and practices, improved emergency services, and retrofitting and replacement of infrastructure elements (Bourrelier et al., 2000; Auld, 2008b; Stevens, 2008; Haasnoot et al., 2009; Hallegatte, 2009; Neumann, 2009; Kwadijk et al., 2010; Wilby and Dessai, 2010).

Strategic environmental assessment approaches, such as those recommended by the Organisation for Economic Cooperation and Development (OECD) and many national environmental assessment agencies, offer an effective means for ensuring that adaptation to climate change and disaster risk management, as well as GHG reduction practices, are mainstreamed into policies and planning for new programs on infrastructure and systems (OECD, 2006; Benson, 2007). Environmental impact assessment approaches can reduce the risks of environmental degradation from a project and reduce future disaster risks from current and changing climate conditions (Benson, 2007). For long-lived infrastructure or networks, studies recommend consideration of likely climate change impacts that will potentially affect the planned useful life of the infrastructure system (e.g., seasonal variability in water flows, temperatures, incidence of extreme weather events) (OECD, 2006; Bosher et al., 2007; Auld, 2008b; Larsen et al., 2008; Neumann, 2009; NRTEE, 2009).

The implementation of adequate national building codes that incorporate up-to-date regionally specific climate data and analyses can improve resilience of infrastructure for many types of weather-related risks (Auld, 2008b; WWC, 2009; Wilby et al., 2009). Typically, infrastructure codes and standards in most countries use historical climate analyses to climate-proof new structures, assuming that the past climate can be extrapolated to represent the future. For example, water-related engineering structures, including both disaster-proofed infrastructure and services infrastructure (e.g., water supply, irrigation and drainage, sewerage, and transportation), are typically designed using analysis of historical rainfall records (Ruth and Coelho, 2007; Auld, 2008b; Haasnoot et al., 2009; Hallegatte, 2009; Wilby and Dessai, 2010). Since infrastructure is built for long life spans and the assumption of climate stationarity will not hold for future climates, it is important that national climate change guidance, tools, and consistent adaptation options be developed to ensure that climate change can be incorporated into infrastructure design (Auld, 2008b; Stevens, 2008; Hallegatte, 2009; Wilby et al., 2009). While some government departments responsible for building regulations and the insurance industry are taking the reality of climate change very seriously, challenges remain about how to incorporate the uncertainty of future climate projections into engineering risk management and into codes and standards, especially for climate elements such as extreme winds and extreme precipitation and their various phases (e.g., short- and long-duration rainfalls, freezing rain, snowpacks) (Sanders and Phillipson, 2003; Auld, 2008b; Haasnoot et al., 2009; Hallegatte, 2009; Kwadijk et al., 2010; Wilby and Dessai, 2010; Lu, 2011). Recent advances in characterizing the uncertainties of climate change projections, in regionalization of climate model outputs, and in the application and mainstreaming of integrated top-down, bottom-up approaches for assessing impacts and adaptation options (Sections 6.3.1 and 6.3.2) will help to ensure that infrastructure and technology can be better adapted to a changing climate. Sections 3.2.3, 3.3, and 3.4 provide further details on scientific advances for the construction, assessment, and communication of climate change projections, including a discussion on recent advances in the development of regionalization or downscaling techniques and approaches used to quantify uncertainties in climate change model outputs.

Some implementation successes are emerging. In one example, discussed in Case Study 9.2.10, the Canadian Standards Association (CSA) and its National Permafrost Working Group developed a Technical Guide, CSA Plus 4011-10, on *Infrastructure in Permafrost: A Guideline for Climate Change Adaptation*, that directly incorporated climate change temperature projections from an ensemble of climate change models. This CSA Guide considered climate change projections of temperature and precipitation and incorporated risks from warming and thawing permafrost to foundations over the planned life spans of the structure (Hayley and Horne, 2008; NRTEE, 2009; CSA, 2010a; Smith et al., 2010; Grosse et al., 2011). The guide suggested possible adaptation options, taking into account the varying levels of risks and the consequences of failure for foundations of structures, whether buildings, water treatment plants, towers, tank farms, tailings ponds, or other infrastructure (NRTEE, 2009; CSA, 2010a; see Case Study 9.2.10). Similarly, working with the Canadian meteorological service, engineering associations, and national water stakeholder associations, the CSA has also developed an initial rainfall Intensity-Duration-Frequency Guideline for water practitioners with adaptation guidance (CSA, 2010b).

In developing countries, structures are often built using prevalent local practices, which may not reflect best practices from disaster risk reduction or adaptation perspectives. These prevalent local practices usually do not include the use of national building standards or adequately account for local climate conditions (Rossetto, 2007). While the perception in some developing countries is that national building codes and standards are too expensive, experience in the implementation of incremental hazard-proof measures in building structures has proven in some countries to be relatively inexpensive and highly beneficial in reducing losses (Rossetto, 2007; ProVention, 2009). In reality, the most expensive components of codes and standards are usually the cost to implement national policies for inspections, knowledge transfer to trades, and national efforts for their uptake and implementation (Rossetto, 2007). Bangladesh, for example, has implemented simple modifications to improve the cyclone resistance of (non-masonry) kutcha or temporary houses, with costs that amounted to only 5% of the construction costs (Lewis and Chisholm, 1996; Rossetto, 2007). Bangladesh is also developing national policies requiring that houses built following disasters include a small section of the replacement house that meets 'climate proofing' standards and acts as a household shelter in the next disaster. In many countries, climate-proofing guidelines and standards are applied to structures that are used as emergency shelters and for structures that form the economic and social lifeline of a society, such as its communications links, hospitals, and transportation networks (Rossetto, 2007).

Many studies advocate that technical and infrastructure solutions are not the only way of adapting to changing climates and that 'soft solutions' such as financial tools, land use planning, and ecosystem conservation or soft engineering approaches are also needed (Adger et al., 2007; Auld, 2008b; Nicholls et al., 2008; Hallegatte, 2009; McEvoy et al., 2010). Land and water use planning, use of bioshields as natural buffers, soft defenses, and green or 'soft engineering' are complementary adaptation options, described further in Section 6.5.2.3 and in Case Studies 9.2.1 and 9.2.8.

6.5.2.2. Human Development and Vulnerability Reduction

Vulnerabilities to climate-related hazards and the options to reduce them vary between and within countries due to factors such as poverty, social positioning, geographic location, gender, age, class, ethnicity, ecosystem condition, community structure, community decisionmaking processes, and political issues (Yodmani, 2001; Yamin et al., 2005; Halsnaes and Traerup, 2009). Overall, studies indicate that the extent of the vulnerability to climate variability and climate change is shaped by both the dependence of the national economy and livelihoods on climate-sensitive natural resources and the resilience or robustness of the country's social institutions to equitable distribution of resources under climate change (Ikeme, 2003; Brooks et al., 2009; Virtanin et al., 2011). The poorest regions are often characterized by vulnerable housing, weak emergency services and infrastructure, and a dependence on agriculture and other natural resources (Ikeme, 2003; Manuel-Navarrete et al., 2007; Reid et al., 2010).

Many vulnerable communities already suffer greater water stress, food insecurity, disease risks, and loss of livelihoods, which have the potential to increase under climate change (Manuel-Navarrete et al., 2007; Brooks et al., 2009; Halsnaes and Traerup, 2009; Virtanin et al., 2011). For example, climate change may increase the risk of waterborne diseases, requiring targeted assistance for health and water sanitation issues (Curriero et al., 2001; Brooks et al., 2009). Small island states and low-lying countries may require support to relocate vulnerable groups to safer locations or other countries, all requiring a complex set of actions at the national and international levels (Manuel-Navarrete et al., 2007; McGranahan et al., 2007). Other studies indicate that resilient housing and safe shelters will remain a key adaptation action to protect vulnerable people from disasters and climate extremes, requiring national guidelines to ensure that new or replacement structures are built with flexibility to accommodate future changes (Ikeme, 2003; Manuel-Navarrete et al., 2007; Rossetto, 2007; Auld, 2008b). Under climate change, it is expected that food security issues among vulnerable populations will become more impacted by climate variability, erratic rainfall, and more frequent extreme events (Ikeme, 2003; IRI, 2006; Brooks et al., 2009; Halsnaes and Traerup, 2009; and regional studies through global partnerships, such as the Consultative Group on International Agricultural Research). When faced with food scarcity, vulnerable populations sometimes adopt maladaptive coping strategies such as overgrazing, deforestation, and unsustainable extraction of water resources that aggravate long-term disaster risks (Brooks et al., 2009; Bunce et al., 2010).

Studies indicate that the greatest losses in suitable agricultural cropland due to climate change are likely to be in Africa, particularly sub-Saharan Africa (Ikeme, 2003; FAO, 2010). Assessing food security issues in this vulnerable area requires consideration of multiple socioeconomic and environmental variables, including climate (Verdin et al., 2005; Virtanin et al., 2011). In sub-Saharan Africa, where large and widely dispersed populations depend on rain-fed agriculture and pastoralism, climate monitoring and forecasting are important inputs to food security analysis and assessments. Since conventional climate and hydro-meteorological networks in these areas are sparse and often report with significant delays, there is a growing need for increased capacity in rainfall observations, forecasting, data management, and modeling applications (Verdin et al., 2005; Heltberg et al., 2009; FAO, 2010). Studies indicate a need for rainfall observation networks to be expanded and to incorporate satellite information; for data management systems to be improved; for tailored forecast information to be disseminated and used by decisionmakers; and for more effective early warning systems that can integrate seasonal forecasts with drought projections as inputs for hazards, food security, and vulnerability analysis (Verdin et al., 2005; Heltberg et al., 2009; FAO, 2010). Other short-term but limited strategies to minimize food security risks include diversifying livelihoods to spread risk, farming in different ecological niches, building social networks, productive safety net and social protection schemes, and risk pooling at the regional or national level to reduce financial exposure (Brooks et al., 2009; Halsnaes and Traerup, 2009; Heltberg et al., 2009; FAO, 2010). Specific longer-term strategies to address the increasing risks, particularly given uncertainties, include land rehabilitation, terracing and reforestation, measures to enhance water catchment and irrigation techniques, improvements to infrastructure quality for better access to markets, and the introduction of drought-resistant crop varieties (Halsnaes and Traerup, 2009; Heltberg et al., 2009).

In the longer term, studies indicate that increasing food security risks under climate change will require higher agricultural productivity, reduced production variability, and agricultural systems that are more resilient to disruptive events (Cline, 2007; Stern, 2007; Halsnaes and Traerup, 2009; FAO, 2010). This implies transformations in the management of natural resources; new climate-smart agriculture policies, practices, and tools; better use of climate science information in assessing risks and vulnerability; and financing for food security (Brooks et al., 2009; Ericksen et al., 2009; FAO, 2010). Other coping strategies may include increased non-farm incomes, migration, government and other financial assistance, microfinance, social protection, other safety nets, and various insurance products (Barrett et al., 2007; Heltberg et al., 2009; FAO, 2010). The Sustainable Livelihoods Approach or Framework has been used internationally for rural and coastal development to holistically describe the variables that impact livelihoods locally and to define the capacity, assets (both natural and social), and policies required for sustainable living, poverty reduction, and recovery from disasters (Brocklesby and Fisher, 2003; Yamin et al., 2005). Sections 2.3, 2.4.3, 2.6.1, 5.2.3, and 5.4, and Case Studies 19.2.1 and 19.2.2, also discuss sustainable livelihood approaches that can be considered in building adaptive capacity and resilience to climate hazards and climate change.

Early identification of populations at risk can enable timely and appropriate actions needed to avert widespread impacts. Reliable and detailed information on the current and future climates and their impacts can play an important role in the recognition of the need to adapt and the successful evolution of effective adaptation strategies (Ikeme, 2003; Verdin et al., 2005; Heltberg et al., 2009; Wilby et al., 2009; and as discussed in Section 6.5.1). Some studies claim that one of the potential barriers for identifying the most vulnerable regions and people in developing countries under future climate change is the limited human resource capacity regionally to downscale global and regional climate projections to a scale suitable to support national-level planning and programming processes (Ikeme, 2003; Verdin et al., 2005; CCCD, 2009; Wilby et al., 2009). Not all of the climate variables of importance for development can be projected and downscaled with confidence, particularly given that many development activities are especially sensitive to changes in climate extremes (Agrawala and van Aalst, 2008). Even when downscaled results are available, their use can be limited by a lack of understanding and interpretation of how these downscaled projections can be translated to highlight vulnerabilities with certainty (Agrawala and van Aalst, 2008; Heltberg et al., 2009). Agrawala and van Aalst (2008) argue that development practitioners and climate scientists should join forces to make climate information more accessible, relevant, and usable.

Because the risks posed by climate change can affect the long-term efficiency with which development resources can be invested and development objectives achieved, studies indicate that it remains important to integrate or mainstream disaster risk management and climate change adaptation into a range of development activities (Agrawala and van Aalst, 2008; Halsnaes and Traerup, 2009; Heltberg et al., 2009; Mitchell et al., 2010a). Lack of awareness within the development community of the many implications of climate change and limitations on resources for implementation are frequently cited reasons for difficulties in mainstreaming adaptation and disaster risk management (Agrawala and van Aalst, 2008; Heltberg et al., 2009; also see Section 6.3.2). Adaptation to climate change and disaster risk management actions can be considered to be successfully mainstreamed when they reduce the vulnerability of susceptible populations to existing climate variability and are also able to strengthen the capacity of the population to prepare for and respond to further changes (Yamin et al., 2005; Manuel-Navarrete et al., 2007; Mertz et al., 2009). Studies indicate that national policies can increase this capacity (Ikeme, 2003; Heltberg et al., 2009). Policies and measures such as the establishment of an LDC fund, Special Climate Fund, Adaptation Fund, climate change Multi-Donor Trust Fund, etc., have all been developed to address the special adaptation and risk reduction issues of vulnerable countries (see Sections 7.4.2 and 7.4.3.3 for more details).

In spite of recommendations to target assistance to the most vulnerable in the developing world, practical 'on the ground' examples have been limited (Yamin et al., 2005; Ayers and Huq, 2009; Heltberg et al., 2009). Nonetheless, some developing countries have implemented successful policies and plans. Nationally, good progress is being made in strengthening some disaster reduction capacities for disaster preparedness and early warning and response systems and in addressing some of the underlying risk drivers in many developing country regions and sectors (Manuel-Navarrete et al., 2007; UNISDR, 2009c). For example, social safety nets and other similar national-level programs, particularly for poverty reduction and attainment of the Millennium Development Goals, have helped the poorest to reduce their exposure to current and future climate hazards (Yamin et al., 2005; Tanner and Mitchell 2008; Heltberg et al., 2009). Some examples of social safety nets are cash transfers to the most vulnerable, versions of weather-indexed crop insurance, employment guarantee schemes, and asset transfers (Yamin et al., 2005; CCCD, 2009; also see Section 6.6.3). A national policy to help the vulnerable build assets should incorporate climate screening in order to remain resilient under a changing climate (UNISDR, 2004; Tanner and Mitchell, 2008; Heltberg et al., 2009). Other measures, such as social pensions that transfer cash from the national level to vulnerable people, provide some buffers against climate hazards (Davies et al., 2008; Heltberg et al., 2009). However, lack of capacity and good governance has remained a major barrier to efficient and effective delivery of assistance to the most vulnerable (Yamin et al., 2005; CCCD, 2009; Heltberg et al., 2009; Warner et al., 2009).

National Adaptation Programme of Actions (NAPA) under the UNFCCC process have helped least-developed countries assess the climate-sensitive sectors and prioritize projects to address the most urgent adaptation issues of the most vulnerable regions, communities, and populations. The NAPA process has proven instrumental in increasing awareness of climate change and its potential impacts in the poorest countries. The proposed adaptation projects under the NAPA usually cover small areas and address a few components within a given sector with a view to addressing urgent and immediate needs. The choice of projects is based on the urgency of the actions as well as cost-effectiveness in cases where delays would increase the costs of later addressing the issue. Assessment of completed NAPAs show different national and regional priority sectors such as health, food security, infrastructure, coastal zone and marine ecosystem, insurance, early warning and disaster management, terrestrial ecosystem, education and capacity building, tourism, energy, water resources, and cross-sectoral areas. The NAPA process forms a good basis for developing medium- and long-term adaptation plans and policies. The capacity within NAPA teams and the subsequent networks that are created are proving very useful in the design of broader national adaptation plans (UNFCCC, 2011a,b).

6.5.2.3. Investing in Natural Capital and Ecosystem-Based Adaptation

Ecosystem-based adaptation, which integrates the use of biodiversity and ecosystem services into an overall adaptation strategy, can be a cost-effective strategy for responding to the effects of weather and climate extremes (SCBD, 2009). It is generally agreed that investment in sustainable ecosystems and environmental management has the potential to also provide improved livelihoods and increased biodiversity conservation (Bouwer, 2006; UNEP, 2006, 2010; McGray et al., 2007; Colls et al., 2009; SCBD, 2009; Sudmeier-Rieux and Ash, 2009; World Bank, 2009).

Healthy, natural or modified, ecosystems (see Section 6.3.1 and Box 6-4) have a critical role to play in reducing risks of climate extremes and disasters (Sidle et al., 1985; Dorren et al., 2004; Phillips and Marden, 2005; Reid and Huq, 2005; UNISDR, 2005, 2007a,b, 2009a,b; Bebi et al., 2009; Colls et al., 2009; SCBD, 2009; Sudmeier-Rieux and Ash, 2009; UNEP, 2009; Lal, 2010). Although the scientific evidence base relating to the role of ecosystem services in reducing the sensitivity of natural systems to weather and climate extremes and reducing vulnerabilities to many disasters is nascent, investment in natural ecosystem management has long been used to reduce risks of disasters (see Box 6-4). Forests, for example, have been used in the Alps and elsewhere as effective risk-reducing measures against avalanches, rockfalls, and landslides since the 1900s (Sidle et al., 1985; Dorren et al., 2004; Phillips and Marden, 2005; Bebi et al., 2009). The damage caused by wildfires, wind erosion, drought, and desertification are reported to have been buffered by forest management, shelterbelts, greenbelts, hedges, and other 'living fences' (ProAct Network, 2008; Dudley et al., 2010). Mangrove replanting has been used as a buffer against cyclones and storm surges, with reports of a 70 to 90% reduction in energy from wind-generated waves in coastal areas (UNEP, 2006) and reduction in the number of deaths from cyclones (Das and Vincent, 2009), depending on the health and extent of the mangroves. Many sectoral examples are provided in Table 6-1 that also provide evidence of the value of ecosystem services in disaster risk reduction and adaption to climate change (see also Section 6.5.2.1).

The extent to which ecosystems support such benefits, though, depends on a complex set of dynamic interactions among ecosystem-related factors, as well as the intensity of the hazard (UNEP, 2006; Sudmeier-Rieux and Ash, 2009) and institutional and governance arrangements (see case studies in Angelsen et al., 2009). Scientific understanding of the relationship between ecosystem structure and function and the reduction of risks associated with weather and climate extremes is limited, though growing.

> **Box 6-4 | Value of Ecosystem Services in Disaster Risk Management: Some Examples**
>
> - In the Maldives, degradation of protective coral reefs necessitated the construction of artificial breakwaters at a cost of US$ 10 million per kilometer (SCBD, 2009).
> - In Vietnam, the Red Cross began planting mangroves in 1994 with the result that, by 2002, some 12,000 hectares of mangroves had cost US$1.1 million for planting but saved annual levee maintenance costs of US$ 7.3 million, shielded inland areas from a significant typhoon in 2000, and restored livelihoods in planting and harvesting shellfish (Reid and Huq, 2005; SCBD, 2009).
> - In the United States, wetlands are estimated to reduce flooding associated with hurricanes at a value of US$ 8,250 per hectare per year, and US$ 23.2 billion a year in storm protection services (Costanza et al., 2008).
> - In Orissa, India, a comparison of the impact of the 1999 super cyclone on 409 villages in two tahsils with and without mangroves showed that villages that had healthy stands of mangroves suffered significantly less loss of lives than those without (or limited areas) healthy mangroves, even though all villages had the benefit of early warnings and accounting for other social and economic variables (Das and Vincent, 2009).

Investment in natural ecosystems also contributes significantly to reduction in GHG emissions, through practices such as those associated with Land Use, Land-Use Change, and Forestry (LULUCF) and through Reduced Carbon Emissions from Deforestation and Forest Degradation (REDD) or REDD+, which additionally includes the value of conservation from sustainable management of forests and enhancement of forest carbon stocks (UNEP, 2006; SCBD, 2009). Mangrove ecosystems, for example, are important for carbon sequestration and storage, containing among the highest carbon pools: 1,060-2,020 t CO_2 ha^{-1} or an annual carbon sequestration of 6.32 t CO_2 ha^{-1} (Murray et al., 2010), as well as providing the buffers against weather and climate extremes, biodiversity values, and livelihood benefits discussed above. Investment in natural ecosystems, through REDD and REDD+ related strategies, can generate alternative sources of income for local communities and provide much needed financial incentives to prevent deforestation (Reid and Huq, 2005; Angelsen et al., 2009; SCBD, 2009; Sudmeier-Rieux and Ash, 2009; Murray et al., 2010), as well as provide additional livelihood benefits from the conservation and restoration of forest ecosystems and the services they support (Longley and Maxwell, 2003; MEA, 2005; SEEDS India, 2008; Sudmeier-Rieux and Ash, 2009; Murray et al., 2010).

Some countries have begun to explicitly consider ecosystem-based solutions for climate change mitigation and/or adaptation to risks

> **Box 6-5 | Some Examples of Ecosystem Based Adaptation Strategies and Disaster Risk Management Interventions Taking into Account the Role of Ecosystem Services**
>
> - Vietnam has applied strategic environmental assessments to land use-planning projects and hydropower development for the Vu Gia-Thu Bon River basin, including climatic disaster risks (OECD, 2009; SCBD, 2009).
> - European countries affected by severe flooding, notably the United Kingdom, The Netherlands, and Germany, have made policy shifts to 'make space for water' by applying more holistic river basin management plans and integrated coastal zone management (DEFRA, 2005; Wood and Van Halsema, 2008; EC, 2009; ONERC 2009).
> - At the regional level, the Caribbean Development Bank has integrated weather and climatic disaster risks into its environmental impact assessments for new development projects (CDB and CARICOM, 2004; UNISDR, 2009c).
> - Under the Amazon Protected Areas Program, Brazil has created a more than 30 million hectare mosaic of biodiversity-rich forests reserve of state, provincial, private, and indigenous land, resulting in a potential reduction in emissions estimated at 1.8 billion tonnes of carbon through avoided deforestation (World Bank, 2009).
> - Swiss Development Cooperation's four-year project in Muminabad, Tajikistan adopted an integrated approach to risk through reforestation and integrated watershed management (SDC, 2008).

associated with weather and climatic extremes as an integral element of national and sectoral development decisions (see Box 6-5).

Ecosystem-based adaptation strategies, often considered as part of 'soft' options, are a widely applicable approach to climate change adaptation because they can be applied at regional, national, and local levels, at both project and programmatic levels, and benefits can be realized over short and long time scales. They can be a more cost-effective adaptation strategy than hard infrastructure and engineering solutions, as also discussed in Section 6.5.2.1, and produce multiple benefits, and are also considerably more accessible to the rural poor than measures based on hard infrastructure and engineering solutions (Sudmeier-Rieux and Ash, 2009). Communities are also able to integrate and maintain traditional and local knowledge and cultural values in their risk reduction efforts (SCBD, 2009).

In the choice of ecosystem-based adaptation options, decisionmakers may at times require making judgements about the tradeoffs between particular climatic risk reduction services and other ecosystem services also valued by humans. Such decisions benefit from information resulting from risk assessments, scenario planning, and adaptive management approaches that recognize and incorporate these potential tradeoffs. This might be the case, for example when deciding to use wetlands for coastal protection that requires emphasis on silt accumulation and stabilization possibly at the expense of wildlife values and recreation (SCBD, 2009), particularly when achieving a full complement of biodiversity values is highly complex and long-term in nature (UNEP, 2006).

However, countries would need to overcome many challenges if they are to be successful in increasing investment in ecosystem-based solutions, including for example:

- Insufficient recognition of the economic and social benefits of ecosystem services under current risk situations, let alone under potential changes in climate extremes and disaster risks (Vignola et al., 2009).
- Lack of interdisciplinary science and implementation capacity for making informed decisions associated with complex and dynamic systems (Leslie and McLeod, 2007; OECD, 2009).
- Ability to estimate economic values of different ecosystem services supported by nature (TEEB, 2009).
- Lack of capacity to undertake careful cost and benefit assessments of alternative strategies to inform choices at the local level. Such assessments could provide the total economic value of the full range of disaster-related ecosystem services, compared with alternative uses of the forested land such as for agriculture (see, e.g., Balmford et al., 2002).
- Where they exist, data on and monitoring of ecosystem status and risk are often dispersed across agencies at various scales and are not always accessible at the sub-national or municipal level where land use-planning decisions are made (UNISDR, 2009a).
- The mismatch in geographic scales and mandates between the administration and responsibilities for disaster reduction, and that of ecosystem extent and functioning, such as in water basins (Leslie and McLeod, 2007; OECD, 2009).

6.5.3. Transferring and Sharing 'Residual' Risks

Not all risk can be reduced, and a residual, often sizeable risk will remain. Mechanisms for sharing and transferring residual risks for households and businesses have been introduced in Section 5.5.2.2 in the context of managing local-level impacts and risks. Chapter 5 also discusses the incentive and disincentive aspects provided by insurance for risk management and adaptation to climate change at the local level. This section sets out the role of national-level institutions, especially governments, in enabling and regulating practices at national scales. It also discusses the need on the part of some governments to transfer their own risks.

Markets offer risk-sharing and transfer solutions, most prominently property and asset insurance for households and businesses, and crop insurance for farmers. Insurance markets are generally segregated and regulated nationally. Existing national insurance systems commonly offer a wide variety of choice in providing protection for property and assets against natural hazards. National insurance systems differentially include hazards, such as storms, hail, floods, earthquake, and also landslides or subsidence. Risks may be covered separately or bundled with a fire policy or covered under an 'all hazards' policy. The contracts differ in the extent of cover offered, as well as indemnity limits, and whether the policies are compulsory, bundled, or voluntary. Importantly, they differ institutionally with regard to the involvement of the public authorities and private insurers and how they allocate liability and responsibility for disaster losses across individual households, businesses, and taxpayers (Schwarze and Wagner, 2004; Aakre et al., 2010).

Yet, insurance coverage is limited and globally only about 20% of the losses from weather-related events have been insured over the period 1980 to 2003 (also see Section 6.2.2). In many instances, insurance providers even in industrialized countries have been reluctant to offer region- or nationwide policies covering flood and other hazards because of the systemic nature of these risks, as well as problems of moral hazard and adverse selection (Froot, 2001; Aakre et al., 2010). In some highly exposed countries, such as The Netherlands for flood risk, insurance is even non-existent and government relief is dispensed in lieu (Botzen et al., 2009). In many developing countries, there is little in terms of insurance for disaster risks, yet novel index-based micro-insurance solutions have been developed and are starting to show results (Hazell and Hess, 2010; see also Sections 5.6.3 and Case Study 9.2.13 on risk financing). Market mechanisms may work less well in developing countries, particularly because there is often limited risk assessment information, limited scope for risk pooling, and little or no supply of insurance instruments. In such circumstances, governments may need to create enabling environments by helping to estimate risk, helping to develop training programs for insurer's staff, and generally promoting awareness among the population at risk (Linnerooth-Bayer et al., 2005; Hoeppe and Gurenko, 2006; Cummins and Mahul, 2009; Hazell and Hess, 2010).

Employing insurance and other risk-financing instruments for helping to manage the vagaries of nature may often involve the building of PPPs in developing and in developed countries in order to tackle market failure, adverse selection, and the sheer non-availability of such instruments (see Aakre et al., 2010). Because of such reasons, there is a role for governments to not only create an enabling environment for private sector engagement, but also to regulate its activities. In the development context, Hazell and Hess (2010) distinguish between protection and promotion models, while acknowledging that in many instances hybrid combinations may contain elements of both. Protection relates to governments helping to protect themselves, individuals, and businesses from destitution and poverty by providing ex-post financial assistance, which, however, is taken out as an ex-ante instrument as insurance before disasters. The promotion model relates to the public sector promoting more stable livelihoods and higher income opportunities by better helping businesses and households access risk financing, including micro-financing.

Private insurers are often not willing to fully underwrite the risks and many countries, including Japan, France, the United States, Norway, and New Zealand, have therefore instituted public-private national insurance systems, where participation of the insured is mandatory or voluntary and single hazards may be insured or comprehensive insurance offered (Linnerooth-Bayer and Mechler, 2007). Further, specific strategies may be employed to increase market penetration of risks that are not easily covered by regular avenues. As one example, in India, pro-poor regulation stipulates that insurers within their regular business segment reserve a certain quota of insurance policies for the poor and thus cross-subsidize fledgling low-income micro-insurance policies (Mechler et al., 2006a).

As well, governments may insure their liabilities through sovereign insurance. Liabilities arise as governments own a large portfolio of public infrastructure and other assets that are exposed to disaster risks. Moreover, most governments accept their role as provider of post-disaster emergency relief and assistance to vulnerable and affected households and businesses. In wealthy countries, government (sovereign) insurance hardly exists at the national level and, in Sweden, insurance for public assets is illegal (Linnerooth-Bayer and Amendola, 2000). On the other hand, states in the United States, Canada, and Australia, although regulated not to incur budget deficits, often carry cover for their public assets (Burby, 1991). As discussed earlier (see Section 6.4.3), this is consistent with the Arrow and Lind Theorem, which suggests that governments can efficiently spread and share risk over their citizens without buying sovereign insurance policies.

Yet, realizing the shortcomings of after-the-event approaches for coping with disaster losses for small, low-income or highly exposed countries with over-stretched tax bases and highly correlated infrastructure risks (OAS, 1991; Pollner, 2000; Mechler, 2004; Cardona, 2006; Linnerooth-Bayer and Mechler, 2007; Mahul and Ghesquiere, 2007), sovereign insurance for public sector assets and relief expenditure has become a recent cornerstone for tackling the substantial and increasing effects of disasters (Mahul and Ghesquiere, 2007). As a general statement, the strategy involves transferring a layer of risks ranging from infrequent risk (such as events with a return period of more than 10 years) up to risks associated with 150-year return periods, beyond which it will become very costly to insure (Cummins and Mahul, 2009). One key element is to define the financial vulnerability indicating the inability to bear losses with a certain return period (Mechler et al., 2010).

Key applications have been implemented in Mexico in 2006, which insured its government emergency relief expenditure, and in the Caribbean with the Caribbean Catastrophe Risk Insurance Facility in 2007 (Ghesquiere, et al., 2006; Cardenas et al., 2007). Like national governments, donor organizations, exposed indirectly through their relief and assistance programs, also have been considering similar transactions; the World Food Programme in 2006, for example, purchased 'humanitarian insurance' for its drought exposure in Ethiopia through index-based reinsurance (see Section 9.2.13). These transactions set innovative and promising precedents in terms of protecting highly exposed developing and transition government portfolios against the risks imposed by disasters.

6.5.4. Managing the Impacts

Even in the rare circumstances where efforts outlined previously are all in place, there still needs to be investment in capacities to manage potential disaster impacts as risk cannot be reduced to zero (Pelling, 2003; Wisner et al., 2004; Coppola, 2007). The scale of the disaster impact should ideally dictate the level and extent of response. Individual household capacities to respond to disasters may be quickly overwhelmed, requiring local resources to be mobilized (del Ninno, 2001). When community-level responses are overwhelmed, regional or central governments are called upon (Coppola, 2007). Some events may overwhelm national government capacities too, and may require mobilization of the international community of humanitarian responders (Fagen, 2008; Harvey, 2009). International responses pose the most complex management challenges for national governments, because of the diversity of actors that are involved and the multiple resources flows that are established (Borton, 1993; Bennett et al., 2006; Ramalingam et al., 2008; ALNAP, 2010a). However, although humanitarian principles call for a proportionate and equitable response, in practice there are a few high-profile disasters that are over-resourced, with many more that are 'forgotten or neglected emergencies' (Slim, 2006). Despite the definition of international or national disasters as those where immediate capacities are overwhelmed, evaluations routinely find that most of the vital life-saving activities happen at the local level, led by households, communities, and civil society (see Sections 5.1 and 5.2; Smillie, 2001; Hilhorst, 2003; ALNAP, 2005; Telford and Cosgrave, 2006).

In terms of how responses are managed nationally, there are different models to consider (ALNAP, 2010b). Many countries now have some standing capacity to manage disaster events (Interworks, 1998) and this should be considered distinct from national systems for managing disaster risk, commonly associated with 'national platforms' detailed in Section 6.4.2. Examples of standing disaster management capacity include the Federal Emergency Management Agency in the United States, Public Safety in Canada, the National Commission for Disaster Reduction in China, the National Disaster Management Authorities in India and Indonesia, National Disaster Management Offices (NDMO) in many Pacific island countries, and the Civil Contingencies Secretariat in the United Kingdom. Comparative analysis of these structures shows that there are a number of common elements (Interworks, 1998; Coppola, 2007). Countries with formal disaster management structures typically operate a system comprised of a National Disaster Committee, which works to provide high-level authority and ministerial coordination, alongside an NDMO to lead the practical implementation of disaster preparedness and response (Interworks, 1998). National Committees are typically composed of representatives from different ministries and departments as well as the Red Cross/Red Crescent. They might also include donor agencies, NGOs, and the private sector. The committee works to coordinate the inputs of different institutions to provide a comprehensive approach to disaster management. NDMOs usually act as the executive arm of the national committee. Focal points for disaster management are usually professional disaster managers. NDMOs may be operational, or in large countries they may provide policy and strategic oversight to decentralized operational entities at federal or local levels. Where formal structures do not exist, national ministerial oversight is provided to the efforts of the NDMO in times of national disasters.

Government ownership of the national disaster management function can vary, with three models evident: it may reside with the presidential or prime ministerial offices; it may sit within a specific ministry; or it may be distributed across ministries (Interworks, 1998). The way in which the international community is engaged in major emergencies is shaped by existing national capabilities and social contracts, with four possible response approaches (Chandran and Jones 2008; ALNAP, 2010b; see Table 6-4). Analysis based on these broad categories helps clarify the ways in which international agencies are mobilized to manage disaster impacts, following from national structure and capabilities.

There may be states where there is an existing or emerging social contract with its citizens, by which the state undertakes to assist and protect them in the face of disasters, and there is a limited role for international agencies, focusing on advocacy and fundraising. By comparison, there are states that have a growing capacity to respond and request international agencies to supplement their effort in specific locally owned ways, through filling gaps in national capacities or resources. Next, there are states that have limited capacity and resources to meet their responsibilities to assist and protect their citizens in the face of disasters, and which request international assistance to cope with the magnitude of a disaster, resulting in a fully fledged international response. Finally, there are states that lack the will to negotiate a resilient social contract, including assisting and protecting their citizens

Table 6-4 | Activities associated with managing the impacts of disasters. Adapted from Coppola (2007) and ALNAP (2010a).

Pre-disaster	Immediate post-disaster	Recovery
• Public education • Awareness raising • Warning and evacuation plans • Pre-positioning of resources and supplies • Last minute alleviation and preparedness measures	• Search and rescue • Emergency medical treatment • Damage and Needs Assessment • Provision of services – water, food, health, shelter, sanitation, social services, security • Resumption of critical infrastructure • Coordination of response • Coordination / Management of development partner support	• Transitional shelter in form of temporary housing or long-term shelter • Demolition of critically damaged structures • Repair of less seriously damaged structures • Clearance, removal, and disposal of debris • Rehabilitation of infrastructure • New construction • Social rehabilitation • 'Building back better' to reduce future risk • Employment schemes • Reimbursement for losses • Reassessment of risks

> **Box 6-6 | National Disaster Preparedness, Prevention, and Management Systems: China and Kenya**
>
> *China*
>
> The Government's disaster management process, developed as National Integrated Disaster Reduction, is a comprehensive system bringing together a number of central and local government sectors and covering the different phases of disasters preparedness, response, and recovery/rehabilitation. China has put in place over 30 laws and regulations regarding disaster management. The Emergency Response Law was adopted on 30 August 2007, as the central legal document governing all disaster-related efforts in China.
>
> Under the related law and regulations, the Government has established an emergency response system consisting of three levels:
> - The National Master Plan for Responding to Public Emergencies – a framework to be used throughout government to ensure public security and cope with public emergency events, including all disaster response activities.
> - Five national thematic disaster response plans that outline the detailed assignment of duties and arrangements for major disaster response categories – disaster relief, flood and drought, earthquakes, geological disasters, and very severe forest fires.
> - Emergency response plans for 15 central government departments and their detailed implementation plans and operation norms (UNESCAP, 2009).
>
> *Kenya*
>
> The government is working toward a national disaster management policy with the intention of preventing disasters and minimizing the disruption they cause through taking steps to reduce risks. The policy will help enhance existing capacities by building resilience to hazard events, building institutional capacity, developing a well-managed disaster response system, reducing vulnerability, and ensuring that disaster policy is integrated with development policy and poverty reduction and takes a multi-sectoral, multi-level approach. The Ministry of State for Special Programmes will be responsible for the coordination of the disaster management policy, will promote integration and coordination of disaster management, and will establish a national institute for disaster research to improve systematic monitoring and promotion of research.
>
> The draft policy published in 2009 stressed the central role of climate change in any future sustainable planned and integrated National Strategy for Disaster Management. It sets out principles for effective disaster management, codes of conduct of different stakeholders, and provides for the establishment of an institutional framework that is legally recognized and embedded within the government structures. It stresses the importance of mobilizing resources to enable the implementation of the policy, with provision of 2% of the annual public budget to a National Disaster Management Fund.
>
> At the time of writing, this policy has not reached Parliament for discussion and approval (MOSSP, 2010).

in times of disaster. These pose significant challenges and involve a combination of direct delivery and advocacy. Across all four categories of response, there are challenges around resources availability, proportionality of distribution, coordination, and leadership (ALNAP, 2010a).

Box 6-6 outlines details of the disaster management systems of two countries, which were chosen to illustrate the different stages of disaster management development that are evident across states.

Although level of response and actors involved can vary considerably between disasters and countries (ALNAP, 2010a), the basic actions taken to manage disaster impacts remain broadly the same across countries, and correspond closely to the different stages of the disaster timeline (see Table 6-4; Coppola, 2007). In general, disaster management employs immediate humanitarian activities, needs assessments, and the delivery of goods and services to meet requirements. The demand for water, food, shelter, sanitation, healthcare, security, and – later on – education, employment, reconstruction, and so on is balanced against available resources (Wisner and Adams, 2002).

Despite the existence of evidence that climate change is not responsible for the vast majority of the increasing trend in disaster losses (see SPM and Section 4.5.3.3), climate change-related disasters are still widely, if incorrectly, seen by particularly the humanitarian community as playing a major role in increasing the overall human impact of disasters. Numerous trends in disaster events are commonly attributed to climate change (IASC, 2009a; IFRC, 2009), and, as such, climate change is often cited as a reason for enhancing both national and international disaster management capacities (HFP, 2007; Oxfam, 2007; IASC, 2009a,b). Consequently, climate change-related considerations are increasingly featuring in literature on disaster management (Barrett et al., 2007;

McGray et al., 2007; Mitchell and Van Aalst, 2008; Venton and La Trobe, 2008; IASC, 2009a). As presented in this report, evidence is available for the influence of climate change on some extreme weather events but not for others (see Chapter 3), and, perhaps because of this, challenges remain in how climate change-related information can be used as a direct guide to decisionmaking in the humanitarian sector (IASC, 2009a).

The challenges of climate change call for institutional changes in approaches to managing disasters that are far from trivial (Salter 1998), with such challenges including more appropriate policies and legislation; decentralization of capacities and resources; greater budgetary allocation; improved capacity building at the local level; and the political will to bridge the divide between disaster risk reduction activities and the humanitarian action associated with managing disasters (Sanderson, 2000; UNISDR, 2005). Recent analyses of the need for greater innovation in international humanitarian responses (Ramalingam et al., 2009) present these shifts as among the most significant and important reforms the international system must undergo.

6.6. Aligning National Disaster Risk Management Systems with the Challenges of Climate Change

As mentioned, climate change presents multidimensional challenges for national systems for managing the risks of climate extremes and disaster risks, including potential changes in the way society views, treats, and responds to risks and projected impacts on hazards, exposure, and vulnerability. As climate change is altering the frequency and magnitude of some extreme events (see Chapter 3) and contributing to trends in exposure and vulnerability (see Chapter 4), the efficacy of national systems of disaster risk management requires review and realignment with the new challenges (UNISDR, 2009c; Mitchell et al., 2010a; Polack, 2010; see FAQ 6.1). Literature suggests that the effectiveness of national systems for managing disaster risk in a changing climate will be improved if they integrate assessments of changing climate extremes and disasters into current investments, strategies, and activities; seek to strengthen the adaptive capacity of all actors; and address the causes of vulnerability and poverty recognizing climate change as one such cause (Schipper, 2009; UNISDR 2009c; Mitchell et al., 2010a). In practice, this might require: (i) new alliances and hybrid organizations across government and potentially across countries; (ii) different actors to join the national system; (iii) new cross-sector relationships; (iv) reallocation of functions, responsibilities, and resources across scales; and (v) new practices (Hedger et al., 2010; Mitchell et al., 2010a; Polack, 2010). As a complement to the available data, information, and knowledge about the impact of climate change and disaster risk presented in Chapters 2, 3, and 4, this section seeks to elaborate the key areas where realignment of national systems could occur – in assessing the effectiveness of disaster risk management in a changing climate (Section 6.6.1), managing uncertainty and adaptive management (Section 6.6.2), in tackling poverty, vulnerability, and their structural causes (Section 6.6.3); and commenting on the practicalities of approaching such changes holistically (Section 6.6.4).

6.6.1. Assessing the Effectiveness of Disaster Risk Management in a Changing Climate

In order to align disaster risk management with the challenges presented by climate change, it is necessary to assess the effectiveness and efficiency of management options in a changing climate based on the best available information, recognizing that that information remains patchy at best. Adopting an economic assessment framework, different approaches have been used to comment on the effectiveness or efficiency of adaptation options. Many climate adaptation studies have focused on the national-level costs of adaptation rather than comparing costs and benefits (i.e., examining the benefits of adaptation or reduced disaster impacts and damage costs) (see Nordhaus, 2006; EEA, 2007; UNFCCC, 2007a; Agrawala and Fankhauser, 2008; World Bank, 2008; ECA, 2009; Parry et al., 2009). National-level adaptation assessments have been conducted, among others, in the European Union, the United Kingdom, Finland, The Netherlands, and Canada, as well as in a number of developing countries using the NAPA approach (UNDP, 2004c; MMM, 2005; DEFRA, 2006; UNFCCC, 2007b; Lemmen et al., 2008; De Bruin et al., 2009a).

Other approaches include assessments of disaster risk management with risk assessment at the core, and focusing on economic efficiency of management responses (see World Bank, 1996; Benson and Twigg, 2004; Mechler, 2004). Using such a rationale, the World Bank, for example, goes as far as suggesting that governments should in many instances prioritize allocating their resources on early warning (such as for floods), critical infrastructure, such as water and electricity lifelines, and supporting environmental buffers such as mangroves, forests, and wetlands, of which the latter should be treated with caution (World Bank and UN, 2010). Another report suggests taking an adaptation cost curve approach to selecting adaptation options (ECA, 2009); this approach organizes adaptation options around their cost-benefit ratios, similar to mitigation cost curves. Interestingly, many of the options considered efficient in this analysis are 'soft' options, such as reviving reefs, using mangroves as barriers, and nourishing beaches.

It is, however, difficult to make conclusive assessments about the effectiveness of disaster risk management in a changing climate, as overall the evidence base used to determine economic efficiency – that is, benefits net of costs of adaptation – remains limited and fragmented (Adger et al., 2007; UNFCCC, 2007a; Agrawala and Fankhauser, 2008). In addition to the rather small number of studies available, there are important limitations of these assessments as well. These relate to the types of hazards examined as well as treatment of extreme events and risk, affecting the robustness of the results. Another key limitation, relevant for this report, is that only very few national level studies assessing economic efficiency of options have focused explicitly on disaster risk, and in most instances the hazards examined have been gradual, such as sea level rise and slower onset impacts, such as drought, on agriculture (see UNFCCC, 2007a; Agrawala and Fankhauser, 2008). Where extreme events and disaster risks have been considered, studies have often adopted deterministic impact metrics, when disaster risk associated with frequency and variability of extreme events can change. Where

FAQ 6.1 | What can a government do to better prepare its people for changing climate-related disaster risks?

In almost all countries, governments create the enabling environment of policies, regulations, institutional arrangements, and coordination mechanisms to guide and support the efforts of all agencies and stakeholders involved in managing disaster risks at different scales. Such risks are increasing and changing because of population growth, migration, climate change, and a range of other factors. National systems for managing disaster risk need to act on these changes in order to build resilience in the short and long term. Accordingly, the following measures can be considered:

- **Generate and communicate robust information about the dynamic nature of disaster risk**: Given the dynamic and changing nature of disaster risks in the context of climate change, regular updates on changes in the level of risk will further strengthen such systems if the information is acted upon. Not possessing information about changing disaster risks or not integrating the information into decisions about longer-term investments can lead to increases in the exposure and vulnerability of people and assets and may increase risk over time. An example could be non-drought-tolerant monoculture agriculture in an area likely to experience increased frequency and/or longer durations of drought conditions, or water harvesting tanks installed in houses or communities that lack the capacity to supply water during longer periods of drought, or roads not raised sufficiently above future projected flood levels. Knowledge about dynamic risks can be generated from scientific observations and models, combined with analysis of patterns of vulnerability and exposure and from the experiences of local communities (see Section 6.5.1).

- **Even without robust information, consider 'no or low regrets' strategies, including ecosystem-based adaptation**: Countries have started to adopt 'no or low regrets' strategies that generate short-term benefits as well as help to prepare for projected changes in disaster risks, even when robust information is not available (see Section 6.3.1). Included in these 'no or low regrets' strategies are ecosystem-based strategies that not only help reduce current vulnerabilities and exposure to hazards under a range of climatic conditions, but also produce other co-benefits such as improved livelihoods and poverty reduction that help reduce vulnerability to projected changes in climate. Table 6-5, a considerably reduced version of Table 6-1, shows a summary of these options. Such 'no or low regrets' practices also tend to include measures to tackle the underlying drivers of disaster risk and are effective irrespective of projected changes in extremes of weather or climate (see Section 6.5.2). Where better information is available, this can be mainstreamed across line ministries and other agencies to shape practices that help to build resilience to projected changes in disaster risk over the longer term. These are highlighted in the right-hand column of Table 6-5.

Table 6-5 | Range of practices to demonstrate comparison between 'no or low regrets' measures and those integrating projected changes in disaster risk.

'No or low regrets' practices with demonstrated evidence of having integrated observed trends in disaster risks to reduce the effects of disasters	Practices that enhance resilience to projected changes in disaster risk
• Effective early warning systems and emergency preparedness (*very high confidence*)	• Crop improvement for drought tolerance and adaptive agricultural practices, including responses to enhanced weather and climate prediction services (*high confidence*)
• Integrated water resource management (*high confidence*)	• Integrated coastal zone management integrating projections of sea level risk and weather/climate extremes (*medium confidence*)
• Rehabilitation of degraded coastal and terrestrial ecosystems (*high confidence*)	• National water policy frameworks and water supply infrastructures, incorporating future climate extremes and demand projections (*medium-high confidence*)
• Robust building codes and standards reflecting knowledge of current disaster risks (*high confidence*)	• Strengthened and enforced building codes, standards for changed climate extremes (*medium confidence*)
• Ecosystem-based/nature-based investments, including ecosystem conservation measures (*high confidence*)	• Advances in human development and poverty reduction, through, for example, social protection, employment, and wealth creation measures, taking future exposure to weather and climate extremes into account (*very high confidence*)
• Micro-insurance, including weather-indexed insurance (*medium confidence*)	
• Vulnerability-reducing measures such as pro-poor economic and human development, through for example improved social services and protection, employment, wealth creation (*very high confidence*)	

Continued next page →

disaster risks have been accounted for, the robustness of future projections of risk is also uncertain (Bouwer, 2010).

Furthermore, many of the economic cost assessments faced key methodological challenges, including the difficulty in estimating economic values of intangible effects of disasters, such as impact on human life, suffering, and ecological services, different rates of time preferences or discounting the future, as well as the difficulties associated with properly accounting for the distribution of costs and benefits across different sectors of society (Parry et al., 2009). Such challenges suggest that the value of tools, such as cost-benefit analysis, for the assessment of economic efficiency, even with risk considerations, may lie in the usefulness of the analytical process rather than the numeric outcomes per se. They suggest that in the context of climate adaptation, such tools may be most usefully employed as a heuristic tool in the context of iterative stakeholder decisionmaking processes (Moench et al., 2007).

- **Use risk-sharing and transfer mechanisms to protect financial security**: To effectively support communities and protect the financial security of the country, governments are increasingly using a range of financial instruments for transferring costs of disaster losses through risk-sharing mechanisms. Key risk transfer instruments include financial insurance, micro-insurance, and micro-financing, investment in social capital, government disaster reserve funds, and intergovernmental risk sharing. The latter two help to provide much needed relief and immediate liquidity after a disaster in regions where individual countries, because of their size and lack of diversity, cannot have viable risk insurance schemes. Such mechanisms can allow for more effective government response, provide some relief of the fiscal burden placed on governments due to disaster impacts, and constitute critical steps in promoting more proactive risk management strategies and responses (see Section 6.5.3).
- **Not all disaster risk can be eliminated, so act to manage residual risk too**: Even with effective disaster risk reduction policies and practices in place, it is impossible to reduce all disaster risks to zero and some residual risks will remain. With disaster risks increasing in many countries, steps could be taken to strengthen governments' ability to effectively manage residual risks effectively, and in doing so will need to strengthen partnerships with other actors and stakeholders to enable quick and effective humanitarian response that includes measures to 'build back better' and build resilience over time (for example, using rapid climate risk assessments to position critical infrastructure or relief camps in safer locations during relief and reconstruction phases). Many governments are also already working to enhance their disaster preparedness and early warning systems, focusing on the accuracy and timeliness of warnings, increasing public awareness, working with communities to ensure messages are communicated and transmitted effectively, and enhancing preparedness measures, such as first aid training, providing swimming lessons, encouraging households to have a disaster plan and an emergency kit, securing and indicating evacuation routes and shelters, and enhancing the skills of relief workers in child protection, for example (see Section 6.5.4).
- **Review resilience-building efforts**: Given competing priorities and development goals, governments are forced to balance resource allocation across development goals. The decision to bear residual losses is always a risk management option due to financial and other constraints. Many governments decide to accept the full risk of very low probability and surprise events, but new information on the impacts of climate change on such events may lead to such decisions being reviewed. Even in such cases where risk reduction and risk transfer is not a viable management option, investments in reducing vulnerability and enhancing early warning, preparedness, and standing capacity for emergency response can lead to positive returns. Furthermore, given uncertainties associated with disasters, efforts to promote flexible institutions, cross-scale learning, improved knowledge and awareness, and redundancies in response systems (in case one part of the system is badly impacted) can all help to promote resilience to very low probability and surprise events. Many governments are also encouraging maintenance and strengthening of social cohesiveness and social networks as a form of insurance enabling families and friends to support each other in times of disasters (see Sections 6.6.2 and 6.6.3).

A limited number of studies have used other tools such as multi-criteria analysis and other variants, which do not rely on just quantitative values, to help in the stakeholder-based adaptation decisionmaking (De Bruin et al., 2009a; Debels et al., 2009; Cardona et al., 2010). Debels et al. (2009) developed a multi-purpose index for a quick evaluation of adaptation practices in terms of proper design, implementation, and post-implementation evaluation and applied it to cases in Latin America. Mechler et al. (2006b) developed a metric for measuring fiscal vulnerability to natural hazards, capturing the relationship between the economic and fiscal losses that a country could experience when a catastrophic event occurs and the availability of funds to address the situation. Cardona et al. (2010), building on this, constructed the Disaster Deficit Index and applied it across a range of Latin American countries to support governmental decisionmaking in disaster risk management over time. De Bruin et al. (2009b) describe a hybrid approach based on qualitative and quantitative assessments of adaptation options for flood risk in The Netherlands. For the qualitative part, stakeholders selected options in terms of their perceived importance, urgency, and other elements. In the quantitative assessment, costs and benefits of key adaptation options are determined. Finally, using priority ranking based on a weighted sum of the qualitative and quantitative criteria suggests that in The Netherlands, for example, an integrated portfolio of nature and water management with risk-based policies has particularly high potential and acceptance for stakeholders. Overall, the assessment of adaptation explicitly considering the risk-based nature of extreme events remains fragmented and incipient, and more work will be necessary to improve the robustness of results and confidence in assessments.

6.6.2. Managing Uncertainties and Adaptive Management in National Systems

Disasters associated with climate extremes are inherently complex, involving socioeconomic as well as environmental and meteorological uncertainty (Hallegatte et al., 2007; see Chapter 3). Population, social, economic, and environmental change all influence the way in which hazards are experienced, through their impact on levels of exposure and on people's sensitivity to hazards (Pielke Jr. et al., 2003; Aldunce et al., 2008). Uncertainty about the magnitude, frequency, and severity of climate extremes is managed, to an extent, through the development of

Box 6-7 | Building Resilience to Disasters in the Cayman Islands

Key aspects that are relevant to building disaster resilience are flexibility, learning, and adaptive governance (Adger et al., 2005; Berkes, 2007), and the Cayman Islands (Tompkins et al., 2008) illustrate how such factors help to successfully manage their disaster risks. For example, in 2004, Hurricane Ivan (which was similar in magnitude to Hurricane Katrina that hit New Orleans in 2005) only caused two fatalities in the island, largely due to the activities of the National Hurricane Committee (NHC), which manages hurricane disaster risk reduction in the Cayman Islands and is responsible for preparedness, response, and recovery. The NHC is a learning-based organization. It learns from its successes, but more importantly from mistakes made. Each year the disaster managers actively assess the previous year's risk management successes and failures. Every year the National Hurricane Plan is revised to incorporate this learning and to ensure that good practices are institutionalized. Evidence of adaptive governance can be observed, for example, in the changing composition of the NHC, its structure, network arrangements, funding allocation, and responsibilities. Policymakers are encouraged to design and to implement new initiatives, to make adjustments, and take motivated actions. Creating such space for experimentation, innovation, learning, and institutional adjustment is crucial for disaster resilience.

predictive models and early warning systems (see Section 3.2.3 and Box 3-2; Section 9.2.11). Early warning systems are also based on models and consequently there is always a probability of their success (or failure) in predicting events accurately, although the failure to heed early warning systems is also a function of social factors, such as perception of risk, trust in the information-providing institution, previous experience of the hazard, degree of social exclusion, and gender (see, e.g., Drabek, 1986, 1999). Enhanced scientific modeling and interdisciplinary approaches to early warning systems can address some of these uncertainties provided good baseline and time series information are available (see Section 3.2.3 and Box 3-2). Even where such information is available, there remain other unresolved questions that influence the outcome of hazards. These relate to the capacity of ecosystems to provide buffering services, and the ability of systems to recover. Management approaches that take these issues into account include adaptive management and resilience, yet these approaches are not without their challenges (also see Section 8.6.3.1).

Adaptive management, as defined in Chapter 8 (Section 8.6.3.1), is "a structured process for improving management policies and practices by systemic learning from the outcomes of implemented strategies, and by taking into account changes in external factors in a proactive manner" (Pahl-Wostl et al., 2009; Pahl-Wostl, 2009). It has come to also mean bringing together interdisciplinary science, experience, and traditional knowledge into decisionmaking through 'learning by doing' by individuals and organizations (Walters, 1997). Decisionmakers, under adaptive management, are expected to be flexible in their approach, and accept new information as it become available, or when new challenges emerge, and not be rigid in their responses. Proponents argue that effective adaptive management contributes to more rapid knowledge acquisition and better information flows between policymakers, and ensures that there is shared understanding of complex problems (Lee, 1993).

In most cases, adaptive management has been implemented at the local or regional scale and there are few examples of its implementation at the national level. Examples of adaptive management abound in ecosystem management (Johnson, 1999; Ladson and Argent, 2000) and in disaster risk management (Thompson and Gaviria, 2004; Tompkins, 2005; see Box 6-7). Nearly 40 years of research, after the seminal paper was published by Holling in 1973, have produced evidence of the impacts of aspects of resilience policy (notably adaptive management) on forests, coral reefs, disasters, and adaptation to climate change; however, most of this has been at the local or ecosystem scale.

One of the main unresolved issues in adaptive management is how to ensure that scientists and engineers tasked with investigating adaptation and disaster risk management processes are able to learn from each other and from practitioners and how this learning can be integrated to inform policy and management practices. In the case of the restoration of the Florida Everglades, a limiting factor to effective management observed was the unwillingness of some parts of society to accept short-term losses for longer-term sustainability of ecosystem services (Kiker et al., 2001). Investment in hurricane preparedness in New Orleans prior to Hurricane Katrina provides a contemporary example of science not being included in disaster risk decisionmaking and planning (Laska, 2004; Congleton, 2006). The Cayman Islands hurricane management, on the other hand, demonstrates a success story in a flexible disaster management committee being prepared to change its strategies and measures from experience, and essentially learning by doing (Box 6-7).

Spare capacity within institutions has been argued to increase the ability of socio-ecological systems to address surprises or external shocks (Folke et al., 2005). McDaniels et al. (2008), in their analysis of hospital resilience to earthquake impacts, agreed with this finding, concluding that key features of resilience include the ability to learn from previous experience, careful management of staff during hazard, daily communication, and willingness of staff to address specific system failures. The latter can be achieved through creating overlapping institutions with shared delivery of services/functions, and providing redundant capacity within these institutions thereby allowing a sharing of the risks (Low et al., 2003). Such redundancy increases the chances of social memory being retained within the institution (Ostrom, 2005). However, if not carefully managed, costs of this approach can include fragmented policy, high transactions costs, duplication, inconsistencies, and inefficiencies (Imperial, 1999).

'Learning by doing' in disaster risk management can only be undertaken effectively if the management institutions are scaled appropriately, where necessary at the local level, or at multiple scales with effective interaction (Gunderson and Holling, 2002; Eriksen et al., 2011). For the management of climate extremes, the appropriate scale is influenced by the magnitude of the hazard and the affected area, including biological diversity. Research suggests that increasing biological diversity of ecosystems allows a greater range of ecosystem responses to hazards, and this increases the resilience of the entire system (Elmqvist et al., 2003). Other research has shown that reducing non-climate stresses on ecosystems can enhance their resilience to climate change. This is the case for coral reefs (Hughes et al., 2003; Hoegh-Guldberg et al., 2008) and rainforests (Malhi et al., 2008). Managing the resources at the appropriate scale, for example, water catchment or coastal zone instead of managing smaller individual tributaries or coastal sub-systems (such as mangroves), is becoming more urgent (Sorensen, 1997; Parkes and Horwitz, 2009).

Climate resilience as a development objective is, however, difficult to implement, particularly as it is unclear as to what resilience means (Folke, 2006). Unless resilience is clearly defined and broadly understood, with measurable indicators designed to fit different local contexts and to show the success, the potential losers from this policy may go unnoticed, causing problems with policy implementation and legitimacy (Eakin et al., 2009). See the Glossary for this report's definition of resilience, and more details regarding uncertainty and resilience related to extreme events in the light of climate change are given in Section 3.2.3, Box 3-2, and Section 8.5.1.

6.6.3. Tackling the Underlying Drivers of Vulnerability

This assessment has found that future trends in exposure, vulnerability, and climate extremes may further alter disaster risk and associated impacts. Future trends in climate extremes will be affected by anthropogenic climate change in addition to natural climate variability, and exposure and vulnerability will be influenced by both climatic and non-climatic factors (SPM; Sections 2.2, 2.3, and 2.5). Accordingly, reducing vulnerability and its underlying drivers is a considered a critical aspect of addressing both observed and projected changes in disaster risk (UNISDR 2009c, 2011b; Figure 6-3). Section 6.5.2.2 discussed the centrality of human development and vulnerability reduction to the goal of disaster risk reduction. As an extension, literature focused on aligning national disaster risk management systems to the challenges posed by climate change and other dynamic drivers of disaster risk places considerable importance on addressing the underlying drivers of vulnerability as one of the most effective 'low or no regrets' measures (see Figure 6-3 and Table 6-5 in FAQ 6.1; Tanner and Mitchell, 2008; Davies et al., 2008; CCCD, 2009; UNISDR, 2009c; Mitchell et al., 2010a). Such underlying drivers of vulnerability include inequitable development; poverty; declining ecosystems; lack of access to power, basic services, and land; and weak governance (Wisner et al., 2004; Schipper 2009; UNISDR 2009c, 2011b). An approach to managing disaster risk in the context of a changing climate highlights that disaster risk management efforts should seek to develop partnerships to tackle vulnerability drivers by focusing on approaches that promote more socially just and economic systems; forge partnerships to ensure the rights and entitlements of people to access basic services, productive assets, and common property resources; empower communities and local authorities to influence the decisions of national governments, NGOs, and international and private sector organizations and to promote accountability and transparency; and promote environmentally sensitive development (Hedger et al., 2010; Mitchell et al., 2010a; Polack, 2010).

To date, strategies for tackling the risks of climate extremes and disasters, in practice, have tended to focus on treating the symptoms of vulnerability, and with it risk, rather than the underlying causes, partly due to disaster risk management still not being a core component of sustainable development (Schipper, 2009). The mid-term review of the HFA indicates that insufficient effort is being made to tackle the conditions that create risk (UNISDR, 2011b), and other studies have found a continued disconnect between disaster risk management and development processes that tackle the structural causes of poverty and vulnerability and between knowledge and implementation at all scales (CCCD, 2009; UNISDR, 2009c). The impacts of climate change, both on disaster risk and on vulnerability and poverty, are viewed by some as a potential force that will help to forge a stronger connection between disaster risk reduction measures and poverty and vulnerability reduction measures, also partly as a result of increased availability of financial resources and renewed political will (Soussan and Burton, 2002; Schipper, 2009; Mitchell et al., 2010a). A recent and growing body of literature has focused on the potential for strengthening the links among particular forms of social protection, disaster risk reduction, and climate change adaptation measures as a way to simultaneously tackle the drivers of vulnerability, poverty, and hence disaster risk (see Section 8.3.1; Davies et al., 2008; Heltberg et al., 2009). With increasing levels of exposure to disaster risk in middle-income countries (see Section 6.1; UNISDR, 2009c, 2011a), reducing vulnerability of poor people and their assets in such locations is becoming a focus for those governments and for CSOs and CBOs (Tanner and Mitchell, 2008).

6.6.4. Approaching Disaster Risk, Adaptation, and Development Holistically

As this chapter has demonstrated, climate change poses diverse and complex challenges for actors in national disaster risk management systems and for disaster risk management policies and practices more broadly. These challenges include changes in the magnitude and frequency of some hazards in some regions, impacts on vulnerability and exposure, new agreements and resource flows, and the potential of climate change to alter value systems and people's perceptions. As Table SPM.1 highlights, it is the complexity resulting from the combination of these factors, in addition to the uncertainty generated, that means national disaster risk management systems and broader national strategies may need to be realigned to maintain and improve their effectiveness. There is *high agreement* but *limited evidence* to suggest that a business-as-usual approach to disaster risk management that fails to take the impacts of climate change into account will become increasingly ineffective. Section 6.6 and other parts of this chapter have assessed evidence on the different elements involved in such a realignment. A selection of these elements is briefly summarized here.

As discussed in Section 6.6.2, there is *high agreement* but *limited evidence* to suggest that flexible and adaptive national systems for disaster risk management, based on the principle of learning by doing, are better suited to managing the challenges posed by changes in exposure, vulnerability, and weather and climate extremes than static and rigid systems (see Section 8.6). This ability to be flexible will be tested by a systems' capacity to act on new knowledge generated by the frequent assessment of dynamic risk needed to capture trends in exposure vulnerability and weather and climate extremes and by information on how the costs and benefits of different response measures change as a result (Section 6.6.1). The accuracy of these assessments will be based on the quality of available data (Section 6.5.2.1). Where such assessments generate uncertainty for decisionmakers, tools such as multi-criteria analysis, scenario planning, and flexible decision paths offer ways of supporting informed action (Section 6.6.1).

There is *high agreement* and *robust evidence* to demonstrate that the mainstreaming of disaster risk management processes into development planning and practice leads to more resilient development pathways. By extension, with climate change and other development processes having an impact on disaster risk, these changes then need to be factored into development and economic planning decisions at different scales. This suggests an ideal national system for managing the risks from climate extremes and disasters would be designed to be fully integrated with economic and social development, environmental, poverty reduction, and humanitarian dimensions to create a holistic approach. The nature of transformational changes in thinking, analysis, planning, approaches, strategies, and actions is the subject of Chapter 8 (particularly Section 8.2.2).

While there is *limited evidence* that some countries have begun to factor climate change into the way disaster risks are assessed and managed (see Sections 6.3 and 6.6.1), few countries appear to have adopted a comprehensive approach – for example, by addressing projected changes in exposure, vulnerability, and extremes as well as adopting a learning-by-doing approach to decisionmaking embedded in the context of national development planning processes. Incremental efforts toward implementing suitable strategies for mainstreaming climate change responses into national development planning and budgetary processes, and climate proofing at the sector and project levels (Sections 6.2 and 6.3) in the context of disaster risk management appear to be the most likely approach adopted by many countries. None of these measures will be easy to implement, as actors and stakeholders at all levels of society are being asked to embrace a dynamic notion of risk as an inherent part of their decisions, and continuously learn and modify policies, decisions, and actions taking into account new traditional and scientific knowledge as it emerges.

The knowledge base for understanding changing climate-related disaster risks and for the way national systems are acting on this understanding through modifying practices, altering the nature of relationships between different actors, and adopting new strategies and policies is fragmented and incomplete. As this chapter has illustrated, incomplete information and knowledge gaps do not need to present blockages to action. As FAQ 6.1 and Section 6.3.1 highlight, there is considerable experience of governments and other actors investing in measures to respond to existing climate variability and disaster risk that can be considered as 'no or low regrets' options when taking into account the uncertainty associated with future climate. However, in conducting this assessment, some knowledge gaps have emerged that, if filled, would aid the creation of enduring national risk management systems for tackling observed and projected disaster risk. These gaps include the need for more research on:

- The extent to which efforts to build disaster risk management capacities at different scales prepare people and organizations for the challenges posed by climate change.
- Whether the current trend of decentralizing disaster risk management functions to sub-national and local governments and communities is effective, given the level of information and capacity requirements, changing risks, and associated uncertainties presented by climate change.
- How the function, roles, and responsibilities of different actors working within national disaster risk management systems are changing, given the impacts of climate change at the national and sub-national level.
- The characteristics of flexibility, learning-by-doing, and adaptive management in the context of national disaster risk management systems in different governance contexts.
- How decisions on disaster risk management interventions are made at different scales if there is limited context specific information.
- The costs and benefits of different risk management interventions if the impacts of climate change and other dynamic drivers of risk are factored in.
- The benefits and tradeoffs of creating integrated programs and policies that seek to manage disaster risk, mitigate GHGs, adapt to climate change, and reduce poverty simultaneously.

References

A digital library of non-journal-based literature cited in this chapter that may not be readily available to the public has been compiled as part of the IPCC review and drafting process, and can be accessed via either the IPCC Secretariat or IPCC Working Group II web sites.

Aakre, S., I. Banaszak, R. Mechler, D. Rübbelke, A. Wreford, and H. Kalirai, 2010: Financial adaptation to disaster risk in the European Union. *Mitigation and Adaptation Strategies for Global Change*, **15(7)**, 721-736.

ActionAid, 2006: *Climate Change, Urban Flooding and the Rights of the Urban Poor in Africa: Key findings from six African cities*. ActionAid International, London, UK and Johannesburg, South Africa.

Adger, W.N., T.P. Hughes, C. Folke, S.R. Carpenter, and J. Rockstrom, 2005: Social-ecological resilience to coastal disasters. *Science*, **5737(309)**, 1036-1039.

Adger, W.N., S. Agrawala, M.M.Q. Mirza, C. Conde, K. O'Brien, J. Pulhin, R. Pulwarty, B. Smit, and K. Takahashi, 2007: Assessment of adaptation practices, options, constraints and capacity. In: *Climate Change 2007. Impacts, Adaptation and Vulnerability. Contribution of Working Group II to the Fourth Assessment Report of the Intergovernmental Panel on Climate Change* [Parry, M.L., O.F. Canziani, J.P. Palutikof, P.J. Van Der Linde, and C.E. Hanson (eds.)]. Cambridge University Press, Cambridge, UK, pp. 717-743.

ADPC, 2007: *Community Self-Resilience and Flood Risk Reduction*. ADB TA 4574-CAM Report, Asian Disaster Preparedness Centre, Asian Development Bank, Manila, Philippines.

African Development Bank Group, 2010: *Nairobi Rivers Rehabilitation and Restoration Program: Sewerage Improvement Project*. Project Appraisal Report, African Development Bank, www.afdb.org/fileadmin/uploads/afdb/Documents/Project-and-Operations/Kenya%20-%20Nairobi%20rivers%20rehabilitation%20and%20restoration%20program%20-%20sewerage%20improvement%20project.pdf. Last accessed on: 26th Sept 2011.

Agrawala, S. and S. Fankhauser, 2008: *Economic Aspects of Adaptation to Climate Change. Costs, benefits and policy instruments*. Organisation for Economic Cooperation and Development, Paris, France.

Agrawala, S. and M. van Aalst, 2008: Adapting development cooperation to adapt to climate change. *Climate Policy*, **8(3)**, 183-193.

Albala-Bertrand, J.M., 1993: *Political Economy of Large Natural Disasters with Special Reference to Developing Countries*. Clarendon Press, Oxford, UK.

Aldunce, P. and M. González, 2009: *Desastres Asociados al Clima en la Agricultura y Medio Rural de Chile*. Universidad de Chile, Santiago, Chile.

Aldunce, P. and A. Leon, 2007: Opportunities for improving disaster management in Chile: A case study. *Disaster Prevention and Management*, **16(1)**, 33-41.

Aldunce, P, C. Neri, and C. Szlafsztein (eds.), 2008: *Hacia la Evaluación de Prácticas de Adaptación ante la Variabilidad y el Cambio Climático*. Nucleo de Meio Ambiente/Universidad Federal do Para, Belém, Brazil.

ALNAP, 2005: *Capacity Building amid Humanitarian Response*. Active Learning Network for Accountability and Performance in Humanitarian Action, London, UK.

ALNAP, 2010a: *The State of the Humanitarian System: assessing performance and progress*. Active Learning Network for Accountability and Performance in Humanitarian Action, London, UK.

ALNAP, 2010b: *The Role of National Governments in International Humanitarian Response to Disasters*. Meeting Background Paper, 26th ALNAP Meeting, Active Learning Network for Accountability and Performance in Humanitarian Action, London, UK, www.alnap.org/pool/files/26-meeting-background-paper.pdf.

Amendola, A., J. Linnerooth-Bayer, N. Okada, and P. Shi, 2008: Towards integrated disaster risk management: case studies and trends from Asia. *Natural Hazards*, **44**, 163-168.

Andersen, T.J., 2001: *Managing Economic Exposures of Natural Disasters. Exploring Alternative Financial Risk Management Opportunities and Instruments*. Inter-American Development Bank, Washington, DC.

Angelsen, A., M. Brockhaus, M. Kanninen, E. Sills, W.D. Sunderlin, and S. Wertz-Kanounnikoff, 2009: *Realising REDD: National strategy and policy options*. Center for International Forestry Research, Indonesia.

Arnell, N., 2004: Climate change and global water resources: SRES emissions and socio-economic scenarios. *Global Environmental Change*, **14(1)**, 31-52.

Arnell, N.W. and E.K. Delaney, 2006: Adapting to climate change: Public water supply in England and Wales. *Climatic Change*, **78**(2-4), 227-255.

Arnold, M., 2008: *The Role of Risk Transfer and Insurance in Disaster Risk Reduction and Climate Change Adaptation*. Policy brief, Commission on Climate Change and Development, Stockholm, Sweden.

Arrow, K. and R. Lind, 1970: Uncertainty and the evaluation of public investment decisions. *The American Economic Review*, **60**, 364-378.

Auffret, P., 2003: *Catastrophe Insurance Market in the Caribbean Region: Market Failures and Recommendations for Public Sector Interventions*. World Bank Policy Research Working Paper 2963, World Bank, Washington DC.

Auld, H., 2008a: Disaster risk reduction under current and changing climate conditions. *World Meteorological Organization Bulletin*, **57(2)**, 118-125.

Auld, H.E., 2008b: Adaptation by design: The impact of changing climate on infrastructure. *Journal of Public Works and Infrastructure*, **3**, 276-288.

Auld, H., D. MacIver, and J. Klaassen, 2004: Heavy rainfall and waterborne disease outbreaks: The Walkerton example. *Journal of Toxicology and Environmental Health, Part A*, **67(20-22)**, 1879-1887.

AusAid, 2009: *Investing in a Safer Future: A Disaster Risk Reduction policy for the Australian aid program*. AusAid, Canberra, Australia, www.ausaid.gov.au/publications/pdf/disasterriskreduction.pdf.

Averyt, K., 2010: Are we successfully adapting science to climate change? *Bulletin of the American Meteorological Society*, **91(6)**, 723-726.

Ayers, J.M. and S. Huq, 2009: The value of linking mitigation and adaptation: A case study of Bangladesh. *Environmental Management*, **43**(5), 753-764.

Balmford, A., A. Bruner, P. Cooper, R. Costanza, S. Farber, R.E. Green, M. Jenkins, P. Jefferiss, V. Jessamy, and J. Madden, 2002: Economic reasons for conserving wild nature. *Science*, **297(5583)**, 950-953.

Barbier, E.B., 2009: *Rethinking the Economic Recovery: A Global Green New Deal*. Report prepared for the Economics and Trade Branch, Division of Technology, Industry and Economics, United Nations Environment Programme (UNEP), Geneva, Switzerland.

Barrett, E., S. Murfitt, and P. Venton, 2007: *Mainstreaming the Environment into Humanitarian Response: An Exploration of Opportunities and Issues*. Environmental Resource Management Limited, London, UK.

Basher, R., 2006: Global early warning systems for natural hazards: Systematic and people-centered. *Philosophical Transactions of the Royal Society A*, **364(1845)**, 2167-2182.

Bebi, P., D. Kulakowski, and C. Rixen, 2009: Snow avalanche disturbances in forest ecosystems – State of research and implications for management. *Forest Ecology and Management*, **257(9)**, 1883-1892.

Bennett, J., W. Bertrand, C. Harkin, S. Samarasinghe, and H. Wickramatillake, 2006: *Coordination of International Humanitarian Assistance in Tsunami Affected Countries*. Tsunami Evaluation Coalition, London, UK.

Benson, C., 2007: *Tools for Mainstreaming Disaster Risk Reduction: Environmental Assessments*. Guidance Note 7, ProVention Consortium, Geneva, Switzerland.

Benson, C. and E. Clay, 2002a: *Understanding the Economic and Financial Impacts of Natural Disasters*. World Bank, Washington, DC.

Benson, C. and E. Clay, 2002b: Disasters, vulnerability, and the global economy. In: *Building Safer Cities: The Future of Disaster Risk* [Kreimer, A., M. Arnold, and A. Carlin (eds.)]. Disaster Risk Management Series No. 3, World Bank, Washington, DC, pp. 3-32.

Benson, C. and E. Clay, 2004: Understanding the *Economic and Financial Impacts of Natural Disasters*. Disaster Risk Management Series 4, World Bank, Washington, DC.

Benson, C. and J. Twigg, 2004: *Measuring Mitigation: Methodologies for Assessing Natural Hazard Risks and the Net Benefits of Mitigation – A Scoping Study*. The International Federation of Red Cross and Red Crescent Societies/The ProVention Consortium, Geneva, Switzerland.

Benson, C., J. Twigg, and M. Myers, 2001: NGO initiatives in risk reduction: An overview. *Disasters*, **25(3)**, 199-215.

Berkes, F., 2007: Understanding uncertainty and reducing vulnerability: Lessons from resilience thinking. *Natural Hazards*, **41(2)**, 283-295.

Berry, P., 2007: *Adaptation Options on Natural Ecosystems*. Report to the Financial and Technical Support Division, United Nations Framework Convention on Climate Change, Bonn, Germany.

Bhavnani, R., M. Owor, S. Vordzorgbe, and F. Bousquet, 2008: *Status of Disaster Risk Reduction in the Sub-Saharan Africa Region*. World Bank, Washington, DC.

Biesbroek, G.R., R.J. Swart, T.R. Carter, C. Cowan, T. Henrichs, H. Mela, M.D. Morcecroft, and D. Rey, 2010: Europe adapts to climate change: Comparing National Adaptation strategies. *Global Environmental Change*, **20(3)**, 440-450.

Birkmann, J., M. Garschagen, F. Kraas, and N. Quang, 2011: Adaptive urban governance: new challenges for the second generation of urban adaptation strategies to climate change. *Sustainability Science*, **5**, 185-206.

Borton, J., 1993: Recent trends in the international relief system. *Disasters*, **17(3)**, 187-201.

Bosher, L., P. Carillo, A. Dainty, J. Glass, and A. Price, 2007: Realising a resilient and sustainable built environment: Towards a strategic agenda for the United Kingdom. *Disasters*, **31**, 236-255.

Botzen, W.J.W., J.C.J.H. Aerts, and J.C.J.M. van den Bergh, 2009: Willingness of homeowners to mitigate climate risk through insurance. *Ecological Economics*, **68(8-9)**, 2265-2277.

Bourrelier, P.H., B.G. Deneufbourg, and B. de Vanssay, 2000: IDNDR objectives: French technical sociological contributions. *Natural Hazards Review*, **1(1)**, 18-26.

Bouwer, L.M., 2006: The benefits of disaster risk reduction and their effects on loss trends. In: *Climate Change and Disaster Losses: Understanding and Attributing Trends and Projections* [Hoppe, P. and R. Pielke (eds.)]. Centre for Science and Technology Policy Research, Tyndall Centre, Munich Re and National Science Foundation, Hamburg, Germany, pp 26-28.

Bouwer, L.M., 2010: *Disasters and Climate Change: Analysis and Methods for Projecting Future Losses from Extreme Weather*. PhD thesis, Vrije Universiteit, Amsterdam, The Netherlands, 141 pp., dare.ubvu.vu.nl/bitstream/1871/16355/1/dissertation.pdf.

Branco, A., J. Suassuna, and S.A. Vainsencher, 2005: Improving access to water resources through rainwater harvesting as a mitigation measure: The case of the Brazilian semi-arid region. *Mitigation and Adaptation Strategies for Global Change*, **10(3)**, 393-409.

Brocklesby, M. and E. Fisher, 2003: Community development in sustainable livelihoods approaches – an introduction. *Community Development Journal*, **38**, 185-198.

Brooks, N. and W.N. Adger, 2005: Adaptation policy frameworks for climate change: developing strategies, policies and measures. In: *Assessing and Enhancing Adaptive Capacity* [Lim, B., E. Spanger-Siegfried, I. Burton, E.L. Malone, and S. Huq (eds.)]. Cambridge University Press, Cambridge, UK, pp. 165-181.

Brooks, N., N. Grist, and K. Brown, 2009: Development futures in the context of climate change: Challenging the present and learning from the past. *Development Policy Review*, **27**, 741-765.

Bunce, M., K. Brown, and S. Rosendo, 2010: Policy misfits, climate change and cross-scale vulnerability in coastal Africa: how development projects undermine resilience. *Environmental Science and Policy*, **13**, 485-497.

Burby, R, 1991. *Sharing Environmental Risks: How to Control Governments' Losses in Natural Disasters*. Westview Press, Boulder, CO.

Butler, C.D. and W. Oluoch-Kosura, 2006: Linking future ecosystem services and future human well-being. *Ecology and Society*, **11(1)**, 30.

Butt, T.A., B.A. Mccarl, J. Angerer, P.T. Dyke, and J.W. Stuth, 2005: The economic and food security implications of climate change in Mali. *Climatic Change*, **68(3)**, 355-378.

CACCA, 2010: *Approaches to Climate Change Adaptation*. Committee on Approaches to Climate Change Adaptation, Global Environment Bureau, Ministry of Environment, Tokyo, Japan, www.env.go.jp/en/earth/cc/adapt_guide.

Campbell, A., V. Kapos, A. Chenery, S.I. Kahn, M. Rashid, J.P.W. Scharlemann, and B. Dickson, 2008: *The Linkages between Biodiversity and Climate Change Mitigation*. UNEP World Conservation Monitoring Center, Cambridge, UK.

Cardenas, V., S. Hochrainer, R. Mechler, G. Pflug, and J. Linnerooth-Bayer, 2007: Sovereign financial disaster risk management: The case of Mexico. *Environmental Hazards*, **7(1)**, 40-53.

Cardona, O., 2006: Measuring vulnerability to natural hazards: Towards disaster resilient societies. In: *A system of indicators for disaster risk management in the Americas* [Birkmann, J. (ed.)]. United Nations University Press, Tokyo, Japan.

Cardona, O.D., M.G. Ordaz, M.C. Marulanda, M.L. Carreno, and A.H. Barbat, 2010: Disaster risk from a macroeconomics perspective: a metric for fiscal vulnerability evaluation. *Disasters*, **34(4)**, 1064-1083.

CARE International, 2008: Community preparedness for emergencies helps poverty. In: *Linking Disaster Risk Reduction and Poverty: Good Practices and Lessons Learnt* [United Nations International Strategy for Disaster Reduction (ed.)]. United Nations International Strategy for Disaster Reduction, Geneva, Switzerland, pp. 6-10.

Carter, T.R., R.N. Jones, X. Lu, S. Bhadwal, C. Conde, L.O. Mearns, B.C. O'Neill, M. Rounsevell, and M.B. Zurek, 2007: New assessment methods and the characterisation of future conditions. In: *Climate Change 2007. Impacts, Adaptation and Vulnerability. Contribution of Working Group II to the Fourth Assessment Report of the Intergovernmental Panel on Climate Change* [Parry, M.L., O.F. Canziani, J.P. Palutikof, P.J. Van Der Linde, and C.E. Hanson (eds.)]. Cambridge University Press, Cambridge, UK, pp.133-171.

Carter, W.N., 1992: *Disaster Management: A Disaster Manager's Handbook*. Asian Development Bank, Manila, Philippines.

Caselli, F. and P. Malthotra, 2004: *Natural Disasters and Growth: From Thought Experiment to Natural Experiment*. International Monetary Fund, Washington, DC.

Cash, D.W. and J. Buizer, 2005: *Knowledge-Action Systems for Seasonal to Interannual Climate Forecasting: Summary of a Workshop*. Roundtable on Science and Technology for Sustainability Policy and Global Affairs Division, National Research Council of the National Academies, The National Academies Press, Washington DC.

CCCD, 2009: *Closing the Gaps: Disaster Risk Reduction and Adaptation to Climate Change in Developing Countries: Final Report*. Commission on Climate Change and Development, Ministry for Foreign Affairs, Stockholm, Sweden.

CDB and CARICOM, 2004: *Sourcebook on the Integration of Natural Hazards into Environmental Impact Assessment (NHIA-EIA Sourcebook)*. Caribbean Development Bank, Bridgetown, Barbados.

Chandran, R. and B. Jones, 2008: *Concepts and Dilemmas of State Building in Fragile Situations: from fragility to resilience*. OECD/DAC Discussion Paper, Organisation for Economic Cooperation and Development, Paris, France.

Charveriat, C., 2000: *Natural disasters in Latin America and the Caribbean: an overview of risk*. Inter-American Development Bank, Washington, DC.

Chhatre, A. and A. Agrawal, 2009: Trade-offs and synergies between carbon storage and livelihood benefits from forest commons. *Proceedings of the National Academy of Sciences*, **106(42)**, 17667-17670.

Christensen, J.H., B. Hewitson, A. Busuioc, A Chen, A. Gao, I. Held, R. Jones, R.K. Koli, W.T. Kwon, R. Laprise, V.M. Rueda, L. Mearns, C.G. Menéndez, J. Räisänen, A. Rinke, A. Sarr, and P. Whetton, 2007: Regional climate predictions. In: *Climate Change 2007: The Physical Science Basis. Contribution of Working Group I to the Fourth Assessment Report of the Intergovernmental Panel on Climate Change* [Solomon, S., D. Qin, M. Manning, Z. Chen, M. Marquis, K.B. Averyt, M. Tignor and H.L. Miller (eds.)]. Cambridge University Press, Cambridge, UK, and New York, NY, pp. 847-940.

Chung, J., 2009: *Situation analysis for strengthening disaster information management systems in Fiji: baseline data preparedness and assessment methodologies*. A report prepared for the Fiji Government and Pacific Disaster Risk Management Partnership, United Nations International Strategy for Disaster Reduction and South Pacific Applied Geoscience Commission, Suva, Fiji.

Cline, W.R., 2007: *Global Warming and Agriculture: Impact Estimates by Country*. Center for Global Development and Peterson Institute for International Economics, Washington, DC.

Coleman, T., 2002: *The Impact of Climate Change on Insurance against Catastrophes*. Insurance Australia Group, Melbourne, Australia.

Colls, A., N. Ash, and N. Ikkala, 2009: *Ecosystem-based Adaptation: A Natural Response to Climate Change*. International Union for Conservation of Nature, Gland, Switzerland.

Congleton, R.D., 2006: The story of Katrina: New Orleans and the political economy of catastrophe. *Public Choice*, **127(1)**, 5-30.

Conway, D. and L. Schipper, 2011: Adaptation to climate change in Africa: Challenges and opportunities identified from Ethiopia. *Global Environmental Change*, **21(1)**, 227-237.

Coppola, D.P, 2007: *Introduction to International Disaster Management*. Butterworth-Heinemann, Oxford, UK.

Costanza, R., O. Pérez-Maqueo, M.L. Martinez, P. Sutton, S.J. Anderson, and K. Mulder, 2008: The value of coastal wetlands for hurricane protection. *Ambio*, **37(4)**, 241-248.

Crabbé, P. and M. Robin, 2006: Institutional adaptation of water resource infrastructures to climate change in eastern Ontario. *Climatic Change*, **78(1)**, 103 -133.

CRGR, 2007a: Lineamientos generales de organización y funcionamiento de la Concertación Regional de Gestión de Riesgos. *Políticas, Prácticas y Gestión del Riesgo*, **1(7)**, 4-7.

CRGR, 2007b: Reforma del Estado y gestión del riesgo de desastres. *Políticas, Prácticas y Gestión del Riesgo*, **2(8)**, 3-11.

CRGR, 2009: Centroamérica frente a los retos del cambio climático. Posicionamiento político de la sociedad civil y pueblos indígenas organizados. *Políticas, Prácticas y Gestión del Riesgo*, **3(12)**, 12-15.

Crowards, T., 2000: *Comparative Vulnerability to Natural Disasters in the Caribbean*. Caribbean Development Bank, Bridgetown, Barbados.

CSA, 2010a: *Infrastructure in Permafrost: A Guideline for Climate Change Adaptation. A Technical Guide.* Canadian Standards Association, Mississauga, Canada.

CSA, 2010b: *Development, interpretation and use of rainfall intensity-duration-frequency (IDF) information: Guideline for Canadian water resources practitioners*. Technical Guide, Canadian Standards Association, Mississauga, Canada.

Cummins, J., and O. Mahul, 2009: *Catastrophe Risk Financing In Developing Countries: Principles For Public Intervention.* World Bank, Washington, DC.

Curriero, F.C., J.A. Patz, J.B. Rose, and S. Lele, 2001: The association between extreme precipitation and waterborne disease outbreaks in the United States, 1948-1994. *American Journal of Public Health*, **91(8)**, 1194-1199.

Das, S. and J.R. Vincent, 2009: Mangroves protected villages and reduced death tolls during Indian super cyclone. *Proceedings of the National Academy of Sciences*, **106(18)**, 7357-7360.

DaSilva, J., B. Garanganga, V. Teveredzi, S.M. Marx, S.J. Mason, and S.J. Connor, 2004: Improving epidemic malaria planning, preparedness and response in Southern Africa. *Malaria Journal*, **3(1)**, 37.

Davies, M., B. Guenther, J. Leavy, T. Mitchell, and T. Tanner, 2008: Adaptive social protection synergies for poverty reduction. *Institute of Development Studies (IDS) Bulletin*, **39(4)**, 105-112.

Davis, K. (ed.), 2004: *Technology Dissemination among Small-Scale Farmers in Meru Central District of Kenya: Impact of Group Participation*. USA: University of Florida.

De Bruin, K., R. Dellink, and S. Agrawala, 2009a: *Economic Aspects of Adaptation to Climate Change: Integrated Assessment Modeling of Adaptation Costs and Benefits.* Environment Working Paper No. 6, Organisation for Economic Cooperation and Development, Paris, France.

De Bruin, K., R.B. Dellink, A. Ruijs, L. Bolwidt, A. van Buuren, J. Graveland, R.S. de Groot, P.J. Kuikman, S. Reinhard, R.P. Roetter, V.C. Tassone, A. Verhagen, and E.C. van Ierland, 2009b: Adapting to climate change in The Netherlands: an inventory of climate adaptation options and ranking of alternatives. *Climatic Change*, **95**, 23-45.

Debels, P., C. Szlafsztein, P. Aldunce, C. Neri, Y. Carvajal, M. Quintero-Angel, A. Celis, A. Bezanilla, and D. Martinez, 2009: IUPA: A tool for the evaluation of the general usefulness of practices for adaptation to climate change and variability. *Natural Hazards*, **50(2)**, 211-233.

DEFRA, 2005: *Making Space for Water. Taking Forward a New Government Strategy for Flood and Coastal Erosion Risk Management in England.* Department for Environment, Food and Rural Affairs, London, UK.

DEFRA, 2006: *The UK's Fourth National Communication under the United Nations Framework Convention on Climate Change*. Department for Environment, Food and Rural Affairs, London, UK.

del Ninno, C., 2001: The 1998 Floods in Bangladesh: Disaster Impacts, Household Coping Strategies, and Response. Research Report 122, International Food Policy Research Institute, Washington, DC.

del Ninno, C., P.A. Dorosh, and L.C. Smith, 2003: Public policy, markets and household coping strategies in Bangladesh; Avoiding a food security crisis following the 1998 floods. *World Development*, **31(7)**, 1221-1238.

Dercon, S., 2005: Risk, insurance, and poverty: A review. In: *Insurance Against Poverty. A study prepared by the World Institute for Development Economics Research of the United Nations University (UNU-WIDER)*[Dercon, S. (ed.)]. Oxford University Press, Oxford, UK, pp. 9-37.

Dessai, S., and M. Hulme, 2007: Assessing the robustness of adaptation decisions to climate change uncertainties: A case study on water resources management in the East of England. *Global Environmental Change*, **17**, 59-72.

DFID, 2005: *Natural Disaster and Disaster Risk Reduction Measures: A Desk Review of Costs and Benefits*. UK Department for International Development, The Stationery Office, London, IK.

DFID, 2011: *Multilateral Aid Review: Ensuring Maximum Value for Money for UK Aid through Multilateral Organisations*. UK Department for International Development, London, UK.

Dilley, M., 2006: Risk identification: A critical component of disaster risk management. *World Meteorological Organisation Bulletin*, **55(1)**, 13-20.

Dolan, A.H. and I.J. Walker, 2006: Understanding vulnerability of coastal communities to climate change related risks. In: *Proceedings of the 8th Int'l Coastal Symposium*. Itajaí, Brazil, 14-18 March 2004. *Journal of Coastal Research*, **SI 39**, 1316-1323.

Dorren, L.K.A., F. Berger, A.C. Imeson, B. Maier, and F. Rey, 2004: Integrity, stability and management of protection forests in the European Alps. *Forest Ecology and Management*, **195(1-2)**, 165-176.

Drabek, T.E., 1986: *Human System Responses To Disaster: An Inventory Of Sociological Finding*. Springer-Verlag New York, NY.

Drabek, T.E., 1999: Understanding disaster warning responses. *The Social Science Journal*, **36**, 515-523.

Dudley, N., S. Stolton, A. Belokurov, L. Krueger, N. Lopoukhine, K. Mackinnon, T. Sandwith, and N. Sekhran, 2010: *Natural Solutions: Protected Areas Helping People Cope with Climate Change*. IUCN, WCPA, TNC, UNDP, WCS, World Bank and WWF, Gland, Switzerland, Washington, DC and New York, NY.

Eakin, H., E.L. Tompkins, D.R. Nelson, and J.M. Anderies, 2009: Adapting to climate change: Thresholds, values, governance. In: *Hidden Costs and Disparate Uncertainties: Trade-Offs Involved in Approaches to Climate Policy* [Adger, W.N., I. Lorenzoni, and L.O.B. Karen (eds.)]. Cambridge University Press, Cambridge, UK, pp. 212-226.

Easterling, W.E., P.K. Aggarwal, P. Batima, K.M. Brander, L. Erda, S.M. Howden, A. Kirilenko, J. Morton, J.-F. Soussana, J. Schmidhuber, and F.N. Tubiello, 2007: Food, fibre and forest products. In: *Climate Change 2007. Impacts, Adaptation and Vulnerability. Contribution of Working Group II to the Fourth Assessment Report of the Intergovernmental Panel on Climate Change* [Parry, M.L., O.F. Canziani, J.P. Palutikof, P.J. Van Der Linde, and C.E. Hanson (eds.)]. Cambridge University Press, Cambridge, UK, pp. 273-313.

Ebi, K.L., 2008: Adaptation costs for climate change-related cases of diarrhoeal disease, malnutrition, and malaria in 2030. *Globalization and Health*, **6**, 9, doi:10.1186/1744-8603.

Ebi, K.L., R.S. Kovats, and B. Menne, 2006. An approach for assessing human health vulnerability and public health interventions to adapt to climate change. *Environmental Health Perspectives*, **114**, 12, doi:10.1289/Ehp.8430.

EC, 2009: Technical Report – 2009 – 040, Guidance Document No. 24, River Basin Management. In: *A Changing Climate, Common Implementation Strategy for the Water Framework*. Directive 2000/60/EC. European Commission, Brussels, Belgium.

ECA, 2009: *Shaping Climate-Resilient Development: A Framework For Decision-Making Study*. Economics of Climate Adaptation Working Group, World Bank, Washington, DC.

ECLAC, 2002: *Handbook for Establishing the Socio-Economic and Environmental Effects of Natural Disasters*. United Nations, ECLAC and World Bank, Mexico City, Mexico.

ECLAC, 2003: *Handbook for Establishing the Socio-Economic and Environmental Effects of Natural Disasters*. United Nations, Economic Commission for Latin America and the Caribbean, and World Bank, Santiago, Chile.

EEA, 2007: *Climate change: The Cost of Inaction and the Cost of Adaptation*. Technical Report, European Environment Agency, Copenhagen, Denmark.

Elmqvist, T., C. Folke, M. Nyström, G. Peterson, J. Bengtsson, B. Walker, and J. Norberg, 2003: Response diversity, ecosystem change, and resilience. *Frontiers in Ecology and the Environment*, **1(9)**, 488-494.

Embrechts, P., C. Klüppelberg, and T. Mikosch, 1997: *Modelling Extremal Events for Insurance And Finance*. Springer, Berlin, Germany.

Ericksen, P., J. Ingram, and D. Liverman, 2009: Food security and global environmental change: emerging challenges. *Environmental Science and Policy*, **12**, 373-377.

Eriksen, S. and L.O. Næss, 2003: Pro-Poor Climate Adaptation: Norwegian Development Cooperation and Climate Change Adaptation – An Assessment of Issues, Strategies and Potential Entry Points. Report 2003:02. Center for International Climate and Energy Research – Oslo (CICERO), Oslo, Norway, 75pp.

Eriksen, S., P. Aldunce, C. Bahinipati, R. D'Almeida, J. Molefe, C. Nhemachena, K. O'Brien, F. Olorunfemi, J. Park, L. Sygna, and K. Ulsrud, 2011: When not every response to climate change is a good one: Identifying principles for sustainable adaptation. *Climate Change and Development*, **3(1)**, 7-20.

Fagen, P., 2008: *Natural Disasters in Latin America and the Caribbean: National, Regional and International Interactions*. HPG Working Paper, Overseas Development Institute, London, UK.

Fankhauser, S., 2010: The costs of adaptation. *Wiley Interdisciplinary Reviews: Climate Change*, **1(1)**, 23-30.

FAO, 2006: *Third session of the sub-committee on aquaculture: Committee on Fisheries* (CoFI). India, New Delhi, 4-8 September. Food and Agriculture Organization, Rome, Italy.

FAO, 2007: *Building Adaptive Capacity to Climate Change: Policies to Sustain Livelihoods and Fisheries*. New Directions in Fisheries – A Series of Policy Briefs on Development Issues, 0816. Food and Agriculture Organization, Rome, Italy.

FAO, 2008a: Climate change adaptation and mitigation in the food and agriculture sector: In: *Technical background document from the expert consultation held on 5 to 7 March 2008* (Paper presented at Climate change, energy and food). Food and Agriculture Organization, Rome, Italy.

FAO, 2008b: *Challenges for Sustainable Land Management for Food Security in Africa*. 25th Regional Conference for Africa, Nairobi, Kenya. Food and Agriculture Organization, Rome, Italy.

FAO, 2008c: *Climate Change for Fisheries and Aquaculture*. (Technical background document from the expert consultation). Food and Agriculture Organization, Rome, Italy.

FAO, 2009: *Seed Security for Food Security in the Light of Climate Change and Soaring Food Prices: Challenges and Opportunities* (Report of the twenty first session of the Committee on Agriculture). Food and Agriculture Organization, Rome, Italy.

FAO, 2010: *"Climate-Smart" Agriculture: Policies, Practices and Financing for Food Security, Adaptation and Mitigation*. Food and Agriculture Organization, Rome, Italy.

Feldman, D.L. and H.M. Ingram, 2009: Making science useful to decision makers: Climate forecasts, water management, and knowledge networks. *Weather, Climate and Society*, **1**, 9-20.

Few, R., H. Osbahr, L.M. Bouwer, D. Viner, and F. Sperling, 2006: *Linking Climate Change Adaptation and Disaster Risk Management for Sustainable Poverty Reduction* (Synthesis report). Vulnerability and Adaptation Resource Group, Washington, DC.

Folke, C., 2006: Resilience: The emergence of a perspective for social-ecological systems analyses. *Global Environmental Change*, **16(3)**, 253-267.

Folke, C., T. Hahn, P. Olsson, and J. Norberg, 2005: Adaptive governance of social-ecological systems. *Annual Review of Environment and Resources*, **30**, 441-473.

Freeman, P.K. and Warner, K., 2001: *Vulnerability of Infrastructure to Climate Variability: How Does this Affect Infrastructure Lending Policies?* World Bank and ProVention Consortium, Washington, DC.

Freeman, P.K, L.A. Martin, J. Linnerooth-Bayer, R. Mechler, G. Pflug, and K. Warner, 2003: *Disaster Risk Management: National Systems for the Comprehensive Management of Disaster Risk and Financial Strategies for Natural Disaster Reconstruction*. Sustainable Development Department, Inter-American Development Bank, Washington, DC.

Froot, K.A., 2001: The market for catastrophe risk: A clinical examination. *Journal of Financial Economics*, **60**, 529-571.

Füssel, H., 2007: Adaptation planning for climate change: Concepts, assessment approaches, and key lessons. *Sustainability Science*, **2(2)**, 265-275.

Gero, A., K. Méheux, and D. Dominey-Howes, 2011: Integrating community based disaster risk reduction and climate change adaptation: examples from the Pacific. *Natural Hazards and Earth System Sciences*, **11**, 101-113.

GFDRR, 2010: *Partnership Charter*. Global Facility for Disaster Risk Reduction, Washington, DC, gfdrr.org/gfdrr/sites/gfdrr.org/files/publication/GFDRR_Partnership_Charter_2010.pdf.

Ghesquiere, F., O. Mahul, M. Forni, and R. Gartley, 2006: *Caribbean Catastrophe Risk Insurance Facility: A solution to the short-term liquidity needs of small island states in the aftermath of natural disasters*. World Bank, Washington, DC,siteresources.worldbank.org/PROJECTS/Resources/Catastrophicriskinsurane facility.pdf.

Glantz, M.H., 2005: Hurricane Katrina rekindles thoughts about fallacies of a so-called "natural" disaster. *Sustainability: Science, Practice and Policy*, **1(2)**, 1-4.

GNDR, 2009: *Clouds but Little Rain. Views from the Frontline: A Local Perspective of Progress towards Implementation of the Hyogo Framework for Action*. Global Network of Civil Society Organisations for Disaster Reduction, Teddington, UK.

Grosse, G., V. Romanovsky, T. Jorgenson, K.W. Anthony, J. Brown, and P.P. Overduin, 2011: Vulnerability and feedbacks of permafrost to climate change. *EOS, Transactions AGU*, **92(9)**, 73-80.

Grossi, P., and H. Kunreuther (eds.), 2005: *Catastrophe Modeling: A New Approach to Managing Risk*. Springer, New York, NY.

GTZ/DKKV, 2007: *National Platforms for Disaster Reduction: Study on current status of disaster reduction, institutional arrangements and potential for national platforms for disaster reduction 3 South and South East Asian Countries*. Gesellschaft für Technische Zusammenarbeit / German Committee for Disaster Risk Reduction, Bonn, Germany.

Guha-Sapir, D. and R. Below, 2002: *Quality and Accuracy of Disaster Data: A Comparative Analysis of Three Global Datasets*. Working document prepared for Disaster Management Facility, World Bank, Washington DC.

Gunderson, L. and C.S. Holling, 2002: *Panarchy Synopsis: Understanding Transformations in Human and Natural Systems*. Island Press, Washington, DC.

Gwimbi, P., 2007: The effectiveness of early warning systems for the reduction of flood disasters: Some experiences from cyclone induced floods in Zimbabwe. *Journal of Sustainable Development in Africa*, **9(4)**, 152-169.

Haasnoot, M., H. Middelkoop, E. van Beek, and W.P.A. van Deursen, 2009: A method to develop sustainable water management strategies for an uncertain future. *Sustainable Development*, **19**, doi:10.1002/sd.438.

Haines, A., R.S. Kovats, D. Campbell-Lendrum, and C. Corvalan, 2006: Climate change and human health: impacts, vulnerability and public health. *Public Health*, **120(7)**, 585-596.

Hallegatte, S., 2009: Strategies to adapt to an uncertain climate change. *Global Environmental Change*, **19(2)**, 240-247.

Hallegatte, S. and M. Ghil, 2007: *Endogenous Business Cycles and the Economic Response to Exogenous Shocks*. Fondazione Eni Enrico Mattei, Milan, Italy.

Hallegatte, S., J.-C. Hourcade, and P. Dumas, 2007: Why economic dynamics matter in assessing climate change damages: Illustration on extreme events. *Ecological Economics*, **62**, 330-340.

Halsnaes, K., and S. Traerup, 2009: Development and climate change: A mainstreaming approach for assessing economic, social, and environmental impacts of adaptation measures. *Environmental Management*, **43**, 765-778.

Hammer, K., N. Arrowsmith, and T. Gladis, 2003: Agrobiodiversity with emphasis on plant genetic resources. *Naturwissenschaften 2003*, **90(6)**, 241-250.

Hammill, A. and T. Tanner, 2010: Climate risk screening and assessment tools: making sense of a crowded field. In: *Meeting of the OECD DAC-EPOC Joint Task Team on Climate Change and Development Co-operation*, Amsterdam, The Netherlands, 12-13 October 2010.

Handmer, J. and S. Dovers, 2007: *Handbook of Disaster and Emergency Policies and Institutions*. Earth Scan, London, UK.

Harris, P. (ed.), 2009: *Climate Change and Foreign Policy: Case Studies from East to West*. Routledge, London, UK.

Harvey, P., 2009: *Towards Good Humanitarian Government: The Role of the Affected State in Disaster Response.* Humanitarian Policy Group, Overseas Development Institute, London, UK.

Hay, J., 2009: *Institutional and Policy Analysis of Disaster Risk Reduction and Climate Change Adaptation in Pacific Island Countries*. A report prepared for the United Nations International Strategy for Disaster Reduction (UNISDR) and the United Nations Development Programme (UNDP). UNISDR and UNDP, Suva, Fiji.

Hayley, D.W. and B. Horne, 2008: Rationalizing climate change for design of structures on permafrost: a Canadian perspective. In: *Proceedings of Ninth International Conference on Permafrost* [Kane, D.L. and K.M. Hinkle (eds.)]. University of Alaska Fairbanks, Fairbanks, AK, **1**, 681-686.

Hazell, P. and U. Hess, 2010: Drought insurance for agricultural development and food security in dryland areas. *Food Security*, **2**, 395-405.

Hedger, M. and J. Cacouris, 2008: *Separate Streams? Adapting Water Resources Management to Climate Change*. Tearfund, Teddington, UK.

Hedger, M., A. Singha, and M. Reddy, 2010: *Building Climate Resilience at State Level: Disaster Risk Management and Rural Livelihoods in Orissa*. Strengthening Climate Resilience Discussion Paper 5, Institute of Development Studies, Brighton, UK.

Heltberg, R., S.L. Jorgensen, and P.B. Siegel, 2008: *Climate Change, Human Vulnerability, and Social Risk Management.* World Bank, Washington, DC.

Heltberg, R., P.B. Siegel, and S.L. Jorgensen, 2009: Addressing human vulnerability to climate change: toward a 'no-regrets' approach. *Global Environmental Change*, **19(1)**, 89-99.

Hess, U. and H. Syroka, 2005: *Weather-based Insurance in Southern Africa. The Case of Malawi*. World Bank, Washington, DC.

HFP, 2007: *Dimensions of Crisis Impacts: Humanitarian Needs by 2015*. Working Paper, Humanitarian Futures Programmes, Kings College London, London, UK.

Hilhorst, D., 2003: Unlocking disaster paradigms: An actor-oriented focus on disaster response. Abstract submitted for Session 3 of the Disaster Research and Social Crisis Network Panels of the *6th European Sociological Conference*, Murcia, Spain 23-26 September 2003.

Hochrainer, S., 2006: *Macroeconomic Risk Management against Natural Disasters*. Deutscher Universita Ts-Verlag, Wiesbaden, Germany.

Hodgson, R. and M. Carter, 1999: Some lessons for a national approach to building for safety in Bangladesh. In: *Natural Disaster Management* [Ingleton, J. (ed.)]. Tudor Rose, Leicester, UK, pp. 160-162.

Hoegh-Guldberg, O., P.J. Mumby, A.J. Hooten, R.S. Steneck, P. Greenfield, E. Gomez, D.R. Harvell, P.F. Sale, A.J. Edwards, K. Caldeira, N. Knowlton, C.M. Eakin, R. Iglesias-Prieto, N. Muthiga, R.H. Bradbury, H. Dubi, and M.E. Hatziolos, 2008: Coral adaptation in the face of climate change. *Science*, **320(5874)**, 315.

Hoeppe, P. and E.N. Gurenko, 2006: Scientific and economic rationales for innovative climate insurance solutions. *Climate Policy*, **6(6)**, 607-620.

Holling, C.S., 1973: Resilience and stability of ecological systems. *Annual Review of Ecology and Systematics*, **4(1)**, 1-23.

Howden, S.M., J.F. Soussana, F.N. Tubiello, N. Chhetri, M. Dunlop, and H. Meinke, 2007: Adapting agriculture to climate change. *Proceedings of the National Academy of Sciences*, **104(50)**, 19691-19696.

Hughes, T.P., A.H. Baird, D.R. Bellwood, M. Card, S.R. Connolly, C. Folke, R. Grosberg, O. Hoegh-Guldberg, J.B.C. Jackson, and J. Kleypas, 2003: Climate change, human impacts, and the resilience of coral reefs. *Science*, **301(5635)**, 929.

IASC, 2009a: *Addressing the Humanitarian Challenges of Climate Change Regional and National Perspectives*. Preliminary Findings from the IASC Regional and National Level Consultations, Inter-Agency Standing Committee, Geneva, Switzerland, www.humanitarianinfo.org/iasc/pageloader.aspx?page=content-news-newsdetails&newsid=134.

IASC, 2009b: *Letter from the IASC Principals to Yvo de Boer, Executive Secretary of the United Nations Framework Convention on Climate Change*. Inter-Agency Standing Committee, Geneva, Switzerland, www.humanitarianinfo.org/iasc/pageloader.aspx?page=content-news-newsdetails&newsid=134.

Ibarraran, M.E., M. Ruth, S. Ahmad, and M. London, 2009: Climate change and natural disasters: Macroeconomic performance and distributional impacts. *Environment, Development and Sustainability*, **11**, 549-569.

ICHARM, 2009: *Global Trends in Water-Related Disasters: An Insight for Policymakers*. International Centre for Water Hazard and Risk Management, World Water Assessment Programme, UNESCO, Paris, France.

IFRC, 2004: *World Disasters Report 2004: Focus on Community Resilience*. Int'l Federation of the Red Cross and Red Crescent Societies, Geneva, Switzerland.

IFRC, 2009: *World Disasters Report: Focus on Early Warning, Early Action*. International Federation of the Red Cross and Red Crescent Societies, Geneva, Switzerland.

IFRC, 2010: *World Disasters Report 2010: Focus on Urban Risk.* International Federation of the Red Cross and Red Crescent Societies, Geneva, Switzerland.

IIED, 2009: *Climate Change and the Urban Poor: Risk and Resilience in 15 of the World's Most Vulnerable Cities.* International Institute for Environment and Development, London, UK.

Ikeme, J., 2003: Climate change adaptational deficiencies in developing countries: The case of sub-Saharan Africa. *Mitigation and Adaptation Strategies for Global Change*, **8**, 29-52.

Imperial, M.T., 1999: Institutional analysis and ecosystem-based management: The institutional analysis and development framework. *Environmental Management*, **24(4)**, 449-465.

Interworks, 1998: *Model for a National Disaster Management Structure, Preparedness Plan, and Supporting Legislation, UNOCHA Disaster Management Training Programme.* UN Office for the Coordination of Humanitarian Affairs, ocha.unog.ch/drptoolkit/PreparednessTools/IL%20Frameworks/Models%20for%20Institutional%20structures%20and%20legislation,%20DMTP,%20InterWorks,1998.pdf.

IRI, 2006: *A Gap Analysis for the Implementation of the Global Climatic Observing System Programme in Africa*. IRI Technical Report No. IRI-TR/06/01, International Research Institute for Climate and Society, Columbia University, New York, NY.

Islam, T., and S. Ferdousi, 2007: Renewable energy development – challenges for Bangladesh. *Energy and Environment*, **18**, 421-430.

Jaspars, S. and D. Maxwell, 2009: *Food Security and Livelihoods Programming in Conflict: A Review*. HPN Network Paper 65. Humanitarian Practice Network, Overseas Development Institute, London, UK.

Johnson, B.L., 1999: Introduction to the special feature: Adaptive management – scientifically sound, socially challenged. *Conservation Ecology*, **3(1)**, 10.

Juhola, S., and L. Westerhoff, 2011: Challenges of adaptation to climate change across multiple scales: a case study of network governance in two European countries. *Environmental Science and Policy*, **14**, 239-247.

Kagiannas, A., D. Askounis, K. Anagnostopoulos, and J. Psarras, 2003: Energy policy assessment of the Euro-Mediterranean Cooperation. *Energy Conversion and Management*, **44(16)**, 2665-2686.

Katz, R. and G.B. Brown, 1992: Extreme events in a changing climate: Variability is more important than averages. *Climatic Change*, **21**, 289-302.

Kay, R. and J. Adler, 2005: *Coastal planning and management.* Third Edition. Routledge, New York, NY.

Kellenberg, D., K. Mobarak, and A. Mushfiq, 2008: Does rising income increase or decrease damage risk from natural disasters? *Journal of Urban Economics*, **63(3)**, 788-802.

Kelman, I., 2008: Relocalising disaster risk reduction for urban resilience. *Urban Design and Planning*, **161(4)**, 197-204.

Kelman, I., J. Lewis, J.C. Gaillard, and J. Mercer, 2011: Participatory action research for dealing with disasters on islands. *Island Studies Journal*, **6**, 59-86.

Kesavan, P.C. and Swaminathan, M.S., 2006: Managing extreme natural disasters in coastal areas. *Philosophical Transactions of the Royal Society A*, **364(1845)**, 2191-2216.

Kiker, C.F., J.W. Milon, and A.W. Hodges, 2001: Adaptive learning for science-based policy: The everglades restoration. *Ecological Economics*, **37(3)**, 403-416.

Kirshen, P., 2007: *Adaptation Options and Costs in Water Supply*. Report to the UNFCCC Secretariat, Financial and Technical Support Division, United Nations Framework Convention on Climate Change, Bonn, Germany.

Klein, R.J.T., R.J. Nicholls, S. Ragoonaden, M. Capobianco, J. Aston, and E.N. Buckley, 2001: Technological options for adaptation to climate change in coastal zones. *Journal of Coastal Research*, **17(3)**, 531-543.

Klein, R.J.T., S. Eriksen, L.O. Næss, A. Hammill, C. Robledo, and K. O'Brien, 2007: Portfolio screening to support the mainstreaming of adaptation to climate change into development. *Climatic Change*, **84**, 23-44.

Klijn, F., M. van Buuren, and S.A.M. van Rooij, 2004: Flood-risk management strategies for an uncertain future: living with Rhine river floods in the Netherlands? *Ambio*, **33**, 141-147.

Krysanova, V., C. Dickens, J. Timmerman, C. Varela-Ortega, M. Schlüter, K. Roest, P. Huntjens, F. Jaspers, H. Buiteveld, E. Moreno, J. de Pedraza Carrera, R. Slámová, M. Martínková, I. Blanco, P. Esteve, K. Pringle, C. Pahl-Wostl, and P. Kabat, 2010: Cross-comparison of climate change adaptation strategies across large river basins in Europe, Africa and Asia. *Water Resources Management*, **24**, 4121-4160.

Kuzak, D., 2004: The application of probabilistic earthquake risk models in managing earthquake insurance risks in Turkey. In: *Catastrophe Risk and Reinsurance: A Country Risk Management Perspective* [Gurenko, E. (ed.)]. Risk Books, London, UK.

Kwadijk, C.J., M. Haasnoot, J.P.M. Hoogvliet, M.C. Marco, A.B.M. Jeuken, N.G.C. van der Oostrom, H.A. Schelfhout, E.H. van Velzen, H. van Waveren, and M.J.M. de Wit, 2010: Using adaptation tipping points to prepare for climate change and sea level rise: a case study in the Netherlands. *Wiley Interdisciplinary Reviews: Climate Change*, **1**, 729-740.

Ladson, A.R. and R.M. Argent, 2000: Adaptive management of environmental flows: Lessons for the Murray-Darling Basin from three large North American rivers. *Australian Journal of Water Resources*, **5(51)**, 89-101.

Lal, P.N., 2010: Vulnerability to natural disasters: an economic analysis of the impact of the 2009 floods on the Fijian sugar belt. *Pacific Economic Bulletin*, **25(2)**, 62-77.

Larsen, P.H., S. Goldsmith, O. Smith, M.L. Wilson, K. Strzepek, P. Chinowsky, and B. Saylor, 2008: Estimating future costs for Alaska public infrastructure at risk from climate change. *Global Environmental Change*, **18(3)**, 442-457.

Laska, S., 2004: What if hurricane Ivan had not missed New Orleans? *Natural Hazards Observer*, **29(2)**, 5-6.

Lavell, A., 1996. Environmental degradation, risks and urban disasters. In: *Issues and Concepts: Towards the Definition of a Research Agenda*. La Red, Lima, Peru, pp. 19-58.

Lavell, A., 1998. Decision making and risk management. In: *Furthering Cooperation in Science and Technology for Caribbean Development*. Port of Spain, Trinidad, 23-25 September 1998.

Lavell, A., 2009: *Unpacking Climate Change Adaptation and Disaster Management: Searching for the Links and Differences: A Conceptual and Epistemological Critique and Proposal*. FLACSO, Bogota, Colombia.

Lavell, A. and E. Franco, 1996: *Estado, sociedad y gestián de loes desastres en america latina: en busqueda del paradigma perdido*. La Red, Bogota, Colombia.

Lee, K., 1993: *Compass and Gyroscope*. Island Press, California.

Lemmen, D., F. Warren, J. Lacroix, and E. Bush, 2008: *From Impacts to Adaptation: Canada in a Changing Climate 2007*. Natural Resources Canada, Ottawa, Canada.

Leslie, H.M. and K.L. Mcleod, 2007: Confronting the challenges of implementing marine ecosystem-based management. *Frontiers in Ecology and the Environment*, **5(10)**, 540-548.

Lewis, J. and M.P. Chisholm, 1996: Cyclone-resistant domestic construction in Bangladesh. In: *Implementing Proceedings of Hazard-Resistant Housing: Proceedings of the First International Housing and Hazards Workshop to Explore Practical Building for Safety Solutions* [Hodgson, R.L.P., S.M. Seraj, and J.R. Choudhury (eds.)]. Dhaka, Bangladesh, 3-5 December 1996.

Linnerooth-Bayer, J. and A. Amendola, 2000: Global change, catastrophic risk and loss spreading. *The Geneva Papers on Risk and Insurance*, **25(2)**, 203-219.

Linnerooth-Bayer, J. and R. Mechler, 2006: Insurance for assisting adaptation to climate change in developing countries: A proposed strategy. *Climate Policy*, **6(6)**, 621-636.

Linnerooth-Bayer, J. and R. Mechler, 2007: Disaster safety nets for developing countries: Extending public-private partnerships. *Environmental Hazards*, **7**, 54-61.

Linnerooth-Bayer, J., R. Mechler, and G. Pflug, 2005: Refocusing disaster aid. *Science*, **309(5737)**, 1044-1046.

Llosa, S. and I. Zodrow, 2011: *Disaster Risk Reduction Legislation as a Basis for Effective Adaptation*. Background Paper prepared for the 2011 Global Assessment Report on Disaster Risk Reduction. United Nations International Strategy for Disaster Reduction, Geneva, Switzerland.

Longley, C. and D. Maxwell, 2003: Livelihoods, chronic conflict and humanitarian response: A review of current approaches. *Natural Resource Perspectives*, **89**, 1-6.

Low, B., E. Ostrom, C. Simon, and J. Wilson, 2003: Navigating social-ecological systems: Building resilience for complexity and change. In: *Redundancy and diversity: Do they influence optimal management?* [Berkes, F., J. Colding, and C. Folke (eds.)]. Cambridge University Press, Cambridge, UK, pp. 83-114.

Lowe, R., 2003: Preparing the built environment for climate change. *Building Research and Information*, **31(3-4)**, 195-199.

Lu, X., 2011: Provision of climate information for adaptation to climate change. *Climate Research*, **47**, 83-94.

Mahul, O. and F. Ghesquiere, 2007: *Sovereign Natural Disaster Insurance for Developing Countries: A Paradigm Shift in Catastrophe Risk Financing*. Policy Research Working Paper 4345, World Bank, Washington, DC.

Malhi, Y., J.T. Roberts, R.A. Betts, T.J. Killeen, W. Li, and C.A. Nobre, 2008: Climate change, deforestation and the fate of the Amazon. *Science*, **319(5860)**, 169-172.

Mansourian, S., A. Belokurov, and P. Stephenson, 2009: The role of forest protected areas in adaptation to climate change. *Unasylva*, **60**, 63-69.

Manuel-Navarrete, D., J. Gomez, and G. Gallopin, 2007: Syndromes of sustainability of development for assessing the vulnerability of coupled human–environmental systems: The case of hydrometeorological disasters in Central America and the Caribbean. *Global Environmental Change*, **17**, 207-217.

Maréchal, K., 2007: The economics of climate change and the change of climate in economics. *Energy Policy*, **35(10)**, 5181-5194.

Mastrandrea, M.D., N.E. Heller, T.L. Root, and S.H. Schneider, 2010: Bridging the gap: linking climate-impacts research with adaptation planning and management. *Climatic Change*, **100(1)**, 87-101.

Masumoto, T., T. Yoshida, and T. Kubota, 2006: An index for evaluating flood-prevention function of paddies. *Paddy and Water Environment*, **4(4)**, 205-210.

Masumoto, T., H. Pham Thanh, and K. Shimizu, 2008: Impact of paddy irrigation levels on floods and water use in the Mekong River Basin. *Hydrological Processes*, **22(9)**, 1321-1328.

McBean, G.A., 2008: Role of prediction in sustainable development and disaster management. *Springer- Hexagon Series on Human and Environmental Security and Peace*, **3**, 929-938.

McCarl, B.A., 2007: *Adaptation Options for Agriculture, Forestry and Fisheries*. A Report to the UNFCCC Secretariat Financial and Technical Support Division, United Nations Framework Convention on Climate Change, Bonn, Germany.

McDaniels, T., S. Chang, D. Cole, J. Mikawoz, and H. Longstaff, 2008: Fostering resilience to extreme events within infrastructure systems: characterizing decision contexts for mitigation and adaptation. *Global Environmental Change: Human and Policy Dimensions*, **18**, 310-318.

McEntire, D.A. and V. Myers, 2004: Preparing communities for disasters: Issues and processes for government readiness. *Disaster Prevention and Management*, **13(2)**, 140-152.

McEvoy, D., P. Matczak, I. Banaszak, and A. Chorynski, 2010: Framing adaptation to climate-related extreme events. *Mitigation and Adaptation Strategies to Global Change*, **15**, 779-795.

McGranahan, G., D. Balk, and B. Anderson, 2007: The rising tide: assessing the risks of climate change and human settlements in low elevation coastal zones. *Environment and Urbanization*, **19**, 17-37.

McGray, H., A. Hammill, R. Bradley, E.L. Schipper, and J.E. Parry, 2007: *Weathering the Storm: Options for Framing Adaptation and Development*. World Resources Institute, Washington, DC.

McKenzie, E., B. Prasad, and A. Kaloumaira, 2005: *Economic Impact of Natural Disasters on Development in the Pacific.* Volume I. SOPAC and USP, Suva, Fiji.

MEA, 2005: *Ecosystems and Human Well-Being: Synthesis*. Millennium Ecosystem Assessment, Island Press, Washington, DC.

Mechler, R., 2004: *Natural Disaster Risk Management and Financing Disaster Losses in Developing Countries.* Verlag für Versicherungswirtschaft, Germany.

Mechler, R., J. Linnerooth-Bayer, and D. Peppiatt, 2006a: *Microinsurance for Natural Disasters in Developing Countries: Benefits, Limitations and Viability*. ProVention Consortium, Geneva, Switzerland.

Mechler, R., J. Linnerooth-Bayer, S. Hochrainer, and G. Pflug, 2006b: Assessing financial vulnerability and coping capacity: The IIASA CATSIM model. In: *Measuring Vulnerability and Coping Capacity to Hazards of Natural Origin. Concepts and Methods* [Birkmann, J. (ed.)]. United Nations University Press, Tokyo, pp. 380-398.

Mechler, R., S. Hochrainer, A. Aaheim, Z. Kundzewicz, N. Lugeri, M. Moriondo, H. Salen, M. Bindi, I. Banaszak, A. Chorynski, E. Genovese, H. Kalirai, J. Linnerooth-Bayer, C. Lavalle, D. Mcevoy, P. Matczak, M. Radziejewski, D. Rübbelke, M.J. Schelhaas, M. Szwed, and A. Wreford, 2010: Making climate change work for us: European perspectives on adaptation and mitigation strategies. In: *A Risk Management Approach for Assessing Adaptation to Changing Flood and Drought Risks in Europe* [Hulme, H.N. (ed.)]. Cambridge University Press, Cambridge, UK, pp. 200-229.

Meehl, G.A., F. Zwiers, J. Evans, T. Knutson, L. Mearns, and P. Whetton, 2000: Trends in extreme weather and climate events: Issues related to modeling extremes in projections of future climate change. *Bulletin of the American Meteorological Society*, **81(3)**, 427-436.

Meinke, H., R. Nelson, P. Kokic, R. Stone, R. Selvaraju, and W. Baethgen, 2006: Actionable climate knowledge: From analysis to synthesis. *Climate Research*, **33(1)**, 101.

Menkhaus, K., 2007: Governance without government in Somalia: Spoilers, state building and the politics of coping. *International Security*, **31(3)**, 74-106.

Mercer, J., D. Dominey-Howes, I. Kelman, and K. Lloyd, 2007: The potential for combining indigenous and western knowledge in reducing vulnerability to environmental hazards in small island developing states. *Environmental Hazards*, **7**, 245-256.

Mertz, O., K. Halsnaes, J. Olesen, and R. Rasmussen, 2009: Adaptation to climate change in developing countries: Special feature. *Environmental Management*, **43**, 743-752.

Miller, F., H. Osbahr, E. Boyd, F. Thomalla, S. Bharwani, G. Ziervogel, B. Walker, J. Birkmann, S. van der Leeuw, J. Rockström, J. Hinkel, T. Downing, C. Folke, and D. Nelson, 2010: Resilience and vulnerability: Complementary or conflicting concepts? *Ecology and Society*, **15(3)**, 11.

Mills, E., 2005: Insurance in a climate of change. *Science*, **309(5737)**, 1040-1044.

Mills, E., 2006: Testimony to the National Association of Insurance Commissioners. The Role of NAIC in responding to climate change. *UCLA Journal of Environmental Law and Policy*, **26(1)**, 129-168.

Mills, E., 2007: Synergisms between climate change mitigation and adaptation: An insurance perspective. *Mitigation and Adaptation Strategies for Global Change*, **12(15)**, 809-842.

Mitchell, T. and S. Maxwell, 2010: *Defining Climate Compatible Development*. Policy Brief, Climate and Development Knowledge Network, London, UK.

Mitchell, T. and M. van Aalst, 2008: *Convergence of Disaster Risk Reduction and Climate Change Adaptation: A Review for DFID*. Department for International Development (DFID), London, UK.

Mitchell, T., M. Ibrahim, K. Harris, M. Hedger, E. Polack, A. Ahmed, N. Hall, K. Hawrylyshyn, K. Nightingale, M. Onyango, M. Adow, and M. Sajjad, 2010a: *Climate Smart Disaster Risk Management. Strengthening Climate Resilience*. Institute of Development Studies, Brighton, UK.

Mitchell, T., M. van Aalst, and P. Villaneuva, 2010b: *Assessing Progress on Integrating Disaster Risk Reduction and Climate Change Adaptation in Development Processes*. Strengthening Climate Resilience Programme, Institute of Development Studies, Brighton, UK.

MMM, 2005: *Finland's National Strategy for Adaptation to Climate Change*. Ministry of Agriculture and Forestry, Helsinki, Finland, www.mmm.fi/en/index/frontpage/ymparisto/ilmastopolitiikka/ilmastomuutos.html.

Moench, M., R. Mechler, and S. Stapleton, 2007: Guidance note on the costs and benefits of disaster risk reduction. In: *UNISDR High Level Platform on Disaster Risk Reduction*, Geneva, Switzerland, 4-7 June 2007. United Nations International Strategy for Disaster Reduction (UNISDR), Geneva, Switzerland.

Moser, C. and D. Satterthwaite, 2008: *Towards Pro-Poor Adaptation to Climate Change in the Urban Centers of Low- and Middle-Income Countries*. International Institute for Environment and Development, London, UK.

MOSSP, 2010: Draft National Policy for Disaster Management in Kenya. Ministry of State of Special Programs, Kenya, www.kecosce.org/downloads/DRAFT_DISASTER_MANAGEMENT_POLICY.pdf.

Muller, M., 2007: Adapting to climate change: Water management for urban resilience. *Environment and Urbanization*, **19(1)**, 99-113.

Munich Re, 2005: *Short Annual Report 2005: Paving the way for opportunities*. Munich Re Group, Munich, Germany.

Munich Re, 2011: *Topics Geo Natural Catastrophes 2010: Analyses, Assessments, Positions*. Munich Re, Munich, Germany.

Murlidharan, T.L. and H.C. Shah, 2001: Catastrophes and macro-economic risk factors: An empirical study. In: *Conference on Integrated Disaster Risk Management: Reducing Socio-Economic Vulnerability*. International Institute for Applied Systems Analysis (IIASA), Laxenburg, Austria.

Murray, B.C., W.A. Jenkins, S. Sifleet, L. Pendleton, and A. Baldera, 2010: *Payment for Blue Carbon: Potential for Protecting Threatened Coastal Habitats*. Nicholas Institute for Environmental Policy Solutions, Duke University, Durham, NC.

Musgrave, R.A., 1959: *The Theory of Public Finance.* McGraw Hill, New York, NY.

National Platform for Kenya, 2009: *Interim National Progress Report on Hyogo Framework for Action*. Prepared for the UN Global Assessment Report 2009, United Nations International Strategy for Disaster Reduction, Geneva, Switzerland, preventionweb.net/files/7432_Kenya.pdf.

Neufeldt, H., A. Wilkes, R.J. Zomer, J. Xu, E. Nang'ole, C. Munster, and F. Place, 2009: *Trees on Farms: Tackling the Triple Challenges of Mitigation, Adaptation and Food Security*. Policy Brief 07, World Agroforestry Centre, Kenya.

Neumann, J., 2009: *Adaptation to Climate Change: Revisiting Infrastructure Norms*. Resources for the Future, Washington, DC.

Nicholls, R.J., 2007: *Adaptation Options for Coastal Areas and Infrastructure: An Analysis for 2030*. Report to the United Nations Framework Convention on Climate Change (UNFCCC), UNFCCC, Bonn, Germany.

Nicholls, R.J., P.P. Wong, V. Burkett, C.D. Woodroffe, and J. Hay, 2008: Climate change and coastal vulnerability assessment: scenarios for integrated assessment. *Sustainability Science*, **3**, 89-102.

Nordhaus, W., 2006: *The Economics of Hurricanes in the United States*. National Bureau of Economic Research, Cambridge, MA.

Noy, I., 2009: The macroeconomic consequences of disasters. *Journal of Development Economics*, **88(2)**, 221-231.

NRC (National Research Council), 2009: *Informing Decisions in a Changing Climate*. Panel on Strategies and Methods for Climate-Related Decision Support, NRC, National Academies Press, Washington, DC.

NRTEE, 2009: *True North: Adapting Infrastructure to Climate Change in Northern Canada*. National Round Table on Environment and Economy, Ottawa, Canada.

O'Brien, G., 2006: UK emergency preparedness – A step in the right direction? *Journal of International Affairs*, **59(2)**, 63-85.

O'Brien, G., 2008: UK emergency preparedness – A holistic local response? *Disaster Prevention and Management*, **17(2)**, 232-243.

O'Brien, G. and P. O'Keefe, 2010: Resilient responses to climate change and variability: A challenge for public policy. *International Journal of Public Policy*, **6(3-4)**, 369-385.

O'Brien, G. and P. Read, 2005: Future UK emergency management: New wine, old skin? *Disaster Prevention and Management*, **14(3)**, 353-361.

O'Brien, G., P. O'Keefe, J. Rose, and B. Wisner, 2006: Climate change and disaster management. *Disasters*, **30(1)**, 64-80.

O'Brien, K., L. Sygna,, R. Leichenko, W.N. Adger, J. Barnett, T. Mitchell, L. Schipper, T. Tanner, C. Vogel, and C. Mortreux, 2008: *Disaster Risk Reduction, Climate Change Adaptation and Human Security*. Report prepared for the Royal Norwegian Ministry of Foreign Affairs by the Global Environmental Change and Human Security (GECHS) Project, GECHS Report 2008:3, Oslo, Norway.

OAS, 1991: *Primer on Natural Hazard Management in Integrated Regional Development Planning*. Organization of American States, Washington, DC.

OECD, 2006: *Applying Strategic Environmental Assessment: Good Practice Guidance For Development Co-Operation*. Organisation for Economic Cooperation and Development, Paris, France.

OECD, 2009: *Policy Guidance on Integrating Climate Change Adaptation into Development Co-operation*. Organisation for Economic Cooperation and Development, Paris, France.

Olhoff, A. and C. Schaer, 2010: *Screening Tools and Guidelines to Support the Mainstreaming of Climate Change Adaptation into Development Assistance – A Stocktaking Report*. United Nations Development Programme, New York, NY.

Olsen, J.R., 2006: Climate change and floodplain management in the United States. *Climatic Change*, **76(3-4)**, 407- 426.

ONERC, 2009: *Climate Change Costs of Impacts and Lines of Adaptation*. A report to the Prime Minister and Parliament, Observatoire National sur les Effets du Réchauffement Climatique, Paris, France.

Ostrom, E., 2005: *Understanding Institutional Diversity*. Princeton University Press, Princeton, NJ.

Otero, R.C. and R.Z. Marti, 1995: The impacts of natural disasters on developing economies: Implications for the international development and disaster community. In: *Disaster Prevention for Sustainable Development: Economic and Policy Issues* [Munasinghe, M. and C. Clarke (eds.)]. World Bank, Washington, DC, pp. 11-40.

Owens, T., T. Hoddinott, and B. Kinsey, 2003: Ex-ante actions and ex-post public responses to drought shocks: Evidence and simulations from Zimbabwe. *World Development*, **31(7)**, 1239-1255.

Oxfam, 2007: *Climate Alarm: Disasters Increase as Climate Change Bites*. Oxfam Briefing Paper 108, Oxfam, Oxford, UK, www.oxfam.org.uk/resources/policy/climate_change/downloads/bp108_weather_alert.pdf.

Oxfam America, 2008: National network for DRR helps curb poverty. In: *Linking Disaster Risk Reduction and poverty: Good practices and Lessons Learnt* [United Nations International Strategy for Disaster Reduction (UNISDR) (ed.)]. UNISDR, Geneva, Switzerland, pp. 11-14.

Pahl-Wostl, C., 2009: A conceptual framework for analysing adaptive capacity and multi-level learning processes in resource governance regimes. *Global Environmental Change*, **19**, 354-365.

Pahl-Wostl, C., J. Sendzimer, and P. Jeffrey, 2009: Resources management in transition. *Ecology and Society*, **14(1)**, 46, www.ecologyandsociety.org/vol14/iss1/art46/.

Parkes, M.W and P. Horwitz, 2009: Water, ecology and health: Ecosystems as settings for promoting health and sustainability. *Health Promotion International*, **24(1)**, 94.

Parry, M., N. Arnell, P. Berry, D. Dodman, S. Fankhauser, C. Hope, S. Kovats, R. NicholLas, D. Satterthwaite, R. Tiffin, and R. Wheeler, 2009: *Assessing the Costs of Adaptation to Climate Change: A review of the UNFCCC and other recent estimates*. International Institute for Environment and Development and Grantham Institute for Climate Change, London, UK.

Patt, A., R.J.T. Klein, and A. de la Vega-Leinert, 2005: Taking the uncertainty in climate-change vulnerability assessment seriously. *Comptes Rendus Geoscience*, **337(4)**, 411-424.

Patz, J.A., M.A. Mcgeehin, S.M. Bernard, K.L. Ebi, P.R. Epstein, A. Grambsch, D.J. Gubler, P. Reither, I. Romieu, J.B. Rose, J.M. Samet, and J. Trtanj, 2000: The potential health impacts of climate variability and change for the United States: Executive Summary of the report of the health sector of the U.S. National Assessment. *Environmental Health Perspectives*, **108(4)**, 367-376.

Patz, J.A., D. Campbell-Lendrum, T. Holloway, and J.A. Foley, 2005: Impact of regional climate change on human health. *Nature*, **438(7066)**, 310-317.

Pelham, L., E. Clay, and T. Braunholz, 2011: *Natural Disasters: What is the Role for Social Safety Nets?* World Bank, Washington, DC.

Pelling, M., 2003: *The Vulnerability of Cities: Natural Disasters and Social Resilience*. Earthscan, London, UK.

Pelling, M. and A. Holloway, 2006: *Legislation for Mainstreaming Disaster Risk Reduction*. Tearfund, Teddington, UK.

Persson, Å. and R.J.T. Klein, 2009: Mainstreaming adaptation to climate change into official development assistance: challenges to foreign policy integration. In: *Climate Change and Foreign Policy: Case Studies from East to West* [P. Harris (ed.)]. Routledge, London, UK, pp. 162-177.

Phillips, C. and M. Marden, 2005: Landslide hazard and risk. In: *Reforestation Schemes to Manage Regional Landslide Risk* [Glade, T., M. Anderson, and M.J. Crozier (eds.)]. John Wiley and Sons Ltd, Sussex, UK, pp. 517-548.

Pielke, Jr., R.A., J. Rubiera, C. Landsea, M.L. Fernández, and R. Klein, 2003: Hurricane vulnerability in Latin America and the Caribbean: Normalized damage and loss potentials. *Natural Hazards Review*, **4**, 101-114.

Pierro, R. and B. Desai, 2008: Climate insurance for the poor: Challenges for targeting and participation. *Institute of Development Studies Bulletin*, **39**, 4.

Polack, E., 2010: *Integrating Climate Change into Regional Disaster Risk Management at the Mekong River Commission*. Strengthening Climate Resilience Discussion Paper 4, Institute of Development Studies, Brighton, UK.

Polackova Brixi, H. and A. Mody, 2002: Dealing with government fiscal risk: An overview. In: *Government at Risk* [Polackova Brixi, H. and A. Schick (eds.)]. World Bank, Washington, DC.

Pollner, J., 2000: *Managing Catastrophic Risks Using Alternative Risk Financing and Insurance Pooling Mechanisms*. World Bank, Washington, DC.

Prabhakar, S.V.R.K., A. Srinivasan, and R. Shaw, 2009: Climate change and local level disaster risk reduction planning: Need, opportunities and challenges. *Mitigation and Adaptation Strategies for Global Change*, **14(1)**, 7-33.

Practical Action Bangladesh, 2008: Risk reduction boosts livelihood security in disaster-prone District. In: Linking Disaster Risk Reduction and Poverty: Good Practices and Lessons Learnt [United Nations International Strategy for Disaster Reduction (UNISDR) (ed.)]. UNISDR, Geneva, Switzerland, pp. 1-5.

Preston, B.L., R.M. Westaway, and E. Yuen, 2011: Climate adaptation planning in practice: an evaluation of adaptation plans from three developed nations. *Mitigation and Adaptation Strategies for Global Change*, **16(4)**, 407-438.

Priest, G.L., 1996. The government, the market, and the problem of catastrophic loss. *Journal of Risk and Uncertainty*, **12(2)**, 219-237.

ProAct Network, 2008: *Environmental Management, Multiple Disaster Risk Reduction and Climate Change Adaptation Benefits for Vulnerable Communities*. ProAct Network, Switzerland.

ProVention, 2009: Cities and resilience. In: *Climate Policy Brief from Cities and Resilience Dialogue*. Bangkok, Thailand, 28-29 September 2009.

Pulwarty, R.S., and K. Redmond, 1997: Climate and salmon restoration in the Columbia River basin: The role and usability of seasonal forecasts. *Bulletin of the American Meteorological Society*, **78(3)**, 381-397.

Pulwarty, R.S., K. Broad, and T. Finan, 2004: El Niño events, forecasts and decision-making. In: *Mapping Vulnerability: Disasters, Development, and People* [Bankoff, G., G. Frerks, and T. Hilhorst (eds.)]. Earthscan, London, UK, pp. 83-98.

Raddatz, C., 2007: Are external shocks responsible for the instability of output in low-income countries? *Journal of Development Economics*, **84**, 155-187.

Rahaman, M. and O. Varis, 2005: Integrated water resources management: Evolution, prospects and future challenges. *Sustainability: Science, Practice and Policy*, **1(1)**, 15-21.

Rahel, F.J. and J.D. Olden, 2008: Assessing the effects of climate change on aquatic invasive species. *Conservation Biology*, **22(3)**, 521-533.

Ramalingam, B., H. Jones, T. Reba, and J. Youngs, 2008: *Exploring the Science of Complexity: Ideas and Implications for International Development and Humanitarian Efforts*. ODI Working Paper 285, Overseas Development Institute, London, UK.

Ramalingam, B., K. Scriven, and C. Foley, 2009: Innovations in International Humanitarian Action. In: *ALNAP 8th Review of Humanitarian Action*. Active Learning Network for Accountability and Performance in Humanitarian Action (ALNAP), London, UK.

Ranger, N., A. Millner, S. Dietz, S. Fankhauser, A. Lopez, and G. Ruta, 2010: *Adaptation in the UK: a Decision Making Process*. Policy Brief, Grantham Research Institute on Climate Change and the Environment and Centre for Climate change Economics and Policy, London, UK.

Ranger, N. and S.-L. Garbett-Shiels, 2011: *How can Decision-Makers in Developing Countries Incorporate Uncertainty about Future Climate Risks into Existing Planning and Policymaking Processes?* Policy Paper, Centre for Climate Change Economics and Policy and Grantham Research Institute on Climate Change and the Environment in collaboration with the World Resources Report, World Resources Institute, Washington, DC, www.worldresourcesreport.org/files/wrr/papers/wrr_ranger_uncertainty.pdf.

Rayner, S., D. Lach, and H. Ingram, 2005: Weather forecasts are for wimps: why water resource managers do not use climate forecasts. *Climatic Change*, **69**(31), 197-227.

Reid, H. and S. Huq, 2005: Tropical forests and adaptation to climate change: In search of synergies. In: *Climate Change – Biodiversity and Livelihood Impacts* [Robledo, C., M. Kanninen, and L. Pedroni, (eds.)]. Center for International Forestry Research, Indonesia.

Reid, H., D. Dodman, R. Janssen, and S. Huq, 2010: Building capacity to cope with climate change in the least developed countries. In: *Changing Climates, Earth Systems and Society* [Dodson, J. (ed.)]. International Year of Planet Earth, Springer, the Netherlands.

Ribeiro, M., C. Losenno, T. Dworak, E. Massey, R. Swart, M. Benzie, and C. Laaser, 2009: *Design of Guidelines for the Elaboration of Regional Climate Change Adaptations Strategies*. Study for the European Commission, Ecologic Institute, Vienna, Austria.

Robledo, C., M. Kanninen, and L. Pedroni (eds.), 2005: *Tropical Forests and Adaptation to Climate Change: In Search of Synergies*. Center for International Forestry Research, Indonesia.

Roeth, H., 2009: *The Development of a Public Partnership Framework and Action Plan for Disaster Risk Reduction in Asia*. United Nations International Strategy for Disaster Reduction, Bangkok, Thailand.

Rogers, D., and V. Tsirkunov, 2010: *Costs and Benefits of Early Warning Systems. Global Assessment Report on Disaster Risk Reduction*. United Nations International Strategy for Disaster Reduction and World Bank, Geneva, Switzerland and Washington, DC.

Romieu, E., T. Welle, S. Schneiderbauer, M. Pelling, and C. Vinchon, 2010: Vulnerability assessment within climate change and natural hazard contexts: revealing gaps and synergies through coastal applications. *Sustainability Science*, **5**, 159-170.

Rosenzweig, C. and F.N. Tubiello, 2007: Adaptation and mitigation strategies in agriculture: An analysis of potential synergies. *Mitigation and Adaptation Strategies for Global Change*, **12(5)**, 855-873.

Rossetto, T., 2007: *Construction Design, Building Standards and Site Selection: Tools for Mainstreaming Disaster Risk Reduction*. ProVention Consortium, Geneva, Switzerland.

Ruth, M., and D. Coelho, 2007: Understanding and managing the complexity of urban systems under climate change. *Climate Policy*, **7**, 317-336.

Sabatier, P.A., 1986: Top-down and bottom-up approaches to implementation research: A critical analysis and suggested synthesis. *Journal of Public Policy*, **6(1)**, 21-48.

Salter, J., 1998. Risk management in the emergency management context. *Australian Journal of Emergency Management*, **13**, 22-28.

Sanders, C.H., and M.C. Phillipson, 2003: UK adaptation strategy and technical measures: The impacts of climate change on buildings. *Building Research and Information*, **31(3-4)**, 210-221.

Sanderson, I., 2000: Cities, disasters and livelihoods. *Environment and Urbanization*, **12(2)**, 93-102.

Satterthwaite, D., 2007. *Adaptation Options for Infrastructure In Developing Countries*. A report to the UNFCCC Financial and Technical Support Division. United Nations Framework Convention on Climate Change, Bonn, Germany.

Satterthwaite, D., S. Huq, M. Pelling, H. Reid, and P.R. Lankao, 2007: *Adapting to Climate Change in Urban Areas*. International Institute for Environment and Development, London, UK.

SCBD, 2009: *Connecting Biodiversity and Climate Change Mitigation and Adaptation*. Report of the second ad hoc technical expert group on biodiversity and climate change, Secretariat of the Convention on Biological Diversity, Montreal, Canada.

Schipper, L., 2009: Meeting at the crossroads? Exploring the linkages between climate change adaptation and disaster risk reduction. *Climate and Development*, **1**, 16-30.

Schipper, L., and M. Pelling, 2006: Disaster risk, climate change and international development: Scope and challenges for integration. *Disasters*, **30**, 19-38.

Schwarze, R. and G. Wagner, 2004. In the aftermath of Dresden: New directions in German flood insurance. *The Geneva Papers on Risk and Insurance*, **29(2)**, 154-168.

Scott, Z. and M. Tarazona, 2011: *Decentralization and Disaster Risk Reduction*. Study on disaster risk reduction, decentralization and political economy analysis for UNDP contribution to the 2011 Global Assessment Report on Disaster Risk Reduction, UNISDR, Geneva, Switzerland.

SDC, 2008: *Natural disaster risk management in Muminabad*. SDC Project Fact Sheet CARITAS, Swiss Development Cooperation, Switzerland.

SEEDS India, 2008: Reducing risk in poor urban areas to protect shelters, hard-won assets and livelihoods. In: *Linking Disaster Risk Reduction and Poverty: Good Practices and Lessons Learnt* [United Nations International Strategy for Disaster Reduction (UNISDR) (ed.)]. UNISDR, Geneva, Switzerland, pp. 20-22.

Sharma, A., 2009: *Progress Review of National Platforms for DRR in the Asia and Pacific Region*. United Nations International Strategy for Disaster Reduction, Bangkok, Thailand.

Shaw, R., Y. Takeuchi, and B. Rouhban, 2009: Education, capacity building and public awareness for disaster reduction. In: *Landslides – Disaster Risk Reduction* [Sassa, P. (ed.)]. Springer-Verlag, Berlin, Germany, pp. 499-515.

Shepherd, G., 2004: *The Ecosystem Approach: Five Steps to Implementation*. World Conservation Union (IUCN), Gland, Switzerland.

Shepherd, G. (ed.), 2008: *The Ecosystem Approach: Learning from Experience*. World Conservation Union (IUCN), Gland, Switzerland.

Shugart, H., R. Sedjo, and B. Sohngen, 2003: *Forests and Global Climate Change: Potential Impacts on U.S. Forest Resources*. Pew Center on Global Climate Change, Arlington, VA.

Sidle, R.C., A.J. Pearce, C.L. O'Loughlin, and A.G. Union, 1985: *Hillslope Stability and Land Use*. American Geophysical Union, Washington, DC.

Skees, J., P. Varangis, D.F. Larson, and P. Siegel, 2005: Can financial markets be tapped to help poor people cope with weather risks? In: *Insurance Against Poverty* [Dercon, S. (ed.)]. Oxford University Press, Oxford, UK, pp. 422-437.

Skidmore, M. and H. Toya, 2002: Do natural disasters promote long run growth? *Economic Inquiry*, **40(4)**, 664-687.

Skoufias, E., 2003: Economic crises and natural disasters: Coping strategies and policy implications. *World Development*, **13(7)**, 1087-1102.

Slim, H., 2006: Global welfare: A realistic expectation for the international humanitarian system? In: *ALNAP Review of Humanitarian Action: Evaluation Utilisation* [Active Learning Network for Accountability and Performance in Humanitarian Action (ed.)]. Overseas Development Institute, London, UK, www.alnap.org/pool/files/ch1-f1.pdf.

Smillie, I., 2001: *Patronage or Partnership: Local capacity building in humanitarian crises*. Kumarian Press, Bloomfield, CT.

Smit, B., O. Pilifosova, I. Burton, B. Challenger, S. Huq, R. Klein, and G. Yohe, 2001: Adaptation to climate change in the context of sustainable development and equity. In: *Climate Change 2007. Impacts, Adaptation and Vulnerability. Contribution of Working Group II to the Fourth Assessment Report of the Intergovernmental Panel on Climate Change* [Parry, M.L., O.F. Canziani, J.P. Palutikof, P.J. Van Der Linde, and C.E. Hanson (eds.)]. Cambridge University Press, Cambridge, UK, pp. 879-906.

Smith, S.L., V.E. Romanovsky, A.G. Lewkowicz, C.R. Burn, M. Allard, G.D. Clow, K. Yoshikawa, and J. Throop, 2010: Thermal state of permafrost in North America: A contribution to the International Polar Year. *Permafrost and Periglacial Processes*, **21**, 117-135.

Sorensen, J., 1997: National and international efforts at Integrated Coastal Management: Definitions, achievements, and lessons. *Coastal Management*, **25**, 3-41.

Soussan, J. and I. Burton, 2002: Adapt and thrive: Combining adaptation to climate change, disaster mitigation, and natural resources management in a new approach to the reduction of vulnerability and poverty. In: *UNDP Expert Group Meeting, Integrating Disaster Reduction and Adaptation to Climate Change*, Havana, Cuba, 17-19 June 2002. United Nations Development Program, Havana, Cuba.

Spence, R., 2004: Risk and regulation: Can improved government action reduce the impacts of natural disasters? *Building Research and Information*, **32(5)**, 391-402.

Sperling, F. and F. Szekely, 2005: *Disaster risk management in a changing climate, vulnerability and adaptation*. Discussion paper prepared for the world conference on disaster reduction – reprint with addendum on conference outcomes. Vulnerability and Adaptation Resource Group, Washington, DC.

Spittlehouse, D.L. and R.B. Stewart, 2003: Adapting to climate change in forest management. *Journal of Ecosystems and Management*, **4**, 7-17.

Stainforth, D.A., T.E. Downing, R. Washington, A. Lopez and M. New, 2007: Issues in the interpretation of climate model ensembles to inform decisions. *Philosophical Transactions of the Royal Society A*, **365**, 2163-2177.

Stern, N.H., 2007: *The Stern Review: The Economics of Climate Change*. Cambridge University Press, Cambridge, UK.

Stevens, L., 2008: *Assessment of Impacts of Climate Change on Australia's Physical Infrastructure*. Academy of Technological Sciences and Engineering, Parksville, Australia.

Stewart, M.G., X. Wang, and M.N. Nguyen, 2011: Climate change impact and risks of concrete infrastructure deterioration. *Engineering Structures*, **33**, 1326-1337.

Sudmeier-Rieux, K. and N. Ash, 2009: *Environmental Guidance Note for Disaster Risk Reduction*. International Union for Conservation of Nature (IUCN), Gland, Switzerland.

Swiss Re, 2000: *Storm over Europe. An underestimated risk*. Swiss Reinsurance Company, Zurich, Switzerland.

Tanner, T.M. and T. Mitchell (eds.), 2008: Poverty in a changing climate. *Institute of Development Studies (IDS) Bulletin*, **39**, 4.

Tearfund, 2008: Livelihood initiatives helps poor women build community resilience. In: *Linking Disaster Risk Reduction and Poverty: Good Practices and Lessons Learnt* [United Nations International Strategy for Disaster Reduction (UNISDR) (ed.)]. UNISDR, Geneva, Switzerland, pp. 45-50

Tearfund, 2011: *Disasters and the Local Church: Guidelines for church leaders in disaster-prone areas*. Tearfund, Teddington, UK.

TEEB, 2009. *The Economics of Ecosystems and Biodiversity for National and International Policy Makers Summary: Responding to the Value of Nature*. The Economics of Ecosystems and Biodiversity, Geneva, Switzerland, teebweb.org/ForPolicymakers/tabid/1019/Default.aspx.

Telford, J. and J. Cosgrave, 2006: *Joint Evaluation of the International Response to the Indian Ocean Tsunami: Synthesis Report*. Tsunami Evaluation Coalition, London, UK.

Thomalla, F., T. Downing, E. Spanger-Siegfried, G. Han, and J. Rockstrom, 2006: Reducing hazard vulnerability: towards a common approach between disaster risk reduction and climate adaptation. *Disasters*, **30(1)**, 39-48.

Thompson, I., B. Mackey, S. McNulty, and A. Mosseler, 2009: *Forest Resilience, Biodiversity, and Climate Change. A synthesis of the biodiversity/resilience/stability relationship in forest ecosystems*. Technical Series no. 43, Secretariat of the Convention on Biological Diversity, Montreal, Canada.

Thompson, M. and I. Gaviria, 2004: *Weathering the storm: Lessons in risk reduction from Cuba*. Oxfam, Boston, MA.

Thorne, C., E. Evans, and E. Penning-Rowsell (eds.), 2006: *Future Flooding And Coastal Erosion Risks*. Thomas Telford, London, UK.

Tompkins, E.L., 2005: Planning for climate change in small islands: Insights from national hurricane preparedness in the Cayman Islands. *Global Environmental Change Part A*, **15(2)**, 139-149.

Tompkins, E.L., M.C. Lemos, and E. Boyd, 2008: A less disastrous disaster: Managing response to climate-driven hazards in the Cayman Islands and NE Brazil. *Global Environmental Change*, **18**, 736-745.

Tompkins, E.L., W.N. Adger, E. Boyd, S. Nicholson-Cole, K. Weatherhead, and N. Arnell, 2010: Observed adaptation to climate change: UK evidence of transition to a well-adapting society. *Global Environmental Change*, **20**, 627-635.

Tschoegl, L., R. Below, and D. Guha-Sapir, 2006: *An Analytical Review of Selected Data Sets on Natural Disasters and Impacts*. Centre for Research on the Epidemiology of Disasters, United Nations Development Programme, Brussels, Belgium.

Tsing, A.L., J.P. Brosius, and C. Zerner, 1999: Assessing community-based natural-resource management. *Ambio*, **28(2)**, 197-198.

Twigg, J., 2004: Project planning. In: *Disaster Risk Reduction: Mitigation and Preparedness in Development and Emergency Programming*. Humanitarian Practice Network, Overseas Development Institute, London, UK, pp. 31-60.

UNDP, 2002: A climate risk management approach to disaster reduction and adaptation to climate change. In: *Proceedings of UNDP Expert Group Meeting*, Havana, Cuba, 19-21 June 2002. United Nations Development Programme, New York, NY.

UNDP, 2004a: *Reducing Disaster Risk: A Challenge for Development*. United Nations Development Programme, New York, NY.

UNDP, 2004b: *A Global Review: UNDP Support for Institutional and Legislative Systems for Disaster Risk Management: Executive Summary*. United Nations Development Programme, New York, NY.

UNDP, 2004c: *Adaptation Policy Frameworks for Climate Change. Developing strategies, policies and measures*. Cambridge University Press, Cambridge, UK.

UNDP, 2009: *Indonesia: Institutional and Legal Systems for Early Warning and Disaster Risk Reduction*. Regional Programme on Capacity Building for Sustainable Recovery and Risk Reduction, United Nations Development Programme, New York, NY.

UNEP, 2006: *In the Front Line: Shoreline Protection and Other Ecosystem Services from Mangroves And Coral Reefs*. United Nations Environment Programme World Conservation Monitoring Centre, Cambridge, UK.

UNEP, 2009. *The Role of Ecosystem Management in Climate Change Adaptation and Disaster Risk Reduction*. Issues paper prepared for the global platform for Disaster Risk Reduction. United Nations Environment Programme, Geneva, Switzerland.

UNEP, 2010: *Using Ecosystems to Address Climate Change – Ecosystem Based Adaptation*. UNEP Regional Seas Information Series, United Nations Environment Programme, Nairobi, Kenya.

UNESCAP, 2009: *Case Study: The National Disaster Management System Of China And Its Response To The Wenchuan Earthquake*. United Nations Economic and Social Commission for Asia and the Pacific, Bangkok, Thailand, www.unescap.org/Idd/Events/Cdrr-2009/Cdr_2e.Pdf.

UNFCCC, 2006a: *Application of Environmentally Sound Technologies for Adaptation to Climate Change*. United Nations Framework Convention on Climate Change, Bonn, Germany.

UNFCCC, 2006b. *Technologies for Adaptation to Climate Change*. United Nations Framework Convention on Climate Change, Bonn, Germany.

UNFCCC, 2007a: *Investment and Financial Flows to Address Climate Change*. United Nations Framework Convention on Climate Change, Bonn, Germany.

UNFCCC, 2007b: *National Adaptation Programmes of Action (NAPAs)*. United Nations Framework Convention on Climate Change, Bonn, Germany.

UNFCCC, 2008a: *Disaster Risk Reduction Strategies and Risk Management Practices: Critical Elements for Adaptation to Climate Change*. United Nations Framework Convention on Climate Change, Bonn, Germany.

UNFCCC, 2008b: *Integrating Practices, Tools and Systems for Climate Risk Assessment and Management and Strategies for Disaster Risk Reduction into National Policies and Programmes*. United Nations Framework Convention on Climate Change, Bonn, Germany.

UNFCCC, 2009a: *Potential Costs and Benefits of Adaptation Options: A Review of Existing Literature*. Technical Paper, United Nations Framework Convention on Climate Change, Bonn, Germany.

UNFCCC, 2011a: *NAPA Priorities Database*. United Nations Framework Convention on Climate Change, Bonn, Germany, unfccc.int/cooperation_support/least_developed_countries_portal/napa_priorities_database/items/4583.php.

UNFCCC, 2011b: *Frequently Asked Questions about LDCs, the LEG and NAPAs*. United Nations Framework Convention on Climate Change, Bonn, Germany, unfccc.int/cooperation_support/least_developed_countries_portal/frequently_asked_questions/items/4743.php.

UNISDR, 2004: *Living with Risk: A Global Review of Disaster Reduction Initiative*. United Nations International Strategy for Disaster Reduction, Geneva, Switzerland.

UNISDR, 2005: *Hyogo Framework for Action 2005–2015: Building the resilience of nations and communities to disasters*. United Nations International Strategy for Disaster Reduction, Geneva, Switzerland.

UNISDR, 2006: *Developing Early Warning Systems: A checklist. The conclusions of the third international conference on early warning*. Germany, Bonn, 27-29. United Nations International Strategy for Disaster Reduction, Geneva, Switzerland.

UNISDR, 2007a: *Building Disaster Resilient Communities: Good Practices and Lessons Learned*. United Nations International Strategy for Disaster Reduction, Geneva, Switzerland.

UNISDR, 2007b: *Words into Action: A Guide for Implementing the Hyogo Framework*. United Nations International Strategy for Disaster Reduction, Geneva, Switzerland.

UNISDR, 2007c: *'Acting with Common Purpose'. Proceedings of the first session of the Global Platform on Disaster Risk Reduction*. Geneva, 5-7 June 2007, United Nations International Strategy for Disaster Reduction, Geneva, Switzerland.

UNISDR, 2008a: *Linking Disaster Risk Reduction and Poverty Reduction: Good Practices and Lessons Learnt*. United Nations International Strategy for Disaster Reduction, Geneva, Switzerland.

UNISDR, 2008b: *Towards National Resilience: Good practices of National Platforms for Disaster Risk Reduction*. United Nations International Strategy for Disaster Reduction, Geneva, Switzerland.

UNISDR, 2009a: *Adaptation to Climate Change by Reducing Disaster Risks: Country Practices and Lessons*. United Nations International Strategy for Disaster Reduction, Geneva, Switzerland.

UNISDR, 2009b: *Applying Disaster Risk Reduction for Climate Change Adaptation: Country Practices and Lessons*. United Nations International Strategy for Disaster Reduction, Geneva, Switzerland.

UNISDR, 2009c: *Risk and Poverty in a Changing Climate: Global Assessment Report on Disaster Risk Reduction*. United Nations International Strategy for Disaster Reduction, Geneva, Switzerland.

UNISDR, 2009d: *Recommendations of National Platforms to the Chair and Participants of the Second Session of the Global Platform for Disaster Risk Reduction*. United Nations International Strategy for Disaster Reduction, Geneva, Switzerland.

UNISDR, 2010: *Early Warning Practices can Save Many Lives: Good Practices and Lessons Learned*. United Nations International Strategy for Disaster Reduction, Geneva, Switzerland.

UNISDR, 2011a: *2011 Global Assessment Report on Disaster Risk Reduction: Revealing Risk, Redefining Development*. United Nations International Strategy for Disaster Reduction, Geneva, Switzerland.

UNISDR, 2011b: *Hyogo Framework for Action 2005-2015: Mid Term Review*. United Nations International Strategy for Disaster Reduction, Geneva, Switzerland.

UNISDR/DKKV/Council of Europe, 2008: *Disaster risk reduction in Europe: Overview of European National Platforms, Hyogo Framework for Action focal points and regional organisations and institutions*. United Nations International Strategy for Disaster Reduction, German Committee for Disaster Risk Reduction, and Council of Europe.

Urwin, K. and A. Jordan, 2008: Does public policy support or undermine climate change adaptation? Exploring policy interplay across different scales of governance. *Global Environmental Change*, **18(1)**, 180-191.

Van Buskirk, R., 2006: Analysis of long-range clean energy investment scenarios for Eritrea, East Africa. *Energy Policy*, **34(14)**, 1807-1817.

Varangis, P., J.R. Skees, and B.J. Barnett, 2002: Weather indexes for developing countries. In: *Climate Risk and the Weather Market* [Dischel, B. (ed.)]. Risk Books, London, UK, pp. 279-294.

Venter, O., W.F. Laurance, T. Iwamura, K.A. Wilson, R.A. Fuller, and H.P. Possingham, 2009: Harnessing carbon payments to protect biodiversity. *Science*, **326(5958)**, 1368.

Venton, P. and S. La Trobe, 2008: *Linking Climate Change Adaptation and Disaster Risk Reduction*. Tearfund, Tweddington, UK.

Verdin, J., C. Funk, G. Senay, and R. Choularton, 2005: Climate science and famine early warning. *Philosophical Transactions of the Royal Society B*, **360**, 2155-2168.

Vignola, R., B. Locatelli, C. Martinez, and P. Imbach, 2009: Ecosystem-based adaptation to climate change: What role for policy-makers, society and scientists? *Mitigation and Adaptation Strategies for Global Change*, **8(14)**, 691-696.

Virtanin, P., E. Palmujpki, and D. Gemechu, 2011: Global climate policies, local institutions and food security in a pastoral society in Ethiopia. *Consilience: The Journal of Sustainable Development*, **5**, 96-118.

Vogel, C., and K. O'Brien, 2006: Who can eat information? Examining the effectiveness of seasonal climate forecasts and regional climate-risk management strategies. *Climate Research*, **33**, 111-122.

Walters, C., 1997. Challenges in adaptive management of riparian and coastal ecosystems. *Conservation Ecology*, **2(1)**, 1.

Wamsler, C., 2004: Managing urban risk: Perceptions of housing and planning as a tool for reducing disaster risk. *Global Built Environment Review*, **4(2)**, 11-28.

Warhurst, A., 2006: *Disaster Prevention: A Role for Business?* Maplecroft and ProVention Consortium, Geneva, Switzerland.

Warner, K., N. Ranger, S. Surminski, M. Arnold, J. Linnnerooth-Bayer, E. Michel-Kerjan, P. Kovacs, and C. Herweijer, 2009: *Adaptation to Climate Change: Linking Disaster Risk Reduction and Insurance*. United Nations International Strategy for Disaster Reduction Secretariat, Geneva, Switzerland.

Weih, M., 2004: Intensive short rotation forestry in boreal climates: Present and future perspectives. *Canadian Journal of Forest Research*, **34(7)**, 1369-1378.

White, P., M. Pelling, K. Sen, D. Seddon, S. Russell, and R. Few, 2004: *Disaster Risk Reduction: A Development Concern. A Scoping Study on Links between Disaster Risk Reduction, Poverty and Development*. Department for International Development, London, UK.

WHO, 2003: *Climate Change and Human Health - Risks and Responses: Summary*. World Health Organization, Geneva, Switzerland.

WHO, 2005: *Health and Climate Change: The "Now and How": A Policy Action Guide*. World Health Organization Regional Office for Europe, Copenhagen, Denmark.

Wickham, F., J. Kinch, and P.N. Lal, 2009: *Institutional Capacity within Melanesian Countries to Effectively Respond to Climate Change Impacts, with a Focus on Vanuatu and the Solomon Islands*. A report prepared for the Bishop Museum and SPREP. Bishop Museum, Honolulu, Hawaii.

Wilby, R.L., 2007: A review of climate change impacts on the built environment. *Built Environment*, **33(1)**, 31-45.

Wilby, R.L., 2009: *Climate for Development in South Asia (ClimDev-SAsia): An inventory of cooperative programmes and sources of climate risk information to support robust adaptation*. Report prepared on behalf of the UK Department for International Development (DFID), DFID, London, UK.

Wilby, R.L. and S. Dessai, 2010: Robust adaptation to climate change. *Weather*, **65(7)**, 180-185.

Wilby, R.L., J. Troni, Y. Biot, L. Tedd, B.C. Hewitson, D.M. Smith, and R.T. Sutton, 2009: A review of climate risk information for adaptation and development planning. *International Journal of Climatology*, **29(9)**, 1193-1215.

Wilches-Chaux, G., 2008: *¿Qu-ENOS pasa? – Guía de LA RED para la gestión radical de riesgos asociados con el fenómeno ENOS*. LA RED, Bogotá, Colombia.

Wisner, B., 2011: Are we there yet? Reflections on integrated disaster risk management after ten years. *Journal of Integrated Disaster Risk Management*, **1**, 1-14.

Wisner, B. and J. Adams, 2002: *Environmental Health in Emergencies and Disasters: A Practical Guide*. World Health Organization, Geneva, Switzerland.

Wisner, B., P. Blaikie, T. Cannon, and I. Davis, 2004: *At Risk: Natural Hazards, People's Vulnerability And Disasters*. 2nd ed. Routledge, London, UK.

WMO, 2004: *Report of the Expert Meeting on Meteorological Information for Locust Control, 18-20 October 2004.* World Meteorological Organization, Geneva, Switzerland.

WMO, 2007: *WMO to Provide Guidance for Heat Health Warning System.* WMO Press Release 781, World Meteorological Organization, Geneva, Switzerland.

Wood, A.P. and Van Halsema, G.E., 2008: *Scoping agriculture, wetland interactions: Towards a sustainable multiple-response strategy.* FAO Water Reports. Rome: Food and Agriculture Organisation (FAO).

World Bank, 1996. *Argentina flood protection project.* Staff Appraisal Report 15354, World Bank, Washington, DC.

World Bank, 2000: *Cities, Seas and Storms: Managing Change in Pacific Island Economies.* World Bank, Washington, DC.

World Bank, 2003: *Protecting New Health Facilities from Natural Disasters: Guidelines for the Promotion of Disaster Mitigation.* World Bank, Washington, DC.

World Bank, 2007: *The Caribbean Catastrophe Risk Insurance Initiative.* Results of preparation work on the design of a Caribbean Catastrophe Risk Insurance Facility, World Bank, Washington, DC.

World Bank, 2008: *The Economics of Adaptation to Climate Change: Methodology Report.* World Bank, Washington, DC.

World Bank, 2009: *Convenient Solutions to an Inconvenient Truth: Ecosystem-Based Approaches to Climate Change.* World Bank, Washington, DC.

World Bank, 2010: *Mainstreaming Adaptation to Climate Change in Agriculture and Natural Resources Management Projects.* Washington D.C.: World Bank.

World Bank and UN, 2010: *Natural Hazards, UnNatural Disasters: The Economics of Effective Prevention.* International Bank for Reconstruction and Development/ World Bank, Washington, DC, 254 pp., gfdrr.org/gfdrr/nhud-home.

World Vision, 2008: Reducing vulnerabilities and poverty through disaster mitigation. In: *Linking Disaster Risk Reduction and Poverty: Good Practices and Lessons Learnt* [United Nations International Strategy for Disaster Reduction (UNISDR) (ed.)]. UNISDR, Geneva, Switzerland, pp. 15-19.

WSSD, 2002: *Report of the World Summit on Sustainable Development*, Johannesburg, South Africa, 26 August – 4 September 2002.

WWAP, 2009: *Water in a Changing World. The United Nations World Water Development Report.* World Water Assessment Programme, Earthscan, London, UK.

WWC, 2009: *Water Supply and Sanitation.* World Water Council, France.

Yamin, F., A. Rahman, and S. Huq, 2005: Vulnerability, adaptation and climate disasters: A conceptual overview. *IDS Bulletin*, **36(4)**, 1-14.

Yodmani, S., 2001: Disaster risk management and vulnerability reduction: Protecting the poor. In: *Proceedings of Asia and Pacific Forum on Poverty: Reforming policies and institutions for poverty reduction.* Manila, Philippines, 5-9 February 2001.

Younger, M., H.R. Morrow-Almeida, S.M. Vindigni, and A.L. Dannenberg, 2008: The built environment, climate change, and health: Opportunities for co-benefits. *American Journal of Preventive Medicine*, **35(5)**, 517-52.

7. Managing the Risks: International Level and Integration across Scales

Coordinating Lead Authors:
Ian Burton (Canada), O. Pauline Dube (Botswana)

Lead Authors:
Diarmid Campbell-Lendrum (Switzerland), Ian Davis (UK), Richard J.T. Klein (Sweden), Joanne Linnerooth-Bayer (USA), Apurva Sanghi (USA), Ferenc Toth (Austria)

Review Editors:
Joy Jacqueline Pereira (Malaysia), Linda Sygna (Norway)

Contributing Authors:
Neil Adger (UK), Thea Dickinson (Canada), Kris Ebi (USA), Md. Tarik ul Islam (Canada / Bangladesh), Clarisse Kehler Siebert (Sweden)

This chapter should be cited as:

Burton, I., O.P. Dube, D. Campbell-Lendrum, I. Davis, R.J.T. Klein, J. Linnerooth-Bayer, A. Sanghi, and F. Toth, 2012: Managing the risks: international level and integration across scales. In: *Managing the Risks of Extreme Events and Disasters to Advance Climate Change Adaptation* [Field, C.B., V. Barros, T.F. Stocker, D. Qin, D.J. Dokken, K.L. Ebi, M.D. Mastrandrea, K.J. Mach, G.-K. Plattner, S.K. Allen, M. Tignor, and P.M. Midgley (eds.)]. A Special Report of Working Groups I and II of the Intergovernmental Panel on Climate Change (IPCC). Cambridge University Press, Cambridge, UK, and New York, NY, USA, pp. 393-435.

Table of Contents

Executive Summary 396

7.1. The International Level of Risk Management 398
7.1.1. Context and Background 398
7.1.2. Related Questions and Chapter Structure 398

7.2. Rationale for International Action 398
7.2.1. Systemic Risks and International Security 399
7.2.2. Economic Efficiency 399
7.2.3. Shared Responsibility 400
7.2.4. Subsidiarity 401
7.2.5. Legal Obligations 401
7.2.5.1. Scope of International Law, Managing Risks, and Adaptation 401
7.2.5.2. International Conventions 402
7.2.5.3. Customary Law and Soft Law Principles 402
7.2.5.4. Non-Legally Binding Instruments 402

7.3. Current International Governance and Institutions 403
7.3.1. The Hyogo Framework for Action 403
7.3.1.1. Evolution and Description 403
7.3.1.2. Status of Implementation 404
7.3.2. The United Nations Framework Convention on Climate Change 406
7.3.2.1. Evolution and Description 406
7.3.2.2. Status of Implementation 406
7.3.3. Current Actors 408
7.3.3.1. International Coordination in Linking Disaster Risk Management and Climate Change Adaptation 408
7.3.3.2. International Technical and Operational Support 409
7.3.3.3. International Finance Institutions and Donors 410

7.4. Options, Constraints, and Opportunities for Disaster Risk Management and Climate Change Adaptation at the International Level 411
7.4.1. International Law 411
7.4.1.1. Limits and Constraints of International Law 411
7.4.1.2. Opportunities for the Application of International Law 412
7.4.2. International Finance 412
7.4.3. Technology Transfer and Cooperation 414
7.4.3.1. Technology and Climate Change Adaptation 414
7.4.3.2. Technologies for Extreme Events 416
7.4.3.3. Financing Technology Transfer 417
7.4.4. Risk Sharing and Transfer 418
7.4.4.1. International Risk Sharing and Transfer 418
7.4.4.2. International Risk-Sharing and Transfer Mechanisms 418
7.4.4.3. Value Added by International Interventions 420
7.4.5. Knowledge Acquisition, Management, and Dissemination 421
7.4.5.1. Knowledge Acquisition 421
7.4.5.2. Knowledge Organization, Sharing, and Dissemination 422

7.5.	**Considerations for Future Policy and Research**	**425**
7.6.	**Integration across Scales**	**426**
7.6.1.	The Status of Integration	426
7.6.2.	Integration at a Spatial Scale	427
7.6.3.	Integration at a Temporal Scale	427
7.6.4.	Integration at a Functional Scale	427
7.6.5.	Toward More Integration	427

References 428

Executive Summary

Increasing global interconnectivity, population, and economic growth, and the mutual interdependence of economic and ecological systems, can serve both to reduce vulnerability and to amplify disaster risks (*high confidence*). Global development pathways are becoming a more important factor in the management of vulnerability and disaster risk. [7.2.1]

The international community has accumulated substantial experience in providing help for disasters and risk management in the context of localized and short-term events associated with climate variability and extremes. Experience in disaster risk management includes both bottom-up and top-down approaches, but most often has developed from disasters considered first as local issues, then at the national level, and only at the international level where needs exceed national capacity, especially in terms of humanitarian assistance and capacity building. [7.2.4]

There are two main mechanisms at the international level that are purpose-built and dedicated to disaster risk management and climate change adaptation. These are the United Nations International Strategy for Disaster Reduction (UNISDR) and the United Nations Framework Convention on Climate Change (UNFCCC), in particular in its adaptation components. This chapter focuses on these two bodies while recognizing that there are many others that have an international role to play. Page limitations require a selective approach and a comprehensive assessment of all relevant bodies is impractical. The UNISDR and the UNFCCC are very different institutions with different mandates and scope and objectives, and with varying strengths and capacities (*high confidence*). Up to the present this fact has made the integration of disaster risk management and climate change adaptation difficult to achieve (*medium confidence*). [7.3] **The evolution of disaster risk management has come from various directions: from the top down where legislation has required safe practice at operational levels and from the local level up to the national and international levels. The evolution of climate change adaptation has been driven primarily by the recognition of the global issue of anthropogenic climate change** (*high confidence*). [7.3]

In addition to the UNISDR and the UNFCCC, other areas of international law and practice are being used to address climate change adaptation and disaster risk management. The relationship between legal aspirations and obligations in these areas of international action and management is complex and neither is well understood or agreed upon (*high confidence*). Other areas include international refugee law, which has been invoked to deal with the displacement of people that might be in part attributed to climate change; human rights law as used by citizens against states for climate change impacting on the enjoyment of human rights; and the attempts to expand existing legal doctrines such as the emerging 'responsibility to protect' doctrine to motive states to act on climate change. Such attempts to use tools from other areas of international law to address climate change adaptation and disaster risk reduction challenges have generally not been successful. [7.2.5]

International action on disaster risk reduction and climate change adaptation can be motivated both by national interests and a concern for the common (global) public good. [7.2] The interdependence of the global economy, the public good, and the transboundary nature of risk management, and the potential of regional risk pooling, can make international cooperation on disaster risk reduction and climate change adaptation more economically efficient than national or sub-national action alone. Notions of solidarity and equity motivate addressing disaster risk reduction and climate change adaptation at the international level in part because developing countries are more vulnerable to physical disasters. [7.2]

Closer integration at the international level of disaster risk reduction and climate change adaptation, and the mainstreaming of both into international development and development assistance, could foster efficiency in the use of available and committed resources and capacity (*high confidence*). [7.4] Neither disaster risk reduction nor climate change adaptation is as well integrated as they could be into current development policies and practices. Both climate change adaptation and disaster risk reduction might benefit from sharing of knowledge and experience in a mutually supportive and synergistic way. Climate change adaptation could be factored into all disaster risk management, and weather-related disasters are becoming an essential component of the adaptation agenda. [7.4]

Opportunities exist to create synergies in international finance for disaster risk management and adaptation to climate change, but these have not yet been fully realized (*high confidence*). International funding for disaster risk reduction remains relatively low as compared to the scale of spending on international humanitarian response. [7.4.2] Governments have committed to mobilize greater amounts of funding for climate change adaptation and this may also help to support the longer-term investments necessary for disaster risk reduction. [7.4.2]

Expanded international financial support for climate change adaptation as specified in the Cancun Agreements of 2010 and the Climate Change Green Fund will facilitate and strengthen disaster risk management (*medium confidence*). The agreements to provide substantial additional finance at the international level for adaptation to climate change have been formulated to include climate- and weather-related disaster risk reduction. There is therefore some prospect that projects and planning for disaster risk reduction and climate change adaptation can increasingly be combined and integrated at the national level (*high confidence*). [7.3.2.2, 7.4]

Technology transfer and cooperation under the United Nations Framework Convention on Climate Change has until recently focused more on the reduction of greenhouse gas emissions than on adaptation (*high confidence*). Technology for disaster risk management, especially to advance and strengthen forecasting and warning systems and emergency response, is promoted through the Hyogo Framework for Action (HFA), but is widely dispersed among many international and national-level organizations and is not closely linked to the UNISDR. Technology transfer and cooperation to advance disaster risk reduction and climate change adaptation are important. Coordination on technology transfer and cooperation between these two fields has been lacking, which has led to fragmented implementation (*high confidence*). [7.4]

International financial institutions, bilateral donors, and other international actors have played a catalytic role in the development of catastrophic risk transfer and other risk-sharing instruments in the more vulnerable countries. **Stronger products and methods for risk sharing and risk transfer are being developed as a relatively new and expanding area of international cooperation to help achieve climate change adaptation and disaster risk reduction (*high confidence*).** [7.4] Established mechanisms include remittances, post-disaster credit, and insurance and reinsurance. Partly in response to concerns about climate change, additional insurance instruments are in various stages of development and expansion including international risk pools and weather index micro-insurance. These processes and products are being developed by international financial institutions as well as by nongovernmental organizations and the private sector. [7.4.4.2]

One lesson from disaster risk reduction and climate change adaptation is that stronger efforts at the international level do not necessarily lead to substantive and rapid results on the ground and at the local level. There is room for improved integration across scales from international to local (*high confidence*). [7.6] The expansion of disaster risk reduction through the International Decade for Natural Disaster Reduction (1990-1999), and the establishment of the UNISDR and the creation and adoption of the HFA have had results that are difficult to specify or to quantify – but which may have contributed to some reduction in morbidity and mortality, while enjoying much less success in the area of economic and property losses. The problems of disaster risk have continued to grow due in large part to the relentless expansion in exposure and vulnerability even as the international management capacity has expanded (*medium confidence*). [7.5, 7.6]

7.1. The International Level of Risk Management

7.1.1. Context and Background

A need to cope with the risks associated with atmospheric processes (floods, droughts, cyclones, and so forth) has always been a fact of human life (Lamb, 1995). In more recent decades, extreme weather events have increasingly come to be associated with large-scale disasters and an increasing level of economic losses (Chapters 2 and 4). Considerable experience has accumulated at the international (as well as local and national) level on ways of coping with or managing the risks.

The same cannot be said for the risks associated with anthropogenic climate change. These are new risks identified as possibilities or probabilities (IPCC 1990, 1996, 2007).

Acceptance of climate change and its growing impacts has led to a stronger emphasis on the need for adaptation, as exemplified, for example, in the Bali Action Plan (adopted at the 13th Session of the Conference of the Parties to the UNFCCC (UNFCCC, 2007a) and the Cancun Agreements of December 2010.

The international community is thus faced with a contrast between a long record of managing disasters and the risks of 'normal' climate extremes, and the new problem of adaptation to anthropogenic climate change and its associated changes in variability and extremes. It has been asked how the comparatively new field of anthropogenic climate change adaptation (CCA) can benefit from the longer experience in disaster risk management (DRM). That question is a major focus of this Special Report.

Climate extremes can have both negative and positive effects. The occurrence of extreme events has raised consciousness of climate change within the public and in policymakers. This can then help to enhance a sense of priority to governmental action in terms of supporting DRM, enhancing adaptation, and promoting mitigation (Adger et al., 2005). An international framework for integration of climate-related DRM and CCA in the development process could provide the potential for reducing exposure and vulnerability (Thomalla et al., 2006; Venton and La Trobe, 2008). Collective efforts at the international level to reduce greenhouse gases are a way to reduce long-term exposure to frequent and more intense climate extremes. International frameworks designed to facilitate adaptation with a deliberate effort to address issues of equity, technology transfer, globalization, and the need to meet the Millennium Development Goals (MDGs) can, when combined with mitigation, lead to reduced vulnerability (Adger et al., 2005; Haines et al., 2006). The 2007/2008 Human Development Report noted that if climate change is not adequately addressed now, 40% of the world's poorest (i.e., 2.6 billion people) will be confined to a future of diminished opportunity (Stern, 2007; Watkins, 2007). The long-term potential to reducing exposure to climate risks lies in sustainable development (O'Brien et al., 2008). Both seek to build resilience through sustainable development (O'Brien et al., 2008).

Some claim that DRM and CCA could be realized through increased awareness and use of synergies and differences, and by the provision of a framework for integration in areas of overlap between the two (Venton and La Trobe, 2008). The World Conference on Disaster Reduction held in Kobe (UNISDR, 2005c), Hyogo Prefecture, Japan in 2005 and the Bali Action Plan both point to the need for incorporation of measures that can reduce climate change impacts within the practice of disaster risk reduction (DRR). Integration of the relevant aspects of DRR and CCA can be facilitated by using the Hyogo Framework for Action (2005-2015) as agreed by 168 governments in Kobe (UNISDR, 2005a).

7.1.2. Related Questions and Chapter Structure

Within the context of the overarching question – how can experience with disaster risk management inform and help with climate change adaptation? – there are a series of other related issues to be addressed in this chapter in order to provide a basis for their closer integration. A first question concerns the rationale for disaster risk management and climate change adaptation at the international level. The issues of systemic risks and international security, economic efficiency, solidarity, and subsidiarity are addressed in Section 7.2.

A second topic concerns the nature and development of institutions and capacity at the international level. This topic is explored in Section 7.3 concentrating on the Hyogo Framework for Action and the United Nations Framework Convention on Climate Change.

A third issue concerns the opportunities for and constraints on disaster risk management and climate change adaptation at the international level. These include the matters of legal, financial, technology, risk transfer, and cooperation, and the creation of knowledge and its management and dissemination. All are addressed in Section 7.4.

Considerations of future policy and research are addressed in Section 7.5.

The challenge of bringing lessons from disaster risk reduction to climate change adaptation takes on a different complexion at different temporal and spatial scales. The question of integration across scales is taken up in Section 7.6.

7.2. Rationale for International Action

This section provides a brief overview of selected concepts and principles that have been invoked to justify (or restrain) financing, assistance, regulation, and other types of international policy interventions for disaster risk management and climate change adaptation. There is no attempt to be comprehensive, and additional principles are discussed in Section 7.2.5. Starting from the reality that risks of extreme weather and risk management interventions cross national borders and transcend single nation policies and procedures, this section discusses the *systemic nature* of these risks and their effects on international security before turning to a discussion of *efficiency, shared responsibility*, and *subsidiarity* as these principles have shaped international discourse, practices, and legal obligations within existing frameworks and conventions.

7.2.1. Systemic Risks and International Security

The term 'systemic risk' refers to risks that are characterized by linkages and interdependencies in a system, where the failure of a single entity or cluster of entities can cause cascading impacts on other interlinked entities. Because of greatly increased international interdependency, shocks occurring in one country can potentially have major and bi-directional systemic impacts on other parts of the world (Kleindorfer, 2009), although the full extent of these impacts is not well documented. Moreover, major interlinked events, such as melting of glaciers, will bring increased levels of hazard to specific areas, and the initial impacts of such changes can extend to second- and third-order impacts (Alexander, 2006). This can apply to the contiguous zones of many countries, such as shared basins with associated flood risks, which calls for transboundary, international mechanisms (Linnerooth-Bayer et al., 2001).

Relationships and connections involving the movement of goods (trade), finance (capital flows and remittances), and people (displaced populations) can also have transboundary impacts as discussed below. Moreover, actions in one country impact another, for example, clearing forests in an upstream riparian country can increase flood risks downstream. Chastened by the unexpected systemic cascading of the 2007-2008 financial crisis, firms with global supply chains are now devoting significant resources to crisis management and disruption risk management (Sheffi, 2005; Harrington and O'Connor, 2009).

A few examples can illustrate the cascading nature of the financial and economic impacts from disaster. Due to Hurricane Katrina in 2005, the International Energy Agency announced a coordinated drawdown of European and Asian oil stocks totaling 60 million barrels (Bamberger and Kumins, 2005), and reportedly oil prices rose not only in the United States but also as far away as Canada and the United Kingdom. Disasters also have an impact on international trade. Using a gravity model across 170 countries (1962-2005), Gassebner et al. (2010) conclude that an additional disaster reduces imports on average by 0.2% and exports by 0.1%. The main conditions determining the impact of disastrous events on trade are the level of democracy and the geographical size of the affected country.

Turning specifically to displaced persons as a cascading impact, estimates of the numbers of current and future migrants due not only to disasters but generally to environmental change are divergent and controversial (Myers, 2001; Christian Aid, 2007). A middle-range estimate puts the figure at 200 million by 2050 (Brown, 2008). Looking only at extreme weather as a cause of migration, a recent report estimates that over 20 million people were displaced due to sudden-onset climate-related disasters in 2008 (OCHA/IDMC, 2009). This report and others, however, acknowledge the difficulty of disentangling the drivers of migration, including climate change risks, rising poverty, spread of infectious diseases, and conflict (Castles, 2002; Myers, 2005; Thomalla et al., 2006; Barnett and Adger 2007; CIENS, 2007; Dun and Gemenne, 2008; Guzmán, 2009; Morrissey, 2009).

As opposed to abrupt displacement due to extreme weather events, mobility and migration can also be an adaptation strategy to gradual climatic change (Barnett and Webber, 2009), which normally leads to slower migration shifts. However, the very poor and vulnerable will in many cases be unable to move (Tacoli, 2009). To the extent that weather extremes contribute to migration, it can result in a huge burden to the destination areas (Barnett and Adger, 2007; Heltberg et al., 2008; Morrissey, 2009; Tacoli, 2009; Warner et al., 2009a). As part of this burden, the conflict potential of migration depends to a significant degree on how the government and people in the transit, destination, or place of return respond. Governance, the degree of political stability, the economy, and whether there is a history of violence are generally important factors (Kolmannskog, 2008).

The international impacts of climate-related disasters can extend beyond financial consequences, international trade, and migration, and affect human security more generally. O'Brien et al. (2008) report on the intricate and systemic linkages between DRR, CCA, and human security, and they emphasize the importance of confronting the societal context, including development levels, governance, inequality, and cultural practices. A further rationale for disaster risk reduction in the face of climate change at the international scale thus places emphasis on ethical issues and the growing connections among people and places in coupled social-ecological systems.

7.2.2. Economic Efficiency

The public policy literature describes situations in which government intervention is justified to address market deficiencies and inefficiencies, a rationale that can also be applied to international interventions. Stern (2007) makes the case that adaptation will not happen autonomously because of inefficiencies in resource allocation brought about by missing and misaligned markets. As a case in point, markets do not allocate resources efficiently in the case of public goods, which are goods that meet two conditions: the consumption of the good by one individual does not reduce availability of the good for consumption by others; and no one can be effectively excluded from using the good. Tompkins and Adger (2005) and Berkhout (2005) discuss how some areas, such as water resources, change from being public to private depending on national regulations and circumstances. Nevertheless, the principles of interdependence and public goods suggested by Stern and others (and which lead to inefficient allocation of resources) are frequently noted in the literature on international responsibility (Stern, 2007; Vernon, 2008; Gupta et al., 2010; World Bank, 2010a).

Early warning systems (as an example of a public good) can depend on regional and international cooperation to make more efficient use of climate data through its exchange. In the field of meteorology, many years of discussion under the auspices of the World Meteorological Organization (WMO) have led to formal agreements on the types of data that are routinely exchanged (WMO, 1995; Basher, 2006). There are similar levels of agreement in other hazard fields, for instance, sharing resources and expertise in managing floods at the river basin scale. As another example of enhanced efficiency through international cooperation, many Caribbean countries have formed a catastrophe insurance pool to reduce reinsurance premiums (see Sections 6.3.3 and 7.4, and Case Study 9.2.13).

7.2.3. Shared Responsibility

It is not only efficiency claims that can be invoked to justify international interventions, but also considerations of shared responsibility and solidarity, especially with those least able to cope with the impacts of extreme events and changes in them due to climate change. This subsection makes reference to selected principles found in the current literature on adaptation to weather-related extremes; there is no attempt to comprehensively assess the moral and ethical literature on this topic.

In the words of the Millennium Declaration that was adopted by 189 nations in September 2000:"We recognize that, in addition to our separate responsibilities to our individual societies, we have a collective responsibility to uphold the principles of human dignity, equality and equity at the global level. Global challenges must be managed in a way that distributes the costs and burdens fairly in accordance with basic principles of equity and social justice. Those who suffer or who benefit least deserve help from those who benefit most" (UNGA, 2000).

In the poorest countries, people have a higher burden in terms of loss of life per event and loss of their assets relative to their income. Based on historical loss data from Munich Re, average fatalities for major disaster events have been approximately 40 times higher in low-income as compared to high-income countries (groupings according to the World Bank), and direct asset losses as a percentage of gross national income have averaged three times greater (Barnett et al., 2008; Linnerooth-Bayer et al., 2010). Changes in frequency, magnitude, and spatial coverage of some climate extremes (see Table 3-1) can result in losses that exceed the capability of many individual countries to manage the risk (Rodriguez et al., 2009). Many have concluded that without significant international assistance the most vulnerable countries will have difficulty in adapting to changes in extreme events and their impacts due to climate change, as well as other impacts of climate change (Agrawala and Fankhauser, 2008; Agrawala and van Aalst, 2008; Klein and Persson, 2008; Klein and Möhner, 2009; Gupta and van de Grijp, 2010; Gupta et al., 2010; World Bank, 2010a). Shared responsibility can take the form of ex-ante interventions to reduce vulnerability and poverty, as well as ex-post disaster response and assistance.

Weather extremes constrain progress toward meeting the MDGs as expressed in the Millennium Declaration, especially the goal of eradicating extreme poverty and hunger (UNDP, 2002; Mirza, 2003; Watkins, 2007; UNISDR, 2009a), which can be interpreted as a direct raison d'être for international intervention in risk management (UNISDR, 2005b; Heltberg et al., 2008). Barrett et al. (2007) have shown that ex-ante risk management strategies on the part of the poor commonly sacrifice expected gains, such as investing in improved seed, to reduce risk of suffering catastrophic loss, a situation perpetuating the 'poverty trap.' The poor can be subject to multiple exposures from climate change and other stresses like geophysical hazards and changing economic conditions (e.g., fluctuating exchange rates) leading to vulnerability to even moderate hazard events (O'Brien and Leichenko, 2000).

Shared responsibility and common human concern have been articulated most effectively with regard to post-disaster humanitarian assistance, and the Millennium Declaration gives specific mention to 'natural' disasters in this context. Section VI (Protecting the Vulnerable) states: "We will spare no effort to ensure that children and all civilian populations that suffer disproportionately the consequences of natural disasters ... are given every assistance and protection so that they can resume normal life as soon as possible." With growing globalization the principle of shared responsibility is further enhanced as offers of disaster relief may provide nations access to new spheres of influence both politically and in terms of new business opportunities. Governments can piggyback a humanitarian effort on top of a for-profit operation involving private companies (Dunfee and Hess, 2000).

Disasters can overwhelm the coping mechanisms of nations, in which case international relief and assistance, as a form of solidarity, are required as a matter of saving lives. Humanitarian assistance will remain essential, but emphasizing disaster response strategies at the expense of proactive integrated approaches to disaster risk reduction can have the effect of perpetuating vulnerability (UNDP, 2002; Bhatt, 2007). For this reason, the DRR and CCA communities are placing great emphasis on pre-disaster investment and planning to redress this balance and reduce overall costs of disaster management (Kreimer and Arnold, 2000; Linnerooth-Bayer et al., 2005). These efforts include encouraging the humanitarian community to become a stronger advocate of DRR and CCA.

Beyond a sense of common human concern, it can be argued that countries contributing most to climate change have an obligation to pay to reduce or compensate losses. This is the principle underlying the 'polluter pays principle.' In addition, it can be claimed that countries have a 'principled' obligation to support those who are most vulnerable and who have made a limited contribution to the creation of the climate change problem. This is the claim underlying the expression of 'common but differentiated responsibilities and respective capabilities' (CBDR), which has emerged as one principle of international environmental law (De Lucia, 2007) and has been explicitly formulated in the context of the 1992 Rio Earth Summit (and subsequently in the Preamble and Article 3 of the UNFCCC). "In view of the different contributions to global environmental degradation, States have common but differentiated responsibilities. The developed countries acknowledge the responsibility that they bear in the international pursuit of sustainable development in view of the pressures their societies place on the global environment and of the technologies and financial resources they command." (Principle 7, the Rio Declaration; UNCED, 1992). The CBDR is discussed further in Section 7.2.5. For purposes here it is important to note that, while the CBDR principle can apply to climate change in general, including incremental change, it is relevant to climate-related disasters only if there is evidence or reason to believe that the disaster would not have occurred or would have been less severe in the absence of climate change.

Another set of literature (e.g., Adger et al., 2009; Caney, 2010) frames equity issues around climate change in terms of 'rights,' namely the right 'not to suffer from dangerous climate change' or 'to avoid dangerous climate change' (Adger, 2004; Caney, 2008). The 'rights' argument, which is highly relevant to international solidarity, can be extended to suggest that individuals and collectives have the right to be protected from risk and disaster imposed by others through the processes that lead to social exclusion, marginality, exposure, and vulnerability. According to this literature, climate change impacts can jeopardize fundamental rights to life and livelihood (such as impacts on disease burden, malnutrition, and food security). Caney (2010, p. 83) also discusses a potential further undeniable right, 'not to be forcibly evicted.' This framing, however, raises a number of difficult issues because of competing fundamental rights (O'Brien et al., 2009).

7.2.4. Subsidiarity

The principle of 'subsidiarity' can be invoked to support a case *against* international intervention. It is best known as articulated in Article 5 of the Treaty of Maastricht on European Union (Maastricht Treaty, 1992). It is based on the concept that centralized governing structures should only take action if deemed more effective or necessary than action at lower levels (Jordan, 2000; Craeynest et al., 2010). The intent is to strengthen accountability and reduce the dangers of making decisions in places remote from their point of application (Gupta and Grubb, 2000). In Europe, the principle of subsidiarity has been interpreted to mean, for example, that international- or national-level involvement in flood protection should only apply to cross-border catchments (Stoiber, 2006). While many regions and river basins are required to develop risk management flood plans, flood protection is considered predominantly a national, and in many countries (e.g., Germany and India), primarily a sub-national (state) responsibility.

The principle also recognizes that multi-level governance requires cooperation between all levels of government (Begg, 2008). As an example of this cooperation, in 2004, the African Union developed a continent-wide African Regional Strategy for Disaster Risk Reduction (African Union, 2010). Below the continental level, disaster management strategies are developed at the regional level (e.g., under the Regional Economic Communities), national level (e.g., National Disaster Management platforms), district level (e.g., District Disaster Management Committees), and local levels (e.g., Village Development Committees). Action at any one level can affect all others in a reflexive fashion.

7.2.5. Legal Obligations

7.2.5.1. Scope of International Law, Managing Risks, and Adaptation

Contemporary international law concerns the coexistence of states in times of war and of peace (19th-century conception of international law, rooted in the Westphalian system), the relationship between a state and citizens (e.g., human rights law), and the cooperation between states and other international actors in order to achieve common goals and address common concerns (e.g., international environmental law). International law, according to the authoritative Article 38 of the Statute of the International Court of Justice, emanates from three primary sources: (1) international conventions, which establish "rules expressly recognized by the ... states," and result from a deliberate process of negotiations; (2) international custom, "as evidence of a general practice accepted as law"; and (3) general principles of law, "recognized by civilized nations" (see also Birnie et al., 2009). This triumvirate of conventional and customary international law, and general principles of law, contains legal norms and obligations that can be used to motivate, justify, and facilitate international cooperation on climate change adaptation, such as contained within the UNFCCC, and in anticipation of and response to natural disasters, such as with the emerging field of international disaster relief law.

In addition to international sources of 'hard law,' 'soft law' principles also exist in the form of non-legally binding resolutions, guidelines, codes of conduct (Chinkin, 1989; Bodansky, 2010), and other non-legally binding instruments adopted by states. Collectively, hard law and soft law provide a framework within which states have obligations (hard law) or commitments (soft law) of relevance to adapting to climate change and disaster risk management. These include obligations to mitigate the effects of drought (United Nations Convention to Combat Desertification), to formulate and implement measures to facilitate adaptation (UNFCCC; see Section 7.3.2), to exercise precaution (Rio Declaration), for international cooperation to protect and promote human rights (OHCHR, 2009, para. 84 et seq.), and to develop national legislation to address disaster risk reduction (HFA; see Section 7.3.1).

At the same time as international law appears to provide a normative framework and to create an obligation to "implement ... measures to facilitate adequate adaptation to climate change" (UNFCCC Article 4.1(b)), the literature suggests that taken together, international legal instruments are not equipped to fully facilitate climate adaptation and to reduce disaster risk. To illustrate, the law of international disaster response, which aims to establish a legal framework for transborder disaster relief and recovery, has been characterized as "dispersed, with gaps of scope, geographic coverage and precision" (Fisher, 2007), with states being "hesitant to negotiate and accept far-reaching treaties that impose legally binding responsibilities with respect to disaster preparedness, protection, and response" (Fidler, 2005). A second example, international refugee law, does not recognize environmental factors as grounds for granting refugee status to those displaced across borders as a result of environmental factors (Kibreab, 1997).

7.2.5.2. International Conventions

Few internationally negotiated treaties deal, at the international level, with managing risk associated with climate extremes or with adaptation to climate change. As the primary treaty to address climate-related risk management at the international level, the UNFCCC commits Parties to facilitate adequate adaptation, to cooperate with planning for extreme weather, and to consider insurance schemes, though at present it is unresolved as to whether this implies international insurance schemes. Specifically, in Article 4.1(b), Parties to the UNFCCC agree to "formulate, implement, publish and regularly update national and, where appropriate, regional programmes containing ... measures to facilitate adequate adaptation to climate change." In Article 4.1(e), Parties agree to "cooperate in preparing for adaptation to the impacts of climate change; develop and elaborate appropriate and integrated plans for coastal zone management, water resources and agriculture, and for the protection and rehabilitation of areas, particularly in Africa, affected by drought and desertification, as well as floods." Article 4.8 of the UNFCCC commits Parties to consider actions "including related to funding, insurance and the transfer of technology" to meet the specific needs and concerns of developing countries. In Article 3.14, UNFCCC's Kyoto Protocol considers the establishment of funding, insurance, and transfer of technology (see also Sections 7.4.2, 7.4.3, and 7.4.4).

In addition to the UNFCCC, Parties to the United Nations Convention to Combat Desertification aim to "combat desertification and mitigate the effects of drought in countries experiencing serious drought and/or desertification ... through effective action at all levels, supported by international cooperation and partnership arrangements" (Article 2).

The Tampere Convention on the Provision of Telecommunication Resources for Disaster Mitigation and Relief Operations is the only contemporary multilateral treaty on the topic of disaster relief (Fidler, 2005). Aiming to reduce regulatory barriers for important equipment for disaster response, and entered into force in 2005, the Tampere Convention's first application has been met with limited success, due primarily to limited membership of many of the most vulnerable states (Fisher, 2007).

7.2.5.3. Customary Law and Soft Law Principles

Customary law and soft law principles, unlike international conventions, emerge from informal processes and do not exist in canonical form (Bodansky, 2010, p. 192 et seq.), though such customary law and soft law principles are often reflected in international treaties. This is the reality of various customs and principles that justify or mandate international action on disaster risk reduction and climate change adaptation. To be established as customary law, two elements are requisite: evidence of generally uniform and continuous state practice (regular behavior), and evidence that this practice is motivated by a sense of legal obligation (*opinio juris*) (Bodansky, 1995). Soft law principles of law, by contrast, are not customary norms and do not reflect behavioral regularities. They are rather an articulation of collective aspiration, important in shaping the "development of international law and negotiations to develop more precise norms" (Bodansky, 2010, p. 200). In practice, the distinction between rules of customary law (reflecting actual practice of states following a legal obligation) and soft law principles is frequently blurred. For instance, the principle of common but differentiated responsibilities and respective capabilities – which would for example suggest that states have differentiated responsibilities in addressing disaster risk and financing adaptation – is increasingly supported by state practice, however *opinio juris* is lacking as it is unclear whether most states consider the principle to be a legal obligation. The principle of common but differentiated responsibilities and respective capabilities might thus fall closer to a general principle than a customary norm. Irrespective of this status, the principle of common but differentiated responsibilities and respective capabilities is nevertheless a principle that states may apply in their international relations, even if it is not a norm of customary international law.

The precautionary principle states that scientific uncertainty does not justify inaction with respect to environmental risks (Trouwborst, 2002), and is articulated in a number of international instruments including Principle 15 of the Rio Declaration, and Article 3 of the UNFCCC. That states have a duty to prevent transboundary harm, provide notice of, and undertake consultations with respect to such potential harms is a soft law norm expressed under international environmental law. The more general duty to cooperate has evolved as a result of the inapplicability of the law of state responsibility to problems of multilateral concern, such as global environmental challenges. The Office of the High Commissioner for Human Rights has noted that "climate change can only be effectively addressed through cooperation of all members of the international community" (OHCHR, 2009). From the duty to cooperate is deduced a duty to notify other states of potential environmental harm. This is reflected in Principles 18 and 19 of the Rio Declaration (a non-legal international instrument), that "States shall immediately notify other States of any natural disasters or other emergencies that are likely to produce sudden harmful effects on the environment of those States" (Rio Principle 18) and "States shall provide prior and timely notification and relevant information to potentially affected States on activities that may have a significant adverse transboundary environmental effect" (Rio Principle 19).

7.2.5.4. Non-Legally Binding Instruments

Many international instruments are non-legal in nature (Raustiala, 2005). This is the case with respect to disaster relief where many of the most significant international instruments are non-binding. Illustrative are the Code of Conduct for the International Red Cross and Red Crescent Movement and Nongovernmental Organizations in Disaster Relief (ICRC, 1995) and the Sphere Project, Humanitarian Charter and Minimum Standards in Disaster Response (Sphere Project, 2004), which focus on the quality of relief developed by the international humanitarian community. These are limited by lack of compliance mechanisms (Fidler, 2005), as well as in their application, as they are the creation of international nongovernmental organizations (NGOs) and are rarely recognized in the policies of national governments. The Guiding Principles on Internal Displacement (Cohen, 1998) articulate principles of disaster prevention and of human vulnerability (Fisher, 2007).

International human rights norms as articulated in the International Bill of Human Rights have also been applied to disaster risk reduction and adaptation to climate change. Notably, the Report of the Office of the High Commission for Human Rights observes that climate change and response measures thereto can have a negative effect on the realization of human rights including rights to life, adequate food, water, health, adequate housing, and self-determination (OHCHR, 2009). These rights could risk being jeopardized when contemplated, for example, in the context of migration induced by extreme weather events. As discussed in Section 7.3.1, the HFA further stipulates key tasks for governments and multi-stakeholder actors; among these is the development of legal frameworks (UNISDR, 2005a, para. 22). The HFA is an international framework, a priority area of which is to ensure that disaster risk reduction is a national priority with an institutional basis for implementation. As to adaptation, the Bali Action Plan agreed to at the 13th Conference of the Parties to the UNFCCC recognizes the need to address consideration of disaster reduction strategies and risk management within adaptation (UNFCCC, 2007a). Adaptation is further addressed in the Cancun Agreements (UNFCCC, 2010c).

7.3. Current International Governance and Institutions

Among the many relevant frameworks and protocols administered by a host of United Nations and other international agencies, the most significant for this Special Report are the HFA, to reduce disaster risk, and the UNFCCC, which includes adaptation to the adverse effects of climate change. Since both DRR and CCA occur within a broader development context and are particularly relevant to the challenges facing developing countries, they are indirectly connected to a third important international framework: the MDGs.

The UNFCCC was adopted in 1992 following one year of negotiations and was further complemented by the Kyoto Protocol adopted in 1997. The Convention came into force in 1994 and the Protocol in 2005. In parallel, the DRR framework was adopted as a nonbinding instrument in 2005 following two years of negotiations and is time bound – 2005 to 2015. The HFA recognizes the relevance of addressing climate change in order to reduce the risk of disasters and, as soon as adopted, the two processes began to work together, collaborating closely in order to synchronize frameworks and approaches so as to create added value to current risk management initiatives. This IPCC Special Report is one example of the initiatives taken by governments. It is one of the first official products of the two communities working within different but related policy frameworks.

This section first introduces the HFA and the UNFCCC, including an overview of their respective objectives, legal nature, and status of implementation. It then presents relevant international actors involved in implementing these two frameworks, as well as a summary of other relevant international policy frameworks and agencies.

7.3.1. The Hyogo Framework for Action

7.3.1.1. Evolution and Description

The first major collective international attempt to reduce disaster impact, particularly within hazard-prone developing countries, took place in 1989, when the United Nations (UN) General Assembly designated the 1990s as the International Decade for Natural Disaster Reduction (IDNDR) (Wisner et al., 2004). About 120 National Committees were established and in 1994, the first World Conference on Natural Disaster Reduction was held in Yokohama, Japan. The conference produced the 'Yokohama Strategy and Plan of Action,' providing policy guidance with a strong technical and scientific focus.

In 2000, the IDNDR was followed by the United Nations International Strategy for Disaster Reduction (UNISDR), which broadened the technical and policy scope of the IDNDR to include increased social action, public commitment, and linkages to sustainable development. The UNISDR system promotes tools and methods to reduce disaster risk while encouraging collaboration between disaster reduction and climate change. The UNISDR Secretariat provides information and guidance on disaster risk reduction and has increasingly widened its focus to embrace adaptation to climate change. The strategy undertakes global reviews of disaster risk and promotes national initiatives to reduce disaster risk. The UNISDR has also promoted the development of National Platforms. A key function is to assist in the compilation, exchange, analysis, and dissemination of good practices and lessons learned in disaster risk reduction (refer to Section 7.4.5).

In January 2005, just three weeks after the Indian Ocean tsunami, the second World Conference on Disaster Reduction was held in Kobe, Japan. 168 governments adopted the Hyogo Framework for Action 2005-2015: Building the Resilience of Nations and Communities to Disasters. The adoption of the framework directly after a devastating tsunami gave the framework high visibility in many countries. The HFA was unanimously endorsed by the UN General Assembly (UNISDR, 2005a). The HFA is not a binding agreement: the governments simply agreed and adopted the framework as a set of recommendations to be utilized voluntarily. In international law it can be described as 'soft law.' Some regard the voluntary nature of the HFA as a useful flexible commitment, largely based on self-regulation and trust, while others regard this as its inherent weakness (Pelling, 2011, p. 44).

The HFA's Strategic Goals include the integration of DRR into sustainable development policies and planning; development and strengthening of institutions, mechanisms, and capacities to build resilience to hazards; and the systematic incorporation of risk reduction approaches into the design and implementation of emergency preparedness, response, and recovery programs (UNISDR, 2005a). The Framework also provides five Priorities for Action:

1) Ensure that DRR is a national and local priority, with a strong institutional basis for implementation
2) Identify, assess, and monitor disaster risks, and enhance early warning
3) Use knowledge, innovation, and education to build a culture of safety and resilience at all levels
4) Reduce the underlying risk factors
5) Strengthen disaster preparedness for effective response at all levels.

The priorities address all hazards with a multi-hazard approach, hence the inclusion of climate change risks and adaptation, but they do not specify the need to factor climate change risks and adaptation into ongoing action. The HFA does identify 'critical tasks' for varied actors, including states who are to "promote the integration of DRR with climate variability and climate change into DRR strategies and adaptation to climate change" (UNISDR 2005a; see also UNISDR, 2009a, 2011a,b; World Bank, 2011a).

7.3.1.2. Status of Implementation

This section will review the various tools that have been used to measure the performance of the HFA in fulfilling its Strategic Goals and Priorities for Action.

The measurement of performance in the implementation of DRR was a matter of considerable debate when the HFA was drafted. The consensus was for the final text not to include targets or indicators of progress, but countries were encouraged to develop their own guidelines to monitor their own progress in reducing their risks. To assist this process, in 2008, UNISDR published guidance notes on 'Indicators of Progress' (UNISDR, 2008). This provided the template for self-assessment that is used in national reports. While there is an obvious value in 'self-assessment' as a learning experience, in the absence of external, objective evaluation, inevitable doubts will always remain concerning such internal reporting on actual progress with DRR and CCA.

The main instruments to encourage HFA applications are the HFA Monitoring Service on PreventionWeb acting mainly as a guidance tool for countries to monitor their own progress in DRR. This is a multi-tier online tool for regional, national, and local progress review. Core Indicators are measured for the five HFA Priorities for Action as noted below, and these are reported with detailed analysis in the Global Assessment Reports (UNISDR, 2009a, 2011a; refer to Section 7.4.5). In addition to these biennial reports, the UNISDR has published a mid-term review of progress in achieving the HFA (UNISDR, 2011b). Further tools to measure progress include the reports to the biennial sessions of the Global Platform for DRR and the regional platforms for DRR and other similar mechanisms. The World Bank and the United Nations Development Programme (UNDP) also utilize the HFA to guide their support to national and local programs on DRR and gradually also for CCA (the HFA is also discussed in Sections 1.3.6 and 6.3.2).

As a result of the adoption of HFA, and the development of performance indicators, global efforts to address DRR have become more systematic. In 2009, the first biennial Global Assessment Report (GAR) on Disaster Risk Reduction was released and in the same year the Global Network of Civil Society Organisations for Disaster Reduction (GNDR) also released a report on the performance of the HFA (GNDR, 2009). The GAR found that since the adoption of the HFA, progress toward decreasing disaster risk is varied across scales. This variation is based on national government agencies self-assessment of progress against the indicators defined by the UNISDR (UNISDR, 2008) and since many of these indicators require a subjective assessment, progress is not directly comparable across countries.

Countries have been making improvements toward increasing capacity, developing institutional systems, and legislation to promote DRR, and early warning systems have been implemented in many areas. However, the Global Assessment Reports (UNISDR, 2009a, 2011a) conclude that progress is still required to mainstream DRR into public investment, development planning, and governance arrangements. During 2010, at the mid-point in the HFA, the UN Secretary General echoed this concern in reporting that "risk reduction is still not hardwired into the 'business processes' of the development sectors, planning ministries and financial institutions" (UNGA, 2010, p. 5).

Further, both the GARs and the GNDR (2009, 2011) noted that at national and international levels, policy and institutional frameworks for climate change adaptation and poverty reduction are not yet synchronized to those for DRR. For example, the 2011 GAR reports on weak coordination and separate management between institutional and program mechanisms (UNISDR, 2011a, p. 150).

The GNDR observed that ecosystem management approaches can provide multiple benefits, including risk reduction, and thus be a central part of DRR strategies. But countries have experienced difficulty in addressing underlying risk drivers (such as food security, social protection, building codes/standards, poverty alleviation, poor urban and local governance, vulnerable rural livelihoods, and ecosystem decline) in a way that leads to a reduction in the risk of damages and economic loss (GNDR, 2009). This Fourth HFA Priority for Action – 'Reduce the Underlying Risk Factors' – remains the greatest challenge to civil society bodies, with all 13 criteria only reaching a rating of 2 on the assessment scale: 'some activity but significant scope for improvements' (GNDR, 2009, pp. 24–26). The

GARs also note this area of weakness, but note that it is possible for countries to address underlying risk drivers using an assortment of mechanisms to increase resilience (e.g., raising awareness, education, training, risk assessments, early warning systems, building safety, micro-insurance in macro-financing schemes) (UNISDR, 2009a, 2011a).

It was also acknowledged in the 2009 GAR that weather-related disaster risk is escalating swiftly, in terms of the regions affected, frequency of events, and losses reported. This frequency relates to occurrence patterns as well as improved reporting of all categories of weather-related hazards. Data was collected from a sample of 12 Asian and Latin American countries: Argentina, Bolivia, Colombia, Costa Rica, Ecuador, the Indian states of Orissa and Tamil Nadu, Iran, Mexico, Nepal, Peru, Sri Lanka, and Venezuela. The report further noted that these increases will magnify the uneven distribution of risk between wealthier and poorer countries (UNISDR, 2009a, p. 11). Furthermore, a conclusion is drawn in the report that climate change is changing the geographical distribution, intensity, and frequency of these weather-related hazards, threatening to exceed the capacities of poorer countries and their communities' abilities to absorb losses and recover from disaster impacts (UNISDR, 2009b). However, the 2011 GAR reported significant progress with a decrease in global mortality risk from tropical cyclones and flooding, with the only exception being South Asia where vulnerability is still increasing (UNISDR, 2011a, p. 28).

The 2009 and 2011 GARs, as well as the discussion they generated in the Global Platforms of 2009, have brought a regional dimension to performance assessment, in an effort to monitor progress.

When evaluating the progress of HFA on each of its five Priorities for Action, the GNDR found that the lowest level of progress across all the five priorities was at the lowest scale in community participation in decisionmaking on DRR (GNDR, 2009). These findings also indicate the need for a stronger link between policy formulation at international and national levels to policy execution at local levels. Rapid progress has been made in the development of comprehensive seasonal and long-term early warning systems (EWS) to anticipate droughts, floods, and tropical storms. These systems have proved to be effective in saving lives and protecting property. In the 2009 GAR, the status of EWS was reviewed (UNISDR, 2009a, Box 5.2 on p. 127). This was based on a detailed progress review of EWS undertaken by WMO (WMO, 2009). Typical examples of the effectiveness of EWS in reducing the impact of cyclones and flooding can be found in Mozambique, where their EWS was first tested in a cyclone in 2007 (Foley, 2007) and in Bangladesh, where the flood and cyclone EWS has been progressively developed over three decades (Paul et al., 2010; also see Case Study 9.2.11).

A key finding concerned the importance of education and sharing knowledge, including indigenous and traditional knowledge, and ensuring easy and systematic access to best practice tools and international standards, tailored to specific sectors (see Section 7.4.5). There is some recognition of the benefits in harmonizing and linking the frameworks and policies for DRM and CCA as core policy and programmatic objectives in national development plans and in support of poverty reduction strategies. DRM policies also need to take account of climate change. Nevertheless, countries are making significant progress in strengthening capacities, institutional systems, and legislation to address deficiencies in disaster preparedness and response (GNDR, 2009; UNISDR, 2009a).

In preparing for the mid-term review of the HFA, the UNISDR secretariat commissioned a desk review of literature to form "a baseline of the disaster risk reduction landscape." Forty-seven key documents were identified, mainly consisting of reports from UNISDR offices and partner organizations: NGOs and international development banks (UNISDR, 2011b).

The HFA Mid-Term Review 2010-2011 raised two important international issues. The first need is to develop accountability mechanisms at all levels to measure the actions taken and progress achieved in DRR. The second need is for the international community to develop a more coherent and integrated approach to support the implementation of the HFA. The review suggests that this will require connected action of the varied international actors (UNISDR, 2011b).

However, it is important to reflect on the reality that all of these methods to review international progress in risk reduction – country progress reports, the 2009 and 2011 GARs, the reports of the GNDR, and the Mid-Term Review of the HFA – are all internally produced reports by the participating agencies with external advisory boards and peer review, but all involving self-assessment. The GNDR's publications are fully independent from the UN and governments, but make no claim to be scientifically accurate assessments. The country HFA reports are online at www.preventionweb.net/english/hyogo/progress/?pid:73&pih:2.

All the above studies attempted to assess HFA performance and, as noted above, none were totally separate from the work or institutions being assessed. Furthermore, none looked specifically at the performance of the lead organization, UNISDR, in comparison with other multilateral bodies. This report came in 2011, when the UK Aid Agency, the Department for International Development (DFID), published a Multilateral Aid Review. The purpose was to ensure maximum value for money for UK aid by examining the performance of 43 multilateral organizations. This peer-reviewed assessment placed the UNISDR in a 43rd-ranked position in an assessment of 43 multilateral organizations (DFID, 2011).

This independent and comparative assessment included an evaluation of UNISDR since its foundation and identified its strength as global coordinator of the three Global Platforms in DRR that have been successful in advocacy and raising awareness. However, the assessment also identified a series of shortcomings in UNISDR. They included its poor performance in international coordination and its focus on national-level responses rather than its global mandate, which is broad rather than specific in focus. Further criticisms include inadequate attention to strategic considerations as well as leadership failures, with the report stating that there was no clear line of sight from UNISDR's mandate, to a strategy, to an implementation plan and that there was an absence of a results-based framework, thus making it difficult to measure results from input to output (DFID, 2011, p. 211).

UNISDR responded to the assessment by noting that the criticisms were also reflected in a UN audit as well as in an external evaluation requested by UNISDR in 2009, and that changes had now been incorporated in a management-reform work program (UNISDR, 2011c).

Whatever method is adopted to monitor progress with risk reduction and climate change adaptation (internal or external, self-assessment or peer review), the implicit problems faced in the measurement of DRR and CCA before a disaster event must be recognized. It is not easy, even with detailed objective scientific measurement, to accurately determine whether a given structural or non-structural measure will actually provide the necessary level of protection to people and property under extreme hazard loads. Structural tests can be carried out and simulation exercises can be usefully conducted to test warning systems or the effectiveness of preparedness, but at best such performance tests can only approximate disaster reality. The ultimate test of DRR and CCA applications will inevitably need to await the impact of the next disaster. But this limitation does not remove the requirement to monitor and measure progress in an objective scientific manner to the upper limits of existing knowledge (Davis, 2004).

7.3.2. The United Nations Framework Convention on Climate Change

7.3.2.1. Evolution and Description

The UNFCCC is a multilateral treaty aimed at addressing climate change. Its ultimate objective as stated in Article 2 is (UN, 1992; see also Oppenheimer and Petsonk, 2005):

> "to achieve … stabilization of greenhouse gas concentrations in the atmosphere at a level that would prevent dangerous anthropogenic interference with the climate system. Such a level should be achieved within a time-frame sufficient to allow ecosystems to adapt naturally to climate change, to ensure that food production is not threatened and to enable economic development to proceed in a sustainable manner."

The UNFCCC was negotiated from February 1991 to May 1992, and opened for signature at the UN Conference on Environment and Development in Rio de Janeiro in June 1992. It entered into force on 21 March 1994, and since 1995 the Conference of the Parties (COP) to the UNFCCC has met in yearly sessions. The rules, institutions, and procedures of the UNFCCC have been described in detail elsewhere (e.g., Yamin and Depledge, 2004; Bodansky, 2005). The development of adaptation as a priority under the UNFCCC has been analyzed by Schipper (2006).

A major thrust of the UNFCCC and subsequent negotiations about its implementation concerns the mitigation of climate change: all policies and measures aimed at reducing the emission of greenhouse gases such as carbon dioxide (CO_2), or at retaining and capturing them in sinks such as forests, oceans, and underground reservoirs. As mentioned by Schipper (2006), adaptation to climate change was initially given little priority, although it is subject to various commitments in the UNFCCC (see Box 7-1). When taken together, these commitments acknowledge the systematic nature of climate change risks and the relevance of the principles of economic efficiency, solidarity, and subsidiarity in adaptation.

The Kyoto Protocol, agreed at COP3 in 1997 and in force since 2005, sets binding targets for 37 industrialized countries and the European Union for reducing greenhouse gas emissions by an average of 5% compared to 1990 over the five year period 2008-2012. Adaptation is all but absent in the Kyoto Protocol, with two exceptions. Article 10(b) specifies that Parties shall formulate, implement, publish, and regularly update national and, where appropriate, regional programs containing measures to mitigate climate change and measures to facilitate adequate adaptation to climate change. Article 12.8, on the Clean Development Mechanism, provides the basis of what later became the Adaptation Fund (see Section 7.4.2).

7.3.2.2. Status of Implementation

There is to date no overall assessment of progress on adaptation under the UNFCCC in the way that the UNISDR has assessed progress under the HFA in the GARs. However, Parties to the UNFCCC are required to submit National Communications on their activities toward implementing the UNFCCC, including adaptation. There is no common reporting template so reports vary widely in content, making aggregation or comparison problematic. The annual sessions of the COP also allow countries to assess their progress toward meeting their commitments under the UNFCCC, and to negotiate and adopt new decisions for further implementation. By June 2011, there were 195 Parties to the UNFCCC: 194 countries and one regional economic integration organization (the European Union).

During the 1990s, adaptation received little attention in the UNFCCC negotiations, reflecting a similarly low level of attention to adaptation from the academic community at the time (Burton et al., 2002). The profile was raised in 2001 with the publication of the IPCC Third Assessment Report, which contained the chapter 'Adaptation to Climate Change in the Context of Sustainable Development and Equity' (Smit et al., 2001). Also in 2001, COP7 adopted a decision (5/CP.7) that outlined a range of activities that would promote adaptation in developing countries, including the preparation of National Adaptation Programmes of Action (NAPAs) by least-developed countries. To this end, COP7 established three funds with which adaptation in developing countries could be supported, namely the Least Developed Countries Fund (LDCF), the Special Climate Change Fund (SCCF), and the Strategic Priority 'Piloting an Operational Approach to Adaptation' (SPA) under the Trust Fund of the Global Environment Facility (GEF). In addition, COP7 took the first steps toward making operational the Adaptation Fund (Huq, 2002;

Box 7-1 | Commitments on Climate Change Adaptation as Included in the UNFCCC

Article 4.1: All Parties, taking into account their common but differentiated responsibilities and their specific national and regional development priorities, objectives, and circumstances, shall:

(b) Formulate, implement, publish, and regularly update national and, where appropriate, regional programs containing measures to mitigate climate change by addressing anthropogenic emissions by sources and removals by sinks of all greenhouse gases not controlled by the Montreal Protocol, and measures to facilitate adequate adaptation to climate change.

(e) Cooperate in preparing for adaptation to the impacts of climate change; develop and elaborate appropriate and integrated plans for coastal zone management, water resources, and agriculture, and for the protection and rehabilitation of areas, particularly in Africa, affected by drought and desertification, as well as floods.

(f) Take climate change considerations into account, to the extent feasible, in their relevant social, economic, and environmental policies and actions, and employ appropriate methods, for example impact assessments, formulated and determined nationally, with a view to minimizing adverse effects on the economy, on public health, and on the quality of the environment, of projects or measures undertaken by them to mitigate or adapt to climate change.

Article 4.4: The developed country Parties and other developed Parties included in Annex II shall also assist the developing country Parties that are particularly vulnerable to the adverse effects of climate change in meeting costs of adaptation to those adverse effects.

Article 4.8: In the implementation of the commitments in this Article, the Parties shall give full consideration to what actions are necessary under the Convention, including actions related to funding, insurance, and the transfer of technology, to meet the specific needs and concerns of developing country Parties [...].

Article 4.9: The Parties shall take full account of the specific needs and special situations of the least developed countries in their actions with regard to funding and transfer of technology.

Dessai, 2003; Mace, 2005). Section 7.4.2 provides more information on the international financing of climate change adaptation.

Since 2001, a number of successive decisions have given increasing priority to climate change adaptation under the UNFCCC. Decision 1/CP.10 built on Decision 5/CP.7; it reiterated the need for support for adaptation in developing countries and started a regional consultation process. Decision 2/CP.11 then established the Nairobi Work Programme on impacts, vulnerability, and adaptation to climate change, which originally ran from 2006 to 2010 – a next phase is currently under consideration, to be decided at COP17 in Durban in 2011. The objective of the Nairobi Work Programme is to assist all Parties, in particular developing countries, (i) to improve their understanding and assessment of impacts, vulnerability, and adaptation to climate change, and (ii) to make informed decisions on practical adaptation actions and measures to respond to climate change on a sound scientific, technical, and socioeconomic basis, taking into account current and future climate change and variability (Decision 2/CP.11). The Nairobi Work Programme is implemented by Parties, intergovernmental and nongovernmental organizations, the private sector, communities, and other stakeholders. Several of the nine work areas of the Nairobi Work Programme are relevant to DRR as well as CCA, in particular 'climate-related risks and extreme events' and 'adaptation planning and practices.'

With Decision 1/CP.13 (also known as the Bali Action Plan), agreed in December 2007, the COP launched "a comprehensive process to enable the full, effective, and sustained implementation of the Convention through long-term cooperative action – now, up to, and beyond 2012 – in order to reach an agreed outcome and adopt a decision at its fifteenth session" in Copenhagen in December 2009 (COP15). The Bali Action Plan gave equal priority to mitigation and adaptation, and identified technology and finance as the key mechanisms for enabling developing countries to respond to climate change (Clémençon, 2008; Ott et al., 2008; Persson et al., 2009). It recognized the need for action to enhance adaptation in five main areas:

1) International cooperation to support urgent implementation of adaptation actions, including through vulnerability assessments, prioritization of actions, financial needs assessments, capacity building, and response strategies, and integration of adaptation actions into sectoral and national planning [...]
2) Risk management and risk reduction strategies, including risk-sharing and transfer mechanisms such as insurance
3) Disaster reduction strategies and means to address loss and damage associated with climate change impacts in developing countries that are particularly vulnerable to the adverse effects of climate change
4) Economic diversification to build resilience
5) Ways to strengthen the catalytic role of the Convention in encouraging multilateral bodies, the public and private sectors, and civil society, building on synergies among activities and processes, as a means to support adaptation in a coherent and integrated manner.

No agreed outcome was reached at COP15, and no comprehensive decision was adopted that included these five issues. Instead, the COP decided to take note of the Copenhagen Accord, a nonbinding document about which there was no consensus among Parties, and which provides considerably less substance on adaptation than the Bali Action Plan (Bodansky, 2010; Grubb, 2010; Klein, 2010). As mentioned in Section 7.4.2, however, the Copenhagen Accord was a milestone toward scaled-up funding for both mitigation and adaptation.

In 2010, Decision 1/CP.16 (part of the Cancun Agreements) established the Cancun Adaptation Framework (Cozier, 2011). It invites all Parties to enhance action on adaptation by undertaking nine activities related to planning, implementation, capacity strengthening, and knowledge development, including "enhancing climate change related disaster risk reduction strategies, taking into consideration the Hyogo Framework for Action where appropriate; early warning systems; risk assessment and management; and sharing and transfer mechanisms such as insurance, at local, national, sub-regional, and regional levels, as appropriate." In addition, Decision 1/CP.16 established (i) a process to enable least-developed countries and other developing countries to formulate and implement national adaptation plans; (ii) an Adaptation Committee that will, among other things, provide technical support, share relevant information, promote synergies, and make recommendations on finance, technology, and capacity building required for further action; and (iii) a work program in order to consider approaches to address loss and damage associated with climate change impacts in developing countries that are particularly vulnerable to the adverse effects of climate change.

Decision 1/CP.16 also established a Technology Mechanism, consisting of a Technology Executive Committee and a Climate Technology Center and Network. The Technology Mechanism should accelerate action at different stages of the technology cycle, including research and development, demonstration, deployment, diffusion, and transfer of technology in support of mitigation and adaptation. Finally, Decision 1/CP.16 established the Green Climate Fund as a new entity operating the financial mechanism of the UNFCCC under Article 11 (see Section 7.4.2).

The unfolding of international adaptation policy under the UNFCCC shows the increasing prominence of adaptation in the negotiations, and the increasing level of detail and concreteness of the relevant COP decisions. It also shows that adaptation under the UNFCCC is increasingly linked with disaster risk reduction, with the Hyogo Framework for Action explicitly mentioned in the Cancun Agreements. Yet, this unfolding, from Decision 5/CP.7 to Decision 1/CP.16, has taken 10 years.

7.3.3. Current Actors

A wide range of actors play a role in DRM and CCA at the international level. This section does not attempt a comprehensive review of all of these, but instead identifies the broad areas in which the international community is providing support at the interface between DRM and CCA, describes some of the main actors under each of these categories, and summarizes, where available, independent assessments of their strengths and weaknesses in performing these roles.

7.3.3.1. International Coordination in Linking Disaster Risk Management and Climate Change Adaptation

Given the wide range of actions and actors that are considered necessary by those involved to carry out DRM and CCA, and to link them to each other, effective international coordination is essential. Overall, there are weaknesses in the current systems; the 2009 Global Assessment Report on Disaster Risk Reduction states that: "Efforts to reduce disaster risk, reduce poverty and adapt to climate change are poorly coordinated" (UNISDR, 2009a).

The main coordination mechanism for DRR, contributing to DRM, is the UNISDR, designed to develop a system of partnerships to support nations and communities to reduce disaster risk. These partners include governments, intergovernmental and nongovernmental organizations, international financial institutions, scientific and technical bodies and specialized networks as well as civil society and the private sector. Among the diverse range of stakeholders across scales, the national governments play the most important roles, including developing national coordination mechanisms; conducting baseline assessments on the status of disaster risk reduction; publishing and updating summaries of national programs; reviewing national progress toward achieving the objectives and priorities of the Hyogo Framework; working to implement relevant international legal instruments; and integrating disaster risk reduction with climate change strategies. Intergovernmental organizations play a supporting role, including, for example, promotion of DRR programs and integration into development planning, and capacity building (UNISDR, 2005b). The fact that the primary roles in planning and implementation are played by national governments, while the UNISDR Secretariat and other intergovernmental organizations provide supporting, monitoring, and information sharing roles at the regional and global level is consistent with the principle of subsidiarity.

UNISDR has made specific efforts to link DRR and CCA, through advocacy of the role of DRR in climate change adaptation, and support for scientific reviews of the linkages (including this report). Two evaluations covering the effectiveness of UNISDR in linking DRR and CCA have recently been published. The UN Special Representative of the Secretary-General for Disaster Risk Reduction and the main donors to UNISDR requested an independent evaluation of the performance of the secretariat, which was published in 2010 (Dalberg, 2010). This review endorsed the overall effectiveness of UNISDR, particularly in advocacy and awareness raising, and in establishing global and regional platforms, and specifically highlights its strong contribution to mainstreaming DRR into climate change policy. However, it also highlights difficulties, including lack of definition of comparative advantage within CCA implementation, and the need to balance the focus and resources spent on DRR in climate change adaptation versus the broader DRR concept. The same review also illustrates challenges in coordination of implementation, particularly the need for effective coordination with UN Country Teams, the World Bank, and other relevant partners at the country level, and in the full implementation and sustainable follow-up of new initiatives. The UK Government also published a review of the performance of the UNISDR Secretariat, alongside other multilateral agencies, in 2011 (DFID, 2011). The review is critical of the overall operational and organizational strengths of the UNISDR, citing a lack of a results-based framework, and weaknesses in strategic direction, coordination focus, and speed of reform. The review does, however, highlight the unique coordinating role of UNISDR, and specifically praises "a good focus on climate change, especially adaptation."

From the CCA side, the main global mechanism to increase understanding and share best practice in CCA is the Nairobi Work Programme (NWP), coordinated by the UNFCCC Secretariat (UNFCCC, 2010a; refer to Section 7.3.2.2). The NWP functions mainly as a forum for interested parties and organizations to specify their own contributions to CCA through 'action pledges,' and for sharing, synthesis, and dissemination of information. Disaster risk reduction is well represented within the NWP, which identifies DRR as one of its 14 specified adaptation delivery activities, with an associated 'call to action' for strengthened work in areas such as linking DRR and CCA, risk mapping, and cost-benefit analysis of adaptation options. Out of the 137 action pledges made by partners, 59 include a component of DRR. Evaluation of the NWP by Parties is only now being carried out, so as yet there is no formal assessment of the degree to which it has supported changes in policy and practice as well as information exchange.

7.3.3.2. International Technical and Operational Support

DRM and CCA are now beginning to be linked not only in international coordination activities, but also in mechanisms for international technical and operational support.

7.3.3.2.1. Climate services for disaster risk reduction and climate change adaptation

National meteorological and hydrological services (NMHSs) are the primary source of meteorological observations and forecasts at time scales relevant to both disaster risk management and climate change adaptation. These national services also constitute the members of the WMO, which serves to set international standards and coordinate among the members, as well as supporting several relevant international programs, including a Disaster Risk Reduction and Service Delivery Branch and a Climate Prediction and Adaptation Branch.

In recent years, a number of studies have identified weaknesses in the way in which the large amount of potentially relevant information that is available from NMHSs at the national and international level is incorporated into development decisions, particularly in the most vulnerable countries. For example a 'gap analysis' of this issue in Africa identified gaps in (i) integrating climate into policy; (ii) integrating climate into practice; (iii) climate services; and (iv) climate data, concluding that "the problem is one of 'market' atrophy: negligible demand coupled with inadequate supply of climate services for development decisions" (IRI, 2006). Studies on specific sectors (e.g., health: Kuhn et al., 2005), or at a local level (Vogel and O'Brien, 2006), conclude that the main deficit is not in generation of data, but in knowledge management. They conclude that this requires more effective mechanisms for decisionmakers to identify their information needs, and to work both with providers of weather and climate information and with institutions working on other dimensions of human and social vulnerability to address these needs.

In response to the need for a comprehensive approach to climate variability and change, and the drive for more demand-driven climate services the, World Climate Conference-3 agreed in 2009 to begin development of a Global Framework on Climate Services (GFCS) (WMO, 2010). This has a goal of "the development and provision of relevant science-based climate information and prediction for climate risk management and adaptation to climate variability and change, throughout the world." The framework therefore explicitly links climate variability (most relevant to DRR), in the context of climate change (most relevant to CCA), and support for risk management decisions (common to both). The GFCS has four major components: a User Interaction Mechanism; a World Climate Services System; Climate Research; and Observation and Monitoring. The initiative will focus on improving access and operational use of climate information, especially in vulnerable, developing countries. The principles and focus of the initiative therefore correspond closely to the objectives of linking DRM and CCA in operational planning across international and smaller scales. In May 2011, the 16th WMO congress committed to "support and facilitate the implementation of the GFCS as a priority of the Organization," including the development of an implementation plan for review and adoption in 2012 (WMO, 2011).

7.3.3.2.2. Technical and operational support from civil society

Some of the largest international civil society organizations involved in disaster risk management and humanitarian response are now beginning to integrate climate change adaptation activities into their operational programs (e.g., CARE International, 2010; Oxfam, 2011). One of the longest established examples of civil society providing technical support to CCA and DRM integration is the Red Cross/Red Crescent Climate Centre. Alongside awareness raising and advocacy, the Centre analyzes forecast information and integrates knowledge of climate risks into Red Cross/Red Crescent strategies, plans, and activities, with a particular focus on implementation at the community level (IFRC, 2011).

The various international civil society organizations working on DRR are now also beginning to coordinate their operational support, and to make explicit links to CCA (UNISDR, 2009a). The GNDR was launched in 2007, and constitutes over 300 organizations across 90 countries. It

Box 7-2 | Disaster Risk Management and Climate Change Adaptation in the Context of International Development

Vulnerability to extreme weather and to climate change is strongly conditioned by socioeconomic development, including income levels and distribution, supportive institutional frameworks, and the capacities of specific sectors. Conversely, the effects of climate change, including through any increase in the frequency of extreme weather events, can also set back economic development (Stern, 2007). Countries that are relatively poor, isolated, and reliant on a narrow range of economic activities are particularly vulnerable to such shocks (UNISDR, 2009a). The objectives of climate change adaptation, disaster risk reduction, and sustainable development are therefore intricately linked, and while the HFA and UNFCCC are the main international frameworks for CCA and DRR, a wider range of other governance and institutional mechanisms have a major influence. These range, for example, from the agreements of the World Trade Organization (affecting development and potentially technology transfer for adaptation; WTO, 2011), to the International Health Regulations (affecting the way that epidemics of climate-sensitive infectious diseases such as cholera are managed across borders; WHO, 2007), to the codes of practice of international humanitarian organizations (such as the Code of Conduct for the International Red Cross and Red Crescent Movement and NGOs in Disaster Relief; ICRC, 1995).

While approaches such as poverty reduction strategies are important in development planning at the national level, arguably the central framework for defining global development objectives is the Millennium Declaration and the associated MDGs. These have been agreed by all members of the United Nations as well as 23 international organizations, with a target date of 2015 (UN, 2011). These are also supported by international aid agreements, such as the Multilateral Debt Relief Initiative to cancel US$ 40 to 55 million dollars' worth of debt (IMF, 2011), and the commitment of economically advanced countries to commit 0.7% of gross national income to overseas development aid (UN ,1970). The eight MDGs break down into 21 quantifiable targets that are measured by 60 indicators (UN, 2011).

Neither DRM nor CCA are explicitly covered in the MDGs. However, they are strongly linked in practice. First, if disasters occur they can set back progress across many of the goals. Second, progress toward the MDGs can help to increase resilience to extreme weather events, and to climate change (Schipper and Pelling, 2006). Linking CCA and DRM with the MDGs is therefore important for the coherence of international development, and the target date of the Hyogo Framework for Action coincides with the intended completion of the MDGs (UNISDR, 2005b).

While there are exceptions, the majority of the LDCs, particularly in sub-Saharan Africa, are currently off track to reach most of the MDGs (UN, 2011). This has been attributed in part to financial, structural, and institutional weaknesses in the affected countries, and also by failure of most developed countries to reach the 0.7% aid target. Failure or delays in reaching the MDGs are therefore likely to be both a cause and a consequence of vulnerability to extreme weather and climate change (UNISDR, 2005b).

has three objectives of (1) influencing DRR public policy formulation (development); (2) increasing public accountability for effective policy administration (implementation); and (3) raising resources and political will for community-based DRR (mobilization). One of the five core strategies of the GNDR is to develop synergies between DRR and climate change to address underlying risk factors (sustainable development), including adapting local-level DRR monitoring infrastructure for climate adaptation, and input to the UNFCCC COP negotiations. Given the recent launch of the initiative there is no evaluation of effectiveness so far.

7.3.3.3. International Finance Institutions and Donors

7.3.3.3.1. Global Environment Facility

The GEF is an independent financial organization established in 1991 that provides grants to developing countries and countries with economies in transition for projects related to biodiversity, climate change, international waters, land degradation, the ozone layer, and persistent organic pollutants. It has become the largest funder of projects to address global environmental challenges and it serves as the financial mechanism for the following conventions:

- Convention on Biological Diversity (CBD)
- United Nations Framework Convention on Climate Change (UNFCCC)
- Stockholm Convention on Persistent Organic Pollutants (POPs)
- UN Convention to Combat Desertification (UNCCD).

The GEF administers the main international funds that have been made available under the UNFCCC for adaptation: the SCCF, which supports adaptation alongside development, technology transfer, capacity building, and sectoral approaches, and the LDCF, which particularly focuses on the development and implementation of NAPAs in the least-developed countries (LDCs). Ten international agencies [UNDP, the United Nations Environment Programme, the World Bank, the Food and

Agriculture Organization (FAO), the Inter-American Development Bank (IADB), the United Nations Industrial Development Organization, the International Fund for Agricultural Development, the European Bank for Reconstruction and Development (EBRD), and the African and Asian Development Banks] implement GEF projects, usually in partnership with national or other international agencies. Following a review of the implementation of the LDCF Fund by the UNFCCC's Subsidiary Body for Implementation, parties to the UNFCCC have requested the GEF, *inter alia*, to speed up the implementation process, update NAPAs, and work with its implementing agencies to improve communication with LDCs (UNFCCC, 2011). The GEF also provides interim secretariat services to the Adaptation Fund, established under the Kyoto Protocol of the UNFCCC, funded mainly through a percentage of the proceeds of the Certified Emission Reductions under the Clean Development Mechanism (Adaptation Fund, 2011a). The Fund finances climate change adaptation projects, including DRR projects, in developing countries (Adaptation Fund, 2011b).

7.3.3.3.2. The World Bank and Regional Development Banks

The major development banks (the African Development Bank, Asian Development Bank, EBRD, IADB, and World Bank Group) manage much of the funding for both climate change and disaster reduction. This includes, for example, the Pilot Program for Climate Resilience, covering a wide remit, including integration of climate risk and resilience into development planning (World Bank, 2009; Climate Funds Update, 2011).

Perhaps the clearest example of the strengths and challenges of international financing for DRM and CCA is provided by the Global Facility for Disaster Reduction and Recovery (GFDRR), managed by the World Bank. This is a partnership of the UNISDR system to support the implementation of the HFA. The GFDRR's mission is to mainstream disaster reduction and climate change adaptation into national policies, plans, and strategies to promote development and achieve the MDGs. The World Bank provides operational services to the GFDRR, on behalf of donors and other partnering stakeholders. The GFDRR supports international collaboration, and provides technical and financial assistance to low- and middle-income countries that are considered to be at high risk from disasters (GFDRR, 2010).

Two independent evaluations of the GFDRR have been conducted (Universalia Management Group, 2010; DFID, 2011). The facility has mobilized significant funds (over US$ 240 million in contributions and pledges from 2006 to 2009). The fund is considered relevant and responsive to stakeholders, and to play a unique role in helping to bridge knowledge, policy, and practice in DRR services, with good coverage of climate change adaptation (Universalia Management Group, 2010). It is also considered to be cost-effective in program implementation (DFID, 2011). However, the resources that have been mobilized through the fund remain much lower than those required, and partnerships, policy integration, and monitoring of results are considered uneven across countries. Despite these challenges, the facility is considered to have achieved important progress, and to be implementing the necessary steps to improve function and to scale up implementation (Universalia Management Group, 2010; DFID, 2011).

7.4. Options, Constraints, and Opportunities for Disaster Risk Management and Climate Change Adaptation at the International Level

7.4.1. International Law

As demonstrated in Section 7.2.5, existing tools and instruments of international law can assist with disaster risk reduction and management and in driving adaptation to climate change, recognizing at the same time that international law is limited in scope and enforceability when applied to addressing these challenges.

7.4.1.1. Limits and Constraints of International Law

Structurally, international law is both facilitated and constrained by the need for explicit or implicit acceptance by nation states, which create and comprise the system. It follows that the relevance of negotiated treaties depends on state consent, while customary law only exists if there is state practice and *opinio juris*. For instance, in the case of the Tampere Convention on the Provision of Telecommunication Resources for Disaster Mitigation and Relief Operations noted in Section 7.2.5, only four of the 25 most disaster-prone states have signed up, limiting its relevance to many of the states that would most benefit from its provisions (Fisher, 2007). The International Bill of Rights, which at face value is highly relevant to disaster risk response and in supporting an obligation to assist with adapting to climate change, does not enjoy universal acceptance. Furthermore, because international law is made by and applicable to states, the many non-state actors relevant to disaster risk reduction and climate change adaptation are not subject to obligations – though as citizens they may benefit from the duty of states.

Some fields of international law provide tools that seem applicable to disaster risk management and/or adaptation to climate change, yet are constrained through inherent limited applicability. International humanitarian law (IHL) enshrined in the 1949 Geneva Conventions enjoys wide applicability due to universal adherence (Lavoyer, 2006; Fisher, 2007), but is limited to situations of armed conflict. In contrast, 'International Disaster Response Law' (IDRL) (see Fisher, 2007), sometimes proposed as a peacetime counterpart to IHL, not only lacks the central regime and universal adhesion of the Geneva Conventions, but further experiences challenges in coordination and monitoring (Fisher, 2007). As a second example, international law has on the one hand been described as "not yet equipped to respond adequately to the diverse causes of climate-induced migration" (Von Doussa et al., 2007; generally Biermann and Boas, 2010), while on the other hand the literature is in disagreement as to *whether* refugee law should provide the instruments to deal with the challenge of migration related to climate change. The application of international refugee law, as codified in the 1951 Convention relating to the Status of Refugees, to those who cross international borders due to climate-induced migration is indeed complex and limited (UNHCR, 2009). Reopening the Convention to expand the term 'refugee,' it is argued, would risk a renegotiation of the Convention and thus potentially result in lower levels of protection for the displaced (Kolmannskog and Myrstad, 2009).

7.4.1.2. Opportunities for the Application of International Law

The potential expansion of the concepts, definitions, and procedures known to international law can also be seen as future opportunity for international law to address the challenges of disaster risk reduction and adaptation to climate change.

Beyond the current international law obligations to mitigate the effects of climate change, facilitate disaster response, and mandate international facilitation of adaptation efforts (see Section 7.2.5), the fact that international law is shaped by nation states and evolves with state practice means that international law may also adapt to future realities. Expanding the interpretation and application of existing international law, and the introduction of new law for disaster response and climate change adaptation, are both plausible in the future.

A controversial candidate field for expanded interpretation is international refugee law. The extant definition of 'refugee' per the Refugee Convention and Protocol is any person who is outside their country of nationality and who, "owing to a well-founded fear of being persecuted" is unable or unwilling to return to their country. Some literature proposes the expansion of 'persecuted' to encompass being subject to environmental disaster or degradation (Warnock, 2007; Kolmannskog and Myrstad, 2009). Comparably, Article 7 of the International Covenant on Civil and Political Rights prohibits torture and "cruel, inhuman, or degrading punishment." Some literature notes the potential expansion of the meaning 'inhuman treatment' to include being left without basic levels of subsistence due to climate change impacts. A step further proposes a new international agreement to share the "emerging burden of climate-induced migration flows" and which "upholds the human rights of the individuals affected" (Von Doussa et al., 2007). The expansion of the definition of refugee remains highly controversial, with many states opposing the use of refugee law to address climate-related, transboundary movement of people.

The emerging legal doctrine of 'responsibility to protect' has also been proposed in application to natural disasters. The emergence of state practice in observing certain responsibilities "before, during, and after natural disasters occur" in the absence of obligations to do so supports an emerging responsibility to protect in the context of natural disaster, and sources of human rights law are to be used in promoting this doctrine (Saechao, 2007).

7.4.2. International Finance

The UNFCCC recognizes that in addition to the need to mitigate emissions of greenhouse gases and adapt to climate change, there is a responsibility on developed countries to support developing countries in this process (see Article 4.4 in Box 7-1). A starting point for the delivery of adaptation finance is the assessment of adaptation finance needs, which have also been interpreted as a proxy for adaptation costs (see Section 4.5). The UNFCCC (2007b) estimated the additional investment and financial flows needed worldwide to be US$ 48 to 171 billion in 2030 (or US$ 60 to 193 billion when also considering current investment needs for ecosystem adaptation). Some US$ 28 to 67 billion of this amount would be needed in developing countries (UNFCCC, 2007b). The largest uncertainty in these estimates is in the cost of adapting infrastructure, which may require anything between US$ 8 and 130 billion in 2030, one-third of which would be for developing countries. The UNFCCC (2007b) also estimated that an additional amount of about US$ 41 billion would be needed for agriculture, water, health, and coastal zone protection, most of which would be used in developing countries. Other studies providing estimates of the annual incremental costs of adaptation in developing countries include those by the World Bank (2006), Stern (2007), Oxfam International (2007), Watkins (2007), and the World Bank (2010b). These estimates are shown in Table 7-1, and discussed in more detail in Parry et al. (2009) and Fankhauser (2010).

While these different estimates highlight the high level of uncertainty, there appears to be consensus that global adaptation costs will total tens of billions of US dollars per year in developing countries. A review by the Organisation for Economic Co-operation and Development (OECD) of the estimates mentioned above found that there is very little quantified information on the costs of adaptation in developing countries, and most studies are constrained to a few sectors within countries (mostly coastal zones and, to a lesser extent, water, agriculture, and health) (Agrawala and Fankhauser, 2008). In addition, these studies assume relatively crude relationships and make strong assumptions, such as perfect foresight and high levels of autonomous adaptation. Almost no cross-sector studies have examined cumulative effects within countries, and only a handful of studies have investigated the wider macroeconomic consequences of impacts or adaptation. Moreover, most of the literature only considers adaptation to average changes in temperature or sea level rise. Little attention has been paid to more

Table 7-1 | Estimated annual adaptation costs and finance needs in developing countries.

	Assessment Year	US$ (Billion)	Time Frame
World Bank	2006	9 - 41	Present
Stern	2006	4 - 37	Present
Oxfam	2007	> 50	Present
UNDP	2007	86 - 109	2015
UNFCCC	2007	28 - 67	2030
World Bank	2010	70 - 100	2010 - 2050

Sources: World Bank, 2006, 2010b; Stern, 2007; Oxfam International, 2007; UNFCCC, 2007b; Watkins, 2007.

abrupt changes in mean conditions or to changes in the frequency and magnitude of extreme events (Agrawala and Fankhauser, 2008).

According to Agrawala and Fankhauser (2008), the consensus on global adaptation costs, even in order of magnitude terms, may therefore be premature. In addition, in most cases the estimates are neither attributed to specific adaptation activities, nor do they articulate the benefits of adaptation investment. Double counting between sectors and scaling up to global levels from very limited (and often local) source material limit utility. At the same time, a point also noted by Parry et al. (2009), many sectors and adaptations have not been included in the estimates.

In addition to these global estimates, total adaptation finance needs can also be assessed by aggregating national estimates, although this is hampered by the absence of a common method to make such estimates, and the fact that they are not available for all countries. The NAPAs (see Section 7.3.2 and Chapter 6), which have now been completed by most LDCs, are the most extensive effort to date to assess adaptation priorities and finance needs in developing countries. The cumulative cost of projects prioritized to respond to urgent and immediate adaptation needs is approximately US$ 1,660 million for the 43 countries that had completed their NAPAs by September 2009 (UNFCCC, 2010b). The divergence from the global estimates mentioned above can be explained by several factors: they cover only 43 LDCs, they include only prioritized projects, and they consider only urgent and immediate adaptation needs, not medium- to long-term needs (Persson et al., 2009).

A challenge for the international community is how to meet the adaptation finance needs that have been identified. The GEF operates the LDCF and SCCF, to provide funding to eligible developing countries to meet the 'additional' or 'incremental' costs of adaptation; the baseline costs of a project or program are borne by the recipient country, by other bilateral or multilateral donors, or both. The LDCF and SCCF rely on voluntary contributions from developed countries. As of May 2010, US$ 315 million had been pledged for adaptation under these two funds (US$ 221 million to the LDCF and US$ 94 million to the SCCF); of this amount, US$ 220 million has been allocated (US$ 135 million from the LDCF and US$ 85 million from the SCCF) (GEF, 2010a). In addition, the GEF has allocated all US$ 50 million it had made available to the SPA (GEF, 2008; see also Klein and Möhner, 2009).

The Adaptation Fund, which became operational in 2009, is operated by a special Adaptation Fund Board. It is the first financial instrument under the UNFCCC and its Kyoto Protocol that is not based solely on voluntary contributions from developed countries. It receives a 2% share of proceeds from project activities under the Clean Development Mechanism (CDM), but can also receive funds from other sources to fund concrete adaptation projects and programs (Persson et al., 2009). The actual amount of money that will be available from the Adaptation Fund depends on the extent to which the CDM is used and on the price of carbon. As of October 2010, the Adaptation Fund had received US$ 202.09 million, of which US$ 130.55 million was generated through CDM activities. Estimates of potential resources available for the Adaptation Fund from 31 October 2010 to 31 December 2012 range from US$ 288.4 million to US$ 401.5 million (Adaptation Fund, 2010).

While the GEF-managed funds have supported adaptation activities in some 80 countries (Persson et al., 2009), there has been criticism, particularly from developing countries, on how the funds are being managed (e.g., Mitchell et al., 2008; Klein and Möhner, 2009; Ministry of Foreign Affairs of Denmark and GEF Evaluation Office, 2009). In addition, concern has been voiced about the predictability and adequacy of funds, and the perceived equity and fairness of decisionmaking (Mace, 2005; Paavola and Adger, 2006; Müller, 2007; Persson et al., 2009). The GEF has acknowledged the criticism and indicated in reports to the COP how it is responding to it (GEF, 2009, 2010b). At the same time, developed countries have raised concern about fiduciary risks in some developing countries, which would need to be addressed through improved accountability and transparency before program-based adaptation can be supported by international finance (Mitchell et al., 2008; GEF, 2010b). The Adaptation Fund has not been operational long enough to allow for such an assessment but the first signals are positive, particularly regarding its governance structure and the option of direct access (Czarnecki and Guilanpour, 2009; Brown et al., 2010; Grasso, 2010).

In addition to the funds operating within the context of the UNFCCC, money for adaptation is provided through several other channels, including developing countries' domestic national, sectoral, and local budgets; bilateral and multilateral development assistance; and private-sector investments. This makes for an adaptation financing landscape that is highly fragmented, resulting in a proliferation not only of funds but also of policies, rules, and procedures (Persson et al., 2009). But despite the proliferation of funds, the amount of money currently available falls substantially short of the adaptation finance needs presented above.

In light of this shortfall, the 2009 Copenhagen Accord was a milestone in international climate finance. It refers to a collective commitment for developed countries to provide "new and additional resources … approaching USD 30 billion" in 'fast start' money for the 2010-2012 period, balanced between adaptation and mitigation, and sets a longer-term collective goal of mobilizing US$ 100 billion per year by 2020 from all sources (public and private, bilateral and multilateral) (Bodansky, 2010). Although the Copenhagen Accord was not adopted by the COP, the collective commitment and longer-term goal are also part of the Cancun Agreements, which the COP adopted a year later. Parties agreed that "scaled-up, new and additional, predictable and adequate funding shall be provided to developing country Parties, taking into account the urgent and immediate needs of developing countries that are particularly vulnerable to the adverse effects of climate change." In the meantime, the High-level Advisory Group on Climate Change Financing, established by the UN Secretary-General, had analyzed the feasibility of mobilizing US$ 100 billion per year by 2020. It concluded that "it is challenging but feasible to meet that goal. Funding will need to come from a wide variety of sources, public and private, bilateral and multilateral, including alternative sources of finance, the scaling up of existing sources, and increased private flows. Grants and highly concessional loans are crucial for adaptation in the most vulnerable developing countries, such as the least developed countries, small island developing States and Africa" (AGF, 2010).

An open question is how climate finance might be linked with other international finance flows. The Bali Action Plan referred to "means to incentivize the implementation of adaptation actions on the basis of sustainable development policies" in its section on the provision of financial resources. The Copenhagen Accord did not discuss the link between adaptation and development, even though the issue of 'mainstreaming' – integrating adaptation to climate change into mainstream development planning and decisionmaking – was much debated in the pre-Copenhagen negotiations on adaptation finance (Persson et al., 2009; Klein, 2010). From an operational perspective, mainstreaming adaptation into development makes common sense: both contribute to enhancing human security, and opportunities to create synergies between the two are increasingly recognized and pursued (Gigli and Agrawala, 2007; Klein et al., 2007; Kok et al., 2008; Gupta and Van de Grijp, 2010). Besides, there is a range of activities that can be seen as contributing to both adaptation and development objectives (McGray et al., 2007).

But from a climate policy perspective, mainstreaming creates a dilemma (Persson and Klein, 2009; Klein, 2010). Financial flows for adaptation and those for development – for example, official development assistance (ODA) – are managed separately. One of the arguments in favor of mainstreamed adaptation is that it makes more efficient use of financial and human resources than adaptation that is designed, implemented, and managed as stand-alone activities (i.e., separately from ongoing development planning and decisionmaking). However, developing countries have expressed the concern that, as a result of donors seeking to create synergies between adaptation and development, finance for adaptation will not be new and additional but in effect will be absorbed into ODA budgets of a fixed size (Michaelowa and Michaelowa, 2007). The concern is fueled by the fact that the amount of money currently available for adaptation falls short of the estimated adaptation finance needs in developing countries. A second, related concern is that mainstreaming could divert any new and additional funds for adaptation into more general development activities, thus limiting the opportunity to evaluate, at least quantitatively, their benefits with respect to climate change specifically (Yamin, 2005). Third, there is concern that donors' use of ODA to pursue mainstreamed adaptation could impose conditionalities on what should be a country-driven process (Gupta et al., 2010).

As mentioned in Section 7.3.2, the Cancun Agreements established the Green Climate Fund as a new entity operating the financial mechanism under Article 11. The Green Climate Fund is not yet operational and it is too early to say how it might address the mainstreaming dilemma, or even how important it will be for climate adaptation in developing countries. All that can be said at this moment is that in the Cancun Agreements, Parties decided that "a significant share of new multilateral funding for adaptation should flow through the Green Climate Fund."

7.4.3. Technology Transfer and Cooperation

7.4.3.1. Technology and Climate Change Adaptation

Technologies receive prominent attention both in adaptation to emerging and future impacts of climate change as well as in mitigating current disasters. The sustainability, operation, and maintenance of technologies can be challenging in many developing countries due to lack of resources, human capacity, and cultural differences. Moreover, technology transfer is complex and requires capacity building as well as a client (technology user) focus as opposed to a developer (technology designer) focus (O'Brien et al., 2007). Intellectual property rights are rarely an issue in the availability and use of technologies for adaptation (Murphy, 2011) but when they are, adequate methods are needed that foster affordable deployment of new technologies but preserve the incentives for technology developers (Doig, 2008). While the importance of transferring technologies from developers/owners to would-be users is widely recognized, the bulk of the literature seems to address the issues at a rather generic level, without going into the details of what technologies for adaptation would need to be transferred in different impact sectors from where to where and via what mechanisms. Institutional, political, technological, economic, information, financial, cultural, legal, and participation and consultation obstacles can hinder the transfer of mitigation and adaptation technologies and concerted efforts are required to overcome those impediments (IEA, 2001). Private-public partnership as a policy instrument could well be a mechanism for transferring the required technologies for adaptation projects (Agrawala and Fankhauser, 2008). In the adaptation literature, publications addressing the transfer of technologies important for reducing vulnerability and increasing the ability to cope with weather-related disasters are even scarcer. This section reviews literature on technologies for adaptation and the issues involved in international technology transfer of such technologies.

The Special Report on *Methodological and Technological Issues in Technology Transfer* by the IPCC defines the term 'technology transfer' as a "broad set of processes covering the flows of know-how, experience and equipment for mitigating and adapting to climate change amongst different stakeholders such as governments, private sector entities, financial institutions, NGOs and research/education institutions" (IPCC, 2000, p. 3). The report uses a broad and inclusive term 'transfer' encompassing diffusion of technologies and technology cooperation across and within countries. It evaluates international as well as domestic technology transfer processes, barriers, and policies. This section focuses on the international aspects.

Adaptation to climate change involves more than merely the application of a particular technology (Klein et al., 2005). Adaptation measures include increasing robustness of infrastructural designs and long-term investments, increasing flexibility of vulnerable managed systems, enhancing adaptability of natural systems, reversing trends that increase vulnerability, and improving societal risk awareness and preparedness. In the case of disasters related to extreme weather events, anticipatory

Box 7-3 | Examples of Technologies for Adaptation in Asia

In Asia, adaptation to climate change, variability, and extreme events at the community level are small scale and concentrate mainly on agriculture, water, and disaster amelioration (Alam et al., 2007). They focus on the livelihood of affected communities, raise awareness to change practices, diversify agriculture, and promote water conservation. For example, Saudi Arabia has already built 215 dams for water storage and 30 desalination plants, passed water protection and conservation laws, and initiated leakage detection and control schemes as well as advanced irrigation water conservation schemes and a system for modified water pumping as part of its climate change adaptation program (Alam et al., 2007). In India, a combination of traditional and innovative technological approaches is used to manage drought risk. Technological management of drought (e.g., development and use of drought tolerant cultivars, shifting cropping seasons in agriculture, flood and drought control techniques in water management) is combined with model-based seasonal and annual to decadal forecasts. Model results are translated into early warning in order to take appropriate drought protection measures (Alam et al., 2007). In China, adaptation technologies have been widely used for flood disaster mitigation (Alam et al., 2007). Another example is related to the Philippines where a typhoon in 1987 completely destroyed over 200,000 homes. The Department of Social Welfare and Development initiated a program of providing typhoon-resistant housing for the population in the most typhoon-prone areas (Diacon, 1992). The so-called Core Shelter houses have typhoon resistant features and can endure wind speeds up to 180 km hr^{-1}. The technology was proved to be successful by providing the required protection and was adopted recently in regions stricken by a landslide (Government of the Philippines, 2008) and typhoons (Government of the Philippines, 2010), partly financed by UNDP.

adaptation is more effective and less costly than emergency measures and retrofitting, and immediate benefits can be gained from better adaptation to climate variability and extreme events. Some factors that determine adaptive capacity of human systems are the level of economic wealth, access to technology, information, knowledge and skills, and existence of institutions, infrastructure, and social capital (Smit et al., 2001; Christoplos et al., 2009).

An extensive list of 'soft' options that are vital to building capacity to cope with climatic hazards with references to publications that either describe the technology in detail or provide examples of its application is available (Klein et al., 2000, 2005). For example, the applications in coastal system adaptation include various types of geospatial information technologies such as mapping and surveying, videography, airborne laser scanning (lidar), satellite and airborne remote sensing, global positioning systems in addition to tide gauges and historical and geological methods. These technologies help formulate adaptation strategies (protection versus retreat), implement the selected strategy (design, construction, and operation), and provide early warning (UNFCCC, 2006a). Another set of examples includes technologies to protect against sea level rise: dikes, levees, floodwalls, seawalls, revetments, bulkheads, groynes, detached breakwaters, floodgates, tidal barriers, and saltwater intrusion barriers among the hard structural options, and periodic beach nourishment, dune restoration and creation, and wetland restoration and creation as examples of soft structural options (Klein et al., 2000, 2005). A combination of these technologies selected on the basis of local conditions constitutes the protection against extreme events in coastal regions. Structural measures are localized solutions and there is a need for localized information such as their environmental and hydrologic impacts. In addition, there are a series of indigenous options (flood and drought management) that might be valuable in regions to be affected by similar events (Klein et al., 2005, p. 19). It is also important to integrate technology transfer efforts for CCA and DRR needs with sustainable development efforts to avoid conflicts and foster synergies between them (Hope Sr., 1996; Sanusi, 2005). Adaptation is normally assumed to be benign for development but Eriksen and Brown (2011) challenge this assumption, arguing that there is emerging evidence that adaptation measures run counter to principles of sustainable development, as both social equality and environmental integrity can be threatened. Placing responses to extreme events into the larger context of other societal and environmental changes will be vital for sustainable development (Yohe et al., 2007; Eriksen et al., 2011).

A report by the UNFCCC (2006a) summarizes the technology needs identified by Parties not included in Annex I to the Convention. Curiously, only one country mentioned 'potential for adaptation' among the commonly used criteria for prioritizing technology needs. Among 30 technologies listed in the report, the technology needs relevant for coping and adapting to weather extremes include, for example, improved drainage, emergency planning, raising buildings and land, and protecting against sea level rise. Many of these are good examples of measures that link DRR and CCA objectives, namely to reduce overall ecological and social vulnerability. Another UNFCCC report (2006b) observes that, unlike those for mitigation, the forms of technology for adaptation are often rather familiar. Many have been used over generations in coping with floods, for example, by building houses on stilts or by cultivating floating vegetable plots. Some other types of technologies draw on new developments in, for example, advanced materials science and satellite remote sensing (see Box 7-3). The UNFCCC report (2006b) provides an overview of the old and new technologies available in adapting to changing environments, including climate change. The Disaster Reduction Hyperbase in Asia is a web-based collection of new and traditional indigenous technologies relevant to DRM that also promotes communication among developing and industrial countries (Kameda, 2007).

7.4.3.2. Technologies for Extreme Events

Approaching the issues of technologies to foster adaptation to extreme weather events and their impacts from the direction of disaster mitigation, Sahu (2009) presents an overview of diverse technologies that might be applied in various stages of disaster management. The list of technologies for adaptation to weather-related extreme events includes early warning and disaster preparedness; search and rescue for disaster survivors; water supply, purification, and treatment; food supply, storage, and safety; energy and electricity supply; medicine and healthcare for disaster victims; disease surveillance; sanitation and waste management; and disaster-resistant housing and construction (Sahu, 2009).

Developing wind-resistant building technologies is crucial for reducing vulnerability to high-wind conditions like storms, hurricanes, and tornadoes. A report by the International Hurricane Research Centre presents hurricane loss reduction devices and techniques (IHRC, 2006). The Wall of Wind testing apparatus (multi-fan systems that generate up to 209 km hr^{-1} winds and include water-injection and debris-propulsion systems with sufficient wind field sizes to test the construction of small single-story buildings) will improve the understanding of the failure mode of buildings and hence lead to technologies and products to mitigate hurricane impacts (Fugate and Crist, 2008).

An absolutely crucial aspect of managing weather extremes both under the present and future climate regime is the ability to forecast and provide early warning. Downscaling projections from global climate models could provide useful information about the changing risks. It is important to note that really useful early warning systems would provide multi-hazard warning and warnings on vulnerability development to the extent it is possible. Satellite and aerial monitoring, meteorological models, and computer tools including geographic information systems (GIS) as well as local and regional communication systems are the most essential technical components. (The focus on technology here does not negate the importance of social and communication aspects of early warning.) The use of GIS in the support of emergency operations in the case of both weather and non-weather disasters is becoming increasingly important in the United States. The benefits of using GIS technologies include informing the public, enabling officials to make smarter decisions, and facilitating first-responder efforts to effectively locate and rescue storm victims (NASCIO, 2006). Lack of locally useable climate change information about projected changes in extreme weather events remains an important constraint in managing weather-related disasters, especially in developing countries. Therefore there is a need to develop regional mechanisms to support in developing and delivering downscaling techniques and tools (see Section 3.2.3 for details on downscaling regional climate models) and transferring them to developing countries.

Space technologies (such as Earth observation, satellite imagery, real-time application of space sensors, mapping) are important in the reduction of disasters, including extreme weather and climate events such as drought, flood, and storms (Rukieh and Koudmani, 2006). These technologies can be particularly helpful in the risk assessment, mitigation, and preparedness phases of disaster management by identifying risk-prone areas, establishing zoning restrictions and escape routes, etc. Space technologies are important for early warning and in managing the effects of disasters. For incorporating the routine use of space technology-based solutions in developing countries, there is a need to increase awareness, build national capacity, and also develop solutions that are customized and appropriate to their needs (Rukieh and Koudmani, 2006). A good example of the application of space technology at international scales and for early warning is the joint initiative of WMO, the National Oceanic and Atmospheric Administration, the US Agency for International Development, and the Hydrologic Research Center on global flash flood guidance. The system uses global data produced by a global center and downscales the global information to regional products that are sent to national entities for further downscaling at the national level and then disseminated to users and communities (WMO, 2007, 2010). It is also important to note that there are existing capabilities within some particularly exposed developing countries (such as India, Bangladesh, China, Philippines) with well-developed remote-sensing capabilities of their own, or existing arrangements with other space agency suppliers.

Support for risk reduction and relief agencies and governments depends, among other factors, on timely availability of information about the scale and nature of these disasters (Holdaway, 2001). Currently, ground-based sources provide most of such information. This could be improved by using input from space-based sensor systems, both for disaster warning and disaster monitoring where the scale of devastation cannot adequately be monitored from ground-based information sources alone. A global space-based monitoring and information system, with the associated ability to provide advanced warning of many types of hazards, can be combined with the latest developments in sensor technology (optical, infrared, radar) including a UK initiative on high-resolution imaging from a microsatellite (Holdaway, 2001). The literature suggests that transferring these technologies and the related know-how will be important for building capacities in CCA and DRR in countries where they are still missing (*medium certainty, limited evidence*).

Microsatellites (low weights and small sizes, just under or well below 500 kg) are seen as an important technology for the detection of and preparation for weather-related hazards in other countries as well. Shimizu (2008) emphasizes the importance of international cooperation in this area. He observes that only a few countries are able to develop large rockets and satellites and launch them from their own territories. Several Asian countries have been cooperating with OECD countries to develop small Earth observation satellites, like DAICHI (Advanced Land Observing Satellite) and WINDS (Wideband Internetworking engineering test and demonstration satellite) that include both optical and microwave sensors. DAICHI operated between 2006 and 2011 based on cooperation of Asian countries with the United States and the European Union and made an important contribution to emergency observations of regions hit by major disasters in this period (JAXA, 2011).

Mitigation of adverse cyclone impacts involves reliable tropical cyclone forecasting and warnings, and efficient ways to convey the information to stakeholders, users, and the general public (Lee et al., 2006). It is important that NMHSs take advantage of the advances in communication technology such as wireless broadband access, Global Positioning System (GPS), and GIS to enhance the relevance and effectiveness of warnings, options, and backup capabilities to disseminate warnings through multiple and diverse channels (Lee et al., 2006). Natural hazards management has advanced to address a major challenge: turning real-time data provided by new technologies (e.g., satellite- and ground-based sensors and instruments) into information products to help people make better decisions about their own safety and prosperity (Groat, 2004).

The literature about technology transfer to foster adaptation to changes in extreme events induced by climate change is very limited. It was necessary to broaden the scope of the literature review and embrace climate change adaptation in general in order to gain lessons about the processes, channels, stakeholders, and barriers of technology transfer. In addition, useful insights were inferred from the literature on technology transfer to support climate change mitigation, disaster risk reduction (prevention, mitigation, and preparedness), and other related areas. The DRR literature on technology development and transfer documents the expanding international cooperation in forecasting and monitoring extreme weather events by collecting and disseminating satellite-based information and the international transfer of know-how to interpret it. There is increasing emphasis on the importance of establishing close linkages across all EWS components ranging from collection of hydro-meteorological data, forecasting how nature will respond (e.g., weather or flood forecasting), to communicating information (or warnings) to decisionmakers (sectoral users or communities) (*medium agreement, limited evidence*).

7.4.3.3. Financing Technology Transfer

Climate change mitigation has been the primary focus of the financing mechanisms and innovative financing in recent years. In contrast, the transfer of technologies for adaptation is hampered by insufficient incentive regimes, increased risks, and high transaction costs (Klein et al., 2005). Yet the lessons from the transfer of mitigation technologies are relevant for adaptation: results of the penetration of energy and industrial technologies in the developing countries depend on many factors ranging from labor skills, market conditions, achieved level of technological development, the reliability of basic services (electricity and water), availability of spare parts, etc. A combination of interrelated socioeconomic, institutional, and governance issues would often determine the success or failure of technology transfer, rather than the technologies themselves (Klein et al., 2005, p. 23). These factors are also important in transferring technologies for adaptation because they determine the feasibility and efficiency of adopting the transferred technologies (e.g., regulations to build and install them, skilled labor, water and electricity to operate them).

UNFCCC (2005) addresses the transfer of environmentally sound technologies for adaptation to climate change: the needs for and the identification and evaluation of technologies for adaptation to climate change, and financing their transfer. Cost is one of the main barriers in technology transfer; therefore innovative financing for the development and transfer of technologies is needed. Potential sources of funding for technology transfer include bilateral activities of Parties, multilateral activities such as the GEF, the World Bank, or regional banks, the SCCF, the LDCF, financial flows generated by Joint Implementation and CDM projects, and the private sector (see also Section 7.3.3.3). The GEF funds for adaptation activities include the SPA trust fund, the LDCF, and the SCCF. In addition, the GEF is providing secretariat services to the Adaptation Fund Board under the Kyoto Protocol (see also Section 7.4.2).

Climate variability is already a major impediment to development and 2% of World Bank funds are devoted to disaster reconstruction and recovery (World Bank, 2008). In order to use available funds efficiently, the World Bank (2009) developed the screening tool ADAPT (Assessment & Design for Adaptation to Climate Change: A Prototype Tool), a software-based tool for assessing development projects for potential sensitivities to climate change. The tool combines climate databases and expert assessments of the threats and opportunities arising from climate variability and change. As of 2011, the knowledge areas covered by the tool cover agriculture and irrigation in India and sub-Saharan Africa and, for all regions, various aspects of biodiversity and natural resources.

Both conventional and innovative options for financing the transfer of technologies for adaptation might be explored. As conventional options, the GEF funds (SPA, LDCF, and SCCF) provide opportunities for accessing financial resources that could be used for deployment, diffusion, and transfer of technologies for adaptation, including initiatives on capacity building, partnerships, and information sharing. Projects identified in technology needs assessments could also be implemented using these financial opportunities. Based on these experiences as well as on special needs of groups of countries such as small island developing states and LDCs, further guidance could be provided to the GEF on funding technologies for adaptation. In addition, there is an opportunity to explore innovative financing mechanisms that can promote, facilitate, and support increased investment in technologies for adaptation (UNFCCC, 2005).

Concerning financing of technological development and transfer, a report by the Expert Group on Technology Transfer (UNFCCC, 2009a) classifies technologies by stage of maturity, the source of financing (public or private sector), and whether they are under or outside the UNFCCC and estimates the financing resources currently available for technology research, development, deployment, diffusion, and transfer. The estimates for financing mitigation technologies are between US$ 70 and 165 billion per year. In the adaptation area, the report claims that research and development is focused on tailoring technologies to specific sites and applications and thus the related expenditures become part of the project costs. Current spending on adaptation projects in developing countries is about US$ 1 billion per year (UNFCCC, 2009a).

The literature clearly shows that the transfer of technologies for adaptation lags behind the transfer of mitigation technologies in terms of the scales of attention and funding. Funding transfer and funding mechanisms for technologies that help reduce vulnerability to climate variability, particularly to weather-related extreme events, appear to be as important for both CCA and DRR (*high confidence*).

7.4.4. Risk Sharing and Transfer

This section examines the current and potential role of the international community – international financial institutions, NGOs, development organizations, private market actors, and the emerging adaptation community – in enabling access to insurance and other financial instruments that share and transfer risks of extreme weather. The international transfer and sharing of risk is an opportunity for individuals and governments of all countries that cannot sufficiently diversify their portfolio of weather risk internally, and especially (as discussed in Section 6.3.3) for governments of vulnerable countries that do not wish to rely on ad hoc and often insufficient post-disaster assistance.

Experience shows that the international community can play a role in enabling individual, national, and international risk-sharing and transfer strategies (*high confidence*). The following discussion identifies successful practices, or value added, as well as constraints on this role.

7.4.4.1. International Risk Sharing and Transfer

Risk transfer (usually with payment) and risk sharing (usually informal with no payment) are recognized by the international community as an integral part of DRM and CCA (see Case Study 9.2.13 for definitions). The 2005 HFA calls on the disaster community "to promote the development of financial risk-sharing mechanisms, particularly insurance and reinsurance against disasters" (UNISDR, 2005a, p. 11). Similarly, the 2007 Bali Action Plan calls for consideration of risk-sharing and transfer mechanisms as a means for enhancing adaptation (UNFCCC, 2007a). The Plan builds on the mandate to consider insurance as set out by Article 4.8 of the UNFCCC and Article 3.14 of the Kyoto Protocol.

Often by necessity risk sharing and transfer are international. Local and national pooling arrangements (discussed in Sections 5.5.2 and 6.3.3) may not be viable for statistically dependent (co-variant) risks that cannot be sufficiently diversified. A single event can cause simultaneous losses to many insured assets, violating the underlying insurance principle of diversification. For this reason, primary insurers, individuals, and governments (particularly in small countries) rely on risk-sharing and transfer instruments that diversify their risks regionally and even globally. A few examples can serve to illustrate international arrangements:

- A government receives international emergency assistance and loans after a major disaster.
- A family locates a relative in a distant country who provides post-disaster relief through remittances.
- After a major disaster, a farm household takes out a loan from an internationally backed micro-lending institution.
- An insurer purchases reinsurance from a private reinsurance company, which spreads these risks to its international shareholders.
- A government issues a catastrophe bond, which transfers risks directly to the international capital markets.
- Many small countries form a catastrophe insurance pool, which diversifies risks and better enables them to purchase reinsurance.

Not only are these financial arrangements international in character, but many are supported by the international development and climate adaptation communities (see, especially, UNISDR, 2005b; UNFCCC, 2009b). At the outset it is important to point out that these instruments cannot stand alone but must be viewed as part of a risk management strategy, for which cost-effective risk reduction is a priority.

7.4.4.2. International Risk-Sharing and Transfer Mechanisms

This section reviews international mechanisms for sharing and transferring risk, including remittances, post-disaster credit, insurance and reinsurance, alternative insurance mechanisms, and regional pooling arrangements.

7.4.4.2.1. Remittances

Remittances – transfers of money from foreign workers or expatriate communities to their home countries – make up a large part of informal risk sharing and transfer, even exceeding official development aid flows. In 2010, the official worldwide flow of remittances was estimated at US$ 325 billion, and unrecorded flows may add another 50% or more. In some cases, remittances can be as large as one-third of the recipient country's gross domestic product (World Bank, 2011b).

A number of studies show that remittances increase substantially following disasters, often exceeding post-disaster donor assistance (Lucas and Stark, 1985; Miller and Paulson, 2007; Yang and Choi, 2007; Mohapatra et al., 2009). Payments can be sent through professional money transfer organizations, but often these channels break down and remittances are carried by hand (Savage and Harvey, 2007). While simple in concept, remittances can be complicated by associated transfer fees. A survey carried out in the United Kingdom found that for an average-sized transfer, the associated costs could vary between 2.5 and 40% (DFID, 2005). Information pertinent to the transfer is often obscure or in an unfamiliar language, and transfers across some borders have been complicated due to initiatives taken by developed nations to counter international money laundering and terrorism financing (Fagen and Bump, 2006). Finally, a major problem is difficulties in communicating with relatives abroad, as well as the high potential of losing necessary documents in a disaster.

The international community has been active in reducing the costs and barriers to post-disaster remittances. DFID, among other development organizations, supports financial inclusion policies including mobile banking and special savings accounts earmarked for disaster recovery that will greatly reduce transaction costs. High-tech proposals for assuring security have included biometric identification cards and retina scanners as forms of identification (DFID, 2005; Pickens et al., 2009).

7.4.4.2.2. Post-disaster credit

One of the most important post-disaster financing mechanisms, credit provides governments and individuals with resources after a disaster, yet with an obligation to repay at a later time. Governments and individuals of highly vulnerable countries, however, can have difficulties borrowing from commercial lenders in the post-disaster context. Since the early 1980s, the World Bank has thus initiated over 500 loans for recovery and reconstruction with a total disbursement of more than US$ 40 billion (World Bank, 2006), and the Asian Development Bank also reports large loans for this purpose (Arriens and Benson, 1999). With the growing importance of pre-disaster planning, a recent innovation on the part of international organizations is to make pre-disaster contingent loan arrangements – for example, the World Bank's catastrophe deferred drawdown option, which disburses quickly after the government declares an emergency (World Bank, 2008).

For micro-finance institutions (MFIs), post-disaster lending has associated risks given increased demand that tempts relaxed loan conditions or even debt pardoning. This risk is particularly acute in vulnerable regions. Recognizing the need for a risk transfer instrument to help MFIs remain solvent in the post-disaster period, the Swiss State Secretariat for Economic Affairs (SECO) and the IADB, as well as private investors, created the Emergency Liquidity Facility (ELF) (UNFCCC, 2008). Located in Costa Rica, ELF provides needed and immediate post-disaster liquidity at low rates to MFIs across the region.

7.4.4.2.3. Insurance and reinsurance

Insurance is an instrument for distributing disaster losses among a pool of at-risk households, farms, businesses, and/or governments, and is the most recognized form of international risk transfer. The insured share of property losses from extreme weather events has risen from a negligible level in the 1950s to approximately 20% of the total in 2007 (Mills, 2007).

Insurance and reinsurance markets attract capital from international investors, making insurance an instrument for transferring disaster risks over the globe. The market is highly international in character, yet uneven in its cover. In the period 2000 to 2005, for example, US insurers purchased reinsurance annually from more than 2,000 different non-US reinsurers (Cummins and Mahul, 2009, p. 115). From 1980 through 2003, insurance covered 4% of total losses from climate-related disasters (estimated at about US$ 1 trillion) in developing countries compared to 40% in high-income countries (Munich Re, 2003).

The international community is playing an active role in enabling insurance in developing countries, particularly by supporting micro- and sovereign (macro) insurance initiatives. The following four examples illustrate this role:

- The World Bank and World Food Programme provided essential technical assistance and support for establishing the Malawi pilot micro-insurance program (see discussion in Section 5.5.2), which provides index-based drought insurance to smallholder farmers (Hess and Syroka, 2005; Suarez et al., 2007).
- The Mongolian government and World Bank support the Mongolian Index-Based Livestock Insurance Program (see Section 5.5.2) by absorbing the losses from very infrequent extreme events (over 30% animal mortality) and providing a contingent debt arrangement to back this commitment, respectively (Skees and Enkh-Amgalan, 2002; Skees et al., 2008).
- The World Food Programme successfully obtained an insurance contract through a Paris-based reinsurer to provide insurance to the Ethiopian government, which assures capital for relief efforts in the case of extreme drought (Hess, 2007).
- The governments of Bermuda, Canada, France, and the United Kingdom, as well as the Caribbean Development Bank and the World Bank, have recently pledged substantial contributions to provide start-up capital for the Caribbean Catastrophe Risk Insurance Facility (discussed in Section 7.4.4.2.5) (Cummins and Mahul, 2009).

These early initiatives, especially micro-insurance schemes, are showing promise in reaching the most vulnerable, but also demonstrate significant challenges to scaling up current operations. Lack of data, regulation, trust, and knowledge about insurance, as well as high transaction costs, are some of the barriers (Hellmuth et al., 2009).

As discussed in Case Study 9.2.13, insurance and other risk transfer instruments can promote DRR and CCA in multiple ways by providing the means to finance recovery, thus reducing long-term losses; adding to knowledge about risks; creating incentives (and imperatives) for risk reduction; and providing the safety net necessary for farms and businesses to take on cost-effective, yet risky, investments that reduce their vulnerability to climate change (Linnerooth-Bayer et al., 2009; Warner et al., 2009b).

7.4.4.2.4. Alternative insurance instruments

Alternative insurance-like instruments, sometimes referred to as risk-linked securities, are financing devices that enable risk to be sold in international capital markets. Given the enormity of these markets, there is a large potential for alternative or non-traditional risk financing, including catastrophic risk (CAT) bonds (explained below, and in Section 6.3.3 and Case Study 9.2.13), industry loss warranties, sidecars (a company purchases a portion or all of an insurance policy to share in the profits and risks), and catastrophic equity puts, all of which are playing an increasingly important role in providing risk finance for large-loss events. A discussion of these instruments goes beyond the scope of this chapter, but it is worth drawing attention to the most prominent risk-linked security, the CAT bond, which is a fully collateralized instrument whereby the investor receives an above-market return when a specific natural hazard event does not occur (e.g., a Category 4 hurricane or greater), but shares the insurer's or government's losses by sacrificing interest or principal following the event if it does occur.

Over 90% of CAT bonds are issued by insurers and reinsurers in developed countries. Although it is still an experimental market, CAT bond placements more than doubled between 2005 and 2006, with a peak at US$ 4.7 billion in 2006 (Cummins and Mahul, 2009), but declining to US$ 3.4 billion in 2009 (Munich Re, 2010).

In 2006 and 2009, the first government-issued disaster relief CAT bond placements were executed by Swiss Re and Deutsche Bank Securities to provide funds to Mexico to insure its catastrophe fund FONDEN against earthquake and (in 2009) hurricane risk, and thus to defray costs of disaster recovery and relief (Cardenas et al., 2007). The World Bank provided technical assistance for these transactions. Although the transaction costs of the Mexican CAT bond were large, and basis risk (the risk that the bond trigger will not be highly correlated with losses) is a further impediment to their success, it is expected that this form of risk transfer will become increasingly attractive especially to highly exposed developing country governments (Lane, 2004). As discussed in Chapter 6, a large number of government treasuries are vulnerable to catastrophic risks, and post-disaster financing strategies generally have high opportunity costs for developing countries.

International and donor organizations have played an important role in another case of sovereign risk transfer (discussed in Section 9.2.13). In 2006, the World Food Programme purchased an index-based insurance instrument to support the Ethiopian government-sponsored Productive Safety Net Programme, which provides immediate cash payments in the case of food emergencies. While this transaction relied on traditional reinsurance instruments, there is current interest in issuing a CAT bond for this same purpose. Tomasini and Van Wassenhove (2009) note the important role that securitized instruments can play in providing backup for humanitarian aid when disasters strike.

7.4.4.2.5. Regional risk pools

Regional catastrophe insurance pools are a promising innovation that can enable highly vulnerable countries, and especially small states, to more affordably transfer their risks internationally. By amalgamating risks across individual countries or regions and accumulating reserves over time, catastrophe insurance pools generate diversification benefits that can eventually reduce insurance premiums. There is also growing empirical evidence that catastrophe insurance pools have been able to diversify intertemporally and thus dampen the volatility of the reinsurance pricing cycle and offer secure premiums to the insured governments (Cummins and Mahul, 2009).

As a recent example (discussed in Section 6.3.3 and Case Study 9.2.13), the Caribbean Catastrophe Risk Insurance Facility (CCRIF) was established in 2007 to provide Caribbean Community governments with an insurance instrument at a significantly lower cost (about 50% reduction) than if they were to purchase insurance separately in the financial markets. Governments of 16 island states contributed resources commensurate with their exposure to earthquakes and hurricanes, and claims will be paid depending on an index for hurricanes (wind speed) and earthquakes (ground shaking). Early cash payments received after an event will help to mitigate the typical post-disaster liquidity crunch (Ghesquiere et al., 2006; World Bank, 2007a,b).

7.4.4.3. Value Added by International Interventions

International financial institutions, donors, and other international actors have played a strongly catalytic role in the development of catastrophic risk-financing solutions in vulnerable countries, most notably by:
- *Exercising convening power*, for example, the World Bank coordinated the development of the CCRIF (Cummins and Mahul, 2009)
- *Supporting public goods* for development of risk market infrastructure, for example, donors might consider funding the weather stations necessary for index-based weather derivatives
- *Providing technical assistance*, for example, the World Food Programme carried out risk assessments and provided other assistance to support the Ethiopian sovereign risk transfer (Hess, 2007), and the World Bank provided technical assistance for the Mexican CAT bond (Cardenas et al., 2007)
- *Enabling markets*, for example, DFID is active in creating the legal and regulatory environment to facilitate access to banking services, which, in turn, greatly expedite remittances (DFID, 2005; Pickens et al., 2009)
- *Financing risk transfer*, as examples, the Bill Gates Foundation subsidizes micro-insurance in Ethiopia (Suarez and Linnerooth-Bayer, 2010); the World Bank provides low-cost capital backing for the Mongolian micro-insurance program (Skees et al., 2008); the Swiss SECO and the IADB provide low-interest credit to the ELF (UNFCCC, 2008); and many countries have contributed to the CCRIF reserve fund (Cummins and Mahul, 2009).

Though only a few of many examples of involvement by the international community in risk-sharing and transfer projects, they show that international financial institutions and development/donor organizations can assist and enable risk-sharing and transfer initiatives in diverse ways, which raises the question of their value added. Largely uncontested is the value of creating the institutional conditions necessary for community-based risk sharing and market-based risk transfer, yet, direct financing, especially of insurance, is controversial. Critics point to the 'economic efficiency principle' discussed in Section 7.2.2, and argue that public and international support, especially in the form of premium subsidies, can distort the price signal and weaken incentives for taking preventive measures, thus perpetuating vulnerability. Supporters point to the 'solidarity principle' discussed in Section 7.2.3 and the important role that solidarity has played in the social systems of the developed world (Linnerooth-Bayer and Mechler, 2008). Other types of assistance, like providing reinsurance to small insurers, can crowd out the (emerging) role of the private market. Finally, critics point out that it may be more efficient to provide the poor with cash grants than to subsidize insurance (Skees, 2001; Gurenko, 2004).

Recognizing these concerns, there may be important and valid reasons for interfering in catastrophe insurance and other risk-financing markets in specific contexts (see discussions by Cummins and Mahul, 2009; Linnerooth-Bayer et al., 2010), especially if:

- The private market is non-existent or embryonic, in which case enabling support (e.g., to improve governance, regulatory institutions, as well as knowledge creation) may be helpful.
- The private market does not function properly, in particular, if premiums greatly exceed the actuarially fair market price due, for example, to limitations on private capital and the uncertainty and ambiguity about the frequency and severity of future losses (Kunreuther and Michel-Kerjan, 2009). In this case economically justified premiums that are lower than those charged by the imperfect private market may be appropriate (Froot, 1999; Cutler and Zeckhauser, 2000).
- The target population cannot afford sufficient insurance coverage, in which case financial support that does not appreciably distort incentives may be called for. The designers of the Mongolian program, for example, argue that subsidizing the 'upper layer' is less price-distorting than subsidizing lower layers of risk because the market may fail to provide insurance for this layer (Skees et. al., 2008).
- The alternative is providing 'free' aid after the disaster happens.

7.4.5. Knowledge Acquisition, Management, and Dissemination

A close integration of DRR and CCA and their mainstreaming into sustainable development agendas for managing risks across scales calls for multiple ways of knowledge acquisition and development, management, sharing, and dissemination at all levels. Knowledge on the level of exposure to hazards and vulnerabilities across temporal and geographical scales (Louhisuo et al., 2007; Heltberg et al., 2008; Kaklauskas et al., 2009); the legal aspects of DRM and CCA; financing mechanisms at different scales; and information on access to appropriate technologies and risk-sharing and transfer mechanisms for disaster risk reduction (see Sections 7.4.1-7.4.4) are key to integrated risk management. Collaboration among scientists of different disciplines, practitioners, policymakers, and the public is pertinent in knowledge acquisition, management, and accessibility (Thomalla et al., 2006). The type, level of detail, and ways of generation and dissemination of knowledge will also vary across scales, that is, from the local level where participatory approaches are used to incorporate indigenous knowledge and build collective ownership of knowledge generated, to national and broader regional to international levels, thus upholding the principle of subsidiarity in the organization, sharing, and dissemination of information on disaster risk management (Marincioni, 2007; Chagutah, 2009).

An internationally agreed mechanism for acquisition, storage and retrieval, and sharing of integrated climate change risk information, knowledge, and experiences is yet to be established (Sobel and Leeson, 2007). Where this has been achieved it is fragmented, assumes a top-down approach, is sometimes carried out by institutions with no clear international mandate, and the quality of the data and its coverage are inadequate. In other cases a huge amount of information is collected but not efficiently used (Zhang et al., 2002; Sobel and Leeson, 2007). Access to data or information under government institutions is often constrained by bureaucracy and consolidating shared information can be hampered by multiple formats and incompatible data sets. The major challenge in achieving coordinated integrated risk management across scales is in establishing clear mechanisms for a networked program to generate and exchange diverse experiences, tools, and information that can enable various DRR and CCA actors at different levels to use different options available for reducing climate risks. Such a mechanism will support efforts to mainstream CCA and DRR into development, for example, in the case of initiatives by UNDP; development organizations such as the World Bank, DFID, and the IADB; the Canadian International Development Agency; the European Commission; and so forth (Benson and Twigg, 2007). Accounting for climate risks within the development context will, among other things, be effectively achieved where appropriate information and knowledge of what is required exist and are known and shared efficiently (Ogallo, 2010).

7.4.5.1. Knowledge Acquisition

Knowledge acquisition by nature is a complex, continuous, nonlinear, and life-long process that spans generations. Knowledge acquisition for DRR and CCA involves acquisition, documentation, and evaluation of knowledge for its authenticity and applicability over time and beyond its point of origin (Rautela, 2005). Knowledge acquisition and documentation has to focus on the shifting emphasis by the HFA from reactive emergency relief to proactive DRR approaches by aiming at strengthening prevention, mitigation, and preparedness and linking with changes in CCA that include greater focus on local scales (refer to Section 7.4.3.2). The Global Spatial Data Infrastructure (GSDI), which aims to coordinate and support the development of Spatial Data Infrastructures worldwide, provides important services for a proactive DRR approach (Köhler and Wächter, 2006). One of the major breakthroughs facilitating the creation of the GSDI has been the development of interoperability standards and technology that form a common foundation for the sharing and interoperability of, for example, geospatial data. However, global geospatial data infrastructure is still largely underutilized for site- and/or application-specific needs (Le Cozannet et al., 2008; Di and Ramapriyan, 2010).

There are huge efforts in DRR- and CCA-related knowledge acquisition, development, and exchange by universities, government agencies, international organizations, and to some extent the private sector, but coordination of these efforts internationally is yet to be achieved (Marincioni, 2007). At the international level, the International Council for Science (ICSU) is the main international body that facilitates and funds efforts to generate global environmental change information that extends into DRR and CCA. ICSU is an NGO with a global membership of national scientific bodies (121 members) and international scientific unions (30 members) that maintain a strong focus on natural sciences (www.icsu.org). However, there have been changes over the years and ICSU now works closely with the International Social Science Council (ISSC). There are four major global environmental change (GEC) research programs facilitated by ICSU: an International Programme of Biodiversity Science (DIVERSITAS), the International Geosphere Biosphere Programme, the International Human Dimensions Programme closely tied to the ISSC, and the World Climate Research Programme. These programs have been supported by a capacity-building and information dissemination wing, the System for Analysis, Research and Training. The four GEC programs have had a significant role in generating the background science that forms the basis for CCA and DRR (Steffen et al., 2004). The link between science and policy within the UN system for CCA is achieved through the IPCC process while for DRR it is through activities of the UNISDR.

There has been growing concern that GEC programs are not integrated and provide fragmented information limited to certain disciplines. This concern led to the establishment of the Earth System Science Partnership aiming to integrate natural and social sciences from the regional to the global scale. However, this has proved inadequate to meet the growing need for integrated information (Leemans et al., 2009). As a result, a major restructuring of the knowledge generation process both at the institutional and science level has been launched by ICSU and the main focus is on increased use of integrated approaches and co-production of knowledge with potential users to deliver regionally and locally relevant information to address environmental risks for sustainable development. These initiatives will influence the process of integration of DRR and CCA and their linkages to development in the future (ICSU, 2010; Reid et al., 2010).

An assessment of climate services for DRR and CCA is given in Section 7.3.3.2. But the generation of climate change information has followed a top-down approach relying on global models to produce broad-scale information with no clear local context and usually with large uncertainties and complex for the public to assimilate hence providing lower incentive for policymakers to act on the risks that are indicated (Weingart et al., 2000; Schipper and Pelling, 2006). Climate change information is primarily provided at long temporal ranges, for example, 2050, which is far beyond the usual five-year attention span of most political governments let alone that of poor people concerned with basic needs. Climate information at all scales is essential for decisionmaking although there are various factors other than climate information that ultimately influence decisionmaking. The IPCC Fifth Assessment Report will cover near-term climate extending to periods earlier than 2050. Efforts to enhance delivery of information at interannual to interdecadal scales will improve assimilation of climate information in risk management (Goddard et al., 2010; Vera et al., 2010). However, expressing impacts, vulnerability, and adaption require description of complex interactions between biophysical characteristics of a risk and socioeconomic factors and relating to factors that usually span far beyond the area experiencing the risk. Communicating these linkages has been a challenge particularly for areas where education levels are low and communication infrastructure is inadequate (Vogel and O'Brien, 2006).

Knowledge acquisition and documentation requires capacity in terms of skilled manpower, infrastructure, and appropriate institutions and funding (Section 7.4.3.1). Long-term research and monitoring with a wide global coverage of different hazards and vulnerabilities is required (Kinzig, 2001). For example, forecasting a hazard is a key aspect of disaster prevention but generating such information comes with a cost. Although weather forecasting through the meteorological networks of WMO is improving, the network of meteorological stations is far from spatially adequate and some have ceased to operate or are not adequately equipped (Ogallo, 2010). Forecasters are challenged to communicate forecasts that are often characterized by large uncertainty but which need to be conveyed in a manner that can be readily understood by policymakers and the public (Vogel and O'Brien, 2006; Carvalho, 2007).

Interdisciplinary generation of information – that is, bridging the traditional divide among the social, natural, behavioral, and engineering sciences – continues to be a great intellectual challenge in climate change risk reduction. The newly formed Integrated Research on Disaster Risk (IRDR) program – co-sponsored by ICSU, ISSC, and UNISDR – aims at applying an integrated approach in understanding natural and human-induced environmental hazards (ICSU, 2008; McBean, 2010). IRDR is intended to address these challenges and gradually provide relevant data, information, and knowledge on vulnerability trends, which are key information for policy- and decisionmakers to formulate integrated policies and measures for DRR and CCA.

7.4.5.2. Knowledge Organization, Sharing, and Dissemination

Exchange of disaster information worldwide has increased tremendously through, for example, mass media and information and communication technologies (ICT). The role of mass media in addressing the broader needs of DRR and CCA as opposed to disaster response is still limited, although various regional initiatives such as the Network of Climate Journalists of the Greater Horn of Africa (NECJOGHA) that involve climate and media experts are being established to improve the situation (Ogallo, 2010). NECJOGHA serves to disseminate integrated information based on, for example, environmental monitoring, climatology, agronomy, public health, and so forth, to the users to enhance sustainable response to climate change. Clearly, multiple strategies for disseminating and sharing knowledge and information are required for different needs at different scales (Glik, 2007; Maitland and Tapia, 2007; Maibach et al., 2008; Saab et al., 2008; see also Chapters 5 and 6). In particular, greater efforts are needed to identify and communicate information on vulnerability development, going beyond and adding to the hazards information, to effectively contribute to reducing risk.

Disaster response and recovery are closely linked to provision of effective communication prior to and throughout the disaster situation (Zhang et al., 2002). Mass media, for example, radio, television, and newspapers are powerful mechanisms for conveying information during and immediately after disasters although they may over-sensationalize issues, which may influence perception of risk and subsequent responses (Vasterman et al., 2005; Glik, 2007). A 'two-step flow' approach where the mass media is combined with interpersonal communication channels has been found to provide a more effective approach to information dissemination (Maibach et al., 2008; Chagutah, 2009; Kaklauskas et al., 2009).

Increased use of ICT such as mobile phones, online blogging websites with interactive functions and links to other web pages and real-time crowd-sourcing electronic commentary, and other forms of web-based social-networked communications such as Twitter, Facebook, etc., represent current tools for timely information dissemination. They facilitate rapid exchange of information, for instance, from the disaster scene to rescuers and/or delivery of vital information to those affected. This is particularly the case where such information is given in an appropriate format and language and facilities to deliver information are accessible (Glik, 2007). There are emerging attempts to develop mobile phone-based disaster response services, for example, that can translate disaster information into different languages (Hasegawa et al., 2005); and use real-time mobile phone-calling data to provide information on location and movement of victims in a disaster area (Madey et al., 2007). Mobile phones are now routinely used to disseminate disaster warning information within industrialized countries and the process is rapidly expanding to developing countries.

Information sharing and dissemination for disaster relief has improved through the establishment of the ReliefWeb site (www.reliefweb.int) by the UN Office for the Coordination of Humanitarian Affairs (OCHA) in 1996. ReliefWeb so far offers the largest Internet-based international disaster information gathering, sharing, and dissemination mechanism (Wolz and Park, 2006; Maitland and Tapia, 2007; Saab et al., 2008). The International Charter (www.disasterscharter.org) provides space data that serve to augment the ReliefWeb. But the OCHA ReliefWeb does not cover preparedness and disaster prevention to fully embrace CCA and DRR compared to the comparatively more recent PreventionWeb (www.preventionweb.net) where disaster risk reduction is covered.

Despite the growing role of mass media and ICT in disaster response, significant improvements are still needed to reduce disaster losses. The full potential of mobile phones and Internet facilities in disaster relief has yet to be exploited. The OCHA ReliefWeb poorly represents local- to national-level humanitarian activities; for example, most of this information is not translated into different languages (Wolz and Park, 2006). There are large sections of the global population who have no access to Internet and other telecommunication services (Samarajiva, 2005) although evidence shows that improved access by disaster workers has overall positive effects on disaster relief (Wolz and Park, 2006). Other initiatives such as RAdio and InterNET (RANET), a satellite broadcast service that combines radio and Internet to communicate hydro-meteorological and climate-related information, are examples of innovative measures being put in place to address the problem of limited access to the Internet in developing countries (Boulahya et al., 2005). Sustainable use of ICT for coordination of information for humanitarian efforts faces challenges of limited resources to mount, maintain, and upgrade these systems (Saab et al., 2008). ICT is also limited to explicit knowledge that is comprised of, for example, documents and data stored in computers but generally lacks tacit knowledge that is based on experience linked to someone's expertise, competence, understanding, professional intuition, and so forth that can be valuable for disaster relief (Kaklauskas et al., 2009). Increased international collaboration on disaster management and also the growing use of interactive web communication facilities provides for the filtering of tacit knowledge.

7.4.5.2.1. Disaster risk reduction and climate change adaptation

In addition to disaster management organizations such as UNISDR, the International Federation of Red Cross and Red Crescent Societies, the Federal Emergency Management Agency, national Red Cross and Red Crescent societies, and so forth, a great deal of knowledge dissemination is accomplished in the academic field. But this knowledge does not translate automatically to the general public. The use of ICTs such as computer networks, digital libraries, satellite communications, remote sensing, grid technology, and GIS for data and information integration for knowledge acquisition and exchange is growing to be important in integrating DRR and CCA (UNISDR, 2005b; Louhisuo et al., 2007; see also Section 7.4.3.2). ICT offers interactive modes of learning that could be of value in distance education and online data sharing and retrieval. For example, the Centre for Research on the Epidemiology of Disasters (CRED) at the Catholic University of Louvain in Belgium maintains the Emergency Events Database (EM-DAT), which has over 18,000 entries on disasters in the world dated from 1900 to present (www.cred.be). The data are recorded on a country-level basis and form a useful resource for disaster preparedness and vulnerability assessments, although information on small-scale disasters is difficult to establish (Tschoegl, 2006). In addition to CRED, a comprehensive database of global natural catastrophe losses is provided by the Munich Re NatCatSERVICE, where nearly 800 events are entered in the database every year; by 2009, the database had more than 25,000 entries with losses spanning from the 1980s, although records for major events go up to 2000 years ago (Schmidt et al., 2009; Zschocke and de Leon, 2010). Because of its strong focus on insured losses, the Munich Re database tends to have less coverage for areas with lower insurance coverage. At a regional level, the DesInventar database in Latin America is an example of a regional database that was developed in 1994 by the Network for Social Studies in Disaster Prevention. The DesInventar database is an inventory of small-, medium-, and greater-impact disasters (www.desinventar.net) and aims to facilitate dialog for risk management between actors, institutions, sectors, and provincial and national governments. This initiative has been extended to the Caribbean, Asia, and Africa by UNDP, while the UNFCCC provides a more local-scale database on local coping strategies (maindb.unfccc.int/public/adaptation).

ICT capabilities in disaster risk reduction also lie in enhancing interaction among individuals and institutions from the national, to regional, to international level, for example, through e-mail, newsgroups, online chats, mailing lists, and web forums (Marincioni, 2007). Attempts have been made, for example, in Japan, to create an integrated disaster risk reduction system where mobile phone communication operates as part of a greater information generation and delivery chain that includes Earth observation data analysis, navigation and web technologies, GIS, and advanced information technology such as grid (Louhisuo et al., 2007). When such innovations are transferred to other regions they contribute to international DRR efforts.

Other initiatives include NetHope International, which combines development and disaster issues into its ICT-centric mandate (Saab et al., 2008). RANET (www.oar.noaa.gov/spotlite/archive/spot_ranet.html), originally developed in Africa for drought and which spread to Asia, Pacific, Central America, and the Caribbean, has a strong community engagement and disseminates comprehensive information from global climate data banks combined with regional and local data and forecasts resulting in spinoffs to food security, agriculture, and health in rural areas (Boulahya et al., 2005). A network of extension agents, development practitioners, and trained members of the community are used in RANET to translate information into local contexts and languages and as a result, RANET is being considered for other educational initiatives such as the Spare Time University to improve access to learning in DRR with benefits for CCA (Glantz, 2007). RANET has been found to reduce vulnerability to climate extremes in different areas in Africa, for example, communication of rainfall forecasts in parts of west Africa assists farmers with decisions on what crop variety to plant and field to use where a choice of fields of different soil type existed, and also where to search for pasture and water for livestock during drought periods. However, RANET faces challenges of unavailability of technical support, follow-up training, power supply, and coordination (Boulahya et al., 2005).

The establishment of the PreventionWeb facility by UNISDR demonstrates the potential of ICT in information sharing for international disaster risk management across scales. PreventionWeb has been evolving since 2006, and was built on the experience of ReliefWeb with the purpose of becoming a single entry point to the full range of global disaster risk reduction information and providing a common platform for institutions to connect, exchange experiences, and share information on DRR, and facilitating integration with CCA and the development process. Updated daily, the PreventionWeb platform contains news, DRR initiatives, event calendars, online discussions, contact directories, policy and reference documents, training events, terminology, country profiles, and fact sheets as well as audio and video content. Hence, while catering primarily to DRR professionals, it also promotes better understanding of disaster risk by non-specialists. PreventionWeb is a response to a need for greater information and knowledge sharing and dissemination advanced in Zhang et al. (2002), Marincioni (2007), Kaklauskas et al. (2009), and others. The web site serves a critical role in supporting the implementation of the HFA where information and knowledge sharing is essential (Zschocke and de Leon, 2010). But the full potential of PreventionWeb has yet to be realized and evaluated since it is a relatively new initiative.

In addition to the PreventionWeb with a DRR focus, the number of web-based resource portals supporting both DRR and CCA has been increasing. These include, among others, ProVention Consortium, which had a DRR and climate focus (www.proventionconsortium.org) but has ceased to operate; the UN Adaptation Learning Mechanism (www.adaptationlearning.net) with links to related online resources and documentation of over 140 countries; Linking Climate Adaptation Network/CBA-X (www.linkingclimateadaptation.org) which has some DRR focus, had over 1,000 members in 2008, and has continued to provide current thinking on climate adaptation and resources and publications for researchers, practitioners, and policy formers; and the WeAdapt/WikiAdapt, an adaptation focus portal (www.weadapt.org) that goes beyond networking and dissemination to cover knowledge integration and other innovative adaptation tools. These portals are relatively new, remain predominantly used by their respective communities, and have also been noted by others to be poorly organized (Mitchell and van Aalst, 2008). Performance of such ICT information resources in disaster risk management could improve with more coordination and integration of CCA, DRR, and the development community.

7.4.5.2.2. Constraints in knowledge sharing and dissemination

For all information tools noted, the quality of information transferred and language used influence their effectiveness. Further, these mechanisms often collapse during a disaster when most needed (Marincioni, 2007; Saab et al., 2008). Some of the new technologies are not easily accessible to the very poor, and even the most innovative tools like RANET show numerous maintenance constraints particularly in remote areas (Boulahya et al., 2005).

There are differences in perception on the role of ICT in the exchange of disaster and hazard risk knowledge as opposed to its role in increased flow of information, with knowledge here defined simply as understanding of information while information refers to organized data (Zhang et al., 2002; Marincioni, 2007). Indications are that, while there is increased circulation of disaster information, this does not always result in increased assimilation of new risk reduction approaches, a factor that is partly attributed to lack of effective sharing although lack of capacity to use/apply the information could be a major factor (Zhang et al., 2002; UNISDR, 2005b).The level of assimilation of ICT technology into disaster risk reduction depends, among other things, on levels of literacy and the working environment including institutional arrangements, hence effectiveness may vary with levels of development (Samarajiva, 2005; Marincioni, 2007; see also Section 7.4.3.2). As a result, the contribution of these relatively new facilities such as PreventionWeb will, among other things, depend on accessibility and assimilation of ICT in the daily operations of institutions across the globe. Evidence shows that information alone is not adequate to address disaster risk reduction; rather, other factors such as availability of resources, effective management structures, and social networks are critical (Glik, 2007; Lemos et al., 2007; Maibach et al., 2008; Chagutah, 2009).

A major constraint in climate change risk management results from the fact that communities working in disaster management, climate change, and development operate separately and this increases vulnerability to climate extremes leading to disasters (Schipper and Pelling, 2006; Lemos et al., 2007). For example, emphasis on humanitarian assistance has been attributed to development agendas that do not adequately integrate risk reduction leading to increased vulnerability (Benson and Twigg, 2007), while development community members are, for example, better equipped with the use of insurance but fail to link this to climate risk reduction thus exposing communities to vulnerability to climate extremes. Similar observations have been made about cities where urban developers have no link with the climate risk management community (Wamsler, 2006). But in fact both the development and climate adaptation communities are concerned with vulnerability to disasters. This could be a common point of focus facilitating collaboration in research, information sharing, and practice as part of global security (Schipper and Pelling, 2006; Lemos et al., 2007).

Communication gaps between professional groups often result from different language styles and jargons. Heltberg et al. (2008) have suggested a need for establishing universally shared basic operational definitions of key terms such as risk, vulnerability, and adaptation across the different actors as a basis for dissemination of knowledge. This has also been noted by others, for example, for better coordination among numerous humanitarian organizations (Saab et al., 2008) and in the FAO guide for disaster risk management (Baas et al., 2008; also see Chapter 1). The move toward establishment of national disaster risk reduction institutions that link to similar regional and international structures by, for example, UNISDR, provides a framework for bringing different stakeholders together including the climate change and development communities at the national level, culminating in greater integration of risk management at the international level. Other efforts include international initiatives to integrate, at the national level, disaster risk reduction with poverty reduction frameworks (Schipper and Pelling, 2006).

In conclusion, there is *high agreement* in the literature indicating that efforts are being made internationally to build information and knowledge bases that support the shift in emphasis by the HFA from reactive emergency relief to proactive DRR (*high confidence*). Conventional media and ICT are major factors in facilitating the required international exchange and dissemination of information on disaster response, CCA, and DRR (*high confidence*). This in turn stimulates generation of new knowledge and will over time lead to greater integration of DRR and CCA, which at the present moment is still limited (*medium confidence*). The limitation of relying heavily on ICT is that there is still a large part of the world where the ICT infrastructure is not adequately developed. There is also *high agreement* in the literature that an increase in the exchange of data and information at the international level on its own is not a complete solution to risk reduction. Resources to generate and supply information and experience in a usable form for each unique case so as to translate this to knowledge and action are a critical dimension in risk reduction (*high confidence*). Further, more attention is required for the international community to identify what information is essential for different stages of climate change risk management, and how it should be captured and used by different actors under different risk reduction scenarios. Data gathering, information, and knowledge acquisition and management for disaster relief has a longer history. The process of building integrated information resource tools that brings together experiences from CCA, DRR, and the development community is still weak, yet these tools hold the promise for reducing vulnerability to disasters in the future (*high confidence*).

7.5. Considerations for Future Policy and Research

How best can experience with disaster risk reduction at the international level be used to help or strengthen climate change adaptation? The characteristics of the DRR regime (as exemplified chiefly by the UNISDR and the Hyogo Framework for Action) and the CCA regime (chiefly the UNFCCC and the IPCC) have been described in detail and assessed to the extent that the literature allows. One frequently made assumption is that the DRR world has much to learn from CCA and vice versa (IPCC, 2009). It is widely proposed in the literature that disaster risk reduction and climate change adaptation should be 'integrated' (Birkmann and von Teichman, 2010).

The call for integration of disaster risk reduction with climate change adaptation goes much further, however (UNISDR, 2009a). It is argued that both disaster risk reduction and climate change adaptation remain outside the mainstream of development activities (UNISDR, 2009a). The United Nations Global Assessment Report on Disaster Risk Reduction calls for "an urgent paradigm shift" in disaster risk reduction to address the underlying risk drivers such as vulnerable rural livelihoods, poor urban governance, and declining ecosystems (UNISDR, 2009a). The report also calls for the harmonization of existing institutional and governance arrangements for disaster risk reduction and climate change adaptation (p. 181), and presents a 20-point plan to reduce risk (pp. 176-177).

These conclusions come from an official UN report (UNISDR, 2009a), and they are widely supported in the scientific literature (O'Brien et al., 2006; Schipper, 2009) as well as in other government reports (DFID, 2005; Birkmann et al., 2009; CCD, 2009) and in the advocacy literature (Venton and La Trobe, 2008). More recently, the widely reviewed ICSU (2010) report (called the Belmont Challenge) on Regional Environmental Change: Human Action and Adaptation, which was commissioned by the major global environmental change research funders to assess the international research capability required to respond to the challenge of delivering knowledge to support human action and adaptation to regional environmental change, concluded by calling for a highly coordinated and collaborative research program to deliver integrated knowledge required to identify and respond to hazards, risks, and vulnerability, and develop mitigation and adaptation strategies. Similarly, ICSU and the ISSC carried out a wide consultative process to rethink the focus and framework of Earth system research. This consultation came out with four Grand Challenges that require a balanced mix of disciplinary and interdisciplinary research to address critical issues at the intersection of Earth systems science and sustainable development (Reid et al., 2010):

- Improve the usefulness of forecasts of future environmental conditions and their consequences for people.
- Develop, enhance, and integrate observation systems to manage global and regional environmental change.
- Determine how to anticipate, avoid, and manage disruptive global environmental change.
- Determine institutional, economic, and behavioral changes to enable effective steps toward global sustainability.

Both the Belmont Challenge and the Grand Challenges are setting an international tone for an integrative approach to challenges such as DRR, CCA, and development. There is no shortage of policy proposals designed to integrate disaster risk reduction and climate change adaptation for their common strengthening and benefit.

Official reports also list many reasons why more movement in this direction has been slow to develop. One constraint is the difficulty of integration across scales, which is addressed in Section 7.6. Two other sets of constraints are described as 'the normative dimension' and 'the knowledge dimension' (Birkmann et al., 2009). The extensive list of challenges and constraints identified includes the following:

- *Normative Dimensions* (adapted from Birkmann et al., 2009)
  - Absence of uniform methods, standards, and procedures in vulnerability and capacity assessment and also in the design, formulation, and implementation of adaptation plans, programs, and projects. Lack of clear norms when applying vulnerability and capacity assessment and when designing and implementing adaptation measures
  - The desire for stability and the tendency to rapidly restore normalcy limit the scope to explore and to take advantage of the opportunity after disaster and recover in an adaptive way by taking account of future climate change. The notion and desire for stability may hamper the chance to take advantage of change and dynamics – after disasters, the chance to use the opportunity and build back in an adaptive way considering future climate change is in most cases not taken – more commonly, infrastructure is rapidly built back to the pre-disaster condition
- *Knowledge Challenges* (adapted from Birkmann et al., 2009)
  - Differences in the form of terminology used – that is, the different terms and definitions framed by both DRR and CCA communities

- Unavailability of information about the concrete effects of climate change at the local level (see Section 7.4.5.1)
- Limited census-based information on relevant census data (social and economic parameters) especially in dynamic areas with, for example, high fluctuations of people and/or economic instability
- Scientific knowledge on climate change acquired by the scientific community has not been translated or trickled down to practitioners or it is communicated in a way that is hard to understand and derive practical knowledge (see Section 7.4.5.2.2)
- Absence or lack of appropriate indicators for assessment that could measure successful adaptation and which could also be incorporated into funding guidelines as well as monitoring and evaluation strategies (ICSU, 2010).

For the purposes of this Special Report, the question has been formulated in terms of what can be learned from the practice of DRR to advance CCA. It is clear from the literature, however, that cooperation between the DRR and CCA communities is a two-way process. This has given rise to questions about how 'integration' in practice at local and national levels might best be facilitated by change at the international level.

7.6. Integration across Scales

7.6.1. The Status of Integration

The literature reflects three different perspectives on the integration of disaster risk reduction and climate change adaptation. One view common among the community of experts and practitioners is that climate change adaptation should be integrated into disaster risk reduction (CCD, 2008a,b,c; Prabhakar et al., 2009, p. 26). It has even been suggested that climate change adaptation is a case of 'reinventing the wheel' (Mercer 2010) since disaster risk reduction covers much of the same ground and is "already well-established within the international development community" (Lewis, 1999; Wisner et al., 2004). Practitioners in disaster risk reduction tend to have the view that climate change is one of a number of factors contributing to vulnerability and disasters (Mercer, 2010), and that therefore climate change adaptation needs to be taken on board.

A second view is adopted by many in the climate change adaptation community. They recognize a diversity of cross-cutting risks that can be associated with the impacts of climate change and consider disaster risk to be one of these (Birkmann and von Teichman, 2010). They conclude that disaster risk reduction should be integrated into climate change adaptation.

A third and perhaps more widespread view is that both disaster risk reduction and climate change adaptation should be more effectively integrated into wider development planning (Glantz, 1999; O'Brien et al., 2006; Lewis, 2007; CCD, 2009; Christoplos et al., 2009; UNISDR, 2009a).

At the practical level there are many steps already underway to bring about such forms of integration (see Chapters 5 and 6). There are numerous hazards and disasters that are not directly linked to climate change but their impacts may serve to increase vulnerability to climate change. Nevertheless, as noted in Section 7.5 there are many obstacles to integration and it is by no means agreed that full integration between disaster risk reduction and climate change adaptation is possible, or desirable.

The potential benefits as well as the obstacles to integration can be examined in terms of three scales: the spatial, the temporal, and the functional (Birkmann and von Teichman, 2010).

7.6.2. Integration at a Spatial Scale

The literature reflects a view that DRR and CCA operate at different spatial scales (Birkmann and von Teichman, 2010) and that therefore their integration in practice has been problematic or impracticable. Disasters are often thought of as events occurring at a specific location whereas climate change is thought of as a global or regional phenomenon. This view is now being modified as the need for locally based climate change adaptation becomes evident (Adger et al., 2005), as the impacts of local disasters are recognized as having more widespread impacts at a larger spatial scale (see Chapters 4 and 6 and Section 7.2.1).

One commonly cited impediment to integration is that climate change projections do not provide precise information at a local scale (see Chapter 3) and that adaptation strategies tend to be designed for entire countries or regions (German Federal Government, 2008; Red Cross and Red Crescent Climate Centre, 2007).

7.6.3. Integration at a Temporal Scale

There is also a perceived difference in the temporal scales of CCA and DRR. The disaster community has traditionally been focused on humanitarian response including relief and reconstruction in the relatively short term. (UNISDR, 2009b), whereas climate change has been recognized as including long-term processes with projections extending from decades to centuries (Chapter 3), which poses problems for development communities usually focusing on a shorter time span. More effective cooperation and integration between the DRR and the CCA practitioners could help to detect, address, and overcome these temporal-scale challenges. This essentially requires the stronger recognition of the risks of climate-related disasters in CCA and the incorporation of longer-term climate change risk factors into DRR.

7.6.4. Integration at a Functional Scale

The functional separation of CCA and DRR institutions, organizations, and mechanisms extends across all three levels of management from local to national to international. At the international level there are weak links between the climate adaptation 'regime' as expressed in the UNFCCC and the leading DRR 'regime' in the form of the UNISDR. The character of the two 'regimes' is radically different, the former having the task of implementing an international agreement and the latter being a UN-wide interagency and advocacy program. The history of the evolution of the two institutional arrangements is markedly different. The disaster field has long been dominated by humanitarian and emergency response measures and has only relatively recently been moving toward a stronger DRR approach (Burton, 2003). Similarly, climate change was initially conceived as an atmospheric pollution issue with greater emphasis on the need to reduce greenhouse gas emissions and has slowly been repositioned, as in the UNFCCC negotiations, as also being a development issue. One consequence of the different evolution has been that the emerging international climate 'regime' (UNFCCC) is linked at the national level to environment ministries, whereas the disaster 'regime' (UNISDR) is linked to emergency planning and preparedness agencies or, in other cases, to the office of President. Neither DRR nor CAA are well linked to economic planning and development agencies (UNISDR, 2009b).

There is also a 'top-down' versus 'bottom-up' distinction (Rayner, 2010). Natural hazards and associated disasters have a long history, and DRR has moved slowly from local to national to international levels in response to the rationale described in Section 7.2. Climate change, on the other hand, came to attention as a result of the work of atmospheric scientists and was first recognized primarily as a global problem, and has subsequently moved down scale as the need for CCA became more apparent and pressing. This shows that the opportunity exists for the two to complement each other, at the international level where DRR has progressed, and at the national and local level to which CCA is moving.

7.6.5. Toward More Integration

The mandate of this Special Report is in part to consider how CCA could be enhanced by learning from the experience of the DRR community, and vice versa. The literature shows a widespread view that the two could both benefit from closer integration with each other and that both would benefit society better if there was more integration into sustainable development (UNISDR, 2009a). Integration in this sense is meant as symbiosis or synthesis rather than formal integration at the institutional level. Integration across scales can be facilitated if integration between DRR and CCA were also to take place at local, national, and international levels. Integration at the international level might help to facilitate integration at national and local levels although the opposite is also possible. This Special Report is itself a prime example of emerging cooperation. It is in line with a wider evolution in the global environmental change science research community whose products serve both disaster risk reduction and climate change adaptation at the international level of management.

References

A digital library of non-journal-based literature cited in this chapter that may not be readily available to the public has been compiled as part of the IPCC review and drafting process, and can be accessed via either the IPCC Secretariat or IPCC Working Group II web sites.

Adaptation Fund, 2010: *Financial Status of the Adaptation Fund Trust Fund.* AFB/EFC.3/7Rev.1, Adaptation Fund, GEF, Washington, DC.

Adaptation Fund, 2011a: *About the Adaptation Fund*. Adaptation Fund, GEF, Washington, DC, adaptation-fund.org/about.

AdaptationFund, 2011b: Funded Projects. Adaptation Fund, GEF, Washington, DC, adaptation-fund.org/funded_projects.

Adger, W.N., 2004: The right to keep cold. *Environment and Planning, A*, **36**, 1711-1715.

Adger, W.N., N.W. Arnell, and E.L. Tompkins, 2005: Successful adaptation to climate change across scales. *Global Environmental Change*, **15**, 77-86.

Adger, W.N., I. Lorenzoni, and K. O'Brien (eds.), 2009: *Adapting to Climate Change: Thresholds, Values, Governance*. Cambridge University Press, Cambridge, UK.

African Union, 2010: Extended Programme of Action for the Implementation of the Africa Regional Strategy for Disaster Risk Reduction (2006-2015). In: *Extended Programme of Action for the Implementation of the Africa Regional Strategy for Disaster Risk Reduction (2006 - 2015) and Declaration of the 2nd African Ministerial Conference on Disaster Risk Reduction 2010*. African Union Commission and United Nations International Strategy for Disaster Reduction, Addis Ababa, Ethiopia, p. 25-54.

AGF, 2010: *Report of the Secretary-General's High-Level Advisory Group on Climate Change Financing*. United Nations, New York, NY, www.un.org/wcm/webdav/site/climatechange/shared/Documents/AGF_reports/AGF%20Report.pdf.

Agrawala, S. and S. Fankhauser (eds.), 2008: *Economic Aspects of Adaptation to Climate Change: Costs, Benefits and Policy Instruments*. OECD, Paris, France.

Agrawala, S. and M. van Aalst, 2008: Adapting development cooperation to adapt to climate change. *Climate Policy*, **8**, 183-193.

Alam, M., A. Rahman, M. Rashid, G. Rabbani, P. Bhandary, S. Bhadwal, M. Lal, and M.H. Soejachmoen, 2007: Impacts, Vulnerability and Adaptation to Climate Change In Asia. Background Paper for the UNFCCC, UNFCCC, Bonn, Germany, unfccc.int/files/adaptation/methodologies_for/vulnerability_and_adaptation/application/pdf/unfccc_asian_workshop_background_paper.pdf.

Alexander, D., 2006: Globalization of disaster: trends, problems and dilemmas. *Journal of International Affairs*, **59**, 1-22.

Arriens, W.T.L. and C. Benson, 1999: Post disaster rehabilitation: The experience of the Asian Development Bank. Paper presented to the *IDNR-ESCAP Regional Meeting for Asia: Risk Reduction and Society in the 21st Century*, Bangkok, Thailand, 23-26 February 1999.

Baas, S., S. Ramasamy, J.D. DePryck, and F. Battista, 2008: *Disaster Risk Management Systems Analysis*. Environment, Climate Change and Bioenergy Division, FAO, Rome, Italy, 68 pp.

Bamberger, R.L. and L. Kumins, 2005: *Oil and Gas: Supply Issues after Katrina*. Congressional Research Service, Library of Congress, Washington, DC.

Barnett, B.J., C.B. Barrett, and J.R. Skees, 2008: Poverty traps and index-based risk transfer products. *World Development*, 36(10), 1766-1785.

Barnett, J. and W.N. Adger, 2007: Climate change, human security and violent conflict. *Political Geography*, **26**, 639-655.

Barnett, J. and M. Webber, 2009: *Accommodating Migration to Promote Adaptation to Climate Change: A Policy Brief*. World Bank and SCCCD, Washington, DC and Stockholm, Sweden.

Barrett, C.B., B.J. Barnett, M.R. Carter, S. Chantarat, J.W. Hansen, A.G. Mude, D.E Osgood, J.R. Skees, C.G. Turvey, and M.N. Ward, 2007: *Poverty Traps and Climate Risk: Limitations and Opportunities of Index-Based Risk Financing*. IRI Technical Report 07-02, International Research Institute for Climate and Society, Palisades, NY, 53 pp.

Basher, R., 2006: Global early warning systems for natural hazards: systematic and people-centred. *Philosophical Transactions of the Royal Society*, **364**, 2167-2182.

Begg, I., 2008: Subsidiarity in regional policy. In: Subsidiarity and economic reform in Europe [Gelauff, G., I. Grilo, and A. Lejour (eds.)]. Springer, Berlin, Germany, pp. 291-310.

Benson, C. and J. Twigg, 2007: *Tools for Mainstreaming Disaster Risk Reduction: Guidance Notes for Development Organisations*. International Federation of Red Cross and Red Crescent Societies/the ProVention Consortium, ProVention Consortium Secretariat, Geneva, Switzerland, 178 pp.

Berkhout, F. 2005: Rationales for adaptation in EU climate change policies. *Climate Policy*, **5**, 377-391.

Bhatt, M.R., 2007: Good practice in local approaches to climate change adaptation and disaster risk management in South Asia: Lessons from Tsunami Evaluation Coalition. In: *Climate Change, Humanitarian Disasters and International Development: Linking vulnerability, risk reduction and response capacity. Research and Policy Workshop*, 27 April 2007. Oslo Center for Interdisciplinary Environmental and Social Research Forskningsparken, Oslo, Norway.

Biermann, F. and I. Boas, 2010: Global adaptation governance: the case of protecting climate refugees. In: *Global Climate Governance Beyond 2012: Architecture, Agency and Adaptation* [Bierman, F., P. Pattberg, and f. Zelli (eds.)]. Cambridge University Press, Cambridge, UK, pp. 255-269.

Birkmann, J. and K. von Teichman, 2010: Integrating disaster risk reduction and climate change adaptation: key challenges – scales, knowledge, and norms. *Sustainability Science*, **5**, 171-184.

Birkmann, J., K. von Teichman, P. Aldunce, C. Bach, N. T. Binh, M. Garschagen, S. Kanwar, N. Setiadi, and L.N. Thach, 2009: Addressing the challenge: Recommendations and quality criteria for linking disaster risk reduction and adaptation to climate change. In: DKKV Publications Series 38 [Birkmann, J., G. Tetzlaff, and K.-O. Zentel (eds.)]. German Committee for Disaster Reduction, Bonn, Germany.

Birnie, P., A. Boyle, and C. Redgwell, 2009: *International Law and the Environment*. 3rd ed. Oxford University Press, Oxford, UK.

Bodansky, D., 1995: Customary (and not so customary) international environmental law. *Indiana Journal of Global Legal Studies*, **3(1)**, 105-119.

Bodansky, D., 2005: The international climate change regime. In: *Perspectives on Climate Change: Science, Economics, Politics, Ethics* [Sinnott-Armstrong, W. and R.B. Howarth (eds.)]. Elsevier, Amsterdam, The Netherlands, pp. 147-180.

Bodansky, D., 2010: *The Art and Craft of International Environmental Law*. Harvard University Press, Cambridge, MA.

Boulahya, M., M.S. Cerda, M. Pratt, and K. Sponberg, 2005: Climate, communications, and innovative technologies: Potential impacts and sustainability of new radio and internet linkages in rural African communities. *Climatic Change*, **70**, 299-310.

Brown, J., N. Bird, and L. Schalatek, 2010: *Direct Access to the Adaptation Fund: Realising the Potential of National Implementing Entities*. Overseas Development Institute, London, UK, 10 pp.

Brown, O., 2008: *Migration and Climate Change*. Research Series No. 31, International Organization for Migration, Geneva, Switzerland.

Burton, I., 2003. Do we have the adaptive capacity to develop and use the adaptive capacity to adapt? In: *Climate Change, Adaptive Capacity and Development* [Smith, J.B., R.J.T. Klein, and S. Huq (eds.)]. Imperial College Press. London, UK, pp 137-161.

Burton, I., S. Huq, B. Lim, O. Pilifosova, and E.L. Schipper, 2002: From impacts assessment to adaptation priorities: the shaping of adaptation policy. *Climate Policy*, **2(2)**, 145-159.

Caney, S., 2008: Human rights, climate change and discounting. *Environmental Politics*, **17**, 536-555.

Caney, S., 2010: Climate change, human rights and moral thresholds. In: *Human Rights and Climate Change* [Humphreys, S. (ed.)]. Cambridge University Press, Cambridge, UK, pp. 69-90.

Cardenas, V., S. Hochrainer, R. Mechler, G. Pflug, and J. Linnerooth-Bayer, 2007: Sovereign financial disaster risk management: the case of Mexico. *Environmental Hazards*, **7**, 40-53.

CARE International, 2010: *Community-Based Adaptation Toolkit*. Cooperative for Assistance and Relief Everywhere (CARE), Chatelaine, Switzerland, www.careclimatechange.org/files/toolkit/CARE_CBA_Toolkit.pdf.

Carvalho, A., 2007: Ideological cultures and media discourses on scientific knowledge: re-reading news on climate change. *Public Understanding of Science*, **16**, 223.

Castles, S., 2002: *Environmental Change and Forced Migration: Making Sense of the Debate*. New Issues in Refugee Research, Working Paper No. 70, United Nations High Commissioner for Refugees, Geneva, Switzerland.

CCD, 2008a: *Incentives and Constraints to Climate Change Adaptation and Disaster Risk Reduction – A Local Perspective*. Commission on Climate Change and Development, Stockholm, Sweden.

CCD, 2008b: *Overview of Adaptation Mainstreaming Initiative*. Commission on Climate Change and Development, Stockholm, Sweden.

CCD, 2008c: *Links between Disaster Risk Reduction, Development and Climate Change*. Commission on Climate Change and Development, Stockholm, Sweden.

CCD, 2009: *Closing the Gaps: Disaster Risk Reduction and Adaptation to Climate Change in Developing Countries*. Commission on Climate Change and Development, Stockholm, Sweden.

Chagutah, T., 2009: Towards improved public awareness for climate related disaster risk reduction in South Africa: A Participatory Development Communication perspective. *Jàmbá: Journal of Disaster Risk Studies*, **2(2)**, 113-126.

Chinkin, C.M., 1989: The challenge of soft law: Development and change in international law. *International and Comparative Law Quarterly*, **38(4)**, 850-866.

Christian Aid, 2007: *Human Tide: The Real Migration Crisis*. Christian Aid, London, UK, www.christianaid.org.uk/Images/human-tide.pdf.

Christoplos, I., S. Anderson, M. Arnold, V. Galaz, M. Hedger, R.J.T. Klein, and K. Le Goulven, 2009: *The Human Dimension of Climate Adaptation: The Importance of Local and Institutional Issues*. Commission on Climate Change and Development, Stockholm, Sweden.

CIENS, 2007: *Climate Change, Humanitarian Disasters and International Development: Linking Vulnerability, Risk Reduction and Response Capacity*. Research and Policy Workshop. 27 April 2007. Oslo Center for Interdisciplinary Environmental and Social Research (CIENS), Forskningsparken, Oslo, Norway.

Clémençon, R., 2008: The Bali Road Map: A first step on the difficult journey to a post-Kyoto Protocol agreement. *Journal of Environment and Development*, **17(1)**, 70-94.

Climate Funds Update, 2011: *Pilot Program for Climate Resilience*. Climate Funds Update, www.climatefundsupdate.org/listing/pilot-program-for-climate-resilience.

Cohen, R., 1998: *The Guiding Principles on Internal Displacement: a New Instrument for International Organizations and NGOs*. Office for the Coordination of Humanitarian Affairs, DC 1-1568, 1 UN Plaza, 10017 NY, New York, USA. Also available on http://www.fmreview.org/FMRpdfs/FMR02/fmr209.pdf.

Cozier, M., 2011: Restoring confidence at the Cancun Climate Change Conference. *Greenhouse Gases: Science and Technology*, **1(1)**, 8-10.

Craeynest, L., L. Gallagher, and C. Sharkey, 2010: *Business as Unusual. Direct Access: Giving power back to the poor?* Caritas Internationalis and CIDSE, Vatican City, Italy and Brussels, Belgium.

Cummins, J.D. and O. Mahul, 2009: *Catastrophe Risk Financing in Developing Countries: Principles for Public Intervention*. World Bank, Washington, DC.

Cutler, D.M. and R. Zeckhauser, 2000: The anatomy of health insurance. In: *Handbook of Health Economics*. Vol. 1 [Culyer, A.J. and J.P. Newhouse (eds.)]. Elsevier Science, Amsterdam, The Netherlands, pp. 563-643.

Czarnecki, R. and K. Guilanpour, 2009: The Adaptation Fund after Poznan. *Carbon and Climate Law Review*, **3(1)**, 79-87.

Dalberg, 2010: *United Nations International Strategy for Disaster Reduction (UNISDR) Secretariat Evaluation: Final Report*.

Davis, I., 2004: The application of performance targets to promote effective earthquake risk reduction strategies. Engineering Paper no 2726 presented at the *Thirteenth World Conference on Earthquakes*, Vancouver, Canada 1-6 August 2004.

De Lucia, V., 2007: Common but differentiated responsibility. In: *The Encyclopedia of Earth* [C.J. Cleveland (ed.)]. Environmental Information Coalition, National Council for Science and the Environment, Washington, DC, www.eoearth.org/article/Common_but_differentiated_responsibility.

Dessai, S., 2003: The Special Climate Change Fund: origins and prioritisation assessment. *Climate Policy*, **3(3)**, 295-302.

DFID, 2005: *Sending Money Home? A Survey of Remittance Products and Services in the United Kingdom*. Department for International Development, London, UK.

DFID, 2011: *The Multilateral Aid Review. Ensuring maximum value for money for UK aid through multilateral organisations*, Annex 6, Assessment Summary ISDR. Criteria for Assessment Table 2, Department for International Development, London, UK, pp 217-218.

Di, L. and H.K. Ramapriyan, 2010: Standards-based data and information systems for Earth observations – An introduction. In: *Standard-Based Data and Information Systems for Earth Observation* [Di, L. and H.K. Ramapriyan (eds.)]. Springer-Verlag, Berlin, Germany, pp. 1-6.

Diacon, D., 1992: Typhoon resistant housing in the Philippines: The Core Shelter Project. *Disasters*, **16(3)**, 266-271.

Doig, A. 2008: *Setting the Bar High at Poznan*. Christian Aid, London, UK, www.christianaid.org.uk/images/poznan-report.pdf.

Dun, O. and F. Gemenne, 2008: Defining environmental migration. *Forced Migration Review*, **31**, 10-11.

Dunfee, T.W. and D. Hess, 2000: The legitimacy of direct corporate humanitarian investment. *Business Ethics Quarterly*, **10(1)**, 95-109.

Eriksen, S. and K. Brown, 2011: Sustainable adaptation to climate change: Prioritising social equity and environmental integrity. *Climate and Development*, **3(1)**, 3-6.

Eriksen, S., P. Aldunce, C.S. Bahinipati, R.D.'A. Martins, J.I. Molefe, C. Nhemachena, K. O'Brien, F. Olorunfemi, J. Park, L. Sygna, and K. Ulstrud, 2011: When not every response to climate change is a good one: identifying principles for sustainable adaptation. *Climate and Development*, **3(1)**, 7-20.

Fagen, P. and M. Bump (ed.), 2006: *Remittances in Conflict and Crises: How Remittances Sustain Livelihoods in War, Crises, and Transitions to Peace*. International Peace Academy, New York, NY.

Fankhauser, S., 2010: The costs of adaptation. *Wiley Interdisciplinary Reviews: Climate Change*, **1(1)**, 23-30.

Fidler, D., 2005: Disaster relief and governance after the Indian tsunami: What role for international law? *Melbourne Journal of International Law*, **6(2)**, 458.

Fisher, D., 2007: The law of international disaster response. In: *Global Legal Challenges: Command of the Commons, Strategic Communications and Natural Disasters* [M.D. Carsten (ed.)]. International Law Studies Vol. 83, US Naval War College, Newport, Rhode Island, pp. 293-320.

Foley, C., 2007: *Mozambique: A Case Study in the Role of the Affected State in Humanitarian Action*. HPG Working Paper, Humanitarian Policy Group and Overseas Development Institute, London, UK.

Froot, K.A. (ed.), 1999: *The Financing of Catastrophe Risk*. The University of Chicago Press, Chicago, IL.

Fugate, W.C. and C. Crist, 2008: *Florida Hurricane Loss Mitigation Program*. Report to the Florida Legislature, Florida Division of Emergency Management, Tallahassee, FL, www.floridadisaster.org/Mitigation/Documents/RCMPAnnualRptSFY07-08-final-secured.pdf.

Gassebner, M., A. Keck, and R. Teh, 2010: Shaken, not stirred: The impact of disasters on international trade. *Review of International Economics*, **18(2)**, 351-368.

GEF, 2008: *Report on the Completion of the Strategic Priority on Adaptation*. GEF/C.34/8, Global Environment Facility, Washington, DC, www.thegef.org/gef/sites/thegef.org/files/documents/C.34.8%20Report%20on%20the%20Completion%20of%20the%20SPA.pdf.

GEF, 2009: *Report of the GEF to the Fifteenth Session of the Conference of the Parties to the United Nations Framework Convention on Climate Change*. FCCC/CP/2009/9, UNFCCC, Bonn, Germany, unfccc.int/resource/docs/2009/cop15/eng/09.pdf.

GEF, 2010a: *Status Report on Least Developed Countries Fund and Special Climate Change Fund*. GEF/LDCF.SCCF.9/Inf.2/Rev.2, GEF, Washington, DC, www.thegef.org/gef/sites/thegef.org/files/documents/Status%20Report%20on%20the%20Climate%20Change%20Funds%20-%20Oct%208,%202010-Rev2.pdf.

GEF, 2010b: *Report of the GEF to the Sixteenth Session of the Conference of the Parties to the United Nations Framework Convention on Climate Change*. FCCC/CP/2010/5, UNFCCC, Bonn, Germany, http://unfccc.int/resource/docs/2010/cop16/eng/05.pdf.

German Federal Government, 2008: *German Strategy for Adaptation to Climate Change*. Adopted by the German Federal Cabinet on 17 December 2008, The Federal Government, Berlin, Germany, www.bmu.de/files/english/pdf/application/pdf/das_gesamt_en_bf.pdf.

GFDRR, 2010: *Partnership Charter*. Global Facility for Disaster Reduction and Recovery, Washington, DC, gfdrr.org/gfdrr/sites/gfdrr.org/files/publication/GFDRR_Partnership_Charter_2010.pdf.

Ghesquiere, F., O. Mahul, M. Forni, and R. Gartley, 2006: *Caribbean Catastrophe Risk Insurance Facility: A solution to the short-term liquidity needs of small island states in the aftermath of natural disasters*. IAT03-13/3, World Bank, Washington, DC.

Gigli, S. and S. Agrawala, 2007: *Stocktaking on Progress on Integrating Adaptation to Climate Change into Development Co-operation Activities*. COM/ENV/EPOC/DCD/DAC(2007)1/FINAL, Environment Directorate and Development Co-operation Directorate, OECD, Paris, France, 74 pp.

Glantz, M.H., 1999: El Niño as a hazard-spawner. In: *The Nature of Hazards: Commemorative Volume to Celebrate Achievements of International Decade for Natural Disaster Reduction* [J. Ingleton (ed.)]. Tudor Rose Publishers, Leicester, UK, pp. 78-79.

Glantz, M.H., 2007: How about a spare-time university? *WMO Bulletin*, **56(2)**, 1-6.

Glik, D.C., 2007: Risk communication for public health emergencies. *Annual Review of Public Health*, **28**, 33-54.

GNDR, 2009 : *Clouds but Little Rain: Views from the Frontline - A local perspective of progress towards implementation of the Hyogo Framework for Action*. Global Network of Civil Society Organisations for Disaster Reduction, London, UK.

GNDR, 2011: *If We do not Join Hands… Summary Report, Views from the Frontline, Local reports of progress on implementing the Hyogo Framework for Action*, Global Network of Civil Society Organisations for Disaster Reduction, London, UK.

Goddard, L., Y. Ait Chellouche, W. Baethgen, M. Dettinger, R. Graham, P. Hayman, M. Kadi, R. Martínez, and H. Meinke, with additional contributions by E. Conrada, 2010: Providing seasonal-to-interannual climate information for risk management and decision-making. *Procedia Environmental Sciences*, **1**, 81-101

Government of the Philippines, 2008: *DSWD leads turnover of core shelter units to landslide victims in Southern Leyte*. Government of the Philippines, Manila, Philippines, www.reliefweb.int/rw/rwb.nsf/db900SID/FBUO-7JSDJY?OpenDocument.

Government of the Philippines, 2010: *Philippines: DSWD, UNDP turn over core shelter project to Sorsogon beneficiaries*. Government of the Philippines, Manila, Philippines, www.reliefweb.int/rw/rwb.nsf/db900sid/MYAI-82W3GG?OpenDocument.

Grasso, M., 2010: An ethical approach to climate adaptation finance. *Global Environmental Change*, **20(1)**, 74-81.

Groat, Ch.G., 2004: Seismographs, sensors, and satellites: better technology for safer communities. *Technology in Society*, **26(2-3)**, 169-179.

Grubb, M., 2010: Copenhagen: back to the future. *Climate Policy*, **10(2)**, 127-130.

Gupta, J. and M. Grubb, 2000: Competence and subsidiarity. In: Climate change and European leadership: a sustainable role for Europe? [Gupta, J. and M. Grubb (eds.)]. Kluwer Academic Publishers, Dordrecht, The Netherlands, 372 pp.

Gupta, J. and N. van de Grijp (eds.), 2010: *Mainstreaming Climate Change in Development Cooperation: Theory, Practice and Implications for the European Union*. Cambridge University Press, Cambridge, UK, 347 pp.

Gupta, J., Å. Persson, L. Olsson, J. Linnerooth-Bayer, N. van der Grijp, A. Jerneck, R.J.T. Klein, M. Thompson, and A.G. Patt, 2010: Mainstreaming climate change in development co-operation policy: conditions for success. In: *Making Climate Change Work for Us: European Perspectives on Adaptation and Mitigation Strategies* [Hulme, M. and H. Neufeldt (eds.)]. Cambridge University Press, Cambridge, UK, pp. 319-339.

Gurenko, E., 2004: *Catastrophe Risk and Reinsurance: A Country Risk Management Perspective*. Risk Books, London, UK.

Guzmán, J.M., 2009: The use of population census data for environmental and climate change analysis. In: *Population Dynamics and Climate Change* [Guzmán, J.M., G. Martine, G. McGranahan, D. Schensul, and C. Tacoli (eds.)]. UNFPA and IIED, New York, NY and London, UK, pp. 192-205.

Haines, A., R. Kovats, D. Campbell-Lendrum, and C. Corvalan, 2006: Climate change and human health: impacts, vulnerability, and mitigation. *The Lancet*, **367(9528)**, 2101-2109.

Harrington, K. and J. O'Connor, 2009: How Cisco succeeds at global risk management. *Supply Chain Management Review*, **July/August 2009**.

Hasegawa, S., K. Sato, S. Matsunuma, M. Miyao, and K. Okamoto, 2005: Multilingual disaster information system: information delivery using graphic text for mobile phones. *AI & Society*, **19**, 265-278.

Hellmuth, M., D. Osgood, U. Hess, A. Moorhead, and H. Bhojwani (eds.), 2009: *Index Insurance and Climate Risk: Prospects for Development and Disaster Management*. Climate and Society No. 2, International Research Institute for Climate and Society, Columbia University, New York, NY, 112 pp.

Heltberg, R., P.B. Siegel, and S.L. Jorgensen, 2008: *Climate Change, Human Vulnerability and Social Risk Management*. World Bank, Washington, DC.

Hess, U., 2007: Risk management framework- the big LEAP in Ethiopia, presentation, Insurance in catastrophe risk.

Hess, U. and J. Syroka, 2005: *Weather-based Insurance in Southern Africa: The Case of Malawi*. Agriculture and Rural Development Discussion Paper 13, World Bank, Washington, DC.

Holdaway, R., 2001: Is space global disaster warning and monitoring now nearing reality? *Space Policy*, **17(2)**, 127-132.

Hope Sr., K.H., 1996: Promoting sustainable community development in developing countries: The role of technology transfer. *Community Development Journal*, **31(3)**, 193-200.

Huq, S., 2002: The Bonn–Marrakech agreements on funding. *Climate Policy*, **2(2)**, 243-246.

ICRC, 1995: *The Code of Conduct for the International Red Cross and Red Crescent Movement and NGOs in Disaster Relief*. International Committee of the Red Cross, Geneva, Switzerland, www.icrc.org/eng/resources/documents/misc/code-of-conduct-290296.htm.

ICSU, 2008: *A Science Plan for Integrated Research on Disaster Risk: addressing the challenge of natural and human induced environmental hazards*. International Council for Science, Paris, France.

ICSU, 2010: *Regional Environmental Change: Human Action and Adaptation*. International Council for Science, Paris, France.

IEA, 2001: *Technology without Borders. Case Studies of Successful Technology Transfer*. OECD/International Energy Agency, Paris, France.

IFRC, 2011: *Red Cross/Red Crescent Climate Centre*. International Federation of Red Cross and Red Crescent Societies, Geneva, Switzerland, www.climatecentre.org.

IHRC, 2006: *News Updates on the Release of Public Hurricane Loss Model (Winter/Spring 2006)*. International Hurricane Research Centre, Florida International University, Miami, FL.

IMF, 2011: *International Monetary Fund Factsheet: The Multilateral Debt Relief Initiative*. International Monetary Fund, Washington, DC, www.imf.org/external/np/exr/facts/pdf/mdri.pdf.

IPCC, 1990: *Climate Change: The IPCC Scientific Assessment*. The First Assessment Report of the Intergovernmental Panel on Climate Change. Cambridge University Press, Cambridge, United Kingdom and New York, NY, USA.

IPCC, 1996: *Climate Change 1995*. The Second Assessment Report of the Intergovernmental Panel on Climate Change. Cambridge University Press, Cambridge, United Kingdom and New York, NY, USA.

IPCC, 2000: *Special Report on Emissions Scenarios*. Cambridge University Press, Cambridge, UK.

IPCC, 2007: *Climate Change 2007*. The Fourth Assessment Report of the Intergovernmental Panel on Climate Change. Cambridge University Press, Cambridge, United Kingdom and New York, NY, USA.

IPCC, 2009. *Scoping Meeting for an IPCC Special Report on Extreme Events and Disasters: Managing the Risks. Proceedings* [Barros, V., et al. (eds.)], Oslo, Norway, 23-26 March 2009.

IRI, 2006: *A Gap Analysis for the Implementation of the Global Climate Observing System Programme in Africa*. IRI Technical Report IRI-TR/06/1, International Research Institute for Climate and Society, Palisades, NY, portal.iri.columbia.edu/portal/server.pt/gateway/PTARGS_0_2_2806_0_0_18/GapAnalysis.pdf.

JAXA, 2011: *DAICHI (ALOS) Operation Completion*. Japan Aerospace Exploration Agency, Tokyo, Japan, www.jaxa.jp/press/2011/05/20110512_daichi_e.html.

Jordan, A., 2000: The politics of multilevel environmental governance: Subsidiarity and environmental policy in the European Union. *Environment and Planning A*, **32(7)**, 1307-1324.

Kaklauskas A., D. Amaratunga, and R. Haigh, 2009: Knowledge model for post-disaster management. *International Journal of Strategic Property Management*, **13**, 117-128.

Kameda, H., 2007: Networking disaster risk reduction technology and knowledge through Disaster Reduction Hyperbase (DRH). In: *Proceedings of the Disaster Reduction Hyperbase (DRH) Contents Meeting*, Kobe, Japan, 12-13 March 2007, drh.edm.bosai.go.jp/Project/Phase2/1Documents/9_EXr.pdf.

Kibreab, G., 1997: Environmental causes and impact of refugee movements: a critique of the current debate. *Disasters*, **21(1)**, 20-38.

Kinzig, A.P., 2001: Bridging disciplinary divides to address environmental and intellectual challenges. *Ecosystems*, **4**, 709-715.

Klein, R.J.T., 2010: Linking adaptation and development finance: a policy dilemma not addressed in Copenhagen. *Climate and Development*, **2(3)**, 203-206.

Klein, R.J.T. and A. Möhner, 2009: Governance limits to effective global financial support for adaptation. In: *Adapting to Climate Change: Thresholds, Values, Governance* [Adger, W.N., I. Lorenzoni, and K.L. O'Brien (eds.)]. Cambridge University Press, Cambridge, UK, pp. 465-475.

Klein, R.J.T. and A. Persson, 2008: *Financing Adaptation to Climate Change: Issues and Priorities*. European Climate Platform (ECP) Report No. 8. An Initiative of Mistra's Climate Policy Research Programme (Clipore) and the Centre for European Policy Studies (CEPS).

Klein, R.J.T., J. Aston, E.N. Buckley, M. Capobianco, N. Mizutani, R.J. Nicholls, P.D. Nunn, and S. Ragoonaden, 2000: Coastal adaptation. In: *Methodological and Technological Issues in Technology Transfer* [Metz, B., O.R. Davidson, J.-W. Martens, S.N.M. van Rooijen, and L. Van Wie McGrory (eds.)]. Cambridge University Press, Cambridge, UK, pp. 349-372.

Klein, R.J.T., W.W. Dougherty, M. Alam, and A. Rahman, 2005: Technology to understand and manage climate risks. *Background Paper for the UNFCCC Seminar on the Development and Transfer of Environmentally Sound Technologies for Adaptation to Climate Change*, Tobago, 14-16 Jun 2005, UNFCCC, Bonn, Germany, unfccc.int/ttclear/pdf/Workshops/Tobago/backgroundPaper_old.pdf.

Klein, R.J.T., S.E.H. Eriksen, L.O. Næss, A. Hammill, T.M. Tanner, C. Robledo, and K.L. O'Brien, 2007: Portfolio screening to support the mainstreaming of adaptation to climate change into development assistance. *Climatic Change*, **84(1)**, 23-44.

Kleindorfer, P.R., 2009: *Climate Change and Insurance: Integrative Principles and Regulatory Risk*. INSEAD Working Paper No. 2009/43/TOM/INSEAD, INSEAD, Fontainebleau, France, ssrn.com/abstract=1456862.

Köhler, P. and J. Wächter, 2006: Towards an open information infrastructure for disaster research and management: Data management and information systems inside DFNK. *Natural Hazards*, **38**, 141-157.

Kok, M., B. Metz, J. Verhagen, and S. van Rooijen, 2008: Integrating development and climate policies: national and international benefits. *Climate Policy*, **8(2)**, 103-118.

Kolmannskog, V., 2008: *Future floods of refugees*. Norwegian Refugee Council, Oslo, Norway.

Kolmannskog, V. and F. Myrstad, 2009: Environmental displacement in European asylum law. *European Journal of Migration and Law*, **11**, 313-326.

Kreimer, A. and M. Arnold, 2000: *Managing Disaster Risk in Emerging Economies*. Disaster Risk Management Series No 2, World Bank, Washington, DC.

Kuhn, K., D. Campbell-Lendrum, A. Haines, and J. Cox, 2005. *Using Climate to Predict Infectious Disease Epidemics*. World Health Organization, Geneva, Switzerland, whqlibdoc.who.int/publications/2005/9241593865.pdf.

Kunreuther, H. and E. Michel-Kerjan, 2009: *At War with the Weather: Managing Large-Scale Risks in a New Era of Catastrophes*. 1st ed. The MIT Press, Cambridge, MA.

Lamb, H., 1995: *Climate, History and a Modern World*. 2nd ed. Routledge, New York, NY.

Lane, M., 2004: The viability and likely pricing of "cat bonds" for developing countries. In: *Catastrophe Risk and Reinsurance: A Country Risk Management Perspective* [Gurenko, E. (ed.)]. Risk Books, London, UK, pp. 239-268.

Lavoyer, J.-P., 2006: International humanitarian law: Should it be reaffirmed, clarified or developed? In: *Issues in International Law and Military Operations* [Jaques, R.B. (ed.)]. International Law Studies Vol. 80, US Naval War College, Newport, Rhode Island, pp. 287-310.

Le Cozannet, G., S. Hosford, J. Douglas, J.-J. Serrano, D. Coraboeuf, and J. Comte, 2008: Connecting hazard analysts and risk managers to sensor information. *Sensors*, **8**, 3932-3937.

Lee, W.-J., R.A. Pielke Jr., and L. Anderson-Berry, 2006: Disaster mitigation, warning systems and societal impact. Rapporteur Report. In: *Sixth International Workshop on Tropical Cyclones*, San José, Costa Rica, 21-30 Nov 2006, WMO, Geneva, Switzerland, severe.worldweather.org/iwtc/document/Topic_5_M_C_Wong.pdf.

Leemans, R., G. Asrar, A. Busalacchi, J. Canadell, J. Ingram, A. Larigauderie, H. Mooney, C. Nobre, A. Patwardhan, M. Rice, F. Schmidt, S. Seitzinger, H. Virji, C. Vorosmarty, and O. Young, 2009: Developing a common strategy for integrative global environmental change research and outreach: the Earth System Science Partnership (ESSP) Strategy paper. *Current Opinion in Environmental Sustainability*, **1**, 4-13.

Lemos, M.C., E. Boyd, E.L. Tompkins, H. Osbahr, and D. Liverman, 2007. Developing adaptation and adapting development. *Ecology and Society*, **12(2)**, 26, www.ecologyandsociety.org/vol12/iss2/art26/.

Lewis, J., 1999: *Development in Disaster-prone Places: Studies of Vulnerability*. Intermediate Technology Publications, London, UK.

Lewis, J., 2007: Climate and disaster reduction. *Tiempo Climate Newswatch*, www.tiempocyberclimate.org/newswatch/comment070217.htm.

Linnerooth-Bayer, J. and R. Mechler, 2008: *Insurance against Losses from Natural Disasters in Developing Countries*. Background paper for United Nations World Economic and Social Survey, United Nations Department of Economic and Social Affairs, New York, NY.

Linnerooth-Bayer, J., R. Loefstedt, and G. Sjostedt (eds.), 2001: *Transboundary Risk Management*. Earthscan Publications, London, UK.

Linnerooth-Bayer, J., R. Mechler, and G. Pflug, 2005: Refocusing disaster aid. *Science*, **309**, 1044-1046.

Linnerooth-Bayer, J., K. Warner, C. Bals, P. Höppe, I. Burton, T. Loster, and A. Haas, 2009: Insurance mechanisms to help developing countries respond to climate change. *The Geneva Papers on Risk and Insurance - Issues and Practice*, **34**, 381-400.

Linnerooth-Bayer, J., C. Bals, and R. Mechler, 2010: Insurance as part of a climate adaptation strategy. In: *Making Climate Change Work for Us: European Perspectives on Adaptation and Mitigation Strategies* [Hulme, M. and H. Neufeldt (eds.)]. Cambridge University Press, Cambridge, UK.

Louhisuo, M., T. Veijonen, J. Ahola, and T. Morohoshi, 2007: A disaster information and monitoring system utilizing earth observation. *Management of Environmental Quality*, **18(3)**, 246-262.

Lucas, R. and O. Stark, 1985: Motivations to remit: evidence from Botswana. *Journal of Political Economy*, **93(5)**, 901-918.

Maastricht Treaty, 1992 : *Provisions Amending the Treaty Establishing the European Economic Community with a view to Establishing the European Community*, www.eurotreaties.com/maastrichtec.pdf.

Mace, M.J., 2005: Funding for adaptation to climate change: UNFCCC and GEF developments since COP-7. *Review of European Community & International Environmental Law*, **14(3)**, 225-246.

Madey, G.R., A. Barabási, N.V. Chawla, M. Gonzalez, D. Hachen, B. Lantz, A. Pawling, T. Schoenhar, G. Szabó, P. Wang, and P. Yan, 2007: Enhanced situational awareness: Application of DDDAS concepts to emergency and disaster management. In: *Computational Science – ICCS 2007: 7th International Conference, Beijing, China, 27-30 May 2007, Proceedings, Part I, LNCS 4487* [Shi, Y. and G.D. Van Albada (eds.)]. Springer-Verlag, Berlin, Germany, pp. 1090-1097.

Maibach, E.W., C. Roser-Renouf, and A. Leiserowitz, 2008: Communication and marketing as climate change–intervention assets. A public health perspective. *American Journal of Preventive Medicine*, **35(5)**, 488-500.

Maitland, C. and A. Tapia, 2007: *Outcomes from the UN OCHA 2002 Symposium & HIN Workshops on Best Practices in Humanitarian Information Management and Exchange*. College of Information Sciences and Technology, Pennsylvania State University, University Park, PA.

Marincioni, F., 2007: Information technologies and the sharing of disaster knowledge: the critical role of professional culture. *Disasters*, **31(4)**, pp. 459-476.

McBean, G.A., 2010: Introduction of a New International Research Program: Integrated Research on Disaster Risk - The Challenge of Natural and Human-Induced Environmental Hazards. In: *Geophysical Hazards: Minimizing Risk, Maximizing Awareness. Part II* [Beer, T. (ed.)]. Springer, Berlin, Germany, pp. 59-69, www.cprm.gov.br/33IGC/1340016.html.

McGray, H., A. Hammill, R. Bradley, E.L. Schipper, and J.-E. Parry, 2007: *Weathering the Storm: Options for Framing Adaptation and Development*. World Resources Institute, Washington, DC, 57 pp.

Mercer, J., 2010. Disaster risk reduction or climate change adaptation: Are we reinventing the wheel? *Journal of International Development*, **22**, 247-264.

Michaelowa, A. and K. Michaelowa, 2007: Climate or development: is ODA diverted from its original purpose? *Climatic Change*, **84(1)**, 5-21.

Miller, D. and A. Paulson, 2007: *Risk Taking and the Quality of Informal Insurance: Gambling and Remittances in Thailand*. Working Paper No. 2007-01, Federal Reserve Bank of Chicago, Chicago, IL, ssrn.com/abstract=956401.

Mills, E., 2007: *From Risk to Opportunity: Insurer Responses to Climate Change*. A Ceres Report, Ceres, Boston, MA.

Ministry of Foreign Affairs of Denmark and GEF Evaluation Office, 2009: *Joint External Evaluation: Operation of the Least Developed Countries Fund for Adaptation to Climate Change*. MFA, Copenhagen, Denmark, 112 pp.

Mirza, M.M.Q., 2003: Climate change and extreme weather events: Can developing countries adapt? *Climate Policy*, **3**, 233-248.

Mitchell, T. and M. van Aalst, 2008: *Convergence of Disaster Risk Reduction and Climate Change Adaptation. A Review for DFID 31st October 2008*. Department for International Development (DFID), London, UK, www.preventionweb.net/files/7853_ConvergenceofDRRandCCA1.pdf.

Mitchell, T., S. Anderson, and S. Huq, 2008: *Principles for Delivering Adaptation Finance*. Institute of Development Studies, University of Sussex, Brighton, UK, 6 pp.

Mohapatra, S., J. George, and D. Ratha, 2009: *Remittances and Natural Disasters: Ex-post Response and Contribution to Ex-ante Preparedness*. Policy Research Working Paper 4972, World Bank, Washington, DC.

Morrissey, J., 2009: *Environmental Change and Forced Migration - A State of the Art Review*. Background Paper, Refugee Studies Centre, Oxford Department of International Development, Queen Elizabeth House, University of Oxford, 48 pp., www.rsc.ox.ac.uk/events/environmental-change-and-migration/EnvChangeandFmReviewWS.pdf.

Müller, B., 2007: *Nairobi 2006: Trust and the Future of Adaptation Funding*. Oxford Institute for Energy Studies, Oxford, UK, 26 pp.

Munich Re, 2003: *TOPICSgeo: Annual Review – Natural Catastrophes 2002*. Munich Reinsurance Group, Geoscience Research, Munich, Germany.

Munich Re, 2010: *Insurance-linked Securities (ILS)*. Market Update Q1 2010, Munich Re, Munich, Germany.

Murphy, B., 2011: *Briefing Paper: Technology for Adapting to Climate Change*. weADAPT, the Collaborative Platform on Climate Adaptation, weadapt.org/knowledge-base/wikiadapt/technology-and-adaptation.

Myers, N., 2001: Environmental refugees: A growing phenomenon of the 21st century. *Philosophical Transactions of the Royal Society B*, **357**, 609-613.

Myers, N., 2005: Environmental refugees: An emergent security issue. In: *Session III – Environment and Migration, 13th Meeting of the OSCE Economic Forum*, Prague, Czech Republic, 23-27 May 2005, pp. 23-27.

NASCIO, 2006: *State of Louisiana GIS Support for Emergency Operations Before, During, and After the Hurricanes of 2005*. National Association of State Chief Information Officers, Lexington, KY, www.nascio.org/awards/nominations/2006Louisiana9.pdf.

O'Brien, G., P. O'Keefe, J. Rose, and B. Wisner, 2006: Climate change and disaster management. *Disasters*, **30(1)**, 64-80.

O'Brien, G., P. O'Keefe, and J. Rose, 2007: Energy, poverty and governance. *International Journal of Environmental Studies*, **64(5)**, 607-618.

O'Brien, K. and R. Leichenko, 2000: Double exposure: assessing the impacts of climate change within the context of economic globalization. *Global Environmental Change*, **10**, 221-232.

O'Brien, K., L. Sygna, R. Leichenko, W.N. Adger, J. Barnett, T. Mitchell, L. Schipper, T. Tanner, C. Vogel, and C. Mortreux, 2008: *Disaster Risk Reduction, Climate Change Adaptation and Human Security*. GECHS Report 2008:3, prepared for the Royal Norwegian Ministry of Foreign Affairs by the Global Environmental Change and Human Security (GECHS) Project.

O'Brien, K., B. Hayward, and F. Berkes, 2009: Rethinking social contracts: building resilience in a changing climate. *Ecology and Society*, **14(2)**, 12, www.ecologyandsociety.org/vol14/iss2/art12/.

OCHA/IDMC, 2009: *Monitoring Disaster Displacement in the Context of Climate Change*. UN Office for the Coordination of Humanitarian Affairs (OCHA) and Internal Displacement Monitoring Center (IDMC), Geneva, Switzerland.

Ogallo, L., 2010: The mainstreaming of climate change and variability information into planning and policy development for Africa. *Procedia Environmental Sciences*, **1**, 405-410.

OHCHR, 2009: *Report of the Office of the United Nations High Commissioner for Human Rights on the Relationship between Climate Change and Human Rights*. UN Doc. A/HRC/10/61, Office of the United Nations High Commissioner for Human Rights, Geneva, Switzerland.

Oppenheimer, M. and A. Petsonk, 2005: Article 2 of the UNFCCC: historical origins, recent interpretations. *Climatic Change*, **73(3)**, 195-226.

Ott, H.E., W. Sterk, and R. Watanabe, 2008: The Bali roadmap: new horizons for global climate policy. *Climate Policy*, **8(1)**, 91-95.

Oxfam, 2011: *Adaptation and Risk Reduction*. Oxfam, Oxford, UK, www.oxfam.org.uk/resources/issues/climatechange/introduction.html#adaptation.

Oxfam International, 2007: *Adapting to Climate Change: What's Needed in Poor Countries, and Who Should Pay*. Oxfam Briefing Paper 104, Oxfam International Secretariat, Oxford, UK, 47 pp.

Paavola, J. and W.N. Adger, 2006: Fair adaptation to climate change. *Ecological Economics*, **56(4)**, 594-609.

Parry, M., N. Arnell, P. Berry, D. Dodman, S. Fankhauser, C. Hope, S. Kovats, R. Nicholls, D. Satterthwaite, R. Tiffin, and T. Wheeler, 2009: *Assessing the Costs of Adaptation to Climate Change: A Review of the UNFCCC and Other Recent Estimates*. International Institute for Environment and Development and Grantham Institute for Climate Change, London, UK, 111 pp.

Paul, B.K., H. Rashid, M. Shahidul Islam, and L. Hunt, 2010: Cyclone evacuation in Bangladesh: Tropical cyclones Gorky (1991) vs. Sidr (2007). *Environmental Hazards*, **9**, 89-101.

Pelling, M., 2011: *Adaptation to Climate Change: From Resilience to Transformation*. Routledge, Abingdon, UK.

Persson, Å. and R.J.T. Klein, 2009: Mainstreaming adaptation to climate change into official development assistance: challenges to foreign policy integration. In: *Climate Change and Foreign Policy: Case Studies from East to West* [Harris, P. (ed.)]. Routledge, London, UK, pp. 162-177.

Persson, Å., R.J.T. Klein, C. Kehler Siebert, A. Atteridge, B. Müller, J. Hoffmaister, M. Lazarus, and T. Takama, 2009: *Adaptation Finance under a Copenhagen Agreed Outcome*. Stockholm Environment Institute, Stockholm, Sweden, 187 pp.

Pickens, M., D. Porteous, and S. Rotman, 2009: *Banking the Poor via G2P Payments*. Focus Note 58, Consultative Group to Assist the Poor, Washington, DC.

Prabhakar, S.V.R.K., S. Anch, and R. Shaw, 2009: Climate change and local level disaster risk reduction planning: need, opportunities and challenges. *Mitigation and Adaptation Strategies for Global Change*, **14**, 7-33.

Raustiala, K., 2005: Form and substance of international agreements. *American Journal of International Law*, 99, 581.

Rautela, P., 2005: Indigenous technical knowledge inputs for effective disaster management in the fragile Himalayan ecosystem. *Disaster Prevention and Management*, **14(2)**, 233-241.

Rayner, S., 2010. How to eat an elephant: a bottom-up approach to climate policy. *Climate Policy*, **10**, 615-621.

Red Cross and Red Crescent Climate Centre, 2007: *Red Cross, Red Crescent Climate Guide*. Red Cross, The Hague, The Netherlands.

Reid, W.V., D. Chen, L. Goldfarb, H. Hackmann, Y.T. Lee, K. Mokhele, E. Ostrom, K. Raivio, J. Rockström, H.J. Schellnhuber, and A. Whyte, 2010: Earth system science for global sustainability: Grand challenges. *Science*, **330**, 916-917.

Rodriguez, J., F. Vos, R. Below, and D. Guha-Sapir, 2009: *Annual Disaster Statistical Review 2008 –The numbers and trends*. Centre for Research on the Epidemiology of Disasters, Université Catholique de Louvain, Louvain, Belgium, www.emdat.be/publications.

Rukieh, M. and M. Koudmani, 2006: *Use of Space Technology for Natural Disaster Detection and Prevention*. General Organisation of Remote Sensing, Damascus, Syria, www.iemss.org/iemss2006/papers/s11/288_RUKIEH_0.pdf.

Saab, D., E. Maldonado, R. Orendovici, L. Ngamassi, A. Gorp, K. Zhao, C. Maitland, and A. Tapia, 2008: Building global bridges: Coordination bodies for improved information sharing among humanitarian relief agencies. In: *Proceedings of the 5th International ISCRAM Conference*, Washington, DC, May 2008 [Fiedrich, F. and B. Van de Walle (eds.)].

Saechao, T.R., 2007: Natural disasters and the responsibility to protect: From chaos to clarity. *Brooklyn Journal of International Law*, **32**, 663.

Sahu, S., 2009: *Guidebook on Technologies for Disaster Preparedness and Mitigation*. Prepared for the Asian and Pacific Centre for Transfer of Technology, New Delhi, India, technology4sme.net/docs/Guidebook%20on%20Technologies%20for%20Disaster%20Preparedness%20&%20Mitigation.pdf.

Samarajiva, R., 2005: Policy Commentary: Mobilizing information and communications technologies for effective disaster warning: lessons from the 2004 tsunami. *New Media & Society*, **7(6)**, 731-747.

Sanusi, Z.A., 2005: *Technology Transfer under Multilateral Environmental Agreements: Analyzing the Synergies*. UNU-IAS Working Paper No. 134, Institute of Advanced Studies, United Nations University, Yokohama, Japan.

Savage, K. and P. Harvey (eds.), 2007: *Remittances during Crises: Implications for Humanitarian Response*. Overseas Development Institute, London, UK.

Schipper, E.L.F., 2006: Conceptual history of adaptation in the UNFCCC process. *Review of European Community & International Environmental Law*, **15(1)**, 82-92.

Schipper, E.L.F., 2009: Meeting at the crossroads?: Exploring the linkages between climate change adaptation and disaster risk reduction. *Climate and Development*, **1**, 16-30.

Schipper, L. and Pelling, M., 2006: Disaster risk, climate change, and international development, scope for, and challenges to integration. *Disasters*, 30 (1) 19-38

Schmidt, S., C. Kemfert, and P. Höppe, 2009: Tropical cyclone losses in the USA and the impact of climate change – A trend analysis based on data from a new approach to adjusting storm losses. *Environmental Impact Assessment Review*, **29**, 359-369.

Sheffi, Y., 2005: *The Resilient Enterprise*. MIT Press, Cambridge, MA.

Shimizu, T., 2008: Disaster Management Satellite System Development and International Cooperation Promotion in Asia. *Science and Technology Trends Quarterly Review*, No. 27/April 2008, 93-108.

Skees, J., 2001: The bad harvest: More crop insurance reform: A good idea gone awry. *Regulation: The CATO Review of Business and Government*, **24**, 16-21.

Skees, J.R. and A. Enkh-Amgalan, 2002: *Examining the feasibility of livestock insurance in Mongolia*. World Bank Working Paper 2886, World Bank, Washington, DC.

Skees, J., B. Barnett, and A. Murphy, 2008: Creating insurance markets for natural disaster risk in lower income countries: the potential role for securitization. *Agricultural Finance Review*, **68**, 151-157.

Smit, B., O. Pilifosova, I. Burton, B. Challenger, S. Huq, R.J.T. Klein, and G. Yohe, 2001: Adaptation to climate change in the context of sustainable development and equity. In: *Climate Change 2007. Impacts, Adaptation and Vulnerability. Contribution of Working Group II to the Fourth Assessment Report of the Intergovernmental Panel on Climate Change* [Parry, M.L., O.F. Canziani, J.P. Palutikof, P.J. Van Der Linde, and C.E. Hanson (eds.)]. Cambridge University Press, Cambridge, UK, pp. 877-912.

Sobel, R.S. and P.T. Leeson, 2007: The use of knowledge in natural-disaster relief management. *The Independent Review*, **11(4)**, 519- 532.

Sphere Project, 2004: *Humanitarian Charter and Minimum Standards in Disaster Response*. The Sphere Project, Geneva, Switzerland.

Steffen, W., A. Sanderson, P.D. Tyson, J. Jäger, P.A. Matson, B. Moore III, F. Oldfield, K. Richardson, H.J. Schellnhuber, B.L. Turner, and R.J. Wasson, 2004: *Global Change and the Earth System: A Planet Under Pressure*. Springer-Verlag, Berlin, Germany.

Stern, N., 2007: *The Economics of Climate Change*: The Stern Review. Cambridge University Press, Cambridge, UK.

Stoiber, E., 2006. Why Europe needs a subsidiarity early-warning mechanism. *Europe's World*, Summer 2006.

Suarez, P. and J. Linnerooth-Bayer, 2010: Micro-insurance for local adaptation. *Wiley Interdisciplinary Reviews: Climate Change*, **1(2)**, 271-278.

Suarez, P., J. Linnerooth-Bayer, and R. Mechler, 2007: *The Feasibility of Risk Financing Schemes for Climate Adaptation: The Case of Malawi*. DEC-Research Group, Infrastructure and Environment Unit, World Bank, Washington, DC.

Tacoli, C., 2009: Crisis or adaptation? Migration and climate change in a context of high mobility. *Environment and Urbanization*, **21**, 513-525.

Thomalla, F., T. Downing, E. Spanger-Siegfried, G. Han, and J. Rockström, 2006: Reducing hazard vulnerability: towards a common approach between disaster risk reduction and climate adaptation. *Disasters*, **30(1)**, 39-48.

Tomasini, R. and L. Van Wassenhove, 2009: *Humanitarian Logistics*. Palgrave Macmillan, London, UK.

Tompkins, E.L. and W.N. Adger, 2005: Defining a response capacity for climate change. *Environmental Science and Policy*, **8**, 562-571.

Trouwborst, A., 2002: *Evolution and Status of the Precautionary Principle in International Law*. Kluwer, Dordrecht, The Netherlands.

Tschoegl, L. (with R. Below and D. Guha-Sapir), 2006: An analytical review of selected data sets on natural disasters and impacts. In: *UNDP/CRED Workshop on Improving Compilation of Reliable Data on Disaster Occurrence and Impact*, CRED, Brussels, Belgium, pp. 2-21.

UN, 1970: *UN General Assembly resolution 2626, The International Development Strategy for the Second United Nations Development Decade (25th session)*. Earthscan, London, UK.

UN, 1992: *United Nations Framework Convention on Climate Change*. United Nations, New York, NY, unfccc.int/resource/docs/convkp/conveng.pdf.

UN, 2011: *Millennium Goals Report 2011*. United Nations, New York, NY, www.un.org/millenniumgoals/pdf/%282011_E%29%20MDG%20Report%202011_Book%20LR.pdf.

UNCED, 1992: *Rio Declaration 1992*. United Nations Conference on Environment and Development, Rio de Janeiro, Brazil.

UNDP, 2002: *A Climate Risk Management Approach to Disaster Reduction and Adaptation to Climate Change*. United Nations Development Programme Expert Group Meeting, Integrating Disaster Reduction with Adaptation to Climate Change, Havana, Cuba, 19-21 June 2002, 24 pp.

UNFCCC, 2005: *Report on the seminar on the development and transfer of technologies for adaptation to climate change. Note by the secretariat*. United Nations Framework Convention on Climate Change, Bonn, Germany, unfccc.int/resource/docs/2005/sbsta/eng/08.pdf.

UNFCCC, 2006a: *Synthesis report on technology needs identified by Parties not included in Annex I to the Convention. Note by the secretariat*. United Nations Framework Convention on Climate Change, Bonn, Germany, unfccc.int/resource/docs/2006/sbsta/eng/inf01.pdf.

UNFCCC, 2006b: *Technologies for adaptation to climate change*. United Nations Framework Convention on Climate Change, Bonn, Germany, unfccc.int/resource/docs/publications/tech_for_adaptation_06.pdf.

UNFCCC, 2007a: *Bali Action Plan*. Report of the Conference of the Parties (COP 13), Bali, adopted by Decision 1/CP.13 of the COP-13, United Nations Framework Convention on Climate Change, Bonn, Germany. unfccc.int/files/meetings/cop_13/application/pdf/cp_bali_action.pdf.

UNFCCC, 2007b: *Investment and Financial Flows to Address Climate Change*. United Nations Framework Convention on Climate Change, Bonn, Germany, 272 pp.

UNFCCC, 2008: *Mechanisms to manage financial risks from direct impacts of climate change in developing countries*. Technical paper, FCCC/TP/2008/9, United Nations Framework Convention on Climate Change, Bonn, Germany.

UNFCCC, 2009a: *Recommendations on future financing options for enhancing the development, deployment, diffusion and transfer of technologies under the Convention*. FCCC/SB/2009/2, United Nations Framework Convention on Climate Change, Bonn, Germany, unfccc.int/resource/docs/2009/sb/eng/02sum.pdf.

UNFCCC, 2009b: *Draft decision CP.15 Copenhagen Accord*. FCCC/CP/2009/L.7, United Nations Framework Convention on Climate Change, Bonn, Germany.

UNFCCC, 2010a: *Nairobi Work Programme*. United Nations Framework Convention on Climate Change, Bonn, Germany, unfccc.int/adaptation/nairobi_work_programme/items/3633.php.

UNFCCC, 2010b: *Potential Costs and Benefits of Adaptation Options: A Review of Existing Literature*. FCCC/TP/2009/2/Rev.1, United Nations Framework Convention on Climate Change, Bonn, Germany, unfccc.int/resource/docs/2009/tp/02r01.pdf.

UNFCCC, 2010c: *Report of the Conference of the Parties on its Sixteenth Session*, held in Cancun from 29 November to 10 December 2010. United Nations Framework Convention on Climate Change, UNFCCC/CP/2010/7/Add.1. Also available on unfccc.int/resource/docs/2010/cop16/eng/07a01.pdf#page=2.

UNFCCC, 2011: *Draft decision -/CP.16: Further guidance for the operation of the Least Developed Countries Fund*. United Nations Framework Convention on Climate Change, Bonn, Germany, unfccc.int/files/meetings/cop_16/conference_documents/application/pdf/20101204_cop16_ldcf.pdf.

UNGA, 2010: *Implementation of the International Strategy for Disaster Reduction*. Report of the Secretary General to the Sixty-fifth session, Ref A/65/388, United Nations General Assembly, New York, NY.

UNGA, 2000: *United Nations Millennium Declaration*. Resolution adopted by the General Assembly, United Nations General Assembly (A/55/L.2), New York, NY.

UNHCR, 2009: *Climate Change, Natural Disasters and Human Displacement: A UNHCR Perspective*. United Nations High Commissioner for Refugees, Geneva, Switzerland, www.unhcr.org/refworld/docid/4a8e4f8b2.html.

UNISDR, 2005a: Hyogo Framework for Action 2005-2015: Building the resilience of nations and communities to disasters. In: *Report of the World Conference on Disaster Risk Reduction, Jan. 2005, Kobe, Japan*, United Nations International Strategy for Disaster Reduction, Geneva, Switzerland, pp. 40-62.

UNISDR, 2005b: *The Link between Millennium Development Goals (MDGs) and Disaster Risk Reduction*. United Nations International Strategy for Disaster Reduction, Geneva, Switzerland, www.unisdr.org/eng/mdgs-drr/link-mdg-drr.htm.

UNISDR, 2005c: *World Conference on Disaster Reduction (WCDR): Proceedings of the Conference held at Kobe Japan*. United Nations International Strategy for Disaster Reduction, Geneva, Switzerland, www.unisdr.org/wcdr/thematic-sessions/WCDR-proceedings-of-the-Conference.pdf.

UNISDR, 2008: *Indicators of Progress: Guidance on Measuring the Reduction of Disaster Risks and the Implementation of the Hyogo Framework of Action*. United Nations International Strategy for Disaster Reduction, Geneva, Switzerland.

UNISDR, 2009a: *Global Assessment Report on Disaster Risk Reduction*. United Nations International Strategy for Disaster Reduction, Geneva, Switzerland, 207 pp., www.preventionweb.net/english/hyogo/gar/report/index.php?id=1130&pid:34&pih.

UNISDR, 2009b: *Second Global Platform on Disaster Risk Reduction, Geneva, June 2009: Concluding Summary by the Platform Chair*. United Nations International Strategy for Disaster Reduction, Geneva, Switzerland.

UNISDR, 2011a: *Revealing Risk, Redefining Development. Global Assessment Report on Disaster Risk Reduction*. United Nations International Strategy for Disaster Reduction, Geneva, Switzerland.

UNISDR, 2011b: *Hyogo Framework for Action 2005-2015: Building the Resilience of Nations and Communities to Disasters, Mid-Term Review 2010-2011*. United Nations International Strategy for Disaster Reduction, Geneva, Switzerland.

UNISDR, 2011c: *Response to the United Kingdom's Department for International Development : Multilateral Aid Review of the UN Secretariat of the International Strategy for Disaster Reduction*. United Nations International Strategy for Disaster Reduction, Geneva, Switzerland.

Universalia Management Group, 2010: *Evaluation of the World Bank Global Facility for Disaster Reduction and Recovery (GFDRR)*. Universalia Management Group, Montreal, Canada, gfdrr.org/docs/GFDRR_EvaluationReportVol-I.pdf.

Vasterman, P., C.J. Yzermans, and A.J.E. Dirkzwager, 2005: The role of the media and media hypes in the aftermath of disasters. *Epidemiologic Reviews*, **27**, 107-114.

Venton, P. and S. La Trobe, 2008: Linking Climate Change Adaptation and Disaster Risk Reduction. Tearfund, Teddington, UK.

Vera, C., M. Barange, O.P. Dube, L. Goddard, D. Griggs, N. Kobysheva, E. Odada, S. Parey, J. Polovina, G. Poveda, B. Seguin and K. Trenberth, 2010: Needs assessment for climate information on decadal timescales and longer. *Procedia Environmental Sciences*, **1**, 275-286.

Vernon, T., 2008: The economic case for pro-poor adaptation: what do we know? *IDS Bulletin*, **39(4)**, 32-41.

Vogel, C. and K. O'Brien, 2006: Who can eat information? Examining the effectiveness of seasonal climate forecasts and regional climate-risk management strategies. *Climate Research*, **33**, 111-122.

Von Doussa, J., A. Corkery, and R. Chartres, 2007: Human rights and climate change. *Australian International Law Journal*, **14**, 161.

Wamsler, C., 2006: Mainstreaming risk reduction in urban planning and housing: a challenge for international aid organizations. *Disasters*, **30(2)**, 151-177.

Warner, K., M. Hamza, A. Oliver-Smith, F. Renaud, and A. Julca, 2009a: Climate change, environmental degradation and migration. *Natural Hazards*, **55(3)**, 689-715.

Warner, K., N. Ranger, S. Surminski, M. Arnold, J. Linnerooth-Bayer, E. Michel-Derjan, P. Kovacs, and C. Herweijer, 2009b: *Adaptation to Climate Change: Linking Disaster Risk Reduction and Insurance*. United Nations International Strategy for Disaster Reduction, Geneva, Switzerland.

Warnock, A., 2007. Small island developing states of the Pacific and climate change: Adaptation and alternatives. *New Zealand Yearbook of International Law*, **4**, 247-286.

Watkins, K. (ed.), 2007: *Fighting Climate Change: Human Solidarity in a Divided World*. Human Development Report 2007/2008 of the United Nations Development Programme, Palgrave Macmillan, Basingstoke, UK and New York, NY, USA, 384 pp.

Weingart, P., A. Engels, and P. Pansegrau, 2000: Risks of communication: discourses on climate change in science, politics, and the mass media. *Public Understanding of Science*, **9(3)**, 261-283.

WHO, 2007: *International Health Regulations*. World Health Organization, Geneva, Switzerland, www.who.int/gb/ebwha/pdf_files/WHA58/WHA58_3-en.pdf.

Wisner, B., P. Blaikie, T. Cannon, and I. Davis, 2004: *At Risk: Natural Hazards, People's Vulnerability and Disasters*. 2nd Ed. Routledge, London, UK.

WMO, 1995: *Resolution 40 (Cg-XII), WMO Policy and Practice for the Exchange of Meteorological and Related Data and Products Including Guidelines on Relationships in Commercial Meteorological Activities*. Twelfth WMO Congress, World Meteorological Organization, Geneva, Switzerland.

WMO, 2007: Commission for Basic Systems Management Group Seventh Session. Final Report. World Meteorological Organization, Geneva, Switzerland.

WMO, 2009: *Thematic progress review on Early Warning Systems*. Background paper for Global Assessment Report on Disaster Risk Reduction, World Meteorological Organization, Geneva, Switzerland.

WMO, 2010: *High-level Taskforce towards the Global Framework for Climate Services (GFCS)*. World Meteorological Organization, Geneva, Switzerland, www.wmo.int/hlt-gfcs/index_en.html.

WMO, 2011: *Resolution 48 (Cg-XVI) - Response to the Report of the High-Level Task Force on the Global Framework for Climate Services*. World Meteorological Organization, Geneva, Switzerland.

Wolz, C., and N. Park, 2006: *Evaluation of ReliefWeb*. Prepared for Office for the Coordination of Humanitarian Affairs, United Nations by Forum One Communications, www.reliefweb.int/rw/lib.nsf/db900SID/LTIO-6VLQJP?OpenDocument.

World Bank, 2006: *Clean Energy and Development: Towards an Investment Framework*. DC2006-0002, World Bank, Washington, DC, 146 pp.

World Bank, 2007a: *A Partnership for Mainstreaming Disaster Mitigation in Poverty Reduction Strategies.* Global Facility for Disaster Reduction and Recovery (GFDRR), World Bank, Washington, DC.

World Bank, 2007b: *The Caribbean catastrophe risk insurance initiative: results of preparation work on the design of a Caribbean Catastrophe Risk Insurance Facility.* World Bank, Washington, DC.

World Bank, 2008: *Catastrophe risk deferred drawdown option (DDO), or CAT DDO.* Background Note, World Bank, Washington, DC.

World Bank, 2009: *Climate Investment Funds: PPCR Programming and Financing Modalities.* World Bank, Washington, DC, siteresources.worldbank.org/INTCC/Resources/ppcrprogrammingdraftvers2april23.pdf.

World Bank, 2010a: *World Development Report 2010: Development and Climate Change.* World Bank, Washington, DC.

World Bank, 2010b: *Economics of Adaptation to Climate Change: Synthesis Report.* World Bank, Washington, DC, 101 pp.

World Bank, 2011a: *Disaster Risk Management Programs for Priority Countries.* World Bank, Global Facility for Disaster Reduction and Recovery (GFDRR), Washington, DC.

World Bank, 2011b: *Migration and Remittances Factbook.* World Bank, Washington, DC.

WTO, 2011: *Legal Texts of the World Trade Organization.* World Trade Organization, Geneva, Switzerland, docsonline.wto.org/gen_browseDetail.asp?preprog=3.

Yamin, F., 2005: The European Union and future climate policy: Is mainstreaming adaptation a distraction or part of the solution? *Climate Policy*, **5(3)**, 349-361.

Yamin, F. and J. Depledge, 2004: *The International Climate Change Regime: A Guide to Rules, Institutions and Procedures.* Cambridge University Press, Cambridge, UK, 699 pp.

Yang, D. and H. Choi, 2007: Are remittances insurance? Evidence from rainfall shocks in the Philippines. *The World Bank Economic Review*, **21(2)**, 219-248.

Yohe, G.W., R.D. Lasco, Q.K. Ahmad, N.W. Arnell, S.J. Cohen, C. Hope, A.C. Janetos, and R.T. Perez, 2007: Perspectives on climate change and sustainable development. In: *Climate Change 2007. Impacts, Adaptation and Vulnerability. Contribution of Working Group II to the Fourth Assessment Report of the Intergovernmental Panel on Climate Change* [Parry, M.L., O.F. Canziani, J.P. Palutikof, P.J. Van Der Linde, and C.E. Hanson (eds.)]. Cambridge University Press, Cambridge, UK, pp. 811-841.

Zhang, D., L. Zhou, and J.F. Nunamaker Jr., 2002: A knowledge management framework for the support of decision making in humanitarian assistance/disaster relief. *Knowledge and Information Systems*, **4**, 370-385.

Zschocke, T. and J.C.V. de Leon, 2010: Towards an ontology for the description of learning resources on disaster risk reduction. In: *Knowledge Management, Information Systems, E-Learning, and Sustainability Research* [Lytras, M.D., P. Ordóñez De Pablos, A. Ziderman, A. Roulstone, H. Maurer, and J.B. Imber (eds.)]. Proceedings of the Third World Summit of the Knowledge Society, WSKS 2010, Corfu, Greece, 22-24 Sep. 2010, Springer-Verlag, Berlin, Germany, pp. 60-74.

8. Toward a Sustainable and Resilient Future

Coordinating Lead Authors:
Karen O'Brien (Norway), Mark Pelling (UK), Anand Patwardhan (India)

Lead Authors:
Stephane Hallegatte (France), Andrew Maskrey (Switzerland), Taikan Oki (Japan), Úrsula Oswald-Spring (Mexico), Thomas Wilbanks (USA), Pius Zebhe Yanda (Tanzania)

Review Editors:
Carlo Giupponi (Italy), Nobuo Mimura (Japan)

Contributing Authors:
Frans Berkhout (Netherlands), Reinette Biggs (South Africa), Hans Günter Brauch (Germany), Katrina Brown (UK), Carl Folke (Sweden), Lisa Harrington (USA), Howard Kunreuther (USA), Carmen Lacambra (Colombia), Robin Leichenko (USA), Reinhard Mechler (Germany), Claudia Pahl-Wostl (Germany), Valentin Przyluski (France), David Satterthwaite (UK), Frank Sperling (Germany), Linda Sygna (Norway), Thomas Tanner (UK), Petra Tschakert (Austria), Kirsten Ulsrud (Norway), Vincent Viguié (France)

This chapter should be cited as:

O'Brien, K., M. Pelling, A. Patwardhan, S. Hallegatte, A. Maskrey, T. Oki, U. Oswald-Spring, T. Wilbanks, and P.Z. Yanda, 2012: Toward a sustainable and resilient future. In: *Managing the Risks of Extreme Events and Disasters to Advance Climate Change Adaptation* [Field, C.B., V. Barros, T.F. Stocker, D. Qin, D.J. Dokken, K.L. Ebi, M.D. Mastrandrea, K.J. Mach, G.-K. Plattner, S.K. Allen, M. Tignor, and P.M. Midgley (eds.)]. A Special Report of Working Groups I and II of the Intergovernmental Panel on Climate Change (IPCC). Cambridge University Press, Cambridge, UK, and New York, NY, USA, pp. 437-486.

Table of Contents

Executive Summary 439

8.1. Introduction 441

8.2. Disaster Risk Management as Adaptation: Relationship to Sustainable Development Planning 443
8.2.1. Concepts of Adaptation, Disaster Risk Reduction, and Sustainable Development and how they are Related 443
8.2.2. Sustainability of Ecosystem Services in the Context of Disaster Risk Management and Climate Change Adaptation 445
8.2.3. The Role of Values and Perceptions in Shaping Response 446
8.2.4. Technology Choices, Availability, and Access 447
8.2.5. Tradeoffs in Decisionmaking: Addressing Multiple Scales and Stressors 448

8.3. Integration of Short- and Long-Term Responses to Extremes 450
8.3.1. Implications of Present-Day Responses for Future Well-Being 450
8.3.2. Barriers to Reconciling Short- and Long-Term Goals 451
8.3.3. Connecting Short- and Long-Term Actions to Promote Resilience 453

8.4. Implications for Access to Resources, Equity, and Sustainable Development 454
8.4.1. Capacities and Resources: Availability and Limitations 454
8.4.2. Local, National, and International Winners and Losers 456
8.4.3. Potential Implications for Human Security 457
8.4.4. Implications for Achieving Relevant International Goals 458

8.5. Interactions among Disaster Risk Management, Adaptation to Climate Change Extremes, and Mitigation of Greenhouse Gas Emissions 458
8.5.1. Thresholds and Tipping Points as Limits to Resilience 458
8.5.2. Adaptation, Mitigation, and Disaster Risk Management Interactions 459
8.5.2.1. Urban 460
8.5.2.2. Rural 461

8.6. Options for Proactive, Long-Term Resilience to Future Climate Extremes 462
8.6.1. Planning for the Future 462
8.6.2. Approaches, Tools, and Integrating Practices 463
8.6.2.1. Improving Analysis and Modeling Tools 464
8.6.2.2. Institutional Approaches 464
8.6.2.3. Transformational Strategies and Actions for Achieving Multiple Objectives 465
8.6.3. Facilitating Transformational Change 466
8.6.3.1. Adaptive Management 467
8.6.3.2. Learning 467
8.6.3.3. Innovation 468
8.6.3.4. Leadership 469

8.7. Synergies between Disaster Risk Management and Climate Change Adaptation for a Resilient and Sustainable Future 469

References 471

Executive Summary

Actions that range from incremental steps to transformational changes are essential for reducing risk from weather and climate extremes (*high agreement, robust evidence*). [8.6, 8.7] Incremental steps aim to improve efficiency within existing technological, governance, and value systems, whereas transformation may involve alterations of fundamental attributes of those systems. The balance between incremental and transformational approaches depends on evolving risk profiles and underlying social and ecological conditions. Disaster risk, climate change impacts, and capacity to cope and adapt are unevenly distributed. Vulnerability is often concentrated in poorer countries or groups, although the wealthy can also be vulnerable to extreme events. Where vulnerability is high and adaptive capacity relatively low, changes in extreme climate and weather events can make it difficult for systems to adapt sustainably without transformational changes. Such transformations, where they are required, are facilitated through increased emphasis on adaptive management, learning, innovation, and leadership.

Evidence indicates that disaster risk management and adaptation policy can be integrated, reinforcing, and supportive – but this requires careful coordination that reaches across domains of policy and practice (*high agreement, medium evidence*). [8.2, 8.3, 8.5, 8.7] Including disaster risk management in resilient and sustainable development pathways is facilitated through integrated, systemic approaches that enhance capacity to cope with, adapt to, and shape unfolding processes of change, while taking into consideration multiple stressors, different prioritized values, and competing policy goals.

Development planning and post-disaster recovery have often prioritized strategic economic sectors and infrastructure over livelihoods and well-being in poor and marginalized communities. This can generate missed opportunities for building local capacity and integrating local development visions into longer-term strategies for disaster risk reduction and adaptation to climate change (*high agreement, robust evidence*). [8.4.1, 8.5.2] A key constraint that limits pathways to post-disaster resilience is the time-bound nature of reconstruction funding. The degradation of ecosystems providing essential services also limits options for future risk management and adaptation actions locally.

Learning processes are central in shaping the capacities and outcomes of resilience in disaster risk management, climate change adaptation, and sustainable development (*high agreement, robust evidence*). [8.6.3, 8.7] An iterative process of monitoring, research, evaluation, learning, and innovation can reduce disaster risks and promote adaptive management in the context of extremes. Technological innovation and access may help achieve resilience, especially when combined with capacity development anchored in local contexts.

Progress toward resilient and sustainable development in the context of changing climate extremes can benefit from questioning assumptions and paradigms, and stimulating innovation to encourage new patterns of response (*medium agreement, robust evidence*). [8.2.5, 8.6.3, 8.7] Successfully addressing disaster risk, climate change, and other stressors often involves embracing broad participation in strategy development, the capacity to combine multiple perspectives, and contrasting ways of organizing social relations.

Multi-hazard risk management approaches provide opportunities to reduce complex and compound hazards in rural and urban contexts (*high agreement, robust evidence*). [8.2.5, 8.5.2, 8.7] Considering multiple types of hazards reduces the likelihood that risk reduction efforts targeted at one type of hazard will increase exposure and vulnerability from other hazards, both in the present and future. Building adaptation into multi-hazard risk management involves consideration of current climate variability and projected changes in climate extremes, which pose different challenges to affected human and natural systems than changes in the means. Where changes in extremes cause greater stresses on human and natural systems, direct impacts may be more unpredictable, increasing associated adaptation challenges.

**The most effective adaptation and disaster risk reduction actions are those that offer development benefits in the relative near term, as well as reductions in vulnerability over the longer term (*high agreement,* *medium evidence*). [8.2.1, 8.3.1, 8.3.2, 8.5.1, 8.6.1] There are tradeoffs between current decisions and long-term goals linked to diverse values, interests, and priorities for the future. Short-term and long-term perspectives on both disaster risk management and adaptation to climate change thus can be difficult to reconcile. Such reconciliation involves overcoming the disconnect between local risk management practices and national institutional and legal frameworks, policy, and planning. Resilience thinking offers some tools for reconciling short- and long-term responses, including integrating different types of knowledge, an emphasis on inclusive governance, and principles of adaptive management. However, limits to resilience are faced when thresholds or tipping points associated with social and/or natural systems are exceeded.

Building a strong foundation for integrating disaster risk management and adaptation to climate change includes making transparent the values and interests that underpin development, including who wins and loses from current policies and practices, and the implications for human security (*high agreement, medium evidence*). [8.2.3, 8.2.4, 8.4.2, 8.4.3, 8.6.1.2] Both disaster risk management and adaptation to climate change share challenges related to (1) reassessing and potentially transforming the goals, functions, and structure of institutions and governance arrangements; (2) creating synergies across temporal and spatial scales; and (3) increasing access to information, technology, resources, and capacity. These challenges are particularly demanding in countries and localities with the highest climate-related risks and weak capacities to manage those risks. Countries with significant capacity and strong risk management records also benefit from addressing these challenges.

Social, economic, and environmental sustainability can be enhanced by disaster risk management and adaptation approaches. A prerequisite for sustainability is addressing the underlying causes of vulnerability, including the structural inequalities that create and sustain poverty and constrain access to resources (*medium agreement, robust evidence*). [8.6.2, 8.7] This involves integrating disaster risk management in other social and economic policy domains, as well as a long-term commitment to managing risk.

The interactions among climate change mitigation, adaptation, and disaster risk management will have a major influence on resilient and sustainable pathways (*high agreement, low evidence*). [8.2.5, 8.5.2, 8.7] Interactions between the goals of mitigation and adaptation in particular will play out locally, but have global consequences.

There are many approaches and pathways to a sustainable and resilient future. Multiple approaches and development pathways can increase resilience to climate extremes (*medium agreement, medium evidence*). [8.2.3, 8.4.1, 8.6.1, 8.7] Choices and outcomes for adaptive actions to climate extremes must reflect divergent capacities and resources and multiple interacting processes. Actions are framed by tradeoffs between competing prioritized values and objectives, and different visions of development that can change over time. Iterative, reflexive approaches allow development pathways to integrate risk management so that diverse policy solutions can be considered, as risk and its measurement, perception, and understanding evolve over time. Choices made today can reduce or exacerbate current or future vulnerability, and facilitate or constrain future responses.

Chapter 8. Toward a Sustainable and Resilient Future

8.1. Introduction

This chapter focuses on the implications of changing climate extremes for development, and considers how disaster risk management and climate change adaptation together can contribute to a sustainable and resilient future. Changes in the frequency, timing, magnitude, and characteristics of extreme events pose challenges to the goals of reducing disaster risk and vulnerability, both in the present and in the future (see Chapter 3). Enhancing the capacity of social-ecological systems to cope with, adapt to, and shape change is central to building sustainable and resilient development pathways in the face of climate change. The concept for social-ecological systems recognizes the interdependence of social and ecological factors in the generation and management of risk, as well as in the pursuit of sustainable development. Despite 20 years on the policy agenda, sustainable development remains contested and elusive (Hopwood et al., 2005). However, within the context of climate change, it is becoming increasingly clear that the sustainability of humans on the Earth is closely linked to resilient social-ecological systems, which is influenced by social institutions, human agency, and human capabilities (Pelling, 2003; Bohle et al., 2009; Adger et al., 2011).

Extremes are translated into impacts by the underlying conditions of exposure and vulnerability associated with development contexts. For example, there is *robust evidence* that institutional arrangements and governance weaknesses can transform extreme events into disasters (Hewitt, 1997; Pelling, 2003; Wisner et al., 2004; Ahrens and Rudolph, 2006). The potential for concatenated global impacts of extreme events continues to grow as the world's economy becomes more interconnected, but in relative terms most impacts will occur in contexts with severe environmental, economic, technological, cultural, and cognitive limits to adaptation (see Section 5.5.3). In relation to extreme events, global risk assessments show that social losses – as well as economic losses as a proportion of livelihood or GDP – are disproportionately concentrated in developing countries, and within these countries in poorer communities and households (UNDP, 2004; UNISDR, 2009, 2011; World Bank, 2010a).

This chapter recognizes that outcomes of changing extreme events depend on responses and approaches to disaster risk reduction and climate change adaptation, both of which are closely linked to development processes. The assessment of literature presented in this chapter shows that changes in extreme events call for greater alignment and integration of climate change responses and sustainable development strategies, and that this alignment depends on greater coherence between short- and long-term objectives. Yet there are different interpretations of development, different preferences and prioritized values and motivations, different visions for the future, and many tradeoffs involved. Research on the resilience of social-ecological systems provides some lessons for addressing the gaps among these objectives. Transformative social, economic, and environmental responses can facilitate disaster risk reduction and adaptation (see Box 8-1). Transformations often include questioning of social values, institutions, and technical practices (Loorbach et al., 2008; Hedrén and Linnér, 2009; Pelling 2010a). A resilient and sustainable future is a choice that involves proactive measures that promote transformations, including adaptive management, learning, innovation, and leadership capacity to manage risks and uncertainty.

In this chapter, we assess a broad literature presenting insights on how diverse understandings and perspectives on disaster risk reduction and climate change adaptation can help to promote a more sustainable and resilient future. Drawing on many of the key messages from earlier chapters, the objective is to assess scientific knowledge on the incremental and transformative changes needed, particularly in relation to integrating disaster risk reduction and climate change adaptation into development policies and pathways. Bringing together experience from a range of disciplines, this chapter identifies proven pathways that can help move from an incremental to an integrative approach that also

Box 8-1 | Transformation in Response to Changing Climate Extremes

Transformation involves fundamental changes in the attributes of a system, including value systems; regulatory, legislative, or bureaucratic regimes; financial institutions; and technological or biophysical systems (see Glossary). This chapter focuses on the transformation of disaster risk management systems in the context of climate extremes, through integration with climate change adaptation strategies and wider systems of human development. This is similar to, yet distinct from, other types of transformation associated with climate change. For example, there have been attempts to understand climate change and development failures by identifying the scope for political (Harvey, 2010), social (Kovats et al., 2005), economic (Jackson, 2009), and value (Leiserowitz et al., 2006) transformation, and so too for disaster risk management (Klein, 2007). Across these cases, observed processes of stasis and change are analogous (often using common language), but actors and objectives are distinct. That said, transformation in wider political, economic, social, and ethical systems can open or close policy space for a more resilient and sustainable form of disaster risk management (Birkland, 2006), just as acts aimed at transformation in managing climate extremes can have implications for wider systems. This is particularly true where contemporary development goals, paths, and hierarchies are identified and addressed as part of the root or proximate causes of vulnerability and risk, that is, when they are seen as part of the solution for building resilient and sustainable futures (Wisner, 2003; Pelling, 2010a). Although there has been some research on how and why social lock-in makes it difficult to move away from established development priorities and trajectories (Pelling and Manuel-Navarrete, 2011), there has been only limited academic work to date on the ways in which wider transformations impact on disaster risk management, and vice versa.

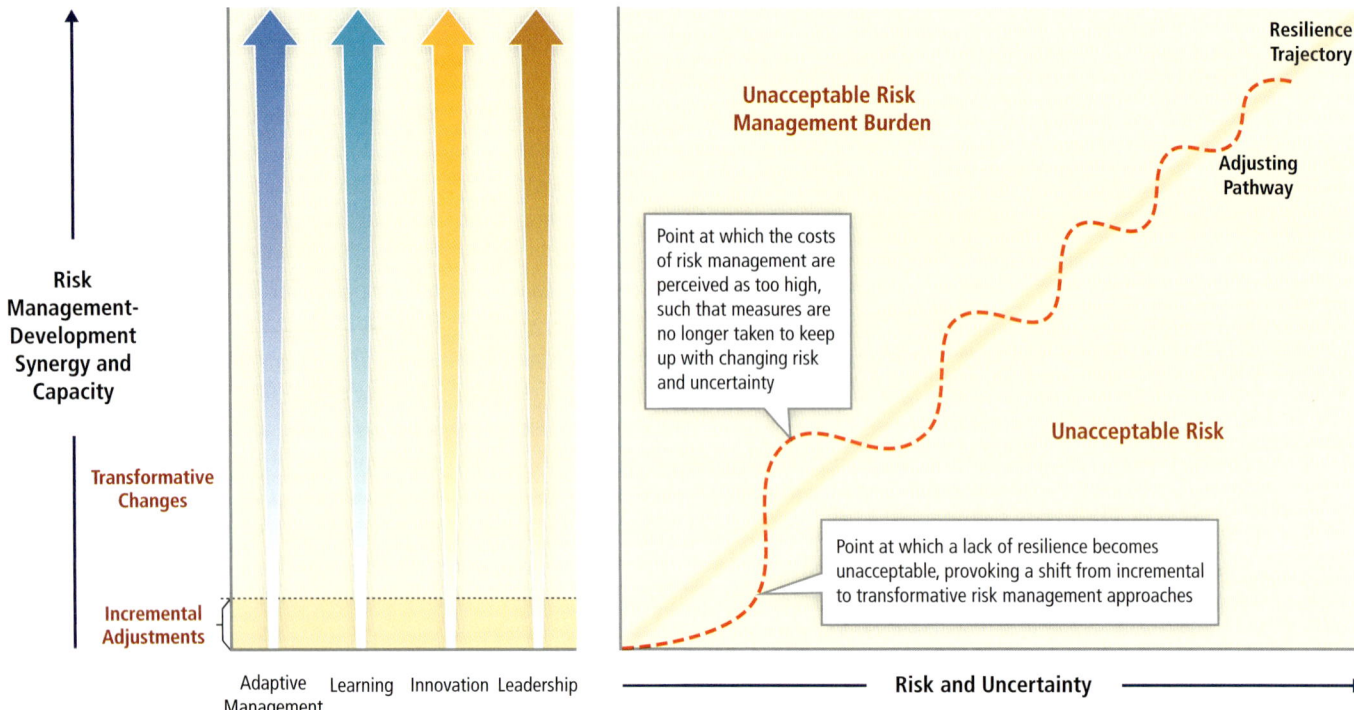

Figure 8-1 | Incremental and transformative pathways to resilience.

embraces transformation – as illustrated in Figure 8-1, which depicts resilience as a moving target that is positioned somewhere between the acceptability of residual risk and the costs of risk management. The target moves as the relationship between risk and uncertainty changes (driven by climate extremes, as well as development trends such as urbanization) in relation to the capacity for risk management (which integrates climate change adaptation, disaster risk management, and development). As risk and uncertainty increase, incremental adjustments in practices may no longer be sufficient to achieve resilience, and at some point the growing resilience gap will provoke a search for transformative solutions. Through enhanced experimentation and learning approaches, climate change adaptation, disaster risk management, and development may provide a pathway for keeping pace with the dynamic drivers and expressions of risk.

After this introduction, this chapter discusses the relationship between disaster risk management, climate change adaptation, and sustainable and resilient development (Section 8.2), highlighting the synergies and conflicts between these objectives and the common obstacles to reaching them (Section 8.2.1) and the specific role of ecosystems and biodiversity (Section 8.2.2). In particular, it emphasizes the importance of values and perceptions (Section 8.2.3) and the role of technologies (Section 8.2.4) in designing sustainability policies. Finally, it highlights the importance of tradeoffs between temporal scales, spatial scales, and multiple stressors (Section 8.2.5).

Focusing on time perspectives, Section 8.3 then discusses options to integrate short- and long-term objectives, by looking at the long-term consequences of present-day responses to disasters (Section 8.3.1), investigating the barriers to integrating short- and long-term responses (Section 8.3.2), and proposing options to overcome these barriers and promote resilience (Section 8.3.3).

Section 8.4 assesses the implications of disaster risk reduction and climate change adaptation for equity and access to resources, and in particular the importance of capacities and resource availability to implement policies for adaptation and disaster risk reduction (Section 8.4.1). It also highlights the existence of losers and winners from disasters and disaster risk reduction and adaptation policies (Section 8.4.2), and the consequences of these distributive effects for human security (Section 8.4.3) and for the possibility to achieve international goals such as the Millennium Development Goals (Section 8.4.4).

Section 8.5 focuses on the specific issue of combining disaster risk management and adaptation with climate change mitigation policies. It starts by stressing the role of thresholds and tipping points as limits to what can be achieved in terms of disaster risk management and adaptation, and thus the importance of considering the three policies together (Section 8.5.1). It then discusses synergies and conflicts between mitigation, adaptation, and disaster risk management in urban and rural areas (Section 8.5.2).

Section 8.6 identifies the tools and options to promote resilience to climate extremes and combine adaptation, disaster risk management, and other policy goals. It first discusses various approaches to planning for the future, including the use of scenarios (Section 8.6.1). It then highlights the existence of a continuum of options to make progress over the short and long term, from incremental to transformational changes (Section 8.6.2). These increasingly ambitious changes include the use of analysis and modeling tools to improve disaster risk management and adaptation (Section 8.6.2.1), the implementation of new institutional tools (Section 8.6.2.2), and transformational strategies to reach multiple objectives (Section 8.6.2.3). Such transformational changes can be facilitated using a combination of approaches (Section 8.6.3), including adaptive management (Section 8.6.3.1), learning (Section 8.6.3.2), innovation (Section 8.6.3.3), and leadership (Section 8.6.3.4). The chapter concludes (Section 8.7) by discussing synergies between disaster risk reduction and climate change adaptation to achieve a resilient and sustainable future.

8.2. Disaster Risk Management as Adaptation: Relationship to Sustainable Development Planning

Earlier chapters discussed the concepts of and relationship between disaster risk management (including disaster risk reduction) and climate change adaptation. The two concepts and practices overlap considerably and are strongly complementary. Disaster risk management considers hazards other than those that are climate-derived, such as earthquakes and volcanoes, while climate change adaptation considers and addresses vulnerabilities related to phenomena that would not normally be classified as discrete disasters, such as gradual changes in precipitation, temperature, or sea level. Examples of hazards that are addressed by both communities include flooding, droughts, and heat waves.

Disaster risk management is increasingly considered as one of the 'frontlines' of adaptation, and perhaps one of the most promising arenas for mainstreaming or integrating climate change adaptation into sustainable development planning (Sperling and Szekely, 2005; G. O'Brien et al., 2006; Schipper and Pelling, 2006; Schipper, 2009). However, it requires modifying development policies, mechanisms, and tools, and identifying and responding to those who gain and lose from living with and creating risk. Contested notions of development and hence differing perspectives on sustainable development planning lead to different conclusions about how disaster risk reduction can contribute to adaptation. This section reviews the definitions of some of the key concepts used in this chapter, and considers the roles that ecosystems services, values and perceptions, technologies, and tradeoffs in decisionmaking can play in influencing sustainable development planning and outcomes. It also considers the tradeoffs that are involved in decisionmaking.

8.2.1. Concepts of Adaptation, Disaster Risk Reduction, and Sustainable Development and how they are Related

Adaptation can be defined as the process of adjustment to actual or expected climate and its effects in order to moderate harm or exploit beneficial opportunities (see Section 1.1.2). Adaptation actions may be undertaken by public or private actors, and can be anticipatory or reactive, and incremental or transformative (Adger et al., 2007; Stafford Smith et al., 2011). In both principle and practice, adaptation is more than a set of discrete measures designed to address climate change; it is an ongoing process that encompasses responses to many factors, including evolving experiences with both vulnerabilities and vulnerability reduction planning and actions, as well as risk perception (Tschakert and Dietrich, 2010; Weber, 2010; Wolf, 2011).

Adaptive capacity underlies action and is defined in this report as the combination of strengths, attributes, and resources available to an individual, community, society, or organization that can be used to prepare for and undertake adaptation. Adaptive capacity can also be described as the capability for innovation and anticipation (Armitage, 2005), the ability to learn from mistakes (Adger, 2003), and the capacity to generate experience in dealing with change (Berkes et al., 2003). Enhancing adaptive capacity under climate change entails paying attention to learning about past, present, and future climate threats, accumulated memory of adaptive strategies, and anticipatory action to prepare for surprises and discontinuities in the climate system (Nelson et al., 2007).

Adaptive capacity is uneven across and within sectors, regions, and countries (K. O'Brien et al., 2006). Although wealthy countries and regions have more resources to direct to adaptation, the availability of financial resources is only one factor determining adaptive capacity (Moss et al., 2010; Ford and Ford, 2011). Other factors include the ability to recognize the importance of the problem in the context of multiple stresses, to identify vulnerable sectors and communities, to translate scientific knowledge into action, and to implement projects and programs (Moser and Ekstrom, 2010). The capacity to adapt is in fact dynamic and influenced by economic and natural resources, social networks, entitlements, institutions and governance, human resources, and technology (Parry et al., 2007). It is particularly important to understand that places with greater wealth are not necessarily less vulnerable to climate impacts and that a socioeconomic system might be as vulnerable as its weakest link (K. O'Brien et al., 2006; Tol and Yohe, 2007). Therefore, even wealthy locations can be severely impacted by extreme events, socially as well as economically, as Europeans experienced during the 2003 heat wave (Salagnac, 2007; see also Case Study 9.2.1).

Current adaptation planning in many countries, regions, and localities involves identification of a wide range of options, although the available knowledge of their costs, benefits, wider consequences, potentials, and limitations is still incomplete (NRC, 2010; see Section 4.5). In many cases, the most attractive adaptation actions are those that offer development benefits in the relatively near term, as well as reductions of vulnerabilities in the longer term (Agrawala, 2005; Klein et al., 2007; McGray et al., 2007; Hallegatte, 2008a; NRC, 2010). This is a lesson already noted, though not always practiced, in disaster preparedness and risk reduction (IFRC, 2002; Pelling, 2010b). An emerging literature discusses adaptation through the lens of sustainability, recognizing that not all adaptation responses are necessarily benign; there are tradeoffs, potentials for negative outcomes, competing interests, different types of knowledge, and winners and losers inherent in adaptation responses (Eriksen and O'Brien, 2007; Ulsrud et al., 2008; Barnett and O'Neill, 2010; Beckman, 2011; Brown, 2011; Eriksen et al., 2011; Gachathi and Eriksen, 2011; Owuor et al., 2011). Sustainable adaptation is defined as a process that addresses the underlying causes of vulnerability and poverty, including ecological fragility; it is considered a way of generating social transformation, or changes in the fundamental attributes of society that contribute to vulnerability (Eriksen and O'Brien, 2007; Eriksen and Brown, 2011).

Disaster risk can be defined in many ways (see Section 1.1.2). In general, however, it is closely associated with the concepts of hazards, exposure, and vulnerability. Hazards are defined in this report as the potential occurrence of a natural or human-induced physical event that may cause negative consequences. Exposure is defined as the presence of people, livelihoods, environmental services and resources, infrastructure, and economic, social, and cultural assets in places that could be adversely affected by climate extremes. Hazards and exposure are changing, not only as the result of climate change, but also due to human activities. For example, hazards associated with floods, landslides, storm surges, and fires can be influenced by declines in ecosystem services that regulate runoff, erosion, etc. The drainage of wetlands, deforestation, the destruction of mangroves, and the changes associated with urban development (such as the impermeability of surfaces and overexploitation of groundwater) are all factors that can modify hazard patterns (Nobre et al., 1991, 2005; MEA, 2005; Nicholls et al., 2008). Consequently, most weather-related hazards now have an anthropogenic element (Cardona, 1999; Lavell, 1999).

Vulnerability has many different (and often conflicting) definitions and interpretations, both across and within the disaster risk and climate communities (see Sections 1.1.2 and 2.2). Vulnerability can increase or decrease over time as a result of both environmental and socioeconomic changes (Blaikie et al., 1994; Leichenko and O'Brien, 2008). In general, improvements in a country's development indicators have been associated with reduced mortality risk, yet an increase in economic loss and insurance claims (UNDP, 2004; Pielke Jr. et al., 2008; Schumacher and Strobl, 2008; ECA, 2009; UNISDR, 2009; World Bank, 2010a). Indeed, recent evidence confirms that, despite increasing exposure, mortality risk from tropical cyclones and floods is now decreasing globally, as well as in heavily exposed regions like Asia (UNISDR, 2011). In contrast, the risk of economic loss is increasing globally because reductions in vulnerability are not compensating for rapid increases in the exposure of economic assets. In the Organisation for Economic Co-operation and Development (OECD) countries, for example, economic losses are increasing at a faster rate than GDP per capita. In other words, the risk of losing wealth in disasters is increasing faster than that wealth is being created (UNISDR, 2011). However, some types of development may increase vulnerability or transfer it between social groups, particularly if development is unequal or degrades ecosystem services (Guojie, 2003). Even where growth is more equitable, vulnerabilities can be generated (e.g., when modern buildings are not constructed to prescribed safety standards) (Hewitt, 1997; Satterthwaite, 2007).

Climate change can magnify many preexisting risks through changes in the frequency, severity, and spatial distribution of weather-related hazards, as well as through increases in vulnerability due to climate impacts (e.g., decreased water availability, decreased agricultural production and food availability, or increased heat stress) (see Section 4.3). Like adaptation, disaster risk reduction may be anticipatory (ensuring that new development does not increase risk) or corrective (reducing existing risk levels) (Lavell, 2009). Given expected population increases in hazard-prone areas, anticipatory disaster risk reduction is fundamental to addressing the risk associated with future climate extremes. At the same time, investments in corrective disaster risk reduction are required to address the accumulation of exposure and susceptibility to existing climate risks, for example, those inherited from past urban planning or rural infrastructure decisions.

Climate change adaptation and disaster risk management (especially disaster risk reduction) are critical elements of long-term sustainability for economies, societies, and environments at all scales (Wilbanks and Kates, 2010). The generally accepted and most widespread definition of sustainable development comes from the Brundtland Commission Report, which defined sustainable development as "development that meets the needs of the present without compromising the ability of future generations to meet their own needs" (WCED, 1987). A number of principles of sustainable development have emerged, including the achievement of a standard of human well-being that meets human needs and provides opportunities for social and economic development; that sustains the life support systems of the planet; that broadens participation in development processes and decisions; and that accelerates the movement of knowledge into action in order to provide a wider range of options for resolving issues (WCED, 1987; Meadowcroft, 1997; NRC, 1999; Swart et al., 2003; MEA, 2005). Because sustainable development means finding pathways that achieve socioeconomic and environmental goals without sacrificing either, it is a concept that is fundamentally political (Wilbanks, 1994).

Discussions of the relationships between sustainable development and climate change have increased over the past decades (Cohen et al., 1998; Yohe et al., 2007; Bizikova et al., 2010). The literature on development has considered how development paths relate to vulnerabilities both to climate change and to climate change policies (e.g., Davis, 2001; Garg et al., 2009), as well as to other hazards. Clearly, some climate change-related environmental shifts are potentially threatening to sustainable development, but they can also help move toward sustainability, especially if the trends or events are severe enough to require significant adjustment of unsustainable development practices or development paths (e.g., the relocation of population or economic activities to less vulnerable areas). In such cases, both disaster risk reduction and climate change adaptation can be important – even essential – contributors to sustainable development.

There are some examples of successful decreases in vulnerability through disaster risk management, but less evidence in relation to climate change adaptation, in part because the ability to attribute observed environmental stresses from and responses to climate change is still limited (Fankhauser et al., 1999; Adger et al., 2007; Repetto, 2008). In terms of disaster risk reduction, a large number of lives have been saved over the last decade due to improved disaster early warning systems (IFRC, 2005), and to increased development and human welfare (UNISDR, 2011). There remains, however, much more that can be done to reduce mortality and counteract growth in the number of people affected by disasters and climate extremes. For example, recent self-assessments of progress by over 100 countries on the objectives of the Hyogo Framework of Action (UNISDR, 2009, 2011) indicate that few developing countries have conducted comprehensive, accurate, and accessible risk assessments, which are a prerequisite for both anticipatory and corrective disaster risk reduction. Furthermore, the assessment shows that few countries are able to quantify their investment in disaster risk reduction. There are numerous ways to evaluate success of disaster risk management or climate adaptation, including gauging the extent to which the goals of a given action (determined in anticipation of a given environmental stressor) are achieved, independent of whether the environmental stressor materializes. Both climate adaptation and disaster risk management can contribute to responses to changes in extreme events due to climate change, yet neither approach alone is sufficient.

Econometric analyses at the national scale have reached different conclusions about the impact of disasters on economic growth, but the balance of evidence suggests a negative impact. Whereas Noy and Nualsri (2007), Noy (2009), Hochrainer (2009), Jaramillo (2009), and Raddatz (2007) suggest that the overall impact on growth is negative, Albala-Bertrand (1993) and Skidmore and Toya (2002) argue that natural disasters have a positive influence on long-term economic growth, often due to both the stimulus effect of reconstruction and the productivity effect. As suggested by Cavallo and Noy (2009) and Loayza et al. (2009), this difference may arise from the different impacts of small and large disasters, the latter having a negative impact on growth and the former enhancing growth. In any case, whether or not disaster losses translate into other social and economic impacts depends on how each individual disaster is managed (Moreno and Cardona, 2011) which in turn is related to capacities and political priorities. At the local scale, Strobl (2011) investigates the impact of hurricane landfall on county-level economic growth in the United States. This analysis shows that a county that is struck by at least one hurricane in a year sees its economic growth reduced on average by 0.79%, and increased by only 0.22% the following year. Noy and Vu (2010) investigate the impact of disasters on economic growth at the province level in Vietnam, and find that lethal disasters decrease economic production while costly disasters increase short-term growth. Rodriguez-Oreggia et al. (2009) focus instead on poverty and the World Bank's Human Development Index at the municipality level in Mexico. They show that municipalities affected by disasters experienced an increase in poverty by 1.5 to 3.6%. Considering these important links between disasters and development, there is a need to consider disaster risk reduction, climate change adaptation, and sustainable development in a consistent and integrated framework (G. O'Brien et al., 2006; Schipper and Pelling, 2006).

8.2.2. Sustainability of Ecosystem Services in the Context of Disaster Risk Management and Climate Change Adaptation

Reducing human pressures on ecosystems and managing natural resources more sustainably can facilitate efforts to mitigate climate change and to reduce vulnerabilities to extreme climate and weather events. The degradation of ecosystems is undermining their capacity to provide ecosystem goods and services upon which human livelihoods and societies depend (MEA, 2005; WWF, 2010), and to withstand disturbances, including climate change. There is evidence that the likelihood of collapse and subsequent regime shifts in ecological and coupled social-ecological systems may be increasing in response to the magnitude, frequency, and duration of climate change and other disturbance events (Folke et al., 2004; MEA, 2005; Woodward, 2010). Large, persistent shifts in ecosystem services not only affect the total level of welfare that people in a community can enjoy, they also impact the welfare distribution between people within and between generations and hence may give rise to new conflicts over resource use and questions on inter-generational equity as a component of sustainable development (Thomas and Twyman, 2005). They could result in domino effects of increased pressure on successive resource systems, as has been suggested in the case of depletion of successive fish stocks (Berkes et al., 2006). However, the thresholds at which ecosystems undergo regime shifts and the points at which these may catalyze social stress remain largely unknown, partly due to variability over space and time (Biggs et al., 2009; Scheffer, 2009).

Ecosystems can act as natural barriers against climate-related hazardous extremes, reducing disaster risk (Conde, 2001; Scholze et al., 2005). For example, mangrove forests are a highly effective natural flood control mechanism that will become increasingly important with sea level rise, and are already used as a coastal defense against extreme climatic and non-climatic events (Adger et al., 2005). The benefits of such ecosystem services are determined by ecosystem health, hazard characteristics, local geomorphology, and the geography and location of the system with respect to the hazard (Lacambra and Zahedi, 2011). In assessing the ecological limits of adaptation to climate change, Peterson (2009) emphasizes that ecosystem regime shifts can occur as the result of extreme climate shocks, but that such shifts depend upon the resilience of the ecosystem, and are influenced by processes operating at multiple scales. In particular, there is evidence that the loss of regulating services (e.g., flood regulation, regulation of soil erosion) erodes ecological resilience (MEA, 2005).

Ecosystems and ecosystem approaches can also facilitate adaptation to changing climatic conditions (Conde, 2001; Scholze et al., 2005). Conservation of water resources and wetlands that provide hydrological sustainability can further aid adaptation by reducing the pressures and impacts on human water supply, while forest conservation for carbon sinks and alternative sources of energy such as biofuels can reduce carbon emissions and have multiple benefits (Reid, 2006), as can coastal defenses and avalanche protection (Silvestri and Kershaw, 2010). In New York, for example, untreated storm water and sewage regularly flood the streets because the aging sewage system is no longer adequate. After heavy rains, overflowing water flows directly into rivers and streams instead of reaching water treatment plants. The US Environmental Protection Agency has estimated that around US$ 300 billion over 20 years would be needed to upgrade sewage infrastructure across the country (UNISDR, 2011). In response, New York City will invest US$ 5.3 billion in green infrastructure on roofs, streets, and sidewalks. This promises multiple benefits: the new green spaces may absorb more rainwater and reduce the burden on the city's sewage system, improve air quality, and reduce water and energy costs. Such changes in the constituents of an ecosystem can be used as levers to enhance the resilience of coupled social-ecological systems (Biggs et al., 2009).

Biodiversity is also important to adaptation. Functionally diverse systems have more scope to adapt to climate change and climate variability than functionally impoverished systems (Elmqvist et al., 2003; Hughes et al., 2003; Lacambra and Zahedi, 2011). A larger gene pool will facilitate the emergence of genotypes that are better adapted to changed climatic conditions. Conservation of biodiversity and maintenance of ecosystem integrity may therefore be a key objective in improving the adaptive capacity of society to cope with climate change extremes (Peterson et al., 1997; Elmqvist et al., 2003; SCBD, 2010).

Strategies that are adopted to reduce climate change through greenhouse gas mitigation can affect biodiversity both negatively and positively (Edenhofer et al., 2011), which in turn influences the capacity to adapt to climate extremes. For example, some bioenergy plantations replace sites with high biodiversity, introduce alien species, and use damaging agrochemicals, which in turn may reduce ecosystem resilience and hence their capacity to respond to extreme events (Foley et al., 2005; Fargione et al., 2009). Large hydropower schemes can cause loss of terrestrial and aquatic biodiversity, inhibit fish migration, and lead to mercury contamination (Montgomery et al., 2000), as well as change watershed sediment dynamics, leading to sediment starvation in coastal areas, which in turn could lead to coastal erosion and make coasts more vulnerable to sea level rise and storm surges (Silvestri and Kershaw, 2010).

The increasing international attention and support for efforts focused on Reducing Emissions from Deforestation and Forest Degradation, maintaining/enhancing carbon stocks, and promoting sustainable forest management (REDD+) is an example of where incentives for the protection and sustainable management of natural resources driven by mitigation concerns also have the potential of generating co-benefits for adaptation. By mediating runoff and reducing flood risk, protecting soil from water and wind erosion, providing climate regulation, and providing migration corridors for species, ecosystem services supplied by forests can increase the resilience to some climatic changes (Locatelli et al., 2010). Primary forests tend to be more resilient to disturbance and environmental changes, such as climate change, than secondary forests and plantations (Thompson et al., 2009). However, forests are also vulnerable to climatic extremes (Nepstad et al., 2007) and the modeled effects of global warming (Vergara and Scholz, 2011). Hence, the role of forest ecosystems in climate change mitigation and adaptation will itself depend on the rate and magnitude of climate change and whether the crossing of ecological tipping points can be avoided.

8.2.3. The Role of Values and Perceptions in Shaping Response

Values and perceptions are important in influencing action on climate change extremes, and they can have significant implications for sustainable development. The disaster risk community has used several points of view for resolving decisions about where to invest limited resources, including considerations of economic rationality and moral obligation (Sen, 2000). Value judgments are embedded in problem framing, solutions, development decisions, and evaluation of outcomes, thus it is important to make them explicit and visible. Values describe what is desirable or preferable, and they can be used to represent the subjective, intangible dimensions of the material and nonmaterial world (O'Brien and Wolf, 2010). They are closely linked to worldviews and beliefs, including perceptions of change and causality (Rohan, 2000; Leiserowitz, 2006; Weber, 2010). Values both inform and are shaped by action, judgment, choice, attitude, evaluation, argument, exhortation, rationalization, and attribution of causality (Rokeach, 1979). However, values do not always clearly translate to particular behaviors (Leiserowitz et al., 2005). Recognizing and reconciling conflicting values increases the need for inclusiveness in decisionmaking and for finding ways to communicate across social and professional boundaries (Rosenberg, 2007; Vogel et al., 2007; Oswald Spring and Brauch, 2011).

Losses from extreme events can have implications beyond objective, measurable impacts such as loss of lives, damage to infrastructure, or economic costs. They can lead to a loss of what matters to individuals, communities, and groups, including the loss of elements of social capital, such as sense of place or of community, identity, or culture. This has long been observed within the disaster risk community (Hewitt, 1997; Mustafa, 2005) and in more recent work in the climate change community (O'Brien, 2009; Adger et al., 2010; Pelling, 2010a). A values-based approach recognizes that socioeconomic systems are continually evolving, driven by innovations, aspirations, and changing values and preferences of the constituents (Simmie and Martin, 2010; Hedlund-de Witt, 2011). This approach raises not only the ethical question of 'whose values count?', but also the important political question of 'who decides?'. These questions have been asked in relation to both disaster risk (Blaikie et al., 1994; Wisner, 2003; Wisner et al., 2004) and climate change (Adger, 2004; Hunt and Taylor, 2009; Adger et al., 2010; O'Brien and Wolf, 2010), and are significant when considering the interaction of climate change and disaster risk, including the complexity of the temporal consequences of policies and decisions (Pelling, 2003).

The probabilistic risk assessments that form the basis for current models of cost-benefit analysis (CBA) rarely take into account the wider consequences that account for a substantial proportion of disaster damage for poorer households and communities (UNISDR, 2004, 2009; Marulanda et al., 2010). These include outcomes such as increased poverty and inequality (Hallegatte, 2006; de la Fuente et al., 2009), health effects (Murray and Lopez, 1996; Grubb et al., 1999; Viscusi and Aldy, 2003), cultural assets and historical building losses (ICOMOS, 1998), and environmental impacts, which are often very difficult to measure in monetary terms. Specific approaches allow accounting for distributive effects in CBA (e.g., distributional-weight CBA, see Harberger, 1978; basic-needs CBA, see Harberger, 1984; or social welfare function built as a sum of individual welfare function that increases nonlinearly with income), but none of them are consensual. Other types of valuation emphasize institutional elements such as the 'moral economy' associated with the collective memory and identities of people living in non-western cultures in many parts of the world (Rist, 2000; Hughes, 2001; Trawick, 2001; Scott, 2003).

Two important philosophical value frameworks have dominated attempts to establish priorities for risk management: human rights and utilitarian approaches. Human rights-based approaches (Wisner, 2003; Gardiner, 2010) emphasize moral obligation to reduce avoidable risk and contain loss, which was recognized in the UN Universal Declaration of Human Rights in 1948: Article 3 provides for the right to "life, liberty and security of person," while Article 25 protects "a standard of living adequate for the health and well-being … in the event of unemployment, sickness, disability, widowhood, or old age or other lack of livelihood in circumstances beyond his [sic] control."

The humanitarian community, and civil society more broadly, has made considerable progress in addressing these aspirations (Kent, 2001), perhaps best exemplified by the Sphere standards. These are a set of self-imposed guidelines for good humanitarian practices that require impartiality in post-disaster actions including shelter management and access to and distribution of relief and reconstruction aid (Sphere, 2004). The ethics and equity dimensions of risk management have also been explored in adaptation through the application of Rawls' theory of justice (Rawls, 1971; Paavola, 2005; Paavola and Adger 2006; Paavola et al., 2006; Grasso, 2009, 2010). From this perspective, priority is given to reducing risk for the most vulnerable, even if this limits the absolute numbers who benefit.

In contrast to focusing on the most excluded or economically poor, utilitarian approaches assume that interpersonal welfare comparisons are possible, and that a social welfare function that summarizes the welfare of a population can be built (Pigou, 1920). Assuming its existence, maximizing this social welfare function reveals where economic benefits of public investments exceed costs. The calculated economic benefits of investing in risk reduction vary, but are often considered significant (see Ghesquiere et al., 2006; World Bank 2010a; UNISDR, 2011). There are, however, extreme difficulties in accounting for the complexity of disaster costs and risk reduction investment benefits (Pelling et al., 2002; Hallegatte and Przyluski, 2010). A key point here is that value frameworks can significantly influence the types of responses to climate and weather extremes.

8.2.4. Technology Choices, Availability, and Access

Technology choices can contribute to both risk reduction and risk enhancement, relative to extreme climate and weather events. As discussed in Section 7.4.3, technologies receive prominent attention in both climate change adaptation and disaster risk reduction. Continuing transitions from one socio-technological state to another frame many aspects of responses to climate change risks. Assessments of roles of technology choices, availability, and access in responding to climate extremes are enmeshed in a wide range of technologies that must be considered within a broad range of development contexts. However, in nearly every case, issues are raised about the balance between risk reduction and risk creation. Technology is a broad concept that embraces a range of areas, including information and communication technologies, roads and infrastructure, food and production technologies, energy systems, and so on. Technology choices can alleviate disaster risk, but they can also significantly increase risks and add to adaptation challenges (Jonkman et al., 2010). For example, some modern energy systems and centralized communication systems are dependent on physical structures that can be vulnerable to storm damage. It has been suggested that relatively centralized high-technology systems are 'brittle,' offering efficiencies under normal conditions but subject to cascading effects in the event of emergencies (Lovins and Lovins, 1982).

In many cases, technologies are considered to be an important part of responses to climate extremes and disaster risk. This includes, for example, attention to physical infrastructure, including how to 'harden' built infrastructure such as bridges or buildings, or natural systems such as hillsides or river channels, such that they are able to withstand higher levels of stress (UNFCCC, 2006; Larsen et al., 2007; CCSP, 2008). Another focus is on technologies that assist with information collection and diffusion, including technologies to monitor possible stresses and vulnerabilities, technologies to communicate with populations and responders in the event of emergencies, and technology applications to disseminate information about possible threats and contingencies – although access to such technologies may be limited in some developing regions. Seasonal climate forecasts based on the results from numerical climate models have been developed in recent decades to provide multi-month forecasts, which can be used to prepare for floods and droughts (Stern and Easterling, 1999). Modern technological development is exploring a wide variety of innovative concepts that may eventually hold promise for disaster risk reduction, for example, through new food production technologies, although ecological, ethical, and human health implications are often as yet unresolved (Altieri and Rosset, 1999).

Attention to technology alternatives and their benefits, costs, potentials, and limitations in cases where disaster risk is created and when risk reduction takes place involve two different time horizons. In the near term, technologies to be considered are those that currently exist or that can be modified relatively quickly. In the longer term, it is possible to consider potentials for new technology development, given identified needs (Wilbanks, 2010). In some circumstances, technologies put in place to reduce short-term risk and vulnerability can increase future

> **FAQ 8.1 | Why is there not a greater emphasis on technology as the solution to climate extremes?**
>
> Technology is an essential part of responses to climate extremes, at least partly because technology choices and uses are so often a part of the problem. Enhancing early warning systems is one example where technology can play an important role in disaster risk management. This example also flags the importance of considering 'hard' (engineering) and 'soft' (social and administrative) technology. Great advances have been made in hard technology around hazard identification, and this has saved many lives. Communicating warnings through the 'soft' technology of institutional reform and communication networks has been less well developed. Both hard and soft technology systems must be responsive to different cultures, environments, and types of governance. Most fundamentally, it is clear that technologies are the product of research and development choices, which reflect particular values, interests, and priorities. The successful transfer of technology is sensitive to local needs, capacities, and development goals. Technologies can have unintended consequences that contribute to maladaptations. For example, some modern agricultural technologies may reduce local biodiversity and constrain future adaptation. Technologies only matter if they are both appropriate and accessible. Technology development and use are necessary for reducing vulnerabilities to climate extremes, both through mitigation and adaptation, but they need to be the right technologies that are deployed in the right ways. This calls for greater reflection on the social, economic, and environmental consequences of technology across both space and time. In many cases, responses to climate extremes can be improved by addressing social vulnerability, rather than focusing exclusively on technological responses.

vulnerability to extreme events or ongoing trends. For example, the use of irrigation has reduced farmer vulnerabilities to low and variable precipitation patterns. However, when the irrigation water is from a nonrenewable source (e.g., the Ogallala-High Plains aquifer system of the United States), the foreseeable reduction in future irrigation opportunities would mean an increase in vulnerability and the risk of increasing crop failures (AAG, 2003; Harrington, 2005).

Similarly, while large dams could mitigate drought and generate electricity, well known costs of social and ecological displacement may be unacceptable (Baghel and Nusser, 2010). Furthermore, unless dams are constructed to accommodate future climate change, they may present new risks to society by encouraging a sense of security that ignores departures from historical experience (Wilbanks and Kates, 2010). In the Mekong region, dikes, dams, drains, and diversions established for flood protection have unexpected consequences for risk over the longer term, because they influence risk-taking behavior (Lebel et al., 2009). In the United States, past building in floodplain areas downstream from dams that have now exceeded their design life has become a major concern; tens of thousands of dams are now considered as having high hazard potential (McCool, 2005; FEMA, 2009; ASCE, 2010).

Investments in physical infrastructure cast long shadows through time, because they tend to assume lifetimes of three to four decades or longer. The gradual modernization of a city's housing stock, transport, or water and sanitation infrastructure takes many decades without targeted planning. If they are maladaptive rather than adaptive, the consequences can be serious. This suggests a reappraisal of technology that might promote more distributed solutions, for example, multiple, smaller dams that can resolve local as well as more distant needs, or widely spread, local energy production (perhaps utilizing micro-solar, wind and water, or geothermal power) that can reduce exposure to secondary impacts from natural disasters when large power generators or power transmission lines are lost during a natural disaster, or when power plants generate secondary disasters after being impacted by a natural hazard, as has happened recently in Japan. The goal of a more distributed and less maximizing development vision has been expressed in Thailand's 'Sufficiency Economy' approach, where local development is judged against its contribution to local, national, and international wealth generation (UNDP, 2007a).

Technology choices, availability, and access depend on more than technology development alone. Unless the technologies, the skills required to use them, and the institutional approaches appropriate to deploy them are effectively transferred from providers to users ('technology transfer'), the effects of technology options, however promising, are minimized (see Section 7.4.3). Challenges in putting science and technology to use for sustainable development have received considerable attention (e.g., Nelson and Winter, 1982; Patel and Pavit, 1995; NRC, 1999; ICSU, 2002; Kristjanson et al., 2009), emphasizing the wide range of contexts that shape both barriers and potentials. If obstacles related to intellectual property rights can be overcome, however, the growing power of the information technology revolution could accelerate technology transfer (linked with local knowledge) in ways that would be very promising (Wilbanks and Wilbanks, 2010).

8.2.5. Tradeoffs in Decisionmaking: Addressing Multiple Scales and Stressors

Sustainable development involves finding pathways that achieve a variety of socioeconomic and environmental goals, without sacrificing any one for the sake of the others. As a result, the relationships between adaptation, disaster risk management, and sustainability are highly political. Successful reconciliation of multiple goals "lies in answers to such questions as who is in control, who sets agendas, who allocates resources, who mediates disputes, and who sets rules of the game"

(Wilbanks, 1994, p. 544). This means that conflicts of interest must be acknowledged and addressed, whether they are between government departments, sectors, or policy arenas, and suggests that simple panaceas are unlikely without tradeoffs in decisionmaking (Brock and Carpenter, 2007).

There is no single or optimal way of adapting to climate change or managing risks, because contexts for risk management vary so widely. For example, risk management decisions can be oriented toward incremental responses to frequent events that are disruptive but perhaps not 'extreme.' Often, tradeoffs between multiple objectives are ambiguous. For example, focusing on and taking actions to protect against frequent events may lead to greater vulnerability to larger and rarer extreme events (Burby, 2006). This is a particular challenge for investing in fixed physical infrastructure. Social investments and risk awareness, including early warning systems, can be strengthened by more frequent low-impact events that maintain risk visibility and allow preparedness for larger, less frequent events (see Case Studies 9.2.11 and 9.2.14). Pielke Jr. et al. (2007) also warn that locating adaptation policy in a narrow risk framework by concentrating only on identifiable anthropogenic risks can distort public policy because vulnerabilities are created through multiple stresses.

As one salient example, during disaster reconstruction, tensions frequently arise between demands for speed of delivery and sustainability of outcome. Response and reconstruction funds tend to be time-limited, often requiring expenditure within 12 months or less from the time of disbursement. This pressure is compounded by multiple agencies working with often limited coordination. Time pressure and competition between agencies tends to promote centralized decisionmaking and the subcontracting of purchasing and project management to non-local commercial actors. Both outcomes save time but miss opportunities to include local people in decisionmaking and learning from the event, with the resulting reconstruction in danger of failing to support local cultural and economic priorities (Berke et al., 1993; Pearce, 2003). At the same time it is important not to romanticize local actors or their viewpoints, which might at times be unsustainable or point to maladaptation, or to accept local voices as representative of all local actors. When successful, participatory reconstruction planning has been shown to build local capacity and leadership, bind communities, and provide mechanisms for information exchange with scientific and external actors (Lyons et al., 2010). As part of any participatory or community-based reconstruction, the importance of a clear conflict resolution strategy has been recognized.

Tradeoffs may also arise through conflicts between economic development and risk management (Kahl, 2003, 2006). The current trend of development in risk-prone areas (e.g., coastal areas in Asia) is driven by socioeconomic benefits yielded by these locations, with many benefits accruing to private investors or governments through tax revenue. For example, export-driven economic growth in Asia favors production close to large ports to reduce transportation time and costs. Consequently, the increase in risk has to be balanced against the socioeconomic gains of development in at-risk areas. Additional construction in at-risk areas is not unacceptable *a priori*, but has to be justified by other benefits, and sometimes complemented by other risk-reducing actions (e.g., early warning and evacuation, improved building norms, specific flood protection). This introduces the possibility for those benefiting financially to offset produced risk through risk reduction mechanisms ranging from fair wages and disaster-resistant housing (to enhance worker resilience) to support for early warning, preparedness, and reconstruction. Such approaches have been considered in some businesses through corporate social responsibility agendas (Twigg, 2001).

One climate change/development tradeoff linked both to timeframes and the magnitude of climate extremes is the future need for risk reduction infrastructure that would require changes in ecologically or historically important areas. For example, when considering additional protection (e.g., dikes and seawalls) in historical centers, aesthetic and cultural elements as well as building costs will be taken into account. Existing planning and design standards to protect cultural heritage or ecological integrity may need to be balanced with the needs of adaptation (Hallegatte et al., 2011a). Difficulties in attributing value to cultural and ecological assets mean that CBAs are not the best tool to approach these types of problems. Multi-criteria decisionmaking tools (Birkmann, 2006) that incorporate a participatory element and can recognize the political, ethical, and philosophical aspects of such decisions can also be useful (Mercer et al., 2008). But the magnitude of emerging climate extremes is an important issue. If climate change is relatively severe, rather than moderate, then the focus on preserving iconic areas is likely to increase, as will the costs.

Another contextual complication that introduces tradeoffs is the fact that impacts of climate change extremes extend across multiple scales. The challenge is to find ways to combine the strengths of addressing multiple scales, rather than having them work against each other (Wilbanks, 2007, 2009). Local scales offer potentials for bottom-up actions that ensure participation, flexibility, and innovation. At the same time, efforts to develop initiatives from the bottom up are often limited by a lack of information, limited resources, and limited awareness of larger-scale driving forces (AAG, 2003). Larger scales offer potentials for top-down actions that assure resource mobilization and cost sharing. Integrating these kinds of assets across scales is often essential for resilience to extremes, but in fact, integration is profoundly impeded by differences in who decides, who pays, and who benefits, and perceptions of scalar effects that often reflect striking ignorance and misunderstanding (Wilbanks, 2007). In recent years, there have been a number of calls for innovative co-management structures that cross scales in order to promote sustainable development (e.g., Bressers and Rosenbaum, 2003; Cash et al., 2006; Campbell et al., 2010).

What might be done to realize potentials for integrating actions at different scales to make them more complementary and reinforcing? Many top-down interventions (from international donor development and disaster response and reconstruction funding to new adaptation fund mechanisms and national programming) may unintentionally discourage local action by imposing bureaucratic conditions for access to financial and other resources (Christoplos et al., 2009). Top-down sustainability initiatives are often preoccupied with *input* metrics, such as criteria for partner selection and justifications (often based on relatively detailed quantitative analyses of such attributes as 'additionality'), rather than on *outcome* metrics, such as whether the results make a demonstrable contribution to sustainability (regarding metrics, see NRC, 2005).

To manage tradeoffs and conflicts in an open, efficient, and transparent way, institutional and legal arrangements are extremely important. The existing literature on legislation for adaptation at the state level is not comprehensive, but those countries studied lack many of the institutional mechanisms and legal frameworks that are important for coordination at the state level (Richardson et al., 2009). This has been found to be the case for Vietnam, Laos, and China (Lin, 2009). In the South Pacific, high exposure to climate change risk has yet to translate into legislative frameworks to support adaptation – with only Fiji, Papua New Guinea, and Western Samoa formulating national climate change regulatory frameworks (Kwa, 2009). Without a supporting and implemented national legislative structure, achieving local disaster reduction and climate change adaptation planning can be complicated (La Trobe and Davis, 2005; Pelling and Holloway, 2006; see also Section 6.4). Still, where local leadership is determined, skillful planning is possible, even without legislation. This has been the experience of Ethekwini Municipality (the local government responsible for the city of Durban, South Africa), which has developed a Municipal Climate Protection Programme with a strong and early focus on adaptation without national-level policy or legal frameworks to guide adaptation planning at the local level (Roberts, 2008, 2010).

One way around the challenges of tradeoffs is to 'bundle' multiple objectives through broader participation in strategy development and action planning, both to identify multiple objectives and to encourage attention to mutual co-benefits. In this sense, both the pathway and outcomes of development planning have scope to shape future social capacity and disaster risk management. Policies and actions to achieve multiple objectives include stakeholder participation, participatory governance (IRGC, 2009), capacity building, and adaptive organizations, including both private and public institutions where there is a considerable knowledge base reflecting both research and practice to use as a starting point (e.g., NRC, 2008). Multi-hazard risk management approaches provide opportunities to reduce complex and compound hazards, both in rural and urban contexts.

8.3. Integration of Short- and Long-Term Responses to Extremes

When considering the linkages between disaster management, climate change adaptation, and development, time scales play an important role. Disaster management increasingly emphasizes vulnerability reduction in addition to the more traditional emergency response and relief measures. This requires addressing underlying exposure and sensitivity in the context of hazards with different frequencies and return periods. As discussed in Chapter 2, there is now a converging focus on vulnerability reduction in the context of disaster risk management and adaptation to climate change (Sperling and Szekely, 2005).

Cross-scale (spatial and temporal) interactions between responses focusing on the short term and those required for long-term adjustment can potentially create both synergies and contradictions among disaster risk reduction, climate change adaptation, and development. This section assesses the literature regarding synergies and tradeoffs between short- and long-term adjustments. First, we consider the implications of present-day responses for future well-being. The barriers to reconciling short- and long-term goals are then assessed. Insights from research on the resilience of social-ecological systems are then considered as a potential means of addressing integration in a long-term perspective.

8.3.1. Implications of Present-Day Responses for Future Well-Being

The implications of present-day responses to both disaster risk and climate change can be either positive or negative for human security and well-being in the long term. Positive implications can include increased resilience, capacity building, broad social benefits from extensive participation in risk management and resilience planning, and the value of multi-hazard planning (see Sections 5.4 and 6.5). Negative implications can include threats to sustainability if the well-being of future generations is not considered; issues related to the economic discounting of future benefits; 'silo effects' of optimizing responses for one system or sector without considering interaction effects with others (see Burby et al., 2001); equity issues regarding who benefits and who pays; and the 'levee effect,' where the adaptive solution to a current risk management problem builds confidence that the problem has been solved, blinding populations to the possibility that conditions may change and make the present adaptation inadequate (Burby, 2006; Burby et al., 2006).

The terms 'coping' and 'adaptation' reflect strategies for adjustments to changing climatic and environmental conditions. In the case of a set of policy choices, both coping and adaptation denote forms of conduct that aim and indeed may achieve modifications in the ways in which society relates to nature, and nature to society (Stehr and von Storch, 2005). As discussed in Section 2.4, coping actions are those that take place when trying to alleviate the impacts or to live with the costs of a specific event. They are usually found during the unfolding of disaster impacts, which can continue for some time after an event – for example, if somebody loses their job or is traumatized. Coping strategies can help to alleviate the immediate impact of a hazard, but may also increase vulnerabilities over the medium to longer term (Swift, 1989; Davies, 1993; Sperling et al., 2008). The different time frames for coping and adaptation can present challenges for risk management. Focusing on short-term responses and coping strategies can limit the scope for adaptation in the long term. For example, drought can force agriculturalists to remove their children from school or delay medical treatment, which may have immediate survival benefits, yet in aggregate undermines the human resources available for long-term adaptation (Norris, 2005; Alderman et al., 2006; Santos, 2007; Sperling et al., 2008).

In both developed and developing countries, a focus on coping with the present is often fueled by the perception that climate change is a long-term issue and that other challenges, including economic growth, food security, water supply (Bradley et al., 2006), sanitation, education, and health care, require more immediate attention (Klein et al., 2005; Adly and Ahmed, 2009; Kameri-Mbote and Kindiki, 2009). Particularly in poor rural contexts, short-term coping may be a tradeoff that increases longer-term risks (UNISDR, 2009; Brauch and Oswald Spring, 2011). Adaptation, on the other hand, is often focused on minimizing potential risk to future losses (Oliver-Smith, 2007). This 'long-term' framing of adaptation can constrain both short-term coping and adaptive capacity, for example, when relocation of settlements to avoid coastal hazards undermines social capital and local livelihoods, limiting household coping and adaptive capacity (Hunter, 2005). There is a large literature and much experience related to slum relocation that is of direct relevance to urban coping and adaptation (Gilbert and Ward, 1984; Davidson et al., 1993; Viratkapan and Perera, 2006). Context is important in discussing tradeoffs between addressing short- and long-term risks, and even in well-governed systems, political expediency will often distort the regulatory process in a way that favors the short term (Platt, 1999).

Disasters can destroy assets and wipe out savings, and can push households into 'poverty traps,' that is, situations where productivity is reduced, making it impossible for households to rebuild their savings and assets (Zimmerman and Carter, 2003; Carter et al., 2007; Dercon and Outes, 2009; López, 2009; van den Berg, 2010). The process by which a series of events generates a vicious spiral of impacts, vulnerability, and risk was first recognized by Chambers (2006), who described it as the ratchet effect of disaster, risk, and vulnerability. These micro-level poverty traps can also be created by health and social impacts of natural disasters: it has been shown that disasters can have long-lasting consequences for psychological health (Norris, 2005), and for child development from reduction in schooling and diminished cognitive abilities (see Alderman et al., 2006; Santos, 2007; Bartlett, 2008).

Where disaster loss is widespread, micro-level poverty traps can aggregate to the regional level. Here, poor regions impacted by disaster are unable to fully recover so that capacity is reduced and vulnerability heightened, making future disasters more likely. Without enough time to rebuild between events, such regions may end up in a state of permanent reconstruction, with resources devoted to repairing and replacing rather than accumulating infrastructure and equipment. This obstacle to capital accumulation and infrastructure development can lead to a permanent disaster-related underdevelopment (Hallegatte et al., 2007; Hallegatte and Dumas, 2008). This can be amplified by other long-term mechanisms, such as changes in risk perception that reduce investments in the affected regions or reduced services that make qualified workers leave the region. These effects have been discussed by Benson and Clay (2004),

and investigated by Noy (2009) and Hochrainer (2009), who found that natural disasters have a negative impact on economic growth and development, especially when direct losses are large. This negative impact is found to be larger when the disaster affects a smaller country, with lower GDP per capita, weaker institutions, lower openness to trade, lower literacy rates, and lower levels of government spending, and when foreign aid and remittances are lower. Such effects have been modeled by Hallegatte et al. (2007) and Hallegatte and Dumas (2008) using a reduced-form economic model that shows that the average GDP impact of natural disasters can be either close to zero if reconstruction capacity is large enough, or very large if reconstruction capacity is too limited, which may be the case in less-developed countries. There are, however, many uncertainties in the ways in which people's spontaneous and organized responses to increasing climate-related hazards feed back to influence long-term adaptive capacity and options. Migration, which can be traumatic for those involved, might lead to enhanced life chances for the children of migrants, building long-term capacities and potentially also contributing to the movement of populations away from places exposed to risk (IOM, 2007, 2009a,b; Ahmed, 2009; Oswald Spring, 2009b; UNDP, 2009).

A broad literature on experiences of community-based and local-level disaster risk reduction indicates options for transiting from short- to longer-term responses, at least in the context of frequently occurring risk manifestations (Lavell, 2009; UNISDR, 2009; Maskrey, 2011). Such approaches, many of which are based on community participation, have progressively moved from addressing disaster preparedness and capacities for emergency management toward addressing the vulnerability of livelihoods, the decline of ecosystems, the lack of social protection, unsafe housing, the improvement of governance, and other underlying risk factors (Bohle, 2009). While managing existing risk will contain loss, addressing underlying risk drivers will contribute to a reduction in future risk to climate extremes.

8.3.2. Barriers to Reconciling Short- and Long-Term Goals

Although there is *robust evidence* in the literature to support disaster risk reduction as a strategy for long-term climate change adaptation, there are numerous barriers to reconciling short- and long-term goals. Many poor countries are very vulnerable to natural hazards but cannot implement the measures that could reduce this vulnerability for financial reasons, or due to a lack of governance capacity or technology. The recent national self-assessments of progress toward achieving the UNISDR Hyogo Framework for Action indicated that some least-developed countries lack the human, institutional, technical, and financial capacities to address even emergency management concerns (UNISDR, 2009). A recently developed index that measures capacities and conditions for risk reduction shows that low- and lower-middle-income countries with weak governance have, with some exceptions, great difficulty addressing underlying drivers of vulnerability. Those at the bottom of the index, such as Haiti, Chad, or Afghanistan, are also experiencing conflict or political instability (UNISDR, 2011). Another obstacle to reconciling short- and long-term goals is access to technology and maintenance of infrastructure. An example is the introduction of water reuse technologies, which have been developed in a few countries, which could bring a great improvement in the management of droughts if they could be disseminated in many developing countries (Metcalf & Eddy, 2005).

Money and technology are not enough to implement efficient disaster risk reduction and adaptation strategies. Indeed, differences in resources cannot explain the differences in exposure and vulnerability among regions (Nicholls et al., 2008). Governance capacities and the inadequacy of and lack of synergy between institutional and legislative arrangements for disaster risk reduction, climate change adaptation, and poverty reduction are also as much a part of the problem as the shortage of resources. Institutional and legal environments and political will are important, as illustrated by the difference in risk management in various regions of the world (Pelling and Holloway, 2006). In many countries disaster risk management and adaptation to climate change measures are overseen by different institutional structures (see Section 1.1.3). This is explained by the historical evolution of both approaches.

Disaster risk management originated from humanitarian assistance efforts, evolving from localized, specific response measures to preventive measures, which seek to address the broader environmental and socioeconomic aspects of vulnerability that are responsible for turning a hazard into a disaster in terms of human and/or economic losses. Within countries, disaster risk management efforts are often coordinated by civil defense agencies, while measures to adapt to climate change are usually developed by environment ministries. Responding to climate change was originally more of a top-down process, where advances in scientific research led to international policy discussions and frameworks. While the different institutional structures may represent an initial coordination challenge, the converging focus on vulnerability reduction represents an opportunity for managing disaster and climate risks more comprehensively within the development context (Sperling and Szekely, 2005). A change in the culture of public administration toward creative partnerships between national and local governments and empowered communities has been found in some cases to dramatically reduce costs (Dodman et al., 2008).

In addition to the barriers described above, there is also a tendency for individuals and groups to focus on the short-run and to ignore low-probability, high-impact events. The following studies discuss some of the psychological and economic barriers shaping how people make decisions under uncertainty:

- *Underestimation of the risk:* Even when individuals are aware of the risks, they often underestimate the likelihood of the event occurring (Smith and McCarty, 2006). This bias can be amplified by natural variability (Pielke Jr. et al., 2008), where there is expert disagreement, and where there is uncertainty. Magat et al. (1987), Camerer and Kunreuther (1989), and Hogarth and Kunreuther (1995), for example, provide considerable empirical evidence that individuals do not seek information on probabilities in making their decisions.

- *Budget constraints*: If there is a high upfront cost associated with investing in adaptation measures, individuals will often focus on short-run financial goals rather than on the potential long-term benefits in the form of reduced risks (Kunreuther et al., 1978; Thaler, 1999).
- *Difficulties in making tradeoffs*: Individuals are also not skilled in making tradeoffs between costs and benefits of these measures, which requires comparing the upfront costs of the measure with the expected discounted benefits in the form of loss reduction over time (Slovic, 1987).
- *Procrastination*: Individuals are observed to defer choosing between ambiguous choices (Tversky and Shafir, 1992; Trope and Liberman, 2003).
- *Samaritan's dilemma*: Anticipated availability of post-disaster support can undermine self-reliance when there are no incentives for risk reduction (Burby et al., 1991).
- *Politician's dilemma*: Time delays between public investment in risk reduction and benefits when hazards are infrequent, and the political invisibility of successful risk reduction can be pressures for a 'not in my term of office' attitude that leads to inaction (Michel-Kerjan, 2008).

Work in West and East Africa has shown that rural communities tend to underestimate external forces that influence their region while overestimating their own response capacity (Enfors et al., 2008; Tschakert et al., 2010). Misjudging external drivers may be explained by the low degree of control people feel they have over these drivers, resulting in reactions that range from powerlessness to denial. Another issue that makes it difficult to reconcile short- and long-term goals arises from the challenges in projecting the long-term climate and corresponding risks (see Section 3.2.3). Examples of this challenge are reflected in the demographic growth of Florida in the 1970s and 1980s, which unfolded during a period of low hurricane activity but may expose larger populations to the risks associated with extreme climate and weather events. Major engineering projects with long lead times from planning to implementation have difficulty factoring in climate change futures, and have instead been planned according to historic hazard risk (Pielke Jr. et al., 2008). Managing natural risks and adapting to climate change requires anticipating how natural hazards will change over the next decades, but uncertainty about climate change and natural variability is a significant obstacle to such anticipation (Reeder et al., 2009).

Climate change is typically viewed as a slow-onset, multigenerational problem. Consequently, individuals, governments, and businesses have been slow to invest in adaptation measures. Research in South Asia shows that in those regions where past development had prioritized short-term gains over long-term resilience, agricultural productivity is in decline because of drought and groundwater depletion, rural indebtedness is increasing, and households are sliding into poverty with particularly insidious consequences for women, who face the brunt of nutritional deprivation as a result (Moench et al., 2003; Moench and Dixit, 2007). Connecting short- and long-term perspectives is thus seen as critical to realizing the synergies between disaster risk management and climate change adaptation.

8.3.3. Connecting Short- and Long-Term Actions to Promote Resilience

The previous section has highlighted the importance of linking short- and long-term responses so that disaster risk reduction and climate change adaptation mutually support each other. A systems approach that emphasizes cross-scale interactions can provide important insights on how to realize synergies between disaster risk management and climate change adaptation. Resilience, a concept fundamentally concerned with how a system, community, or individual can deal with disturbance and surprise, increasingly frames contemporary thinking about sustainable futures in the context of climate change and disasters (Folke, 2006; Walker and Salt, 2006; Brand and Jax, 2007; Bahadur et al., 2010). It has developed as a fusion of ideas from several bodies of literature: ecosystem stability (e.g., Holling, 1973; Gunderson, 2009), engineering robust infrastructures (e.g., Tierney and Bruneau, 2007), the behavioral sciences (Norris, 2010), psychology (e.g., Lee et al., 2009), disaster risk reduction (e.g., Cutter et al., 2008), vulnerabilities to hazards (Moser, 2009), and urban and regional development (e.g., Simmie and Martin, 2010). In the context of this report, resilience refers to a system's capacity to anticipate and reduce, cope with, and respond to and recover from external disruptions (see Sections 1.1.2.1 and 1.3.2). Resilience perspectives can be used as an approach for understanding the dynamics of human-environmental systems and how they respond to a range of different perturbations (Carpenter et al., 2001; Walker et al., 2004).

'Resilience thinking' (Walker and Salt, 2006) may provide a useful framework to understand the interactions between climate change and other challenges, and in reconciling and evaluating tradeoffs between short- and longer-term goals in devising response strategies. Approaches that focus on resilience emphasize the need to manage for change, to see change as an intrinsic part of any system, social or otherwise, and to 'expect the unexpected.' Resilience thinking goes beyond the conventional engineering systems' emphasis on capacity to control and absorb external shocks in systems assumed to be stable (Folke, 2006). For social-ecological systems (examined as a set of interactions between people and the ecosystems they depend on), resilience involves three properties: the amount of change a system can undergo and retain the same structure and functions; the degree to which it can reorganize; and the degree to which it can build capacity to learn and adapt (Folke, 2006). Resilience can also be considered a dynamic process linked to human agency, as expressed in the ability to deal with hazards or disturbance, to engage with uncertainty and future changes, to adapt, cope, learn, and innovate, and to develop leadership capacity (Bohle et al., 2009; Obrist et al., 2010).

Resilience approaches offer four key contributions for living with extremes: first, in providing a holistic framework to evaluate hazards in coupled social-ecological systems; second, in putting emphasis on the capacities to deal with hazard or disturbance; third, in helping to explore options for dealing with uncertainty and future changes; and fourth, in identifying enabling factors to create proactive responses (Berkes, 2007; Obrist et al., 2010). The concept of resilience is already being applied as a guiding principle to disaster risk reduction and adaptation issues, as well as to examine specific responses to climate change in different developed and developing country contexts (e.g., Cutter et al., 2008). Eakin and Webbe (2008) use a resilience framework to show that the interplay between individual and collective adaptation can be related to wider system sustainability. Goldstein (2009) uses resilience concepts to strengthen communicative planning approaches to dealing with surprise. Linnenluecke and Griffiths (2010) use a resilience framework to explore organizational adaptation to climate change and weather extremes, and suggest that organizations may need to develop multiple capabilities and response approaches in response to changing extremes. Nelson et al. (2007) have shown how resilience thinking can enhance analyses of adaptation to climate change: as adaptive actions affect not only the intended beneficiaries but have repercussions for other regions and times, adaptation is part of a path-dependent trajectory of change. Resilience also considers a distinction between incremental adjustments and system transformation, which may broaden the expanse of adaptation and also provide space for agency (Nelson et al., 2007). Resilience approaches can be seen as complementary to agent-based analyses of climate change responses that emphasize processes of negotiation and decisionmaking, as they can provide insights into the systems-wide implications. Adger et al. (2011) show that dealing with specific risks without taking into account the nature of system resilience can lead to responses that potentially undermine long-term resilience.

Recent work on resilience and governance has focused on communication of science between actors and depth of inclusiveness in decisionmaking as key determinants of the character of resilience. In support of these approaches it is argued that inclusive governance facilitates better flexibility and provides additional benefit from the decentralization of power. On the down side, greater participation can lead to loose institutional arrangements that may be captured and distorted by existing vested interests (Adger et al., 2005; Plummer and Armitage, 2007). Still, the balance of argument (and existing centrality of institutional arrangements) calls for a greater emphasis to be placed on the inclusion of local and lay voices and of diverse stakeholders in shaping agendas for resilience through adaptation and adaptive management (Nelson et al., 2007). Striking the right balance between top-down command-and-control approaches, which offer stability over the short term but reduced long-term resilience, and more flexible, adaptive forms of risk management is the core practical challenge that disaster risk management brings to climate change adaptation under conditions of climatic extremes and projected increases in disaster risk and impacts (Sperling and Szekely, 2005).

Resilience thinking is not without its critiques (Nelson, 2009; Pelling, 2010a). Shortcomings include the downplaying of human agency in systems approaches and difficulty in including analysis of power in explanations of change, which combine to effectively promote stability rather than flexibility, that is, maintaining the status quo and thus serving particular interests rather than supporting adaptive management, social learning, or inclusive decisionmaking. One challenge to enhancing resilience of desired system states is to identify how responses to any single stressor influence the larger, interconnected social-ecological system, including the system's ability to absorb perturbations or shocks, its ability to adapt to current and future changes, and its ability to learn and create new types or directions of change. Responses to one stressor alone may inadvertently undermine the capacity to address other stressors, both in the present and future. For example, coastal towns in eastern England, experiencing worsening coastal erosion exacerbated by sea level rise, are taking their own action against immediate erosion in order to protect livelihoods and homes, affecting sediments and erosion rates down the coast (Milligan et al., 2009). While such actions to protect the coast are effective in the short term, in the long term, investing to 'hold the line' may diminish capital resources for other adaptations and hence reduce adaptive capacity to future sea level rise. Thus, dealing with specific risks without a full accounting of the nature of system resilience can lead to responses that can potentially undermine long-term resilience. Despite an increasing emphasis on managing for resilience (Walker et al., 2002; Lebel et al., 2006), the resilience lens alone may not sufficiently illuminate how to enhance agency and move from the understanding of complex dynamics to transformational action.

8.4. Implications for Access to Resources, Equity, and Sustainable Development

The previous section assessed the links between short- and long-term responses to climate extremes. This section takes the idea of links further. It explores the relationships between climate change adaptation, disaster risk management, and mitigation, and larger issues related to equity, access to resources, environmental and ecosystem protection, and related development processes. This draws out the importance of governance in determining the relationship between disaster risk and underlying processes of unequal socioeconomic development and environmental injustices (Maskrey, 1994; Sacoby et al., 2010). The section discusses issues related to capacity and equity, the existence of winners and losers from disaster and disaster management policy, and opportunities for contributing to wider development goals including the enhancing of human security.

8.4.1. Capacities and Resources: Availability and Limitations

The capacity to manage risks and adapt to change is unevenly distributed within and across nations, regions, communities, and households (Hewitt, 1983; Wisner et al., 2004; Beck, 2007). The literature on how these capacities contribute to disaster risk management and climate change adaptation emphasizes the role of economic, financial, social, cultural, human, and natural capital, and of institutional context (see

Sections 1.4 and 2.4). When the poor are impacted by disasters, limited resources are quickly expended in coping actions that can further undermine household sustainability in the long run, reducing capital and increasing hazard exposure or vulnerability. In these vicious cycles of decline, households tend first to expend savings and then, if pressures continue, to withdraw members from non-productive activities such as school, and finally to sell productive assets. As households begin to collapse, individuals may be forced to migrate or in some cases enter into culturally inappropriate, dangerous, or illegal livelihoods such as the sex industry (Mgbako and Smith, 2010; Ferris, 2011). This poverty and vulnerability trap means that recovery to pre-disaster levels of well-being becomes increasingly difficult (Burton et al., 1993; Adger, 1996; Wisner et al., 2004; Chambers, 2006).

Children, the elderly, and women stand out as more vulnerable to extreme climate and weather events. The vulnerability of children and their capacity to respond to climate change and disasters is discussed in Box 8-2 (see also Section 5.5.1 and Case Study 9.2.14). Among the elderly, increasing numbers will become exposed to climate change impacts in the coming decades, particularly in OECD countries where populations are aging most rapidly. By 2050, it is estimated that one in three people will be older than 60 years in OECD countries, as well as one in five at the global scale (UN, 2002). The elderly are made additionally vulnerable to climate change-related hazards by characteristics that also increase vulnerability to other social and environmental hazards (thus compounding overall vulnerability): deterioration of health, personal lifestyles, social isolation, poverty, and inadequate access to health and social infrastructures (OECD, 2006). Gender impacts vulnerability in many ways. In the 1991 cyclone in Bangladesh, the death toll among women was reportedly five times higher than among men (UNDP, 2007b). Cultural as well as physiological factors are widely cited for the over-representation of female deaths from flooding. Gender inequality extends into female-headed households to compound the vulnerability of dependent children or elderly (Cannon, 2002; UNISDR, 2008; Oxfam, 2010). Inequality has many other important faces: race, caste, religious affiliation, and physical disability, all of which help determine individual and household vulnerability, and they cross-cut gender and age effects. Importantly, the social construction of vulnerability through these characteristics highlights the ways in which vulnerability changes over time – in this case with changes in family structure and access to services in response to economic cycles and political and cultural trends evolving as the climate changes with potentially compounding effects (Leichenko and O'Brien, 2008).

Studies also show that female-headed households more often borrow food and cash than rich and male-headed households during difficult times. This coping strategy is considered to be a dangerous one as the households concerned will have to return the food or cash soon after harvests, leaving them more vulnerable as they have less food or cash to last them the season and to be prepared if disaster strikes (Young and Jaspars, 1995). This may leave households in a cycle of poverty from one season to the next. Literature shows that this outcome is linked to unequal access by women to resources, land, and public and privately

Box 8-2 | Children, Extremes, and Equity in a Changing Climate

The linkages between children and extreme events have been addressed through two principal lenses.

1. Differentiated Impacts and Vulnerability

The literature estimates that 66.5 million children are affected annually by disasters (Penrose and Takaki, 2006). Research on disaster impacts among children focuses on short- and long-term physical and psychological health impacts (Norris et al., 2002; Bunyavanich et al., 2003; del Ninno and Lindberg, 2005; Balaban, 2006; Waterson, 2006; Bartlett, 2008). Vulnerability to these impacts in part is due to the less-developed physical and mental state of children, and therefore differential capacities to cope with deprivation and stress in times of disaster (Cutter, 1995; Bartlett, 2008; Peek, 2008).

Most literature points toward higher mortality and morbidity rates among children due to climate stresses and extreme events (Cutter, 1995; Bunyavanich et al., 2003; Telford et al., 2006; Waterson, 2006; Bartlett, 2008; Costello et al., 2009). This is especially acute in developing countries, where climate-sensitive health outcomes such as malnutrition, diarrhea, and malaria are already common and coping capacities are lowest (Haines et al., 2006), although research in the United States found relatively low child mortality from disasters and considerable differences across age groups for different types of hazard (Zahran et al., 2008).

Recent studies conducted in Bolivia, Indonesia, Mexico, Mozambique, Nepal, the Philippines, and Vietnam provide evidence of how extensive (low impact/high frequency) disasters negatively affect children's education, health, and access to services such as water and sanitation, an issue of critical importance given the importance of primary education for human and long-term economic development. In areas in Bolivia that experienced the greatest incidence of extensive disasters, the gender gap in primary education achievement widened, preschool enrollment rates decreased, and dropout rates increased. Equivalent areas in Nepal and Vietnam saw, respectively, reduced primary enrollment rates and a drop in the total number of children in primary education. Extensive disasters also led to an increased incidence of diarrhea in children under five years of age in Bolivia, an increased proportion of malnourished children under three in Nepal, an increased infant mortality rate in Vietnam, and an increase in the incidence of babies born with low birth weight in Mozambique. This study also found evidence of negative impacts in terms of access to water and sanitation in Mexico and Vietnam (UNISDR, 2011).

These studies underpin the need for resources for child protection during and after disaster events (Last, 1994; Jabry, 2003; Bartlett, 2008; Lauten and Lietz, 2008; Weissbecker et al., 2008). These include protection from abuse, especially during displacement, social safety nets to guard against withdrawal from school due to domestic or livelihood duties, and dealing with psychological and physical health issues (Norris et al., 2002; Keenan et al., 2004; Evans and Oehler-Stinnett, 2006; Waterson, 2006; Bartlett, 2008; Davies et al., 2008; Lauten and Lietz, 2008; Peek, 2008).

2. Children's Agency and Resource Access

Rather than just vulnerable victims requiring protection, children also have a critical role to play in tackling extreme events in the context of climate change (Tanner, 2010). There is also increasing attention on child-centered approaches to preventing, preparing for, coping with, and adapting to extreme events (Peek, 2008; Tanner, 2010).

While often centered on disaster preparedness and climate change programs in education and schools (Wisner, 2006; Bangay and Blum, 2010), more recent work emphasizes the latent capacity of children to participate directly in disaster risk reduction or adaptation supported through child-centered programs (Back et al., 2009; Tanner et al., 2009). This emphasis acknowledges the unique risk perceptions and risk communication processes of children, and their capacity to act as agents of change before, during, *and* after disaster events (see collections of case studies in Peek, 2008; Back et al., 2009; and Tanner, 2010). Examples demonstrate the ability to reduce risk behavior at household and community scales, but also to mobilize adults and external policy actors to change wider determinants of risk and vulnerability (Mitchell et al., 2008; Tanner et al., 2009).

provided services (Agarwal, 1991; Thomas-Slayter et al., 1995; Nemarundwe, 2003; Njuki et al., 2008). But women are also often the majority holders of social capital and the mainstay of social movements and local collective action, providing important mechanisms for household and local risk reduction and potentially transformative resilience. Important here are local saving groups and microcredit/micro-finance groups, some of which extend to micro-insurance. In a review of micro-finance for disaster risk reduction and response in South Asia, Chakrabarti and Bhatt (2006) identify numerous initiatives, including those that build on extensive social networks and connect to the formal financial services sector.

Demographic and sociological diversity is also difficult to capture in decisionmaking, where non-scientific knowledge is less easily incorporated into formal decisionmaking. The importance of culture, including traditional knowledge, in shaping strategies for adaptation is recognized (Heyd and Brooks, 2009; ISET, 2010). There is a long tradition of seeking to identify and bring such knowledge into planned disaster risk management in urban and rural contexts through participatory and community-based disaster risk management (Bruner et al., 2001; Fearnside, 2001; Pelling, 2007; Mercer et al., 2008) and tools such as participatory geographic information systems that explicitly seek to bring scientific and local knowledge together (Tran et al., 2009). Both development planning and post-disaster recovery have tended to prioritize strategic economic sectors and infrastructure over local livelihoods and poor communities (Maskrey, 1989, 1996). However, this represents a missed opportunity for building local capacity and including local visions for the future in planning the transition from reconstruction to development – opportunities that could increase long-term sustainability (Christoplos, 2006).

8.4.2. Local, National, and International Winners and Losers

While climate-related disasters cannot always be prevented, the scale of loss and its social and geographical distribution do differ significantly, determined by the characteristics of those at risk and overarching structures of governance including the legacy of preceding development paths for social institutions, economies, and physical assets (Oliver-Smith, 1994). But some people also benefit from disasters. These may be organizations or individuals who benefit economically from reconstruction or response (West and Lenze, 1994; Hallegatte, 2008b), through supplying materials, equipment, and services – often at a premium price generated by local scarcity and inflationary pressures (Benson and Clay, 2004) or as a result of poorly managed tendering processes (Klein, 2007). Areas not impacted by disaster can also experience economic benefits, for example, in the Caribbean where hurricanes have caused international tourist flows to be redirected (Pelling and Uitto, 2001). Political actors can also benefit by demonstrating strong post-disaster leadership, at times even when past political decisions have contributed to generating disaster risk (Olson and Gawronski, 2003; Le Billon and Waizenegger, 2007; Gaillard et al., 2008). The same can be said for climate change,

with very unequal consequences in various regions of the world and various economic sectors and social categories (Adger et al., 2003; O'Brien et al., 2004; Tol et al., 2004). Less directly, those who have benefited from policies and processes, such as expansion of commercial agriculture or logging, can also be described as benefiting from decisions that have generated vulnerability and prefigured disaster for others. Such costs and benefits are often separated geographically and temporally, making any efforts at distributional equity challenging. For example, in the case of Hurricane Mitch, which killed more than 10,000 people and caused as much as US$ 8.5 billion in damages, deforestation and rapid urban growth are often cited among the key causes of the disaster-related losses (Alves, 2002; Pielke Jr. et al., 2003), with those benefiting from such development including distant speculators.

Analyses of winners and losers associated with climate change and discrete hazards need differentiation. In almost every circumstance, what one part of society views as a win can be viewed by another part as a loss. In examining possible responses to risks of climate extremes, it is essential to recognize that possible impacts interact with vested interests of different locations, sectors, and population groups in very different ways. In virtually every case, the question is: benefits for whom? Who says that this course of action is best for society as a whole? What compensation is offered for those who are losers? In particular, who is listening to views of those parts of society that have less political power and influence?

While individual events can be assessed as a snapshot of winners and losers, climate change as an ongoing process has no final state. Over time, it may produce different distributions of winners and losers, for example, as areas experience positive and then negative consequences of changes in temperature or precipitation. Whether or not a particular place produces winners or losers from an extreme event or a combination of climate extremes and other driving forces also depends on perceptions. These may be shaped by the recovery process, but are strongly influenced by prioritized values (Quarantelli, 1984, 1995; O'Brien, 2009; O'Brien and Wolf, 2010). In considering winners and losers from extreme climate and weather events, and also from the outcomes of policies directed at reducing disaster risk or responding to climate change, it is thus vital to recognize the subjective understanding of winners and losers.

Much depends upon an individual, group, or society's dominant values, perspectives, and access to information. While some regard winners and losers as a natural and inevitable outcome of ecological changes and/or economic development, others suggest that winners and losers are deliberately generated by unequal political and social conditions (O'Brien and Leichenko, 2003; Wisner, 2003). Lurking behind discourses about winners and losers are issues of liability and compensation for losses: that is, if a population or an area experiences severe losses due to an extreme event (at least partly) attributed to climate change, can fault be prescribed? Does responsibility lie with those who have generated local environmental change through settling a hazard-exposed area, those who have promoted or permitted such settlement, or those who have failed to mitigate local hazard or global environmental change?

Issues of equity, justice, and compensation are emerging in climate change adaptation, but few have begun to deal with questions of liability for disaster risk production beyond the local scale (Kent, 2001; Mitchell, 2001; Wisner, 2001; O'Brien et al., 2010b). It seems that efforts to assign responsibility will emerge as an issue for both governments and courts at a range of scales (Farber, 2007).

8.4.3. Potential Implications for Human Security

Changes in extreme climate and weather events threaten human security, and both disaster risk management and climate change adaptation represent strategies that can improve human security while also avoiding disasters. Human security addresses the combined but related challenges of upholding human rights, meeting basic human needs, and reducing social and environmental vulnerability (UNDP, 1994; Sen, 2003; Bogardi and Brauch, 2005; Brauch 2005a,b, 2009; Fuentes and Brauch, 2009). It also emphasizes equity, ethics, and reflexivity in decisionmaking and a critical questioning and contestation of the drivers of climate change (O'Brien et al., 2010b) and local impacts (Pelling, 2010a). Human security is realized through the capacity of individuals and communities to respond to threats to their environmental, social, and human rights (GECHS, 2000; Barnett et al., 2010). A number of studies have assessed the relationship between climate change and human security, demonstrating that the linkages are often both complex and context-dependent (Barnett, 2003; Barnett and Adger, 2007; Brauch et al., 2008, 2009, 2011; Buhaug et al., 2008; O'Brien et al., 2010a). Among the most likely human security threats are impacts felt through damage to health, food, water, or soil conditions (Oswald Spring, 2009a, 2011b).

Among the most widely discussed humanitarian and human security issues related to climate change are the possibilities of increased migration and/or violent conflict resulting from the biophysical or ecological disruptions associated with climate change (Reuveny, 2007; O'Brien et al., 2008; Raleigh et al., 2008; Warner et al., 2010). There are indications that migration conditions followed disasters in the distant past, as well as in current situations (see, e.g., Le Roy Ladurie, 1971; Kinzig et al., 2006; Peeples et al., 2006). Migration is a key coping mechanism for poor rural households, not only in extreme circumstances (e.g., during a prolonged drought, as with the 20th-century US Dustbowl period and Sahelian droughts) but also as a means of diversifying and increasing income (Harrington et al., 2009; Oswald Spring, 2011a; Scheffran, 2011). The opportunities that population movement opens for risk reduction are seen in international remittance flows from richer to poorer countries. These are estimated to have exceeded US$ 318 billion in 2007, of which developing countries received US$ 240 billion (World Bank, 2008).

Disasters linked to extreme events often lead to forced displacement of people, as well as provoke voluntary migration among the less poor. The relationship between climate risk and displacement is a complex one and there are numerous factors that affect migration (UNDP, 2009). Nonetheless, recent research suggests that adverse environmental impacts associated with climate change have the potential to trigger displacement of an increased number of people (Kolmannskog, 2008; Feng et al., 2010). Studies suggest that most migration will take place internally within individual countries; that in most cases when climatic extremes occur in developing countries they will not lead to net out-migration because people tend to return to re-establish their lives after a disaster; and that while long-term environmental changes may cause more permanent migration this will also tend to be internal (Piguet, 2008; UNDP, 2009). More negatively, forced land abandonment is stressful for migrants whose culture and sense of identity are affected (Mortreux and Barnett, 2009; Brauch and Oswald Spring, 2011). The social dislocation provoked by migration can lead to a breakdown in traditional rural institutions and associated coping mechanisms, for example, in the erosion of traditional community-based water management committees in central and west Asia (Birkenholtz, 2008). Local collective coping and adaptive capacity can also be limited by increases in the number of female-headed households as men migrate (Oswald Spring, 1991, 2009a).

Attention has been mainly focused on population displacement associated with large disasters. Pakistan's 2010 floods have to date left an estimated 6 million people in need of shelter; India's 2008 floods also uprooted roughly 6 million people; Hurricane Katrina displaced more than half a million people in the United States; and Cyclone Nargis uprooted 800,000 people in Myanmar and South Asia. However, the compound effect of smaller, more frequent events can also contribute to displacement. Hazards such as floods, although often causing relatively low mortality, destroy many houses and hence cause considerable displacement. Between 1970 and 2009 in Colombia, for example, 24 disaster loss reports detailed floods that killed fewer than 10 people but destroyed more than 500 houses. In total, around 26,500 houses were destroyed, potentially displacing more than 130,000 people. In the Indian state of Orissa, 265 floods with similar low mortality rates destroyed more than half a million houses. It is estimated that such extensive disasters account for an additional 19% displacement of people who are typically less visible than those displaced in larger events that attract international media and humanitarian assistance (UNISDR, 2011).

Despite the opportunities to enhance development, disaster response is often better at meeting basic needs than securing or extending human rights. Indeed, the political neutrality that underpins the humanitarian imperative makes any overt actions to promote human rights by humanitarian actors difficult. In this way, disaster response and reconstruction can to only a partial extent claim to enhance human security (Pelling and Dill, 2009). Work at the boundaries between humanitarian and development actors, new partnerships, the involvement of government, and meaningful local participation are all emerging as ways to resolve this challenge. One successful case has been the reconstruction process in Aceh, Indonesia, following the Indian Ocean tsunami, where collaboration between government and local political interests facilitated by international humanitarian actions on the ground and through political-level peace-building efforts have increased rights locally, contained armed conflict, and provided an economic recovery plan (Le Billon and Waizenegger, 2007; Gaillard et al., 2008; Törnquist et al., 2010).

Coping with the new and unprecedented threats to human security posed by climate change has raised questions about whether existing geopolitics and geo-strategies have become obsolete (Dalby, 2009). The concepts, strategies, policies, and measures of the geopolitical and strategic toolkits of the past as well as the short-term interests dominating responses to climate change have been increasingly questioned, while the potential for unprecedented disasters has led to a consideration of the security implications of climate change (UNSC, 2007; EU, 2008; UNGA, 2009; UNSG, 2009). Concerns range from increased needs for humanitarian assistance to concerns over environmental migration, emergent diseases for humans or in food chains, potentials for conflict between nations or localities over resources, and potential for political/governmental destabilization due to climate-related stresses in combination with other stresses, along with efforts to assign blame (Ahmed 2009; Brauch and Oswald Spring, 2011).

Climate change is generally regarded to act as a threat multiplier for instability in some of the most volatile regions of the world (CNA Corporation, 2007). Even in stable polities, adaptation planning that seeks long-term resilience is confronted by political instability directly after disasters (Drury and Olson, 1998; Olson, 2000; UNDP, 2004; Pelling and Dill, 2009). When disasters strike across national boundaries or within areas of conflict, they can provide a space for rapprochement, but effects are usually short-lived unless the underlying political and social conditions are addressed (Kelman and Koukis, 2000; Kelman, 2003). New interest in disaster and climate change as a security concern has brought in lessons from international law (Ammer et al., 2010) and security policy (Campbell et al., 2007) on planning for relatively low-probability, high-consequence futures. Although during times of stress it is easy for polities to drift toward militarization and authoritarianism for managing disaster risk (Albala-Bertrand, 1993), there are alternatives, such as inclusive governance, that can meet the goals of sustainable development and human security over the long term (Olson and Gawronski, 2003; Brauch, 2009; Pelling and Dill, 2009; Bauer, 2011).

8.4.4. Implications for Achieving Relevant International Goals

Addressing – or failing to address – disaster risk reduction and climate change adaptation can influence the success of international goals, particularly those linked to development. Successive reports have noted the potential for climate change to derail the Millennium Development Goals (MDGs). In 2003, the Asian Development Bank and nine other development organizations first highlighted that climate change may impact on progress toward the MDGs and in particular constrain progress beyond 2015, underlying the importance of managing climate risks within and across development sectors. The UK Department for International Development (DFID, 2004, 2006) and UNDP (2004) show how each of the MDGs is dependent on some aspect of disaster risk for success. Disaster impacts on the MDG targets are both direct and indirect. For example, MDG 1 (to eradicate extreme poverty and hunger) is impacted directly by damage to productive and reproductive assets of the poor and less poor (who may remain in poverty or slip into poverty as a result of disaster loss), and indirectly affected by negative macroeconomic impacts. The 2007 UN Human Development Report noted that enhanced adaptation is required to protect the poor, with climate change potentially acting as a brake on development beyond 2015.

The UNISDR Hyogo Framework for Action 2005-2015: Building the Resilience of Nations and Communities to Disasters (HFA) explicitly recognizes that climate variability and change are important contributors to disaster risk and includes strong support for better linking disaster management and climate change adaptation efforts (Sperling and Szekely, 2005; see also Section 7.3.1). The HFA priorities for action have proven foresightful in including resilience explicitly as a component. Priority Three calls for "knowledge, innovation and education to build a culture of safety and resilience at all levels." This provides a strong justification for international actors to support investment in institutional and human capacity for national and local resilience building, and one that does not require the addition of new international agreements to start work. Frameworks for such action exist, for example, in Common Country Assessments, United Nations Development Assistance Frameworks, National Adaptation Plans of Action, and Poverty Reduction Strategies, but limited progress has been made on this to date (DFID, 2004). Some have proposed integrating climate change and disaster risk management into any equivalent of the MDGs post-2015 (Tribe and Lafon, 2009; Gostin et al., 2011).

8.5. Interactions among Disaster Risk Management, Adaptation to Climate Change Extremes, and Mitigation of Greenhouse Gas Emissions

8.5.1. Thresholds and Tipping Points as Limits to Resilience

Recent literature suggests that climate change could trigger large-scale, system-level regime shifts that could significantly alter climatic and socioeconomic conditions (MEA, 2005; Lenton et al., 2008; Hallegatte et al., 2010; see Section 3.1.7). Examples of potential system changes include dieback of the Amazon rainforest, decay of the Greenland ice sheet, and changes in the Indian summer monsoon (Lenton et al., 2008). At smaller scales, also, climate change is exacerbating well-established examples of environmental regime shifts, such as freshwater eutrophication (Carpenter, 2003), shifts to algae-dominated coral reefs (Hughes et al., 2003), and woody encroachment of savannas (Midgley and Bond, 2001). The abruptness and persistence of such changes in social and ecological systems, coupled with the fact that they can be difficult and sometimes impossible to reverse, means that they can have substantial impacts on human well-being (Scheffer et al., 2001; MEA, 2005; Scheffer, 2009). The notion of regimes shifting once thresholds or tipping points are crossed contrasts with discussions of climate thresholds (see Section 3.1.1) and traditional thinking about gradual, linear, and more predictable changes in ecological and social-ecological systems, and emphasizes the possibility of multiple futures determined by the crossing of critical thresholds (Levin, 1998; Gunderson and Holling, 2002). Similar discussion on socioeconomic systems has, for example, identified profitability limits in economic activities as critical thresholds that can bring about sudden collapse and regime change (Schlenker and Roberts, 2006; OECD, 2007).

The metaphor of tipping points, or points at which a system shifts from one state to another, can also be applied to disaster events. Disasters themselves are threshold-breaching events, where coping capacities of communities are overwhelmed (e.g., Blaikie et al., 1994; Sperling et al., 2008). Disasters may lead to secondary hazards, for example, when the impacts from one disaster breach the coping capacities of related systems, as when hurricane impacts trigger landslides or when different disasters produce concatenated impacts over time (Biggs et al., 2011). For example, losses associated with droughts and fires during the 1997/1998 El Niño-Southern Oscillation event in Central America increased landslide and flood hazard during Hurricane Mitch in 1998 (Villagrán de León, 2011). Critical social thresholds may be crossed as disaster impacts spread across society. Disaster response is as much about containing such losses as assisting those hurt by the initial disaster impact.

For the poor, life and health are immediately at risk; for those living in societies that take measures to protect infrastructure and economic and physical assets, the lives and health of the population are less at risk. However, a threshold can be crossed when hazards exceed anticipated limits, are novel or unexpected in a specific risk management domain (Beniston, 2004; Schär et al., 2004; Salagnac, 2007), or when vulnerability has increased or resilience decreased due to spillover from market and other shocks (Wisner, 2003).

In 2010, for example, western Russia experienced the hottest summer since the beginning of systematic weather data recording 130 years ago. Lack of rainfall in early 2010 and July temperatures almost 8°C above the long-term average led to parched fields, forests, and peatlands that posed a high wildfire risk. During and after the wildfires, Russia's mortality rate increased by 18%. In August alone, 41,300 more people died as compared to August 2009, due to both the extreme heat and smoke pollution. Social and economic change had greatly increased the risk posed by wildfires. Traditional agricultural livelihoods have declined, accompanied by out-migration and reduced management of surrounding forests, arguably exacerbated by the decentralization of national management and increased exploitation by the private sector (UNISDR, 2011).

The recognition of nonlinearity and the importance of thresholds as limiting points for existing systems has led climate scientists to increase their attention to the 'tails' of impact probability density functions (Weitzman, 2009). This is in contrast to the disasters research community, which, after focusing on major extremes, is now recognizing the importance of small or local disasters and the secondary disasters that make up concatenated events (UNDP, 2004; UNISDR, 2009). Both lenses are valuable for a comprehensive understanding of the interaction of disaster impact with development.

Tipping points in natural and human systems are more likely to arise with relatively severe and/or rapid climate change than with moderate levels and rates (Wilbanks et al., 2007). Because of this, less success with climate change mitigation implies greater challenges for adaptation and disaster risk management. Not only does adaptation need to consider incremental change in hazard and vulnerability, but the possibility of threshold breaching, systems-wide changes. The nonlinear changes associated with breaching thresholds may exceed adaptation capacity to avoid serious disruptions. Examples of ecological system changes of this kind and social impacts include the disappearance of glaciers currently feeding urban and agricultural water supply (Orlove, 2009), effects of climate change on traditional livelihoods for the sustainability of indigenous cultures (Turner and Clifton, 2009), widespread loss of corals in acidifying oceans and fisher livelihoods (Reaser et al., 2000), and profitability limits for important economic activities like agriculture, fisheries, and tourism. When socioeconomic systems are already under stress (e.g., fisheries in many countries), sustainability thresholds may be more easily passed. Responses to potential thresholds or tipping points include efforts to improve the information available to decisionmakers, for example, through monitoring systems to provide early warning of an impending system collapse (Biggs et al., 2009; Scheffer et al., 2009), but also initiatives researching the balance of risks associated with geoengineering (Royal Society, 2009) aiming to avoid such tipping points (Virgoe, 2002; Kiehl, 2006).

8.5.2. Adaptation, Mitigation, and Disaster Risk Management Interactions

As indicated above, the extent to which future adaptation and disaster risk reduction action will be required is likely to be dependent on the extent and rapidity with which climate change mitigation actions may be taken and resulting risk unfolds for any given development context. This section assesses the ways in which mitigation, disaster risk reduction, and adaptation interact with development in urban and rural contexts.

In many instances, climate change mitigation and adaptation may be synergistic, such as land use planning to reduce transport-related energy consumption and limit exposure to floods, or building codes to reduce heating energy consumption and enhance robustness to heat waves (McEvoy et al., 2006). There is an emerging literature exploring the linkages between climate change mitigation and adaptation, and the possibility of approaches that address both objectives simultaneously (Wilbanks and Sathaye, 2007; Hallegatte, 2009; Bizikova et al., 2010; Wilbanks, 2010; Yohe and Leichenko, 2010). In this section we enlarge the scope of the interactions to include disaster risk management. This builds on experience within the disaster management community that has recently sought to integrate risk modeling (UNDP, 2004; UNISDR, 2009, 2011) and planning to consider multi-hazard contexts. An important lesson from this is that avoiding superficial integration means seeking out and addressing shared root causes of exposure and vulnerability to hazards, and not just addressing the surface expressions of risk (Wisner, 2011). The extent of adaptation required will depend on the climate change mitigation efforts undertaken, and it is possible that these requirements could increase drastically if levels of climate change exceed systemic thresholds – whether in geophysical or socioeconomic systems.

Practical integration of climate change mitigation and adaption into a development context is complicated because of a differential distribution in costs and benefits (e.g., mitigation benefits are distributed and accrue globally; adaptation benefits, like disaster risk management, are often easier to measure locally). In addition, the research and policy discourses of these three policy domains are quite separated and in areas technically unrelated, and the constituencies and decisionmakers are often different (Wilbanks et al., 2007). In many cases, the challenge of bringing the entire range of issues and options into focus – seeking synergies and avoiding conflicts – is most likely to occur in discussions of climate change responses and development objectives in particular places: localities and small regions where compliance with national or international mitigation agendas provides a logic for local action (Wilbanks, 2003). The following subsections present the urban and rural contexts as examples.

8.5.2.1. Urban

In an increasingly urbanized world, global sustainability in the context of a changing climate will depend on achieving sustainable and climate-resilient cities. Urban spatial form is critical for energy consumption, emission patterns, and disaster risk management (Desplat et al., 2009), and it influences where and how residents live and the modes of transport that they use. Urban planning is a tool that can be used to pursue climate change mitigation, adaptation, and disaster risk reduction as part of the everyday development process (Newman and Kenworthy, 1989; Bento et al., 2005; Handy et al., 2005; Ewing and Rong, 2008; Grazi et al., 2008; Brownstone and Golob, 2009; Glaeser and Kahn, 2010). Urban form also influences the spatial and social inequalities that largely shape vulnerability, coping, and adaptive capacity (Pelling, 2003; Gusdorf et al., 2008; Leichenko and Solecki, 2008). The historical failure of urban planning in most developing country cities has had tremendous environmental and social consequences (UN-HABITAT, 2009; World Bank, 2010a). Also in richer countries, where planning is not comprehensive, maladaptation can take place rather than synergistic risk reduction, for example, where urban heat wave risk management results in increased private air conditioning without decarbonized energy available (Lindley et al., 2006). Similarly, a denser city may reduce greenhouse gas emissions but increase heat wave vulnerability (Hamin and Gurran, 2009). However, since urban forms influence both greenhouse gas emissions and vulnerability (McEvoy et al., 2006), scope for synergistic planning and action can also be found. For example, managing car use may contribute to decreased greenhouse gas emissions, but also lower local particulate pollution and reduce the health impacts of urban heat waves (Dennekamp and Carey, 2010).

As yet there is only limited evidence that opportunities for synergistic planning offered by urbanization are being realized, especially for those most marginalized and vulnerable. More typically, urbanization compounds environmental problems. As countries urbanize, the risks associated with economic asset loss tend to increase through rapid growth in infrastructure and productive and social assets, while mortality risk tends to decrease (Birkmann, 2006). As cities grow, they also modify the surrounding rural environment, and consequently may generate a significant proportion of the hazard to which they are also exposed. For example, as areas of hinterland are paved over, runoff increases during storms, greatly magnifying flood hazards (Mitchell, 1999; Pelling, 1999). As mangroves are destroyed in coastal cities, storm-surge hazard can increase (Hardoy et al., 2001). Likewise, within urban areas (though often beyond the reach of urban planning), the expansion of informal settlements can lead to increased local population exposure to landslide and flood hazards (Satterthwaite, 1997; UNDP, 2004). Global risk models indicate that expansion of urban risk is primarily due to rapidly increasing exposure, which outpaces improvements in capacity to reduce vulnerability (such as through improvements legislating and applying building standards and land use planning), at least in rapidly growing low- and middle-income nations (UNISDR, 2009, 2011). As a consequence, risk is becoming increasingly urbanized (Mitchell, 1999; Pelling, 2003; Leichenko and O'Brien, 2008). There are dramatic differences, nonetheless, between developed and developing countries. In most developed countries (and increasingly in a number of cities in middle-income countries, e.g., Bogota, Mexico City), risk-reducing capacities exist that can manage increases in exposure. In contrast, in much of the developing world (and particularly in the poorest least-developed countries where large proportions of the urban population live in unplanned settlements) such capacities are greatly restricted, while population growth drives exposure. Financial and technical constraints matter for risk management, but differences in wealth alone do not explain differences in risk reduction investments, which also depend on risk perceptions and political choice (e.g., Satterthwaite, 1998; Hardoy et al., 2001; Hanson et al., 2011).

Urban planning can be a vehicle for synergy, but it takes time to produce significant effects. Synergy in planning requires anticipation of future climate change, taking into account how climate will change over many decades, the uncertainty of this information, the vulnerability of urban systems, and the capacity of social agents. The Asian Cities Climate Change Resilience Network found that catalyzing city-level actors to assess these plans are essential, rather than depending on external experts or national agencies to prepare urban plans (Tyler et al., 2010). Built forms are difficult to change because they exhibit strong inertia and irreversibility: when a low-density city is created, transforming it into a high-density city is a long, expensive, and difficult process (Gusdorf et al., 2008). This point is crucial in the world's most rapidly growing cities, where urban forms of the future are being decided based on actions taken in the present, and where current trends indicate that low-density, automobile-dependent forms of suburban settlement are rapidly expanding (Solecki and Leichenko, 2006). Some work has started to investigate these aspects of climate change adaptation and mitigation (Newman et al., 1996). At the same time, there are specific opportunities when cities enter periods of large-scale transformation. This is happening in Delhi, Mumbai, and other cities in India as private capital redevelops low-income city neighborhoods into commercial districts and middle- and high-income housing areas with associated low-income housing. There is rare scope here to promote disaster risk reduction, climate change adaptation, and mitigation alongside existing demands for market profitability and social justice in urban and building design. There are also growing numbers of large-scale slum/informal settlement upgrading programs that aim to improve housing and living conditions for low-income households (Boonyabancha, 2005; Satterthwaite, 2010).

Disaster reconstruction also creates opportunities for synergistic development planning. For example, reconstruction after the 2005 Hurricane Katrina disaster in New Orleans, Louisiana, included rebuilding to Green Building Council 'Leadership in Energy and Environmental Design' (LEED) standards (USGBC, 2010). Similarly, in May 2007, Greensburg, Kansas, was virtually destroyed by a tornado and LEED standards have been applied (Harrington, 2010). Echoing the tradeoffs between speed and sustainability presented in Section 8.2.5, the actions in Greensburg have also slowed rebuilding of the town, leading in this instance to an erosion in community and associated aspects of resilience in the short run, while attempting to create a model 'green' community in the long run.

In short, despite the many opportunities for building synergy into urban development planning and practice, examples of success are not plentiful. Lack of synergy is more the norm, to take just one example of urbanization in central Dhaka, Bangladesh. These flood-prone areas had until recently been occupied by natural water bodies and drains, vital to the regulation of floods. The Dhaka Metropolitan Development Plan restricts development in many of these areas, but despite the Plan, infilling continues with both private- and public-sector projects. Destruction of retention ponds and drains increases risks of flooding and building in the drained wetlands generates new risks of liquefaction following earthquakes (UNISDR, 2011).

8.5.2.2. Rural

Rural areas are the primary site for climate change mitigation. Rural areas have considerable experience in disaster risk management and more recently in climate change adaptation (UNDP, 2007b). Nonetheless, as for urban areas, the evidence base is limited for consciously synergistic development projects and policies that consider climate change mitigation, adaptation, and disaster management together. There are, however, several important opportunities where climate change mitigation and adaptation or risk management have shown scope for integration and opportunities are being explored, for example in agroforestry (Verchot et al., 2007).

Any scope for synergy needs to be seen within the context of contemporary development pressures (Goklany, 2007). For small farms in particular, pressures are strong for diversification into non-farm activities, where such opportunities exist, but strong support is needed to enable transitions in economic activity (Roshetko et al., 2007). Climate change affects the range of choices available, for example, in low-lying coastal zones where saltwater intrusion and coastal flooding are already making traditional agriculture marginal and leading to the adoption of saltwater tolerant crops or a shift from agriculture to aquaculture (Adger, 2000). While urban areas have expanded in size and influence, the majority of the poor continue to reside in rural areas in many countries, particularly in Africa, and are among the most resource-scare and capacity-limited population groups (UNDP, 2009). For populations that may also be isolated from markets and communication networks, even small increases in the frequency or severity of hazard can cause local livelihoods to collapse, though recent developments in communication technology may bridge this gap (Aker and Mbiti, 2010). Where political and economic systems disrupt food distribution and market functioning, vulnerability to food insecurity escalates (Misselhorn, 2005).

Hard choices also have to be made between expanding rural populations or economies and natural capital. Too often, local natural assets are exploited not by local actors to build local capacities but by external agents, such that resources are extracted with little benefit accruing locally. The balance and implementation of controls on natural resource exploitation is both a potential damper on current capacity building and a critical mechanism for ensuring long-term sustainability of rural livelihoods and ecosystem services (Chouvy and Laniel, 2007). Non-farm income now represents a substantial proportion of total income for many rural households and can, in turn, increase resilience to weather- and climate-related shocks (Brklacich et al., 1997; Smithers and Smit, 1997; Wandel and Smit, 2000). The implications of these transitions for local rural risk, and how far they may provide scope for mitigation, has not been fully explored in the literature.

While urban sites offer opportunities for mitigation through diversified (household) production and energy conservation, rural areas are a focus for concentrated low- or no-carbon energy production ranging from hydroelectric power (HEP) to solar and wind farms, biofuel crops, and carbon sink functions associated with forestry in particular and REDD+ projects. These investments can have significant local impacts on disaster risk through changes in land use and land cover that may influence hydrology, or through economic effects and consequences for livelihoods. There is scope for synergy, for example, through small HEP/flood or water conservation dams, and some have gone as far to say that this joined-up approach is part of a transformed development policy for meeting combined energy and water demands in vulnerable rural communities, most particularly in sub-Saharan Africa (Foster and Briceño-Garmendia, 2011). Some impacts can even go beyond local places. Recent impacts of biofuel production on rural livelihoods and global food security indicate the interdependence of vulnerability in rural and urban systems, and the care required in transformations of this kind where impacts can quickly spread and be amplified through global markets (Dufey, 2006; de Fraiture et al., 2008).

Flows of investment, remittances, migration, and material transfers through trade and also in the movement of resources (water, food, waste, and energy) intimately connect rural and urban economies and societies, and the local with the global, such that the sustainability of one will influence the other. The existence of multiple, intersecting stressors in rural and urban contexts draws attention to the importance of addressing the underlying drivers of risk as a means of both disaster risk management and adaptation, and of promoting climate change mitigation.

8.6. Options for Proactive, Long-Term Resilience to Future Climate Extremes

Considering the broad challenges described previously, it is important to assess the range of existing planning tools and the ways they are used, who uses them, and how they interact or change over time. Pursuing sustainable and resilient development pathways requires integrated and ambitious policy that is science-based and knowledge-driven, and that is capable of addressing issues of heterogeneity and scale. The latter issues are particularly vexing, as the consequences of, and responses to, extreme climate and weather events are local, but these responses need to be supported and enabled by actions at regional, national, and global scales.

This section first considers the challenges of planning for the future, then assesses the literature pertaining to tools and practices that can help address these issues. As the preceding sections in this chapter and other chapters in the report have argued, achieving a sustainable and resilient future draws attention to the need for both incremental and transformational changes. Based on an assessment of the literature, the final section discusses why such changes may involve a combination of adaptive management, learning, innovation, and leadership.

8.6.1. Planning for the Future

Disaster risk management and climate change adaptation are fundamentally about planning for an uncertain future, a process that involves combining one's own aspirations (individual and collective) with perspectives on what is to come (Stevenson, 2008). Planning for the future is challenging when the stakes are high, values disputed, and decisions urgent, and these factors often create tensions among different visions of development. Typically, decisionmakers (representing households, local or national governments, international institutions, etc.) look to the future partly by remembering the past (e.g., projections of the near future are often derived from recent experiences with extreme events) and partly by projecting how the future might be different (using forecasts, scenarios, visioning processes, or story lines – either formal or informal) (Miller, 2007). Projections further into the future are necessarily shrouded in larger uncertainties. The most common approach for addressing these uncertainties is to develop multiple visions of the future (quantitative scenarios or narrative storylines), that in early years can be compared with actual directions of change (Boulanger et al., 2006a,b; Moss et al., 2010).

Scenario development has become an established research tool both in the natural sciences (e.g., Nakicenovic et al., 2000; Lobell et al., 2008) and in the social sciences (e.g., Wack, 1985; Davis, 1998; Robinson, 2003; Galer, 2004; Kahane, 2004; Rosegrant et al., 2011). Scenarios can be based on different spatial (e.g., global, national, and local) and temporal scales (e.g., from a few years to several decades or centuries). The challenges for integrated disaster risk management and climate change adaptation scenarios are to generate climate data that can be downscaled to at least regional and sub-national scales, while extending disaster risk projections to longer time scales (see Gaffin et al., 2004; Theobald, 2005; Bengtsson et al., 2006; van Vuuren et al., 2006; Grübler et al., 2007; Moss et al., 2010; Hallegatte et al., 2011a).

Scenario development has traditionally been carried out in a sequential manner (Moss et al., 2010). For example, a first step in developing climate change scenarios has typically involved structural projections of key determinants of greenhouse gas emissions (e.g., population changes, urbanization, etc.). These have been used to estimate concentrations and radiative forcing from emissions, leading to climate projections that can be used in impacts research. One difficulty in using climate scenarios for disaster risk management and climate change adaptation has been the uncertainties associated with extreme climate and weather events, including the behavior of local climates (see Section 3.2.3). Future socioeconomic changes (e.g., demography, population preferences, technologies) are also highly uncertain, thus scenarios must consider a wide range of possible futures to design adaptation strategies and analyze tradeoffs (e.g., Hall, 2007; Lempert, 2007; Lempert and Collins, 2007; WGBU, 2008; Dessai et al., 2009a,b; Hallegatte, 2009). Alternative approaches have focused first on scenarios of radiative forcing, followed by an analysis of the combinations of economic, technological, demographic, policy, and institutional factors that can influence such trajectories (Moss et al., 2010). Other approaches are based on robust decisionmaking (e.g., Groves and Lempert, 2007; Lempert and Collins, 2007; Groves et al., 2008); information gap analysis (Hine and Hall, 2010); or on the search for co-benefits, no regrets strategies, flexibility, and reversibility (e.g., Fankhauser et al., 1999; Goodess et al., 2007; Hallegatte, 2009).

Scenario development requires substantial climate, social, environmental, and economic data, which are not equally available or accessible for all parts of the world. Qualitative scenarios can also be produced based on expert judgment (e.g., Delphi exercises) or on storylines designed through consultative processes. Such scenarios often reflect different mindsets or worldviews that represent contrasting visions of the future.

To adapt to changing climate and weather extremes, difficult choices may become increasingly necessary. In many locations, for example, adapting to scenarios of reduced water availability may involve increased investments in water infrastructure to provide enough irrigation to maintain existing agricultural production, or a shift from current production to less water-consuming crops (see Rosenzweig et al., 2004; ONERC, 2009; Gao and Hu, 2011). In considering adaptation to future flood risk in the Thames Estuary, the UK Environment Agency

(2009) applied four scenarios over three time periods to flood management. Through a wide consultation process, it was determined that improving the current infrastructure continues to be the preferred strategy until 2070, when construction of an outer barrage may become justifiable, especially as economic and climate change conditions change over time.

Evaluating choices among different options depends on how the stakeholders view the region in coming decades, and on adaptation decisions that are informed by political processes. One scenario approach that explicitly acknowledges both social and environmental uncertainties entails identification of flexible adaptation pathways for managing the future risks associated with climate change (Yohe and Leichenko, 2010). Based on principles of risk management (which emphasize the importance of diversification and risk-spreading mechanisms in order to improve social and/or private welfare in situations of profound uncertainty), this approach can be used to identify a sequence of adaptation strategies that are designed to keep society at or below acceptable levels of risk. These strategies, which policymakers, stakeholders, and experts develop and implement, are expected to evolve over time as knowledge of climate change and associated climate hazards progresses. The flexible adaptation or adaptive management approach that underpins this also stresses the connections between adaptation and mitigation of climate change, recognizing that climate change mitigation will be needed in order to sustain society at or below an acceptable level of risk (Yohe and Leichenko, 2010).

In contrast to predictive scenarios and risk management approaches, exploratory and normative approaches can be used to develop scenarios that represent desirable alternative futures. This is particularly important in the case of sustainability, where the most likely futures may not be the most desirable (Robinson, 2003), and where poverty, inequity, and injustice are recognized by many as incompatible with sustainable development (Redclift 1987, 1992; St. Clair, 2010). Pathways that require considerable transformation to reach sustainable futures of this kind can be supported by backcasting techniques. The process of backcasting involves developing normative scenarios that explore the feasibility and implications of achieving certain desired outcomes (Robinson, 2003; Carlsson-Kanyama et al., 2008). It is concerned with how desirable futures can be attained, focusing on policy measures that would be required to reach such conditions. Participatory backcasting, which involves local stakeholders in visionary activities related to sustainable development, can also open deliberative opportunities and inclusiveness in decision framing and making. Where visioning is repeated it can also open possibilities for tracking development and learning processes that make up adaptive strategies for disaster risk management, based on the explicit acknowledgement of the beliefs, values, and preferences of citizens (Robinson, 2003). Changing attitudes and core beliefs, including those on climate change, its causes, and consequences, is a slow process (Volkery and Ribeiro, 2009).

Adding an anticipatory dimension to planning for the future is critical for striving toward transformational actions in the face of multiple and dynamic uncertainties. The literature on anticipatory action learning provides some experience on what this might look like (Stevenson, 2002; Kelleher, 2005). The framing and negotiation of decisionmaking and policy is made inclusive and reflexive through multiple rounds of stakeholder engagement to explore meanings of what different futures may involve, reflect upon unavoidable tradeoffs and the winners and losers, and establish confidence to creatively adapt to new challenges (Inayatullah, 2006). This type of learning stresses the skills, knowledge, and visions of those at risk and aims to support leadership from even the most vulnerable. A combination of local- and global-scale scenarios that link storylines developed at several organizational levels (Biggs et al., 2007), personalizing narratives to create a sense of ownership (Frittaion et al., 2010), and providing safe and repeated learning spaces (Kesby, 2005) can enhance learning.

While scenarios, projections, and forecasts are all useful and important inputs for planning, actual planning and decisionmaking is a complex socio-political process involving different stakeholders and interacting agents. Although much progress has been made by employing scenario building and narrative creation to explore uncertainties, surprises, extreme events, and tipping points, the transition from envisioning to planning, policymaking, and implementation remains poorly understood (Lempert, 2007). Similarly, more widespread uptake of even scientifically highly robust scenarios may be hampered by conflicting understandings of and practical approaches to uncertainty, different scalar needs, and lack of training among users (Gawith et al., 2009). Experiences in scenario building emphasize their usefulness for raising awareness on climate change (Gawith et al., 2009). However, to move from framing public debates to policymaking and implementation, useful scenario building requires procedural stability, permanent yet flexible institutional and governance structures that build trust, and experience to take advantage of new insights for effective and fair risk management (Volkery and Ribeiro, 2009).

Developing the capacity for adaptive learning to accommodate complexity and uncertainty requires exploratory and imaginative visions for the future that support choices and can accommodate multiple values and aspirations (Miller, 2007). Disaster risk management and climate change adaptation, and synergies between the two, can contribute toward planning for a sustainable and resilient future, but this involves expanding the diversity of futures that are considered and identifying those that are desirable, as well as the short- and long-term values and actions that are consistent with them (Lempert, 2007).

8.6.2. Approaches, Tools, and Integrating Practices

As discussed above, scenarios, narrative storylines (Tschakert and Dietrich, 2010) and simulations (Nicholls et al., 2007) can help to project and facilitate discussion of possible futures. This section considers the tools that are available for helping decisionmakers and planners think about and plan for the future in the context of extreme climate and weather events. Past experiences with enhancing resilience to climate extremes include examples of both specific decision support tools and the governance and institutional contexts in which these tools are used and subsequent decisions are made (OECD, 2009b; Burch et al., 2010; Whitehead et al., 2010). Tools include those that enable information gathering, monitoring, analysis, and assessment; simulate threats; develop projections of possible impacts; and explore implications for response. Effective approaches combine understandings of potential stresses from climate extremes, along with possible tipping points for affected social and physical systems, with monitoring systems for tracking changes and identifying emerging threats in time for adaptive responses. This is, of course, challenging, requiring methodologies that can be open to both quantitative and qualitative data and their analysis, including participatory deliberation (NRC, 2010).

Institutional innovations aimed at improving the availability of disaster information to decisionmakers include the creation of national or regional institutions to manage and distribute disaster risk information (von Hesse et al., 2008; Corfee-Morlot et al., 2011), bringing together previously fragmented efforts centered in national meteorological, geological, oceanographic, and other agencies. The World Meteorological Organization and partner organizations have proposed the creation of a Global Framework for Climate Services, a collaborative effort to help the global community to better adapt to climate variability and change by developing and incorporating science-based climate information and prediction into planning, policy, and practice (WCC-3, 2009). New open-source tools for comprehensive probabilistic risk assessment are beginning to offer ways of compiling information at different scales and from different institutions (OECD, 2009b). A growing number of countries are also systematically recording disaster loss and impacts at the local level (DesInventar, 2010) and developing mechanisms to use such information to inform and guide public investment decisions (Comunidad Andina, 2007, 2009; von Hesse et al., 2008) and national planning. Unfortunately, there is as yet only limited experience with the integrated deployment of such tools and institutional approaches, especially in ways that cross scales of risk management strategy development and decisionmaking, and very limited evaluations of such deployment.

8.6.2.1. Improving Analysis and Modeling Tools

Various tools can be used to design environmental and climate policies. Among them, integrated environment-energy-economy models produce long-term projections taking into account demographic, technological, and economic trends (e.g., Edenhofer et al., 2006; Clarke and Weyant, 2009). These models can be used to assess the consequences of various policies. However, most such models are at spatial and temporal scales that do not resolve specific climate extremes or disasters (Hallegatte et al., 2007). At higher spatial resolution, numerical models (e.g., input-output models, calculable general equilibrium models) can help to assess disaster consequences and, therefore, balance the cost of disaster risk management actions and their benefits (Rose et al., 1997; Gordon et al., 1998; Okuyama, 2004; Rose and Liao, 2005; Tsuchiya et al., 2007; Hallegatte, 2008b). In particular, they can compare the cost of responding to disasters with the cost of preventing disasters. Since disasters have intangible consequences (e.g., loss of lives, ecosystem losses, cultural heritage losses, distributional consequences) that are difficult to measure in economic terms, the quantitative models are necessary but not sufficient to determine desirable policies and disaster risk management actions. Whether incorporated in models or used in other forms of analysis, CBA is useful to compare costs and benefits; but when intangibles play a large role and when no consensus can be reached on how to value these intangibles, other decisionmaking tools and approaches are needed. Multi-criteria decisionmaking (Birkmann, 2006), robust decisionmaking (e.g., Lempert 2007; Lempert and Collins, 2007), transition management approaches (e.g., Kemp et al., 2007; Loorbach, 2010), and group-process analytic-deliberative approaches (Mercer et al., 2008) are examples of such alternative decisionmaking methodologies.

Also necessary are indicators to measure the successes and failures of policies. For example, climate change adaptation policies often target the enhancement of adaptive capacity. The effects and outcomes of policies are often measured using classical economic indicators such as GDP. The limits of such indicators are well known, and have been summarized in several recent reports (e.g., CMEPSP, 2009; OECD, 2009a). To measure progress toward a resilient and sustainable future, one needs to include additional components, such as measures of stocks, other capital types (natural capital, human capital, social capital), distribution issues, and welfare factors (health, education, etc.). Many alternative indicators have been proposed in the literature, but no consensus exists. Examples of these alternative indicators include the Human Development Index, the Genuine Progress Indicator, the Index of Sustainable Economic Welfare, the Ecological Footprint, the normalized GDP, and various indicators of vulnerability and adaptive capacity (Costanza, 2000; Yohe and Tol, 2002; Lawn, 2003; Costanza et al., 2004; Eriksen and Kelly, 2006; Jones and Klenow, 2010).

8.6.2.2. Institutional Approaches

Among the most successful disaster risk management and adaptation efforts have been those that have facilitated the development of partnerships between local leaders and other stakeholders, including extra-local governments (Bicknell et al, 2009; Pelling and Wisner, 2009; Gero et al, 2011). This allows local strength and priorities to surface in disaster risk management, while acknowledging also that communities (including local government) have limited resources and strategic scope and alone cannot always address the underlying drivers of risk (Bhattamishra and Barrett, 2010). Local programs are now increasingly moving from a focus on strengthening disaster preparedness and response to reducing both local hazard levels and vulnerability (e.g., through slope stabilization, flood control measures, improvements in drainage, etc.) (Lavell, 2009; UNISDR, 2009; Reyos, 2010). Most of the cases where sustainable local processes have emerged are where national governments have decentralized both responsibilities and resources to the local level, and where local governments have become more accountable to their citizens, as for example in cities in Colombia such as Manizales (Velásquez, 1998, 2005). In Bangladesh and Cuba, successes in disaster preparedness and response leading to drastic reduction in mortality due to tropical cyclones, built on solid local organization, have relied on sustained support from the national level (Haque and Blair, 1992; Bern et al., 1993; Ahmed et al., 1999; Chowdhury, 2002; Kossin et al., 2007; Elsner et al., 2008; Karim and Mimura, 2008; Knutson et al., 2010; World Bank, 2010b). A growing number of examples now exist of community-driven approaches that are supported by local and national governments as well as by international agencies, through mechanisms such as social funds (Bhattamishra and Barrett, 2010).

Risk transfer instruments, such as insurance, reinsurance, insurance pools, catastrophe bonds, micro-insurance, and other mechanisms, shift economic risk from one party to another and thus provide compensation in exchange for a payment, often a premium (ex-post effect) (see Sections 5.6.3, 6.5.3, and 7.4, and Case Study 9.2.13). In addition, these mechanisms can also help to anticipate and reduce (economic) risk as they reduce volatility and increase economic resilience at the household, national, and regional levels (Linnerooth-Bayer et al., 2005). As one example, with such insurance, drought-exposed farmers in Malawi have been able to access improved seeds for higher yielding and higher risk crops, thus helping them to make a leap ahead in terms of generating higher incomes and the adoption of higher return technologies (World Bank, 2005; Hazell and Hess, 2010). However, many obstacles to such schemes still exist, particularly in low-income and many middle-income countries, including the absence of comprehensive risk assessments and required data, legal frameworks, and the necessary infrastructure, and probably more experience is required to determine the contexts in which they can be effective (Linnerooth-Bayer and Mechler, 2007; Cummins and Mahul, 2009; Mahul and Stutley, 2010).

Disaster risk management and adaptation can also be addressed through the enhancement of generic adaptive capacity alongside hazard-specific response strategies (IFRC, 2010). This capacity includes access to information, the skills and resources needed to reflect upon and apply new knowledge, and institutions to support inclusive decisionmaking. These are cornerstones of both sustainability and resilience. While uncertainty may make it difficult for decisionmakers to commit funds for hazard-specific risk reduction actions, these barriers do not prevent investment in generic foundations of resilient and sustainable societies (Pelling, 2010a). Importantly, from such foundations, local actors may be able to make better-informed choices on how to manage risk in their own lives, certainly over the short and medium terms. For instance, federations formed by slum dwellers have become active in identifying and acting on disaster risk within their settlements and seeking partnerships with local governments to make this more effective and larger scale (IFRC, 2010).

Changes in systems and structures may call for new ways of thinking about social contracts, which describe the balance of rights and responsibilities between different parties. Social contracts that are suitable for technical problems can be limiting and insufficient for addressing adaptive challenges (Heifetz, 2010). Pelling and Dill (2009) describe the ways that current social contracts are tested when disasters occur, and how disasters may open up a space for social transformation, or catalyze transformative pathways building on pre-disaster trajectories. O'Brien et al. (2009) consider how resilience thinking can contribute to new debates about social contracts in a changing climate, drawing attention to tradeoffs among social groups and ecosystems, and to the rights of and responsibilities toward distant others and future generations.

8.6.2.3. Transformational Strategies and Actions for Achieving Multiple Objectives

If extreme climate and weather events increase significantly in coming decades, climate change adaptation and disaster risk management are likely to require not only incremental changes, but also *transformative* changes in systems and institutions. Transformation can be defined as a fundamental qualitative change, or a change in composition or structure that is often associated with changes in perspectives or initial conditions (see Box 8-1). It often involves a change in paradigm and may include shifts in perception and meaning, changes in underlying norms and values, reconfiguration of social networks and patterns of interaction, changes in power structures, and the introduction of new institutional arrangements and regulatory frameworks (Folke et al., 2009, 2010; Pahl-Wostl, 2009; Smith and Stirling, 2010).

Although transformational policies and measures may be deliberately invoked as a strategy to reduce disaster risk and adapt to climate change, in many cases such strategies are precipitated by an extreme event, sometimes referred to as a 'focusing event' (Birkland, 1996). However, whether an extreme event leads to any change at all is unclear, as processes of policy change are often subtle and complex and linked to learning processes (Birkland, 2006). Exploring the relationship between systematic learning processes and small disasters, Voss and Wagner (2010) find that a failure to learn is the most common prerequisite for future disasters. There are, however, many dimensions to learning (e.g., cognitive, normative, and relational; see Huitema et al., 2010), and learning may be a necessary but insufficient condition for initiating transformational change.

Understanding processes of deliberate change and change management can provide insights on societal responses to extreme climate and weather events. Traditional approaches to managing change successfully in businesses and organizations focus on a series of defined steps (Harvard Business Essentials, 2003). Kotter (1996), for example, identifies an eight-step process for promoting change: (1) create a sense of urgency; (2) pull together the guiding team; (3) develop the change vision and strategy; (4) communicate for understanding and buy in; (5) empower others to act; (6) produce short-term wins; (7) don't let up; and (8) create a new culture. Kotter (1995) also identifies eight errors that are often made when leading change, including, for example, allowing too much complacency, failing to create a sufficiently powerful guiding coalition, and underestimating the power of a sound vision. It is also

> **FAQ 8.2 | Are transformational changes desirable and even possible, and if so, who will lead them?**
>
> Transformation in and of itself is not always desirable. It is a complex process that involves changes at the personal, cultural, institutional, and systems levels. Transformation can imply the loss of the familiar, which can create a sense of disequilibrium and uncertainty. In some cases, notable changes in the nature, form, or appearance of a system or process may be inconsistent with the values and preferences of some groups. Transformation can thus be perceived as threatening by some and instrumental by others, as the potential for real or perceived winners and losers at different scales stimulates social unease or tension. Desirable or not, it is important to recognize that transformations are now occurring at an unprecedented rate and scale, influenced by globalization, social and technological development, and environmental change. Climate change itself represents a system-scale transformation that will have widespread consequences for ecology and society, including through changes in climate extremes. Responses to climate change and changes in disaster risk can be both incremental and transformational. Transformational responses are not always radical or monumental – sometimes they simply involve a questioning of assumptions or viewing a problem from a new perspective. Transformational responses are not only possible, but they can be facilitated through learning processes, especially reflexive learning that explores blind spots in current thinking and approaches to disaster risk management and climate change adaptation. However, because there are risks and barriers, transformation also calls for leadership – not only from authority figures who hold positions and power, but from individuals and groups who are able to connect present-day actions with their values, and with a collective vision for a sustainable and resilient future. Considering the balance between incremental and transformative adjustments flags the importance of scale: first, because of the opportunities for enhancing leadership capacity that come from greater involvement of those locally at risk or undertaking adaptive experimentation for risk management; and second, because of the potential for transformation, incremental change, or stability at one systems level or sector (e.g., administrative, social, technical) to provoke or restrict adjustments in other systems and scales. Inter-scale and inter-sectoral communication therefore become important tools for managing adaptive disaster risk management.

important to recognize that many change initiatives create uncertainty and disequilibria, and are considered disruptive or disorienting (Heifetz et al., 2009). Furthermore, vested interests seldom choose transformation, particularly when there is much to lose from change (Christensen, 1997). As discussed in Section 8.5.2, there are winners and losers not only from extreme climate and weather events, but also from responses. Consequently, fundamental change is often resisted by the people that it affects the most (Kotter, 1996; Kegan and Lahey, 2009). Helping people, groups, organizations, and governments to manage the resulting disequilibria is seen as essential to successful transformation.

Many of the recent approaches to change and transformation focus on learning organizations, and the importance of changing individual and collective mindsets or mental models (Senge, 1990; Heifetz et al., 2009; Kegan and Lahey, 2009; Scharmer, 2009). This transformational change literature distinguishes between technical problems that can be addressed through management based on existing organizational and institutional structures and cultural norms, and adaptive challenges that require a change in mindsets, including changes in assumptions, beliefs, priorities, and loyalties (Heifetz et al., 2009; Kegan and Lahey, 2009). Treating disaster risk management and climate change adaptation as technical problems may focus attention only on improving technologies, reforming institutions, or managing displaced populations, whereas viewing them as an adaptive challenge shifts attention toward gaps between values and behaviors (e.g., values that promote human security versus policies or behaviors that undermine health and livelihoods), beliefs (e.g., a belief that disasters are inevitable or that adaptation will occur autonomously), and competing commitments (e.g., a commitment to maintaining aid dependency or preserving social hierarchies). Although most problems have both technical and adaptive elements, treating an adaptive challenge only as a technical problem limits successful outcomes (Heifetz et al., 2009).

Transformative changes that move society towards the path of openness and adaptability depend not only on changes in mindsets, but also on changes in systems and structures. Case studies of social-ecological systems suggest that there are three phases involved in systems transformations. The first phase includes being prepared, or preparing the system, for change. The second phase calls for navigating the transition by making use of a sudden crisis as an opportunity for change, whether the crisis is real or perceived. The third phase involves building resilience of the new system (Olsson et al., 2004; Chapin et al., 2010). Traditional management approaches emphasize the reduction of uncertainties, with the expectation that this will lead to systems that can be predicted and controlled. However, in the case of climate change, future projections of climate variables and extremes will contain uncertainty (see Section 3.2.3). Consequently, there is a need for management approaches that are adaptive and robust in the presence of large and irreducible uncertainties.

8.6.3. Facilitating Transformational Change

Adapting to climate and weather extremes associated with rapid and severe climate change, such as a warming beyond 4°C within this century, without transformational policy and social change will be difficult: if not chosen through proactive policies, forced transformations and crises are likely to result (New et al., 2011). Adaptation that is transformative marks a shift from emphasizing finite projects with linear trajectories and readily identifiable, discrete strategies and outcomes (Schipper, 2007) toward an approach that includes adaptive management, learning, innovation, and leadership, among other elements. These aspects of adjustment are increasingly seen as being embedded in ongoing socio-cultural and institutional learning processes. This can be observed in the many adaptation projects that emphasize learning about risks, evaluating response options, experimenting with and rectifying options, exchanging information, and making tradeoffs based on public values using reversible and adjustable strategies (McGray et al., 2007; Leary et al., 2008; Hallegatte, 2009; Hallegatte et al., 2011b).

Transformational adaptations are likely to be enabled by a number of factors. Some of the factors arise from external drivers such as focal events that catalyze attention to vulnerabilities or the presence of other sources of stress that also encourage considerations of major changes. Supportive social contexts such as the availability of understandable and socially acceptable options, access to resources for action, and the presence of incentives may also be important. Other factors are related to effective institutions and organizations, including those described in the following subsections.

8.6.3.1. Adaptive Management

In general terms, adaptive management can be defined as a structured process for improving management policies and practices by systemic learning from the outcomes of implemented strategies, and by taking into account changes in external factors in a proactive manner (Pahl-Wostl et al., 2007; Pahl-Wostl, 2009; Section 6.6.2). Principles of adaptive management can contribute to a more process-oriented approach to disaster risk management, and have already shown some success in promoting sustainable natural resource management under conditions of uncertainty (Medema et al., 2008). Adaptive management is often associated with 'adaptive' organizations that are not locked into rigid agendas and practices, such that they can consider new information, new challenges, and new ways of operating (Berkhout et al., 2006; Pelling et al., 2007). Organizations that can monitor environmental, economic, and social conditions and changes, respond to shifting policies and leadership changes, and take advantage of opportunities for innovative interventions are a key to resilience, especially with respect to conceivable but long-term and/or relatively low-probability events. Those social systems that appear most adept at adapting are able to integrate formal organizational roles with cross-cutting informal social spaces for learning, experimentation, communication, and for trust-based and speedy disaster response that is nonetheless accountable to beneficiaries (Pelling et al., 2007).

Adaptive management is a challenge for those organizations that perceive reputational risk from experimentation and the knowledge that some local experiments may fail (Fernandez-Gimenez et al., 2008). Where this approach works best, outcomes have gone beyond specific management goals to include trust-building among stakeholders – a resource that is fundamental to any policy environment facing an uncertain future, and which also has benefits for quality of life and market competitiveness (Fernandez-Gimenez et al., 2008). It requires revisiting the relationship between the state and local actors concerning facilitation of innovation, particularly when experiments go wrong. Investing in experimentation and innovation necessarily requires some tolerance for projects that may not be productive or cost effective, or at least not in the short term or under existing risk conditions. However, it is exactly the existence of this diversity of outcomes that makes societies fit to adapt once risk conditions change, particularly in unexpected and nonlinear directions.

8.6.3.2. Learning

The dynamic notion of adaptation calls for learning as an iterative process in order to build resilience and enhance adaptive capacity now, rather than targeting adaptation in the distant future (see Section 1.4.2). Social and collective learning includes support for joint problem solving, power sharing, and iterative reflection (Berkes, 2009). The need to take into account the arrival of new information in the design of response strategies has also been mentioned for mitigation policies (Ha Duong et al., 1997; Ambrosi et al., 2003). Adaptive management is an incremental and iterative learning-by-doing process, whereby participants make sense of system changes, engage in actions, and finally reflect on changes and actions. Lessons from learning theories, including experiential learning (Kolb, 1984) and transformative learning (Mezirow, 1995), stress the importance of learning-by-doing in concrete learning cycles, problem-solving actions, and the reinterpretation of meanings and values associated with learning activities.

Learning is a key component for living with uncertainty and extreme events, and is nurtured by building the right kind of social/institutional space for learning and experimentation that allows for competing worldviews, knowledge systems, and values, and facilitates innovative and creative adaptation (Thomas and Twyman, 2005; Armitage et al., 2008; Moser, 2009; Pettengell, 2010). Examples include promoting shared platforms for dialogs and participatory vulnerability assessments that include a wide range of stakeholders (see ISET, 2010). It is equally important to acknowledge that abrupt and surprising changes may surpass existing skills and memory (Batterbury, 2008). Adaptation projects have demonstrated that fostering adaptive capacity and managing uncertainty in real-time, by adjusting as new information, techniques, or conditions emerge, especially among populations exposed to multiple risks and stressors, is more effective than more narrowly designed planning approaches that target a given impact and are dependent on particular future climate information (McGray et al., 2007; Pettengell, 2010). In the humanitarian sector, institutionalized processes of learning have contributed to leadership innovation (see Box 8-3).

Action research and learning provide a powerful complement to resilience thinking, as they focus explicitly on iterative or cyclical learning through reflection of successes and failures in experimental

> **Box 8-3 | Institutionalized Research and Learning in the Humanitarian Sector**
>
> An important attribute of the humanitarian sector is its readiness to learn. Research and learning unfolds at multiple levels, including sector-wide reviews of performance and practice such as those undertaken by the Active Learning Network for Accountability and Performance in Humanitarian Action. Research and learning is also structured around the internal needs of organizations (e.g., Red Cross and Red Crescent Societies) or the outcomes of individual events (e.g., the landmark report on humanitarian sector practice following the Indian Ocean tsunami (Telford et al., 2006). Organizations have different methodologies, target audiences, and frames of reference, making cross-sector learning difficult (Amin and Goldstein, 2008), but they all have led to practical and procedural changes. Less well-developed is active experimentation in the field of practice, with a view of proactive learning (Corbacioglu and Kapucu, 2006). This is difficult in the humanitarian sector, where stakes are high and rapid action has typically made it difficult to implement learning-while-doing experiments. Where experimentation may be more observable, for example, in disaster prevention and risk reduction or reconstruction activities, there are significant gaps in documentation that have slowed down the transferring of learning outcomes between organizations. Hierarchical models of governance have fostered a lack of cooperation and generated competition between agencies within the humanitarian and development sectors, partly explaining why there is more learning based on the sharing of experience inside organizations than across sectors (Kapucu, 2009). But the increasing scale and diversity of risk associated with climate change, and compounded by other development trends such as growing global inequality and urbanization, puts more pressure on donors to promote cross-sector communication of productive innovations and of the research and experimentation such innovation builds upon.

action, transfer of knowledge between learning cycles, and the next learning loop that will lead to new types of action (List, 2006; Ramos, 2006). Referring to the learning processes described in Section 1.4.2, critical reflection is paramount to triple-loop learning; it also constitutes the key pillar of double-loop learning, or the questioning of what works and why that is fundamental to shifts in reasoning and behavior (Kolb and Fry, 1975; Argyris and Schön, 1978; Keen et al., 2005). Allowing time for reflection in this iterative learning process is important because it provides the necessary space to develop and test theories and strategies under ever-changing conditions. It is through such learning processes that individual and collective empowerment can emerge and potentially be scaled up to trigger transformation (Kesby, 2005).

8.6.3.3. Innovation

The transformation of society toward sustainability and resilience involves both social innovations and technological innovations – incremental as well as radical. Innovation can refer to non-material changes related to knowledge, cognition, communication, or intelligence, or it can refer to any kind of material resources. In some cases, small adjustments in practices or technologies may represent innovative steps toward sustainability, while in other cases there is a strong need for more radical transformations. Some of the literature on innovation focuses on ensuring economic competitiveness for firms in an increasingly globalized economy (Fløysand and Jakobsen, 2010), and some concentrates on the relationship between environment on the one hand and the competitiveness of firms on the other (Mol and Sonnenfeld, 2000). In addition, there is a body of social science literature on innovation that has emerged during the last 15 years, motivated by the need for transforming society as a whole in more sustainable directions. Recent literature has brought out new ideas and frameworks for understanding and managing technology and innovation-driven transitions, such as the Multi Level Perspective (MLP) (Rip and Kemp, 1998; Geels, 2002; Geels and Schot, 2007; Markard and Truffer, 2008). Combining insights from evolutionary theory and sociology of technology, MLP conceptualizes major transformative change as the product of interrelated processes occurring at the three levels of niches, regimes, and landscape. The model emphasizes the incremental nature of innovation in socio-technical regimes. Transitions – that is, shifts from one stable socio-technical regime to another – occur when regimes are destabilized through landscape pressures, which provide breakthrough opportunities for niche innovations.

In this field of research, there is a strong focus on systems innovation and transformation of socio-technical systems, with the potential of facilitating transitions from established systems for transport, energy supply, agriculture, housing, etc., to alternative, sustainable systems (Geels, 2002; Hoogma et al., 2002; Smith et al., 2005; Raven et al., 2010). The systems innovation literature analyzes the emergence and dynamics of large-scale, long-term socio-technical transformations (Kemp et al., 1998).

Though not directly dependent on changes in technology, technological and social innovations are often closely interrelated, not the least in that they involve changes in social practices, institutions, cultural values, knowledge systems, and technologies (Rohracher, 2008). Box 8-4 describes such innovation in water management. A central, basic insight established within this research is that social and technological change is an interactive process of co-development between technology and society (Kemp, 1994; Hoogma et al., 2002; Rohracher, 2008). Throughout history, new socio-technical systems have emerged and replaced old ones in so-called technological revolutions, and an important characteristic of such transitions is the interactions and conflicts between new, emerging systems and established and dominating socio-technical regimes, with strong actors defending business as usual (Kemp, 1994; Perez, 2002).

Box 8-4 | Innovation and Transformation in Water Management

The impacts of climate change in many regions are predominantly linked to the water system, in particular through increased exposure to floods and droughts (Lehner et al., 2006; Smith and Barchiesi, 2009; see Section 2.5). Considering water as a key structuring element or guiding principle for landscape management and land use planning requires technology, integrated systems thinking, and the art of thinking in terms of attractiveness and mutual influence, or even mutual consent, between different authorities, experts, interest groups, and the public. One of the most pronounced changes can be observed in The Netherlands, where the government has requested a radical rethinking of water management in general and flood management in particular. The resulting policy stream, initiated through the 'Room for the River' (Ruimte voor de Rivier) policy, has strongly influenced other areas of government policy. Greater emphasis is now given to the integration of water management and spatial planning, with the regulating services provided by landscapes with natural flooding regimes being highly valued. This requires a revision of land use practices and reflects a gradual movement toward integrated landscape planning, whereby water is recognized as a natural, structural element. The societal debate about the plans to build in deep-lying polders and other hydrologically unfavorable spots, and new ideas on floating cities, indicate a considerable social engagement of both public and private parties with the issue of sustainable landscapes and water management. However, although such innovative ideas have been adopted in policy, they take time to implement, as there is considerable social resistance (Wolsink, 2006).

8.6.3.4. Leadership

Leadership can be critical for disaster risk management and climate change adaptation, particularly in initiating processes and sustaining them over time (Moser and Ekstrom, 2010). Change processes are shaped both by the action of individual champions (as well as by those resisting change) and their interactions with organizations, institutional structures, and systems. Leadership can be a driver of change, providing direction and motivating others to follow, thus the promotion of leaders by institutions is considered an important component of adaptive capacity (Gupta et al., 2010), although knowledge about how to create and enable leadership remains elusive. Leadership and leaders often do not develop independent of the institutional context, which includes institutional rules, resources, and organizational culture (Kingdon, 1995).

Leaders who facilitate transformation have the capacity to understand and communicate a wide set of technical, social, and political perspectives related to a particular issue or problem. They are also able to reframe meanings, overcome contradictions, synthesize information, and create new alliances that transform knowledge into action (Folke et al., 2009). Leadership also involves diagnosing the kinds of losses that some people, groups, organizations, or governments may experience through transformative change, such as the loss of status, wealth, security, loyalty, or competency, not to mention loved ones (Heifetz, 2010). Leaders help individuals and groups to take action to mobilize 'adaptive work' in their communities, such that they and others can thrive in a changing world by managing risk and creating alternative development pathways or engaging and directing people during times of choice and change (Heifetz, 2010).

8.7. Synergies between Disaster Risk Management and Climate Change Adaptation for a Resilient and Sustainable Future

Drawing on the assessment presented in this chapter, it becomes clear that there are many potential synergies between disaster risk management and climate change adaptation that can contribute to social, economic, and environmental sustainability and a resilient future. There is, however, no single approach, framework, or pathway to achieve this. Responding to a diversity of extremes in the present and under varying social and environmental conditions can contribute to future resilience in situations of uncertainty. Nonetheless, some important contributing factors have been identified and discussed in this chapter, and are confirmed by the wider literature (e.g., Lemos et al., 2007; Tompkins et al., 2008; Pelling, 2010a; Wisner, 2011). Eight critical factors stand out as important:

1) A capacity to reconcile short- and long-term goals
2) A willingness to reconcile diverse expressions of risk in multi-hazard and multi-stressor contexts
3) The integration of disaster risk reduction and climate change adaptation into other social and economic policy processes
4) Innovative, reflexive, and transformative leadership (at all levels)
5) Adaptive, responsive, and accountable governance
6) Support for flexibility, innovation, and learning, locally and across sectors
7) The ability to identify and address the root causes of vulnerability
8) A long-term commitment to managing risk and uncertainty and promoting risk-based thinking.

Lessons learned in climate change adaptation and disaster risk management illustrate that managing uncertainty through adaptive management, anticipatory learning, and innovation can lead to more flexible, dynamic, and efficient information flows and adaptation plans, while creating openings for transformational action. Reducing vulnerability has been identified in many contemporary disaster studies as the most important prerequisite for a resilient and sustainable future. Research has consistently found that for long-term sustainability, disaster risk management is most impactful when combined with structural reforms that address underlying causes of vulnerability and the structural

> **FAQ 8.3 | What practical steps can we take to move toward a sustainable and resilient future?**
>
> The disruptions caused by disaster events often reveal development failures. They also provide an opportunity for reconsidering development through reconstruction and disaster risk reduction. Practical steps can address both the root causes of risk found in development relations, including enhancing human rights, gender equity, and environmental integrity, and more proximate causes expressed most commonly through a need for extending land and property rights, access to critical services and basic needs, including social safety nets and insurance mechanisms, and transparent decisionmaking, especially at the local level. Identifying the drivers of hazard and vulnerability in ways that empower both those at risk and risk managers to take action is key. This can be done best where local and scientific knowledge is combined in the generation of risk maps or risk management plans. Greater use of local knowledge when coupled with local capacity can initiate enhanced accountability in integrated risk decisionmaking that helps to break unsustainable development relations.
>
> The uncertainty that comes with climatic variability and extremes reinforces arguments for better coordination and accountability within governance hierarchies and across sectors, as well as between generations and for non-human species in development. Local, national, and international actors bring different strengths and tools to questions of environmental change and its relationship to trends in human development. While offering a range of specific practical measures, both local and national approaches to risk management can better meet the flexibility demands of adaptation and resilience when they have strong, accountable leadership, and are enhanced by systematic experimentation and support for innovation in the development of tools as part of planned adaptive risk management approaches. International actors can help by providing an institutional framework to support experimentation, innovation, and flexibility. This can be part of national and local strategies to move development away from incentives that promote short-term gain and toward those that promote longer-term sustainability and flexibility.

inequalities that create and sustain poverty and constrain access to resources (Hewitt, 1983; Wisner et al., 2004; Lemos et al., 2007; Collins, 2009; Pelling, 2010b).

Engaging possible and desirable futures and options for decisionmaking fosters knowledge generation essential for adaptive risk management as well as iterative change processes. Zooming in on uncertain elements and their potential impacts (e.g., changes in rainfall and variability) and identifying factors that currently limit adaptive capacity (e.g., marginalization, lack of access to resources, or information gaps) allows for more robust decisionmaking that also integrates local contexts (asset portfolios, spreading and managing risks) with the climate context (current trends, likely futures, and uncertainties) to identify the most feasible, appropriate, and equitable response strategies, policies, and external interventions (Pettengell, 2010). Creating space and recognizing a diversity of voices often means reframing what counts as knowledge, engaging uncertainties, nourishing the capacity for narrative imagination, and articulating agency and strategic adaptive responses in the face of already experienced changes and to anticipate and prepare for future disturbances and shocks (Tschakert and Dietrich, 2010).

Challenges remain with respect to anticipating low-probability, high-impact events and potentially catastrophic tipping points that represent futures too undesirable to imagine, especially under circumstances where exposure and vulnerability are high and adaptive capacity low (Volkery and Ribeiro, 2009). At a practical level, there are many gaps and barriers to realizing synergies for integration to foster a sustainable and resilient future. For example, overcoming the current disconnect between local risk management practices and national institutional and legal frameworks, policy, and planning can be considered key to reconciling short- and long-term goals for vulnerability reduction. Even where capacity is present, it can take effort to shift into more critical, learning modes of governance (Corbacioglu and Kapucu, 2006). Moreover, anticipating vulnerabilities as well as feasible and fair actions may also reveal limits of adaptation and risk management, and thus raise the potential need for transformation. Because transformative changes open up questions about the values and priorities shaping development and risk futures, who wins and loses, and the balance of tradeoffs, decisions about when and where to facilitate transformative change and to whose benefit are inherently normative and political. Transformation cannot be approached without understanding related ethical and governance dimensions. At the same time, incremental changes, in supporting many aspects of business-as-usual, also possess implicit ethical and normative aspects. At heart, it is perhaps in failing to fully reveal and question these normative positions that current disaster risk management practice and policy has remained outside of development planning and policy processes, inhabiting a long acknowledged, but still present 'disasters archipelago' in the policy world (Hewitt, 1983, p. 12).

Disasters often require urgent action and represent a time when everyday processes for decisionmaking are disrupted. Although it is a useful approach in responding to emergency events and disaster relief, such top-down command and control frameworks work less well in disaster risk reduction and this is likely to be the case too in integrated adaptive risk management. In such systems, it is often the most vulnerable to hazards that are left out of decisionmaking processes (Pelling, 2003, 2007; Cutter, 2006; Mercer et al., 2008), whether it is within households (where the knowledge of women, children, or the elderly may not be recognized), within communities (where divisions among social groups may hinder learning), or within nations (where marginalized groups may not be heard, and where social division and political power influence the development and adaptation agenda). Disaster periods are frequently the times when the development visions and aspirations for the future of those most affected are not recognized. This reflects a widespread limitation on the quality and comprehensiveness of local participation in disaster risk reduction and its integration into everyday development planning. Instead, the humanitarian imperative, limited-term reconstruction budgets, and an understandable desire for rapid action over deliberation means that too often international social movements and humanitarian nongovernmental organizations, government agencies, and local relief organizations impose their own values and visions, often with the best of intentions. It is also important to recognize the potential for some people or groups to prevent sustainable decisions by employing their veto power or lobbying against reforms or regulations based on short-term political or economic interests (Klein, 2007). The distribution of power in society and who has the responsibility or right to shape the future through decisionmaking today is thus significant, and includes the role of international as well as national and local actors. Within the international humanitarian community, efforts such as the Sphere Standards and the Humanitarian Accountability Partnership are steps toward addressing this challenge.

Actions to reduce disaster risk and responses to climate change invariably involve tradeoffs with other societal goals, and conflicts related to different values and visions for the future. Innovative and successful solutions that combine multiple perspectives, differing worldviews, and contrasting ways of organizing social relations have been described by Verweij et al. (2006) as 'clumsy solutions.' Such solutions, they argue, depend on institutions in which all perspectives are heard and responded to, and where the quality of interactions among competing viewpoints foster creative alternatives. Drawing on the development ethics literature, St. Clair (2010) notes that when conflict and broad-based debate arise, alternatives often flourish and many potential spaces for action can be created, tapping into people's innovation and capacity to cope, adapt, and build resilience. Pelling (2010a) stresses the importance of social learning for transitional or transformational adaptation, and points out that it requires a high level of trust, a willingness to experiment and accept the possibility of failure in processes of learning and innovation, transparency of values, and active engagement of civil society. Committing to such a learning process is, as Tschakert and Dietrich (2010, p. 17) argue, preferable to alternatives because "learning by shock is neither an empowering nor an ethically defensible pathway."

The conjuncture of hazard and vulnerability, realized through disasters, forces coping and adaptation on individuals and society. Climate change and ongoing development place more people and assets at risk. Noteworthy progress in disaster risk management has been made, especially through the action of early warning on reducing mortality, but underlying vulnerability remains high (as indicated by increasing numbers of people affected and economic losses from disaster) and demographic and economic development trends continue to raise the stakes and present a choice: risk can be denied or faced, and adaptation can be forced or chosen. A reduction in the disaster risks associated with climate and weather extremes is therefore a question of political choice that involves addressing issues of equity, rights, and participation at all levels.

References

A digital library of non-journal-based literature cited in this chapter that may not be readily available to the public has been compiled as part of the IPCC review and drafting process, and can be accessed via either the IPCC Secretariat or IPCC Working Group II web sites.

AAG, 2003: *Global Change and Local Places: Estimating, Understanding, and Reducing Greenhouse Gases*. Association of American Geographers, Cambridge University Press, Cambridge, UK, 290 pp.

Adger, W.N., 1996: *Approaches to Vulnerability to Climate Change*. CSERGE Working Paper #GEC 96-05, University of East Anglia, Norwich, UK, 66 pp.

Adger, W.N., 2000: Social and ecological resilience: are they related? *Progress in Human Geography*, **24(3)**, 347-364.

Adger, W.N., 2003: Social capital, collective action, and adaptation to climate change. *Economic Geography*, **79(4)**, 387-404.

Adger, W.N., 2004: The right to keep cold. *Environment and Planning A*, **36(10)**, 1711-1715.

Adger, W.N., S. Huq, K. Brown, D. Conway, and M. Hulme, 2003: Adaptation to climate change in the developing world. *Progress in Development Studies*, **3(3)**, 179-195.

Adger, W.N., T.P. Hughes, C. Folke, S.R. Carpenter, and J. Rockstrom, 2005: Social-ecological resilience to coastal disasters. *Science*, **309**, 1036-1039.

Adger, W.N., S. Agrawala, M.M.Q. Mirza, C. Conde, K. O'Brien, J. Pulhin, R. Pulwarty, B. Smit, and K. Takahashi, 2007: Assessment of adaptation practices, options, constraints and capacity. In: *Climate Change 2007: Impacts, Adaptation and Vulnerability. Contribution of Working Group II to the Fourth Assessment Report of the Intergovernmental Panel on Climate Change* [Parry, M.L., O.F. Canziani, J.P. Palutikof, P.J. van der Linden and C.E. Hanson (eds.)], Cambridge University Press, Cambridge, UK, pp. 717-743.

Adger, W.N., J. Barnett, and H. Ellemor, 2010: Unique and valued places. In: *Climate Change Science and Policy* [Schneider, S. H., A. Rosencranz, M. Mastrandrea, and K. Kuntz-Duriseti (eds.)]. Island Press, Washington, DC, pp. 131-138.

Adger, W.N., K. Brown, D. Nelson, F. Berkes, H. Eakin, C. Folke, K. Galvin, L. Gunderson, M. Goulden, K. O'Brien, J. Ruitenbeek, and E. Tompkins, 2011: Resilience implications of policy responses to climate change. *WIRES Climate Change*, doi:10.1002/WCC.133.

Adly, E. and T. Ahmed, 2009: Water and food security in the river Nile basin: Perspectives of the government and NGOs in Egypt. In: *Facing Global Environmental Change: Environmental, Human, Energy, Food, Health and Water Security Concepts* [Brauch, H.G., N.C. Behera, P. Kameri-Mbote, J. Grin, Ú. Oswald Spring, B. Chourou, C. Mesjasz, and H. Krummenacher (eds.)]. Springer, Berlin, Germany, pp. 645-654.

Agarwal, B., 1991: Social security and the family: Coping with seasonality and calamity in rural India. In: *Social Security in Developing Countries* [Ahmad, E., J. Dreze, J. Hills, and A. Sen (eds.)]. Clarendon Press, Oxford, UK, pp. 171-244.

Agrawala, S. (ed.), 2005: *Bridge over Troubled Waters: Linking Climate Change and Development*. Organisation for Economic Co-operation and Development, Paris, France, 153 pp.

Ahmed, A.U., M. Alam, and A.A. Rahman, 1999: Adaptation to climate change in Bangladesh: future outlook. In: *Vulnerability and Adaptation to Climate Change for Bangladesh* [Huq, S., M. Asaduzzaman, Z. Karim, and F. Mahtab (eds.)]. J. Kluwer Academic Publishers, Dordrecht, The Netherlands, pp. 125-143.

Ahmed, I., 2009: Environmental refugees and environmental distressed migration as a security challenge for India and Bangladesh. In: *Facing Global Environmental Change: Environmental, Human, Energy, Food, Health and Water Security Concepts* [Brauch, H.G., Ú. Oswald Spring, J. Grin, C. Mesjasz, P. Kameri-Mbote, N. Chadha Behera, B. Chourou, and H. Krummenacher (eds.)]. Springer, Berlin, Germany, pp. 295-308.

Ahrens, J. and P.M. Rudolph, 2006: The importance of governance in risk reduction and disaster management. *Journal of Contingencies and Crisis Management*, **14(4)**, 207-220.

Aker, J.C. and I.M. Mbiti, 2010: Mobile phones and economic development in Africa. *The Journal of Economic Perspectives*, **24(3)**, 207-232.

Albala-Bertrand, J.M., 1993: *The Political Economy of Large Natural Disasters with Special Reference to Developing Countries*. Clarendon Press, Oxford, UK, 259 pp.

Alderman, H., J. Hodditnott, and B. Kinsey, 2006: Long-term consequences of early childhood malnutrition. *Oxford Economic Papers*, **58(3)**, 450-474.

Altieri, M.A. and P. Rosset, 1999: Ten reasons why biotechnology will not ensure food security, protect the environment, and reduce poverty in the developing world. *AgBioForum*, **2(3/4)**, 155-162.

Alves, D., 2002: An analysis of the geographical patterns of deforestation in Brazilian Amazonia in the 1991–1996 period. In: *Patterns and Processes of Land Use and Forest Change in the Amazon* [Wood, C. and R. Porro (eds.)]. University of Florida Press, Gainesville, Florida, pp. 95-105.

Ambrosi, P., J.-C. Hourcade, S. Hallegatte, P. Lecocq, P. Dumas, and M. Ha Duong, 2003: Optimal control models and elicitation of attitudes towards climate damages. *Environmental Modeling and Assessment*, **8(3)**, 133-147.

Amin, S. and M. Goldstein (eds.), 2008: *Data against Natural Disasters: Establishing Effective Systems for Relief, Recovery and Reconstruction*. World Bank, Washington, DC, 342 pp.

Ammer, M., M. Nowak, L. Stadlmayr, and G. Hafner, 2010: *Legal Status and Legal Treatment of Environmental Refugees*. Federal Environment Ministry, Dessau-Roßlau, Germany, 16 pp.

Argyris, C. and D. Schön, 1978: *Organizational Learning: A Theory of Action Perspective*. Addison Wesley, Reading, MA.

Armitage, D., 2005: Adaptive capacity and community-based natural resource management. *Environmental Management*, **35(6)**, 703-715.

Armitage, D., M. Marschke, and R. Plummer, 2008: Adaptive co-management and the paradox of learning. *Global Environmental Change*, **18(1)**, 86-98.

ASCE, 2010: *Report Card for America's Infrastructure: Dams*. American Society of Civil Engineers, Reston, VA.

Back, E., C. Cameron, and T.M. Tanner, 2009: *Children and Disaster Risk Reduction: Taking Stock and Moving Forward*. Children in a changing climate report, Institute of Development Studies, Brighton, UK, 44 pp.

Baghel, R. and M. Nusser, 2010: Discussing large dams in Asia after the World Commission on Dams: is a political ecology approach the way forward? *Water Alternatives*, **3(2)**, 231-248.

Bahadur, A.V., M. Ibrahim, and T. Tanner, 2010: *The Resilience Renaissance? Unpacking of Resilience for Tackling Climate Change and Disasters*. Strengthening Climate Resilience Discussion Paper, 1, Institute of Development Studies, Brighton, UK, 43 pp.

Balaban, V., 2006: Psychological assessment of children in disasters and emergencies. *Disasters*, **30(2)**, 178-198.

Bangay, C. and N. Blum, 2010: Education responses to climate change and quality: two parts of the same agenda? *International Journal of Educational Development*, **30(4)**, 359-368.

Barnett, J., 2003: Security and climate change. *Global Environmental Change*, **13(1)**, 7-17.

Barnett, J. and W.N. Adger, 2007: Climate change, human security and violent conflict. *Political Geography*, **26(6)**, 639-655.

Barnett, J. and S. O'Neill, 2010: Maladaptation. *Global Environmental Change*, **20**, 211-213.

Barnett, J., R. Matthew, and K. O'Brien, 2010: Global environmental change and human security: An introduction. In: *Global Environmental Change and Human Security* [Matthew, R.A., B. McDonald, J. Barnett, and K. O'Brien (eds.)]. MIT Press, Cambridge, MA, pp. 3-32.

Bartlett, S., 2008: After the tsunami in Cooks Nagar: The challenges of participatory rebuilding. *Children, Youth and Environments*, **18(1)**, 470-484.

Batterbury, S., 2008: Anthropology and global warming: the need for environmental engagement. *Australian Journal of Anthropology*, **19(1)**, 62-68.

Bauer, S., 2011: Stormy weather: International security in the shadow of climate change. In: *Coping with Global Environmental Change, Disaster and Security* [Brauch, H.G., Ú. Oswald Spring, C. Mesjasz, J. Grin, P. Kameri-Mbote, B. Chourou, P. Dunay, and J. Birkmann (eds.)]. Springer, Berlin, Germany, pp. 719-734.

Beck, U., 2007: *World at Risk*. Polity, Cambridge, UK, 269 pp.

Beckman, M., 2011: Converging and conflicting interests in adaptation to env. change in Central Vietnam. *Climate and Development*, **3(1)**, 32-41.

Bengtsson, M., Y. Shen, and T. Oki, 2006: A SRES-based gridded global population dataset for 1990–2100. *Population and Environment*, **28(2)**, 113-131.

Beniston, M., 2004: The 2003 heat wave in Europe: a shape of things to come? An analysis based on Swiss climatological data and model simulations. *Geophysical Research Letters*, **31(2)**, L02202.

Benson, C. and E. Clay, 2004: *Understanding the Economic and Financial Impact of Natural Disasters*. International Bank for Reconstruction and Development, Disaster Risk Management Series 4, World Bank, Washington, DC, 134 pp.

Bento, A.M., M.L. Cropper, A.M. Mobarak, and K. Vinha, 2005: The effects of urban spatial structure on travel demand in the United States. *Review of Economics and Statistics*, **87(3)**, 466-478.

Berke, P.R., J. Kartez, and D. Wenger, 1993: Recovery after disaster: achieving sustainable development, mitigation and equity. *Disasters*, **17(2)**, 93-109.

Berkes, F., 2007: Understanding uncertainty and reducing vulnerability: Lessons from resilience thinking. *Natural Hazards*, **41(2)**, 283-295.

Berkes, F., 2009: Evolution of co-management: role of knowledge generation, bridging organizations and social learning. *Journal of Environmental Management*, **90(5)**, 1692-1702.

Berkes, F., J. Colding, and C. Folke (eds.), 2003: *Navigating Social-ecological Systems: Building Resilience for Complexity and Change*. Cambridge University Press, Cambridge, UK, 398 pp.

Berkes, F., T.P. Hughes, R.S. Stenech, J.A. Wilson, D.R. Bellwood, B. Crona, C. Folke, L.H. Gunderson, H.M. Leslie, J. Norberg, M. Nyström, P. Olsson, H. Österblom, M. Scheffer, and B. Worm, 2006: Globalization, roving bandits, and marine resources. *Science*, **311(5767)**, 1557-1558.

Berkhout, F., J. Hertin, and D.M. Gann, 2006: Learning to adapt: Organisational adaptation to climate change impacts. *Climatic Change*, **78(1)**, 135-156.

Bern, C., J. Sniezek, G.M. Mathbor, M.S. Siddiqi, C. Ronsmans, A.M. Chowdhury, A.E. Chowdhury, K. Islam, M. Bennish, E. Noji, and R.I. Glass, 1993: Risk factors for mortality in the Bangladesh Cyclone 1991. *Bulletin of the World Health Organization*, **71(1)**, 73-78.

Bhattamishra, R. and C.B. Barrett, 2010: Community-based risk management arrangements: A review. *World Development*, **38(7)**, 923-932.

Bicknell, J., D. Dodman, and D. Satterthwaite (eds.), 2009: *Adapting Cities to Climate Change: Understanding and Addressing the Development Challenges*. Earthscan Publications, London, p. 384.

Biggs, D., R. Biggs, V. Dakos, R.J. Scholes, and M.L. Schoon, 2011: One crisis after another: Are we entering an era of concatenated global shocks? *Ecology and Society*, **16(2)**, 27.

Biggs, R., C. Raudsepp-Hearne, C. Atkinson-Palombo, E. Bohensky, E. Boyd, G. Cundill, H. Fox, S. Ingram, K. Kok, S. Spehar, M. Tengö, D. Timmer, and M. Zurek, 2007: Linking futures across scales: a dialog on multiscale scenarios. *Ecology and Society*, **12(1)**, 17.

Biggs, R., S.R. Carpenter, and W.A. Brock, 2009: Turning back from the brink: detecting an impending regime shift in time to avert it. *Proceedings of the National Academy of Sciences*, **106(3)**, 826-831.

Birkenholtz, T., 2008: Irrigated landscapes, produced scarcity and adaptive social institutions in Rajasthan, India. *Annals of the Association of American Geographers*, **99(1)**, 118-137.

Birkland, T.A., 1996: Natural disasters as focusing events: Policy communities and political response. *International Journal of Mass Emergencies and Disasters*, **14(2)**, 221-243.

Birkland, T.A., 2006: *Lessons of Disaster: Policy Change after Catastrophic Events*. Georgetown University Press, Washington, DC, 224 pp.

Birkmann, J., (ed.), 2006: *Measuring Vulnerability to Natural Hazard: Towards Disaster Resilient Societies*. UNU Press, Tokyo, Japan, 400 pp.

Bizikova, L., S. Burch, S. Cohen, and J. Robinson, 2010: Linking sustainable development with climate change adaptation and mitigation. In: *Climate Change, Ethics and Human Security* [O'Brien, K., A. St. Clair, and B. Kristoffersen (eds.)]. Cambridge University Press, Cambridge, UK, pp. 157-179.

Blaikie, P., T. Cannon, I. Davis and B. Wisner, 1994: *At Risk: Natural Hazards, People's Vulnerability and Disaster*. 1st ed. Routledge, London, UK, 286 pp.

Bogardi, J. and H.G. Brauch, 2005: Global environmental change: A challenge for human security – Defining and conceptualizing the environmental dimension of human security. In: *UNEO – Towards an International Environmental Organisation - Approaches to a Sustainable Reform of Global Environmental Governance* [Rechkemmer, A. (ed.)]. Nomos, Baden-Baden, Germany, pp. 85-109.

Bohle, H.G., 2009: Sustainable livelihood security. Evolution and application. In: *Facing Global Environmental Change: Environmental, Human, Energy, Food, Health and Water Security Concepts* [Brauch, H.G., Ú. Oswald Spring, J. Grin, C. Mesjasz, P. Kameri-Mbote, N. Chadha Behera, B. Chourou, and H. Krummenacher (eds.)]. Springer, Berlin, Germany, pp. 521-528.

Bohle, H.G., B. Etzold, and M. Keck, 2009: Resilience as agency. *IHDP Update*, **2**, 8-13.

Boonyabancha, S., 2005: Baan Mankong; going to scale with 'slum' and squatter upgrading in Thailand. *Environment and Urbanization*, **2(2)**, 309-329.

Boulanger, J.-P., F. Martinez, and E.C. Segura, 2006a: Projection of future climate change conditions using IPCC simulations, neural networks and Bayesian statistics. Part 1. Temperature mean state and seasonal cycle in South America. *Climate Dynamics*, **27**, 233-259.

Boulanger, J.-P., F. Martinez, and E.C. Segura, 2006b: Projection of future climate change conditions using IPCC simulations, neural networks and Bayesian statistics. Part 2. Precipitation mean state and seasonal cycle in South America. *Climate Dynamics*, **28**, 255-271.

Bradley, R.S., M. Vuille, H. Diaz, and W. Vergara, 2006: Threats to water supplies in the tropical Andes. *Science*, **312**, 1755-1756.

Brand, F.S. and K. Jax, 2007: Focusing the meaning(s) of resilience: resilience as a descriptive concept and a boundary object. *Ecology and Society*, **12(1)**, 23.

Brauch, H.G., 2005a: *Environment and Human Security. Towards Freedom from Hazard Impact*. InterSecTions No. 2, United Nations University–Institute for Environment and Human Security, Bonn, Germany, 60 pp.

Brauch, H.G., 2005b: *Threats, Challenges, Vulnerabilities and Risks in Environmental and Human Security*. Source No. 1, United Nations University–Institute for Environment and Human Security, Bonn, Germany, 100 pp.

Brauch, H.G., 2009: Introduction: facing global environmental change and sectorialization of security. In: *Facing Global Environmental Change: Environmental, Human, Energy, Food, Health and Water Security Concepts* [Brauch, H.G., Ú. Oswald Spring, J. Grin, C. Mesjasz, P. Kameri-Mbote, N. Chadha Behera, B. Chourou, and H. Krummenacher (eds.)]. Springer, Berlin, Germany, pp. 21-42.

Brauch, H.G. and Ú. Oswald Spring, 2011: Introduction: coping with global environmental change in the anthropocene. In: *Coping with Global Environmental Change, Disaster and Security* [Brauch, H.G., Ú. Oswald Spring, C. Mesjasz, J. Grin, P. Kameri-Mbote, B. Chourou, P. Dunay, and J. Birkmann (eds.)]. Berlin, Springer, Germany, pp. 31-60.

Brauch, H.G., Ú. Oswald Spring, C. Mesjasz, J. Grin, P. Dunay, N. Chadha Behera, B. Chourou, P. Kameri-Mbote, and P.H. Liotta (eds.), 2008: *Globalization and Environmental Challenges: Reconceptualizing Security in the 21 st Century*. Springer, Berlin, Germany, 1148 pp.

Brauch, H.G., Ú. Oswald Spring, J. Grin, C. Mesjasz, P. Kameri-Mbote, N. Chadha Behera, B. Chourou, and H.Krummenacher (eds.), 2009: *Facing Global Environmental Change: Environmental, Human, Energy, Food, Health and Water Security Concepts*. Springer, Berlin, Germany, 1588 pp.

Brauch, H.G., Ú. Oswald Spring, C. Mesjasz, J. Grin, P. Kameri-Mbote, B. Chourou, P. Dunay, and J. Birkmann (eds.), 2011: *Coping with Global Environmental Change, Disasters and Security – Threats, Challenges, Vulnerabilities and Risks*. Springer, Berlin, Germany, 1815 pp.

Bressers, H.T.A. and W.A. Rosenbaum (eds), 2003: *Achieving Sustainable Development: the Challenge of Governance Across Social Scales*. Praeger Publishers, Westport, CT.

Brklacich, M., D. McNabb, C. Bryant, and J. Dumanski, 1997: Adaptability of agriculture systems to global climate change: A Renfrew county, Ontario, Canada pilot study. In: *Agricultural Restructuring and Sustainability: A Geographical Perspective* [Ilbery, B., Q. Chiotti, and T. Rickard (eds.)]. CABI, Wallingford, UK, pp. 351-364.

Brock, W.A. and S.R. Carpenter, 2007: Panaceas and diversification of environmental policy. *Proceedings of the National Academy of Sciences*, **104(39)**, 15206-15211.

Brown, K., 2011: Sustainable adaptation: an oxymoron? *Climate and Development*, **3(1)**, 21-31.

Brownstone, D. and T.F. Golob, 2009: The impact of residential density on vehicle usage and energy consumption. *Journal of Urban Economics*, **65(1)**, 91-98.

Bruner, A.G., R.E. Gullison, R.E. Rice, and G.A.B. da Fonseca, 2001: Effectiveness of parks in protecting tropical biodiversity. *Science*, **291**, 125-128.

Buhaug, H., N.P. Gleditsch, and O.M. Theisen, 2008: *Implications of Climate Change for Armed Conflict*. World Bank, Washington DC, 52 pp.

Bunyavanich, S., C.P. Landrigan, A.J. McMichael, and P.R. Epstein, 2003: The impact of climate change on child health. *Ambulatory Pediatrics*, **3(1)**,44-52.

Burby, R.J., 2006: Hurricane Katrina and the paradoxes of government disaster policy: bringing about wise governmental decisions for hazardous areas. *The Annals of the American Academy of Political and Social Science*, **604(1)**, 171-191.

Burby, R.J., B.A. Cigler, S.P. French, E.J. Kaiser, J. Kartez, D. Roenigk, D. Weist, and D. Whittington, 1991: *Sharing Environmental Risks: How to Control Governments' Losses in Natural Disasters*. Westview, Boulder, CO, 280 pp.

Burby, R.J., A.C. Nelson, D. Parker, and J. Handmer, 2001: Urban containment policy and exposure to natural hazards: is there a connection? *Journal of Environmental Planning and Management*, **44(4)**, 475-490.

Burby, R.J., A.C. Nelson, and T.W. Sanchez, 2006: The problem of containment and the promise of planning. In: *Rebuilding Urban Places After Disaster, Lessons from Hurricane Katrina* [Birch, E.L. and S.M. Wachter (eds.)]. University of Pennsylvania Press, Philadelphia, PA, pp. 47-65.

Burch, S., S.R.J. Sheppard, A. Shaw, and O. Flanders, 2010: Planning for climate change in a flood-prone community: municipal barriers to policy action and the use of visualizations as decision-support tools. *Journal of Flood Risk Management*, **3(2)**, 126-139.

Burton, I., R.W. Kates, and G.F. White: 1993: *The Environment as Hazard*. 2nd ed. Guildford Press, London, UK, 290 pp.

Camerer, C. and H. Kunreuther, 1989: Decision processes for low probability events: policy implications. *Journal of Policy Analysis and Management*, **8(4)**, 565-592.

Campbell, B.M., J.A. Sayer, and B. Walker, 2010: Navigating trade-offs: working for conservation and development outcomes. *Ecology and Society*, **15(2)**, 16.

Campbell, K.M., J. Gulledge, J.R. McNeill, J. Podesta, P. Ogden, L. Fuerth, R.J. Woolsey, A.T.J. Lennon, J. Smith, R. Weitz, and D. Mix, 2007: *The Age of Consequences: The Foreign Policy and National Security Implications of Global Climate Change*. Center for Strategic and International Studies, Washington, DC, 124 pp.

Cannon, T., 2002: Gender and climate hazards in Bangladesh. *Gender and Development*, **10(2)**, 45-50.

Cardona, O., 1999: Environmental management and disaster prevention: Two related topics. In: *Cities at Risk: Environmental Degradation, Urban Risks and Disaster in Latin America* [Fernandez, M.A. (ed.)]. La Red/USAID, Quito, Ecuador, pp. 79-102.

Carlsson-Kanyama, A., K.H. Dreborg, H.C. Moll, and D. Padovan, 2008: Participative backcasting: a tool for involving stakeholders in local sustainability planning. *Futures*, **40(1)**, 34-46.

Carpenter, S.R., 2003: *Regime Shifts in Lake Ecosystems: Pattern and Variation*. Excellence in Ecology Series, 15, Ecology Institute, Oldendorf/Luhe, Germany.

Carpenter, S.R., B.H. Walker, J.M. Anderies, and N. Abel, 2001: From metaphor to measurement: resilience of what to what? *Ecosystems*, **4(8)**, 765-781.

Carter, M., P.D. Little, T. Mogues, and W. Negatu, 2007: Poverty traps and natural disasters in Ethiopia and Honduras. *World Development*, **35(5)**, 835-856.

Cash, D.W., W. Adger, F. Berkes, P. Garden, L. Lebel, P. Olsson, L. Pritchard, and O. Young, 2006: Scale and cross-scale dynamics: governance and information in a multilevel world. *Ecology and Society*, **11(2)**, 8.

Cavallo, E. and I. Noy, 2009: *The Economics of Natural Disasters: A Survey*. IDB Working Paper Series 124, Inter-American Development Bank, Washington, DC, 50 pp.

CCSP, 2008: *Weather and Climate Extremes in a Changing Climate. Regions of Focus: North America, Hawaii, Caribbean, and U.S. Pacific Islands* [Karl, T.R., G.A. Meehl, C.D. Miller, S.J. Hassol, A.M. Waple, and W.L. Murray (eds.)]. A report by the U.S. Climate Change Science Program and the Subcommittee on Global Change Research, Department of Commerce, NOAA National Climatic Data Center, Washington, DC, 164 pp.

Chakrabarti, P.G.D. and M.R. Bhatt (eds.), 2006: *Micro-finance and Disaster Risk Reduction*. Knowledge World, New Delhi, India.

Chambers, R., 2006: Vulnerability, coping and policy. *IDS Bulletin*, **37(4)**, 33-40.

Chapin III, F.S., S.R. Carpenter, G.P. Kofinas, C. Folke, N. Abel, W.C. Clark, P. Olsson, D.M. Stafford Smith, B.H. Walker, O.R. Young, F. Berkes, R. Biggs, J.M. Grove, R.L. Naylor, E. Pinkerton, W. Steffen, and F.J. Swanson, 2010: Ecosystem stewardship: sustainability strategies for a rapidly changing planet. *Trends in Ecology and Evolution*, **25(4)**, 241-249.

Chouvy, P.-A. and L.R. Laniel, 2007: Agricultural drug economies: cause or alternative to intra-state conflicts? *Crime, Law, Social Change*, **48**, 133-150.

Chowdhury, K.M.M.H., 2002: Cyclone preparedness and management in Bangladesh. In: *Improvement of Early Warning System and Responses in Bangladesh Towards Total Disaster Risk Management Approach*. Bangladesh Public Administration Training Centre, Dhaka, Bangladesh, pp. 97-106.

Christensen, C.M., 1997: *The Innovator's Dilemma: When New Technologies Cause Great Firms to Fail*. Harvard Business School Press, Boston, MA, 256 pp.

Christoplos, I., 2006: *Links between Relief, Rehabilitation and Development in the Tsunami Response*. Tsunami Evaluation Coalition, London, UK, 102 pp.

Christoplos, I., S. Anderson, M. Arnold, V. Galaz, M. Hedger, R.J.T. Klein, and K. Le Goulven, 2009: *The Human Dimension of Climate Adaptation: The Importance of Local and Institutional Issues*. Commission on Climate Change and Development, Edita Sverige AB, Stockholm, Sweden, 38 pp.

Clarke, L. and J. Weyant, 2009: Introduction to the EMF 22 special issue on climate change control scenarios. *Energy Economics*, **31(2)**, 63.

CMEPSP, 2009: *Report of the Commission on the Measurement of Economic Performance and Social Progress*. Commission on the Measurement of Economic Performance and Social Progress, France, 292 pp.

CNA Corporation, 2007: *National Security and the Threat of Climate Change*. CNA Corporation, Alexandria, VA, 63 pp.

Cohen, S., D. Demeritt, J. Robinson, and D. Rothman, 1998: Climate change and sustainable development: towards dialogue. *Global Environmental Change*, **8(4)**, 341-371.

Collins, A.E., 2009: *Disaster and Development*. Routledge Perspectives on Development, Routledge, London, UK.

Comunidad Andina, 2007: *Agenda Ambiental Andina 2006–2010* (Andean Environmental Agenda 2006–2010). Secretary General, San Isidro, Perú, 12 pp.

Comunidad Andina, 2009: *Articulando la gestión del riesgo y la adaptación al cambio climático en el sector agropecuario* (Articulating risk assessment and adaptation to climate change in agriculture and livestock). Secretary General, San Isidro, Perú, 128 pp.

Conde, J.E., 2001: The Orinoco River Delta Venezuela. In: *Coastal Marine Ecosystems of Latin America* [Seeliger, U. and B. Kjerfve (eds.)]. Springer, Berlin, Germany, pp. 61-70.

Corbacioglu, S. and N. Kapucu, 2006: Organisational learning and self adaptation in dynamic disaster environments. *Disasters*, **30(2)**, 212-233.

Corfee-Morlot, J., I. Cochran, P.-J. Teasdale, and S. Hallegatte, 2011: Multilevel governance and deliberative practice to support local climate action. *Climatic Change*, **104(1)**, 169-197.

Costanza, R., 2000: The dynamics of the ecological footprint concept. *Ecological Economics*, **32**, 341-345.

Costanza, R., J. Erickson, K. Fligger, A. Adams, C. Adams, G. Altschuler, S. Balter, B. Fisher, J. Hike, J. Kelly, T. Kerr, M. McCauley, K. Montone, M. Rauch, K. Schmiedeskamp, D. Saxton, L. Sparacino, W. Tusinski, and L. Williams, 2004: Estimates of the Genuine Progress Indicator (GPI) for Vermont, Chittenden County and Burlington, from 1950 to 2000. *Ecological Economics*, **51(1-2)**, 139-155.

Costello, A., M. Abbas, A. Allen, S. Ball, S. Bell, R. Bellamy, S. Friel, N. Groce, A. Johnson, M. Kett, M. Lee, C. Levy, M. Maslin, D. McCoy, B. McGuire, H. Montgomery, D. Napier, C. Pagel, J. Patel, J. Antonio, P. de Oliveira, N. Redclift, H. Rees, D. Rogger, J. Scott, J. Stephenson, J. Twigg, J. Wolff, and C. Patterson, 2009: Managing the health effects of climate change. *The Lancet*, **373**, 1693-1733.

Cummins, J.D. and O. Mahul, 2009: *Catastrophe Risk Financing in Developing Countries: Principles for Public Intervention*. World Bank, Washington, DC, 268 pp.

Cutter, S.L., 1995: The forgotten casualties: women, children, and environmental change. *Global Environmental Change*, **5(3)**, 181-194.

Cutter, S.L., 2006: *Hazards, Vulnerability and Environmental Justice*. Earthscan, London, UK, 418 pp.

Cutter, S.L., L. Barnes, M. Berry, C. Burton, E. Evans, E. Tate, and J. Webb, 2008: A place-based model for understanding community resilience to natural disasters. *Global Environmental Change*, **18(4)**, 598-606.

Dalby, S., 2009: *Security and Environmental Change*. Polity, Cambridge, UK, 199 pp.

Davidson, F., M. Zaaijer, M. Peltenburg, and M. Rodell, 1993: *Relocation and Resettlement Manual: A Guide to Managing and Planning Relocation*. Institute for Housing and Urban Development Studies, Rotterdam, The Netherlands, 69 pp.

Davies, M., B. Guenther, J. Leavy, T. Mitchell, and T.M. Tanner, 2008: Adaptive social protection: synergies for poverty reduction. *IDS Bulletin*, **39(4)**, 105-112.

Davies, S., 1993: Are coping strategies a cop-out? *IDS Bulletin*, **24(4)**, 60-72.

Davis, G., 1998: *Creating Scenarios for Your Company's Future*. The 1998 Conference on Corporate Environmental, Health, and Safety Excellence, Bringing Sustainable Development Down to Earth, New York, USA, 5 pp.

Davis, M., 2001: *Late Victorian Holocausts. El Niño Famines and the Making of the Third World*. Verso, London, UK, 464 pp.

de Fraiture, C., M. Giordano, and Y. Liao, 2008: Biofuels and implications for agricultural water use: blue impacts of green energy. *Water Policy*, **10(S1)**, 67-81.

de la Fuente, A., A. Revi, F. Lopez-Calva, J. Serje, F. Ramirez, C. Rosales, A. Velasquez, and S. Dercan, 2009: Deconstructing disaster: Risk patterns and poverty trends at the local level. In: *Global Assessment Report on Disaster Risk Reduction*. United Nations International Strategy for Disaster Reduction, Geneva, Switzerland, pp. 59-85.

del Ninno, C. and M. Lindberg, 2005: Treading water: the long-term impact of the 1998 flood on nutrition in Bangladesh. *Economics and Human Biology*, **3(1)**, 67-96.

Dennekamp, M. and M. Carey, 2010: Air quality and chronic disease: why action on climate change is also good for health. *New South Wales Public Health Bulletin*, **21(6)**, 115-121.

Dercon, S. and I. Outes, 2009: *Income Dynamics in Rural India: Testing for Poverty Traps and Multiple Equilibria*. Background paper for the U.N.-World Bank Assessment on the Economics of Disaster Risk Reduction, Washington, DC, 40 pp.

DesInventar, 2010: online.desinventar.org/desinventar/.

Desplat, J., J-L. Salagnac, R. Kounkou, A. Lemonsu, M. Colombert, M. Lauffenburger, and V. Masson, 2009: Multidisciplinary study of the impacts of climate change on the scale of Paris. In: *Proceedings of the Seventh International Conference on Urban Climate*, Yokohama, Japan, June 2009, 5 pp.

Dessai, S., M. Hulme, R. Lempert, and R. Pielke, Jr., 2009a: Climate prediction: a limit to adaptation? In: *Adapting to Climate Change: Thresholds, Values, Governance* [Adger, W.N., I. Lorenzoni, and K.L. O'Brien (eds.)]. Cambridge University Press, Cambridge, UK, pp. 64-78.

Dessai, S., M. Hulme, R. Lempert, and R. Pielke, Jr., 2009b: Do we need better predictions to adapt to a changing climate? *Eos*, **90(13)**, 111-112.

DFID, 2004: *Disaster Risk Reduction: a Development Concern*. Department for International Development, London, UK, 74 pp.

DFID, 2006: *Eliminating World Poverty: Making Governance Work for the Poor*. A White Paper on International Development, Department for International Development, London, UK, 132 pp.

Dodman, D., J. Hardoy, and D. Satterthwaite, 2008: *Urban Development and Intensive and Extensive Risk*. Contribution to the Global Assessment Report on Disaster Risk Reduction (2009), International Institute for Environment and Development, London, UK, 67 pp.

Drury, A.C. and R.S. Olson, 1998: Disasters and political unrest: an empirical investigation. *Journal of Contingencies and Crisis Management*, **6(3)**, 153-161.

Dufey, A., 2006: *Biofuels Production, Trade and Sustainable Development: Emerging Issues*. Sustainable Markets Discussion Paper 2, International Institute for Environment and Development, London, UK, 62 pp.

Eakin, H. and M. Webbe, 2008: Linking local vulnerability to system sustainability in a resilience framework: Two cases from Latin America. *Climatic Change*, **93**, 355-377.

ECA, 2009: *Shaping Climate-Resilient Development: A Framework for Decision-Making*. A report of the Economics of Climate Adaptation Working Group. ClimateWorks Foundation, Global Environment Facility, European Commission, McKinsey and Company, The Rockefeller Foundation, Standard Chartered Bank, and Swiss Re, 164 pp.

Edenhofer, O., K. Lessmann, C. Kemfert, M. Grubb, and J. Köhler, 2006: Induced technological change: exploring its implications for the economics of atmospheric stabilization: synthesis report from innovation modeling comparison project. *The Energy Journal*, **27(S1)**, 57-107.

Edenhofer, O., R. Pichs-Madruga, Y. Sokona, K. Seyboth, P. Matschoss, S. Kadner, T. Zwickel, P. Eickemeier, G. Hansen, S. Schlömer, and C. von Stechow (eds.), 2011: *IPCC Special Report on Renewable Energy Sources and Climate Change Mitigation*. Cambridge University Press, Cambridge, UK, 1544 pp.

Elmqvist, T., C. Folker, M. Nyström, G. Peterson, J. Bengtsson, B. Walker, and J. Norberg, 2003: Response diversity, ecosystem change, and resilience. *Frontiers in Ecology and the Environment*, **1(9)**, 488-494.

Elsner, J.B., J.P. Kossin, and T.H. Jagger, 2008: The increasing intensity of the strongest tropical cyclones. *Nature*, **455**, 92-95.

Enfors, E. I., L.J. Gordon, G.D. Peterson, and D. Bossio, 2008: Making investments in dryland development work: participatory scenario planning in the Makanya catchment, Tanzania. *Ecology and Society*, **13(2)**, 42.

Eriksen, S. and K. Brown, 2011: Sustainable adaptation to climate change. *Climate and Development*, **3(1)**, 3-6.

Eriksen, S.H. and P.M. Kelly, 2006: Developing credible vulnerability indicators for climate adaptation policy assessment. *Mitigation and Adaptation Strategies for Global Change*, **12(4)**, 495-524.

Eriksen, S. and K. O'Brien, 2007: Vulnerability, poverty and the need for sustainable adaptation measures. *Climate Policy*, **7(4)**, 337-352.

Eriksen, S., P. Aldunce, C.S. Bahinipati, R. Martins, J.I. Molefe, C. Nhemachena, K. O'Brien, F. Olorunfemi, J. Park, L. Sygna, and K. Ulsrud, 2011: When not every response to climate change is a good one: identifying principles for sustainable adaptation. *Climate and Development*, **3(1)**, 7-20.

EU, 2008: *Climate Change and International Security*. Paper from the High Representative and the European Commission to the European Council, European Commission, Brussels, Belgium, 11 pp.

Evans, L. and J. Oehler-Stinnett, 2006: Children and natural disasters: a primer for school psychologists. *School Psychology International*, **27(1)**, 33-55.

Ewing, R. and F. Rong, 2008: The impact of urban form on US residential energy use. *Housing Policy Debate*, **19(1)**, 1-30.

Fankhauser, S., J.B. Smith, and R.S.J. Tol, 1999: Weathering climate change: some simple rules to guide adaptation decisions. *Ecological Economics*, **30(1)**, 67-78.

Farber, D., 2007: Basic compensation for victims of climate change. *University of Pennsylvania Law Review*, **155(6)**, 1605-1656.

Fargione, J.E., T.R. Cooper, D.J. Flaspohler, J. Hill, C. Lehman, T. McCoy, S. Mcleod, E.J. Nelson, K.S. Oberhauser, and D. Tilman, 2009: Bioenergy and wildlife: threats and opportunities for grassland conservation. *BioScience*, **59(9)**, 767-777.

Fearnside, P.M., 2001: Status of South American natural ecosystems. In: *Encyclopedia of Biodiversity* [Levin, S.A. (ed.)]. Academic Press, San Diego, CA, pp. 345-359.

FEMA, 2009: *Dam Safety in the United States: A Progress Report on the National Dam Safety Program Fiscal Years 2006 and 2007*. Federal Emergency Management Agency, Washington, DC, 60 pp.

Feng, S., A.B. Krueger, and M. Oppenheimer, 2010: Linkages among climate change, crop yields and Mexico-US cross-border migration. *Proceedings of the National Academy of Sciences*, **107(32)**, 14257-14262.

Fernandez-Gimenez, M.E., H.L. Ballard, and V.E. Sturtevant, 2008: Adaptive management and social learning in collaborative and community-based monitoring: a study of five community-based forestry organizations in the western USA. *Ecology and Society*, **13(2)**, 4.

Ferris, E., 2011: *Protecting Civilians in Disasters and Conflicts*. Brookings Policy Brief 182, Brookings Institute, Washington, DC, 4 pp.

Fløysand, A. and S.-E. Jakobsen, 2010: The complexity of innovation: A relational turn. *Progress in Human Geography*, **35(1)**, 328-344.

Foley, J.A., R. DeFries, G.P. Asner, C. Barford, G. Bonan, S.R. Carpenter, F.S. Chapin, M.T. Coel, G.C. Daily, H.K. Gibbs, T. Helkowski, E.A. Howard, C.J. Kucharik, C. Monfreda, J.A. Patz, I.C. Prentice, N. Ramankutty, and P.K. Snyder, 2005: Global consequences of land use. *Science*, **309**, 570-574.

Folke, C., 2006: Resilience: The emergence of a perspective for social-ecological systems analyses. *Global Environmental Change*, **16(3)**, 253-267.

Folke, C., S. Carpenter, B. Walker, M. Scheffer, T. Elmqvist, L. Gunderson, and C.S. Holling, 2004: Regime shifts, resilience, and biodiversity in ecosystem management. *Annual Review of Ecology, Evolution and Systematics*, **35**, 557-581.

Folke, C., F.S. Chapin III, and P. Olsson, 2009: Transformations in ecosystem stewardship. In: *Principles of Ecosystem Stewardship: Resilience-Based Natural Resource Management in a Changing World* [Chapin III, F.S., G.P. Kofinas, and C. Folke (eds.)]. Springer, New York, NY, pp. 103-125.

Folke, C., S.R. Carpenter, B.H. Walker, M. Scheffer, F.S. Chapin III, and J. Rockström, 2010: Resilience thinking: integrating resilience, adaptability and transformability. *Ecology and Society*, **15(4)**, 20.

Ford, J.D. and L.B. Ford (eds.), 2011: *Climate Change Adaptation in Developed Nations: From Theory to Practice*. Springer, Dordrecht, The Netherlands, 295 pp.

Foster, V. and C. Briceño-Garmendia (eds.), 2011: *Africa's Infrastructure. A Time for Transformation*. World Bank and French Development Agency, Washington, DC, 44 pp.

Frittaion, C., P.N. Duinker, and J. Grant, 2010: Suspending disbelief: Influencing engagement in scenarios of forest futures. *Technological Forecasting and Social Change*, **78(3)**, 421-436.

Fuentes, C.J. and H.G. Brauch, 2009: The human security network: A global North-South coalition. In: *Facing Global Environmental Change: Environmental, Human, Energy, Food, Health and Water Security Concepts* [Brauch, H.G., Ú. Oswald Spring, J. Grin, C. Mesjasz, P. Kameri-Mbote, N. Chadha Behera, B. Chourou, and H. Krummenacher (eds.)]. Springer, Berlin, Germany, pp. 991-1001.

Gachathi, F. and S. Eriksen, 2011: Gums and resins: the potential for supporting sustainable adaptation in Kenya's drylands. *Climate and Development*, **3(1)**, 59-70.

Gaffin, S, X. Xing, and G. Yetman, 2004: Downscaling and geospatial gridding of socio-economic projections from the IPCC special report on emissions scenarios (SRES). *Global Environmental Change*, **14**, 105-123.

Gaillard, J.C., E. Clavé, and I. Kelman, 2008: Wave of peace? Tsunami disaster diplomacy in Aceh, Indonesia. *Geoforum*, **39(1)**, 511-526.

Galer, G., 2004: Preparing the ground? Scenarios and political change in South Africa. *Development*, **47(4)**, 26-34.

Gao, Z. and Y. Hu, 2011: Coping with population growth, climate change, water scarcity and growing food demand in China in the 21st century. In: *Coping with Global Environmental Change, Disaster and Security* [Brauch, H.G., Ú. Oswald Spring, C. Mesjasz, J. Grin, P. Kameri-Mbote, B. Chourou, P. Dunay, and J. Birkmann (eds.)] Springer, Berlin, Germany, pp. 957-968.

Gardiner, S. 2010: Climate change: a global test for contemporary political institutions and theories. In: *Climate Change, Ethics and Human Security* [O'Brien, K., A. Lera St. Clair, and B. Kristoffersen (eds.)]. Cambridge University Press, Cambridge, UK, pp. 131-153.

Garg, A., R.C. Dhiman, S. Bhattacharya, and P.R. Shukla, 2009: Development, malaria and adaptation to climate change: A case study from India. *Environmental Management Journal*, **43(5)**, 779-789.

Gawith, M., R. Street, R. Westaway, and A. Steynor, 2009: Application of the UKCIP02 climate change scenarios: reflections and lessons learnt. *Global Environmental Change*, **19(1)**, 113-121.

GECHS, 2000: *Science Plan: Global Environmental Change and Human Security Project*. IHDP Report 11, International Human Dimensions Programme, Bonn, Germany, 60 pp.

Geels, F.W., 2002: Technological transitions as evolutionary reconfiguration processes: A multi-level perspective and a case-study. *Research Policy*, **31(8/9)**, 1257-1274.

Geels, F.W. and J. W. Schot, 2007: Typology of sociotechnical transition pathways. *Research Policy*, **36(3)**, 399-417.

Gero, A., K. Méheux, and D. Dominey-Howes, 2011: Integrating community based disaster risk reduction and climate change adaptation: examples from the Pacific. *Natural Hazards and Earth Systems Science*, **11**, 101-113.

Ghesquiere, F., L. Jamin, and O. Mahul, 2006: *Earthquake Vulnerability Reduction Program in Colombia: A Probabilistic Cost-Benefit Analysis*. Policy Research Working Paper 3939, World Bank, Washington, DC, 22 pp.

Gilbert, A. and P. Ward, 1984: Community participation in upgrading irregular settlements: the community response. *World Development*, **12(9)**, 913-922.

Glaeser, E.L. and M.E. Kahn, 2010: The greenness of cities: Carbon dioxide emissions and urban development. *Journal of Urban Economics*, **67(3)**, 404-418.

Goklany, I.M., 2007: Integrated strategies to reduce vulnerability and advance adaptation, mitigation, and sustainable development. *Mitigation and Adaptation Strategies for Global Change*, **12(5)**, 755-786.

Goldstein, B.E., 2009: Resilience to surprises through communicative planning. *Ecology and Society*, **14(2)**, 33.

Goodess, C.M., J.W. Hall, M. Best, R. Betts, L. Cabantous, P.D. Jones, C.G. Kilsby, A. Pearman, and C.J. Wallace, 2007: Climate scenarios and decision making under uncertainty. *Built Environment*, **33(1)**, 10-30.

Gordon, P., H. Richardson, and B. Davis, 1998: Transport related impacts of the Northridge earthquake. *Journal of Transportation and Statistics*, **1**, 21-36.

Gostin, L.O., E.A. Friedman, G. Ooms, T. Gebauer, N. Gupta, D. Sridhar, W. Chenguang, J-A. Røttingen, and D. Sanders, 2011: The joint action and learning initiative: towards a global agreement on national and global responsibilities for health. *PLoS Medicine*, **8(5)**, e1001031.

Grasso, M., 2009: An ethical approach to climate adaptation finance. *Global Environmental Change*, **20(1)**, 74-81.

Grasso, M., 2010: *Justice in Funding Adaptation Under the International Climate Change Regime*. Springer, Dordrecht, The Netherlands, 184 pp.

Grazi, F., J.C.J.M. van den Bergh, and J. N van Ommeren, 2008: An empirical analysis of urban form, transport, and global warming. *The Energy Journal*, **29(4)**, 97-122.

Groves, D.G. and R.J. Lempert, 2007: A new analytic method for finding policy-relevant scenarios. *Global Environmental Change*, **17(1)**, 73-85.

Groves, D.G., D. Knopman, R.J. Lempert, S.H. Berry, and L. Wainfan, 2008: *Presenting Uncertainty about Climate Change to Water-Resource Managers. A Summary of Workshops with the Inland Empire Utilities Agency*. Technical Reports 505, Rand Corporation, Santa Monica, CA, 100 pp.

Grubb, M., C. Vrolijk, and D. Brack, 1999: *The Kyoto Protocol: A Guide and Assessment*. Royal Institute of International Affairs and Earthscan, London, UK, 342 pp.

Grübler, A., B. O'Neill, K.Riahi, V. Chirkov, A. Goujon, P. Kolp, I. Prommer, S. Scherbov, and E. Slentoe, 2007: Regional, national, and spatially explicit scenarios of demographic and economic change based on SRES. *Technological Forecasting and Social Change*, **74(7)**, 980-1029.

Gunderson, L., 2009: *Comparing Ecological and Human Community Resilience*. CARRI Research Paper 5, Community and Regional Resilience Initiative, National Security Directorate, Oak Ridge, TN, 35 pp.

Gunderson, L.H. and C.S. Holling (eds.), 2002: *Panarchy: Understanding Transformations in Human and Natural Systems*. Island Press, Washington, DC, 450 pp.

Guojie, C., 2003: Ecological reconstruction of the upper reaches of the Yangtze River. In: *Natural Disasters and Development in a Globalizing World* [Pelling, M. (ed.)]. Routledge, London, UK, pp. 214-230.

Gupta, J., C. Termeer, J. Klostermann, S. Meijerink, M. van den Brink, P. Jong, S. Nooteboom, and E. Mbergsma, 2010: The adaptive capacity wheel: a method to assess the inherent characteristics of institutions to enable the adaptive capacity of society. *Environmental Science and Policy*, **13(6)**, 459-471.

Gusdorf, F., S. Hallegatte, and A. Lahellec, 2008: Time and space matter: how urban transitions create inequality. *Global Environment Change*, **18(4)**, 708-719.

Ha Duong, M., M.J. Grubb, and J.-C. Hourcade, 1997: Influence of socioeconomic inertia and uncertainty on optimal CO_2-emission abatement. *Nature*, **390**, 270-274.

Haines, A., R.S. Kovats, D. Campbell-Lendrum, and C. Corvalan, 2006: Climate change and human health: impacts, vulnerability and public health. *Public Health*, **120**, 585-596.

Hall, J.W., 2007: Probabilistic climate scenarios may misrepresent uncertainty and lead to bad adaptation decisions. *Hydrological Processes*, **21(8)**, 1127-1129.

Hallegatte, S., 2006: *A Cost-Benefit Analysis of the New Orleans Flood Protection System*. Regulatory Analysis 2, American Enterprise Institute–Brookings Joint Center, Washington, DC, 18 pp.

Hallegatte, S., 2008a: An adaptive regional input-output model and its application to the assessment of the economic cost of Katrina. *Risk Analysis*, **28(3)**, 779-799.

Hallegatte, S., 2008b: A Note on Including Climate Change Adaptation in an International Scheme. Idées pour le Débat 18, Institut du Développement Durable et des Relations Internationals, Paris, France, 15 pp.

Hallegatte, S., 2009: Strategies to adapt to an uncertain climate change. *Global Environmental Change*, **19(2)**, 240-247.

Hallegatte, S. and P. Dumas, 2008: Can natural disasters have positive consequences? Investigating the role of embodied technical change. *Ecological Economics*, **68(3)**, 777-786.

Hallegatte, S. and V. Przyluski, 2010: *The Economics of Natural Disasters: Concepts and Methods*. Policy Research Working Paper 5507, World Bank, Washington, DC, 29 pp.

Hallegatte, S., J.-C. Hourcade, and P. Dumas, 2007: Why economic dynamics matter in assessing climate change damages: illustration on extreme events. *Ecological Economics*, **62(2)**, 330-340.

Hallegatte, S., P. Dumas, and J.-C. Hourcade, 2010: *A Note on the Economic Cost of Climate Change and the Rationale to Limit it Below 2°C*. Background Paper to the 2010 World Development Report, Policy Research Working Paper 5179, World Bank, Washington, DC, 19 pp.

Hallegatte, S., F. Henriet, and J. Corfee-Morlot, 2011a: The economics of climate change impacts and policy benefits at city scale: A conceptual framework. *Climatic Change*, **104(1)**, 51-87.

Hallegatte, S., F. Lecocq, and C. De Perthuis, 2011b: *Designing Climate Change Adaptation Policies: An Economic Framework*. Policy Research Working Paper 5568, World Bank, Washington, DC, 41 pp.

Hamin, E.M. and N. Gurran, 2009: Urban form and climate change: Balancing adaptation and mitigation in the U.S. and Australia. *Habitat International*, **33(3)**, 238-245.

Handy, S., X. Cao, and P. Mokhtarian, 2005: Correlation or causality between the built environment and travel behavior? Evidence from Northern California. *Transportation Research Part D: Transport and Environment*, **10(6)**, 427-444.

Hanson, S., R. Nicholls, N. Ranger, S. Hallegatte, J. Corfee-Morlot, C. Herweijer, and J. Chateau, 2011: A global ranking of port cities with high exposure to climate extremes. *Climatic Change*, **104(1)**, 89-111.

Haque, C.E. and D. Blair, 1992: Vulnerability to tropical cyclones: evidence from the April 1991 cyclone in coastal Bangladesh. *Disasters*, **16(3)**, 217-229.

Harberger, A.C., 1978: On the use of distributional weights in social cost benefit analysis. *Journal of Political Economy*, **86(2)**, 87-120.

Harberger, A.C., 1984: Basic needs versus distributional weights in social cost-benefit analysis. *Economic Development and Cultural Change*, **32(3)**, 455-474.

Hardoy, J.E., D. Mitlin, and D. Satterthwaite, 2001: *Environmental Problems in an Urbanizing World*. Earthscan, London, UK.

Harrington, L.M.B., 2005: Vulnerability and sustainability concerns for the US High Plains. In: *Rural Change and Sustainability: Agriculture, the Environment and Communities* [Essex, S.J., A.W. Gilg, R.B. Yarwood, J. Smithers, and R. Wilson (eds.)]. CABI, Wallingford, UK, pp. 169-184.

Harrington, L.M.B., 2010: The U.S. Great Plains, change, and place development. In: *The Next Rural Economies: Constructing Rural Place in Global Economies* [Halseth, G., S. Markey, and D. Bruce (eds.)]. CABI, Wallingford, UK, pp. 32-44.

Harrington, L.M.B., M. Lu, and J.A. Harrington, Jr., 2009: Fossil water and agriculture in southwestern Kansas. In: *Sustainable Communities on a Sustainable Planet: The Human-Environment Regional Observatory Project* [Yarnal, B., C. Polsky, and J. O'Brien (eds.)]. Cambridge University Press, Cambridge, UK, pp. 269-291.

Harvard Business Essentials, 2003: *Managing Change and Transition*. Harvard Business School Press, Boston, MA.

Harvey, D., 2010: *The Enigma of Capital: And the Crisis of Capitalism*. Profile, London, UK.

Hazell, P. and U. Hess, 2010: Drought insurance for agricultural development and food security in dryland areas. *Food Security*, **2(4)**, 395-405.

Hedlund-de Witt, A., 2011: The rising culture and worldview of contemporary spirituality: A sociological study of potentials and pitfalls for sustainable development. *Ecological Economics*, **70**, 1057-1065.

Hedrén, J. and B.O. Linnér, 2009: Utopian thought and the politics of sustainable development. *Futures*, **41(4)**, 210-219.

Heifetz, R., 2010: Leadership. In: *Political and Civic Leadership: A Reference Handbook* [Couto, R.A. (ed.)]. Sage Publications, London, UK, pp. 12-23.

Heifetz, R., A. Grashow, and M. Linsky, 2009: *The Practice of Adaptive Leadership: Tools and Tactics for Changing Your Organization and the World*. Harvard Business Press, Boston, MA.

Hewitt, K. (ed.), 1983: *Interpretations of Calamity*. Allen and Unwin, London, UK, 304 pp.

Hewitt, K., 1997: *Regions of Risk: Geographical Introduction to Disasters*. Longman, London, UK, 389 pp.

Heyd, T. and N. Brooks, 2009: Exploring cultural dimensions of adaptation to climate change. In: *Adapting to Climate Change: Thresholds, Values, Governance* [Adger, W.N., I. Lorenzoni, and K. O'Brien (eds.)]. Cambridge University Press, Cambridge, UK, pp. 269-282.

Hine, D. and J.W. Hall, 2010: Information gap analysis of flood model uncertainties and regional frequency analysis. *Water Resources Research*, **46**, W01514.

Hochrainer, S., 2009: *Assessing Macroeconomic Impacts of Natural Disasters: Are There Any?* Policy Research Working Paper 4968, World Bank, Washington, DC, 43 pp.

Hogarth, R. and H. Kunreuther, 1995: Decision making under ignorance: arguing with yourself. *Journal of Risk and Uncertainty*, **10(1)**, 15-36.

Holling, C.S., 1973: Resilience and stability of ecological systems. *Annual Review of Ecology and Systematics*, **4**, 1-23.

Hoogma, R., R. Kemp, J. Schot, and B. Truffer, 2002: *Experimenting for Sustainable Transport: The Approach of Strategic Niche Management*. Routledge, London, New York.

Hopwood, B., M. Mellor, and G. O'Brien, 2005: Sustainable development: mapping different approaches. *Sustainable Development*, **13(1)**, 38-52.

Hughes, D., 2001: Cadastral politics: The making of community-based resource management in Zimbabwe and Mozambique. *Development and Change*, **32(4)**, 741-768.

Hughes, T., A.H. Baird, D.R. Bellwood, M. Card, S.R. Connolly, C. Folke, R. Grosberg, O. Hoegh-Guldberg, J.B.C. Jackson, J. Kleypas, J.M. Lough, P. Marshall, M. Nystro, S.R. Palumbi, J.M. Pandolfi, B. Rosen, and J. Roughgarden, 2003: Climate change, human impacts, and the resilience of coral reefs. *Science*, **301**, 929-933.

Huitema, D., C. Cornelisse, and B. Ottow. 2010. Is the jury still out? Toward greater insight in policy learning in participatory decision processes – the case of Dutch citizens' juries on water management in the Rhine Basin. *Ecology and Society*, **15(1)**, 16.

Hunt, A. and T. Taylor, 2009: Values and cost-benefit analysis: economic efficiency criteria in adaptation. In: *Adapting to Climate Change: Thresholds, Values, Governance* [Adger, W.N., I. Lorenzoni, and K.L. O'Brien (eds.)]. Cambridge University Press, Cambridge, UK, pp. 197-211.

Hunter, L.M., 2005: Migration and environmental hazards. *Population and Environment*, **26(4)**, 273-302.

ICOMOS, 1998: *Report on Economics of Conservation: An Appraisal of Theories, Principles and Methods.* International Economics Committee, International Council on Monuments and Sites, Paris, France, 123 pp.

ICSU, 2002: *Science and Technology for Sustainable Development*. Consensus Report and Background Document, Mexico City Synthesis Conference, May 2002. ICSU Series on Science for Sustainable Development 9, International Council for Science, Paris, France, 37 pp.

IFRC, 2002: *World Disasters Report: Focus on Reducing Risk*. International Federation of Red Cross and Red Crescent Societies, Eurospan, London, UK, 239 pp.

IFRC, 2005: *World Disasters Report: Focus on Information in Disasters*. International Federation of Red Cross and Red Crescent Societies, Eurospan, London, UK, 246 pp.

IFRC, 2010: *World Disasters Report 2010: Focus on Urban Risk*. International Federation of Red Cross and Red Crescent Societies, Geneva, Switzerland, 220 pp.

Inayatullah, S., 2006: Anticipatory action learning: Theory and practice. *Futures*, **38(6)**, 656-666.

IOM, 2007: *Discussion Note: Migration and the Environment*. MC/ING/288, 94th Session, International Organization for Migration, Geneva, Switzerland, 9 pp.

IOM, 2009a: *Migration, Climate Change and the Environment*. IOM Policy Brief, International Organization for Migration, Geneva, Switzerland, 9 pp.

IOM, 2009b: *Compendium of IOM's Activities on Migration, Climate Change and the Environment.* International Organization for Migration, Geneva, Switzerland, 311 pp.

IRGC, 2009: *Risk Governance Deficits: An Analysis and an Illustration of the Most Common Deficits in Risk Governance*. International Risk Governance Council, Geneva, Switzerland, 92 pp.

ISET, 2010: *The Shared Learning Dialogue: Building Stakeholder Capacity and Engagement for Resilience Action*. Climate Resilience in Concept and Practice, Working Paper 1, Institute for Social and Environmental Transition, Boulder, CO, 29 pp.

Jabry, A. (ed.), 2003: *Children in Disasters: After the Cameras Have Gone*. Plan UK, London, UK, 68 pp.

Jackson, T., 2009: *Prosperity without Growth: Economics for a Finite Planet*. Earthscan, London, UK, 278 pp.

Jaramillo, C.R.H., 2009: *Do Natural Disasters have Long-Term Effects on Growth?* Manuscript, Universidad de los Andes, Bogota, Colombia, 44 pp.

Jones, C. and P. Klenow, 2010: *Beyond GDP? Welfare across Countries and Time*. NBER Working Paper No. 16352, National Bureau of Economic Research, Cambridge, MA, 52 pp.

Jonkman, S.N., A. Lentz, and J.K. Vrijling, 2010: A general approach for the estimation of loss of life due to natural and technological disasters. *Reliability Engineering and System Safety*, **95(11)**, 1123-1133.

Kahane, A., 2004: Colombia: speaking up. *Development*, **47(4)**, 95-98.

Kahl, C.H., 2003: The political ecology of violence: lessons for the Mediterranean. In: *Security and Environment in the Mediterranean. Conceptualising Security and Environmental Conflicts* [Brauch, H.G., P.H. Liotta, A. Marquina, P.F. Rogers, and M. El-Sayed Selim (eds.)]. Springer, Berlin, Germany, pp. 465-476.

Kahl, C.H., 2006: *States, Scarcity and Civil Strife in the Developing World*. Princeton University Press, Princeton, NJ, 352 pp.

Kameri-Mbote, P. and K. Kindiki, 2009: Water and food security in the Nile River Basin: perspectives of governments and NGOs of upstream countries. In: *Facing Global Environmental Change: Environmental, Human, Energy, Food, Health and Water Security Concepts* [Brauch, H.G., Ú. Oswald Spring, J. Grin, C. Mesjasz, P. Kameri-Mbote, N. Chadha Behera, B. Chourou, and H. Krummenacher (eds.)]. Springer, Berlin, Germany, pp. 655-664.

Kapucu, N., 2009: Interorganizational coordination in complex environments of disasters: the evolution of intergovernmental disaster response systems. *Journal of Homeland Security and Emergency Management*, **6(1)**, 47.

Karim, M.F. and N. Mimura, 2008: Impacts of climate change and sea level rise on cyclonic storm surge floods in Bangladesh. *Global Environmental Change*, **18(3)**, 490-500.

Keen, M., V.A. Brown, and R. Dyball, 2005: *Social Learning in Environmental Management: Towards a Sustainable Future*. Earthscan, London, UK, 270 pp.

Keenan, H.T., S.W. Marshall, M.A. Nocera, and D.K. Runyan, 2004: Increased incidence of inflicted traumatic brain injury in children after a natural disaster. *American Journal of Preventive Medicine*, **26(3)**, 189-193.

Kegan, R. and L.L. Lahey, 2009: *Immunity to Change: How to Overcome It and Unlock the Potential in Yourself and Your Organization*. Harvard Business Press, Boston, 341 pp.

Kelleher, A., 2005: A personal philosophy of anticipatory action-learning. *Journal of Future Studies*, **10(1)**, 85-90.

Kelman, I., 2003: Beyond disaster, beyond diplomacy. In: *Natural Disasters and Development in a Globalizing World* [Pelling, M. (ed.)]. Routledge, London, UK, pp. 110-123.

Kelman, I. and T. Koukis, 2000: Disaster diplomacy. *Cambridge Review of International Affairs*, **14(1)**, 214-294.

Kemp, R., 1994: Technology and the transition to environmental sustainability. The problems of technological regime shifts. *Futures*, **26(10)**, 1023-1046.

Kemp, R., J. Schot, and R. Hoogma, 1998: Regime shifts to sustainability through processes of niche formation: The approach of strategic niche management. *Technology Analysis & Strategic Management*, **10(2)**, 175-195.

Kemp, R., D. Loorbach, and R. Rotmans, 2007: Transition management as a model to manage processes of co-evolution towards sustainable development. *International Journal of Sustainable Development and World Ecology*, **14(1)**, 78-91.

Kent, G., 2001: The human right to disaster mitigation and relief. *Environmental Hazards*, **3(3-4)**, 137-138.

Kesby, M., 2005: Retheorizing empowerment-through-participation as a performance in space: beyond tyranny to transformation. *Signs: Journal of Women in Culture and Society*, **30(4)**, 2037-2065.

Kiehl, J.T., 2006: Geoengineering climate change: Treating the symptom over the cause? *Climatic Change*, **77(3-4)**, 227-228.

Kingdon, J.W., 1995: *Agendas, Alternatives and Public Policies*. 2nd ed. HarperCollins, New York, NY, 273 pp.

Kinzig, A.P., P. Ryan, M. Etienne, H. Allison, T. Elmqvist, and B.H. Walker, 2006: Resilience and regime shifts: assessing cascading effects. *Ecology and Society*, **11(1)**, 20.

Klein, N., 2007: *The Shock Doctrine: The Rise of Disaster Capitalism*. Penguin, London, UK, 701 pp.

Klein, R.J.T., E.L.F. Schipper, and S. Dessai, 2005: Integrating mitigation and adaptation into climate and development policy: three research questions. *Environmental Science & Policy*, **8(6)**, 579-588.

Klein, R.J.T., S.E.H. Eriksen, L.O. Næss, A. Hammill, T.M. Tanner, C. Robledo, and K.L. O'Brien, 2007: Portfolio screening to support the mainstreaming of adaptation to climate change into development assistance. *Climatic Change*, **84(1)**, 23-44.

Knutson, T.R., J.L. McBride, J. Chan, K. Emanuel, G. Holland, C. Landsea, I. Held, J.P. Kossin, A.K. Srivastava, and M. Sugi, 2010: Tropical cyclones and climate change. *Nature Geoscience*, **3**, 157-163.

Kolb, D.A., 1984: *Experiential Learning*. Englewood Cliffs, Prentice Hall, NJ, 256 pp.

Kolb, D.A. and R. Fry, 1975: Toward an applied theory of experiential learning. In: *Theories of Group Process* [Cooper, C. (ed.)]. John Wiley, London, UK, pp. 33-57.

Kolmannskog, V., 2008: *Future Floods of Refugees: A Comment on Climate Change, Conflict and Forced Migration*. Report by the Norwegian Refugee Council, Oslo, Norway, 44 pp.

Kossin, J.P., K.R. Knapp, D.J. Vimont, R.J. Murnane, and B.A. Harper, 2007: A globally consistent reanalysis of hurricane variability and trends. *Geophysical Research Letters*, **34**, L04815.

Kotter, J.P., 1995: Leading change: Why transformation efforts fail. *Harvard Business Review OnPoint*, **March-April**, 1-10.

Kotter, J.P., 1996: *Leading Change*. Harvard Business School Press, Boston, MA, 200 pp.

Kovats, S., D. Campbell-Lendrum, and F. Matthies, 2005: Climate change and human health: estimating avoidable deaths and disease. *Risk Analysis*, **25(6)**, 1409-1418.

Kristjanson, P., R.S. Reid, N. Dickson, W.C. Clark, D. Romney, R. Puskur, S. MacMillan, and D. Grace, 2009: International agricultural research knowledge with action for sustainable development. *Proceedings of the National Academy of Sciences*, **106(13)**, 5047-5052.

Kunreuther, H., R. Ginsberg, L. Miller, P. Sagi, P. Slovic, B. Borkan, and N. Katz, 1978: *Disaster Insurance Protection: Public Policy Lessons*. John Wiley and Sons, New York, NY.

Kwa, E., 2009: Climate change and indigenous peoples in the South Pacific: the need for regional and local strategies. In: *Climate Law and Developing Countries: Legal and Policy Challenges for the World Economy* [Richardson, B.J., Y. le Bouthillier, H. McLeod-Kilmurray, and S. Wood (eds.)]. Edward Elgar, Cheltenham, UK, pp. 102-124.

La Trobe, S. and I. Davis, 2005: *Mainstreaming Disaster Risk Reduction: A Tool for Development Organizations*. Tearfund, London, 20 pp.

Lacambra, C. and K. Zahedi, 2011: Climate change, natural hazards and coastal ecosystems in Latin-America: A framework for analysis. In: *Coping with Global Environmental Change, Disaster and Security* [Brauch, H.G., Ú. Oswald Spring, C. Mesjasz, J. Grin, P. Kameri-Mbote, B. Chourou, P. Dunay, and J. Birkmann (eds.)]. Springer, Berlin, Germany, pp. 585-602.

Larsen, P., S. Goldsmith, O. Smith, M. Wilson, K. Strzepek, P. Chinowsky, and B. Saylor, 2007: *Estimating Future Costs for Alaska Public Infrastructure at Risk from Climate Change*. Institute of Social and Economic Research, University of Alaska, Anchorage, AK, 108 pp.

Last, M., 1994: Putting children first. *Disasters*, **18(3)**, 192-202.

Lauten, A.W. and K. Lietz, 2008: A look at the standards gap: comparing child protection responses in the aftermath of hurricane Katrina and the Indian Ocean tsunami. *Children, Youth and Environments*, **18(1)**, 158-201.

Lavell, A., 1999: *Desastres en América Latina: Avences Teóricos y Prácticos: 1990-1999*. Anuario Social y Politico de América Latina y el Caribe, FLACSO-Nueva Sociedad, Caracas, Venezuela, 34 pp.

Lavell, A., 2009: *Unpacking Climate Change Adaptation and Disaster Management: Searching for the Links and Differences: A Conceptual and Epistemological Critique and Proposal*. FLACSO, Caracas, Venezuela.

Lawn, P., 2003: A theoretical foundation to support the Index of Sustainable Economic Welfare (ISEW), Genuine Progress Indicator (GPI), and other related indexes. *Ecological Economics*, **44**, 105-118.

Le Billon, P. and A. Waizenegger, 2007: Peace in the wake of disaster? Secessionist conflicts and the 2004 Indian Ocean tsunami. *Transactions of the Institute of British Geographers*, **32(3)**, 411-427.

Le Roy Ladurie, E., 1971: *Times of Feast, Times of Famine: A History of Climate Since the Year 1000*. Doubleday, New York, NY.

Leary, N., J. Adequwon, V. Barros, I. Burton, J. Kukarni, and R. Lasco (eds.), 2008: *Climate Change and Adaptation*. Earthscan, London, UK.

Lebel, L., J.M. Anderies, B. Campbell, C. Folke, S. Hatfield-Dodds, T.P. Hughes, and J. Wilson, 2006: Governance and the capacity to manage resilience in regional social-ecological systems. *Ecology and Society*, **11(1)**, 19.

Lebel, L., B.T. Sinh, P. Garden, S. Seng, L.A. Tuan, and D. Van Truc, 2009: The promise of flood protection: dykes and dams, drains and diversions. In: *Contested Waterscapes in the Mekong Region: Hydropower, Livelihoods and Governance* [Molle, F., T. Foran, and M. Kakonen (eds.)]. Earthscan Publications, London, UK, pp. 283-306.

Lee, E.K.O., C. Shen, and T.V. Tran, 2009: Coping with Hurricane Katrina: psychological distress and resilience among African Americans evacuees. *Journal of Black Psychology*, **35(1)**, 5-23.

Lehner, B., P. Döll, J. Alcamo, T. Henrichs, and F. Kaspar, 2006: Estimating the impact of global change on flood and drought risks in Europe: A continental, integrated analysis. *Climatic Change*, **75(3)**, 273-299.

Leichenko, R.M. and K.L. O'Brien, 2008: *Environmental Change and Globalization: Double Exposures*. Oxford University Press, New York, NY, 192 pp.

Leichenko, R. and W. Solecki, 2008: Consumption, inequity, and environmental justice: the making of new metropolitan landscapes in developing countries. *Society and Natural Resources*, **21(7)**, 611-624.

Leiserowitz, A., 2006: Climate change risk perception and policy preferences: the role of affect, imagery, and values. *Climatic Change*, **77(1-2)**, 45-72.

Leiserowitz, A.A., R.W. Kates, and T.M. Parris. 2005: Do global attitudes and behaviors support sustainable development? *Environment*, **47(9)**, 22-38.

Leiserowitz, A.A., R. Kates, and T. M. Parris. 2006: Sustainability values, attitudes, and behaviors: A review of multinational and global trends. *Annual Review of Environment and Resources*, **31**, 413-444.

Lemos, M.C., E. Boyd, E.L. Tompkins, H. Osbahr, and D. Liverman, 2007: Developing adaptation and adapting development. *Ecology and Society*, **12(2)**, 26.

Lempert, R., 2007: Can scenarios help policymakers be both bold and careful? In: *Blindside: How to Anticipate Forcing Events and Wild Cards in Global Politics* [Fukuyama, F. (ed.)]. Brookings Institution Press, Washington, DC, pp. 109-179.

Lempert, R.J. and M.T. Collins, 2007: Managing the risk of uncertain thresholds responses: comparison of robust, optimum, and precautionary approaches. *Risk Analysis*, **27(4)**, 1009-1026.

Lenton, T.M., H. Held, E. Kriegler, J.W. Hall, W. Lucht, S. Rahmstorf, and H.J. Schellnhuber, 2008: Tipping elements in the Earth's climate system. *Proceedings of the National Academy of Sciences*, **105(6)**, 1786-1793.

Levin, S.A., 1998: Ecosystems and the biosphere as complex adaptive systems. *Ecosystems*, **1(5)**, 431-436.

Lin, J., 2009: Supporting adaptation in developing countries at the national and global levels. In: *Climate Law and Developing Countries: Legal and Policy Challenges for the World Economy* [Richardson, B.J., Y. le Bouthillier, H. McLeod-Kilmurray, and S. Wood (eds.)]. Edward Elgar, Cheltenham, UK, pp. 127-150.

Lindley, S., D. McEvoy, and J. Handley, 2006: Adaptation and mitigation in urban areas: synergies and conflicts. *Municipal Engineer*, **159(4)**, 185-191.

Linnenluecke, M. and A. Griffiths, 2010: Beyond adaptation: resilience for business in light of climate change and weather extremes. *Business & Society*, **49(3)**, 477-511.

Linnerooth-Bayer, J. and R. Mechler, 2007: Disaster safety nets for developing countries: extending public-private partnerships. *Environmental Hazards*, **7(1)**, 54-61.

Linnerooth-Bayer, J., R. Mechler, and G. Pflug, 2005: Refocusing Disaster Aid. *Science*, **309(5737)**, 1044-1046.

List, D., 2006: Action Research cycles for multiple futures perspectives. *Futures*, **38(6)**, 673-684.

Loayza, N., E. Olaberria, J.Rigolini, and L. Christiansen, 2009: *Natural Disasters and Growth – Going Beyond the Averages.* World Bank Policy Research Working Paper 4980, World Bank, Washington, DC, 42 pp.

Lobell, D.B., M.B. Burke, C. Tebaldi, M.D. Mastrandrea, W.P. Falcon, and R.L. Naylor, 2008: Prioritizing climate change adaptation needs for food security in 2030. *Science*, **319(5863)**, 607-610.

Locatelli, B., V. Evans, A. Wardell, A. Andrade, and R. Vignola, 2010: *Forests and Climate Change in Latin America Linking Adaptation and Mitigation in Projects and Policies*. CIFOR Infobrief No. 31, Center for International Forestry Research, Bogor, Indonesia, 8 pp.

Loorbach, D., 2010: Transition management for sustainable development: a prescriptive, complexity-based governance framework. *Governance*, **23(1)**, 161-183.

Loorbach, D., R. van der Brugge, and M. Taanman, 2008: Governance in the energy transition: practice of transition management in the Netherlands. *International Journal of Environmental Technology and Management*, **9(2/3)**, 294-315.

López, R., 2009: *Natural Disasters and the Dynamics of Intangible Assets*. Policy Research Working Paper 4874, Background paper for the UN–World Bank Assessment on the Economics of Disaster Risk Reduction, World Bank, Washington, DC, 79 pp.

Lovins, A. and H. Lovins, 1982: *Brittle Power: Energy Strategy for National Security.* Brick House, Andover, MA, 486 pp.

Lyons, M., T. Schilderman, and C. Boano (eds.), 2010: *Building Back Better: Delivering People-Centred Reconstruction at Scale*. Practical Action Publishing, Rugby, UK, 380 pp.

Magat, W., K.W. Viscusi, and J. Huber, 1987: Risk-dollar tradeoffs, risk perceptions, and consumer behaviour. In: *Learning About Risk* [Viscusi, W. and W. Magat (eds.)]. Harvard University Press, Cambridge, MA, pp. 83-97.

Mahul, O. and C.J. Stutley, 2010: *Government Support to Agricultural Insurance: Challenges and Options for Developing Countries*. World Bank, Washington, DC, 19 pp.

Markard, J, and B. Truffer, 2008: Technological innovation systems and the multi-level perspective: towards an integrated framework. *Research Policy*, **37(4)**, 596-615.

Marulanda, M.C., O.D. Cardona, and A.H. Barbat, 2010: Revealing the socioeconomic impact of small disasters in Colombia using the DesInventar database. *Disasters*, **34(2)**, 552-570.

Maskrey, A., 1989: *Disaster Mitigation: A Community Based Approach*. Oxfam, Oxford, UK, 114 pp.

Maskrey, A., 1994: Comunidad y desastres en América Latina: Estrategias de Intervención. In: *Viviendo en Riesgo: Comunidades vulnerables y prevención de desastres en América Latina* [Lavell, A. (ed.)]. Red de Estudios Sociales en Prevención de Desastres en América Latina, Bogota, Colombia, pp. 14-38.

Maskrey, A. (ed.), 1996: *Terremotos en el tropico humedo:La gestion de los desastres del Alto Mayo, Peru (1990-1991), Limon, Costa Rica (1991) y Atrato Medio, Colombia (1992)*. LA RED, Tercer Mundo, Bogota, Colombia, 256 pp.

Maskrey, A., 2011: Revisiting community-based disaster risk management. *Environmental Hazards*, **10(1)**, 42-52.

McCool, D., 2005: The river commons: a new era in U.S. water policy. *Texas Law Review*, **83(7)**, 1903-1927.

McEvoy, D., S. Lindley, and J. Handley, 2006: Adaptation and mitigation in urban areas: synergies and conflict. *Municipal Engineer*, **159(4)**, 185-191.

McGray, H., A. Hammill, R. Bradley, E.L. Schipper, and J.-E. Parry, 2007: *Weathering the Storm: Options for Framing Adaptation and Development*. World Resources Institute, Washington, DC, 57 pp.

MEA, 2005: *Ecosystems and Human Well-being A Framework for Assessment*. Millennium Ecosystem Assessment Series, Island Press, Washington, DC, 266 pp.

Meadowcroft, J., 1997: Planning for sustainable development: insights from the literatures of political science. *European Journal of Political Research*, **31(4)**, 427-454.

Medema, W., B.S. McIntosh, and P.J. Jeffrey, 2008: From premise to practice: a critical assessment of integrated water resources management and adaptive management approaches in the water sector. *Ecology and Society*, **13(2)**, 29.

Mercer, J., I. Kelman, K. Lloyd, and S. Suchet-Pearson, 2008: Reflections on use of participatory research for disaster risk reduction. *Area*, **40(2)**, 172-183.

Metcalf & Eddy, 2005: *Water Reuse: Issues, Technologies and Applications*. Mcgraw-Hill, New York, NY, 1570 pp.

Mezirow, J., 1995: Transformation theory in adult learning. In: *In Defense of the Life World* [Welton, M.R. (ed.)]. State University of New York Press, Albany, NY, pp. 39-70.

Mgbako, C. and L.A. Smith, 2010: Sex work and human rights in Africa. *Fordham International Law Journal*, **33(4)**, 1178-1220.

Michel-Kerjan, E., 2008: Disasters and public policy: Can market lessons help address government failures? In: *Proceedings of the 99th National Tax Association Conference*, Boston, MA, 16-18 Nov. 2006, 15 pp.

Midgley, J. J. and W. J. Bond, 2001: A synthesis of the demography of African acacias. *Journal of Tropical Ecology*, **17(6)**, 871-886.

Miller, R., 2007: Futures literacy: A hybrid strategic scenario method. *Futures*, **39(4)**, 341-362.

Milligan, J., T. O'Riordan, S.A. Nicholson-Cole, and A.R. Watkinson, 2009: Nature conservation for future sustainable shorelines: lessons from seeking to involve the public. *Land Use Policy*, **26(2)**, 203-213.

Misselhorn, A.A., 2005: What drives food insecurity in southern Africa? A meta-analysis of household economy studies. *Global Environmental Change A*, **15(1)**, 33-43.

Mitchell, J.K., 1999: *Crucibles of Hazard: Mega-Cities and Disasters in Transition*. UNU Press, New York, NY, 535 pp.

Mitchell, J.K., 2001: Policy forum: human rights to disaster assistance and mitigation. *Environmental Hazards*, **3(3-4)**, 123-124.

Mitchell, T., K. Haynes, N. Hall, W. Choong, and K. Oven, 2008: The role of children and youth in communicating disaster risk. *Children, Youth and Environments*, **18(1)**, 254-279.

Moench, M. and A. Dixit, 2007: *Working with the Winds of Change: Toward Strategies for Responding to the Risks Associated with Climate Change and other Hazards*. Institute for Social and Environmental Transition, Kathmandu, Nepal, 296 pp.

Moench, M., A. Dixit, S. Janakarajan, M.S. Rathore, and S. Mudrakartha, 2003: *The Fluid Mosaic: Water Governance in the Context of Variability, Uncertainty and Change*. Institute for Social and Environmental Transition, Boulder, CO, 71 pp.

Mol, A.P.J. and D.A. Sonnenfeld (eds.), 2000: *Ecological Modernization around the World*. Frank Cass Publishers, Portland, OR, 312 pp.

Montgomery, S., M. Lucotte, and I. Rheault, 2000: Temporal and spatial influences of flooding on dissolved mercury in boreal reservoirs. *The Science of the Total Environment*, **260(1-3)**, 147-157.

Moreno, A. and O.D. Cardona, 2011: *Efectos de los desastres naturales sobre el crecimiento, el desempleo, la inflación y la distribución del ingreso: Una evaluación de los casos de Colombia y México*. Background Paper prepared for the 2011 Global Assessment Report on Disaster Risk Reduction, United Nations International Strategy for Disaster Reduction, Geneva, Switzerland, 30 pp.

Mortreux, C. and J. Barnett, 2009: Climate change, migration and adaptation in Funafuti, Tuvalu. *Global Environmental Change*, **19(1)**, 105-112.

Moser, S., 2009: Now more than ever: The need for more socially relevant research on vulnerability and adaptation to climate change. *Applied Geography*, **30(4)**, 464-474.

Moser, S. and J.A. Ekstrom, 2010: A framework to diagnose barriers to climate change adaptation. *Proceedings of the National Academy of Sciences*, **107(51)**, 22026-22031.

Moss, R.H., J.A. Edmonds, K.A. Hibbard, M.R. Manning, S.K. Rose, D.P. van Vuuren, T.R. Carter, S. Emori, M. Kainuma, T. Kram, G.A. Meehl, J.F.B. Mitchell, N. Nakicenovic, K. Riahi, S.J. Smith, R.J. Stouffer, A.M. Thomson, J.P. Weyant and T.J. Wilbanks, 2010: A new paradigm for the next generation of climate change scenarios. *Nature*, **463**, 747-756.

Murray, C. and A. Lopez (eds.), 1996: *The Global Burden of Disease*. Harvard University Press, Cambridge, MA, 1022 pp.

Mustafa, D., 2005: The production of an urban hazardscape in Pakistan: modernity, vulnerability and the range of choice. *The Annals of the Association of American Geographers*, **95(3)**, 566-586.

Nakicenovic, N., J. Alcamo, G. Davis, B. de Vries, J. Fenhann, S. Gaffin, K. Gregory, A. Grübler, T.Y. Jung, T. Kram, E.L. La Rovere, L. Michaelis, S. Mori, T. Morita, W. Pepper, H. Pitcher, L. Price, K. Riahi, A. Roehrl, H-H. Rogner, A. Sankovski, M. Schlesinger, P. Shukla, S. Smith, R. Swart, S. van Rooijen, N. Victor, and Z. Dadi, 2000: *Special Report on Emissions Scenarios: A Special Report of Working Group III of the Intergovernmental Panel on Climate Change*. Cambridge University Press, Cambridge, UK, 599 pp.

Nelson, D.R., 2009: Conclusions: transforming the world. In: *Adapting to Climate Change: Thresholds, Values, Governance* [Adger, W.N., I. Lorenzoni, and K.L. O'Brien (eds.)]. Cambridge University Press, Cambridge, UK, pp. 491-500.

Nelson, D.R., W.N. Adger, and K. Brown, 2007: Adaptation to environmental change: contributions of a resilience framework. *Annual Review of Environment and Resources*, **32**, 395-419.

Nelson, R. and S. Winter, 1982: *An Evolutionary Theory of Economic Change*. Harvard University Press, Cambridge, MA, 454 pp.

Nemarundwe, N., 2003: *Negotiating Resource Access; Institutional Arrangement for Woodlands and Water Use in Southern Zimbabwe*. Doctoral Thesis, University of Uppsala, Sweden, 279 pp.

Nepstad, D.C., I.M. Tohver, D. Ray, P. Moutinho, and G. Cardinot, 2007: Mortality of large trees and lianas following experimental drought in an Amazon forest. *Ecology*, **88(9)**, 2259-2269.

New, M., D. Liverman, H. Schroeder, and K. Anderson, 2011: Four degrees and beyond: the potential for a global temperature increase of four degrees and its implications. *Philosophical Transactions of the Royal Society A*, **369**, 6-19.

Newman, P. and J.R. Kenworthy, 1989: *Cities and Automobile Dependence: A Sourcebook*. Gower Publishing Company, Farnham, UK, 388 pp.

Newman, P.W.G., B. Birrell, D. Holmes, C. Mathers, P. Newton, G. Oakley, A. O'Connor, B. Walker, A. Spessa, and D. Tait, 1996: Human settlements. In: *Australian State of the Environment Report* [Beeton, R.J.S., K.I> Buckley, G.J. Jones, D. Morgan, R.E. Reichelt, and D. Trewin (eds.)]. Department of Environment, Sport and Territories, Canberra, pp. 7-18.

Nicholls, R., A. Watkinson, M. Mokrech, S. Hanson, J. Richards, J. Wright, S. Jude, S. Nicholson-Cole, M. Walkden, J. Hall, R. Dawson, P. Stansby, G.K. Jacoub, M. Rounsvell, C. Fontaine, L. Acosta, J. Lowe, J. Wolf, J. Leake, and M. Dickson, 2007: Integrated coastal simulation to support shoreline management planning. In: *Proceedings of the 42nd Flood and Coastal Management Conference*, York, UK, 3-5 July 2007, 11 pp.

Nicholls, R.J., S. Hanson, C. Herweijer, N. Patmore, S. Hallegatte, J. Chateau, and R. Muir-Wood, 2008: *Ranking Port Cities with High Exposure and Vulnerability to Climate Extremes: Exposure Estimates*. Environment Working Papers 1, Organisation for Economic Co-operation and Development, Paris, France, 61 pp.

Njuki, J.M., M.T. Mapila, S. Zingore, and R. Delve, 2008: The dynamics of social capital in influencing use of soil management options in the Chinyanja Triangle of southern Africa. *Ecology and Society*, **13(2)**, 9.

Nobre, C.A., P. Sellers, and J. Shukla, 1991: Amazonian deforestation model and regional climate change. *Journal of Climate*, **4**, 957-988.

Nobre, C.A., E.D. Assad, and M.D. Oyama, 2005: Mudança ambiental no Brasil: o impacto do aquecimento global nos ecossistemas da Amazônia e na agricultura. *Scientific American Brasil*, **Special Issue: A Terra na Estufa**, 70-75.

Norris, F.H., 2005: *Range, Magnitude, and Duration of the Effects of Disasters on Mental Health: Review Update 2005*. Dartmouth Medical School and National Center for PTSD, Hanover, NH, 23 pp.

Norris, F., 2010: *Behavioral Science Perspectives on Resilience*. CARRI Research Paper 10, Community and Regional Resilience Institute, Oak Ridge, TN, 50 pp.

Norris, F.H., M. J. Friedman, P.J. Watson, C.M. Byrne, E. Diaz, and K. Kaniasty, 2002: 60,000 disaster victims speak: Part I. An empirical review of the empirical literature, 1981-2001. *Psychiatry*, **65(3)**, 207-239.

Noy, I., 2009: The macroeconomic consequences of disasters. *Journal of Development Economics*, **88(2)**, 221-231.

Noy, I. and A. Nualsri, 2007: *What do Exogenous Shocks Tell Us about Growth Theories?* Working Paper 07-28, University of Hawaii at Manoa, Department of Economics, Manoa, HI, 31 pp.

Noy, I. and T.B. Vu, 2010: The economics of natural disasters in a developing country: The case of Vietnam. *Journal of Asian Economics*, **21(4)**, 345-354.

NRC, 1999: *Our Common Journey: A Transition Towards Sustainability*. National Research Council, National Academies Press, Washington, DC, 384 pp.

NRC, 2005: *Thinking Strategically: The Appropriate Use of Metrics for the Climate Change Science Program*. National Research Council, National Academies Press, Washington, DC, 162 pp.

NRC, 2008: *Disaster Risk Management in an Age of Climate Change*. A Summary of the April 3, 2008 Workshop of the Disasters Roundtable [Anderson, W.A. (ed.)], National Research Council, National Academies Press, Washington, DC, 18 pp.

NRC, 2010: *Adapting to the Impacts of Climate Change*. Panel on Adapting to Impacts of Climate Change, National Research Council, National Academies Press, Washington, DC, 292 pp.

O'Brien, G., P. O'Keefe, J. Rose, and B. Wisner, 2006: Climate change and disaster management. *Disasters*, **30(1)**, 64-80.

O'Brien, K., 2009: Do values subjectively define the limits to climate change adaptation? In: *Adapting to Climate Change: Thresholds, Values, Governance* [Adger, W.N., I. Lorenzoni, and K. O'Brien (eds.)]. Cambridge University Press, Cambridge, UK, pp. 164-180.

O'Brien, K. and R. Leichenko, 2003: Winners and losers in the context of global change. *Annals of the Association of American Geographers*, **93(1)**, 89-103.

O'Brien, K. and J. Wolf. 2010: A values-based approach to vulnerability and adaptation to climate change. *Wiley Interdisciplinary Reviews: Climate Change*, **1(2)**, 232-242.

O'Brien, K.L., R. Leichenko, U. Kelkar, H. Venema, G. Aandahl, H. Tompkins, A. Javed, S. Bhadwal, S. Barg, L. Nygaard, and J. West, 2004: Mapping vulnerability to multiple stressors: climate change and globalization in India. *Global Environmental Change*, **14(4)**, 303-313.

O'Brien, K. L., S. Eriksen, L. Sygna, and L.O. Næss, 2006: Questioning complacency: climate change impacts, vulnerability, and adaptation in Norway. *Ambio*, **35(2)**, 50-56.

O'Brien, K., L. Sygna, R. Leichenko, W.N. Adger, J. Barnett, T. Mitchell, L. Schipper, T. Tanner, C. Vogel, and C. Mortreux, 2008: *Disaster Risk Reduction, Climate Change Adaptation and Human Security*. A Commissioned Report for the Norwegian Ministry of Foreign Affairs, GECHS Report 2008:3, University of Oslo, Oslo, Norway, 76 pp.

O'Brien, K., B. Hayward, and F. Berkes, 2009: Rethinking social contracts: building resilience in a changing climate. *Ecology and Society*, **14(2)**, 12.

O'Brien, K., A. St Clair, and B. Kristoffersen (eds.), 2010a: *Climate Change, Ethics and Human Security*. Cambridge University Press, Cambridge, UK, 246 pp.

O'Brien, K., A. St Clair, and B. Kristoffersen, 2010b: The framing of climate change: Why it matters. In: *Climate Change, Ethics and Human Security* [O'Brien, K., A. St Clair, and B. Kristoffersen (eds.)]. Cambridge University Press, Cambridge, UK, pp. 3-22.

Obrist, B., C. Pfeiffer, and R. Henley, 2010: Multi-layered social resilience: a new approach in mitigation research. *Progress in Development Studies*, **10(4)**, 283-293.

OECD, 2006: *Declaration On Integrating Climate Change Adaptation into Development Cooperation*. Organisation for Economic Co-operation and Development, Paris, France, 7 pp.

OECD, 2007: *Climate Change in the European Alps: Adapting Winter Tourism and Natural Hazards Management* [Agrawala, S. (ed.)]. Organisation for Econonomic Co-operation and Development, Paris, France.

OECD, 2009a: *Measuring and Fostering the Progress of Societies*. Document prepared for OECD Meeting of the Council at Ministerial Level, 27-28 May 2010. C/MIN(2010)13, 12 May 2010, Organisation for Econonomic Co-operation and Development, Paris, France, 17 pp.

OECD, 2009b: *Innovation in Country Risk Management*. Organisation for Economic Co-operation and Development, Paris, France, 47 pp.

Okuyama, Y., 2004: Modeling spatial economic impacts of an earthquake: Input-output approaches. *Disaster Prevention and Management*, **13(4)**, 297-306.

Oliver-Smith, A., 1994: Peru's five hundred year earthquake: vulnerability in historical context. In: *Disasters, Development and Environment* [Varley, A. (ed.)]. John Wiley and Sons, London, UK, pp. 31-48.

Oliver-Smith, A., 2007: Communities after catastrophes: reconstructing the material, reconstituting the social. In: *Community Building in the Twenty First Century* [Hyland, S.E. (ed.)]. School of American Research Press, Santa Fe, NM, pp. 49-75.

Olson, R.S., 2000: Toward a politics of disaster losses, values, agendas, and blame. *International Journal of Mass Emergencies and Disasters*, **18(2)**, 265-287.

Olson, R.S. and V. Gawronski, 2003: Disasters as 'critical junctures' Managua, Nicaragua 1972 and Mexico City 1985. *International Journal of Mass Emergencies and Disasters*, **21(1)**, 5-35.

Olsson, P., C. Folke, and T. Hahn, 2004: Social–ecological transformation for ecosystem management: the development of adaptive co-management of a wetland landscape in southern Sweden. *Ecology and Society*, **9(4)**, 2.

ONERC, 2009: *Climate Change: Costs of Impacts and Lines of Adaptation*. Report to the Prime Minister and Parliament, Observatoire National sur les Effets du Réchauffement Climatique, Paris, France, 136 pp.

Orlove, B., 2009: Glacier retreat: reviewing the limits of human adaptation to climate change. *Environment: Science and Policy for Sustainable Development*, **51(3)**, 22-34.

Oswald Spring, Ú., 1991: *Estrategias de Supervivencia en la Ciudad de México*. Centro Regional de Investigaciones Multidisciplinarias, Universidad Nacional Autónoma de México, Cuernavaca, México, 219 pp.

Oswald Spring, Ú., 2009a: A HUGE gender security approach: towards human, gender and environmental security. In: *Facing Global Environmental Change: Environmental, Human, Energy, Food, Health and Water Security Concepts* [Brauch, H.G., Ú. Oswald Spring, J. Grin, C. Mesjasz, P. Kameri-Mbote, N. Chadha Behera, B. Chourou, and H. Krummenacher (eds.)]. Springer, Berlin, Germany, pp.1165-1190.

Oswald Spring, Ú., 2009b: Food as a new human and livelihood security challenge. In: *Facing Global Environmental Change: Environmental, Human, Energy, Food, Health and Water Security Concepts* [Brauch, H.G., Ú. Oswald Spring, J. Grin, C. Mesjasz, P. Kameri-Mbote, N. Chadha Behera, B. Chourou, and H. Krummenacher (eds.)]. Springer, Berlin, Germany, pp. 471-500.

Oswald Spring, Ú., 2011a: Environmentally-forced migration in rural areas: security risks and threats in Mexico. In: *Climate Change, Human Security and Violent Conflict: Challenges for Societal Stability* [Scheffran, J., M. Brzoska, H.G. Brauch, P.M. Link, and J. Schilling (eds.)]. Hexagon Series on Human and Environmental Security and Peace 8, Springer, Heidelberg, Germany (in press).

Oswald Spring, Ú. (ed.), 2011b: *Retos de la Investigación del Agua en México*. Centro Regional de Investigaciones Multidisciplinarias, Universidad Nacional Autónoma de México / Scientific Network on Water – National Council on Science and Technology, Cuernavaca, México.

Oswald Spring, Ú. and H.G. Brauch, 2011: Coping with global environmental change – sustainability revolution and sustainable peace. In: *Coping with Global Environmental Change, Disaster and Security* [Brauch, H.G., Ú. Oswald Spring, C. Mesjasz, J. Grin, P. Kameri-Mbote, B. Chourou, P. Dunay, and J. Birkmann (eds.)]. Springer, Berlin, Germany, pp. 1487-1504.

Owuor, B., W. Mauta, and S. Eriksen, 2011: Strengthening sustainable adaptation: examining interactions between pastoral and agropastoral groups in dryland Kenya. *Climate and Development*, **3(1)**, 42-58.

Oxfam, 2010: *Gender, Disaster Risk Reduction and Climate Change Adaptation, A Learning Companion*. Oxfam GB, Oxford, UK, 15 pp.

Paavola, J., 2005: Seeking justice: international environmental governance and climate change. *Globalizations*, **2(3)**, 309-322.

Paavola, J. and W.N. Adger, 2006: Fair adaptation to climate change. *Ecological Economics*, **56(1)**, 594-609.

Paavola, J., W.N. Adger, and S. Huq, 2006: Multifaceted justice in adaptation to climate change. In: *Fairness in Adaptation to Climate Change* [Adger, W.N., J. Paavola, S. Huq, and M.J. Mace (eds.)]. MIT Press, Cambridge, MA, 263-277.

Pahl-Wostl, C., 2009: A conceptual framework for analysing adaptive capacity and multi-level learning processes in resource governance regimes. *Global Environmental Change*, **19(3)**, 354-365.

Pahl-Wostl, C., M. Craps, A. Dewulf, E. Mostert, D. Tabara, and T.Taillieu, 2007: Social learning and water resources management. *Ecology and Society*, **12(2)**, 5.

Parry, M.L., O.F. Canziani, J.P. Palutikof and Co-authors, 2007: Technical summary. In: *Climate Change 2007: Impacts, Adaptation and Vulnerability. Contribution of Working Group II to the Fourth Assessment Report of the Intergovernmental Panel on Climate Change* [Parry, M.L., O.F. Canziani, J.P. Palutikof, P.J. van der Linden, and C.E. Hanson (eds.)]. Cambridge University Press, Cambridge, UK, pp. 23-78.

Patel, P. and K. Pavit, 1995: Patterns of technological activity: their measurement and interpretation. In: *Handbook of Economics of Innovation and Technological Change* [Stoneman, P. (ed.)]. Wiley-Blackwell, Oxford, UK, pp. 14-51.

Pearce, L., 2003: Disaster management and community planning, and public participation: how to achieve sustainable hazard mitigation. *Natural Hazards*, **28(2-3)**, 211-228.

Peek, L., 2008: Children and disasters: understanding vulnerability, developing capacities and promoting resilience – an introduction. *Children, Youth and Environments*, **18(1)**, 1-29.

Peeples, M.A., C.M. Barton, and S. Schmich, 2006: Resilience lost: intersecting land use and landscape dynamics in the prehistoric southwestern United States. *Ecology and Society*, **11(2)**, 22.

Pelling, M., 1999: The political ecology of flood hazard in urban Guyana. *Geoforum*, **30**, 249-261.

Pelling, M., 2003: *Vulnerability of Cities, Natural Disasters and Social Resilience*. Earthscan, London, UK, 212 pp.

Pelling, M., 2007: Learning from others: scope and challenges for participatory disaster risk assessment. *Disasters*, **31(4)**, 373-385.

Pelling, M., 2010a: *Adaptation to Climate Change: From Resilience to Transformation*. Taylor and Francis, London, UK, 224 pp.

Pelling, M., 2010b: *Systematisation Review of Urban Disaster Risk Reduction Projects in the Caribbean Region*. Oxfam, Oxford, UK, 90 pp.

Pelling, M. and K. Dill, 2009: Disaster politics: tipping points for change in the adaptation of socio-political regimes. *Progress in Human Geography*, **34(1)**, 21-37.

Pelling, M. and A. Holloway, 2006: *Legislation for Mainstreaming Disaster Risk Reduction*. Tearfund, London, UK, 36 pp.

Pelling, M. and D. Manuel-Navarrete, 2011: From resilience to transformation: the adaptive cycle in two Mexican urban centers. *Ecology and Society*, **16(2)**, 11.

Pelling, M. and J.I. Uitto, 2001: Small island developing states: natural disaster vulnerability and global change. *Environmental Hazards: Global Environmental Change B*, **3(2)**, 49-62.

Pelling, M. and B. Wisner (eds.), 2009: *Disaster Risk Reduction: Cases from Urban Africa*. Earthscan, London, UK.

Pelling, M., A. Özerdem, and S. Barakat, 2002: The macro-economic impact of disasters. *Progress in Development Studies*, **2(4)**, 283-305.

Pelling, M., C. High, J. Dearing, and D. Smith, 2007: Shadow spaces for social learning: a relational understanding of adaptive capacity to climate change within organizations. *Environment and Planning A*, **40(4)**, 867-884.

Penrose, A. and M. Takaki, 2006: Children's rights in emergencies and disasters. *The Lancet*, **367**, 698-699.

Perez, C., 2002: *Technological Revolutions and Financial Capital. The Dynamics of Bubbles and Golden Ages*. Edward Elgar, Cheltenham, UK, 198 pp.

Peterson, G., 2009: Ecological limits of adaptation to climate change. In: *Adapting to Climate Change: Thresholds, Values, Governance* [Adger, W.N., I. Lorenzoni, and K. O'Brien (eds.)]. Cambridge University Press, Cambridge, UK, pp. 25-41.

Peterson, G., G.A. De Leo, J.J. Hellmann, M.A. Janssen, A. Kinzig, J.R. Malcolm, K.L. O'Brien, S.E. Pope, D.S. Rothman, E. Shevliakova, and R.R.T Tinch, 1997: Uncertainty, climate change, and adaptive management. *Conservation Ecology*, **1(2)**, 4.

Pettengell, C., 2010: *Climate Change Adaptation: Enabling People Living in Poverty to Adapt*. Oxfam International Research Report, Oxfam, Oxford, UK, 48 pp.

Pielke, Jr., R., J. Rubiera, C. Landsea, M. Fernández, and R. Klein, 2003: Hurricane vulnerability in Latin America and the Caribbean: Normalized damage and loss potentials. *Natural Hazards Review*, **4(3)**, 101-114.

Pielke, Jr., R., G. Prins, S. Rayner, and D. Sarewitz, 2007: Climate change 2007: lifting the taboo on adaptation. *Nature*, **445**, 597-598.

Pielke, Jr., R.A., J. Gratz, C.W. Landsea, D. Collins, M. Saunders, and R. Musulin, 2008: Normalized hurricane damages in the United States: 1900-2005. *Natural Hazards Review*, **9(1)**, 29-42.

Pigou, A.C., 1920: The *Economics of Welfare*. Macmillan, London, UK, 976 pp.

Piguet, E., 2008: *Climate Change and Forced Migration*. Research Paper No 153, Evaluation and Policy Analysis Unit, UN High Commissioner for Refugees, Geneva, Switzerland, 15 pp.

Platt, R.H., 1999: *Disasters and Democracy: the Politics of Extreme Natural Events*. Island Press, Washington, DC, 320 pp.

Plummer, R. and D.R. Armitage, 2007: Charting the new territory of adaptive co-management: a Delphi study. *Ecology and Society*, **12(2)**, 10.

Quarantelli, E.L., 1984: Perceptions and reactions to emergency warnings of sudden hazards. *Ekistics*, **309**, 511-515.

Quarantelli, E.L., 1995: *What is a Disaster?* Routledge, London, UK, 312 pp.

Raddatz, C., 2007: Are external shocks responsible for the instability of output in low-income countries? *Journal of Development Economics*, **84(1)**, 155-187.

Raleigh, C., L. Jordan, and I. Salehyan, 2008: *Assessing the Impact of Climate Change on Migration and Conflict*. World Bank, Washington DC, 57 pp.

Ramos, J.M., 2006: Dimensions in the confluences of futures studies and action research. *Futures*, **38(6)**, 642-655.

Raven, R.P.J.M., S. van den Bosch, and R. Weterings, 2010: Transitions and strategic niche management: towards a competence kit for practitioners. *International Journal of Technology Management*, **51(1)**, 57-74.

Rawls, J., 1971: *A Theory of Justice*. Harvard University Press, Cambridge, MA, 607 pp.

Reaser, J.K., R. Pomerance, and P.O. Thomas, 2000: Coral bleaching and global climate change: scientific findings and policy recommendations. *Conservation Biology*, **14(5)**, 1500-1511.

Redclift, M., 1987: *Sustainable Development: Exploring the Contradictions*. Methuen, London, UK, 221 pp.

Redclift, M., 1992: The meaning of sustainable development. *Geoforum*, **25(3)**, 395-403.

Reeder, T., J. Wicks, L. Lovell, and O. Tarrant, 2009: Protecting London from tidal flooding: limits to engineering adaptation. In: *Adapting to Climate Change: Thresholds, Values, Governance* [Adger, W.N., I. Lorenzoni, and K. O'Brien (eds.)]. Cambridge University Press, Cambridge, UK, pp. 54-78.

Reid, H., 2006: Climatic change and biodiversity in Europe. *Conservation Society*, **4(1)**, 84-101.

Repetto, R., 2008: *The Climate Crisis and the Adaptation Myth*. Working Paper 13, Yale School of Forestry and Environmental Studies, New Haven, CT, 24 pp.

Reuveny, R., 2007: Climate change induced migration and violent conflict. *Political Geography*, **26(6)**, 656-673.

Reyos, J., 2010: *Community-driven Disaster Intervention: Experiences of the Homeless People's Federation in the Philippines*. Homeless Peoples Federation Philippines, Philippines Action for Community-led Shelter Initiatives, and International Institute for Environment and Development, Manila, Philippines and London, UK, 70 pp.

Richardson, B.J., Y. Le Bouthillier, H. McLeod-Kilmurray, and S. Wood (eds.), 2009: *Climate Law and Developing Countries: Legal and Policy Challenges for the World Economy*. Edward Elgar, Cheltenham, UK, 448 pp.

Rip, A. and R. Kemp, 1998: Technological change. In: *Human Choice and Climate Change* [Rayner, S. and L. Malone (eds.)]. Batelle Press, Washington, DC, pp. 327-399.

Rist, S., 2000: Linking ethics and market – Campesino economic strategies in the Bolivian Andes. *Mountain Research and Development*, **20(4)**, 310-315.

Roberts, D., 2008: Thinking globally, acting locally – institutionalizing climate change at the local government level in Durban, South Africa. *Environment and Urbanization*, **20(2)**, 521-537.

Roberts, D., 2010: Prioritizing climate change adaptation and local level resilience in Durban, South Africa. *Environment and Urbanization*, **22(2)**, 397-413.

Robinson, J., 2003: Future subjunctive: backcasting as social learning. *Futures*, **35(8)**, 839-856.

Rodriguez-Oreggia, E., A. de la Fuente, R. de la Torre, H. Moreno, and C. Rodriguez, 2009: *The Impact of Natural Disasters on Human Development and Poverty at the Municipal Level in Mexico*. CID Working Paper 43, Center for International Development at Harvard University, Cambridge, MA, 35 pp.

Rohan, M.J., 2000: A rose by any name? The values construct. *Personal and Social Psychology Review*, **4(3)**, 255-277.

Rohracher, H., 2008: Energy systems in transition: contributions from social sciences. *International Journal of Environmental Technology and Management*, **9(2-3)**, 144-161.

Rokeach, M., (ed.), 1979: *Understanding Human Values: Individual and Societal*. The Free Press, New York, NY, 322 pp.

Rose, A. and S.-Y. Liao, 2005: Modeling regional economic resilience to disasters: A computable general equilibrium analysis of water service disruptions. *Journal of Regional Science*, **45(1)**, 75-112.

Rose, A., J. Benavides, S.E. Chang, P. Szczesniak, and D. Lim, 1997: The regional economic impact of an earthquake: Direct and indirect effects of electricity lifeline disruptions. *Journal of Regional Science*, **37(3)**, 437-458.

Rosegrant, M.W., T. Zhu, S. Msangi, and T. Sulser, 2011: Global scenarios for biofuels: impacts and implications. *Applied Economic Perspectives and Policy*, **30(3)**, 595-505.

Rosenberg, S.W. (ed.), 2007: *Deliberation, Participation and Democracy: Can the People Govern?* Palgrave Macmillan, New York, NY, 322 pp.

Rosenzweig, C., K.M. Strzepek, D.C. Major, A. Iglesias, D.N. Yates, A. McCluskey, and D. Hillel, 2004: Water resources for agriculture in a changing climate: international case studies. *Global Environmental Change*, **14(4)**, 345-360.

Roshetko, J.M., R.D. Lasco, and M.S.D. Angeles, 2007: Smallholder agroforestry systems for carbon storage. *Mitigation and Adaptation Strategies for Global Change*, **12**, 219-242.

Royal Society, 2009: *Geoengineering The Climate: Science, Governance and Uncertainty*. The Royal Society, London, UK, 98 pp.

Sacoby, M., R.R. Wilson, J. Lesley, and E. Williams, 2010: Climate change, environmental justice, and vulnerability: an exploratory spatial analysis. *Environmental Justice*, **3(1)**, 13-19.

Salagnac, J.-L., 2007: Lessons from the 2003 heat wave: a French perspective. *Building Research and Information*, **35(4)**, 450-457.

Santos, I., 2007: *Disentangling the Effects of Natural Disasters on Children: 2001 Earthquakes in El Salvador*. Doctoral Dissertation, Kennedy School of Government, Harvard University, Cambridge, MA.

Satterthwaite, D., 1997: Sustainable cities or cities that contribute to sustainable development? *Urban Studies*, **34(10)**, 1167-1691.

Satterthwaite, D., 1998: Meeting the challenge of urban disasters. In: *World Disasters Report 1998*. International Federation of Red Cross and Red Crescent Societies, Oxford University Press, Oxford, UK, pp. 9-19.

Satterthwaite, D. (ed.), 2007: *The Transition to a Predominantly Urban World and its Underpinnings*. International Institute for Environment and Development, London, UK, 91 pp.

Satterthwaite, D., 2010: Slum upgrading. *Economic and Political Weekly*, **45(10)**, 12-16.

Schär, C., P.L. Vidale, D. Lüthi, C. Frei, C.Häberli, M.A. Liniger, and C. Appenzeller, 2004: The role of increasing temperature variability in European summer heatwaves. *Nature*, **427**, 332-336.

Scharmer, C.O., 2009: *Theory U: Leading from the Future as it Emerges*. The Society for Organizational Learning, Cambridge, MA, 533 pp.

Scheffer, M., 2009. *Critical Transitions in Nature and Society*. Princeton University Press, Princeton, NJ, 400 pp.

Scheffer, M., S.R. Carpenter, J.A. Foley, C. Folke, and B. Walker, 2001: Catastrophic shifts in ecosystems. *Nature*, **413**, 591-596.

Scheffer, M., J. Bascompte, W.A. Brock, V. Brovkin, S.R. Carpenter, V. Dakos, H. Held, E.H. van Nes, M. Rietkerk, and G. Sugihara, 2009: Early-warning signals for critical transitions. *Nature*, **461**, 53-59.

Scheffran, J., 2011: Security risks of climate change: vulnerabilities, threats, conflicts and strategies. In: *Coping with Global Environmental Change, Disaster and Security* [Brauch, H.G., Ú. Oswald Spring, C. Mesjasz, J. Grin, P. Kameri-Mbote, B. Chourou, P. Dunay, and J. Birkmann (eds.)]. Springer, Berlin, Germany, pp. 735-756.

Schipper, E.L.F., 2007: *Climate Change Adaptation and Development: Exploring the Linkages*. Working Paper 107, Tyndall Centre for Climate Change Research, Norwich, UK, 20 pp.

Schipper, L., 2009: Meeting at the crossroads? Exploring the linkages between climate change adaptation and disaster risk reduction. *Climate and Development*, **1(1)**, 16-30.

Schipper, L. and M. Pelling, 2006: Disaster risk, climate change and international development: scope and challenges for integration. *Disasters*, **30(1)**, 19-38.

Schlenker, W. and M.J. Roberts, 2006: Nonlinear effects of weather on corn yields. *Review of Agricultural Economics*, **28(3)**, 391-398.

Scholze, M., W. Knorr, N.W. Arnell, and I.C. Prentice, 2005: A climate change risk analysis for world ecosystems. *Proceedings of the National Academy of Sciences*, **103**, 13116-13120.

Schumacher, I. and E. Strobl, 2008: *Economic Development and Losses due to Natural Disasters: The Role of Risk*. Cahiers 2008-32, Département d'Economie, Ecole Polytechnique, Palaiseau, France, 39 pp.

SCBD, 2010: *Global Biodiversity Outlook 3*. Secretariat of the Convention on Biological Diversity, Montréal, Canada.

Scott, J.C., 2003: *The Moral Economy of the Peasant: Rebellion and Subsistence in Southeast Asia*. Yale University Press, New Haven, CT, 246 pp.

Sen, A, 2000: *Un Nouveau Modèle Économique. Développement, Justice, Liberté*. Editions Odile Jacob, Paris, France, 356 pp.

Sen, A, 2003: Human security now. *Soka Gakkai International Quarterly*, **33**, 1-7.

Senge, P.M., 1990: *The Fifth Discipline: The Art and Practice of the Learning Organization*. Currency/Doubleday, New York, NY, p. 445.

Silvestri, S. and F. Kershaw (eds.), 2010: *Framing the flow: Innovative Approaches to Understand, Protect and Value Ecosystem Services across Linked Habitats*. United Nations Environment Programme World Conservation Monitoring Centre, Cambridge, UK, 66 pp.

Simmie, J. and R. Martin. 2010: The economic resilience of regions: towards an evolutionary approach. *Economy and Society*, **3(1)**, 27-43.

Skidmore, M. and H. Toya, 2002: Do natural disasters promote long-run growth? *Economic Inquiry*, **40(4)**, 664-687.

Slovic, P., 1987: Perception of risk. *Science*, **236**, 280-285.

Smith, A. and A. Stirling, 2010: The politics of social–ecological resilience and sustainable sociotechnical transitions. *Ecology and Society*, **15(1)**, 11.

Smith, A., A. Stirling, and F. Berkhout, 2005: The governance of sustainable socio-technical transitions: a quasi-evolutionary model. *Research Policy*, **34(10)**, 1491-1510.

Smith, D.M. and S. Barchiesi, 2009: *Environment as Infrastructure: Resilience to Climate Change impacts on Water through Investments in Nature*. International Union for Conservation of Nature, Gland, Switzerland, 13 pp.

Smith, S.K. and C. McCarty, 2006: *Florida's 2004 Hurricane Season: Demographic Response and Recovery*. Bureau of Economic and Business Research, University of Florida, Gainesville, FL, 40 pp.

Smithers, J. and B. Smit, 1997: Agricultural system response to environmental stress. In: *Agricultural Restructuring and Sustainability: A Geographical Perspective* [Ilbery, B., Q. Chiotti, and T. Rickard (eds.)]. CABI, Wallingford, UK, pp. 167-183.

Solecki, W. and R. Leichenko, 2006: Urbanization and the metropolitan environment: lessons from New York and Shanghai. *Environment*, **48(4)**, 8-23.

Sperling, F. and F. Szekely, 2005: *Disaster Risk Management in Changing Climate*. Discussion Paper prepared for the World Conference on Disaster Risk Reduction on behalf of the Vulnerability and Adaptation Resource Group (VARG). Reprint with Addendum on Conference Outcomes, VARG, Washington, DC, 45 pp.

Sperling, F., C. Valdivia, R. Quiroz, R. Valdivia, L. Angulo, A. Seimon, and I. Noble, 2008: *Transitioning to Climate Resilient Development: Perspectives from Communities in Peru*. World Bank Environment Department Papers 115, World Bank, Washington, DC, 103 pp.

Sphere, 2004: *Humanitarian Charter and Minimum Standards in Humanitarian Response*. The Sphere Project, Belmont Press Ltd., Northampton, UK, 402 pp.

St. Clair, A., 2010: Global poverty: towards the responsibility to protect. In: *Climate Change, Ethics and Human Security* [O'Brien, K., A. St. Clair, and B. Kristoffersen (eds.)]. Cambridge University Press, Cambridge, UK, pp. 180-198.

Stafford Smith, M., L. Horrocks, A. Harvey, and C. Hamilton, 2011: Rethinking adaptation for a 4°C world. *Philosophical Transactions of the Royal Society A*, **369**, 196-216.

Stehr, N. and H. Von Storch, 2005: Introduction to papers on mitigation and adaptation strategies for climate change: protecting nature from society or protecting society from nature? *Environmental Science Policy*, **8(6)**, 537-540.

Stern, P. and W. Easterling (eds.), 1999: *Making Use of Seasonal Climate Forecasts*. National Academies Press, Washington, DC, 175 pp.

Stevenson, T., 2002: Anticipatory action learning: conversations about the future. *Futures*, **34(5)**, 417- 425.

Stevenson, T., 2008: Enacting the vision for sustainable development. *Futures*, **41(4)**, 246-252.

Strobl, E., 2011: The economic growth impact of hurricanes: evidence from US coastal counties. *The Review of Economics and Statistics*, **93(2)**, 575-589.

Swart, R., J. Robinson, and S. Cohen, 2003: Climate change and sustainable development: expanding the options. *Climate Policy*, **391**, 810-849.

Swift, J., 1989: Why are rural people vulnerable to famine? *IDS Bulletin*, **20(2)**, 8-15.

Tanner, T.M., 2010: Shifting the narrative: child-led responses to climate change and disasters in El Salvador and the Philippines. *Children and Society*, **24(4)**, 339-351.

Tanner, T.M., M. Garcia, J. Lazcano, F. Molina, G. Molina, G. Rodríguez, B. Tribunalo, and F. Seballos, 2009: Children's participation in community-based disaster risk reduction and adaptation to climate change. *Participatory Learning and Action*, **60**, 54-64.

Telford, J., J. Cosgrave, and R. Houghton, 2006: *Joint Evaluation of the International Response to the Indian Ocean Tsunami: Synthesis Report*. Tsunami Evaluation Coalition, London, UK, 178 pp.

Thaler, R., 1999: Mental accounting matters. *Journal of Behavioral Decision Making*, **12**, 183-206.

Theobald, D., 2005: Landscape patterns of exurban growth in the USA from 1980 to 2020. *Ecology and Society*, **10(1)**, 32.

Thomas, D.S.G. and C. Twyman, 2005: Equity and justice in climate change adaptation amongst natural-resource dependent societies. *Global Environmental Change Part A*, **15(2)**, 115-124.

Thomas-Slayter, B., R. Polestico, A.L. Esser, O. Taylor, and E. Mutua, 1995: *A Manual for Socio-Economic and Gender Analysis: Responding to the Development Challenge.* Clark ECOGEN Publication, Worcester, MA, 278 pp.

Thompson, I., B. Mackey, S. McNulty, and A. Mosseler, 2009: *Forest Resilience, Biodiversity, and Climate Change. A Synthesis of the Biodiversity/Resilience/Stability Relationship in Forest Ecosystems.* Technical Series no. 43, Secretariat of the Convention on Biological Diversity, Montreal, Canada, 67 pp.

Tierney, K. and M. Bruneau, 2007: Conceptualizing and measuring resilience: A key to disaster loss reduction. *TR News 250*, **May-June**, 14-17.

Tol, R. and G. Yohe, 2007: The weakest link hypothesis for adaptive capacity: An empirical test. *Global Environmental Change*, **17(2)**, 218-227.

Tol, R.S.J., T.E. Downing, O.J. Kuik, and J.B. Smith: 2004. Distributional aspects of climate change impacts. *Global Environmental Change*, **14(3)**, 259-272.

Tompkins, E.L., M.C. Lemos, and E. Boyd, 2008: A less disastrous disaster: Managing response to climate-driven hazards in the Cayman Islands and NE Brazil. *Global Environmental Change*, **18(4)**, 736-745.

Törnquist, O., S.A. Prasetyo, and T. Birks (eds.), 2010: *Aceh: The Role of Democracy for Peace and Reconstruction*. PCD Press Indonesia, Jogjakarta, Indonesia, 431 pp.

Tran, P., R. Shaw, G. Chantry, and J. Norton, 2009: GIS and local knowledge in disaster management: a case study of flood risk mapping in Viet Nam. *Disasters*, **33(1)**, 152-169.

Trawick, P., 2001: The moral economy of water: equity and antiquity in the Andean commons. *American Anthropologist*, **103(2)**, 361-379.

Tribe, M. and A. Lafon, 2009: *After 2015: Promoting Pro-Poor Policy after the MDGS: The Plenary Presentations and Discussion*. Report on the EADI – DSA – IDS– ActionAid – DIFID High Level Policy Forum, June 2009, 17 pp.

Trope, Y. and N. Liberman, 2003: Temporal construal. *Psychological Review*, **110(3)**, 403-421.

Tschakert, P. and K.A. Dietrich, 2010: Anticipatory learning for climate change adaptation and resilience. *Ecology and Society*, **15(2)**, 11.

Tschakert, P., R. Sagoe, G. Darko, and S.N. Codjoe, 2010: Floods in the Sahel: an analysis of anomalies, memory, and anticipatory learning. *Climatic Change*, **103(3-4)**, 471-502.

Tsuchiya, S., H. Tatano, and N. Okada, 2007: Economic loss assessment due to railroad and highway disruptions. *Economic Systems Research*, **19(2)**, 147–162.

Turner, N.J. and H. Clifton, 2009: It's so different today: climate change and indigenous lifeways in British Columbia. *Global Environmental Change*, **19(2)**, 180-190.

Tversky, A. and E. Shafir, 1992: Choice under conflict: the dynamics of deferred decision. *Psychological Science*, **3(6)**, 358-361.

Twigg, J., 2001: *Corporate Social Responsibility and Disaster Reduction: A Global Overview*. Benfield Greig Hazard Research Centre, University College London, UK, 84 pp.

Tyler, S., S.O. Reed, K. Macclune, and S. Chopde, 2010: *Planning for Urban Climate Resilience: Framework and Examples from the Asian Cities Climate Change Resilience Network (ACCCRN)*. Climate Resilience in Concept and Practice Working Paper Series 3, Institute for Social and Environmental Transition, Boulder, CO, 168 pp.

UK Environment Agency, 2009: The Thames estuary is changing. In: *Managing Flood Risk through London and the Thames Estuary*. TE2100 Plan Consultation Document, Environment Agency, London, UK, pp. 20-23.

Ulsrud, K., L. Sygna, and K.L. O'Brien, 2008: *More than Rain, Identifying Sustainable Pathways for Climate Adaptation and Poverty Reduction*. Report prepared for the Development Fund, Oslo, Norway, 68 pp.

UN, 2002: *World Population Ageing 1950 – 2050. Executive Summary*. United Nations, Department of Economic and Social Affairs, Population Division, New York, NY.

UN-HABITAT, 2009: *Planning Sustainable Cities: Global Report on Human Settlements 2009*. Earthscan Publications Ltd., London, UK, 306 pp.

UNDP, 1994: *Human Development Report 1994: New Dimensions of Human Security*. United Nations Development Programme, Oxford University Press, New York, NY.

UNDP, 2004: *Reducing Disaster Risk: A Challenge for Development*. United Nations Development Programme, Oxford University Press, New York, NY, 161 pp.

UNDP, 2007a: *Thailand Human Development Report 2007: Sufficiency Economy and Human Development*. United Nations Development Programme, Bangkok, Thailand, 152 pp.

UNDP, 2007b: *Human Development Report 2007/2008: Fighting Climate Change: Human Solidarity in a Divided World*. United Nations Development Programme, Palgrave Macmillan, New York, NY, 399 pp.

UNDP, 2009: *Human Development Report 2009: Overcoming Barriers: Human Mobility and Development*. United Nations Development Programme, Oxford University Press, Oxford, UK.

UNFCCC, 2006: *Technologies for Adaptation to Climate Change*. Adaptation, Technology and Science Programme of the United Nations Framework Convention on Climate Change Secretariat, Bonn, Germany, 40 pp.

UNGA, 2009: *General Assembly, Expressing Deep Concern, Invites Major United Nations Organs: to Intensify Efforts in Addressing Security Implications of Climate Change*. GA/10830, 3 June 2009, United Nations General Assembly, New York, NY.

UNISDR, 2004: *Living with Risk. A Global Review of Disaster Reduction Initiatives*. United Nations International Strategy for Disaster Reduction Secretariat, Geneva, Switzerland, 429 pp.

UNISDR, 2008: *Gender Perspectives: Integrating Disaster Risk Reduction into Climate Change Adaptation. Good practices and Lessons Learned*. United Nations International Strategy for Disaster Reduction, Geneva, Switzerland, 87 pp.

UNISDR, 2009: *Global Assessment Report on Disaster Risk Reduction*. United Nations International Strategy for Disaster Reduction Secretariat, Geneva, Switzerland, 207 pp.

UNISDR, 2011: *Global Assessment Report on Disaster Risk Reduction*. United Nations International Strategy for Disaster Reduction, Geneva, Switzerland.

UNSC, 2007: *Security Council Holds First-Ever Debate on Impact of Climate Change on Peace, Security, Hearing over 50 Speakers*. UN Security Council, 5663rd Meeting, 17 April 2007. UN Security Council, New York, NY.

UNSG, 2009: *Climate Change and its Possible Security Implications*. Report of the Secretary-General. United Nations, New York, NY.

USGBC, 2010: *USGBC Celebrates Five Years of Green Building, Economic & Educational Progress in New Orleans*. Press Release, US Green Building Council, Washington, DC.

van den Berg, M., 2010: Household income strategies and natural disasters: dynamic livelihoods in rural Nicaragua. *Ecological Economics*, **69(3)**, 592-602.

van Vuuren, D.P., P.L. Lucas, and H. Hilderink, 2006: Downscaling drivers of global environmental change scenarios: enabling use of the IPCC-SRES scenarios at the national and grid level. *Global Environmental Change*, **17(1)**, 114-130.

Velásquez, L.S., 1998: Agenda 21: A form of joint environmental management in Manizales, Colombia. *Environment and Urbanization*, **10(2)**, 9-36.

Velásquez, L.S., 2005: The bioplan: decreasing poverty in Manizales, Colombia, through shared environmental management. In: *Reducing Poverty and Sustaining the Environment* [Bass, S., H. Reid, D. Satterthwaite, and P. Steele (eds.)]. Earthscan Publications, London, UK, pp. 44-72.

Verchot, L.V., M. van Noordijk, S. Kandji, T. Tomich, C. Ong, A. Albrecht, J. Mackensen, C. Bantilan, K.V. Anupama, and C. Palm, 2007: Climate change: linking adaptation and mitigation through agroforestry. *Mitigation and Adaptation Strategies for Global Change*, **12(5)**, 901-918.

Vergara, W. and S.M. Scholz (eds.), 2011: *Assessment of the Risk of Amazon Dieback*. World Bank, Washington, DC, 99 pp.

Verweij, M., M. Douglas, R. Ellis, C. Engel, F. Hendriks, S. Lohmann, S. Ney, S. Rayner, and M. Thompson, 2006: Clumsy solutions for a complex world: the case of climate change. *Public Administration*, **84(4)**, 817-843.

Villagrán de León, J.C., 2011: Risks in Central America: bringing them under control. In: *Coping with Global Environmental Change, Disaster and Security* [Brauch, H.G., Ú. Oswald Spring, C. Mesjasz, J. Grin, P. Kameri-Mbote, B. Chourou, P. Dunay, and J. Birkmann (eds.)]. Springer, Berlin, Germany, pp. 1147-1158.

Viratkapan, V. and R. Perera, 2006: Slum relocation projects in Bangkok: what has contributed to their success or failure? *Habitat International*, **30(1)**, 157-174.

Virgoe, J., 2002: International governance of a possible geoengineering intervention to combat climate change. *Climatic Change*, **95(1-2)**, 103-119.

Viscusi, W.K. and J. Aldy, 2003: The value of a statistical life: a critical review of market estimates throughout the world. *Journal of Risk and Uncertainty*, **27(1)**, 5-76.

Vogel, C., S.C. Moser, R.E. Kasperson, and G. Dabelko. 2007: Linking vulnerability, adaptation, and resilience science to practice: pathways, players, and partnerships. *Global Environmental Change*, **17(3-4)**, 349-364.

Volkery, A. and T. Ribeiro, 2009: Scenario planning in public policy: understanding use, impacts and the role of institutional context factors. *Technological Forecasting and Social Change*, **76(9)**, 1198-1207.

von Hesse, M., J. Kamiche, and C. de la Torre, 2008: *Contribución Temática de America Latina al Informe Bienal de Evaluación Mundial Sobre la Reducción de Riesgo 2009*. Contribution to the GTC-UNDP Background Paper prepared for the 2009 Global Assessment Report on Disaster Risk Reduction, United Nations International Strategy for Disaster Reduction, Geneva, Switzerland, 131 pp.

Voss, M. and K. Wagner, 2010: Learning from (small) disasters. *Natural Hazards*, **55**, 657-669.

Wack, P., 1985: Scenarios: shooting the rapids. *Harvard Business Review*, **63(6)**, 139-150.

Walker, B. and D. Salt, 2006: *Resilience Thinking: Sustaining Ecosystems and People in a Changing World*. Island Press, Washington, DC, 176 pp.

Walker, B., S. Carpenter, J. Anderies, N. Abel, G.S. Cumming, M. Janssen, L. Lebel, J. Norberg, G.D. Peterson, and R. Pritchard, 2002: Resilience management in social-ecological systems: a working hypothesis for a participatory approach. *Conservation Ecology*, **6(1)**, 14.

Walker, B., C.S. Holling, S.R. Carpenter, and A. Kinzig, 2004: Resilience, adaptability and transformability in social–ecological systems. *Ecology and Society*, **9(2)**, 5.

Wandel, J. and B. Smit, 2000: Agricultural risk management in light of climate variability and change. In: *Agricultural and Environmental Sustainability in the New Countryside* [Millward, H., K. Beesley, B. Ilbery, and L. Harrington (eds.)]. Hignell Printing Limited, Winnipeg, Canada, pp. 30-39.

Warner, K., M. Hamza, A. Oliver-Smith, F. Renaud, and A. Julca, 2010: Climate change, environmental degradation and migration. *Natural Hazards*, **55(3)**, 689-715.

Waterson, T., 2006: Climate change – the greatest crisis for children? *Journal of Tropical Pediatrics*, **52(6)**, 383-385.

WCC-3, 2009: *Global Framework for Climate Services*. Background paper prepared by WMO secretariat, World Climate Conference 3, World Meteorological Organization, Geneva, Switzerland, 6 pp.

WCED, 1987: *Our Common Future*. World Commission on Environment and Development, Oxford University Press, Oxford, UK.

Weber, E.U., 2010: What shapes perceptions of climate change? *Wiley Interdisciplinary Reviews: Climate Change*, **1(3)**, 332-342.

Weissbecker, I., S.E. Sephton, M.B. Martin, and D.M. Simpson, 2008: Psychological and physiological correlates of stress in children exposed to disaster: review of current research and recommendations for intervention. *Children, Youth and Environments*, **18(1)**, 30-70.

Weitzman, M.L., 2009: On modeling and interpreting the economics of catastrophic climate change. *Review of Economics and Statistics*, **91(1)**, 1-19.

West, C.T. and D.G. Lenze, 1994: Modeling the regional impact of natural disasters and recovery: a general framework and an application to hurricane Andrew. *International Regional Science Review*, **17(2)**, 121-150.

WGBU, 2008: *Climate Change as a Security Risk*. German Advisory Council on Climate Change (WGBU), Earthscan, London, UK, 271 pp.

Whitehead, J., R. Bacon, G.Carbone, K. Dow, J. Thigpen, and D. Tufford, 2010: Using climate extension to assist coastal decision-makers with climate adaptation. In: *22nd International Conference of the Coastal Society, Shifting Shorelines: Adapting to the Future*, Wilmington, NC, 13-16 June 2010.

Wilbanks, T.J., 1994: "Sustainable development" in geographic context. *Annals of the Association of American Geographers*, **84(4)**, 541-557.

Wilbanks, T., 2003: Integrating climate change and sustainable development in a place-based context. *Climate Policy*, **3(1)**, 147-154.

Wilbanks, T., 2007: Scale and sustainability. *Climate Policy*, **7(4)**, 278-287.

Wilbanks, T., 2009: *How Geographic Scale Matters in Seeking Community Resilience*. CARRI Research Report 7, Community and Resilience Research Institute, Oak Ridge, TN, 28 pp.

Wilbanks, T., 2010: Research and development priorities for climate change mitigation and adaptation. In: *Dealing with Climate Change: Setting a Global Agenda for Mitigation and Adaptation* [Pachauri, R. (ed.)]. The Energy and Resources Institute, New Delhi, India, pp. 77-99.

Wilbanks, T. and R. Kates, 2010: Beyond adapting to climate change: embedding adaptation in responses to multiple threats and stresses. *Annals of the Association of American Geographers*, **100(4)**, 719-728.

Wilbanks, T. and J. Sathaye, 2007: Integrating mitigation and adaptation as responses to climate change: a synthesis. *Mitigation and Adaptation Strategies for Global Change*, **12(5)**, 957-962.

Wilbanks, J. and T. Wilbanks, 2010: Science, open communication, and sustainable development. *Sustainability*, **2(4)**, 993-1015.

Wilbanks, T., P. Leiby, R. Perlack, J.T. Ensminger, and S.B. Wright, 2007: Toward an integrated analysis of mitigation and adaptation: Some preliminary findings. *Mitigation and Adaptation Strategies for Global Change*, **12(5)**, 713-725.

Wisner, B., 2001: Disasters: what the United Nations and its world can do. *Environmental Hazards*, **3(3-4)**, 125-127.

Wisner, B., 2003: Changes in capitalism and global shifts in the distribution of hazard and vulnerability. In: *Natural Disasters and Development in a Globalizing World* [Pelling, M. (ed.)]. Routledge, London, UK, pp. 43-56.

Wisner, B., 2006: *Let Our Children Teach Us! A Review of the Role of Education and Knowledge in Disaster Risk Reduction*. Books for Change, Bangalore, India, 148 pp.

Wisner, B., 2011: Are we there yet? Reflections on integrated disaster risk management after ten years. *Journal of Integrated Disaster Risk Management*, **1(1)**, doi:10.5595/idrim.2011.0015.

Wisner, B., P. Blaikie, T. Cannon, and I. Davis, 2004: *At Risk: Natural Hazards, People's Vulnerability and Disaster*. 2nd ed. Routledge, London, UK, 472 pp.

Wolf, J., 2011: Climate change adaptation as a social process. In: *Climate Change Adaptation in Developed Nations: From Theory to Practice* [Ford, J. and L. Ford Berrang (eds.)]. Springer, Dordrecht, The Netherlands, pp. 21-32.

Wolsink, M., 2006: River basin approach and integrated water management: Governance pitfalls for the Dutch Space-Water-Adjustment Management Principle. *Geoforum*, **37(4)**, 473-487.

Woodward, G. (ed.), 2010: *Advances in Ecological Research 42: Ecological Networks*. Academic Press, London, UK, 454 pp.

World Bank, 2005: *Managing Agricultural Production Risk*. World Bank, Washington, DC, 129 pp.

World Bank, 2008: *World Development Report 2009: Reshaping Economic Geography*. World Bank, Washington, DC.

World Bank, 2010a: *Natural Hazards, Unnatural Disasters, The Economics of Effective Prevention*. World Bank and United Nations, Washington, DC, 254 pp.

World Bank, 2010b: *Vulnerability of Bangladesh to Cyclones in a Changing Climate. Potential Damages and Adaptation Cost*. Policy Research Working Paper 5280. World Bank, Washington, DC, 54 pp.

WWF, 2010: *Living Planet Report*. World Wide Fund for Nature, Gland, Switzerland, 119 pp.

Yohe, G. and R. Leichenko, 2010: Adopting a risk-based approach. *Annals of the New York Academy of Sciences*, **1196(1)**, 29-40.

Yohe, G. and R.S.J. Tol, 2002: Indicators for social and economic coping capacity – moving toward a working definition of adaptive capacity. *Global Environmental Change*, **12**, 25-40.

Yohe, G.W., R.D. Lasco, Q.K. Ahmad, N.W. Arnell, S.J. Cohen, C. Hope, A.C. Janetos, and R.T. Perez, 2007: Perspectives on climate change and sustainability. In: *Climate Change 2007: Impacts, Adaptation and Vulnerability. Contribution of Working Group II to the Fourth Assessment Report of the Intergovernmental Panel on Climate Change* [Parry, M.L., O.F. Canziani, J.P. Palutikof, P.J. Van Der Linde, and C.E. Hanson (eds.)]. Cambridge University Press, Cambridge, UK, pp. 811-841.

Young, H. and S. Jaspars, 1995: *Nutrition Matters. People Food and Famine.* Intermediate Technology Publications, London, UK, 151 pp.

Zahran, S., L. Peek, and S.D. Brody, 2008: Youth mortality by forces of nature. *Children, Youth and Environments*, **18(1)**, 371-388.

Zimmerman, F.J. and M.R. Carter, 2003: Asset smoothing, consumption smoothing and the reproduction of inequality under risk and subsistence constraints. *Journal of Development Economics*, **71(2)**, 233-260.

9. Case Studies

Coordinating Lead Authors:
Virginia Murray (UK), Gordon McBean (Canada), Mihir Bhatt (India)

Lead Authors:
Sergey Borsch (Russian Federation), Tae Sung Cheong (Republic of Korea), Wadid Fawzy Erian (Egypt), Silvia Llosa (Peru), Farrokh Nadim (Norway), Mario Nunez (Argentina), Ravsal Oyun (Mongolia), Avelino G. Suarez (Cuba)

Review Editors:
John Hay (New Zealand), Mai Trong Nhuan (Vietnam), Jose Moreno (Spain)

Contributing Authors:
Peter Berry (Canada), Harriet Caldin (UK), Diarmid Campbell-Lendrum (UK / WHO), Catriona Carmichael (UK), Anita Cooper (UK), Cherif Diop (Senegal), Justin Ginnetti (USA), Delphine Grynzspan (France), Clare Heaviside (UK), Jeremy Hess (USA), James Kossin (USA), Paul Kovacs (Canada), Sari Kovats (UK), Irene Kreis (Netherlands), Reza Lahidji (France), Joanne Linnerooth-Bayer (USA), Felipe Lucio (Mozambique), Simon Mason (USA), Sabrina McCormick (USA), Reinhard Mechler (Germany), Bettina Menne (Germany / WHO), Soojeong Myeong (Republic of Korea), Arona Ngari (Cook Islands), Neville Nicholls (Australia), Ursula Oswald Spring (Mexico), Pascal Peduzzi (Switzerland), Rosa Perez (Philippines), Caroline Rodgers (Canada), Hannah Rowlatt (UK), Sohel Saikat (UK), Sonia Seneviratne (Switzerland), Addis Taye (UK), Richard Thornton (Australia), Sotiris Vardoulakis (UK), Koko Warner (Germany), Irina Zodrow (Switzerland / UNISDR)

This chapter should be cited as:
Murray, V., G. McBean, M. Bhatt, S. Borsch, T.S. Cheong, W.F. Erian, S. Llosa, F. Nadim, M. Nunez, R. Oyun, and A.G. Suarez, 2012: Case studies. In: *Managing the Risks of Extreme Events and Disasters to Advance Climate Change Adaptation* [Field, C.B., V. Barros, T.F. Stocker, D. Qin, D.J. Dokken, K.L. Ebi, M.D. Mastrandrea, K.J. Mach, G.-K. Plattner, S.K. Allen, M. Tignor, and P.M. Midgley (eds.)]. A Special Report of Working Groups I and II of the Intergovernmental Panel on Climate Change (IPCC). Cambridge University Press, Cambridge, UK, and New York, NY, USA, pp. 487-542.

Table of Contents

Executive Summary 489

9.1.   Introduction 490

9.2.   Case Studies 492
9.2.1.   European Heat Waves of 2003 and 2006 492
9.2.2.   Response to Disaster Induced by Hot Weather and Wildfires 496
9.2.3.   Managing the Adverse Consequences of Drought 498
9.2.4.   Recent *Dzud* Disasters in Mongolia 500
9.2.5.   Cyclones: Enabling Policies and Responsive Institutions for Community Action 502
9.2.6.   Managing the Adverse Consequences of Floods 505
9.2.7.   Disastrous Epidemic Disease: The Case of Cholera 507
9.2.8.   Coastal Megacities: The Case of Mumbai 510
9.2.9.   Small Island Developing States: The Challenge of Adaptation 512
9.2.10. Changing Cold Climate Vulnerabilities: Northern Canada 514
9.2.11. Early Warning Systems: Adapting to Reduce Impacts 517
9.2.12. Effective Legislation for Multilevel Governance of Disaster Risk Reduction and Adaptation 519
9.2.13. Risk Transfer: The Role of Insurance and Other Instruments
         in Disaster Risk Management and Climate Change Adaptation in Developing Countries 522
9.2.14. Education, Training, and Public Awareness Initiatives for Disaster Risk Reduction and Adaptation 526

9.3.   Synthesis of Lessons Identified from Case Studies 529

References 530

Executive Summary

Case studies contribute more focused analyses which, in the context of human loss and damage, demonstrate the effectiveness of response strategies and prevention measures and identify lessons about success in disaster risk reduction and climate change adaptation. The case studies were chosen to complement and be consistent with the information in the preceding chapters, and to demonstrate aspects of the key messages in the Summary for Policymakers and the Hyogo Framework for Action Priorities.

The case studies were grouped to examine types of extreme events, vulnerable regions, and methodological approaches. For the extreme event examples, the first two case studies pertain to events of extreme temperature with moisture deficiencies in Europe and Australia and their impacts including on health. These are followed by case studies on drought in Syria and *dzud*, cold-dry conditions in Mongolia. Tropical cyclones in Bangladesh, Myanmar, and Mesoamerica, and then floods in Mozambique are discussed in the context of community actions. The last of the extreme events case studies is about disastrous epidemic disease, using the case of cholera in Zimbabwe, as the example.

The case studies chosen to reflect vulnerable regions demonstrate how a changing climate provides significant concerns for people, societies, and their infrastructure. These are: Mumbai as an example of a coastal megacity; the Republic of the Marshall Islands, as an example of small island developing states with special challenges for adaptation; and Canada's northern regions as an example of cold climate vulnerabilities focusing on infrastructures.

Four types of methodologies or approaches to disaster risk reduction (DRR) and climate change adaptation (CCA) are presented. Early warning systems; effective legislation; risk transfer in developing countries; and education, training, and public awareness initiatives are the approaches demonstrated. The case studies demonstrate that current disaster risk management (DRM) and CCA policies and measures have not been sufficient to avoid, fully prepare for, and respond to extreme weather and climate events, but these examples demonstrate progress.

A common factor was the need for greater information on risks before the events occur, that is, early warnings. The implementation of early warning systems does reduce loss of lives and, to a lesser extent, damage to property and was identified by all the extreme event case studies (heat waves, wildfires, drought, *dzud*, cyclones, floods, and epidemic disease) as key to reducing impacts from extreme events. A need for improving international cooperation and investments in forecasting was recognized in some of the case studies but equally the need for regional and local early warning systems was heavily emphasized, particularly in developing countries.

A further common factor identified overall was that it is better to invest in preventative-based DRR plans, strategies, and tools for adaptation than in response to extreme events. Greater investments in proactive hazard and vulnerability reduction measures, as well as development of capacities to respond and recover from the events were demonstrated to have benefits. Specific examples for planning for extreme events included increased emphasis on drought preparedness; planning for urban heat waves; and tropical cyclone DRM strategies and plans in coastal regions that anticipate these events. However, as illustrated by the small island developing states case study, it was also identified that DRR planning approaches continue to receive less emphasis than disaster relief and recovery.

One recurring theme is the value of investments in knowledge and information, including observational and monitoring systems, for cyclones, floods, droughts, heat waves, and other events from early warnings to clearer understanding of health and livelihood impacts. In all cases, the point is made that with greater information available it would be possible to know the risks better and ensure that response strategies were adequate to face the coming threat. Research improves our knowledge, especially when it integrates the natural, social, health, and engineering sciences and their applications. The case studies have reviewed past events and identified lessons which could be considered for the future. Preparedness through DDR and DRM can help to adapt to climate change and these case studies offer examples of measures that could be taken to reduce the damage that is inflicted as a result of extreme events. Investment in increasing knowledge and warning systems, adaptation techniques, and tools and preventive measures will cost money now but will save money and lives in the future.

9.1. Introduction

In this chapter, case studies are used as examples of how to gain a better understanding of the risks posed by extreme weather and climate-related events while identifying lessons and best practices from past responses to such occurrences. Using the information in Chapters 1 to 8, it was possible to focus on particular examples to reflect the needs of the whole Special Report. The chosen case studies are illustrative of an important range of disaster risk reduction, disaster risk management, and climate change adaptation issues. They are grouped to examine representative types of extreme events, vulnerable regions, and methodological approaches.

For the extreme event examples, the first two case studies pertain to extreme temperature with moisture deficiencies: the European heat waves of 2003 and 2006 and response to disaster induced by hot weather and wildfires in Australia. Managing the adverse consequences of drought is the third case study, with the focus on Syrian droughts. The combination of drought and cold is examined through two recent *dzud* disasters in Mongolia, 1999-2002 and 2009-2010. Tropical cyclones in Bangladesh, Myanmar, and Mesoamerica are used as examples of how a difference can be made via enabling policies and responsive institutions for community action. The next case study shifts the geographical focus to floods in Mozambique in 2000 and 2007. The last of the extreme events case studies is about disastrous epidemic disease, using the case of cholera in Zimbabwe, as the example.

The case studies chosen to reflect a few vulnerable regions all demonstrate how a changing climate provides significant concerns for people, societies, and their infrastructure. The case of Mumbai is used as an example of a coastal megacity and its risks. Small island developing states have special challenges for adaptation, with the Republic of the Marshall Islands being the case study focus. Cold climate vulnerabilities, particularly the infrastructure in Canada's northern regions, provide the final vulnerable region case study.

Following examples of extreme events and vulnerable regions, this chapter presents case study examination of four types of methodologies or approaches to DRR and CCA. Early warning systems provide the opportunity for adaptive responses to reduce impacts. Effective legislation to provide multi-level governance is another way of reducing impacts. The case study on risk transfer examines the role of insurance and other instruments in developing countries. The final case study is on education, training, and public awareness initiatives. This selection provides a good basis of information and serves as an indicator of the resources needed for future DRR and CCA. Additionally, it allows good practices to be identified and lessons to be extracted.

The case studies provide the opportunity for connecting with common elements across the other chapters. Each case study is presented in a consistent way to enable better comparison of approaches. After an introduction, authors provide background to the event, vulnerable region, or methodology. Then the description of the events, vulnerability, or strategy is given as appropriate. Next is the discussion of interventions, followed by the outcomes and/or consequences. Each case study concludes with a discussion of lessons identified. These case studies relate to the key messages of the Summary for Policymakers and also to the Hyogo Framework for Action Priorities (see Table 9-1).

Case studies are widely used in many disciplines including health care (Keen and Packwood, 1995; McWhinney, 2001), social science (Flyvbjerg, 2004), engineering, and education (Verschuren, 2003). In addition, case studies have been found to be useful in previous IPCC Assessment Reports, including the 2007 Working Group II report (Parry et al., 2007). Case studies offer records of innovative or good practices. Specific problems or issues experienced can be documented as well as the actions taken to overcome these. Case studies can validate our understanding and encourage re-evaluation and learning. It is apparent that (i) case studies capture the complexity of disaster risk and disaster situations; (ii) case studies appeal to a broad audience; and (iii) case studies should be fully utilized to provide lessons identified for DRR and DRM for adaptation to climate change (Grynszpan et al., 2011). Several projects have identified lessons from case studies (Kulling et al., 2010). The Forensic Investigation of Disaster (Burton, 2011; FORIN, 2011) Project of the Integrated Research on Disaster Risk (ICSU, 2008) program has developed a methodology and template for future case study investigations to provide a basis for future policy analysis and literature for assessments. The FORIN template lays out the elements: (a) critical cause analysis; (b) meta-analysis; (c) longitudinal analysis; and (d) scenarios of disasters.

Case studies included in this chapter have been extracted from a variety of literature sources from many disciplines. As a result, an integrated approach examining scientific, social, health, and economic aspects of disasters was used where appropriate and included different spatial and temporal scales, as needed. The specialized insights they provide can be useful in evaluating some current disaster response practices.

This chapter addresses events whose impacts were felt in many dimensions. A single event can produce effects that are felt on local, regional, national, and international levels. These effects could have been the direct result of the event itself, from the response to the event, or through indirect impacts such as a reduction in food production or a decrease in available resources. In addition to the spatial scales, this chapter also addresses temporal scales, which vary widely in both event-related impacts and responses. However, the way effects are felt is additionally influenced by social, health, and economic factors. The resilience of a society and its economic capacity to allay the impact of a disaster and cope with the after effects has significant ramifications for the community concerned (UNISDR, 2008a). Developing countries with fewer resources, experts, equipment, and infrastructure have been shown to be particularly at risk (see Chapter 5). Developed nations are usually better equipped with technical, financial, and institutional support to enable better adaptive planning including preventive measures and/or quick and effective responses (Gagnon-Lebrun and Agrawala, 2006). However, they still remain at risk of high-impact events, as exemplified by the European heat wave of 2003 and by Hurricane Katrina (Parry et al., 2007).

Table 9-1 | Matrix demonstrating the connectivity between the case studies (9.2.1-9.2.14) and the Summary for Policymakers messages. Those with the strongest relationship are shown. Connectivity between the case studies and the Hyogo Framework for Action Priority Areas (UNISDR, 2005b) are also shown.

	Key Message	9.2.1 Heat-waves	9.2.2 Hot weather and wildfires	9.2.3 Drought	9.2.4 Dzud	9.2.5 Cyclones	9.2.6 Floods	9.2.7 Epidemic Disease	9.2.8 Mega-cities	9.2.9 SIDS	9.2.10 Cold Climate	9.2.11 EWS	9.2.12 Legislation	9.2.13 Risk Transfer	9.2.14 Education
A. Context	Exposure and vulnerability are key determinants of disaster risk.	●		●		●	●								
	A changing climate leads to changes in the frequency, intensity, spatial extent, duration, and timing of extreme weather and climate events, and can result in unprecedented extreme weather and climate events.	●	●		●										
B. Observations of Exposure, Vulnerability, Climate Extremes, Impacts, and Disaster Losses	Exposure and vulnerability are dynamic, varying across temporal and spatial scales, and depend on economic, social, geographic, demographic, cultural, institutional, governance, and environmental factors.								●				●		
	Settlement patterns, urbanization, and changes in socioeconomic status have all influenced observed trends in exposure and vulnerability to climate extremes.					●			●	●					
C. Disaster Risk Management and Adaptation to Climate Change: Past Experience with Climate Extremes	Trends in exposure and vulnerability are major drivers of changes in disaster risk.	●		●				●							
	Inequalities influence local coping and adaptive capacity, and pose disaster risk management and adaptation challenges from the local to national levels.		●			●			●					●	
	Humanitarian relief is often required when disaster risk reduction measures are absent or inadequate.	●		●	●		●	●							
	Post-disaster recovery and reconstruction provide an opportunity for reducing weather and climate-related disaster risk and for improving adaptive capacity.					●	●						●		●
	Risk sharing and transfer mechanisms at local, national, regional, and global scales can increase resilience to climate extremes.													●	
	Attention to the temporal and spatial dynamics of exposure and vulnerability is particularly important given that the design and implementation of adaptation and disaster risk management strategies and policies can reduce risk in the short term, but may increase exposure and vulnerability over the longer term.	●								●					
	Closer integration of disaster risk management and climate change adaptation, along with the incorporation of both into local, subnational, national, and international development policies and practices, could provide benefits at all scales.								●	●		●	●		
D. Future Climate Extremes, Impacts, and Disaster Losses	Models project substantial warming in temperature extremes by the end of the 21st century.	●	●								●				
	It is *likely* that the frequency of heavy precipitation or the proportion of total rainfall from heavy falls will increase in the 21st century over many areas of the globe.						●								
	There is *medium confidence* that droughts will intensify in the 21st century in some seasons and areas, due to reduced precipitation and/or increased evapotranspiration.			●											

Continued next page →

Table 9-1 (continued)

		9.2.1 Heat-waves	9.2.2 Hot weather and wildfires	9.2.3 Drought	9.2.4 Dzud	9.2.5 Cyclones	9.2.6 Floods	9.2.7 Epidemic Disease	9.2.8 Mega-cities	9.2.9 SIDS	9.2.10 Cold Climate	9.2.11 EWS	9.2.12 Legislation	9.2.13 Risk Transfer	9.2.14 Education
E. Managing Changing Risk of Climate Extremes and Disasters	Measures that provide benefits under current climate and a range of future climate change scenarios, called low-regrets measures, are available starting points for addressing projected trends in exposure, vulnerability, and climate extremes. They have the potential to offer benefits now and lay the foundation for addressing projected changes.													•	
	Effective risk management generally involves a portfolio of actions to reduce and transfer risk and to respond to events and disasters, as opposed to a singular focus on any one action or type of action.					•					•		•		
	Multi-hazard risk management approaches provide opportunities to reduce complex and compound hazards.		•												
	Integration of local knowledge with additional scientific and technical knowledge can improve disaster risk reduction and climate change adaptation.			•	•	•	•				•			•	•
	Appropriate and timely risk communication is critical for effective adaptation and disaster risk management.		•					•	•	•		•			
Hyogo Framework for Action – Priorities for Action	1: Ensure that disaster risk reduction is a national and a local priority with a strong institutional basis for implementation.	•										•	•		•
	2: Identify, assess and monitor disaster risks and enhance early warning.		•								•	•		•	•
	3: Use knowledge, innovation and education to build a culture of safety and resilience at all levels.												•	•	•
	4: Reduce the underlying risk factors.			•										•	
	5: Strengthen disaster preparedness for effective response at all levels.				•	•			•	•		•	•		

Most importantly, this chapter highlights the complexities of disasters in order to encourage effective solutions that address these complexities rather than just one issue or another. The lessons of this chapter provide examples of experience that can help develop strategies to adapt to climate change.

9.2. Case Studies

9.2.1. European Heat Waves of 2003 and 2006

9.2.1.1. Introduction

Extreme heat is a prevalent public health concern throughout the temperate regions of the world and extreme heat events have been encountered recently in North America, Asia, Africa, Australia, and Europe. It is *very likely* that the length, frequency, and/or intensity of warm spells, including heat waves, will continue to increase over most land areas (Section 3.3.1). As with other types of hazards, extreme heat can have disastrous consequences, particularly for the most vulnerable populations. Risk from extreme heat is a function of hazard severity and population exposure and vulnerability. Extreme heat events do not necessarily translate into extreme impacts if vulnerability is low. It is important, therefore, to consider factors that contribute to hazard exposure and population vulnerability. Recent literature has identified a host of factors that can amplify or dampen hazard exposure. Experience with past heat waves and public health interventions suggest that it is possible to manipulate many of these variables to reduce both exposure and vulnerability and thereby limit the impacts of extreme heat events. This case study, which compares the European heat wave of 2003 with 2006, demonstrates developments in disaster risk management and adaptation to climate change.

9.2.1.2. Background/Context

Extreme heat is a prevalent public health concern throughout the temperate regions of the world (Kovats and Hajat, 2008), in part because heat-related extreme events are projected to result in increased mortality (Peng et al., 2010). Extreme heat events have been encountered recently in North America (Hawkins-Bell and Rankin, 1994; Klinenberg, 2002), Asia (Kumar, 1998; Kalsi and Pareek, 2001; Srivastava et al., 2007), Africa (NASA, 2008), Australia (DSE, 2008b), and Europe (Robine et al., 2008; Founda and Giannakopoulos, 2009). This concern may also be present in non-temperate regions, but there is little research on this.

As with other types of hazards, extreme heat events can have disastrous consequences, partly due to increases in exposure and particular types of vulnerabilities. However, it is important to note that reducing the impacts of extreme heat events linked to climate change will necessitate further actions, some of which may be resource intensive and further exacerbate climate change.

9.2.1.2.1. Vulnerabilities to heat waves

Physiological: Several factors influence vulnerability to heat-related illness and death. Most of the research related to such vulnerability is derived from experiences in industrialized nations. Several physiological factors, such as age, gender, body mass index, and preexisting health conditions, play a role in the body's ability to respond to heat stress. Older persons, babies, and young children have a number of physiological and social risk factors that place them at elevated risk, such as decreased ability to thermoregulate (the ability to maintain temperature within the narrow optimal physiologic range; Havenith, 2001). Preexisting chronic disease – more common in the elderly – also impairs compensatory responses to sustained high temperatures (Havenith, 2001; Shimoda, 2003). Older adults tend to have suppressed thirst impulse resulting in dehydration and increased risk of heat-related illness. In addition, multiple diseases and/or drug treatments increase the risk of dehydration (Hodgkinson et al., 2003; Ebi and Meehl, 2007).

Social: A wide range of socioeconomic factors are associated with increased vulnerability (see Sections 2.3 and 2.5). Areas with high crime rates, low social capital, and socially isolated individuals had increased vulnerability during the Chicago heat wave in 1995 (Klinenberg, 2002). People in areas of low socioeconomic status are generally at higher risk of heat-related morbidity and mortality due to higher prevalence of chronic diseases – from cardiovascular diseases such as hypertension to pulmonary disease, such as chronic obstructive pulmonary disease and asthma (Smoyer et al., 2000; Sheridan, 2003). Minorities and communities of low socioeconomic status are also frequently situated in higher heat stress neighborhoods (Harlan et al., 2006). Protective measures are often less available for those of lower socioeconomic status, and even if air conditioning, for example, is available, some of the most vulnerable populations will choose not to use it out of concern over the cost (O'Neill et al., 2009). Other groups, like the homeless and outdoor workers, are particularly vulnerable because of their living situation and being more acutely exposed to heat hazards (Yip et al., 2008). Older persons may also often be isolated and living alone, and this may increase vulnerability (Naughton et al., 2002; Semenza, 2005).

9.2.1.2.2. Impact of urban infrastructure

Addressing vulnerabilities in urban areas will benefit those at risk. Around half of the world's population live in urban areas at present, and by 2050, this figure is expected to rise to about 70% (UN, 2008). Cities across the world are expected to absorb most of the population growth over the next four decades, as well as continuing to attract migrants from rural areas (UN, 2008). In the context of a heat-related extreme event, certain infrastructural factors can either amplify or reduce vulnerability of exposed populations. The built environment is important since local heat production affects the urban thermal budget (from internal combustion engines, air conditioners, and other activities). Other factors also play a role in determining local temperatures, including surface reflectivity or albedo, the percent of vegetative cover, and thermal conductivity of building materials. The urban heat island effect, caused by increased absorption of infrared radiation by buildings and pavement, lack of shading, evapotranspiration by vegetation, and increased local heat production, can significantly increase temperatures in the urban core by several degrees Celsius, raising the likelihood of hazardous heat exposure for urban residents (Clarke, 1972; Shimoda, 2003). Street canyons where building surfaces absorb heat and affect air flow are also areas where heat hazards may be more severe (Santamouris et al., 1999; Louka et al., 2002). The restricted air flow within street canyons may also cause accumulation of traffic-related air pollutants (Vardoulakis et al., 2003).

Research has also identified that, at least in the North American and European cities where the phenomenon has been studied, these factors can have a significant impact on the magnitude of heat hazards on a neighborhood level (Harlan et al., 2006). One study in France has shown that higher mortality rates occurred in neighborhoods in Paris that were characterized by higher outdoor temperatures (Cadot et al., 2007). High temperatures can also affect transport networks when heat damages roads and rail tracks. Within cities, outdoor temperatures can vary significantly (Akbari and Konopacki, 2004), resulting in the need to focus preventive strategies on localized characteristics.

Systems of power generation and transmission partly explain vulnerability since electricity supply underpins air conditioning and refrigeration – a significant adaptation strategy particularly in developed countries, but one that is also at increased risk of failure during a heat wave (Sailor and Pavlova, 2003). It is expected that demand for electricity to power air conditioning and refrigeration units will increase with rising ambient temperatures. Areas with lower power capacities face increased risk of disruptions to generating resources and transmission under excessive heat events.

In addition to increased demand, there can be a risk of reduced output from power generating plants (UNEP, 2004). The ability of inland thermal power plants, both conventional and nuclear, to cool their generators is restricted by rising river temperatures. Additionally, fluctuating levels of water availability will affect energy outputs of hydropower complexes. During the summer of 2003 in France, six power plants were shut down and others had to control their output (Parry et al., 2007).

9.2.1.2.3. Heat waves and air pollution

Concentrations of air pollutants such as particulate matter and ozone are often elevated during heat waves due to anticyclonic weather conditions, increased temperatures, and light winds. Photochemical production of ozone and emissions of biogenic ozone precursors increase during hot, sunny weather, and light winds do little to disperse the build-up of air pollution. Air pollution has well-established acute effects on health, particularly associated with respiratory and cardiovascular illness, and can result in increased mortality and morbidity (WHO, 2006a). Background ozone levels in the Northern Hemisphere have doubled since preindustrial times (Volz and Kley, 1988) and increased in many urban areas over the last few decades (Vingarzan, 2004). Air quality standards and regulations are helping to improve air quality, although particles and ozone are still present in many areas at levels that may cause harm to human health, particularly during heat waves (Royal Society, 2008; EEA, 2011). The effects of climate change (particularly temperature increases) together with a steady increase in background hemispheric ozone levels is reducing the efficacy of measures to control ozone precursor emissions in the future (Derwent et al., 2006). The increased frequency of heat waves in the future will probably lead to more frequent air pollution episodes (Stott et al., 2004; Jones et al., 2008).

9.2.1.3. Description of Events

9.2.1.3.1. European heat wave of 2003

During the first two weeks of August 2003, temperatures in Europe soared far above historical norms. The heat wave stretched across much of Western Europe, but France was particularly affected (InVS, 2003). Maximum temperatures recorded in Paris remained mostly in the range of 35 to 40°C between 4 and 12 August, while minimum temperatures recorded by the same weather station remained almost continuously above 23°C between 7 and 14 August (Météo France, 2003). The European heat wave had significant health impacts (Lagadec, 2004). Initial estimates were of costs exceeding €13 billion, with a death toll of over 30,000 across Europe (UNEP, 2004). It has been estimated that mortality over the entire summer could have reached about 70,000 (Robine et al., 2008) with approximately 14,800 excess deaths in France alone (Pirard et al., 2005). The severity, duration, geographic scope, and impact of the event were unprecedented in recorded European history (Grynszpan, 2003; Kosatsky, 2005; Fouillet et al., 2006) and put the event in the exceptional company of the deadly Beijing heat wave of 1743, which killed at least 11,000, and possibly many more (Levick, 1859; Bouchama, 2004; Lagadec, 2004; Pirard et al., 2005; Robine et al., 2008).

During the heat wave of August 2003, air pollution levels were high across much of Europe, especially surface ozone (EEA, 2003). A rapid assessment was performed for the United Kingdom after the heat wave, using published exposure-response coefficients for ozone and PM_{10} (particulate matter with an aerodynamic diameter of up to 10 μm). The assessment associated 21 to 38% of the total 2,045 excess deaths in the United Kingdom in August 2003 to elevated ambient ozone and PM_{10} concentrations (Stedman, 2004). The task of separating health effects of heat and air pollution is complex; however, statistical and epidemiological studies in France also concluded that air pollution was a factor associated with detrimental health effects during August 2003 (Dear et al., 2005; Filleul et al., 2006).

9.2.1.3.2. European heat wave of 2006

Three years later, between 10 and 28 July 2006, Europe experienced another major heat wave. In France, it ranked second only to the one in 2003 as the most severe heat wave since 1950 (Météo France, 2006; Fouillet et al., 2008). The 2006 heat wave was longer in duration than that of 2003, but was less intense and covered less geographical area (Météo France, 2006). Ozone levels were high across much of southern and northwestern Europe in July 2006, with concentrations reaching levels only exceeded in 2003 to date (EEA, 2007). Across France, recorded maximum temperatures soared to 39 to 40°C, while minimum recorded temperatures reached 19 to 23°C (compared with 23 to 25°C in 2003) (Météo France, 2006). Based on a historical model, the temperatures were expected to cause around 6,452 excess deaths in France alone, yet only around 2,065 excess deaths were recorded (Fouillet et al., 2008).

9.2.1.4. Interventions

Efforts to minimize the public health impact for the heat wave in 2003 were hampered by denial of the event's seriousness and the inability of many institutions to instigate emergency-level responses (Lagadec, 2004). Afterwards several European countries quickly initiated plans to prepare for future events (WHO, 2006b). France, the country hit hardest, developed a national heat wave plan, surveillance activities, clinical treatment guidelines for heat-related illness, identification of vulnerable populations, infrastructure improvements, and home visiting plans for future heat waves (Laaidi et al., 2004).

9.2.1.5. Outcomes/Consequences

The difference in impact between the heat waves in 2003 and 2006 may be at least partly attributed to the difference in the intensity and geographic scope of the hazard. It has been considered that in France at least, some decrease in 2006 mortality may also be attributed to increased awareness of the ill effects of a heat wave, the preventive measures instituted after the 2003 heat wave, and the heat health watch system set up in 2004 (Fouillet et al., 2008). While the mortality reduction may demonstrate the efficacy of public health measures, the persistent excess mortality highlights the need for optimizing existing public health measures such as warning and watch systems (Hajat et al., 2010), health communication with vulnerable populations (McCormick, 2010a), vulnerability mapping (Reid et al., 2009), and heat wave response plans (Bernard and McGeehin, 2004). It also highlights the need for other, novel measures such as modification of the urban form to reduce exposure (Bernard and McGeehin, 2004; O'Neill et al., 2009; Reid et al., 2009; Hajat et al., 2010; Silva et al., 2010). Thus the outcomes from the two European heat waves of 2003 and 2006 are extensive and are considered below. They include public health approaches to reducing exposure, assessing heat mortality, communication and education, and adapting the urban infrastructure.

9.2.1.5.1. Public health approaches to reducing exposure

A common public health approach to reducing exposure is the Heat Warning System (HWS) or Heat Action Response System. The four components of the latter include an alert protocol, community response plan, communication plan, and evaluation plan (Health Canada, 2010). The HWS is represented by the multiple dimensions of the EuroHeat plan, such as a lead agency to coordinate the alert, an alert system, an information outreach plan, long-term infrastructural planning, and preparedness actions for the health care system (WHO, 2007). The European Network of Meteorological Services has created Meteoalarm as a way to coordinate warnings and to differentiate them across regions (Bartzokas et al., 2010). There are a range of approaches used to trigger alerts and a range of response measures implemented once an alert has been triggered. In some cases, departments of emergency management lead the endeavor, while in others public health-related agencies are most responsible (McCormick, 2010b).

As yet, there is not much evidence on the efficacy of heat warning systems. A few studies have identified an effect of heat preparedness programming. For example, the use of emergency medical services during heat wave events dropped by 49% in Milwaukee, Wisconsin, between 1995 and 1999; an outcome that may be partly due to heat preparedness programming or to differences between the two heat waves (Weisskopf et al., 2002). Evidence has also indicated that interventions in Philadelphia, Pennsylvania, are *likely* to have reduced mortality rates by 2.6 lives per day during heat events (Ebi et al., 2004). An Italian intervention program found that caretaking in the home resulted in decreased hospitalizations due to heat (Marinacci et al., 2009). However, for all these studies, it is not clear whether the observed reductions were due to the interventions. Questions remain about the levels of effectiveness in many circumstances (Cadot et al., 2007).

Heat preparedness plans vary around the world. Philadelphia, Pennsylvania – one of the first US cities to begin a heat preparedness plan – has a ten-part program that integrates a 'block captain' system where local leaders are asked to notify community members of dangerous heat (Sheridan, 2006; McCormick, 2010b). Programs like the Philadelphia program that utilize social networks have the capacity to shape behavior since networks can facilitate the sharing of expertise and resources across stakeholders; however, in some cases the influence of social networks contributes to vulnerability (Crabbé and Robin, 2006). Other heat warning systems, such as that in Melbourne, Australia, are based solely on alerting the public to weather conditions that threaten older populations (Nicholls et al., 2008). Addressing social factors in preparedness promises to be critical for the protection of vulnerable populations. This includes incorporating communities themselves into understanding and responding to extreme events. It is important that top-down measures imposed by health practitioners account for community-level needs and experiences in order to be more successful. Greater attention to and support of community-based measures in preventing heat mortality can be more specific to local context, such that participation is broader (Semenza et al., 2007). Such programs can best address the social determinants of health outcomes.

9.2.1.5.2. Assessing heat mortality

Assessing excess mortality is the most widely used means of assessing the health impact of heat-related extreme events. Mortality represents only the 'tip of the iceberg' of heat-related health effects; however, it is more widely and accurately reported than morbidity, which explains its appeal as a data source. Nonetheless, assessing heat mortality presents particular challenges. Accurately assessing heat-related mortality faces challenges of differences in contextual variations (Hémon and Jougla, 2004; Poumadere et al., 2005), and coroner's categorization of deaths (Nixdorf-Miller et al., 2006). For example, there are a number of estimates of mortality for the European heat wave that vary depending on geographic and temporal ranges, methodological approaches, and risks considered (Assemblée Nationale, 2004). The different types of analyses used to assess heat mortality, such as certified heat deaths and heat-related mortality measured as an excess of total mortality over a given time period, are important distinctions in assessing who is affected by the heat (Kovats and Hajat, 2008). Learning from past and other countries' experience, a common understanding of definitions of heat waves and excess mortality, and the ability to streamline death certification in the context of an extreme event could improve the ease and quality of mortality reporting.

9.2.1.5.3. Communication and education

One particularly difficult aspect of heat preparedness is communicating risk. In many locations populations are unaware of their risk and heat wave warning systems go largely unheeded (Luber and McGeehin, 2008). Some evidence has even shown that top-down educational messages do not result in appropriate resultant actions (Semenza et al., 2008). The receipt of information is not sufficient to generate new behaviors or the development of new social norms. Even when information is distributed through pamphlets and media outlets, behavior of at-risk populations often does not change and those targeted by such interventions have suggested that community-based organizations be involved in order to build on existing capacity and provide assistance (Abrahamson et al., 2008). Older people, in particular, engage better with prevention campaigns that allow them to maintain independence and do not focus on their age, as many heat warning programs do (Hughes et al., 2008). More generally, research shows that communication about heat preparedness centered on engaging with communities results in increased awareness compared with top-down messages (Smoyer-Tomic and Rainham, 2001).

9.2.1.5.4. Adapting the urban infrastructure

Several types of infrastructural measures can be taken to prevent negative outcomes of heat-related extreme events. Models suggest that significant reductions in heat-related illness would result from land use modifications that increase albedo, proportion of vegetative cover, thermal conductivity, and emissivity in urban areas (Yip et al., 2008; Silva et al., 2010). Reducing energy consumption in buildings can improve resilience, since localized systems are less dependent on vulnerable energy infrastructure. In addition, by better insulating residential dwellings, people would suffer less effect from heat hazards. Financial incentives have been tested in some countries as a means to increase energy efficiency by supporting those who are insulating their homes. Urban greening can also reduce temperatures, protecting local populations and reducing energy demands (Akbari et al., 2001).

9.2.1.6. Lessons Identified

With climate change, heat waves are *very likely* to increase in frequency and severity in many parts of the world (Section 3.3.1). Smarter urban planning, improvements in existing housing stock and critical infrastructure, along with effective public health measures will assist in facilitating climate change adaptation.

Through understanding local conditions and experiences and current and projected risks, it will be possible to develop strategies for improving heat preparedness in the context of climate change. The specificity of heat risks to particular sub-populations can facilitate appropriate interventions and preparedness.

Communication and education strategies are most effective when they are community-based, offer the opportunity for changing social norms, and facilitate the building of community capacity.

Infrastructural considerations are critical to reducing urban vulnerability to extreme heat events. Effective preparedness includes building techniques that reduce energy consumption and the expansion of green space.

Heat wave preparedness programs may be able to prevent heat mortality; however, testing and development is required to assess the most effective approaches.

Further research is needed on the efficacy of existing plans, how to improve preparedness that specifically focuses on vulnerable groups, and how to best communicate heat risks across diverse groups. There are also methodological difficulties in describing individual vulnerability that need further exploration.

9.2.2. Response to Disaster Induced by Hot Weather and Wildfires

9.2.2.1. Introduction

Climate change is expected to increase global temperatures and change rainfall patterns (Christensen et al., 2007). These climatic changes will increase the risk of temperature- and precipitation-related extreme weather and climate events. The relative effects will vary by regions and localities (Sections 3.3.1, 3.3.2, and 3.5.1). In general, an increase in mean temperature, and a decrease in mean precipitation can contribute to increased fire risk (Flannigan et al., 2009). When in combination with severe droughts and heat waves, which are also expected to increase in many fire regions (Sections 3.3.1 and 3.5.1), fires can become catastrophic (Bradstock et al., 2009). Wildfires occur in many regions of the world, and due to their extreme nature, authorities and the public in general are acquainted with such extreme situations, and plans have been enacted to mitigate them. However, at times, the nature of fire challenges these plans and disasters emerge. This case study uses the example from Victoria, Australia, in 2009. The goal is to present hot weather and wildland fire hazards and their effects and potential impacts, and to provide an overview of experience to learn to manage these extreme risks, as well as key lessons for the future.

9.2.2.2. Background

Wildfire risk occurs in many regions of the globe; however embodying this risk in a single and practical universal index is difficult. The relationships between weather and wildfires have been studied for many areas of the world; in some, weather is the dominant factor in ignitions, while in others, human activities are the major cause of ignition, but weather and environmental factors mainly determine the area burned (Bradstock et al., 2009). Wildfire behavior is also modified by forest and land management and fire suppression (Allen et al., 2002; Noss et al., 2006). Wildfires do not burn at random in the landscape (Nunes et al., 2005), and occur at particular topographic locations or distances from towns or roads (Mouillot et al., 2003; Badia-Perpinyà and Pallares-Barbera, 2006; Syphard et al., 2009). The intensity and rate of spread of a wildfire is dependent on the amount, moisture content, and arrangement of fine dead fuel, the wind speed near the burning zone, and the terrain and slope where it is burning. Wildfire risk is a combination of all factors that affect the inception, spread, and difficulty of fire control and damage potential (Tolhurst, 2010).

9.2.2.3. Description of Events

An episode of extreme heat waves began in South Australia on 25 January 2009. Two days later they had become more widespread over southeast Australia. The exceptional heat wave was caused by a slow-moving high-pressure system that settled over the Tasman Sea, in combination with an intense tropical low located off the northwest

Australian coast and a monsoon trough over northern Australia. This produced ideal conditions for hot tropical air to be directed over southeastern Australia (National Climate Centre, 2009).

In Melbourne the temperature was above 43°C for three consecutive days (28 to 30 January 2009), reaching a peak of 45.1°C on 30 January 2009. This was the second-highest temperature on record. The extremely high day and night temperatures combined to produce a record high daily mean temperature of 35.4°C on 30 January (Victorian Government, 2009). The 2008 winter season was characterized by below-average precipitation across much of Victoria. While November and December 2008 experienced average and above average rainfall, respectively, in January and February the rainfall was substantially below average (Parliament of Victoria, 2010a). During the 12 years between 1998 and 2007, Victoria experienced warmer than average temperatures and a 14% decline in average rainfall (DSE, 2008a). In central Victoria the 12-year rainfall totals were approximately 10 to 20% below the 1961 to 1990 average (Australian Government, 2009).

This heat wave had a substantial impact on the health of Victorians, particularly the elderly (National Climate Centre, 2009; Parliament of Victoria, 2009). A 25% increase in total emergency cases and a 46% increase over the three hottest days were reported for the week of the heat wave. Emergency departments reported a 12% overall increase in presentations, with a greater proportion of acutely ill patients and a 37% increase in patients 75 years or older (Victorian Government, 2009). Attribution of mortality to a heat wave can be difficult, as deaths tend to occur from exacerbations of chronic medical conditions as well as direct heat-related illness; this is particularly so for the frail and elderly (Kovats and Hajat, 2008). However, excess mortality can provide a measure of the impact of a heat wave. With respect to total all-cause mortality, there were 374 excess deaths with a 62% increase in total all-cause mortality. The total number of deaths during the four days of the heat wave was 980, compared to a mean of 606 for the previous five years. Reported deaths in people 65 years and older more than doubled compared to the same period in 2008 (Victorian Government, 2009).

On 7 February 2009, the temperatures spiked again. The Forest Fire Danger Index, which is calculated using variables such as temperature, precipitation, wind speed, and relative humidity (Hennessy et al., 2005), this time reached unprecedented levels, higher than the fire weather conditions experienced on Black Friday in 1939 and Ash Wednesday in 1983 (National Climate Centre, 2009) – the two previous worse fire disasters in Victoria.

By the early afternoon of 7 February, wind speeds were reaching their peak, resulting in a power line breaking just outside the town of Kilmore, sparking a wildfire that would later generate extensive pyrocumulus cloud and become one of the largest, deadliest, and most intense firestorms ever experienced in Australia's history (Parliament of Victoria, 2010a). The majority of fire activity occurred between midday and midnight on 7 February, when wind speeds and temperature were at their highest and humidity at its lowest. A major wind change occurred late afternoon across the fire ground, turning the northeastern flank into a new wide fire front, catching many people by surprise. This was one of several hundred fires that started on this day, most of which were quickly controlled; however, a number went on to become major fires resulting in much loss of life. The worst 15 of these were examined in detail by the Victorian Bushfires Royal Commission (Parliament of Victoria, 2010a). A total of 173 people died as a result of the Black Saturday bushfires (Parliament of Victoria, 2010a). They also destroyed almost 430,000 ha of forests, crops, and pasture, and 61 businesses (Parliament of Victoria, 2009). The Victorian Bushfires Royal Commission conservatively values the cost of the 2009 fire at AUS$ 4.4 billion (Parliament of Victoria, 2010a).

9.2.2.4. Interventions

The Victorian Government had identified the requirement to respond to predicted heat events in the Sustainability Action Statement and Action Plan (released in 2006 and revised in January 2009), which committed to a Victorian Heat Wave Plan involving communities and local governments. As a part of this strategy, the Victorian Government has established the heat wave early warning system for metropolitan Melbourne and is undertaking similar work for regional Victoria. The government is also developing a toolkit to assist local councils in preparing for a heat wave response that could be integrated with existing local government public health and/or emergency management plans (Victorian Government, 2009).

The 'Prepare, Stay and Defend, or Leave Early' (SDLE) approach instructs that residents decide well before a fire whether they will choose to leave when a fire threatens but is not yet in the area, or whether they will stay and actively defend their property during the fire. SDLE also requires residents to make appropriate preparations in advance for either staying or leaving. Prior to 7 February 2009, the Victorian State Government devoted unprecedented efforts and resources to informing the community regarding fire risks. The campaign clearly had benefits, but there were a number of weaknesses and failures with Victoria's information and warning systems (Bushfire CRC, 2009; Parliament of Victoria, 2010b).

Another key focus during the wildfire season is protecting the reservoirs, especially the Upper Yarra and Thomson catchments that provide the majority of Melbourne's water supply (Melbourne Water, 2009a). During the February 2009 fires, billions of liters of water were moved from affected reservoirs to other safe reservoirs to protect Melbourne's drinking water from contamination with ash and debris (Melbourne Water, 2009b).

The Victorian Bushfires Royal Commission made wide-ranging recommendations about the way fire is managed in Victoria. These have included proposals to replace all single-wire power lines in Victoria, and new building regulations for bushfire-prone areas (Parliament of Victoria, 2010c).

9.2.2.5. Outcomes/Consequences

Following the findings from the various inquiries into the 2009 Victorian Bushfires, which found failings in assumptions, policies, and implementation, a number of far-reaching recommendations were developed (Parliament of Victoria, 2010c). National responses have been adopted through the National Emergency Management Committee, including: (i) revised bushfire safety policies to enhance the roles of warning and personal responsibility; (ii) increased fuel reduction burning on public lands; (iii) community refuges established in high-risk areas; (iv) improved coordination and communication between fire organizations; (v) modifying the 'Prepare, stay and defend, or leave early' approach (now 'Prepare, act, survive') to recognize the need for voluntary evacuations on extreme fire days; and (vi) further ongoing investment in bushfire research, including a national research center.

9.2.2.6 Lessons Identified

Australia has recognized the need for strengthening risk management capacities through measures including: (i) prior public campaigns for risk awareness; (ii) enhanced information and warning systems; (iii) translation of messages of awareness and preparedness into universal action; (iv) sharing responsibility between government and the people; (v) development of integrated plans; and (vi) greater investment in risk mitigation and adaptation actions.

Predicted changes in future climate will only exacerbate the impact of other factors through increased likelihood of extreme fire danger days (Hennessy et al., 2005). We are already seeing the impact of many factors on wildfires and heat waves, for example, demographic and land use changes. In the future, a better understanding of the interplay of all the causal factors is required. The Victorian Bushfires Royal Commission stated "It would be a mistake to treat Black Saturday as a 'one off' event. With populations at the rural-urban interface growing and the impact of climate change, the risks associated with bushfire are likely to increase" (Parliament of Victoria, 2010c).

9.2.3. Managing the Adverse Consequences of Drought

9.2.3.1. Introduction

Water is a critical resource throughout the world (Kundzewicz et al., 2007). Drought can increase competition for scarce resources, cause population displacements and migrations, and exacerbate ethnic tensions and the likelihood of conflicts (Barnett and Adger, 2007; Reuveny, 2007; UNISDR, 2011a). Mediterranean countries are prone to droughts that can heavily impact agricultural production, cause economic losses, affect rural livelihoods, and may lead to urban migration (UNISDR, 2011b). This case study focuses on Syria as one of the countries that has been affected by drought in recent years (2007-2010) (Erian et al., 2011).

9.2.3.2. Background

The Eastern Mediterranean region is subject to frequent soil moisture droughts, and in areas where annual rainfall ranges between 120-150 and 400 mm, rain-fed crops are strongly affected (Erian et al., 2011). During the period from 1960 to 2006, a severe decrease in annual rainfall has been documented in some major cities in Syria. These reductions were related to decreases in spring and winter rainfall (Skaff and Masbate, 2010). The negative trend in precipitation in Syria during the past century and the beginning of the 21st century is of a similar magnitude to that predicted by most general circulation models for the Mediterranean region in the coming decades (Giannakopoulos et al., 2009).

9.2.3.3. Description of Events

Syria is considered to be a dry and semi-arid country (FAO and NAPC, 2010). Three-quarters of the cultivated land depends on rainfall and the annual rate is less than 350 mm in more than 90% of the overall area (FAO, 2009; FAO and NAPC, 2010). Syria has a total population of 22 million people, 47% of whom live in rural areas (UN, 2011). The National Programme for Food Security in the Syrian Arab Republic reported that in the national economy of Syria, the agricultural and rural sector is vital, but with the occurrence of frequent droughts, this sector is less certain of maintaining its contribution of about 20 to 25% of GDP and employment of about 38 to 47% of the work force (UNRCS and SARPCMSPC, 2005; FAO and NAPC, 2010).

The prolonged drought, that in 2011 was in its fourth consecutive year, has affected 1.3 million people, and the loss of the 2008 harvest has accelerated migration to urban areas and increased levels of extreme poverty (UN, 2009, 2011; Sowers et al., 2011). During the 2008-2009 winter grain-growing season there were significant losses of both rain-fed and irrigated winter grain crops (USDA, 2008a). This was exacerbated by abnormally hot spring temperatures (USDA, 2008a). Wheat production decreased from 4,041 kt in 2007 to 2,139 kt in 2008, an almost 50% reduction (SARPMETT, 2010). Of the farmers who depended on rain-fed production, most suffered complete or near-total loss of crops (FAO, 2009). Approximately 70% of the 200,000 affected farmers in the rain-fed areas have produced minimal to no yields because seeds were not planted due to poor soil moisture conditions or failed germination (USDA, 2008b; FAO, 2009).

Herders in the region were reported to have lost around 80% of their livestock due to barren grasslands, and animal feed costs rose by 75%, forcing sales at 60 to 70% below cost (FAO, 2008). Many farmers and herders sold off productive assets, eroding their source of livelihoods, with only few small-scale herders retaining a few animals, possibly as few as 3 to 10% (FAO, 2009).

Drought has affected the livelihoods of small-scale farmers and herders, threatening food security and having negative consequences for entire families living in affected areas (FAO, 2009; UN, 2009). It is estimated that 1.3 million people have been affected by drought with up to 800,000 (75,641 households) being severely affected (FAO, 2009; UN, 2009). Of those severely affected, around 20% (160,000 people) are considered to be highly vulnerable, including female-headed households, pregnant women, children 14 and under, those with illness, the elderly, and the disabled (UN, 2009).

The United Nations (UN) estimates that a large number of the severely affected population are living below the poverty line (US$ 1/person/day) (UN, 2009). When combined with an increase in the price of food and basic resources, this reduced income has resulted in negative consequences for whole households (FAO, 2009). Many cannot afford basic supplies or food, which has led to a reduction in their food intake, the selling of assets, a rise in the rate of borrowing money, the degradation of land, urban migration, and children leaving school (FAO, 2009; UN, 2009; Solh, 2010). The UN assessment mission stated that the reasons for removing children from school included financial hardship, increased costs of transport, migration to cities, and the requirement for children to work to earn extra income for families (UN, 2009). Consequently, due to poor food consumption, the rates of malnutrition have risen between 2007 and 2008, with the Food and Agriculture Organization (FAO) estimating a doubling of malnutrition cases among pregnant women and children under five (FAO, 2008). Due to inadequate consumption of micro- and macronutrients in the most-affected households, it has been estimated that the average diet contains less than 15% of the recommended daily fat intake and 50% of the advised energy and protein requirements (UN, 2009).

One of the most visible effects of the drought was the large migration of between 40,000 and 60,000 families from the affected areas (UN, 2009; Solh, 2010; Sowers and Weinthal, 2010). In June 2009, it was estimated that 36,000 households had migrated from the Al-Hasakah Governorate (200,000-300,000 persons) to the urban centers of Damascus, Dara'a, Hama, and Aleppo (UN, 2009; Solh, 2010). Temporary settlements and camps were required, creating further strains on resources and public services, that had already been attempting to support approximately 1 million Iraqi refugees (UN, 2009; Solh, 2010). In addition, migration leads to worse health, educational, and social indicators among the migrant population (IOM, 2008; Solh, 2010).

Deficits in water resources exceeding 3.5 billion cubic meters have arisen in recent years due to growing water demands and drought (FAO and NAPC, 2010; SARPMETT, 2010). Interventions by a project further upstream to control the flow of the Euphrates and Tigris rivers have been initiated and these have had a significant impact on water variability downstream in Iraq and Syria, which, added to the severe drought, have caused these rivers to flow at well-below normal levels (USDA, 2008a; Daoudy, 2009; Sowers et al., 2011).

9.2.3.4. Interventions

The UN Syria Drought Response Plan was published in 2009. It was designed to address the emergency needs of, and to prevent further impact on, the 300,000 people most affected by protracted drought (FAO, 2009). The Response Plan identified as its strategic priorities the rapid provision of humanitarian assistance, the strengthening of resilience to future drought and climate change, and assisting in the return process and ensuring socioeconomic stability among the worst-affected groups (UN, 2009). Syria also welcomed international assistance provided to the drought-affected population through multilateral channels (Solh, 2010). Various loans to those affected, including farmers and women entrepreneurs, were provided (UN, 2009).

9.2.3.5. Outcomes/Consequences

A combination of actions including food and agriculture assistance, supplemented by water and health interventions, and measures aimed at increasing drought resilience, were identified as required to allow affected populations to remain in their villages and restart agricultural production (UN, 2009). Ongoing interventions with the aim of reducing vulnerability and increasing resilience to drought were summarized by the UN Syrian Drought Response Plan (UN, 2009) and FAO (FAO, 2009). These interventions were aimed at providing support through four main approaches: (1) the rapid distribution of wheat, barley, and legume seeds to 18,000 households in the affected areas, potentially assisting 144,000 people; (2) sustaining the remaining asset base of the approximately 20,000 herders by providing animal feed and limited sheep restocking to approximately 1,000 herders; (3) the development of a drought early warning system to facilitate the government taking early actions before serious and significant losses occur and to develop this to ensure sustainability; and (4) building capability to implement the national drought strategy by developing and addressing all stages of the disaster management cycle (FAO, 2009). Conservation agriculture (which has been defined as no-tillage, direct drilling/seeding, and drilling/seeding through a vegetative cover) is considered to be a way forward for sustainable land use (Stewart et al., 2008; Lalani 2011). However, how to take this forward has caused considerable debate (Stewart et al., 2008).

9.2.3.6. Lessons Identified

The need for the UN Syrian Drought Response Plan was identified and has facilitated the understanding of the work programs and links to the interventions listed in Case Study 9.2.3.4 (UN, 2009). Other response strategies that have been considered include:
- Development of capacities to identify, assess, and monitor drought risks through national and local multi-hazard risk assessment; building systems to monitor, archive, and disseminate data (Lalani, 2011), taking into account decentralization of resources, community participation, and regional early warning systems and networks (UNISDR, 2011a).
- Integrating activities in the national strategy for CCA and DRR, including drought risk loss insurance; improved water use efficiency; adopting and adapting existing water harvesting techniques; integrating use of surface and groundwater; upgrading irrigation

practices at both the farm level and on the delivery side; developing crops tolerant to salinity and heat stress; changing cropping patterns; altering the timing or location of cropping activities; diversifying production systems into higher value and more efficient water use options; and capacity building of relevant stakeholders in vulnerable national and local areas (Abou Hadid, 2009; El-Quosy, 2009).

- Building resilience through knowledge, advocacy, research, and training by making information on drought risk accessible (UNISDR, 2007a), and having any adaptation measures be developed as part of, and closely integrated into, overall and country-specific development programs and strategies that should be understood as a 'shared responsibility' (Easterling et al., 2007). This could be achieved through educational material and training to enhance public awareness (UN, 2009).

9.2.4. Recent *Dzud* Disasters in Mongolia

9.2.4.1. Introduction

This case study introduces *dzud* disaster: the impacts, intervention measures, and efforts toward efficient response using the example of two events that occurred in Mongolia in 1999-2002 and 2009-2010. Mongolia is a country of greatly variable, highly arid and semi-arid climate, with an extensive livestock sector dependent upon access to grasslands (Batima and Dagvadorj, 2000; Dagvadorj et al., 2010; Marin, 2010). The Mongolian term *dzud* denotes unusually extreme weather conditions that result in the death of a significant number of livestock over large areas of the country (Morinaga et al., 2003; Oyun, 2004). Thus, the term implies both exposure to such combinations of extreme weather conditions but also the impacts thereof (Marin, 2010).

9.2.4.2. Background

The climate of Mongolia is continental with sharply defined seasons, high annual and diurnal temperature fluctuations, and low rainfall (Batima and Dagvadorj, 2000). Summer rainfall seldom exceeds 380 mm in the mountains and is less than 50 mm in the desert areas (Dagvadorj et al., 2010). *Dzud* is a compound hazard (see Section 3.1.3 for discussion of compound events) occurring in this cold dry climate, and encompasses drought, heavy snowfall, extreme cold, and windstorms. It can last all year round and can cause mass livestock mortality and dramatic socioeconomic impacts – including unemployment, poverty, and mass migration from rural to urban areas, giving rise to heavy pressure on infrastructure and social and ecosystem services (Batima and Dagvadorj, 2000; Batjargal et al., 2001; Oyun, 2004; AIACC, 2006; Dagvadorj et.al., 2010).

There are several types of *dzud*. If there is heavy snowfall, the event is known as a *white dzud*, conversely if no snow falls, a *black dzud* occurs, which results in a lack of drinking water for herds (Morinaga et al., 2003; Dagvadorj et al., 2010). The trampling of plants by passing livestock migrating to better pasture or too high a grazing pressure leads to a *hoof dzud*, and a warm spell after heavy snowfall resulting in an icy crust cover on short grass blocking livestock grazing causes an *iron dzud* (Batjargal et al., 2001; Marin, 2008). Livestock have been the mainstay of Mongolian agriculture and the basis of its economy and culture for millennia (Mearns, 2004; Goodland et al., 2009). This sector is likely to continue to be the single most important sector of the economy in terms of employment (Mearns, 2004; Goodland et al., 2009).

In the last decades, *dzuds* occurred in 1944-1945, 1954-1955, 1956-1957, 1967-1968, 1976-1977, 1986-1987, 1993-1994, and 1996-1997, with further *dzuds* discussed below (Morinaga et al., 2003; Sternberg, 2010). The *dzud* of 1944-1945 was a record for the 20th century with mortality of one-third of Mongolia's total livestock (Batjargal et al., 2001). The 2009-2010 *dzud* caused similarly high animal mortality (NSO, 2011). The large losses of animals in *dzud* events demonstrates that Mongolia as a whole has low capacity to combat natural disaster (Batjargal et al., 2001). These potential losses are considered to be beyond the financial capacity of the government and the domestic insurance market (Goodland et al., 2009).

9.2.4.3. Description of Events – *Dzud* of 1999-2002 and of 2009-2010

Dzud disasters occurred in 1999-2002 and 2009-2010, causing social and economic impacts. These disasters occurred as a result of environmental and human-induced factors. The environmental factors included drought resulting in very limited pasture grass and hay with additional damage to pasture by rodents and insects (Batjargal et al., 2001; Begzsuren et al., 2004; Saizen et al., 2010). Human factors included budgetary issues for preparedness in both government and households, inadequate pasture management and coordination, and lack of experience of new and/or young herders (Batjargal et al., 2001). Climatic factors contributing to both *dzuds* were summer drought followed by extreme cold and snowfall in winter. However the autumn of 1999-2000 brought heavy snowfall and unusual warmth with ice cover, while the winter and spring of 2009-2010 also brought windstorms. Summer drought was a more significant contributor to the 1999-2000 *dzud* (Batjargal et al., 2001), while winter cold was more extreme in the 2009-2010 *dzud*.

9.2.4.3.1. *Dzud* of 1999-2002

The *dzud* began with summer drought followed by heavy snowfall and unusual warmth with ice cover in the autumn and extreme cold and snowfall in the winter. The sequence of events was as follows (Batjargal et al., 2001):

- **Drought**: In the summer of 1999, 70% of the country suffered drought. Air temperature reached 41 to 43°C, exceeding its highest value recorded at meteorological stations since the 1960s. The condition persisted for a month and grasslands dried up. As a result, animals were unfit for the winter, with insufficient haymaking for winter preparedness.

- **Iron dzud**: Autumn brought early snowfall and snow depth reached 30 to 40 cm, even 70 to 80 cm in some places. Heavy snowfall exceeding climatic means was recorded in October. Moreover, a warming in November and December by 1.7 to 3.5°C above the climatic mean resulted in snow cover compaction and high density, reaching 0.37 g cm^{-3}, and ice cover formation, both of which blocked livestock pasturing.
- **White dzud**: In January, air temperature dropped down to -40 to -50°C over the western and northern regions of the country. The monthly average air temperature was lower than climatic means by 2 to 6°C. The cold condition persisted for two months. Abundant snowfall resulted with 80% of the country's territory being covered in snow of 24 to 46 cm depth.
- **Black dzud**: Lack of snowfall in the Gobi region and Great Lake depression caused water shortages for animals.
- **Hoof dzud**: The improper pasture management led to unplanned concentration of a great number of livestock in a few counties in the middle and south Gobi provinces that were not affected by drought and snowfall.

Animals were weakened as a result of long-lasting climatic hardship and forage shortage of this *dzud* (Batjargal et al., 2001). After three years of *dzuds* that occurred in sequence, the country had lost nearly one-third (approximately 12 million) of its livestock and Mongolia's national gross agricultural output in 2003 decreased by 40% compared to that in 1999 (Mearns, 2004; Oyun, 2004; AIACC, 2006; Lise et al., 2006; Saizen et al., 2010). It was reported that in 1998 there were an estimated 190,000 herding households but as a result of the *dzud*, 11,000 families lost all their livestock (Lise et al., 2006). Thus the *dzud* had severe impacts on the population and their livelihoods, including unemployment, poverty, and negative health impacts (Batjargal et al., 2001; Oyun, 2004; AIACC, 2006; Morris, 2011).

9.2.4.3.2. Dzud of 2009-2010

In the summer of 2009, Mongolia suffered drought conditions, restricting haymaking and foraging (UNDP Mongolia, 2010; Morris, 2011). Rainfall at the end of November became a sheet of ice, and, in late December, 19 of 21 provinces recorded temperatures below -40°C; this was followed by heavy and continuous snowfall in January and February 2010 (Sternberg, 2010; UNDP Mongolia, 2010). Over 50% of all the country herders' households and their livestock were affected by the *dzud* (Sternberg, 2010). By April, 75,000 herder families had lost all or more than half their livestock (Sternberg, 2010).

9.2.4.4. Interventions

9.2.4.4.1. Dzud of 1999-2002

The government of Mongolia issued the order for intensification of winter preparedness in August 1999, but allocated funding for its implementation in January 2000, by which time significant animal mortality had already occurred (Batjargal et al., 2001). The government then appealed to its citizens and international organizations for assistance and relief, including distribution of money, fodder, medicine, clothes, flour, rice, high energy and high protein biscuits for children, health and veterinary services, medical equipment, and vegetable seeds (Batjargal et al., 2001). Capacity-building activities through mass media campaigns were also carried out, focused on providing advice on methods of care and feeding for weak animals (Batjargal et al., 2001).

Herders rely upon traditional informal coping mechanisms and ad hoc support from government and international agencies (Mahul and Skees, 2005). For affected areas, after immediate relief, the main longer term support has conventionally been through restocking programs (Mahul and Skees, 2005). Evaluation has shown that these can be expensive, relatively inefficient, and fail to provide the right incentives for herders (Mahul and Skees, 2005). Restocking in areas with drought, poor pasture condition, and unfit animals can actually increase livestock vulnerability in the following year (Mahul and Skees, 2005) as a result of greater competition for scarce resources.

The government has prioritized the livestock sector with parliament-approved state policy (MGH, 2003) and, with support from donors, responded to *dzud* disasters with reforms that include greater flexibility in pasture land tenure, coupled with increased investment in rural infrastructure and services (Mahul and Skees, 2005). However, livestock sector reforms and approaches have not yet proved sufficient to cope with catastrophic weather events (Mahul and Skees, 2005). Although the State Reserves Agency is working to reduce the effects of *dzud*, catastrophic livestock mortality persists (Mahul and Skees, 2005).

9.2.4.4.2. Dzud of 2009-2010

At a local level: The National Climate Risk Management Strategy and Action Plan (MMS, 2009) sets a goal of building climate resilience at the community level through reducing risk and facilitating adaptation by: (i) improving access to water through region-specific activities such as rainwater harvesting and creation of water pools from precipitation and flood waters, for use for animals, pastureland, and crop irrigation purposes; (ii) improving the quality of livestock by introducing local selective breeds with higher productivity and more resilient to climate impacts; (iii) strengthened veterinarian services to reduce animal diseases and parasites and cross-border epidemic infections; and (iv) using traditional herding knowledge and techniques for adjusting animal types and herd structure to make appropriate for the carrying capacity of the pastureland and pastoral migration patterns. The formation of herders' community groups and establishment of pasture co-management teams (Ykhanbai et al., 2004), along with better community-based disaster risk management, could also facilitate effective DRR and CCA (Baigalmaa, 2010).

At a national level: Mongolia's millennium development priorities clearly state an aim to adapt to climate change and desertification and to implement strategies to minimize negative impacts (Mijiddorj, 2008; UNDP Mongolia, 2009a). The recent national CCA report outlines government strategy priorities as: (i) education and awareness campaigns among the decisionmakers, rural community, herders, and the general public; (ii) technology and information transfer to farmers and herdsmen; (iii) research and technology to ensure the development of agriculture that could successfully deal with various environmental problems; and (iv) improved coordination of stakeholders' activities based on research, inventory, and monitoring findings (Dagvadorj et al., 2010). The management of risk in the livestock sector requires a combination of approaches. Traditional herding and pastoral risk reduction practices can better prepare herders for moderate weather events. For country-wide *dzud* events, however, high levels of livestock mortality are often unavoidable, even for the most experienced herders, and pasture resource and herd management must be complemented by risk-financing mechanisms that provide herders with instant liquidity in the aftermath of a disaster (Goodland et al., 2009).

At an international level: As Mongolia is a country extremely prone to natural disasters, addressing climate change risks is a priority in Mongolia. In 2009, the Mongolian Government undertook the project 'Strengthening the Disaster Mitigation and Management Systems in Mongolia' under the National Emergency Management Agency (UNDP Mongolia, 2009b; Sternberg, 2010).

9.2.4.5. Consequences

The most critical consequences of *dzud* are increased poverty and mass migration from rural to urban and from remote to central regions (Oyun, 2004; Dagvadorj et al., 2010). According to national statistics there has been a continuous increase in poverty over the last decade (NSO, 2011).

In response to the climatic hardship, a growing proportion of the rural population has migrated to urban areas and the central region (Dagvadorj et al., 2010; UNDP Mongolia, 2010). Livestock-herding families are forced to migrate because of poverty caused by loss of livestock from catastrophic weather events (Sternberg, 2010). Besides poverty, there are reasons why members of herding families may wish to leave the livestock sector, including obtaining a better education for their children and access to health care (Mahul and Skees, 2005). Many migrants travel from Western Mongolia to the capital city Ulaanbaatar (Saizen et al., 2010; Sternberg, 2010).

9.2.4.6. Lessons Identified

Current policies and measures are mainly limited to post-disaster government relief and restocking activities with donors' funding and individual herder's traditional knowledge and practices (Batjargal et al., 2001; AIACC, 2006). These can be insufficient to avoid, prepare for, and respond to a *dzud* (Goodland et al., 2009). Various practices have been identified as effective for DRR, and could further contribute to promote CCA. These include localized seasonal climate prediction and improvement of early warning (Morinaga et al., 2003; MMS, 2009), risk-insuring systems (Skees and Enkh-Amgalan, 2002; Mahul and Skees, 2005), and policy (Batjargal et al., 2001; AIACC, 2006; Goodland et al., 2009).

At present, adaptation occurs through increased mobility of herders in search of better pasture for their animals in *dzud* disasters (Batjargal et al., 2001), and as a response to changed rain patterns occurring over small areas, which the herders call 'silk embroidery rain' (Marin, 2010). Livelihood diversification to create resilient livelihoods for herders has also been seen as being effective for building climate resilience (Mahul and Skees, 2005; Borgford-Parnell, 2009; MMS, 2009, Dagvadorj et al., 2010).

9.2.5. Cyclones: Enabling Policies and Responsive Institutions for Community Action

9.2.5.1. Introduction

Tropical cyclones, also called typhoons and hurricanes, are powerful storms generated over tropical and subtropical waters. Their extremely strong winds damage buildings, infrastructure and other assets; the torrential rains often cause floods and landslides; and high waves and storm surge often lead to extensive coastal flooding and erosion – all of which have major impacts on people. Tropical cyclones are typically classified in terms of their intensity, based on measurements or estimates of near-surface wind speed (sometimes categorized on a scale of 1 to 5 according to the Saffir-Simpson scale). The strongest storms (Categories 3, 4, and 5) are comparatively rare but are generally responsible for the majority of damage (Section 3.4.4).

The focus of this case study is the comparison between the response to Indian Ocean cyclones in Bangladesh (Sidr in 2007) and in Myanmar (Nargis in 2008) in the context of the developments in preparedness and response in Bangladesh resulting from experiences with Cyclone Bhola in 1970, Gorky in 1991, and other events. To provide a more global context, the impacts and responses to 2005 Hurricanes Stan and Wilma in Central America and Mexico are also discussed. These clearly demonstrate that climate change adaptation efforts can be effective in limiting the impacts from extreme tropical cyclone events by use of disaster risk reduction methods.

Changes in tropical cyclone activity due to anthropogenic influences are discussed in Section 3.4.4. There is *low confidence* that any observed long-term increases in tropical cyclone activity are robust, after accounting for past changes in observing capabilities. The uncertainties in the historical tropical cyclone records, the incomplete understanding of the physical mechanisms linking tropical cyclone metrics to climate change, and the degree of tropical cyclone variability provide only *low confidence* for the attribution of any detectable changes in tropical cyclone activity to anthropogenic influences. There is *low confidence* in projections of changes in tropical cyclone genesis, location, tracks, duration, or areas of impact. Based on the level of consistency among models, and physical reasoning, it is *likely* that tropical cyclone-related rainfall rates will increase with greenhouse warming. It is *likely* that the global frequency of tropical cyclones will either decrease or remain essentially unchanged. An increase in mean tropical cyclone maximum wind speed is *likely*, although increases may not occur in all tropical regions. While it is *likely* that overall global frequency will either decrease or remain essentially unchanged, it is *more likely than not* that the frequency of the most intense storms will increase substantially in some ocean basins. Although there is evidence that surface sea temperature (SST) in the tropics has increased due to increasing greenhouse gases, the increasing SST does not yet have a fully understood physical link to increasingly strong tropical cyclones (Section 3.4.4).

9.2.5.2. Indian Ocean Cyclones

Although only 15% of world tropical cyclones occur in the North Indian Ocean (Reale et al., 2009), Bangladesh and India account for 86% of mortality from tropical cyclones (UNISDR, 2009c). The 2011 Global Assessment Report (UNISDR, 2011b) provides strong evidence that weather-related mortality risk is highly concentrated in countries with low GDP and weak governance. Many of the countries exposed to tropical cyclones in the North Indian Ocean are characterized by high population density and vulnerability and low GDP.

9.2.5.2.1. Description of events – Indian Ocean cyclones

In 2007, Cyclone Sidr made landfall in Bangladesh on 15 November and caused over 3,400 fatalities (Paul, 2009). Cyclone Nargis hit Myanmar on 2 May 2008 and caused over 138,000 fatalities (CRED, 2009; Yokoi and Takayabu, 2010), making it the eighth deadliest cyclone ever recorded (Fritz et al., 2009). Sidr and Nargis were both Category 4 cyclones of similar severity; affecting coastal areas with a comparable number of people exposed (see Table 9-2). Although Bangladesh and Myanmar both are considered least developed countries (Giuliani and Peduzzi, 2011), these two comparable events had vastly different impacts. The reasons for the differences follow.

9.2.5.2.2. Interventions – Indian Ocean cyclones

Bangladesh has a significant history of large-scale disasters (e.g., Cyclones Bhola in 1970 and Gorky in 1991; see Table 9-2). The Government of Bangladesh has made serious efforts aimed at DRR from tropical cyclones. It has worked in partnership with donors, nongovernmental organizations (NGOs), humanitarian organizations and, most importantly, with coastal communities themselves (Paul, 2009).

First, they constructed multi-storied cyclone shelters with capacity for 500 to 2,500 people (Paul and Rahman, 2006) that were built in coastal regions, providing safe refuge from storm surges for coastal populations. Also, killas (raised earthen platforms), which accommodate 300 to 400 livestock, have been constructed in cyclone-prone areas to safeguard livestock from storm surges (Paul, 2009).

Second, there has been a continued effort to improve forecasting and warning capacity in Bangladesh. A Storm Warning Center has been established in the Meteorological Department. System capacity has been enhanced to alert a wide range of user agencies with early warnings and special bulletins, soon after the formation of tropical depressions in the Bay of Bengal. Periodic training and drilling practices are conducted at the local level for cyclone preparedness program (CPP) volunteers for effective dissemination of cyclone warnings and for raising awareness among the population in vulnerable communities.

Third, the coastal volunteer network (established under the CPP) has proved to be effective in disseminating cyclone warnings among the coastal communities. These enable time-critical actions on the ground, including safe evacuation of vulnerable populations to cyclone shelters (Paul, 2009). With more than sevenfold increase in cyclone shelters and twofold increase in volunteers from 1991 to 2007, 3 million people were safely evacuated prior to landfall of Sidr in 2007 (Government of Bangladesh, 2008).

In addition, a coastal reforestation program, including planting in the Sundarbans, was initiated in Bangladesh in the late 1960s, covering riverine coastal belt and abandoned embankments (Saenger and Siddiqi, 1993). Sidr made landfall on the western coast of Bangladesh, which is

Table 9-2 | Key data for extreme cyclones in Bangladesh, Myanmar, and Mexico. Sources: García et al., 2006; National Hurricane Center, 2006; Government of Bangladesh, 2008; Karim and Mimura, 2008; Webster, 2008; CRED, 2009; Paul, 2009; Giuliani and Peduzzi, 2011.

Cyclone Event	Year	Storm Surge (m)	Maximum Wind Speed (km h^{-1})	Category (Saffir-Simpson)	Number of Affected People (approximate in millions)	Mortality (approximate)	Damages (US$ billion)
Bhola	1970	6 – 9	223	3	1	300,000 - 500,000	Unknown
Gorky	1991	6 – 7.5	260	4	15.4	138,000	1.8
Sidr	2007	5 – 6	245	4	8 – 10	4,200	2.3
Nargis	2008	~ 4	235	4	2 – 8	138,000	4.0
Stan[a]	2005	Negligible	130	1	3 – 8	1,726	3.9
Wilma	2005	12.8	295	5	10[b]	62 (8 in Mexico)	29 (7.5 in Mexico)

Notes:
[a] Most of damage and mortality caused by landslides and river flooding.
[b] Affecting Jamaica, Bahamas, Haiti, Cayman Islands, Belize, Honduras, El Salvador, Nicaragua, Honduras, Yucatán Peninsula (Mexico), and Florida (USA).

lined by the world's largest mangrove forest, the Sundarbans. This region is the least-populated coastal area in the country and has been part of a major reforestation effort in recent years (Hossain et al., 2008). The Sundarbans provided an effective attenuation buffer during Sidr, greatly reducing the impact of the storm surge (Government of Bangladesh, 2008).

In contrast to Bangladesh, Myanmar has very little experience with previous powerful tropical cyclones. The landfall of Nargis was the first time in recorded history that Myanmar experienced a cyclone of such a magnitude and severity (Lateef, 2009) and little warning was provided.

9.2.5.2.3. Outcomes – Indian Ocean cyclones

Despite Nargis being both slightly less powerful and affecting fewer people than Sidr, it resulted in human losses that were much higher. Bangladesh and Myanmar are both very poor countries with low levels of HDI (World Bank, 2011a). The relatively small differences in poverty and development cannot explain the discrepancy in the impacts of Sidr and Nargis. However, the governance indicators developed by the World Bank (Kaufmann et al., 2010) suggest significant differences between Bangladesh and Myanmar in the quality of governance, notably in voice and accountability, rule of law, regulatory quality, and government effectiveness. Low quality of governance, and especially voice and accountability, has been highlighted as a major vulnerability component for human mortality due to tropical cyclones (Peduzzi et al., 2009).

9.2.5.3. Mesoamerican Hurricanes

9.2.5.3.1. Description of events – Mesoamerican hurricanes

Central America and Mexico (Mesoamerica) are heavily affected by strong tropical storms. From 1-13 October 2005, Hurricane Stan affected the Atlantic coast of Central America and the Yucatan Peninsula in Mexico. Stan was a relatively weak storm that only briefly reached hurricane status. It was associated with a larger non-tropical storm system that resulted in torrential rains and caused debris flows, rockslides, and widespread flooding. Guatemala reported more than 1,500 fatalities and thousands of missing people. El Salvador reported 69 fatalities while Mexico reported 36 (CRED, 2009). Wilma hit one week later (19-24 October). It was an intense cyclone in the Atlantic (National Hurricane Center, 2006; Table 9-2), with winds reaching a speed of 297 km h^{-1}. Wilma caused 12 fatalities in Haiti, 8 in Mexico, and 4 in the United States. Most residents in western Cuba, and tourists and local inhabitants in the Yucatan Peninsula in Mexico, were evacuated (CRED, 2009).

9.2.5.3.2. Interventions – Mesoamerican hurricanes

While Stan mainly affected the poor indigenous regions of Guatemala, El Salvador, and Chiapas, Wilma affected the international beach resort of Cancun. A joint study of Mexico's response to the hurricanes funded by the World Bank and conducted through the Economic Commission for Latin America and the Caribbean (ECLAC, 2006) and its Commission for Latin American and the Caribbean and the Mexican National Center for Prevention of Disasters (García et al., 2006) showed that Stan caused about US$ 2.2 billion damage in that country, 65% of which were direct losses and 35% due to future impacts on agricultural production. About 70% of these damages were reported in the state of Chiapas (Oswald Spring, 2011), representing 5% of the GDP of the state.

Comparing the management of the two hurricanes by the Mexican authorities, in the same month and year, highlights important issues in DRM. Evacuation of areas in Mexico affected by Stan only started during the emergency phase, when floods in 98 rivers had already affected 800 communities. 100,000 people fled from the mountain regions to improvised shelters – mostly schools – and 'guest families' (Oswald Spring, 2010). In comparison, following the early alert for Wilma, people were evacuated from dangerous places, most tourists were moved to safe areas, and local inhabitants and remaining tourists were taken to shelters (García et al., 2006). Before the hurricane hit the coast, heavy machines and emergency groups were mobilized in the region, to reestablish water, electricity, communications, and health services immediately after the event. After the disaster, all ministries were involved in reopening the airport and tourist facilities as quickly as possible. By December, most hotels were operating, and the sand lost from the beaches had been reestablished (Oswald Spring, 2011).

9.2.5.3.3. Outcomes – Mesoamerican hurricanes

Comparing government responses to these two hurricanes in the same month, it is possible to note vastly different official actions in terms of early warning, evacuation, and reconstruction (Oswald Spring, 2011). The federal institutions in charge of DRM functioned well during Hurricane Wilma. A massive recovery support strategy restored almost all services and hotels in Cancun within two months, with a significant portion of costs being covered by insurance companies (García et al., 2006). The government response to Stan left the poor indigenous population with limited advice, insufficient disaster relief, and scant reconstruction support, especially among the most marginal groups (Oswald Spring, 2011).

9.2.5.4. Lessons Identified

Comparative studies of disaster risk management practices for tropical cyclones demonstrate that choices and outcomes for response to climatic extreme events are triggered by multiple interacting processes, and competing priorities. Indigenous, poor, and illiterate people have low resilience, limited resources, and are highly exposed without early warnings and DRM. Government response to similar extreme events may be quite different in neighboring countries, or even within the same country.

Tropical cyclone DRM strategies in coastal regions that create protective measures, and anticipate and plan for extreme events, along with continuing changes in vulnerability and in causal processes, increase the resilience of potentially exposed communities. International cooperation and investment in the following measures are essential in improving the capacity of developing nations in coping with extreme tropical cyclone events:
- Improvement of forecasting capacity and implementation of improved early warning systems (including evacuation plans and infrastructure)
- Protection of healthy ecosystems
- Post-disaster support service to dispersed communities
- Transparent management of recovery funds directly with the victims.

Awareness, early warnings and evacuation, hurricane experience, disaster funds, and specialized bodies reduce the impact of tropical cyclones on socially vulnerable people. Good governance and participation of people at risk in the decisionmaking process may overcome conflicting governmental priorities. Disaster risk management is most effectively pursued by understanding the diverse ways in which social processes contribute to the creation, management, and reduction of disaster risk with the involvement of people at risk. A development planning perspective that includes disaster risk management as an integral part of the development framework is the key to a coherent strategy for the reduction of risk associated with extreme weather events.

9.2.6. Managing the Adverse Consequences of Floods

9.2.6.1. Introduction

Floods are a major natural hazard in many regions of the world (Ahern et al., 2005). Averaged over 2001 to 2010, floods and other hydrological events accounted for over 50% of the disasters (Guha-Sapir et al., 2011); for example, it was reported that, in 2007, flooding worldwide accounted for four of the top five deadliest natural disasters (Subbarao et al., 2008). Currently about 800 million people live in flood-prone areas and about 10% are annually exposed to floods (Chapter 4; Peduzzi et al., 2011; UNISDR, 2011b). Causes of floods are varied, but may occur as a result of heavy, persistent, and sustained rainfall or as a result of coastal flooding (Ahern et al., 2005; see also Section 3.5.2). Flooding impacts are wide ranging, potentially interrupting food and water supplies, affecting economic development, and causing acute as well as subsequent long-term health impacts (Ahern et al., 2005; Subbarao et al., 2008). It is important to study flooding events to develop or enhance reliable approaches to risk reduction as well as systems for forecasting and informing the population, in order to help minimize negative consequences (ICSU, 2008). This case study examines the impacts on the population and economy of Mozambique from the 2000 and 2007 flooding events.

Effective functioning of DRR and DRM programs at all levels can help to reduce the risks from extreme events including floods (UNISDR, 2005a). These programs operate best with a combination of local, national, and international strategies (Hellmuth et al., 2007; UNISDR, 2011a). A variety of strategies have been used to reduce the impact of floods. For example, dams and sea walls prevent flooding of coastal areas but are expensive and difficult to maintain and these facilities can be breached (ProAct Network, 2008). Furthermore, urban drainage systems are recognized as an important tool to reduce urban flood risk, but less than half (46%) of low-income countries have invested in drainage infrastructure in flood-prone areas (UNISDR, 2011b). Timely flood warnings in many countries have been developed as part of DRR and DRM programs (Case Study 9.2.11).

The Global Assessment Report (UNISDR, 2011b) reported that the 2000 floods in Mozambique are one of the four examples of large disasters that have highlighted DRM capacity gaps that have led to institutional and legislative changes.

9.2.6.2. Background

Mozambique has high socioeconomic vulnerability with approximately 50% of its population of 21 million living below the poverty line (see Sections 2.3 and 2.5; WMO, 2011a; World Bank, 2011b). Its development has been restricted by previous civil war and conflict with neighboring South Africa. Further examples of its vulnerability include rising HIV/AIDS rates, an almost 70% female illiteracy rate, and most of the population depending on subsistence farming (Hellmuth et al., 2007; World Bank, 2011b).

Geographic position and climatic factors contribute to Mozambique's high physical vulnerability. Mozambique has a 2,700-km coastline and the whole country and neighboring countries are subjected to cyclones and resultant flooding (Hellmuth et al., 2007; WMO, 2011a; World Bank, 2011b). Nine of the 11 rivers in Mozambique are transboundary (Hellmuth et al., 2007) making its location downstream more susceptible to rainfall events across a large region such that increases in river levels and flows in neighboring countries can result in or exacerbate floods. Therefore the development and operating of early warning and flood control systems in Mozambique depend on a close collaboration with other countries of the Southern Africa Development Community and its protocol on shared watercourse systems (SADC, 2000).

The World Bank (2005a) reported that Mozambique experienced 12 major floods, 9 major droughts, and 4 major cyclone events between 1965 and 1998. In 1999, a national government policy on disaster management was articulated and a National Institute for Disaster Management (NIDM), with an emphasis on coordination rather than delivery, created (World Bank, 2005a).

9.2.6.3. Description of Events – 2000 Floods in Mozambique

In February 2000, catastrophic floods caused the loss of more than 700 lives with over half a million people losing their homes, and more than

4.5 million affected (Mirza, 2003; Hellmuth et al., 2007; WMO, 2011a; World Bank, 2011b).

The flooding was the result of a cascade of events. It started with above-average rainfall in southern Mozambique and adjacent countries from October to December 1999 (Hellmuth et al., 2007). Exacerbating the situation was the series of cyclones Astride, Connie, Eline, and Gloria with the main impact coming from cyclone Eline (UNESC, 2000; Asante et al., 2007; Hellmuth et al., 2007). Cyclone Eline, after tracking over 7,000 km west across the tropical south Indian Ocean (Reason and Keibel, 2004), made landfall on 22 February 2000, crossing the Mozambique coastline and moving over the headwater basins of the Limpopo River, making a critical situation worse.

The rainfall that occurred over Mozambique and the northeastern parts of South Africa and Zimbabwe was exceptional; record flooding ensued downstream on the Limpopo and Zambezi rivers (Carmo Vaz, 2000; Kadomura, 2005), and in parts of the Sabie catchment the return period was in excess of 200 years (Smithers et al., 2001).

As a result of the floods it was reported that many small towns and villages remained under water for approximately two months (Hellmuth et al., 2007). Access roads were rendered impassable with railways, bridges, water management systems (including water intake and treatment plants), and more than 600 primary schools damaged or destroyed (UNESC, 2000; Dyson and van Heerden, 2001; Reason and Keibel, 2004). The UN World Food Programme reported that Mozambique lost 167,000 ha of agricultural land (FAO and WFP, 2000). Dams were overwhelmed; for example, the total inflow to Massingir reservoir between January and March was approximately eight times the storage capacity of the reservoir at that time (Carmo Vaz, 2000).

Although floodwaters can wash away breeding sites and, hence, reduce mosquito-borne disease transmission (Sidley, 2000), the collection of emergency clinic data and interviews of 62 families found that the incidence of malaria was reported as increasing by a factor of 1.5 to 2.0. Diarrhea also increased by a factor of 2 to 4 (Kondo et al., 2002).

The government declared an emergency, mobilized its disaster response mechanisms, and made appeals for assistance from other countries (Hellmuth et al., 2007).The enormous material damage and human losses during the floods in Mozambique in 2000 were associated with the following problems:
- *Institutional problems*: It was only in 1999 that the National Policy on Disaster Management in Mozambique began to shift from a reactive to a proactive approach, with an aim to develop a culture of prevention (Asante et al., 2005; Hellmuth et al., 2007).
- *Technological problems*: In 2000, in Mozambique, there were problems with the installation and maintenance of *in situ* gauging equipment due to financial constraints. In addition, the hydrological and precipitation gauges were washed away and many key stations were destroyed, leaving Mozambican water authorities with no source of information on the actual magnitude of floodwaters (Dyson and van Heerden, 2001; Smithers et al., 2001; Asante et al., 2005).
- *Financial problems*: The UN Economic and Social Council (UNESC, 2000) reported that the Government of Mozambique responded to the emergency despite limited means, but due to the extensive international financial support requested help in its coordination from the UN. The World Bank estimates that the direct losses as a result of the 2000 floods amounted to US$ 273 million (UNESC, 2000).

9.2.6.4. Interventions

After the catastrophic floods in 2000, the Government of Mozambique took a range of measures to improve the effectiveness of disaster risk management. In 2001, an Action Plan for the Reduction of Absolute Poverty (PARPA I) was adopted (Republic of Mozambique, 2001); and this was revised for the period 2006 to 2009 (PARPA II) (Republic of Mozambique, 2006a,b; Foley, 2007). In 2006, the government also adopted a Master Plan, which provides a comprehensive strategy for dealing with Mozambique's vulnerability to natural disasters (Republic of Mozambique, 2006a).

After the 2000 floods, Mozambique implemented intensive programs to move people to safe areas (World Bank, 2005a). Since the 2000 floods, a large resettlement program for communities affected by the floods and tropical cyclones was initiated, with about 59,000 families resettled although a lack of funds for improved livelihoods has reduced the success of this program (WMO, 2011a).

Success and effectiveness of warnings depend not only on the accuracy of the forecast, but also their delivery in adequate time before the disaster to put in place prevention strategies. From November 2006 to November 2007 the Severe Weather Forecasting Demonstration Project, conducted by the World Meteorological Organization in southeastern Africa, tested a new concept for capacity building and this service contributed to the forecasting and warnings about Cyclone Favio in February 2007 (Poolman et al., 2008). The demonstration phase was found to be valuable, and the implementation phase – with training, supported with efficient and effective forecasting and warning of tropical cyclones in developing countries – continues (WMO, 2011b).

Besides high-level alerting it is important that a warning is received by each person in the disaster zone, in an easily understandable way (UNISDR, 2010). In 2005 and 2006 the German Agency for Technical Cooperation developed a simple but effective early warning system along the River Búzi (Bollin et al., 2005; Loster and Wolf, 2007). This warning system was adapted to the specific needs and skills of the people. The village officials receive daily precipitation and water levels at strategic points along the Búzi River basin. If precipitation is particularly heavy or the river reaches critical levels, this information is passed on by radio and blue, yellow, or red flags are raised depending on the flood alert level (Bollin et al., 2005; UNISDR, 2010).

9.2.6.5. Outcomes/Consequences – 2007 Floods in Mozambique

Seven years after the catastrophic floods of 2000, similar flooding occurred in Mozambique, but the country was prepared to a greater extent than before. Between December 2006 and February 2007, heavy rains across northern and central Mozambique together with a severe downpour in neighboring countries led to flooding in the Zambezi River basin (IFRC, 2007). Additional flooding was caused by the approach of tropical cyclone Favio, which struck the Búzi area at the end of February 2007 (Poolman et al., 2008). During the flood period on the southern coast of Mozambique, 29 people were killed, 285,000 people affected, and approximately 140,000 displaced (Kienberger, 2007; World Bank, 2011b). The heavy rains and floods damaged health centers, public buildings, drug stocks, and medical equipment and affected safe water and sanitation facilities (UNOCHA, 2007). In total, the floods and cyclone caused approximately US$ 71 million of damage to local infrastructure and destroyed 277,000 ha of crops (USAID, 2007).

During the course of January 2007, it became clear that there was an imminent threat of severe flooding in the Zambezi River basin valley (Foley, 2007). A multinational flood warning covering Zambia, Malawi, and Mozambique was issued on 26 January 2007. With forecasts and warnings increasing over the next week, NIDM increased the flood warning until a 'Red Alert' was issued (UNISDR, 2010). This was a test of the earlier work undertaken by Bollin et al. (2005); when the rivers rose rapidly, it was reported that approximately 12,800 people who were at risk had been well prepared by prior training (Loster and Wolf, 2007). The district's disaster mitigation committee had alerted threatened villages two days previously (blue-flag alert) and now with a red-flag alert announced evacuations, which were completed in less than two days, with approximately 2,300 going to accommodation centers (Loster and Wolf , 2007).

In the emergency period, NIDM, with local and international partner organizations, established networks with local centers to coordinate the emergency operations. The International Federation of Red Cross and Red Crescent Societies and its local partners, the US Agency for International Development, and other organizations worked to distribute basic goods, food, and medical assistance during the emergency period (IFRC, 2007; USAID, 2007).

A resettlement program, although a policy of last resort, to move inhabitants from flood-prone areas to safer areas was initiated (Stal, 2011; WMO, 2011a). Resettlement is not an easy option. Although brick-built housing was provided in flood-safe areas with new (or nearby) schools and health facilities, these have not been as well received as intended as these flood-safe resettlements suffer from water scarcity and drought, and growing crops is therefore difficult (Stal, 2011).

The floods of 2000 and 2007 along with other natural hazards are considered to have undone years of development efforts (Sietz et al., 2008) and to have undermined national efforts in realizing Mozambique's poverty reduction strategy (IMF, 2011).

9.2.6.6. Lessons Identified

This comparison of the two floods events that occurred in Mozambique in 2000 and 2007 shows:
- Floods, as one of the most dangerous natural phenomena, are a real threat to the sustainable development of nations (Ahern et al., 2005; Guha-Sapir et al., 2011)
- The consequences of floods depend on the long-term adaptation to extremes of climate, and associated hydrologic extremes require further understanding. After the 2000 floods in Mozambique, national and international organizations updated their strategies to include disaster preparedness, risk management, and contingency and response capacities according to the lessons of catastrophic floods. The Government of Mozambique introduced new DRM structures between 2000 and 2007, illustrating the flexibility needed to accommodate the scientific and communication systems that need to be in place to adapt to a climate change-driven disaster and that this can be done in liaison with and with guidance from external agencies. Realization of the new program of DRM led to a reduction in consequences from the floods in 2007 (Republic of Mozambique, 2006a).
- Experience in Mozambique shows that creation and development of effective and steadily functioning systems of hydrological monitoring and early warning systems at a local, regional, and national level as key components of DRM allowing more realistic warnings of flooding threats (WMO, 2011a).
- The implementation of resettlement programs in periodically flooded areas in 2007 has reduced flood damage, but these measures are not easy to implement (WMO, 2011a).
- Limited available resources are one of the most important problems for both disaster preparedness and disaster response. The extreme poverty of the people makes them highly vulnerable to floods and other natural disasters, despite the best efforts of the government to protect them (World Bank, 2011b).
- The example of Mozambique shows that climate change adaptation needs to be achieved through the understanding of vulnerability in all sectors (social, infrastructure, production, and environmental) and this knowledge needs to be used for the formulation of preparedness and response mechanisms (Sietz et al., 2008).

9.2.7. Disastrous Epidemic Disease: The Case of Cholera

9.2.7.1. Introduction

Weather and climate have a wide range of health impacts and play a role in the ecology of many infectious diseases (Patz et al., 2000). The relationships between health and weather, climate variability, and long-term climate change are complex and often indirect (McMichael et al., 2006). As with other impacts explored in this report, not all extreme health impacts associated with weather and climate result from extreme events; some result instead from less dramatic events unfolding in the context of high population vulnerability. In such cases, impacts are typically indirect and are mediated by a constellation of factors, as opposed to the direct health impacts of severe weather, for example, traumatic injuries resulting directly from exposure to kinetic energy associated with storms (Noji, 2000).

Commonly, underdeveloped health and other infrastructure, poverty, political instability, and ecosystem disruptions interact with weather to impact health adversely, sometimes to a disastrous degree (Myers and Patz, 2009). For example, cholera is an infectious disease that is perpetuated by poverty and associated factors, though outbreaks are commonly associated with rainy season onset. Research in the last decade has demonstrated that cholera is also sensitive to climate variability (Rodó et al., 2002; Koelle et al., 2005a; Constantin de Magny et al., 2007). Assuming persistence of these vulnerability factors, cholera outbreaks may become more widespread as the climate continues to change (Lipp et al., 2002) due to the projected *likely* increase in frequency of heavy precipitation over many areas of the globe, and tropical regions in particular (see Table 3-1). Insights into the disease's ecology, however, including its climate sensitivity, may one day inform early warning systems and other interventions that could blunt its disastrous impact. Equally, if not more important, poverty reduction and improvements in engineering, critical infrastructure, and political stability and transparency can reduce vulnerability among exposed populations to the degree that cholera could be contained.

9.2.7.2. Background

Cholera has a long history as a human scourge. The world is in the midst of the seventh global pandemic, which began in Indonesia in 1961 and is distinguished by continued prevalence of the El Tor strain of the *Vibrio cholerae* bacterium (Zuckerman et al., 2007; WHO, 2010). Primarily driven by poor sanitation, cholera cases are concentrated in areas burdened by poverty, inadequate sanitation, and poor governance. Between 1995 and 2005, the heaviest burden was in Africa, where poverty, water source contamination, heavy rainfall and floods, and population displacement were the primary risk factors (Griffith et al., 2006).

V. cholerae is flexible and ecologically opportunistic, enabling it to cause epidemic disease in a wide range of settings and in response to climate forcings (Koelle et al., 2005b). Weather, particularly seasonal rains, has long been recognized as a risk factor for cholera epidemics.

Cholera is one of a handful of diseases whose incidence has been directly associated with climate variability and long-term climate change (Rodó et al., 2002). One driver of cholera's presence and pathogenicity is the El Niño-Southern Oscillation (ENSO), which brings higher temperatures, more intense precipitation, and enhanced cholera transmission. ENSO has been associated with cholera outbreaks in coastal and inland regions of Africa (Constantin de Magny et al., 2007), South Asia (Constantin de Magny et al., 2007), and South America (Gil et al., 2004). There is concern that climate change will work synergistically with poverty and poor sanitation to increase cholera risk.

As with other disasters, the risk of disastrous cholera epidemics can be deconstructed into hazard probability, exposure probability, and population vulnerability, which can be further broken down into population susceptibility and adaptive capacity. As noted in the introduction, some disastrous cholera epidemics are not associated with discrete extreme weather events, but extreme impacts are triggered instead by exposure to a less dramatic weather event in the context of high population vulnerability. We focus on factors affecting exposure and vulnerability in general, then apply this discussion to the Zimbabwe cholera epidemic that began in 2008.

9.2.7.2.1. Exposure

Cholera epidemics occur when susceptible human hosts are brought into contact with toxigenic strains of *V. cholerae* serogroup O1 or serogroup O139. A host of ecological factors affect *V. cholerae*'s environmental prevalence and pathogenicity (Colwell, 2002) and the likelihood of human exposure (Koelle, 2009). In coastal regions, there is a commensal relationship between *V. cholerae*, plankton, and algae (Colwell, 1996). Cholera bacteria are attracted to the chitin of zooplankton exoskeletons, which provide them with stability and protect them from predators. The zooplankton feed on algae, which bloom in response to increasing sunlight and warmer temperatures. When there are algal blooms in the Bay of Bengal, the zooplankton prosper and cholera populations grow, increasing the likelihood of human exposure. Precipitation levels, sea surface temperature, salinity, and factors affecting members of the marine and estuarine ecosystem, such as algae and copepods, affect exposure probability (Huq et al., 2005). Many of these factors appear to be similar across regions, although their relative importance varies, such as the association of *V. cholerae* with chitin (Pruzzo et al., 2008) and the importance of precipitation and sea level (Emch et al., 2008). For example, marine and estuarine sources were the source of pathogenic *V. cholerae* strains responsible for cholera epidemics in Mexico in recent El Niño years (Lizarraga-Partida et al., 2009).

Other variables are associated with increased likelihood of exposure, including conflict (Bompangue et al., 2009), population displacement, crowding (Shultz et al., 2009), and political instability (Shikanga et al., 2009). Many of these factors are actually mediated by the more conventional cholera risk factors of poor sanitation and lack of access to improved water sources and sewage treatment.

9.2.7.2.2. Population susceptibility

Population susceptibility includes both physiological factors that increase the likelihood of infection after cholera exposure, as well as social and structural factors that drive the likelihood of a severe, persistent epidemic once exposure has occurred. Physiologic factors that affect cholera risk or severity include malnutrition and co-infection with intestinal parasites (Harris et al., 2009) or the bacterium *Helicobacter pylori*. Infections are more severe for people with blood group O, for children, and for those with increased health-related vulnerability. Waxing and waning immunity as a result of prior exposure has a significant impact on population vulnerability to cholera over long periods (Koelle et al., 2005b).

While physiologic susceptibility is important, social and economic drivers of population susceptibility persistently seem to drive epidemic risk. Poverty is a strong predictor of risk on a population basis (Ackers et al., 1998; Talavera and Perez, 2009), and political factors, as illustrated by the Zimbabwe epidemic, are often important drivers of epidemic severity and persistence once exposure occurs. Many recent severe epidemics exhibit population susceptibility dynamics similar to Zimbabwe, including in other poor communities (Hashizume et al., 2008), in the aftermath of political unrest (Shikanga et al., 2009), and following population displacement (Bompangue et al., 2009).

9.2.7.2.3. Adaptive capacity

Cholera outbreaks are familiar sequelae of complex emergencies. The DRM community has much experience with prevention efforts to reduce the likelihood of cholera epidemics, containing them once they occur, and reducing the associated morbidity and mortality among the infected. Best practices include guidelines for water treatment and sanitation and for population-based surveillance (Sphere Project, 2004).

9.2.7.3. Description of Event

Zimbabwe has had cholera outbreaks every year since 1998, with the 2008 epidemic the worst the world had seen in two decades, affecting approximately 92,000 people and killing over 4,000 (Mason, 2009). The outbreak began on 20 August 2008, slightly lagging the onset of seasonal rains, in Chitungwiza city, just south of the capital Harare (WHO, 2008a). In the initial stages, several districts were affected. In October, the epidemic exploded in Harare's Budiriro suburb and soon spread to include much of the country, persisting well into June 2009 and ultimately seeding outbreaks in several other countries. Weather appears to have been crucial in the outbreak, as recurrent point-source contamination of drinking water sources (WHO, 2008a) was almost certainly amplified by the onset of the rainy season (Luque Fernandez et al., 2009). In addition to its size, this epidemic was distinguished by its urban focus and relatively high case fatality rate (CFR; the proportion of infected people who die) ranging from 4 to 5% (Mason, 2009). Most outbreaks have CFRs below 1% (Alajo et al., 2006). Underlying structural vulnerability with shortages of medicines, equipment, and staff at health facilities throughout the country compounded the effects of the cholera epidemic (WHO, 2008b).

9.2.7.4. Intervention

There are several risk management considerations for preventing cholera outbreaks and minimizing the likelihood that an outbreak becomes a disastrous epidemic (Sack et al., 2006). Public health has a wide range of interventions for preventing and containing outbreaks, and several other potentially effective interventions are in development (Bhattacharya et al., 2009). As is the case in managing all climate-sensitive risks, the role of institutional learning is becoming ever more important in reducing the risk of cholera and other epidemic disease as the climate shifts.

9.2.7.4.1. Conventional public health strategies

The conventional public health strategies for reducing cholera risk include a range of primary, secondary, and tertiary prevention strategies (Holmgren, 1981).

Primary prevention, or prevention of contact between a hazardous exposure and susceptible host, includes promoting access to clean water and reducing the likelihood of population displacement; secondary prevention, or prevention of symptom development in an exposed host, includes vaccination; and tertiary prevention, or containment of symptoms and prevention of complications once disease is manifest, includes dehydration treatment with oral rehydration therapy.

9.2.7.4.2. Newer developments

Enhanced understanding of cholera ecology has enabled development of predictive models that perform relatively well (Matsuda et al., 2008) and fostered hope that early warning systems based on remotely sensed trends in sea surface temperature, algal growth, and other ecological drivers of cholera risk can help reduce risks of epidemic disease, particularly in coastal regions (Mendelsohn and Dawson, 2008). Strategies to reduce physiologic susceptibility through vaccination have shown promise (Calain et al., 2004; Chaignat et al., 2008; Lopez et al., 2008; Sur et al., 2009) and mass vaccination campaigns have potential to interrupt epidemics (WHO, 2006c), and may be cost effective in resource-poor regions or for displaced populations where provision of sanitation and other services has proven difficult (Jeuland and Whittington, 2009). Current World Health Organization policy on cholera vaccination holds that vaccination should be used in conjunction with other control strategies in endemic areas and be considered for populations at risk for epidemic disease, and that cholera immunization is a temporizing measure while more permanent sanitation improvements can be pursued (WHO, 2010). Ultimately, given the strong association with poverty, continued focus on development may ultimately have the largest impact on reducing cholera risk.

9.2.7.5. Outcomes

Managing the risk of climate-sensitive disease, like risk management of other climate-sensitive outcomes, will necessarily become more iterative and adaptive as climate change shifts the hazard landscape and heightens vulnerability in certain populations. Learning is an important component of this iterative process (see Sections 1.4 and 8.6.3.2).

There are multiple opportunities for learning to enhance risk management related to epidemic disease. First, while reactive containment processes can be essential for identifying and containing outbreaks, this approach often glosses over root causes in an effort to return to the status quo. As the World Health Organization states, "Current responses to cholera outbreaks are reactive, taking the form of a more or less well-organized emergency response," and prevention is lacking (WHO, 2006c). Without losing the focus on containment, institutional learning could incorporate strategies to address root causes, reducing the likelihood of future outbreaks. This includes continued efforts to better understand cholera's human ecology to explore deeper assumptions, structures, and policy decisions that shape how risks are constructed. In the case of cholera, such exploration has opened the possibility of devising warning systems and other novel risk management strategies. Another equally important conclusion – one that experts on climate's role in driving cholera risk have emphasized (Pascual et al., 2002) – is that poverty and political instability are the fundamental drivers of cholera risk, and that emphasis on development and justice are risk management interventions as well.

9.2.7.6. Lessons Identified

The 2008 cholera epidemic epitomized the complex interactions between weather events and population vulnerability that can interact to produce disastrous epidemic disease. Recent studies of cholera, including its basic and human ecology, demonstrate the potential for early warning and potential points of leverage that may be useful for interventions to contain future epidemics. The key messages from this work include:
- Variability in precipitation and temperature can affect important epidemic diseases such as cholera both through direct effects on the transmission cycle, but also potentially through indirect effects, for example through problems arising from inadequate basic water and sanitation services.
- If other determinants remained constant, climate change would be expected to increase risk by increasing exposure likelihood – through increased variability in precipitation and gradually rising temperatures and by increasing population vulnerability.
- The health impacts of cholera epidemics are strongly mediated through individual characteristics such as age and immunity, and population-level social determinants, such as poverty, governance, and infrastructure.
- Experience from multiple cholera epidemics demonstrates that non-climatic factors can either exacerbate or override the effects of weather or other infection hazards.
- The processes of DRM and preventive public health are closely linked, and largely synonymous. Strengthening and integrating these measures, alongside economic development, should increase resilience against the health effects of extreme weather and gradual climate change.

9.2.8. Coastal Megacities: The Case of Mumbai

9.2.8.1. Introduction

In July 2005, Mumbai, India, was struck by an exceptional storm (Revi, 2005). In one 24-hour span alone, the city received 94 cm of rain, and the storm left more than 1,000 dead, mostly in slum settlements (De Sherbinin et al., 2007; Sharma and Tomar, 2010). A week of heavy rain disrupted water, sewer, drainage, road, rail, air transport, power, and telecommunications systems (Revi, 2005). As a result of this 'synchronous failure,' Mumbai-based automated teller machine banking systems ceased working across much of the country, and the Bombay and National Stock Exchanges were temporarily forced to close (Revi, 2005; UNISDR, 2011b). This demonstrates that within megacities, risk and loss are both concentrated and also spread through networks of critical infrastructure as well as connected economic and other systems.

9.2.8.2. Background

At present, Mumbai is the city with the largest population exposed to coastal flooding – estimated at 2,787,000 currently, and projected to increase to more than 11 million people exposed by 2070 (Hanson et al., 2011). During that same period, exposed assets are expected to increase from US$ 46.2 billion to nearly US$ 1.6 trillion (Hanson et al., 2011).

Mumbai's significant, and increasing, exposure of people and assets – both within the urban fabric but also outside, connected to the city's functions through networks of critical infrastructure, financial, and resource flows – will be affected by changes in climate means and climate extremes (Nicholls et al., 2007; Revi, 2008; Fuchs et al., 2011; Ranger et al., 2011). It is difficult to associate a single extreme event with climate change, but it may be possible to discuss the changed probability of an event's occurrence in relation to a particular cause, such as global warming (see FAQ 3.2). For the Indian monsoon, for example, extreme rain events have an increasing trend between 1901 and 2005, with the trend being stronger since 1950 (see Section 3.4.1).

9.2.8.3. Description of Vulnerability

Attributing causes of changes in monsoons is difficult due to substantial differences between models, and the observed maximum rainfall on India's west coast, where Mumbai is located, is poorly simulated by many models (see Section 3.4.1). That being said, increases in precipitation are projected for the Asian monsoon, along with increased interannual seasonally averaged precipitation variability (see Section 3.4.1).

Furthermore, extreme sea levels can be expected to change in the future as a result of mean sea level rise and changes in atmospheric storminess, and it is *very likely* that sea level rise will contribute to increases in extreme sea levels in the future (see Section 3.5.3).

The development failures that have led to an accumulation of disaster risk in Mumbai and allowed its transmission beyond the urban core are common to many other large urban centers. The IPCC Fourth Assessment Report (AR4) stated with *very high confidence* that the impact of climate change on coasts is exacerbated by increasing human-induced pressures, with subsequent studies being consistent with this assessment (see Section 3.5.5).

The AR4 also reported with *very high confidence* that coasts will be exposed to increasing risks, including coastal erosion, over coming decades due to climate change and sea level rise, both of which will be exacerbated by increasing human-induced pressures (see Section 3.5.5).

The July 2005 flooding in Mumbai underscores the fact that coastal megacities are already at risk due to climate-related hazards (De Sherbinin et al., 2007; McGranahan et al., 2007). Refuse and debris commonly clog storm drains, causing flooding even on the higher ground in Mumbai's slums, and landslides are another threat to squatter communities that are near or on the few hillsides in the city (De Sherbinin et al., 2007). Urban poor populations often experience increased rates of infectious disease after flood events, and after the July 2005 floods the prevalence of leptospirosis rose eightfold in Mumbai (Maskey et al., 2006; Kovats and Akhtar, 2008).

To the present, drivers of flood risk have been largely driven by socioeconomic processes and factors, such as poverty, ecosystem degradation, and poorly governed rapid urbanization (Revi, 2005, 2008; De Sherbinin et al., 2007; Huq et al., 2007; UNISDR, 2009c, 2011b; Hanson et al., 2011; Ranger et al., 2011). These processes are interrelated, and within these cities, vulnerability is concentrated in the poorest neighborhoods, which often lack access to sanitation, health care, and transportation infrastructure, and whose homes and possessions are unprotected by insurance (Revi, 2005; De Sherbinin et al., 2007; UNISDR, 2009c; Ranger et al., 2011).

Slum settlements are often located in sites with high levels of risk due to environmental and social factors. For example, they are often located in floodplains or on steep slopes, which means their residents suffer from a considerable degree of physical exposure and social vulnerability to losses from flood events (Huq et al., 2007; McGranahan et al., 2007; Chatterjee, 2010).

Mumbai is one of many coastal megacities that have been built in part on reclaimed land, a process that increases flood risks to low-lying areas where slums are frequently located (Chatterjee, 2010). Its slums do not benefit from structural flood protection measures and are located in low-lying areas close to marshes and other marginal places and are frequently flooded during monsoon season, especially when heavy rainfall occurs during high tides (McGranahan et al., 2007; Chatterjee, 2010). A rise in sea level of 50 cm, together with storm surges, would render uninhabitable the coastal and low-lying areas (De Sherbinin et al., 2007) where many of Mumbai's informal settlements are currently located.

9.2.8.4. Outcomes/Consequences

India's 2001 census indicated that in Mumbai 5,823,510 people (48.9% of the population) lived in slums (Government of India, 2001). In 2005, the global slum population was nearly 1 billion, and it is projected to reach 1.3 to 1.4 billion by 2020, mostly concentrated in cities in developing countries (UN-HABITAT, 2006). In addition to Mumbai, Hanson et al. (2011) found that the following cities will have the greatest population exposure to coastal flooding in 2070: Kolkata, Dhaka, Guangzhou, Ho Chi Minh City, Shanghai, Bangkok, Rangoon, Miami, and Hai Phòng. Many of these cities are already characterized by significant population and asset exposure to coastal flooding, and all but Miami are located in developing countries in Asia.

Africa does not have a large share of the world's biggest coastal cities but most of its largest cities are on the coast and large sections of their population are at risk from flooding (Awuor et al., 2008; Adelekan, 2010). Compared to Asia, Europe, and the Americas, a greater percentage of Africa's population lives in coastal cities of 100,000 to 5 million people, which is noteworthy because Africa's medium-to-large cities tend to be poor and many are growing at much higher rates than cities on the other continents (McGranahan et al., 2007).

The amount of vulnerability concentrated within these cities will define their risks, and in the absence of adaptation there is *high confidence* that locations currently experiencing adverse effects, such as coastal erosion and inundation, will continue to do so in the future (see Section 3.5.5).

However, there is a certain limit to adaptation given that these cities are fixed in place and some degree of exposure to hazards is 'locked in' due to the unlikelihood of relocation (Hanson et al., 2011). For example, India's large infrastructure investments, which have facilitated Mumbai's rapid growth, were built to last 50 to 150 years (Revi, 2008). This forecloses some adaptation and DRR strategies, such as risk avoidance. Furthermore, all large coastal cities are centers of high population density, infrastructure, investments, networking, and information (McGranahan et al., 2007; Chatterjee, 2010). This concentration and connectivity make them important sources of innovation and economic growth, especially in developing countries where these ingredients may be absent elsewhere. This underscores the importance of governance and economic relations, including insurance and more general basic needs of health and education, in allowing urban systems and those at risk to build resilience if they cannot avoid hazard.

9.2.8.5. Lessons Identified

Measures to reduce exposure to existing weather-related hazards can also serve as means of adapting to climate change (McGranahan et al., 2007; UNISDR, 2009c, 2011b; Chapters 1 and 2). At the time of the 2005 flood, Mumbai lacked the capacity to address a complex portfolio of (interrelated) risks (De Sherbinin et al., 2007; Revi, 2008), and its multi-hazard risk plan from 2000 was not well implemented (Revi, 2005). Risk protection in most other megacities in developing countries was also found to be more informal than robust (Hanson et al., 2011). Multi-hazard risk models, based on probabilistic analysis, can help governments better reduce risks and facilitate better management of and preparedness for risks that cannot be reduced cost effectively (Revi, 2008; Ranger et al., 2011; UNISDR, 2011b).

Given that up to US$ 35 trillion (approximately 9% of projected global GDP) may be exposed to climate-related hazards in port cities by 2070 (in purchasing power parity, 2001 US$) (Hanson et al., 2011), managing – and reducing – these risks represents a high-leverage policy area for adaptation. The scale of economic assets at risk is impressive and to this must be added the livelihoods and health of the poor that may be disproportionately impacted by disaster events but have partial visibility in macroeconomic assessments.

The need to adapt is especially acute in developing countries in Asia given that 14 of the top 20 urban agglomerations projected to have the greatest exposure of assets in 2070 are in developing countries in this region (Hanson et al., 2011). This suggests that scaled-up financing for adaptation may be needed to safeguard the residents and economic activity in these cities to a level comparable to that of other coastal megacities that face similar population and asset exposure, such as New York or Tokyo. Two critical distinctions are the degree of poverty and the less complete reach of local government in those cities most at risk.

Despite efforts to assess the impacts of climate change at the city scale, analysis of the economic impacts of climate change at this scale has received relatively little attention to date (Hallegatte and Corfee-Morlot, 2011). In developing countries, the sometimes incomplete understanding of climate risks and the limited institutional capacity have meant that analysis of climate change impacts at the city scale has generally considered only flood risks and not yet assessed additional potential impacts (Hunt and Watkiss, 2011). A standardized, multi-hazard impact analysis at the city scale would be useful and facilitate comparisons between cities (Hunt and Watkiss, 2011).

9.2.9. Small Island Developing States: The Challenge of Adaptation

9.2.9.1. Introduction

Small Island Developing States (SIDS) are defined as those that are small island nations, have low-lying coastal zones, and share development challenges (UNCTAD, 2004). Strengthening of SIDS technical capacities to enable resilience building has been recommended (UNESC, 2011).

This case study explores the critical vulnerabilities of the Republic of the Marshall Islands (RMI). Additional data from the Maldives, also highly vulnerable to sea level rise and extreme weather events and where the tsunami caused significant damage, and Grenada, which is a country with a small open economy vulnerable to external shocks and natural disasters, are used to develop the full context of the limits of adaptation. Specifically, the RMI highlights the availability of fresh water as a major concern. "There is strong evidence that under most climate change scenarios, water resources in small islands are likely to be seriously compromised (*very high confidence*)" (Mimura et al., 2007).

9.2.9.2. Background

SIDS can be particularly vulnerable to hazards and face difficulties when responding to disasters (TDB, 2007). SIDS share similar development challenges including small but growing populations, economic dependence on international funders, and lack of resources (e.g., freshwater, land) (World Bank, 2005b; UNFCCC, 2007a). The IPCC (Mimura et al., 2007) concluded that "small islands, whether located in the tropics or higher latitudes, have characteristics which make them especially vulnerable to the effects of climate change, sea level rise, and extreme events (*very high confidence*)." Many SIDS share vulnerabilities with high levels of poverty and are reported to suffer serious environmental degradation and to have weak human and institutional capacities for land management that is integrated and sustainable (GEF, 2006). The range of physical resources available to states influences their options to cope with disasters and the relatively restricted economic diversity intrinsic to SIDS minimizes their capacity to respond in emergencies with measures such as shelter or evacuation (Boruff and Cutter, 2007). Hence SIDS are among the most vulnerable states to the impacts of climate change (UNFCCC, 2007b). As of 2010, 38 UN member nations and 14 non-UN Members/Associate Members of the Regional Commissions were classified as SIDS (UN-OHRLLS, 2011).

The RMI provides an example of critical vulnerabilities. It is made up of five islands and 29 atolls that are spread across more than 1.9 million square kilometers of Pacific Ocean (World Bank, 2006a). The country has a population of 64,522, approximately two-thirds of which is concentrated in urban areas on just two atolls (UNDESA, 2010; World Bank, 2011b). The other one-third live on the even more remote outer islands and atolls (World Bank, 2006a). Even the main inhabited islands remain extremely isolated; the nearest major port is over 4,500 km from Majuro, the capital atoll (World Bank, 2005b).

The Maldives and Grenada both provide other examples of vulnerability to extreme events and disasters and climate change adaptation needs:
- The Maldives consist of 1,192 small islands. 80% are 1 m or less above sea level (Quarless, 2007), of which only three islands have a surface area of more than 500 ha (De Comarmond and Payet, 2010). These characteristics make them highly vulnerable to damage from sea level rise and extreme weather events. The economic and survival challenges of the people of the Maldives were evident after the 2004 tsunami caused damage equivalent to 62% of national GDP (World Bank, 2005c). As of 2009, the country still faced a deficit of more than US$ 150 million for reconstruction.

- Such devastation in a SIDS might be countered with further disaster preparation and efforts to maintain emergency funds to rebuild their economies (De Comarmond and Payet, 2010).
- Grenada is a small tri-island state in the Eastern Caribbean with a population of 102,000 and a per capita GDP of US$ 4,601 in 2004 (IMF, 2011). It is a small open economy, vulnerable to external shocks and natural disasters as seen by the effects of Hurricane Ivan, which created large fiscal and balance of payments financing needs in 2004, and Hurricane Emily, which struck in 2005 (IMF, 2004, 2006). Hurricane Ivan brought major disruption to an economic recovery process, eventually costing the island an estimated US$ 3 billion (Boruff and Cutter, 2007). It is projected that Ivan reduced the country's forecasted growth rate from 5.7% to -1.4% (Quarless, 2007). Hurricane Emily followed 10 months later, virtually completing the trail of destruction started by Ivan. The impact was seen in every sector of the economy. Capital stock was severely damaged and employment was significantly affected (UNDP, 2006).

9.2.9.3. Description of Vulnerability

Many SIDS face specific disadvantages associated with their small size, insularity, remoteness, and susceptibility to natural hazards. SIDS are particularly vulnerable to climate change because their key economic sectors such as agriculture, fisheries, and tourism are all susceptible (Barnett and Adger, 2003; Read, 2010) (a more extensive discussion is provided in Chapter 4, especially Sections 4.2.1, 4.4.4, 4.5.2, and 4.5.3). The hazards of extreme weather events are coupled with other long-term climate change impacts, especially sea level rise (see Box 3-4). Low-lying atoll communities, such as the Maldives and Cook Islands, are especially vulnerable (Ebi et al., 2006; Woodroffe, 2008; Kelman and West, 2009) and are expected to lose significant portions of land (Mimura et al., 2007). Small island states and particularly atoll countries may experience erosion, inundation, and saline intrusion resulting in ecosystem disruption, decreased agricultural productivity, changes in disease patterns, economic losses, and population displacement – all of which reinforce vulnerability to extreme weather events (Pernetta, 1992; Nurse and Sem, 2001; Mimura et al., 2007).

SIDS suffer higher relative economic losses from natural hazards and are less resilient to those losses so that one extreme event may have the effect of countering years of development gains (UN, 2005; Kelman, 2010). The distances between many SIDS and economic centers make their populations among the most isolated in the world (World Bank, 2005b).

Underdevelopment and susceptibility to disasters are mutually reinforcing: disasters not only cause heavy losses to capital assets, but also disrupt production and the flow of goods and services in the affected economy, resulting in a loss of earnings (Pelling et al., 2002). In both the short and the long term, those impacts can have sharp repercussions on the economic development of a country, affecting GDP, public finances, and foreign trade, thus increasing levels of poverty and public debt (Mirza, 2003; Ahrens and Rudolph, 2006). Climate change threatens to exacerbate existing vulnerabilities and hinder socioeconomic development (UNFCCC, 2007b).

The RMI faces major climate-related natural hazards including sea level rise, tropical storms or typhoons with associated storm surges, and drought. These hazards should be considered within the context of additional hazards and challenges such as ecosystem degradation, pollution of the marine environment, and coastal erosion as well as food security. The RMI faces physical and economic challenges that amplify the population's vulnerability to climate hazards, including high population density, high levels of poverty, low elevation, and fragile freshwater resources (World Bank, 2011b). The Global Facility for Disaster Reduction and Recovery report concludes that the hazard that poses the most threat is sea level rise, as the highest point on RMI is only 10 m above sea level (World Bank, 2011b). Consequently, multilateral donors considered the RMI 'high risk' and the Global Facility for Disaster Reduction and Recovery has identified it as a priority country for assistance (World Bank, 2011b).

9.2.9.4. Outcomes/Consequences

A range of both local and donor-supported actions have endeavored to build resilience among SIDS. The example of the RMI shows the benefits that risk reduction and climate change adaptation efforts may offer other island states.

Freshwater availability is a major concern for many SIDS (Quarless, 2007), including the RMI. Since SIDS are especially vulnerable to extreme weather events, their water supplies face rapid salinization due to seawater intrusion and contamination (PSIDS, 2009). According to one study, countries such as the RMI lack the financial and technical resources to implement seawater desalination for their populations (UNDESA, 2011). Some disaster and climate risk management gains may come from simple technology (UNDESA, 2010). New scavenger technology for wells has been introduced (UNESC, 2004) as one of the ways forward. Simple abstraction of freshwater from thin groundwater lenses (a typical practice in oceanic atolls) often results in upward coning of saltwater, which, in turn, causes contamination of the water supplies. The RMI has benefited from its use of new, pioneering technology to limit the effects of extreme weather events on its water supply (UNDESA, 2011). The improvement of climate sensitivity knowledge, particularly in the context of risk management, is key to adaptation to climate change. Climate and disaster risk are closely entwined and, for example, resilience to drought and resilience to climate change both stand to be enhanced through a single targeted program.

In addition to project-oriented development assistance, the RMI receives substantial financial assistance from the United States through a Compact of Free Association (Nuclear Claims Tribunal, Republic of the Marshall Islands, 2007). Grants and budget support provided under Compact I over the period of 1987 to 2001 totaled an average of over 30% of GDP, not including any other form of bilateral assistance (World

Bank, 2005b). The RMI stands out among other lower middle income countries, receiving average aid per capita of US$ 1,183, compared with the average of US$ 8 for other lower middle income countries (World Bank, 2005b). This assistance, buttressed by national disaster management policies dating back to the RMI's independence in 1986 and including the Global Facility for Disaster Reduction and Recovery's role in assessing the RMI's systems and noting existing gaps for future development partner projects, has resulted in a range of national and regional disaster and climate risk management initiatives (World Bank, 2009, 2011b).

9.2.9.5. Lessons Identified

The physical, social, and economic characteristics by which SIDS and developing countries are defined (education, income, and health, for example) increase their vulnerability to extreme climate events. Experiences from the Marshall Islands, the Maldives, and Grenada indicate that limited freshwater supplies and inadequate drainage infrastructure are key vulnerability factors. These examples also indicate an important difference between risk of frequent smaller hazards and catastrophic risk of infrequent but extreme events.

The cases of Grenada and the Maldives demonstrate the high relative financial impact that a hazard can have on a small island state. For the RMI, financial support from donors has enabled a range of risk management programs. Although the importance of disaster risk reduction strategies is apparent, preventive approaches continue to receive less emphasis than disaster relief and recovery (Davies et al., 2008). Considering the range of challenges facing policymakers in some SIDS, preventive climate adaptation policies can seem marginal compared with pressing issues of poverty, affordable energy, affordable food, transportation, health care, and economic development.

National policymaking in this context remains a major challenge and availability of funding for preventive action – such as disaster and climate risk management – may continue to be limited for many countries (Ahmad and Ahmed, 2002; Jegillos, 2003; Huq et al., 2006; Yohe et al., 2007). Although most developing countries participate in various international protocols and conventions relating to climate change and sustainable development and most have adopted national environmental conservation and natural disaster management policies (Yohe et al., 2007), the policy agendas of many developing countries do not yet fully address all aspects of climate change (Beg et al., 2002).

9.2.10. Changing Cold Climate Vulnerabilities: Northern Canada

9.2.10.1. Introduction

In cold climate regions all over the world, climate change is occurring more rapidly than over most of the globe (Anisimov et al., 2007). These changes have implications for the built environment. The vulnerability of residents of the Canadian North is complex and dynamic. In addition to the increasing risks from extreme weather events, there are climate impacts upon travel, food security, and infrastructural integrity, which in turn affect many other aspects of everyday life (Pearce et al., 2009; Ford et al., 2010). Additionally, the relative isolation of these northern communities makes exposure to climate-related risk more difficult to adapt to, thus increasing their level of vulnerability (Ford and Pearce, 2010). This case study will examine the increased vulnerabilities in regions of the Canadian North due to climate change's effect on infrastructure through changes in permafrost thaw and snow loading. The study illustrates existing and projected risks and governmental responses to them at the municipal, provincial/territorial, and national levels. Canada has three territories: the Yukon (YT); the Northwest Territories (NWT); and Nunavut (NU); this study deals with all three and, to a much lesser extent, the northern regions of the provinces, such as Nunavik in northern Quebec. Though both permafrost thaw and changing snow loads are slowly progressing events, as opposed to one-time extreme events, their impacts can result in disasters. Future protection relies upon risk reduction and adaptation. Sections 3.3.1 and 3.5.7 discuss changes in cold extremes and other climate variables at high latitudes.

9.2.10.2. Background

Over the past few decades, the northern regions of Canada have experienced a rate of warming about twice that of the rest of the world (McBean et al., 2005; Field et al., 2007; Furgal and Prowse, 2008). In northern Canada, winter temperatures are expected to rise by between 3.5 and 12.5°C by 2080, with smaller changes projected for spring and summer; in more southerly regions of northern Canada, temperatures could warm to be above freezing for much longer periods (Furgal and Prowse, 2008). For example, whereas it was estimated that the Northwest Passage would be navigable for ice-strengthened cargo ships in 2050 (Instanes et al., 2005), it has already been navigable in 2007 (Barber et al., 2008). Recent studies have suggested that some communities in northern Canada will be vulnerable to the accelerated rate of climate change (Ford and Smit, 2004; Ford and Furgal, 2009).

Higher temperatures have several implications for infrastructure that plays an important role in maintaining the social and economic functions of a community (CSA, 2010). Permafrost thaw and changing snow loads have the potential to affect the structural stability of essential infrastructure (Nelson et al., 2002; Couture and Pollard, 2007). Design standards in northern Canada were based on permafrost and snow load levels of a previous climate regime (CSA, 2010). Adaptation is essential to avoid higher operational and maintenance costs for structures and to ensure that the designed long lifespan of each structure remains viable (Allard et al., 2002). Addressing these impacts of climate change is a complex task. Naturally each structure will be differently affected and the resulting damage can exacerbate existing weaknesses and create new vulnerabilities. For example, although increasing snow loads alone can have negative impacts on infrastructure, the fact that many buildings have been structurally weakened by permafrost thaw adds to the damage potential during any snow event (CSA, 2010).

9.2.10.3. Description of Vulnerability

9.2.10.3.1. Permafrost thaw

Permafrost thaw is one of the leading factors increasing climate-related vulnerability. Permafrost is by definition dependent on a sub-zero temperature to maintain its state (CSA, 2010; NRCAN, 2011a). With a changing climate, it is difficult to predict where permafrost is most likely to thaw, but about half of Canada's permafrost zones are sensitive to small, short-term increases in temperature, compromising the ability of the ground to support infrastructure (Nielson, 2007; NRTEE, 2009; CSA, 2010). The rate of thaw (and hence implications for infrastructure stability) is also dependent on soil type within the permafrost zone (Nielson, 2007). Areas that have ice-rich soil are much more likely to be affected than those with a lower ice-soil ratio or those that are underlain by bedrock (Nelson et al., 2002). Municipalities in discontinuous or sporadic permafrost zones may feel the impacts of a warming climate more intensely since the permafrost is thinner than it would be in continuous zones where ice has built up over time (Nelson et al., 2002).

Though some infrastructure maintenance will always be required, climate-related permafrost thaw will increase the needs for infrastructure maintenance and the rate of damage that is inflicted (Allard et al., 2002). Permafrost thaw affects different types of infrastructure in radically different ways. In northern Canada, municipalities have experienced many different climate-related impacts on physical infrastructure including the following (Infrastructure Canada, 2006; Nielson, 2007; NRTEE, 2009):

- Nunavik, in northern Quebec, reported that local roads and airport runways have suffered from severe erosion, heaving, buckling, and splitting (Nielson, 2007; Fortier et al., 2011).
- In Iqaluit, in Nunavut, 59 houses have required foundation repair and/or restoration and buildings with shallow foundation systems have been identified as needing attention in the near future (Nielson, 2007). In Inuvik, in the Northwest Territories, a recent study estimated that 75% of the buildings in the municipality would experience structural damage (Bastedo, 2007) depending on the rate of permafrost thaw.
- The Tibbitt to Contwoyto winter road (Northwest Territories) experienced climate-related closures in 2006, remaining open for only 42 days compared to 76 in 2005 (Bastedo, 2007). This resulted in residents and businesses having to airlift materials to their communities instead. In particular, the Diavik Diamond Mine was forced to spend millions of dollars flying in materials (Bastedo, 2007; Governments of Northwest Territories, Nunavut, and Yukon, 2010).
- The Northwest Territories reported that the airport runway in Yellowknife required extensive retrofitting when the permafrost below it began to thaw (Infrastructure Canada, 2006).

The impacts of permafrost thaw on infrastructure have implications for the health, economic livelihood, and safety of northern Canadian communities. The costs of repairing and installing technologies to adapt to climate change in existing infrastructure can range from several million to many billions of dollars, depending on the extent of the damage and the type of infrastructure that is at risk (Infrastructure Canada, 2006). Lessons from municipalities in the United States have proven that these costs can be large. For instance, while the Yukon had financial difficulties with CDN\$ 4,000 km^{-1} yr^{-1} costs related to permafrost damage to highways, Alaska is experiencing costs of up to CDN\$ 30,000 km^{-1} yr^{-1} for an annual cost of over CDN\$ 6 million over a 200-km stretch (Governments of Northwest Territories, Nunavut, and Yukon, 2010). In the future, as infrastructure needs to be replaced, costs will multiply rapidly (Larsen et al., 2008).

9.2.10.3.2. Snow loading

In most northern Canadian communities, buildings and roadways are built using historical snow load standards (Nielson, 2007; Auld, 2008). This makes them particularly vulnerable to climate change since snow loads are expected to increase with higher levels of winter precipitation (Christensen et al., 2007; NRTEE, 2009). Already in the Northwest Territories, 10% of public access buildings have been retrofitted since 2004 to address critical structural malfunctions. An additional 12% of buildings are on high alert for snow load-related roof collapse (Auld et al., 2010). In Inuvik, NWT, a local school suffered a complete roof collapse under a particularly heavy snowfall (Bastedo, 2007). As permafrost continues to thaw, resulting in a loss of overall structural integrity, greater impacts will be linked to the increase in snow loads as previously weakened or infirm structures topple under larger or heavier snowfalls.

9.2.10.4. Outcomes

In response to these vulnerabilities, government and community leaders have put emphasis on action and preparedness (Government of Northwest Territories, 2008; Governments of Northwest Territories, Nunavut, and Yukon, 2010). The social impacts of relocating communities or complete restoration after a major disaster, as well as the financial costs, provide a strong deterrent to complacency and relocation will be utilized where necessary as a last resort (USARC, 2003). Though each government tier, from federal to municipal levels, tackles the issue from a different angle, their approaches are proving complementary as is demonstrated below. This section explores adaptation efforts from each level of government and the contribution they make to adaptive capacity in northern Canadian communities.

9.2.10.4.1. Federal level

The Canadian government contributes to numerous adaptation efforts at different levels and through various programs (Lemmen et al., 2008).

Some federal-level climate change adaptation programs are reactive; for example, at the most basic level, the federal government is responsible for the provision of assistance after a disaster or in order to relocate structures and communities (Henstra and McBean, 2005). Other programs are more proactive, designed to prevent disasters from occurring; for example, climate change is currently being incorporated into the 2015 version of the National Building Code (Environment Canada, 2010), which would help ensure that future infrastructure is built to a more appropriate standard and that adaptive measures are incorporated into the design and building of any new infrastructure. This could also help ensure that adaptation measures are implemented in a uniform way across the country.

In addition, several federal-level departments have programs specially designed to prevent damage from climate-related impacts. As part of the Climate Change Adaptation Program offered by the Aboriginal Affairs and Northern Development Canada, the Assisting Northerners in Assessing Key Vulnerabilities and Opportunities helps to support aboriginal and northern communities, organizations and territories in addressing the urgent climate-related risks (INAC, 2010). For example, the program offers risk assessments for existing infrastructure, water quality, and management programs and helps to identify new infrastructure designs to reduce risk from climate change (INAC, 2010).

Similarly, the Regional Adaptation Collaborative (RAC) funding provided by Natural Resources Canada was designed to assist communities that are adapting to climate change (NRCAN, 2011b). The Northern RAC initiatives are focused on identifying vulnerabilities in the mining sector. Permafrost thaw and snow loading are examples of factors that the program will examine (NRCAN, 2011b).

Another adaptation initiative that has come from the federal level is the site selection guidelines developed by the Canadian Standards Association (CSA, 2010). Though voluntary, this set of guidelines encourages engineers, land use planners, and developers to consider environmental factors including the rate of permafrost thaw and type of soil when building (CSA, 2010). Additionally, it strongly encourages the use of projections and models in the site selection process, instead of relying on extrapolated weather trends (CSA, 2010).

Similarly, federal-level design requirements such as the Canadian environmental assessment process are required to account for climate change in the design phase of significant new projects such as tailings containment, water retention, pipelines, or roads (Furgal and Prowse, 2008). Facilitating use of the guidelines and environmental assessment requirements are proactive responses that aim to prevent future permafrost-related damage to infrastructure.

9.2.10.4.2. Provincial/territorial level

The territorial governments are contributing to the protection of infrastructure in several ways, including conducting and funding research to identify vulnerable areas and populations (INAC, 2010). The Yukon transportation department has undertaken several adaptation initiatives including the design and implementation of road embankments to minimize melting; construction of granular blankets on ice-rich slopes to provide for stability and to prevent major slope failure; and the installation of culverts in thawed streambeds (Government of Yukon, 2010). Ground-penetrating radar and resistivity to assess permafrost conditions underground are being used in Nunavik, Quebec (Fortier et al., 2011). To protect existing permafrost, light-colored pavement on roadways is being used to reflect greater amounts of sunlight and prevent heat absorption (Walsh et al., 2009). Collaborations with federal-level departments to address community infrastructure resilience are being conducted with, for example, the Nunavut Climate Change Partnership, which involves the Government of Nunavut, Natural Resources Canada, Aboriginal Affairs and Northern Development Canada, and the Canadian Institute of Planners (NRCAN, 2011c). These programs help communities to develop action plans that detail suitable options for addressing issues related to climate change. The Yukon government is providing funding for municipalities to develop their own climate change adaptation plans through the Northern Strategy Trust Fund to the Northern Climate ExChange (Government of Yukon, 2009).

About 85 flat-loop thermosyphons, a sort of ground-source heat pump, which extract heat from the ground (through convection) during the winter and reduce thawing, have been constructed in territorial-owned buildings including schools and hospitals, prisons, and visitor centers in Nunavut, the Northwest Territories, and the Yukon (Holubec, 2008; CSA, 2010). The installation of thermosyphon technology is not, in itself, a long-term strategy but merely prolongs the lifetime of most infrastructures (CSA, 2010). Finally, screw jack foundations, a technology that helps to stabilize vulnerable foundations and has been used to prevent damage due to permafrost thaw and related shifting of house foundations, have been implemented in new buildings built by the Northwest Territories Housing Corporation (Government of Northwest Territories, 2008).

9.2.10.4.3. Municipal level

The municipal level is often most involved in building adaptive capacity and implementing adaptation strategies (Black et al., 2010) because municipal governments feel the effects of damaged infrastructure more keenly than higher levels of government (Richardson, 2010). Municipalities, community groups, and businesses all over the three territories have contributed in many ways. Some examples include:
- Urban planning and design are being used to reduce exposure to wind and snowdrifts as well as minimize heat loss from buildings in Iqaluit, NU (NRCAN, 2010).
- Insulated lining was placed underneath a 100-m section of runway to prevent damage from permafrost thaw in Yellowknife, NWT (Infrastructure Canada, 2006).
- Ice-rich soil under important infrastructure has been replaced with gravel and heat-absorbing pavement in Yellowknife, NWT (Bastedo, 2007).

- Wind deflection fins are being used to prevent snow loading on roofs and obstructions around exits in NWT (Waechter, 2005).
- In Tuktoyaktuk, NWT, important buildings, including the police station and a school at risk of severe damage or loss have been moved inland (Governments of Northwest Territories, Nunavut, and Yukon, 2010) and concrete mats bound together with chains are being used to limit erosion (Johnson et al., 2003).
- Shims or pillars to elevate buildings are being used to make them less vulnerable to permafrost thaw (USARC, 2003).
- Construction of new bridges and all-weather roads to replace ice roads that are no longer stable is underway (Infrastructure Canada, 2006).

9.2.10.5. Lessons Identified

Northern Canada can be considered a vulnerable region given the expected climate-related risks. As the climate continues to warm in the North, infrastructure in many remote communities will become more vulnerable as well.

More research, especially into vulnerabilities in northern regions of the globe, and the identification of adaptation options for established communities would be of benefit for adaptation. Additionally, while governmental programs and support are available, a significant portion of it has been devoted to adaptation planning and strategizing. An important issue is the funding needed to help northern Canadian communities implement adaptation actions.

Finally, codes and standards are an integral part of addressing climate impacts on infrastructure. Given the importance of this task, building codes in vulnerable regions need more review and attention to protect communities. An evaluation and monitoring program that focuses on codes and structures as well as adaptation options is noticeably lacking. Despite the complexity of these risks, however, a concerted effort from three tiers of government and community can work to reduce the vulnerability of infrastructure and northern communities.

9.2.11. Early Warning Systems: Adapting to Reduce Impacts

9.2.11.1. Introduction

It is recognized that vulnerability and exposure can never be reduced to zero but risk can be reduced by effective systems for early warning of extreme events that may occur in the near- through to longer-term future (Broad and Agrawala, 2000; Da Silva et al., 2004; Haile, 2005; Patt et al., 2005; Hansen et al., 2011). This sense of 'seeing the future' by understanding current and projected risks is essential to effectively prepare for, respond to, and recover from extreme events and disasters. It is important to recognize that a changing climate poses additional uncertainty and therefore early warning systems can contribute to climate-smart disaster risk management. Effective disaster risk management in a changing climate is facilitated by strong coordination within and between sectors to realize adaptation potentials through assessing vulnerabilities and taking anticipatory actions (Choularton, 2007; Braman et al., 2010).

9.2.11.2. Background

The Hyogo Framework for Action (UNISDR, 2010; Chapter 7) stresses that early warning systems should be "people centered" and that warnings need to be "timely and understandable to those at risk" and need to "take into account the demographic, gender, cultural, and livelihood characteristics of the target audiences." "Guidance on how to act upon warnings" should be included. An early warning system is thus considerably more than just a forecast of an impending hazard.

In 2006, the United Nations International Strategy for Disaster Reduction completed a global survey of early warning systems. The executive summary opened with the statement that "If an effective tsunami early warning system had been in place in the Indian Ocean region on 26 December 2004, thousands of lives would have been saved. ... Effective early warning systems not only save lives but also help protect livelihoods and national development gains" (Basher, 2006; UN, 2006). Improved early warning systems have contributed to reductions in deaths, injuries, and livelihood losses over the last 30 years (IFRC, 2009). Early warning systems are important at local (Chapter 5), national (Chapter 6), and international scales (Chapter 7). Toward the achievement of sustainable development, early warning systems provide important information for decisionmaking and in avoiding tipping points (Chapter 8).

9.2.11.3. Description of Strategy of Early Warning Systems

Early warning systems are to alert and inform citizens and governments of changes on time scales of minutes to hours for immediate threats requiring urgent evasive action; weeks for more advanced preparedness; and seasons and decades for climate variations and changes (Brunet et al., 2010). To date most early warning systems have been based on weather predictions, which provide short-term warnings often with sufficient lead time and accuracy to take evasive action. However, the range of actions that can be taken is limited. Weather predictions often provide less than 24 hours notice of an impending extreme weather event and options in resource-poor areas may not extend beyond the emergency evacuations of people (Chapter 5). Thus although lives may be saved, livelihoods can be destroyed, especially those of the poorest communities.

While most of the successfully implemented early warning systems to date have focused on shorter time scales, for example, for tornadoes (Doswell et al., 1993), benefits of improved predictions on sub-seasonal to seasonal scales are being addressed (Nicholls, 2001; Brunet et al., 2010; Webster et al., 2010). While hazardous atmospheric events can develop in a matter of minutes (in the case of tornadoes), it can be across seasons and decades that occurrence of extremes can change climatically (McBean, 2000). Since planning for hazardous events involves decisions across a full range of time scales, 'An Earth-system Prediction Initiative for the 21st Century' covering all scales has been proposed (Shapiro et al., 2007, 2010).

With the rapid growth in the number of humanitarian disasters, the disaster risk management community has become attentive to changes in extreme events possibly attributed to climate change including floods, droughts, heat waves, and storms that cause the most frequent and economically damaging disasters (Gall et al., 2009; Munich Re, 2010; Vos et al., 2010). Early warning systems provide an adaptation option to minimize damaging impacts resulting from projected severe events. Such systems also provide a mechanism to increase public knowledge and awareness of natural risks and may foster improved policy- and decisionmaking at various levels.

Important developments in recent years in the area of sub-seasonal and seasonal-to-interannual prediction have led to significant improvements in predictions of weather and climate extremes (Nicholls, 2001; Simmons and Hollingsworth, 2002; Kharin and Zwiers, 2003; Medina-Cetina and Nadim, 2008). Some of these improvements, such as the use of soil moisture initialization for weather and (sub-) seasonal prediction (Koster et al., 2010), have potential for applications in transitional zones between wet and dry climates, and in particular in mid-latitudes (Koster et al., 2004). Such applications may potentially be relevant for projections of temperature extremes and droughts (Lawrimore et al., 2007; Schubert et al., 2008; Koster et al., 2010). Decadal and longer time scale predictions are improving and could form the basis for early warning systems in the future (Meehl et al., 2007, 2009; Palmer et al., 2008; Shukla et al., 2009, 2010).

Developing resiliency to weather and climate involves developing resiliency to its variability on a continuum of time scales, and in an ideal world, early warnings would be available across this continuum (Chapters 1 and 2; McBean, 2000; Hellmuth et al., 2011). However, investments in developing such resiliency are usually primarily informed by information only over the expected lifetime of the investment, especially among poorer communities. For the decision of which crops to grow next season, some consideration may be given to longer-term strategies but the more pressing concern is likely to be the expected climate over the next season. Indeed, there is little point in preparing to survive the impacts of possible disasters a century in the future if one is not equipped to survive more immediate threats. Thus, within the disaster risk management community, preparedness for climate change must involve preparedness for climate variability (Chapters 3 and 4).

Improving prediction methods remains an active area of research and it is hoped significant further progress will be reached in coming years (Brunet et al., 2010; Shapiro et al., 2010). However for such predictions to be of use to end users, improved communication will be required to develop indices appropriate for specific regional impacts. A better awareness of such issues in the climate modeling community through greater feedback from the disaster risk management community (and other user communities) may lead to the development of additional applications for weather and climate hazard predictions. Prediction systems, if carefully targeted and sufficiently accurate, can be useful tools for reducing the risks related to climate and weather extremes (Patt et al., 2005; Goddard et al., 2010).

Despite an inevitable focus on shorter-term survival and hence interest in shorter-term hazard warnings, the longer time scales cannot be ignored if reliable predictions are to be made. Changing greenhouse gas concentrations are important even for seasonal forecasting, because including realistic greenhouse gas concentrations can significantly improve forecast skill (Doblas-Reyes et al., 2006; Liniger et al., 2007). Similarly, adaptation tools traditionally based on long-term records (e.g., stream flow measurements over 50 to 100 years) coupled with the assumption that the climate is not changing may lead to incorrect conclusions about the best adaptation strategy to follow (Milly et al., 2008). Thus reliable prediction and successful adaptation both need a perspective that includes consideration of short to long time scales (days to decades).

While there are potential benefits of early warning systems (NRC, 2003; Shapiro et al., 2007) that span a continuum of time scales, for much of the disaster risk management community the idea of preparedness based on predictions is a new concept. Most communities have largely operated in a reactive mode, either to disasters that have already occurred or in emergency preparedness for an imminent disaster predicted with high confidence (Chapter 5). The possibility of using weather and climate predictions longer than a few days to provide advanced warning of extreme conditions has only been a recent development (Brunet et al., 2010; Shapiro et al., 2010). Despite over a decade of operational seasonal predictions in many parts of the globe, examples of the use of such information by the disaster risk management community are scarce, due to the uncertainty of predictions and comprehension of their implications (Patt et al., 2005; Meinke et al., 2006; Hansen et al., 2011). Most seasonal rainfall predictions, for example, are presented as probabilities that total rainfall over the coming few (typically three) months will be amongst the highest or lowest third of rainfall totals as measured over a historical period and these are averaged over large areas (typically tens of thousands of square kilometers). Not only are the probabilities lacking in precision but the target variable – seasonal rainfall total – does not necessarily map well onto flood occurrence. Although higher-than-normal seasonal rainfall will often be associated with a higher risk of floods, it is possible for the seasonal rainfall total to be unusually high yet no flooding occurs. Alternatively, the total may be unusually low, yet flooding might occur because of the occurrence of an isolated heavy rainfall event (Chapter 3). Thus even when seasonal predictions are understood properly, it may not be obvious how to utilize them. These problems emphasize the need for the development of tools to translate such information into quantities directly relevant to end users. Better communication between modeling centers and end users is needed (Chapters 5 and 6). Where targeted applications have been developed, some success has been reported (e.g., for malaria prediction) (Thomson et al., 2006; Jones et al., 2007). Nonetheless there may be additional obstacles such as policy constraints that restrict the range of possible actions.

9.2.11.4. Interventions

There are many examples of interventions of early warning systems outlined in the other case studies of this chapter and also in Chapters 5, 6, and 7. As a part of their strategy of reducing risk, the Victorian Government in Australia has established the heat wave early warning system for metropolitan Melbourne and is undertaking similar work for regional Victoria (see Case Study 9.2.2). A Storm Warning Center and associated coastal volunteer network has been established in Bangladesh and has been proven effective (Case Study 9.2.5). The absence of a storm warning system in Myanmar contributed to the tragedy of that event (Case Study 9.2.5). The benefits of early warning systems are also discussed with respect to floods (Case Study 9.2.6), heat waves (Case Study 9.2.1), epidemic disease (Case Study 9.2.7), and drought (Case Study 9.2.3).

9.2.11.5. Outcomes

There have been examples of major benefits of early warning systems (Einstein and Sousa, 2007). Assessments of community capacity to respond to cyclone warnings have been performed for India (Sharma et al., 2009), Florida (Smith and McCarty, 2009), New Orleans (Burnside et al., 2007), New South Wales, Australia (Cretikos et al., 2008), and China (Wang et al., 2008). Predictions of landfall for tropical cyclones are important (Davis et al., 2008). In Bangladesh (Case Study 9.2.5; Paul, 2009), the implementation of an early warning system enabled people to evacuate a hazardous area promptly (Paul and Dutt, 2010; Stein et al., 2010). If forecasts are frequently incorrect, the response of people is affected (Chapter 5; Dow and Cutter, 1998). Public health impacts of hazards also depend on the preparedness of the local community (Vogt and Sapir, 2009) and this can be improved by early warnings. However, accurate predictions alone are insufficient for a successful early warning system, as is demonstrated by the case in the United Kingdom – a country that regularly experiences flooding (Parker et al., 2009). Severe damage and health problems followed flooding in 2007 due to insufficiently clear warning communication, issued too late and inadequately coordinated, so that people, local government, and support services were unprepared (UNISDR, 2009c). Heat-health warnings (Case Study 9.2.1) have proved more effective (Fouillet et al., 2008; Hajat et al., 2010; Michelozzi et al., 2010; Rubio et al., 2010) although improvements are still needed (Kalkstein and Sheridan, 2007).

Notwithstanding the difficulties outlined for use of seasonal predictions in disaster risk management, the successful use of such predictions has been possible (IRI, 2011). Since all preparative actions have some direct cost and it is impractical to be always prepared for all eventualities, seasonal predictions can help to choose priorities from a list of actions.

9.2.11.6. Lessons Identified

Early warning systems for extreme weather- or climate-related events, such as heat waves, floods, and storms have been implemented to provide warnings on time scales of hours to days. The skill of warnings beyond a few days ahead is improving as seasonal predictions are now demonstrating benefits for drought, floods, and other phenomena, and decadal forecasts of increased numbers of intense precipitation events and heat waves are now being factored into planning decisions (NRC, 2003; Lazo et al., 2009; Goddard et al., 2010). It is expected that early warning systems will enable the implementation of DRR and CCA. Early warning systems rely on the ability of people to factor information on the future into plans and strategies and need to be coupled with education programs, legislative initiatives, and scientific demonstrations of the skill and cost value benefits of these systems.

9.2.12. Effective Legislation for Multilevel Governance of Disaster Risk Reduction and Adaptation

9.2.12.1. Introduction

This case study, through focus on South Africa's disaster risk management law and comparable legal arrangements in other states, such as the Philippines and Colombia, explores critical provisions for effective legislation. South Africa's legislation has served as a model for others (Pelling and Holloway, 2006; Van Niekerk, 2011) because it focuses on prevention, decentralizes DRR governance, mandates the integration of DRR into development planning, and requires stakeholder inclusiveness. Implementation has proven challenging, however, particularly at the local level (NDMC, 2007, 2010; Visser and Van Niekerk, 2009; Botha et al., 2011; Van Niekerk, 2011) as is the case for most states (GNDR, 2009; UNISDR, 2011b). Through analysis of South Africa's legislation and the difficulties that it faces in implementation, this study provides relevant information to other governments as they assess whether their own national legislation to reduce and manage disaster risk is adequate for adapting to climate change.

9.2.12.2. Background

A legal framework establishes legal authority for programs and organizations that relate to hazards, risk, and risk management. These laws may dictate – or encourage – policies, practices, processes, the assignment of authorities and responsibilities to individuals and/or institutions, and the creation of institutions or mechanisms for coordination or collaborative action among institutions (Mattingly, 2002). Law can be used to provide penalties and incentives by enforcing standards, to empower existing agencies or establish new bodies with new responsibilities, and to assign budget lines (Pelling and Holloway, 2006). In short, legislation enables and promotes sustainable engagement, helps to avoid disjointed action at various levels, and provides recourse for society when things go wrong.

Most states have some form of disaster risk management legislation or are in the process of enacting it (UNDP 2005; UNISDR 2005b). In 2011, 48 countries reported substantial achievements in developing national policy and legislation; importantly, almost half are low or lower-middle income countries (UNISDR, 2011b). An increasing number of countries have been adopting or updating existing legislation modeled on Hyogo Framework for Action principles. Countries with new or updated laws include India and Sri Lanka in 2005; El Salvador, Saint Lucia, Saint Vincent, and the Grenadines in 2006; Anguilla (United Kingdom) and Gambia in 2007; Indonesia in 2008; Egypt and the Philippines in 2009; and Zambia and Papua New Guinea in 2010 (UNISDR, 2011b). As yet some of the new laws addressing disaster risk have not been harmonized with preexisting legislative frameworks in relevant sectors, such as water, agriculture, and energy (UNISDR, 2011b). Although these national legislations for disaster management do not necessarily include a disaster risk reduction orientation (Pelling and Holloway 2006), the evidence suggests a global paradigm shift from the former responsive approach to disaster management toward more long-term, sustainable preventive action (Britton, 2006; Benson, 2009). India, Pakistan, Indonesia, South Africa, and several Central American states have enacted such paradigm-changing amendments to disaster management legislation and, in Ecuador, the "notion of risk-focused disaster management was rooted directly into its new constitution adopted in 2008" (IFRC, 2011).

In the case of South Africa, the country was impacted by floods and droughts, and there was a high motivation for change in the post-Apartheid era (Pelling and Holloway, 2006; NDMC, 2007), which starting in 1994 led to legislative reform for disaster risk reduction. A "Green Paper" first solicited public input and debate, and a second "White Paper" translated responses into policy options for further technical and administrative deliberations. These documents are noteworthy for their consultative approach and their emphasis on disaster risk reduction rather than traditional response (Pelling and Holloway, 2006; NDMC, 2007). Thereafter the Government passed three disaster management bills that culminated in the promulgation of the Disaster Management Act No. 57 of 2002 and of the National Disaster Management Framework in 2005 (Pelling and Holloway, 2006; NDMC, 2007).

9.2.12.3. Description of Strategy

South Africa's 2002 Disaster Management Act and its National Disaster Management Policy Framework of 2005 (Republic of South Africa, 2002, 2005) are noteworthy because they were among the first to focus on prevention, decentralize DRR governance, mandate the integration of DRR into development planning, and require stakeholder inclusiveness.

The Act and Framework define the hierarchical institutional structure that governs disaster risk reduction at national, provincial, and municipal levels. They effectively decentralize DRR by mandating each level of government to create:
- A disaster risk management framework – a policy focused on the prevention and mitigation of risk
- A disaster risk management center – *inter alia*, to promote an integrated and coordinated system of management, to integrate DRR into development plans, to maintain disaster risk management information, to monitor implementation, and to build capacity
- A disaster risk management advisory forum – for government and civil society stakeholders in DRR to coordinate their actions
- An interdepartmental disaster risk management committee – for government departments to coordinate or integrate activities for DRR, to compile disaster risk management plans, and to provide interdepartmental accountability (Republic of South Africa, 2002, 2005; Van Niekerk, 2006).

The Act further details each entity's responsibilities. South Africa's legislation makes a legal connection between disaster risk reduction and development planning. Some other countries adopting this approach include Comoros, Djibouti, Ethiopia, Hungary, Ivory Coast, Mauritius, Romania, and Uganda (Pelling and Holloway, 2006).

The Act requires that municipalities include risk management plans in their integrated development plans (Republic of South Africa, 2002; Van Niekerk, 2006). Municipal-level requirements are supported with a mandate for provincial governments to ensure that their disaster risk management plans "form an integral part of development planning" and for the National Centre of Disaster Management to develop guidelines for the integration of plans and strategies into development plans (Republic of South Africa, 2002).

Closely related to the ability to influence development planning is the authority to lead coordinated government action for DRR across government agencies. The interdepartmental committees mandated by the Act for each level provide the opportunity to communicate plans and develop strategies across ministries and departments, avoiding unilateral action that may increase risk. The forums established by the Act similarly give voice to additional stakeholders to participate in DRR decisionmaking. South Africa, Colombia, and the Philippines' DRR laws, for example, include provisions for the involvement of NGOs, traditional leaders, volunteers, community members, and the private sector in disaster risk reduction.

9.2.12.4. Outcomes

Implementation of South Africa's benchmark legislative provisions has proven challenging. Many district municipalities have not yet established the disaster management centers required by the Act or these are not yet functioning adequately (Van Riet and Diedericks, 2010; Botha et al., 2011). The majority of local municipalities (which are subdivisions of district municipalities) have not yet established advisory forums although it should be noted that the Act does not require their creation at this level (Botha et al., 2011). A greater percentage of metropolitan districts have established advisory forums. Similarly, interdepartmental committees, which facilitate cross-sectoral governmental collaboration and the integration of DRR into development planning, have also not yet been established in a majority of municipalities. Although municipalities reported good progress in integrating their disaster management plans into integrated development plans (Botha et al., 2011), such plans as yet contain little evidence of integration (Van Niekerk, 2011).

Provincial and municipal levels attribute the lack of progress in implementing the Act and Framework to inadequate resources for start-up costs for municipalities as well as for the continuous operations of disaster risk reduction projects (Visser and Van Niekerk, 2009; NDMC, 2010). There is also the continuing need for resources for response recovery and rehabilitation activities. Reasons for the lack of funding include a lack of clarity of the Act on funding sources and confusion regarding the processes to access various sources of funding (Visser and Van Niekerk, 2009). Although the mechanisms to obtain funding exist, not all municipalities and provinces are using them, hence the perception of inadequate funding persists (Van Niekerk, 2011). In some cases there appears to be an absolute lack of funds, such as in district municipalities, which are more rural and less densely populated, and have a narrower tax base to fund DRR (Van Riet and Diedericks, 2010). This situation is similar to that in other countries, such as Colombia, where more than 80% of municipalities are able to assign only 20% of their own non-earmarked resources to risk reduction and disaster response. Because the law does not stipulate percentages and amounts, municipalities allocate minimal sums for disaster risk reduction (MIJ, 2009) given competing infrastructure and social spending needs (Cardona and Yamín, 2007).

South Africa and Colombia's experiences are replayed around the world. Governments informed in their 2011 Hyogo Framework progress reports that the lack of efficient and appropriate budget allocations remains one of the major challenges for effective disaster risk reduction legislation (UNISDR, 2011a). Even in countries in which funding for disaster risk management is mandated by law, actual resource allocation for disaster risk reduction remains low and is concentrated in preparedness and response (UNDP, 2007). The Philippines has new legislation that attempts to address these issues. The Philippines new Disaster Risk Reduction and Management Act 10121 renames the Local Calamity Fund as the Local Disaster Risk Reduction and Management Fund and stipulates that no less than 5% shall be set aside for risk management and preparedness (Republic of the Philippines, 2010). Further, to carry out the provisions of the Act, the Commission allocated one billion pesos or US$ 21.5 million (Republic of the Philippines, 2010). Unspent money will remain in the fund to promote risk reduction and disaster preparedness. The adequacy of this provision has yet to be tested as the Act is recent and implementation has yet to begin.

In South Africa, all relevant national departments have yet to undertake required DRR activities or identified sectoral focal points; consequently, the advisory committee at the national level is not yet functioning optimally (Van Niekerk, 2011). Similarly, at provincial and municipal levels, departmental representatives are absent or too junior to make decisions at meetings, reflecting lack of understanding about their department's role in DRR and about DRR generally (Van Riet and Diedericks, 2010).

Moreover, as mentioned above, between 55 and 73% of municipalities have not established a committee, which Botha et al. (2011) point out hampers local government's ability to implement integrated multi-sectoral DRM.

The Philippines' Climate Change Act, enacted in 2009, addresses the challenge of inter-sectoral government collaboration by creating a commission to be chaired by the president and attached to that office, thus ensuring highest-level political support for collaborative implementation of the law (Republic of the Philippines, 2009). The commission is composed of the secretaries of all relevant departments as well as the Secretary of the Department of National Defense, as Chair of the National Disaster Coordinating Council, and representatives from the disaster risk reduction community. The main functions of the Commission are to "ensure the mainstreaming of climate change, in synergy with disaster risk reduction, into the national, sectoral, and local development plans and programs" and to create a panel of technical experts, "consisting of practitioners in disciplines that are related to climate change, including disaster risk reduction" (Republic of the Philippines, 2009).

Implementing the multi-sectoral DRM envisioned by South Africa's legislation may be hindered by the placement of the National Disaster Management Centre within a line ministry (the Department of Cooperative Governance) (Van Niekerk, 2011). Sub-national levels have likewise placed their centers within sectors with insufficient political authority; consequently, local municipal and district levels rate current interdepartmental collaboration as low (Botha et al., 2011). This placement allows other departments to disregard DRM, as the National Disaster Management Centre cannot enforce punitive measures (Van Niekerk, 2011).

Similar to South Africa's arrangements, the Philippines' highest policymaking and coordinating body for disaster risk management, the National Disaster Risk Reduction and Management Council (formerly called the National Disaster Coordinating Council), sits within the Department of National Defense. As such, it is focused on disaster preparedness and response and does not have sustainable development and poverty reduction responsibilities. The Philippines' new Disaster Risk Reduction and Management Act of 2010 attempts to redress this issue by including experts from all relevant fields as members of the Council and expressly defining its mandate on mainstreaming disaster risk reduction into sustainable development and poverty reduction strategies, policies, plans, and budgets at all levels (Republic of the Philippines, 2010).

Positioning DRR institutions within the highest levels of government has proven effective because this position often determines the amount of political authority of the national disaster risk management body (UNDP, 2007; UNISDR, 2009a). National disaster risk management offices attached to prime ministers' offices usually can take initiatives affecting line ministries, while their colleagues operating at the sub-ministerial level often face administrative bottlenecks (UNDP, 2007). High-level support is particularly important to enable disaster risk reduction legislation to provide a framework for strategies to build risk reduction into development and reconstruction (Pelling and Holloway, 2006).

9.2.12.5. Lessons Identified

The main lesson that emerges from this case study is that carefully crafted legislation buttresses DRR activities, thus avoiding a gap between the law's vision and its implementation. The experiences of South Africa and the Philippines in implementing their DRR legislation (as described by Visser and Van Niekerk, 2009; NDMC, 2010; Van Riet and Diedericks, 2010; Botha et al., 2011; Van Niekerk, 2011) and the literature on DRR legislation (Mattingly, 2002; Britton, 2006; Pelling and Holloway, 2006; UNDP, 2007; Benson, 2009; UNISDR, 2009c) point to the following elements of effective legislation and implementation:
- The law allocates adequate funding for implementation at all levels with clarity about the generation of funds and procedures for accessing resources at every administrative level.
- The institutional arrangements provide both access to power for facilitating implementation and opportunities to 'mainstream' disaster risk reduction and adaptation into development plans.
- The law includes provisions that increase accountability and enable coordination and implementation, that is, the clear identification of roles and responsibilities and access to participate in decisionmaking.

An additional element is the need for periodic assessment and revision to ensure that legislation for disaster risk reduction and adaptation is dynamic and relevant (Llosa and Zodrow, 2011). For instance, the Philippines' Disaster Risk Reduction Management (DRRM) Act calls for the development of a framework to guide disaster risk reduction and management efforts to be reviewed "on a five-year interval, or as may be deemed necessary, in order to ensure its relevance to the times" (Republic of the Philippines, 2010). The DRRM Act also calls for the development of assessments on hazards and risks brought about by climate change (Republic of the Philippines, 2010). Likewise, the Philippines Climate Change Act calls for the framework strategy that will guide climate change planning, research and development, extension, and monitoring of activities to be reviewed every three years or as necessary (Republic of the Philippines, 2009). Similarly, the United Kingdom's Climate Change Act establishes the preparation of a report informing parliament on risks of current and predicted impact of climate change no later than five years after the previous report (United Kingdom, 2008). Thus an additional element for effective DRR-adaptation legislation may be that the law be based on up-to-date risk assessment and mandates periodic reassessment as risks evolve and knowledge of climate change impacts improves.

Developing and enacting legislation takes considerable time and political capital. It took South Africa and the Philippines about a decade to enact comprehensive disaster risk reduction frameworks. Linking the development of disaster risk reduction legislation to the politically prominent climate change discussion could substantially increase the sense of urgency and thus speed of parliamentary processes (Llosa and Zodrow, 2011).

Another method for hastening the legislative process would be to first assess the adequacy of existing disaster risk reduction legislation and strengthen these laws rather than starting a wholly new drafting and negotiations process for adaptation that may create a parallel legal and operational system (Llosa and Zodrow, 2011). As frequently reported (e.g., UNDP, 2007; UNISDR, 2009c), an overload of laws and regulations without a coherent and comprehensive framework, clear competencies, and budget allocations hinders the effective implementation of disaster risk reduction legislation.

9.2.13. Risk Transfer: The Role of Insurance and Other Instruments in Disaster Risk Management and Climate Change Adaptation in Developing Countries

9.2.13.1. Introduction

The human and economic toll from disasters can be greatly amplified by the long-term loss in incomes, health, education, and other forms of capital resulting from the inability of communities to restore infrastructure, housing, sanitary conditions, and livelihoods in a timely way (Mechler, 2004; Mills, 2005). By providing timely financial assistance following extreme event shocks, insurance and other risk-transfer instruments contribute to DRR by reducing the medium- and long-term consequences of disasters. These instruments are widespread in developed countries, and are gradually becoming part of disaster management in developing countries, where novel micro-insurance programs are helping to put cash into the hands of affected poor households so they can begin rebuilding livelihoods (Bhatt et al., 2010). These mechanisms can also contribute to reducing vulnerability and advancing development even before disasters strike by providing the requisite security for farmers and firms to undertake higher-return, yet more risky investments in the face of pervasive risk. Governments also engage in risk transfer. Investors can be encouraged to invest in a country if there is evidence that the government has reduced its risks (Gurenko, 2004).

9.2.13.2. Background

This case study focuses on instruments for risk transfer in order to manage catastrophe risk in developing countries (see also Sections 5.5.2, 6.5.3, and 7.4.4). Table 9-3 provides an overview of financial instruments and arrangements, including risk transfer, as they are employed by households, farmers, small- and medium-sized enterprises (SMEs), and governments, as well as international organizations and donors. Typically, losses are reimbursed on an ad hoc basis after disasters strike through appeals to solidarity, for example, from neighbors, governments, and international donors. Households and other agents also rely on savings and credit, and many governments set aside national or sub-national level reserve funds. Alternatively, agents can engage in risk transfer (the shaded cells

Table 9-3 | Examples of risk financing mechanisms (shaded cells) at different scales. Source: adapted from Linnerooth-Bayer and Mechler, 2009.

	Local Households, Farmers, SMEs	**National** Governments	**International** Development organizations, donors, NGOs
Solidarity	Help from neighbors and local organizations	Government post-disaster assistance; government guarantees/bailouts	Bilateral and multilateral assistance, regional solidarity funds
Informal risk transfer (sharing)	Kinship and other reciprocity obligations, semi-formal micro-finance, rotating savings and credit arrangements, remittances		
Savings, credit, and storage (inter-temporal risk spreading)	Savings; micro-savings; fungible assets; food storage; money lenders; micro-credit	Reserve funds; domestic bonds	Contingent credit; emergency liquidity funds
Insurance instruments	Property insurance; crop and livestock insurance; micro-insurance	National insurance programs; sovereign risk transfer	Re-insurance; regional catastrophe insurance pools
Alternative risk transfer	Weather derivatives	Catastrophe bonds	Catastrophe bonds; risk swaps, options, and loss warranties

in Table 9-3), which is defined by UNISDR as "the process of formally or informally shifting the financial consequences of particular risks from one party to another whereby a household, community, enterprise or state authority will obtain resources from the other party after a disaster occurs, in exchange for ongoing or compensatory social or financial benefits provided to that other party" (UNISDR, 2009b). Risk sharing can be considered synonymous with risk transfer, although the latter is often used to connote more informal forms of shifting risk without explicit compensation or payment, for example, mutual non-market arrangements among family or community. Insurance is the best known form of market risk transfer; yet, risks can be transferred with many formal and informal instruments as described in the following section.

Traditional channels for financing disaster relief and recovery, although in many cases less costly than risk transfer, have in the past proved to be inadequate for managing large-scale weather-related events in highly vulnerable countries (Cohen and Sebstad, 2003; Cardenas et al., 2007; Barnett et al., 2008). In poor countries, households and businesses usually do not have the resources to purchase commercial insurance to cover their risks with the additional difficulty that in many developing countries the commercial insurance providers do not exist. If there is no support from family or government, disasters can lead to a worsening of poverty in the absence of insurance. The victims then must either obtain high-interest loans (or default on existing loans), sell their important and valued assets and livestock, or engage in low-risk, low-yield farming to lessen their exposure to extreme events (Varangis et al., 2002). In recognition of these issues and to reduce the overall costs of disasters, investments in disaster risk reduction and proactive risk transfer are strongly encouraged by governments, the insurance sector, and the donor community (Kreimer and Arnold, 2000; Gurenko, 2004; Linnerooth-Bayer et al., 2005).

9.2.13.3. Description of Strategy – Catastrophe Risk Transfer Mechanisms and Instruments

As shown by the shaded cells in Table 9-3, risk transfer includes a range of pre-disaster mechanisms and instruments (Cummins and Mahul, 2009), the most important of which are briefly described below:
- **Informal mutual arrangements** involve pre-agreed non-market exchanges of post-disaster support (informal risk sharing).
- **Insurance** is a "well-known form of risk transfer, where coverage of a risk is obtained from an insurer in exchange for ongoing premiums paid to the insurer" (UNISDR, 2009b). A contractual transaction based on a premium is used to guarantee financial protection against potentially large loss; contracts typically cover losses to property, productive assets, commercial facilities, crops and livestock, public infrastructure (sovereign insurance), and business interruption.
- **Micro-insurance**, based on the same principles as insurance, is aimed, most often, at lower-income individuals who cannot afford traditional insurance and hence the premiums are lower but also the coverage may be restricted. In some cases, the individuals are unable to access more traditional insurance (Mechler et al., 2006). Often it is provided in innovative partnerships involving communities, NGOs, self-help groups, rural development banks, insurers, government authorities, and donors.
- **Alternative risk transfer** denotes a range of arrangements that hedge risk (Mechler et al., 2006). These include catastrophe bonds, which are instruments where the investor receives an above-market return when a pre-specified catastrophe does not occur within a specified time interval. However, the investor sacrifices interest or part of the principal following the event.
- **Weather derivatives** typically take the form of a parametric (indexed-based) transaction, where payment is made if a chosen weather index, such as 5-day rainfall amounts, exceeds some predetermined threshold.
- **Contingent credit** (also called deferred drawdown option) is a prearranged loan contingent on a specified event; it can be provided by the insurance industry to other insurers, or by international financial institutions to governments.
- **Risk pools** aggregate risks regionally (or nationally) allowing individual risk holders to spread their risk geographically.

9.2.13.4. Interventions – Examples of Local, National, and International Risk Transfer for Developing Countries

Development organizations working together with communities, governments, insurers, and NGOs have initiated or supported many recent pilot programs offering risk transfer solutions in developing countries. Three examples at the local, national, and international scales are briefly discussed below.

9.2.13.4.1. Covering local risks: index-based micro-insurance for crop risks in India

Micro-insurance to cover, for example, life and health is widespread in developing countries, but applications for catastrophic risks to crops and property are in the beginning phases (see Morelli et al., 2010 for a review, and Loster and Reinhard, 2010 for a focus on micro-insurance and climate change). Typically a micro-insurance company, often operating on a not-for-profit basis, evolves from an organization that has developed insurance products for a community. Most are based on the expectation that the pool of participants will provide payments that cover the costs incurred, including expected damage claims (which are generally low because of infrequent and small claims), administrative costs (which are reduced through group contracts or linking contracts to loans), taxes, and regulatory fees. Many depend on the support of government subsidies and international development organizations and participation of NGOs (Mechler et al., 2006).

An innovative insurance program set up in India in 2003 covers non-irrigated crops in the state of Andhra Pradesh against the risk of insufficient rainfall during key times during the cropping season. The index-based policies are offered by a commercial insurer and marketed to growers through microfinance banks. In contrast to conventional insurance, which is written against actual losses, this index-based (parametric) insurance is written against a physical or economic trigger, in this case rainfall measured by a local rain gauge. The scheme owes its existence to technical assistance provided by the World Bank (Hess and Syroka, 2005). Schemes replicating this approach are currently targeting 700,000 exposed farmers in India (Cummins and Mahul, 2009).

One advantage of index-based insurance is the substantial decrease in transaction costs due to eliminating the need for expensive post-event claims handling, which has impeded the development of insurance mechanisms in developing countries (Varangis et al., 2002). A disadvantage is basis risk, which is the lack of correlation of the trigger with the loss incurred. If the rainfall measured at the weather station is sufficient, but for isolated farmers insufficient, they will not receive compensation for crop losses. Similar schemes are implemented or underway, for instance, in Malawi, Ukraine, Peru, Thailand, and Ethiopia (Hellmuth et al., 2009). A blueprint for insuring farmers in developing countries who face threats to their livelihoods from adverse weather has been developed (World Bank, 2005d). Overcoming major institutional and other barriers must be done in order for these programs to achieve this target (Hellmuth et al., 2009).

Weather insurance and especially index-based contracts contribute, in at least two ways, to climate change adaptation and disaster risk reduction. Since farmers will receive payment based on rainfall and thus have an incentive to plant weather-resistant crops, indexed contracts eliminate moral hazard, which is defined as the disincentive for risk prevention provided by the false perception of security when purchasing insurance coverage. Second, an insurance contract renders high-risk farmers more creditworthy, which enables them to access loans for agricultural inputs. This was illustrated in the pilot program in Malawi, where farmers purchased index-based drought insurance linked to loans to cover costs of hybrid seed, with the result that their productivity was doubled (Linnerooth-Bayer et al., 2009). Increased productivity decreases vulnerability to weather extremes, thus contributing to climate change adaptation (to the extent that risks of weather extremes are increased by climate change). In another innovative micro-insurance project in Ethiopia, farmers can pay their premiums by providing labor on risk-reducing projects (Suarez and Linnerooth-Bayer, 2010).

9.2.13.4.2. Covering national risks: the Ethiopian weather derivative

The World Food Programme (WFP), to supplement and partly replace its traditional food-aid approach to famine, has recently supported the Ethiopian government-sponsored Productive Safety Net Programme (PSNP). The WFP is now insuring it against extreme drought (World Bank, 2006b). When there is a food emergency, the PSNP is able to provide immediate cash payments that may be sufficient to save lives even in the case of very severe droughts (Hess et al., 2006). However, these payments may not be sufficient to restore livelihoods (World Bank, 2006b). To provide extra capital in the case of extreme drought, an index-based contract, sometimes referred to as a weather derivative, was designed by the WFP. The amount of capital is based on contractually specified catastrophic shortfalls in precipitation based on the Ethiopia Drought Index (EDI). The EDI depends on rainfall amounts that were measured at 26 weather stations that represent the various agricultural areas of Ethiopia. In 2006, the WFP successfully obtained an insurance contract based on the EDI through an international reinsurer (Hess et al., 2006). A drawback of this arrangement, in contrast to the micro-insurance programs in India and Malawi, is that it perpetuates dependence on post-drought government assistance with accompanying moral hazard.

9.2.13.4.3. Intergovernmental risk sharing: the Caribbean Catastrophe Risk Insurance Facility (CCRIF)

The world's first regional catastrophe insurance pool was launched in 2007 in the Caribbean region; this is the Caribbean Catastrophe Risk Insurance Facility (discussed in Section 7.4.4). Sixteen participating governments secured insurance protection against costs associated with catastrophes such as hurricanes and earthquakes (Ghesquiere et al., 2006; World Bank, 2007). Several of the participating countries represent the countries experiencing the greatest economic losses from disasters in the last few decades, when measured as a share of GDP (CCRIF, 2010).

The aim of the Caribbean facility is to provide immediate liquidity to cover part of the costs that participating governments expect to incur while they provide relief and assistance for recovery and rehabilitation. Because it does not cover all costs, CCRIF provides an incentive for governments to invest in risk reduction and other risk transfer tools. The cost of participation is based on estimates of the respective countries' risk (measured as probability and cost). The advantage of pooling is that due to diversifying risk it greatly reduces the costs of reinsurance compared to the price each government would have paid individually. Funding for the program, although mainly the responsibility of participating countries, has been supported by a donor conference hosted by the World Bank.

Insofar as weather extremes are increased by climate change, the CCRIF contributes directly to disaster risk reduction and climate change adaptation. By providing post-event capital it enables governments to restore critical infrastructure so important for reducing the long-term human and economic impacts from hurricanes. Experience with CCRIF also shows the importance of designing programs that reflect the needs of the participating countries. Finally, it demonstrates how international assistance can support disaster management in tandem with national responsibility.

9.2.13.4.4. Outcomes – the role of risk transfer for advancing disaster risk reduction and climate change adaptation

As these examples illustrate, risk-transfer instruments and especially insurance can promote disaster risk reduction and climate change adaptation by enabling recovery and productive activities. By providing means to finance relief, recovery of livelihoods, and reconstruction, insurance reduces long-term indirect losses – even human losses – that do not show up in the disaster statistics. Risk transfer arrangements thus directly lead to the reduction of post-event losses from extreme weather events, what is commonly viewed as adaptation. Moreover, insured households and businesses can plan with more certainty, and because of the safety net provided by insurance, they can take on cost-effective, yet risky, investments. This ultimately reduces vulnerability to weather extremes and by so doing contributes to climate change adaptation.

Experience in developed countries has demonstrated additional ways in which insurance and other risk-transfer instruments have promoted DRR and CCA as listed below:
- Because risk-transfer instruments require detailed analysis of risk, they can both raise awareness and provide valuable information for its response and reduction; for example, in some developed countries insurers with other partners have made flood and other hazard maps publicly available (Botzen et al., 2009; Warner et al., 2009). Potential challenges include the technical difficulties related to risk assessment, dissemination of appropriate information, and overcoming education and language barriers in some areas.
- By pricing risk, insurance can provide incentives for investments and behavior that reduce vulnerability and exposure, especially if premium discounts are awarded. Differential premium pricing has been effective in discouraging construction in high-risk areas; for example, UK insurers price flood policies according to risk zones, but insurers are reluctant to award premium discounts for other types of mitigation measures, such as reinforcing windows and doors to protect against hurricanes (Kunreuther and Roth, 1998; Kunreuther and Michel-Kerjan, 2009). The incentive effect of actuarial risk pricing should be weighed against the benefits of increasing insurance penetration to those unable to afford risk-based premiums. The positive incentives provided by insurance should also not overshadow the potential for negative incentives or moral hazard.
- Insurers and other providers can make risk reduction a contractual stipulation, for example, by requiring fire safety measures as a condition for insuring a home or business (Surminski, 2010). The US National Flood Insurance Program requires communities to reduce risks as a condition for offering subsidized policies to their residents (Kunreuther and Roth, 1998; Linnerooth-Bayer et al., 2007). It was noted above that the WFP might require risk-reducing activities as a condition for its support for weather derivatives.
- Providers can partner with government and communities to establish appropriate regulatory frameworks and promote, for instance, land use planning, building codes, emergency response, and other types of risk-reducing policies. Ungern-Sternberg (2003) has shown that Swiss cantons having public monopolies that provide disaster insurance outperform cantons with private systems in reducing risks and premiums, mainly because the public monopolies have better access to land use planning institutions, fire departments, and other public authorities engaged in risk reduction. In many countries, insurers have co-financed research institutes and disaster management centers, and in other cases, have partnered with government to achieve changes in the planning system and investment in public protection measures (Surminski, 2010).

9.2.13.5. Lessons Identified

Governments, households, and businesses can experience liquidity gaps limiting their ability to recover from disasters (*high confidence*). There is *robust evidence* to suggest that risk-transfer instruments can help reduce this gap, thus enabling recovery.

There are a range of risk-transfer instruments, where insurance is the most common. With support from the international community, risk transfer is becoming a reality in developing countries at the local, national, and international scales, but the future is still uncertain. Index-based contracts greatly reduce transaction costs and moral hazard (*medium confidence*); while more costly than many traditional financing measures, insurance has benefits both before disasters (by enabling productive investment) and after disasters (by enabling reconstruction and recovery) (*medium confidence*). Insurance and other forms of risk transfer can be linked to disaster risk reduction and climate change adaptation by enabling recovery, reducing vulnerability, and providing knowledge and incentives for reducing risk (*medium confidence*).

9.2.14. Education, Training, and Public Awareness Initiatives for Disaster Risk Reduction and Adaptation

9.2.14.1. Introduction

Disasters can be substantially reduced if people are well informed and motivated to prevent risk and to build their own resilience (UNISDR, 2005b). Disaster risk reduction education is broad in scope: it encompasses primary and secondary schooling, training courses, academic programs, and professional trades and skills training (UNISDR, 2004), community-based assessment, public discourse involving the media, awareness campaigns, exhibits, memorials, and special events (Wisner, 2006). Given the breadth of the topic, this case study illustrates just a few practices in primary school education, training programs, and awareness-raising campaigns in various countries.

9.2.14.2. Background

The Hyogo Framework calls on states to "use knowledge, innovation, and education to build a culture of safety and resilience at all levels" (UNISDR, 2005b). States report minor progress in implementation, however (UNISDR, 2009c). Challenges noted include the lack of capacity among educators and trainers, difficulties in addressing needs in poor urban and rural areas, the lack of validation of methodologies and tools, and little exchange of experiences. On the positive side, the 2006-2007 international campaign "Disaster Risk Reduction Begins at School" (UNISDR, 2006) raised awareness of the importance of education with 55 governments undertaking awareness-raising activities and 22 governments reporting success in making schools safer (e.g., 175 schools developed disaster plans in Gujarat, India) by developing educational and training materials, introducing school drills, and implementing DRR teacher trainings (UNISDR, 2008b). Furthermore, the implementation scheme of the United Nations Decade of Education for Sustainable Development 2005-2014 seeks to improve the knowledge base on disaster reduction as one of the keys to sustainable development.

A related emerging trend is to engage children in disaster risk reduction and adaptation, as children are increasingly understood as effective agents of change (Mitchell et al., 2009). Children's inclusion also increases the likelihood that they will maintain their own DRR and adaptation learning (Back et al., 2009). A report from five NGOs (Twigg and Bottomley, 2011) states that their DRR work with children and young people involves risk identification and action planning for preparedness; training of school teachers and students; DRR curriculum development; youth-led prevention and risk reduction actions, such as mangrove and tree conservation; awareness raising (e.g., through peer-to-peer community exchanges and children's theater); and "lobbying and networking in promoting and supporting children's voice and action."

Effective DRR education initiatives seek to elicit behavioral change not only by imparting knowledge of natural hazards but also by engaging people in identifying and reducing risk in their surroundings. In formal education, disaster risk education should not be confined within the school but promoted to family and community (Shaw et al., 2004). Lectures can create knowledge, particularly if presented with visual aids and followed up with conversation with other students. Yet it is family, community, and self learning, coupled with school education, that transform knowledge into behavioral change (Shaw et al., 2004).

9.2.14.3. Description of Strategies

9.2.14.3.1. School curriculum

States are increasingly incorporating DRR in the curriculum (UNISDR, 2009c) and have set targets for so doing in all school curricula by 2015 (UNISDR, 2009c). Initiatives to integrate the teaching of climate change and DRR are also emerging, such as the described Philippines program. Importantly, the new Philippines disaster risk reduction and climate change laws mandate the inclusion of DRR and climate change, respectively, in school curricula; the following example predates these laws, however.

The Asian Disaster Preparedness Centre and UN Development Programme, with the National Disaster Coordinating Council and support from the European Commission Humanitarian Aid and Civil Protection, assisted the Ministry of Education in the Philippines, Cambodia, and Lao People's Democratic Republic to integrate disaster risk reduction into the secondary school curriculum. Each country team developed its own draft module, adapting it to local needs.

The Philippines added climate change and volcanic hazards into its disaster risk reduction curriculum. The relevant lessons addressed 'what is climate change,' they then asked 'what is its impact,' and finally 'how can you reduce climate change impact?' Other lessons focused on the climate system, typhoons, heat waves, and landslides, among other related topics (Luna et al., 2008). The Philippines' final disaster risk reduction module was integrated into 12 lessons in science and 16 lessons in social studies for the first year of secondary school (Grade 7) (Luna et al., 2008). Each lesson includes group activities, questions to be asked of the students, the topics that the teacher should cover in the lecture, and a learning activity in which students apply knowledge gained and methodology for evaluation of learning by the students (Luna et al., 2008). The project reports that it reached 1,020 students, including 548 girls, who learned about disaster risk reduction and climate change. Twenty-three teachers participated in the four-day orientation session. An additional 75 teachers and personnel were trained to train others and replicate the experience across the country (Luna et al., 2008).

9.14.2.3.2. Training for disaster risk reduction and adaptation

In order to effectively include disaster risk reduction and adaptation in the curriculum, teachers require (initial and in-service) training on the substantive matter as well as the pedagogical tools (hands-on, experiential learning) to elicit change (Shiwaku et al., 2006; Wisner, 2006). Education program proponents might have to overcome teachers' resistance to incorporate yet another topic into overburdened curricula. To enlist teachers' cooperation, developing a partnership with the ministry of education and school principals can be helpful (UNISDR, 2007b; World Bank et al., 2009). The following program in Indonesia and the evaluation results from Nepal demonstrate the importance of engaging teachers for effective education. The subsequent example from Nepal, Pakistan, and India focuses on training builders through extensive hands-on components in which new techniques are demonstrated and participants practice these techniques under expert guidance (World Bank et al., 2009).

The Disaster Awareness in Primary Schools project, which provides teacher training, was launched in Indonesia in 2005 with German support and is ongoing. By 2007, through this project, 2,200 school teachers had received DRR training. Project implementers found that existing teaching methods were not conducive to active learning. Students listened to teacher presentations, recited facts committed to memory, and were not encouraged to understand concepts and processes. The training took teachers' capabilities into account by emphasizing the importance of clarity and perseverance in delivering lessons so as to avoid passing on faulty life-threatening information (e.g., regarding evacuation routes). Scientific language was avoided and visual aids and activities encouraged. Teachers were asked to take careful notes and to participate in practical activities such as first-aid courses, thus modeling proactive learning. Continuity with the teachers' traditional teaching methods was maintained by writing training modules in narrative form and following the established lesson plan model. Moreover, to avoid further burdening teachers' heavy lesson requirements and schedules, the modules were designed to be integrated into many subjects, such as language and physical education, and to require minimum preparation (UNISDR, 2007b).

In Nepal, Kyoto University researchers evaluated the knowledge and perceptions of 130 teachers in 40 schools, most of whom were imparting disaster education (Shiwaku et al., 2006). Through responses to a survey, the researchers found that the content of the disaster risk education being imparted depended on the awareness of individual teachers. Teachers focused lessons on the effects of disasters with which they could relate from personal experience. The researchers concluded that teacher training is the most important step to improve disaster risk education in Nepal. Most social studies teachers reported a need for teacher training but the survey analysis recommended that training programs be designed to integrate DRR into any subject rather than taught in special classes (Shiwaku et al., 2006).

The National Society for Earthquake Technology in Nepal conducted large-scale training for masons, carpenters, bar benders, and construction supervisors in 2007 over a five-month period to impart risk-resilient construction practices and materials. Participants from Kathmandu and five other municipalities formed working groups to train other professionals. As the project was successful, a mason-exchange program was designed with the Indian NGO Seeds. Nepali masons were sent to Gujarat, India, to mentor local masons in the theory and practice of safer construction. Also in India, the government of Uttar Pradesh trained two junior engineers in the rural engineering service in each district to carry out supervisory inspection functions and delegated the construction management to school principals and village education committees. Similarly, the Department of Education of the Philippines mandated principals to take charge of the management of the repair and/or construction of typhoon-resistant classrooms after the 2006 typhoons. Assessment, design, and inspection functions were provided by the Department's engineers, who also assist with auditing procurement (World Bank et al., 2009).

9.2.14.3.3. Raising public awareness

In addition to the insights on the psychological and sociological aspects of risk perception, risk reduction education has benefited from lessons in social marketing. These include involving the community and customizing for audiences using cultural indicators to create ownership; incorporating local community perspectives and aggressively involving community leaders; enabling two-way communications and speaking with one voice on messages (particularly if partners are involved); and evaluating and measuring performance (Frew, 2002).

According to the UNISDR Hyogo Framework Mid-Term Review (UNISDR, 2011a), few DRR campaigns have translated into public action and greater accountability. However, successful examples include Central America and the Caribbean, where the media played an important role, including through radio soap operas. The UNISDR review also found a high level of risk acceptance, even among communities demonstrating heightened risk awareness. In some cultures, the spreading of alarming or negative news – such as information on disaster risks – is frowned upon (UNISDR, 2011a). The following examples from Brazil, Japan, and the Kashmir region illustrate good practice in raising awareness for risk reduction.

Between 2007 and 2009, the Brazilian Santa Catarina State Civil Defence Department, with the support of the Executive Secretariat and the state university, undertook a public awareness initiative to reduce social vulnerability to disasters induced by natural phenomena and human action (SCSCDD, 2008a,b). During the two-year initiative, 2,000 educational kits were distributed free of charge to 1,324 primary schools. Students also participated in a competition of drawings and slogans that were made into a 2010 calendar. As the project's goal was public awareness of risk, the project jointly launched a communications network in partnership with media and social networks to promote better dissemination of risk and disasters (SCSCDD, 2008a,b). The initiative also focused on the most vulnerable populations. A pilot project for 16 communities precariously perched on a hill prone to landslides featured a 44-hour course on risk reduction. Community participants elaborated risk maps and reduction strategies, which they had to put to use immediately. Shortly into the course, heavy rains battered the state, triggering a state of emergency; 10 houses in the pilot project area had to be removed and over 50 remained at risk. The participants' risk reduction plans highlighted the removal of garbage and large rocks as well as the building of barriers. The plans also identified public entities for partnership and the costs for services required. The training closed with a workshop on climate change and with the community leaders' presentation of the major risk reduction lessons learned (SCSCDD, 2008c). On international disaster risk reduction day, representatives of the community, Civil Defence, and other public entities visited the most at-risk areas of the hill community, planted trees, installed signs pointing out risky areas and practices, distributed educational pamphlets, and discussed risk. One of the topics of discussion was improper refuse disposal and the consequent blocking of drains, causing flooding (SCSCDD, 2008d).

In 2004, typhoons resulted in flooding in urban areas of Saijo City (Ehime Prefecture of Shikoku Island, Japan). There were also landslides in the mountains. As a result, a public awareness campaign was implemented. Saijo City, a small city with semi-rural mountainous areas, faces challenges in disaster risk reduction that are relatively unique. In Japan, young people have a tendency to leave smaller communities and move to larger cities. The result is that Japanese smaller towns have older than the national average populations. Since younger, able-bodied people are important for community systems of mutual aid and emergency preparedness, there is a special challenge. Saijo City has an urban plain, semi-rural and isolated villages on hills and mountains, and a coastal area and, hence, is spread over a mix of geographic terrains (Yoshida et al., 2009; ICTILO et al., 2010); this brings another challenge. In 2005, the Saijo City Government launched a risk awareness program to meet both of these challenges through a program targeted at school children. The project for 12-year olds has a 'mountain-watching' focus for the mountainside and a 'town-watching' focus for the urban area (ICTILO et al., 2010). The students are taken, accompanied by teachers, forest workers, local residents, and municipal officials, on risk-education field trips. In the mountains, the young urban dwellers meet with the elderly and they learn together about the risks the city faces. Part of the process is to remember the lessons learned from the 2004 typhoons. Additionally, a 'mountain and town watching' handbook has been developed, a teachers' association for disaster education was formed, a kids' disaster prevention club started, and a disaster prevention forum for children was set up (Yoshida et al., 2009; ICTILO et al., 2010). This is an example of a local government both conceiving and implementing the program. The city government led a multi-stakeholder and community-based disaster risk awareness initiative that then became self-sustaining. Professionals from disaster reduction and education departments were provided through government support. The government also funds the town and mountain watching and puts on an annual forum (ICTILO et al., 2010).

The Centre for Environment Education (CEE) Himalaya is undertaking a disaster risk reduction and climate change education campaign in 2,000 schools and 50 Kashmir villages in the Himalayas. In the schools, teachers and students are involved in vulnerability and risk mapping through rapid visual risk assessment and in preparing a disaster management plan for their school. Disaster response teams formed in selected schools have been trained in life-saving skills and safe evacuation (CEE Himalaya, 2009).

CEE Himalaya celebrated International Mountain Day 2009 with educators by conducting a week-long series of events on climate change adaptation and disaster risk reduction. About 150 participants including teachers and officials of the Department of Education, Ganderbal, participated in these events (CEE Himalaya, 2009). Participants worked together to identify climate change impacts in the local context, particularly in terms of water availability, variation in microclimate, impact on agriculture/horticulture and other livelihoods, and vulnerability to natural disasters. The concept of School Disaster Management Plans (SDMPs) was introduced. Participants actively prepared SDMPs for their schools through group exercises, and discussed their opinions about village contingency plans (CEE Himalaya, 2009). Some of the observations on impacts of climate change in the area discussed by participants included the melting, shrinking, and even disappearance of some glaciers and the drying up of several wetlands and perennial springs. Heavy deforestation, decline and extinction of wildlife, heavy soil erosion, siltation of water bodies, fall in crop yields, and reduced availability of fodder and other non-timber forest produce were some of the other related issues discussed (CEE Himalaya, 2009). Participants watched documentaries about climate change and played the Urdu version of 'Riskland: Let's Learn to Prevent Disasters.' They received educational kits on disaster risk reduction and on climate change, translated and adapted for Kashmir (CEE Himalaya, 2009).

9.2.14.4. Lessons Identified

The main lesson that can be drawn from the various initiatives described above is that effective DRR education does not occur in a silo. As the examples from Japan, Brazil, and the Himalayas illustrate, successful programs actively engage participants and their wider communities to elicit risk-reducing behavioral change (Shaw et al., 2004; Wisner, 2006; Bonifacio et al., 2010). Lessons on actively engaging participants include:

- Assessing community risk, discussing risk with others, and joining a risk-reducing activity in school or community forums provide opportunities for active learning. Engaging children and community members in vulnerability and capacity assessments has been found to be effective in disaster risk reduction and adaptation programs (Twigg and Bottomley, 2011; see Himalaya example).
- Interactive lectures with visual aids can be effective in building knowledge (Shaw et al., 2004; see teacher training in Indonesia example) and should be followed up with discussion with peers and family – and action – beyond the classroom (Shaw et al., 2004; Wisner, 2006).

Additional lessons of good practice illustrated above include:
- Integrating climate change information into DRR education and integrating both into various subject matters is simple and effective.

The Philippines example shows that such integration is underway, and the teacher training in Indonesia example concludes that such integration can be helpful in avoiding overburdening full curricula.
- Training of teachers and professionals in all relevant sectors can have a positive multiplier effect. As the Nepalese teachers' evaluation example shows, teacher training is critical to address risk self-perception and ensure that teachers pass on appropriate DRR knowledge. The training of builders example in Nepal, India, and the Philippines illustrates the successful dissemination of DRR methods and tools within a critical sector across borders.

As well as providing further examples of current adaptation and DRR initiatives, a United Nations Framework Convention on Climate Change synthesis report of initiatives undertaken by Nairobi Work Programme partners concludes that the integration of activities relating to education, training, and awareness-raising into relevant ongoing processes and practices is key to the long-term success of such activities (UNFCCC, 2010).

9.3. Synthesis of Lessons Identified from Case Studies

This chapter examined case studies of extreme climate events, vulnerable regions, and methodological management approaches in order to glean lessons and good practices. Case studies are provided to add context and value to this report. They contribute to a focused analysis and convey, in part, the reality of an event: the description of how certain extreme events develop; the extent of human loss and financial damage; the response strategies and interventions; the DRR, DRM, and CCA measures and their effect on the overall outcomes; and cultural or region-specific factors that may influence the outcome. Most importantly, case studies provide a medium through which to learn practical lessons about successes in DRR that are applicable for adaptation to climate change. The lessons identified will prove useful at various levels from the individual to national and international organizations as people try to respond to extreme events and disasters and adapt to climate change.

The case studies highlight several recurring themes and lessons.

A common factor was the need for greater amounts of useful information on risks before the events occur, including early warnings. The implementation of early warning systems does reduce loss of lives and to a lesser extent damage to property. Early warning was identified by all the extreme event case studies – heat waves, wildfires, drought, *dzud*, cyclones, floods, and epidemics – as key to reducing the impacts from extreme events. A need for improving international cooperation and investments in forecasting was recognized in some of the case studies, but equally the need for regional and local early warning systems was heavily emphasized, particularly in developing countries.

A further common factor identified overall was that it is better to invest in preventive-based DRR plans, strategies, and tools for adaptation than in response to extreme events. Greater investments in proactive hazard and vulnerability reduction measures, as well as development of capacities to respond and recover from the events were demonstrated to have benefits. Specific examples for planning for extreme events included increased emphasis on drought preparedness; planning for urban heat waves; and tropical cyclone DRM strategies and plans in coastal regions that anticipate these events. However, as illustrated by the SIDS case study, it was also identified that DRR planning approaches continue to receive less emphasis than disaster relief and recovery.

It was also identified that DRM and preventive public health are closely linked and largely synonymous. Strengthening and integrating these measures, along with economic development, should increase resilience against the health effects of extreme weather and facilitate adaptation to climate change. Extreme weather events and population vulnerability can interact to produce disastrous epidemic disease through direct effects on the transmission cycle and also potentially through indirect effects, such as population displacement.

Another lesson is that in order to implement a successful DRR or CCA strategy, legal and regulatory frameworks are beneficial in ensuring direction, coordination, and effective use of funds. The case studies are helpful in this endeavor as effective and implemented legislation can create a framework for governance of disaster risks. While this type of approach is mainly for national governments and the ways in which they devolve responsibilities to local administrations, there is an important message for international governance and institutions as well. Frameworks that facilitate cooperation with other countries to attain better analysis of the risks will allow institutions to modify their focus with changing risks and therefore maintain their effectiveness. This cooperation could be at the local through national to international levels. Here and in other ways, civil society has an important role.

Insurance and other forms of risk transfer can be linked to disaster risk reduction and climate change adaptation by providing knowledge and incentives for reducing risk, reducing vulnerability, and enabling recovery.

A lesson identified by many case studies was that effective DRR education contributes to reduce risks and losses, and is most effective when it is not done in isolation, but concurs with other policies. The integration of activities relating to education, training, and awareness-raising into relevant ongoing processes and practices is important for the long-term success of DRR and DRM activities. Investing in knowledge at primary to higher education levels produces significant DRR and DRM benefits.

Research improves our knowledge, especially when it includes integration of natural, social, health, and engineering sciences and their applications. In all cases, the point was made that with greater information available it would be possible to better understand the risks and to ensure that response strategies were adequate to face the risks. It further poses a set of questions to guide the investigations.

The case studies have reviewed past events and identified lessons that could be considered for the future. Preparedness through DDR and DRM can help to adapt for climate change and these case studies offer examples of measures that could be taken to reduce the damage that is inflicted as a result of extreme events. Investment in increasing knowledge and warning systems, adaptation techniques and tools, and preventive measures will cost money now but they will save money and lives in the future.

References

A digital library of non-journal-based literature cited in this chapter that may not be readily available to the public has been compiled as part of the IPCC review and drafting process, and can be accessed via either the IPCC Secretariat or IPCC Working Group II web sites.

Abou Hadid, A.F., 2009: Food Production. In: *Arab Environment: Climate Change. Impact of Climate Change on Arab Countries* [Tolba, M.K. and N.W. Saab (eds.)]. Arab Forum for Environment and Development, Beirut, Lebanon, pp. 63-74, www.afedonline.org/afedreport09/Full%20English%20Report.pdf.

Abrahamson, V., J. Wolf, B. Fenn, S. Kovats, P. Wilkinson, W.N. Adger, and R. Raine, 2008: Perceptions of heatwave risks to health: interview-based study of older people in London and Norwich, UK. *Journal of Public Health*, **31(1)**, 119-126.

Ackers, M.-L., R.E. Quick, C.J. Drasbek, L. Hutwagner, and R.V. Tauxe, 1998: Are there national risk factors for epidemic cholera? The correlation between socioeconomic and demographic indices and cholera incidence in Latin America. *International Journal of Epidemiology*, **27(2)**, 330-334.

Adelekan, I.O., 2010: Vulnerability of poor urban coastal communities to flooding in Lagos, Nigeria. *Environment and Urbanization*, **22(2)**, 433-450.

Ahern, M, R.S. Kovats, P. Wilkinson, R. Few, and F. Matthies, 2005: Global health impacts of floods: Epidemiologic evidence. *Epidemiologic Reviews*, **27**, 36-46.

Ahmad, Q.K. and A.U. Ahmed (eds.), 2002: *Bangladesh: citizen's perspective on sustainable development.* Bangladesh Unnayan Parishad, Dhaka, Bangladesh, 181 pp.

Ahrens, J., and P.M. Rudolph, 2006: The importance of governance in risk reduction and disaster management. *Journal of Contingencies and Crisis Management*, **14(4)**, 207-220.

AIACC, 2006: *Climate Change Vulnerability and Adaptation in the Livestock Sector of Mongolia.* A Final Report Submitted by P. Batima to Assessments of Impacts and Adaptations to Climate Change (AIACC), Project No. AS 06, International START Secretariat, Washington, DC, 84 pp.

Akbari, H. and S. Konopacki, 2004: Energy effects of heat-island reduction strategies in Toronto, Canada. *Energy*, **29(2)**, 191-210.

Akbari, H., M. Pomerantza, and H. Tahaa, 2001: Cool surfaces and shade trees to reduce energy use and improve air quality in urban areas. *Solar Energy*, **70(3)**, 295-310.

Alajo, S.O., J. Nakavuma, and J. Erume, 2006: Cholera in endemic districts in Uganda during El Nino rains: 2002-2003. *African Health Sciences*, **6(2)**, 93-97.

Allard, M., R. Fortier, C. Duguay, and N. Barrette 2002: A trend of fast climate warming in Northern Quebec since 1993: impacts on permafrost and man-made infrastructures. *Eos Transactions, American Geophysical Union*, **83(47)**, F258.

Allen, C.D., M. Savage, D. Falk, K.F. Suckling, T.W. Swetnam, T. Schulke, P.B. Stacey, P. Morgan, M. Hoffman, and J. Klingel, 2002: Ecological restoration of southwestern ponderosa pine ecosystems: a broad perspective. *Ecological Applications*, **12**, 1418-1433.

Anisimov, O.A., D.G. Vaughan, T.V. Callaghan, C. Furgal, H. Marchant, T.D. Prowse, H. Vilhjálmsson, and J.E. Walsh. 2007: Polar regions (Arctic and Antarctic). In: *Climate Change 2007. Impacts, Adaptation and Vulnerability. Contribution of Working Group II to the Fourth Assessment Report of the Intergovernmental Panel on Climate Change* [Parry, M.L., O.F. Canziani, J.P. Palutikof, P.J. Van Der Linde, and C.E. Hanson (eds.)]. Cambridge University Press, Cambridge, UK, pp. 653-685.

Asante, K.O., J.S. Famiglietti, and J.P. Verdin, 2005: Current and future applications of remote sensing for routine monitoring of surface water controllers. In: Proceedings Pecora 16 Conf., Sioux Falls, SD, 23-27 October 2005, pp. 1-9.

Asante, K.O., R.D. Macuacua, G.A. Artan, R.W. Lietzow, and J.P. Verdin, 2007: Developing a flood monitoring system from remotely sensed data for the Limpopo Basin. *IEEE Transactions on Geoscience and Remote Sensing*, **45(6)**, 1709-1714.

Assemblée Nationale, 2004: *Rapport de la Commission D'Enquête sur les Conséquences Sanitaires et Sociales de la Canicule.* Assemblée Nationale, France, vols. 1 and 2.

Auld, H.E., 2008: Adaptation by design: the impact of changing climate on infrastructure. *Journal of Public Works and Infrastructure*, **3**, 276-288.

Auld, H., J. Waller, S. Eng, J. Klaassen, R. Morris, S. Fernandez, V. Cheng, and D. MacIver, 2010: *The Changing Climate and National Building Codes and Standards.* Environment Canada, Ottawa, Canada, ams.confex.com/ams/pdfpapers/174517.pdf.

Australian Government, 2009: Metrological aspects of the 7 February 2009 Victorian fires, an overview. Bureau of Meteorology Report for the 2009 Victorian Bushfires Royal Commission. www.royalcommission.vic.gov.au/getdoc/f1eaba2f-414f-4b24-bcda-d81ff680a061/WIT.013.001.0012.pdf.

Awuor, C.B., V.A. Orindi, and A. Adwerah, 2008: Climate change and coastal cities: The case of Mombasa, Kenya. *Environment and Urbanization*, **20(1)**, 231-242.

Back, E., C. Cameron, and T. Tanner, 2009: *Children and Disaster Risk Reduction: Taking stock and moving forward.* Children in a Changing Climate, Institute of Development Studies and UNICEF, Brighton, UK, 42 pp.

Badia-Perpinyà, A. and M. Pallares-Barbera, 2006: Spatial distribution of ignitions in Mediterranean periurban and rural areas: the case of Catalonia. *International Journal of Wildland Fire*, **15**, 187-196.

Baigalmaa, P., 2010: Community base disaster risk management, Self Sharpen – 2, guideline. In: *Strengthening disaster management system in Mongolia, MON/08/305 project* [Boldbaatar, S. (ed.)]. Phase III, United Nations Development Programme, New York, NY, 137 pp., www.un-mongolia.mn/publication/ UNDP_gamshig_04-14.pdf.

Barber, D., J. Lukovich, J. Keogak, S. Baryluk, L. Fortier, and G. Henry, 2008: Changing climate of the arctic. *Arctic*, **61(1)**, 7-26.

Barnett, B.J., C.B. Barrett, and J.R. Skees, 2008: Poverty traps and index-based risk transfer products. *World Development*, **36(10)**, 1766-1785.

Barnett, J. and W.N. Adger, 2003: Climate dangers and atoll countries. *Climatic Change*, **61**, 321-337.

Barnett, J. and W.N. Adger, 2007: Climate change, human security and violent conflict. *Political Geography*, **26(6)**, 639-655.

Bartzokas, A., J. Azzopardi, L. Bertotti, A. Buzzi, L. Cavaleri, D. Conte, S. Davolio, S. Dietrich, A. Drago, O. Drofa, A. Gkikas, V. Kotroni, K. Lagouvardos, C.J. Lolis, S. Michaelides, M. Miglietta, A. Mugnai, S. Music, K. Nikolaides, F. Porcù, K. Savvidou, and M. I. Tsirogianni, 2010: The RISKMED project: philosophy, methods and products. *Natural Hazards and Earth System Sciences*, **10**, 1393-1401.

Basher, R., 2006: Global early warning systems for natural hazards: Systematic and people-centered. *Philosophical Transactions of the Royal Society A*, **364(1845)**, 2167-2182.

Bastedo, J., 2007: *On the Frontlines of Climate Change: What's Really Happening in the Northwest Territories.* Government of the Northwest Territories, Yellowknife, Canada, sen.parl.gc.ca/nsibbeston/Final%20for%20WEB%20or%20Email.pdf.

Batima, P. and D. Dagvadorj (eds.), 2000: *Climate change and its Impacts in Mongolia.* Mongolian National Agency for Meteorology, Hydrology and Environmental Monitoring (NAMHEM). JEMR, Ulaanbaatar, Mongolia, 227 pp., ISBN 99929-70-35-9.

Batjargal, Z., R. Oyun, S. Sangidansranjav, and N. Togtokh, 2001: *Lessons Learned from Dzud 1999-2000* [Batjargal, Z., S. Sangidansranjav, R. Oyun, and N.Togtokh (eds.)]. Case study funded by UNDP, conducted by joint team of National Agency of Meteorology, Hydrology and Environmental Monitoring, Civil Defense Agency, Ministry of Agriculture and JEMR, Ulaanbaatar, Mongolia, 347 pp., ISBN 99929-70-54-7.

Beg, N., J.C. Morlot, O. Davidson, Y. Afrane-Okesse, L. Tyani, F. Denton, Y.Sokona, J.P. Thomas, E.L. La Rovere, J.K. Parikh, K. Parikh, and A.A. Rahman, 2002: Linkages between climate change and sustainable development. *Climate Policy*, **2**, 129-144.

Begzsuren, S., J.E. Ellis, D.S. Ojima, M.B. Coughenour, T. Chuluun, 2004: Livestock responses to droughts and severe winter weather in the Gobi Three Beauty National Park, Mongolia. *Journal of Arid Environments*, **59(4)**, 785-796.

Benson, C., 2009: *Mainstreaming Disaster Risk Reduction into Development: Challenges and Experience in the Philippines*. International Federation of Red Cross and Red Crescent Societies and Provention Consortium, Geneva, Switzerland, 60 pp.

Bernard, S.M. and M.A. McGeehin, 2004: Municipal heat wave response plans. *American Journal of Public Health*, **94(9)**, 1520-1522.

Bhatt, M., T. Reynolds, and M. Pandya, 2010: Disaster insurance for the poor: All India Disaster Mitigation Institute and Afat Vimo. In: Microinsurance – An innovative tool for risk and disaster management [Morelli, E., G.A. Onnis, W.J. Ammann, and C. Sutter (eds.)]. Global Risk Forum, Davos, Switzerland, pp. 341-353.

Bhattacharya, S., R. Black, L. Bourgeois, J. Clemens, A. Cravioto, J.L. Deen, G. Dougan, R. Glass, R.F. Grais, M. Greco, I. Gust, J. Holmgren, S. Kariuki, P.-H. Lambert, M.A. Liu, I. Longini, G.B. Nair, R. Norrby, G.J.V. Nossal, P. Ogra, P. Sansonetti, L. von Seidlein, F. Songane, A.-M. Svennerholm, D. Steele, and R. Walker, 2009: The cholera crisis in Africa, *Science*, **324(5929)**, 885.

Black, R.A., J.P. Bruce, and I.D.M. Egener, 2010: *Managing the Risks of Climate Change: A Guide for Arctic and Northern Communities*. Centre for Indigenous Environmental Resources, Winnipeg, Canada.

Bollin, C., N. Lamade, and J. Ferguson, 2005: *Disaster Risk Management along the Rio Búzi*. Case Study on the Background, Concept and Implementation of Disaster Risk, Management in the Context of the GTZ-Programme for Rural Development (PRODER), German Gesellschaft für Technische Zusammenarbeit, Governance and Democracy Division, Eschborn, Germany, 26 pp.

Bompangue, D., P. Giraudoux, M. Piarroux, G. Mutombo, R. Shamavu, B. Sudre, A. Mutombo, V. Mondonge, and R. Piarroux, 2009: Cholera epidemics, war and disasters around Goma and Lake Kivu: an eight-year survey. *PLoS Neglected Tropical Diseases*, **3(5)**, e436.

Bonifacio, A.C., Y. Takeuchi, and R. Shaw, 2010: Mainstreaming climate change adaptation and disaster risk reduction through school education: perspectives and challenges. In: *Climate Change Adaptation and Disaster Risk Reduction: Issues and Challenges* [Shaw, R., J. Pulhin, and J. Pereira (eds.)]. Emerald Group Publishing, Bingley, UK, pp.143-169.

Borgford-Parnell, N., 2009: Mongolia: a case for economic diversification in the face of a changing climate. *Sustainable Development Law & Policy*, **9(2)**, 54-56.

Boruff, B. and S. Cutter, 2007: The environmental vulnerability of Caribbean island nations. *The Geographical Review*, **97(1)**, 24-45.

Botha, D., D. Van Niekerk, G. Wentink, C. Coetzee, K. Forbes, Y. Maartens, E. Annandale, T. Tshona, and E. Raju, 2011: *Disaster Risk Management Status Assessment at Municipalities in South Africa*. South African Local Government Association, Pretoria, South Africa, 108 pp.

Botzen, W.J.W., J.C.J.H. Aerts, and J.C.J.M. van den Bergh, 2009: Willingness of homeowners to mitigate climate risk through insurance. *Ecological Economics*, **68(8-9)**, 2265-2277, doi:10.1016/j.ecolecon.2009.02.019.

Bouchama, A., 2004: The 2003 European heat wave. *Intensive Care Medicine*, **30(1)**, 1-3.

Bradstock, R.A., J.S. Cohn, A.M. Gill, M. Bedward, and C. Lucas, 2009: Prediction and probability of large fires in the Sydney Region of South-eastern Australia using fire weather. *International Journal of Wildland Fire*, **18**, 932-943.

Braman, L., P. Suarez, and M.K. van Aalst, 2010: Climate change adaptation: integrating climate science into humanitarian work. *International Review of the Red Cross*, **92(879)**.

Britton, N.R., 2006: Getting the foundation right: in pursuit of effective disaster legislation for the Philippines. In: *Proceedings of the Second Asian Conference on Earthquake Engineering*, Manila, Philippines, March 2006, 15 pp.

Broad, K. and S. Agrawala, 2000: The Ethiopia food crisis: uses and limits of climate forecasts. *Science*, **289(5485)**, 1693-1694.

Brunet, G., M. Shapiro, B. Hoskins, M. Moncrieff, R. Dole, G.N. Kiladis, B. Kirtman, A. Lorenc, B. Mills, R. Morss, S. Polavarapu, D. Rogers, J. Schaake, and J. Shukla, 2010: Collaboration of the weather and climate communities to advance subseasonal-to-seasonal prediction. *Bulletin of the American Meteorological Society*, **91**, 1397-1406.

Burnside, R., D.S. Miller, and J.D. Rivera, 2007: The impact of information and risk perception on the hurricane evacuation decision-making of greater New Orleans residents. *Sociological Spectrum*, **27(6)**, 727-740.

Burton, I, 2011: Forensic disaster investigations in depth: A new case study model. *Environment Magazine*, **52(5)**, 36-41.

Bushfire CRC, 2009: *Household Mail Survey*. Bushfire Co-operative Research Centre report, Bushfire CRC, East Melbourne, Australia, www.bushfirecrc.com/managed/ resource/mail-survey-report-10-1-10-rt-2.pdf.

Cadot, E., V.G. Rodwin, and A. Spiral, 2007: In the heat of the summer: lessons from the heat waves in Paris. *Journal of Urban Health*, **84(4)**, 466-468.

Calain, P., J.P. Chaine, E. Johnson, M.L. Hawley, M.J. O'Leary, H. Oshitani, and C.L. Chaignat, 2004: Can oral cholera vaccination play a role in controlling a cholera outbreak? *Vaccine*, **22(19)**, 2444-2451.

Cardenas, V., S. Hochrainer, R. Mechler, G. Pflug, and J. Linnerooth-Bayer, 2007: Sovereign Financial Disaster Risk Management: The Case of Mexico. *Environmental Hazards*, **7**, 40-53.

Cardona, O.D. and L.E. Yamín, 2007: *Información para la gestión de riesgo de desastres. Estudio de caso de cinco países: Colombia*. United Nations, Inter-American Development Bank, and Economic Commission for Latin America and the Caribbean (ECLAC), 2007, Mexico City, Mexico, 25 pp.

Carmo Vaz, Á., 2000: Coping with floods – The experience of Mozambique. In: *1st WARFSA/WaterNet Symposium: Sustainable Use of Water Resources*, Maputo, Mozambique, 1-2 November 2000, www.bvsde.paho.org/bvsacd/cd46/coping.pdf.

CCRIF, 2010: *Annual Report 2009-2010*. Caribbean Catastrophe Risk Insurance Facility, Grand Cayman, Cayman Islands.

CEE Himalaya, 2009: Disaster risk reduction in the mountains. *Ceenario*, **24**, December 16-31, Centre for Environment Education, Ahmedabad, India, 4 pp.

Chaignat, C.L., V. Monti, J. Soepardi, G. Petersen, E. Sorensen, J. Narain, and M.P. Kieny, 2008: Cholera in disasters: do vaccines prompt new hopes? *Expert Review of Vaccines*, **7(4)**, 431-435.

Chatterjee, M., 2010: Slum dwellers response to flooding events in the megacities of India. *Mitigation and Adaptation Strategies for Global Change*, **15**, 337-353.

Choularton, R., 2007: Contingency planning and humanitarian action: a review of practice. *Humanitarian Practice Network*, **59**.

Christensen, J.H., B. Hewitson, A. Busuioc, A. Chen, X. Gao, I. Held, R. Jones, R.K. Kolli, W.-T. Kwon, R. Laprise, V. Magaña Rueda, L. Mearns, C.G. Menéndez, J. Räisänen, A. Rinke, A. Sarr, and P. Whetton, 2007: Regional climate projections. In: *Climate Change 2007: The Physical Science Basis. Contribution of Working Group I to the Fourth Assessment Report of the Intergovernmental Panel on Climate Change* [Solomon, S., D. Qin, M. Manning, Z. Chen, M. Marquis, K.B. Averyt, M. Tignor and H.L. Miller (eds.)]. Cambridge University Press, Cambridge, UK, and New York, NY, pp. 847-940.

Clarke, J., 1972: Some effects of the urban structure on heat mortality, *Environmental Research*, **5**, 93-104.

Cohen, M. and J. Sebstad, 2003: *Reducing Vulnerability: the Demand for Microinsurance*. MicroSave-Africa, Nairobi, Kenya.

Colwell, R., 1996: Global climate and infectious disease: the cholera paradigm. *Science*, **274**, 2025-2031.

Colwell, R.R., 2002: A voyage of discovery: cholera, climate and complexity. *Environmental Microbiology*, **4(2)**, 67-69.

Constantin de Magny, G., J.F. Guegan, M. Petit, and B. Cazelles, 2007: Regional-scale climate-variability synchrony of cholera epidemics in West Africa. *BMC Infectious Diseases*, **7**, 20.

Couture, N.J. and W.H. Pollard, 2007: Modelling geomorphic response to climatic change. *Climatic Change*, **85(3-4)**, 407-431.

Crabbé, P. and M. Robin, 2006: Institutional adaptation of water resource infrastructures to climate change in Eastern Ontario. *Climatic Change*, **78**, 103-133.

CRED, 2009: *EM-DAT, Emergency Events Database*. Centre for Research on the Epidemiology of Disasters (CRED), Université Catholique de Louvain, Brussels, Belgium, www.emdat.be/.

Cretikos, M., K. Eastwood, C. Dalton, T. Merritt, F. Tuyl, L. Winn, and D. Durrheim, 2008: Household disaster preparedness and information sources: rapid cluster survey after a storm in New South Wales, Australia. *BMC Public Health*, **8**, 195.

CSA, 2010: *Technical Guide – Infrastructure in Permafrost: A Guideline for Climate Change Adaptation*. Special Publication PLUS 4011-10, Canadian Standards Association, Mississauga, Canada.

Cummins, D. and O. Mahul, 2009: *Catastrophe Risk Financing in Developing Countries: Principles for Public Intervention*. World Bank, Washington, DC.

Da Silva, J., B. Garanganga, V. Teveredzi, S.M. Marx, S.J. Mason, and S.J. Connor, 2004: Improving epidemic malaria planning, preparedness and response in Southern Africa. *Malaria Journal*, **3**, 37.

Dagvadorj, D., L., Natsagdorj, J. Dorjpurev, and B. Namkhainyam, 2010: *Mongolia Assessment Report on Climate Change 2009*. Second publication with corrections, Ministry of Nature, Environment and Tourism, Ulaanbaatar, Mongolia, 226 pp.

Daoudy, M., 2009: Asymmetric power : Negotiating water in the Euphrates and Tigris. *International Negotiation*, **14**, 361-391.

Davies, M., B. Guenther, J. Leavy, T. Mitchell, and T. Tanner, 2008: 'Adaptive social protection': synergies for poverty reduction. *IDS Bulletin*, **39(4)**, 105-112.

Davis, C., W. Wang, S.S. Chen, Y. Chen, K. Corbosiero, M. DeMaria, J. Dudhia, G. Holland, J. Klemp, J. Michalakes, H. Reeves, R. Rotunno, C. Snyder, and Q. Xiao, 2008: Prediction of land falling hurricanes with the advanced hurricane WRF model. *Monthly Weather Review*, **126**, 1990-2005.

De Comarmond, A. and R. Payet, 2010: Small Island Developing States: incubators of innovative adaptation and sustainable technologies? In: *Coastal Zones and Climate Change* [Michel, D. and A. Pandya (eds.)]. The Henry L. Stimson Center, Washington, DC, www.stimson.org/images/uploads/research-pdfs/Alain.pdf.

De Sherbinin, A., A. Schiller, and A. Pulsipher, 2007: The vulnerability of global cities to climate hazards. *Environment and Urbanization*, **19(1)**, 39-64.

Dear, K., G. Ranmuthugala, T. Kjellström, C. Skinner, and I. Hanigan, 2005: Effects of temperature and ozone on daily mortality during the August 2003 heat wave in France. *Archives of Environmental and Occupational Health*, **60**, 205-212.

Derwent, R.G., P.G. Simmonds, S. O'Doherty, D.S. Stevenson, W.J. Collins, M.G. Sanderson, C.E. Johnson, F. Dentener, J. Cofala, R. Mechler and M. Amann, 2006: External influences on Europe's air quality: baseline methane, carbon monoxide and ozone from 1990 to 2030 at Mace Head, Ireland. *Atmospheric Environment*, **40**, 844-855.

Doblas-Reyes, F.J., R. Hagedorn, T.N. Palmer, and J.J. Morcrette, 2006: Impact of increasing greenhouse gas concentrations in seasonal ensemble forecasts. *Geophysical Research Letters*, **33**, L07708.

Doswell III, C.A., S.J. Weiss, and R.H. Johns, 1993: Tornado forecasting: a review. In: *The Tornado: Its Structure, Dynamics, Prediction, and Hazards* [Church, C., D. Burgess, C. Doswell, and R. Davies-Jones (eds.)]. Geophysical Monograph 79, American Geophysical Union, Washington, DC, pp. 557-571.

Dow, K. and S.L. Cutter, 1998: Crying wolf: repeat responses to hurricane evacuation orders. *Coastal Management*, **26(4)**, 237-252.

DSE, 2008a: *Climate Change in Port Phillip and Westernport*. Department of Sustainability and Environment, East Melbourne, Victoria, Australia.

DSE, 2008b: *Victoria's State of the Forests Report 2008*. Victorian Government, Department of Sustainability and Environment, Melbourne, Australia.

Dyson, L.L. and J. van Heerden, 2001: The heavy rainfall and floods over the northeastern Interior of South Africa during February 2000. *South African Journal of Science*, **97**, 80-86.

Easterling, W.E., P.K. Aggarwal, P. Batima, K.M. Brander, L. Erda, S.M. Howden, A. Kirilenko, J. Morton, J.-F. Soussana, J. Schmidhuber, and F.N. Tubiello, 2007: Food, fibre and forest products. In: *Climate Change 2007. Impacts, Adaptation and Vulnerability. Contribution of Working Group II to the Fourth Assessment Report of the Intergovernmental Panel on Climate Change* [Parry, M.L., O.F. Canziani, J.P. Palutikof, P.J. Van Der Linde, and C.E. Hanson (eds.)]. Cambridge University Press, Cambridge, UK, pp. 273-313.

Ebi, K.L. and G.A. Meehl, 2007: *The Heat Is On: Climate Change And Heatwaves in the Midwest*. Pew Center on Global Climate Change, Arlington, VA, 14 pp.

Ebi, K.L., T.J. Teisberg, L.S. Kalkstein, L. Robinson, and R.F. Weiher, 2004: Heat watch/warning systems save lives: estimated costs and benefits for Philadelphia 1995-1998. *Bulletin of the American Meteorological Society*, **85**, 1067-1073.

Ebi, K., N. Lewis, and C. Corvalan, 2006: Climate variability and change and their potential health effects in small island states: Information for adaptation planning in the health sector. *Environmental Health Perspectives*, **114(12)**, 1957-1963.

ECLAC, 2006: *Características e impacto socioeconómico de los huracanes "Stan" y "Wilma" en la República*. Economic Commission for Latin America and the Caribbean, Santiago, Chile, biblioteca.cepal.org/record=b1096781~S0.

EEA, 2003: *Air Pollution by Ozone in Europe in Summer 2003*. Topic report 3/2003, European Environment Agency, Copenhagen, Denmark.

EEA, 2007: *Air Pollution by Ozone in Europe in Summer 2006*. Technical report 5/2007, European Environment Agency, Copenhagen, Denmark.

EEA, 2011: *Air Pollution by Ozone across Europe during Summer 2010*. Technical report 6/2011, European Environment Agency, Copenhagen, Denmark.

Einstein, H.H. and R. Sousa, 2007: Warning systems for natural threats. *Georisk: Assessment and Management of Risk for Engineered Systems and Geohazards*, **1(1)**, 3 -20.

El-Quosy, D., 2009: Fresh Water. In: *Arab Environment: Climate Change. Impact of Climate Change on Arab Countries* [Tolba, M.K. and N.W. Saab (eds.)]. Arab Forum for Environment and Development, Beirut, Lebanon, pp. 75-86, www.afedonline.org/afedreport09/Full%20English%20Report.pdf.

Emch, M., C. Feldacker, M. Yunus, P.K. Streatfield, V. DinhThiem, D.G. Canh, and M. Ali, 2008: Local environmental predictors of cholera in Bangladesh and Vietnam. *American Journal of Tropical Medicine & Hygiene*, **78(5)**, 823-832.

Environment Canada, 2010: *Climate Information to Inform New Codes and Standards*. Environment Canada, Ottawa, Canada, www.ec.gc.ca/sc-cs/default.asp?lang=En&n=20CD1ADB-1.

Erian, W., B. Katlan, and O. Babah, 2011: *Drought Vulnerability in the Arab Region, Case Study – Drought in Syria*. Arab Center for the Studies of Arid Zones and Dry Lands and United Nations International Strategy for Disaster Reduction, Geneva, Switzerland, www.preventionweb.net/english/hyogo/gar/2011/en/bgdocs/Erian_Katlan_&_Babah_2010.pdf.

FAO, 2008: *FAO's role in the 2008 Syria drought appeal*. Food and Agriculture Organization, Rome, Italy, www.faoiraq.org/images/word/The%20FAO%20proposed%20%20program%20under%20this%20Syria%20l.pdf.

FAO, 2009: *FAO's role in the Syria Drought Response Plan 2009*. Food And Agriculture Organization, Rome, Italy, www.fao.org/fileadmin/templates/tc/tce/pdf/app_syriadrought2009.pdf.

FAO and NAPC, 2010: *National Programme for Food Security in the Syrian Arab Republic*. Food and Agriculture Organization and National Agriculture Policy Centre, Rome Italy and Damascus, Syria, www.napcsyr.org/events/ws/f_s_program/NPFS_final_en.pdf.

FAO and WFP, 2000: *FAO/WFP Crop and Food Supply Assessment Mission to Mozambique*. Food and Agriculture Organization and World Food Programme Special Report, FAO, Rome, Italy.

Field, C.B., L.D. Mortsch, M. Brklacich, D.L. Forbes, P. Kovacs, J.A. Patz, S.W. Running, and M.J. Scott, 2007: North America. In: *Climate Change 2007. Impacts, Adaptation and Vulnerability. Contribution of Working Group II to the Fourth Assessment Report of the Intergovernmental Panel on Climate Change* [Parry, M.L., O.F. Canziani, J.P. Palutikof, P.J. Van Der Linde, and C.E. Hanson (eds.)]. Cambridge University Press, Cambridge, UK, pp. 617-652.

Filleul, L., S. Cassadou, S. Médina, P. Fabres, A. Lefranc, D. Eilstein, A. Le Tertre, L. Pascal, B. Chardon, M. Blanchard, C. Declercq, J.F. Jusot, H. Prouvost, and M. Ledrans, 2006: The relation between temperature, ozone, and mortality in nine French cities during the heat wave of 2003. *Environmental Health Perspectives*, **114**, 1344-1347.

Flannigan, M.D., M.A. Krawchuk, W.J. de Groot, B.M. Wotton, and L.M. Gowman, 2009: Implications of changing climate for global wildland fire. *International Journal of Wildland Fire*, **18**, 483-507.

Flyvbjerg, B., 2004: Five misunderstandings about case-study research. In: *Qualitative Research Practice* [Seale, C., G. Gobo, J.F. Gubrium, D. Silverman (eds.)]. Sage Publications, London, UK, pp. 420-434.

Foley, C., 2007: *Mozambique: A Case Study In the Role of the Affected State in Humanitarian Action*. HPG Working Paper, Humanitarian Policy Group, Overseas Development Institute, London, UK, www.odi.org.uk/resources/download/ 2557.pdf.

Ford, J.D. and C. Furgal, 2009: Foreword to the special issue: climate change adaptation and vulnerability in the arctic, *Polar Research*, **28**, 1-9.

Ford, J.D. and T. Pearce, 2010: What we know, do not know, and need to know about climate change vulnerability in the western Canadian Arctic: A systematic literature review. *Environmental Research Letters*, **5**, 014008.

Ford, J. and B. Smit, 2004: A framework for assessing the vulnerability of communities in the Canadian Arctic to risks associated with climate change. *Arctic*, **57(4)**, 389-400.

Ford, J.D., T. Pearce, F. Duerden, C. Furgal, and B. Smit, 2010: Climate change policy responses for Canada's Inuit population: the importance of and opportunities for adaptation. *Global Environmental Change*, **20**,177-191.

FORIN, 2011: *Forensic Disaster Investigations – The FORIN Project*. Integrated Research on Disaster Risk, Beijing, China, www.irdrinternational.org/about-irdr/scientific-committee/working-group/forensic-investigations/.

Fortier, R., A-.M. LeBlanc, and W. Yu, 2011: Impacts of permafrost degradation on a road embankment at Umiujaq in Nunavik (Quebec), Canada. *Canadian Geotechnical Journal*, **48**, 720-740, doi:10.1139/T10-101.

Fouillet, A., G. Rey, F. Laurent, G. Pavillon, S. Bellec, C. Guihenneuc-Jouyaux, J. Clavel, E. Jougla, and D. Hemon, 2006: Excess mortality related to the August 2003 heat wave in France. *International Archives of Occupational & Environmental Health*, **80(1)**, 16-24.

Fouillet, A.R., V. Wagner, K. Laaidi, P. Empereur-Bissonnet, A. Le Tertre, P. Frayssinet, P. Bessemoulin, F. Laurent, P. De Crouy-Chanel, E. Jougla, and D. Hemon, 2008: Has the impact of heat waves on mortality changed in France since the European heat wave of summer 2003? A study of the 2006 heat wave. *International Journal of Epidemiology*, **37(2)**, 309-317.

Founda, D. and C. Giannakopoulos, 2009: The exceptionally hot summer of 2007 in Athens, Greece – A typical summer in the future climate? *Global and Planetary Change*, **67(3-4)**, 227-236.

Frew, S.L., 2002: *Public Awareness and Social Marketing*. Proceedings of a workshop, Asian Disaster Preparedness Center Regional Workshop on Best Practices in Disaster Management, Bangkok, Thailand, pp. 381-393.

Fritz, H.M., C.D. Blount, S. Thwin, M.K. Thu, and N. Chan, 2009: Cyclone Nargis storm surge in Myanmar. *Nature Geoscience*, **2**, 448-449.

Fuchs, R., M. Conran, and E. Louis, 2011: Climate change and Asia's coastal urban cities. *Environment and Urbanization Asia*, **2(1)**, 13-28.

Furgal, C. and T. Prowse, 2008: Northern Canada. In: *From Impacts to Adaptation: Canada in a Changing Climate 2007* [Lemmen, D., F. Warren, E. Bush, and J. Lacroix (eds.)]. Natural Resources Canada, Ottawa, Canada, pp. 57-118.

Gagnon-Lebrun, F. and S. Agrawala, 2006: *Progress on Adaptation to Climate Change in Developed Countries: An Analysis of Broad Trends*. OECD, Paris, France, 11 pp.

Gall, M., K.A. Borden, and S.L. Cutter, 2009: When do losses count? Six fallacies of natural hazards loss data. *Bulletin of the American Meteorological Society*, **90**, 799-809.

García, A., R.M.C. Norland, and K. Méndez Estrada, 2006: *Características e impacto socioeconómico de los huracanes "Stan" y "Wilma" en la República Mexicana en 2005*. Secretaria de Gobernación, Centro Nacional de Prevención de Desastres, Mexico.

GEF, 2006: *Sustainable Land Management in Least Developed Countries and Small Island Developing States*. Global Environment Facility, Washington, DC, www.cndwebzine.hcp.ma/cnd_sii/IMG/pdf/SLM_Support_in_LDCs_SIDS.pdf.

Ghesquiere, F., O. Mahul, M. Forni, and R. Gartley, 2006: *Caribbean Catastrophe Risk Insurance Facility: A Solution to the Short-Term Liquidity Needs of Small Island States in the Aftermath of Natural Disasters*. IAT03-13/3, World Bank, Washington, DC.

Giannakopoulos, C., P. Le Sager, M. Bindi, M. Moriondo, E. Kostopoulou, and C.M. Goodess, 2009: Climatic changes and associated impacts in the Mediterranean resulting from a 2°C global warming. *Global and Planetary Change*, **68(3)**, 209-224.

Gil, A.I., V.R. Louis, I.N. Rivera, E. Lipp, A. Huq, C.F. Lanata, D.N. Taylor, E. Russek-Cohen, N. Choopun, R.B. Sack, and R.R. Colwell, 2004: Occurrence and distribution of *Vibrio cholerae* in the coastal environment of Peru. *Environmental Microbiology*, **6(7)**, 699-706.

Giuliani, G. and P. Peduzzi, 2011: The PREVIEW Global Risk Data Platform: a geoportal to serve and share global data on risk to natural hazards. *Natural Hazards and Earth System Sciences*, **11**, 53-66, doi:10.5194/nhess-11-53-2011.

GNDR, 2009: *Clouds but Little Rain - Views from the Frontline: A Local Perspective of Progress towards Implementation of the Hyogo Framework for Action*. Global Network of Civil Society Organisations for Disaster Reduction, Teddington, UK, 59 pp.

Goddard, L., Y. Aichellouche, W. Baethgen, M. Dettinger, R. Graham, P. Hayman, M. Kadi, R. Martinez, and H. Meinke, 2010: Providing seasonal-to-interannual climate information for risk management and decision-making. *Procedia Environmental Sciences*, **1**, 81-101.

Goodland, A., D. Sheehy, and T. Shine, 2009: *Mongolia Livestock Sector Study, Volume I – Synthesis Report*. Sustainable Development Department, East Asia and Pacific Region, World Bank, Washington, DC.

Government of Bangladesh, 2008: *Cyclone Sidr in Bangladesh: Damage, Loss and Needs Assessment for Disaster Recovery and Reconstruction*. Government of Bangladesh, Dhaka, Bangladesh.

Government of India, 2001: *Census of India: (Provisional) Slum Population in Million Plus Cities (Municipal Corporations): Part A*. Government of India, Ministry of Home Affairs, Office of the Registrar General and Census Commissioner, New Delhi, India, censusindia.gov.in/Tables_Published/Admin_Units/Admin_links/slum1_m_plus.html.

Government of Northwest Territories, 2008: *NWT Climate Change Impacts and Adaptation Report*. Government of Northwest Territories, Yellowknife, Canada, www.enr.gov.nt.ca/_live/documents/content/NWT_Climate_Change_Impacts_and_Adaptation_Report.pdf.

Government of Yukon, 2009: *Yukon Government Climate Change Action Plan*. Government of Yukon, Whitehorse, Canada, www.env.gov.yk.ca/pdf/YG_Climate_Change_Action_Plan.pdf.

Government of Yukon, 2010: *Shawak and Permafrost*. Government of Yukon, Whitehorse, Canada, www.hpw.gov.yk.ca/trans/engineering/765.html.

Governments of Northwest Territories, Nunavut, and Yukon. 2010: *A Northern Vision: A Stronger North and a Better Canada Pan-Territorial Adaptation Strategy*. Governments of Northwest Territories, Nunavut, and Yukon, Canada, www.anorthernvision.ca/strategy/.

Griffith, D.C., L.A. Kelly-Hope, and M.A. Miller, 2006: Review of reported cholera outbreaks worldwide, 1995-2005. *American Journal of Tropical Medicine & Hygiene*, **75(5)**, 973-977.

Grynszpan, D., 2003: Lessons from the French heatwave. *Lancet*, **362**, 1169-1170.

Grynszpan, D., V. Murray, and S. Llossa, 2011: The value of case studies in disaster assessment. *Prehospital and Disaster Medicine*, doi:10.1017/S1049023X11006406.

Guha-Sapir, D., F. Vos, and R. Below, with S. Ponserre, 2011: *Annual Disaster Statistical Review 2010: The Numbers and Trends*. Centre for Research on the Epidemiology of Disasters, Université Catholique de Louvain, Brussels, Belgium, 42 pp. www.cred.be/sites/default/files/ADSR_2010.pdf.

Gurenko, E., 2004: *Catastrophe Risk and Reinsurance: A Country Risk Management Perspective*. Risk Books, London, UK.

Haile, M., 2005: Weather patterns, food security and humanitarian response in sub-Saharan Africa. *Philosophical Transactions of the Royal Society B*, **360(1463)**, 2169-2182.

Hajat, S., S.C. Sheridan, M. Allen, M. Pascal, K. Laaidi, A. Yagouti, U. Bickis, A.Tobias, D. Bourque, B.G. Armstrong, and T. Kosatsky, 2010: Heat-health warning systems: a comparison of the predictive capacity of different approaches to identifying dangerously hot days. *American Journal of Public Health*, **100(6)**, 1137-1144.

Hallegatte, S. and J. Corfee-Morlot, 2011: Understanding climate change impacts, vulnerability and adaptation at city scale: an introduction. *Climatic Change*, **104**, 1-12.

Hansen, J.W., S.J. Mason, L. Sun, and A. Tall, 2011: Review of seasonal climate forecasting for agriculture in sub-Saharan Africa. *Experimental Agriculture*, **47**, 205-240.

Hanson, S., R. Nicholls, N. Ranger, S. Hallegatte, J. Corfee-Morlot, and C. Herweijer, 2011: A global ranking of port cities with high exposure to climate extremes. *Climatic Change*, **104**, 89-111.

Harlan, S.L., A.J. Brazel, L. Prashad, W.L. Stefanov, and L. Larsen, 2006: Neighborhood microclimates and vulnerability to heat stress. *Social Science and Medicine*, **63**, 2847-2863.

Harris, J.B., M.J. Podolsky, T.R. Bhuiyan, F. Chowdhury, A.I. Khan, R.C. Larocque, T. Logvinenko, J. Kendall, A. S. Faruque, C.R. Nagler, E.T. Ryan, F. Qadri, and S.B. Calderwood, 2009: Immunologic responses to *Vibrio cholerae* in patients co-infected with intestinal parasites in Bangladesh. *PLoS Neglected Tropical Diseases*, **3(3)**, e403.

Hashizume, M., Y. Wagatsuma, A.S. Faruque, T. Hayashi, P.R. Hunter, B. Armstrong, and D.A. Sack, 2008: Factors determining vulnerability to diarrhoea during and after severe floods in Bangladesh. *Journal of Water & Health*, **6(3)**, 323-332.

Havenith, G., 2001: Individualized model of human thermoregulation for the simulation of heat stress response. *Journal of Applied Physiology*, **90(5)**, 1943-1954.

Hawkins-Bell, L. and J.T. Rankin, 1994: Heat-related deaths – Philadelphia and United States, 1993–1994. *Morbidity and Mortality Weekly Report*, **43(25)**, 453-455.

Health Canada, 2010: *Communicating the Health Risks of Extreme Heat Events: Toolkit for Public Health and Emergency Management Officials*. Health Canada, Ottawa, Canada.

Hellmuth, M.E., A. Moorhead, M.C. Thomson, and J. Williams (eds.), 2007: *Climate Risk Management in Africa: Learning from Practice*. International Research Institute for Climate and Society, Columbia University, New York, NY.

Hellmuth, M.E., D.E. Osgood, U. Hess, A. Moorhead, and H. Bhojwani (eds.), 2009: *Index Insurance and Climate Risk: Prospects for Development and Disaster Management*. International Research Institute for Climate and Society, Columbia University, New York, NY.

Hellmuth, M., S.J. Mason, C. Vaughan, M.K. van Aalst, and R. Choularton (eds.), 2011: *A Better Climate for Disaster Risk Management*. International Research Institute for Climate and Society, Columbia University, New York, NY, 118 pp.

Hémon, D. and E. Jougla, 2004: The heat wave in France in August 2003. *Revue d'Epidémiologie et de Sante Publique*, **52(3-5)**.

Hennessy, K., C. Lucas, N. Nicholls, J. Bathols, R. Suppiah, and J. Ricketts, 2005: *Climate Change Impacts on Fire-Weather In South-East Australia*. CSIRO Marine and Atmospheric Research, Aspendale, Australia.

Henstra, D. and G.A. McBean, 2005: Canadian Disaster Management Policy: Moving Toward a Paradigm Shift? *Canadian Public Policy*, **31**, 303-318.

Hess, U. and H. Syroka, 2005: *Weather-based Insurance in Southern Africa. The Case of Malawi*. World Bank, Washington, DC.

Hess, U., W. Wiseman, and T. Robertson, 2006: *Ethiopia: Integrated Risk Financing to Protect Livelihoods and Foster Development*. Discussion Paper, UN World Food Programme, Rome, Italy.

Hodgkinson, R.D., J. Evans, and J. Wood, 2003: Maintaining oral hydration in older adults: a systematic review. *International Journal of Nursing Practice*, **9**, 19-28.

Holmgren, J., 1981: Actions of cholera toxin and the prevention and treatment of cholera. *Nature*, **292(5822)**, 413-417.

Holubec, I., 2008: *Flat Loop Thermosyphon Foundations in Warm Permafrost*. Government of Northwest Territories, Yellowknife, Canada.

Hossain, M.Z., M.T. Islam, T. Sakai, and M. Ishida, 2008: Impact of tropical cyclones on rural infrastructures in Bangladesh. *Agricultural Engineering International: the CIGR Ejournal*, **X**, Invited Overview No. 2, 13 pp.

Hughes, K., E. Van Beurden, E.G. Eakin, L.M. Barnett, E. Patterson, J. Backhouse, S. Jones, D. Hauser, J.R. Beard, and B. Newman, 2008: Older persons' perception of risk of falling: implications for fall-prevention campaigns. *American Journal of Public Health*, **98(2)**, 351-357.

Hunt, A., and P. Watkiss, 2011: Climate change impacts and adaptation in cities: a review of the literature. *Climatic Change*, **104**, 13-49.

Huq, A., R.B. Sack, A. Nizam, I.M. Longini, G.B. Nair, A. Ali, J.G. Morris, Jr., M.N. Khan, A.K. Siddique, M. Yunus, M.J. Albert, D.A. Sack, and R.R. Colwell, 2005: Critical factors influencing the occurrence of *Vibrio cholerae* in the environment of Bangladesh. *Applied & Environmental Microbiology*, **71(8)**, 4645-4654.

Huq, S., H. Reid and L.A. Murray, 2006: *Climate Change and Development Links*. Gatekeeper Series 123, IIED, London, UK, 24 pp.

Huq, S., S. Kovats, H. Reid, and D. Satterthwaite, 2007: Editorial: Reducing risks to cities from disasters and climate change. *Environment and Urbanization*, **19(1)**, 3-15.

ICSU, 2008: *A Science Plan for Integrated Research on Disaster Risk: Addressing the Challenge Of Natural And Human-Induced Environmental Hazards*. International Council for Science, Paris, France, www.irdrinternational.org/reports.

ICTILO, UNDP, and UNISDR, 2010: *Local Governments and Disaster Risk Reduction: Good Practices and Lessons Learned*. International Training Centre of the International Labour Organization, UN Development Programme, and United Nations International Strategy for Disaster Reduction, Geneva, Switzerland, 86 pp.

IFRC, 2007: *Mozambique Floods and Cyclones*. DREF Bulletin No. MDRMZ002, Update no. 1, Glide no. FL-2006-000198-MOZ, International Federation of the Red Cross and Red Crescent Societies, Geneva, Switzerland, 4 pp., www.ifrc.org/docs/appeals/07/MDRMZ00201.pdf.

IFRC, 2009: World Disasters Report 2009 – Focus on Early Warning and Early Action. International Federation of the Red Cross and Red Crescent Societies, Geneva, Switzerland, 204 pp., www.ifrc.org/publicat/wdr2009/summaries.asp.

IFRC, 2011: Desk review on trends in the promotion of community-based disaster risk reduction legislation. In: *Global Assessment Report on Disaster Risk Reduction: Revealing Risk, Redefining Development*. United Nations International Strategy for Disaster Reduction, Geneva, Switzerland.

IMF, 2004: *Grenada: Use of Fund Resources-Request for Emergency Assistance-Staff Report; Press Release on the Executive Board Discussion; and Statement by the Executive Director for Grenada*. International Monetary Fund, Washington, DC, www.imf.org/external/pubs/ft/scr/2004/cr04405.pdf.

IMF, 2006: *Grenada: Request for a Three-Year Arrangement under the Poverty Reduction and Growth Facility – Staff Report; and Press Release on the Executive Board Discussion*. International Monetary Fund, Washington, DC, www.imf.org/external/pubs/ft/scr/2006/cr06277.pdf.

IMF, 2011: *World Economic Outlook Database*. International Monetary Fund, Washington, DC, www.imf.org/external/pubs/ft/weo/2011/01/weodata/index.aspx.

INAC, 2010: *Report on Adaptation to Climate Change Activities in Arctic Canada*. Indian and Northern Affairs Canada, Gatineau, Canada.

Infrastructure Canada, 2006: *Adapting Infrastructure to Climate Change in Canada's Cities and Communities: A Literature Review*. Infrastructure Canada, Ottawa, Canada, cbtadaptation.squarespace.com/storage/CdnInfrastructure Adaptation-LiteratureReview.pdf.

Instanes, A., O.A. Anisimov, L. Brigham, D. Goering, L.N. Khrustalev, B. Ladanyi, and J.O. Larsen, 2005: Infrastructure: buildings, support systems, and industrial facilities. In: *Arctic Climate Impact Assessment*. Cambridge University Press, Cambridge, UK, pp. 907-944.

InVS, 2003: *Impact sanitaire de la vague de chaleur en France survenue en août 2003. Rapport d'étape (Health impact of the heatwave which took place in France in August 2003. Progress Report)*. Institut de Veille Sanitaire (InVS), Département Santé Environnement, Saint-Maurice, France.

IOM, 2008: *Migration and Climate Change*. IOM Research Series No. 31, International Organization for Migration, Geneva, Switzerland, www.migrationdrc.org/publications/resource_guides/Migration_and_Climate_Change/MRS-31.pdf.

IRI, 2011: *Progress Report to National Oceanographic and Atmospheric Administration*. International Research Institute for Climate and Society, Columbia University, New York, NY, portal.iri.columbia.edu/portal/server.pt/gateway/PTARGS_0_2_7713_0_0_18/NOAAReport2011_FINAL_strip.pdf.

Jegillos, S.R., 2003: Methodology. In: *Sustainability in Grass-Roots Initiatives: Focus on Community Based Disaster Management* [Shaw, R. and K. Okazaki (eds.)]. United Nations Centre for Regional Development, Disaster Management Planning Hyogo Office, Hyogo, Japan, pp. 19-28.

Jeuland, M. and D. Whittington, 2009: Cost-benefit comparisons of investments in improved water supply and cholera vaccination programs. *Vaccine*, **27(23)**, 3109-3120.

Johnson, K., S. Solomon, D. Berry, and P. Graham, 2003: *Erosion Progression and Adaptation Strategy in a Northern Coastal Community*. Swets & Zeitlinger, Lisse, The Netherlands.

Jones, A.E., U.U. Wort, A.P. Morse, I.M. Hastings, and A.S. Gagnon, 2007: Climate prediction of El Niño malaria epidemics in north-west Tanzania. *Malaria Journal*, **6**, 162.

Jones, G.S., P.A. Stott, and N. Christidis, 2008: Human contribution to rapidly increasing frequency of very warm Northern Hemisphere summers. *Journal of Geophysical Research*, **113**, D02109.

Kadomura, H., 2005: Climate anomalies and extreme events in Africa in 2003, including heavy rains and floods that occurred during northern hemisphere summer. *African Study Monographs*, **S30**, 165-181.

Kalkstein, A.J. and S.C. Sheridan, 2007: The social impacts of the heat-health watch/warning system in Phoenix, Arizona: assessing the perceived risk and response of the public. *International Journal of Biometeorology*, **52(1)**, 43-55.

Kalsi, S.R. and R.S. Pareek, 2001: Hottest April of the 20th century over north-west and central India. *Current Science*, **80(7)**, 867-872.

Karim, M.F. and N. Mimura, 2008: Impacts of climate change and sea level rise on cyclonic storm surge floods in Bangladesh. *Global Environmental Change*, **18(3)**, 490-500.

Kaufmann, D., A. Kraay, and M. Mastruzzi, 2010: *The Worldwide Governance Indicators: Methodology and Analytical Issues*. Policy Research Working Paper No. 5430, World Bank, papers.ssrn.com/sol3/ papers.cfm?abstract_id=1682130.

Kelman, I., 2010: Policy arena: introduction to climate, disasters and international development. *Journal of International Development*, **22**, 208-217.

Kelman, I. and J. West, 2009: Climate change and small island developing states: a critical review. *Ecological and Environmental Anthropology*, **5(1)**, 1-16.

Keen, J. and T. Packwood, 1995. Qualitative research: Case study evaluation. *BMJ*, **311(1)**, 444-446.

Kharin, V.V. and F.W. Zwiers, 2003: Improved seasonal probability forecasts. *Journal of Climate*, **16**, 1684-1701.

Kienberger, S., 2007: Assessing the vulnerability to natural hazards on the provincial/community level in Mozambique: the contribution of GIScience and remote sensing. In: *The 3rd International Symposium on Geo-information for Disaster Management*, Joint CIG/ISPRS Conference, Toronto, Canada, 23-25 May 2007.

Klinenberg, E., 2002: *Heatwave: A Social Autopsy of Disaster in Chicago*. University of Chicago Press, Chicago, IL.

Koelle, K., 2009: The impact of climate on the disease dynamics of cholera. *Clinical Microbiology & Infection*, **15(S1)**, 29-31.

Koelle, K., M. Pascual, and M. Yunus, 2005a: Pathogen adaptation to seasonal forcing and climate change. *Proc of the Royal Society of London B*, **272(1566)**, 971-977.

Koelle, K., X. Rodo, M. Pascual, M. Yunus, and G. Mostafa, 2005b: Refractory periods and climate forcing in cholera dynamics. *Nature*, **436(7051)**, 696-700.

Kondo, H., N. Seo, T. Yasuda, M. Hasizume, Y. Koido, N. Ninomiya, and Y. Yamamoto, 2002: Post-flood infectious diseases in Mozambique. *Prehospital and Disaster Medicine*, **17(3)**, 126-133, pdm.medicine.wisc.edu/Volume_17/issue_3/kondo.pdf.

Kosatsky, T., 2005: The 2003 European heat waves. *Euro Surveillance*, **10**, 148-149.

Koster, R.D., M.J. Suarez, P. Liu, U. Jambor, A. Berg, M. Kistler, R. Reichle, M. Rodell, and J. Famiglietti, 2004: Realistic initialization of land surface states: impacts on subseasonal forecast skill. *Journal of Hydrometeorology*, **5**, 1049-1063.

Koster, R.D., S.P.P. Mahanama, T.J. Yamada, G. Balsamo, A.A. Berg, M. Boisserie, P.A. Dirmeyer, F.J. Doblas-Reyes, G. Drewitt, C.T. Gordon, Z. Guo, J.-H. Jeong, D.M. Lawrence, W.-S. Lee, Z. Li, L. Luo, S. Malyshev, W.J. Merryfield, S.I. Seneviratne, T. Stanelle, B.J.J.M. van den Hurk, F. Vitart, and E.F. Wood, 2010: Contribution of land surface initialization to subseasonal forecast skill: first results from a multi-model experiment. *Geophysical Research Letters*, **37**, L02402.

Kovats, R.S. and S. Hajat, 2008: Heat stress and public health: a critical review. *Annual Review of Public Health*, **29**, 41-55.

Kovats, S. and R. Akhtar, 2008: Climate, climate change and human health in Asian cities. *Environment and Urbanization*, **20(1)**, 165-175.

Kreimer, A. and M. Arnold, 2000: World Bank's role in reducing impacts of disasters. *Natural Hazards Review*, **1(1)**, 37-42.

Kulling, P., M. Birnbaum, V. Murray, and G. Rockenschaub, 2010: Guidelines for reports on health crises and critical health events. *Prehospital and Disaster Medicine*, **25(4)**, 377-383.

Kumar, S., 1998: India's heat wave and rains result in massive death toll. *Lancet*, **351**, 1869.

Kundzewicz, Z.W., L.J. Mata, N.W. Arnell, P. Döll, P. Kabat, B. Jiménez, K.A. Miller, T. Oki, Z. Sen, and I.A. Shiklomanov, 2007: Freshwater resources and their management. In: *Climate Change 2007. Impacts, Adaptation and Vulnerability. Contribution of Working Group II to the Fourth Assessment Report of the Intergovernmental Panel on Climate Change* [Parry, M.L., O.F. Canziani, J.P. Palutikof, P.J. Van Der Linde, and C.E. Hanson (eds.)]. Cambridge University Press, Cambridge, UK, 173-210.

Kunreuther, H. and E. Michel-Kerjan, 2009: *At War with the Weather: Managing Large-Scale Risks in a New Era of Catastrophes*. MIT Press, Cambridge, MA.

Kunreuther, H. and R.J. Roth, 1998: *Paying the Price: The Status and Role of Insurance Against Natural Disasters in the United States*. Joseph Henry Press, Washington, DC.

Laaidi, K., M. Pascal, M. Ledrans, A. Le Tertre, S. Medina, C. Caserio, J.C. Cohen, J. Manach, P. Beaudeau, and P. Empereue-Bissonet, 2004: *Le système français d'alerte canicule et santé (SACS 2004): Un dispositif intégéré au Plan National Canicule (The French Heatwave Warning System and Health: A System Integrated into the National Heatwave Plan)*. Institut de Veille Sanitaire, Saint-Maurice, France.

Lagadec, P., 2004: Understanding the French 2003 heat wave experience: beyond the heat, a multi-layered challenge. *Journal of Contingencies and Crisis Management*, **12**, 160-169.

Lalani, B., 2011: Drought mitigation: Assessing technological trade-offs and challenges posed by policy solutions; A case study of Salamieh District, Syria. *World Applied Sciences Journal*, **13(4)**, 936-946.

Larsen, P.H., S. Goldsmith, O. Smith, M. Wilon, K. Strzepek, P. Chinowsky, and B. Saylor, 2008: Estimating future costs for Alaska public infrastructure at risk from climate change. *Global Environmental Change*, **18**, 442-457.

Lateef, F., 2009: Cyclone Nargis and Myanmar: a wake up call. *Journal of Emergencies, Trauma, and Shock*, **2(2)**, 106-113.

Lawrimore, J., R. Heim, T. Owen, M. Svoboda, V. Davydova, D. Chobanik, R. Rippey, and D. Lecomte, 2007: The North American drought monitor and a global drought early warning system. In: *The Full Picture*. Group on Earth Observations, Tudor Rose, London, UK, pp. 142-144.

Lazo, J.K., R.E. Morss, and J.L. Demuth, 2009: 300 billion served: Sources, perceptions, uses, and values of weather forecasts. *Bulletin of the American Meteorological Society*, **90(6)**, 785-798.

Lemmen, D.S., F.J. Warren, J. Lacroix, and E. Bush (eds.), 2008: *From Impacts to Adaptation: Canada in a Changing Climate 2007*. Government of Canada, Ottawa, Canada, 448 pp.

Létard, V., H. Flandre, and S. Lepeltier, 2004: *La France et les Français face à la canicule: les leçons d'une crise (How France and French people faced a heatwave: lessons from a crisis)*. Report No. 195 (2003-2004) to the Sénat, Government of France, Paris, France, 391 pp.

Levick J.J., 1859: Remarks on sunstroke. *American Journal of the Medical Sciences*, **73**, 40-42.

Liniger, M.A., H. Mathis, C. Appenzeller, and F.J. Doblas-Reyes, 2007: Realistic greenhouse gas forcing and seasonal forecasts. *Geophysical Research Letters*, **34**, L04705.

Linnerooth-Bayer, J. and R. Mechler, 2009: Insurance against Losses from Natural Disasters in Developing Countries. DESA Working Paper No. 85, United Nations Department of Economic and Social Affairs, New York, NY, USA, 37 pp.

Linnerooth-Bayer, J., R. Mechler, and G. Pflug, 2005: Refocusing disaster aid. *Science*, **309**, 1044-1046.

Linnerooth-Bayer, J., R. Mechler, and M.J. Mace, 2007: Insurance for assisting adaptation to climate change in developing countries. A proposed strategy. *Climate Policy*, **6**, 621-636.

Linnerooth-Bayer, J., P. Suarez, M. Victor and R. Mechler, 2009: Drought insurance for subsistence farmers in Malawi. *Natural Hazards Observer*, **33(5)**, 5-9.

Lipp, E.K., A. Huq, and R.R. Colwell, 2002: Effects of global climate on infectious disease: the cholera model. *Clinical Microbiology Reviews*, **15**, 757-770.

Lise, W., S. Hess, and B. Purev, 2006: Pastureland degradation and poverty among herders in Mongolia: data analysis and game estimation. *Ecological Economics*, **58**, 350-364.

Lizarraga-Partida, M.L., E. Mendez-Gomez, A.M. Rivas-Montano, E. Vargas-Hernandez, A. Portillo-Lopez, A.R. Gonzalez-Ramirez, A. Huq, and R.R. Colwell, 2009: Association of *Vibrio cholerae* with plankton in coastal areas of Mexico. *Environmental Microbiology*, **11(1)**, 201-208.

Llosa, S. and I. Zodrow, 2011: Disaster risk reduction legislation as a basis for effective adaptation. In: *Global Assessment Report on Disaster Risk Reduction: Revealing Risk, Redefining Development*. United Nations International Strategy for Disaster Reduction, Geneva, Switzerland.

Lopez, A.L., J.D. Clemens, J. Deen, and L. Jodar, 2008: Cholera vaccines for the developing world. *Human Vaccines*, **4(2)**, 165-169.

Loster, T. and D. Reinhard, 2010: Microinsurance and climate change. In: Microinsurance- An innovative tool for risk and disaster management [Morelli, E., G.A. Onnis, W.J. Amman, and C. Sutter (eds.)]. Global Risk Forum, Davos, Switzerland, pp. 39-42.

Loster, T. and A. Wolf, 2007: *IntoAction. Flood warning system in Mozambique. Completion of Búzi project*. Munich Re Foundation, Munich, Germany, 16 pp.

Louka, P., G. Vachon, J.-F. Sini, P.G. Mestayer, and J.-M. Rosant, 2002: Thermal effects on the airflow in a street canyon – Nantes '99 experimental results and model simulation. *Water, Air and Soil Pollution: Focus*, **2(5-6)**, 351-364.

Luber, G. and M. McGeehin, 2008: Climate change and extreme heat events. *American Journal of Preventive Medicine*, **35(5)**, 429-435.

Luna, E.M., M.L.P. Bautista, and M.P. de Guzman, 2008: *Mainstreaming Disaster Risk Reduction in the Education Sector in the Philippines: Integrating Disaster Risk Reduction in the School Curriculum, Impacts of Disasters on the Education Sector, School Construction Current Practices and Improvements Needed*. Centre for Disaster Preparedness, Quezon City, the Philippines, 160 pp.

Luque Fernandez, M.A., A. Bauernfeind, J.D. Jimenez, C.L. Gil, N. El Omeiri, and D.H. Guibert, 2009: Influence of temperature and rainfall on the evolution of cholera epidemics in Lusaka, Zambia, 2003-2006: analysis of a time series. *Transactions of the Royal Society of Tropical Medicine & Hygiene*, **103(2)**, 137-143.

Mahul, O. and J. Skees, 2005: *Managing Agricultural Risk at the Country Level: The Case of Index-Based Livestock Insurance in Mongolia*. World Bank, Washington, DC, 35 pp.

Marin, A., 2008: Between cash cows and golden calves. Adaptations of Mongolian pastoralism in the 'age of the market.' *Nomadic Peoples*, **12(2)**, 75-101.

Marin, A., 2010: Riders under storms: Contributions of nomadic herders' observations to analysing climate change in Mongolia. *Global Environmental Change – Human and Policy Dimensions*, **20(1)**, 162-176.

Marinacci, C., M. Marino, E. Ferracin, L. Fubini, L. Gilardi, P. Visentin, E. Cadum, and G. Costa, 2009: Testing of interventions for prevention of heat wave related deaths: results among frail elderly and methodological problems. *Epidemiologia e Prevenzione*, **33(3)**, 96-103.

Maskey, M., J.S. Shastri, K. Saraswathi, R. Surpam, and N. Vaidya, 2006: Leptospirosis in Mumbai: post-deluge outbreak 2005. *Indian Journal of Medical Microbiology*, **24(S4)**, 337-338.

Mason, P.R., 2009: Zimbabwe experiences the worst epidemic of cholera in Africa. *Journal of Infection in Developing Countries*, **3(2)**, 148-151.

Matsuda, F., S. Ishimura, Y. Wagatsuma, T. Higashi, T. Hayashi, A.S. Faruque, D.A. Sack, M. Nishibuchi, and A.S.G. Faruque, 2008: Prediction of epidemic cholera due to *Vibrio cholerae* O1 in children younger than 10 years using climate data in Bangladesh. *Epidemiology & Infection*, **136(1)**, 73-79.

Mattingly, S., 2002: *Policy, legal and institutional arrangements*. Proceedings of Regional Workshop on Best Practices in Disaster Mitigation, Asian Disaster Preparedness Centre, Bangkok, Thailand, 51 pp.

McBean, G.A., 2000. *Forecasting in the 21st Century*. Publication 916, World Meteorological Organization, Geneva, Switzerland, 18 pp.

McBean, G., G. Alekseev, D. Chen, E. Førland, J. Fyfe, P.Y. Groisman, R. King, H. Melling, R. Vose, and P.H. Whitfield, 2005: Arctic climate: past and present. In: *Arctic Climate Impact Assessment*. Cambridge University Press, Cambridge, UK, pp. 22-60.

McCormick, S., 2010a: Dying of the heat: diagnostic debates, calculations of risk, and actions to advance preparedness. *Environment and Planning A*, **42**, 1513-1518.

McCormick, S., 2010b: Hot or not? Obstacles to the emergence of climate-induced illness movements. In: *Social Movements and Health Care in the United States* [Zald, M., J. Banaszak-Holl, and S. Levitsky (eds.)]. Oxford University Press, Oxford, UK.

McGranahan, G., D. Balk, and B. Anderson, 2007: The rising tide: assessing the risks of climate change and human settlements in low elevation coastal zones. *Environment and Urbanization*, **19(1)**, 17-37.

McMichael, A.J., R.E. Woodruff, and S. Hales, 2006: Climate change and human health: present and future risks. *Lancet*, **367(9513)**, 859-869.

McWhinney, I.R., 2001: The value of case studies. *European Journal of General Practice*, **7(3)**, 88-89.

Mearns, R., 2004: Sustaining livelihoods on Mongolia's pastoral commons: Insights from a participatory poverty assessment. *Development and Change*, **35(1)**, 107-139.

Mechler, R., 2004: *Natural Disaster Risk Management and Financing Disaster Losses in Developing Countries*. Verlag für Versicherungswirtschaft, Karlsruhe, Germany.

Mechler, R., J. Linnerooth-Bayer, and D. Peppiatt, 2006: *Disaster Insurance for the Poor? A Review of Microinsurance for Natural Disaster Risks in Developing Countries*. ProVention Consortium and IIASA, Geneva, Switzerland and Laxenburg, Austria.

Medina-Cetina, Z. and F. Nadim, 2008: Stochastic design of an early warning system. *Georisk: Assessment and Management of Risk for Engineered Systems and Geohazards*, **2(4)**, 223-236.

Meehl, G.A., T.F. Stocker, W.D. Collins, P. Friedlingstein, A.T. Gaye, J.M. Gregory, A. Kitoh, R. Knutti, J.M. Murphy, A. Noda, S.C.B. Raper, I.G. Watterson, A.J. Weaver, and Z.-C. Zhao, 2007: Global climate projections. In: *Climate Change 2007: The Physical Science Basis. Contribution of Working Group I to the Fourth Assessment Report of the Intergovernmental Panel on Climate Change* [Solomon, S., D. Qin, M. Manning, Z. Chen, M. Marquis, K.B. Averyt, M. Tignor and H.L. Miller (eds.)]. Cambridge University Press, Cambridge, UK, and New York, NY, pp. 747-845.

Meehl, G.A., L. Goddard, J. Murphy, R.J. Stouffer, G. Boer, G. Danabasoglu, K. Dixon, M.A. Giorgetta, A.M. Greene, E. Hawkins, G. Hegerl, D. Karoly, N. Keenlyside, M. Kimoto, B. Kirtman, A. Navarra, R. Pulwarty, D. Smith, D. Stammer, and T. Stockdale, 2009: Decadal prediction: can it be skillful? *Bulletin of the American Meteorological Society*, **90**, 1467-1485.

Meinke, H., R. Nelson, P. Kokic, R. Stone, R. Selvaraju, and W. Baethgen, 2006: Actionable climate knowledge – from analysis to synthesis. *Climate Research*, **33**, 101-110.

Melbourne Water, 2009a: *Water Report*. Melbourne Water, Melbourne, Australia, www.melbournewater.com.au/content/water_storages/water_report/water_report.asp.

Melbourne Water, 2009b: Bushfires in Catchments. Melbourne Water, Melbourne, Australia, www.melbournewater.com.au/content/water_storages/bushfires_in_catchments/bushfires_in_catchments.asp.

Mendelsohn, J. and T. Dawson, 2008: Climate and cholera in KwaZulu-Natal, South Africa: the role of environmental factors and implications for epidemic preparedness. *International Journal of Hygiene & Environmental Health*, **211(1-2)**, 156-162.

Météo France, 2003: *Retour sur la canicule d'août 2003 - Bilan de la canicule d'août 2003 adressé par Météo-France au ministère de l'Equipement, des Transports, du Logement et de la Mer le 20 août 2003 (Analysis of the August 2003 heatwave presented by Météo-France to the Ministry for Equipement, Transport, Housing and Sea on 20 August 2003)*. Météo France, Paris, France, france.meteofrance.com/france/actu/bilan/archives/2003/canicule?page_id=10035&document_id=4523&portlet_id=40336.

Météo France, 2006: *Retour sur la canicule de juillet 2006 (Analysis of the July 2006 heatwave)*. Météo France, Paris, France, france.meteofrance.com/france/actu/bilan/archives/2006/canicule?page_id=10043&document_id=4516&portlet_id=40408.

MGH, 2003: *State Policy for Food and Agriculture*. Annex 29 of resolution of Mongolia Great Hural (MGH), Ulaanbaatar, Mongolia, mofa.gov.mn/livestock/index.php?option=com_content&view=article&id=73%3A2010-09-06-08-07-14&catid=52%3A2010-07-02-06-32-10&Itemid=59&lang=en.

Michelozzi, P., F.K. de' Donato, A.M. Bargagli, D. D'Ippoliti, M. de Sario, C. Marino, P. Schifano, G. Cappai, M. Leone, U. Kirchmayer, M. Ventura, M. di Gennaro, M. Leonardi, F. Oleari, A. de Martino, and C.A. Perucci, 2010: Surveillance of summer mortality and preparedness to reduce the health impact of heat waves in Italy. *International Journal of Environmental Research and Public Health*, 7(5), 2256-2273.

MIJ, 2009: *Informe Nacional del Progreso en la Implementación del Marco de Acción de Hyogo*. Report to UNISDR HFA Monitor, Colombia Ministerio del Interior y de Justicia, Dirección de Prevención y Atención de Desastres, Bogota, Colombia.

Mijiddorj, R., 2008: *Global Warming and Challenge to Desert*. Mongolia University of Science and Technology, Ulaanbaatar, Mongolia, 62 pp.

Mills, E., 2005: Insurance in a climate of change. *Science*, **309**, 1040-1044.

Milly, P.C.D., J. Betancourt, M. Falkenmark, R.M. Hirsch, Z.W. Kundzewicz, D.P. Lettenmaier, and R.J. Stouffer, 2008: Stationarity is dead: whither water management? *Science*, **319**, 573-574.

Mimura, N., L. Nurse, R.F. McLean, J. Agard, L. Briguglio, P. Lefale, R. Payet, and G. Sem, 2007: Small islands. In: *Climate Change 2007. Impacts, Adaptation and Vulnerability. Contribution of Working Group II to the Fourth Assessment Report of the Intergovernmental Panel on Climate Change* [Parry, M.L., O.F. Canziani, J.P. Palutikof, P.J. Van Der Linde, and C.E. Hanson (eds.)]. Cambridge University Press, Cambridge, UK, pp. 687-716.

Mirza, M.M.Q., 2003: Climate change and extreme weather events: can developing countries adapt? *Climate Policy*, **3**, 233-248.

Mitchell, T., T. Tanner, and K. Haynes, 2009: *Children as Agents of Change for Disaster Risk Reduction: Lessons from El Salvador and the Philippines*. Working Paper 1, Children in a Changing Climate, IDS, Brighton, UK, 48 pp.

MMS, 2009: *National Climate Risk Management Strategy and Action Plan (Draft)* [by expert team headed by Togtokh, N. and D. Tuvdendorj]. National Emergency Management Authority, Mongolia Meteorological Society (MMS), United Nations Development Programme, Ulaanbaatar, Mongolia.

Morelli, E., G.A. Onnis, W.J. Amman, and C. Sutter (eds.), 2010: Microinsurance – An innovative tool for risk and disaster management. Global Risk Forum, Davos, Switzerland, 360 pp.

Morinaga, Y., S. Tian, and M. Shinoda, 2003: Winter snow anomaly and atmospheric circulation in Mongolia. *International Journal of Climatology*, **23**, 1627-1636.

Morris, L., 2011: Death and suffering in the land of Genghis Khan. *CMAJ*, 183(5), doi:10.1503/cmaj.109-3796.

Mouillot, F., J.P. Ratte, R. Joffre, J.M. Moreno, and S. Rambal, 2003: Some determinants of the spatio-temporal fire cycle in a Mediterranean landscape (Corsica, France). *Landscape Ecology*, **18**, 665-674.

Munich Re, 2010: Topics Geo: Natural catastrophes 2009 – Analyses, assessments, positions. Munich Re, Munich, Germany, www.munichre.com.

Myers, S.S. and J.A. Patz, 2009: Emerging threats to human health from global environmental change. *Annual Review of Environment and Resources*, 34(1), 223-252.

NASA, 2008: *Heatwave in Northern Africa and Southern Europe*. Earth Observatory, NASA Goddard Space Flight Center, Greenbelt, MD, earthobservatory.nasa.gov/NaturalHazards/view.php?id=15257.

National Climate Centre, 2009: *The exceptional January-February 2009 heatwave in southeastern Australia*. Special Climate Statement 17, Bureau of Meteorology, Melbourne, Australia, www.bom.gov.au/climate/current/statements/scs17c.pdf.

National Hurricane Center, 2006: *Dennis, Katrina, Rita, Stan, and Wilma "Retired" from List of Storm Names*. National Oceanic and Atmospheric Administration, Washington, DC, www.noaanews.noaa.gov/stories2006/s2607.htm.

Naughton, M.P., A. Henderson, M.C. Mirabelli, R. Kaiser, J.L. Wilhelm, S.M. Kieszak, C.H. Rubin, and M.A. McGeehin, 2002: Heat-related mortality during a 1999 heat wave in Chicago. *American Journal of Preventive Medicine*, **22(4)**, 221-227.

NDMC, 2007: *Inaugural Annual Report 2006-2007*. National Disaster Management Centre, Provincial and Local Government Department, Pretoria, South Africa, 172 pp.

NDMC, 2010: *National Education, Training and Research Needs and Resources Analysis (NETaRNRA)*. Consolidated report compiled by the National Disaster Management Centre, South African Department of Cooperative Governance and Traditional Affairs, Pretoria, South Africa, 23 pp.

Nelson, F., O. Anisimov, and N. Shiklomanov, 2002: Climate change and hazard zonation in the circum-arctic permafrost regions. *Natural Hazards*, **26**, 203-225.

Nicholls, N., 2001: Atmospheric and climatic hazards: improved monitoring and prediction for disaster mitigation. *Natural Hazards*, **23**, 137-155.

Nicholls, N., C. Skinner, M. Loughnan, and N. Tapper, 2008: A simple heat alert system for Melbourne, Australia. *International Journal of Biometeorology*, **52(5)**, 375-384.

Nicholls, R.J., P.P. Wong, V.R. Burkett, J.O. Codignotto, J.E. Hay, R.F. McLean, S. Ragoonaden, and C.D. Woodroffe, 2007: Coastal systems and low-lying areas. In: *Climate Change 2007. Impacts, Adaptation and Vulnerability. Contribution of Working Group II to the Fourth Assessment Report of the Intergovernmental Panel on Climate Change* [Parry, M.L., O.F. Canziani, J.P. Palutikof, P.J. Van Der Linden, and C.E. Hanson (eds.)]. Cambridge University Press, Cambridge, UK, 315-356.

Nielson, D., 2007: *The City of Iqaluit's Climate Change Impacts, Infrastructure Risks and Adaptive Capacity Project*. City of Iqaluit, Canada.

Nixdorf-Miller, A., D.M. Hunsaker, and J.C. Hunsaker III, 2006: Hypothermia and hyperthermia medicolegal investigation of morbidity and mortality from exposure to environmental temperature extremes. *Archives of Pathology & Laboratory Medicine*, **130**, 1297-1304.

Noji, E.K., 2000: The public health consequences of disasters. *Prehospital and Disaster Medicine*, **15(4)**, 147-157.

Noss, R.F., J.F. Franklin, W.L. Baker, T. Schoennagel, and P.B. Moyle, 2006: Managing fire-prone forests in the western United States. *Frontiers in Ecology and Environment*, **4**, 481-487.

NRC, 2003: *Fair Weather: Effective Partnerships in Weather and Climate Services*. National Research Council, National Academies Press, Washington, DC.

NRCAN, 2010: *Adaptation case studies: Iqaluit's Sustainable Subdivision*. Natural Resources Canada, Ottawa, Canada, adaptation.nrcan.gc.ca/case/iqaluit_e.php.

NRCAN, 2011a: *Geological Survey of Canada*. Natural Resources Canada, Ottawa, Canada, gsc.nrcan.gc.ca/permafrost/whatis_e.php.

NRCAN, 2011b: *Regional Adaptation Collaboratives: The Nunavut Mining Sector and Climate Change: Risks, Vulnerabilities and Good Environmental Practices Project*. Natural Resources Canada, adaptation.nrcan.gc.ca/collab/no_e.php.

NRCAN, 2011c: *Nunavut Climate Change Partnership*. Natural Resources Canada, Ottawa, Canada, www.nrcan.gc.ca/com/elements/issues/49/panpcn-eng.php.

NRTEE, 2009: *True North: Adapting Infrastructure to Climate Change in Northern Canada*. National Round Table on the Environment and the Economy of Canada, Ottawa, Canada, 160 pp.

NSO, 2011: *Mongolian Statistical Yearbook, 2010*. National Statistical Office of Mongolia (NSO), Ulaanbaatar, Mongolia, 463 pp.

Nuclear Claims Tribunal, Republic of the Marshall Islands, 2007: *Agreement between the Government of the United States and the Government of the Marshall Islands for the Implementation of Section 177 of the Compact of Free Association*. www.nuclearclaimstribunal.com/177text.htm.

Nunes, M.C.S., M.J. Vasconcelos, J.M.C. Pereira, N. Dasgupta, R.J. Alldredge, and F.C. Rego, 2005: Land cover type and fire in Portugal: do fires burn land cover selectively? *Landscape Ecology*, **20**, 661-673.

Nurse, L. and G. Sem, 2001: Small island states. In: *Climate Change 2001: Impacts, Adaptation, and Vulnerability. Contribution of Working Group II to the Third Assessment Report of the Intergovernmental Panel on Climate Change* [McCarthy, J.J., O.F. Canziani, N.A. Leary, D.J. Dokken, and K.S. White (eds.)]. Cambridge University Press, Cambridge, UK, 834-875.

O'Neill, M.S., R. Carter, J.H. Kish, C.J. Gronlund, J.L. White-Newsome, X. Manarolla, A. Zanobetti, and J.D. Schwartz, 2009: Preventing heat-related morbidity and mortality: new approaches in a changing climate. *Maturitas*, **64(2)**, 98-103.

Oswald Spring, Ú., 2011: Social vulnerability, discrimination, and resilience-building in disaster risk reduction. In: *Coping with Global Environmental Change, Disasters and Security – Threats, Challenges, Vulnerabilities and Risks* [Brauch, H.G., U. Oswald Spring, C. Mesjasz, J. Grin, P. Kameri-Mbote, B. Chourou, P. Dunay, and J. Birkmann (eds.)]. Springer Verlag, Berlin, Germany, pp. 1169-1188.

Oyun, R. (ed.), 2004: *Impact of Current Climate Hazards on the Livelihoods of the Herders' Households*. United Nations Development Programme, Poverty Research Group, JEMR Consulting, Ulaanbaatar, Mongolia, 87 pp.

Palmer, T.N., F.J. Doblas-Reyes, A. Weisheimer, and M.J. Rodwell, 2008: Toward seamless prediction: calibration of climate change projections using seasonal forecasts. *Bulletin of the American Meteorological Society*, **89**, 459-470.

Parker, D.J., S.J. Priest and S.M. Tapsell, 2009: Understanding and enhancing the public's behavioural response to flood warning information. *Meteorological Applications*, **16**, 103-114.

Parliament of Victoria, 2009: *2009 Victorian Bushfires Royal Commission – Interim Report*. No. 225 – Session 2006–09, Government Printer for the State of Victoria, Melbourne, Australia.

Parliament of Victoria, 2010a: *2009 Victorian Bushfires Royal Commission, Final Report, Volume 1*. Government Printer for the State of Victoria, Melbourne, Australia, www.royalcommission.vic.gov.au/Commission-Reports/Final-Report/Volume-1.

Parliament of Victoria, 2010b: *2009 Victorian Bushfires Royal Commission, Final Report, Volume 2*. Government Printer for the State of Victoria, Melbourne, Australia, www.royalcommission.vic.gov.au/Commission-Reports/Final-Report/Volume-2.

Parliament of Victoria, 2010c: *2009 Victorian Bushfires Royal Commission, Summary Report, Volume 2*. Government Printer for the State of Victoria, Melbourne, Australia, www.royalcommission.vic.gov.au/Commission-Reports/Final-Report/Summary.

Parry, M.L., O.F. Canziani, J.P. Palutikof, P.J. van der Linden, and C.E. Hanson (eds.), 2007: *Climate Change 2007: Impacts, Adaptation and Vulnerability. Contribution of Working Group II to the Fourth Assessment Report of the Intergovernmental Panel on Climate Change*. Cambridge University Press, Cambridge, UK, 1000 pp.

Pascual, M., M.J. Bouma, and A.P. Dobson, 2002: Cholera and climate: revisiting the quantitative evidence. *Microbes & Infection*, **4(2)**, 237-245.

Patt, A., P. Suarez, and C. Gwata, 2005: Effects of seasonal climate forecasts and participatory workshops among subsistence farmers in Zimbabwe. *Proceedings of the National Academy of Sciences*, **102(35)**, 12623-12628.

Patz, J., D. Engelberg, and J. Last, 2000: The effects of changing weather on public health. *Annual Review of Public Health*, **21**, 271-307.

Paul, A. and M. Rahman, 2006: Cyclone mitigation perspectives in the islands in Bangladesh: a case study of Swandip and Hatia Islands. *Coastal Management*, **34(2)**, 199-215.

Paul, B.K., 2009: Why relatively fewer people died? The case of Bangladesh cyclone Sidr. *Natural Hazards*, **50**, 289-304.

Paul, B.K. and S. Dutt, 2010: Hazard warnings and responses to evacuation orders: the case of Bangladesh's cyclone Sidr. *Geographical Review*, **100(3)**, 336-355.

Pearce, T., J. Ford, G. Laidler, B. Smit, F. Duerden, M. Allarut, M. Andrachuk, S. Baryluk, A. Dialla, P. Elee, A. Goose, T. Ikummaq, E. Joamie, F. Kataoyak, E. Loring, S. Meakin, S. Nickels, K. Shappa, J. Shirley, and J. Wandel, 2009: Community collaboration and climate change research in the Canadian Arctic. *Polar Research*, **28**, 10-27.

Peduzzi, P., B. Chatenoux, H. Dao, C. Herold, and G. Giuliani, 2011: Preview Global Risk Data Platform. UNEP/GRID and UNISDR, Geneva, Switzerland, preview.grid.unep.ch/index.php?preview=tools&cat=1&lang=eng.

Peduzzi, P., U. Deichmann, A. Maskrey, F.A. Nadim, H. Dao, B. Chatenoux, C. Herold, A. Debono, G. Giuliani, and S. Kluser, 2009: Global disaster risk: patterns, trends and drivers. In: *Global Assessment Report on Disaster Risk Reduction*. United Nations International Strategy for Disaster Reduction, Geneva, Switzerland, pp. 17-57.

Pelling, M. and A. Holloway, 2006: *Legislation for Mainstreaming Disaster Risk Reduction*. Tearfund, Teddington, UK, 36 pp.

Pelling, M., A. Ozerdem, and S. Barakat, 2002: The macro-economic impact of disasters. *Progress in Development Studies*, **2(4)**, 283-305.

Peng, R.D., J.F. Bobb, C. Tebaldi, L. McDaniel, M.L. Bell, and F. Dominici, 2010: Toward a quantitative estimate of future heat wave mortality under global climate change. *Environmental Health Perspectives*, **119(5)**, 701-706.

Pernetta, J.C., 1992: Impacts of climate change and sea-level rise on small island states: National and international responses. *Global Environmental Change*, **2(1)**, 19-31.

Pirard, P., S. Vandentorren, M. Pascal, K. Laaidi, A. Le Tertre, S. Cassadou, and M. Ledrans, 2005: Summary of the mortality impact assessment of the 2003 heat wave in France. *Euro Surveillance*, **10(7)**, pii=554.

Poolman, E., H. Chikoore, and F. Lucio, 2008: Public benefits of the Severe Weather Forecasting Demonstration Project in southeastern Africa. *Meteo World*, **Dec 2008**, www.wmo.int/pages/publications/meteoworld/archive/dec08/swfdp_en.html.

Poumadere, M., C. Mays, S. LeMer, and R. Blong, 2005: The 2003 heat wave in France: dangerous climate change here and now. *Risk Analysis*, **25**, 1483-1494.

ProAct Network, 2008: The Role of Environmental Management and eco-engineering in Disaster Risk Reduction and Climate Change Adaptation. ProAct Network, Gaia Group, Finnish Ministry of Environment, and the United Nations International Strategy for Disaster Reduction Secretariat, Switzerland, www.preventionweb.net/english/professional/publications/v.php?id=4148.

Pruzzo, C., L. Vezzulli, and R.R. Colwell, 2008: Global impact of *Vibrio cholerae* interactions with chitin. *Environmental Microbiology*, **10(6)**, 1400-1410.

PSIDS, 2009: *Fiji, Marshall Islands, Micronesia (Federated States of), Nauru, Palau, Papua New Guinea, Samoa, Solomon Islands, Tonga, Tuvalu, Vanuatu: Views on the Possible Security Implications of Climate Change to be included in the report of the Secretary-General to the 64th Session of the United Nations General Assembly*, Pacific Small Island Developing States, Permanent Mission of the Republic of Nauru to the United Nations, New York, NY, www.un.org/esa/dsd/resources/res_pdfs/ga-64/cc-inputs/PSIDS_CCIS.pdf.

Quarless, D., 2007: Addressing the vulnerability of SIDS. *Natural Resources Forum*, **31(2007)**, 99-101.

Ranger, N., S. Hallegatte, S. Bhattacharya, M. Bachu, S. Priya, K. Dhore, F. Rafique, P. Mathur, N. Naville, F. Henriet, C. Herweijer, S. Pohit, and J. Corfee-Morlot, 2011: An assessment of the potential impact of climate change on flood risk in Mumbai. *Climatic Change*, **104**, 139-167.

Read, R., 2010: *Economic Vulnerability and Resilience in Small Island Developing State*. Global Platform on Climate Change, Trade and Sustainable Energy, International Centre for Trade and Sustainable Development, Geneva, Switzerland.

Reale, O., W.K. Lau, J. Susskind, R. Rosenberg, E. Brin, E. Liu, L.P. Riishojgaard, M. Fuentes, and R. Rosenberg, 2009: AIRS impact on the analysis and forecast track of tropical cyclone Nargis in a global data assimilation and forecasting system. *Geophysical Research Letters*, **36**, L06812, doi:10.1029/2008GL037122.

Reason, C.J.C. and A. Keibel, 2004: Tropical Cyclone Fline and its unusual penetration and impacts over the Southern African mainland. *Weather and Forecasting*, **19**, 789-805.

Reid, C.E., M.S. O'Neill, C.J. Gronlund, S.J. Brines, D.G. Brown, A.V. Diez-Roux, and J. Schwartz, 2009: Mapping community determinants of heat vulnerability, *Environmmental Health Perspectives*, **117**, 1730-1736.

Republic of Mozambique, 2001: *The National Action Plan for the Reduction of Absolute Poverty, 2001-2005*. Government of Mozambique, Maputo, Mozambique.

Republic of Mozambique, 2006a: *Master Plan for Prevention and Mitigation of Natural Disasters*. Council of Ministers, Republic of Mozambique, Maputo, Mozambique.

Republic of Mozambique, 2006b: *Action Plan for the Reduction of Absolute Poverty, 2006-2009 (PARPA II)*. Republic of Mozambique, Maputo, Mozambique.

Republic of South Africa, 2002: *Disaster Management Act, 2002*. Act No. 57 of 2002. Government Printer, Pretoria, South Africa.

Republic of South Africa, 2005: *National Disaster Management Policy Framework*. Government Printer, Pretoria, South Africa.

Republic of the Philippines, 2009: *Climate Change Act of 2009*. Republic Act 9729, Fourteenth Congress of the Philippines, Manila, Philippines.

Republic of the Philippines, 2010: *National Disaster Risk Reduction and Management Act*. Republic Act 10121, Fourteenth Congress of the Philippines, Manila, Philippines.

Reuveny, R., 2007: Climate change-induced migration and violent conflict. *Political Geography*, **26(6)**, 656-673.

Revi, A., 2005: Lessons from the deluge: Priorities for multi-hazard risk mitigation. *Economic and Political Weekly*, **40(36)**, 3911-3916.

Revi, A., 2008: Climate change risk: an adaptation and mitigation agenda for Indian cities. *Environment and Urbanization*, **20(1)**, 207-229.

Richardson, G.R.A., 2010: *Adapting to Climate Change: An Introduction for Canadian Municipalities*. Natural Resources Canada, Ottawa, Canada, 40 pp.

Robine, J.M., S.L. Cheung, S. Le Roy, H. Van Oyen, C. Griffiths, J.P. Michel, and F.R. Herrmann, 2008: Death toll exceeded 70,000 in Europe during the summer of 2003. *Comptes Rendus Biologies*, **331(2)**, 171-178.

Rodó, X., M. Pascual, G. Fuchs, and A. Faruque, 2002: ENSO and cholera: a nonstationary link related to climate change? *Proceedings of the National Academy of Sciences*, **99**, 12901-12906.

Royal Society, 2008: *Ground-Level Ozone in the 21st Century: Future Trends, Impacts and Policy Implications*. RS Policy document 15/08, Royal Society, London, UK.

Rubio, J.C.M., I.J.M Pérez, J.J. Criado-Álvarez, C. Linares, and J.D. Jiménez, 2010: Heat health warning systems. Possibilities of improvement. *Revista Espanola de Salud Publica*, **84(2)**, 137-149.

Sack, D.A., R.B. Sack, and C.-L. Chaignat, 2006: Getting serious about cholera. *New England Journal of Medicine*, **355(7)**, 649-651.

SADC, 2000: *Revised Protocol on Shared Watercourses*. Southern African Development Community, Gabarone, Botswana.

Saenger, P. and N.A. Siddiqi, 1993: Land from the sea: the mangrove afforestation program of Bangladesh. *Ocean and Coastal Management*, **20(1)**, 23-39.

Sailor, D.J. and A.A. Pavlova, 2003: Air conditioning market saturation and long-term response of residential cooling energy demand to climate change. *Energy*, **28(9)**, 941-951.

Saizen, I., A. Maekawa, and N. Yamamura, 2010: Spatial analysis of time-series changes in livestock distribution by detection of local spatial associations in Mongolia. *Applied Geography*, **30**, 639-649.

Santamouris, M., N. Papanikolaou, I. Koronakis, I. Livada, and D. Assimakopoulos, 1999: Thermal and airflow characteristics in a deep pedestrian canyon, under hot weather conditions. *Atmospheric Environment*, **33**, 4503-4521.

SARPMETT, 2010: *Syrian Economy Bulletin*. Syrian Arab Republic Prime Ministry Economic Technical Team, Damascus, Syria, zunia.org/uploads/media/knowledge/ Syrian%20Economy%20Bulletin%20EN1284201253.pdf.

Schubert, S.D., M.J. Suarez, P.J. Pegion, R.D. Koster, and J.T. Bacmeister, 2008: Potential predictability of long-term drought and pluvial conditions in the United States Great Plains. *Journal of Climate*, **21**, 802-816.

SCSCDD, 2008a: Percepção de risco, a descoberta de um novo olhar. Santa Catarina State Civil Defence Department, Florianópolis, Brazil. *EIRD Informa*, **15**, 2 pp.

SCSCDD, 2008b: *O Projeto: Percepção de risco, a descoberta de um novo olhar*. Santa Catarina State Civil Defence Department, Florianópolis, Brazil, 2 pp., www.percepcaoderisco.sc.gov.br/?ver=projeto.

SCSCDD, 2008c: Finalizada capacitação sobre percepção de risco para comunidades em Florianópolis. *Notícias*,18 December, Santa Catarina State Civil Defence Department, Florianópolis, Brazil, 1 pp., www.percepcaoderisco.sc.gov.br/?ver=noticia-completa&noticia=49.

SCSCDD, 2008d: Morro da penitenciária tem ação comunitária sobre riscos locais. *Notícias*, 9 October, Santa Catarina State Civil Defence Department, Florianópolis, Brazil, 1 pp.

Semenza, J.C., 2005: Building healthy cities: A focus on interventions. In: *Handbook of Urban Health* [Galea, S. and D. Vlahov (eds.)]. Springer Science and Business Media, New York, NY.

Semenza, J.C., T. March, and B. Bontempo, 2007: Community-initiated urban development: an ecological intervention. *Journal of Urban Health*, **84(1)**, 8-20.

Semenza, J., D.J. Wilson, J. Parra, B.D. Bontempo, M. Hart, D.J. Sailor, and L.A. George, 2008: Public perception and behavior change in relationship to hot weather and air pollution. *Environmental Research*, **107(3)**, 401-411.

Shapiro, M.A., J. Shukla, B. Hoskins, K. Trenberth, M. Beland, G. Brasseur, M. Wallace, G. McBean, J. Caughey, D. Rogers, G. Brunet, L. Barrie, A. Henderson-Sellers, D. Burridge, T. Nakazawa, M. Miller, P. Bougeault, R. Anthes, Z. Toth, and T. Palmer, 2007: The socio-economic and environmental benefits of a revolution in weather, climate and Earth-system analysis and prediction. In: *The Full Picture*, Group on Earth Observations, Tudor Rose, London, UK, pp. 137-139.

Shapiro, M.A., J. Shukla, G. Brunet, C. Nobre, M. Béland, R. Dole, K. Trenberth, R. Anthes, G. Asrar, L. Barrie, P. Bougeault, G. Brasseur, D. Burridge, A. Busalacchi, J. Caughey, D. Chen, J. Church, T. Enomoto, B. Hoskins, Ø. Hov, A. Laing, H. Le Treut, J. Marotzke, G. McBean, G. Meehl, M. Miller, B. Mills, J. Mitchell, M. Moncrieff, T. Nakazawa, H. Olafsson, T. Palmer, D. Parsons, D. Rogers, A. Simmons, A. Troccoli, Z. Toth, L. Uccellini, C. Velden, and J.M. Wallace, 2010: An earth-system prediction initiative for the 21st century. *Bulletin of the American Meteorological Society*, **91**, 1377-1388.

Sharma, D. and S. Tomar, 2010: Mainstreaming climate change adaptation in Indian cities. *Environment and Urbanization*, **22(2)**, 451-465.

Sharma, U., A. Patwardhan, and D. Parthasarathy, 2009: Assessing adaptive capacity to tropical cyclones in the east coast of India: a pilot study of public response to cyclone warning information. *Climatic Change*, **94(1-2)**, 189-209.

Shaw, R., K. Shiwaku, H. Kobayashi, and M. Kobayashi, 2004: Linking experience, education, perception and earthquake preparedness. *Disaster Prevention and Management*, **13(1)**, 39-49.

Sheridan, S.C., 2003: Heat, mortality, and level of urbanisation. *Climate Research*, **24**, 255-265.

Sheridan, S.C., 2006: A survey of public perception and response to heat warnings across four North American cities: an evaluation of municipal effectiveness. *International Journal of Biometeorology*, **52**, 3-15.

Shikanga, O.T., D. Mutonga, M. Abade, S. Amwayi, M. Ope, H. Limo, E.D. Mintz, R.E. Quick, R.F. Breiman, and D.R. Feikin, 2009: High mortality in a cholera outbreak in western Kenya after post-election violence in 2008. *American Journal of Tropical Medicine & Hygiene*, **81(6)**, 1085-1090.

Shimoda, Y., 2003: Adaptation measures for climate change and the urban heat island in Japan's built environment. *Building Research & Information*, **31(3-4)**, 222-230.

Shiwaku, K., R. Shaw, C. Kandel, S.N. Shrestha, and A.M. Dixit, 2006: Promotion of disaster education in Nepal: role of teachers as change agents. *International Journal of Mass Emergency and Disaster*, **24(3)**, 403-420.

Shukla, J., R. Hagedorn, B. Hoskins, J. Kinter, J. Marotzke, M. Miller, T. Palmer, and J. Slingo, 2009: Revolution in climate prediction is both necessary and possible: a declaration at the world modelling summit for climate prediction. *Bulletin of the American Meteorological Society*, **90**, 16-19.

Shukla, J., T.N. Palmer, R. Hagedorn, B. Hoskins, J. Kinter, J. Marotzke, M. Miller, and J. Slingo, 2010: Towards a new generation of world climate research and computing facilities. *Bulletin of the American Meteorological Society*, **91**, 1407-1412.

Shultz, A., J.O. Omollo, H. Burke, M. Qassim, J.B. Ochieng, M. Weinberg, D.R. Feikin, and R.F. Breiman, 2009: Cholera outbreak in Kenyan refugee camp: risk factors for illness and importance of sanitation. *American Journal of Tropical Medicine & Hygiene*, **80(4)**, 640-645.

Sidley, P., 2000; Malaria epidemic expected in Mozambique. *BMJ*, **320**, 669.

Sietz, D., M. Boschütz, R.J.T. Klein, and A. Lotsch, 2008: *Mainstreaming Climate Adaptation into Development Assistance in Mozambique: Institutional Barriers and Opportunities*. Policy Research Working Paper 4711, World Bank, Washington, DC.

Silva, H.R., P.E. Phelan, and J.S. Golden, 2010: Modeling effects of urban heat island mitigation strategies on heat-related morbidity: a case study for Phoenix, Arizona, USA. *International Journal of Biometeorology*, **54**, 13-22.

Simmons, A.J. and A. Hollingsworth, 2002: Some aspects of the improvement in skill of numerical weather prediction. *Quarterly Journal of the Royal Meteorological Society*, **128**, 647-677.

Skaff, M. and Sh. Masbate, 2010: Precipitation change, and its potential effects on vegetation and crop productivity in Syrian Al Jazerah Region. *The Arab Journal for Arid Environments*, **3(2)**, 71-78.

Skees, R. and A. Enkh-Amgalan, 2002: *Examining the Feasibility of Livestock Insurance in Mongolia*. Policy Research Working Paper 2886, World Bank, Washington, DC.

Smith, S.K. and C. McCarty, 2009: Fleeing the storm(s): an examination of evacuation behavior during Florida's 2004 hurricane season. *Demography*, **46(1)**, 127-145.

Smithers, J.C., R.E. Schulze, A. Pike, and G.P.W. Jewitt, 2001: A hydrological perspective of the February 2000 floods: a case study in the Sabie River Catchment. *Water SA*, 27(3), 25-31.

Smoyer, K.E., D.G. Rainham, and J.N. Hewko, 2000: Heat stress-related mortality in five cities in Southern Ontario: 1980-1996. *International Journal of Biometeorology*, **44(4)**, 190-197.

Smoyer-Tomic, K.E. and D.G.C. Rainham, 2001: Beating the heat: Development and evaluation of a Canadian hot weather health-response plan. *Environmental Health Perspectives*, **109(12)**, 1241-1248.

Solh, M., 2010: Tackling the drought in Syria. *Nature Middle East*, doi:10.1038/nmiddleeast.2010.206.

Sowers, J. and E. Weinthal, 2010: *Climate Change Adaptation in the Middle East and North Africa: Challenges and Opportunities*. Working Paper No. 2, Dubai Initiative, Dubai School of Government, Dubai, United Arab Emirates, belfercenter.ksg.harvard.edu/files/Sowers-Weinthal_DI_Working-Paper-2.PDF.

Sowers, J., A. Vengosh, and E. Weinthal, 2011: Climate change, water resources, and the politics of adaptation in the Middle East and North Africa. *Climatic Change*, **104(3-4)**, 599-627.

Sphere Project, 2004: *Humanitarian Charter and Minimum Standards in Disaster Response*. The Sphere Project, Geneva, Switzerland.

Srivastava, A,K, M.M. Dandekar, S.R. Kshirsagar, and S.K. Dikshit, 2007: Is summer becoming more uncomfortable at Indian Cities? *Mausam*, **58(3)**, 335-344.

Stal, M., 2011. Flooding and relocation: The Zambezi River Valley in Mozambique. International Migration, 49(S1), e125-e145.

Stedman, J.R., 2004: The predicted number of air pollution related deaths in the UK during the August 2003 heatwave. *Atmospheric Environment*, **38**, 1087-1090.

Stein, R.M., L. Dueñas-Osorio, and D. Subramanian, 2010: Who evacuates when hurricanes approach? The role of risk, information, and location. *Social Science Quarterly*, **91(3)**, 816-834.

Sternberg, T., 2010: Unravelling Mongolia's extreme winter disaster of 2010. *Nomadic Peoples*, **14(1)**, 72-86.

Stewart, B.I., A.F. Asfary, A. Belloum, K. Steiner, and T. Friedrich, 2008: *Conservation Agriculture for Sustainable Land Management to Improve the Livelihood of People in Dry Areas*. Arab Center for the Studies of Arid Zones and Dry Lands and the German Agency for Technical Cooperation, Damascus, Syria and Eschborn, Germany, www.fao.org/ag/ca/doc/CA%20Workshop%20procedding%2008-08-08.pdf.

Stott, P.A., D.A. Stone, and M.R. Allen, 2004: Human contribution to the European heat wave of 2003. *Nature*, **432**, 610-614.

Suarez, P. and J. Linnerooth-Bayer, 2010: Micro-insurance for local adaptation. *Wiley Interdisciplinary Reviews: Climate Change*, **1(2)**, 271-278.

Subbarao, I., N.A. Bostick, and J.J. James, 2008: Applying yesterday's lessons to today's crisis: Improving the utilization of recovery services following catastrophic flooding. *Disaster Medicine and Public Health Preparedness*, 2(3), 132-133.

Sur, D., A.L. Lopez, S. Kanungo, A. Paisley, B. Manna, M. Ali, S.K. Niyogi, J.K. Park, B. Sarkar, M.K. Puri, D.R. Kim, J.L. Deen, J. Holmgren, R. Carbis, R. Rao, T.V. Nguyen, A. Donner, N.K. Ganguly, G.B. Nair, S.K. Bhattacharya, and J.D. Clemens, 2009: Efficacy and safety of a modified killed-whole-cell oral cholera vaccine in India: an interim analysis of a cluster-randomised, double-blind, placebo-controlled trial. *Lancet*, **374(9702)**, 1694-1702.

Surminski, S., 2010: *Adapting to the Extreme Weather Impacts of Climate Change – How Can the Insurance Industry Help?* Climate Wise, London, UK.

Syphard, A.D., V.C. Radeloff, T.J. Hawbaker, and S.I. Stewart, 2009: Conservation threats due to human-caused increases in fire frequency in Mediterranean-climate ecosystems. *Conservation Biology*, **23(3)**, 758-769.

Talavera, A. and E.M. Perez, 2009: Is cholera disease associated with poverty. *Journal of Infection in Developing Countries*, **3(6)**, 408-411.

TDB, 2007: *"Structurally weak, vulnerable and small economies": Who are they? What can UNCTAD do for them?* Trade and Development Board, United Nations Conference on Trade and Development, Geneva, Switzerland, www.unctad.org/ en/docs/tdb54crp4_en.pdf.

Thomson, M.C., F.J. Doblas-Reyes, S.J. Mason, R. Hagedorn, S.J. Connor, T. Phindela, A.P. Morse, and T.N. Palmer, 2006: Malaria early warnings based on seasonal climate forecasts from multi-model ensembles. *Nature*, **439**, 576-579.

Tolhurst, K., 2010: *Report on Fire Danger Ratings and Public Warning*. The University of Melbourne, Melbourne, Australia.

Twigg, J. and H. Bottomley, 2011: *Disaster Risk Reduction NGO Inter-Agency Group Learning Review*. Action Aid, Christian Aid, Plan International, Practical Action, and Tearfund, London, UK, 40 pp.

UN, 2005: *Report of International Meeting to Review the Implementation of the Programme of Action for the Sustainable Development of Small Island Developing States*, Port Louis, Mauritius, 10-14 January 2005, United Nations, New York, NY, www.sidsnet.org/docshare/other/20050622163242_English.pdf.

UN, 2006: *Global Survey of Early Warning Systems*. Prepared by UN International Strategy for Disaster Reduction for the United Nations, United Nations, Geneva, Switzerland, 46 pp.

UN, 2008: *World Urbanization Prospects – The 2007 Revision*. United Nations Department of Economic and Social Affairs, Population Division, New York, NY, www.un.org/esa/population/publications/wup2007/2007WUP_Highlights_web.pdf.

UN, 2009: *Syria Drought Response Plan*. United Nations, New York, NY, reliefweb.int/sites/reliefweb.int/files/resources/2A1DC3EA365E87FB8525760F0051E91A-Full_Report.pdf.

UN, 2011: *General Assembly Report of the Special Rapporteur on the right to food, Olivier De Schutter. Mission to the Syrian Arab Republic*. United Nations, New York, NY, www2.ohchr.org/english/bodies/hrcouncil/docs/16session/A.HRC.16.49.Add.2_en.pdf.

UNCTAD, 2004: *Is a Special Treatment of Small Island Developing States Possible?* United Nations Conference on Trade and Development, New York, NY, www.unctad.org/en/docs/ldc20041_en.pdf.

UNDESA, 2010: *Trends in Sustainable Development: Small Island Developing States*. Division for Sustainable Development, United Nations Department of Economic and Social Affairs, New York, NY, www.cbd.int/islands/doc/sids-trends-report-v4-en.pdf.

UNDESA, 2011: *Case Study Detail Record, Marshall Islands: Groundwater Supply on the Marshall Islands*. Division for Sustainable Development, United Nations Department of Economic and Social Affairs, New York, NY, webapps01.un.org/dsd/caseStudy/public/displayDetailsAction.do?code=326.

UNDP, 2005: *A Global Review: Support to Institutional and Legislative Systems for Disaster Risk Management*. United Nations Development Programme, New York, NY.

UNDP, 2006: *Country Programme Action Plan. 2006 – 2009*. United Nations Development Programme, New York, NY, www.bb.undp.org/uploads/file/pdfs/general/UNDP-GRN%20CPAP%202006-2009.pdf.

UNDP, 2007: *A Global Review: UNDP Support to Institutional and Legislative Systems for Disaster Risk Management*. United Nations Development Programme, Bureau for Crisis Prevention and Recovery, New York, NY.

UNDP Mongolia, 2009a: *The Millennium Development Goals Implementation, Third National Report*. UNDP, Ulaanbaatar, Mongolia, 142 pp.

UNDP Mongolia, 2009b: *Moving Forward to Reduce Climate Change Risks*. United Nations Development Programme, Ulaanbaatar, Mongolia, www.undp.mn/news-snrm-230409.html.

UNDP Mongolia, 2010: *Project Report: Support to the Dzud in Mongolia/early recovery*. United Nations Development Programme, Ulaanbaatar, Mongolia, www.undp.mn/pprojects.html.

UNEP, 2004: *Impacts of Summer 2003 Heat Wave in Europe*. Environment Alert Bulletin 2, United Nations Environment Programme, Geneva, Switzerland, www.grid.unep.ch/product/publication/download/ew_heat_wave.en.pdf.

UNESC, 2000: *Assistance to Mozambique following the devastating floods: Report of the Secretary-General*. A/55/123-E/2000/89, United Nations General Assembly Economic and Social Council, New York, NY, www.un.org/documents/ga/docs/55/a55123.pdf.

UNESC, 2004: *Freshwater management: progress in meeting the goals, targets and commitments of Agenda 21, the Programme for the Further Implementation of Agenda 21, and the Johannesburg Plan of Implementation*. Report of the Secretary-General, A/CONF.207/3, United Nations Economic and Social Council, New York, NY, www.preventionweb.net/files/resolutions/N0423881.pdf.

UNESC, 2011: *Integrated Analysis of United Nations System Support to Small Island Developing States*. Report of the Secretary-General, E/2011/110, United Nations Economic and Social Council, New York, NY, www.un.org/ga/search/view_doc.asp?symbol=E/2011/110&referer=http://sids-l.iisd.org/news/advance-report-of-secretary-general-calls-for-improved-msi-delivery/?referrer=linkages-iisdrs&Lang=E.

UNFCCC, 2007a: *Vulnerability and Adaptation To Climate Change In Small Island Developing States*. United Nations Framework Convention on Climate Change, Bonn, Germany, unfccc.int/files/adaptation/adverse_effects_and_response_measures_art_48/application/pdf/200702_sids_adaptation_bg.pdf.

UNFCCC, 2007b: *Climate Change: Impacts, Vulnerabilities and Adaptation in Developing Countries*, United Nations Framework Convention on Climate Change, Bonn, Germany, unfccc.int/resource/docs/publications/impacts.pdf.

UNFCCC, 2010: *Action On the Ground: A Synthesis of Activities in the Areas of Education, Training and Awareness-Raising For Adaptation*. Nairobi Work Programme on Impacts, Vulnerability and Adaptation to Climate Change, United Nations Framework Convention on Climate Change, Bonn, Germany, 116 pp.

Ungern-Sternberg, T., 2003: State Intervention on the Market for Natural Damage Insurance in Europe. CESifo Working Paper 1067, CESifo Group, Munich, Germany.

UN-HABITAT, 2006: *State of the World's Cities 2006/2007*. United Nations Human Settlements Programme, Nairobi, Kenya.

UNISDR, 2004: *Living with Risk: A Global Review of Disaster Reduction Initiatives*. United Nations International Strategy for Disaster Reduction, Geneva, Switzerland, 429 pp.

UNISDR, 2005a: Review of the Yokohama Strategy and Plan of Action for a Safer World. Conference document A/CONF.206/L.1. In: *Proceedings of the World Conference on Disaster Reduction*, Kobe, Hyogo, Japan, 18-22 January 2005, United Nations International Strategy for Disaster Reduction, Geneva, Switzerland, 23 pp.

UNISDR, 2005b: *Hyogo Framework for Action: Building the Resilience of Nations and Communities to Disasters, 2005-2015*. United Nations International Strategy for Disaster Reduction, Geneva, Switzerland.

UNISDR, 2006: *Disaster Risk Reduction Begins at School, 2006-2007 International Disaster Risk Reduction Campaign*. United Nations International Strategy for Disaster Reduction, Geneva, Switzerland, www.unisdr.org/eng/public_aware/world_camp/2006-2007/wdrc-2006-2007.htm.

UNISDR, 2007a: *Drought Risk Reduction Framework and Practices: Contributing to the Implementation of the Hyogo Framework for Action*. United Nations International Strategy for Disaster Reduction, Geneva, Switzerland, 98 pp.

UNISDR, 2007b: *Towards a Culture of Prevention: Disaster Risk Reduction Begins at School; Good Practices and Lessons Learned*. United Nations International Strategy for Disaster Reduction, Geneva, Switzerland, 143 pp.

UNISDR, 2008a: *Climate Change and Disaster Risk Reduction*. United Nations International Strategy for Disaster Reduction, Geneva, Switzerland.

UNISDR, 2008b: *Annual Report 2007: The Secretariat of the International Strategy for Disaster Reduction*. United Nations International Strategy for Disaster Reduction, Geneva, Switzerland, 68 pp.

UNISDR, 2009a: *Outcome Document: Chair's Summary of the Second Session of the Global Platform for Disaster Risk Reduction*. Proceedings of a conference, Global Platform for Disaster Risk Reduction, 16-19 June 2009, United Nations International Strategy for Disaster Reduction, Geneva, Switzerland, 4 pp.

UNISDR, 2009b: *Terminology on Disaster Risk Reduction, 2009*. United Nations International Strategy for Disaster Reduction, Geneva, Switzerland, www.unisdr.org/we/inform/terminology.

UNISDR, 2009c: Global Assessment Report on Disaster Risk Reduction. International Strategy for Disaster Reduction, Geneva, Switzerland, 207 pp.

UNISDR, 2010: *Early Warning Practices can Save Many Lives: Good Practices and Lessons Learned*. United Nations Secretariat of the International Strategy for Disaster Reduction, UN International Strategy on Disaster Risk Reduction, Bonn, Germany, 67 pp., www.unisdr.org/files/15254_EWSBBLLfinalweb.pdf.

UNISDR, 2011a: *Mid-Term Review 2010-2011: Hyogo Framework for Action 2005-2015: Building the Resilience of Nations and Communities to Disasters*. United Nations International Strategy for Disaster Reduction, Geneva, Switzerland, 107 pp.

UNISDR, 2011b: *Global Assessment Report on Disaster Risk Reduction: Revealing Risk, Redefining Development*. United Nations International Strategy for Disaster Reduction, Geneva, Switzerland, www.preventionweb.net/english/hyogo/gar/2011/en/home/index.html.

United Kingdom, 2008: *Climate Change Act 2008*. The Stationary Office Limited, London, UK.

UNOCHA, 2007: *Mozambique 2007 Flash Appeal, Executive Summary*. United Nations Office for the Coordination of Human Affairs, New York, NY, ochaonline.un.org/humanitarianappeal/webpage.asp?Page=1558.

UN-OHRLLS, 2011: *List of Small Island Developing States (UN Members)*. United Nations Office of the High Representative for the Least Developed Countries, Landlocked Developing Countries and the Small Island Developing States, New York, NY, www.un.org/special-rep/ohrlls/sid/list.htm.

UNRCS and SARPCMSPC, 2005. *Second National Report on the Millennium Development Goals (MDGs) in the Syrian Arab Republic 2005*. United Nations Resident Coordinator System and Syrian Arab Republic Presidency of the Council of Ministers State Planning Commission, Damascus, Syria, planipolis.iiep.unesco.org/upload/Syrian%20AR/SyriaMDGSecondReportEnglish.pdf.

USAID, 2007: *Mozambique – Floods and Cyclone*. US Agency for International Development, Washington, DC, www.usaid.gov/our_work/humanitarian_assistance/disaster_assistance/countries/mozambique/template/fs_sr/mozambique_fl_fs01_03-22-2007.pdf.

USARC, 2003: *Climate Change, Permafrost and Impacts on Civil Infrastructure: Permafrost Task Force Report*. United States Arctic Research Commission, Washington, DC.

USDA, 2008a: Middle East & Central Asia: Continued Drought in 2009/10 Threatens Greater Food Grain Shortages. US Department of Agriculture http://www.pecad.fas.usda.gov/highlights/2008/09/mideast_cenasia_drought/

USDA, 2008b: *Syria: Wheat Production in 2008/09 Declines Owing to Season-Long Drought*. US Department of Agriculture, Washington, DC, www.pecad.fas.usda.gov/highlights/2008/05/Syria_may2008.htm.

Van Niekerk, D., 2006: Disaster risk management in South Africa: The function and the activity – towards an integrated approach. *Politeia*, **25(2)**, 95-115.

Van Niekerk, D., 2011: *Concept Paper: The South African Disaster Risk Management Policy and Legislation – A critique*. African Centre for Disaster Studies, Potchefstroom, South Africa, acds.co.za/uploads/research_reports/SA_law_2011.pdf.

Van Riet, G. and M. Diedericks, 2010: The placement of the disaster management function within district, metropolitan and provincial government structures in South Africa. *Administratio Publica*, **18(4)**, 155-173.

Varangis, P., J.R. Skees, and B.J. Barnett, 2002: Weather indexes for developing countries. In: *Climate Risk and the Weather Market* [Dischel, B. (ed.)]. Risk Books, London, UK, pp. 279-294.

Vardoulakis, S., B.E.A. Fisher, K. Pericleous, and N. Gonzalez-Flesca, 2003: Modelling air quality in street canyons: a review. *Atmospheric Environment*, **37**, 155-182.

Verschuren, P.J.M., 2003: Case study as a research strategy: Some ambiguities and opportunities. *International Journal of Social Research Methodology: Theory and Practice*, **6(2)**, 121-139.

Victorian Government, 2009: *January 2009 Heatwave in Victoria: an Assessment of Health Impacts*. Victorian Government, Department of Human Services, Melbourne, Australia.

Vingarzan, R., 2004: A review of surface ozone background levels and trends. *Atmospheric Environment*, **38**, 3431-3442.

Visser, R. and D. Van Niekerk, 2009: *A Funding Model for the Disaster Risk Management Function of Municipalities*. National Disaster Management Centre, South African Department of Cooperative Governance and Traditional Affairs, Pretoria, South Africa.

Vogt, F. and D.G. Sapir, 2009: Cyclone Nargis in Myanmar: lessons for public health preparedness for cyclones. *American Journal of Disaster Medicine*, **4(5)**, 273-278.

Volz, A. and D. Kley, 1988: Evaluation of the Montsouris series of ozone measurements made in the nineteenth century. *Nature*, **332**, 240-242.

Vos, F., J. Rodriguez, R. Below, and D. Guha-Sapir, 2010: *Annual Disaster Statistical Review 2009: The Numbers and Trends*. Centre for Research on the Epidemiology of Disasters, Brussels, Belgium, www.cred.be/sites/default/files/ADSR_2009.pdf.

Waechter, B., 2005: *Working with Nature: Wind Deflection Fins in Action*. Rowan Williams Davies & Irwin Inc., Guelph, Canada, www.rwdi.com/cms/publications/16/t05.pdf.

Walsh, R., S. Orban, R. Walker, J. Coates, J. Croteau, D. Stone, and T. Strynadks, 2009: *Front Street Paving Project, Dawson City, Yukon: Adapting to Climate Change in a National Historic District*. Yukon Highways and Public Works, Whitehorse, Canada, www.colascanada.ca/uploads/colascanada/File/expertise/EnvironmentalAchievementYukonPaper.pdf.

Wang, W.-G., X.-R.Wang, Y.-L. Xu, Y.-H. Duan, and J.-Y. Li, 2008: Characteristics of super-typhoon "SaoMai" and benefits of forecast service. *Journal of Natural Disasters*, **17(3)**, 106-111.

Warner, K., N. Ranger, S. Surminski, M. Arnold, J. Linnnerooth-Bayer, E. Michel-Kerjan, P. Kovacs, and C. Herweijer, 2009: *Adaptation to Climate Change: Linking Disaster Risk Reduction and Insurance*. United Nations International Strategy for Disaster Reduction, Geneva, Switzerland, 30pp.

Webster, P.J., 2008: Myanmar's deadly daffodil. *Nature Geoscience*, **1**, 488-490.

Webster, P.J., J. Jian. T.M. Hopson, C.D. Hoyos, P.A. Agudelo, H.-R. Chang, J.A Curry, R.I. Grossman, T.N. Palmer, and A.R. Subbiah, 2010: Extended-range probabilistic forecasts of Ganges and Brahmaputra floods in Bangladesh. *Bulletin of the American Meteorological Society*, **91**, 1493-1514.

Weisskopf, M.A., H.A. Anderson, S. Foldy, L.P. Hanrahan, K. Blair, T.J. Torok, and P.D. Rumm, 2002: Heat wave morbidity and mortality, Milwaukee, Wis, 1999 vs 1995: An improved response? *American Journal of Public Health*, **92(5)**, 830-833.

WHO, 2006a: *Air Quality Guidelines: Global Update 2005, Particulate Matter, Ozone, Nitrogen Dioxide And Sulphur Dioxide*. World Health Organization Regional Office for Europe, Copenhagen, Denmark.

WHO, 2006b: *First meeting of the project: Improving Public Health Responses to Extreme Weather/Heat-waves*. EuroHEAT Report on a WHOMeeting in Rome, Italy, 20–22 June 2005, World Health Organization Regional Office for Europe, Copenhagen, Denmark.

WHO, 2006c: Cholera 2005. *Weekly Epidemiologic Record*, **81**, 297-308.

WHO, 2007: *Improving Public Health Responses to Extreme Weather/Heat-Waves – EuroHEAT, Technical Summary*. World Health Organization Regional Office for Europe, Copenhagen, Denmark.

WHO, 2008a: *Cholera in Zimbabwe: Epidemiological Bulletin Number 1*. World Health Organization, Harare, Zimbabwe.

WHO, 2008b: *Health System Problems Aggravate Cholera Outbreak in Zimbabwe*. News Release, World Health Organization, Geneva, Switzerland, www.who.int/mediacentre/news/releases/2008/pr49/en/index.html.

WHO, 2010: Cholera vaccines: WHO position paper. *Weekly Epidemiologic Record*, **85(13)**, 117-128.

Wisner, B., 2006: *Let our Children Teach Us! A Review of the Role of Education and Knowledge in Disaster Risk Reduction*. UNISDR System Thematic Cluster/Platform on Knowledge and Education, Geneva, Switzerland, 135 pp.

WMO, 2011a: *Climate Knowledge for Action: A Global Framework for Climate Services – Empowering the Most Vulnerable*. The Report of the High-Level Taskforce for the Global Framework for Climate Services, WMO No. 1065, World Meteorological Organization, Geneva, Switzerland.

WMO, 2011b: *Tropical Cyclone Programme. Report to Plenary on item 4.3*. World Meteorological Organization, Geneva, Switzerland, www.hydrometeoindustry.org/Reports2011/CONGRESS2011/d04-3_TROPICAL_CYCLONE.pdf.

Woodroffe, C., 2008: Reef-island topography and the vulnerability of atolls to sea-level rise. *Global and Planetary Change*, **247(1-2)**, 159-177.

World Bank, 2005a: *Learning Lessons from Disaster Recovery: The Case of Mozambique*. Disaster Risk Management Working Paper Series No. 12, World Bank, Washington, DC.

World Bank, 2005b: *Regional Engagement Framework FY 2006-2009 for Pacific Islands*. Report No. 32261-EAP, World Bank, Washington, DC, siteresources.worldbank.org/INTPACIFICISLANDS/Resources/PI-Strategy-05.pdf.

World Bank, 2005c: *Republic of Maldives, Tsunami Impact and Recovery*. Joint assessment, World Bank, Asian Development Bank, and UN System, Washington, DC, www.adb.org/Documents/Reports/Tsunami/joint-needs-assessment.pdf .

World Bank, 2005d: *Managing Agricultural Production Risk*. World Bank, Washington, DC, siteresources.worldbank.org/INTARD/Resources/Managing_Ag_Risk_FINAL.pdf.

World Bank, 2006a: *Opportunities to Improve Social Services in the Republic of the Marshall Islands*. Summary Report, World Bank, Washington, DC, www-wds.worldbank.org/external/default/WDSContentServer/WDSP/IB/2007/05/17/000310607_20070517100419/Rendered/PDF/397780EAP0P0791Development01PUBLIC1.pdf.

World Bank, 2006b: *Ethiopia – Second Productive Safety Net APL Project*. The World Bank, Washington, DC, www-wds.worldbank.org/external/default/main?pagePK=64193027&piPK=64187937&theSitePK=523679&menuPK=64187510&searchMenuPK=64187283&siteName=WDS&entityID=000020953_20070117142245.

World Bank, 2007: *The Caribbean Catastrophe Risk Insurance Initiative. Results of Preparation Work on the Design of a Caribbean Catastrophe Risk Insurance Facility*. World Bank, Washington, DC.

World Bank, 2009: *Reducing the Risk of Disasters and Climate Variability in the Pacific Islands. Republic of the Marshall Islands Country Assessment*. World Bank, Washington, DC, siteresources.worldbank.org/INTPACIFICISLANDS/Resources/MARSHALL_ASSESSMENT_f.pdf .

World Bank, 2011a: *World Development Indicators Database*. World Bank, Washington, DC, data.worldbank.org/data-catalog/world-development-indicators.

World Bank, 2011b: *Disaster Risk Management Programs for Priority Countries*. Global Facility for Disaster Risk Reduction, Washington, DC.

World Bank, Global Facility for Disaster Reduction and Recovery, Inter-Agency Network for Education in Emergencies (INEE), and UNISDR, 2009: *Guidance Notes on Safer School Construction*. Global Facility for Disaster Reduction and Recovery and INEE Secretariat, Washington, DC and New York, NY, 133 pp.

Yip, F.Y., W.D. Flanders, A. Wolkin, D. Engelthaler, W. Humble, A. Neri, L. Lewis, L. Backer, and C. Rubin, 2008: The impact of excess heat events in Maricopa County, Arizona: 2000–2005. *International Journal of Biometeorology*, **52(8)**, 765-772.

Ykhanbai, H., E.U. Beket, R. Vernooy, and J. Graham, 2004: Reversing grassland degradation and improving herders' livelihoods in the Altai Mountains of Mongolia. *Mountain Research and Development*, **24(2)**, 96-100.

Yohe, G.W., R.D. Lasco, Q.K. Ahmad, N.W. Arnell, S.J. Cohen, C. Hope, A.C. Janetos. and R.T. Perez, 2007: Perspectives on climate change and sustainability. In: *Climate Change 2007. Impacts, Adaptation and Vulnerability. Contribution of Working Group II to the Fourth Assessment Report of the Intergovernmental Panel on Climate Change* [Parry, M.L., O.F. Canziani, J.P. Palutikof, P.J. Van Der Linde, and C.E. Hanson (eds.)]. Cambridge University Press, Cambridge, UK, 811-841.

Yokoi, S. and Y.N. Takayabu, 2010: Environmental and external factors in the genesis of tropical cyclone Nargis in April 2008 over the Bay of Bengal. *Journal of the Meteorological Society of Japan*, **88**, 425-435.

Yoshida, Y., Y. Takeuchi, and R. Shaw, 2009: Town watching as a useful tool in urban risk reduction in Saijo. *Community, Environment and Disaster Risk Management*, **1**, 189-205.

Zuckerman, J., L. Rombo, and A. Fisch, 2007: The true burden and risk of cholera: implications for prevention and control. *Lancet Infectious Diseases*, **7**, 521-530.

IV
Annexes I to IV

Authors and Expert Reviewers

This annex should be cited as:
IPCC, 2012: Authors and expert reviewers annex. In: *Managing the Risks of Extreme Events and Disasters to Advance Climate Change Adaptation* [Field, C.B., V. Barros, T.F. Stocker, D. Qin, D.J. Dokken, K.L. Ebi, M.D. Mastrandrea, K.J. Mach, G.-K. Plattner, S.K. Allen, M. Tignor, and P.M. Midgley (eds.)]. A Special Report of Working Groups I and II of the Intergovernmental Panel on Climate Change (IPCC). Cambridge University Press, Cambridge, UK, and New York, NY, USA, pp. 545-553.

Authors and Expert Reviewers

Argentina
Vicente Barros, CIMA/Universidad de Buenos Aires
Ines Camilloni, CIMA/Universidad de Buenos Aires
Hernan Carlino, Universidad Torcuato Di Tella
Mario Nunez, CIMA/Universidad de Buenos Aires
Matilde Rusticucci, Universidad de Buenos Aires
Haris Eduardo Sanahuja, Senior Consultant
Pablo Suarez, Boston University, Red Cross/Red Crescent Climate Centre
Carolina Vera, CIMA/Universidad de Buenos Aires

Australia
Jonathan Abrahams, World Health Organization
Lisa Alexander, The University of New South Wales
Julie Arblaster, National Center for Atmospheric Research, Australian Bureau of Meteorology
Jon Barnett, Melbourne University
Ian Carruthers, National Climate Change Adaptation Research Facility
Bob Cechet, Geoscience Australia
Lynda Chambers, Australian Bureau of Meteorology
John Church, Commonwealth Scientific and Industrial Research Organization
Paul Della-Marta, Partner Reinsurance Company
Amy Dumbrell, Australian Government Department of Climate Change and Energy Efficiency
Jill Edwards, Australasian Fire and Emergency Service Authorities Council
Ailie Gallant, University of Melbourne
John Handmer, Centre for Risk and Community Safety, RMIT University
Mark Hemer, Commonwealth Scientific and Industrial Research Organization
Adriana Keating, RMIT University
Monique Ladds, RMIT University
Padma Narsey Lal, International Union for Conservation of Nature-Oceania
Yun Li, Commonwealth Scientific and Industrial Research Organisation
Kathleen McInnes, Commonwealth Scientific and Industrial Research Organization, Marine and Atmospheric Research
Neville Nicholls, Monash University, School of Geography and Environmental Science
Lauren Amy Rickards, University of Melbourne
Anthony Swirepik, Australia Government Department of Climate Change and Energy Efficiency
Frank Thomalla, Macquarie University
Richard Thornton, Bushfire Cooperative Research Centre
Blair Trewin, Australian Bureau of Meteorology
Anya M. Waite, University of Western Australia
Xiaoming Wang, Commonwealth Scientifc and Industrial Research Organisation
Penny Whetton, Commonwealth Scientific and Industrial Research Organisation
Joshua Whittaker, RMIT University

Austria
Stefan Hochrainer, International Institute for Applied System Analysis
Helmut Hojesky, Bundesministerium fur Land-und Forstwirtschaft, Umwelt und Wasserwirtschaft
Klaus Radunsky, Umweltbundesamt GmbH
Petra Tschakert, Pennsylvania State University

Bangladesh
Md. Siarjul Islam, North South University
Tarik ul Islam, United Nations Development Programme-Bangladesh
Alimullah Miyan, South Asian Disaster Management Centre, International University of Business, Agriculture, and Technology
Ainun Nishat, BRAC University
Ataur Rahman, Centre for Global Environmental Culture, International University of Business, Agriculture, and Technology

Belgium
Johan Bogaert, Flemish Government
Lieven Bydekerke, VITO – Flemish Institute for Technological Research
Cathy Clerbaux, Universite Libre de Bruxelles and CNRS France
Luc Feyen, Joint Research Centre, European Commission
Leen Gorissen, VITO – Flemish Institute for Technological Research
Julien Hoyaux, Agence Wallonne de l'Air et du Climat
Philippe Marbaix, Université Catholique de Louvain
Anne Mouchet, Universite de Liege
Andrea Tilche, European Commission Directorate
Hans van de Vyvere, Royal Meteorological Institute of Belgium
Jean-Pascal van Ypersele, Université Catholique de Louvain
Martine Vanderstraeten, Belgian Federal Science Policy
Patrick Willems, Katholieke Universiteit Leuven

Botswana
Pauline Dube, University of Botswana

Brazil
Andre Odenbreit Carvalho, Environmental Policy and Sustainable Development
Jose Marengo, National Institute for Space Research, Earth System Science Centre
Jose Domingos Gonzales Miguez, Tecnologia Commission for Global Climate Change
Carlos Nobre, Ministry of Science and Technology
Vinicius Rocha, Operador Nacional do Sistema Elétrico
Maria Assuncao Silva Dias, University of Sao Paulo

Canada
Heather Auld, Environment Canada
Anik Beaudoin, Foreign Affairs and International Trade Canada
Peter Berry, Climate Change and Health Office, Health Canada
Marie Boehm, Agriculture and Agri-Food Canada
Roy Brooke, United Nations
Ross Brown, Environment Canada at Ouranos
Stephen Burridge, Foreign Affairs and International Trade Canada
Ian Burton, University of Toronto
Elizabeth Bush, Environment Canada
John Clague, Simon Fraser University
J. Graham Cogley, Trent University
Neil Comer, Environment Canada
John Cooper, Health Canada
Thea Dickinson, Burton Dickinson Consulting
Karen Dodds, Environment Canada
Jimena Eyzaguirre, National Round Table on the Environment and the Economy
Donald Forbes, Natural Resources Canada
Jim Frehs, Health Canada
Nathan P. Gillett, Environment Canada
Sunling Gong, Environment Canada
Christian Gour, Foreign Affairs and International Trade Canada
Patrick Hebert, Foreign Affairs and International Trade Canada
Ole Hendrickson, Environment Canada
Matt Jones, Environment Canada
Dan Jutzi, Environment Canada
Viatcheslav V. Kharin, Environment Canada
Grace Koshida, Environment Canada
Paul Kovacs, Institute for Catastrophic Loss Reduction
Beth Lavender, Foreign Affairs and International Trade Canada
Donald Lemmen, Natural Resources Canada
Guilong Li, Environment Canada
Brad Little, Environment Canada
John Loder, Fisheries and Oceans Canada
Heather Low, Foreign Affairs and International Trade Canada
Katie Lundy, Environment Canada
Gordon McBean, Institute for Catastrophic Loss Reduction
Brian Mills, Environment Canada
Seung-Ki Min, Environment Canada

Monirul Mirza, Environment Canada
Niall O'Dea, Natural Resources Canada
Michael Ott, Fisheries and Oceans Canada
William Perrie, Fisheries and Oceans Canada
Paul Pestieau, Environment Canada
Caroline Rodgers, Clean Air Partnership, Toronto
Slobodan Simonovic, University of Western Ontario
Sharon Smith, Natural Resources Canada
Ronald Stewart, University of Manitoba
John Stone, Carleton University
Roger B. Street, UK Climate Impacts Programme
Kit Szeto, Environment Canada
Richard Tarasofsky, Foreign Affairs and International Trade Canada
Martin Tremblay, Indian and Northern Affairs Canada
Liette Vasseur, Brock University
Elizabeth Walsh, Natural Resources Canada
Xiaolan Wang, Environment Canada
Bin Yu, Environment Canada
Xuebin Zhang, Environment Canada
Francis Zwiers, Pacific Climate Impacts Consortium, University of Victoria

Chile
Paulina Aldunce, University of Chile
Daniel Barrera, Ministry of Agriculture
Gonzalo Leon, Ministry of Environment
Alejandro León, Universidad de Chile
Sebastian Vicuna, Pontificia Universidad Catolica de Chile

China
Xing Chen, Nanjing University
Ying Chen, Chinese Academy of Social Sciences
Qin Dahe, China Meteorological Administration
Shibo Fang, Chinese Academy of Meteorological Sciences
Qiang Feng, Chinese Academy of Sciences
Ge Gao, National Climate Center, China Meteorological Administration
Daoyi Gong, State Key Laboratory of Earth Surface Processes and Resource Ecology
Anhong Guo, Agrometeorological Center of National Meteorological Centre
Tong Jiang, China Meteorological Administration
Jianping Li, Institute of Atmospheric Physics, Chinese Academy of Sciences
Jing Li, Beijing Normal University
Ning Li, Beijing Normal University
Erda Lin, Chinese Academy of Agricultural Sciences
Hongbin Liu, China Meteorological Administration
Xinhua Liu, Severe Weather Prediction Center of National Meteorological Center
Houquan Lu, National Meteorological Center
Xianfu Lu, United Nations Framework Convention on Climate Change
Yali Luo, Chinese Academy of Sciences
Wenjun Ma, Shanghai Jiao Tong University
Huixin Meng, Institute for Urban and Environment Studies
Jiahua Pan, Chinese Academy of Social Sciences
Gubo Qi, College of Humanities and Development Studies
Fumin Ren, China Meteorological Administration
Guoyu Ren, Chinese Academy of Social Sciences
WU Shaohong, Institute of Geographical Sciences, Chinese Academy of Sciences
Shangbai Shi, Chinese Academy of Social Sciences
Ying Sun, China Meteorological Administration
Changke Wang, China Meteorological Administration
Dongming Wang, Policy Study
Jianwu Wang, Chinese Academy of Social Sciences
Ming Wang, Academy of Disaster Reduction and Emergency Management
Xiaoyi Wang, Institute of Sociology
Yongguang Wang, China Meteorological Administration
Fengying Wei, China Meteorological Administration
Jet-Chau Wen, National Yunlin University of Science and Technology
Xiangyang Wu, Research Centre for Sustainable Development
Liyong Xie, Shenyang Agricultural University
Wei Xu, State Key Laboratory of Earth Surface Processes and Resource Ecology
Yinlong Xu, Chinese Academy of Agricultural Sciences
Zheng Yan, Institute of Urban and Environmental Studies
Saini Yang, Beijing Normal University
Tao Ye, State Key Lab of Earth Surface Processes and Resource Ecology
Yi Yuan, National Disaster Reduction Center of China
Yin Yunhe, Institute of Geographical Sciences, Chinese Academy of Sciences
Panmao Zhai, Chinese Academy of Meteorological Sciences
Chenyi Zhang, China Meteorological Administration
Zhao Zhang, State Key Laboratory of Earth Surface Processes and Resource Ecology
Zong-Ci Zhao, National Climate Center
Guangsheng Zhou, China Meteorological Administration
Hongjian Zhou, Nationals Disaster Reduction Center of China
Tao Zhou, State Key Laboratory of Earth Surface Processes and Resource Ecology
Tianjun Zhou, Chinese Academy of Sciences, Institute of Atmospheric Physics
Furong Zhu, Ningxia Economic Research Center
Xukai Zou, China Meteorological Administration

Colombia
Ana Campos-Garcia, Consultant
Omar-Dario Cardona, Universidad Nacional de Colombia
Carmen Lacambra, Cambridge Coastal Research Unit
Pedro Simon Lamprea Quiroga, Colombian Institute of Hydrology, Meteorology, and Environmental Studies
Walter Vergara, Inter-American Development Bank
Gustavo Wilches-Chaux, Universidad Andina Simon Bolivar

Cook Islands
Arona Ngari, Cook Islands Meteorological Service

Costa Rica
Allan Lavell, Programme for the Social Study of Risk and Disaster
Roberto Villalobos Flores, Instituto Meteorologico Nacional

Cuba
Raul J. Garrido Vazquez, Ministry Science, Technology and Environment
Tomas Gutierrez Perez, Instituto de Meteorologia
Avelino G. Suarez, Institute of Ecology and Systematic, Cuban Environmental Agency

Cyprus
Silas Michaelides, Ministry of Agriculture, Natural Resources, and Environment

Denmark
Kirsten Halsnaes, Risø DTU
Torkil Jonch Clausen, DHI
Anne Mette Jorgensen, Centre Danish Meteorological
Karen G. Villholth, GEUS, Geological Survey of Denmark and Greenland

Egypt
Fatma El Mallah, League of Arab States
Wadid Fawzy Erian, Arab Center for the Studies of Arid Zones and Dry Lands
Amal Saad-Hussein, National Research Centre
Adel Yasseen, Ain Shams University, Institute of Environmental Research and Studies

El Salvador
Luis Ernesto Romano, Centro Humboldt Nicaragua

Finland
Timothy Carter, Finnish Environment Institute
Hilppa Gregow, Finnish Meteorological Institute

Simo Haanpää, Aalto University
Pirkko Heikinheimo, Prime Minister's Office of Finland
Susanna Kankaanpää, HSY Helsinki Region Environmental Services Authority
Sanna Luhtala, Ministry of Agriculture and Forestry
Markku Niinioja , Ministry for Foreign Affairs
Hannu Raitio, Finnish Forest Research Institute
Karoliina Saarniharo, United Nations Framework Convention on Climate Change
Kristiina Säntti, Finnish Meteorological Institute
Heikki Tuomenvirta, Finnish Meteorological Institute
Elina Vapaavuori, Finnish Forest Research Institute
Hanna Virta, Finnish Meteorological Institute

France

Franck Arnaud, Ministry of Ecology, Sustainable Development, Transport, and Housing
Slimane Bekki, Institut Pierre Simon Laplace, Laboratoire Atmospheres, Milieux, Observations Spatiales
Nicolas Beriot, Ministry of Ecology, Sustainable Development, Transport, and Housing
Olivier Bommelaer, Ministry of Ecology, Sustainable Development, Transport, and Housing
Olivier Boucher, Met Office Hadley Centre
Paul-Henri Bourrelier, French Association for Disaster Risk Reduction
Jean-Marie Carriere, Meteo-France
Fabrice Chauvin, Meteo-France
Chantal Claud, Institute Pierre-Simon, Laboratoire de Meteorologie Dynamique
Fabio D'Andrea, Institute Pierre-Simon, Laboratoire de Meteorologie Dynamique
Sylvie de Smedt, Ministry of Ecology, Sustainable Development, Transport and Housing
Henri Decamps, Centre National de la Recherche Scientifique
Pascale Delecluse, Meteo-France
Michel Deque, Meteo-France
Pierre-Yves Dupuy, Service Hydrographique et Oceanographique de la Marine
Nicolas Eckert, Cemagref
Rene Feunteun, French Association for Disaster Risk Reduction
JC Gaillard, The University of Auckland
Francois Gerard, French Association for Disaster Risk Reduction
Marc Gillet, Meteo-France
Pascal Girot, IUCN
Frederic Grelot, Cemagref
Delphine Grynszpan, UK Health Protection Agency
Eric Guilyardi, Institut Pierre Simon Laplace, Laboratoire D'Oceanographie et du Climat
Stephane Hallegatte, CIRED and Meteo-France
Sylvie Joussaume, Paris Consortium on Climate
Reza Lahidji, Groupement de Recherches en Gestion a HEC
Michel Lang, Cemagref
Goneri Le Cozannet, Bureau de Recherches Geologiques et Minieres
Jean-Michel Le Quentrec, French Association for Disaster Risk Reduction
Antoine Leblois, CIRED
Alexandre Magnan, Institute fior Sustainable Development and International Relations
Eric Martin, Meteo-France
Olivier Mestre, Meteo-France
David Meunier, Ministry of Ecology, Sustainable Development, Transport, and Housing
Hormoz Modaressi, Bureau of Geological and Mining Research
Roland Nussbaum, Mission Risques Naturels
Sylvie Parey, EDF-France
Cedric Peinturier, Ministry of Ecology, Sustainable Development, Transport, and Housing
Michel Petit, Conseil General de L'industrie, de L'energie et des Technologies
Pierre Picard, Ecole Polytechnique
Serge Planton, Méto-France
Valentin Przyluski, Centre International de Recherche sur l'Environnement et le Developpement
Jean-Luc Salagnac, Centre Scientifique et Technique de Batiment
Bernard Seguin, INRA
Karen Sudmeier-Reiux, University of Lausanne, IUCN Commission on Ecosystems Management
Nicolas Taillefer, Centre Scientifique et Technique de Batiment
Jean-Philippe Torterotot, Cemagref
Robert Vautard, Institut Pierre Simon Laplace, Laboratoire des Sciences du Climat et de l'Environnement
Jean-Philippe Vidal, Cemagref
Vincent Viguié, Centre International de Recherche sur l'Environnement et le Développement
Pascal Yiou, Laboratoire des Sciences du Climat et de l'Environnement

Germany

Gotelind Alber, Women for Climate Justice
Christoph Bals, Germanwatch
Hubertus Bardt, Cologne Institute for Economic Research
Andreas Baumgärtner, Project Management Agency of DLR, German IPCC Coordination Office
Paul Becker, German Weather Service
Joern Birkmann, UN University Institute for Environment and Human Security
Hans-Georg Bohle, University of Bonn
Hans Gunter Brauch, Freie Universitaet Berlin
Michael Bründl, WSL Institute for Snow and Avalanche Research SLF
Achim Daschkeit, German Federal Environment Agency
Carmen de Jong, University of Savoy
Thomas Deutschländer, German Meteorological Service
Paul Dostal, Project Management Agency of DLR
Dirk Engelbart, Federal Ministry of Transport, Building and Urban Development
Eberhard Faust, Munich Reinsurance Company
Roland Fendler, German Federal Environment Agency
Tobias Fuchs, German Weather Service
Hans-Martin Fuessel, European Environment Agency
Stefan Goessling-Reisemann, University of Bremen
Robert Grassmann, Deutsche Welthungerhilfe e.V.
Edeltraud Guenther, Technische Universität Dresden
Josef Haider, KfW Development Bank
Angela Michiko Hama, United Nations International Strategy for Disaster Reduction
Sven Harmeling, Germanwatch
Fred Fokko Hattermann, Potsdam Institute for Climate Impact Research
Gabriele Hegerl, University of Edinburgh
Hans-Joachim Herrmann, German Federal Environment Agency
Anne Holsten, Potsdam Institute of Climate Impact Research
Anke Jentsch, University of Koblenz-Landau
Marcus Kaplan, German Development Institute
Christina Koppe-Schaller, Deutscher Wetterdienst
Christoph Kottmeier, Karlsruhe Institute of Technology
Frank Kreienkamp, Climate and Environment Consulting Potsdam GmbH
Christian Kuhlicke, Helmhotz Centre for Environmental Research
Birgit Kuna, Project Management Agency of DLR
Nana Künkel, German Agency for International Development
Michael Kunz, Karlsruhe Institute of Technology
Ole Langniss, Fichtner GmbH & Co KG
Rocio Lichte, United Nations Framework Convention on Climate Change
Petra Mahrenholz, German Federal Environment Agency
Reinhard Mechler, International Institute for Applied Systems Analysis, Vienna University of Economics
Bettina Menne, World Health Organization, Regional Office for Europe
Annette Mohner, United Nations Framework Convention on Climate Change
Guido Mücke, German Federal Environment Agency
Christian L.C. Müller, Federal Ministry for the Environment, Nature Conservation, and Nuclear Safety
Claudia Pahl-Wostl, Institute of Environmental Systems Research, University of Osnabruck
Gertrude Penn-Bressel, German Federal Environment Agency
Jurgen Pohl, University of Bonn
Joerg Rapp, Deutscher Wetterdienst

Annex I

Markus Reichstein, Max-Planck Institute for Biogeochemistry
Joachim Rock, Johann Heinrich von Thuenen-Institute
Benno Rothstein, University of Applied Forest Sciences Rottenburg
Peter Rottach, Diakonie Katastrophenhilfe Consultant
Julia Rufin, Federal Minitry for the Environment, Nature Conservation, and Nuclear Safety
Evelina Santa, Federal Ministry of Education and Research
Philipp Schmidt-Thome, Geological Survey of Finland
Gudrun Schütze, German Federal Environment Agency
Reimund Schwarze, Helmholtz Center for Environmental Research
Joachim H. Spangenberg, Sustainable Europe Research Institute
Frank Sperling, World Wildlife Fund, Norway
Jochen Stuck, Project Management Agency of DLR
Swenja Surminski, Association of British Insurers
Christiane Textor, Project Management Agency of DLR, German IPCC Coordination Office
Annegret Thieken, University of Potsdam
Uwe Ulbrich, Freie Universitat Berlin
Martine Vatterodt, Federal Ministry for Economic Cooperation and Development
Monika Vees, German Federal Environment Agency
Gottfried von Gemminingen, Federal Ministry for Economic Cooperation and Development
Hans von Storch, GKSS Research Center
Martin Voss, Katastrophenforschungsstelle Berlin
Koko Warner, United Nations University, Institute for Environment and Human Security
Juergen Weichselgartner, GKSS Research Center
Johanna Wolf, Memorial University of Newfoundland
Sabine Wurzler, North Rhine Westphalia State Environment Agency
Karl-Otto Zentel, Deutsches Komitee Katastrophenvorsorge e.V.

Ghana
Seth Vordzorgbe, United Nations Development Programme

Greece
Christina Anagnostopoulou, Aristotle University of Thessaloniki
Helena Flocas, University of Athens
Panagiota Galiatsatou, Aristotle University of Thessaloniki
Antonis Koussis, National Observatory of Athens
Aristeidis Koutrouli, Technical University of Crete
Athanasios Louka, University of Thessaly
Petroula Louka, Hellenic National Meteorological Service
Dimitrios Melas, Aristotle University of Thessaloniki
Panayotis Prinos, Aristotle University of Thessaloniki
Ioannis Tsanis, Technical University of Crete
Adonis Velegrakis, University of the Aegean
Christos Zerefos, Academy of Athens

Guatemala
Edwin Castellanos, Universidad del Valle de Guatemala

Hungary
Joseph Feiler, Ministry of National Development
Ferenc L. Tóth, International Atomic Energy Agency

Iceland
Halldor Bjornsson, Icelandic Meteorological Office
HalldorSigrun Karlsdottir, Icelandic Meteorological Office
Arni Snorrason, Icelandic Meteorological Office

India
Unnikrishnan Alakkat, National Institute of Oceanography
Subbiah Arjunapermal, Asian Disaster Preparedness Center
Suruchi Bhadwal, The Energy and Resources Institute
Mihir Bhatt, India Disaster Mitigation Institute
Amit Garg, Indian Institute of Management Ahmedabad
B.N. Goswami, Indian Institute for Tropical Meteorology
Manu Gupta, SEEDS
Umesh Haritashya, University of Dayton
Ritesh Kumar, Wetlands International - South Asia
Pradeep Mujumdar, Indian Institute of Science
Anand Patwardhan, Indian Institute of Technology Bombay
Apurva Sanghi, The World Bank
Akhilesh Surjan, United Nations University

Indonesia
Edvin Aldrian, Badan Meteorologi Klimatologi dan Geofisika

Iran
Rahman Davtalab, Ministry of Energy
Saeid Eslamian, Isfahan University of Technology
Mahnaz Khazaee, Atmospheric Science and Meteorological Research Center
Mohammad Rahimi, Semnan University
Fatemeh Rahimzadeh, Atmospheric Science and Meteorological Research Center
Saviz Sehat Kashani, Atmospheric Sciences and Meteorological Research Center

Ireland
Ian Bryceson, Norwegian University of Life Sciences
Noel Casserly, Department of the Environment

Italy
Marina Baldi, National Research Council, Institute of Biometeorology
Roberto Bertolini, World Health Organization
Francesco Bosello, Fondazione Eni Enrico Mattei, Milan University
Stefano Bovo, ARPA Piemonte
Carlo Giupponi, University Ca' Foscari of Venice and Euro-Mediterranean Centre for Climate Change
Georg Kaser, University of Innsbruck
Valentina Pavan, ARPA Emilia-Romagna
Roberto Ranzi, University of Brescia
Carlo Scaramella, World Food Programme
Rodica Tomozeiu, ARPA Emilia-Romagna

Japan
Shiho Asano, Forestry and Forest Products Research Institute
Fumiaki Fujibe, Meteorological Research Institute
Koji Fujita, Nagoya University
Masahiro Hashizume, Institute of Tropical Medicine, Nagasaki University
Yasushi Honda, University of Tsukuba
Shinjiro Kanae, Tokyo Institute of Technology
Takehiro Kano, Division Ministry of Foreign Affairs
Miwa Kato, UNFCCC Secretariat
Hiroyasu Kawai, Port and Airport Research Institute
So Kazama, Tohoku University
Akio Kitoh, Meteorological Research Institute
Masahide Kondo, University of Tsukuba
Kazuo Kurihara, Meteorological Research Institute
Shoji Kusunoki, Meteorological Research Institute
Takao Masumoto, National Institute for Rural Engineering, National Agriculture and Food Research Organization
Nobuo Mimura, Ibaraki University
Hisayoshi Morisugi, Nihon University
Toshiyuki Nakaegawa, Meteorological Research Institute
Elichi Nakakita, Kyoto University
Motoki Nishimori, National Institute for Agro-Environmental Sciences
Taikan Oki, University of Tokyo
Rajib Shaw, Kyoto University

Hideo Shiogama, National Institute for Environmental Studies
Yasuto Tachikawa, Kyoto University
Kiyoshi Takahashi, National Institute for Environmental Studies
Izuru Takayabu, Meteorological Research Institute
Kuniyoshi Takeuchi, International Centre for Water Hazard and Risk Management
Tadashi Tanaka, University of Tsukuba
Makoto Tani, Kyoto University
Tsugihiro Watanabe, Research Institute for Humanity and Nature
Hiroya Yamano, National Institute for Environmental Studies

Kenya
Peter Ambenje, Kenya Meteorological Department
Samwel Marigi, Kenya Meteorological Department
Charles Mutai, Ministry of Environment and Mineral Resources
Christopher Oludhe, University of Nairobi, Department of Meteorology

Latvia
Olga Vilima, United Nations International Strategy for Disaster Reduction

Malaysia
Joy Jacqueline Pereira, Universiti Kebangsaan Malaysia
Salmah Zakaria, United Nations Economic and Social Commission

Mauritania
Gueladio Cisse, Swiss Tropical and Public Health Institute

Mexico
Victor Cardenas, Climate Change and Natural Disaster Risk Management
Tereza Cavazos, El Centro de Investigación Científica y de Educación Superior de Ensenada
Carolina Neri, Universidad Nacional Autonoma de Mexico
Ursula Oswald-Spring, Universidad Nacional Autónoma de México
Ricardo Zapata-Marti, United Nations Economic Commission for Latin America and the Caribbean

Mongolia
Ravsal Oyun, JEMR Consulting Company

Morocco
Abdalah Mokssit, Direction de la Météorologie Nationale

Mozambique
Felipe Lucio, Global Framework for Climate Services Office, WMO

New Zealand
Reid Basher, Secretariat of the High-Level Taskforce on the Global Framework for Climate Services
Leonard Brown, Ministry for the Environment - Manatu Mo Te Taiao
John Campbell, University of Waikato
John Hay, University of the South Pacific
Glenn McGregor, University of Auckland
Matthew McKinnon, DARA
Helen Plume, Ministry for the Environment - Manatu Mo Te Taiao
David Wratt, National Institute of Water and Atmospheric Research

Niger
Abdelkrim Ben Mohamed, University of Niamey

Norway
Torgrim Asphjell, Climate and Pollution Agency
Rasmus Benestad, The Norwegian Meteorological Institute
Tor A. Benjaminsen, Norwegian University of Life Sciences
Elzbieta Maria Bitner-Gregersen, Det Norske Veritas AS
Oyvind Christophersen, Climate and Pollution Agency
Solveig Crompton, Ministry of the Environment
Linda Dalen, Norwegian Directorate for Nature Management
Lars Ingolf Eide, Det Norske Veritas
Siri Eriksen, Norwegian University of Life Sciences
Christoffer Grønstad, Climate and Pollution Agency
Hege Haugland, Climate and Pollution Agency
Hege Hisdal, Norwegian Water Resources and Energy Directorate
Dag O. Høgvold, Directorate for Civil Protection and Emergency Planning
Linn Bryhn Jacobsen, Climate and Pollution Agency
Vikram Kolmannskog, Norwegian Refugee Council
Ole-Kristian Kvissel, Climate and Pollution Agency
Farrokh Nadim, International Centre for Geohazards
Lars Otto Naess, Institute of Development Studies
Karen O'Brien, University of Oslo
Ellen Øseth, Norwegian Polar Institute
Marit Viktoria Pettersen, Ministry of Foreign Affairs
Asgeir Sorteberg, University of Bergen
Linda Sygna, University of Oslo
Kirsten Ulsrud, University of Oslo
Vigdis Vestreng, Climate and Pollution Agency

Pakistan
Muhammad Mohsin Iqbal, Global Change Impact Studies Centre
Jawed Ali Khan, Ministry of Environment
Maira Zahur, Women for Climate Justice

Palestinian National Authority
Nedal Katbeh-Bader, Environment Quality Authority

Peru
Eduardo Calvo, Universidad Nacional Mayor de San Marcos
Encinas Carla, Intercooperation
Silvia Llosa, United Nations International Strategy for Disaster Reduction

Philippines
Imelda Abarquez, Oxfam Hong Kong
Sanny Jegillos, United Nations Development Programme
Rosa Perez, Manila Observatory

Poland
Janusz Filipiak, Institute of Meteorology and Water Management
Zbigniew Kundzewicz, Polish Academy of Sciences
Zbigniew Ustrnul, Institute of Meteorology and Water Management, Jagiellonian University
Joanna Wibig, University of Lodz

Republic of Korea
So-Min Cheong, University of Kansas
Tae Sung Cheong, National Emergency Management Agency
Soojeong Myeong, Korea Environment Institute

Republic of Maldives
Amjad Abdulla, IPCC Vice Chair WG II, Climate Change Energy Department, Ministry of Housing and Environment

Romania
Roxana Bojariu, National Meteorological Administration
Sorin Cheval, National Meteorological Administration

Russian Federation
E. M. Akentyeva, Main Geophysical Observatory
Sergey Borsch, Hydromet Center of Russia

N. V. Kobysheva, Main Geophysical Observatory
Boris Porfiriev, Institute for Economic Forecasting, Russian Academy of Sciences
Vladimir Semenov, A.M. Obukhov Institute of Atmospheric Physics
Boris Sherstyukov, All Russian Research Institute of Hydrometeorological Information World Data Center

Senegal
Cherif Diop, Senegalese Meteorological Agency

South Africa
Reinette (Oonsie) Biggs, Stockholm Resilience Centre, Stockholm University
Bruce Glavovic, Massey University
Bettina Koelle, Indigo Development and Change
Noel Oettle, Environmental Monitoring Group
Coleen Vogel, University of Witwatersrand
Gina Ziervogel, University of Cape Town

Spain
Enric Aguilar, Universitat Rovira i Virgili
Gerardo Benito, Spanish Council for Scientific Research
Jorge Bonnet Fernandez Trujillo, Government of the Canary Islands
Francisco Garcia Novo, University of Seville
José Manuel Gutiérrez, Consejo Superior de Investigaciones Científicas
Ana Iglesias, Universidad Politecnica de Madrid
José Antonio López-Díaz, Agencia Estatal de Meteorología
Concepcion Martinez-Lope, Spanish Bureau for Climate Change
José Moreno, University of Castilla-La Mancha
Francisco Pascual, Spanish Bureau for Climate Change
Jose Ramon Picatoste-Ruggeroni, Spanish Bureau for Climate Change
Ernesto Rodriguez-Camino, Spanish Meteorological Agency
Sabater Sergi, University Girona

Sudan
Ismail Fadl El Moula Mohamed, Sudan Meteorological Authority
Balgis Osman-Elasha, African Development Bank

Sweden
Cecilia Alfredsson, Swedish Civil Contingencies Agency
Lars Barring, Swedish Meteorological and Hydrological Institute
Sten Bergstrom, Swedish Meteorological and Hydrological Institute
Pelle Boberg, Swedish Meteorological and Hydrological Institute
Henrik Carlsen, Swedish Defence Research Agency
Carl Folke, The Beijer Institute, Stockholm University
Clarisse Kehler Siebert, Stockholm Environment Institute
Carina Keskitalo, Umea University
Richard Klein, Stockholm Environment Institute
Georg Lindgren, Lund University
Elin Lovendahl, Swedish Meteorological and Hydrological Institute
Barbro Naslund-Landenmark, Swedish Civil Contingencies Agency
Carin Nilsson, Swedish Meteorological and Hydrological Institute
Ulrika Postgard, Swedish Civil Contingencies Agency
Markku Rummukainen, Swedish Meteorological and Hydrological Institute
Johan Schaar, Ministry of Foreign Affairs
Lisa Schipper, Stockholm Environment Institute
Ake Svensson, Swedish Civil Contingencies Agency

Switzerland
Simon Allen, IPCC WGI Technical Support Unit
Walter J. Ammann, Global Risk Forum GRF Davos
Neville Ash, United Nations Environment Programme
Stefan Brönnimann, University of Bern
Carlo Casty, Partner Reinsurance Company
Nicole Clot, Intercooperation

Paul Della-Marta, Partner Reinsurance Company
Andreas Fischlin, Swiss Federal Institute of Technology, Systems Ecology
Markus Gerber, University of Bern, Climate and Environmental Physics
Peter Greminger, Federal Office for Environment
Christian Huggel, University of Zurich
Matthias Huss, University of Fribourg
Daniel Kull, International Federation of Red Cross and Red Crescent Societies
Juerg Luterbacher, Justus Liebig University
Joy Muller, International Federation of Red Cross and Red Crescent Societies
Urs Neu, Swiss Academy of Sciences
Boris Orlowsky, Swiss Federal Institute of Technology Zurich
Pascal Peduzzi, United Nations Environment Programme
Gian-Kasper Plattner, IPCC WGI Technical Support Unit
Dieter Rickenmann, Swiss Federal Research Institute WSL
Stephan Rist, Centre for Development and Environment, University of Bern
Jose Romero, Federal Office for the Environment
Sonia Seneviratne, Swiss Federal Institute of Technology Zurich
Andreas Spiegel, Swiss Re
Thomas Stocker, University of Bern
Philippe Thalmann, EPFL Swiss Federal Institute of Technology Lausanne
Heinz Wanner, University of Bern
Andre Wehrli, European Environment Agency
Heini Wernli, Swiss Federal Institute of Technology Zurich
Irina Zodrow, United Nations International Strategy for Disaster Reduction

Tanzania
Emmanuel Mpeta, Tanzania Meteorological Agency
Khamaldin Daud Mutabazi, Sokoine University of Agriculture
Pius Zebhe Yanda, University of Dar es Salaam

Thailand
Seree Supratid, Rangsit University

The Netherlands
Frans Berkhout, Vrije University
Laurens Bouwer, Institute for Environmental Studies, Vrije University
Hein W. Haak, The Royal Netherlands Meteorological Institute
Albert Klein Tank, The Royal Netherlands Meteorological Institute
Irene Kreis, Health Protection Agency
Adriaan Perrels, Finnish Meteorological Institute
Maarten van Aalst, Red Cross Red Crescent Climate Centre
Bart van den Hurk, The Royal Netherlands Meteorological Institute
Henny A.J. van Lanen, Wageningen University
Geert Jan van Oldenborgh, The Royal Netherlands Meteorological Institute
Jeroen Warner, Wageningen University

Trinidad and Tobago
Veronica Belgrave, Ministry of Planning, Housing, and the Environment

Turkey
Salahattin Incecik, Istanbul Technical University

United Kingdom
Neil Adger, Tyndall Centre, University of East Anglia
Alex Arnall, University of Reading
Nigel Arnell, University of Reading
Victoria Bell, Centre for Ecology and Hydrology
Enrico Biffis, Imperial College, London
Katrina Brown, University of East Anglia
Simon Brown, Met Office Hadley Centre
Sal Burgess, UK Department for Environment, Food, and Rural Affairs
Harriet Caldin, Health Protection Agency
Diarmid Campbell-Lendrum, World Health Organization

Authors and Expert Reviewers — Annex I

Catriona Carmichael, Health Protection Agency
Amanda Charles, UK Government Office for Science
Declan Conway, University of East Anglia
Tim Conway, UK Department for International Development
Anita Cooper, Health Protection Agency
Geoff Darch, Atkins Consultants and University of East Anglia
Ian Davis, Cranfield University
Ken De Souza, UK Department for International Development
Andrew Dlugolecki, Climatic Research UnitUniversity of East Anglia
Maureen Fordham, Northumbria University
Tim Forsyth, London School of Economics and Political Science
Clare Goodess, University of East Anglia Climatic Research Unit
Jim Hall, Newcastle University
Lucy Hayes, UK Department of Energy and Climate Change
Clare Heaviside, Centre for Radiation, Chemical and Environmental Hazards, Health Protection Agency
Debbie Hillier, Oxfam International
Robert Hodgson, University of Exeter
Sari Kovats, London School of Hygiene and Tropical Medicine
Bo Lim, United Nations Development Programme
Andrew Maskrey, UN International Strategy for Disaster Reduction
Michael McCall, Universidad Nacional Autonoma de Mexico
William McGuire, University College London
Thomas Mitchell, Overseas Development Institute
John Morton, Natural Resources Institute, University of Greenwich
Alessandro Moscuzza, UK Department for International Development
Robert Muir-Wood, Risk Management Solutions
Virginia Murray, Health Protection Agency
Katherine Nightingale, Christian Aid
Geoff O'Brien, Northumbria University
Phil O'Keefe, Northumbria University
Jean Palutikof, Griffith University
Mark Pelling, King's College London
Emily Polack, Institute of Development Studies
Ben Ramalingam, Overseas Development Institute
Nicola Ranger, London School of Economics
John Rees, Natural Environment Research Council, UK
Hannah Rowlatt, Health Protection Agency and University of Sheffield
Sohel Saikat, Health Protection Agency
David Satterthwaite, International Institute for Environment & Development
Chris Sear, UK Department of Energy and Climate Change
A. Simmons, European Centre for Medium-Range Weather Forecasts
Robert Siveter, International Petroleum Industry Environmental Conservation Association
David Smith, University of the West Indies
Stephen Smith, UK Committee on Climate Change
David Stephenson, University of Exeter
Peter Stott, Met Office Hadley Centre
Robert Sykes, International Petroleum Industry Environmental Conservation Association
Thomas Tanner, Institute of Development Studies
Addis Taye, Health Protection Services
Emma Tompkins, University of Southampton, Highfield Campus
John Twigg, University College London
Sotiris Vardoulakis, Health Protection Agency
Emma Visman, Humanitarian Futures Programme, King's College, London
Tim Waites, UK Department for International Development
David Warrilow, UK Department of Energy and Climate Change
Paul Watkiss, Paul Watkiss Associates
Robert Wilby, University of Loughborough
Michelle Winthrop, UK Department for International Development
Philip Woodworth, National Oceanography Centre
Ronald Young, Young International Ltd / Knowledge Associates International Ltd

United States of America

David Allen, US Global Change Research Program
Tom Armstrong, US Global Change Research Program
Jeff Arnold, US Army Corps of Engineers
Margaret Arnold, The World Bank
Bilal Ayyub, University of Maryland
Donald Ballantyne, MMI Engineering
Ko Barrett, National Oceanic and Atmospheric Administration
Stephen Bender, Organization of American States (retired)
Lisa M. Butler-Harrington, The Wharton School, Kansas State University
JoAnn Carmin, Massachusetts Institute of Technology
Edward Carr, US Administration for International Development
DeWayne Cecil, National Oceanic and Atmospheric Administration
Christina Chan, US Department of State
David Cleaves, US Forest Service
Thomas Cronin, US Geological Survey
Susan Cutter, University of South Carolina
Kirstin Dow, University of South Carolina
David Easterling, National Oceanic and Atmospheric Administration, National Climatic Data Center
Kristie Ebi, IPCC WGII Technical Support Unit
Barbara Ellis, Center for Disease Control
Christopher Emrich, University of South Carolina
Sandy Eslinger, National Oceanic and Atmospheric Administration
Ross Faith, US Subcommittee on Disaster Reduction
Christopher Field, Carnegie Institution for Science
Stephen Gill, National Oceanic and Atmospheric Administration
Justin Ginnetti, United Nations International Strategy for Disaster Reduction
William Gutowski, Iowa State University
D.E. (Ed) Harrison, National Oceanic and Atmospheric Administration, Pacific Marine Environmental Laboratory
Jerry Hatfield, US Department of Agriculture
Robert Heilmayr, Emmett Interdisciplinary Program for Environment and Resources, Stanford University
Molly Hellmuth, International Research Institute for Climate and Society
Jeremy Hess, Centers for Disease Control and Prevention
Robert Hirsch, US Geological Survey
Robert Jarrett, US Geological Survey
Terry Jeggle, University of Pittsburgh
Mark Keim, Centers for Disease Control and Prevention
Paul Knappenberger, New Hope Environmental Services
Thomas Knutson, National Oceanic and Atmospheric Administration
James Kossin, National Oceanic and Atmospheric Administration, National Climatic Data Center
Howard Kunreuther, The Wharton School, University of Pennsylvania
David Lea, US Department of State
Arthur Lee, Chevron Services Company
Robin Leichenko, Rutgers University
Maria Carmen Lemos, University of Michigan
Robert Lempert, RAND Corporation
David Levinson, US Forest Service
Joanne Linnerooth-Bayer, International Institute for Applied Systems Analysis
Peter Liotta, Independent Scholar
Chris Little, Woodrow Wilson School of Public and International Affairs, Princeton University
David Lobell, Stanford University
Pat Longstaff, Syracuse University
Alexander Lotsch, The World Bank
Michael MacCracken, Climate Institute
Katharine Mach, IPCC WGII Technical Support Unit
Simon Mason, Columbia University
Michael Mastrandrea, IPCC WGII Technical Support Unit
Sabrina McCormick, US Environmental Protection Agency

Linda Mearns, National Center for Atmospheric Research
Jerry Meehl, National Center for Atmospheric Research
Chris Milly, US Geological Survey
James Mitchell, Rutgers University
Marcus Moench, Institute for Social and Environmental Transition
Susanne Moser, Susanne Moser Research and Consulting
Meredith Muth, National Oceanic and Atmospheric Administration
Robert J Naiman, University of Washington
Robert O'Connor, National Science Foundation
Ian O'Donnell, Asian Development Bank
Michael Oppenheimer, Princeton University
Jacob Park, Green Mountain College
Roger Pielke Jr., University of Colorado
David Pierce, Scripps Institution of Oceanography
Mark Powell, National Oceanic and Atmospheric Administration
Michael Prather, University of California, Irvine
Roger Pulwarty, National Oceanic and Atmospheric Administration
David Reidmiller, US Department of State
Dian Seidel, National Oceanic and Atmospheric Administration
Emil Simiu, National Institute of Standards and Technology
Anthony-Oliver Smith, Emeritus, University of Florida
Joel Smith, Stratus Consulting
Susan Solomon, National Oceanic and Atmospheric Administration
Doreen Stabinsky, College of the Atlantic
Amanda Staudt, National Wildlife Federation
Ronald Stouffer, National Oceanic and Atmospheric Administration
Trigg Talley, US Department of State
Wassila Thiaw, National Oceanic and Atmospheric Administration
John Tiefenbacher, Texas State University
Sezin Tokar, US Agency for International Development
Kevin E. Trenberth, National Center for Atmospheric Research
Thomas Wagner, National Aeronautics and Space Administration
Robert Webb, National Oceanic and Atmospheric Administration

Elke Weber, Columbia University
Michael Wehner, Lawrence Berkeley National Laboratory
Jason Westrich, University of Georgia, Odum School of Ecology
Thomas Wilbanks, Oak Ridge National Laboratory
Benjamin Wisner, Aon-Benfield UCL Hazard Research Centre, University College London
Richard Wright, American Society of Civil Engineers
Donald Wuebbles, University of Illinois
Tingjun Zhang, University of Colorado National Snow and Ice Data Center

Venezuela
Maria Teresa Abogado, Ministry of People's Powers for Foreign Affairs
Jose Azuaje, Ministry of People's Powers for Foreign Affairs
Salvano Briceno, United Nations
Claudia Salerno Caldera, Ministry of People's Powers for Foreign Affairs
Isabel Di Carlo Quero, Ministry of People's Powers for Foreign Affairs
Rafael Hernandez, Ministry of People's Powers for Foreign Affairs
Federico Lagarde, Ministry of People's Powers for Foreign Affairs
Alejandro Linayo, Research Center on Disaster Risk Reduction
Luis Jose Mata, International Monetary Fund
Yessica Pereira, Ministry of People's Powers for Foreign Affairs
Reina Perez, Ministry of People's Powers for Foreign Affairs
Rafael Rebolledo, Ministry of People's Powers for Foreign Affairs
Dirk Thielen, Ministry of People's Powers for Foreign Affairs

Vietnam
Mai Trong Nhuan, Vietnam National University
Bach Tan Sinh, National Institute for Science and Technology Policy and Strategy Studies

Zambia
Raban Chanda, University of Botswana

ANNEX II

Glossary of Terms

This annex should be cited as:
IPCC, 2012: Glossary of terms. In: *Managing the Risks of Extreme Events and Disasters to Advance Climate Change Adaptation* [Field, C.B., V. Barros, T.F. Stocker, D. Qin, D.J. Dokken, K.L. Ebi, M.D. Mastrandrea, K.J. Mach, G.-K. Plattner, S.K. Allen, M. Tignor, and P.M. Midgley (eds.)]. A Special Report of Working Groups I and II of the Intergovernmental Panel on Climate Change (IPCC). Cambridge University Press, Cambridge, UK, and New York, NY, USA, pp. 555-564.

Glossary of Terms

Abrupt climate change
The nonlinearity of the climate system may lead to abrupt climate change, sometimes called rapid climate change, abrupt events, or even surprises. The term abrupt often refers to time scales faster than the typical time scale of the responsible forcing. However, not all abrupt climate changes need be externally forced. Some changes may be truly unexpected, resulting from a strong, rapidly changing forcing of a nonlinear system.

Adaptation
In human systems, the process of adjustment to actual or expected climate and its effects, in order to moderate harm or exploit beneficial opportunities. In natural systems, the process of adjustment to actual climate and its effects; human intervention may facilitate adjustment to expected climate.

Adaptation assessment
The practice of identifying options to adapt to climate change and evaluating them in terms of criteria such as availability, benefits, costs, effectiveness, efficiency, and feasibility.

Adaptive capacity
The combination of the strengths, attributes, and resources available to an individual, community, society, or organization that can be used to prepare for and undertake actions to reduce adverse impacts, moderate harm, or exploit beneficial opportunities.

Aerosols
A collection of airborne solid or liquid particles, with a typical size between 0.01 and 10 µm, that reside in the atmosphere for at least several hours. Aerosols may be of either natural or anthropogenic origin. Aerosols may influence climate in several ways: directly through scattering and absorbing radiation, and indirectly by acting as cloud condensation nuclei or modifying the optical properties and lifetime of clouds.

Albedo
The fraction of solar radiation reflected by a surface or object, often expressed as a percentage. Snow-covered surfaces have a high albedo, the surface albedo of soils ranges from high to low, and vegetation-covered surfaces and oceans have a low albedo. The Earth's planetary albedo varies mainly through varying cloudiness, snow, ice, leaf area, and land cover changes.

Anthropogenic
Resulting from or produced by human beings.

Anthropogenic emissions
Emissions of greenhouse gases, greenhouse gas precursors, and aerosols associated with human activities. These activities include the burning of fossil fuels, deforestation, land use changes, livestock, fertilization, etc., that result in a net increase in emissions.

Atlantic Multi-decadal Oscillation (AMO)
A multi-decadal (65- to 75-year) fluctuation in the North Atlantic, in which sea surface temperatures showed warm phases during roughly 1860 to 1880 and 1930 to 1960 and cool phases during 1905 to 1925 and 1970 to 1990 with a range of the order of 0.4°C.

Atmosphere
The gaseous envelope surrounding the Earth. The dry atmosphere consists almost entirely of nitrogen (78.1% volume mixing ratio) and oxygen (20.9% volume mixing ratio), together with a number of trace gases, such as argon (0.93% volume mixing ratio), helium, and radiatively active greenhouse gases such as carbon dioxide (0.035% volume mixing ratio) and ozone. In addition, the atmosphere contains the greenhouse gas water vapor, whose amounts are highly variable but typically around 1% volume mixing ratio. The atmosphere also contains clouds and aerosols.

Available potential energy
That portion of the total potential energy that may be converted to kinetic energy in an adiabatically enclosed system.

Baseline/reference
The baseline (or reference) is the state against which change is measured. It might be a 'current baseline,' in which case it represents observable, present-day conditions. It might also be a 'future baseline,' which is a projected future set of conditions excluding the driving factor of interest. Alternative interpretations of the reference conditions can give rise to multiple baselines.

Capacity
The combination of all the strengths, attributes, and resources available to an individual, community, society, or organization, which can be used to achieve established goals.

Carbon cycle
The term used to describe the flow of carbon (in various forms, e.g., as carbon dioxide) through the atmosphere, ocean, terrestrial biosphere, and lithosphere.

Carbon dioxide (CO_2)
A naturally occurring gas fixed by photosynthesis into organic matter. A byproduct of fossil fuel combustion and biomass burning, it is also emitted from land use changes and other industrial processes. It is the principal anthropogenic greenhouse gas that affects the Earth's radiative balance. It is the reference gas against which other greenhouse gases are measured, thus having a Global Warming Potential of 1.

Catchment
An area that collects and drains precipitation.

Clausius-Clapeyron relationship (or equation)
The differential equation relating the pressure of a substance (usually

water vapor) to temperature in a system in which two phases of the substance (water) are in equilibrium.

Climate
Climate in a narrow sense is usually defined as the average weather, or more rigorously, as the statistical description in terms of the mean and variability of relevant quantities over a period of time ranging from months to thousands or millions of years. The classical period for averaging these variables is 30 years, as defined by the World Meteorological Organization. The relevant quantities are most often surface variables such as temperature, precipitation, and wind. Climate in a wider sense is the state, including a statistical description, of the climate system. In various chapters in this report different averaging periods, such as a period of 20 years, are also used.

Climate change
A change in the state of the climate that can be identified (e.g., by using statistical tests) by changes in the mean and/or the variability of its properties and that persists for an extended period, typically decades or longer. Climate change may be due to natural internal processes or external forcings, or to persistent anthropogenic changes in the composition of the atmosphere or in land use.[1] See also Climate variability and Detection and attribution.

Climate extreme (extreme weather or climate event)
The occurrence of a value of a weather or climate variable above (or below) a threshold value near the upper (or lower) ends of the range of observed values of the variable. For simplicity, both extreme weather events and extreme climate events are referred to collectively as 'climate extremes.' The full definition is provided in Section 3.1.2.

Climate feedback
An interaction mechanism between processes in the climate system is called a climate feedback when the result of an initial process triggers changes in a second process that in turn influences the initial one. A positive feedback intensifies the original process, and a negative feedback reduces it.

Climate model
A numerical representation of the climate system that is based on the physical, chemical, and biological properties of its components, their interactions, and feedback processes, and that accounts for all or some of its known properties. The climate system can be represented by models of varying complexity, that is, for any one component or combination of components a spectrum or hierarchy of models can be identified, differing in such aspects as the number of spatial dimensions, the extent to which physical, chemical, or biological processes are explicitly represented, or the level at which empirical parameterizations are involved. Coupled Atmosphere-Ocean Global Climate Models (AOGCMs), also referred to as Atmosphere-Ocean General Circulation Models, provide a representation of the climate system that is near the most comprehensive end of the spectrum currently available. There is an evolution toward more complex models with interactive chemistry and biology. Climate models are applied as a research tool to study and simulate the climate, and for operational purposes, including monthly, seasonal, and interannual climate predictions.

Climate projection
A projection of the response of the climate system to emissions or concentration scenarios of greenhouse gases and aerosols, or radiative forcing scenarios, often based upon simulations by climate models. Climate projections are distinguished from climate predictions in order to emphasize that climate projections depend upon the emission/concentration/radiative-forcing scenario used, which are based on assumptions concerning, e.g., future socioeconomic and technological developments that may or may not be realized and are therefore subject to substantial uncertainty.

Climate scenario
A plausible and often simplified representation of the future climate, based on an internally consistent set of climatological relationships that has been constructed for explicit use in investigating the potential consequences of anthropogenic climate change, often serving as input to impact models. Climate projections often serve as the raw material for constructing climate scenarios, but climate scenarios usually require additional information such as about the observed current climate.

Climate system
The climate system is the highly complex system consisting of five major components: the atmosphere, the oceans, the cryosphere, the land surface, the biosphere, and the interactions between them. The climate system evolves in time under the influence of its own internal dynamics and because of external forcings such as volcanic eruptions, solar variations, and anthropogenic forcings such as the changing composition of the atmosphere and land use change.

Climate threshold
A critical limit within the climate system that induces a non-linear response to a given forcing. See also Abrupt climate change.

Climate variability
Climate variability refers to variations in the mean state and other statistics (such as standard deviations, the occurrence of extremes, etc.) of the climate at all spatial and temporal scales beyond that of individual weather events. Variability may be due to natural internal processes within the climate system (internal variability), or to variations in natural

---

[1] This definition differs from that in the United Nations Framework Convention on Climate Change (UNFCCC), where climate change is defined as: "a change of climate which is attributed directly or indirectly to human activity that alters the composition of the global atmosphere and which is in addition to natural climate variability observed over comparable time periods." The UNFCCC thus makes a distinction between climate change attributable to human activities altering the atmospheric composition, and climate variability attributable to natural causes.

or anthropogenic external forcing (external variability). See also Climate change.

Cold days/cold nights
Days where maximum temperature, or nights where minimum temperature, falls below the 10th percentile, where the respective temperature distributions are generally defined with respect to the 1961-1990 reference period.

Community-based disaster risk management
See Local disaster risk management.

Confidence
Confidence in the validity of a finding, based on the type, amount, quality, and consistency of evidence and on the degree of agreement. Confidence is expressed qualitatively.

Control run
A model run carried out to provide a 'baseline' for comparison with climate change experiments. The control run uses constant values for the radiative forcing due to greenhouse gases and anthropogenic aerosols appropriate to pre-industrial conditions.

Convection
Vertical motion driven by buoyancy forces arising from static instability, usually caused by near-surface cooling or increases in salinity in the case of the ocean and near-surface warming in the case of the atmosphere. At the location of convection, the horizontal scale is approximately the same as the vertical scale, as opposed to the large contrast between these scales in the general circulation. The net vertical mass transport is usually much smaller than the upward and downward exchange.

Coping
The use of available skills, resources, and opportunities to address, manage, and overcome adverse conditions, with the aim of achieving basic functioning in the short to medium term.

Coping capacity
The ability of people, organizations, and systems, using available skills, resources, and opportunities, to address, manage, and overcome adverse conditions.

Detection and attribution
Climate varies continually on all time scales. Detection of climate change is the process of demonstrating that climate has changed in some defined statistical sense, without providing a reason for that change. Attribution of causes of climate change is the process of establishing the most likely causes for the detected change with some defined level of confidence.

Diabatic
A process in which external heat is gained or lost by the system.

Disaster
Severe alterations in the normal functioning of a community or a society due to hazardous physical events interacting with vulnerable social conditions, leading to widespread adverse human, material, economic, or environmental effects that require immediate emergency response to satisfy critical human needs and that may require external support for recovery.

Disaster management
Social processes for designing, implementing, and evaluating strategies, policies, and measures that promote and improve disaster preparedness, response, and recovery practices at different organizational and societal levels.

Disaster risk
The likelihood over a specified time period of severe alterations in the normal functioning of a community or a society due to hazardous physical events interacting with vulnerable social conditions, leading to widespread adverse human, material, economic, or environmental effects that require immediate emergency response to satisfy critical human needs and that may require external support for recovery.

Disaster risk management (DRM)
Processes for designing, implementing, and evaluating strategies, policies, and measures to improve the understanding of disaster risk, foster disaster risk reduction and transfer, and promote continuous improvement in disaster preparedness, response, and recovery practices, with the explicit purpose of increasing human security, well-being, quality of life, and sustainable development.

Disaster risk reduction (DRR)
Denotes both a policy goal or objective, and the strategic and instrumental measures employed for anticipating future disaster risk; reducing existing exposure, hazard, or vulnerability; and improving resilience.

Diurnal temperature range
The difference between the maximum and minimum temperature during a 24-hour period.

Downscaling
Downscaling is a method that derives local- to regional-scale (up to 100 km) information from larger-scale models or data analyses. The full definition is provided in Section 3.2.3.

Drought
A period of abnormally dry weather long enough to cause a serious hydrological imbalance. Drought is a relative term (see Box 3-3), therefore any discussion in terms of precipitation deficit must refer to the particular precipitation-related activity that is under discussion. For example, shortage of precipitation during the growing season impinges on crop production or ecosystem function in general (due to soil moisture

drought, also termed agricultural drought), and during the runoff and percolation season primarily affects water supplies (hydrological drought). Storage changes in soil moisture and groundwater are also affected by increases in actual evapotranspiration in addition to reductions in precipitation. A period with an abnormal precipitation deficit is defined as a meteorological drought. A megadrought is a very lengthy and pervasive drought, lasting much longer than normal, usually a decade or more.

Early warning system
The set of capacities needed to generate and disseminate timely and meaningful warning information to enable individuals, communities, and organizations threatened by a hazard to prepare and to act appropriately and in sufficient time to reduce the possibility of harm or loss.

El Niño-Southern Oscillation (ENSO)
The term El Niño was initially used to describe a warm-water current that periodically flows along the coast of Ecuador and Peru, disrupting the local fishery. It has since become identified with a basin-wide warming of the tropical Pacific Ocean east of the dateline. This oceanic event is associated with a fluctuation of a global-scale tropical and subtropical surface pressure pattern called the Southern Oscillation. This coupled atmosphere-ocean phenomenon, with preferred time scales of 2 to about 7 years, is collectively known as the El Niño-Southern Oscillation. It is often measured by the surface pressure anomaly difference between Darwin and Tahiti and the sea surface temperatures in the central and eastern equatorial Pacific. During an ENSO event, the prevailing trade winds weaken, reducing upwelling and altering ocean currents such that the sea surface temperatures warm, further weakening the trade winds. This event has a great impact on the wind, sea surface temperature, and precipitation patterns in the tropical Pacific. It has climatic effects throughout the Pacific region and in many other parts of the world, through global teleconnections. The cold phase of ENSO is called La Niña.

Emissions scenario
A plausible representation of the future development of emissions of substances that are potentially radiatively active (e.g., greenhouse gases, aerosols), based on a coherent and internally consistent set of assumptions about driving forces (such as technological change, demographic and socioeconomic development) and their key relationships. Concentration scenarios, derived from emissions scenarios, are used as input to a climate model to compute climate projections. In the IPCC 1992 Supplementary Report, a set of emissions scenarios was presented, which were used as a basis for the climate projections in the IPCC Second Assessment Report. These emissions scenarios are referred to as the IS92 scenarios. In the IPCC Special Report on Emissions Scenarios, new emissions scenarios, the so-called SRES scenarios, were published. SRES scenarios (e.g., A1B, A1FI, A2, B1, B2) are used as a basis for some of the climate projections shown in Chapter 3 of this report.

Ensemble
A group of parallel model simulations used for climate projections. Variation of the results across the ensemble members gives an estimate of uncertainty. Ensembles made with the same model but different initial conditions only characterize the uncertainty associated with internal climate variability, whereas multi-model ensembles including simulations by several models also include the impact of model differences. Perturbed parameter ensembles, in which model parameters are varied in a systematic manner, aim to produce a more objective estimate of modeling uncertainty than is possible with traditional multi-model ensembles.

Evapotranspiration
The combined process of evaporation from the Earth's surface and transpiration from vegetation.

Exposure
The presence of people; livelihoods; environmental services and resources; infrastructure; or economic, social, or cultural assets in places that could be adversely affected.

External forcing
External forcing refers to a forcing agent outside the climate system causing a change in the climate system. Volcanic eruptions, solar variations, and anthropogenic changes in the composition of the atmosphere and land use change are external forcings.

Extratropical cyclone
Any cyclonic-scale storm that is not a tropical cyclone. Usually refers to a middle- or high-latitude migratory storm system formed in regions of large horizontal temperature variations. Sometimes called extratropical storm or extratropical low.

Extreme coastal high water (also referred to as extreme sea level)
Extreme coastal high water depends on average sea level, tides, and regional weather systems. Extreme coastal high water events are usually defined in terms of the higher percentiles (e.g., 90th to 99.9th) of a distribution of hourly values of observed sea level at a station for a given reference period.

Extreme weather or climate event
See Climate extreme.

Famine
Scarcity of food over an extended period and over a large geographical area, such as a country. Famines may be triggered by extreme climate events such as drought or floods, but can also be caused by disease, war, or other factors.

Flood
The overflowing of the normal confines of a stream or other body of water, or the accumulation of water over areas that are not normally submerged. Floods include river (fluvial) floods, flash floods, urban floods, pluvial floods, sewer floods, coastal floods, and glacial lake outburst floods.

Frozen ground
Soil or rock in which part or all of the pore water is frozen. Perennially frozen ground is called permafrost. Ground that freezes and thaws annually is called seasonally frozen ground.

Glacial lake outburst flood (GLOF)
Flood associated with outburst of glacial lake. Glacial lake outburst floods are typically a result of cumulative developments and occur (i) only once (e.g., full breach failure of moraine-dammed lakes), (ii) for the first time (e.g., new formation and outburst of glacial lakes), and/or (iii) repeatedly (e.g., ice-dammed lakes with drainage cycles, or ice fall).

Glacier
A mass of land ice that flows downhill under gravity (through internal deformation and/or sliding at the base) and is constrained by internal stress and friction at the base and sides. A glacier is maintained by accumulation of snow at high altitudes, balanced by melting at low altitudes or discharge into the sea.

Global climate model (also referred to as general circulation model, both abbreviated as GCM)
See Climate model.

Global surface temperature
The global surface temperature is an estimate of the global mean surface air temperature. However, for changes over time, only anomalies, as departures from a climatology, are used, most commonly based on the area-weighted global average of the sea surface temperature anomaly and land surface air temperature anomaly.

Governance
The way government is understood has changed in response to social, economic, and technological changes over recent decades. There is a corresponding shift from government defined strictly by the nation-state to a more inclusive concept of governance, recognizing the contributions of various levels of government (global, international, regional, local) and the roles of the private sector, of nongovernmental actors, and of civil society.

Greenhouse effect
Greenhouse gases effectively absorb thermal infrared radiation, emitted by the Earth's surface, by the atmosphere itself due to the same gases, and by clouds. Atmospheric radiation is emitted to all sides, including downward to the Earth's surface. Thus, greenhouse gases trap heat within the surface-troposphere system. This is called the greenhouse effect. Thermal infrared radiation in the troposphere is strongly coupled to the temperature of the atmosphere at the altitude at which it is emitted. In the troposphere, the temperature generally decreases with height. Effectively, infrared radiation emitted to space originates from an altitude with a temperature of, on average, -19°C, in balance with the net incoming solar radiation, whereas the Earth's surface is kept at a much higher temperature of, on average, 14°C. An increase in the concentration of greenhouse gases leads to an increased infrared opacity of the atmosphere and therefore to an effective radiation into space from a higher altitude at a lower temperature. This causes a radiative forcing that leads to an enhancement of the greenhouse effect, the so-called enhanced greenhouse effect.

Greenhouse gas
Greenhouse gases are those gaseous constituents of the atmosphere, both natural and anthropogenic, which absorb and emit radiation at specific wavelengths within the spectrum of thermal infrared radiation emitted by the Earth's surface, by the atmosphere itself, and by clouds. This property causes the greenhouse effect. Water vapor (H_2O), carbon dioxide (CO_2), nitrous oxide (N_2O), methane (CH_4), and ozone (O_3) are the primary greenhouse gases in the Earth's atmosphere. Moreover, there are a number of entirely human-made greenhouse gases in the atmosphere, such as the halocarbons and other chlorine- and bromine-containing substances, dealt with under the Montreal Protocol. Besides CO_2, N_2O, and CH_4, the Kyoto Protocol deals with the greenhouse gases sulfur hexafluoride (SF_6), hydrofluorocarbons (HFCs), and perfluorocarbons (PFCs).

Hazard
The potential occurrence of a natural or human-induced physical event that may cause loss of life, injury, or other health impacts, as well as damage and loss to property, infrastructure, livelihoods, service provision, and environmental resources.

Heat wave (also referred to as extreme heat event)
A period of abnormally hot weather. Heat waves and warm spells have various and in some cases overlapping definitions. See also Warm spell.

Holocene
The Holocene geological epoch is the latter of two Quaternary epochs, extending from about 11.6 thousand years before present to and including the present.

Human security
Human security can be said to have two main aspects. It means, first, safety from such chronic threats as hunger, disease, and repression. And second, it means protection from sudden and hurtful disruptions in the patterns of daily life – whether in homes, in jobs, or in communities. Such threats can exist at all levels of national income and development.

Hydrological cycle (also referred to as water cycle)
The cycle in which water evaporates from the oceans and the land surface, is carried over the Earth in atmospheric circulation as water vapor, condenses to form clouds, precipitates again as rain or snow, is intercepted by trees and vegetation, provides runoff on the land surface, infiltrates into soils, recharges groundwater, and/or discharges into streams and flows out into the oceans, and ultimately evaporates again from the oceans or land surface. The various systems involved in the hydrological cycle are usually referred to as hydrological systems.

Impacts
Effects on natural and human systems. In this report, the term 'impacts' is used to refer to the effects on natural and human systems of physical events, of disasters, and of climate change.

Indian Ocean Dipole (IOD)
Large-scale interannual variability of sea surface temperature in the Indian Ocean. This pattern manifests through a zonal gradient of tropical sea surface temperature, which in one extreme phase in boreal autumn shows cooling off Sumatra and warming off Somalia in the west, combined with anomalous easterlies along the equator.

Insurance/reinsurance
A family of financial instruments for sharing and transferring risk among a pool of at-risk households, businesses, and/or governments. See Risk transfer.

Landslide
A mass of material that has moved downhill by gravity, often assisted by water when the material is saturated. The movement of soil, rock, or debris down a slope can occur rapidly, or may involve slow, gradual failure.

Land surface air temperature
The air temperature as measured in well-ventilated screens over land at 1.5 to 2 m above the ground.

Land use and land use change
Land use refers to the total of arrangements, activities, and inputs undertaken in a certain land cover type (a set of human actions). The term land use is also used in the sense of the social and economic purposes for which land is managed (e.g., grazing, timber extraction, and conservation). Land use change refers to a change in the use or management of land by humans, which may lead to a change in land cover. Land cover and land use change may have an impact on the surface albedo, evapotranspiration, sources and sinks of greenhouse gases, or other properties of the climate system and may thus have radiative forcing and/or other impacts on climate, locally or globally.

Lapse rate
The rate of change of an atmospheric variable, usually temperature, with height. The lapse rate is considered positive when the variable decreases with height.

Latent heat flux
The flux of heat from the Earth's surface to the atmosphere that is associated with evaporation or condensation of water vapor at the surface; a component of the surface energy budget.

Likelihood
A probabilistic estimate of the occurrence of a single event or of an outcome, for example, a climate parameter, observed trend, or projected change lying in a given range. Likelihood may be based on statistical or modeling analyses, elicitation of expert views, or other quantitative analyses.

Local disaster risk management (LDRM)
The process in which local actors (citizens, communities, government, non-profit organizations, institutions, and businesses) engage in and have ownership of the identification, analysis, evaluation, monitoring, and treatment of disaster risk and disasters, through measures that reduce or anticipate hazard, exposure, or vulnerability; transfer risk; improve disaster response and recovery; and promote an overall increase in capacities. LDRM normally requires coordination with and support from external actors at the regional, national, or international levels. Community-based disaster risk management is a subset of LDRM where community members and organizations are in the center of decisionmaking.

Mass movement
Mass movement in the context of mountainous phenomena refers to different types of mass transport processes including landslides, avalanches, rock fall, or debris flows.

Mean sea level
Sea level measured by a tide gauge with respect to the land upon which it is situated. Mean sea level is normally defined as the average relative sea level over a period, such as a month or a year, long enough to average out transients such as waves and tides. See Sea level change.

Meridional overturning circulation (MOC)
Meridional (north-south) overturning circulation in the ocean quantified by zonal (east-west) sums of mass transports in depth or density layers. In the North Atlantic, away from the subpolar regions, the MOC (which is in principle an observable quantity) is often identified with the thermohaline circulation, which is a conceptual interpretation. However, it must be borne in mind that MOC can also include shallower, wind-driven overturning cells such as occur in the upper ocean in the tropics and subtropics, in which warm (less dense) waters moving poleward are transformed to slightly denser waters and subducted equatorward at deeper levels.

Mitigation (of disaster risk and disaster)
The lessening of the potential adverse impacts of physical hazards (including those that are human-induced) through actions that reduce hazard, exposure, and vulnerability.

Mitigation (of climate change)
A human intervention to reduce the sources or enhance the sinks of greenhouse gases.

Modes of climate variability
Natural variability of the climate system, in particular on seasonal and longer time scales, predominantly occurs with preferred spatial patterns

and time scales, through the dynamical characteristics of the atmospheric circulation and through interactions with the land and ocean surfaces. Such patterns are often called regimes, modes, or teleconnections. Examples are the North Atlantic Oscillation (NAO), the Pacific-North American pattern (PNA), the El Niño-Southern Oscillation (ENSO), the Northern Annular Mode (NAM; previously called the Arctic Oscillation, AO), and the Southern Annular Mode (SAM; previously called the Antarctic Oscillation, AAO).

Monsoon
A monsoon is a tropical and subtropical seasonal reversal in both the surface winds and associated precipitation, caused by differential heating between a continental-scale land mass and the adjacent ocean. Monsoon rains occur mainly over land in summer.

Nonlinearity
A process is called nonlinear when there is no simple proportional relation between cause and effect. The climate system contains many such nonlinear processes, resulting in a system with a potentially very complex behavior. Such complexity may lead to abrupt climate change. See also Predictability.

North Atlantic Oscillation (NAO)
The North Atlantic Oscillation consists of opposing variations in barometric pressure near Iceland and near the Azores. It therefore corresponds to fluctuations in the strength of the main westerly winds across the Atlantic into Europe, and thus to fluctuations in the embedded cyclones with their associated frontal systems.

Northern Annular Mode (NAM)
A winter fluctuation in the amplitude of a pattern characterized by low surface pressure in the Arctic and strong mid-latitude westerlies. NAM has links with the northern polar vortex into the stratosphere. Its pattern has a bias to the North Atlantic and has a large correlation with the North Atlantic Oscillation.

Pacific Decadal Oscillation (PDO)
The pattern and time series of the first empirical orthogonal function of sea surface temperature over the North Pacific north of 20°N. PDO broadened to cover the whole Pacific Basin is known as the Interdecadal Pacific Oscillation (IPO). The PDO and IPO exhibit virtually identical temporal evolution.

Parameterization
In climate models, this term refers to the technique of representing processes that cannot be explicitly resolved at the spatial or temporal resolution of the model (sub-grid scale processes) by relationships between model-resolved larger-scale flow and the area- or time-averaged effect of such sub-grid scale processes.

Percentile
A percentile is a value on a scale of 100 that indicates the percentage of the data set values that is equal to or below it. The percentile is often used to estimate the extremes of a distribution. For example, the 90th (10th) percentile may be used to refer to the threshold for the upper (lower) extremes.

Permafrost
Ground (soil or rock and included ice and organic material) that remains at or below 0°C for at least 2 consecutive years.

Predictability
The extent to which future states of a system may be predicted based on knowledge of current and past states of the system.

Probability density function (PDF)
A probability density function is a function that indicates the relative chances of occurrence of different outcomes of a variable. The function integrates to unity over the domain for which it is defined and has the property that the integral over a sub-domain equals the probability that the outcome of the variable lies within that sub-domain. For example, the probability that a temperature anomaly defined in a particular way is greater than zero is obtained from its PDF by integrating the PDF over all possible temperature anomalies greater than zero. Probability density functions that describe two or more variables simultaneously are similarly defined.

Projection
A projection is a potential future evolution of a quantity or set of quantities, often computed with the aid of a model. Projections are distinguished from predictions in order to emphasize that projections involve assumptions concerning, for example, future socioeconomic and technological developments that may or may not be realized, and are therefore subject to substantial uncertainty. See also Climate projection and Climate prediction.

Proxy climate indicator
A proxy climate indicator is a local record that is interpreted, using physical and biophysical principles, to represent some combination of climate-related variations back in time. Climate-related data derived in this way are referred to as proxy data. Examples of proxies include pollen analysis, tree ring records, characteristics of corals, and various data derived from ice cores. The term 'proxy' can also be used to refer to indirect estimates of present-day conditions, for example, in the absence of observations.

Radiative forcing
Radiative forcing is the change in the net, downward minus upward, irradiance (expressed in W m^{-2}) at the tropopause due to a change in an external driver of climate change, such as, for example, a change in the concentration of carbon dioxide or the output of the Sun. Radiative forcing is computed with all tropospheric properties held fixed at their unperturbed values, and after allowing for stratospheric temperatures, if perturbed, to readjust to radiative-dynamical equilibrium. Radiative

forcing is called instantaneous if no change in stratospheric temperature is accounted for. For the purposes of this report, radiative forcing is further defined as the change relative to the year 1750 and, unless otherwise noted, refers to a global and annual average value. Radiative forcing is not to be confused with cloud radiative forcing, a similar terminology for describing an unrelated measure of the impact of clouds on the irradiance at the top of the atmosphere.

Reanalysis
Reanalyses are atmospheric and oceanic analyses of temperature, wind, current, and other meteorological and oceanographic quantities, created by processing past meteorological and oceanographic data using fixed state-of-the-art weather forecasting models and data assimilation techniques. Using fixed data assimilation avoids effects from the changing analysis system that occur in operational analyses. Although continuity is improved, global reanalyses still suffer from changing coverage and biases in the observing systems.

Relative sea level
See Mean sea level.

Resilience
The ability of a system and its component parts to anticipate, absorb, accommodate, or recover from the effects of a hazardous event in a timely and efficient manner, including through ensuring the preservation, restoration, or improvement of its essential basic structures and functions.

Return period
An estimate of the average time interval between occurrences of an event (e.g., flood or extreme rainfall) of (or below/above) a defined size or intensity.

Return value
The highest (or, alternatively, lowest) value of a given variable, on average occurring once in a given period of time (e.g., in 10 years).

Risk transfer
The process of formally or informally shifting the financial consequences of particular risks from one party to another whereby a household, community, enterprise, or state authority will obtain resources from the other party after a disaster occurs, in exchange for ongoing or compensatory social or financial benefits provided to that other party.

Runoff
That part of precipitation that does not evaporate and is not transpired, but flows through the ground or over the ground surface and returns to bodies of water. See Hydrological cycle.

Scenario
A plausible and often simplified description of how the future may develop based on a coherent and internally consistent set of assumptions about driving forces and key relationships. Scenarios may be derived from projections, but are often based on additional information from other sources, sometimes combined with a narrative storyline. See also Climate scenario and Emissions scenario.

Sea level change
Changes in sea level, globally or locally, due to (i) changes in the shape of the ocean basins, (ii) changes in the total mass and distribution of water and land ice, (iii) changes in water density, and (iv) changes in ocean circulation. Sea level changes induced by changes in water density are called steric. Density changes induced by temperature changes only are called thermosteric, while density changes induced by salinity changes are called halosteric. See also Mean sea level.

Sea surface temperature (SST)
The sea surface temperature is the temperature of the subsurface bulk temperature in the top few meters of the ocean, measured by ships, buoys, and drifters. From ships, measurements of water samples in buckets were mostly switched in the 1940s to samples from engine intake water. Satellite measurements of skin temperature (uppermost layer; a fraction of a millimeter thick) in the infrared or the top centimeter or so in the microwave are also used, but must be adjusted to be compatible with the bulk temperature.

Sensible heat flux
The flux of heat from the Earth's surface to the atmosphere that is not associated with phase changes of water; a component of the surface energy budget.

Significant wave height
The average height of the highest one-third of the wave heights (trough to peak) from sea and swell occurring in a particular time period.

Soil moisture
Water stored in or at the land surface and available for evapotranspiration.

Southern Annular Mode (SAM)
The fluctuation of a pattern like the Northern Annular Mode, but in the Southern Hemisphere.

SRES scenarios
See Emissions scenario.

Storm surge
The temporary increase, at a particular locality, in the height of the sea due to extreme meteorological conditions (low atmospheric pressure and/or strong winds). The storm surge is defined as being the excess above the level expected from the tidal variation alone at that time and place.

Storm tracks
Originally, a term referring to the tracks of individual cyclonic weather systems, but now often generalized to refer to the regions where the

main tracks of extratropical disturbances occur as sequences of low (cyclonic) and high (anticyclonic) pressure systems.

Streamflow
Water flow within a river channel, for example, expressed in $m^3\ s^{-1}$. A synonym for river discharge.

Subsidiarity
The principle that decisions of government (other things being equal) are best made and implemented, if possible, at the lowest most decentralized level closest to the citizen. Subsidiarity is designed to strengthen accountability and reduce the dangers of making decisions in places remote from their point of application. The principle does not necessarily limit or constrain the action of higher orders of government, it merely counsels against the unnecessary assumption of responsibilities at a higher level.

Surface temperature
See Global surface temperature, Land surface air temperature, and Sea surface temperature.

Sustainable development
Development that meets the needs of the present without compromising the ability of future generations to meet their own needs.

Transpiration
The evaporation of water vapor from the surfaces of leaves through stomata.

Transformation
The altering of fundamental attributes of a system (including value systems; regulatory, legislative, or bureaucratic regimes; financial institutions; and technological or biological systems).

Tropical cyclone
The general term for a strong, cyclonic-scale disturbance that originates over tropical oceans. Distinguished from weaker systems (often named tropical disturbances or depressions) by exceeding a threshold wind speed. A tropical storm is a tropical cyclone with one-minute average surface winds between 18 and 32 $m\ s^{-1}$. Beyond 32 $m\ s^{-1}$, a tropical cyclone is called a hurricane, typhoon, or cyclone, depending on geographic location.

Uncertainty
An expression of the degree to which a value or relationship is unknown. Uncertainty can result from lack of information or from disagreement about what is known or even knowable. Uncertainty may originate from many sources, such as quantifiable errors in the data, ambiguously defined concepts or terminology, or uncertain projections of human behavior. Uncertainty can therefore be represented by quantitative measures, for example, a range of values calculated by various models, or by qualitative statements, for example, reflecting the judgment of a team of experts. See also Likelihood and Confidence.

Urban heat island
The relative warmth of a city compared with surrounding rural areas, associated with changes in runoff, the concrete jungle effects on heat retention, changes in surface albedo, changes in pollution and aerosols, and so on.

Vulnerability
The propensity or predisposition to be adversely affected.

Warm days/warm nights
Days where maximum temperature, or nights where minimum temperature, exceeds the 90th percentile, where the respective temperature distributions are generally defined with respect to the 1961-1990 reference period.

Warm spell
A period of abnormally warm weather. Heat waves and warm spells have various and in some cases overlapping definitions. See also Heat wave.

ANNEX III

Acronyms

Acronyms

AAO	Antarctic Oscillation
ADAPT	Assessment & Design for Adaptation to Climate Change: A Prototype Tool
AMO	Atlantic Multi-decadal Oscillation
AO	Arctic Oscillation
AR5	Fifth Assessment Report
CAPE	Convective Available Potential Energy
CAT	catastrophic risk
CBA	cost-benefit analysis or community-based adaptation
CBD	Convention on Biological Diversity
CBDR	common but differentiated responsibilities and respective capabilities
CBO	community-based organization
CCA	climate change adaptation
CCRIF	Caribbean Catastrophe Risk Insurance Facility
CCSP	Climate Change Science Program (US)
CDD	Consecutive Dry Days
CDM	Clean Development Mechanism
CEE	Centre for Environment Education
CEI	Climate Extremes Index
C-ERA-40	Corrected ERA-40 reanalysis
CFR	case fatality rate
CH_4	methane
CMIP3	Coupled Model Intercomparison Project 3
CO_2	carbon dioxide
COP	Conference of the Parties
CPP	cyclone preparedness program
CRED	Centre for Research on the Epidemiology of Disasters
CSA	Canadian Standards Association
CSO	civil society organization
CSR	corporate social responsibility
DDI	Disaster Deficit Index
DFID	Department for International Development (UK)
DJF	December-January-February
DRM	disaster risk management
DRR	disaster risk reduction
DRRM	disaster risk reduction management
EbA	ecosystem-based adaptation
EBRD	European Bank for Reconstruction and Development
EDI	Ethiopia Drought Index
ELF	Emergency Liquidity Facility
EM-DAT	Emergency Events Database
ENSO	El Niño-Southern Oscillation
ERA-40	European Centre for Medium Range Weather Forecasts 40-year reanalysis
EVT	extreme value theory
EWS	early warning system
FAO	Food and Agriculture Organization
FONDEN	Fund for Natural Disasters
GAR	Global Assessment Report on Disaster Risk Reduction
GCM	global climate model
GDP	gross domestic product
GEC	global environmental change
GEF	Global Environment Facility
GFCS	Global Framework on Climate Services
GFDRR	Global Facility for Disaster Reduction and Recovery
GHG	greenhouse gas
GIS	geographic information system
GLOF	glacial lake outburst flood
GNCSODR	Global Network of Civil Society Organisations for Disaster Reduction
GPS	Global Positioning System
GSDI	Global Spatial Data Infrastructure
H_2O	water
HARS	Heat Action Response System
HDI	Human Development Index
HEP	hydroelectric power
HFA	Hyogo Framework for Action
HFC	hydrofluorocarbon
HWDI	Heat Wave Duration Index
HWS	Heat Warning System
IADB	Inter-American Development Bank
IAM	integrated assessment model
ICSU	International Council for Science
ICT	information and communication technology
ICZM	integrated coastal zone management
IDMC	Internal Displacement Monitoring Centre
IDNDR	International Decade for Natural Disaster Reduction
IDP	internally displaced person
IDRL	International Disaster Response Law
IHL	international humanitarian law
IOD	Indian Ocean Dipole
IPO	Inter-decadal Pacific Oscillation
IRDR	Integrated Research on Disaster Risk program
ISSC	International Social Science Council
ITCZ	Inter-Tropical Convergence Zone
IWRM	integrated water resource management
JJA	June-July-August
LA RED	Red de Estudios Sociales en Prevención de Desastres en América Latina
LDC	least-developed country
LDCF	Least Developed Countries Fund
LDRM	local disaster risk management
LEED	Leadership in Energy and Environmental Design
LIDAR	Light Detection and Ranging
MDGs	Millennium Development Goals
MFI	micro-finance institution
MJO	Madden-Julian Oscillation
MLP	multi-level perspective
MME	Multi-Model Ensemble (CMIP3)
MOC	meridional overturning circulation
MPBI	Indonesian Society for Disaster Management
MSLP	mean sea level pressure
N_2O	nitrous oxide

NAM	Northern Annular Mode	SDLE	Prepare, Stay and Defend, or Leave Early
NAO	North Atlantic Oscillation	SDMP	School Disaster Management Plans
NAPA	National Adaptation Programme of Action	SECO	Swiss State Secretariat for Economic Affairs
NaTech	Natural Hazard Triggering a Technological Disaster	SF_6	sulfur hexafluoride
NDMO	National Disaster Management Office	SHELDUS	Spatial Hazard Events and Losses Database for the United States
NECJOGHA	Network of Climate Journalists of the Greater Horn of Africa	SIDS	small island developing states
NGO	nongovernmental organization	SIS	small island states
NHC	National Hurricane Committee	SMA	soil moisture anomaly
NIDM	National Disaster Management Institute	SMEs	small- and medium-sized enterprises
NMHS	national meteorological and hydrological service	SOI	Southern Oscillation Index
NTR	non-tide residuals	SPA	Strategic Priority 'Piloting an Operational Approach to Adaptation'
NU	Nunavut		
NWP	Nairobi Work Programme	SPEI	Standardized Precipitation-Evapotranspiration Index
NWT	Northwest Territories	SPI	Standard Precipitation Index
O_3	ozone	SRES	Special Report on Emissions Scenarios
OCHA	United Nations Office for the Coordination of Humanitarian Affairs	SST	sea surface temperature
		SWH	significant wave height
ODA	official development assistance	UN	United Nations
OECD	Organisation for Economic Co-operation and Development	UNCCD	United Nations Convention to Combat Desertification
		UNDP	United Nations Development Programme
OFDA	Office of Foreign Disaster Assistance	UNFCCC	United Nations Framework Convention on Climate Change
OLR	outgoing longwave radiation		
PAR	pressure and release	UNISDR	United Nations International Strategy for Disaster Reduction
PDF	probability density function		
PDO	Pacific Decadal Oscillation	WDSI	Warm Spell Duration Index
PDSI	Palmer Drought Severity Index	WFP	World Food Programme
PESETA	Projection of Economic impacts of climate change in Sectors of the European Union based on boTtom-up Analysis	WHO	World Health Organization
		WMO	World Meteorological Organization
		YT	Yukon Territory
PFC	perfluorocarbon		
PICs	Pacific Island Countries and Territories		
PNA	Pacific North American pattern		
POPs	persistent organic pollutants		
PPEA	Precipitation Potential Evaporation Anomaly		
PPP	public-private partnership		
Pr	precipitation		
PSNP	Productive Safety Net Programme		
PTSD	post-traumatic stress disorder		
PVI	Prevalent Vulnerability Index		
RAC	Regional Adaptation Collaborative		
RANET	RAdio and InterNET		
RCM	regional climate model		
REDD	reduced carbon emissions from deforestation and forest degradation		
REDD+	reduced carbon emissions from deforestation and forest degradation, maintaining/enhancing carbon stocks, and promoting sustainable forest management		
RMI	Republic of the Marshall Islands		
SAM	Southern Annular Mode		
SAMS	South American Monsoon System		
SCCF	Special Climate Change Fund		

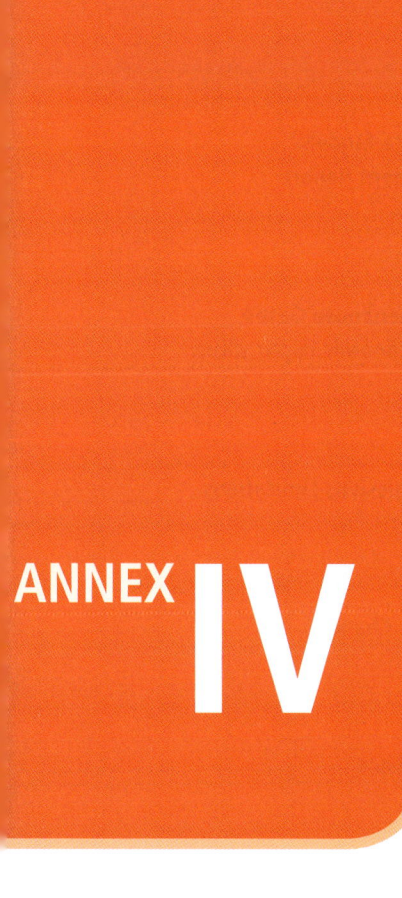

ANNEX IV

List of Major IPCC Reports

List of Major IPCC Reports

Climate Change: The IPCC Scientific Assessment
Report of the IPCC Scientific Assessment Working Group
1990

Climate Change: The IPCC Impacts Assessment
Report of the IPCC Impacts Assessment Working Group
1990

Climate Change: The IPCC Response Strategies
Report of the IPCC Response Strategies Working Group
1990

Climate Change 1992: The Supplementary Report to the IPCC Scientific Assessment
Report of the IPCC Scientific Assessment Working Group
1992

Climate Change 1992: The Supplementary Report to the IPCC Impacts Assessment
Report of the IPCC Impacts Assessment Working Group
1992

Climate Change: The IPCC 1990 and 1992 Assessments – IPCC First Assessment Report Overview and Policymaker Summaries, and 1992 IPCC Supplement
1992

Climate Change 1994: Radiative Forcing of Climate Change and an Evaluation of the IPCC IS92 Emission Scenarios
IPCC Special Report
1994

Climate Change 1995: The Science of Climate Change
Contribution of Working Group I
to the IPCC Second Assessment Report
1996

Climate Change 1995: Impacts, Adaptations, and Mitigation of Climate Change: Scientific-Technical Analyses
Contribution of Working Group II
to the IPCC Second Assessment Report
1996

Climate Change 1995: Economic and Social Dimensions of Climate Change
Contribution of Working Group III
to the IPCC Second Assessment Report
1996

Climate Change 1995: IPCC Second Assessment Synthesis of Scientific-Technical Information Relevant to Interpreting Article 2 of the UN Framework Convention on Climate Change
1996

Technologies, Policies, and Measures for Mitigating Climate Change
IPCC Technical Paper I
1996

An Introduction to Simple Climate Models used in the IPCC Second Assessment Report
IPCC Technical Paper II
1997

Stabilization of Atmospheric Greenhouse Gases: Physical, Biological, and Socio-Economic Implications
IPCC Technical Paper III
1997

Implications of Proposed CO_2 Emissions Limitations
IPCC Technical Paper IV
1997

The Regional Impacts of Climate Change
IPCC Special Report
1998

Aviation and the Global Atmosphere
IPCC Special Report
1999

Methodological and Technological Issues in Technology Transfer
IPCC Special Report
2000

Land Use, Land-Use Change, and Forestry
IPCC Special Report
2000

Emissions Scenarios
IPCC Special Report
2000

Climate Change 2001: The Scientific Basis
Contribution of Working Group I
to the IPCC Third Assessment Report
2001

Climate Change 2001: Impacts, Adaptation, and Vulnerability
Contribution of Working Group II
to the IPCC Third Assessment Report
2001

Climate Change 2001: Mitigation
Contribution of Working Group III
to the IPCC Third Assessment Report
2001

Annex IV — List of Major IPCC Reports

Climate Change 2001: IPCC Third Assessment Synthesis Report
2001

Climate Change and Biodiversity
IPCC Technical Paper V
2002

**Safeguarding the Ozone Layer and the Global Climate System:
Issues Related to Hydrofluorocarbons and Perfluorocarbons**
IPCC Special Report
2005

Carbon Dioxide Capture and Storage
IPCC Special Report
2005

Climate Change 2007: The Physical Science Basis
Contribution of Working Group I
to the IPCC Fourth Assessment Report
2007

Climate Change 2007: Impacts, Adaptation, and Vulnerability
Contribution of Working Group II
to the IPCC Fourth Assessment Report
2007

Climate Change 2007: Mitigation of Climate Change
Contribution of Working Group III
to the IPCC Fourth Assessment Report
2007

Climate Change 2007: Synthesis Report
2008

Climate Change and Water
IPCC Technical Paper VI
2008

Renewable Energy Sources and Climate Change Mitigation
IPCC Special Report
2011

**Managing the Risks of Extreme Events and Disasters
to Advance Climate Change Adaptation**
IPCC Special Report
2012

Enquiries: IPCC Secretariat, c/o World Meteorological Organization, 7 bis, Avenue de la Paix, Case Postale 2300, CH - 1211 Geneva 2, Switzerland

Index

Index | Key

Terms defined in the glossary are marked with an asterisk(*). Bold page numbers indicate page spans for entire chapters. Italicized page numbers denote tables, figures, and boxed material.

A

Abrupt climate change[*], 122, 458
Access
    to resources, 454-456
    to technology, 447-448
ADAPT (Assessment & Design for Adaptation Tool), 417
Adaptation[*], *6*, 36, 443
    barriers to, 451-453
    community-based, 295, *300*, 321, *322*
    continuum/process, 323, *324*
    coping and, 50-56
    dimensions of, *51*
    DRM and, 37-39, 47-48, 398, 408-409, 443-462
    DRM as, 443-450
    DRR and, 237, 396-397, 423-424, 443-446
    ecosystem-based, 343, *352*, 357, 370-371, *371*, 445-446
    effectiveness and tradeoffs, 439-440, 448-450
Adaptation, inequalities, 10, 294, 313-317
Adaptation[*]
    integration with DRM, 11, 28, 50, 355-357, 396-397, 425-427, 439, 450-454, 469-471
    interactions with DRM and mitigation, 20, 440, 458-462
    international level, 396-397, 411
    learning and, 27-28, 50, 53-56
    local knowledge and, 293, 311-312
    local level, 293-295, 319-320, *319*
    mainstreaming, 348, 355-357
    maladaptative actions, 55, 368
    management options, 16-20, *18-19*
    past experience and, 10-11, 38-39
    portfolio of approaches, 17, 295
    robust, 56
    sector-based, *352-354*, 357
    sustainability and, 444
    synergies with DRM, 48-50, 68, 357, 469-471
    technology and, 414-415, *415*
    UNFCCC commitments on, *407*
Adaptation costs, 264-269, 412-414
    assessment of, 273-274, *273*, *274*
    in developing countries, 412, *412*
    evaluation of, 236, 266-269
    funding for, 406-407, 411
Adaptation deficit, 265
Adaptation Fund, 406-407, 411, 413
Adaptation limits, 319-320, *319*
Adaptation options, *352-354*
Adaptation planning, 349-357, 443-444
Adaptive capacity[*], 6-7, 33, 72-76, *73*, 443
    local, 294, 308-313, 315
    post-disaster recovery/reconstruction and, 293, 439
Adaptive management, 343, 377-379, 467
Adjustments, 450-454

Afforestation, *352*, 355
Africa, 253-254
    adaptation costs, 274, *274*
    agricultural impacts, 246-247, 253
    drought, *19*, 170, 171, 172, 174, 175, 253
    floods, 176, 253-254, 297
    food security, 368
    monsoons, 153, 154
    precipitation, 142, 143, *145-146*, 147, *193*, *199*, 253-254
    sand/dust storms, 190, 254
    temperature, 134, *135*, *139-140*, 141, *193*, *199*
Africa Union, 401
Age/aging, 234-235, 314, *315*
Agriculture
    crop insurance, 323
    DRM/adaptation options, *352*
    food security and, 368-369
    impacts, 235, 246-247, 272-273
    regional impacts, 253, 259
    temperature impacts, 247, 255
    vulnerability/exposure, 235
Alaska, *191*, *197*
    drought, 170
    precipitation, *145-146*, *191*, *197*
    temperature, *135*, *139-140*, *191*, *197*
Alpine regions, *251*
Amazon region, *194*, *200*, *371*
Antarctica, 261-263
    ice sheet, 179, 188
Anthropogenic influences
    changes in extremes, 9, 112, 125-126, 268
    changes in precipitation, 143-144, 149
    changes in temperature, 135-136, 141
    climate change, 40
Aral Sea basin, 239
Arctic region, 261-263
Arid environments, 143, 149, *167*, 190
    case studies, 498-502
Asia, 254-255
    adaptation costs, 273, *274*
    adaptation technologies, 415, *415*
    Disaster Reduction Hyperbase, 415
    drought, 171, 255
    floods, 176, 177, 254-255
    monsoons, 154
    precipitation, 142-143, *145-146*, 147-148, *194-195*, *201-202*
    sand/dust storms, 190
    temperature, 135, *135*, *139-140*, 141, *194-195*, *201-202*, 255
    wildfires, 255
Assets, 315-316
Assistance
    humanitarian, 10, 293, 299-300
    international, 400
Atlantic Multi-decadal Oscillation (AMO)[*], 153, 157
Atolls, 263-264, *263*
    community relocation, 300-301
Attribution
    of changes in extremes, 112, *119-120*, 125-128, 268
    of changes in precipitation, 143-144
    of changes temperature, 135-136, 141
    of extreme events, 40
    human-induced changes, 9, 12, 125-126
    of impacts, 9, 268-269
    multi-step, 127-128
    single-step, 127-128

Auckland, 260, 261
Australia, 260-261
    case study (Victoria), 496-498
    coastal impacts, 183, 185, 186, 261
    drought, 170, 171, 172, 174, 261-262
    heat waves, 496-498
    land use change, 261
    precipitation, 142, 143, *145-146*, 147, 148-149, *195*, *202*, 261
    temperature, 134, 135, *135*, *139-140*, 141, *195*, *202*
    waves, 182
    wildfire, *239*, 261, 496-498
    wind, 150
Avalanches, 187-188, 189
    See also Landslides

B

Bali Action Plan, 398, 403, 407, 414
Bangladesh
    cyclones, 502-505, *503*
    disaster preparedness, 305, 465
    early warning system, 405
    floods, 254, 461
Barriers to adaptation, 451-453
Baseline/reference[*], 117, 364
Belmont Challenge, 425-426
Bilateral/multilateral agencies, 348-349
Biodiversity, 343, 370, *371*, 410, 446
Bottom-up mechanisms/approach, 266-267, *346*, 350-351, *350*, 396, 427
Brazil, *194*, *200*
    dams, *304*
    drought, 172, 174, 175
    precipitation, *145-146*, 148, *194*, *200*
    public awareness initiative, 527-528
    temperature, *139-140*, *194*, *200*
Building codes, 305, 317, *353*
    LEED standards, 461
    national, 347, 367-368
Building measures, 304-305
    wind-resistant building, 416

C

Cambodia, 526
Canada, 258-260
    case study, 514-517
    coastal impacts, 183
    CSA (Infrastructure in Permafrost) guide, 367
    drought, 170
    floods, 176, 177
    forest fires, 252, 259
    precipitation, 142, 143, *145-146*, 148, *191*, *197*
    temperature, *135*, 138, *139-140*, *191*, *197*
    wind, 150
Cancun Agreements, 397, 398, 403, 408, 413, 414
Capabilities, 33
Capacity[*], 33, 72-76, *73*
    availability and limitations, 454-456
Capacity building, 33, 308-313, 415
Capacity needs, 74-76
Carbon dioxide (CO_2)[*], 244
Carbon sequestration, 307, 370
Carbon sink/source, 244
Caribbean, *184*, 263
    adaptation costs, 273, *274*
    adaptation strategies, *371*
    coastal impacts, 183
    hurricanes, *19*
    insurance/risk pool, 372, 400, 419, 524-525

Index

resilience building, *378*
tourism impacts, 251, *251*
Case studies, 487-542
synthesis of, *491-492*, 529-530
Caste/class, 454
Catastrophic risk bonds, 419-420, 465
Catchments[*], 41, 113, 177-178, 242
Cayman Islands, *378*
Central America, 255-256
disaster losses, 256
drought, 172, 175
hurricanes, *503*, 504-505
precipitation, 142, *145-146*, *191*, *196-197*, *201*
temperature, *135*, *139-140*, *191*, *196-197*, *201*
Central Asia, 195
Children, 314, 454, *455*
China
adaptation technologies, 415
drought, 174
floods, 254-255
national disaster management, *374*
precipitation, 143
sand/dust storms, 190
temperature, *135*
tropical cyclones, 254
vulnerability/exposure, 347-348
wind, 150, 152
Cholera, 316, *410*, 507-510
Circulation (Walker/Hadley), 149, 151, 154
Cities, 248, 317, 460-461
DRM/adaptation options, *353*
megacities, 294, 317, 510-512
See also Urban areas
Civil society organizations, 348, 404, 409-410
Clausius-Clapeyron relationship[*], 126, 143-144
Clean development mechanisms, 411, 413
Climate[*]
global mean, 121-122
is it becoming more extreme? (FAQ), *124-125*
Climate change[*], **25-64**, 444
abrupt, 122
attribution of impacts to, 9, 268-269
changes in extremes and, 7, 111, 115, *127*
concepts/definitions, 30-37
defined, *5*, 29
DRM and, 27-28, 37-39, 375-380, *376-377*
DRM challenges, 46-47
Climate change adaptation. *See* Adaptation
Climate Change Green Fund, 397
Climate change mitigation. *See* Mitigation
Climate events
categories of, 115
disasters and, 115-118
interactions of, 238-239, *239*
Climate extremes[*], **65-290**
changes in, 8-9, **109-230**
context, 4-7, *4*
costs of, 264-274
definition and analysis, *5*, 111, *116-117*, 237
factors and confidence in, 111-112, 120-121
impacts on humans and ecosystems, **231-290**
impacts on physical environment, 8-9, **167-190**
indices of, *116-117*, *125*
managing changing risks, 16-20
methods and requirements, 122-133
natural and socioeconomic systems, 237-239
observed changes, 7-9, 111-112, *119-120*, 133-152
past experience with, 10-11
phenomena related to, *119*, 152-166

projected changes, 11-16, 112-114, *119-120*, 133-152
regional and global climate, 121-122
regionally based impacts, 252-264
unprecedented, 7, 111
vulnerability/exposure and, **65-108**, 239-264
See also Extreme events
Climate feedback[*], 112, 118-120
Climate information, 421-422
Climate models[*], 13, 112-113, 128-133, 147
ensembles, 131-133
modeling tools, 464
planning approach, 350-351, *350*
See also Projections
Climate modes[*], 113, 155-158
observed and projected changes, 15-16, *119*
uncertainty in projections of, 113
See also El Niño-Southern Oscillation
Climate projections[*], *119-120*
Climate scenarios[*]. *See* Scenarios
Climate services, 409, 422
Climate variability, 7, 115, 155-158
Coastal erosion, 113, 182-183, 185, 186
Coastal floods, 259-260
Coastal impacts, 182-186
case study, 510-512
DRM/adaptation options, *352*
exposure and, *249*, 258
extreme high water, 9, 15, *18*, 113, 178, 182
observed changes, *120*, 183
projected changes, 113, *120*, 183-186
regional impacts, 254-255, 256-257, 259-260, 263
sea level and, 182, 183, *184*, 185
waves and, 180-182
Coastal inundation, *18*, 113, 182-183, *184-185*, 185, 248, *249*
Coastal settlements, 7-8, 235, 248, *249*, 258, 460
megacities (case study), 510-512
Cold climate regions (case study), 514-517
Cold days/cold nights[*], 8, *116*, 134, 135, *137*, *138*, 141
regional projections, *191-202*
Cold spell, 134
Collective action, 309-310, *309*, *321*
Colombia, 519-522
Communication, 302-304
gaps, 425
national systems, 349, *376*
of risk, 17, 67, 95, 294, 302-304, *303*, *376*
technologies, 422-423
Community-based adaptation, 295, *300*, 321, *322*
Community-based DRM[*], 308-313
Community-based DRR, 310, 321
Community-based organizations, 348
Complex Bayesian framework, 133
Complex systems, 46, 48, 53
Complexity, 27, 42-44, 53
Compound events, 118
Conference of the Parties (COP), 406
Confidence[*], 8, *21*, 112, 120-121, *132*
Conflict/warfare, 297
Contingent credit, 523
Contingent liabilities, 361, *361*
Cooperation, 401, 402
Coordination (across scales and sectors), 342, 356, 358-360, *360*, 439
international, 408-409
national, 358-360, *360*
Copenhagen Accord, 408, 413

Coping[*], 33, 50-56, 450-451
dimensions of, *51*
local, 294, 298-301
Coping capacity[*], 6-7, 51-53, 72-76, *73*
Coping range, 52-53
Coral reefs, *185*, 263
Corporate social responsibility, 347
Cost-benefit analysis, 267-268
Costa Rica, 252
Costs, 264-274
adaptation costs, 264-265, 266-269, 273-274, *273*, *274*
assessment of, 269-274
damage costs, 264
direct costs, 264, 266, 317
DRM costs, 317-319
economic costs, 267
evaluating (estimating), 236, 266-269
financial costs, 267
framing of, 264-265
global and regional costs, 269-271, *270*, *271*, *272*
indirect costs, 264, 266, 317
intangibles, 264, 266, 317
investment costs, 267
local DRM costs, 317-319
methods for evaluating, 266-269
observed increase in, 270-271, *271*
potential damage costs, 264
regional costs, 256, 270, *270*
residual damage costs, 264
top-down vs bottom-up approach, 266-267
uncertainty and, 274
Crop insurance, 524
Cultural dimensions, 7, 84-85
Cultural heritage loss, 317, *318*, 319
Cultural norms and values, 84-85, 309, 310-311
Culture of safety, 362-366
Cumulative impacts, 6-7, 38, 67, 69
Customary law, 402
Cyclones, 41, 158-166
case study, 502-505
extratropical[*], *119*, 163-166
economic losses, 272, *272*
observed changes, 163-166
poleward shift of, 8, 164-166
projected changes, 13, 272, *272*
tropical[*], *119*, 150, 151, 158-163, 502-505
economic losses, 235, 254, 271, *272*
exposure to, *240*
forecasting/warnings, 416-417
impacts, 248
observed changes, 159-161, 163
projected changes, 13, 161-163, 271, *272*
regional impacts, 254

D

Dams, 176, *304*, 415
Databases, 364, 415, 423-424, 464
Debt, *93*
Decentralization, 28, 46, 312-313, 360, 464
Decisionmaking, 67, 342
decentralized, 28
local, 305-306, 308, 310
maps of, 325
support tools, 463-466
tradeoffs in, 448
Definitions, *5*
See also Glossary
Deforestation, 238, 252, 446

575

Index

Demographic changes, 80, 234-235
Desertification, 190, 402
DesInventar database, 423-424
Developed countries
    economic/disaster losses in, 9, 234, 265
    mainstreaming, 356-357
    urban capacities, 460
    vulnerability, 78, 265
Developing countries
    adaptation costs, 412, *412*
    adaptation funding, 406-407, 411, 413
    disaster preparedness, 369
    economic/disaster losses in, 9, 234, 265, 269-270, 400
    infrastructure, 367-368
    insurance (case study), 522-525
    macroeconomic issues, 344, 412
    mainstreaming, 357
    urban capacities, 460
    vulnerability, 77-78, 265, 269, 441
Development, 10, 11, 27
    disaster risk and, 10, 27-28, 265-266
    extreme events/impacts and, 265-266
    international, *410*
    land use changes and, 238
    planning, 439, 443-444, 460-461
    post-disaster, 293, 439
    setbacks, *410*
    skewed, exposure/vulnerability and, 10, 67, 70
    sustainability and, 20, **437-486**
    values and interests in, 440, 446-447
    vulnerability and, 67, 70, 78-80, 265
    See also Sustainable development
Development pathways
    disaster risk and, 27, 293
    global, 10, 396
    vulnerability/exposure and, 78-80
Dhaka, 252, 461, 511
Diarrhea, 252, 297, 506
Dikes/levees, 11, *52*, 55, 68, 305
Direct losses, 264, 266, 317
Disaster[*], 27, 31, 39-44
    cycle, 35
    defined, *5*, 31, 237
    humanitarian relief for, 10, 293, 299-300
    impacts of, 31, 32, 42
    weather/climate related to, 115-122
Disaster costs, 264-274, *270*, *271*, *272*
Disaster databases, 364, 423-424
Disaster Deficit Index (DDI), *93*
Disaster losses, 9, 11-16
    See also Economic losses/impacts
Disaster management[*], 35
Disaster mitigation, 36
Disaster preparedness, 36, 364-366, *376-377*
Disaster prevention, 36, 69
Disaster Reduction Hyperbase, 415
Disaster risk[*], 10, **25-64**
    concept/relationships, 31-32, *31*, 44, 444
    continuum, 35, 69
    defined, *5*
    development and, 27-28
    future, anticipating and responding to, 302
    increase in, 29, 405
    risk accumulation, 95-96
Disaster risk management (DRM)[*], 16-20, **291-435**
    adaptation and, 37-39, 47-48, 295, 408-409, 443-462

allocation of efforts, 44-50
bottom up mechanisms, 396
challenges and opportunities, 46-47, 302, 313
climate change and, 27-28, 46-47, *49*, 375-380, *376-377*
climate variability and, 297
community participation, 28, 295, *300*
context, 4-7, *4*
coordination, 342, 356
corrective, 36
costs of, 317-319
cycle, 35
defined, *5*, 34
effective strategies, 17, 27, 67, 342, 439-440
effectiveness, assessment of, 375-377
implementation, national level, 341-342
information and, 421-425
integration with adaptation, 11, 28, 50, 355-357, 396-397, 425-427, 439, 450-454, 469-471
international level, **393-435**
legislation, 342, 358, *359*, 519-522
local knowledge and, 293, 311-312
local level, **291-338**
mainstreaming, 348, 355-357, 380
national level, **339-392**
options/strategies, *6*, 16-20, *18-19*, 320-323, 439-440
past experience, 10-11
policies and options, *352-354*, *361*
prospective, 36
public policies/components, 89-90
sector-based strategies, *352-354*, 357
short- and long-term responses, 450-454
synergies with adaptation, 36, 48-50, 68, 357, 469-471
top down mechanisms, 396
tradeoffs in, 20, 439-440, 448-450
See also Risk management
Disaster risk reduction (DRR)[*], 34, 366-371
    as adaptation, 237, 443-446
    adaptation and, 423-424
    community/local knowledge and, 28
    ecosystem-based solutions, 343
    integration with adaptation, 396-397
    international level, 396-397, 403-425
    investment in, 529
    post-disaster recovery/reconstruction, 293, 301, *315*, 439
    short-term strategies, 305-306
    sustainable land management, 293
    technical and operational support, 409 410
Discounting, 268, 319, 376
Disease, 252, 259, *410*, 507-510
Disease vectors, 252, 253, 259, 316, 506
Displaced people/refugees, 80-81, 238, 457
    international refugee law, 396, 402, 412
Displacement, 80-81, 300-301
Distribution
    of climate variables, 40, *41*, 121, 130
    probability distribution, 40, *41*, *116-117*, 117, 121
Diversity, cultural, 84-85, 456
    See also Biodiversity
Downscaling[*], 129-131, 133, 147, 416
Drivers
    of capacity, 76
    of disaster risk, 10
    of economic losses, 16, 235
    underlying, need to address, 20, 425, 440, *470*
    of vulnerability, 70-72, 379

Drought[*], 167-175
    agricultural impacts, 247, 252
    attribution of, 171-172, 174-175
    case study, 498-500
    defined, 167-169
    drivers of, *168*
    ecosystem impacts, 246, 252
    impacts, 242
    indicators/indices, *168-169*, 242, *245*
    management/adaptation options, *19*
    megadroughts, paleoclimatic, 170
    observed changes, 8, *19*, *119*, 170-171, 174
    projected changes, *19*, 113-114, *119*, 172-175, *173*, *191-195*, 242
    regional impacts, 253, 255, 256, 259, 260-261, 264
    regional projections, *191-195*
    socioeconomic impacts, 238-239
    technological management, *415*
Dust storms, 190, 254
Dynamical downscaling, 129-130, 133, 147
Dzud, 500-502

E

Early warning systems[*], 303-304, 342, 364-366, 405, 416-417
    case study, 517-519
    Heat Warning System, 495
    strategy of, 517-519
Economic costs of disasters, 264-274
    See also Costs; Economic losses/impacts
Economic efficiency, 399-400, 420
Economic growth, 445
Economic losses/impacts, 9, 234, 264-274
    assessment of, 269-273
    attribution of, 268-269
    developed vs developing countries, 234, 265
    direct, indirect, intangible, 264, 266, 317
    Disaster Deficit Index (DDI), *93*
    drivers of, 16, 235
    exposure and, 9, 16, 234-235, 269, 273
    global and regional, 269-271, *270*, *271*, *272*, 441
    percent of GDP, 234, 270, 344, 441
    Probable Maximum Loss (PML), *93*
    projected changes, 16
    top-down vs. bottom-up approach, 266-267
    types of, 264
    See also Costs
Economic vulnerability, 86-87
Ecosystem-based adaptation, 343, *352*, 357, 370-371, *371*, 445-446
Ecosystem management and restoration, 294, 306-307, 315, 404
Ecosystem services, 370, *370*, 445-446
Ecosystems, 231-290
    DRM and adaptation, 343, *352*, 370-371
    impacts, 244-246
    local management/protection, 294, 306-307, 315
    reproduction, climate extremes and, 238
Education, 364-366, 526-529
    case study, 526-529
    vulnerability/exposure, 81-82
Efficiency, 399-400, 420
El Niño-Southern Oscillation (ENSO)[*], 155-157, 255, 459
    observed changes, *119*, 155-156
    projected changes, 15-16, 113, *119*, 156-157
El Salvador, 504, 520
Elderly, 454
Electrical networks, 257, 307

576

EM-DAT database, 364
Emissions scenarios[*], 11-12, 112-113
    precipitation projections, *145-146*, 147-149
    SRES, 112-113, 136-137, *137-140*, 141
    temperature projections, 137-138, *139-140*, 141
Empowerment, 310, 316
Energy systems, *353*
Ensembles, 131-133
Entitlements, 315-316
Environment, physical, **109-230**
    impacts on, 167-190
Environmental dimensions of vulnerability, 76-80
Environmental justice, 320
Epidemic disease, 507-510
Equity, 320, 401, 454-456
Ethiopia, 365, 420, 524
Ethnicity. *See* Race/ethnicity
Europe, 256-258
    adaptation costs, *274*
    case study (heat waves), 492-496
    coastal impacts, 183, 185, 186, 256-257
    drought, 170-171, 172-174, 175, 242, *243*, 246, 256
    ecosystem-based adaptation, *371*
    floods, 176, 177, 256-257, 258
    heat waves, *19*, 133, *192*, 256, *257*, 492-496
    landslides, 258
    precipitation, 142, *145-146*, 148, 149, *192*, *198-199*
    snow, 258
    temperature, 133-135, *135*, *139-140*, 141, *192*, *198-199*
    waves, 181, 182
    wildfires, 256
    wind, 150, 151, 152, 257
Evapotranspiration[*], 113-114, 118, 167, *167-169*
Ex post vs *ex ante* **actions**, 74, 90
Exposure[*], 4-7, 32, **65-108**
    concept/definition, 5, 32, 69, 237, 444
    development and, 10, 67, 70
    dimensions and trends, 76-89
    economic losses and, 9, 16, 234-235, 269, 273
    interactions of, 238-239, *239*
    management options, *18-19*
    observations of, 7-9, *18-19*
    regionally based aspects, 252-264
    risk and, 69
    scales and factors in, 67, 237
    system- and sector-based aspects, 239-252
    See also Vulnerability
External forcing, 5, 29
Extratropical cyclones. *See* Cyclones
Extreme events[*], 39-44
    anthropogenic influences and, 9, 112, 125-126, 268
    case studies, 492-507
    climate change and, 40, *127*
    compound events, 118
    comprehensive/integral/holistic focus, 38
    context, 4-7
    costs, 264-274
    defined, 30, 40-41, 111, *116-117*
    extreme weather vs climate, 117
    factors in, 4, 111
    impacts on humans and ecosystems, **231-290**
    impacts on physical environment, **109-230**
    observed changes, 111-112, *119-120*, 133-152
    physical aspects, 38, 40-41
    projected changes, 112-113, *119-120*, 133-152
    risk management, 16-20
    short- and long-term responses to, 450-454

    technologies for, 416-417
    vulnerability/exposure and, **65-108**, 239-264
    See also Climate extremes; Extremes; Impacts
Extreme impacts, 27, 41-44, 237
Extreme indices, *116-117*
Extreme value theory, *116-117*
Extremes
    changing climate and, 40, *124-125*, *127*
    defined, *5*, 40
    diversity of, 40-41
    methods and requirements, 122-133
    observed changes, *119-120*, 133-152
    projected changes, *119-120*, 133-152
    traditional adjustment to, 43-44
    See also Climate extremes; Extreme events

F

Fairness, 320
Faith-based organizations, *296*, *346*, 348
FAQs (frequently asked questions)
    climate: is it becoming more extreme?, *124-125*
    DRM strategies: in changing climate, *49*
    government: preparedness measures, *376-377*
    local context: importance of, *298*
    local level: adaptation limits, *319*
    local level: cost estimation, *318*
    local level: lessons learned, *300*
    relationship: climate change and individual extreme events, *127*
    relationship: extreme events and disasters, *33*
    resilience: practical steps for, *470*
    technology: emphasis on, *448*
    transformational changes, *466*
Fatality rates, 9, 234, 344, 400
Feedbacks, 112, 118-120
Finance/budgeting, *523*
    adaptation in developing countries, 406-407, 411, 413
    international, 17, 357, 397, 410-411, 412-414
    national, 360-362, *523*
    technology transfer, 417
    See also Costs
Fire. *See* Forest fires; Wildfire
Fisheries, *352*
Flexibility, 343, 355, 380
Flood control, 11, *52*, 55, 68, 305
Floods[*], 41, 175-178
    case studies, 505-507, 510-512
    climate change and, *245*
    costs, 256-257, *272*
    ecosystem impacts, 246
    exposure and, *241*, *245*
    flash floods, *18*, 175
    floodplains, 77
    fluvial (river) floods, 113, 175, 178, 242-244
    glacial lake outburst floods[*], 114, 175, 186, 258
    observed changes, 8, *119*, 175-177, 178
    projected changes, 13, 113, *119*, 177-178
    regional impacts, 253-254, 256-257, 258, 259-260, 261, 262
FONDEN, *362*
Food security, 235, 246-247, 368-370
    DRM/adaptation options, *352*
Forest fires, 252, 256, 259
    See also Wildfire
Forestry sector, 235, 446
    costs, *273*
    deforestation, 238, 252, 446
    DRM/adaptation options, *352*

Frameworks, 403-411
    See also UNFCCC; UNISDR
France, 176, 185
Frequently asked questions. *See* FAQs
Funds for adaptation, 406-407, 411

G

Garifuna women, *82*
GDP, 234, 270, 344, 441
Gender, 91, 313, *315*
Geographic information systems (GIS), 416
Glaciers[*], 186-189
    glacial lake outburst floods (GLOFs)[*], 114, 175, 186, 258
    glacier melting, 15, 114, 188
Global Assessment Reports (GARs), 404-405, 408, 425
Global climate models[*], 128-133
Global costs
    of adaptation, 273-274, *273*
    of disasters, 269-271, *270*, *271*, *272*
Global Environment Facility (GEF), 410-411, 413, 417
Global Facility for Disaster Reduction and Recovery (GFDRR), 411
Global Framework on Climate Services (GFCS), 409, 464
Global interdependence, 10, 396, 399
Global mean climate, 121-122
Global Network of Civil Society Organisations for Disaster Reduction (GNDR), 404
Global Platforms, 404, 405
Governance[*]
    case study, 519-522
    international, 403-411
    local, 312-313
    national, 341, 346-347
    preparedness and, *376-377*
    risk governance framework, 27, 44
    risk-neutral approach, 360-361
    sub-national governments, 346-347
    vulnerability and, 85-86
Grand Challenges, 426
Green Climate Fund, 408, 414
Greenhouse gases[*], 9, 112, *124*, 126
    ENSO and, 156
    mitigation, 446, 458-462
Greenland, *191*, *197*
    ice sheet, 179, 188
    precipitation, *145-146*, *191*, *197*
    temperature, *135*, *139-140*, *191*, *197*
Grenada, 512-514
Gross domestic product. *See* GDP
Groundwater, 42, *167-168*, 242
    salinization of, 250, 256
Growing season, 247
Guatemala, 504

H

Hadley/Walker circulation, 149, 151, 154
Hail, 141-142, *143*, 148-149
Hazards[*], 31, 69, 444
    'creeping', 365
    multi-hazard risk management, 17, 439, 450
    preparedness and, 364-366
Health and well-being, 251-252
    DRM/adaptation options, *354*
    impacts, 235, 251-252, 259, 316
    inequalities in, 316

Index

regional impacts, 259
vulnerability, 82-84
Heat waves[*]
    case study, 492-496
    ecosystem impacts, 244-246
    human impacts, 248, 251-252, 316
    observed changes, 8, 133, 134
    projected changes, 15, 114, *191-202*
    regional impacts, 253, 256, 258-259
        Australia, *239*, 496-498
        Europe, *19*, 256, *257*, 492-496
    regional projections, *191-202*
    vulnerability to, 234
High-income countries, 9, 234, 270, 343, 344, 347
High-latitude changes, 189-190
Holistic approaches, 379-380
Honduras, *82*, 314
Housing options, *352-354*
Human development index, 266
Human health. *See* Health and well-being
Human rights, 401, 403, 447, 457
Human rights law, 396, 401
Human security[*], 293, 297-298, 440, 457-458
Human settlements, 7-8, 247-251, 316-317
    coastal, 7-8, 235, 248, *249*, 258, 460, 510-512
    informal, 237, 247, 460
    vulnerability/exposure, 234-235, 247-251
    vulnerability reduction, 368-370
    See also Cities; Urban areas
Human systems, impacts on, **231-290**
Humanitarian/disaster relief, 10, 293, 299-300, 452
Humanitarian sector, research in, *468*
Hurricanes, *19*, 181, 260
    economic losses, *19*, 256, 260
    ecosystem impacts, 246
    Katrina, 9, 55, 158, 260, 308, *315*, 457
    local response to, 297
    management/adaptation options, *19*
    Mesoamerican (case study), 504-505
    storm surge, 158
Hydroclimatic extremes, 41
Hyogo Framework for Action, 398, 403-406
    national systems and, 341-342, 344
    Strategic Goals, 404

I

Ice cover, 262
Ice sheets, 179, 188
Iceland, *191*
Impacts[*], 8-9, 16, **109-290**, 441
    cascading, 399
    costs/economic losses, 234-236, 264-274
    cumulative, 6-7, 38, 67, 69
    databases, 364
    extreme, 6, 27, 41-44, 237
    extreme and non-extreme events, 6, 234
    global, 399, 441
    human systems and ecosystems, 16, **231-290**
    international, 399
    local, 293
    management of, 373-375
    physical environment, 8-9, *119-120*, 167-190
    projected, 11-16
    regionally-based, 252-264
    resistance to, 38
    sector-based, 235, 239-252
    social, 42-43
    system-based, 239-252
    trends in, 271-273, *272*

uncertainty and, 239
vulnerability/exposure and, 10, 67, 234-235, 238-239, 271
Impacts-first approach, 350-351, *350*
Incremental change, 20, 439, *442*
Index-based contracts, 323
Indexes (indices), *92-93*, *116-117*, *125*, *168-169*
India
    adaptation technologies, *415*
    floods, 254, 457
    wetlands, *370*
Indian Ocean, 181, *184*
    cyclones, 503-504, *503*
    tsunami (2004), 307, 457
Indian Ocean Dipole (IOD)[*], 157-158
Indigenous peoples, 80-81, *82*, 84-85, 298, 311-312
Indirect losses, 264, 266
Individual scale, 38, *39*
Indonesia, 359, 457, 526
Inequalities, 10, 294, 313-317, 405, 447, 454
Informal settlements, 237, 247, 460
    flash floods and, *18*
Information, 342, 356, 363-364, *365*, 529
    access to, 82
    acquisition, 421-422
    international level support, 409, 421-425
    on scale/nature of events, 416
    sharing and dissemination, 422-425
    See also Communication
Information and communication technologies (ICT), 422-423
Information gap, 409
Infrastructure, 16, 248-250, 366-368
    adaptation costs, 412
    DRM/adaptation options, *353*, 366-368
    education, 81-82
    permafrost and, 367, 515-517
    risk reduction, 304-306
    technology and, 447, 448
    tipping points, 366
    transport, 235, 248-249, 249, *250*, 348
    vulnerability, 235
Infrastructure thresholds, 366
Innovation, 20, 294, 439, 447, 468, *469*
Institutional approaches, 464-465
Institutional/governance dimensions, 85-86, 294, 308-313
Insurance[*], 294-295, 343, 371-373, 465, 523
    case study, 522-525
    crop insurance, 524
    distribution of, 343
    index-based/index-linked, 323
    international programs, 419-420
    local level, 322-323
    micro-insurance, 322-323, 420, 523, 524
    national systems, 343, 346, 360, 371-373
    regional pools, 372, 400, 419, 420, 523, 524-525
Intangibles, 264, 266, 317, 446
Integrated models, 131-133
Integrated systems approach, 27, 38, 50
Integration
    across scales, 17, 397, 426-427
    of DRM and adaptation, 11, 28, 50, 355-357, 396-397, 425-427, 439, 450-454, 469-471
    international level, 396, 425-427
    short- and long-term responses, 450-454
Intellectual property rights, 414, 448
Interdependence, 10, 396, 399
International Bill of Rights, 411

International conventions, 402
International Decade for Natural Disaster Reduction (IDNDR), 397, 403
International development, *410*
International Disaster Response Law, 411
International finance, 357, 397, 410-411
International humanitarian law, 411
International institutions, 348
    See also UNFCCC; UNISDR
International instruments, 402-403
International law, 396, 401-403, 411-412
    hard law, 401
    limits and constraints, 411-412
    soft law, 401, 402, 403-404
International refugee law, 396, 402, 412
International risk management, **393-435**
    adaptation and, 396-397
    context, 398
    current actors, 408-411
    current governance and institutions, 403-411
    economic efficiency, 399-400
    finance, 17, 357, 397, 410-411, 412-414
    future policy and research, 425-426
    integration across scales, 426-427
    knowledge/information, 421-425
    options, constraints, and opportunities, 411-425
    rationale for, 398-403
    risk sharing and transfer, 418-421
    shared responsibility, 400-401
    subsidiarity, 401
    systemic risks, 399
    technology transfer and cooperation, 414-418
International technical support, 409-410
Irrigation, 448
Island states. *See* Small island states

J

Japan
    disaster risk reduction, 424
    precipitation, 148
    public awareness campaign, 528
    storm surge, 254
    waves, 182
Judgments about risk, *45*, 46-47
Justice, 320
    See also Equity

K

Katrina (hurricane), 55, 158, 308, 457
    economic losses, 9, 260
    international impacts, 399
    recovery and reconstruction, *315*
Kenya, *360*, *374*
Knowledge acquisition, 421-422
Knowledge/information sharing, 422-425, 526-529
Kyoto Protocol, 402, 403, 406, 411

L

Land use/land use change, 69, 238, 293
Land use, land use change, and forestry (LULUCF), 370
Land use planning, 306-307
Landslides[*], 41, 114, *120*, 187-189, 255
    regional impacts, 258
Leadership, 469
Learning, 27-29, 53-56, 439, 467-468
    humanitarian sector, *468*
    'learning by doing', 378-379, 380
    learning loops, 53-54, 56

Index

transformation and, 324
See also Case studies
Least Developed Countries Fund (LDCF), 357, 406, 413, 417
Legislation, 358, *359*
    case study, 519-522
Lessons learned, 469-470, 529-530
    local level, 295, *300*
    See also Case studies
Levees, 55, 254-255, 305
Liabilities, 361, *361*
Light Detection and Ranging (LIDAR) data, 186
Likelihood[*], *21*, 112, 120-121
Limpopo River/Basin, 253, 309, 506
Livelihoods, 314-315
    agricultural sector, 246
    ecosystem management and restoration, 294
    tourism sector, 251, *251*
    vulnerability and, 87
Livestock, 237, 253, 259
Local adaptation, 293-295
    limits to, 319-320, *319*
Local coping, 294, 298-301
    differences/inequalities in, 313-317
Local decisionmaking, 305-306, 308, 310, 325
Local disaster risk management (LDRM)[*], 291-338
    anticipating future risk, 302-307
    capacity building, 308-313
    challenges and opportunities, 295, 313-320
    community participation, 28, 295, *300*, 321, *322*
    context and, 298, *298*
    costs, 295, 317-319, *318*
    current coping, 298-301
    differences/inequalities in, 10, 294, 313-317
    disaster relief and assistance, 299-300
    gaps in information, 323-325
    importance of local level, 296-298
    insurance, 294-295, 322-323
    lack of data, 295
    lessons learned, 295, *300*
    links to global/national levels, *296*
    risk communication, 294, 302-304, *303*
    risk sharing, 321-323
    strategies, 293, 320-323
Local government, 312-313
Local knowledge, 17, 293, 311-312
Local-level institutions/planning, 294, 308-313
Local recovery and reconstruction, 301, *315*
Localized social networks/norms, 310-311
Loss of life/fatalities, 9, 234, 344, 400
Low-income countries, 9, 234, 343, 344, 347
Low regrets strategies, 16-17, 56, 342, 351, *352-354*, *376*

M

Maastricht Treaty, 401
Mainstreaming, 348, 355-357, 380, 411
Maladaptive actions, 54, 55, 368, *448*
Malaria, 252, 253, 316
Maldives, *370*, 512-514
Mangroves, 343, 370, *370*, 450
Mass media, 422-423
Media coverage, 422-423
Mediterranean region, *192*, *198-199*
    drought, 172-174, 175, 256, 498-500
    precipitation, 142, *145-146*, *192*, *198-199*
    temperature, 133-134, *135*, 138, *139-140*, *192*, *198-199*

tourism, 251, *251*
waves, 182
wind, 150
Mega-deltas, 254
Megacities, 294, 317, 510-512
Melbourne, *239*, 497
Mental health impacts, 252, 316
Meridional overturning circulation (MOC)[*], 122
Mesoamerican hurricanes, *503*, 504-505
Mexico, 258-260
    drought, 170, 172, 174, 175, 259
    fund for disasters (FONDEN), *362*
    hurricanes, *503*, 504-505
    insurance, 372
    precipitation, 142, *145-146*, *191*, *197*
    temperature, *135*, *139-140*, *191*, *197*
Micro-finance, 419
Micro-insurance, 322-323, 420, 523, 524
Microsatellites, 416
Migration, 16, 80-81, 293, 300-301, 457
    as adaptive, 237, 300, 399
    destination area issues, 399
Millennium Declaration, 400, *410*
Millennium Development Goals (MDGs), 369, 398, 400, *410*, 458
Mitigation (climate change), 36, 446
    interactions of, 20, 458-462
Mitigation (disasters and disaster risk)[*], 36
Models. *See* Climate models
Modes. *See* Climate modes
Moisture content
    atmospheric, 126
    soil, 118-119, 167-175, *167-169*
Mongolia, 419, 500-502
Monsoons[*], *119*, 152-155
Moral economy, 309, *309*, 447
Mountain environments, 8, 15, 114, 186-189, 248
Mozambique, 253, 297, 405, 505-507
Mudslides, 158, 255
    See also Landslides
Multidisciplinary management, 36, 37
Mumbai, 461, 510-512
Myanmar, 502-505, *503*

N

Nairobi floods, *18*
Nairobi Work Programme, 407, 409, 529
Namibia, 253-254
National adaptation plans, 349-355, 356, 370
National Adaptation Programme of Actions (NAPA), 369-370, 406
National building standards, 347, 367-368
National Platforms, 359, *360*
National systems, 11, 36, **339-392**, *346*
    adaptation options, *352-354*, *361*
    adaptive management, 377-379
    aligning with climate change, 342, 351-355, 375-380
    communication, 349, *376*
    coordination, 358-360, *360*
    culture of safety, 362-366
    disaster risk management, 341-343, *352-354*, *361*
    disaster risk reduction, 366-371
    finance and budget, 360-362
    flexibility in, 343, 355, 380
    holistic approaches, 379-380
    impact management, 373-375, *374*
    implementation of DRM, 341-342
    insurance, 343, 346, 371-373

legislation and compliance, 358, *359*
    planning and policies, 349-357, *352-354*
    practices, methods, and tools, 362-375
    preparedness, 364-366, *376-377*
    public goods, 341, 363
    risk assessments, 380
    risk pooling, 343
    risk transfer, 355, 371-373, *376*
    sector-based risk management, *352-354*, 357
    strategies, 357-362, *376-377*
    systems and actors, 345-349, *346*
    top-down vs bottom-up approaches, 350-351, *350*
    uncertainties, managing, 377-379
Natural physical environment. *See* Physical environment
Nepal, 526
Netherlands, *52*, 257
    flood impacts, 272-273, *272*
    flood management, *469*
Networks, 309-311, *309*, 404
New Orleans, 55, 158, 260, 308, *315*
New York, 303
New Zealand, 260-261
    drought, 172, 261
    precipitation, 143, *145-146*, *195*, *202*, 261
    temperature, *139-140*, *195*, *202*
    wildfire, 261
No regrets/low regrets options, 16-17, 56, 351, *352-354*, *376*
Nongovernmental organizations (NGOs), 313, 403
Normative dimension, 426
Norms, 310-311
North America, 258-260
    adaptation costs, *274*
    drought, 170, 174, 175, 259
    floods, 177, 259-260
    forest fires, 252
    heat waves, 258-259
    hurricanes, *19*, 260, *315*
    monsoons, 153, 154
    precipitation, 142, *145-146*, 149, *191*, *196-197*
    temperature, 134, 135, *135*, *139-140*, 141, *191*, *196-197*
    wildfire, 259
North Atlantic Oscillation (NAO)[*], *119*, 149, 157-158
Northern Annular Mode (NAM)[*], 150
Northwest Territories, 514-517
Nunavut, 514-517

O

Observed changes, 7-9, *18-19*, 111-112, *119-120*, 133-152
    coastal impacts, *120*, 183
    drought, *19*, *119*, 170-171, 174
    ENSO, *119*, 155-156
    extratropical cyclones, 163-166
    floods, *119*, 175-177, 178
    glaciers and mountains, 186-188
    heat waves, *19*
    methods for analyzing, 122-125
    monsoons, 152-153
    permafrost, *120*, 186-188, 189-190
    sand/dust storms, 190
    sea level, *120*, 178-179
    temperature, *119*, 133-136, 141
    tropical cyclones, 159-161, 163
    wind, *119*, 149-150

Index

Ocean acidification, *185*, 261
Oceania, 260-261
Open oceans, 261

P

Pacific Decadal Oscillation (PDO)[*], 153, 157
Pacific Island Countries, 183, 263-264
Pacific Ocean, 181, *184*, 263
Pakistan, 457
Palmer Drought Severity Index (PDSI), *168-169*
Permafrost[*], 189-190
    case study, 514-517
    infrastructure and, 367
    observed changes, *120*, 186-188, 189-190
    projected changes, 15, 114, *120*
    regional impacts, 263
Perturbed-physics ensembles, 131-133
Philadelphia, 495
Philippines, 239, 519-522, 526
Physical environment, **109-230**
    impacts on, *119-120*, 167-190
Planning, 10, 462-469
    adaptation planning, 349-357, 443-444
    community participation in, 28
    development, 439
    local level, 308-313
    national level, 349-357, 380
    risk reduction, 69
    urban planning, 460-461
Polar regions, 261-263
Politician's dilemma, 452
Polluter pays principle, 400
Population growth, 237, 238
Population movements. *See* Migration
Post-disaster credit, 419
Post-disaster recovery and reconstruction, 10, 293, 301, *315*, 439, 457
Post-traumatic stress disorder, 252
Poverty, 20, 314, 344
    rural, 238
    vulnerability/exposure and, 87, 247, 400, 441
Poverty reduction strategies, *410*
Poverty traps, 451, 452
Precautionary principle, 402
Precipitation, 8, 141-149
    attribution of changes, 143-144, 149
    observed changes, 41, *119*, 142-144, 149
    projected changes, 13, *14*, *15*, 113, *119*, 144-149, *144-146*, *191-202*
    rainfall, 113, *119*, 142-144, 148, 149
    regional impacts, 253-254, 255
    regional projections, *191-202*
    snowfall, 141, 189
    uncertainties, 148-149
    See also Drought; Floods
Preparedness, 36, 364-366, 369, *376-377*, 517-519
Prevention, 36, 69
    See also Disaster risk reduction
PreventionWeb, 404, 405, 424
Private sector organizations, 347-348
Probabilistic risk analysis, 42, *43*, 44-45, 446
Probability distributions, 7, *7*, 40, *41*, 116-117, *117*, 121
Probability of occurrence, 40, *41*
    See also Return period; Return value
Projected changes, 11-16, *18-19*, 112-114, *119-120*, 133-152
    coastal impacts, 113, *120*, 183-186
    cyclones, extratropical, 13, *272*

cyclones, tropical, 161-163, 271, *272*
drought, 19, 113-114, *119*, 172-175, *173*
ENSO, 113, *119*, 156-157
floods, 13, 113, *119*, 177-178
glaciers and mountains, 189
heat waves, 15, *19*, 114
monsoons, 153-157
permafrost, 114, *120*, 190
precipitation, 13, *14*, *15*, 113, 144-149, *191-202*
sand/dust storms, 190
sea level, 15, *120*, 179-180
temperature, *12*, 13, 112-113, *119*, 133-152, *137*, *138*, *191-202*
wind, 13, 113, *119*, 151-152, *151*, 248
Projections[*], 11-16, *119-120*, 462-463
    likelihood/confidence in, 112, 120-121
    time scale of, 112
    uncertainty and, 112, 130-133
    See also Climate models; Scenarios
Property, 293
Property rights, 306
Psychological/mental health, 252, 316
Public awareness, 364-366, 526-529
Public goods, 341, 363, 399-400
Public health, 83, 529
    heat waves and, 492-493
    infectious diseases, 252, 253, 259, 316, 507-510
Public-private partnerships, 343, 347

R

Race/ethnicity, 84, *315*
RANET, 423, 424
Rationing, 307
Reconstruction, 10, 301, 457
Recovery and reconstruction, 10, 293, 301, *315*, 439, 457
    funding for, 417
Red Cross/Red Crescent, 348, 359, 403, 409, *410*
Refugees, 238, 412
    international refugee law, 396, 402, 412
Regime shifts (ecosystems), 122, 170, 445, 448
Regional climate models, 129-133
Regional climate projections, 121-122, *191-202*
    cyclones and floods, *240-241*
    precipitation, *145-146*, 147-149, *191-202*
    temperature, *135*, 138, *139-140*, 141, *191-202*
Regional costs/economic losses, 256, 270, *270*
    adaptation costs, 273-274, *274*
Regional development banks, 411
Regional impacts, 252-264
Regional risk pools, 372, 400, 419, *420*, 524-525
Regrets/low regrets, 16-17, 56, 342, 351, *352-354*, *376*
Reinsurance, 323, *362*, 364, 524-525
ReliefWeb, 423
Relocation, 293, 300-301
Remittances, 418-419
Republic of Korea, 201, 246
Republic of the Marshall Islands, 512-514
Resilience[*], 20, 34, **437-486**
    access to resources, 454-458
    adaptation and, 443-450
    building, *376*, 378
    concept/definition, *5*, 34, 238
    disaster risk management and, 443-450, 458-462
    equity and, 454-456
    integration of policies, 11, 439, 450-454, 469-471
    interactions among processes, *324*, 440, 458-462
    long-term options, 462-469

    multi-hazard risk management, 439, 450
    multiple approaches and pathways, 17, 440
    national/government actions, *376*
    pathways to, *442*
    planning and proactive actions, 462-469
    practical steps for, *470*
    questioning assumptions and, 20, 439
    risk and uncertainty and, *442*
    risk sharing/transfer and, 10-11, 465
    short- and long-term responses, 450-454
    sustainable development, 454-458
    synergies of DRM and adaptation, 469-471
    thresholds and tipping points, 458-459
    transformation and, *442*
    vulnerability and, 72
Resilience thinking, 453-454
Resistance, 38
Resources
    access to, 454-456
    scarcity of, *297*, 316
    storage and rationing, 307
Responsibility
    common but differentiated, 400
    corporate, 347
    responsibility to protect, 396, 412
    shared, 400-401
Return period[*], 46, 113
    precipitation projections, 142, *146*, 147, *198*
    temperature projections, *135*, 137, *140*
Return value[*], *117*, 121
    precipitation projections, 141, *145-146*, 147, *196*
    temperature projections, *135*, *139-140*, 141, *196*
Rights, 401, 403, 447
    International Bill of Rights, 411
Rio Earth Summit, 400-401
Risk, **65-108**, 444
    acceptability of, *361*, 362
    factors in, 69-70
    judgments about, *45*, 46-47
    social construction of, 36-37, 45
    systemic, 399
    unacceptable, *442*
    underestimation of, 452
Risk acceptance threshold, *361*
Risk accumulation, 95-96
Risk assessment, 27, 67, 90-95
    challenges in, 27, 46
    local level, 294
    methods, 274
    uncertainty and, 274
Risk awareness, 364-366
Risk communication, 17, 67, 95, 294, 302-304, *303*, *376*
Risk governance framework, 27, 44
Risk identification, 90, 369
Risk-linked securities, 419-420
Risk management, 16-20, 34, *376*
    effective, 17, 67, 439
    integrated systems approach, 27
    international, **393-435**
    iterative, 47-48
    local, **291-338**
    multi-hazard approaches, 17, 439, 450
    national, **339-392**
    sector-based, *352-354*, 357
    short-term strategies, 305-306
    See also Disaster risk management
Risk-neutral approach, 360-361
Risk pooling, 11, 343, 372, 400, 419, 523, 524-525

Index

Risk sharing, 10-11, 397, *523*
    international level, 397, 418-421
    local level, 321-323
    national level, 371-373, *376*
Risk transfer[*], 10-11, 35, 294-295, 371-373, 465
    case study, 522-525
    international level, 397, 419-421, *523*
    local level, 294-295, 321-323, *523*
    national and sector-based programs, *352-354*
    national systems, 341, 355, 371-373, *376*, *523*
River discharge/runoff, 176-178
River regulation, 176, 238
    dams, 176, *304*, *415*
Robustness, 56
Runoff[*], 177, 238, 262
Rural areas, 79-80, 461-462
    continuum with urban areas, 298
Russian Federation, 262

S

Safety, culture of, 362-366
Saffir-Simpson scale, 149, 158, 162, 502
Sahara, *193*, *199*
    drought, 174
    precipitation, *145-146*, *193*, *199*
    sand/dust storms, 190
    temperature, *139-140*, *193*, *199*
Sahel, 171, 190
Saijo City, 528
Saltwater intrusion, 114, *184-185*
Samaritan's dilemma, 452
Sand and dust storms, 190, 254
Satellite-based technologies, 416, 423
Saudi Arabia, *415*
Scale, 111-112, *132*
    coordination/linking across, 358-360, *360*
    exposure and, 237
    individual, 38, *39*
    integration across scales, 397, 426-427
    multiple scales, 448-450
    small- and medium-scale, 38
    territorial, *39*
    timescales, 88-89, 112
Scenario development, 462-463
Scenarios, 11-12, 112-113
    precipitation projections, *145-146*, 147-149
    SRES, 112-113, 136-137, *137-140*, 141
    temperature projections, 137-138, *139-140*, 141
Scenarios-impacts approach, 350-351, *350*
School curricula, 526
Science and technology, 89
Sea ice, 122, 261, 262
Sea level, 178-180
    change[*], *120*
    coastal impacts, *18*, 182, 183, *184*, 185
    extreme, *18*, 178-180
    mean[*], 178
    observed changes, *120*, 178-179
    projected changes, 15, *120*, 179-180
    small island states and, *184*, 263
Sea level rise, 113, 178-179
    impacts, *18*, *184*, 248, 263
    observed changes, 178-179
    projected changes, 15, 113, 179-180
Sea surface temperature[*], *185*, 261
Sector-based organizations, 357
Sectors, *235*, 239-252
    DRM/adaptation, *352-354*, 357
    See also specific sectors

Settlement patterns, 7-8, 78-80, 234, 235, 247-251
    See also Human settlements
Shared responsibility, 400-401
Shelter-in-place, 308-309
Simulations, 462-463
Skewed development processes, 10, 67, 70
Slums. *See* Informal settlements
Small island states, *18*, *184-185*, 263-264
    case study, 512-514
    coastal impacts, 183
    drought, 264
    impacts, 15, 183, 248, 251
    relocation issues, 300-301, 368
    sea level and, *184*, 263
    small island developing states (SIDS), *18*, 512-514
    tourism impacts, 251, *251*
    vulnerability/exposure, 114, *184*, 235, 248, 263, *263*
Snow load, 514, 515
Snowmelt, 113, *119*, 175-178
Snowpack, 176-177
Social capital, 310, 315
Social construction, 36-37, 45, 70
Social groups, vulnerability, 81
Social impacts, 42-43
Social justice, 320
Social networks, 310-311, 315
    web-based, 423
Social norms, 310
Socio-ecological systems, 441
Socio-political networks, 309, *309*
Socioeconomic change/inequities, 10, 238
Socioeconomic conditions, 16, 234, 235, 238, *410*
Socioeconomic systems, 237-239
Soft law, 401, 402, 403-404
Soft vs hard engineering, *448*
Soil moisture[*], 118-119, 124, 167-175, *167-169*, *173*
    soil moisture anomalies (SMA), 15, *169*, 172, *173*
    soil moisture drought, *168-169*, 171-172
Solidarity, 400-401, 420-421, *523*
South Africa, *193*, *199*
    case study (DRM legislation), 519-522
    collective action in, *322*
    precipitation, *145-146*, 147, *193*, *199*
    temperature, *139-140*, *193*, *199*
South America, 255-256
    adaptation costs, 273, *274*
    drought, 171, 174
    floods, 176
    precipitation, 142, *145-146*, 148, *194*, *200*, 255
    temperature, 134, 135, *135*, 138, *139-140*, 141, *194*, *200*
    waves, 181
    wildfires, 255-256
South Asia, *195*, *201*
Southern Annular Mode (SAM)[*], 150, 157-158
Space-based monitoring, 416
Special Climate Change Fund (SCCF), 406
SRES emissions scenarios, 112-113, 136-137, *137-140*, 141
Stationarity/non-stationarity, 46, 131
Statistical downscaling, 129-131, 147
Storm surge[*], 158, 179-180
    regional impacts, 254, 260, 261
Storm tracks[*], poleward shift in, 8, 164-166
Streamflow[*], 176-177
Structural measures, 293, 304-306, 415
Sub-Saharan Africa
    adaptation costs, 273, *274*
    floods, 176

    food security, 368
    MDGs, *410*
Subsidiarity[*], 401
Surprises, 122
Susceptibility, 72, 183
Sustainability, 20, **437-486**
    adaptation and, 444
    DRM and, 440, 443-450, 458-462
    of ecosystem services, 445-446
    incremental steps, 439, *442*
    interactions of mitigation, adaptation, and DRM, 20, 440, 458-462
    multiple approaches, 440
    practical steps for, *470*
    prerequisites for, 440
    synergies of DRM and adaptation, 469-471
    thresholds and tipping points, 20, 458-459
    transformational change, 20, 439, 465-469
    See also Resilience
Sustainable development[*], 20, 439-440
    climate change and, 444
    DRM and, 440, 443-446
    questioning assumptions in, 20, 439
    technology and, 414-418
Sustainable land management, 293
Synergies, 48-50, 68, 357, 397, 469-471
Syria, 498-500
Systemic risks, 399

T

Tampere Convention, 402, 411
Tanzania, *39*
Teacher training, 526-527
Technology, 89, 366-368, 447-448
    adaptation and, 414-415, *415*
    emphasis on (FAQ), 448
    extreme events and, 416-417
    hard vs soft, *448*
    international technical support, 409-410
Technology choices, 447-448
Technology Mechanism, 408
Technology transfer, 397, 414-418, 448
    financing for, 417
Temperature, 133-141
    agricultural impacts, 247
    attribution of changes, 135-136, 141
    case studies, 492-498, 514-517
    extremes, *135*, *137*, *138*, 141
    extremes, projected, 13, 112-113, *119*, 133-152, *137*, *138*
    global surface[*], 126
    observed changes, *119*, 133-136, 141
    polar regions, 262-263
    projected changes, *12*, 13, *119*, *135*, 136-141, *137-140*, *191-202*
    regional projections, *191-202*
    sea surface[*], *185*, 261
    summary, 141
    uncertainties, 141
    See also Heat waves
Thailand, 448
Thermohaline circulation, 122
Three Mile Island, *45*
Thresholds, 20, 458-459
    abrupt climate change, 122
    absolute, 116
    extreme events, 116
    infrastructure, 366
    risk acceptance, *361*

581

Index

tipping points, 122, 351, 458-459
vulnerability/thresholds approach, 350-351, *350*
Thunderstorms, 123, 141, 143
Tibetan plateau, *195, 202*
    precipitation, 143, *145-146, 195, 202*
    temperature, *139-140, 195, 202*
    wind, 150
Timescales, 88-89, 112
Tipping points, 122, 351, 366, 458-459
Top-down mechanisms/approach, 266, *346*, 350-351, *350*, 396, 427
Tornadoes, 151, 152
Tourism, 235, 250-251, *251*
Tradeoffs, 20, 440, 448-450
Traditional behaviors, 84-85
Traditional knowledge, 311-312, 456
Transformation[*], *324*, 441, 465-469
    context, 441-442
    defined, *5*
    local framework, 323
    pathways, *442*
Transformational change, 20, 439, *442*, 465-469, *466*
    facilitating, 466-469
Transpiration[*], 113-114, 118, 167, *167-169*
Transportation sector, 83, 88, 235, 246, 259, *353*
    infrastructure, 235, 248-249, *250*, 348
Tropical cyclones. See Cyclones
Tsunami (2004), 307, 457

U

Uncertainty[*], *21*, 47-48, 112, 130-133, *132*
UNFCCC, 344, 396, 406-408, 411-412
    adoption of, 403
    commitments on adaptation, *407*
    finance and, 412-413
    technology transfer, 397
UNISDR, 396, 397, 403, 408-409
    evaluation of, 405-406
United Kingdom
    coastal impacts, 185, 186
    floods, 176, 177
    precipitation, 142, 148
United Nations Development Programme (UNDP), 404
United Nations Framework Convention on Climate Change. See UNFCCC
United Nations International Strategy for Disaster Reduction. See UNISDR
United States, 258-260
    drought, 170, 174, 259
    floods, 176
    heat waves, 258-259
    hurricanes, *19*, 55, 158, 260, 308, *315*, 457
    precipitation, 142, 143, 148
    temperature, 134, 138
    waves, 181, 182
    wetlands, *370*
    wind, 152
Urban areas, 460-461
    continuum with rural areas, 298
    growth in, 294, 317, 366
    infrastructure, 250, 366-367
    vulnerability/exposure, 78-79, 238, 247-248, 317
    See also Cities

Urban floods, 175
Urban heat islands[*], 235, 248, 493
Urban planning, 460-461
Urbanization, 294, 317, 460
    exposure/vulnerability and, 67, 78, 234, 248, 294
    migration and, 238
Utilitarian approach, 447

V

Values and perceptions, 443, 446-447
Vanuatu, *309*
Variability
    climate, 7, 15, 115, 157
    climate modes, 15-16, 155-158
Vectors of disease, 252, 253, 259, 316, 506
Victoria, Australia, 496-498
Vietnam, 317, 358, *370, 371*
Violent conflict, *297*, 457
Volcanoes, 188-189
Vulnerability[*], 4-7, 65-108
    age/aging and, 234-235
    capacity and, 72-74
    concept/definition, *5*, 32-33, 69-70, 237-238, 444
    conceptual frameworks, 71-72
    dimensions and trends, 76-89
    drivers of, 70-72, 379
    economic dimensions, 86-87
    environmental dimensions, 76-80
    extreme and nonextreme events and, 67
    interactions and integration, 87-88, 238-239, *239*
    management options, 18-19
    observations of, 7-9, *18-19*
    Prevalent Vulnerability Index (PVI), *92*
    regionally based aspects, *92-93*, 252-264
    risk and, 69-70
    risk identification and assessment, 89-95
    scales and factors in, 67, 69-72
    sector-based aspects, 235, 239-252
    skewed development and, 10, 67, 70
    social dimensions, 80-86
    socioeconomic systems and, 237-239, *410*
    system-based aspects, 239-252
    timing and timescales, 88-89
    underlying causes, need to address, 20, 425, 440, *470*
    urbanization and, 67, 78, 234, 248, 294
Vulnerability assessments, 320-321
Vulnerability thresholds approach, 350-351, *350*

W

Warm days/warm nights[*], 8, 134-135, 137-138, *137, 138*, 141
    regional projections, *191-202*
Warm spells[*], 134, 141, *191-202*
Warning systems, 303-304, 364-366, 416-417
Water management, 16, *469*
Water scarcity, *167*, 238-239, 255
Water sector, 241-244
    DRM/adaptation options, *353*
    impacts, 235, 241-244
    vulnerability/exposure, 235, 241-244
    water supply system, 242
Watersheds. *See* Catchments
Waves/wave height, 180-182

WeAdapt/WikiAdapt, 424
Wealth, 314
Weather derivatives, 523, 524
Weather events
    categories of, 115
    disasters and, 115-118
Weather extremes
    defined, 40, *116-117*
    impacts on humans and ecosystems, **231-290**
    impacts on physical environment, **109-230**
    observed changes, *119-120*, 133-152
    phenomena related to, *119*, 152-166
    projected changes, *119-120*, 133-152
    vs climate extremes, 117
    See also Climate extremes; Extreme events; Impacts
Weather forecasting, 416-417
Weather indices, *116-117, 125, 168-169*
Wetland reduction, 238, 444, 461
Wetland services, 248, 255, 370, *370*, 445
Wildfire
    Australia, *239*, 496-498
    case study (Victoria), 496-498
    regional impacts, 255-256, 259, 261
    See also Forest fires
Win-win options, *352-354*, 355, 357
Wind, 149-152
    impacts, 248
    observed changes, *119*, 149-150
    projected changes, 13, 113, *119*, 151-152, *151*, 248
    regional impacts, 257
    technologies for, 416
Winners and losers, 456-457
Women, 81, 313, 454-456
    Garifuna women, *82*
    proactive role (Honduras), *314*
World Bank, 361, 404, 409, 411, 417, 419, 420
World Climate Conference-3 (2009), 409
World Conference on Disaster Reduction (Kobe), 398, 403
World Food Programme, 419, 420, 524
World Trade Organization, *410*

Y

Yangtze Basin, 147, 148, 176
Yucatan Peninsula, *503*, 504
Yukon, 514-517

Z

Zambezi River/Basin, 253, 506, 507
Zimbabwe, 506, 508-509